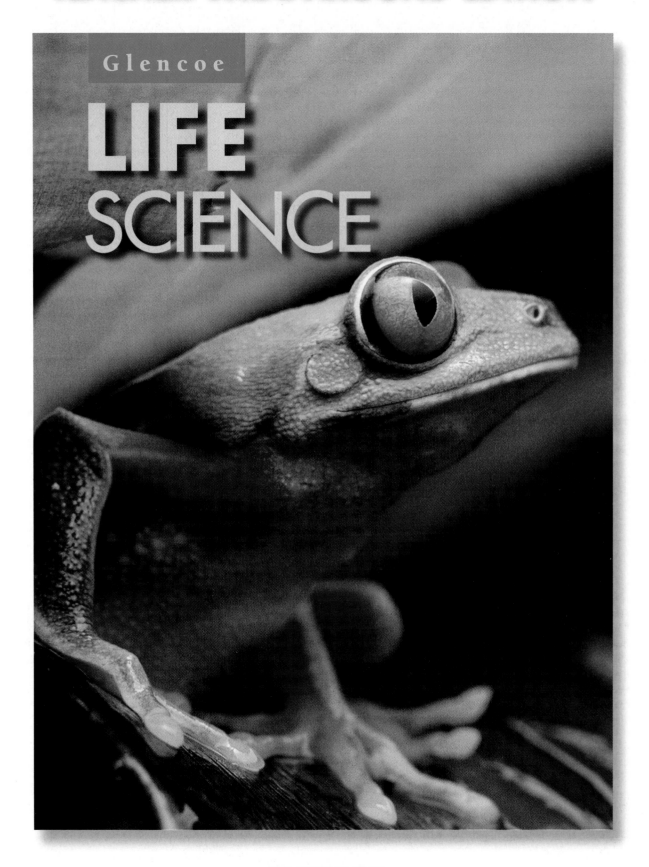

Glencoe

LIFE
SCIENCE

GLENCOE
McGraw-Hill

New York, New York   Columbus, Ohio   Woodland Hills, California   Peoria, Illinois

GLENCOE Life Science

Student Edition

Teacher Wraparound Edition

Laboratory Manual: SE

Laboratory Manual: TE

Study Guide: SE

Study Guide: TE

Teaching Transparency Package

Section Focus Transparency Package

Science Integration Transparency Package

Assessment

Performance Assessment

Computer Test Bank

MindJogger Videoquiz

English/Spanish Audiocassettes

Interactive Videodisc Program

Glencoe/McGraw-Hill

A Division of The **McGraw·Hill** Companies

Send all inquiries to:

Glencoe/McGraw-Hill
936 Eastwind Drive
Westerville, OH 43081
ISBN 0-02-827776-7

Printed in the United States of America.
3 4 5 6 7 8 9 027/043 04 03 02 01 00 99

Teacher Guide

About Our Authors/Table of Contents 4T

New Directions 7T
National Standards 7T
Responding to Changes in
    Science Education 10T
Themes and Scope and Sequence 12T

Program Resources 14T

Developing Thinking Skills 26T
Constructivism 26T
Developing and Applying Thinking
    Skills 27T
Thinking Skills Map 30T
Flex Your Brain 31T
Concept Maps 32T

Planning Your Course 34T

Strategies 38T
Using Technology in the Classroom 38T
Using the Internet 39T
Cultural Diversity 40T
Meeting Individual Needs 41T
Assessment 44T
Tech-Prep Education 47T
Cooperative Learning 48T

Activities in the Classroom 49T
Managing Activities in the
    Science Classroom 49T
Laboratory Safety 50T
Chemical Storage and Disposal 51T
Using a Graphing Calculator 52T
Activity Materials 54T

Equipment List 55T

References 62T
Supplier Addresses 62T

About Our Authors

ALTON BIGGS is a biology instructor at Allen High School, Allen, Texas. Mr. Biggs received his B.S. in natural sciences and an M.S. in biology from East Texas State University. He has completed additional graduate work at Ball State University in genetics, and the University of Texas at Dallas in environmental science. He was a Resident in Science and Technology at Oak Ridge National Laboratory in 1986, a GTE fellow in 1988, and a Genentech Access Excellence Fellow in 1995. Among the teaching awards he has received are Texas Outstanding Biology Teacher in 1982 and 1995; Presidential Science Teacher Award Finalist in 1986; Teacher of the Year Award, Allen Independent School District in 1987; and Texas Teacher of the Year Finalist in 1988. Mr. Biggs is past president of NABT and the founding president of TABT.

LUCY DANIEL, a North Carolina science teacher in Rutherford County, has 37 years of experience teaching biology, life science, physical science, and science education. Ms. Daniel holds a B.S. degree from the University of North Carolina at Greensboro, an M.A.S.E. from Western Carolina University, and an Ed.D. from East Carolina University. She received the Presidential Award for Excellence in Science and Mathematics Teaching in 1984. She has also received the Governor's Business Award for Excellence in Science and Mathematics Education, the Austin D. Bond Award for Distinguished Service to Science Education from East Carolina University, and the North Carolina Science Teachers Award for Distinguished Service to Science Education.

ED ORTLEB is a science consultant for the St. Louis, Missouri Public Schools. He holds an A.B. in education from Harris Teachers College and an M.A. and Advanced Graduate Certificate in science education from Washington University, St. Louis. Mr. Ortleb is a lifetime member of NSTA, having served as its president in 1978-79. He has received numerous awards for his service to science education.

Table of Contents

Student Edition

UNIT 1	Life	2
1	**EXPLORING LIFE**	**4**
1-1	Living Things	6
1-2	Where does life come from?	10
1-3	What is science?	14
1-4	Science and Society: Technology and the Dairy Industry	24
2	**THE STRUCTURE OF VIRUSES AND CELLS**	**30**
2-1	Viruses	32
2-2	Science and Society: AIDS Vaccine?	38
2-3	Cells: The Units of Life	40
2-4	Cell Organization	45
3	**CELL PROCESSES**	**62**
3-1	Chemistry of Living Things	64
3-2	Cell Transport	71
3-3	Energy in Cells	78
3-4	Science and Society: Energy from Biomass	84
4	**CELL REPRODUCTION**	**90**
4-1	Cell Growth and Division	92
4-2	Sexual Reproduction and Meiosis	99
4-3	DNA	104

Table of Contents

4-4   Science and Society:
Techniques in Cloning   112

UNIT 2   HEREDITY AND EVOLUTION   120

5   HEREDITY   122
5-1   What is genetics?   124
5-2   Genetics Since Mendel   132
5-3   Human Genetics   137
5-4   Science and Society:
The Human Genome Project   144

6   EVOLUTION   150
6-1   Mechanisms of Evolution   152
6-2   Evidence for Evolution   162
6-3   Primate Evolution   171
6-4   Science and Society:
Extinction and Evolution   176

UNIT 3   Diversity of Life   184

7   CLASSIFYING LIVING THINGS   186
7-1   What is classification?   188
7-2   Modern Classification   193
7-3   Science and Society:
Diminishing Diversity   198
7-4   Identifying Organisms   200

8   BACTERIA   208
8-1   Two Kingdoms of Bacteria   210
8-2   Bacteria in Your Life   216
8-3   Science and Society:
Fighting Tuberculosis   222

9   PROTISTS AND FUNGI   228
9-1   Kingdom Protista   230
9-2   Kingdom Fungi   241
9-3   Science and Society:
Monitoring Red Tides   248

UNIT 4   Plants   256

10   INTRODUCTION TO PLANTS   258
10-1   Characteristics of Plants   260
10-2   Seedless Plants   267

10-3   Science and Society:
Cleaner Coal   278

11   THE SEED PLANTS   284
11-1   Seed Plants   286
11-2   Parts of Complex Plants   292
11-3   Seed Plant Reproduction   296
11-4   Science and Society:
Medical Treasures   306

12   PLANT PROCESSES   312
12-1   Photosynthesis and
Respiration   314
12-2   Plant Responses   324
12-3   Science and Society:
Transgenic Crops   331

UNIT 5   Animals   340

13   INTRODUCTION TO ANIMALS   342
13-1   What is an animal?   344
13-2   Science and Society:
Building Artificial Coral Reefs   348
13-3   Sponges and Cnidarians   350
13-4   Flatworms and
Roundworms   357

**14   MOLLUSKS, WORMS, ARTHROPODS,
AND ECHINODERMS   368**
14-1   Mollusks   370
14-2   Segmented Worms   375
14-3   Arthropods   382
14-4   Science and Society:
Pesticides and Insects   390
14-5   Echinoderms   392

15   FISH, AMPHIBIANS, AND REPTILES   398
15-1   Fish   400
15-2   Amphibians   409
15-3   Science and Society:
Amphibians and
Ultraviolet Light   414
15-4   Reptiles   416

16   BIRDS AND MAMMALS   424
16-1   Birds   426
16-2   Mammals   435

Table of Contents

16-3 Science and Society:
California Sea Otters 446

17 ANIMAL BEHAVIOR 452
17-1 Types of Behavior 454
17-2 Behavioral Adaptations 462
17-3 Science and Society:
Zoos and Captive Breeding 470

UNIT 6 Ecology 478

18 LIFE AND THE ENVIRONMENT 480
18-1 The Living Environment and the Nonliving Environment 482
18-2 Interactions Among Living Organisms 488
18-3 Matter and Energy 496
18-4 Science and Society:
Bringing Back the Wolves 502

19 ECOSYSTEMS 508
19-1 How Ecosystems Change 510
19-2 Land Environments 516
19-3 Science and Society:
Protecting Antarctica 528
19-4 Water Environments 530

20 RESOURCES AND THE ENVIRONMENT 538
20-1 Natural Resources 540
20-2 Science and Society:
Recycling 548
20-3 Conservation and Wildlife Protection 550
20-4 Maintaining a Healthy Environment 557

UNIT 7 The Human Body 570

21 BONES, MUSCLES, AND SKIN 572
21-1 The Skeletal System 574
21-2 Science and Society:
Biomaterial—Speeding Bone Fracture Recovery 582
21-3 The Muscular System 584
21-4 Skin 591

22 NUTRIENTS AND DIGESTION 598
22-1 Nutrition 600
22-2 Science and Society:
Nutrients Combat Cancer 610
22-3 Your Digestive System 612

23 THE CIRCULATORY SYSTEM 624
23-1 Circulation 626
23-2 Science and Society:
"Growing" Heart Valves 634
23-3 Blood 636
23-4 Your Lymphatic System 644

24 RESPIRATION AND EXCRETION 650
24-1 Your Respiratory System 652
24-2 Science and Society:
Restricting Cigarette Advertising 664
24-3 Your Urinary System 666

25 THE NERVOUS AND ENDOCRINE SYSTEMS 676
25-1 Your Nervous System 678
25-2 Science and Society:
Care and Treatment of Alzheimer's Patients 686
25-3 The Senses 688
25-4 Your Endocrine System 695

26 REPRODUCTION AND GROWTH 702
26-1 Human Reproduction 704
26-2 Fertilization to Birth 710
26-3 Development After Birth 716
26-4 Science and Society:
A New View of the Old 722

27 IMMUNITY 728
27-1 Disease 730
27-2 Your Immune System 737
27-3 Science and Society:
Should a harmful virus be destroyed? 744
27-4 Noncommunicable Disease 746

Glencoe Life Science
The National Science Education Standards

The *National Science Education Standards*, published by the National Research Council and representing the contribution of thousands of educators and scientists, offer a comprehensive vision of a scientifically literate society. The standards not only describe what students should know but also offer guidelines for science teaching and assessment. Bill Aldridge, a Glencoe author, helped originate this standards initiative and served on its Chair's Advisory Committee. If you are using, or plan to use, the standards to guide changes in your science curriculum, you can be assured that *Glencoe Life Science* aligns with the *National Science Education Standards*.

Glencoe Life Science is an example of how Glencoe's commitment to effective science education is changing the materials used in science classrooms today. More than just a collection of facts in a textbook, *Glencoe Life Science* is a program that provides numerous opportunities for students, teachers, and school districts to meet the *National Science Education Standards*.

Content Standards
The accompanying table shows the close alignment between *Glencoe Life Science* and the content standards. The integration included in *Glencoe Life Science* allows students to discover concepts within each of the content standards, giving them opportunities to make connections between the science disciplines. Our hands-on activities and inquiry-based lessons reinforce the science processes emphasized in the standards.

Teaching Standards
Alignment with the *National Science Education Standards* requires much more than alignment with the outcomes in the content standards. The way in which concepts are presented is critical to effective learning. The teaching

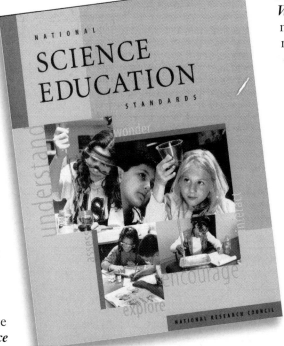

standards within the *National Science Education Standards* recommend an inquiry-based program facilitated and guided by teachers. *Glencoe Life Science* provides such opportunities through activities and discussions that allow students to discover by inquiry critical concepts and apply the knowledge they've constructed to their own lives. Throughout the program, students are building critical skills that will be available to them for lifelong learning. *The Teacher Wraparound Edition* helps you make the most of every instructional moment. It offers an abundance of effective strategies and suggestions for guiding students as they explore science.

Assessment Standards
The assessment standards are supported by many of the components that make up the *Glencoe Life Science* program. *The Teacher Wraparound Edition* and *Teacher Classroom Resources* provide multiple chances to assess students' understanding of important concepts as well as their ability to perform a wide range of skills. Ideas for portfolios, performance activities, written reports, and other assessment activities accompany every lesson. Rubrics and performance task assessment lists can be found in Glencoe's Professional Series booklet *Performance Assessment in the Science Classroom.*

Program Coordination
The scope of the content standards requires students to meet the outcomes over the course of their education. The correlation on the following pages demonstrates the close alignment of this course of *Glencoe Life Science* with the content standards.

Correlation of Glencoe Life Science *to the National Science Standards, Content Standards, Grades 5–8*

Content Standard	Page Numbers
(UCP) UNIFYING CONCEPTS AND PROCESSES	
1. Systems, order, and organization	6-9, 14-23, 32-33, 43-44, 45-55, 64-70, 71-76, 77-81, 92-98, 99-103, 104-109, 132-136, 138, 144-145, 188-192, 193-197, 200-203, 210-211, 230-239, 241-245, 260-261, 264-265, 267-275, 286-288, 292-295, 296-299, 314-323, 324-330, 344-347, 348-349, 350-355, 357-363, 370-373, 375-381, 382-388, 392-394, 400-405, 409-412, 416-420, 426-434, 435-445, 454-459, 462-464, 466-469, 483-487, 488-495, 530-533, 574-579, 584-588, 591-594, 600-608, 612-617, 626-632, 636-641, 644-645, 652-654, 656-661, 666-671, 678-684, 688-694, 695-697, 704-708, 710-715, 730-735, 737-740
2. Evidence, models, and explanation	10-13, 17, 41-42, 48, 97, 104-107, 108, 124-130, 132-136, 140, 157, 160, 167, 198-199, 414-415, 496-501
3. Change, constancy, and measurement	22-23, 67, 79-80, 92-97, 124-130, 153-155, 198-199, 245, 270, 348-349, 467-468, 510-513, 540-546, 584-588, 600-608, 652-654, 716-720
4. Evolution and equilibrium	7, 11, 72-73, 79-80, 152-161, 162-170, 171-175, 176-177, 198-199, 216, 230, 261, 265, 270, 289, 322, 348-349, 352, 356, 381, 402, 410, 417, 434, 445, 462, 510-513, 516-527
5. Form and function	38-39, 48, 52, 76-77, 99, 104-109, 211-212, 292-295, 301, 314, 315, 348-349, 351, 355, 358, 371, 375, 377, 382, 392-393, 401, 404, 419, 427-428, 430-433, 435-437, 440-444, 634-635, 657, 667-668, 679-680
(A) SCIENCE AS INQUIRY	
1. Abilities necessary to do scientific inquiry	11, 14-23, 39, 44, 50, 53, 56-57, 67, 68, 72, 80, 81, 82-83, 97, 98, 100, 101, 102, 108, 109, 112-113, 125, 127, 130, 133, 134-135, 138, 155, 156, 157, 158-159, 163, 165, 189, 190-191, 192, 201, 212, 213, 214, 217, 220-221, 223, 224, 233, 237, 244, 245, 246, 268, 271, 272-273, 277, 291, 297, 301, 302, 304-305, 318, 321, 323, 327, 328-329, 345, 346, 352, 356, 358, 360-361, 372, 378-379, 381, 385, 388, 393, 403, 404, 406-407, 410, 411, 420, 427, 428, 430, 436, 438-439, 441, 445, 455, 458, 460-461, 464, 466, 468, 469, 490-491, 494, 495, 499, 500, 514-515, 517, 518, 520, 522, 523, 525, 532, 541, 545, 546, 552-553, 555, 558, 559, 564, 577, 579, 580-581, 586, 587, 592, 593, 594, 602, 606, 607, 611, 615, 618-619, 628, 630, 632, 640, 641, 642-643, 660, 662-663, 667, 669, 670, 671, 682, 683, 690, 691, 692-693, 696, 708, 712, 713, 714, 717, 718, 719, 733, 734, 739, 742-743, 748, 749
2. Understandings about scientific inquiry	10-13, 14-23, 41-44, 53, 56-57, 77, 82-83, 110-111, 112-113, 125-130, 131, 134-135, 158-159, 167, 190-191, 204, 215, 220-221, 240, 246, 272-273, 280, 300, 304-305, 316, 328-329, 354, 360-361, 378-379, 389, 406-407, 413, 429, 438-439, 460-461, 472, 485, 490-491, 514-515, 519, 552-553, 561, 580-581, 589, 609, 618-619, 633, 642-643, 662-663, 685, 692-693, 709, 721, 736, 742-743
(B) PHYSICAL SCIENCE	
1. Properties and changes of properties in matter	41-42, 64-70, 78-81, 102, 105, 174, 262, 290, 317, 318-322, 351, 466, 499-501, 511, 541, 560, 588-590, 600-608, 613, 639, 657, 689, 705, 732, 747
2. Motions and forces	348-349, 372-373, 430, 585, 631, 657
3. Transfer of energy	49-50, 78-81, 84-85, 165, 318-322, 496-499, 516-517, 542-546, 603
(C) LIFE SCIENCE	
1. Structure and function in living systems	32-37, 43-44, 45-55, 71-76, 78-81, 99-103, 104-108, 142-143, 156-157, 171, 188-192, 193-196, 200-203, 210-214, 216-218, 288-289, 292-295, 296-303, 314-323, 344-347, 348-349, 351, 370-373, 375-381, 382-388, 392-394, 400-405, 409-412, 416-420, 426-434, 435-445, 574-579, 584-588, 591-594, 612-617, 626-632, 636-641, 644-645, 652-661, 666-671, 678-684, 688-694, 695-697, 704-708, 710-715
2. Reproduction and heredity	7, 34-35, 38-39, 92-98, 99-103, 104-109, 112-113, 125-130, 132-136, 137-143, 168-170, 212, 242-244, 263, 268-269, 276-277, 296-303, 352, 355, 358, 377, 384-385, 393, 403, 412, 437, 440-441, 454, 462-463, 704-708, 710-715
3. Regulation and behavior	7, 47, 71-76, 78-81, 93-95, 107-108, 214, 248-249, 314-315, 324-330, 350-354, 358, 371-373, 375-377, 383-388, 393-394, 404, 410-411, 420, 426, 454-459, 462-464, 466-469, 613-617, 626-631, 657, 666-669, 679-683, 686-687, 696-697, 707-708, 716-720, 730-735, 737-740
4. Populations and ecosystems	132, 176-177, 198-199, 214, 216, 241, 245, 248-249, 270, 348-349, 414-415, 446-447, 470-471, 482-487, 488-495, 496-499, 502-503, 510-513, 520-527, 530-533, 548, 550-556
5. Diversity and adaptations of organisms	45-55, 97-98, 142-143, 156-157, 168-170, 188-191, 195, 210-212, 214, 230-239, 241-245, 262-263, 270, 292-295, 296-303, 306, 346-347, 348-349, 371, 372-373, 377, 385, 401, 404, 409, 416-417, 426-428, 430-431, 435-437, 462-469, 510-513, 527

(D) EARTH AND SPACE SCIENCE	
1. Structure of the Earth system	13, 70, 163, 203, 214, 245, 270, 356, 483-484, 516-524, 526, 528-529, 530-533, 543-545, 562-564
2. Earth's history	11, 13, 163-166, 176, 275, 417
3. Earth in the solar system	516-517
(E) SCIENCE AND TECHNOLOGY	
1. Abilities of technological design	194, 279, 333, 419
2. Understandings about science and technology	12, 24-25, 41, 42, 54, 75, 84-85, 107, 112-113, 142-143, 144-145, 167, 173, 194, 218, 243, 248-249, 276, 278-279, 293, 319, 331-333, 362, 384, 419, 433, 467, 493, 523, 545, 582-583, 592, 606, 610-611, 629, 634-635, 670, 683, 712, 714
(F) SCIENCE IN PERSONAL AND SOCIAL PERSPECTIVES	
1. Personal health	24-25, 36, 58, 114, 222-223, 241, 306-307, 380, 582-583, 594, 600-608, 610-611, 617, 632, 638-641, 645, 659-661, 664-665, 670-671, 684, 686-687, 722-723, 730-735, 737-740
2. Populations, resources, and environments	156-157, 176-177, 198-199, 348-349, 446-447, 489, 502-503, 528-529, 550-556, 557-564
3. Natural hazards	58, 112-113, 156, 176-177, 217-218, 241, 373
4. Risks and benefits	24-25, 38-39, 58, 222-223, 306, 390-391, 640, 659-661, 664-665, 744-745
5. Science and technology in society	12, 24-25, 54, 75, 84-85, 107, 112-113, 142-143, 144-145, 222-223, 248-249, 276, 278-279, 293, 331-333, 348-349, 362, 390-391, 433, 467, 493, 523, 545, 549, 582-583, 592, 606, 610-611, 629, 634-635, 670, 683, 712, 714
(G) HISTORY AND NATURE OF SCIENCE	
1. Science as a human endeavor	10-13, 36-37, 40-44, 58, 104-105, 112-113, 114, 125-130, 152-161, 188-192, 219, 250, 306-307, 308, 334, 364, 390-391, 404, 408, 446-447, 448, 456-457, 458, 534, 620, 664, 698, 724, 730, 731, 732, 744-745
2. Nature of science	10-13, 36-37, 40-44, 58, 104-105, 125-130, 152-161, 188-192
3. History of science	10-13, 36-37, 40-44, 58, 104-105, 125-130, 152-155, 188-192, 219, 404, 664, 730, 731, 732, 741

Responding to Changes in
Science Education

The need for new directions in science education

By today's projections, seven out of every ten American jobs will be related to science, mathematics, or electronics by the year 2000. And according to the experts, if students haven't grasped the fundamentals—they probably won't go further in science and may not have a future in a global job market. Studies also reveal that students are avoiding taking "advanced" science classes.

The time for action is now!

In the past decade, educators, public policy makers, corporate America, and parents have recognized the need for reform in science education. These groups have united in a call to action to solve this national problem. As a result, three important projects have published reports to point the way for America:

- **Project on National Science Standards ...** by the National Research Council.
- **Benchmarks for Science Literacy ...** by the American Association for the Advancement of Science.
- **Scope, Sequence, and Coordination of Secondary School Science ...** by the National Science Teachers Association (NSTA).

Together, these reports spell out unified guiding principles for new directions in U.S. science education. *Glencoe Life Science* is based on these guiding principles:

- developing scientific, technological, and mathematical literacy in all students.
- educating students to use scientific principles and processes appropriately in making personal decisions.
- helping students to experience the richness and excitement of knowing about and understanding the natural world.
- teaching students how to engage intelligently in public discourse and debate about matters of scientific and technological concern.

GLENCOE LIFE SCIENCE answers the challenge!

At Glencoe Publishing, we believe that *Glencoe Life Science* will help you bring science reform to the front lines—the classrooms of America. But more importantly, we believe it will help students succeed in science so that they will want to continue learning about science through high school and into adulthood.

In response to the goals of science curriculum reform, *Glencoe Life Science* promotes:

- solid, accurate content,
- a thematic orientation,
- hands-on activities that promote a constructivist approach through more student-planned activities,
- frequent practice of science process skills,
- integration of science concepts across the curriculum,
- numerous opportunities for performance assessment, and
- decision making and problem solving.

GLENCOE LIFE SCIENCE is loaded with activities.

You'll choose from dozens of activities in the *Student Edition.* These activities are easy to set up and manage and will allow you to teach using a hands-on, inquiry-based approach to learning.

MiniLABs and **Activities** involve students in learning and applying scientific methods to practice thinking skills and construct science concepts. They provide an engaging, diverse, active program. **MiniLABs** require a minimum of equipment, and students may take responsibility for organization and execution.

Activities develop and reinforce or restructure concepts as well as develop the ability to use process skills. Activity formats are structured to guide students to make their own discoveries. Students collect real evidence and are encouraged through open-ended questions to reflect and reformulate their ideas based on this evidence.

In each chapter there is one open-ended Activity called **Design Your Own Experiment** that gives students a broad topic to explore and guides them with leading questions to the point where they can determine the direction of the investigation themselves. Students are asked to brainstorm hypotheses, make a decision to investigate one that can be tested, plan procedures, and in the end, think about why their hypotheses were supported or not.

GLENCOE LIFE SCIENCE integrates science and math.

Mathematics is a tool that all students, regardless of their career goals will use throughout their lives. *Glencoe Life Science* provides opportunities to hone mathematics skills while learning about the natural world. **Using Math, Practice Problems, MiniLABs,** and **Activities** offer numerous options for practicing math, including making and using tables and graphs, measuring in SI, and calculating. **Across the Curriculum** strategies in the *Teacher Wraparound Edition* provide additional connections between science and mathematics.

GLENCOE LIFE SCIENCE integrates the sciences for understanding.

No subject exists in isolation. At appropriate points in *Glencoe Life Science*, attention is called to other sciences in the *Student Edition*, in the *Teacher Wraparound Edition*, and in the *Teacher Classroom Resources*.

Connect to... margin features relate basic questions in physics, chemistry, Earth science, and life science to one another. Suggested answers to these questions are contained in the *Teacher Wraparound Edition*. Opportunities for **Integration** of science topics are highlighted throughout the *Student Edition*. Further explanation of how to weave the **Integration** topic into your lesson is provided in the *Teacher Wraparound Edition*. Look for this logo for ideas on how to make the sciences connect for your students. *Science Integration Activities* are full-page activities in the *Teacher Classroom Resources* that relate physics, chemistry, Earth science, and life science to each other. *Science Integration Transparencies* also relate the sciences to one another using full-color illustrations that can be used with an overhead projector. These opportunities for integrating the sciences will help your students discover the science behind things they see every day.

It's time for new directions in science education.

The need for new directions in science education has been established by the experts. America's students must prepare themselves for the high-tech jobs of the future. We at Glencoe believe that *Glencoe Life Science* answers this challenge. We believe *Glencoe Life Science* will assist you better in preparing your students for a lifetime of science learning.

Themes & Scope & Sequence

Our society is becoming more aware of the interrelationship of the disciplines of science. It is also necessary to recognize the precarious nature of the stability of some systems and the ease with which this stability can be disturbed so that the system changes. For most people, then, the ideas that unify the sciences and make connections between them are the most important.

Themes provide a construct for unifying the sciences. Woven throughout *Glencoe Life Science,* themes integrate facts and concepts. Like story lines, themes provide "big ideas" that link the science disciplines. Themes are an important part of any teaching strategy because they help students see the importance of truly understanding concepts rather than simply memorizing isolated facts. While there are many possible themes around which to unify science, we have chosen four: Energy, Systems and Interactions, Scale and Structure, and Stability and Change.

Energy

Energy is a central concept of the physical sciences that pervades the biological and Earth sciences. In physical terms, energy is the ability of an object to change itself or its surroundings—the capacity to do work. In chemical terms, it forms the basis of reactions between compounds. In biological terms, it gives living systems the ability to maintain themselves, to grow, and to reproduce. Energy sources are crucial in the interactions among science, technology, and society.

Systems and Interactions

A system can be incredibly tiny, such as an atom's nucleus and electrons; extremely complex, such as an ecosystem; or unbelievably large, as the stars in a galaxy. By defining the boundaries of the system, one can study the interactions among its parts. The interactions may be a force of attraction between the positively charged nucleus and negatively charged electron. In an ecosystem, on the other hand, the interactions may be between the predator and its prey, or among the plants and animals.

Scale and Structure

Used as a theme, "structure" emphasizes the relationship among different structures. "Scale" defines the focus of the relationship. As the focus is shifted from a system to its components, the properties of the structure may remain constant. In other systems, an ecosystem for example, which includes a change in scale from interactions between prey and predator to the interactions among systems inside an animal, the structure changes drastically. In *Glencoe Life Science,* the authors have tried to stress how we know what we know and why we believe it to be so. Thus, explanations remain on the macroscopic level until students have the background needed to understand how the microscopic structure operates.

Stability and Change

A system that is stable is constant. Often the stability is the result of a system being in equilibrium. If a system is not stable, it undergoes change. Changes in an unstable system may be characterized as trends (position of falling objects), cycles (the motion of planets around the sun), or irregular changes (radioactive decay).

Theme Development

These four major themes are developed within the student material and discussed throughout the *Teacher Wraparound Edition.* Each chapter of *Glencoe Life Science* incorporates at least one of these themes. These themes are interwoven throughout each level and are developed as appropriate to the topic presented.

The *Teacher Wraparound Edition* includes a **Theme Connection** section for each unit opener. This section provides a framework for integrating concepts discussed in the unit opener. Each chapter opener includes a **Theme Connection** section to explain the chapter's themes and to point out the major chapter concepts supporting those themes. Throughout the chapters, **Theme Connections** show specifically how a topic in the *Student Edition* relates to the themes.

Theme connections stressed in each chapter of *Glencoe Life Science* are indicated in the chart on page 13T.

GLENCOE LIFE SCIENCE Chapters

Themes

	Scale and Structure	Energy	Stability and Change	Systems and Interactions
1 Exploring Life				✔
2 The Structure of Viruses and Cells	✔			
3 Cell Processes			✔	
4 Cell Reproduction			✔	
5 Heredity			✔	
6 Evolution			✔	
7 Classifying Living Things			✔	
8 Bacteria				✔
9 Protists and Fungi				✔
10 Introduction to Plants			✔	
11 The Seed Plants				✔
12 Plant Processes				✔
13 Introduction to Animals			✔	
14 Mollusks, Worms, Arthropods, and Echinoderms	✔			
15 Fish, Amphibians, and Reptiles			✔	
16 Birds and Mammals			✔	
17 Animal Behavior			✔	
18 Life and the Environment				✔
19 Ecosystems			✔	
20 Resources and the Environment		✔		
21 Bones, Muscles, and Skin		✔		
22 Nutrients and Digestion		✔		
23 The Circulatory System			✔	
24 Respiration and Excretion		✔		
25 The Nervous and Endocrine Systems			✔	
26 Reproduction and Growth		✔		
27 Immunity				✔

Resources for Different Needs

In addition to the wide array of instructional options provided in the student and teacher editions, *Glencoe Life Science* also offers an extensive list of support materials and program resources. Some of these materials offer alternative ways of enriching or extending your science program, others provide tools for reinforcing and evaluating student learning, while still others will help you directly in delivering instruction. You won't have time to use them all, but the ones you use will help you save the time you have.

Hands-On Activities

If you want more hands-on options, the *Lab Manual* offers you one or more additional labs per chapter. Each lab is complete with setup diagrams, data tables, and space for student responses.

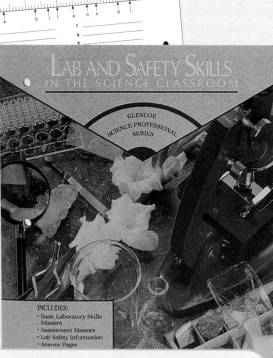

Reinforce basic lab and safety skills, including graphing and measurement, with activities from the *Lab and Safety Skills* book.

NAME DATE CLASS

Chapter 15
MINILAB Text Page 404

How does a fish adjust to different depths?
Young children often wear air-filled water wings while they are learning to swim. Boats carry life jackets for all passengers, in case the boat should capsize. In this activity, you will find out what these safety devices have in common with a fish's swim bladder. Find out what happens when a fish fills or empties its swim bladder.

Materials
• balloons
• bowl
• water

Procedure
1. Fill a balloon with air.
2. Place it in a bowl of water.
3. Fill another balloon partially with water, then blow air into it until it is the same size as the balloon filled only with air.
4. Place the second balloon in the bowl of water.

Analysis
1. Compare the balloon filled with air to the balloon with air and water. Where is each balloon in the water?

2. What structure do these balloons model?

3. Compare where in the water (on the surface, below the surface) a fish would be with swim bladders similar to the two balloons.

92 Copyright © Glencoe/McGraw-Hill, a division of The McGraw-Hill Companies, Inc.

NAME DATE CLASS

Chapter 15
ACTIVITY 15-1 Text Page
• Water Temperature and
the Respiration Rate of F

Design Your Own Experiment

Lab Preview
1. What safety symbols are associated with this activity?

2. How do fish take oxygen from water?

Last summer it was very hot, with few storms. One day after many sunny windless days, you noticed that a lot of dead fish were floating on the surface of your neighbor's pond. What caused these fish t

PREPARATION

Problem
How does water temperature affect the respiration rate of fish?

Form a Hypothesis
Fish obtain oxygen from the water. State a hypothesis about how water temperature affects the respiration rate of fish.

Objectives
• Design and carry out an experiment to measure the effect of water temperature on the rate of respiration in fish.
• Observe the breathing rate of fish.

Possible Materials
• goldfish
• aquarium water
• 1 small fishnet
• 600-mL beakers
• container of ice water
• stirring rod
• thermometer
• aquarium

Safety Precautions
Protect your clothing. Use t
transfer fish into beakers.

88

> Each of the **Activities** and **MiniLABs** in the student text is also available in reproducible master form in the *Activity Worksheets* book.

ACTIVITY WORKSHEETS

Glencoe
LIFE SCIENCE

SCIENCE INTEGRATION ACTIVITIES

Glencoe
LIFE SCIENCE

Includes:
• Twenty-seven activities that integrate the physical sciences with other sciences
• Answers to questions

NAME DATE CLASS

Chapter 11
INTEGRATION • Using Growth Rings As
Indicators of Climate

Trees are plants with woody stems. These stems are called trunks. As shown in Figure 1, three types of vascular tissue—xylem, cambium, and phloem—make up the stem of a tree. Of these three tissues, xylem—the tissue that transports water and minerals up from the roots—occupies the most space.

If a tree is cut across its stem, a series of rings formed from alternating bands of lighter colored and darker colored wood can be seen. These rings are called growth rings, because each ring represents one year in the growth of the tree. Each growth ring is made up of both a dark and a light area.

Growth rings result from differences in the rate at which xylem is produced at different times of the year. For example, spring wood forms from xylem cells that are large and wide. Such xylem cells are produced in the greatest amounts during the spring, when growing conditions such as plentiful water are favorable to growth. When conditions become drier, summer wood is formed from the xylem cells. Summer wood is formed from xylem cells that are narrower and have thicker walls than the xylem that forms spring wood. Summer wood carries less water than spring wood and forms when conditions are drier.

Trees grow the most when environmental conditions such as rainfall and temperature are the most favorable. Thus, the size of the growth rings give a clue about the climate conditions of an area over time. During years of abundant rainfall, when plenty of water is available, growth rings are wide. In contrast, the growth rings of a tree are narrower during dry periods, such as during a drought.

Materials
• metric ruler

Procedure
1. In your Science Journal, copy the table from the Data and Observations on page 24.

Copyright © Glencoe

> ***Science Integration Activities*** are laboratory activities that relate Earth and physical science to specific life science chapters.

Reinforcement and Enrichment

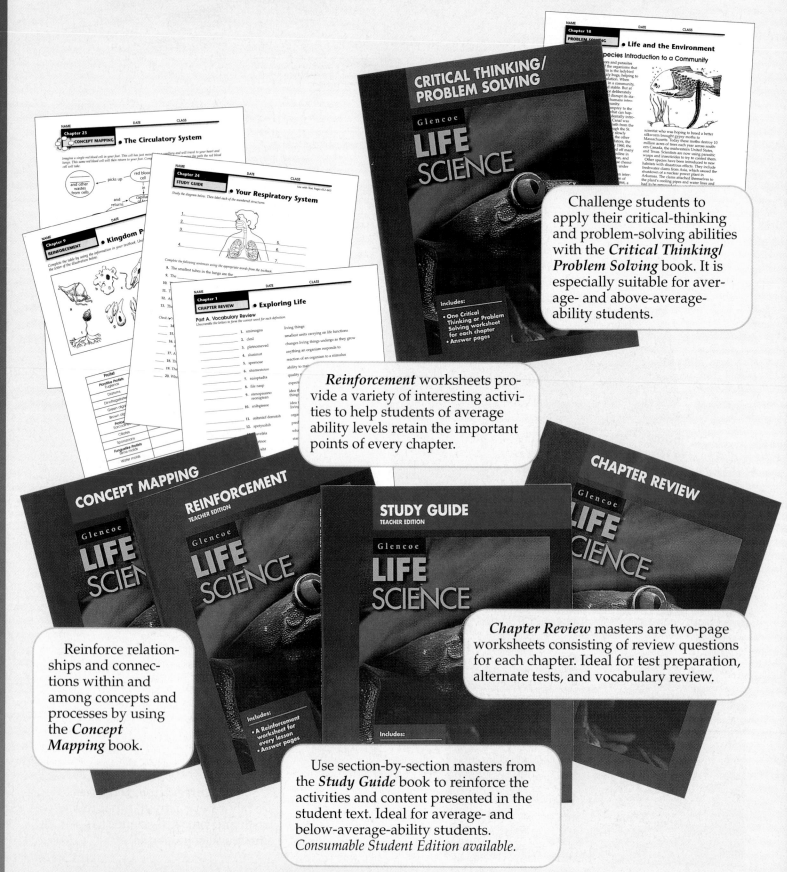

Challenge students to apply their critical-thinking and problem-solving abilities with the *Critical Thinking/Problem Solving* book. It is especially suitable for average- and above-average-ability students.

Reinforcement worksheets provide a variety of interesting activities to help students of average ability levels retain the important points of every chapter.

Reinforce relationships and connections within and among concepts and processes by using the *Concept Mapping* book.

Chapter Review masters are two-page worksheets consisting of review questions for each chapter. Ideal for test preparation, alternate tests, and vocabulary review.

Use section-by-section masters from the *Study Guide* book to reinforce the activities and content presented in the student text. Ideal for average- and below-average-ability students.
Consumable Student Edition available.

Enrichment worksheets challenge your students of above average ability to design, interpret, and research scientific topics based on the text in each lesson.

Science and Society Integration worksheets encourage further involvement with science as it applies to society.

The *Cross-Curricular Integration* book includes interdisciplinary worksheets that emphasize learning by doing and provide valuable insight into the connection life science has with other disciplines.

Technology Integration masters explore recent developments in science and technology or explain how familiar machines, tools, or systems work.

Explore past and present contributions to science of individuals and societies of various cultural backgrounds through the readings and activities in the *Multicultural Connections* book.

Assessment

Assess student learning and performance through a variety of questioning strategies and formats in the *Assessment* book. Test items cover activity procedures and analysis as well as science concepts within the four pages of reproducible masters available for each chapter.

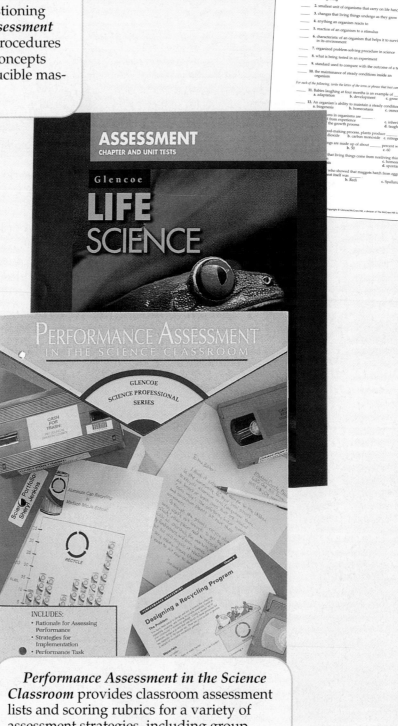

Performance Assessment in the Science Classroom provides classroom assessment lists and scoring rubrics for a variety of assessment strategies, including group assessment, oral presentations, modeling, and writing in science.

The *Performance Assessment* book contains skill assessments and summative assessments that tie together major concepts of each chapter.

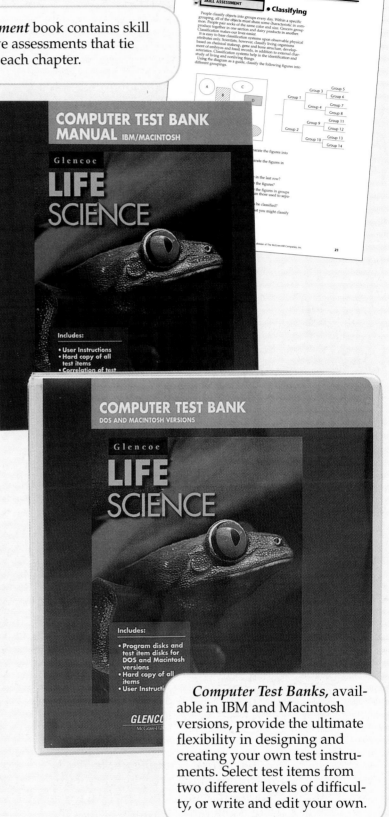

COMPUTER TEST BANK
MANUAL IBM/MACINTOSH

Glencoe
LIFE
SCIENCE

Includes:
• User Instructions
• Hard copy of all test items
• Correlation of test

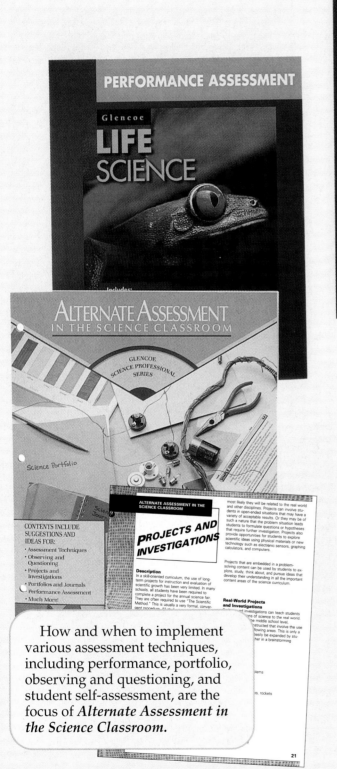

PERFORMANCE ASSESSMENT

Glencoe
LIFE
SCIENCE

Includes:

ALTERNATE ASSESSMENT
IN THE SCIENCE CLASSROOM

GLENCOE
SCIENCE PROFESSIONAL
SERIES

CONTENTS INCLUDE
SUGGESTIONS AND
IDEAS FOR:
• Assessment Techniques
• Observing and Questioning
• Projects and Investigations
• Portfolios and Journals
• Performance Assessment
• Much More!

COMPUTER TEST BANK
DOS AND MACINTOSH VERSIONS

Glencoe
LIFE
SCIENCE

Includes:
• Program disks and test item disks for DOS and Macintosh versions
• Hard copy of all items
• User Instructi

GLENCO
McGraw-Hill

How and when to implement various assessment techniques, including performance, portfolio, observing and questioning, and student self-assessment, are the focus of *Alternate Assessment in the Science Classroom.*

Computer Test Banks, available in IBM and Macintosh versions, provide the ultimate flexibility in designing and creating your own test instruments. Select test items from two different levels of difficulty, or write and edit your own.

Teaching Resources

Chapter-by-chapter *Lesson Plans* will help you organize your lessons more efficiently.

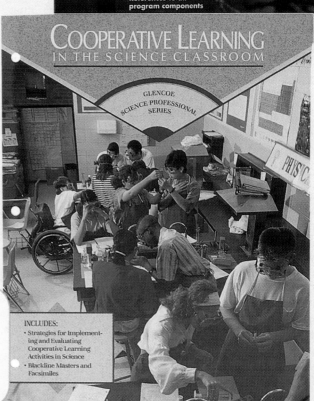

Cooperative Learning in the Science Classroom contains background information, strategies, and practical tips for using cooperative learning techniques whenever you do activities from *Glencoe Life Science.*

Enhance your presentation of science concepts with the color transparency packages. The *Teaching Transparency Package* includes two full-color transparencies per chapter and a **Study Guide** booklet with reproducible student worksheets and blackline reproductions for each of the program's full-color transparencies. The *Section Focus Transparency Package* includes one full-color transparency for each section of the student text and a **Study Guide** booklet with reproducible student worksheets. The *Science Integration Transparency Package* includes one full-color transparency per chapter and a **Study Guide** booklet with student worksheets.

Help your Spanish-speaking students get more out of your science lessons by reproducing pages from the *Spanish Resources* book. In addition to a complete English-Spanish glossary, the book contains translations of all objectives, key terms, activities, and main ideas for each chapter of the student text.

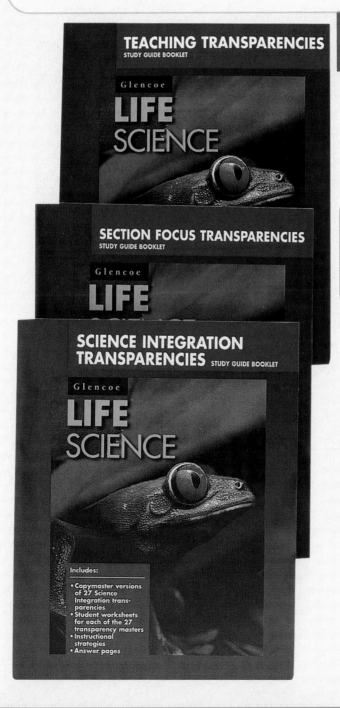

TEACHING TRANSPARENCIES
STUDY GUIDE BOOKLET
Glencoe
LIFE SCIENCE

SECTION FOCUS TRANSPARENCIES
STUDY GUIDE BOOKLET
Glencoe
LIFE

SCIENCE INTEGRATION TRANSPARENCIES STUDY GUIDE BOOKLET
Glencoe
LIFE SCIENCE

Includes:
• Copymaster versions of 27 Science Integration transparencies
• Student worksheets for each of the 27 transparency masters
• Instructional strategies
• Answer pages

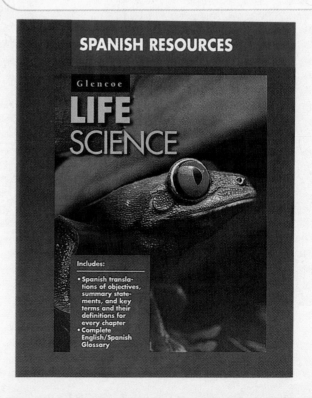

SPANISH RESOURCES
Glencoe
LIFE SCIENCE

Includes:
• Spanish translations of objectives, summary statements, and key terms and their definitions for every chapter
• Complete English/Spanish Glossary

The Technology Connection

Glencoe Life Science offers a wide selection of video and audio products to stimulate and challenge your students.

Interactive videos that provide a fun way for your students to review chapter concepts make up the *MindJogger Videoquiz* series.

The *Glencoe Interactive Videodisc Program* includes interactive lessons that illustrate and reinforce major life science concepts.

The *English/Spanish Audiocassettes* summarize the content of the Student Edition in both English and Spanish.

The *Secret of Life* videotapes produced by WGBH in Boston can demonstrate the myriad ways in which our expanding knowledge of DNA and heredity have affected all areas of biology and created many ethical dilemmas.

Lab Partner Software provides spreadsheet and graphing tools to record and display data from lab activities.

The *Infinite Voyage* (videotapes and videodiscs) is 20 programs from the award-winning PBS series, covering physical sciences, biology, ecology and environmental education, archaeology, and ancient history. An *Instructor's Guide* accompanies each videotape.

Program Resources

Glencoe is proud to offer quality educational technology from the *National Geographic Society.* These interactive learning tools remove the walls from your classroom and transport your students to the far corners of the world. Award-winning titles from National Geographic give students exciting, easy-to-use resources. With the click of a mouse, students can explore the world—and all that's in it—through video, photographs, sound, and text. Also available from Glencoe is the *Science and Technology Videodisc Series (STVS),* based on the Mr. Wizard television program. Both National Geographic Society and STVS products have been correlated to each chapter of *Glencoe Life Science.*

NATIONAL GEOGRAPHIC SOCIETY

- STV (Single-Topic Videodiscs)
    - Atmosphere
    - Water
    - Human Body Volumes 1, 2, & 3
    - Plants
    - The Cell
    - Rain Forest
    - Solar System
    - Restless Earth
    - Biodiversity

National Geographic Society
- GTV (Group-Topic Videodiscs)
    Planetary Manager

The *Science and Technology Videodisc Series* is a seven-disc series that contains more than 280 full-motion video reports. Reports cover a broad spectrum of topics relating to current research in various science fields, innovations in technology, and science and society issues. In addition to reinforcing science concepts, the videoreports are ideal for illustrating science methods, laboratory techniques, and careers in science.

National Geographic Society
Newton's Apple (Videodisc)
  Physical Sciences
  Life Sciences

National Geographic Society
CD-ROM
  NGS Picture Show

Constructivism

Strategies suggested in *Glencoe Life Science* support a constructivist approach to science education. The role of the teacher is to provide an atmosphere in which students design and direct activities. To develop the idea that science investigation is not made up of closed-end questions, the teacher should ask guiding questions and be prepared to help his or her students draw meaningful conclusions when their results do not match predictions. Through the numerous activities, cooperative learning opportunities, and a variety of critical thinking exercises in *Glencoe Life Science,* you can feel comfortable taking a constructivist approach to science in your classroom.

Activities

A constructivist approach to science is rooted in an activities-based plan. Students must be provided with sensorimotor experiences as a base for developing abstract ideas. *Glencoe Life Science* utilizes a variety of learning-by-doing opportunities. **MiniLABs** allow students to consider questions about the concepts to come, make observations, and share prior knowledge. **MiniLABs** require a minimum of equipment, and students may take responsibility for organization and execution.

Design Your Own Activities develop and reinforce or restructure concepts as well as develop the ability to use process skills. **Design Your Own Activity** formats guide students to make their own discoveries. Students collect real evidence and are encouraged through open-ended questions to reflect and reformulate their ideas based on this evidence.

Cooperative Learning

Cooperative learning adds the element of social interaction to science learning. Group learning allows students to verbalize ideas, and encourages the reflection that leads to active construction of concepts. It allows students to recognize the inconsistencies in their own perspectives and the strengths of others'. By presenting the idea that there is no one, "ready-made" answer, all students may gain the courage to try to find a viable solution. **Cooperative Learning** strategies appear in the *Teacher Wraparound Edition* margins whenever appropriate.

And More...

Flex Your Brain—a self-directed, critical-thinking matrix is introduced in Chapter 1, "The Nature of Science." This activity, referenced wherever appropriate in the **Teacher Wraparound Edition** margins, assists students in identifying what they already know about a subject, then in developing independent strategies to investigate further.

Students are encouraged to discover the pleasure of solving a problem through a variety of features. The **Science and Society** features in each chapter invite students to confront real-life problems. **Explore the Issue** or **Explore the Technology** questions encourage students to reflect on issues related to technology and society. The **Skill Handbook** gives specific examples to guide students through the steps of developing thinking and process skills. **Developing Skills** and **Thinking Critically** sections of the **Chapter Review** allow the teacher to assess and reward successful thinking skills.

Developing and Applying
Thinking Skills

Science is not just a collection of facts for students to memorize. Rather, it is a process of applying those observations and intuitions to situations and problems, formulating hypotheses, and drawing conclusions. This interaction of the thinking process with the content of science is the core of science and should be the focus of science study. Students, much like scientists, will plan and conduct research or experiments based on observations, evaluate their findings, and draw conclusions to explain their results. This process then begins again, using the new information for further investigation.

Basic Process Skills

Observing

The most basic process is observing. Through observation—seeing, hearing, touching, smelling, tasting—the student begins to acquire information about an object or event. Observation allows a student to gather information regarding size, shape, texture, or quantity of an object or event.

Classifying

One of the simplest ways to organize information gathered through observation is by classifying. Classifying involves the sorting of objects according to similarities or differences.

Using Numbers

Numbers are used to quantify information, including variables and measurements. Quantified information is useful for making comparisons and classifying data or objects.

Communicating

Communicating information is an important part of science. Once all the information is gathered, it is necessary to organize the observations so that the findings can be considered and shared by others. Information can be presented in tables, charts, a variety of graphs, or models.

Developing Thinking Skills

Measuring

Measuring is a way of quantifying information in order to classify or order it by magnitude, such as area, length, volume, or mass. Measuring can involve the use of instruments and the necessary skills to manipulate them.

Inferring

Inferences are logical conclusions based on observations and are made after careful evaluation of all the available facts or data. Inferences are a means of explaining or interpreting observations and are based on making judgments. Therefore, inferences can be invalid.

Predicting

Predicting involves suggesting future events based on observations and inferences about current events. Reliable predictions are based on making accurate observations and measurements and interpreting them accurately.

Using Space/Time Relationships

This process skill involves describing the spatial relationships of objects and how those relationships change with time. For example, a student may be required to describe the motion, direction, symmetry, or shape of an object or objects.

Sequencing

Sequencing involves arranging objects or events in a particular order and may imply a hierarchy or a chronology. Developing a sequence involves the skill of identifying relationships among objects or events. A sequence may be in the form of a numbered list or a series of objects or ideas with directional arrows.

Comparing and Contrasting

Comparing is a way of identifying similarities among objects or events, while contrasting identifies their differences. Comparing and contrasting provides a way of describing and evaluating what is known about something.

Recognizing Cause and Effect

Recognizing cause and effect involves observing actions or events and making logical inferences about why they occur. Recognizing cause and effect can lead to further investigation to isolate a specific cause of a particular event.

Interpreting Scientific Illustrations

Illustrations provide examples that clarify difficult concepts or give additional information about the topic being studied. Interpreting illustrations involves processing visual information into a conceptual construct.

Integrated Process Skills

Interpreting Data

Interpreting data involves synthesizing information (collected data) to make generalizations about the problem under study and apply those generalizations to new problems. Interpreting data may therefore involve many of the other process skills, such as predicting, inferring, classifying, and using numbers.

Defining Operationally

Forming operational definitions involves stating the meaning of something in terms of its function. Operational definitions are formed through observation.

Experimenting

Experimenting involves gathering data to test a hypothesis. For data to be reliable, the experiment must be conducted under controlled conditions with a limited number of variables.

Controlling Variables

Controlling variables requires understanding and regulating all the possible variables that might affect the outcome of an experiment. Failure to control variables can lead to unreliable results.

Formulating Hypotheses

Formulating a hypothesis involves making observations followed by some kind of inference as to what those observations mean. The hypothesis is then tested via experimentation to determine its validity.

Formulating Models

A model is a way to concretely represent abstract ideas or relationships. Models can also be used to predict the outcome of future events or relationships. Models may be expressed physically, verbally, or mentally.

Concept Mapping

Concept Maps are visual representations or graphic organizers of relationships among concepts. Concept mapping may involve sequencing temporal events in which a final outcome is observed (events chain), or in which the last event relates back to the beginning of the sequence (cycle map). Hierarchies can be represented by network tree concept maps. Central and supporting ideas that are nonhierarchical can be represented by spider concept maps.

Making and Using Tables

Making tables involves understanding cause and effect and relationships among ideas and representing those relationships in a tabular format.

Making and Using Graphs

Making graphs involves interpreting data and representing that data in a visual format.

Interaction of content and process

Glencoe Life Science encourages the interaction between science content and thinking processes. We've known for a long time that hands-on activities are a way of providing a bridge between science content and student comprehension. *Glencoe Life Science* encourages the interaction between content and thinking processes by offering literally hundreds of hands-on activities that are easy to set up and do. In the student text, the **MiniLABs** require students to make observations and collect and record a variety of data. Full-page **Activities** connect hands-on experiences to the content information.

At the end of each chapter, students use thinking processes as they complete **Developing Skills**, **Thinking Critically**, and **Performance Assessment** questions.

Skill Handbook

The **Skill Handbook** provides the student with another opportunity to practice the thinking processes relevant to the material they are studying. The **Skill Handbook** also provides examples of the processes that students may refer to as they do the **Skillbuilders**, **Activities**, and **MiniLABs**.

Thinking Skills in the *Teacher Wraparound Edition*

These processes are also featured throughout the *Teacher Wraparound Edition.* In the margins are suggestions for students to write in a journal. Keeping a journal encourages students to communicate their ideas, a key process in science.

Flex Your Brain is a self-directed activity designed to assist students in developing thinking processes as they investigate content areas. Suggestions for using this decision-making matrix are in the margins of the *Teacher Wraparound Edition* whenever appropriate for the topic. Further discussion of **Flex Your Brain** can be found in the *Activity Worksheets* booklet of the *Teacher Classroom Resources.*

Thinking Skills Map

On page 30T, you will find a Thinking Skills Map that indicates how frequently thinking skills are encouraged and developed in *Glencoe Life Science.*

Developing Thinking Skills

Thinking Skills

Chapters

Thinking Skills	1	2	3	4	5	6	7	8	9	10	11	12	13	14	15	16	17	18	19	20	21	22	23	24	25	26	27
ORGANIZING INFORMATION																											
Communicating	✓	✓	✓	✓	✓	✓	✓	✓	✓		✓	✓	✓	✓	✓		✓	✓	✓		✓	✓	✓	✓	✓		✓
Classifying		✓				✓	✓	✓	✓		✓		✓		✓	✓	✓	✓	✓	✓		✓	✓				
Sequencing				✓		✓							✓	✓	✓		✓		✓		✓			✓	✓	✓	
Concept Mapping	✓	✓	✓	✓	✓	✓	✓	✓	✓		✓	✓	✓	✓	✓	✓	✓	✓	✓	✓	✓	✓	✓	✓	✓	✓	✓
Making and Using Tables	✓		✓	✓	✓	✓	✓		✓	✓	✓	✓				✓	✓	✓	✓	✓		✓	✓	✓			✓
Making and Using Graphs	✓					✓	✓		✓	✓						✓		✓		✓		✓			✓	✓	✓
THINKING CRITICALLY																											
Observing and Inferring	✓	✓	✓	✓	✓	✓	✓	✓		✓	✓	✓	✓	✓	✓	✓	✓	✓	✓	✓	✓	✓	✓	✓	✓		✓
Comparing and Contrasting	✓	✓	✓	✓			✓		✓	✓	✓	✓	✓	✓	✓	✓		✓	✓	✓	✓	✓	✓	✓	✓	✓	✓
Recognizing Cause and Effect	✓		✓				✓	✓		✓	✓			✓	✓	✓	✓		✓		✓		✓				✓
Forming Operational Definitions		✓	✓	✓		✓	✓	✓			✓	✓		✓	✓	✓	✓	✓									
PRACTICING SCIENTIFIC PROCESSES																											
Forming a Hypothesis	✓		✓	✓	✓	✓		✓	✓	✓	✓	✓	✓	✓	✓		✓	✓	✓	✓	✓	✓	✓	✓	✓		✓
Designing an Experiment to Test a Hypothesis	✓	✓	✓		✓	✓	✓	✓	✓	✓	✓			✓	✓	✓		✓			✓	✓	✓	✓	✓		✓
Separating and Controlling Variables	✓		✓		✓	✓		✓	✓	✓	✓			✓	✓	✓	✓	✓			✓	✓		✓	✓		✓
Interpreting Data	✓	✓	✓	✓	✓	✓	✓	✓	✓	✓	✓	✓	✓	✓	✓	✓	✓	✓	✓	✓	✓	✓	✓	✓	✓	✓	✓
REPRESENTING AND APPLYING DATA																											
Interpreting Scientific Illustrations		✓			✓		✓				✓	✓		✓								✓		✓			
Making Models	✓			✓	✓		✓	✓											✓	✓		✓					✓
Measuring in SI	✓			✓		✓	✓		✓		✓	✓			✓	✓		✓		✓	✓	✓			✓		
Predicting	✓	✓	✓		✓				✓				✓					✓			✓			✓	✓		✓
Using Numbers					✓	✓	✓	✓	✓			✓	✓		✓			✓			✓			✓	✓	✓	✓

Flex Your Brain

A key element in the coverage of problem-solving and critical-thinking skills in *Glencoe Life Science* is a critical-thinking matrix called **Flex Your Brain.**

Flex Your Brain provides students with an opportunity to explore a topic in an organized, self-checking way, and then identify how they arrived at their responses during each step of their investigation. The activity incorporates many of the skills of critical thinking. It helps students to consider their own thinking and learn about thinking from their peers.

Where is Flex Your Brain found?

Chapter 1, "Exploring Life," is an introduction to the topics of critical thinking and problem solving. **Flex Your Brain** accompanies the text section in the introductory chapter. A worksheet for **Flex Your Brain** appears on page 5 of the *Activity Worksheets* book of the *Teacher Resources.* This version provides spaces for students to write in their responses.

In the *Teacher Wraparound Edition,* suggested topics are given in each chapter for the use of *Flex Your Brain.* You can photocopy the worksheet master from the *Teacher Resources.*

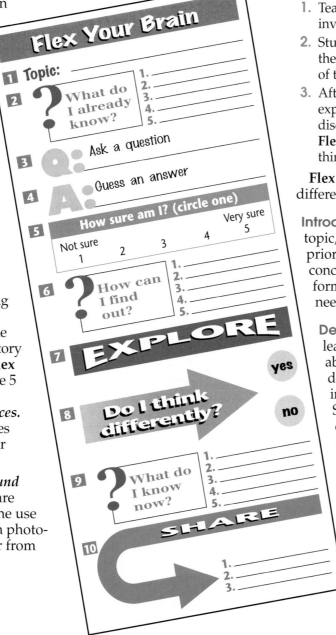

Using Flex Your Brain

Flex Your Brain can be used as a whole-class activity or in cooperative groups, but is primarily designed to be used by individual students within the class. There are three basic steps.

1. Teachers assign a class topic to be investigated using **Flex Your Brain.**
2. Students use **Flex Your Brain** to guide them in their individual explorations of the topic.
3. After students have completed their explorations, teachers guide them in a discussion of their experiences with **Flex Your Brain,** bridging content and thinking processes.

Flex Your Brain can be used at many different points in the lesson plan.

Introduction: Ideal for introducing a topic, **Flex Your Brain** elicits students' prior knowledge and identifies misconceptions, enabling the teacher to formulate plans specific to student needs.

Development: Flex Your Brain leads students to find out more about a topic on their own, and develops their research skills while increasing their knowledge. Students actually pose their own questions to explore, making their investigations relevant to their personal interests and concerns.

Review and Extension: Flex Your Brain allows teachers to check student understanding while allowing students to explore aspects of the topic that go beyond the material presented in class.

Concept Maps

In science, concept maps make abstract information concrete and useful, improve retention of information, and show students that thought has shape.

Concept maps are visual representations or graphic organizers of relationships among particular concepts. Concept maps can be generated by individual students, small groups, or an entire class. *Glencoe Life Science* develops and reinforces four types of concept maps—the **network tree, events chain, cycle concept map,** and **spider concept map**—that are most applicable to studying science. Examples of the four types and their applications are shown on this page and page 33T. Students can learn how to construct each of these types of concept maps by referring to the **Skill Handbook.**

Building Concept Mapping Skills

The **Developing Skills** section of the **Chapter Review** provides opportunities for practicing concept mapping. A variety of concept mapping approaches is used in almost every chapter. Students may be directed to make a specific type of concept map and be provided the terms to use. At other times, students may be given only general guidelines. For example, concept terms to be used may be provided and students required to select the appropriate model to apply, or vice versa. Finally, students may be asked to provide both the terms and type of concept map to explain relationships among concepts. When students are given this flexibility, it is important for you to recognize that, while sample answers are provided, student responses may vary. Look for the conceptual strength of student responses, not absolute accuracy. You'll notice that most network tree maps provide connecting words that explain the relationships between concepts. We recommend that you not require all students to supply these words, but many students may be challenged by this aspect.

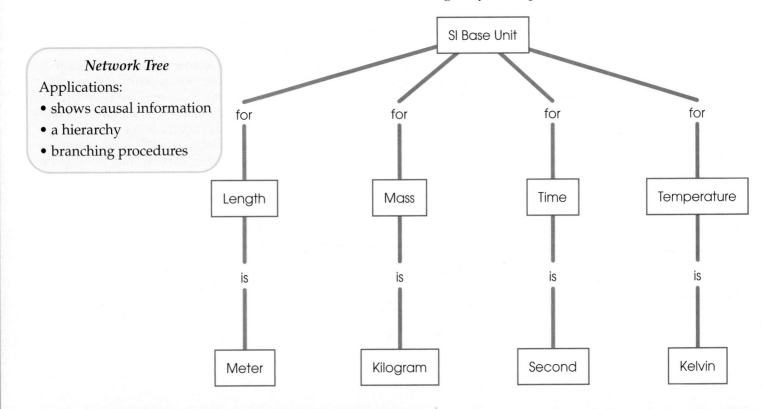

Network Tree

Applications:
- shows causal information
- a hierarchy
- branching procedures

Concept Mapping Booklet

The *Concept Mapping* book of the *Teacher Classroom Resources,* too, provides a developmental approach for students to practice concept mapping.

As a teaching strategy, generating concept maps can be used to preview a chapter's content by visually relating the concepts to be learned and allowing the students to read with purpose. Using concept maps for previewing is especially useful when there are many new key science terms for students to learn. As a review strategy, constructing concept maps reinforces main ideas and clarifies their relationships. Construction of concept maps using cooperative learning strategies as described in this Teacher Guide will allow students to practice both interpersonal and process skills.

Cycle Concept Map

Application:
- shows how a series of events interact to produce a set of results again and again

Initiating Event

Determine the problem.

Make a hypothesis.

Test your hypothesis.

Analyze the results.

Draw conclusions.

Events Chain

Applications:
- describes the stages of a process
- the steps in a linear procedure
- a sequence of events

Spider Concept Map

Applications:
- nonhierarchical, except within a category
- unparallel categories
- brainstorming

CO_2 + H_2O Sunlight/Chlorophyll

Respiration — Energy released — Energy stored — Photosysthesis

O_2 + $C_6H_{12}O_6$

Burning of fossil fuels

Release of sulfur and nitrogen oxides into air

Causes

Lower pH in waterways

Death of fish and other aquatic organisms

Death of trees in evergreen forests

Problems

Acid Rain

Less reliance on fossil fuels

Mass transit

Nuclear power

Possible Solutions

Coal scrubbers

Planning Your Course

Glencoe Life Science provides flexibility in the selection of topics and content that allows teachers to adapt the text to the needs of individual students and classes. In this regard, the teacher is in the best position to decide what topics to present, the pace at which to cover the content, and what material to give the most emphasis. To assist the teacher in planning the course, a planning guide has been provided.

Glencoe Life Science may be used in a full-year, two-semester course that is comprised of 180 periods of approximately 45 minutes each. This type of schedule is represented in the table under the heading of Single-Class Scheduling. As an alternative, the table also outlines a plan for block scheduling, which is described in the next section.

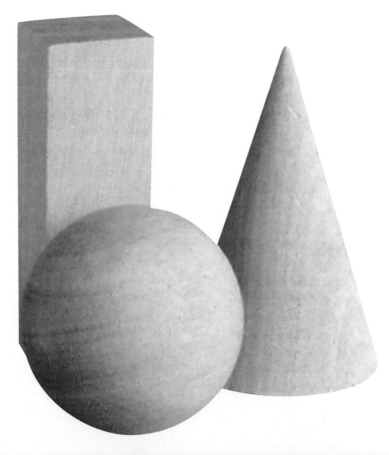

Block Scheduling

To build flexibility into the curriculum, many schools are introducing a block scheduling approach. Block scheduling often involves covering the same information in fewer days with longer class periods. Thus, a course that lasts an entire year may take only one semester under a block scheduling approach. This type of approach allows curriculum supervisors and teachers to tailor the curriculum to meet students' needs while achieving local and/or state curriculum goals. Long, concentrated periods of study can facilitate the learning of complex material. Furthermore, students may be able to take a wider variety of coursework under a block scheduling plan than under a traditional full-year plan, thus enriching their high-school experience and giving them a broader foundation for college-level work.

If you follow a block schedule, you may want to consider either combining lessons or eliminating certain topics and spending more time on the topics you do cover. *Glencoe Life Science* provides the flexibility that allows you to tailor the program to your needs. *Glencoe Life Science* also provides a wide variety of support materials that will help you and your students whether you follow a block schedule or full-year schedule. A block schedule may enable you to complete some of the longer **Activities** in one class period. Have students fill out the accompanying *Activity Worksheets* from the *Teacher Classroom Resources* as they complete the activities. The *Teacher Classroom Resources* provide a wealth of materials that provide classroom management support.

In the table shown here, it is assumed that for block scheduling the course will be taught for one semester and include 90 periods of approximately 90 minutes each.

Please remember that the planning guide is provided as an aid in planning the best course for your students. You should use the planning guide in relation to your curriculum and the ability levels of the classes you teach, the materials available for activities, and the time allotted for teaching.

Planning Guide for GLENCOE LIFE SCIENCE

Unit	Chapter/Section	Single-Class (180 days)	Block (90 days)
UNIT 1	Life	26	17
1	EXPLORING LIFE	5	3
	1-1 Living Things	1	$\frac{1}{2}$
	1-2 Where does life come from?	1	$\frac{1}{2}$
	1-3 What is science?	1	1
	1-4 Science and Society: Technology and the Dairy Industry	1	$\frac{1}{2}$
	Chapter Review	1	$\frac{1}{2}$
2	THE STRUCTURE OF VIRUSES AND CELLS	6	4
	2-1 Viruses	1	1
	2-2 Science and Society: AIDS Vaccine?	1	$\frac{1}{2}$
	2-3 Cells: The Units of Life	1	1
	2-4 Cell Organization	2	1
	Chapter Review	1	$\frac{1}{2}$
3	CELL PROCESSES	7	5
	3-1 Chemistry of Living Things	$1\frac{1}{2}$	1
	3-2 Cell Transport	$2\frac{1}{2}$	2
	3-3 Energy in Cells	$1\frac{1}{2}$	1
	3-4 Science and Society: Energy from Biomass	$\frac{1}{2}$	$\frac{1}{2}$
	Chapter Review	1	$\frac{1}{2}$
4	CELL REPRODUCTION	8	5
	4-1 Cell Growth and Division	2	1
	4-2 Sexual Reproduction and Meiosis	3	2
	4-3 DNA	$1\frac{1}{2}$	1
	4-4 Science and Society: Techniques in Cloning	$\frac{1}{2}$	$\frac{1}{2}$
	Chapter Review	1	$\frac{1}{2}$

Planning Guide for GLENCOE LIFE SCIENCE

Unit	Chapter/Section	Single-Class (180 days)	Block (90 days)
UNIT 2	Heredity and Evolution	14	9
5	HEREDITY	8	5
	5-1 What is genetics?	1	$\frac{1}{2}$
	5-2 Genetics Since Mendel	3	2
	5-3 Human Genetics	2	$1\frac{1}{2}$
	5-4 Science and Society: The Human Genome Project	1	$\frac{1}{2}$
	Chapter Review	1	$\frac{1}{2}$
6	EVOLUTION	6	4
	6-1 Mechanisms of Evolution	2	2
	6-2 Evidence for Evolution	1	$\frac{1}{2}$
	6-3 Primate Evolution	1	$\frac{1}{2}$
	6-4 Science and Society: Extinction and Evolution	1	$\frac{1}{2}$
	Chapter Review	1	$\frac{1}{2}$
UNIT 3	Diversity of Life	16	8
7	CLASSIFYING LIVING THINGS	6	3
	7-1 What is classification?	1	–
	7-2 Modern Classification	2	$1\frac{1}{2}$
	7-3 Science and Society: Diminishing Diversity	1	$\frac{1}{2}$
	7-4 Identifying Organisms	1	$\frac{1}{2}$
	Chapter Review	1	$\frac{1}{2}$
8	BACTERIA	5	2
	8-1 Two Kingdoms of Bacteria	2	1
	8-2 Bacteria in Your Life	1	$\frac{1}{2}$
	8-3 Science and Society: Fighting Tuberculosis	1	–
	Chapter Review	1	$\frac{1}{2}$
9	PROTISTS AND FUNGI	5	3
	9-1 Kingdom Protista	$1\frac{1}{2}$	1
	9-2 Kingdom Fungi	$1\frac{1}{2}$	1
	9-3 Science and Society: Monitoring Red Tides	1	$\frac{1}{2}$
	Chapter Review	1	$\frac{1}{2}$

Planning Guide

Unit	Chapter/Section	Single-Class (180 days)	Block (90 days)
UNIT 4	**Plants**	20	13
10	**INTRODUCTION TO PLANTS**	6	4
	10-1 Characteristics of Plants	2	1
	10-2 Seedless Plants	2	2
	10-3 Science and Society: Cleaner Coal	1	$\frac{1}{2}$
	Chapter Review	1	$\frac{1}{2}$
11	**THE SEED PLANTS**	8	5
	11-1 Seed Plants	2	1
	11-2 Parts of Complex Plants	$1\frac{1}{2}$	1
	11-3 Seed Plant Reproduction	$2\frac{1}{2}$	2
	11-4 Science and Society: Medical Treasures	1	$\frac{1}{2}$
	Chapter Review	1	$\frac{1}{2}$
12	**PLANT PROCESSES**	6	4
	12-1 Photosynthesis and Respiration	3	2
	12-2 Plant Responses	1	1
	12-3 Science and Society: Transgenic Crops	1	$\frac{1}{2}$
	Chapter Review	1	$\frac{1}{2}$
UNIT 5	**Animals**	32	18
13	**INTRODUCTION TO ANIMALS**	7	5
	13-1 What is an animal?	$\frac{1}{2}$	$\frac{1}{2}$
	13-2 Science and Society: Building Artificial Coral Reefs	1	$\frac{1}{2}$
	13-3 Sponges and Cnidarians	2	$1\frac{1}{2}$
	13-4 Flatworms and Roundworms	$2\frac{1}{2}$	2
	Chapter Review	1	$\frac{1}{2}$
14	**MOLLUSKS, WORMS, ARTHROPODS, AND ECHINODERMS**	8	5
	14-1 Mollusks	$1\frac{1}{2}$	1
	14-2 Segmented Worms	1	$\frac{1}{2}$
	14-3 Arthropods	$2\frac{1}{2}$	2

Unit	Chapter/Section	Single-Class (180 days)	Block (90 days)
	14-4 Science and Society: Pesticides and Insects	1	$\frac{1}{2}$
	14-5 Echinoderms	1	$\frac{1}{2}$
	Chapter Review	1	$\frac{1}{2}$
15	**FISH, AMPHIBIANS, AND REPTILES**	6	4
	15-1 Fish	1	1
	15-2 Amphibians	$1\frac{1}{2}$	1
	15-3 Science and Society: Amphibians and Ultraviolet Light	1	$\frac{1}{2}$
	15-4 Reptiles	$1\frac{1}{2}$	1
	Chapter Review	1	$\frac{1}{2}$
16	**BIRDS AND MAMMALS**	6	4
	16-1 Birds	2	$1\frac{1}{2}$
	16-2 Mammals	2	$1\frac{1}{2}$
	16-3 Science and Society: California Sea Otters	1	$\frac{1}{2}$
	Chapter Review	1	$\frac{1}{2}$
17	**ANIMAL BEHAVIOR**	5	
	17-1 Types of Behavior	1	
	17-2 Behavioral Adaptations	2	
	17-3 Science and Society: Zoos and Captive Breeding	1	
	Chapter Review	1	
UNIT 6	**Ecology**	21	15
18	**LIFE AND THE ENVIRONMENT**	7	5
	18-1 The Living Environment and the Nonliving Environment	$1\frac{1}{2}$	1
	18-2 Interactions Among Living Organisms	$1\frac{1}{2}$	1
	18-3 Matter and Energy	2	2
	18-4 Science and Society: Bringing Back the Wolves	1	$\frac{1}{2}$
	Chapter Review	1	$\frac{1}{2}$

Planning Guide for GLENCOE LIFE SCIENCE

Unit	Chapter/Section	Single-Class (180 days)	Block (90 days)
19	ECOSYSTEMS	7	5
19-1	How Ecosystems Change	1	1
19-2	Land Environments	2	$1\frac{1}{2}$
19-3	Science and Society: Protecting Antarctica	1	$\frac{1}{2}$
19-4	Water Environments	2	$1\frac{1}{2}$
	Chapter Review	1	$\frac{1}{2}$
20	RESOURCES AND THE ENVIRONMENT	7	5
20-1	Natural Resources	2	2
20-2	Science and Society: Recycling	1	$\frac{1}{2}$
20-3	Conservation and Wildlife Protection	$1\frac{1}{2}$	1
20-4	Maintaining a Healthy Environment	$1\frac{1}{2}$	1
	Chapter Review	1	$\frac{1}{2}$
UNIT 7	The Human Body	51	10
21	BONES, MUSCLES, AND SKIN	7	
21-1	The Skeletal System	2	
21-2	Science and Society: Biomaterial—Speeding Bone Fracture Recovery	1	
21-3	The Muscular System	2	
21-4	Skin	1	
	Chapter Review	1	
22	NUTRIENTS AND DIGESTION	6	
22-1	Nutrition	2	
22-2	Science and Society: Nutrients Combat Cancer	1	
22-3	Your Digestive System	2	
	Chapter Review	1	
23	THE CIRCULATORY SYSTEM	8	
23-1	Circulation	2	
23-2	Science and Society: "Growing" Heart Valves	1	

Planning Guide for GLENCOE LIFE SCIENCE

Unit	Chapter/Section	Single-Class (180 days)	Block (90 days)
23-3	Blood	2	
23-4	Your Lymphatic System	2	
	Chapter Review	1	
24	RESPIRATION AND EXCRETION	7	
24-1	Your Respiratory System	$2\frac{1}{2}$	
24-2	Science and Society: Restricting Cigarette Advertising	1	
24-3	Your Urinary System	$2\frac{1}{2}$	
	Chapter Review	1	
25	THE NERVOUS AND ENDOCRINE SYSTEMS	8	
25-1	Your Nervous System	2	
25-2	Science and Society: Care and Treatment of Alzheimer's Patients	1	
25-3	The Senses	2	
25-4	Your Endocrine System	2	
	Chapter Review	1	
26	REPRODUCTION AND GROWTH	8	5
26-1	Human Reproduction	2	$1\frac{1}{2}$
26-2	Fertilization to Birth	2	$1\frac{1}{2}$
26-3	Development After Birth	2	1
26-4	Science and Society: A New View of the Old	1	$\frac{1}{2}$
	Chapter Review	1	$\frac{1}{2}$
27	IMMUNITY	7	5
27-1	Disease	1	1
27-2	Your Immune System	2	2
27-3	Science and Society: Should a harmful virus be destroyed?	1	$\frac{1}{2}$
27-4	Noncommunicable Disease	2	1
	Chapter Review	1	$\frac{1}{2}$

Using Technology in the Classroom

Technology helps you adapt your teaching methods to the needs of your students. Glencoe classroom technology products provide many pathways to help you match students' different learning styles. To make your lesson planning easier, all of the technology products listed below are correlated to the student text.

Videodiscs

Glencoe's *Integrated Science Videodiscs* are designed to be used interactively in the classroom. Bar codes in this book allow you to step through the programs and pause to discuss and answer on-screen questions. Bar codes for the following videodiscs also appear throughout this Teacher Edition: *Mr. Wizard's Science and Technology Videodisc, Infinite Voyage, Newton's Apple,* and *National Geographic Society.*

MindJogger Videoquiz

A videoquiz for each chapter can be used to assess prior knowledge or review content before an exam. Student teams work cooperatively to answer three rounds of questions posed by the video host. Each round requires higher-level thinking skills than the previous round.

Glencoe Interactive CD-ROMs

The *Glencoe Life Science* CD-ROM is correlated to the Student Edition.

You can:
- use a CD-ROM interaction as a whole-class presentation;
- allow 2-3 small groups to use the CD-ROMs at once;
- rotate student groups through a single computer station;
- place the materials in a computer lab; or
- use CD-ROMs as a library resource.

NATIONAL GEOGRAPHIC SOCIETY

Glencoe also offers three National Geographic Society CD-ROMs. These image-rich, reference CD-ROMs are faithful to the Society's long history of excellence in science teaching and journalism.

Computer Test Bank

Glencoe's Test Generator for Macintosh and for DOS makes creating, editing, and printing tests quick and easy. You also can edit questions or add your own favorite questions and graphics.

English/Spanish Audiocassettes

Audio chapter summaries in English and in Spanish are a way for auditory learners, lower-level readers, and LEP students to review key chapter concepts. Students can listen individually during class or check out tapes and use them at home. You may find them useful for reviewing the chapter as you plan lessons.

Glencoe Software Support Hotline   **1-800-437-3715**

Should you encounter any difficulty when setting up or running Glencoe software, contact the Software Support Center at Glencoe Publishing between 8:30 A.M. and 4:30 P.M. Eastern Time.

Using the Internet

If you're already familiar with the Internet, skip to the sites listed at the bottom of this page. If you need some tips on how to get started, keep reading.

The Internet is an enormous reference library and a communication tool. You can use it to retrieve information quickly from computers around the globe. Like any good reference, it has an index so you can locate the right piece of information. An Internet index entry is called a *Universal Resource Locator*, or URL. Here's an example:

http://www.glencoe.com/intro/index.html

The first part of the URL tells the computer how to display the information. The second part, after the double slash, names the organization and the computer where the information is stored. The part after the first single slash tells the computer which directory to search and which file to retrieve. File locations change frequently. If you can't find what you're looking for, use the first part of the address only, for example, http://www.glencoe.com/ from the URL shown, and follow links to what you need.

The World Wide Web

The World Wide Web (WWW), a subset of the Internet, began in 1992. Unlike regular text files, Web files can have links to other text files, images, and sound files. By clicking on a link, you can see or hear the linked information.

How do I get access?

To use the Internet, you need a computer, a modem, a telephone line, and a connection to the Internet. If your school doesn't have a connection, contact your local public library or a university; they often give free access to students and educators.

CAUTION: Contents may shift!

The sites referenced in Glencoe's Internet Connections are not under Glencoe's control. Therefore, Glencoe can make no representation concerning the content of these sites. Extreme care has been taken to list only reputable links by using educational and government sites whenever possible. Internet searches have been used that return only sites that contain no content apparently intended for mature audiences.

Where to Start

The brief list of science Internet sites below may prove useful. You can also find Internet addresses throughout this book correlated to selected features.

Useful tools for searching the Internet include:
http://www.yahoo.com/search.html
http://www.altavista.digital.com and
http://www.msn.com/access/
    allinone/hv1

Science Internet Site	Description
http://ericir.syr.edu/	**Ask ERIC**, an ask-the-expert service for K-12 teachers
http://www.enc.org/	**The Eisenhower National Clearinghouse for Math and Science**, instructional materials
http://medinfo.wustl.edu/~ysp/MSN/	**The Mad Scientist Information Network**, an ask-the-expert service for science students

For more information about Glencoe Technology, call Customer Service at **1-800-334-7344.**

Cultural Diversity

American classrooms reflect the rich and diverse cultural heritage of the American people. Students come from different ethnic backgrounds and different cultural experiences into a common classroom that must assist all of them in learning. The diversity itself is an important focus of the learning experience.

Diversity can be repressed, creating a hostile environment; ignored, creating an indifferent environment; or appreciated, creating a receptive and productive environment. Responding to diversity and approaching it as a part of every curriculum is challenging to a teacher, experienced or not. The goal of science is understanding. The goal of multicultural education is to promote the understanding of how people from different cultures approach and solve the basic problems all humans have in living and learning.

Cultural Diversity in Science

Every culture has utilized science to explore fundamental questions, meet challenges, and address human needs. The growth of knowledge in science has advanced along culturally diverse roots. No single culture has a monopoly on the development of scientific knowledge. The history of science is rich with the contributions and accomplishments of men and women from diverse cultural backgrounds. A brief look at some of the milestones in science reveals this diversity. Cultural groups on all continents have devised methods for measuring time and space. For example, the Mayan civilization, which flourished in what is now southern Mexico, Guatemala, and Belize, developed sophisticated and accurate mathematical systems and calendars. Lewis Latimer, an African-American draftsman and inventor, contributed to the development of the electric light bulb and drew the plans that resulted in the 1876 patent of the telephone. Today, individuals from diverse cultural and ethnic backgrounds continue to make significant contributions to science. The **People and Science** feature in the *Student Edition* of *Glencoe Life Science* highlights some of these individuals and their contributions.

The scientific enterprise is a human enterprise that utilizes science processes (observing and inferring, classifying, and so on) to discover, explore, and explain the environment. Consequently, students must experience and understand that all people, regardless of the cultural orientation, have made major contributions to science. *Glencoe Life Science* addresses this issue. In addition to the individuals highlighted in the *Student Edition,* the **Cultural Diversity** sections of the *Teacher Wraparound Edition* provide information about people and groups who have traditionally been misrepresented or omitted. The intent is to build awareness and appreciation for the global community in which we all live.

Goals of Multicultural Education

The *Glencoe Life Science* **Teacher Classroom Resources** include a **Multicultural Connections** booklet that offers additional opportunities to integrate multicultural materials into the curriculum. By providing these opportunities, *Glencoe Life Science* is helping to meet four major goals of multicultural education:

1. promoting the strength and value of cultural diversity
2. promoting human rights and respect for those who are different from oneself
3. promoting social justice and equal opportunity for all people
4. promoting equity in the distribution of power among groups

These goals are accomplished when students see others like themselves in positive settings in their textbooks. They also are accomplished when students realize that major contributions in science are the result of the work of hundreds of individuals from diverse backgrounds. *Glencoe Life Science* has been designed to enable you, the science teacher, to achieve these goals.

Books that provide additional information on multicultural education are

Atwater, Mary, et al. *Multicultural Education: Inclusion of All.* Athens, Georgia: University of Georgia Press, 1994.

Banks, James A. (with Cherry A. McGee Banks). *Multicultural Education: Issues and Perspectives.* Boston: Allyn and Bacon, 1989.

Selin, Helaine. *Science Across Cultures: An Annotated Bibliography of Books on Non-Western Science, Technology, and Medicine.* New York: Garland, 1992.

Meeting Individual Needs

Each student brings his or her own unique set of abilities, perceptions, and needs into the classroom. It is important that the teacher try to make the classroom environment as receptive to these differences as possible and to ensure a good learning environment for all students.

It is important to recognize that individual learning styles are different and that learning style does not reflect a student's ability level. While some students learn primarily through visual or auditory senses, others are kinesthetic learners and do best if they have hands-on exploratory interaction with materials. Some students work best alone and others learn best in a group environment. While some students seek to understand the "big picture" in order to deal with specifics, others need to understand the details first in order to put the whole concept together.

In an effort to provide all students with a positive science experience, this text offers a variety of ways for students to interact with materials so that they can utilize their preferred method of learning the concepts. The variety of approaches allows students to become familiar with other learning approaches, as well.

Ability Levels

The activities are broken down into three levels to accommodate all student ability levels. *Glencoe Life Science Teacher Wraparound Edition* designates the activities as follows:

L1 activities are basic activities designed to be within the ability range of all students. These activities reinforce the concepts presented.

L2 activities are application activities designed for students who have mastered the concepts presented. These activities give students an opportunity for practical application of the concepts presented.

L3 activities are challenging activities designed for the students who are able to go beyond the basic concepts presented. These activities allow students to expand their perspectives on the basic concepts presented.

The chart on pages 42T-43T gives tips you may find useful in structuring the learning environment in your classroom to meet students' special needs. In addition, **Inclusion Strategies** that address special needs are provided in the bottom margins of the *Teacher Wraparound Edition.*

Limited English Proficiency

In providing for the student with limited English proficiency, the focus needs to be on overcoming a language barrier. Once again, it is important not to confuse ability in speaking/reading English with academic ability or "intelligence." In general, the best method for dealing with LEP, variations in learning styles, and ability levels is to provide all students with a variety of ways to learn, apply, and be assessed on the concepts. Look for this symbol **LEP** in the teacher margin for specific strategies for students with limited English proficiency.

Learning Styles

We at Glencoe believe it is our responsibility to provide you with a program that allows you to apply diverse instructional strategies to a population of students with diverse learning styles. Several learning styles are emphasized in the *Teacher Wraparound Edition:* Kinesthetic, Visual-Spatial, Logical-Mathematical, Interpersonal, Intrapersonal, Linguistic, and Auditory-Musical. A student with a kinesthetic style learns from touch, movement, and manipulating objects. Visual-spatial learners respond to images and illustrations. Using numbers and reasoning are characteristics of the logical-mathematical learner. A student with an interpersonal style has confidence in social settings, while a student with an intrapersonal style may prefer to learn on his or her own. Linguistic learning involves the use and understanding of words. Finally, auditory-musical learning involves listening to the spoken word and to tones and rhythms.

Any student may display any or all of these styles depending on the learning situation. The *Student Edition* and *Teacher Wraparound Edition* provide a number of strategies for encouraging students with diverse learning styles. These include **Using Math, Activities, MiniLABs,** and **Science Journal** features. The *Teacher Wraparound Edition* contains **Visual Learning** and **Science Journal,** as well as a number of other strategies. Look for this logo **LS** that identifies strategies for different learning styles.

Meeting Individual Needs

	Description	Sources of Help/Information
Learning Disabled	All learning-disabled students have an academic problem in one or more areas, such as academic learning, language, perception, social-emotional adjustment, memory, or attention.	*Journal of Learning Disabilities* *Learning Disability Quarterly*
Behaviorally Disordered	Children with behavior disorders deviate from standards or expectations of behavior and impair the functioning of others and themselves. These children may also be gifted or learning-disabled.	*Exceptional Children* *Journal of Special Education*
Physically Challenged	Children who are physically disabled fall into two categories—those with orthopedic impairments and those with other health impairments. Orthopedically impaired children have the use of one or more limbs severely restricted, so the use of wheelchairs, crutches, or braces may be necessary. Children with other health impairments may require the use of respirators or other medical equipment.	Batshaw, M.L. and M.Y. Perset. *Children with Handicaps: A Medical Primer.* Baltimore: Paul H. Brooks, 1981. Hale, G. (Ed.). *The Source Book for the Disabled.* New York: Holt, Rinehart & Winston, 1982. *Teaching Exceptional Children*
Visually Impaired	Children who are visually disabled have partial or total loss of sight. Individuals with visual impairments are not significantly different from their sighted peers in ability range or personality. However, blindness may affect cognitive, motor, and social development, especially if early intervention is lacking.	*Journal of Visual Impairment and Blindness* *Education of Visually Handicapped* American Foundation for the Blind
Hearing Impaired	Children who are hearing impaired have partial or total loss of hearing. Individuals with hearing impairments are not significantly different from their hearing peers in ability range or personality. However, the chronic condition of deafness may affect cognitive, motor, and social development if early intervention is lacking. Speech development also is often affected.	*American Annals of the Deaf* *Journal of Speech and Hearing Research* *Sign Language Studies*
Limited English Proficiency	Multicultural and/or bilingual children often speak English as a second language or not at all. The customs and behavior of people in the majority culture may be confusing for some of these students. Cultural values may inhibit some of these students from full participation.	*Teaching English as a Second Language Reporter* R.L. Jones (Ed.). *Mainstreaming and the Minority Child.* Reston, VA: Council for Exceptional Children, 1976.
Gifted	Although no formal definition exists, these students can be described as having above-average ability, task commitment, and creativity. Gifted students rank in the top five percent of their class. They usually finish work more quickly than other students, and are capable of divergent thinking.	*Journal for the Education of the Gifted* *Gifted Child Quarterly* *Gifted Creative/Talented*

Tips for Instruction
With careful planning, the needs of all students can be met in the science classroom.

1. Provide support and structure; clearly specify rules, assignments, and duties.
2. Practice skills frequently. Use games and drills to help maintain student interest.
3. Allow students to record answers on tape and allow extra time to complete tests and assignments.
4. Provide outlines or tape lecture material.
5. Pair students with peer helpers, and provide class time for pair interaction.

---

1. Provide a clearly structured environment with regard to scheduling, rules, room arrangement, and safety.
2. Clearly outline objectives and how you will help students obtain objectives.
3. Reinforce appropriate behavior and model it for students.
4. Do not expect immediate success. Instead, work for long-term improvement.
5. Balance individual needs with group requirements.

---

1. Openly discuss with the student any uncertainties you have about when to offer aid.
2. Ask parents or therapists and students what special devices or procedures are needed, and if any special safety precautions need to be taken.
3. Allow physically disabled students to do everything their peers do, including participating in field trips, special events, and projects.
4. Help nondisabled students and adults understand physically disabled students.

---

1. Help the student become independent. Modify assignments as needed.
2. Teach classmates how to serve as guides.
3. Limit unnecessary noise in the classroom.
4. Provide tactile models whenever possible.
5. Describe people and events as they occur in the classroom.
6. Provide taped lectures and reading assignments.
7. Team the student with a sighted peer for laboratory work.

The *Student Edition* and *Teacher Wraparound Edition* contain a variety of strategies for addressing the individual learning styles of students:

- Both structured and open-ended **Activities**
- Hands-on **MiniLABS**
- **Enrichment** activities
- **Visual Learning** strategies
- **Across the Curriculum** strategies

The following Glencoe products also can help you tailor your instruction to meet the individual needs of your students:

- **Study Guide**
- **Teaching Transparencies**
- **Section Focus Transparencies**
- **Concept Mapping**
- **MindJogger Videoquiz**
- **English/Spanish Audiocassettes**
- **Critical Thinking/Problem Solving**
- **Cooperative Learning Resource Guide**
- Various multimedia products, including **Interactive Videodisc Program, The Secret of Life Series, The Infinite Voyage Series, Science and Technology Videodisc Series, National Geographic Society Series**

---

1. Seat students where they can see your lip movements easily, and avoid visual distractions.
2. Avoid standing with your back to the window or light source.
3. Using an overhead projector allows you to maintain eye contact while writing.
4. Seat students where they can see speakers.
5. Write all assignments on the board, or hand out written instructions.
6. If the student has a manual interpreter, allow both student and interpreter to select the most favorable seating arrangements.

---

1. Remember, students' ability to speak English does not reflect their academic ability.
2. Try to incorporate the student's cultural experience into your instruction. The help of a bilingual aide may be effective.
3. Include information about different cultures in your curriculum to help build students' self-image. Avoid cultural stereotypes.
4. Encourage students to share their cultures in the classroom.

---

1. Make arrangements for students to take selected subjects early and to work on independent projects.
2. Let students express themselves in art forms such as drawing, creative writing, or acting.
3. Make public services available through a catalog of resources, such as agencies providing free and inexpensive materials, community services and programs, and people in the community with specific expertise.
4. Ask "what if" questions to develop high-level thinking skills. Establish an environment safe for risk taking.
5. Emphasize concepts, theories, ideas, relationships, and generalizations.

Assessment

What criteria do you use to assess your students as they progress through a course? Do you rely on formal tests and quizzes? To assess students' achievement in science, you need to measure not only their knowledge of the subject matter, but also their ability to handle apparatus; to organize, predict, record, and interpret data; to design experiments; and to communicate orally and in writing. *Glencoe Life Science* has been designed to provide you with a variety of assessment tools, both formal and informal, to help you develop a clearer picture of your students' progress.

Performance Assessment

Performance assessments are becoming more common in today's schools. Science curricula are being revised to prepare students to cope with change and with futures that will depend on their abilities to think, learn, and solve problems. Although learning fundamental concepts will always be important in the science curriculum, the concepts alone are no longer sufficient in a student's scientific education.

Defining Performance Assessment

Performance assessment is based on judging the quality of a student's response to a performance task. A performance task is constructed to require the use of important concepts with supporting information, work habits important to science, and one or more of the elements of scientific literacy. The performance task attempts to put the student in a "real-world" context so that the class learning can be put to authentic uses.

Performance Assessment Is Also a Learning Activity

Performance assessment is designed to improve the student. While a traditional test is designed to take a snapshot of what the student knows, the performance task involves the student in work that actually makes the learning more meaningful and builds on the student's knowledge and skills. As a student is engaged in a performance task with performance assessment lists and examples of excellent work, both learning and assessment are occurring.

Performance Task Assessment Lists

Performance task assessment lists break the assessment criteria into several well-defined categories. Possible points for each category are assigned by the teacher. Both the teacher and the student assess the work and assign the number of points earned. The teacher is scoring not only the quality of the product, but also the quality of the student's self-assessment.

Besides using the performance task assessment list to help guide his or her work and then to self-assess it, the student needs to study examples of excellent work. These examples could come from published sources, or they could be students' work. These examples include the same type of product for different topics, but they would all be rated excellent using the assessment list for that product type.

Performance Assessment in *GLENCOE LIFE SCIENCE*

Many performance task assessment lists can be found in Glencoe's *Performance Assessment in the Science Classroom.* These lists were developed for the summative and skill performance tasks in the *Performance Assessment* book that accompanies the *Glencoe Life Science* program. The *Performance Assessment* book contains a mix of skill assessments and summative assessments that tie together major concepts of each chapter. The lists can be used to support **MiniLABs, Activities,** and the **Performance**

Assessment problems in the **Chapter Review.** Glencoe's *Alternate Assessment in the Science Classroom* provides additional background and examples of performance assessment. Activity sheets in the *Activity Worksheets* book provide yet another vehicle for formal assessment of student products. The **MindJogger Videoquiz** series offers interactive videos that provide a fun way for your students to review chapter concepts. You can extend the use of the videoquizzes by implementing them in a testing situation. Questions are at three difficulty levels: basic, intermediate, and advanced.

Assessment and Grading

Assessment is giving the learner feedback on the individual elements of his or her performance or product. Therefore, assessment provides specific information on strengths and weaknesses, and allows the student to set targets for improvement. Grading is an act to evaluate the overall quality of the performance or product according to some norms of quality. From the performance task assessment list, the student can see the quality of the pieces. The points on an assessment list can be summed and an overall grade awarded. The grade alone is not very helpful to the student, but it does describe overall quality of the performance or product. The information from both the assessment list and the grade should be reported to the student, parents, and other audiences.

Assessing Student Work with Rubrics

A rubric is a set of descriptions of the quality of a process and/or a product. The set of descriptions includes a continuum of quality from excellent to poor. Rubrics for various types of assessment products are in the Glencoe Professional Development Series booklet: *Performance Assessment in the Science Classroom.*

When to Use the Performance Task Assessment List and When to Use the Rubric

Performance task assessment lists are used for all or most performance tasks. If a grade is also necessary, it can be derived from the total points earned on the assessment list. Rubrics are used much less often. Their best use is to help students periodically assess the overall quality of their work. After making a series of products, the student is asked how he or she is doing overall on one of these types of products. With reference to the standards of quality set for that product at that grade level, the student can decide where on the continuum of quality his or her work fits. The student is asked to assign a rubric score and explain why the score was chosen. Because the student has used a performance task assessment list to examine the elements of each of the products, he or she can justify the rubric score. Experience with performance task assessment lists comes before use of a rubric.

Group Performance Assessment

Recent research has shown that a cooperative learning environment improves student learning outcomes for students of all ability levels. *Glencoe Life Science* provides many opportunities for cooperative learning and, as a result, many opportunities to observe group work processes and products. *Glencoe Life Science: Cooperative Learning Resource Guide* provides strategies and resources for implementing and evaluating group activities. In cooperative group assessment, all members of the group contribute to the work process and the products it produces. For example, if a mixed-ability, four-member laboratory work group conducts an activity, you can use a rating scale or checklist to assess the quality of both group interaction and work skills. An example, along with information about evaluating cooperative work, is provided in the booklet *Alternate Assessment in the Science Classroom.*

The Science Journal

A science journal is intended to help the student organize his or her thinking. It is not a lecture or laboratory notebook. It is a place for students to make their thinking explicit in drawings and writing. It is the place to explore what makes science fun and what makes it hard.

Portfolios: Putting It All Together

The portfolio should help the student see the "big picture" of how he or she is performing in gaining knowledge and skills and how effective his or her work habits are. The portfolio is a way for students to see how the individual performance tasks done during the year fit into a pattern that reveals the overall quality of their learning. The process of assembling the portfolio should be both integrative (of process and content) and reflective. The performance portfolio is not a complete collection of all worksheets and other assignments for a grading period. At its best, the portfolio should include integrated performance products that show growth in concept attainment and skill development.

The Portfolio: Criteria for Success

To be successful, a science portfolio should

1. improve the student's performance in science.
2. promote the student's skills of self-assessment and goal setting.
3. promote a sense of ownership and pride of accomplishment in the student.
4. be a reasonable amount of work for the teacher.
5. be highly valued by the teacher receiving the portfolio the next year.
6. be useful in parent conferences.

The Portfolio: Its Contents

Evidence of the student's growth in the following five categories should be included.

1. Range of thinking and creativity
2. Use of scientific method
3. Inventions and models
4. Connections between science and other subjects
5. Readings in science

For each of the five categories, students should make indexes that describe what they have selected for their portfolios and why. From their portfolio selections, have students pick items that show the quality of their work habits. Finally, have each student write a letter to next year's science teacher introducing him- or herself to the teacher and explaining how the work in this section shows what the student can do.

Opportunities for using Science Journals and Portfolios

Glencoe Life Science presents a wealth of opportunities for performance portfolio development. Each chapter in the student text contains enrichment activities and connections with life, society, and literature. Each of the student activities results in a product. A mixture of these products can be used to document student growth during the grading period. Descriptions, examples, and assessment criteria for portfolios are discussed in *Alternate Assessment in the Science Classroom.* Glencoe's *Performance Assessment in the Science Classroom* contains even more information on using science journals and making portfolios. Performance task assessment lists and rubrics for both journals and portfolios are found there.

Content Assessment

While new and exciting performance skill assessments are emerging, paper-and-pencil tests are still a mainstay of student evaluation. Students must learn to conceptualize, process, and prepare for traditional content assessments. Presently and in the foreseeable future, students will be required to pass pencil-and-paper tests to exit high school and to enter college, trade schools, and other training programs.

Glencoe Life Science contains numerous strategies and formative checkpoints for evaluating student progress toward mastery of science concepts. Throughout the chapters in the student text, **Section Wrap-up** questions and application tasks are presented. This spaced review process helps build learning bridges that allow all students to confidently progress from one lesson to the next.

For formal review that precedes the written content assessment, *Glencoe Life Science* presents a three-page **Chapter Review** at the end of each chapter. By evaluating the student responses to this extensive review, you can determine whether any substantial reteaching is needed.

For the formal content assessment, a two-page review and a four-page chapter test are provided for each chapter in the *Review and Assessment* book. Using the review in a whole-class session, you can correct any misperceptions and provide closure for the text. If your individual assessment plan requires a test that differs from the chapter test in the resource package, customized tests can be easily produced using the *Computer Test Bank.*

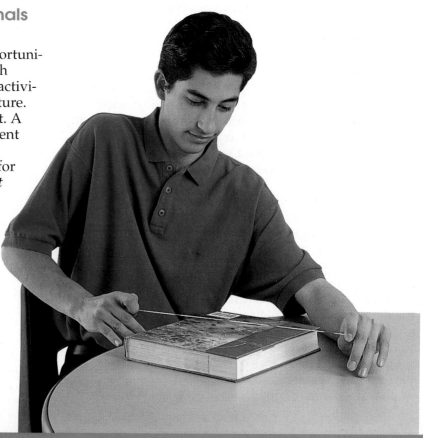

Tech-Prep Education

What Is Tech Prep?

Tech prep is a rigorous and focused program of study that aims to create a workforce in the United States that is technically literate. It is designed to prepare students enrolled in a general curriculum for the demands of further education or for employment by providing them with essential academic and technical foundations, along with problem-solving, group-process, and lifelong-learning skills. These goals can be achieved by integrating vocational study with higher-level academic study.

What Are the Characteristics of the Tech Prep Curriculum?

In 1990, Congress passed the Carl D. Perkins Vocational and Applied Technology Act to set aside funds for the development and administration of Tech Prep programs. The criteria outlined in Title III of the Perkins Act specify that tech prep programs take place during the last two years of high school, followed by two years of post-secondary occupational education, and that this education culminate in a certificate or associate degree. The Secretary's Commission on Achieving Necessary Skills (SCANS), an arm of the United States Department of Labor, published a report in June 1991 that outlined several competencies that characterize successful workers. The Tech Prep curriculum seeks to address these competencies, which include the

- ability to use resources productively.
- ability to use interpersonal skills effectively, including fostering teamwork, teaching others, serving customers, leading, negotiating, and working well with individuals from culturally diverse backgrounds.
- ability to acquire, evaluate, interpret, and communicate data and information.
- ability to understand social, organizational, and technological systems.
- ability to apply technology to specific tasks.

How Does Tech Prep Relate to the Middle School Curriculum?

Tech prep provides an alternative to the college-prep course of study that leads to a career goal.

Because of this, tech prep is available to all students whether they are taking a college prep, vocational, or general course of study. The middle school years provide an opportunity to identify those students who might benefit from a tech-prep curriculum once they reach high school. They also provide an opportunity to introduce unfamiliar students to technological applications leading to career opportunities, or to provide practice for students who already have some knowledge of how technology is used. Young children are fascinated by the technology that surrounds them, and they are especially quick to learn how to use and apply it. Take advantage of their enthusiasm by preparing them for a possible tech-prep education when they are in middle school.

How Does *GLENCOE PHYSICAL SCIENCE* Address Tech Prep Issues?

Glencoe Physical Science helps you develop scientific and technological literacy in your students through a variety of performance-based activities that emphasize problem solving, critical thinking skills, and teamwork. In the **Student Edition, Problem Solving, Using Technology,** and **Design Your Own Experiment** provide opportunities for practical applications of concepts. **Using Computers** and **Using Math** features give additional practice in computer and math applications. **Unit Projects** provide ideas for activities that can be used as extended projects or for science fairs.

Collecting weather data using computers

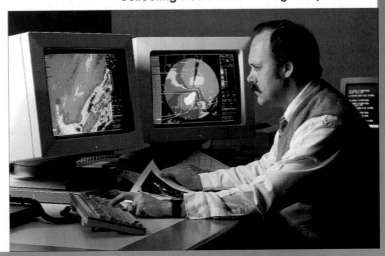

Cooperative Learning

What Is Cooperative Learning?

In cooperative learning, students work together in small groups to learn academic material and interpersonal skills. Group members each learn that they are responsible for accomplishing an assigned group task as well as for learning the material. Cooperative learning fosters academic, personal, and social success for all students.

Recent research shows that cooperative learning results in

- development of positive attitudes toward science and toward school.
- lower drop-out rates for at-risk students.
- building respect for others regardless of race, ethnic origin, or sex.
- increased sensitivity to and tolerance of diverse perspectives.

Establishing a Cooperative Classroom

Cooperative groups in the high school usually contain from two to five students. Heterogeneous groups that represent a mixture of abilities, genders, and ethnicity expose students to ideas different from their own and help them learn to work with different people.

Initially, cooperative learning groups should work together for only a day or two. After the students are more experienced, they can work with a group for longer periods of time. It is important to keep groups together long enough for each group to experience success and to change groups often enough that students have the opportunity to work with a variety of students.

Students must understand that they are responsible for group members learning the material. Before beginning, discuss the basic rules for effective cooperative learning: (1) listen while others are speaking, (2) respect other people and their ideas, (3) stay on tasks, and (4) be responsible for your own actions.

The *Teacher Wraparound Edition* uses the code **COOP LEARN** at the end of activities and teaching ideas where cooperative learning strategies are useful. For additional help, refer again to these pages of background information on cooperative learning.

Using Cooperative Learning Strategies

The *Cooperative Learning Resource Guide* of the *Teacher Classroom Resources* provides help for selecting cooperative learning strategies, as well as methods for troubleshooting and evaluation.

During cooperative learning activities, monitor the functioning of groups. Praise group cooperation and good use of interpersonal skills. When students are having trouble with the task, clarify the assignment, reteach, or provide background as needed. Answer questions only when no students in the group can.

Evaluating Cooperative Learning

At the close of the lesson, have groups share their products or summarize the assignment. You can evaluate group performance during a lesson by frequently asking questions to group members picked at random or having each group take a quiz together. You might have all students write papers and then choose one at random to grade. Assess individual learning by your traditional methods.

Managing Activities

In the Science Classroom

Glencoe Life Science engages students in a variety of experiences to provide all students with an opportunity to learn by doing. The many hands-on activities throughout *Glencoe Life Science* require simple, common materials, making them easy to set up and manage in the classroom.

MiniLAB

MiniLABs are intended to be short and occur many times throughout the text. The integration of these activities with the core material provides for thorough development and reinforcement of concepts.

Design Your Own Experiment

What makes science exciting to students is working in the lab, observing natural phenomena, and tackling concrete problems that challenge them to find their own answers. *Glencoe Life Science* provides more than the same "cookbook" activities you've seen hundreds of times before. Students work cooperatively to develop their own experimental designs in **Design Your Own Experiments.** They discover firsthand that developing procedures for studying a problem is not as hard as they thought it might be. Watch your students grow in confidence and ability as they progress through the self-directed **Design Your Own Experiments.**

Preparing Students for Open-Ended Lab Experiences

To prepare students for the **Design Your Own Experiments,** you should follow the guidelines in the *Teacher Wraparound Edition,* especially in the sections titled Possible Procedures and Teaching Strategies. In these sections, you will be given information about what demonstrations to do and what questions to ask students so that they will be able to design their experiment. Your introduction to a **Design Your Own Experiment** will be very different from traditional activity introductions in that it will be designed to focus students on the problem and stimulate their thinking without giving them directions for how to set up their experiment. Different groups of students will develop alternative hypotheses and alternative procedures. Check their procedures before they begin. In contrast to some "cookbook" activities, there may not be just one right answer. Finally, students should be encouraged to use questions that come up during the activity in designing a new experiment.

If you feel that your students are not prepared to begin designing experiments, you may want to practice some paper-and-pencil lab designs using the following format: Select a simple scenario and ask students to design an experiment that will test a hypothesis they make about the problem presented in the reading. Give them a set of questions that will lead them to an appropriate experimental design.

Laboratory Safety

Safety is of prime importance in every classroom. However, the need for safety is even greater when science is taught. The activities in *Glencoe Life Science* are designed to minimize dangers in the laboratory. Even so, there are no guarantees against accidents. Careful planning and preparation as well as being aware of hazards can keep accidents to a minimum. Numerous books and pamphlets are available on laboratory safety with detailed instructions on preventing accidents. In addition, the *Glencoe Life Science* program provides safety guidelines in several forms. The *Lab and Safety Skills* booklet contains detailed guidelines, in addition to masters you can use to test students' lab and safety skills. The *Student Edition* and *Teacher Wraparound Edition* provide safety precautions and symbols designed to alert students to possible dangers. Know the rules of safety and what common violations occur. Know the **Safety Symbols** used in this book. A safety symbol chart is on page 759. Know where emergency equipment is stored and how to use it. Practice good laboratory housekeeping and management to ensure the safety of your students.

It is most important to use safe laboratory techniques when handling all chemicals. Many substances may appear harmless but are, in fact, toxic, corrosive, or very reactive. Always check with the manufacturer or with Flinn Scientific, Inc., (312) 879-6900. Chemicals should never be ingested. Be sure to use proper techniques to smell solutions or other agents. Always wear safety goggles and an apron. The following general cautions should be used.

1. Poisonous/corrosive liquid and/or vapor. Use in the fume hood. Examples: *acetic acid, hydrochloric acid, ammonia hydroxide, nitric acid.*

2. Poisonous and corrosive to eyes, lungs, and skin. Examples: *acids, limewater, iron(III) chloride, bases, silver nitrate, iodine, potassium permanganate.*

3. Poisonous if swallowed, inhaled, or absorbed through the skin. Examples: *glacial acetic acid, copper compounds, barium chloride, lead compounds, chromium compounds, lithium compounds, cobalt(II) chloride, silver compounds.*

4. Always add acids to water, never the reverse.

5. When sulfuric acid or sodium hydroxide is added to water, a large amount of thermal energy is released. Sodium metal reacts violently with water. Use extra care when handling any of these substances.

Unless otherwise specified, solutions are prepared by adding the solid to a small amount of distilled water and then diluting with water to the volume listed. For example, to make a $0.1M$ solution of aluminum sulfate, dissolve 34.2 g of $Al_2(SO_4)_3$ in a small amount of distilled water and dilute to a liter with water. If you use a hydrate that is different from the one specified in a particular preparation, you will need to adjust the amount of the hydrate to obtain the required concentration.

Chemical
Storage & Disposal

General Guidelines

Be sure to store all chemicals properly. The following are guidelines commonly used. Your school, city, county, or state may have additional requirements for handling chemicals. It is the responsibility of each teacher to become informed as to what rules or guidelines are in effect in his or her area.

1. Separate chemicals by reaction type. Strong acids should be stored together. Likewise, strong bases should be stored together and should be separated from acids. Oxidants should be stored away from easily oxidized materials, and so on.

2. Be sure all chemicals are stored in labeled containers indicating contents, concentration, source, date purchased (or prepared), any precautions for handling and storage, and expiration date.

3. Dispose of any outdated or waste chemicals properly according to accepted disposal procedures.

4. Do not store chemicals above eye level.

5. Wood shelving is preferable to metal. All shelving should be firmly attached to the wall and should have antiroll edges.

6. Store only those chemicals that you plan to use.

7. Hazardous chemicals require special storage containers and conditions. Be sure to know what those chemicals are and the accepted practices for your area. Some substances must even be stored outside the building.

8. When working with chemicals or preparing solutions, observe the same general safety precautions that you would expect from students. These include wearing an apron and goggles. Wear gloves and use the fume hood when necessary. Students will want to do as you do whether they admit it or not.

9. If you are a new teacher in a particular laboratory, it is your responsibility to survey the chemicals stored there and to be sure they are stored properly or disposed of. Consult the rules and laws in your area concerning what chemicals can be kept in your classroom. For disposal, consult up-to-date disposal information from the state and federal governments.

Disposal of Chemicals

Local, state, and federal laws regulate the proper disposal of chemicals. These laws should be consulted before chemical disposal is attempted. Although most substances encountered in high school biology can be flushed down the drain with plenty of water, it is not safe to assume that this is always true. It is recommended that teachers who use chemicals consult the following books from the National Research Council:

Prudent Practices for Handling Hazardous Chemicals in Laboratories. Washington, DC: National Academy Press, 1981.

Prudent Practices for Disposal of Chemicals from Laboratories. Washington, DC: National Academy Press, 1983.

These books are useful and still in print, although they are several years old. Current laws in your area would, of course, supersede the information in these books.

Disclaimer

Glencoe Publishing Company makes no claims to the completeness of this discussion of laboratory safety and chemical storage. The material presented is not all-inclusive, nor does it address all of the hazards associated with handling, storage, and disposal of chemicals, or with laboratory management.

Using a Graphing Calculator

What is it?

What does it do?

How is it going to help me learn math?

These are just a few of the questions many students ask themselves when they first see a graphing calculator. Some students may think, "Oh, no! Do we *have* to use one?" while others may think, "All right! We get to use these neat calculators!" There are as many thoughts and feelings about graphing calculators as there are students, but one thing is for sure: a graphing calculator can help you learn mathematics and science.

So what is a graphing calculator? Very simply, it is a calculator that draws graphs. This means that it will do all of the things that a "regular" scientific calculator will do, *plus* it will draw graphs of equations. This capability is helpful as you analyze graphs.

But a graphing calculator can do more than just calculate and draw graphs. For example, you can program it or work with data to make statistical graphs and perform computations. If you need to generate random numbers, you can do that on the graphing calculator. You can even draw and manipulate geometric figures on some graphing calculators. It's really a very powerful tool—so powerful that it is often called a pocket computer.

As you may have noticed, graphing calculators have some keys that other calculators do not. These instructions are for the Texas Instruments TI-82 and TI-83 calculators. Most of the keystrokes for the TI-82 and TI-83 are the same. The keys on the bottom half of the TI-82 and TI-83 are those found on scientific calculators. The keys located just below the screen are the graphing keys. You will also notice the up, down, left, and right arrow keys. These allow you to move the cursor around on the screen, to "trace" graphs that have been plotted, and to choose items from the menus. The other keys located on the top half of the calculator access the special features such as statistical computations and programming features.

A few of the keystrokes that can save you time when using the graphing calculator are listed below.

- The commands above the calculator keys are accessed with the [2nd] or [ALPHA] key. On a TI-82, the [2nd] key and the commands accessed by it are blue and the [ALPHA] key and its commands are gray. On a TI-83, the [2nd] key and its commands are yellow and the [ALPHA] and its commands are green.

- [2nd] [ENTRY] copies the previous calculation so you can edit and use it again.

- Pressing [ON] while the calculator is graphing stops the calculator from completing the graph.

- [2nd] [QUIT] will return you to the home (or text) screen.

- [2nd] [A-LOCK] locks the [ALPHA] key, which is like pressing "shift lock" or "caps locks" on a typewriter or computer. The result is that all letters will be typed and you do not have to repeatedly press the [ALPHA] key. (This is handy for programming.) Stop typing letters by pressing [ALPHA] again.

- [2nd] [OFF] turns the calculator off.

Some commonly used mathematical functions are shown in the table below. As with any scientific calculator, the graphing calculator observes the order of operations.

Mathematical Operation	Examples	Keys	Display
evaluate expressions	Find 2 + 5.	2 [+] 5 [ENTER]	2+5 7
exponents	Find 3^5.	3 [^] 5 [ENTER]	3^5 243
multiplication	Evaluate 3(9.1 + 0.8).	3 [×] [(] 9.1 [+] .8 [)] [ENTER]	3(9.1+.8) 29.7
roots	Find $\sqrt{14}$.	[2nd] [√] 14 [ENTER]	√ 14 3.741657387
opposites	Enter −3.	[(−)] 3	− 3

Graphing on the TI-82 or TI-83

Before graphing, we must instruct the calculator how to set up the axes in the coordinate plane. To do this, we define a **viewing window.** The viewing window for a graph is the portion of the coordinate grid that is displayed on the graphics screen of the calculator. The viewing window is written as [left, right] by [bottom, top] or [Xmin, Xmax] by [Ymin, Ymax]. A viewing window of [−10, 10] by [−10, 10] is called the **standard viewing window** and is a good viewing window to start with to graph an equation. The standard viewing window can be easily obtained by pressing [ZOOM] 6. Try this. Move the arrow keys around and observe what happens. You are seeing a portion of the coordinate plane that includes the region from −10 to 10 on the x-axis and from −10 to 10 on the y-axis. Move the cursor, and you can see the coordinates of the point for the position of the cursor.

Any viewing window can be set manually by pressing the [WINDOW] key. The window screen will appear and display the current settings for your viewing window. Using the arrow and [ENTER] keys, move the cursor to edit the window settings. Xscl and Yscl refer to the x-scale and y-scale. This is the number of units between tick marks on the x- and y-axes. Xscl=1 means that there will be a tick mark for every unit of one along the x-axis.

Programming on the TI-82 or TI-83

The TI-82 and TI-83 have programming features that allow us to write and execute a series of commands to perform tasks that may be too complex or cumbersome to perform otherwise. Each program is given a name. Commands begin with a colon (:), which the calculator enters automatically, followed by an expression or an instruction. Most of the features of the calculator are accessible from program mode.

When you press [PRGM], you see three menus: EXEC, EDIT, and NEW. EXEC allows you to execute a stored program, EDIT allows you to edit or change a program, and NEW allows you to create a program. The following tips will help you as you enter and run programs on the TI-82.

- To begin entering a new program, press [PRGM] [▶] [▶] [ENTER].

- After a program is entered, press [2nd] [QUIT] to exit the program mode and return to the home screen.

- To execute a program, press [PRGM]. Then use the down arrow key to locate the program name and press [ENTER], or press the number or letter next to the program name.

- If you wish to edit a program, press [PRGM] [▶] and choose the program from the menu.

- To immediately re-execute a program after it is run, press [ENTER] when "Done" appears on the screen.

- To stop a program during execution, press [ON].

While a calculator cannot do everything, it can make some things easier. To prepare for whatever lies ahead, you should try to learn as much as you can. The future will definitely involve technology, and using a graphing calculator is a good start toward becoming familiar with technology. Who knows? Maybe one day you will be designing the next satellite, building the next skyscraper, or helping students learn mathematics and science with the aid of a graphing calculator!

Activity Materials

Get the materials you need quickly and easily!

Glencoe and Science Kit, Inc. have teamed up to make materials selection for *Glencoe Life Science* easier with an activity materials folder. This folder contains two convenient ways to order materials and equipment for the program: the **Activity Plan Checklist** and the **Activity Materials List** master.

Call Science Kit at 1-800-828-7777 to get your folder.

Materials Support Provided By:

Science Kit® & Boreal® Laboratories

Your Classroom Resource

777 East Park Drive

Tonawanda, NY 14150-6784

800-828-7777 Fax 716-874-9572

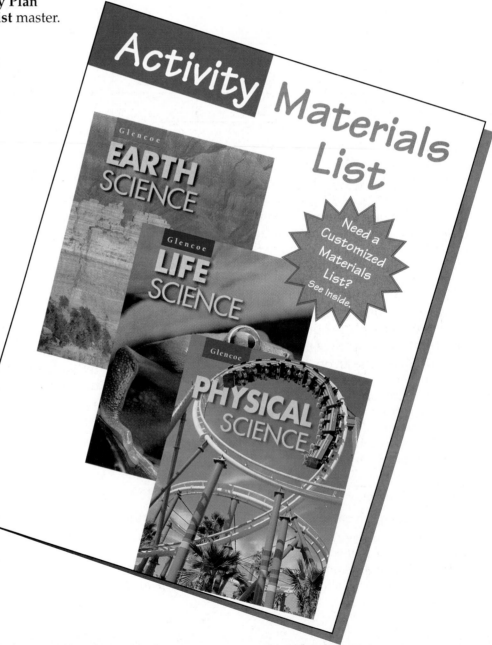

Non-Consumables

Item	Activity	MiniLAB	Explore
Apron	1-1, 3-1, 3-2, 8-1, 8-2, 9-2, 10-1, 10-2, 11-1, 11-2, 12-2, 13-2, 14-1, 16-2, 17-1, 20-1, 22-1, 22-2, 24-1, 24-2, 27-2		
Aquarium with appropriate supplies (pump, filter, glass cover, gravel, etc.)	15-1, 15-2, 17-1		
Balance, pan	3-2, 16-2	22, 268, 607	
Beaker, 100-mL	18-1	318, 559	5
Beaker, 150-mL	11-2	48, 212	
Beaker, 250-mL	2-2, 18-1	70, 500, 539, 607	63
Beaker, 400-mL	24-1		
Beaker, 500-mL		268	
Beaker, 600-mL	15-1, 16-2		
Beans, red (100), white (100)	5-1		
Binoculars	19-2		
Books, on childhood immunizations		718	
Books, on human fetal development		713	
Books, on local plants and animals	19-2		
Books, on pond life	19-1		
Books, on seed plants		301	
Bowl(s), large		404	
Bowl(s), small, identical (2)		428	63
Box(es), cardboard, shoe, with lid	6-2		
Brush or comb, each student's personal		436	
Cellophane, red		466	
Chairs, identical (2)	23-1		
Clamshells			369
Clothespin, snap-type			573
Coffee can, small with lid	17-2		
Container(s), 500-mL capacity, with lids (2)	3-1		
Container(s), large (to hold ants)	17-2		
Container(s), widemouthed, 0.5-L capacity (3)	1-1		
Cookie sheet		607	
Coverslip(s)	2-1, 2-2, 9-1, 9-2, 10-2, 12-1, 19-1	212, 352, 436	209
Dish, shallow, glass	13-1, 17-1	41, 352	31, 209
Dishpans, plastic, identical (2)		628	
Dissecting equipment, pan or cutting board	21-2		
Dissecting equipment, probes, needle	10-2, 21-2		
Dropper(s)	2-1, 2-2, 9-1, 10-2, 13-1, 13-2, 19-1, 22-2	321, 352, 358, 630, 733	729
Feathers, contour and down	16-1		
Fishnet, small	15-1, 15-2, 17-1	532	
Flashlight		48, 466	
Forceps	2-1, 2-2, 10-2, 11-1, 12-1, 27-2		
Glass(es), identical (2)		615	
Glass(es), tall	22-2	271	
Glass(es), small, filled with tap water		22	285
Glass(es), transparent, empty		41	
Goggles, safety	1-1, 2-2, 3-1, 3-2, 4-2, 8-1, 8-2, 9-2, 10-1, 10-2, 11-2, 12-2, 14-1, 16-2, 17-1, 22-1, 22-2, 24-1, 24-2	318, 321	

Equipment List

Non-Consumables

Item	Activity	MiniLAB	Explore
Graduated cylinder, 10-mL	10-1, 22-1	212	
Graduated cylinder, 100-mL	22-2	268	
Graduated cylinder, 250-mL	3-1, 11-2, 16-2, 24-1		
Gravel, aquarium (washed)	15-2		
Gravel, or coarse sand, pea-sized gravel	17-1		425
Hand lens	1-1, 6-1, 7-1, 10-1, 10-2, 16-1, 17-2, 18-1, 19-2, 24-2	44, 133, 297, 352, 495, 532, 667	31, 187, 285, 369
Hand lens or microscope	20-1, 21-1, 21-2		
Heat source	16-2	500	
Hole punch			151
Immunization records of each student		718	
Incubator	8-2		
Insect collection (optional)			187
Jar(s), baby food, identical (4)	27-2		
Jar(s), large (1-L capacity), with lid(s)	18-1, 19-1		
Jar(s), small glass		385	
Light source, bright		318	
Light source, 100-watt	13-2, 20-2		
Light source, or sunny windowsill	18-2		
Marking pencil	22-1, 22-2, 27-1, 27-2		
Measuring tape, metric	19-2		
Meterstick	25-1	690	481, 599
Metric ruler	5-2, 6-1, 7-1, 10-1, 17-1, 25-2, 27-2	677, 713	5, 31
Microscope	2-1, 2-2, 4-1, 8-1, 9-1, 9-2, 10-2, 12-1, 13-2, 19-1, 21-2, 23-2	212, 352, 436	209
Microscope, stereoscopic	13-1, 13-2, 15-2	358	
Microscope slide(s)	2-1, 2-2, 9-1, 9-2, 10-2, 12-1, 19-1	212, 352, 436	209
Mortar and pestle		67	425
Muscle diagram		587	
Needles	9-2		
Paintbrush, small	13-2		
Paper clip(s), small		22	
Pen, felt-tipped, black			5
Penny	25-1		703
Penny, (100)	6-2		
Pie plate(s), small aluminum	20-1	466	
Plant trays or plastic cups	11-2		
Plastic tubing, 15 cm in length, 1 cm diameter		630	
Plexiglas®, clear, 30 cm × 30 cm square			453
Rain gauge		559	
Refrigerator	18-2	219	
Ring stand	12-2		
Rock, small		22	
Rubber bands	16-2, 20-2	385, 466	
Safety pins, linked together (15)		458	
Saucepan, large		219	
Scale, metric	18-1		
Scalpel	11-1, 21-1	271, 385, 667	91, 285
Scissors	2-2, 4-2, 16-1, 27-2	97, 108, 352, 428, 656	5

Item	Activity	MiniLAB	Explore
Seedling-warming cables	11-2		
Seeds, different kinds (10)	7-1		
Shoe box		410	
Shoe, rubber-soled			399
Sphygmomanometer	23-1		
Splint, wooden	1-1		
Spoon(s), measuring		615	
Spoon(s), metal		22	
Spoon(s), or trowel	16-2, 17-2, 18-2	219	
Spray bottle	10-1		539
Stethoscope	23-1		
Stirring rod	3-2, 14-2, 15-1		
Stopwatch or clock with second hand	16-2, 20-2, 22-2, 24-1	70, 212, 268, 318, 321, 393, 593, 628, 690	63, 625, 651
Stove		219	
Table	23-1		
Teaspoon		219, 268	
Test tube(s)	16-2	67, 321	
Test-tube rack	3-2, 22-1, 22-2	11	
Test tube(s), 16 mm × 150 mm (4)	3-2	733	
Test tube(s), identical	22-1, 22-2	11	
Thermometer(s), Celsius	11-2, 15-1, 16-2, 17-1, 19-2, 20-2		
Thermos bottle with lid, clean, 1-quart capacity		219	
Tub, small plastic		495	
Tub, small plastic, or aluminum foil baking pan			539
Tuning fork			399
Tweezers	9-2		
Watch glass	15-2	358, 500	
World atlas		520	509

Consumables

Item	Activity	MiniLAB	Explore
Alcohol, rubbing	23-1, 27-1		
Ant food (honey, sugar, bread crumbs)	17-2		
Apple(s), fresh (6), rotting (1)	27-1		
Aromatic foods, colognes, or household products		691	
Bag(s), lunch, paper (2)	5-1		
Bag(s), plastic, food storage, large		593	313
Bag(s), plastic, self-sealing	16-2, 27-1	532	91
Balloon, small round (2)		404	
Banana, ripe		385	
Bean(s), dried		212	
Bean(s), lima, or other large seed			285
Beef bouillon, sterile		11	
Beef, ground (raw)	14-2		
Bleach		352	

Equipment List

Consumables

Item	Activity	MiniLAB	Explore
Bones, beef cut in half and lengthwise; turkey leg cut in half and lengthwise	21-1		
Bones, beef sliced crosswise; beef, whole; chicken sliced crosswise; chicken, whole	16-2		
Bottles, clear, plastic, 2-L, soda with tops cut off and labels removed (2)	20-2		
Celery stalks, celery stalks with leaves		271, 607	
Cheesecloth		268, 385	
Chocolate chip cookie or nut-filled brownie		541	
Clay, modeling		201	
Cotton cloth, lightweight, 30 cm × 30 cm piece		428	
Coffee filters			5
Container, small paper		162	
Cotton balls	23-1, 27-1	11	729
Cotton-tipped swabs	8-2		
Cup(s), 150-mL capacity		628	
Cup(s), small paper (3)	18-1		
Cup(s), small plastic		532	
Disinfectant, hydrogen peroxide, mouthwash, and rubbing alcohol	27-2		
Egg, chicken	3-1		31
Filter paper	27-2		
Foil, aluminum	18-2		
Food coloring, green, red		70, 271	
Food samples, 4 different kinds		733	
"Fossil" such as small seashell, key, or leaf		162	
Gelatin, small box, unflavored	22-2	48	
Glue, white, all-purpose, or tape	25-2	97	
Guppy food	17-1		
Gymnosperm cone		297	
Ice cubes		500	
Index cards, small (5)	25-2		599
Ink pad with washable ink		133	
Juices, each in a dropping bottle, orange, pineapple, apple, lemon, tomato, cranberry, carrot, lime, mixed vegetable	22-1		
Labels	1-1, 3-1, 18-2, 20-1, 27-1, 27-2		63
Labels, or marking pencil		70	
Lettuce leaves, boiled	15-2		
Lettuce leaves, in dish of water	12-1		
Litmus paper, blue		11	
Liver, beef	13-2		
Mesh, fine wire	14-1, 17-2		
Milk, 1 quart		67, 219	
Mushrooms, edible		242	229
Newsprint		41, 485	
Newsprint, classified ads			151
Oil, cooking		615	
Oil, mineral		733	
Paper, construction, different colors	4-2, 17-2	97, 108	151
Paper, graph	5-2, 6-1, 6-2, 19-2, 26-2	734	
Paper, large sheets (2)		656	

Item	Activity	MiniLAB	Explore
Paper, notebook	19-2		703
Paper, plain white	7-1, 24-2	133, 242, 297	151, 453
Paper, tissue, colored		97	
Paper plate, small		541	
Paper towel(s)	9-2, 12-2, 27-1	297, 466	63, 91
Pebbles, small	14-1		
Pencil(s), colored	23-2, 26-2		
Pencil(s), with eraser	10-2		
Peppermint food flavoring			729
Petri dish(es)	12-2, 20-1		
Petri dish(es), sterile (5)	27-2		
Petri dish(es), sterile, with nutrient agar	8-2, 9-2		
Petroleum jelly or baby oil		428	
pH indicator paper and chart		559	
Pinto beans that have been soaked overnight in water			91
Plant pots, clay	20-1		
Plant pots, plastic, 15-cm	14-1, 18-2		
Plaster of paris		162	
Plastic food wrap	20-2		
Poster board, large sheet, small sheet	4-2	97	
Potato, raw, identical slices (2)			63
Salt, table	1-1, 11-2, 12-1		63
Sand, moist			539
Sand, or gravel	20-1		
Sandpaper, small piece	27-1		
Seeds	11-2		
Seeds, bean, corn	12-2, 18-2		
Seeds, cracked corn, sunflower seeds, nuts, or other seeds			425
Shortening, solid	16-2	630	
Soap, antibacterial	14-2, 16-2, 18-2, 19-1, 20-1, 27-1		
Soap, dishwashing, liquid		615	
Sodium bicarbonate	3-2		
Soil, for ants	17-2		
Soil, freshly dug, 3 kinds	18-1		
Soil, potting	11-2, 14-1, 18-2, 20-1, 20-2		539
Sponge, small moist	17-2		
Straw(s), plastic, drinking	24-1	318	
String	12-2		
Sugar solution		321	
Syrup, white table	3-1		
Tape, masking	20-2	593	599
Tape, transparent	4-2, 8-2, 9-2, 12-2, 27-2	656	
Thread		97	
Toothpick(s)	13-1, 25-2	97, 541, 630	209
Turkey leg, cooked	21-2		
Twist tie or string		268	
Vinegar, white	3-1		
Wastes, biodegradable and nonbiodegradable	20-1		
Water, aquarium	15-1		
Water, distilled	1-1, 10-1, 11-2	212, 559	
Water, lake or pond	15-2		
Water, pH of 5, 6, and 8	10-1		
Water, pond, sterile	9-2, 19-1		

Equipment List

Consumables

Item	Activity	MiniLAB	Explore
Yarn, different colors		97	
Yeast solution		321	
Yogurt, plain			209
Yogurt, starter		219	

Chemical Supplies

Item	Activity	MiniLAB	Explore
Bromothymol blue solution	24-1	321	
Bromothymol blue solution, in dropping bottle	3-2		
Hydrochloric acid, dilute	22-2		
Indophenol solution	22-1		
Iodine solution with dropper	11-1, 24-2		
Methylene blue with dropper		212, 733	209
Pepsin powder	22-2		
Phenol red solution		318	
Rennin tablet		67	

Preserved Specimens

Item	Activity	MiniLAB	Explore
Anabaena, prepared slide	8-1		
Ferns with gametophytes and sporophytes, prepared slide	10-2		
Gloeocapsa, prepared slide	8-1		
Human blood and blood of two other vertebrates (fish, amphibian, reptile, or bird), prepared slides	23-2		
Human cheek cells, prepared slide	2-1		
Liverworts, prepared slide	10-2		
Mosses, prepared slide	10-2		
Muscles (smooth, skeletal, and cardiac), prepared slides	21-2		
Nitrogen-fixing bacteria, prepared slide		495	
Onion root tip, prepared slide	4-1		
Slime mold, prepared slide	9-1		
Tapeworm, prepared slide	13-2		
Whitefish embryo, prepared slide	4-1		

Living Organisms

Item	Activity	MiniLAB	Explore
Amoeba culture	9-1		
Ants (20)	17-2		
Brine shrimp eggs	1-1		
Crayfish in small aquarium	14-2		
Daphnia or brine shrimp	13-1		
Dicot flowers and seeds	11-1		
Earthworm(s)	14-1	466	
Elodea plant	2-1		
Elodea, small pieces	3-2		
Elodea, sprig		318	
Euglena culture	9-1		
Ferns with gametophytes and sporophytes	10-2		
Frog egg masses	15-2		
Frog, in large container		410	
Goldfish	15-1		
Grantia sponge		352	
Guppies	17-1		
Houseplants, small	14-1		
Hydra culture	13-1		
Kidney, large		667	
Legumes, freshly picked (alfalfa, beans, clover, vetch)		495	
Lichens, various kinds	9-2		
Liverworts	10-2		
Micrasterias culture	9-1		
Monocot flowers and seeds	11-1		
Moss plants	10-1, 10-2		539
Nonlegumes, freshly picked		495	
Nostoc and *Oscillatoria* living cultures	8-1		
Onion plant with roots and leaves	2-2		
Paramecium culture	9-1		
Physarum polycephaalum culture or materials for growing this slime mold		237	
Planarian culture	13-2	358	
Plant(s), freshwater	15-2, 17-1		
Plant(s), small, potted			313
Plant(s), with leaves, flower(s), and seeds	6-1		
Pond debris	19-1	532	
Snails, freshwater	17-1		
Sphagnum moss	14-1	268	
Spirogyra culture	9-1		

Supplier Addresses

Audio-Visual Suppliers

Ambrose Video Publishing, Inc.
Exclusive Distributor of
Time-Life Videos
28 West 44th Street, Suite 2100
New York, NY 10036
(800) 526-4663

Barr Media Group
12801 Schabarum Avenue
Irwindale, CA 91706
(800) 234-7878

Center for the Humanities
Box 1000
90 S. Bedford Road
Communications Park
Mount Kisco, NY 10549
(800) 431-1242

Clearvue, Inc.
6465 N. Avondale Avenue
Chicago, IL 60631
(800) 253-2788

Coronet/MTI Film and Video
4350 Equity Drive
Columbus, OH 43228
(800) 777-2400

Education Development Center
55 Chapel Street
Newton, MA 02158
(800) 225-4276

Educational Activities, Inc.
P.O. Box 392
Freeport, NY 11520
(800) 645-3739

Encyclopaedia Britannica Educational Corp.
310 South Michigan Avenue
Chicago, IL 60604
(800) 323-1229

Films, Inc.
5547 North Ravenswood Ave.
Chicago, IL 60640
(800) 343-4312

Handel Film Corp.
8787 Shoreham Drive
Los Angeles, CA 90069
(800) 395-8990

Hawkhill Associates
125 East Gilman Street
P.O. Box 1029
Madison, WI 53701
(800) 422-4295

International Film Bureau
Suite 450
332 South Michigan Avenue
Chicago, IL 60604-4382
(800) 432-2241

JCE Software
Department of Chemistry
University of
Wisconsin-Madison
1101 University Avenue
Madison, WI 53706
(608) 263-2837

Mindscape, Inc.
Educational Division
88 Rowland Way
Novato, CA 94945
(800) 866-5967

National Geographic Society
Educational Services
1145 17th Street, N.W.
Washington, DC 20036-4688
(800) 647-5463

Optical Data Corp.
30 Technology Drive
Warren, NJ 07059
(800) 248-8478

Pyramid Films
P.O. Box 1048
Santa Monica, CA 90406
(800) 421-2304

TVOntario
1140 Kildare Farm Road, Ste 308
Cary, NC 27511
(800) 331-9566

Videodiscovery
1700 Westlake Ave. North, Ste 600
Seattle, WA 98109-3012
(800) 548-3472

Ward's Natural Science Intl., Inc.
Source of CHEM Study Films
P.O. Box 92912
Rochester, NY 14692
(800) 962-2660

University Rental Services

Indiana University
Instructional Support Services
Bloomington, IN 47405-5901
(800) 552-8620

Kent State University
Audio-Visual Services
330 University Library
Kent, OH 44242
(800) 338-5718

SUNY College at Buffalo
Butler Library Media Services
Bonita J. Percival
1300 Elmwood Avenue
Buffalo, NY 14222-1095
(716) 878-6682

Washington State University
Instructional Support Services
Pullman, WA 99164-5604
(509) 335-7336

Software Suppliers

Agency for Instructional Technology (AIT)
Box A
Bloomington, IN 47402-0120
(800) 457-4509

American Chemical Society
1155 16th Street NW
Washington, DC 20036
(800) 227-5558

Bergwall Productions, Inc.
540 Baltimore Pike
P.O. Box 2400
Chadds Ford, PA 19317
(800) 645-3565

Carolina Biological Supply Co.
2700 York Road
Burlington, NC 27215
(800) 334-5551

Cross Educational Software
P.O. Box 1536
508 E. Kentucky Avenue
Ruston, LA 71270
(800) 768-1969

Educational Images, Ltd.
P.O. Box 3456, West Side
Elmira, NY 14905
(800) 527-4264

Educational Materials and Equipment Co. (EME)
P.O. Box 2805
Danbury, CT 06813-2805
(800) 848-2050

Focus Media, Inc.
63 Commercial Avenue
Garden City, NY 11530
(516) 248-1888

IBM Educational Systems
Department PC
4111 Northside Parkway
Atlanta, GA 30327
(800) 426-9920

J and S Software
P.O. Box 1276
Port Washington, NY 11050
(800) 767-8696

Merlan Scientific, Ltd.
247 Armstrong Avenue
Georgetown, Ontario,
Canada L7G 4X6
(800) 387-2474

Minnesota Educational Computing Corp. (MECC)
6160 Summit Drive
Minneapolis, MN 55430
(800) 685-6322

Project Seraphim
Department of Chemistry
University of
Wisconsin-Madison
1101 University Avenue
Madison, WI 53706
(608) 263-2837

Queue, Inc.
338 Commerce Drive
Fairfield, CT 06432
(800) 232-2224

Sunburst Communications
39 Washington Avenue
Pleasantville, NY 10570
(800) 431-1934

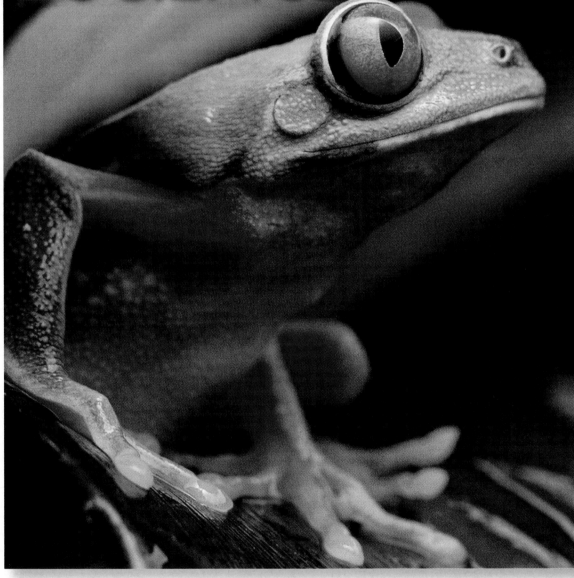

Glencoe
LIFE
SCIENCE

GLENCOE
McGraw-Hill

New York, New York   Columbus, Ohio   Woodland Hills, California   Peoria, Illinois

A GLENCOE PROGRAM

Glencoe Life Science

Student Edition
Teacher Wraparound Edition
Study Guide SE and TE
Reinforcement SE and TE
Enrichment SE and TE
Concept Mapping
Critical Thinking/Problem Solving
Activity Worksheets
Chapter Review
Chapter Review Software
Lab Manual SE and TE
Science Integration Activities
Cross-Curricular Integration
Science and Society Integration

Technology Integration
Multicultural Connections
Performance Assessment
Assessment
Lesson Plans
Spanish Resources
MindJogger Videoquizzes and
　　Teacher Guide
English/Spanish Audiocassettes
CD-ROM Multimedia System
Interactive Videodisc Program
Computer Testbank—
　　DOS and Macintosh Versions

Transparency Packages:
　　Teaching Transparencies
　　Section Focus Transparencies
　　Science Integration Transparencies

The Glencoe Science Professional Development Series
　　Performance Assessment in the Science Classroom
　　Lab and Safety Skills in the Science Classroom
　　Cooperative Learning in the Science Classroom
　　Alternate Assessment in the Science Classroom
　　Exploring Environmental Issues

Cover Photograph: Red-eyed tree frog *Agalychnis sp.* by Stephen Dalton

Glencoe/McGraw-Hill
A Division of The **McGraw·Hill** Companies

Send all inquiries to:
Glencoe/McGraw-Hill
936 Eastwind Drive
Westerville, OH 43081

ISBN 0-02-827777-5
Printed in the United States of America.

4 5 6 7 8 9 10 071/043 06 05 04 03 02 01 00 99

Authors

Lucy Daniel, a North Carolina science teacher in Rutherford County, has 37 years of teaching experience in biology, life science, physical science, and science education. She holds a B.S. degree from the University of North Carolina at Greensboro, an M.A.S.E. from Western Carolina University at Cullowhee, and an Ed.D. from East Carolina University, Greenville, North Carolina. She was the recipient of the Presidential Award for Excellence in Science and Mathematics Teaching in 1984. She has also received the Governor's Business Award for Excellence in Science and Mathematics Education, the Austin D. Bond Award for Distinguished Service to Science Education from East Carolina University, and the North Carolina Science Teachers Award for Distinguished Service to Science Education. She is coauthor of Glencoe's *Biology: An Everyday Experience.*

Ed Ortleb is a science consultant for the St. Louis, Missouri, Public Schools. He holds an A.B. in Education/Biology from Harris Teachers College, and an M.A. and Advanced Graduate Certificate in Science Education from Washington University, St. Louis. Mr. Ortleb is a life member of NSTA and served as its president in 1978-79. He has received numerous awards for his service to science education. He is a coauthor of Merrill Publishing Company's *Science Connections* and Glencoe's *Science Interactions.*

Alton Biggs is Biology Instructor at Allen High School, Allen, Texas. Mr. Biggs received his B.S. in Natural Sciences and an M.S. in Biology from East Texas State University. He has completed additional graduate work at Ball State University in genetics, and The University of Texas at Dallas in environmental science. He was a Resident in Science and Technology at Oak Ridge National Laboratory in 1986, a GTE fellow in 1988, and a Genentech Access Excellence Fellow in 1995. Among the teaching awards that he has received are Texas Outstanding Biology Teacher in 1982 and 1995, Teacher of the Year for Allen ISD in 1987, and Texas Teacher of the Year Finalist in 1988. Mr. Biggs is a Past President and Honorary Member of NABT and TABT. He is coauthor of Glencoe's *Biology: The Dynamics of Life.*

Contributing Writers

Linda Barr
Freelance Writer
Westerville, Ohio

Dan Blaustein
Science Teacher and Author
Evanston, Illinois

Pam Bliss
Freelance Science Writer
Grayslake, Illinois

Mary Dylewski
Freelance Science Writer
Houston, Texas

Helen Frensch
Freelance Writer
Santa Barbara, California

Steve Glazer
Freelance Writer
Champaign, Illinois

Rebecca Johnson
Science Writer and Author
Sioux Falls, South Dakota

Devi Mathieu
Freelance Science Writer
Sebastopol, California

Nancy Ross-Flanigan
Syndicated Science Reporter
Detroit, Michigan

Patricia West
Freelance Writer
Oakland, California

Consultants

David M. Armstrong, Ph.D.
Department of Environmental, Population,
and Organismic Biology
University of Colorado
Boulder, Colorado

Mary D. Coyne, Ph.D.
Professor of Biological Sciences
Wellesley College
Wellesley, Massachusetts

Richard E. Duhrkopf
Assistant Professor of Biology
Baylor University
Waco, Texas

Leland G. Johnson, Ph.D.
Lofthus Distinguished Professor of Biology
Augustana College
Sioux Falls, South Dakota

Robert Klips, Ph.D.
Visiting Assistant Professor of Science
Denison University
Granville, Ohio

Richard Storey, Ph.D.
Chairman and Professor of Biology
Colorado College
Colorado Springs, Colorado

Multicultural

Karen Muir, Ph.D.
Lead Instructor
Department of Social and
Behavioral Sciences
Columbus State Community
College
Columbus, Ohio

Reading

Elizabeth L. Gray, Ph.D.
Graduate Education Program
Otterbein College
Westerville, Ohio
Middle School Reading
Specialist
Heath City Schools
Heath, Ohio

Safety

Douglas K. Mandt
Science Education Consultant
Sumner, Washington

Reviewers

James Andrews
Herndon Magnet School
Belcher, Louisiana

Grant Critchfield
Texas Association for Environmental
    Education
Salado, Texas

Rosa G. Davis
Farragut Middle School
Knoxville, Tennessee

Richard Dwayne Gardner
Van Buren School
Keosauqua, Iowa

Kathy Giammittorio
Blessed Sacrament School
Alexandria, Virginia

Fleurette Hershman
Montclair College Preparation School
Van Nuys, California

Doug A. Lawrence
Les Bois Junior High School
Boise, Indiana

Sherri Marchant
Austin Junior High School
Irving, Texas

Lillian J. Pater
St. Mary School
Cincinnati, Ohio

Patsye D. Peebles
University Lab School
Baton Rouge, Louisiana

James Ristine
Bermudian Springs Middle School
York Springs, Pennsylvania

G. Stuart Rogers
Tioga Central School
Tioga Center, New York

William V. Soranno
Union High School
Union, New Jersey

Terry Thompson
Richmond Middle School
Hanover, New Hampshire

Contents

UNIT 1

Life   2

Chapter 1   **Exploring Life**   **4**

1-1   Living Things   6
1-2   Where does life come from?   10
1-3   What is science?   14
    Activity 1-1/Design Your Own Experiment:
      Using Scientific Methods   18
1-4   Science and Society—Issue:
      Technology and the Dairy Industry   24

Chapter 2   **The Structure of**
                 Viruses and Cells   **30**

2-1   Viruses   32
2-2   Science and Society—Technology:
      AIDS Vaccine?   38
2-3   Cells: The Units of Life   40
2-4   Cell Organization   45
    Activity 2-1/Comparing Plant
      and Animal Cells   53
    Activity 2-2/Design Your Own Experiment:
      Observing the Parts of an Organism   56

Chapter 3   **Cell Processes**   **62**

3-1   Chemistry of Living Things   64
3-2   Cell Transport   71
    Activity 3-1/Observing Osmosis   77
3-3   Energy in Cells   78
    Activity 3-2/Design Your Own Experiment:
      Photosynthesis and Respiration   82
3-4   Science and Society—Technology:
      Energy from Biomass   84

Contents

Chapter 4   **Cell Reproduction**   **90**

4-1   Cell Growth and Division   92

Activity 4-1/Mitosis in Plant and
Animal Cells   96

4-2   Sexual Reproduction and Meiosis   99

4-3   DNA   104

Activity 4-2/Design Your Own Experiment:
Making a Model of DNA   110

4-4   Science and Society—Technology:
Techniques in Cloning   112

Unit 1 Project   In Search of Cells   118

UNIT 2

Heredity and Evolution   120

Chapter 5   **Heredity**   **122**

5-1   What is genetics?   124

Activity 5-1/Expected and
Observed Results   131

5-2   Genetics Since Mendel   132

Activity 5-2/Design Your Own Experiment:
Determining Polygenic Inheritance   134

5-3   Human Genetics   137

5-4   Science and Society—Technology:
The Human Genome Project   144

Contents

Chapter 6   **Evolution**   **150**
  6-1   Mechanisms of Evolution   152
      Activity 6-1/Design Your Own Experiment:
      Recognizing Variation in a Population   158
  6-2   Evidence for Evolution   162
      Activity 6-2/A Radioactive Dating Model   167
  6-3   Primate Evolution   171
  6-4   Science and Society—Issue:
      Extinction and Evolution   176
Unit 2 Project   Keep Those Toes A-Tapping!   182

Unit 3 Diversity of Life   184

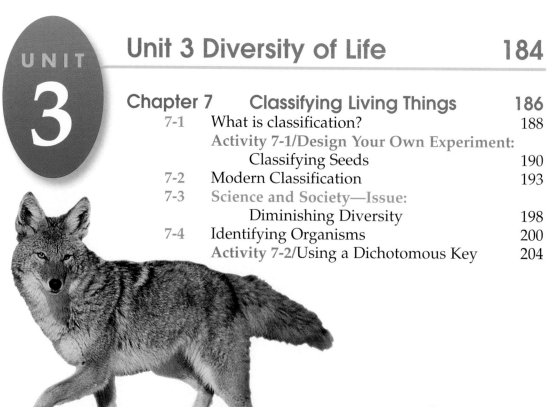

UNIT 3

Chapter 7   **Classifying Living Things**   **186**
  7-1   What is classification?   188
      Activity 7-1/Design Your Own Experiment:
      Classifying Seeds   190
  7-2   Modern Classification   193
  7-3   Science and Society—Issue:
      Diminishing Diversity   198
  7-4   Identifying Organisms   200
      Activity 7-2/Using a Dichotomous Key   204

Contents

Chapter 8   **Bacteria**   **208**

8-1   Two Kingdoms of Bacteria   210

    Activity 8-1/Observing Cyanobacteria   215

8-2   Bacteria in Your Life   216

    Activity 8-2/Design Your Own Experiment:
    Where are the most bacteria found?   220

8-3   Science and Society—Issue:
    Fighting Tuberculosis   222

Chapter 9   **Protists and Fungi**   **228**

9-1   Kingdom Protista   230

    Activity 9-1/Comparing Algae
    and Protozoans   240

9-2   Kingdom Fungi   241

    Activity 9-2/Design Your Own Experiment:
    Investigating Lichens   246

9-3   Science and Society—Technology:
    Monitoring Red Tides   248

Unit 3 Project   Don't Toss It! Compost It!   254

Unit 4 Plants   256

UNIT 4

Chapter 10   **Introduction to Plants**   **258**

10-1   Characteristics of Plants   260

10-2   Seedless Plants   267

    Activity 10-1/Design Your Own Experiment:
    The Effects of Acid Rain   272

10-3   Science and Society—Technology:
    Cleaner Coal   278

    Activity 10-2/Comparing Mosses,
    Liverworts, and Ferns   280

Contents

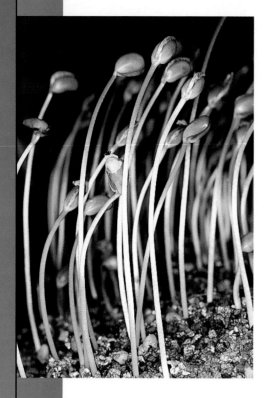

Chapter 11   The Seed Plants   284

11-1   Seed Plants   286
11-2   Parts of Complex Plants   292
11-3   Seed Plant Reproduction   296
        Activity 11-1/Comparing Flowers
        and Seeds   300
        Activity 11-2/Design Your Own Experiment:
        Germination Rate of Seeds   304
11-4   Science and Society—Issue:
        Medical Treasures   306

Chapter 12   Plant Processes   312

12-1   Photosynthesis and Respiration   314
        Activity 12-1/Stomata in Leaves   316
12-2   Plant Responses   324
        Activity 12-2/Design Your Own Experiment:
        Plant Tropisms   328
12-3   Science and Society—Technology:
        Transgenic Crops   331
Unit 4 Project   Homes Built from Plants   338

Unit 5 Animals   340

Chapter 13   Introduction to Animals   342

13-1   What is an animal?   344
13-2   Science and Society—Technology:
        Building Artificial Coral Reefs   348
13-3   Sponges and Cnidarians   350
        Activity 13-1/Observing a Cnidarian   354
13-4   Flatworms and Roundworms   357
        Activity 13-2/Design Your Own Experiment:
        Comparing Free-Living and Parasitic
        Worms   360

Contents

Chapter 14   Mollusks, Worms, Arthropods,
              and Echinoderms   368
  14-1   Mollusks   370
  14-2   Segmented Worms   375
         Activity 14-1/Design Your Own Experiment:
              Earthworms and Plant Growth 378
  14-3   Arthropods   382
         Activity 14-2/Observing a Crayfish 389
  14-4   Science and Society—Issue:
              Pesticides and Insects   390
  14-5   Echinoderms   392

Chapter 15   Fish, Amphibians,
              and Reptiles   398
  15-1   Fish   400
         Activity 15-1/Design Your Own
         Experiment:
              Water Temperature and the
              Respiration Rate of Fish   406
  15-2   Amphibians   409
         Activity 15-2/Metamorphosis
              in Frogs   413
  15-3   Science and Society—Technology:
              Amphibians and Ultraviolet
              Light   414
  15-4   Reptiles   416

Chapter 16   Birds and Mammals   424
  16-1   Birds   426
         Activity 16-1/Observing Contour and
              Down Feathers   429
  16-2   Mammals   435
         Activity 16-2/Design Your Own Experiment:
              Whale Insulation   438
  16-3   Science and Society—Issue:
              California Sea Otters   446

Contents

Chapter 17   Animal Behavior   452
  17-1   Types of Behavior   454
      Activity 17-1/Design Your Own Experiment:
        Investigating Conditioning   460
  17-2   Behavioral Adaptations   462
  17-3   Science and Society—Issue:
      Zoos and Captive Breeding   470
      Activity 17-2/Observing Social Behavior
      in Ants   472
  Unit 5 Project   Was it a reptile? Was it a bird?
      Was it really warm?   476

Unit 6 Ecology   478

Chapter 18   Life and the Environment   480
  18-1   The Living Environment and the
      Nonliving Environment   482
      Activity 18-1/Soil Composition   485
  18-2   Interactions Among Living Organisms   488
      Activity 18-2/Design Your Own Experiment:
        Identifying a Limiting Factor   490
  18-3   Matter and Energy   496
  18-4   Science and Society—Issue:
      Bringing Back the Wolves   502

Contents

Chapter 19   Ecosystems   508

19-1   How Ecosystems Change   510
         Activity 19-1/Design Your Own Experiment:
                 Observing Succession   514
19-2   Land Environments   516
         Activity 19-2/Studying a Land Environment 519
19-3   Science and Society—Issue:
                 Protecting Antarctica   528
19-4   Water Environments   530

**Chapter 20   Resources and the
                         Environment   538**

20-1   Natural Resources   540
20-2   Science and Society—Technology:
                 Recycling   548
20-3   Conservation and Wildlife Protection   550
         Activity 20-1/Design Your Own Experiment:
                 Investigating Biodegradable
                 Substances   552
20-4   Maintaining a Healthy Environment   557
         Activity 20-2/Modeling the Greenhouse
                 Effect   561
Unit 6 Project   Ecology on the Internet   568

Contents

Unit 7 The Human Body   570

Chapter 21   **Bones, Muscles, and Skin**   **572**
21-1   The Skeletal System   574
     Activity 21-1/Design Your Own Experiment:
     Observing Bones   580
21-2   Science and Society—Technology:
     Biomaterial—Speeding Bone
     Fracture Recovery   582
21-3   The Muscular System   584
     Activity 21-2/Observing Muscle   589
21-4   Skin   591

Chapter 22   **Nutrients and Digestion**   **598**
22-1   Nutrition   600
     Activity 22-1/Identifying Vitamin C Content   609
22-2   Science and Society—Technology:
     Nutrients Combat Cancer   610
22-3   Your Digestive System   612
     Activity 22-2/Design Your Own Experiment:
     Protein Digestion   618

Chapter 23   **The Circulatory System**   **624**
23-1   Circulation   626
     Activity 23-1/Taking Blood Pressure   633
23-2   Science and Society—Technology:
     "Growing" Heart Valves   634
23-3   Blood   636
     Activity 23-2/Design Your
     Own Experiment:
     Comparing Blood Cells   642
23-4   Your Lymphatic System   644

Contents

Chapter 24   Respiration and Excretion   650

24-1   Your Respiratory System   652
        Activity 24-1/Design Your Own Experiment:
        The Effects of Exercise on Respiration   662
24-2   **Science and Society—Issue:**
        Restricting Cigarette Advertising   664
24-3   Your Urinary System   666
        Activity 24-2/Sweat Glands in the Skin   672

**Chapter 25   The Nervous and Endocrine
                Systems   676**

25-1   Your Nervous System   678
        Activity 25-1/Reaction Time   685
25-2   **Science and Society—Issue:**
        Care and Treatment of Alzheimer's
        Patients   686
25-3   The Senses   688
        Activity 25-2/Design Your Own Experiment:
        Investigating Skin Sensitivity   692
25-4   Your Endocrine System   695

Contents

Chapter 26   Reproduction and Growth   702

26-1   Human Reproduction   704
        Activity 26-1/Interpreting Diagrams   709
26-2   Fertilization to Birth   710
26-3   Development After Birth   716
        Activity 26-2/Average Growth Rate
            in Humans   721
26-4   Science and Society—Technology:
            A New View of the Old   722

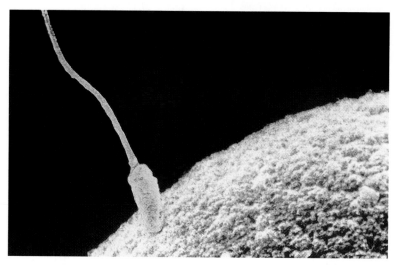

Chapter 27   Immunity   728

27-1   Disease   730
        Activity 27-1/Microorganisms and Disease   736
27-2   Your Immune System   737
        Activity 27-2/Design Your Own Experiment:
            Stopping Growth of Microorganisms   742
27-3   Science and Society—Issue:
            Should a harmful virus be destroyed?   744
27-4   Noncommunicable Disease   746
Unit 7 Project   On Your Mark! Get Set! Go!   754

Contents

Appendices

Appendix A Care and Use of a Microscope   756

Appendix B SI/Metric to English Conversions   757

Appendix C Safety in the Classroom   758

TRIPLE BEAM BALANCE
700 SERIES U.S. PAT. NO. 2,705,439 800 SERIES
2610g CAPACITY 5 lb 2 oz

Safety Symbols   759

Appendix D   Classification   760

Skill Handbook   765

Glossary   785

Spanish Glossary   799

Index   815

Activities

1-1 Design Your Own Experiment: Using Scientific Methods   18

2-1 Comparing Plant and Animal Cells   53

2-2 Design Your Own Experiment: Observing the Parts of an Organism   56

3-1 Observing Osmosis   77

3-2 Design Your Own Experiment: Photosynthesis and Respiration   82

4-1 Mitosis in Plant and Animal Cells   96

4-2 Design Your Own Experiment: Making a Model of DNA   110

5-1 Expected and Observed Results   131

5-2 Design Your Own Experiment: Determining Polygenic Inheritance   134

6-1 Design Your Own Experiment: Recognizing Variation in a Population   158

6-2 A Radioactive Dating Model   167

7-1 Design Your Own Experiment: Classifying Seeds   190

7-2 Using a Dichotomous Key   204

8-1 Observing Cyanobacteria   215

8-2 Design Your Own Experiment: Where are the most bacteria found?   220

9-1 Comparing Algae and Protozoans   240

9-2 Design Your Own Experiment: Investigating Lichens   246

10-1 Design Your Own Experiment: The Effects of Acid Rain   272

10-2 Comparing Mosses, Liverworts, and Ferns   280

11-1 Comparing Flowers and Seeds   300

11-2 Design Your Own Experiment: Germination Rate of Seeds   304

12-1 Stomata in Leaves   316

12-2 Design Your Own Experiment: Plant Tropisms   328

13-1 Observing a Cnidarian   354

13-2 Design Your Own Experiment: Comparing Free-Living and Parasitic Worms   360

Activities

14-1 Design Your Own Experiment: Earthworms and Plant Growth   378

14-2 Observing a Crayfish   389

15-1 Design Your Own Experiment: Water Temperature and the Respiration Rate of Fish   406

15-2 Metamorphosis in Frogs   413

16-1 Observing Contour and Down Feathers   429

16-2 Design Your Own Experiment: Whale Insulation   438

17-1 Design Your Own Experiment: Investigating Conditioning   460

17-2 Observing Social Behavior in Ants   472

18-1 Soil Composition   485

18-2 Design Your Own Experiment: Identifying a Limiting Factor   490

19-1 Design Your Own Experiment: Observing Succession   514

19-2 Studying a Land Environment   519

20-1 Design Your Own Experiment: Investigating Biodegradable Substances   552

20-2 Modeling the Greenhouse Effect   561

21-1 Design Your Own Experiment: Observing Bones   580

21-2 Observing Muscle   589

22-1 Identifying Vitamin C Content   609

22-2 Design Your Own Experiment: Protein Digestion   618

23-1 Taking Blood Pressure   633

23-2 Design Your Own Experiment: Comparing Blood Cells   642

24-1 Design Your Own Experiment: The Effects of Exercise on Respiration   662

24-2 Sweat Glands in the Skin   672

25-1 Reaction Time   685

25-2 Design Your Own Experiment: Investigating Skin Sensitivity   692

26-1 Interpreting Diagrams   709

26-2 Average Growth Rate in Humans   721

27-1 Microorganisms and Disease   736

27-2 Design Your Own Experiment: Stopping Growth of Microorganisms   742

MiniLABs

1-1 Can you repeat Pasteur's experiment?   11

1-2 How are things measured?   22

2-1 How are objects magnified?   41

2-2 How can you model cytoplasm?   48

3-1 How do enzymes work?   68

3-2 How does temperature affect the rate of diffusion of molecules?   72

4-1 How does one cell become two?   97

4-2 Comparing RNA and DNA   108

5-1 What are some common traits?   125

5-2 What are the distinguishing features of fingerprints?   133

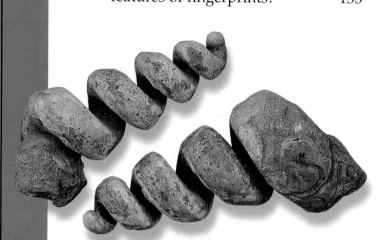

6-1 How does evolution occur?   157

6-2 How are fossils made?   163

7-1 How are organisms named?   201

8-1 How quickly do bacteria grow?   212

8-2 How can you make yogurt?   217

9-1 What do slime molds look like?   237

9-2 How are spore prints made?   244

10-1 How much water can a moss hold?   268

10-2 How does water move in a plant?   271

11-1 What's in a pinecone?   297

11-2 How do plants disperse seeds?   301

12-1 How can the use of carbon dioxide by plants be shown?   318

12-2 How can respiration in yeast be shown?   321

13-1 What are the adaptations of a sponge?   352

13-2 How do planarians move?   358

14-1 What type of metamorphosis do fruit flies undergo?   385

14-2 How do tube feet open clamshells?   393

15-1 How does a fish adjust to different depths?   404

15-2 What are the adaptations of a frog?   410

MiniLABs

16-1 How does oil affect the feathers of a bird? 428

16-2 What are the characteristics of hair? 436

17-1 How does insight help you solve problems? 458

17-2 How do earthworms respond to red light? 466

18-1 How do nitrogen-fixing bacteria affect plants? 494

18-2 How does temperature affect the water cycle? 500

19-1 Is there tundra or taiga in South America? 520

19-2 Freshwater Environments 532

20-1 How does mining for mineral deposits affect the environment? 541

20-2 Is acid rain falling on your town? 559

21-1 How do muscle pairs work? 587

21-2 Do you always sweat? 593

22-1 How much water? 607

22-2 How are fats emulsified? 615

23-1 How hard does your heart work? 628

23-2 What happens in a clogged artery? 630

24-1 How do alveoli change surface area? 656

24-2 What do kidneys look like? 667

25-1 How is balance maintained? 690

25-2 Who has a better sense of smell? 691

26-1 How does an embryo develop? 713

26-2 What immunizations are given to babies and young children? 718

27-1 Are bacteria present in food? 733

27-2 How fast do bacteria reproduce? 734

Explore Activities

1 Find out how scientific research begins. 5
2 Observe a large cell. 31
3 Compare the effects of plain water and salt water on potato slices. 63
4 Observe seeds. 91
5 Observe differences in human hairlines. 123
6 Model evolution by exploring camouflage coloring. 151
7 Use distinguishing features to classify organisms. 187

8 Find out what bacteria look like. 209
9 Dissect a mushroom to find out how it is put together. 229
10 Determine what plant parts are edible. 259
11 Dissect a seed to find out what is inside. 285
12 Observe the relationship between plants and water. 313
13 Identify which organisms are animals. 343
14 Observe a scallop shell. 369
15 Model snake hearing. 399
16 Model how a bird's gizzard works. 425
17 Investigate a simple response. 453
18 Determine the space available to an individual. 481

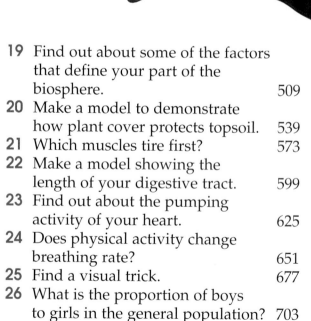

19 Find out about some of the factors that define your part of the biosphere. 509
20 Make a model to demonstrate how plant cover protects topsoil. 539
21 Which muscles tire first? 573
22 Make a model showing the length of your digestive tract. 599
23 Find out about the pumping activity of your heart. 625
24 Does physical activity change breathing rate? 651
25 Find a visual trick. 677
26 What is the proportion of boys to girls in the general population? 703
27 How are disease-causing organisms spread? 729

Problem Solving

Life Scientists and Problem Solving	16
Surface Area and Volume of Living Cells	50
Permeability	80
Chromosome Numbers and Meiosis	101
Boy or Girl?	139
Why isn't Earth covered with pumpkins and pike?	155
Classifying Unknown Organisms	189
Controlling Bacterial Growth	213
Puzzled About Slime	238
What is in nature's medicine chest?	264
Are seed company claims true?	302
The Forgotten Plant	327
Identifying Animals	346
How do tube worms eat?	377
Red-Spotted Newts and Parasites	411
Bats and Sound Waves	441
Abilities of Homing Pigeons	468
Changes in Antarctic Food Webs	499
Saving the Rain Forests	525
Indoor Air Pollution	558
Shape Influences Strength	577
A Menu for Endurance	602
The Baby Exchange	640
Frederick's Dilemma	669
Why am I so tired?	696
When to Test	712
Fighting TB	739

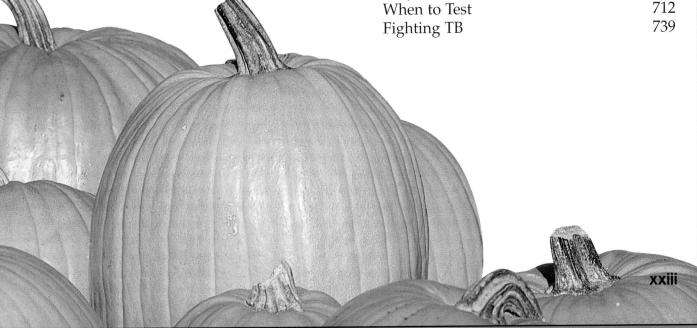

Using Technology

Fission Track Dating   12
Replacement Skin   54
Slowing Cystic Fibrosis   75
DNA Testing   107
Karyotyping   141
Bringing Bones to Life   173
Beyond Appearances   194
Lyme Disease Vaccine   218
Confocal Microscopy   243
The Ozone Connection   276

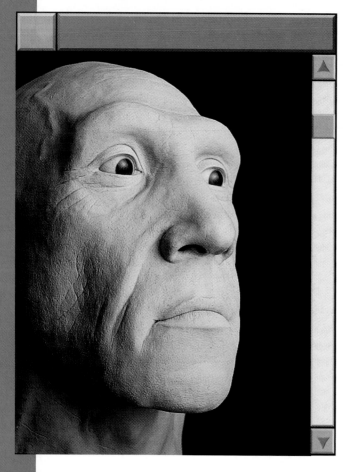

Skinning Your Fruit   293
Pollution Cleanup   319
A Hookworm Vaccine   362
A Fly in My Wound   384
Treating Snakebites   419
Saving the Nene   433
Artificial Insecticides   467
Monitoring Mayflies   493
Ecosystems in Space?   523
Geothermal Power   545
Robot Skin   592
Bioengineered Foods   606
A New Heart Pump   629
Kidney Stones   670
Watching the Brain   683
Prenatal Surgery   714
Shots for Good Health   749

Skill Builders

ORGANIZING INFORMATION
Sequencing: 98, 363, 388, 405, 412, 513, 588, 661, 708

Classifying: 161, 203, 295, 445, 501

Outlining: 720

THINKING CRITICALLY
Observing and Inferring: 13, 130, 192, 373, 394, 459, 487, 617

Comparing and Contrasting: 9, 55, 81, 136, 245, 291, 323, 330, 356, 533, 556, 564, 579, 645, 697, 740

Recognizing Cause and Effect: 735

DESIGNING AN EXPERIMENT
Measuring in SI: 219

Hypothesizing: 70, 265, 303

Using Variables, Constants, and Controls: 469

GRAPHICS
Concept Mapping: 23, 37, 44, 76, 109, 143, 175, 196, 214, 277, 347, 420, 434, 527, 546, 594, 632, 671, 684

Making and Using Tables: 103, 170, 239, 495, 608, 641, 694, 750

Making and Using Graphs: 715

Interpreting Scientific Illustrations: 381

People and Science

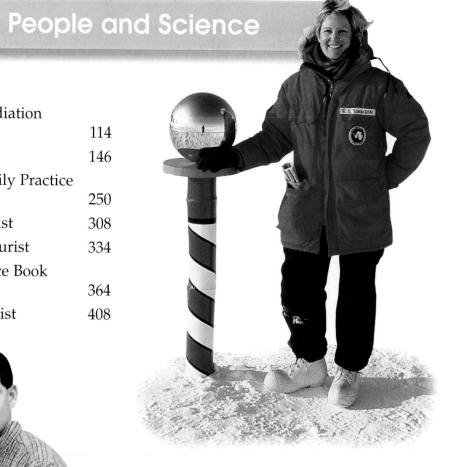

Dr. Alfred L. Goldson, Radiation
    Oncologist   114
Sean B. Carroll, Geneticist   146
Dr. Regina Benjamin, Family Practice
    Physician   250
Mark Plotkin, Ethnobotanist   308
Flora Ninomiya, Horticulturist   334
Rebecca L. Johnson, Science Book
    Author   364
Nancy Aguilar, Ichthyologist   408

Tania Zenteno-Savin, Marine
    Biologist   448
Alfred Gonzalez, Salmon Habitat
    Researcher   534
Georgia Larson, Clinical Dietitian   620
Maureen Diggins, Researcher   698
Carol Ambrose, Nurse Midwife   724

Science Connections

Science and History

Viruses Through History   58
Clarifying Classification   197
Ethics and the Hippocratic Oath   741

Science and Literature

The Lives of a Cell   86
Spoon River Anthology   224
Honeybees   465
Never Cry Wolf   504
Frankenstein   590
Fantastic Voyage   646

Science and Art

The Art of Bev Doolittle:
   A Study in Observation   26
Stone Age Art   178
The Art of Beatrix Potter   266
Mollusks and the Art of the
   South Pacific Islanders   374
Trash Art   547
Breath Control in Singers   655

Life

In Unit 1, students explore the features, needs, and origins of viruses and living things using scientific methods and problem solving. The importance of the development of microscopes to the study of cells is discussed. Students relate the functions of cells, the processes of cells, and the reproduction of cells to simple and complex organisms. The unit concludes with students learning that cells grow and divide by mitosis and meiosis.

CONTENTS

Chapter 1
Exploring Life   4
1-1   Living Things
1-2   Where does life come from?
1-3   What is science?
1-4   Science and Society: Technology and the Dairy Industry

Chapter 2
The Structure of Viruses and Cells   30
2-1   Viruses
2-2   Science and Society: AIDS Vaccine?
2-3   Cells: The Units of Life
2-4   Cell Organization

Chapter 3
Cell Processes   62
3-1   Chemistry of Living Things
3-2   Cell Transport
3-3   Energy in Cells
3-4   Science and Society: Energy from Biomass

Chapter 4
Cell Reproduction   90
4-1   Cell Growth and Division
4-2   Sexual Reproduction and Meiosis
4-3   DNA
4-4   Science and Society: Using Transplanted Cells to Treat Diabetes

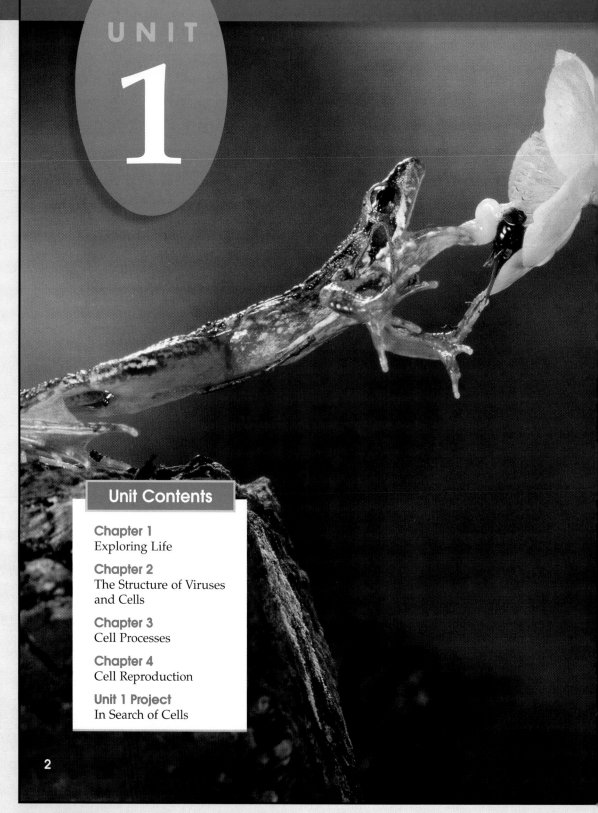

UNIT 1

Unit Contents

Chapter 1
Exploring Life

Chapter 2
The Structure of Viruses and Cells

Chapter 3
Cell Processes

Chapter 4
Cell Reproduction

Unit 1 Project
In Search of Cells

2

Science at Home

What is alive? Have students make lists of all the living things in their homes (pets, family members, plants, etc.). Have them make lists of things that were once living (dried flowers, preserved animals). Ask students to compare and contrast the living and once-living things. *Once-living things no longer move, respond, use energy, adapt, or reproduce.* **L1** **LEP**

Life

What's Happening Here?

A fly lands on a flower. Nearby, a frog sits motionless. Suddenly, the fly is snapped up by the frog's long, sticky tongue. Why does a frog eat flies? Like all living things, the frog requires food and energy to develop, grow, and reproduce. The Venus's-flytrap in the smaller photograph produces its own food, but it, too, traps flies to obtain nutrients. All living things need food to produce the energy needed to live. In this Unit, you will learn about these and the other characteristics that all living things share.

Science Journal

Animals obtain food in many different and sometimes unusual ways. The angler fish wiggles a small wormlike structure on its head to attract other fish. Smaller fish that feed on worms get the surprise of their life when they become a meal instead of getting a meal. Find three other animals that use unusual methods to obtain their food. Describe these methods in your Science Journal.

3

Theme Connection

Systems and Interactions The survival of any organism depends on the effective interaction of all its specialized structures. Ask students to identify ways in which they interact with their environment and explain how the interactions help them carry out their life processes. *They use their senses to avoid danger; they eat plants and animals from the environment to obtain energy.*

Cultural Diversity

Charles Drew, Medical Pioneer When blood is taken from donors, it is separated into red cells and plasma. The red cells spoil rapidly, but the plasma does not. Dr. Charles Drew, an African-American surgeon, discovered that patients recovered from blood loss whether they were given plasma or whole blood. Ask students what effect Dr. Drew's discovery had on blood storage and transportation. [L2]

INTRODUCING THE UNIT

What's Happening Here?

Have students look at the photos and read the text. Ask them how the frog and the Venus's flytrap are similar. *Both are living.* Then ask them what two features of living things are shown in the photo. *using food for energy and responding to other organisms in the environment*

Background

All living things, regardless of their size, need the energy supplied by food. Green plants use the sun's energy, carbon dioxide, water, and minerals from the soil to make the food used by most organisms.

Previewing the Chapters

- In the first unit, students should be encouraged to look through the chapters to acquaint themselves with features of the text. Have students find examples of the following features: Problem Solving, Connect to, Design Your Own Experiment, Skill Builder, Key Terms, and Section Wrap-up.

Tying to Previous Knowledge

Discuss with students the building blocks of living and nonliving things. Elicit that in all living things, the basic unit of organization is the cell. Ask: **Do all living things move? What are some of the ways in which they move? Why can movement alone not be used to classify living things? How do plants move?**

Chapter Organizer

Section	Objectives/Standards	Activities/Features
Chapter Opener		**Explore Activity:** Find out how scientific research begins. p. 5
1-1 **Living Things** (1 session, ½ block)*	1. **Identify** the features of living things in an organism. 2. **Recognize** the needs of living things and explain how these needs are met. National Content Standards: (5-8) UCP1, UCP4, C2, C3	**Skill Builder:** Comparing and Contrasting, p. 9 **Using Computers:** Database, p. 9
1-2 **Where does life come from?** (1 session, ½ block)*	1. **Summarize** the results of Redi's and Spallanzani's experiments. 2. **Explain** how Pasteur's experiments disproved the theory of spontaneous generation. 3. **State** the theory of biogenesis. National Content Standards: (5-8) UCP2, UCP4, A1, A2, D1, D2, E2, F5, G1, G2, G3	**MiniLAB:** Can you repeat Pasteur's experiment? p. 11 **Using Technology:** Fission Track Dating, p. 12 **Connect to Earth Science,** p. 13 **Skill Builder:** Observing and Inferring, p. 13 **Using Math,** p. 13
1-3 **What is science?** (1 session, 1 block)*	1. **Describe** methods scientists use to solve problems. 2. **Identify** and use the SI units of length, volume, mass, and temperature. National Content Standards: (5-8) UCP1, UCP2, UCP3, A1, A2	**Problem Solving:** Life Scientists and Problem Solving, p. 16 **Activity 1-1:** Using Scientific Methods, pp. 18-19 **MiniLAB:** How are things measured? p. 22 **Skill Builder:** Concept Mapping, p. 23 **Using Math,** p. 23
1-4 **Science and Society** Issue: Technology and the Dairy Industry (1 session, ½ block)*	1. **Recognize** how our lives are affected by technology. 2. **Determine** whether products resulting from the use of technology should be labeled. National Content Standards: (5-8) E2, F1, F4, F5	**Explore the Issue,** p. 25 **Science and Art:** The Art of Bev Doolittle: A Study in Observation, p. 26 **Science Journal,** p. 26

* A complete Planning Guide that includes block scheduling is provided on pages 32T-35T.

Activity Materials

Explore	Activities	MiniLABs
page 5 coffee filters, scissors, metric ruler, black felt-tipped pen, tap water, small beaker	pages 18-19 3 widemouthed 0.5-L containers, brine shrimp eggs, wooden splint, 500 mL distilled water, 500 mL weak salt solution, 500 mL strong salt solution, 3 labels, hand lens, apron, safety goggles	page 11 2 identical test tubes containing sterile beef bouillon closed off with cotton stoppers, test-tube rack, blue litmus paper page 22 pan balance, small paper clip, small glass of tap water, metal spoon, small rock

Need Materials? Call Science Kit (1-800-828-7777).

Teacher Classroom Resources

Reproducible Masters	Transparencies	Teaching Resources

Study Guide, p. 5
Reinforcement, p. 5
Enrichment, p. 5
Cross-Curricular Integration, p. 5
	Section Focus Transparency 1,
	  A Nod Toward Light
		Glencoe Life Science Interactive
		  Videodisc
		Life Science CD-ROM
		Spanish Resources
		English/Spanish Audiocassettes
		Cooperative Learning Resource
		  Guide
		Lab Partner
		Lab and Safety Skills
		Lesson Plans

Study Guide, p. 6
Reinforcement, p. 6
Enrichment, p. 6
Activity Woksheets, p. 9
Critical Thinking/Problem Solving, p. 5
	Section Focus Transparency 2,
	  Spark of Life
	Science Integration Transparency 1,
	  Earth's Calendar	

		Assessment Resources

Study Guide, p. 7; Reinforcement, p. 7
Enrichment, p. 7
Activity Worksheets, pp. 7-8, 10
Lab Manual 1, The Scientific Method
Lab Manual 2, Using the Scientific
  Method
Science Integration Activity 1,
  Measuring SI Units
Concept Mapping, p. 7
Technology Integration, p. 5
Multicultural Connections, p. 5
	Section Focus Transparency 3,
	  UFOs: Facts or Fictions?
	Teaching Transparency 1,
	  Flex Your Brain
	Teaching Transparency 2,
	  SI/Metric to English Conversions
		Chapter Review, pp. 5-6
		Assessment, pp. 5-8
		Performance Assessment in the
		  Science Classroom (PASC)
		MindJogger Videoquiz
		Alternate Assessment in the
		  Science Classroom
		Performance Assessment
		Chapter Review Software
		Computer Test Bank

Study Guide, p. 8 Reinforcement, p. 8 Enrichment, p. 8 Science and Society Integration, p. 5	Section Focus Transparency 4, Super Tomato	

Key to Teaching Strategies

The following designations will help you decide which activities are appropriate for your students.

L1 Level 1 activities should be within the ability range of all students, including those with learning difficulties.

L2 Level 2 activities should be within the ability range of the average to above-average student.

L3 Level 3 activities are designed for the ability range of above-average students.

LEP LEP activities should be within the ability range of Limited English Proficiency students.

LS These activities are designed to address different learning styles.

COOP LEARN Cooperative Learning activities are designed for small group work.

P These strategies represent student products that can be placed into a best-work portfolio.

GLENCOE TECHNOLOGY

The following multimedia resources are available from Glencoe.

**Science and Technology
Videodisc Series
(STVS)**
Plants & Simple Organisms
  Selective Breeding in
  Cows

Animals
  Temperature Regulation
  in Dogs
**National Geographic
Society Series**
STV: The Cell
Life Science CD-ROM

Teacher Classroom Resources

This is a representation of key blackline masters available in the Teacher Classroom Resources.

Teaching Aids

Section Focus Transparencies

Science Integration Transparencies

Teaching Transparencies

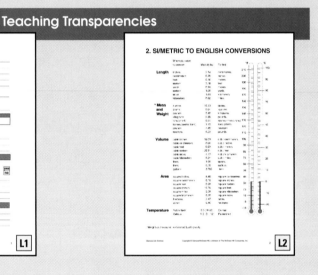

Meeting Different Ability Levels

Study Guide

Reinforcement

Enrichment Worksheets

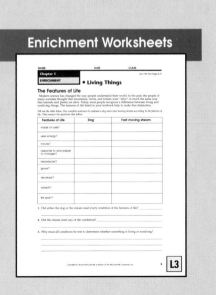

Hands-On Activities

Science Integration Activity

Lab Manual

Assessment

Performance Assessment

Enrichment and Application

Critical Thinking/ Problem Solving

Cross-Curricular Integration

Science and Society Integration

Technology Integration

Multicultural Connections

Concept Mapping

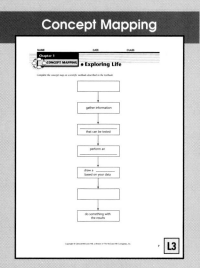

Chapter 1

Exploring Life

CHAPTER OVERVIEW

Section 1-1 Section 1-1 identifies and gives examples of characteristics of living things. Energy and the materials needed for life are discussed.

Section 1-2 This section reviews the scientists and their experiments that led to the understanding of the concept of biogenesis. The work of Oparin and its importance is discussed.

Section 1-3 Section 1-3 investigates what science is, scientific methods, critical thinking, and theories and laws. The use of SI units and safety in science are explained.

Section 1-4 Science and Society This section explores the impact of technology on the dairy industry and its consumers. The Explore the Issue feature asks students whether milk containing a recombinant hormone should be labeled.

Chapter Vocabulary

organism	biogenesis
cell	scientific
stimulus	methods
response	hypothesis
adaptation	variable
homeostasis	control
development	theory
life span	law
spontaneous	technology
generation	

Theme Connection

Systems and Interactions Four unifying themes occur throughout this book: stability and change, systems and interactions, scale and structure, and energy. In this section, the theme is systems and interactions. Interactions between an organism and its environment are necessary for life.

Previewing the Chapter

Section 1-1   Living Things
▶ Features of Life
▶ Basic Needs of Living Things

Section 1-2   Where does life come from?
▶ Life Comes from Life
▶ Life's Origins

Section 1-3   What is science?
▶ The Work of Science
▶ Solving Problems
▶ Theories and Laws
▶ Critical Thinking
▶ Measuring with Scientific Units
▶ Safety First

Section 1-4   Science and Society
Issue: Technology and the Dairy Industry

4

Learning Styles Look for the following logo for strategies that emphasize different learning modalities. **LS**

Kinesthetic	Inclusion Strategies, p. 7
Visual-Spatial	Explore, p. 5; Demonstration, p. 7; Visual Learning, pp. 7, 8, 15
Interpersonal	MiniLAB, p. 11; Brainstorming, p. 15; Activity 1-1, pp. 18-19
Intrapersonal	Demonstration, p. 15
Logical-Mathematical	MiniLAB, p. 22
Linguistic	Integrating the Sciences, p. 11; Across the Curriculum, p. 17; Science and Art, p. 26

Exploring Life

How do birds stay up in the air? Why do antibiotics cure you of some infections but not others? If you have questions such as these about the world in which you live, you think like a scientist. Scientists observe what goes on around them. They ask questions about their observations and try to find answers to their questions. In the following activity, you will observe and try to answer some questions. In the chapter that follows, you will learn about the features of living things and the methods that are used in science to study living things.

EXPLORE ACTIVITY

Find out how scientific research begins.

1. Cut a strip of coffee filter paper 3 cm by 12 cm.
2. Use a black felt-tip pen to draw a line across the paper 2 cm from the end. Describe the line.
3. Put 1 cm of water in a beaker. Put the filter paper strip in the beaker so the end is in the water and the black line is above the water.

Observe: In your Science Journal, predict what will happen to the black line. Describe what you observe as the paper absorbs the water. Does the black line change? If so, how?

Previewing Science Skills

▶ In the **Skill Builders,** you will compare and contrast, observe and infer, and map concepts.

▶ In the **Activity,** you will hypothesize, observe and infer, design an experiment, and predict.

▶ In the **MiniLABs,** you will observe and infer and measure in SI.

5

Assessment Planner

Portfolio
Refer to page 27 for suggested items that students might select for their portfolios.

Performance Assessment
See page 27 for additional Performance Assessment options.
Skill Builders, pp. 9, 13, 23
MiniLABs, pp. 11, 22
Activity 1-1, pp. 18-19

Content Assessment
Section Wrap-ups, pp. 9, 13, 23, 25
Chapter Review, pp. 27-29
Mini Quizzes, pp. 9, 13

Group Assessment
Opportunities for group assessment occur with Cooperative Learning Strategies and Flex Your Brain Activities.

EXPLORE ACTIVITY

Purpose
Visual-Spatial Use the Explore Activity to introduce students to observing in science. Explain that they will be developing skills used in science by performing and designing experiments. L1
COOP LEARN

Preparation
Obtain coffee filter paper, beakers, and water-soluble felt-tip pens.

Materials
2 or 3 filter strips (3 cm × 12 cm), 1 black water-soluble felt-tip pen, 1 beaker or glass, water for every two students

Teaching Strategies
• Have students make sure the black line does not go under the water. If this happens, the colors will not separate as the water moves up the paper.
• When the water stops moving up the paper, have students remove the paper strips.

Observe
Some students may predict that the black ink will move up the paper. Others may predict that the colors in the black ink will separate out.

Assessment
Performance Have students repeat the experiment using a different type of felt-tip pen. Different types of black ink will separate into different colors. Use the Performance Task Assessment List for Carrying Out a Strategy and Collecting Data in **PASC,** p. 25. P

Prepare

Section Background

- All the characteristics of living things are needed for survival of the individual, except for reproduction. Reproduction is needed for the survival of the species.

- Energy is the ability to do work. Work is the result of a force moving an object through a distance. The amount of work done in any action is calculated by the formula $W = F \times d$, where F is the force exerted, and d is the distance an object is moved.

1 Motivate

Bellringer

 Before presenting the lesson, display **Section Focus Transparency 1** on the overhead projector. Assign the accompanying **Focus Activity** worksheet. L1 LEP

Tying to Previous Knowledge

Students will have ideas about characteristics and needs of all living organisms. Ask what all living things have in common and write the answers on the chalkboard.

GLENCOE TECHNOLOGY

CD-ROM

Life Science CD-ROM

Have students perform the interactive exploration for Chapter 1 to reinforce important chapter concepts and thinking processes.

1•1 Living Things

Science Words

- organism
- cell
- stimulus
- response
- adaptation
- homeostasis
- development
- life span

Objectives

- Identify the features of living things in an organism.
- Recognize the needs of living things and explain how these needs are met.

Figure 1-1
Organisms are organized to perform life functions. *What life functions does this dog carry out?*

Features of Life

This book is about life—what it is and how it is maintained. Have you ever tried to explain what it means to be alive?

Imagine that it's a hot summer day. You can't wait to get out of your stuffy house or apartment. With your dog, you head off to a nearby stream. At first, you practice skipping stones across the water, while the dog runs along the bank barking. Then you sit down on some sand on the bank under a large tree. Dragonflies skim over the water, and the noise of the insects and birds is all you can hear. The dog settles down on the bank under a tree. It sure is great to be alive. But what does this mean? Living things such as you and your dog and the tree are **organisms.** Organisms have certain features that rocks and water don't have. What makes you and your dog different from water and rocks?

Organization

You have some things in common with water and rocks in the stream because you are made up of chemicals as they are. However, if someone were to look at you or your dog or the tree under a microscope, it would be clear that each of you is made of units called cells. Water and rocks are not. **Cells** are the smallest units of organisms that carry on the functions of life. Whether an organism is made up of one or many cells, it is organized to perform particular functions. Among the functions that an organism performs are reproduction, adjusting to its surroundings, and growth and development. **Figure 1-1** illustrates an organism made of many cells that are organized to perform life functions.

Program Resources

 Reproducible Masters
Study Guide, p. 5 L1
Reinforcement, p. 5 L1
Enrichment, p. 5 L3
Cross-Curricular Integration, p. 5

 Transparencies
Section Focus Transparency 1 L1

Reproduction

Your dog came from a litter of puppies. You might have looked at its parents to see what it would be like when full grown. Organisms produce new individuals that are usually like the parent organisms. Oak trees produce acorns that become oak trees. Canada geese, as shown in **Figure 1-2,** lay eggs that develop into young Canada geese. All living things reproduce.

Adjusting to Surroundings

Does your dog come running every time it hears a can opener? Anything that an organism responds to, such as sound, is a **stimulus.** The reaction of an organism to a stimulus, in this case probably tail wagging and barking, is a **response.** Organisms are able to respond to stimuli, as illustrated in **Figure 1-3,** and make adjustments to changes in their surroundings. Any characteristic an organism has that makes it better able to survive in its surroundings is an **adaptation.** Adaptations are inherited. They are not merely responses the organism makes to an immediate need. Your dog's coat is an adaptation. In winter, the dog's fur grows and his body stays warmer. In summer, dogs pant to release excess body heat. These adaptations allow dogs to adapt to changes in their surroundings. Plants have adaptations too, such as the thick bark of trees or seeds with tiny hooks.

Organisms also adjust to internal conditions. Maintaining steady conditions inside an organism, no matter what is going on outside the organism, is called homeostasis. **Homeostasis** is the regulation of an organism's internal environment to maintain conditions that allow it to live. Controls in an organism work constantly to bring it back to normal after it has been stimulated.

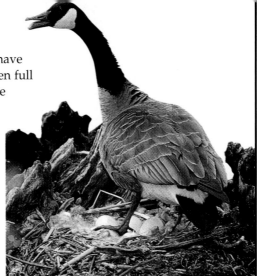

Figure 1-2

Every bird lays eggs to reproduce. *What kind of bird will hatch from these eggs?*

Figure 1-3

Cats respond to a stimulus involving food even when they are not hungry. *Why does a cat crouch down and walk slowly when it stalks a fish?*

1-1 Living Things   **7**

---

2 Teach

Demonstration

[LS] **Visual-Spatial** Place a petri dish half-filled with water on the overhead projector. Add two or three drops of Duco cement (cover the container so students will not see what it is). Have students observe. Add two or three small pieces of pencil shavings. **Ask students what characteristics of living things they see.** *Students should be able to observe movement and ingestion.* Discuss how nonliving things may appear to be living. To be classified as living, a thing must have all of the characteristics of life. **LEP**

Content Background

Explain that responses in plants are usually less obvious than in animals. However, mimosa plants show rapid response to touch. Obtain mimosa plants to demonstrate the rapid response.

GLENCOE TECHNOLOGY

⊙ **Videodisc**
STVS: Animals
Disc 5, Side 2
Temperature Regulation in Dogs (Ch. 19)

Use the **Cross-Curricular Integration,** p. 5, as you teach this lesson.

7

Students may think that growth is just an increase in size, as an icicle adds more water to increase in size. Explain that living things grow because their cells increase in size and divide.

**NATIONAL
GEOGRAPHIC
SOCIETY**

 Videodisc

STV: The Cell

Mother cat and kittens

46941

Healthy and wilted plants

46947

3 Assess

**Check for
Understanding**

 FLEX Your Brain

Use the Flex Your Brain activity to have students explore THE BASIC NEEDS OF LIVING THINGS.

📁 **Activity Worksheets,** p. 5

Reteach

Write the characteristics of life on 3" × 5" cards. Have students draw a card and give an example of the characteristic.

Extension

📁 For students who have mastered this section, use the **Reinforcement** and **Enrichment** masters.

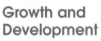

Figure 1-4

A dog's growth and development is complete in about two years. *How long does it take a human to reach adulthood?*

Figure 1-5

This tree gets energy from the sun. *How does this fungus get the energy it requires?*

8   Chapter 1 Exploring Life

Growth and Development

When a dog is born, it is small. With care and feeding, it grows larger. Most living things increase in size. They grow. All the changes organisms undergo as they grow are called **development.** Dogs can't see or walk when they are first born. But in eight or nine days, their eyes open, and their legs become strong enough to hold them up. A dog develops into an adult in about two years, as you can see in **Figure 1-4.** Adult dogs live for up to 18 more years and then they die. The length of time an organism is expected to live is its **life span.** For some, the life span is short. Millions of mayflies live only one day, but some bristlecone pine trees have been alive more than 4500 years. Your life span is about 80 years.

Basic Needs of Living Things

Sitting on the bank of the stream, you and your dog are probably not aware that you interact with everything else in your surroundings. However, all organisms take part in many interacting cycles with each other and with all the nonliving things around them.

Energy

All organisms need energy and raw materials. The main source of energy for living organisms is the sun. Green plants use the sun's energy, along with carbon dioxide, water, and minerals from the soil, to make food in the form of a sugar. The food is made in green leaves. This process also releases oxygen from water. Oxygen is used by most organisms to convert energy from food for other purposes. Some organisms get the energy they need either directly, by eating or breaking down plants, or indirectly, by eating organisms that eat plants. **Figure 1-5** shows an organism that breaks down other living things for the energy it needs.

Visual Learning

Figure 1-4 How long does it take a human to reach adulthood? *Answers may vary; 18 to 21 years may be the most common answer.*
Figure 1-5 How does this fungus get the energy it requires? *by breaking down cells of the tree* **LEP** **LS**

Water, Oxygen, and Minerals

Nonliving things that organisms need are water, oxygen, and minerals. Oxygen, water, carbon dioxide, and other chemicals are used, returned to the environment, and used over again, as they have been since life began.

Water is especially important to living things, as you can see in **Figure 1-6.** Your dog might live two or three weeks without food, but he would die in a short time without water. Living things are made up of about 70 percent water, and they need to maintain that level. Many substances in nature dissolve in water. Blood and the sap of trees are mostly water. Many organisms are born in or live in water all their lives.

Now you can answer the question posed on page 6. How are you and your dog different from the water and rocks? To be considered alive, an organism must have certain features. It must be organized to perform life functions, be able to reproduce, adjust to its surroundings, and grow and develop. Organisms need energy, water, oxygen, and minerals. Do rocks and water have these properties?

Figure 1-6

Every living thing needs water.

Section Wrap-up

Review

1. What is the main source of energy used by most organisms?

2. What are the needs of organisms, and how are these needs supplied?

3. **Think Critically:** Why is homeostasis important to the existence of organisms?

Skill Builder
Comparing and Contrasting

Compare and contrast the features of a goldfish and the flame of a burning candle. If you need help, refer to Comparing and Contrasting in the **Skill Handbook.**

Using Computers

Database Use references to find the life span of humans, horses, cows, dogs, cats, turtles, elephants, penguins, porpoises, spiders, and other animals. Use your computer to make a database. Then graph the life spans from least to greatest.

Skill Builder
The goldfish is made of cells, moves, responds, uses energy, adjusts to changes in the environment, grows and develops, adapts, and has a life span. The candle flame moves, responds to wind in the environment, uses energy from the candle, gives off heat energy, produces waste products—carbon dioxide and water—and uses oxygen to burn.

✓ Assessment

Process Have students heat water in a beaker and dissolve as much sugar in the beaker as they can. Tie a string around a stirring rod. Place the rod across the top of the beaker so the string is immersed. Leave the beaker until sugar crystals begin to grow. Ask students why these crystals are not alive. Use the Performance Task Assessment List for Making Observations and Inferences in **PASC,** p. 17. [P]

---

4 Close

•MINI•QUIZ•

Use the Mini Quiz to check students' recall of chapter content.

1. **A living thing is a(n)** _____ . *organism*

2. **The smallest units of organisms that carry on the functions of life are** _____ . *cells*

3. **Anything an organism responds to is a(n)** _____ . *stimulus*

4. **The ability to maintain steady conditions no matter what is going on inside or outside an organism is** _____ . *homeostasis*

Section Wrap-up

Review

1. the sun

2. They need energy and raw materials. Energy comes from food or the sun; raw materials come from water, oxygen, and minerals.

3. **Think Critically** Without self-regulation, the organism would not be able to function and would die.

Using Computers

Life spans will vary in different references. Some examples are: humans, 74 years; horses, 30 years; cows, 24 years; dogs, 15 years; cats, 12 years; turtles, 125 years; elephants, 60 years; penguins, 23 years; porpoises, 21 years; shrews, 1 1/2 years; snakes, 11 years; spiders, 3 years.

Prepare

Section Background

- Redi believed that spontaneous generation was the only way to explain galls, swollen growths on plant stems and leaves. After Redi's death, one of his students showed that the gall was made by a wasp larva hatched from an egg inside the plant tissue.

- Most biologists now accept the idea of original spontaneous generation at the molecular level. This is considered to have happened under quite different environmental conditions from those that exist today.

Preplanning

Prepare sterile beef bouillon for the MiniLAB. Obtain test tubes and cotton.

1 Motivate

Bellringer

Before presenting the lesson, display **Section Focus Transparency 2** on the overhead projector. Assign the accompanying **Focus Activity** worksheet. [L1] [LEP]

Tying to Previous Knowledge

Review the characteristics of life and the needs of living things from the previous section and connect those ideas with concepts of the origins of life in this section.

1•2 Where does life come from?

Science Words

spontaneous generation
biogenesis

Objectives

- Summarize the results of Redi's and Spallanzani's experiments.
- Explain how Pasteur's experiments disproved the theory of spontaneous generation.
- State the theory of biogenesis.

Life Comes from Life

Have you ever taken a walk after a thunderstorm and found earthworms all over the sidewalk? People have noticed that earthworms are often found on the ground after rainstorms. It's no wonder that people used to think the earthworms had fallen from the sky when it rained. It was a logical conclusion based on repeated experience. But was it true? Jan Baptist van Helmont wrote a recipe for making mice by placing grain in a corner and covering it with rags. For much of history, people believed that living things came from nonliving matter, an idea called the theory of **spontaneous generation.**

Controlled Experiments

People also believed that maggots grew from decaying meat. In 1668, Francesco Redi, an Italian doctor, conducted one of the first controlled experiments in science. A controlled experiment is one in which you change just one thing to find out the result. Redi put meat into two jars. He covered one jar with a cloth but left the other jar open. The two jars and their contents were the same, except for one thing: the cover on one jar. He showed that maggots hatched from eggs that flies had laid on meat, and not from the meat itself. Redi's experiment is shown in **Figure 1-7.**

In the late 1700s, Lazzaro Spallanzani designed an experiment to show that tiny organisms came from other tiny organisms in the air. He sterilized broth in two flasks, sealed one, and left the other one open to the air. The open flask became cloudy with organisms. The sealed flask developed no organisms. Spallanzani's experiment showed that organisms from the air entered one flask and not the other because it was sealed.

Egg Pupa Larva Adult

Figure 1-7

In Redi's experiment, maggots developed on snake meat that flies could reach (A), but not on meat that was protected from the flies (B).

Program Resources

Reproducible Masters

Study Guide, p. 6 [L1]
Reinforcement, p. 6 [L1]
Enrichment, p. 6 [L3]
Activity Worksheets, pp. 5, 9 [L1]
Critical Thinking/Problem Solving, p. 5

Transparencies

Section Focus Transparency 2 [L1]
Science Integration Transparency 1

Broth is boiled. Air and microbes are driven out.

Microbes are trapped in the neck.

One year or more: broth remains without growth.

Flask is tilted.

Shortly, microbes develop in broth.

Figure 1-8

Pasteur's experiment disproved the theory of spontaneous generation.

Biogenesis

It was not until the mid-1800s that Louis Pasteur, a French chemist, showed conclusively that living things do not come from nonliving materials. In the experiment illustrated in **Figure 1-8,** Pasteur boiled broth in flasks with long, curved necks. The broth became contaminated only when dust that had collected in the curved neck of one flask was allowed to mix with the broth. The work of Redi, Spallanzani, Pasteur, and others provided enough evidence finally to disprove the theory of spontaneous generation. It was replaced with **biogenesis,** the theory that living things come only from other living things.

Life's Origins

If living things can come only from other living things, how then did life on Earth begin? Scientists hypothesize that about 5 billion years ago, the solar system was a whirling mass of gas and dust. The sun and planets formed from this mass. Earth is thought to be about 4.6 billion years old. Rocks found in Australia, which are more than 3.5 billion years old, contain fossils of once-living organisms.

Oparin's Hypothesis

One hypothesis on the origin of life was proposed by Alexander I. Oparin, a Russian scientist. Oparin suggested that the atmosphere of early Earth was made up of gases similar to ammonia, hydrogen, methane, and water vapor. No free oxygen was present as it is today. Energy from lightning and ultraviolet rays from the sun helped these early gases to combine. The gases formed the chemical compounds of which living things are made. Oparin suggested that as the compounds were formed, they fell into hot seas. Over a period of time, the chemical compounds in the seas formed new and more complex compounds. Eventually, the complex compounds were able to copy themselves and make use of other chemicals for energy and food.

MiniLAB

Can you repeat Pasteur's experiment?

Procedure

1. Obtain two test tubes of sterile beef bouillon from your teacher. Remove the cotton stopper from one of the tubes.
2. Place the test tubes in a warm, dark place.
3. At the end of two weeks, observe both test tubes. Hold both tubes of bouillon up to the light to see if the bouillon is cloudy.
4. Record the smell and color of broth in both test tubes.

Analysis

1. Which test tube turned cloudy?
2. Why didn't the other test tube turn cloudy?
3. Which test tube had a bad odor? Can you explain why?

 INTEGRATION
Earth Science

Integrating the Sciences

Astronomy Have students use reference books to learn about planetary features such as temperature and relative distance from the sun, and relate these characteristics to the presence of life on Earth. L1 LS
P

2 Teach

MiniLAB

Purpose

LS **Interpersonal** Students will conduct an experiment similar to Pasteur's. L1
COOP LEARN

Materials

2 test tubes of sterile beef bouillon stoppered with cotton, blue litmus paper

Teaching Strategies

Dissolve one beef bouillon cube in a cup of water for four test tubes of broth. Seal the test tubes with cotton. Place the test tubes in a pan of water and heat until the water has boiled for 30 minutes. Broth in the unsealed tube will become cloudy and turn blue litmus red. Blue litmus turns red in acid solution. Organic acids are produced by the bacteria during growth.

Analysis

1. the open one
2. It was sealed and no organisms could enter.
3. The unsealed test tube; it had microorganisms that produced gases.

📁 **Activity Worksheets,** pp. 5, 9

✓ Assessment

Oral Have students make a hypothesis about what microorganisms need for growth based on this experiment. Use the Performance Task Assessment List for Formulating a Hypothesis in **PASC,** p. 21. P

📁 Use **Critical Thinking/ Problem Solving,** p. 5, as you teach this lesson.

11

INTEGRATION
Earth Science

See page 11.

Oparin and Haldane independently hypothesized that organic compounds were created in Earth's primitive atmosphere and became organized to form cells. In 1953, Miller and Urey showed that organic compounds, including amino acids, could be synthesized in a simulation of Earth's primitive atmosphere using sparks instead of lightning. P

USING TECHNOLOGY

Use your local library to obtain more information on fission track dating.

Think Critically
From minerals, rocks, and fossils, the ages of layers of Earth can be determined and then scientists know when different events occurred.

3 Assess

Check for Understanding

FLEX Your Brain

Use the Flex Your Brain activity to have students explore LIFE'S ORIGINS.

Reteach

Write a description of each of the experiments mentioned in this section on 3" × 5" cards. Have a student draw a card, read it, and describe the conclusions of the experiment.

Extension

For students who have mastered this section, use the **Reinforcement** and **Enrichment** masters.

USING TECHNOLOGY

Fission Track Dating

How do scientists know how old rocks are? One possible way to figure out the absolute age of a rock is to use radioactive isotopes of elements. Recently, a simpler way to date some rocks has been found.

Fission Leaves Tracks

Fission track dating relies on the spontaneous fission, or splitting, of radioactive atoms in some minerals. As these atoms break apart, they emit high-energy particles. The high-energy particles tear through the crystals of the mineral, creating tracks. This process occurs over a long time at a constant rate. A sample of the mineral can be soaked in an acid solution for a short time to enlarge the tracks. Then they can be seen using an ordinary microscope. Geologists count the tracks to learn how many radioactive atoms have already split. Then they bombard the sample with neutrons so the remaining radioactive atoms will split. They count the tracks again, and the total count of tracks tells how many radioactive atoms existed when the mineral was formed. This method is used to give accurate dates for rocks that formed from about 40 000 to 1 million years ago.

Think Critically:
Why is it helpful to know the age of minerals?

Fission Tracks

Miller's Experiment

Ever since Oparin formed his hypothesis, other scientists have been testing it. In 1953, an American scientist, Stanley L. Miller, set up an experiment using the chemicals suggested in Oparin's hypothesis. Miller's experiment is illustrated in **Figure 1-9.** Electrical sparks were sent through the mixture of chemicals. At the end of a week, amino acids, sugars, and other organic compounds had formed. This showed that substances present in living things *could* come from nonliving materials in the environment. It did not prove that life was formed in this way.

Evidence suggests that life was formed from nonliving matter sometime between 4.6 billion and 3.5 billion years ago. However, scientists are still investigating where the first life came from.

High-voltage source

Electrode

Condenser for cooling

Stopcock for removing samples

Mixture of methane, ammonia, water vapor, and hydrogen

Solution of organic compounds

Figure 1-9

Miller's experiment was a model of what many scientists think Earth's early atmosphere was like. Volcanic eruptions like the one to the left remind us of what early Earth was probably like.

Section Wrap-up

Review

1. Compare the theory of spontaneous generation with the theory of biogenesis.

2. What was most important about Redi's and Spallanzani's experiments?

3. Think Critically: Explain how it is possible for some of Pasteur's flasks to be uncontaminated after 100 years.

Skill Builder
Observing and Inferring

People believed that Spallanzani had destroyed some "vital force" in the broth. Did he? Based on the experiment, what could they have inferred about where organisms come from? If you need help, refer to Observing and Inferring in the **Skill Handbook.**

USING MATH

Earth's age is estimated at 4.6 billion years old. Life began 3.5 billion years ago. Life has been present for what percent of Earth's existence?

Skill Builder

In a sense, he did. By boiling the broth, he destroyed the microorganisms that were in it at that time. People at that time did not understand that to grow and reproduce in the broth, the organisms had to be reintroduced.

✓ Assessment

Performance Observe the results of Redi's experiment on page 10 and write your interpretations. *The jars that were left open to the air attracted flies, and maggots appeared on the meat. Flies could not get to the meat in the jars that were covered, and no maggots appeared.* Use the Performance Task Assessment List for Making Observations and Inferences in **PASC,** p. 17. **P**

Scientists hypothesize that Earth's oceans originally formed when water vapor was released into the atmosphere from numerous volcanic eruptions. Once it cooled, rain fell to fill Earth's low areas. *Identify* other materials that are produced by volcanoes.

gases including carbon dioxide, plus compounds of silica, sulfur, and nitrogen

4 Close

•MINI•QUIZ•

Use the Mini Quiz to check students' recall of chapter content.

1. The idea that living things came from nonliving matter is called the theory of _____ . *spontaneous generation*

2. The first person known to conduct a controlled experiment was _____ . *Redi*

3. The theory that living things come only from other living things is _____ . *biogenesis*

Section Wrap-up

Review

1. spontaneous generation—living things come from nonliving matter; biogenesis—living things come only from other living things of the same kind

2. Redi conducted one of the first controlled experiments; Spallanzani's was an attempt at a controlled experiment. Both tried to demonstrate that life comes from life.

3. Think Critically If sterilized, stoppered flasks remain sealed, they would still be uncontaminated.

USING MATH

76%

Prepare

Section Background

- All scientific understanding is based on observation.
- Scientists use many methods of investigation. They approach problems from a variety of perspectives and structure their research based on available data, tools, and time. Observations often generate new hypotheses that lead to additional testing.
- The SI system was developed in France in the 1790s. SI stands for the French Système International d'Unités.

Preplanning

- To prepare for Activity 1-1, obtain glass containers, brine shrimp eggs, distilled water, weak salt solution, strong salt solution, labels, hand lenses, and wooden splints.
- For the MiniLAB, set up several balances and obtain a variety of small objects, including paper clips, small glasses, spoons, and rocks.

1 Motivate

Bellringer

Before presenting the lesson, display **Section Focus Transparency 3** on the overhead projector. Assign the accompanying **Focus Activity** worksheet. L1 LEP

Tying to Previous Knowledge

Students will be familiar with everyday problem solving. Pose questions such as, **How did you decide what to wear to school today?** Relate solving everyday problems to scientific methods, theories, and laws.

1•3 What is science?

Science Words

> **scientific methods**
> **hypothesis**
> **variable**
> **control**
> **theory**
> **law**

Objectives

- Describe methods scientists use to solve problems.
- Identify and use the SI units of length, volume, mass, and temperature.

Figure 1-10

Life scientists often identify unknown organisms. *Are these organisms plants or animals?*

A *Lithops*

B Sea anemone

The Work of Science

In this textbook, you'll be exposed to many branches of science. Chemistry is the study of elements and matter. Physics is the study of how energy and matter are related. Earth science is the study of the structure of Earth and the forces that affect it. All of these sciences relate to life science, the study of living things.

Life Science

Imagine you are a life scientist. How would you identify the organisms in **Figure 1-10**? There are millions of different animals, plants, and other organisms on Earth. Life scientists who are zoologists study animals. Botanists study plants. What do you think a bacteriologist studies? Besides studying particular organisms, life scientists are also interested in the relationships among organisms and their environment. Scientists who study the interrelationships among organisms and their environment are ecologists. Other life scientists are interested in genetics. Genetics explains how traits of organisms are passed from generation to generation. Life scientists, like all scientists, learn by asking questions.

C Kelp

Solving Problems

Do scientists always find answers to their questions? No. Do they always do experiments in laboratories? Not always. There is no set method that all scientists use to solve problems. Each problem is different. Solving any problem requires organization. In science, this organization often takes the form of a series of procedures called **scientific methods.**

Program Resources

Reproducible Masters
Study Guide, p. 7 L1
Reinforcement, p. 7 L1
Enrichment, p. 7 L3
Activity Worksheets, pp. 5, 7-8, 10 L1
Science Integration Activities, pp. 1-2
Technology Integration, pp. 5-6
Concept Mapping, p. 7

Multicultural Connections, pp. 5-6
Lab Manual 1, 2

Transparencies
Section Focus Transparency 3 L1
Teaching Transparencies 1, 2

State the Problem

Figure 1-11 shows an order in which scientific methods might be used. You begin by observing something that you cannot explain. Suppose you awake one morning with a sore throat. It hurts all day, so you go to the doctor after school. The first step in a scientific method is to state the problem. A scientist can't begin to solve a problem until it's clearly stated. You tell the doctor that you have a sore throat.

Gather Information

The second step is to gather information about the problem. Tell the doctor everything you've observed about your sore throat—when it began hurting and how it feels now. The doctor makes her own observations. She takes your temperature and examines your throat.

Form a Hypothesis

The next step is to ask a question and form a hypothesis. A **hypothesis** is a prediction that can be tested. Based on her experience, the doctor hypothesizes that you have strep throat. She knows she can test her hypothesis right there in her laboratory. A hypothesis always has to be something you can test.

Perform an Experiment

To answer her question, the doctor will test her hypothesis by performing an experiment, as illustrated in **Figure 1-12**. In an experiment, a series of steps is followed that tests a hypothesis using controlled conditions. The doctor takes a throat culture from you. She has a petri dish with a growth medium, which she marks "experimental." She gently rubs the surface of the medium in the experimental dish with your throat culture. The dish marked experimental is the variable. A **variable** is the factor tested in an experiment. The doctor also has a sample of the organism that is known to cause strep throat. She will use this sample as a control to compare with your culture. A **control** is the standard used to compare with the outcome of a test.

Figure 1-11

The steps in solving a problem in science are a series of procedures called scientific methods.

Figure 1-12

Performing an experiment is an important step in solving a problem scientifically. Here the doctor takes a throat culture to find out whether or not the student's throat infection is caused by the streptococcus bacterium. Some strands of streptococcus cells are also shown.

15

2 Teach

Brainstorming

LS **Interpersonal** Divide the class into groups of four. Ask the question: **What do you think life scientists do, and where do you think they work?** Have each group decide upon its own answer to the question. Have the groups present their conclusions to the class. When all groups have presented their ideas, discuss what most students perceive a life scientist to be. Dispel stereotyping.
COOP LEARN

Demonstration

LS **Intrapersonal** Fill a 4-L glass jar with pond water. Place it in a lighted area. Have students hypothesize how the water will appear in ten days. Collect the written hypotheses, and at the end of ten days, have students check their hypotheses against the conditions of the jar. **L1**

Inquiry Questions

• **How does a dog get fleas? Why do people get sore throats? How does a gardener get rid of weeds? How does a model ship get inside a bottle? How many ways can you divide a pizza?** *List suggestions on the chalkboard.*

• **Why should an experiment have only one variable?** *If an experiment had more than one variable, the scientist would not know which variable caused the result.*

Visual Learning

Figure 1-10 Are these organisms plants or animals? *sea anemones—animals; kelp—protists; Lithops—plants* **LEP**
LS

Problem Solving

Solve the Problem

1. The mealworm changed into a pupa, the pupa became a beetle, and the beetle laid eggs that hatched into mealworms (larvae).
2. Answers will vary; generally two or three weeks. Larvae hatch soon after the eggs are laid.
3. Eggs are probably in the flour when it is packaged at the mill.

Think Critically

No; mealworm eggs are packaged in the flour.

Content Background

- Check scientific journals out of the library and display them. Explain that before scientists begin an experiment, they search for all the information that has already been written on a subject.
- Explain that one of the most important characteristics of science is that conclusions have to be based on evidence known as data.
- Help students understand that inferring is going beyond what they actually observe.
- A scientist must be careful that his or her opinions and emotions do not influence observations. An opinion that influences an observation is a bias. A scientist's observations should be free of bias.

Inquiry Question

Why should the results of an experiment be published or shared? *Other scientists will know about the results of the experiment and can verify them or use them to develop new hypotheses.*

Problem Solving

Life Scientists and Problem Solving

Sometimes when you open a package of flour, you find mealworms in the flour. How do the mealworms get into the flour when it has been tightly sealed? To find out, place about 5 cm of wheat bran in the bottom of a 1-L glass jar. Add a slice of potato and a few strips of newspaper. Place a mealworm, a mealworm pupa, and a mealworm beetle in the jar. Observe the jar two or three times a week for four weeks.

Solve the Problem:

1. Describe the life cycle of the mealworm.
2. How long did it take to have new mealworms?
3. How do mealworms get into flour?

Think Critically:

Your mother threw out flour that had mealworms in it. She bought a new bag of flour at the grocery store. As soon as she got home, she poured the flour into a plastic container and closed it with a tight-fitting lid. Will this prevent mealworms from appearing in the flour? Explain your answer.

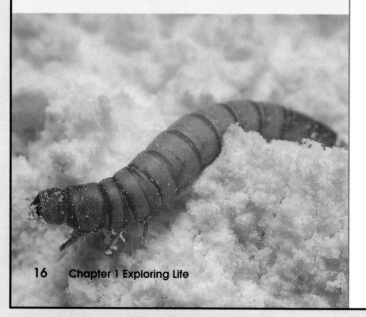

Inclusion Strategies

Learning Disabled Students can design a bulletin board showing the steps in scientific methods.

Analyze Data

The doctor tells you that she will call you in 24 hours with the test results. The next day, a lab technician identifies the growth in the experimental petri dish. The data are recorded in your chart. The doctor uses the data to draw her conclusions. A conclusion is a logical answer to a question based on data and observations of the test materials.

The next step in the search for an answer is to accept or reject the hypothesis. The doctor calls you and tells you that the experimental dish shows that your sore throat was caused by another kind of bacterium. You don't have strep throat. The doctor has had to reject her original hypothesis. Your doctor must now revise her hypothesis to include a different cause of your sore throat.

Report Results

The last step in solving a problem scientifically is to do something with the results. Your doctor may prescribe an antibiotic for you to take. An antibiotic is a drug that kills bacteria. Your doctor also may use the data in a paper she is preparing on different kinds of sore throats found in adolescents. She may have begun to notice some common factors in her data. Your sore throat may be part of a trend of sore throats caused by a new bacterium. Your doctor wants to publish her paper to find out whether other doctors have noticed the same thing. Sharing results is part of the scientific method.

Theories and Laws

Observing something you cannot explain and performing an experiment that allows you to make some conclusions results either in support or in rejection of your hypothesis. When experimentation results in the same conclusions over and over again, you may develop a theory that explains your conclusions. A **theory** is an explanation of things or events based on many observations. A theory is not a guess or someone's opinion, nor is a theory a vague idea. Hypotheses that have been tested over and over again and cannot be shown to be false support theories. You have already read about the theory of spontaneous generation. Theories can be changed as new data uncover new information.

Figure 1-13
Research in life science may involve equipment such as a petri dish. Data are records of observations. Data are recorded in a journal.

Laws

Large amounts of data in science often show a trend. A scientific **law** based on these repeating data tells us how nature works. A law is a reliable description of nature based on many observations. In physical science, you might learn about the law of gravity. Laws may change as more information becomes known.

Scientific methods help answer questions. Your questions may be as simple as "Where did I leave my house key?" or as complex as "What can we do about air pollution?" Will these methods guarantee that you will get an answer? Not always. Often they just lead you to more questions, but that is the work of science.

USING MATH

List several mathematical concepts that you might use as you study life science.

USING MATH

Averaging, ranges, measuring angles, and calculating percentages are a few examples.

Across the Curriculum

Language Arts Have students research a major concept in life science and the person who made the discovery. Some scientists who have made notable discoveries are: Francesco Redi, William Harvey, Alexander Fleming, Barbara McClintock, and George Washington Carver. [L1] [LS]

Content Background

- Carefully review and emphasize the relationship between *theory* and *hypothesis*.
- Emphasize that hypotheses and theories may be changed or dropped as new information is obtained.

Brainstorming

This is a good time to have students begin science fair projects. You may want to have a brainstorming session on ideas for projects. In each project, the scientific method should be followed for both the actual project and the report of the project.

GLENCOE TECHNOLOGY

 Videotape

The Secret of Life
Use the videotape *On the Brink: Portraits of Modern Science* to illustrate that controlled experimentation is the backbone of the process called science.

NATIONAL GEOGRAPHIC SOCIETY

Videodisc
STV: The Cell
Viewing the Cell
Unit 1
Introduction

00373-03727
Early Pioneers

03758-04908
Modern Microscopes

07212-11798

17

Activity 1-1

PREPARATION

Purpose

IS **Interpersonal** Design and carry out an experiment using scientific methods to infer why brine shrimp live in the ocean. **L1** **COOP LEARN**

Process Skills

observing and inferring, comparing and contrasting, recognizing cause and effect, interpreting data, hypothesizing, communicating, making and using tables, making and using graphs, designing an experiment, separating and controlling variables, making models, measuring in SI

Time

50 minutes during first day of the experiment, then 5 minutes a day for 3 days, finally 30 minutes to summarize

Materials

- Purchase brine shrimp from a pet store or a biological supply house.

  To prepare the solutions used, follow these directions:

- Dechlorinated water: Allow tap water to stand for 48 hours.

- Weak salt solution: Add 20 mL noniodized salt to 4 L dechlorinated water. Stir until dissolved.

- Strong salt solution: Add 75 mL noniodized salt to 4 L dechlorinated water. Stir until dissolved.

  Do not place too many brine shrimp eggs in each container. Brine shrimp are orange-colored and swim with a jerking motion. To maintain the brine shrimp, add a small amount of yeast to the container two or three times a week.

Activity 1-1

Design Your Own Experiment
Using Scientific Methods

You are to use scientific methods to determine how salt affects the growth of brine shrimp. Brine shrimp are tiny organisms that live in the ocean. How can you find out why they live where they do?

PREPARATION

Problem

How can you use scientific methods to determine how salt affects the hatching and growth of brine shrimp?

Form a Hypothesis

Based on your observations, state a hypothesis about how salt affects the hatching and growth of brine shrimp.

Objectives

Design and carry out an experiment using scientific methods to infer why brine shrimp live in the ocean.

Possible Materials

- 3 widemouthed 0.5-L containers
- brine shrimp eggs
- wooden splint
- 500 mL distilled water
- 500 mL weak salt solution
- 500 mL strong salt solution
- 3 labels
- hand lens

Safety Precautions

Protect clothing and eyes. Be careful when working with live organisms.

18 Chapter 1 Exploring Life

Alternate Materials Pint mayonnaise jars may be used for the containers in this activity.

Safety Precautions

Caution students to use care when working with live animals.

Possible Hypotheses

Most student hypotheses will reflect that brine shrimp will grow best in a strong salt solution.

📁 **Activity Worksheets**, pp. 5, 7-8

📁 As an alternative to this activity, use **Lab Manual 1** or **2**.

PLAN THE EXPERIMENT

1. As a group, agree upon and write out the hypothesis statement.
2. As a group, list the steps that you need to take to test your hypothesis. Be specific; describe exactly what you will do at each step. List your materials.
3. How will you know whether brine shrimp hatch and grow?
4. Decide what data you need to collect during the experiment. Design a data table in your Science Journal to record your data.
5. Identify the steps in solving a problem that each of your actions represents. For example, what action have you taken that represents stating the problem or gathering information? Make sure you include all the steps needed to reach a conclusion.

Check the Plan
1. Read over your entire experiment to make sure that all steps are in logical order.
2. Identify any constants, variables, and controls of the experiment.
3. Can you explain what variable you are testing?
4. Decide whether you need to run your tests more than once. Can your data be summarized in a graph?
5. *Make sure your teacher approves your plan before you proceed.*

DO THE EXPERIMENT

1. Carry out the experiment as planned.
2. While the experiment is going on, write down any observations that you make and complete the data table in your Science Journal.

Analyze and Apply
1. **Compare** your results with those of other groups.

2. Was your hypothesis supported by your data?
3. **Describe** how the physical conditions in the container in which the brine shrimp hatched and grew best are similar to those in the ocean.
4. How did you use scientific methods to solve this problem?

Go Further

Design an experiment to find out whether there is a limit to the amount of salt that brine shrimp can live with.

Possible Procedures
The same amount of brine shrimp eggs can be added to the three solutions and observed.

Teaching Strategies
Tying to Previous Knowledge Most students will know that the ocean is salty and will be aware that brine shrimp are saltwater organisms.

DO THE EXPERIMENT

Expected Outcome
Most results will reflect that the brine shrimp grew best in the strong salt solution.

Analyze and Apply
1. Student results should be consistent with those of other groups.
2. The hypothesis that brine shrimp hatch best in a salt solution was supported.
3. The container had salt, water, and oxygen and received light.
4. Students formed a hypothesis, performed an experiment with a variable and a control, recorded data, and drew conclusions.

Go Further

Experimental design should follow the activity's general format.

Assessment

Performance Have students design an experiment to determine how ocean currents affect brine shrimp. Use the Performance Task Assessment List for Designing an Experiment in **PASC**, p. 23. **P**

Discussion

Discuss inductive and deductive reasoning aspects in determining why the tulips discussed in the text did not come up. In the process of inductive reasoning, one generalizes from many experiences. After generalizing about the planting of tulips, one can reach a general idea of how tulips should be planted. Then one can eliminate the factors in planting that caused the tulips not to come up. This elimination is an example of deductive reasoning.

Content Background

To acquaint students with the Flex Your Brain strategy, divide the class into small groups and have them explore a topic using the Flex Your Brain worksheet. Students should have some familiarity with the topic chosen. Possible topics might include cacti, volcanoes, robots, or topics from a list the class brainstorms.

📁 Use these program resources as you teach this lesson.

Technology Integration, pp. 5-6

Multicultural Connections, pp. 5-6

Concept Mapping, p. 7

🗄 Use **Teaching Transparency 2** as you teach this lesson.

Figure 1-14

Tulips must be planted deeply enough in soil so they don't freeze over the winter. Tulips require a period of cold in order to grow each spring.

Critical Thinking

Whether or not you become a scientist, you are going to solve problems all your life. Most of the problems you encounter will be solved in your head by a process of sorting through ideas to see what will or will not work. Suppose you would like to have lots of tulips in your garden this spring. You plant tulip bulbs early in the spring, but then no tulips emerge later on. Even though you planted the bulbs the correct depth in the soil, you must have done something wrong. After thinking a bit, you realize you forgot one important thing: tulip bulbs must be planted in the fall, not the spring! Tulips need to go through a period of cold weather before they will grow. How did you figure out that lack of a period of cold weather was the problem? Without being aware of it, you probably used a form of problem solving called critical thinking.

Using Thinking Skills

Critical thinking is a process that uses certain skills to solve problems. For example, you identified the problem by mentally comparing the tulip bulbs that didn't grow with other bulbs you have planted. First, you separated important information from unimportant information. For instance, you may have realized that tulip bulbs must be planted fairly deep in the soil. You know you planted them deep enough. What else did you do? After looking at the planting directions again, you concluded that you planted the bulbs too late in the year.

Flex Your Brain

"Flex Your Brain" is an activity that will help you think about and examine your way of thinking. It takes you through steps of exploration from what you already know and believe, to new conclusions and awareness. Then, it encourages you to review and talk about the steps you took.

"Flex Your Brain" will help you improve your critical thinking skills. "Flex Your Brain" is found on the next page.

Science Journal

Humane Animal Treatment Have students find out about humane treatment of animals used in the laboratory for scientific testing and write a report. Pamphlets and books describing humane treatment may be obtained from the local humane society. L1

Flex Your Brain

1 Topic: _____

2 ❓ **What do I already know?**
1. _____
2. _____
3. _____
4. _____
5. _____

3 **Q:** Ask a question
_____

4 **A:** Guess an answer
_____

5 **How sure am I? (circle one)**

Not sure				Very sure
1	2	3	4	5

6 ❓ **How can I find out?**
1. _____
2. _____
3. _____
4. _____
5. _____

7 **EXPLORE**

8 **Do I think differently?** ➡ yes / no

9 ❓ **What do I know now?**
1. _____
2. _____
3. _____
4. _____
5. _____

10 **SHARE**
1. _____
2. _____
3. _____

---

1 Fill in the topic.

2 Jot down what you already know about the topic.

3 Using what you already know (step 2), form a question about the topic. Are you unsure about one of the items you listed? Do you want to know more? Do you want to know what, how, or why? Write down your question.

4 Guess an answer to your question. In the next few steps, you will be exploring the reasonableness of your answer. Write down your guesses.

5 Circle the number in the box that matches how sure you are of your answer in step 4. This is your chance to rate your confidence in what you've done so far and, later, to see how your level of sureness affects your thinking.

6 How can you find out more about your topic? You might want to read a book, ask an expert, or do an experiment. Write down ways you can use to find out more.

7 Make a plan to explore your answer. Use the resources you listed in step 6. Then, carry out your plan.

8 Now that you've explored, go back to your answer in step 4. Would you answer differently?

9 Considering what you learned in your exploration, answer your question again, adding new things you've learned. You may completely change your answer.

10 It's important to be able to talk about thinking. Choose three people to tell about how you arrived at your response in every step. For example, don't just read what you wrote down in step 2. Try to share how you thought of those things.

Use the Flex Your Brain activity to have students explore PROBLEMS THAT LIFE SCIENTISTS ARE CURRENTLY TRYING TO SOLVE.

An example of how students might fill in the Flex Your Brain blanks:

1. Pollution
2. 1. Pollution harms wildlife
   2. Pollution affects the air, soil, and water
3. Which country pollutes the most?
4. United States
5. 1 2 3 4 5
6. 1. Read magazine articles
   2. Write to environmental organizations
7. Students may work cooperatively.
8. Choice depends on step 7.
9. Answers should reveal that students have learned something new.
10. Have students present their results.

Students should be encouraged to keep a record of their Flex Your Brain exercise in their Science Journals so that they can submit the exercises in their portfolios later.

📁 A worksheet for completing Flex Your Brain activities is found on page 5 of the **Activity Worksheets** booklet.

🕹 Use **Teaching Transparency 1** as you teach this lesson.

MiniLAB

How are things measured?
Use a balance to determine the mass of items.

Procedure
1. Obtain a pan balance. Follow the instructions given by your teacher for using the balance.
2. Separately measure the mass of the following items: a small paper clip, a small glass, the same glass with water, a metal spoon, and a rock.
3. Record your data in a table in your Science Journal.

Analysis
1. Are these items best measured in kilograms, milligrams, or grams?
2. Why did you choose the unit you did?

Measuring with Scientific Units

Think about how many things you use every day that are measured. The thermostat keeps the air in your home at a certain temperature. Meters in your house measure water, electricity, and gas usage. Your food comes in pounds, ounces, and liters, and you step on a scale to check your weight.

International System of Units, SI

Scientists use a system of measurement to make observations. Scientists around the world have agreed to use the International System of Units, or SI. SI is based on certain metric units. Using the same system gives scientists a common language. They can understand each other's research and compare results. Most of the units you will use in science are shown in **Table 1-1.** Because you are used to using the English system of pounds, ounces, and inches, a chart has been included in Appendix B to help you convert these units to SI.

SI is based on units of ten. It is easy to use because calculations are made by multiplying or dividing by ten. Prefixes are used with units to change them to larger or smaller units. **Figure 1-15** shows equipment used for measuring in SI.

You will use these measurements to work in the laboratory this year. Working in the laboratory will help you understand the life science concepts in your textbook.

Safety First

Having the chance to do science can be much more interesting than reading about it. Some of the scientific equipment that you will use is the same that scientists use in the field or

Table 1-1

Common SI Measurements			
Measurement	**Unit**	**Symbol**	**Equal to**
Length	1 millimeter	mm	0.001 (1/1000) m
	1 centimeter	cm	0.01 (1/100) m
	1 meter	m	1
	1 kilometer	km	1000 m
Volume	1 milliliter	mL	1 cm^3
	1 liter	L	1000 mL
Mass	1 gram	g	1000 mg
	1 kilogram	kg	1000 g
	1 tonne	t	1000 kg = 1 metric ton

Figure 1-15

Some of the equipment used by scientists is shown here.

A A metric ruler or a meter-stick is used to measure length.

B Mass is the amount of matter in an object. Mass is measured with a balance.

C Scientists often use the Celsius scale to measure temperature. On the Celsius scale, water freezes at 0°C and boils at 100°C at sea level.

D The amount of space occupied by an object is its volume. Volume is found using a graduated cylinder.

in their laboratories. Of course, safety is of great importance in a laboratory. Most injuries are due to burns from heated objects or splatters and broken glass. The symbols shown below are used throughout your text to alert you to situations that require your special attention. A description of each symbol can be found in the Safety Appendix. Following safety rules will protect you and others from injury and help you get more from your lab experiences.

Section Wrap-up

Review

1. How does a theory differ from a law?

2. What SI units are used to measure length, volume, and mass?

3. **Think Critically:** Identify the scientific methods that were used in the Explore Activity on page 5.

Skill Builder
Concept Mapping
Make an events chain concept map showing the steps you took to perform the experiment in Activity 1-1. If you need help, refer to Concept Mapping in the **Skill Handbook.**

USING MATH

Sometimes temperature is measured in Fahrenheit degrees. In the Fahrenheit temperature scale, water freezes at 32° and boils at 212°. Use the conversion chart in Appendix B to answer the following question. Normal body temperature is 98.6°F. What is normal body temperature in Celsius degrees?

1-3 What is science? **23**

3 Assess

Check for Understanding

? FLEX Your Brain

Use the Flex Your Brain activity to have students explore PROBLEM SOLVING.

Activity Worksheets, p. 5

Reteach

Give students a label that names one of the steps in the scientific method. After all steps (labels) have been assigned, ask students to line up in the correct order.

Extension

For students who have mastered this section, use the **Reinforcement** and **Enrichment** masters.

4 Close

Have students find out what each safety symbol refers to by looking at Appendix C on page 759.

Section Wrap-up

Review

1. A theory is an explanation based on many observations. A law is a reliable description of how nature works, also based on many observations.

2. meter, cubic meter, kilogram, respectively

3. **Think Critically** Students may originally make a hypothesis that the black line is black. Then they perform an experiment and make observations. Because the line of ink separates into two or more colors, they may revise their hypotheses.

USING MATH

37°

Skill Builder
Student concept maps should indicate steps similar to those shown in **Figure 1-11** on page 15.

✓ Assessment

Performance Assess students' abilities to measure in SI by having them measure ten items at home. Use the Performance Task Assessment List for Using Math in Science in PASC, p. 29. **P**

SCIENCE & SOCIETY
Section 1•4

Prepare

Section Background

Although much technology is advantageous, some of it is harmful to our environment and to human health.

1 Motivate

Bellringer

 Before presenting the lesson, display **Section Focus Transparency 4** on the overhead projector. Assign the accompanying **Focus Activity** worksheet. **L1** **LEP**

2 Teach

Teacher F.Y.I.

Bovine somatotrophin (BST) is a genetically engineered cow hormone that increases a cow's milk production by 10 to 15%. The FDA approved BST in 1993 for use in dairy cows.

GLENCOE TECHNOLOGY

Videodisc

STVS: Plants and Simple Organisms
Disc 4, Side 1
Selective Breeding in Cows (Ch. 22)

3 Assess

Check for Understanding

Have students answer questions 1 and 2 in the Section Wrap-up.

SCIENCE & SOCIETY

ISSUE:
1•4 Technology and the Dairy Industry

Science Words
technology

Objectives
- Recognize how our lives are affected by technology.
- Determine whether products resulting from the use of technology should be labeled.

Living with Technology

Scientific methods often result in new information. **Technology** is the use of scientific knowledge to solve everyday problems or improve the quality of life. Sanitation, the antibiotic penicillin, fluoridated water, and irradiated food are examples of technologies that have improved the quality of our lives. Years of research in the life sciences have enabled people to benefit directly from science. For example, farmers today are able to produce great quantities of healthful food at reasonable prices in many countries of the world as a result of agricultural research.

Are all of these advances in the life sciences beneficial? There are some problems due to unforeseen side effects. The technology involved in the development of pesticides, food preservatives, and artificial sweeteners for example has contributed to higher rates of cancer.

Figure 1-16

Dairy cows produce milk that many humans drink. Many people want milk that is free of added chemicals.

Dairy Farmers and New Technology

Recently, the Food and Drug Administration (FDA) approved for use a new drug that helps cows produce more milk. The drug is a genetically engineered version of a cow hormone called bovine somatotrophin (BST). How does BST work? BST increases milk production by causing the cow to use more of her available nutrients for milk production and less for body weight gain.

Why would dairy farmers want to use BST? Dairy farmers, like all farmers, are in business to make money. Farmers who use BST will increase the milk production from their cows without needing to buy and feed more cows. However, cows that are given BST tend to have an increase in mastitis, an infection in cow udders. Mastitis is usually treated with antibiotics. Treating mastitis increases the farmer's costs, but not as much as buying and feeding additional animals would.

24 Chapter 1 Exploring Life

Program Resources

 Reproducible Masters
Study Guide, p. 8 **L1**
Reinforcement, p. 8 **L1**
Enrichment, p. 8 **L3**
Science and Society Integration, p. 5

 Transparencies
Section Focus Transparency 4

2 Points of View

Consumers Wary of BST in Milk

Consumers worry that hormones and antibiotics given to cows will end up in their milk. Antibiotics in milk, for example, could result in the development of bacteria in humans that are resistant to antibiotics used to combat diseases such as strep throat. Consumers believe that dairy products produced from cows that have been given BST should be labeled. The consumer could then exercise other options such as paying more for milk without BST.

Dairy Farmers Want Increased Milk Production

Milk processors are concerned about consumer reaction to the publicity that has been given BST and fear that sales will suffer even more if milk products are labeled. Individual farmers must remain competitive, and projections indicate that those farmers who use BST to increase milk production will show more profit. For this reason, some farmers do not want milk products labeled.

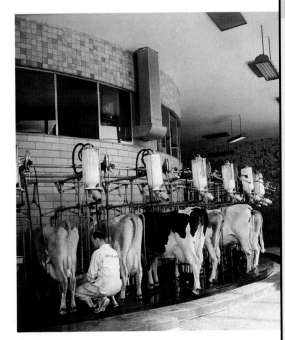

Figure 1-17

To produce milk at competitive prices, farmers must invest in new machinery.

Section Wrap-up

Review

1. Explain how antibiotics such as penicillin have improved human lives.

2. Why do dairy farmers want to give BST to their cows?

3. Should consumers be concerned about milk from cows given BST?

*inter*NET CONNECTION The FDA's Center for Veterinary Medicine is responsible for approving the safety of all drugs used in animals that produce meat, milk, or eggs. Visit the Chapter 1 Internet Connection at Glencoe Online Science, **www.glencoe.com/sec/science/life,** for a link to more information about drugs in milk.

SCIENCE & SOCIETY

1-4 Technology and the Dairy Industry   **25**

Reteach

Have students list the countries that they think are technologically advanced and those they believe are not. First review the two lists, then discuss the effects of technology on the advanced countries and the advantages enjoyed by countries that have never developed such technologies.

Extension

For students who have mastered this section, use the **Reinforcement** and **Enrichment** masters.

4 Close

In a short paragraph, have students define *technology* and name five ways in which it is used in their homes.

Section Wrap-up

Review

1. Antibiotics have improved human lives by destroying harmful bacteria in human bodies.

2. It increases milk production by causing the cow to use more of her available nutrients for milk production and less for body weight gain.

3. BST affects cells. It is contained in the milk of cows that have received it and could affect the cells of humans who drink the milk.

Explore the Issue

The FDA could check milk that is not labeled to make certain that it does not contain BST. Consumers cannot avoid dairy products that contain BST unless they are labeled.

Science & ART

Source

Maclay, Elise. *The Art of Bev Doolittle*. New York: Bantam Books, 1990.

Biography

Bev Doolittle was born and raised in a large family in Southern California. A sister always managed to color their coloring books, so Bev turned to drawing her own pictures. She won an art award in a contest sponsored by the San Gabriel Historical Society at age 15. She also won a Saturday Scholarship at the Art Center College of Design in Los Angeles. After graduation from high school, she enrolled in the College of Design full time. She married an artist, Jay Doolittle, from the same school.

Both Bev and Jay Doolittle are dedicated artists. They both love the outdoors and wildlife. Even trips that they have taken to the Grand Canyon, Yosemite, Sequoia National Park, Alaska, and Africa were times for gathering ideas and photos for future paintings. They have one son, Jayson Bevson Doolittle, who was born in 1981 and also paints.

As an outgrowth of her love of the wilderness, Ms. Doolittle's fascination for the culture of Native Americans developed over the years. At first, she was interested in their daily lives because she wanted to make her paintings realistic and accurate. As she matured as an artist, she sought to express the mystical aspects of the Plains Indians' culture. She has received many letters from Native Americans, praising her ability to see things as they do.

Science Journal

Use the Bev Doolittle painting as the topic for a poem or an essay you will write in your Science Journal. Describe the feelings that the painting evokes in you.

The Art of Bev Doolittle: A Study in Observation

In this chapter, you learned about the role of observation in the scientific method. Bev Doolittle is an artist whose paintings may stimulate your powers of observation. Here is what she wants her paintings to do:

"I like to take the viewer's attention and move it around through a painting so that you see one thing first, then a second, then a third, each time seeing something new."

Hidden Images, Hidden Meanings

Doolittle seems to enjoy teasing your powers of observation. She upsets your usual way of looking at a picture. Sometimes she does this by painting hidden images that require a second or third look for you to realize where she is leading. To test this, find the hidden image in her painting *The Season of the Eagle*, finished in 1988.

At first, you see a party of Native American hunters taking advantage of the retreating snow to track down prey. Closer observation reveals that the artist has used the snow to paint the silhouette of a soaring eagle. Even trickier is her device of having half the silhouette as a reflection in the lake. Besides hidden images, the artist also includes hidden meanings in the painting. She implies the Native American belief that all creatures are brothers by having the eagle accompany the hunters. Because the eagle is a symbol of hope, its silhouette in the snow may be the promise of happy hunting for the group.

Camouflage

Many of Bev Doolittle's paintings are devoted to revealing the surprises of camouflage. She deliberately tries to misdirect your eyes so you will look again and again to find all that she has so skillfully laid out in her work.

Teaching Strategies

LS **Linguistic** Have students share their poems or essays about the painting.

- Ask students how they think the artist might have conceived the idea for *The Season of the Eagle*. Mention that a great deal of thought and planning were involved.

- Encourage students to contribute ideas of how a study of Bev Doolittle's paintings might help their scientific skills.

Chapter 1 Review

Chapter 1
Review

Summary

1-1: Living Things
1. Organisms are made up of cells, use energy, move, respond, adjust, reproduce, grow and develop, and adapt to their environment.
2. Organisms need energy and materials.

1-2: Where does life come from?
1. Redi and Spallanzani tried to disprove the idea that living things come from nonliving matter.
2. Pasteur's experiment proved biogenesis, the theory that life comes from life.
3. Scientists try to explain how life began on Earth. Oparin's hypothesis, which says that energy combined early gases to form the chemical compounds of life, has been tested.

1-3: What is science?
1. Scientists investigate observations that are made about living and nonliving things with the help of problem-solving techniques.
2. Scientists use SI measurements to gather measurable data.
3. Safe laboratory practices help you to learn more about science.

1-4: Science and Society: Technology and the Dairy Industry
1. Technology is the use of scientific knowledge to solve everyday problems.

2. A cow hormone called BST is FDA approved for dairy cows. Its use poses feeding and management problems for farmers and consumers alike.

Key Science Words

a. adaptation
b. biogenesis
c. cell
d. control
e. development
f. homeostasis
g. hypothesis
h. law
i. life span
j. organism
k. response
l. scientific methods
m. spontaneous generation
n. stimulus
o. technology
p. theory
q. variable

Reviewing Vocabulary

Match each phrase with the correct term from the list of Key Science Words.

1. the length of time an organism is expected to live
2. changes undergone during growth
3. a change in the environment that brings about a response
4. living thing
5. the use of science for solving problems or improving life
6. theory that nonliving things produce living things
7. organized steps used to solve a problem
8. a prediction that can be tested

Summary

Have students read the summary statements to review the major concepts of the chapter.

Reviewing Vocabulary

1. i	**6.** m
2. e	**7.** l
3. n	**8.** g
4. j	**9.** c
5. o	**10.** d

✓ Assessment

Portfolio Encourage students to place in their portfolios one or two items of what they consider to be their best work. Examples include:
- Assessment, p. 9
- Integrating the Sciences, p. 11
- Across the Curriculum, p. 22 P

Performance Additional performance assessments may be found in **Performance Assessment** and **Science Integration Activities.** Performance Task Assessment Lists and rubrics for evaluating these activities can be found in Glencoe's **Performance Assessment in the Science Classroom.**

GLENCOE TECHNOLOGY

 MindJogger Videoquiz

Chapter 1 Have students work in groups as they play the Videoquiz game to review key chapter concepts.

Chapter 1 Review

Checking Concepts

1. b	**6.** d
2. c	**7.** a
3. a	**8.** b
4. d	**9.** b
5. c	**10.** b

Understanding Concepts

11. Growth is the process of becoming larger, such as becoming taller; development is a change in structure that brings forth new abilities, such as learning to walk.

12. Technology often enables scientists to see new problems to be solved.

13. Response is a reaction to a stimulus; an adaptation is an inherited trait that helps an organism to survive.

14. Energy is needed for all life functions: growth, movement, and so on. Without energy, these processes could not take place and organisms would eventually die.

15. Add a certain amount of water to the graduated cylinder. Record the volume of water in milliliters. Add the small wooden block to the graduated cylinder. Record the volume in milliliters. The difference between the first reading and the second reading is the mass of the wooden block in grams. The wooden block has displaced that amount of water.

Thinking Critically

16. Scientists can compare and repeat experiments; they have a common tool for measurement.

17. A bird is made up of cells; a bird uses energy to fly and breathe; a bird displays movement by flying or hopping; a bird jumps at a loud sound or re-

9. unit of life
10. a standard for comparing the result of an experiment

Checking Concepts

Choose the word or phrase that completes the sentence.

1. An infant cutting teeth is an example of _____.
   - a. growth
   - b. development
   - c. respiration
   - d. adaptation
2. A bright light that causes you to shut your eyes is a _____.
   - a. need
   - b. response
   - c. stimulus
   - d. variable
3. The source of most energy for life is _____.
   - a. the sun
   - b. oxygen
   - c. water
   - d. carbon dioxide
4. Living things are made up of about 70 percent _____.
   - a. oxygen
   - b. carbon dioxide
   - c. minerals
   - d. water
5. _____ disproved the theory of spontaneous generation.
   - a. Oparin
   - b. Spallanzani
   - c. Pasteur
   - d. Redi
6. Scientists think that _____ was missing from Earth's early atmosphere.
   - a. ammonia
   - b. hydrogen
   - c. methane
   - d. oxygen
7. _____ are inherited characteristics that may help an organism to survive.
   - a. Adaptations
   - b. Stimuli
   - c. Raw materials
   - d. Theories
8. The _____ is/are the part of the experiment that is tested.
   - a. conclusion
   - b. variable
   - c. control
   - d. data

9. A(n) _____ is a prediction that has to be testable.
   - a. experiment
   - b. hypothesis
   - c. theory
   - d. law
10. The SI unit used to measure liquids is the _____.
    - a. meter
    - b. liter
    - c. gram
    - d. degree

Understanding Concepts

Answer the following questions in your Science Journal using complete sentences.

11. Use an example to explain the difference between growth and development.
12. How does technology help science?
13. What is the difference between a response and an adaptation?
14. Explain why living things need energy.
15. How can a graduated cylinder marked in milliliters be used to determine the volume of a small wooden block?

Thinking Critically

16. How does SI benefit scientists in different parts of the world?
17. Using a bird as an example, explain how it has all the characteristics of living things.
18. If a plant had no carbon dioxide available, would this affect animal life?
19. Show how Pasteur correctly used scientific methods to disprove the theory of spontaneous generation.
20. How are stimulus and response related to homeostasis?

sponds to food; a bird maintains a constant body temperature; birds reproduce young birds; birds grow larger and develop the ability to fly; a bird is adapted with certain structures that enable it to fly; different species of birds live for one or more years.

18. No carbon dioxide would result in death of all life because it is needed for photosynthesis.

19. Pasteur's hypothesis was that air is necessary for an organism's survival and that air contains small particles of life. He had a control and a variable. The special flask allowed air in. Based on his results, he showed that no organisms grew in a curved-neck flask.

20. A stimulus momentarily causes an organism to respond. For example, you are thirsty (stimulus); you drink water to quench the thirst (response). As a result of responding to the stimulus by getting a drink of water, your body's homeostasis is maintained.

Developing Skills

If you need help, refer to the **Skill Handbook.**

21. **Designing an Experiment:** Design an experiment to test the effects of plant food on growing plants. Identify scientific methods in your experiment.

22. **Concept Mapping:** Use the following terms to complete an events chain concept map showing the order in which you might use scientific methods: *collect data, perform an experiment, state your hypothesis.*

State the problem
↓
State hypothesis
↓
Perform an experiment
↓
Collect data
↓
Publish your results

23. **Using Variables, Constants, and Controls:** You place one lima bean plant under a green light, another under a red light, and a third under a blue light. You measure their growth for four weeks to determine which color is most favorable to plant growth. What are the variables in this experiment? What is the hypothesis in this experiment? How could the experiment be improved?

24. **Observing and Inferring:** What evidence must scientists have to prove that cancer can be caused by cigarette smoking?

25. **Classifying:** Which type of measurement would you use for each of the following? Classify as meter, gram, cubic meter, degree, or liter.
    a. distance run
    b. how hot a pan of water is
    c. how much juice is in a glass
    d. your mass

Performance Assessment

1. **Bulletin Board:** Interview people in your community whose jobs require a knowledge of life science. Make a Life Science Careers bulletin board. Summarize each person's job and what he or she had to study to prepare for that job.

2. **Writing in Science:** Find out what kind of organism each of the following scientists studies: mycologist, ornithologist, entomologist, ichthyologist, and paleobotanist. Write a job description for each of these scientists and share your report with the class.

3. **Letter:** Most countries in the world now use SI measurements in their everyday practices. Write a letter to your Representatives or Senators asking why the United States has not converted to this system. Share your responses with your class.

Developing Skills

21. **Designing an Experiment** An experiment should contain a hypothesis, observations, variables, a control, and an interpretation of results.

22. **Concept Mapping** See student page.

23. **Using Variables, Constants, and Controls** The variables are the different colors of light and whatever effects they may produce. A hypothesis might be that green light will cause most favorable plant growth. This experiment could be improved by having fewer variables, more plants, and white light as a control.

24. **Observing and Inferring** Scientists research the histories of cancer patients who have smoked; laboratory tests show tumors that have been caused by cigarette smoke.

25. **Classifying** a. meter (*kilo-*), b. degree, c. liter (*milli-*), d. grams (*kilo-*)

Performance Assessment

1. The bulletin board could include farmers, produce clerks, florists, health care workers, and life science teachers. Use the Performance Task Assessment List for Bulletin Board in **PASC,** p. 59. **P**

2. mycologist—fungi; ornithologist—birds; entomologist—insects; ichthyologist—fish; paleobotanist—fossil plants. Use the Performance Task Assessment List for Writing in Science in **PASC,** p. 87. **P**

3. Check letters before they are mailed to make sure congressional representatives are addressed appropriately. Use the Performance Task Assessment List for Letter in **PASC,** p. 67. **P**

Assessment Resources

 Reproducible Masters

Chapter Review, pp. 5-6
Assessment, pp. 5-8
Performance Assessment, p. 15

Glencoe Technology

- **Chapter Review Software**
- **Computer Test Bank**
- **MindJogger Videoquiz**

Chapter Organizer

Section	Objectives/Standards	Activities/Features
Chapter Opener		Explore Activity: Observe a large cell. p. 31
2-1 Viruses (1 session, 1 block)*	1. **Describe** the structure of a virus and explain how viruses reproduce and cause disease. 2. **Explain** the benefits of vaccines. 3. **Describe** some helpful uses of viruses. National Content Standards: (5-8) UCP1, C1, C2, F1, G1, G2, G3	Skill Builder: Concept Mapping, p. 37 Using Computers: Spreadsheet, p. 37
2-2 Science and Society Technology: AIDS Vaccine? (1 session, ½ block)*	1. **Determine** how HIV affects the immune system. 2. **Recognize** how a potential AIDS vaccine can prevent the disease. National Content Standards: (5-8) UCP5, A1, C2, F4	Using Math, p. 39 Explore the Technology, p. 39
2-3 Cells: The Units of Life (1 session, 1 block)*	1. **Discuss** the history leading to the cell theory. 2. **Explain** the difference between the compound light microscope and the electron microscope. 3. **Explain** the importance of the cell theory. National Content Standards: (5-8) UCP1, UCP2, A1, A2, B1, C1, E2, G1, G2, G3	MiniLAB: How are objects magnified? p. 41 Connect to Physics, p. 42 Skill Builder: Concept Mapping, p. 44 Using Math, p. 44 MiniLAB: How can you model cytoplasm? p. 48 Problem Solving: Surface Area and Volume of Living Cells, p. 50 Activity 2-1, p. 53
2-4 Cell Organization (2 sessions, 1 block)*	1. **Diagram** a plant cell and an animal cell; identify the parts and the function of each part. 2. **Describe** the importance of the nucleus in the cell. 3. **Explain** the differences among tissues, organs, and organ systems. National Content Standards: (5-8) UCP1, UCP2, UCP5, A1, A2, B3, C1, C3, C5, E2, F1, F3, F4, F5, G1, G2, G3	Using Technology: Replacement Skin, p. 54 Skill Builder: Comparing and Contrasting, p. 55 Science Journal, p. 55 Activity 2-2, pp. 56-57 Science and History, Viruses Through History, p. 58 Science Journal, p. 58

* A complete Planning Guide that includes block scheduling is provided on pages 32T-35T.

Activity Materials

Explore	Activities	MiniLABs
page 31 chicken egg, dish, metric ruler, hand lens	page 53 microscope, microscope slide, coverslip, forceps, dropper, *Elodea* plant, prepared slide of human cheek cells, tap water pages 56-57 onion plant with roots and leaves, microscope, 3 microscope slides, 3 coverslips, 250-mL beaker, tap water, dropper, forceps, scissors, safety goggles	page 41 empty glass, glass bowl, tap water, magnifying glass, sheet of newspaper page 48 150-mL beaker, tap water, unflavored gelatin, flashlight

Need Materials? Call Science Kit (1-800-828-7777).

Chapter 2 The Structure of Viruses and Cells

Teacher Classroom Resources

Reproducible Masters	Transparencies	Teaching Resources
Study Guide, p. 9 **Reinforcement,** p. 9 **Enrichment,** p. 9 **Critical Thinking/Problem Solving,** p. 6 **Cross-Curricular Integration,** p. 6 **Multicultural Connections,** p. 7	**Section Focus Transparency 5,** Fighting the Flu	**Glencoe Life Science Interactive** **Videodisc** **Life Science CD-ROM** **Spanish Resources** **English/Spanish Audiocassettes** **Cooperative Learning Resource** **Guide** **Lab Partner** **Lab and Safety Skills** **Lesson Plans**
Study Guide, p. 10 **Reinforcement,** p. 10 **Enrichment,** p. 10	**Section Focus Transparency 6,** The AIDS War	
		Assessment Resources
Study Guide, p. 11 **Reinforcement,** p. 11 **Enrichment,** p. 11 **Activity Worksheets,** p. 15 **Lab Manual 3,** The Microscope **Science Integration Activity 2,** The Magnifying Power of Lenses **Science and Society Integration,** p. 6	**Section Focus Transparency 7,** Portrait of a Virus **Science Integration Transparency 2,** Micro-Optics **Teaching Transparency 3,** The Microscope	**Chapter Review,** pp. 7-8 **Assessment,** pp. 9-12 **Performance Assessment in the** Science Classroom (PASC) **MindJogger Videoquiz** **Alternate Assessment in the** Science Classroom **Performance Assessment** **Chapter Review Software** **Computer Test Bank**
Study Guide, p. 12 **Reinforcement,** p. 12 **Enrichment,** p. 12 **Activity Worksheets,** pp. 11-12, 13-14, 16 **Lab Manual 4,** Observing Cells **Concept Mapping,** p. 9	**Section Focus Transparency 8,** Cells Are Factories **Teaching Transparency 4,** Animal and Plant Cells	

Key to Teaching Strategies

The following designations will help you decide which activities are appropriate for your students.

L1 Level 1 activities should be within the ability range of all students, including those with learning difficulties.

L2 Level 2 activities should be within the ability range of the average to above-average student.

L3 Level 3 activities are designed for the ability range of above-average students.

LEP LEP activities should be within the ability range of Limited English Proficiency students.

LS These activities are designed to address different learning styles.

COOP LEARN Cooperative Learning activities are designed for small group work.

P These strategies represent student products that can be placed into a best-work portfolio.

GLENCOE TECHNOLOGY

The following multimedia resources are available from Glencoe.

Science and Technology
Videodisc Series
(STVS)
Plants & Simple Organisms
Mapping a Virus
The Infinite Voyage
Series
Miracles by Design

National Geographic
Society Series
Newton's Apple: Life
Sciences
STV: The Cell
STV: Human Body Vol. 3
Life Science CD-ROM

Teacher Classroom Resources

This is a representation of key blackline masters available in the Teacher Classroom Resources.

Teaching Aids

Section Focus Transparencies

Science Integration Transparencies

Teaching Transparencies

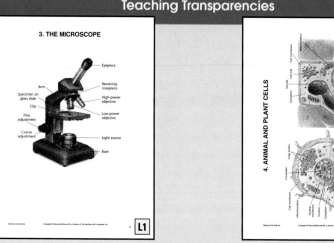

Meeting Different Ability Levels

Study Guide

Reinforcement

Enrichment Worksheets

Chapter 2 The Structure of Viruses and Cells

Hands-On Activities

Science Integration Activity

The Magnifying Power of Lenses

L1

Lab Manual

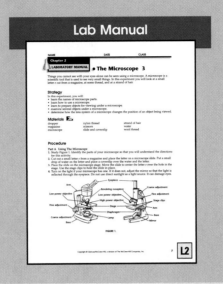

The Microscope 3

L2

Assessment

Performance Assessment

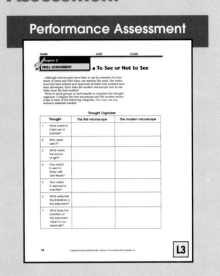

To See or Not to See

L3

Enrichment and Application

Critical Thinking/Problem Solving

The Structure of Viruses and Cells

L2

Cross-Curricular Integration

The Structure of Viruses and Cells

L1

Science and Society Integration

The Structure of Viruses and Cells

L2

Multicultural Connections

Garlic Gives Doctors a Strong Smell of Success

L2

Concept Mapping

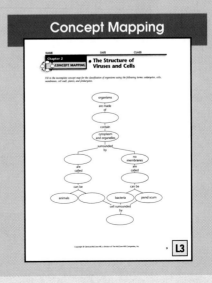

The Structure of Viruses and Cells

L3

The Structure of Viruses and Cells

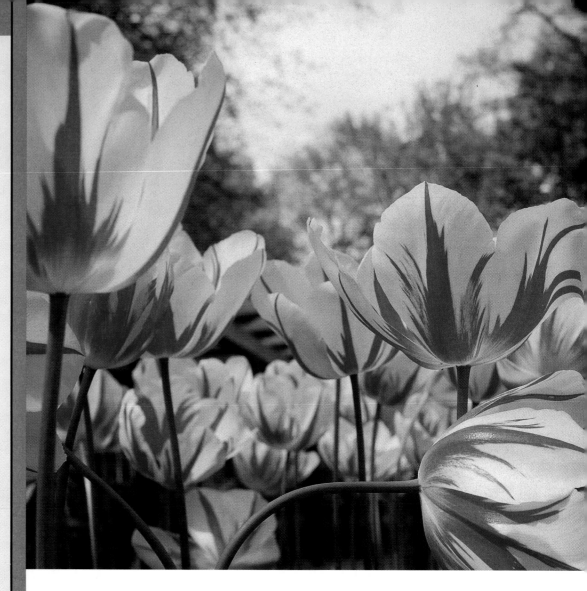

CHAPTER OVERVIEW

Section 2-1 Viral structure, classification, and reproduction are described. How vaccines are useful in preventing viral diseases is discussed. The use of viruses in DNA technology is given as an example of a positive use of viruses.

Section 2-2 Science and Society How HIV mutates to cause AIDS and the potential for developing a vaccine are presented.

Section 2-3 The importance of the development of microscopes in the discovery of cells is discussed. The cell theory is presented as a major theory in life science.

Section 2-4 Prokaryotic and eukaryotic cells and the parts and functions common to all cells are discussed. Animal, plant, and bacterial cells are compared. How cells are organized to form tissues, organs, systems, and organisms is presented.

Chapter Vocabulary

virus	cytoplasm
host cell	organelle
vaccine	endoplasmic
AIDS	reticulum
compound	ribosome
light	Golgi body
microscope	mitochondria
electron	lysosome
microscope	cell wall
cell theory	chloroplast
cell membrane	tissue
nucleus	organ
chromatin	

Theme Connection

Scale and Structure The scale and structure of viruses and cells are compared and the parts of which they are composed are described.

30

Previewing the Chapter

Section 2-1   Viruses
  ▶ Characteristics of Viruses
  ▶ Reproduction of Viruses
  ▶ Viral Diseases
  ▶ Helpful Viruses

Section 2-2   Science and Society
  Technology: AIDS Vaccine?

Section 2-3   Cells: The Units of Life
  ▶ The Microscope
  ▶ The Cell Theory

Section 2-4   Cell Organization
  ▶ An Overview of Cells
  ▶ Animal Cells
  ▶ Plant and Bacterial Cells
  ▶ Organizing Cells

30

Learning Styles

Look for the following logo for strategies that emphasize different learning modalities. **LS**

Kinesthetic   Explore, p. 31; Making a Model, p. 34; Activity, p. 47; MiniLAB, p. 48

Visual-Spatial   Visual Learning, pp. 33, 36, 42, 49, 51, 52; MiniLAB, p. 41; Activity, p. 46; Demonstration, pp. 47, 51, 52; Inclusion Strategies, p. 52; Activity 2-1, p. 53

Interpersonal   Activity 2-2, pp. 56-57

Logical-Mathematical Across the Curriculum, p. 34; Using Math, p. 44

Linguistic   Inclusion Strategies, p. 33; Using Science Words, p. 35; Reteach, p. 36; Across the Curriculum, p. 46

The Structure of Viruses and Cells

Tulips come in many varieties: solid colors, multiple colors, and streaked. The streaked color pattern in the tulips on the opposite page was originally the result of a viral infection. Obtaining this streaked color pattern using a virus is now illegal because such viral infections can destroy entire tulip populations. Viruses that caused such changes attacked individual cells in these flowers. Viruses are much smaller than most cells. In the following activity, find out about a cell that can be seen without a microscope. Then in the chapter that follows, learn about the structure of both viruses and cells.

EXPLORE ACTIVITY

Observe a large cell.
1. Break a chicken egg into a dish. Observe the yolk.
2. Estimate the diameter of the yolk and then measure its diameter.
3. With a hand lens, observe the yolk closely. The largest known living cell is the yolk of bird eggs—not the white, just the yolk.

Observe: Describe and measure the size of the cell of a chicken egg in your Science Journal.

Previewing Science Skills

▶ In the **Skill Builders**, you will map concepts, compare and contrast, and interpret scientific illustrations.

▶ In the **Activities**, you will observe, compare and contrast, diagram, and conclude.

▶ In the **MiniLABs**, you will observe, identify, compare and contrast, and infer.

31

Assessment Planner

Portfolio
Refer to page 59 for suggested items that students might select for their portfolios.

Performance Assessment
See page 59 for additional Performance Assessment options.
Skill Builders, pp. 37, 44, 55
MiniLABs, pp. 41, 48
Activities 2-1, p. 53; 2-2, pp. 56-57

Content Assessment
Assessment—Oral, pp. 31, 37, 48
Section Reviews, pp. 37, 39, 44, 55
Chapter Review, pp. 59-61
Mini Quizzes, pp. 37, 44, 55

Group Assessment
Opportunities for group assessment occur with Cooperative Learning Strategies and Flex Your Brain Activities.

EXPLORE ACTIVITY

Purpose
Ⓚ **Kinesthetic** Use the Explore Activity to introduce students to cells. Explain that they will be learning about the size of viruses and cells, cell parts, and how cells were discovered. L1
LEP   COOP LEARN

Preparation
Purchase one egg for every two students in the class.

Materials
Provide one flat dish, one egg, one metric ruler, and one hand lens for every two students. The eggs can be used again in your next class.

Teaching Strategies
• Have each student measure the diameter of the yolk and look at the egg with the hand lens.
• Have students measure in centimeters. Review the metric units of length from Chapter 1 and refer to Appendix B on p. 757.
• Students should wash their hands after they have completed the activity and returned the eggs to you.

Observe
Students will observe that the egg is from 4 to 5 cm in diameter.

✔ Assessment

Oral Have students compare their results with their partner's and list reasons why different cells are different sizes. Use the Performance Task Assessment List for Analyzing the Data in **PASC**, p. 27. Ⓟ

Prepare

Section Background

- Viruses were first identified as the cause of disease in 1892.
- In the 1980s, the smallpox virus became the first infectious agent to be eliminated from the human population.

1 Motivate

Bellringer

 Before presenting the lesson, display **Section Focus Transparency 5** on the overhead projector. Assign the accompanying **Focus Activity** worksheet. **L1** **LEP**

Tying to Previous Knowledge

Have students recall viral diseases they or members of their families may have had. Ask what the symptoms of viral diseases are and what living things are affected by viruses.

NATIONAL GEOGRAPHIC SOCIETY

 Videodisc

Newton's Apple: Life Sciences
HIV and AIDS
Chapter 3, Side B
What is a virus?

31094-31814
How small is a virus?

31814-33217

2•1 Viruses

Science Words

virus
host cell
vaccine

Objectives

- Describe the structure of a virus and explain how viruses reproduce and cause disease.
- Explain the benefits of vaccines.
- Describe some helpful uses of viruses.

Characteristics of Viruses

Imagine something that doesn't grow, respond, or eat, yet reproduces. This something which appears to be neither living nor nonliving is a virus. A **virus** is a nonliving particle consisting of a core of hereditary material surrounded by a protein coat. Viruses can reproduce only inside a living cell. Viruses are so small that scientists must use extremely powerful microscopes to see them.

Viruses are unlike living things. Some viruses can be made into crystals and stored in a jar on a shelf for years. Then, if they enter an organism, they may quickly reproduce and cause new infections, often causing damage to the organism. Viruses, therefore, have a major potential for impact on the living world.

20 000×

A The polio virus is many-sided and looks like a small crystal when magnified 20 000 times.

Figure 2-1

Viruses, from a Latin word for *poison*, come in a variety of shapes and are responsible for many diseases.

B The AIDS virus, HIV, is spherical in shape and is studded with projections. *How might this structure enable the virus to infect cells?*

46 000×

32   Chapter 2 The Structure of Viruses and Cells

Program Resources

 Reproducible Masters
Study Guide, p. 9 **L1**
Reinforcement, p. 9 **L1**
Enrichment, p. 9 **L3**
Cross-Curricular Integration, p. 6
Multicultural Connections, p. 7
Critical Thinking/Problem Solving, p. 6

 Transparencies
Section Focus Transparency 5 **L1**

36 700×   215 000×

Figure 2-2

Many viruses are named for the disease they cause or where they were first found. The rabies virus, shown on the left, is named for the disease it causes in animals. The adenovirus, shown on the right, is so named because it infects adenoid tissue.

Classification of Viruses

Viruses, as illustrated in **Figure 2-1,** are classified by their shape, the kind of hereditary material they have, the kind of organism that they infect, and their method of reproduction. The protein covering of viruses gives them their shapes. As shown in **Figure 2-2,** viruses are often named for the diseases they cause, such as the polio virus, or the organ or tissue they infect.

C The tobacco-mosaic virus is rod-shaped and has a coat of proteins spiraling around a single strand of hereditary material. It causes tobacco plants to become stunted and the leaves to become discolored and blotchy.

220 000×

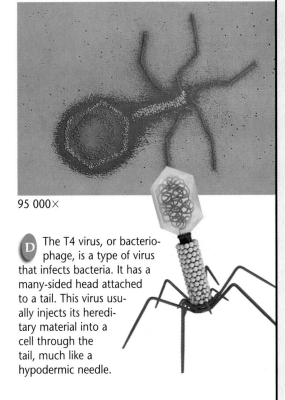

95 000×

D The T4 virus, or bacteriophage, is a type of virus that infects bacteria. It has a many-sided head attached to a tail. This virus usually injects its hereditary material into a cell through the tail, much like a hypodermic needle.

2-1 Viruses   **33**

Reproduction of Viruses

When most people hear the word *virus,* they relate it to a cold sore, a cold, and HIV, the virus that causes AIDS. Mumps, measles, and chicken pox are also diseases caused by viruses. A virus has to be inside a living cell to reproduce. The cell in which a virus reproduces is called a **host cell.** Once a virus is in a host cell, the virus can act in two ways. It can either be active, as shown below, or it can become latent and be an inactive part of the cell for a while.

Active Viruses

If a virus enters a cell and becomes active right away, it causes the cell to make new viruses, which destroy the host cell. Follow the steps in **Figure 2-3** as it outlines the steps an active virus takes to reproduce itself inside a bacterial cell.

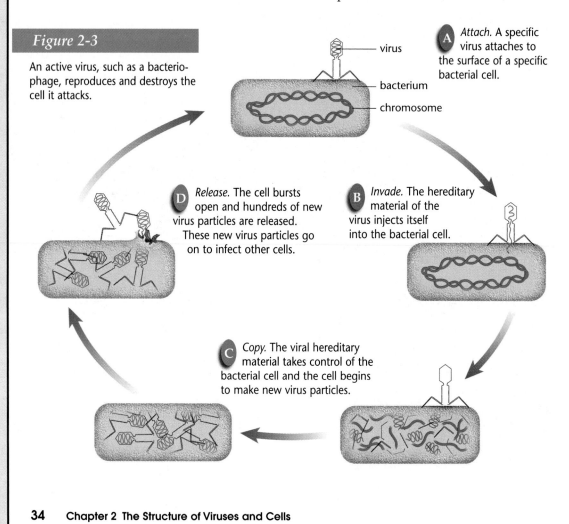

Figure 2-3

An active virus, such as a bacteriophage, reproduces and destroys the cell it attacks.

A *Attach.* A specific virus attaches to the surface of a specific bacterial cell.
— virus
— bacterium
— chromosome

B *Invade.* The hereditary material of the virus injects itself into the bacterial cell.

C *Copy.* The viral hereditary material takes control of the bacterial cell and the cell begins to make new virus particles.

D *Release.* The cell bursts open and hundreds of new virus particles are released. These new virus particles go on to infect other cells.

34 Chapter 2 The Structure of Viruses and Cells

Across the Curriculum

Math Viruses range in size from 0.02 to 0.3 *nm.* A common unit of measure for viruses is the nanometer (nm), which is 1000 times smaller than a micrometer and 1 million times smaller than a millimeter. A typical virus would be 20 nanometers in size. **What part of a millimeter would that be?** *0.00002 mm* **A virus of 0.3 micrometers would be how many nanometers?** *0.0003 nm*

Latent Viruses

Some viruses may be latent viruses. A latent virus enters a cell and becomes part of the cell's hereditary material without immediately destroying the cell or making new viruses. Latent viruses may appear to "hide" inside host cells for many years. Then, at any time, the virus becomes active. Follow **Figure 2-4** as it outlines the reproduction cycle of latent viruses.

Figure 2-4

A latent virus may not immediately destroy the cell it attacks.

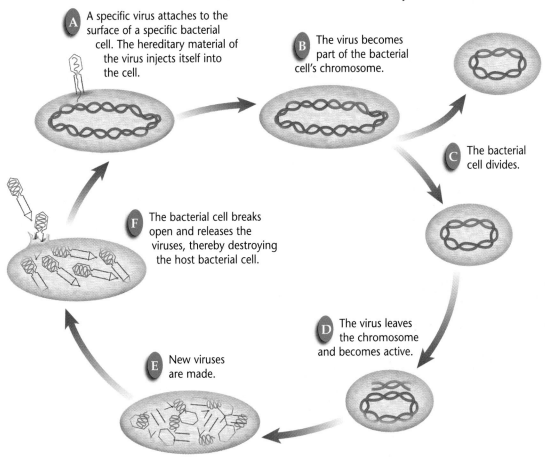

A A specific virus attaches to the surface of a specific bacterial cell. The hereditary material of the virus injects itself into the cell.

B The virus becomes part of the bacterial cell's chromosome.

C The bacterial cell divides.

D The virus leaves the chromosome and becomes active.

E New viruses are made.

F The bacterial cell breaks open and releases the viruses, thereby destroying the host bacterial cell.

If you have ever had a cold sore, you've been infected by a virus that has gone from the latent phase into the active phase. The appearance of a cold sore on your lip is a sign that the virus is active and destroying cells in your lip. Stress, or too much sun or cold, may cause a virus to become active. When the cold sore disappears, the virus has become latent again. The virus is still in your body cells, but you just don't realize it.

Content Background

Some scientists hypothesize that viruses were once part of the hereditary material of a cellular organism. A fragment of hereditary material that includes a recognition site for an enzyme that initiates DNA replication can reproduce on its own in a cell. The DNA enzyme binds to the recognition site on the DNA fragment on viral DNA and begins to copy the DNA. This is why a virus can reproduce inside a cell. The cell recognizes it as if it were its own DNA. (The enzyme is DNA polymerase.)

Discussion

In what ways are latent viruses unlike active viruses? *Latent viruses can go a long time inside the cell without causing the cell to produce new viruses or die.*

Using Science Words

LS **Linguistic** Have students look up the terms *lysis, virion,* and *virology.* Write a paragraph about the relevance of each to the study of viruses. **L1**

NATIONAL GEOGRAPHIC SOCIETY

Videodisc

STV: Human Body Vol. 3
Immune System
Unit 1, Side 1
Cell bursting, releasing cold virus

48789
Cold virus entering a cell (art)

48787
Cold virus entering a cell (art)

48788

Community Connection

Tissue Culture Virologists often grow viruses in human tissue culture. Invite a technician from a hospital laboratory or a commercial laboratory to talk with the students about tissue culture.

Table 2-1

Viral Diseases in Humans		
Disease	**Affected Body Part**	**Vaccine**
AIDS	Immune system	No
Chicken pox	Skin	Yes
Common cold	Respiratory system	No
Influenza, or flu	Respiratory system	Yes
Measles	Skin	Yes
Mumps	Salivary tissue	Yes
Polio	Nervous system	Yes
Rabies	Nervous system	Yes
Smallpox	Skin	Yes

Figure 2-5

Modern vaccination procedures help prevent many childhood and adult diseases, including chicken pox. *What other common vaccinations are given?*

Can Chickenpox BE AVOIDED?

Today, medical research has developed a vaccine that can help prevent chickenpox in children, teens, and adults.

Viral Diseases

Viruses may cause disease in plants, animals, fungi, bacteria, and protists. There are no antibiotic medications to *cure* viral diseases. But some viral diseases can be *prevented* by vaccines. A **vaccine** is made from damaged virus particles that can't cause disease anymore. Vaccines have been developed for measles, polio, chicken pox, and for bacterial diseases.

Vaccines

Edward Jenner, an English doctor, is credited with developing a vaccine in 1796. Jenner developed a vaccine for smallpox, a disease that was greatly feared, even into the twentieth century. Jenner noticed that people who milked cows and came down with a disease called cowpox didn't get smallpox. He prepared a vaccine from the sores of milkmaids who had cowpox. When injected into healthy people, the cowpox vaccine seemed to protect them from smallpox. Did Jenner know he was fighting a virus? No. At that time, no one understood what caused disease or how the body fought disease.

Vaccinations are an important step in maintaining health. **Figure 2-5** shows a young child receiving a new vaccine against chicken pox by receiving a shot. **Table 2-1** lists many viral diseases found in humans. The existence of vaccines available to fight such infections is also indicated.

Studies Using Viruses

Most of what you hear about viruses makes you think that viruses always act in a harmful way. However, there are some cases where, through research, scientists are discovering uses for viruses that may make them helpful.

One experimental method, called gene therapy, involves substituting correctly coded hereditary material for a cell's incorrect hereditary material. Correctly coded hereditary material is enclosed in a virus. The virus may then "infect" defective cells, taking the new strand of hereditary material into such cells to replace the incorrect hereditary material.

Using gene therapy, scientists hope to help people with genetic disorders. For example, some people have the genetic disorder sickle-cell anemia. Because of a defective gene, hemoglobin in their red blood cells does not release oxygen when it reaches body tissues. With the help of a virus, a repaired gene was allowed to infect blood cells in a mouse, and the mouse blood cells began to produce the correct substance. Researchers are hoping to use similar techniques for humans with sickle-cell disease and cancer.

Section Wrap-up

Review

1. Describe the structure of viruses and explain how viruses reproduce.

2. How are vaccines beneficial?

3. Explain how some viruses may be helpful in the process of gene therapy.

4. **Think Critically:** Explain why a doctor might not give you any medication if you had a viral disease.

Skill Builder
Concept Mapping
Make an events chain concept map to show what happens when a latent virus becomes active. Refer to Concept Mapping in the **Skill Handbook.**

Using Computers

Spreadsheet Enter the following data in a spreadsheet and make a line graph. How does temperature affect viruses?
At 36.9°C, there are 1.0 million viruses; at 37.2°C, 1.0 million; at 37.5°C, 0.5 million; at 37.8°C, 0.25 million; at 38.3°C, 0.10 million; and at 38.9°C, 0.05 million.

Skill Builder
Map should show the following order: latent virus enters cell; virus becomes part of cell's DNA; cell divides; virus reproduced as part of cell division; virus becomes active; virus forms new virus particles; cell destroyed.

Assessment

Oral How do viruses invade a cell? *The virus injects its hereditary material into the cell, which becomes infected.* In what ways are latent viruses unlike active viruses? *Latent viruses can go a long time inside the cell without causing the cell to produce new viruses or die.* Use the Performance Task Assessment List for Making Observations and Inferences in **PASC,** p. 17. P

---

4 Close

Section Wrap-up

Review

1. A virus is a particle of hereditary material covered by protein. Viruses reproduce by entering a host cell, causing the cell to make new viruses. The new viruses are released and the host cell is destroyed.

2. Vaccines, when properly administered, help prevent many childhood and adult diseases.

3. They may be used to transfer correctly coded DNA into a cell.

4. **Think Critically** There are no medicines to cure a viral infection.

Using Computers

Viruses cannot survive as well at higher temperatures, and therefore a fever helps to kill them.

Effect of Temperature Upon Viruses

Prepare

Section Background
- The AIDS virus was first identified in 1983 at the Pasteur Institute in Paris after being observed for the first time in 1978.
- The immune system attacks invading microorganisms and prevents them from growing inside the body.

1 Motivate

Bellringer
Before presenting the lesson, display **Section Focus Transparency 6** on the overhead projector. Assign the accompanying **Focus Activity** worksheet. L1 LEP

2 Teach

USING MATH
$$\frac{1\,025\,073}{426\,022} \times 100\% = 241\%$$

3 Assess

Check for Understanding
Have students answer questions 1 and 2 in the Section Wrap-up.

Reteach
Point out that AIDS is at epidemic stages in many parts of the world. **Ask why AIDS is spreading so fast.** *unprotected sexual intercourse and injecting drugs with previously used needles*

2•2 AIDS Vaccine?

Science Words
AIDS

Objectives
- Determine how HIV affects the immune system.
- Recognize how a potential AIDS vaccine can prevent the disease.

Figure 2-6

In this electron micrograph, HIV particles (colored blue) are attacking a larger white blood cell. White blood cells are an important part of your body's immune system.

From HIV to AIDS

Most persons remain healthy because their bodies have natural defenses against disease. These defenses, called the immune system, attack invading viruses and organisms that cause disease and keep them from reproducing inside the body. HIV, the human immunodeficiency virus shown in **Figure 2-6**, attacks the body's immune system. HIV causes a disease called **AIDS,** or acquired immune deficiency syndrome. AIDS allows the body to be infected by many organisms that cause disease. Eventually, death results from other infections, rather than from HIV itself.

The War

Recent research indicates that HIV replicates wildly inside cells, mutating often and destroying many cells of the immune system. This growth is usually kept under control for many years by the body's defensive response. But when the variety of HIV mutants finally becomes overwhelming, the immune defense system collapses. When the defense system can no longer keep up with the viruses, a condition known as full-blown AIDS develops. Most HIV-infected persons develop AIDS over the course of 10 years or more. Some patients, however, are diagnosed within two years of infection while others do not have AIDS for 15 years or more.

According to this new *warring* theory, this delay in the onset of AIDS may result from fierce but equally matched competition between the rapidly changing or mutating HIV virus and the body's defense systems.

The Mutating Enemy

In studying the HIV life cycle, researchers have concluded that the virus is adapted to mutate despite the pressures exerted by the host immune system. Its genetic makeup changes constantly. A high

38 Chapter 2 The Structure of Viruses and Cells

Program Resources

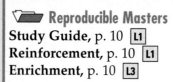
Reproducible Masters
Study Guide, p. 10 L1
Reinforcement, p. 10 L1
Enrichment, p. 10 L3

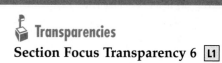
Transparencies
Section Focus Transparency 6 L1

mutation rate of the HIV virus increases the probability that some genetic change will give rise to an advantageous trait. This pattern of replication makes HIV the most variable virus known.

A Potential AIDS Vaccine

A canarypox virus is being made as a potential AIDS vaccine by researchers at Johns Hopkins University. The vaccine is made from a vaccine used in France to inoculate canaries. This new vaccine is a modified form of an earlier vaccine that caused the immune system to produce antibodies against HIV in adults.

The modified virus can enter human cells just as other viruses do. However, this virus cannot live and multiply in human cells. Instead, the modified viruses manufacture and package HIV proteins into tiny virus-like particles. These tiny particles have an empty center where the hereditary material of real viruses is located. The particles look like the whole AIDS virus but they do not have all the hereditary materials HIV needs to cause the disease. The immune system is triggered to produce antibodies by these viruses lacking hereditary material. Further research toward developing an anti-AIDS virus vaccine is ongoing among medical scientists.

USING MATH

In 1990, 426 022 cases of AIDS worldwide had been reported to the World Health Organization. By the end of 1994, the number of known AIDS cases had risen to 1 025 073. What percent increase does this represent?

Section Wrap-up

Review

1. According to this new *warring* theory, why are some persons diagnosed with AIDS within two years of HIV infection when it takes others 10 years or more to get full-blown AIDS?

2. How does the potential AIDS vaccine stop the production of HIV viruses?

Explore the Technology

Is it more important to develop a vaccine for AIDS or to develop anti-viral drugs and medicines? Until there is a cure for AIDS, what seems to be the best approach for slowing the spread of the disease? Prepare a report of your findings.

SCIENCE & SOCIETY

NATIONAL GEOGRAPHIC SOCIETY

Videodisc

Newton's Apple: Life Sciences
HIV and AIDS
Chapter 3, Side B
HIV compared to other viruses

30632-31056

4 Close

Discuss the preceding video material with students.

Section Wrap-up

Review

1. The competition between the virus and the immune system is an even match until a variety of mutants cause the immune system to collapse.

2. by triggering the immune system to produce antibodies

Explore the Technology

Medicines can do little to cure viral diseases. Antibiotics kill bacteria by inhibiting enzymes or biosynthetic processes specific to the bacteria. Viruses have few or no enzymes of their own; they use those of their hosts. The best approach for slowing the spread of AIDS is educating people about the practices that transmit HIV, such as unprotected sex and sharing needles.

Prepare

Section Background

A beam of electrons in an electron microscope is similar to the light used in a light microscope. Magnets in the electron microscope are like lenses in a light microscope. The magnets focus the electron beam on a fluorescent screen that produces an image of a specimen.

Preplanning

Assemble materials for the MiniLAB—drinking glass, clear glass bowl, water, magnifying glass, newspaper pages.

1 Motivate

Bellringer

 Before presenting the lesson, display **Section Focus Transparency 7** on the overhead projector. Assign the accompanying **Focus Activity** worksheet. L1 LEP

Tying to Previous Knowledge

Students may have used magnifying glasses and binoculars. Explain that a microscope magnifies in the same way.

INTEGRATION
Physics

Microscopes—both simple and compound—use convex lenses to enlarge images of objects. Microscopes have magnifying power and resolving power. Resolving power is the ability to see two objects as separate and distinct.

Science Words

compound light microscope
electron microscope
cell theory

Objectives

- Discuss the history leading to the cell theory.
- Explain the difference between the compound light microscope and the electron microscope.
- Explain the importance of the cell theory.

INTEGRATION
Physics

Ivy cells
500×

40

2•3 Cells: The Units of Life

The Microscope

You have learned that viruses are small particles that need cells to reproduce themselves. What are cells? Unlike a virus, cells are the smallest units that carry out all of the activities of life in organisms. Just as you cannot see viruses without a microscope, you cannot see cells when you look at most plants and animals. You need a device, such as a magnifying glass or microscope, to see most cells.

Trying to see separate cells in a large plant, like the one in **Figure 2-7**, is like trying to see individual bricks in a wall from three blocks away. If you start to walk toward the wall, it becomes easier to see individual bricks. When you get right up to the wall, you can see each brick in detail and many of the small features of each brick. A microscope performs a similar function. A microscope has one or more lenses that make an enlarged image of an object. Through these lenses, you are brought closer to the leaf, and you see the individual cells that carry on life processes.

Early Microscopes

Microscopes are simple or compound, depending on how many lenses they contain. A simple microscope is similar to a

Figure 2-7

Individual plant cells become visible when the leaves are looked at with a microscope.

Program Resources

Reproducible Masters
Study Guide, p. 11 L1
Reinforcement, p. 11 L1
Enrichment, p. 11 L3
Science and Society Integration, p. 6
Activity Worksheets, p. 15 L1
Science Integration Activities, pp. 3, 4
Lab Manual 3

Transparencies
Section Focus Transparency 7 L1
Science Integration Transparency 2
Teaching Transparency 3

Eyepiece

Revolving nosepiece

Arm

High-power objective

Specimen on glass slide

Low-power objective

Clip

Fine adjustment

Coarse adjustment

Light source

Base

Figure 2-8
A compound light microscope magnifies organisms or parts of organisms, making details of structures visible.

magnifying glass. It has only one lens. In 1590, a Dutch maker of reading glasses, Zacharias Janssen, put two magnifying glasses together in a tube. The result was the first crude compound microscope. By combining two lenses he got an image that was larger than an image made by only one lens. These early compound microscopes weren't satisfactory, however. The lenses would make a large image, but it wasn't always sharp or clear.

In the mid 1600s, Anton Van Leeuwenhoek, another Dutch scientist, made a simple microscope with a tiny glass bead for a lens. With it, he reported seeing things in pond water that no one had ever imagined before. His microscope could magnify up to 270 times. Another way to say this is that his microscope could make an image of an object 270 times larger than its actual size. Today we would say his lens had a power of 270×.

The Compound Light Microscope

The microscope you will use in studying life science is a compound light microscope like the one in **Figure 2-8**. A **compound light microscope** lets light pass through an object and then through two or more lenses. Convex lenses are used as magnifying lenses. The lenses enlarge the image and bend the light toward your eye. It has an eyepiece lens and an objective

MiniLAB

How are objects magnified?

Early scientists used various tools to view objects. Discover what tools you can use to magnify type in a newspaper.

Procedure

1. Look at a newspaper through both the curved side and the flat bottom of an empty glass.
2. Look at the newspaper through a glass bowl filled with water and then with a magnifying glass.

Analysis

1. In your Science Journal, compare how well you can see the newspaper through each of the objects.
2. What did early scientists learn by using such tools?

2 Teach

MiniLAB

Purpose

Visual-Spatial Students will discover objects that can be used to magnify.
L1 LEP

Materials

drinking glass, clear glass bowl, water, magnifying glass, newspaper pages

Teaching Strategies

Try this activity with the specific glasses to be used by students. Determine amount of water needed beforehand.

Analysis

1. Each of the objects magnifies the newsprint.
2. Early scientists learned that cells exist and began to infer that living things were made up of cells.

Assessment

Performance Place a sheet of waxed paper over a word on the newsprint. Then place a drop of water on the waxed paper over the word. Explain what you see. Is the drop concave or convex? *The word appears magnified because the water drop is curved as if it were a concave lens.* Use the Performance Task Assessment List for Making Observations and Inferences in **PASC**, p. 17. P

Activity Worksheets, pp. 5, 15

Use **Science Integration Activities,** pp. 3-4, as you teach this lesson.

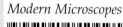
lens. An eyepiece lens usually has a power of 10×. An objective lens may have a power of 43×. Together, they have a total magnification of 430× (10× times 43×). Some compound microscopes have more powerful lenses that magnify an object up to 2000 times (2000×) its original size.

The Stereo Microscope

Your classroom may have a stereoscopic (stereo) light microscope that gives you a three-dimensional view of an object. This type of microscope has an objective and an ocular for each eye. Stereo microscopes are used to look at thick structures that light can't pass through, such as whole insects or leaves, or your fingertips.

Electron Microscopes

Things that are too small to be seen with a light microscope can be viewed with an electron microscope. Instead of using lenses to bend beams of light, an **electron microscope** uses a magnetic field to bend beams of electrons. Electron microscopes can magnify images up to 1 000 000 times. **Figure 2-9** shows the kind of detail that can be seen with an electron microscope.

There are several kinds of electron microscopes. One is the transmission electron microscope (TEM), which is used to study parts inside a cell. The object has to be sliced very thin and placed in a vacuum. **Figure 2-9A** shows a flea as seen through a TEM. There is no air in a vacuum. As a

Figure 2-9

Electron microscopes show details that cannot be seen using a compound light microscope.

A Transmission electron microscopes (TEM) provide images that show great detail. This TEM shows a thin slice of a flea.

B Scanning electron microscope (SEM) transmissions show great detail of the surface of an organism. This is an SEM of a flea. *Could the flea be alive?*

19×

63×

result, only dead cells and tissues can be observed this way. A scanning electron microscope (SEM) is used to see the surfaces of whole objects. With an SEM, you can view and photograph living cells. **Figure 2-9B** shows a flea through an SEM. From the time of Van Leeuwenhoek until the present, the microscope has been a valuable tool for studying cells. You will see how it was used to develop the cell theory.

The Cell Theory

During the centuries when Europeans were exploring new lands, scientists were busy observing everything they could about the smaller world around them. Curiosity made them look through their microscopes and lenses at mud from ponds and drops of rainwater. They examined blood and scrapings from their own teeth.

Cells weren't discovered until the microscope was improved. In 1665, Robert Hooke, an English scientist, made a very thin slice of cork and looked at it under his microscope. To Hooke, the cork seemed to be made up of little empty boxes, which he called *cells.* The photograph of cork cells reproduced in **Figure 2-10** is similar to those observed by Robert Hooke more than 300 years ago. Actually, Hooke was not aware of the importance of what he was seeing.

100×

Figure 2-10

Robert Hooke made drawings of cork cells very similar to those shown here.

Development of the Cell Theory

In 1838, Matthias Schleiden, a German scientist, used a microscope to study plant parts. He concluded that all plants were made of cells. Just a year later, another

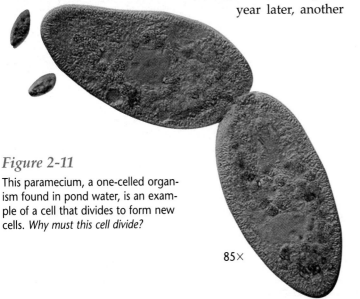

Figure 2-11

This paramecium, a one-celled organism found in pond water, is an example of a cell that divides to form new cells. *Why must this cell divide?*

85×

2-3 Cells: The Units of Life   **43**

Content Background

Explain that Hooke did not see cells; he saw the cell walls of dead cells from the bark of the cork oak tree. The cork tree is a live oak. It is green year-round and grows abundantly in Portugal and Spain.

NATIONAL GEOGRAPHIC SOCIETY

Videodisc

STV: The Cell
Viewing the Cell
Unit 1
Cell Theorists

04948-07166
Parts of the Cell
Unit 2
Scientist at scanning electron microscope

46956
Scientist at electron microscope

47014

3 Assess

Check for Understanding

? FLEX Your Brain

Use the Flex Your Brain activity to have students explore THE CELL THEORY.

📁 **Activity Worksheets,** p. 5

📁 Use **Science and Society Integration,** p. 6, as you teach this lesson.

Across the Curriculum

Art Discuss the use of art before the camera was invented. Have students compare Hooke's drawing of cells with photographs throughout the chapter. Have students write their opinions of the advantages and disadvantages of using artwork and photography. L2 P

Place a large drawing of a compound microscope on the bulletin board. Write the functions of each part on 3" × 5" cards. Have students select a card, name the part, and point out the part on the drawing.

Extension

📁 For students who have mastered this section, use the **Reinforcement** and **Enrichment** masters.

4 Close

•MINI•QUIZ•

Use the Mini Quiz to check students' recall of chapter content.

1. _____ was credited with discovering cells. *Robert Hooke*
2. A(n) _____ allows light to pass through an object and two or more lenses. *compound light microscope*
3. The _____ uses a magnetic field to bend electrons. *electron microscope*

Section Wrap-up

Review

1. Cells were too small to be seen without the microscope.
2. It is the basis for scientists who study the parts of cells, how they are organized, and how they reproduce and change through time.
3. **Think Critically** Accept all reasonable answers. Dead cells will show cell parts differently than will living cells.

USING MATH

📐 **Logical-Mathematical** The low-power magnification is 80 (8 × 10) and the high-power magnification is 320 (8 × 40).

Figure 2-12
The observations and conclusions of many scientists became known as the cell theory. The major ideas of the cell theory are:

The Cell Theory

1. All organisms are made up of one or more cells.
2. Cells are the basic units of structure and function in all organisms.
3. All cells come from cells that already exist.

German scientist, Theodor Schwann, after observing many different animal cells, concluded that all animals were made up of cells. Together, they became convinced that all living things were made of cells.

About 15 years later, a German doctor, Rudolph Virchow, hypothesized that new cells don't form on their own. Instead, cells divide to form new cells. This was a startling idea. Remember that, at that time, people thought earthworms fell from the sky when it rained. They thought that life came about spontaneously. What Virchow said was that every cell that is or has ever been came from a cell that already existed. **Figure 2-11** on page 43 shows how a cell divides to produce a new cell.

The cell theory is one of the major theories in science. It is not based on the hypotheses and observations of only one person, but is the result of the discoveries of many scientists. Today it serves as the basis for scientists who study the parts of cells, how cells are organized, and how cells and organisms reproduce and change through time. **Figure 2-12** summarizes the **cell theory.**

Section Wrap-up

Review

1. Explain why the invention of the microscope was important in the study of cells.
2. Why is the cell theory important?
3. **Think Critically:** Why would it be better to be able to see the details of living cells than dead cells?

Skill Builder
Concept Mapping
Using a network tree concept map, show the differences between compound light microscopes and electron microscopes. If you need help, refer to Concept Mapping in the **Skill Handbook.**

USING MATH

Calculate the total low power and high power magnification of a microscope that has an 8× eyepiece, a 10× objective, and a 40× high power objective.

Skill Builder

✓ Assessment

Performance Use this Skill Builder to assess students' concept mapping skills as they map microscopes. Have students write a paragraph describing the limitations of each microscope as they understand them. Use the Performance Task Assessment List for Concept Map in **PASC**, p. 89. P

Cell Organization 2•4

An Overview of Cells

In contrast to the dry cork boxes that Hooke saw, living cells are dynamic and have several things in common. They all have a membrane and a gel-like material called cytoplasm inside the membrane. In addition, they all have something that controls the life of the cell. This control center is either a nucleus or less organized nuclear material.

Prokaryotic and Eukaryotic Cells

Scientists have found that there are two basic types of cells. Cells that have no membrane around their nuclear material are prokaryotic cells. Bacteria and cells that form pond scum, like those in **Figure 2-13,** are prokaryotic cells. A eukaryotic cell has a nucleus with a membrane around it. The animal and plant cells in this chapter are all eukaryotic cells.

Figure 2-13

Pond scum is made up of prokaryotic cells.

Spyrogyra 200×

Science Words

- cell membrane
- nucleus
- chromatin
- cytoplasm
- organelle
- endoplasmic reticulum
- ribosome
- Golgi body
- mitochondria
- lysosome
- cell wall
- chloroplast
- tissue
- organ

Objectives

- Diagram a plant cell and an animal cell; identify the parts and the function of each part.
- Describe the importance of the nucleus in the cell.
- Explain the differences among tissues, organs, and organ systems.

Prepare

Section Background

- The branch of biology that deals with the formation, structure, and function of cells is cytology.
- Cells range in size from 2 m long, as in the nerve cells that extend down a giraffe's leg, to 0.2 micrometer in diameter, as in some bacteria.
- Not all cells have nuclei. The red blood cells of mammals have nuclei when they are first formed in bone marrow. They lose their nuclei before entering circulation.

Preplanning

- Obtain *Elodea* and prepared cheek cell slides for Activity 2-1.
- Obtain unflavored gelatin for the MiniLAB and onion plants and iodine stain for Activity 2-2.

1 Motivate

Bellringer

Before presenting the lesson, display **Section Focus Transparency 8** on the overhead projector. Assign the accompanying **Focus Activity** worksheet. L1 LEP

Tying to Previous Knowledge

Explain that it is in the cell that the features of life discussed in Chapter 1 are carried out.

Program Resources

📁 **Reproducible Masters**
Study Guide, p. 12 L1
Reinforcement, p. 12 L1
Enrichment, p. 12 L3
Activity Worksheets, pp. 11-12, 13-14, 16 L1
Lab Manual 4
Concept Mapping, p. 9

📽 **Transparencies**
Section Focus Transparency 8 L1
Teaching Transparency 4

2 Teach

Activity

 Visual-Spatial Make a bulletin board showing unlabeled parts of an animal cell and a plant cell. As each cell part is studied, have a student place its label beside it on the bulletin board. L1

Discussion

Ask students what color cells are. Living tissues are, for the most part, colorless. Photographs of cells have been color enhanced, and prepared slides are generally stained. Ask students why stains are used.

NATIONAL GEOGRAPHIC SOCIETY

Videodisc

STV: The Cell
Parts of the Cell
Unit 2
Cell diagram (1)

46988

Cell diagram (2)

46989

Use **Teaching Transparency 4** as you teach this lesson.

Use Lab 4 in the **Lab Manual** as you teach this lesson.

Animal Cells

Each cell in your body is constantly active and has a specific job to do. The activities in your cells might be compared to a business that operates 24 hours a day making dozens of different products. A business operates inside a building. A cell is similar. It functions within a structure called the cell membrane. Materials that are needed to make specific products are brought into the building. Finished products are often moved out. Similarly, nutrients are absorbed into the cell and waste products are released.

A The parts of the animal cell are labeled in the illustration to the right.

1875×

Figure 2-14

Animal cells are typical eukaryotic cells. Refer to this diagram as you read about cell parts and their jobs.

Cell membrane • Chromatin • Golgi bodies • Mitochondrion • Nuclear membrane • Nucleus • Cytoplasm • Lysosome • Endoplasmic reticulum (ER) • Ribosome • Nucleolus • Vacuole

B The cell membrane is made up of a double layer of fats with some proteins scattered throughout.

Across the Curriculum

Language Arts Greek and Latin words are used in naming cell parts. Help students pronounce the names of cell parts in the animal cell diagram. Have students make a list of cell parts, and use a dictionary to find the origins of the words and their meanings. LS

Cell Membrane

The **cell membrane** is a structure that forms the outer boundary of the cell and allows only certain materials to move into and out of the cell. The membrane, as shown in **Figure 2-14,** is flexible. It is made up of a double layer of fats with some proteins scattered throughout. The cell membrane helps to maintain a chemical balance between materials inside and outside the cell. Food and oxygen move into the cell through the membrane. Waste products also leave through the membrane. In Chapter 3, you will learn that many types of substances enter and leave the cell in different ways.

Nucleus

The largest organelle in the cytoplasm of a eukaryotic cell is the **nucleus,** a structure that directs all the activities of the cell. The nucleus, shown in **Figure 2-15,** is like a manager who directs everyday business for a company and passes on information to new cells. A nucleus is separated from the cytoplasm by a nuclear envelope. Materials enter and leave through openings in the membrane. The nucleus contains genetic blueprints for the operations of the cell in the form of long strands called **chromatin,** a form of hereditary material. Chromatin is made up of proteins and DNA, the chemical that controls the activities of the cell. When a cell begins to divide, the strands of chromatin thicken and take on the form of chromosomes, which are easier to see. A structure called a nucleolus, which is involved in making proteins, is also found in the nucleus.

Figure 2-15

The nucleus of a cell is surrounded by a double membrane. DNA in the nucleus controls the activities in the cell.

20 500×

Nucleolus
Nucleus
Pores
Chromatin
Nuclear membrane

2-4 Cell Organization **47**

Activity

Kinesthetic Use three-dimensional models of a plant cell and an animal cell to help students identify cell parts and learn their functions. **L1**

Demonstration

Visual-Spatial Use a microprojector to show the nuclei in prepared slides of frog blood and onion root tip.

Inquiry Question

What would happen if the nucleus of a cell became damaged in some way? *The cell would no longer function correctly because the nucleus controls all of the activities of the cell.*

Videodisc
STV: The Cell
Parts of the Cell
Unit 2
Cell Membrane

12694-15290
Nucleus

15330-19100

Use **Concept Mapping,** p. 9, as you teach this lesson.

Integrating the Sciences

Earth Science The cell membrane has been described as two layers of fats with large proteins scattered among the fat molecules. In some ways, it is similar to an ocean with icebergs. Icebergs are large pieces of ice that move randomly through the ocean. **How does this analogy help you visualize the behavior of large protein molecules in cells?** *The large proteins move about through the fat layers.*

Using an Analogy

Use an analogy for organelles, pointing out that just as different parts of a machine perform different functions in making a product, so each organelle performs a different function in the cell.

MiniLAB

Purpose

☐ **Kinesthetic** Students will make a model of cytoplasm. **L1** **LEP**

Materials

100-mL beaker, unflavored gelatin, water, flashlight, stirring rod

Teaching Strategies

- Tell students that a model is used by scientists to explain a concept that is not easy to understand.
- Have students stir the gelatin well before shining the light on the beaker.

Analysis

1. Particles represent organelles suspended in the cytoplasm.
2. A model is used to help explain a concept that is not easy to see or understand.

◥Assessment

Oral Have students compare the chemical composition of cytoplasm with that of gelatin. *Both are mixtures (suspensions) composed mainly of water. Cytoplasm is 80 percent water.* Use the Performance Task Assessment List for Making Observations and Inferences in **PASC**, p. 17. **P**

📁 **Activity Worksheets,** pp. 5, 16

MiniLAB

How can you model cytoplasm?

Cytoplasm is a mixture of proteins, fats, carbohydrates, salts, water, and many other chemicals. These particles are suspended in the liquid part of this mixture. Make a model of cytoplasm.

Procedure

1. Fill a beaker with 100 mL water.
2. Add unflavored gelatin and stir.
3. Shine a flashlight through the beaker.

Analysis

1. Describe what you see.
2. How does a model help you understand what a real thing looks like?

Cytoplasm

Cytoplasm is the gel-like material inside the cell membrane and outside the nucleus. Cytoplasm contains a large amount of water and many chemicals and structures that carry out the life processes in the cell. Unlike a gelatin dessert, however, cytoplasm constantly moves or streams.

The structures within the cytoplasm of eukaryotic cells are **organelles.** Each one has a specific job or jobs. Some organelles break down food. Others move wastes to be expelled from the cell. Still others store materials. Most organelles are surrounded by membranes.

Organelles in the Cytoplasm

The **endoplasmic reticulum (ER),** illustrated in **Figure 2-16,** is a folded membrane that moves materials around in the cell. The ER extends from the nucleus to the cell membrane and takes up quite a lot of space in some cells. The ER is like a system of conveyor belts in a business or tunnels along which materials move from one place to another.

One chemical that takes part in nearly every cell activity is protein. Proteins are found in cell membranes. Other proteins are needed for chemical reactions that take place in the cytoplasm. Cells make their own proteins on small structures in the cytoplasm called **ribosomes.** Ribosomes receive directions

Figure 2-16

Endoplasmic reticulum (ER) is a complex series of membranes in the cytoplasm of the cell. If ribosomes are present on the endoplasmic reticulum, then the endoplasmic reticulum is referred to as *rough ER.* Proteins are made on these ribosomes. *What would smooth ER look like?*

63 548×

Ribosomes

Cultural Diversity

Ernest Everett Just Just, an African-American biologist in the early 1900s, studied cells and how they function. His research showed that all parts of the cell influence its activities, not just the nucleus, as scientists then believed. This idea changed scientific opinion concerning the basis of life. Discuss how Ernest Just's research is important to the study of cells today.

Figure 2-17

Golgi bodies package materials and move them to the outside of the cell. *Why are materials removed from the cell?*

62 000×

from the hereditary material in the nucleus on how and when to make specific proteins. Some ribosomes are scattered in the cytoplasm. Others are attached to the ER. Ribosome parts are made in the nucleolus.

In a business, products are made, packaged, and moved to loading docks to be carried away. In cells, structures called **Golgi bodies** are stacks of membrane-covered sacs that package and move proteins to the outside of the cell. Golgi bodies, as illustrated in **Figure 2-17,** are the packaging and secreting organelles of the cell. When something is secreted, it is given off by the cell.

Cells require a continuous supply of energy. **Mitochondria,** pictured in **Figure 2-18,** are organelles where food molecules

Figure 2-18

Mitochondria are known as the "powerhouses of the cell." In these organelles, food molecules are broken down, and energy is released. *What types of cells may contain many mitochondria?*

62 000×

Outer membrane

Folded inner membrane

2-4 Cell Organization **49**

Problem Solving

Solve the Problem

2. Surface areas of cubes:
   1-cm cube = 6 cm²
   2-cm cube = 24 cm²
   4-cm cube = 96 cm²

3. Volumes of cubes:
   1-cm cube = 1 mL
   2-cm cube = 8 mL
   4-cm cube = 64 mL

4. Surface area-to-volume ratios:
   1-cm cube = 6
   2-cm cube = 3
   4-cm cube = 1.5

Think Critically

1. it decreases
2. Shape does not affect surface area.
3. The cell can divide.

GRAPHING CALCULATOR

Have students use graphing calculators to graph $Y = 6/X$, where Y is the surface area-to-volume ratio, X is the length of one side of a cube, and 6 is a conversion factor. Have them use the trace function to check the surface area-to-volume ratios asked for in Solve the Problem. L3

NATIONAL GEOGRAPHIC SOCIETY

 Videodisc

STV: The Cell
Parts of the Cell
Unit 2
Lysosomes identified

46938
Other Organelles

19135-23966

50

Surface Area and Volume of Living Cells

The cells in mice, elephants, and humans are about the same size. More than ten trillion cells make up the human body. If 1000 of these cells were lined up, they would total less than 2 centimeters in length—only about the width of a thumbnail.

Why are most cells microscopic? In order to survive a cell must take in nutrients and remove wastes. Substances move into and out of a cell by passing through the membrane that covers the surface of the cell. This fact limits the size to which a cell can grow. Why?

Solve the Problem:

1. Assume that a cell is like a cube.
2. Find the surface area of a cube illustrated below. Surface area is width × length × six (There are six faces in a cube).
3. Calculate the volume of each cube. The volume of a cube is length × width × height.
4. Find the surface area to volume ratio of each cube by dividing the surface area by the total volume.

Think Critically:

1. What happens to the ratio of the surface area to volume as the size of the cell increases?
2. Does the shape of the cell affect the surface area?
3. How can a cell that is too large solve its low surface area to volume ratio?

1 cm   2 cm   4 cm

---

are broken down and energy is released. This energy is then stored in other molecules that can power cell reactions easily. Just as a power plant supplies energy to a business, mitochondria release energy for the cell. Some types of cells are more active than others. Muscle cells, which are always undergoing some type of movement, have large numbers of mitochondria. Why would active cells have more or larger mitochondria?

An active cell also constantly produces waste products. In the cytoplasm, organelles called **lysosomes** contain chemicals that digest wastes and worn-out cell parts as well as break down food. When a cell dies, chemicals in the lysosomes act to quickly break down the cell. In a healthy cell, the membrane around the lysosome keeps the chemicals in the organelle itself from breaking down the cell.

Many businesses have warehouses for storing products until they are sold. Vacuoles are fluid-filled temporary storage areas in cells. They may store water, food, and other materials needed by cells while other vacuoles store waste products. **Figure 2-19** shows that a vacuole may take up most of the space in some plant cells. In animal cells, vacuoles are small.

Plant and Bacterial Cells

The major difference between an animal cell and a plant cell is that plant cells have cell walls. **Figure 2-19** shows the different parts of a plant cell. The **cell wall** is a rigid structure outside the cell membrane that supports and protects the plant cell. It is made of bundles of tough cellulose fibers and other materials made by the cell.

Plant cells also differ from animal cells because they can make their own food. **Chloroplasts** are organelles in plant cells in which light energy is changed into chemical energy in the form of a sugar. One of the chemicals in chloroplasts that traps light energy is chlorophyll, a green pigment.

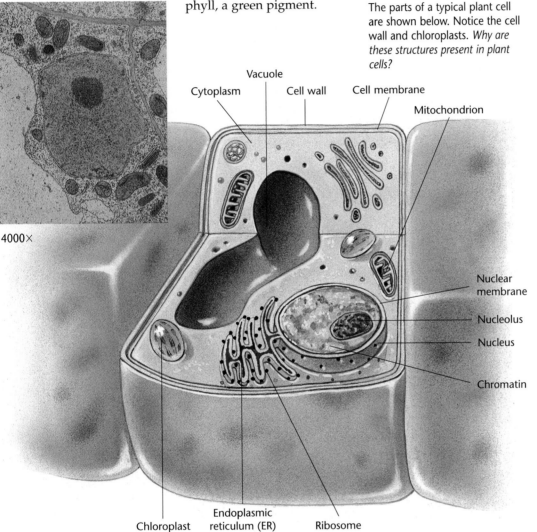

4000×

Figure 2-19

The parts of a typical plant cell are shown below. Notice the cell wall and chloroplasts. *Why are these structures present in plant cells?*

Vacuole

Cytoplasm

Cell wall

Cell membrane

Mitochondrion

Nuclear membrane

Nucleolus

Nucleus

Chromatin

Chloroplast

Endoplasmic reticulum (ER)

Ribosome

2-4 Cell Organization　　**51**

Visual Learning

Figure 2-19 Why are these structures present in plant cells? *Cell walls support and protect plant parts, and chloroplasts are needed to produce food.* **LEP** **LS**

Demonstration

LS **Visual-Spatial** Use a model of a plant cell to point out a vacuole, chloroplasts, and cell walls. **LEP**

Inquiry Question

How do vacuoles in a plant cell differ from those in an animal cell? *Plant cells usually have one large vacuole, and animal cells may have several smaller ones or none that are noticeable.*

NATIONAL GEOGRAPHIC SOCIETY

Videodisc

STV: The Cell
Parts of the Cell
Unit 2
Plant Cells

25786-27234
Cell wall of onion cell

47010
Chloroplasts

47011
Chloroplasts within plant cells

46944
Plant cells with chlorophyll

47012

Across the Curriculum

Health Fiber is an important part of a healthful diet. It comes from the cell walls of plants. Make a list of healthful foods that contain fiber. Lists could include fruits, grains, and leafy vegetables. **L1**

Integrating the Sciences

Chemistry The cell is the smallest unit of living matter. **What is the smallest unit of nonliving matter?** *atom*

Bacterial Cells

In contrast to plant and animal cells, bacterial cells are prokaryotic cells. Bacteria don't have membrane-covered organelles, as shown in **Figure 2-20**. A bacterial cell has a cell wall and cytoplasm, but it has only a single chromosome. There are no nuclei in bacteria, but they do contain ribosomes.

All cells have similar parts. Animal and plant cells have organelles. Bacterial cells have no membrane-covered structures.

Figure 2-20

A bacterial cell is a one-celled prokaryotic organism. It has no membrane around its nuclear material. This type of bacterial cell lives in human intestines.

17 650×

How Cells Differ

Cells come in a variety of sizes. A single nerve cell in your leg may be a meter in length. A human egg cell, on the other hand, is no bigger than a dot on this i. Going a step further, a human red blood cell is about one-tenth of the size of a human egg cell.

The shape of a cell may also tell you something about the job the cell does. The nerve cell in **Figure 2-21** with its fine extensions sends impulses through your body. Look at its shape in contrast to the white blood cell, which can change shape. Some cells in plant stems are long and hollow with holes. They transport food and water through the plant. Human red blood cells, on the other hand, are disk-shaped and have to be small and flexible enough to move through tiny blood vessels.

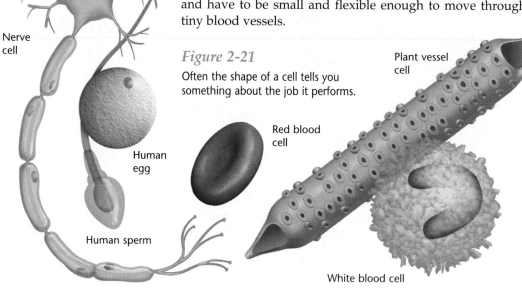

Nerve cell

Human egg

Human sperm

Plant vessel cell

Red blood cell

White blood cell

Figure 2-21

Often the shape of a cell tells you something about the job it performs.

Activity 2-1

Comparing Plant and Animal Cells

If you were to compare a goldfish to a rose bush, you would find the two to be different. However, when the individual cells of these organisms are compared, will they be as different? Try this activity to see how plant and animal cells compare.

Problem
How do plant and animal cells compare?

Materials
- microscope
- 1 microscope slide
- 1 coverslip
- forceps
- dropper
- *Elodea* plant
- prepared slide of human cheek cells

Procedure
1. Copy the data table in your Science Journal. Record whether you observed the cell parts listed.
2. Follow the directions in Appendix A on page 756 for use of low and high power objectives on your microscope and for making a wet-mount slide.
3. Using forceps, make a wet-mount slide of a young leaf from the tip of an *Elodea* plant.
4. Observe the leaf on low power. Focus on the top layer of cells. Draw what you see. Carefully focus down through the top layer of cells to observe the layers of cells.
5. Focus on one cell. Observe the movement of chloroplasts along the cell membrane.
6. Observe the cell nucleus. It looks like a clear ball. Look at the nucleus on high power.
7. Make a drawing of the *Elodea* cell. Label the cell wall, cytoplasm, chloroplasts, and nucleus. Return to low power and remove the slide.

8. Place a prepared slide of cheek cells on the microscope stage. Locate the cells under low power.
9. Switch to high power and observe the cell nucleus. Draw and label the cell membrane, cytoplasm, and nucleus.

Analyze
1. What parts of the *Elodea* cell were you able to identify?
2. How many cell layers could you see in the *Elodea* leaf?
3. What parts of the cheek cell were easy to identify?

Conclude and Apply
4. **Compare and contrast** the shape of the cheek cell and the *Elodea* cell.
5. What can you conclude about the differences between plant and animal cells?

Data and Observations

Sample Data

Cell Part	*Elodea*	Cheek
cytoplasm	X	X
nucleus	X	X
chloroplasts	X	
cell wall	X	
cell membrane	present, not visible	X

2-4 Cell Organization 53

Activity 2-1

3 Assess

Check for Understanding

Use a poster of a plant cell, animal cell, and bacterial cell to have students identify the parts and describe the functions of each part.

Reteach

Have each student write an answer to the following question: **Based on what you know about cell structure, which do you think are more closely related: plant and bacterial cells, animal and plant cells, or animal and bacterial cells? Why?** *animal and plant cells because they contain more common parts*

Extension

For students who have mastered this section, use the **Reinforcement** and **Enrichment** masters.

USING TECHNOLOGY

Replacement Skin

Third-degree burns destroy both the outer and inner layers of skin cells. This can be dangerous as well as painful because it leaves the body without protection against harmful microorganisms. Surgeons have used the skin of dead individuals to temporarily treat such burns since 1881. The body's immune system usually rejects this skin in two weeks, so the skin must continually be replaced until the burn heals.

A New Product

A new, permanent replacement for the outer skin is a product also made from the skin of dead individuals, but with one important difference. All the skin cells are removed because they are the parts that are destroyed by the body's defense system. The remaining tissue of the outer layer of skin has substances that the defense system does not attack.

In healthy skin, cells called fibroblasts continually help regenerate the outer layer of skin. The protein in this new skin replacement product signals fibroblasts from other parts of the body to replace its missing fibroblasts. The fibroblasts reconstruct the skin using this permanent skin replacement as a pattern.

Thinking Critically:

How does this permanent skin-replacement product speed up the healing process?

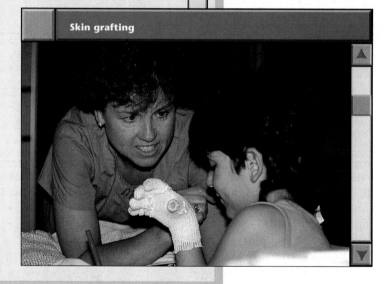

Skin grafting

Community Connection

Burn Units Invite a technician from a local burn unit to speak to the class. Have students prepare questions about treating burn patients. Questions may request information regarding lengths of treatments and technologies used during treatments, as well as common causes of burns.

Organizing Cells

A one-celled organism performs all its life functions by itself. Cells in a many-celled organism, however, do not work alone. Instead, each cell depends on other cells in some way as the organism carries out its functions. This interaction helps the whole organism stay alive.

In **Figure 2-22** you can see a single plant cell. You also see a group of the same type of plant cells that together form a tissue on the outside of a plant leaf.

In many-celled organisms, cells are organized into **tissues,** which are groups of similar cells that do the same sort of work. Each tissue cell carries on all the functions needed to keep it alive. The cells in a tissue may look alike.

Different tissues are further organized into organs. An **organ** is a structure made up of different types of tissues that work together to do a particular job.

Cell

Tissue

Organ

Figure 2-22

In a many-celled organism, different types of tissues are organized into organs and systems.

Your heart is an organ made up of muscle, nerve, and blood tissues. Several different tissues make up a plant leaf, an organ of the plant in which food is made.

A group of organs working together to do a certain job is an organ system. Your heart and blood vessels make up your cardiovascular system. What plant systems can you think of?

In a many-celled organism, several systems work together. Roots, stems, and leaves in a plant work together to keep the plant alive. Name three systems in your body.

Each cell in a many-celled organism carries on its own life functions. Although cells in an organism may differ in appearance and function, all the cells work together to keep the organism alive.

Organ system

Organism

Section Wrap-up

Review

1. Explain the importance of the cell nucleus in the life of a cell.

2. Give an example of an organ system in an animal and name the parts that make up the organ system.

3. **Think Critically:** How is the cell of a one-celled organism different from the cells in many-celled organisms?

Skill Builder
Comparing and Contrasting

Organize information about cell organelles in a table. Use this information to diagram, compare, and contrast animal cells, plant cells, and bacterial cells. If you need help, refer to Comparing and Contrasting in the **Skill Handbook.**

Science Journal

Your textbook compared cell functions to that of a business. In your Science Journal, write an essay that explains how a cell is like your school or town.

2-4 Cell Organization   **55**

✔ Assessment

Performance To further assess students' abilities to compare and contrast different types of cells, have them write a statement in their Science Journals comparing **Figures 2-14** and **2-19.** Use the Performance Task Assessment List for Science Journal in **PASC,** p. 103. **P**

•MINI•QUIZ•

Use the Mini Quiz to check students' recall of chapter content.

1. The _____ is a structure that forms the outer boundary of the cell and allows only certain material to move in and out. *cell membrane*

2. _____ are packaging and secreting organelles of the cell. *Golgi bodies*

3. The _____ of a cell is related to its function. *shape*

4. A group of tissues working together to do a certain job is a(n) _____ . *organ*

Section Wrap-up

Review

1. It directs the activities of the cell.

2. the circulatory system: heart, blood vessels, blood

3. **Think Critically** One-celled organisms perform all life functions alone. In a many-celled organism, different cell types are interdependent.

Science Journal

A cell has many parts, just as a town or school has many parts. Different people supply the needed services in a town or school. Each person has a special job. Each building has a special function. All these things working together make the town or school function properly. The different parts of a cell have special jobs, and each part helps the cell carry out its life processes.

PREPARATION

Purpose

IS **Interpersonal** Design and carry out an experiment to show the parts that make up an organism. L1 **COOP LEARN**

Process Skills

observing, inferring, comparing and contrasting, predicting, communicating, interpreting data, and designing an experiment

Time

60 minutes

Materials

onion plant with roots and leaves, microscope, microscope slides (3), coverslips (3), 250-mL beaker with water, dropper, forceps, scissors

- Small green onions can be purchased in bunches at the grocery store.
- If the piece of onion bulb is placed in water, the onion skin tissue will be easier to place on the slide.
- Forceps can be used to remove thin layers from the roots, bulb, and leaves. You may wish to use a scalpel to cut a root in half for each student group.

Safety Precautions

Caution students to use extreme care when working with the microscope and microscope slides. Use care with scissors.

Think Critically

Parts of an organism include cells, tissues, organs, and organ systems. Most one-celled organisms are able to function on their own, while the cells of many-celled organisms are dependent upon one another in order to function properly.

📁 **Activity Worksheets,** pp. 5, 13, 14

Activity 2-2

Design Your Own Experiment

Observing the Parts of an Organism

How are the parts of your body different? How are they alike? How are the parts of plants different? Are the cells in an organism alike?

PREPARATION

Problem

Observe an organism and determine what parts it has. How do the cells in each of these parts differ?

Think Critically

List the parts of an organism and explain how they work together. How are many-celled organisms different from one-celled organisms?

Objectives

- Observe the similarities and differences in various parts of an organism.
- Design an experiment to observe differences in the cells of each part of an organism.

Possible Materials

- onion plant with roots and leaves
- microscope
- microscope slides (3)
- coverslips (3)
- 250-mL beaker with water
- dropper
- forceps
- iodine stain
- scissors

Safety Precautions

Protect clothing, and be careful using sharp objects, such as scissors.

PLAN THE EXPERIMENT

1. As a group planning your experiment, be sure to consider how you can examine the different levels of organization of the onion.
2. To examine the cellular level of the onion, make a wet-mount slide of the membrane. Adding a drop of iodine stain to the onion before adding the coverslip will enable you to see the structures more clearly.
3. As a group, list the steps that you need. Be very specific, describing exactly what you will do at each step. List your materials.

4. If you need a data table, design one in your Science Journal, so that it is ready to use as your group collects data.

Check Your Plan

1. Read over your entire experiment to make sure that all steps are in logical order.
2. Identify any constants, variables, and controls in the experiment.
3. Will the data be summarized in a table?
4. *Make sure your teacher approves your plan before you proceed.*

DO THE EXPERIMENT

1. Carry out the experiment as planned.
2. While the experiment is going on, write down any observations that you make and complete the data table in your Science Journal.

Analyze and Apply

1. **Compare** your results with those of other groups.
2. What parts did you divide the onion plant into and what are they called?
3. What did you observe about the cells of each part?

4. What are the parts of an onion plant and how do they work together?

Go Further

Predict whether similar structures would be found in any plant. How would you check your predictions?

2-4 Cell Organization 57

Go Further

Students will probably predict that similar structures would be found in any plant. They could check their predictions by making wet mount slides of tissues found in another plant and observing them under the microscope.

Assessment

Performance Have students observe another organism—plant or animal—or a photograph of one and hypothesize what parts it has, and how the cells in each part will be different. Use the Performance Task Assessment List for Making Observations and Inferences in **PASC**, p. 17. **P**

PLAN THE EXPERIMENT

Possible Procedures

One possible procedure is to observe the onion plant and then use the scissors to cut it into parts, prepare wet-mount slides of each of the parts, and observe the slides under the microscope.

Teaching Strategies

Make sure that students record and analyze their observations accurately. Encourage them to draw what they see.

DO THE EXPERIMENT

Expected Outcome

Most students will observe that the parts of an onion plant are cells, tissues, organs, and organ systems.

Analyze and Apply

1. Student results should be similar to those of other groups.
2. Roots, bulb, and leaves. These parts are organ systems. The thick layers are tissues made from cells.
3. All the cells of each part were oblong shaped, those from the leaf were green, and the ones from the roots and bulb were not green.
4. The onion plant has organs, and each organ is composed of tissues made of cells that have the same function. The root cells absorb water, the stem cells (bulb) transport water and food, and the leaf cells make food for the onion plant. **P**

Source

Facklam, Howard, and Margery Facklam. *Viruses.* New York: Henry Holt, 1994.

Preston, Richard. *The Hot Zone.* New York: Random House, 1994.

Background

- Different strains of the same virus can infect humans, birds, and pigs. As the virus moves from host to host, mutations often occur, weakening the virus—or making it more deadly. When two strains of the same virus infect one organism, they can combine and rearrange their DNA into a unique strain of the virus. In fact, new virus strains are likely when people are in close contact with waterfowl and pigs, which seem to serve as ready hosts.

- Mutations of flu viruses cause epidemics because people have developed no antibodies for the new strain. During the flu epidemics of 1957 and 1968, flu deaths in the United States increased from 20 000 in an average year to 90 000.

- As the appearance of *Ebola reston* proves, the United States is not immune to rare viruses. The Hantaan virus, which first appeared in ancient China, was brought to the United States in the 1950s by soldiers returning from the Korean War. The virus apparently remained dormant until 1993, when several Navajos in the same community came down with a mysterious disease. Eventually, scientists discovered that these people had contact with deer mice, which were serving as hosts for the ancient Hantaan virus.

![inter**NET** CONNECTION]

Scientists have determined that Marburg virus, Ebola Zaire, and Ebola Reston all belong to the virus family Filoviridae. Visit the Chapter 2 Internet Connection at Glencoe Online Science, **www.glencoe.com/sec/science/life**, for a link to more information about the Marburg and Ebola viruses.

Viruses Through History

Smallpox is just one of many viruses that have been difficult to control and eliminate. As these viruses multiply rapidly inside living cells, they also mutate. Many new viruses develop in tropical regions, perhaps because so many species of plants and animals live there. This wide range of living cells increases the probability for viruses to multiply and mutate under favorable conditions.

The first case of AIDS was reported in Africa in 1959. Ten years later, again in Africa, scientists identified the first case of Lassa fever, also caused by a virus. Yet even after a new or rare virus is identified, there may be no vaccine or cure.

Ebola Viruses

This is true for the HIV virus and for the Ebola virus, named for the Ebola river in Africa. In 1976, the first known outbreak of Ebola killed half of the people in a Sudan village; 60 percent of those who were infected died. Two months later, a more deadly form of Ebola swept through 50 villages in nearby Zaire. It killed more than 430 people, 90 percent of those infected. Both times, the virus was controlled by isolating its victims.

In 1989, monkeys imported from the Philippines brought the Ebola virus to a research laboratory in Reston, just outside Washington, DC. Scientists gradually figured out, to their horror, why the monkeys were dying. Although all 500 monkeys died from the virus or were destroyed to prevent its spread, the scientists were relieved to discover that the Ebola virus had mutated again. The Reston strain is fatal to monkeys, but harmless to humans.

Predicting the Future

Nevertheless, the most deadly strain—so far—of Ebola reappeared in 1995 in Zaire, killing more than 230 people. Some experts believe that new and rare viruses will appear more frequently as people have more contact with animals and insects in tropical regions. Crowded conditions help spread these viruses, and global travel can carry them far from their origins. Scientists are trying to guess where and how these mutating viruses will next appear—and hoping to be able to develop new treatments and vaccines.

Teaching Strategies

- Ask students why they think *Ebola* outbreaks have been controlled fairly quickly so far. Can we count on this continuing? Why?

- Discuss reasons why people living in the United States might feel safe from new and rare viruses and reasons why we should still be concerned.

Bibliography

- Greenberg, Lorna. *AIDS: How It Works in the Body.* New York: Watts, 1992.
- Jaret, Peter. "Viruses." *National Geographic,* July 1994.

Summary

2-1: Viruses

1. A virus is a structure containing hereditary material surrounded by a protein coat. A virus can reproduce only inside a living cell.
2. Vaccines prevent viral infections.

2-2: Science and Society: AIDS Vaccine?

1. AIDS destroys the body's ability to combat disease. Death results from other infections.
2. Potential vaccines for AIDS are being researched.

2-3: Cells: The Units of Life

1. Janssen made the first compound microscope in 1590. Leeuwenhoek used fine lenses to make accurate observations. Hooke, Schleiden, and Schwann all drew their conclusions about cells with the help of the microscope.
2. Compound light microscopes use light and lenses to make images. Electron microscopes bend beams of electrons in a magnetic field.
3. According to the cell theory, the cell is the basic unit of life. Organisms are made of one or more cells, and all cells come from other cells.

2-4: Cell Organization

1. There are differences among animal, plant, and bacterial cells. Plant cells have cell walls and often have chloroplasts. Bacteria have no membrane-covered organelles.

2. Cell functions are performed by organelles under control of DNA in the nucleus.
3. Most many-celled organisms are organized into tissues, organs, and organ systems that perform specific jobs to keep an organism alive.

Key Science Words

a. AIDS	k. Golgi body
b. cell membrane	l. host cell
c. cell theory	m. lysosome
d. cell wall	n. mitochondria
e. chloroplast	o. nucleus
f. chromatin	p. organ
g. compound light microscope	q. organelle
	r. ribosome
h. cytoplasm	s. tissue
i. electron microscope	t. vaccine
	u. virus
j. endoplasmic reticulum	

Reviewing Vocabulary

Match each phrase with the correct term from the list of Key Science Words.

1. hereditary material in the nucleus
2. directs cell activities
3. where energy is generated for the cell
4. structure within cytoplasm of eukaryotic cells
5. contains all the same kind of cell

Summary

Have students read the summary statements to review the major concepts of the chapter.

Reviewing Vocabulary

1. f	**6.** r
2. o	**7.** h
3. n	**8.** e
4. q	**9.** k
5. s	**10.** t

✓Assessment

Portfolio Encourage students to place in their portfolios one or two items of what they consider to be their best work. Examples include:

- Inclusion Strategies, p. 33
- Across the Curriculum discussion notes, p. 43
- Activity 2-2, data, answers, predictions, p. 57 P

Performance Additional performance assessments may be found in **Performance Assessment** and **Science Integration Activities.** Performance Task Assessment Lists and rubrics for evaluating these activities can be found in Glencoe's **Performance Assessment in the Science Classroom.**

GLENCOE TECHNOLOGY

 MindJogger Videoquiz

Chapter 2 Have students work in groups as they play the Videoquiz game to review key chapter concepts.

Chapter 2 Review

Checking Concepts

1. c	**6.** b
2. a	**7.** b
3. c	**8.** c
4. a	**9.** a
5. c	**10.** a

Understanding Concepts

11. Improved lenses in compound microscopes allowed individual parts of cells to be seen more clearly. Stereomicroscopes allow for three-dimensional viewing; greater magnification is possible with electron microscopes.

12. Answers may include that lysosomes might respond only to old or worn-out cell parts and foreign substances.

13. Answers might include the observation that if something is made up of cells, it is an organism and that cells have to come from cells that already exist.

14. The virus is adapted to mutate despite the pressures exerted by the host immune system.

15. The AIDS virus leaves the body unable to fight infections. As a result, people with AIDS get many secondary infections, which could lead to death.

Thinking Critically

16. Once a virus infects a cell, it uses the cell to produce more viruses. The immune system has to work to get rid of viruses because there are no drugs that can kill viruses.

17. A stereomicroscope would enable you to view a large part of the living organism, as well as looking closely at the mold.

---

6. where proteins are made
7. gel-like substance inside cell membrane
8. traps light energy in plant cells
9. packages and secretes substances
10. substance that can prevent but not cure a disease

Checking Concepts

Choose the word or phrase that completes the sentence.

1. _____ is an example of a viral disease.
   - a. Tuberculosis
   - c. Small pox
   - b. Anthrax
   - d. Tetanus
2. A microscope that uses lenses and objectives to magnify is the _____.
   - a. compound light microscope
   - b. scanning electron microscope
   - c. transmission electron microscope
   - d. atomic force microscope
3. A microscope that magnifies parts inside a cell up to 1 000 000 times or more is the _____.
   - a. compound light microscope
   - b. stereoscopic microscope
   - c. transmission electron microscope
   - d. scanning electron microscope
4. The scientist who gave the name *cells* to structures he viewed was _____.
   - a. Hooke
   - c. Schleiden
   - b. Schwann
   - d. Virchow
5. A _____ is an organelle that can destroy old cell parts.
   - a. chloroplast
   - c. lysosome
   - b. vacuole
   - d. cell wall
6. A structure that allows only certain things to pass in and out of the cell is the _____.
   - a. cytoplasm
   - c. cell wall
   - b. cell membrane
   - d. nuclear envelope

7. Structures in the cytoplasm of the eukaryotic cell are called _____.
   - a. organs
   - c. organ systems
   - b. organelles
   - d. tissues
8. Materials can move around inside the cell through the folded membranes of _____.
   - a. chromatin
   - c. endoplasmic reticulum
   - b. cytoplasm
   - d. Golgi bodies
9. A bacterial cell has _____.
   - a. a cell wall
   - c. mitochondria
   - b. lysosomes
   - d. a nucleus
10. Groups of different tissues form an _____, such as the heart.
   - a. organ
   - c. organ system
   - b. organelle
   - d. organism

Understanding Concepts

Answer the following questions in your Science Journal using complete sentences.

11. How has the development of different microscopes helped scientists study cells?
12. Suggest a reason why a lysosome can exist inside a living cell without destroying the cell.
13. Explain why the cell theory is important.
14. Why is HIV sometimes considered to be the most variable virus known?
15. Why do most AIDS victims die from secondary infections?

Thinking Critically

16. Why is it difficult to get rid of viruses?
17. What type of microscope would be best to use to look at a piece of moldy bread? Give reasons to support your choice.

60   Chapter 2 The Structure of Viruses and Cells

---

18. No light energy can be changed into food or chemical energy, and therefore the plant cell will die.

19. No proteins could be made and the animal cell would die.

20. Look for an organized nucleus. If there is no nucleus, it is a bacterial cell. If there is a nucleus surrounded by a membrane, then the cell is eukaryotic. If there is also a cell wall and chloroplasts, it is a plant cell. A cell with a nucleus but no cell wall or chloroplasts is an animal cell.

Developing Skills

21. Interpreting Scientific Illustrations The long shape of nerve cells allows them to carry messages to other structures; hollow plant stem cells allow for transport of water; human blood cells are small enough to pass through vessels; sperm cells have tails that allow them to move easily.

22. Concept Map Refer to student page.

18. What would happen to a plant cell that suddenly lost its chloroplasts?
19. What would happen to an animal cell if it didn't have ribosomes?
20. How would you decide whether an unknown cell was an animal cell, a plant cell, or a bacterial cell?

Developing Skills

If you need help, refer to the **Skill Handbook.**

21. **Interpreting Scientific Illustrations:** Use the illustrations of cells on page 52 to describe how the shape of each cell may be related to its function.
22. **Concept Mapping:** Complete the following concept map of the basic units of life.

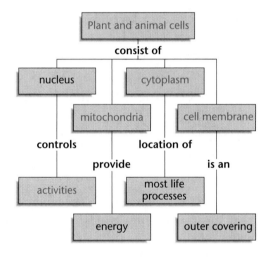

23. **Observing and Inferring:** Infer the effects of a damaged cell membrane on a cell.

24. **Making and Using Tables:** Make a table similar to the one below that lists the names of the cell structure(s) that are involved in each of the functions listed. Structures may be listed in more than one column.

Cell Function and Structure

Function	Structure(s)
Protection	
Movement of materials	Answers provided in the teacher margin
Breaking down substances	
Making substances	
Storing substances	

25. **Comparing and Contrasting:** In a table, compare and contrast a virus, a bacterial cell, and a eukaryotic cell according to the structures they contain.

Performance Assessment

1. **Model:** Use materials that resemble a cell and cell organelles to make a model of either a plant cell or an animal cell. Make a key to the cell parts to explain your model.
2. **Poster:** Draw and label a time line illustrating the history of vaccination. Look up the term *variola* in a library reference and read about early attempts to protect people from disease, especially smallpox.
3. **Scientific Drawing:** Observe cells in a flower petal or tomato skin. Write a report based on your observations. Illustrate your report.

Assessment Resources

 Reproducible Masters
Chapter Review, pp. 7, 8
Assessment, pp. 9-12
Performance Assessment, p. 16

Glencoe Technology
- **Chapter Review Software**
- **Computer Test Bank**
- **MindJogger Videoquiz**

---

23. **Observing and Inferring** There would be a lack of control of what could enter or leave the cell, causing cell damage and, eventually, death.

24. **Making and Using Tables**

Function	Structures
Protection	cell membrane; cell wall
Movement of materials	Golgi bodies; ER; cytoplasm
Breaking down substances	lysosomes; mitochondria
Making substances	ER; ribosomes; nucleus
Storing substances	Golgi bodies; ER; vacuoles

25. **Comparing and Contrasting**

Virus or cell type	Structures	Nucleus
Virus	nucleic acid and protein coat	No
Bacteria	cell wall, cytoplasm, ribosomes,	No
Eukaryote cell	cell membranes, cytoplasm, organelles, DNA in a nucleus	Yes

Performance Assessment

1. The model should have all the parts shown on page 46 or 51. Use the Performance Task Assessment List for Model in **PASC,** p. 51. **P**
2. Begin the timeline with the work of Edward Jenner in 1796. Students should research *vaccination* to fill in the time line. Use the Performance Task Assessment List for Poster in **PASC,** p. 73. **P**
3. Scientific drawing and observations should be clear and logical. Include labels. Use the Performance Task Assessment List for Scientific Drawing in **PASC,** p. 55. **P**

Chapter Organizer

Section	Objectives/Standards	Activities/Features
Chapter Opener		**Explore Activity:** Compare the effects of plain water and salt water on potato slices. p. 63
3-1 Chemistry of Living Things (1½ sessions, 1 block)*	1. **Describe** differences among atoms, elements, molecules, and compounds. 2. **Recognize** the relationship between chemistry and life science. 3. **Compare** inorganic and organic compounds. National Content Standards: (5-8) UCP1, UCP3, A1, B1, D1	**Using Math,** p. 67 **MiniLAB:** How do enzymes work? p. 68 **Skill Builder:** Hypothesize, p. 70 **Using Computers:** Spreadsheet, p. 70
3-2 Cell Transport (2½ sessions, 2 blocks)*	1. **Explain** the function of a selectively permeable membrane. 2. **Describe** the processes of diffusion and osmosis. 3. **Compare** and **contrast** passive transport and active transport and give examples of each. National Content Standards: (5-8) UCP1, UCP4, UCP5, A1, A2, C1, C3, E2, F5	**MiniLAB:** How does temperature affect the rate of diffusion of molecules? p. 72 **Connect to Physics,** p. 73 **Using Technology:** Slowing Cystic Fibrosis, p. 75 **Skill Builder:** Concept Mapping, p. 76 **Science Journal,** p. 76 **Activity 3-1:** Observing Osmosis, p. 77
3-3 Energy in Cells (1½ sessions, 1 block)*	1. **Explain** the difference between producers and consumers. 2. **Compare** and **contrast** the processes of photosynthesis and respiration. 3. **Describe** how cells get energy from glucose through the process of fermentation. National Content Standards: (5-8) UCP1, UCP3, UCP4, A1, A2, B1, B3, C1, C3	**Problem Solving:** Permeability, p. 80 **Skill Builder:** Comparing and Contrasting, p. 81 **Using Math,** p. 81 **Activity 3-2:** Photosynthesis and Respiration, pp. 82-83
3-4 Science and Society Technology: Energy from Biomass (½ session, ½ block)*	1. **Describe** how photosynthesis and fermentation are related to biomass. 2. **Identify** useful biomass materials. 3. **Compare** the advantages and disadvantages of using biomass as an energy source. National Content Standards: (5-8) B3, E2, F5	**Explore the Technology,** p. 85 **Science and Literature:** The Lives of a Cell, p. 8⟨ **Science Journal,** p. 86.

* A complete Planning Guide that includes block scheduling is provided on pages 32T-35T.

Activity Materials

Explore	Activities	MiniLABs
page 63 250-mL beaker; tap water; 2 identical, small bowls; 15 g table salt; labels; 2 identical slices of potato; clock or stopwatch; paper towels	page 77 raw egg, two 500-mL containers with covers, 250 mL vinegar, 250 mL white table syrup, 250-mL graduated cylinder, labels, safety goggles, apron pages 82-83 four 16 × 150-mm test tubes, test-tube rack, stirring rod, balance with masses, sodium bicarbonate, bromothymol blue solution in dropping bottle, aged tap water, 2 pieces of *Elodea*, safety goggles, apron	page 68 test tube, milk, mortar and pestle, rennin tablet page 72 two 250-mL beakers, cold water, hot tap water, labels or marking pencil, green food coloring, clock or stopwatch

Need Materials? Call Science Kit (1-800-828-7777).

Teacher Classroom Resources

Reproducible Masters	Transparencies	Teaching Resources
Study Guide, p. 13 **Reinforcement**, p. 13 **Enrichment**, p. 13 **Activity Worksheets**, p. 21 **Lab Manual 5**, Chemical Changes **Science Integration Activity 3**, Types of Mixtures **Cross-Curricular Integration**, p. 7 **Science and Society Integration**, p. 7	**Section Focus Transparency 9**, Chemicals of Life **Science Integration Transparency 3**, Earth's Elements and the Human Body	**Glencoe Life Science Interactive Videodisc** **Life Science CD-ROM** **Spanish Resources** **English/Spanish Audiocassettes** **Cooperative Learning Resource Guide** **Lab Partner** **Lab and Safety Skills** **Lesson Plans**
Study Guide, p. 14 **Reinforcement**, p. 14 **Enrichment**, p. 14 **Activity Worksheets**, pp. 17-18, 22 **Lab Manual 6**, Diffusion **Critical Thinking/Problem Solving**, p. 7 **Multicultural Connections**, p. 9	**Section Focus Tansparency 10**, Skin Deep **Teaching Transparency 5**, Selective Permeabililty	**Assessment Resources** **Chapter Review**, pp. 9-10 **Assessment**, pp. 13-16 **Performance Assessment in the Science Classroom (PASC)** **MindJogger Videoquiz** **Alternate Assessment in the Science Classroom** **Performance Assessment** **Chapter Review Software** **Computer Test Bank**
Study Guide, p. 15 **Reinforcement**, p. 15 **Enrichment**, p. 15 **Activity Worksheets**, pp. 19-20 **Lab Manual 7**, Cell Respiration **Concept Mapping**, p. 11	**Section Focus Transparency 11**, Consumer's Delight **Teaching Transparency 6**, Photosynthesis/Respiration Cycle	
Study Guide, p. 16 **Reinforcement**, p. 16 **Enrichment**, p. 16	**Section Focus Transparency 12**, Waste Not Want Not	

Key to Teaching Strategies

The following designations will help you decide which activities are appropriate for your students.

L1 Level 1 activities should be within the ability range of all students, including those with learning difficulties.

L2 Level 2 activities should be within the ability range of the average to above-average student.

L3 Level 3 activities are designed for the ability range of above-average students.

LEP LEP activities should be within the ability range of Limited English Proficiency students.

LS These activities are designed to address different learning styles.

COOP LEARN Cooperative Learning activities are designed for small group work.

P These strategies represent student products that can be placed into a best-work portfolio.

GLENCOE TECHNOLOGY

The following multimedia resources are available from Glencoe.

Science and Technology Videodisc Series (STVS)
Chemistry
  Images of Atoms
Human Biology
  Detecting Cystic Fibrosis
The Infinite Voyage Series
Unseen Worlds

National Geographic Society Series
Newton's Apple: Life Sciences
STV: The Cell
STV: Plants
Glencoe Life Science Interactive Videodisc
Photosynthesis and Cellular Respiration
Life Science CD-ROM

Teacher Classroom Resources

This is a representation of key blackline masters available in the Teacher Classroom Resources.

Teaching Aids

Section Focus Transparencies

Science Integration Transparencies

Teaching Transparencies

Meeting Different Ability Levels

Study Guide

Reinforcement

Enrichment Worksheets

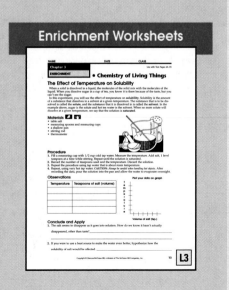

Hands-On Activities

Science Integration Activity

Lab Manual

Assessment

Performance Assessment

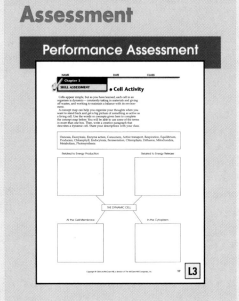

Enrichment and Application

Critical Thinking/ Problem Solving

Cross-Curricular Integration

Science and Society Integration

Multicultural Connections

Concept Mapping

Cell Processes

CHAPTER OVERVIEW

Section 3-1 This section provides a basic explanation of the chemistry needed to understand life processes. Atoms make up elements. Elements are joined by chemical bonds to form compounds. Compounds are classified as organic or inorganic.

Section 3-2 The processes of diffusion, osmosis, and active transport and how they move molecules in and out of cells are discussed.

Section 3-3 The processes of photosynthesis, respiration, and fermentation and their roles in providing energy for cells of living organisms are described.

Section 3-4 Science and Society Biomass fuels are identified. Advantages and disadvantages of using biomass as an energy source are discussed.

Chapter Vocabulary

mixture	active
organic	transport
compound	endocytosis
enzyme	exocytosis
inorganic	metabolism
compound	producer
diffusion	consumer
equilibrium	fermentation
osmosis	biomass
passive	gasohol
transport	biogas

Theme Connection

Stability and Change Living things function as a result of chemical reactions that take place in cells. The equilibrium maintained by cells results from their selectively permeable membranes. This is critical to the life of cells and the organism as a whole.

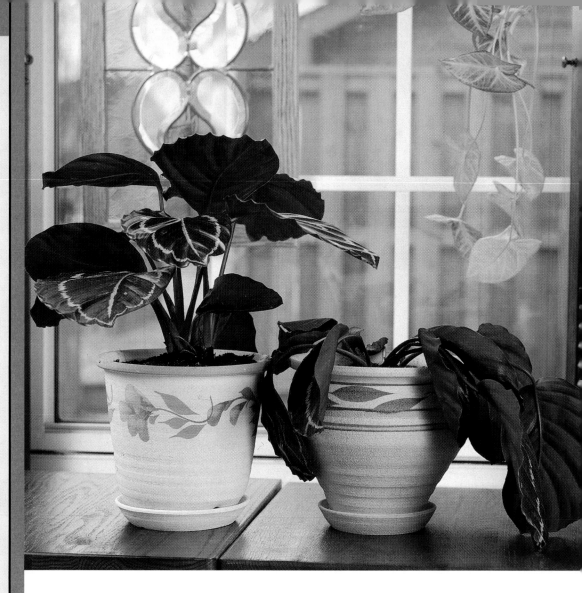

Previewing the Chapter

Section 3-1   Chemistry of Living Things
 ▶ The Nature of Matter
 ▶ Organic Compounds

Section 3-2   Cell Transport
 ▶ Control of Materials by Cells
 ▶ Types of Transport

Section 3-3   Energy in Cells
 ▶ Trapping Energy for Life
 ▶ Releasing Energy for Life

Section 3-4   Science and Society Technology: Energy from Biomass

62

Learning Styles Look for the following logo for strategies that emphasize different learning modalities. [LS]

Kinesthetic	Explore, p. 63; Demonstration, p. 74
Visual-Spatial	MiniLAB, pp. 68, 72; Across the Curriculum, p. 69; Reteach, p. 75; Activity, p. 79
Interpersonal	Activity 3-2, pp. 82-83; Science & Literature, p. 86
Intrapersonal	Science & Literature, p. 86
Logical-Mathematical	Tying to Previous Knowledge, p. 64; Using Math, pp. 68, 81; Using Computers, p. 70; Activity 3-1, p. 77
Linguistic	Integrating the Sciences, p. 68; Using Science Words, p. 69; Cultural Diversity, p. 72; Science Journal, p. 76; Across the Curriculum, p. 79; Science & Literature, p. 86

Cell Processes

You may have forgotten to water a plant in the past, only to be reminded to when the plant begins wilting. After watering, the plant looks healthier and straightens up. What caused the plant to straighten? How long does it take for a wilted plant to return to normal? Why does this happen? In the following activity, find out about water entering and leaving plant cells. Then in the chapter that follows, learn about the movement of water across cell membranes.

EXPLORE ACTIVITY

Compare the effects of plain water and salt water on potato slices.

1. Pour 250 mL of water into each of two small bowls.
2. First, label one bowl *salt water.* Then stir 15 g of salt into the bowl labeled "salt water."
3. Place two slices of raw potato into each bowl. In your Science Journal, predict how the potato slices will be affected.
4. After 20 minutes, take the slices from each of the bowls and examine them. Describe the slices.

Observe: Describe in your Science Journal how your observations relate to the plants in the photograph on page 62.

Previewing Science Skills

▶ In the **Skill Builders**, you will **hypothesize**, **map concepts**, and **compare and contrast**.

▶ In the **Activities**, you will **measure**, **use tables**, **conclude**, and **infer**.

▶ In the **MiniLABs**, you will **observe**, **infer**, and **predict**.

63

Assessment Planner

Portfolio
Refer to page 87 for suggested items that students might select for their portfolios.

Performance Assessment
See page 87 for additional Performance Assessment options.
Skill Builders, pp. 70, 76, 81
MiniLABs, pp. 68, 72
Activities 3-1, p. 77; 3-2, pp. 82-83

Content Assessment
Section Wrap-ups, pp. 70, 76, 81, 85
Chapter Review, pp. 87-89
Mini Quizzes, pp. 70, 76, 81

Group Assessment
Opportunities for group assessment occur with Cooperative Learning Strategies and Flex Your Brain Activities.

EXPLORE ACTIVITY

Purpose
LS **Kinesthetic** Use the Explore Activity to show students that water moves into and out of potato cells. Explain that they will be learning about atoms, molecules, and compounds and that these are the materials that move into and out of cells.
L1 **LEP** **COOP LEARN**

Preparation
Purchase potatoes for the activity.

Materials
2 bowls or baby food jars, stirring rod, 6 slices of potato (0.5 cm thick)

Teaching Strategies
• Have the water at room temperature so the salt will dissolve more readily.
• The amount of salt is not critical as long as the salt concentration is high.

Observe
Students should observe that water moves out of the potato cells in salt water just as it has moved out of the cells of the wilted plant. In plain water, water moves into the potato cells just as it does in the plant that is not wilted.

✓ Assessment

Oral **When a plant wilts, why does it become rigid again after it has been watered?** *Water diffuses out of a plant into the soil when it wilts, making the cells shrink. Adding water to the soil causes water to diffuse back into the plant.* Use the Performance Task Assessment List for Making Observations and Inferences in **PASC,** p. 17.
P

Prepare

Section Background

- Presently, there are 112 known elements. Ninety-one occur naturally, and the others have been made in the laboratory. About 25 of the 91 elements are found in living matter. Carbon, hydrogen, oxygen, and nitrogen are the elements most abundant in living organisms.

- A diatomic molecule is made of two atoms. Hydrogen (H_2), oxygen (O_2), and nitrogen (N_2) are diatomic.

- It was once believed that organic compounds could form only within living organisms. However, scientists have been able to make organic compounds for many years. Plastics and synthetic fibers are examples.

Preplanning

Obtain rennin tablets from a pharmacist for the MiniLAB.

1 Motivate

Bellringer

Before presenting the lesson, display **Section Focus Transparency 9** on the overhead projector. Assign the accompanying **Focus Activity** worksheet. L1 LEP

Tying to Previous Knowledge

LM **Logical-Mathematical** Bring in labels from foods and cleaning products. Many foods and soaps contain sodium chloride (NaCl). Sodium hydroxide (lye, NaOH) is in drain cleaners. Have students use the periodic table and determine what elements are in the compounds. L1

64

3•1 Chemistry of Living Things

Science Words
mixture
organic compound
enzyme
inorganic compound

Objectives
- Describe differences among atoms, elements, molecules, and compounds.
- Recognize the relationship between chemistry and life science.
- Compare inorganic and organic compounds.

Figure 3-1

An oxygen atom model shows the placement of electrons, protons, and neutrons.

 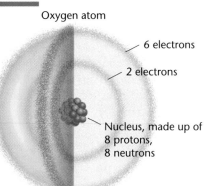

Oxygen atom

6 electrons

2 electrons

Nucleus, made up of 8 protons, 8 neutrons

INTEGRATION
Chemistry

Figure 3-2

When fireflies blink and shine in the dark, energy is released in the form of light as chemical bonds are broken and new ones are formed.

64 Chapter 3 Cell Processes

The Nature of Matter

Do you know what makes up Earth? If you ask a scientist, he or she will probably say "matter and energy." Matter is anything that has mass and takes up space. How many things can you think of that fit that category?

Atoms

Everything that can be measured is made up of matter. You are made of matter and so are the fireflies pictured on this page. Even the air that they fly through is made up of matter. Matter exists in the form of small units called atoms. **Figure 3-1** shows a model of an oxygen atom. At the center of the atom is a nucleus, which contains two kinds of particles, protons and neutrons. A proton has a positive charge, and neutrons have no charge. Electrons are particles found outside the nucleus. They are negatively charged.

What about the energy that the scientist mentioned? Energy in matter is locked in chemical bonds that hold atoms together. To transfer this energy to another molecule or to do work, the atoms that are bound together have to be rearranged or broken apart in a chemical reaction. As illustrated in **Figure 3-2**, chemical reactions take place in living organisms.

Program Resources

Reproducible Masters
Study Guide, p. 13 L1
Reinforcement, p. 13 L1
Enrichment, p. 13 L3
Cross-Curricular Integration, p. 7
Activity Worksheets, p. 21 L1
Lab Manual 5
Science Integration Activities, pp. 5-6
Science and Society Integration, p. 7

Transparencies
Section Focus Transparency 9 L1
Science Integration Transparency 3

Table 3-1

Elements That Compose the Human Body		
Symbol	**Element**	**Percent**
O	Oxygen	65.0
C	Carbon	18.5
H	Hydrogen	9.5
N	Nitrogen	3.3
Ca	Calcium	1.5
P	Phosphorus	1.0
K	Potassium	0.4
S	Sulfur	0.3
Na	Sodium	0.2
Cl	Chlorine	0.2
Mg	Magnesium	0.1

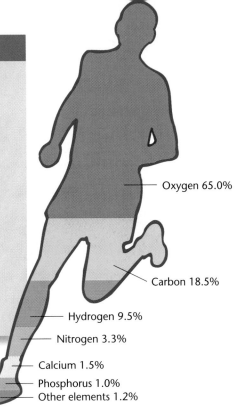

Oxygen 65.0%

Carbon 18.5%

Hydrogen 9.5%

Nitrogen 3.3%

Calcium 1.5%

Phosphorus 1.0%

Other elements 1.2%

Elements

When something is made up of only one kind of atom, it is called an element. An element can't be broken down into any simpler form. The element oxygen (O_2) is made up of only oxygen atoms, and hydrogen (H_2) is made up of only hydrogen atoms. Each element has its own symbol. More than 90 elements occur naturally on Earth. Everything, including you, is made of one or a combination of these elements. **Table 3-1** lists elements as they occur in the human body. What are the two most common elements in your body?

How Atoms Combine

Suppose you've just eaten an apple. Did you know that the sugar in the apple contains the elements carbon, oxygen, and hydrogen? Water, also contained in the apple, contains both oxygen and hydrogen. Notice that the apple and water both contain oxygen. Yet in one case oxygen is part of a colorless, tasteless liquid—water. Oxygen is also part of a structure that's hard and colorful—the apple. How can the same element, made from the same type of atom, help

Figure 3-3

An apple is made up of carbon, oxygen, and hydrogen.

3-1 Chemistry of Living Things **65**

Figure 3-4

Table salt, or NaCl, is a compound that is commonly added to food items such as popcorn and french fries.

make up two materials that are so different? The atoms of elements combine chemically to form new substances called compounds. One compound is shown in **Figure 3-4.** When elements form a compound, the properties of the individual elements change. How does this happen?

One way that atoms combine is by sharing their outermost electrons. The combined atoms form a molecule. For example, two atoms of hydrogen can share electrons with one atom of oxygen to form one molecule of water. Water is a compound composed of molecules. A compound is a type of matter that has properties different from the properties of each of the elements in it. Glucose, the major form of sugar used by animal cells for energy, is a compound. It is made of the elements carbon, hydrogen, and oxygen, and its chemical formula is $C_6H_{12}O_6$.

The properties of hydrogen and oxygen are changed when they combine to form water as illustrated in **Figure 3-5.** Under normal conditions on Earth, you will find the elements oxygen and hydrogen only as gases. Yet water can be a liquid or a solid as well as a gas. When hydrogen and oxygen combine, a change occurs and a new substance forms.

Figure 3-5

The chemical formula for water is H_2O.

Ions

You know that some atoms combine by sharing electrons. But atoms also combine because they've become positively or negatively charged.

As you discovered earlier, atoms are usually neutral—they have no overall electric charge. Under certain conditions, however, atoms can lose or gain electrons. When an atom loses electrons, it has more protons than electrons so the atom is positively charged. When an atom gains electrons, it has more electrons than protons so the atom is negatively charged. Electrically charged atoms are called ions.

Ions are attracted to one another when they have opposite charges. Oppositely charged ions join to form electrically neutral compounds. Table salt forms in this way. A sodium (Na) atom loses an electron and becomes a positively charged ion. Then it comes close to a negatively charged chlorine (Cl) ion. They are attracted to each other and form the compound NaCl as illustrated in **Figure 3-6.**

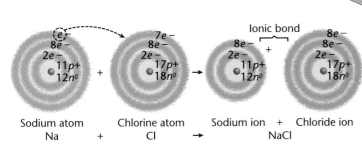

Figure 3-6

A sodium atom (Na) contains 11 electrons, with only one electron in the outermost energy level. A chlorine atom (Cl) has 17 electrons, with the outermost energy level needing one electron to fill it. When sodium and chlorine combine, the sodium atom loses one electron and the chlorine atom gains it. The chlorine atom becomes stable with a complete outer energy level. It has a negative charge. The sodium atom is also stable; it has lost one electron from its outer level and now has a positive charge.

Mixtures

Many times, two substances can be mixed together without combining chemically. A **mixture** is a combination of substances in which individual substances retain their own properties. For example, if you mix sugar and salt together, neither substance would change, nor would they combine chemically. Mixtures can be solids, liquids, or gases.

Figure 3-7

Sugar molecules and tea molecules spread out evenly in water to make iced tea, a mixture.

Inquiry Question

What might happen when a negatively charged ion comes in contact with a positively charged ion? *The two ions may combine to form an electrically neutral compound.*

Content Background

The two most common types of bonds in compounds are covalent bonds and ionic bonds. In covalent bonds, the outermost electrons are shared by the atoms in the compounds. In compounds with ionic bonds, ions are held together by the strong attraction of their opposite charges.

USING MATH

Logical-Mathematical
2 molecules × 2 atoms = 4 atoms hydrogen

Integrating the Sciences

Environmental Science Have students research the damage done to the environment by certain common compounds such as chlorofluorocarbons (CFCs), which are used as refrigerants. CFCs are responsible for damaging Earth's ozone layer. **L1**

Figure 3-8

Solutions are mixtures that are important to living things.

A Seawater is an example of a solution.

MiniLAB

How do enzymes work?

Procedure

1. Fill a test tube with milk.
2. Use a mortar and pestle to crush a rennin tablet.
3. Add the rennin tablet to the milk and let the milk stand undisturbed during your class period.
4. Observe what happens to the milk.

Analysis

1. What can you conclude about the milk?
2. On what part of the milk did the enzyme act?
3. Infer how the action of rennin on milk might be helpful.

B When a test tube holding a suspension of whole blood is left standing, left, the blood cells and the plasma separate, right.

Solutions and Suspensions

What is so important about molecules? In living organisms, cell membranes, cytoplasm, and other substances are all in the form of molecules. During its lifetime, a cell will put together and break apart many molecules for growth, for repair, and for energy. Many of these molecules are found dissolved in the cytoplasm, forming a solution. A solution is a mixture in which one or more substances mix evenly with other substances. When you dissolve salt in water, you get a salt solution. You've probably noticed the taste of salt when you perspire. That's because your cells are bathed in a salt solution. Living organisms also contain suspensions. A suspension is a mixture in which substances are evenly spread through a liquid or gas. Unlike solutions, the particles in a suspension settle out over time. Blood, shown in **Figure 3-8B,** is an example of a suspension. If a test tube of blood is left undisturbed, the red blood cells and white blood cells will gradually settle out. Blood in your body is moved constantly by the pumping action of your heart. Therefore, the cells don't settle out.

Organic Compounds

Compounds in living organisms are classified as either organic or inorganic. Most compounds containing carbon are **organic compounds.** Organic compounds make up foods and cell membranes. **Table 3-2** compares the four groups of organic compounds that make up all living things. These are carbohydrates, lipids, proteins, and nucleic acids.

Carbohydrates are organic compounds made up of carbon, hydrogen, and oxygen. Carbohydrates supply energy to power cell processes. But lipids, commonly called fats and oils, are organic compounds that store and release even larger amounts of energy than carbohydrates.

Table 3-2

Organic Compounds That Make Up Life				
	Carbohydrate	**Lipid**	**Protein**	**Nucleic Acid**
Composition	organic compound made up of carbon, hydrogen, and oxygen	organic compound made up of mostly carbon, hydrogen, and oxygen	organic compound; amino acids; proteins contain carbon, hydrogen, oxygen, and nitrogen	large organic molecule made of carbon, hydrogen, oxygen, nitrogen, and phosphorus
Examples	sugars, starch, and cellulose	fats, oils, and waxes	enzymes	DNA, RNA
Function	break down to release energy; energy is used to power cell processes	store and release larger amounts of energy than even carbohydrates	used to build cell parts	carries hereditary information
Location	cell membranes	cell membranes are made up of two layers of lipids	scattered throughout cell membranes	DNA in chromosomes, mitochondria, chloroplasts, nucleus

Proteins are the building blocks of many structural components of organisms. They are scattered throughout cell membranes and are made up of smaller molecules called amino acids. Certain proteins called **enzymes** speed up chemical reactions in cells without being changed. Enzymes take part in nearly all the chemical reactions in cells.

Nucleic acids are large organic molecules that store important information in cells in the form of a code. One nucleic acid, DNA, is found in the nucleus, in chromosomes, in mitochondria, and in chloroplasts. It carries information that directs each cell's activities. A second nucleic acid, RNA, carries information for making proteins and enzymes in cytoplasm where these substances are built.

Using Science Words

IS **Linguistic** Ask students the common meaning of the word *organic* in organic gardening, organic fertilizer, and organic food.

Revealing Preconceptions

Students may think that all lipids (fats) are harmful. Explain that fats have many useful functions in living organisms, especially in cell membranes.

3 Assess

Check for Understanding

? **FLEX Your Brain**

Use the Flex Your Brain activity to have students explore ORGANIC COMPOUNDS. P

📁 **Activity Worksheets,** p. 5

Reteach

Display a number of carbon compounds, such as soap, vinegar, alcohol, starch, and chalk. Ask why these are inorganic compounds.

Extension

📁 For students who have mastered this section, use the **Reinforcement** and **Enrichment** masters.

Across the Curriculum

Nutrition Assign the nutrients proteins, carbohydrates, and lipids to small groups in the class. Have them report on the functions of each nutrient and make a collage of photographs of food high in each nutrient. L1 **COOP LEARN** IS

4 Close

Section Wrap-up

Review

1. Atoms are the basic units of matter. Molecules are made up of atoms.

2. Most compounds that contain carbon are organic compounds.

3. **Think Critically** Salt and sugar will dissolve in water, forming a solution. Sand will form a suspension.

Using Computers

LS **Logical-Mathematical** The circle graph for the human body should show 65% oxygen, 18.5% carbon, 9.5% hydrogen, 3.3% nitrogen, 1.5% calcium, and 2.0% other elements. The circle graph for Earth's crust should show 46.6% oxygen, 27.7% silicon, 8.1% aluminum, 5.0% iron, 3.6% calcium, 2.8% sodium, 2.6% potassium, 2.1% magnesium, and 1.5% other elements.

Figure 3-9
All organisms are dependent upon water for life.

Inorganic Compounds

Most **inorganic compounds** are made from elements other than carbon. Water, an inorganic compound shown in **Figure 3-9,** makes up a large part of living matter. It is, therefore, one of the most important compounds in living things. In order for substances to be used in cells, they have to be dissolved in water. Nutrients and waste materials are carried throughout your body in solution form.

Section Wrap-up

Review

1. How are atoms different from molecules?

2. How do organic and inorganic compounds differ?

3. **Think Critically:** If you mix salt, sand, and sugar with water in a small jar, will the result be a suspension, a solution, or both?

Skill Builder
Hypothesizing
Cut part of an apple into thin slices. Place the slices in the sun near a window. Hypothesize what will happen. Observe and record your observations. What can you infer about the chemical composition of living things? Refer to Hypothesizing in the **Skill Handbook.**

Using Computers

Spreadsheet Find the percentage of elements that make up Earth's crust. Enter these data into a spreadsheet. Use this spreadsheet and **Table 3-1** to make circle graphs of the percentage of elements that make up the human body and Earth's crust. Only identify the five elements with the highest percentages.

Skill Builder
Water is an important element in living things.

Assessment

Oral Would a cell placed in distilled water lose or gain water? *It would gain water because the distilled water is pure water.* **P**

Cell Transport   3•2

Control of Materials by Cells

Cells obtain food, oxygen, and other substances from their environment. They also release waste materials into this environment. If a cell has a membrane around it, how do these things move into and out of the cell? How does the cell control what enters and leaves?

Have you ever bought marbles in a mesh bag? Although the mesh bag has holes in it, the marbles stay inside because they are larger than the holes. If you put sand in with the marbles, the sand will fall right through the holes because sand grains are smaller than the holes, as illustrated in **Figure 3-10.** The marbles and sand are models for molecules. The property of a cell membrane that allows some materials to pass through while keeping others out is known as selective permeability. The mesh bag is said to be selectively permeable based on hole size because it allows some things to pass through it but not others.

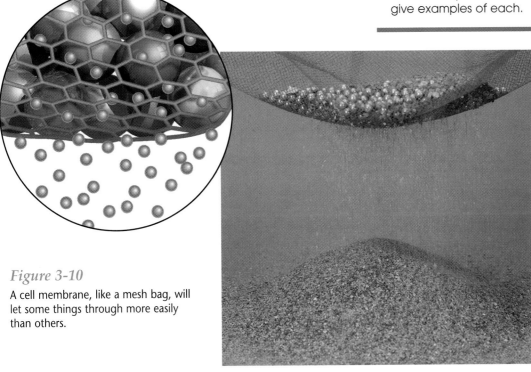

Figure 3-10
A cell membrane, like a mesh bag, will let some things through more easily than others.

Science Words
- **diffusion**
- **equilibrium**
- **osmosis**
- **passive transport**
- **active transport**
- **endocytosis**
- **exocytosis**

Objectives
- Explain the function of a selectively permeable membrane.
- Describe the processes of diffusion and osmosis.
- Compare and contrast passive transport and active transport and give examples of each.

Prepare

Section Background
- Membrane permeability depends not only on size of molecules but on other properties, such as electrical charge and molecular shape.
- Molecular motion is directly proportional to temperature; therefore, passive transport is temperature dependent.

Preplanning
- Obtain vinegar, syrup, and eggs for Activity 3-1.
- Obtain food coloring for the MiniLAB.

1 Motivate

Bellringer
Before presenting the lesson, display **Section Focus Transparency 10** on the overhead projector. Assign the accompanying **Focus Activity** worksheet. L1 LEP

Tying to Previous Knowledge
Review the parts of the cell in Chapter 2. The cell membrane maintains the balance between the cell and its environment.

3-2 Cell Transport   **71**

Program Resources

📁 Reproducible Masters
Study Guide, p. 14 L1
Reinforcement, p. 14 L1
Enrichment, p. 14 L3
Activity Worksheets, pp. 17-18, 22 L1
Critical Thinking/Problem Solving, p. 7
Multicultural Connections, p. 9
Lab Manual 6

🖳 Transparencies
Teaching Transparency 5
Section Focus Transparency 10 L1

MiniLAB

Purpose

Ⓛ Visual-Spatial Students will investigate the effect of temperature on the rate of diffusion of molecules. **L1**

LEP **COOP LEARN**

Materials

2 beakers, hot water, cold water, food coloring, dropper

Teaching Strategies

• Have students observe the length of time it takes for the food coloring to evenly diffuse through both the cold water and the hot water.

• Caution students not to move the beakers once the water is in the beakers.

Analysis

1. The food coloring spreads more rapidly in hot water.

2. Heat increases the rate of diffusion.

✓ Assessment

Performance To further assess students' understanding of the effect of temperature on diffusion, have them repeat the activity using ice water instead of hot water. Use the Performance Task Assessment List for Carrying Out a Strategy and Collecting Data in **PASC**, p. 25. **P**

📂 **Activity Worksheets,** pp. 5, 22

📂 Use Lab 6 in the **Lab Manual** as you teach this lesson.

Figure 3-11

A few blue crystals were added to a beaker of water.

Ⓐ In a few minutes, the blue color formed by the ions of the dissolving compound began to diffuse throughout the water.

Ⓑ After two days, diffusion has occurred to a greater extent.

Ⓒ Sometimes it will take months or even years for the ions to diffuse completely throughout the water if it is left undisturbed and covered.

MiniLAB

How does temperature affect the rate of diffusion of molecules?

Procedure

1. Prepare and label two beakers with equal amounts of cold and hot tap water. **CAUTION:** *Do not spill hot water on your skin.*

2. Add one drop of food coloring to each beaker. Release the drop just above the water's surface to avoid splashing.

3. Observe the beakers and record your observations. Repeat your observations after 10 minutes.

Analysis

1. Describe what happens when food coloring is added to each beaker.

2. How does temperature affect the rate at which this process occurs?

Diffusion

Molecules in solids, liquids, and gases move about. Molecules move constantly. As they move, they move from places where they are crowded together into places where there are fewer of them. This movement of molecules is known as **diffusion.** Diffusion is the net movement of molecules from an area where there are many to an area where there are few. Diffusion results from the random movement of particles.

Particles diffuse in liquids and in gases. **Figure 3-11** illustrates the diffusion of a blue compound in a beaker of water. You experience diffusion when someone opens a bottle of perfume in a closed room.

Drop a sugar cube into a glass of water and taste the water. Then taste it again in three or four hours, the next day, and the next week. At first, the sugar molecules are concentrated near the sugar cube in the bottom of the glass. Then slowly, the sugar molecules diffuse throughout the water until they are more evenly distributed. When the molecules of a substance are spread evenly throughout a space, a state called **equilibrium** occurs. But molecules don't stop moving when equilibrium is reached. They continue to move, and equilibrium is maintained.

Cultural Diversity

Preserving Food Native Americans preserved foods by drying and salting. Have students research the use of dried and salted foods in other cultures. For example, the French developed a vegetable dehydrator in 1795. **L1** **Ⓛ**

Osmosis—The Diffusion of Water

You may remember that water makes up a large part of living matter. The diffusion of water through a cell membrane is called **osmosis.** Osmosis is important to cells because most cells are surrounded by water molecules and they contain water molecules. Water molecules move from where they are in relatively larger numbers to where they are in relatively smaller numbers. When the transfer of water molecules into and out of the cell reaches the same rate, a state of equilibrium is reached, as illustrated in **Figure 3-12**. If cells aren't surrounded by relatively pure water, they will lose water. This is why water left the potato cells in the Explore Activity at the beginning of the chapter. Because there were relatively fewer water molecules in the salt solution around the potato cells than in the potato cells, water tended to move out of the cells and dilute the salt solution. The cell membranes pulled away from the cell walls, and the cells became limp. When the potato slices were taken out of the salt water and put in pure water, the tendency was for the water around the cells to move back into the cells. The cells expanded and became firm again.

CONNECT TO

PHYSICS

R epeat the Explore Activity on page 63. Using what you now know about osmosis and diffusion, *infer* what took place in the potato left in salt water.

CONNECT TO

PHYSICS

Answer: Water diffused out of the potato slice into the salt water.

Discussion

Discuss the differences between diffusion and osmosis. Osmosis is diffusion of water across a cell membrane.

Inquiry Question

Roads are sometimes salted to melt ice. **How does this affect the plants along the roadside?** *It sometimes kills the plants because the water moves out of the plant cells.*

Figure 3-12

Cells respond differently when the concentration of water inside the cell is equal to, greater than, or less than the water concentrations in the environment.

B Equilibrium exists when water leaves and enters the cells at the same rate.

A Wilting occurs when more water leaves the cells than enters them.

Types of Transport

Cells take in a variety of substances. Water, water-soluble materials, lipids, and lipid-soluble materials pass through the cell membrane by diffusion. The movement of particles across the cell membrane by diffusion is called **passive transport** because the cell uses no energy to move the materials. Other substances, such as glucose molecules, are so large that they enter the cell only with the help of transport protein molecules embedded in the cell membrane. This process is known as facilitated diffusion. Transport proteins allow needed substances or wastes to move through the cell membrane. This is a passive transport process also, as no energy is used by the cell.

Active Transport

When materials require energy to move through a cell membrane, **active transport** takes place. Transport proteins are, therefore, also involved in active transport.

Sometimes cells are required to move substances from where there are small amounts to where there are larger amounts. This process is the opposite of diffusion. Active transport requires energy. An example of this type of active transport can be seen when plant root cells take in minerals. Root cells require minerals from the soil, and there are already more minerals in the root cells than around the root. The cells use energy to move additional minerals into the root cells.

In active transport, a transport protein binds with the particle to be transported. Energy from the cell is used to move the particle to the other side of the membrane as seen in **Figure 3-13.** When the particle is released, the transport protein is ready to move another particle.

Figure 3-13

Plant root cells use active transport to take in minerals from the surrounding soil. Because there are already more minerals in the root cells than in the water around the roots, the root cells use carrier molecules and energy to move additional minerals into the root cells.

Community Connection

Lawn Care If fertilizer is used on lawns and plants and it doesn't rain, plants often die. Have students investigate the use of fertilizer on lawns and plants by interviewing farmers or lawn-maintenance workers. **L1**

Endocytosis and Exocytosis

Some substances are too large to pass through the cell membrane by passive or active transport. Large protein molecules and bacteria enter the cell by becoming enclosed in a part of the cell membrane that folds in to form an almost complete sphere. The sphere pinches off, and the resulting vacuole with its contents goes into the cytoplasm in a process called **endocytosis.** Phagocytosis is a type of endocytosis in which large particles move into a cell. Some one-celled organisms, such as amoebas, ingest food by phagocytosis. Pinocytosis is a type of endocytosis in which fluids move into a cell. The details of endocytosis are not yet fully understood.

USING TECHNOLOGY

Slowing Cystic Fibrosis

Cystic fibrosis (CF) is a hereditary disease in which large amounts of thick mucus build up in the body. The organs most affected by this mucus are the lungs, pancreas, and liver. Common problems with this disease are difficulty breathing and in digesting food.

The symptoms of this disease are caused by a defective cell membrane protein that normally pumps sodium chloride out of cells. Sodium chloride accumulates in cells. This causes the cells to take up water by osmosis from the mucus that surrounds the cells. The result is a buildup of thick mucus. The accumulation of mucus can block the ducts of major organs as well as cause other problems.

Treatments

Antibiotics to treat infections, and special diets and enzyme pills to help with digestion are common treatments for persons with CF. Recent studies have shown that ibuprofen, a common pain and fever-reducing medication, can slow the lung deterioration caused by CF. This treatment is most effective in patients under the age of 13. This drug should be taken only under a doctor's supervision.

Think Critically:

How might the pancreas, an enzyme-secreting organ, be affected in people with CF?

Ibuprofen

3-2 Cell Transport   **75**

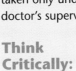

Cultural Diversity

Cultural Diffusion Another meaning of *diffusion* is the spread of cultural elements from one group of people to another by contact with each other. Ask students for examples of cultural diffusion. Examples may include the invasion of India by Muslim Turks during the late 1100s, the colonization of the Americas by Spain in the 1500s, and the imperialist expansion into Africa by Europe in the 1800s.

USING TECHNOLOGY

For more information on cystic fibrosis, contact a local physician or public health nurse.

Think Critically

The pancreas may be susceptible to infection due to the blockage of the ducts that secrete enzymes into the digestive tract.

GLENCOE TECHNOLOGY

Videodisc

STVS: Human Biology
Disc 7, Side 2
Detecting Cystic Fibrosis
(Ch. 17)

3 Assess

Check for Understanding

? FLEX Your Brain

Use the Flex Your Brain activity to have students explore ACTIVE TRANSPORT.
L1

 Activity Worksheets, p. 5

Reteach

Visual-Spatial Place several carrot strips in beakers of plain water, distilled water, and salt water. Ask students to explain their observations in terms of water movement.
LEP

Extension

For students who have mastered this section, use the **Reinforcement** and **Enrichment** masters.

4 Close

•MINI•QUIZ•

Use the Mini Quiz to check students' recall of chapter content.

1. **If the movement of materials through the cell membrane does not require the use of energy, it is called _____ .** *passive transport*

2. **The passive transport of water through a membrane by diffusion is _____ .** *osmosis*

Section Wrap-up

Review

1. They allow some things to pass through but not others.

2. In both, molecules move from where there are many to where there are few. Osmosis is the diffusion of water through a selectively permeable membrane.

3. **Think Critically** Water in the environment will diffuse into the fruits and vegetables and keep them from drying out.

Science Journal

LS Linguistic Students could compare active transport with the task of getting a bicycle to the top of a hill. Energy must be used to get to the top. In the same way, the cell uses energy to move substances from areas of low concentration to areas of high concentration. Energy is not used to ride the bicycle down a hill. Energy is not used to move substances from areas of high concentration to areas of low concentration in cells (passive transport).

Figure 3-14

Substances that are too large to pass through cell membranes by passive or active transport enter and leave cells by endocytosis and exocytosis.

A A white blood cell uses endocytosis to engulf a bacterial cell.

B In exocytosis, substances in small sacs are released at the cell membrane.

Cell membrane

In the opposite way, wastes in vacuoles or proteins packaged by Golgi bodies move to the cell membrane. The package fuses with the cell membrane, and its contents are released from the cell in a process called **exocytosis. Figure 3-14** illustrates both endocytosis and exocytosis.

Section Wrap-up

Review

1. In what way are cell membranes selectively permeable?

2. Compare and contrast osmosis and diffusion.

3. **Think Critically:** Why are fresh fruits and vegetables sprinkled with water in produce markets?

Skill Builder
Concept Mapping

Make a network tree concept map to use as a study guide to help you tell the difference between passive transport and active transport. Begin with the phrase *Transport through membranes.* If you need help, refer to Concept Mapping in the **Skill Handbook.**

Science Journal

In your Science Journal, compare passive and active transport. Find your own analogies to describe these processes.

Skill Builder

Transport through membrane
- Active — requires — Energy — moves — Large molecules, Minerals
- Passive — requires — No energy — examples — Diffusion, Osmosis

Assessment

Oral Would a cell placed in syrup lose or gain water? Explain your answer. *It would lose water because water molecules move by diffusion from where there is a large concentration to where there is a lesser concentration.* Use the Performance Task Assessment List for Making Observations and Inferences in **PASC,** p. 17. **P**

Observing Osmosis

It is difficult to see osmosis occurring in cells because of the small size of most cells. However, there are a few cells that can be seen without the aid of a microscope. Try this activity to see how osmosis occurs in a large cell.

Problem

How does osmosis occur in an egg cell?

Materials

- raw egg
- 500-mL containers with covers (2)
- 250 mL vinegar
- 250 mL corn syrup
- 250-mL graduated cylinder
- 2 labels, A and B

Procedure

1. Copy the data table and use it to record your observations.
2. Measure 250 mL of vinegar into container A. Add 250 mL of syrup to container B. Place the egg into container A. Cover both containers.
3. Observe the egg for two days. Record the appearance of the egg in a data table.
4. At the end of two days, use a spoon to remove the egg from container A. Rinse the egg and place it into container B.
5. Observe the egg the next day. Record its appearance.
6. Measure the remaining liquid in containers A and B. Record the amounts.

Analyze

1. What did you **observe** after 30 minutes?
2. What did you **observe** after two days?
3. What did you **observe** after three days?

Conclude and Apply

4. What caused the movement of water in container A and container B?
5. Calculate the amount of water that moved in containers A and B.
6. **Infer** what part of the egg controlled what moved into and out of the cell.

Data and Observations

Sample Data

Time	30 min	2 days	3 days
Vinegar volume	250 mL	220 mL	
Observations		Egg becomes larger	
Table syrup volume		250 mL	290 mL
Observations		Egg becomes small and rubbery	

the vinegar in container A. There was more water inside the egg in container B than in the syrup, so the water diffused from the egg into the syrup.

5. Answers may vary. About 30 mL water entered the egg from the vinegar in container A, and about 40 mL water left the egg when it was placed in the table syrup in container B.

6. cell membrane

 Assessment

Performance Place 10 dried beans in water and let them remain overnight. Explain your results. (The dried beans will take in water and increase in size.) Use the Performance Task Assessment List for Carrying Out a Strategy and Collecting Data in **PASC**, p. 25. **P**

Purpose

LS **Logical-Mathematical** Students will observe the diffusion of water through the membrane of an egg and measure the amount of water that diffuses. **L1**

COOP LEARN

Process Skills

measuring, observing and inferring, communicating, recognizing cause and effect, forming operational definitions, making models

Time

50 minutes to set up, 5 minutes each day to observe, 15 minutes to summarize

Activity Worksheets, pp. 5, 17-18

Teaching Strategies

Alternate Materials 500-mL jars with lids can be used for the activity.

Troubleshooting Thick, white table syrup gives better results than thin syrup.

- If eggshells have not completely dissolved at the end of two days, leave them in the vinegar an extra day.
- Caution students to handle the egg very carefully after the shell has dissolved to avoid breaking the membrane.
- The bubbles that form around the eggs in the vinegar are caused by the reaction of calcium carbonate of the eggshell with the acetic acid in the vinegar.

Answers to Questions

1. Bubbles formed around the egg.
2. The eggshell had dissolved.
3. The egg in the syrup decreased in size.
4. Water molecules diffuse from where there are large amounts to where there are smaller amounts. There was more water outside the egg in

Prepare

Section Background

- In addition to glucose, all other organic materials originate from photosynthesis.
- Respiration releases 18 times more energy from a food molecule than fermentation does.

Preplanning

Obtain *Elodea* for Activity 3-2.

1 Motivate

Bellringer

 Before presenting the lesson, display **Section Focus Transparency 11** on the overhead projector. Assign the accompanying **Focus Activity** worksheet. [L1] [LEP]

Tying to Previous Knowledge

Display pictures of persons using energy—playing sports, gardening, and so on. Discuss the source of their energy (food from green plants that had captured the sun's energy).

NATIONAL GEOGRAPHIC SOCIETY

⊚ Videodisc

Newton's Apple: Life Sciences
Plant Growth
Chapter 2, Side A
Four ingredients of photosynthesis

20103-20376

3•3 Energy in Cells

Science Words

metabolism
producer
consumer
fermentation

Objectives

- Explain the difference between producers and consumers.
- Compare and contrast the processes of photosynthesis and respiration.
- Describe how cells get energy from glucose through the process of fermentation.

Trapping Energy for Life

Think of all the energy used in a basketball game. Where do the players get all that energy? The simplest answer is, "from the food they eat." Cells take chemical energy stored in food and change it into other forms of energy that can be used in metabolism. **Metabolism** is the total of all chemical activities of an organism that enable it to stay alive, grow, and reproduce.

Living things are divided into two groups based on how they obtain their food energy. These two groups are producers and consumers. Organisms that make their own food, such as the plants pictured in **Figure 3-15**, are called **producers.** Organisms that can't make their own food are **consumers.**

Photosynthesis

Producers change light energy into chemical energy in a process called photosynthesis. During photosynthesis, the energy from sunlight is used to make a sugar from carbon dioxide (CO_2) and water. In this process, illustrated in **Figure 3-16,** oxygen is also given off. Plants and other producers use pigments, such as the green one called chlorophyll found in chloroplasts, to capture light energy. Therefore, photosynthesis in green plants takes place in the chloroplasts. Producers use some of the sugars they make during photosynthesis for energy. The rest is stored as carbohydrates or lipids.

Do you eat vegetables? Have you ever watched sheep graze? People and sheep are both consumers. Consumers eat producers such as vegetables and take in the stored energy. Consumers also eat other consumers in order to get energy. These relationships form a food chain. Producers are at the beginning of every food chain.

Figure 3-15

Green plants are producers because they use sunlight, carbon dioxide, and water to produce chemical energy in the form of glucose.

78   Chapter 3 Cell Processes

Program Resources

 Reproducible Masters

Study Guide, p. 15 [L1]
Reinforcement, p. 15 [L1]
Enrichment, p. 15 [L3]
Activity Worksheets, pp. 19-20 [L1]
Lab Manual 7
Concept Mapping, p. 11

Transparencies

Teaching Transparency 6
Section Focus Transparency 11 [L1]

Figure 3-16

Plants use light energy from the sun for photosynthesis.

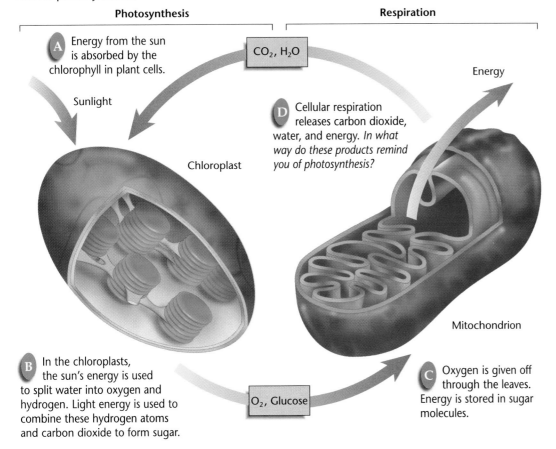

Photosynthesis

Respiration

A Energy from the sun is absorbed by the chlorophyll in plant cells.

Sunlight

CO₂, H₂O

Energy

D Cellular respiration releases carbon dioxide, water, and energy. *In what way do these products remind you of photosynthesis?*

Chloroplast

Mitochondrion

B In the chloroplasts, the sun's energy is used to split water into oxygen and hydrogen. Light energy is used to combine these hydrogen atoms and carbon dioxide to form sugar.

O₂, Glucose

C Oxygen is given off through the leaves. Energy is stored in sugar molecules.

Photosynthesis: $6CO_2 + 6H_2O + \text{light energy} \longrightarrow C_6H_{12}O_6 + 6O_2$

carbon dioxide   water   chlorophyll   sugar   oxygen

Respiration: $C_6H_{12}O_6 + 6O_2 \longrightarrow 6CO_2 + 6H_2O + \text{energy}$

sugar   oxygen   carbon dioxide   water

Releasing Energy for Life

Whether an organism is a producer or a consumer, it has to have some way to release energy from food in order to make use of it. To do this, both producers and consumers break down food in their cells in a process called cellular respiration.

3-3 Energy in Cells   **79**

Visual Learning

Figure 3-16 Use **Figure 3-16** to compare the processes of photosynthesis and respiration. **LEP** **LS**

Across the Curriculum

Language Arts *Chlorophyll* comes from the Greek words *chloros* and *phyllon*. Ask students what they think these words mean (*chloros* = green, *phyllon* = leaf). *Photosynthesis* comes from the Greek words *photos* and *syntithenai* (*photos* = light, *syntithenai* = putting together). **L1** **LS**

---

2 Teach

Revealing Preconceptions

Some students may believe that plants take in food from the soil in which they live. Plants obtain a variety of minerals from the soil—calcium, potassium, magnesium, sulfur, phosphorus, nitrogen, iron—by absorption through their roots, but these materials are not used as food. Plants produce their own food by the process of photosynthesis.

Activity

LS **Visual-Spatial** Fill a jar with water and submerge a freshly picked leaf in the water. Observe the surface of the leaf. **What forms on the leaf?** *bubbles* **Why?** *The leaf is giving off oxygen.* **L1** **LEP**

NATIONAL GEOGRAPHIC SOCIETY

Videodisc

Newton's Apple: Life Sciences
Plant Growth
Chapter 2, Side A
The process of photosynthesis

22935-23542
Question: How do plants grow?

18929
How a leaf uses light, water, and carbon dioxide

23258

Problem Solving

Solve the Problem

1. The latex balloon can be compared with the cell membrane.
2. Helium escapes through the tiny pores in the latex balloons, which are semipermeable.

Think Critically

1. Mylar balloons are less permeable to helium than latex balloons.
2. The water would stay in the balloon. Latex is not permeable to water.

3 Assess

Check for Understanding

Use the Flex Your Brain activity to have students explore PHOTOSYNTHESIS.

Activity Worksheets, p. 5

Reteach

Have students identify those organisms in the classroom aquarium in which photosynthesis takes place and those in which respiration takes place.

Extension

For students who have mastered this section, use the **Reinforcement** and **Enrichment** masters.

Use **Teaching Transparency 6** as you teach this lesson.

Use Lab 7 in the **Lab Manual** as you teach this lesson.

Problem Solving

Permeability

You are probably familiar with helium-filled balloons that are often used to celebrate birthdays or other special occasions. After a day or two, these balloons end up on the floor. A latex balloon filled with helium will usually float in the air because helium is lighter than the mixture of gases in air. So why did the balloon deflate?

Solve the Problem:

1. **With what cell part can the latex balloon be compared?**
2. **Why do the balloons fall to the floor?**

Think Critically:

1. Compared to latex balloons, why do balloons made out of a silvery polyester film seem to last forever?
2. If you filled the latex balloons with water, what would happen? Explain.

Inside most animal cells, glucose, an organic sugar, is the food that is broken down. The breakdown process takes place within mitochondria and uses the oxygen that you take in as you breathe.

What happens in respiration?

During respiration, food molecules are broken down to release stored energy. The energy is released in a series of steps. Carbon dioxide and water are given off as waste products. Some of the energy released in respiration is used to produce high-energy molecules, and some of the energy is lost as heat.

$$C_6H_{12}O_6 + 6O_2$$
sugar   oxygen

↓

$$6CO_2 + 6H_2O + energy$$
carbon   water
dioxide

Fermentation

Sometimes, however, during periods of strenuous activity, muscle cells run low on oxygen. The muscles begin to burn and sting. You begin to breathe harder and faster in an effort to supply the needed oxygen. But your body has another method to continue supplying small amounts of energy. When oxygen levels are low, the muscle cells begin to convert energy from glucose by fermentation. **Fermentation** is a form of respiration that converts energy from glucose when oxygen is insufficient. Lesser amounts of energy are produced by fermentation. In addition, an organic compound called lactic acid is formed. It is lactic acid that causes the muscles to burn and to be sore and stiff. Fermentation takes place in the cytoplasm of cells.

Yeast and some bacteria use alcohol fermentation to release energy and CO_2. **Figure 3-17** discusses the role of yeast in making the types of bread shown in the photograph.

Photosynthesis and respiration are alike in that both involve complex sets of reactions. Both occur in specific organelles, involve energy, and require enzymes. In many ways, photosynthesis is the opposite of respiration. The end products of respiration are the starting materials of photosynthesis. In summary, producers capture the sun's energy and store it in the form of food. Consumers eat producers. Cells use the glucose made by producers for energy.

Figure 3-17

Yeast is an organism that breaks down the glucose in bread dough to produce carbon dioxide and alcohol. The bubbles of carbon dioxide that are released cause the dough to rise while the alcohol is released into the air.

Section Wrap-up

Review

1. Explain the difference between producers and consumers.

2. Explain how the energy used by all living things on Earth can be traced back to sunlight.

3. **Think Critically:** How can indoor plants help to purify the air in a room with cigarette smoke?

Skill Builder
Comparing and Contrasting
Make a table that compares and contrasts respiration and fermentation. If you need help, refer to Comparing and Contrasting in the **Skill Handbook.**

USING MATH

Calculate the minimum number of carbon and oxygen atoms that are required for cellular respiration. Then, calculate the minimum number of carbon and oxygen atoms that are given off in respiration.

3-3 Energy in Cells   **81**

4 Close

• MINI • QUIZ •

Use the Mini Quiz to check students' recall of chapter content.

1. **The total of all the activities of an organism that enable it to live, grow, and reproduce is _____ .** *metabolism*

2. **Organisms that can make their own food are _____ .** *producers*

3. **The process in which energy is released without the use of oxygen is _____ .** *fermentation*

Section Wrap-up

Review

1. Producers use sunlight to make food. Consumers obtain energy by eating producers or the food producers make.

2. Producers change light from the sun into chemical energy used by all living things.

3. **Think Critically** Plants can remove carbon monoxide found in cigarette smoke from the air by using it in photosynthesis and releasing oxygen.

USING MATH

LS **Logical-Mathematical** Six carbon atoms and 18 oxygen atoms are both required and given off in cellular respiration.

Skill Builder

	Ferm.	**Resp.**
Start	Glucose	O_2, glucose
O_2	No	Yes
Energy released	small amount	large amount

P

✓ Assessment

Portfolio Use this Skill Builder to assess students' abilities to compare and contrast photosynthesis and respiration. Use the Performance Task Assessment List for Making Observations and Inferences in **PASC,** p. 17.

P

Activity 3-2

Purpose

LS **Interpersonal** Students will observe photosynthesis and respiration in green plants and infer whether the processes occur in light or darkness. **L1** **COOP LEARN**

Process Skills

measuring, observing and inferring, communicating, comparing and contrasting, recognizing cause and effect, forming a hypothesis, designing an experiment, separating and controlling variables, interpreting data

Time

50 minutes

Materials

Elodea should be kept in the dark for two days before the activity. Use sharp scissors to make a clean diagonal cut at the bottom of each stem.

Alternate Materials Baby food jars with lids may be used for test tubes.

Safety Precautions

Caution students to use extreme care when working with chemicals.

Possible Hypothesis

Most student hypotheses will reflect that plants carry on both photosynthesis and respiration.

📁 **Activity Worksheets,** pp. 5, 19-20

Activity 3-2

Design Your Own Experiment
Photosynthesis and Respiration

Things to consider in designing this experiment are the reactants and products of photosynthesis and respiration. Sodium hydrogen carbonate releases carbon dioxide when mixed with water. Carbon dioxide forms carbonic acid when released in water.

PREPARATION

Problem

Do plants carry on both photosynthesis and respiration?

Form a Hypothesis

Based on what you have learned about photosynthesis and respiration, state a hypothesis about whether green plants carry on both photosynthesis and respiration.

Objectives

- Observe green plants placed in the light and dark.
- Design an experiment that tests the effect of a control in determining whether green plants carry on both photosynthesis and respiration.

Possible Materials

- 16 × 150 mm test tubes (4)
- test-tube rack
- stirring rod
- balance
- sodium hydrogen carbonate
- bromothymol blue solution in dropping bottle
- aged tap water
- pieces of *Elodea* (2)

Safety Precautions

Protect clothing and eyes and be careful using chemicals. Do not get chemicals on your skin.

PLAN THE EXPERIMENT

1. As a group, agree upon and write out the hypothesis statement.
2. As a group, test the steps that you need to take to test your hypothesis. Be specific, describing exactly what you will do at each step. List your materials.
3. A color change of green to yellow upon adding 5 drops of bromothymol blue solution to each test tube indicates the presence of an acid.
4. Be sure to use the same size pieces of *Elodea* in all of your trials. Consider the effects of bright light and darkness on photosynthesis and respiration.
5. If you need a data table, design one in your Science Journal, so that it is ready to use as your group collects data.

Check the Plan

1. Read over your entire experiment to make sure that all steps are in logical order.
2. Identify any constants, variables, and controls of the experiment.
3. *Make sure your teacher approves your plan before you proceed.*

DO THE EXPERIMENT

1. Carry out the experiment as planned.
2. While the experiment is going on, write down any observations that you make and complete the data table in your Science Journal.

Analyze and Apply

1. **Compare** your results with those of other groups.

2. Which product of respiration could you detect in this experiment? How?
3. Which gas is necessary for photosynthesis to occur? How could you detect its use?
4. **Interpret** the results of your experiment.

Go Further

Add 5 mL of yeast suspension to a test tube and add 5 mL of water to another. Add 5 mL of sucrose solution to each. Stir each tube and add 5 mL of bromothymol blue to each. What do your results indicate? What is this process called?

Go Further

Yeast gives off carbon dioxide. This process is called fermentation.

 Assessment

Oral How are fermentation and respiration similar? *Both processes release energy.* Use the Performance Task Assessment List for Making Observations and Inferences in **PASC**, p. 17. **P**

PLAN THE EXPERIMENT

Possible Procedures
One possible procedure is to pour water into four test tubes until each is three-fourths full. Add 1 g of sodium bicarbonate to each test tube. Sodium bicarbonate releases carbon dioxide when mixed with water. Add 5 drops of bromothymol blue to each test tube. Bromothymol blue turns from green to yellow in the presence of an acid. Cut two 10-cm pieces of *Elodea* and place each in a test tube. Stopper the test tubes and record the color in each test tube. Place a test tube and a control (without *Elodea*) in the light and the other test tube with a control in the dark for 30 minutes, then observe and record the colors.

Teaching Strategies
Tying to Previous Knowledge Most students are aware that green plants use sunlight to make food and that photosynthesis will occur in the tube in the light.

DO THE EXPERIMENT

Expected Outcome
Most results will reflect that plants used carbon dioxide in the light and gave off carbon dioxide in the dark.

Analyze and Apply

1. Student results will be consistent with those of other groups.
2. Carbon dioxide—bromothymol blue changes to a deep yellow.
3. Carbon dioxide—as the carbon dioxide was used up in the water, the water turned blue.
4. A green plant carries on photosynthesis in the presence of light.

Prepare

Section Background

Point out to students that biomass is the primary source of energy for about half of the world's population, primarily in developing countries.

1 Motivate

Bellringer

 Before presenting the lesson, display **Section Focus Transparency 12** on the overhead projector. Assign the accompanying **Focus Activity** worksheet. L1 LEP

2 Teach

NATIONAL GEOGRAPHIC SOCIETY

 Videodisc

STV: Plants
What Is a Plant?
Unit 1, Side 1
Leaves and Flowers

06785-11684
How Plants Are Used
Unit 3, Side 2
Electricity and Fuel

15317-19429

TECHNOLOGY:
3•4 Energy from Biomass

Science Words

biomass
gasohol
biogas

Objectives

- Describe how photosynthesis and fermentation are related to biomass.
- Identify useful biomass materials.
- Compare the advantages and disadvantages of using biomass as an energy source.

Figure 3-18

Brazil has a national program to substitute crop-based ethanol for imported petroleum. In 1989, 12 billion L of ethanol replaced almost 200 000 barrels of gasoline a day in approximately 5 million Brazilian cars and trucks.

Useful Biomass

Do you have a fireplace or a wood-burning stove in your home? Wood fires have been a source of heating and cooking for thousands of years. Organic material from plants or animals that is used for energy is called **biomass.** Useful biomass includes materials such as wood; fast-growing plant and algal crops; wood chips, sawdust, and other leftover materials from the timber industry; crop wastes; and municipal wastes. The chemical energy found in this organic material can be traced back to radiant energy from the sun.

Biomass fuel can be a solid, a liquid, or a gas. Solid biomass such as firewood, charcoal, animal dung, and peat supplies a large portion of the world's energy today.

Liquid Fuels

In the process of alcoholic fermentation, microorganisms convert the sugar in sugarcane, corn, and grain into ethanol. Ethanol can be burned directly in automobile engines adapted to its use, or it can be mixed with gasoline to be used in any gasoline engine as shown in **Figure 3-18.** A mixture of gasoline and ethanol is **gasohol.**

Methanol, or wood alcohol, can also be used as a fuel for automobiles. It burns at a lower temperature than gasoline or diesel. Its use could eliminate heavy radiators, making automobiles more efficient. Alcohol production from biomass, though economical, is not energy efficient. A large amount of energy is required to grow and harvest the crop, and much of the original energy in biomass is lost during its conversion to alcohol.

Other Sources of Energy

It is possible to change animal wastes into biogas. **Biogas** is a mixture of gases that can be stored and transported like natural gas. When burned, it produces fewer pollutants than either coal or biomass. In China, biogas digesters use micro-

Program Resources

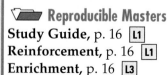 **Reproducible Masters**
Study Guide, p. 16 L1
Reinforcement, p. 16 L1
Enrichment, p. 16 L3

Transparencies
Section Focus Transparency 12 L1

organisms to decompose household and agricultural wastes to produce biogas that is used for heating and cooking. The solid remains are removed and used as fertilizer. Nonfood crops, as illustrated in **Figure 3-19,** can be grown to produce liquid fuel.

Advantages and Disadvantages

Biomass reduces dependence on fossil fuels and makes use of wastes, which reduces the waste-disposal problem. In southern California, cow manure is burned in special furnaces to generate electricity for thousands of homes. Biomass does not pollute the atmosphere with carbon dioxide as long as the plant matter burned equals the plant matter produced each year. It has a high net energy efficiency when it is collected and burned close to the source of production. Burning some forms of biomass, such as urban refuse, reduces the need for land disposal.

There are some problems associated with the use of biomass, especially from plants. Plant production requires land and water. The use of land for energy crops competes with the growing of food crops, so shifting the balance toward energy may decrease food production. Improper management of croplands and forests could lead to soil erosion, sedimentation, destruction of reservoirs, and flooding. Removing crop and forestry wastes may reduce soil nutrients. The use of crop and forest wastes for biomass must be carefully managed.

Figure 3-19

Nonfood crops can be grown to produce liquid fuel.

A A desert shrub, *Euphorbia lathyri*, found in Mexico and the southwestern United States, produces an oil substance that can be refined to make fuel.

B Sunflower oil can also be used instead of diesel fuel.

Section Wrap-up

Review

1. How are photosynthesis and fermentation related to biomass production?

2. What are some sources of biomass?

*inter*NET CONNECTION Visit the Chapter 3 Internet Connection at Glencoe Online Science, **www.glencoe.com/sec/science/life,** for a link to more information about energy from biomass.

SCIENCE & SOCIETY

Integrating the Sciences

Earth Science Wind energy is an enormous resource, abundant in certain parts of every continent. Tapping the world's windiest places could provide 13 times the electricity now produced worldwide. Wind energy is clean and renewable, and is safe to operate. On a typical wind farm, only 10% of the land is used for roads and windmills. The remaining land can be used for grazing or planting.

3 Assess

Check for Understanding

Have students answer questions 1 and 2 in the Section Wrap-up.

Reteach

Have students determine how the use of biomass can help reduce our dependence on nonrenewable energy sources.

Extension

For students who have mastered this section, use the **Reinforcement** and **Enrichment** masters.

4 Close

Have students prepare a list of sources of biomass energy. Ask why alcohol production from biomass is not energy efficient.

Section Wrap-up

Review

1. The chemical energy in biomass can be traced back to radiant energy from the sun, which was used by photosynthetic organisms to form the organic molecules of biomass. Sugar in plants is converted to alcohol in the process of fermentation.

2. Sources of biomass are wood, fast-growing plant and algal crops, residues from the timber industry, crop wastes, animal wastes, and municipal wastes.

Explore the Technology

Animals eat plants and the wastes come from plants.

Source

Thomas, Lewis. *The Lives of a Cell: Notes of a Biology Watcher.* Viking, 1974.

Biography

Lewis Thomas was born in New York. He attended Princeton University and Harvard Medical School. In his long medical and teaching career, he served as a professor of pediatric research at the University of Minnesota, chairman of the departments of pathology and medicine as well as dean at the New York University-Bellevue Medical Center, chairman of pathology as well as dean at Yale Medical School, and president of Memorial Sloan-Kettering Cancer Center in New York City. Thomas won the National Book Award in 1974 for the book from which this excerpt was taken.

Teaching Strategies

LS **Linguistic** Have students discuss the points of comparison and contrast that Earth and a cell have. They may want to chart the points as a way of graphically organizing them.

LS **Linguistic** Invite students to share the essays they wrote in their Science Journals comparing a cell to something else. Small groups can discuss the points of comparison as well as contrast in each essay.

LS **Interpersonal** Lewis Thomas writes of thinking about Earth as he is "driving through a hilly, wooded part of southern New England." Ask students if they have a favorite spot where they go to reflect and think about the wonders of life. If so, invite them to share and describe this spot.

The Lives of a Cell
by Lewis Thomas

In this chapter, you have studied many of the processes that go on in the cells of living things. When you think of how microscopic most cells are, their activity is amazing. Read how one award-winning essayist, Lewis Thomas, put the wonder of cells in a special perspective:

I have been trying to think of the earth as a kind of organism, but it is no go. I cannot think of it this way. It is too big, too complex, with too many working parts lacking visible connections. The other night, driving through a hilly, wooded part of southern New England, I wondered about this. If not like an organism, what is [the earth] like, what is it most like? Then, satisfactorily for the moment, it came to me: it is most like a single cell.

Science Journal

What would you compare a cell to? In your Science Journal, write an essay that compares a cell with something else. Give reasons for your comparison.

For Lewis Thomas, the immense Earth can best be compared to a single microscopic cell. Think of all the "parts" to Earth and all the related activity that goes on within and on the planet. Do you agree with Thomas that it is much like a cell? How can a cell be viewed as "too complex, with too many working parts lacking visible connections"?

LS **Intrapersonal** Share with students the scientific and medical background of the author, Lewis Thomas. Ask whether it surprises them that a scientist might also be a writer. Discuss the special perspective that a scientist can bring to thinking and writing about the world.

Other Works

Other works by Lewis Thomas include *The Medusa and the Snail: More Notes of a Biology Watcher* (Viking, 1979) and *The Fragile Species: Notes of an Earth Watcher* (Simon & Schuster, 1992).

Chapter 3 Review

Summary

3-1: Chemistry of Living Things

1. Everything is made of matter, which is composed of atoms arranged in elements, molecules, and compounds.
2. Both inorganic and organic molecules are important to living things.
3. Carbohydrates, lipids, proteins, and nucleic acids are organic compounds. Most inorganic compounds do not contain carbon.

3-2: Cell Transport

1. The cell membrane controls what molecules can pass through it.
2. Molecules move by diffusion from areas of greater numbers to areas of lesser numbers. In osmosis, water diffuses through a selectively permeable membrane.
3. The cell expends energy to move molecules by active transport, but not by passive transport.

3-3: Energy in Cells

1. Producers make energy in the form of food, which consumers eat.
2. Green plants use light energy to make chemical energy in the form of sugars during photosynthesis. In respiration, sugars are broken down and energy is released.
3. Some yeasts, bacteria, and oxygen-deprived cells carry out fermentation to release small amounts of energy from glucose without using oxygen.

3-4: Science and Society: Energy from Biomass

1. Biomass has potential as a renewable energy source.
2. Biomass can be burned directly or converted to gas or liquid fuels.

Key Science Words

a. active transport
b. biogas
c. biomass
d. consumer
e. diffusion
f. endocytosis
g. enzyme
h. equilibrium
i. exocytosis
j. fermentation
k. gasohol
l. inorganic compound
m. metabolism
n. mixture
o. organic compound
p. osmosis
q. passive transport
r. producer

Reviewing Vocabulary

Match each phrase with the correct term from the list of Key Science Words.

1. organism that makes its own food
2. the movement of water molecules through a selectively permeable membrane
3. movement of molecules that requires energy
4. protein that speeds up a chemical reaction
5. organic material that is used for energy
6. made from elements other than carbon

Chapter 3

Review

Summary

Have students read the summary statements to review the major concepts of the chapter.

Reviewing Vocabulary

1. r
2. p
3. a
4. g
5. c
6. l
7. h
8. i
9. j
10. k

Assessment

Portfolio Encourage students to place in their portfolios one or two items of what they consider to be their best work. Examples include:

- Flex Your Brain, p. 69
- Skillbuilder concept map, p. 76
- Skillbuilder table, p. 81 P

Performance Additional performance assessments may be found in **Performance Assessment** and **Science Integration Activities.** Performance Task Assessment Lists and rubrics for evaluating these activities can be found in Glencoe's **Performance Assessment in the Science Classroom.**

GLENCOE TECHNOLOGY

 Videodisc

Glencoe Life Science Interactive Videodisc
Side 1, Lesson 4
Photosynthesis and Cellular Respiration

37736 to 49167

 MindJogger Videoquiz

Chapter 3 Have students work in groups as they play the Videoquiz game to review key chapter concepts.

Checking Concepts

1. b	**6.** c
2. b	**7.** d
3. a	**8.** a
4. b	**9.** c
5. d	**10.** d

Understanding Concepts

11. Diffusion is the movement of molecules from an area of higher concentration to one of lower concentration. Osmosis is the diffusion of water molecules through a selectively permeable membrane.

12. Accept all reasonable responses. One answer might be that respiration takes place in mitochondria and energy is released. Muscle cells utilize more energy than other cells.

13. Membranes are selectively permeable. Molecules pass through depending upon size.

14. Water molecules diffuse from outside to inside until there are 17 molecules inside the cell as well as outside.

15. In osmosis, water will diffuse out of the celery stalk, where there is a higher concentration of water, into the salt water, where there is a lower concentration of water. Consequently, the celery loses turgor and becomes limp and rubbery.

Thinking Critically

16. The red blood cell would burst because so many water molecules would move into the cell. The water would move from an area of greater concentration (outside the cell) to an area of lesser concentration (inside the cell) in an effort to reach equilibrium.

---

7. when molecules are spread evenly throughout a space
8. release of a substance from a small sac at the membrane
9. the converting of glucose to another energy form without oxygen
10. mixture of gasoline and ethanol

Checking Concepts

Choose the word or phrase that completes the sentence.

1. Energy is used to move molecules by _____.
  a. diffusion   b. active transport
  c. osmosis   d. passive transport
2. Bacteria are taken into cells by _____.
  a. osmosis   c. exocytosis
  b. endocytosis   d. diffusion
3. An example of an organic compound is _____.
  a. $C_6H_{12}O_6$ c. H_2O
  b. NO_2 d. O_2
4. _____ are examples of carbohydrates.
  a. Enzymes   c. Waxes
  b. Sugars   d. Proteins
5. Organic compounds in the chromosomes are _____.
  a. carbohydrates   c. lipids
  b. water molecules   d. nucleic acids
6. The organic molecule that releases the greatest amount of energy is _____.
  a. carbohydrate   c. lipid
  b. water   d. nucleic acid
7. Water is a(n) _____ molecule.
  a. organic   c. carbohydrate
  b. lipid   d. inorganic
8. _____ occurs when molecules are evenly distributed through a solid, liquid, or gas.
  a. Equilibrium   c. Fermentation
  b. Metabolism   d. Cellular respiration

---

9. _____ are organisms that can't make their own food.
  a. Biodegradables   c. Consumers
  b. Producers   d. Enzymes
10. Chlorophyll is needed for _____.
  a. fermentation
  b. cellular respiration
  c. diffusion
  d. photosynthesis

Understanding Concepts

Answer the following questions in your Science Journal using complete sentences.

11. Compare and contrast diffusion and osmosis.
12. Why might there be many more mitochondria in muscle cells than in other types of cells?
13. Explain how some substances, but not others, can pass through the cell membrane.
14. Describe how a cell reaches equilibrium with its environment if there are ten water molecules inside and 24 molecules outside.
15. Describe the process that causes a stalk of celery to become limp and rubbery when placed in salt water.

Thinking Critically

16. If you could place a single red blood cell in distilled water, what do you think you would see happen to the cell? Explain.
17. In snowy states, salt is used to melt ice on the roads. Explain what happens to many roadside plants as a result.

---

17. Plants die as water molecules move out of the cells.
18. Sugar molecules move faster in hot water due to the increased energy in the heated water.
19. Consumers would also die; they depend on producers for food.
20. The enzymes break down the bonds of the proteins; this makes the meat more tender.

22.

18. Explain why sugar dissolves faster in hot tea than in iced tea.
19. Explain what would happen to the consumers in a lake if all the producers died.
20. Meat tenderizers contain enzymes. How do these enzymes affect protein in meat?

Developing Skills

If you need help, refer to the **Skill Handbook**.

21. **Interpreting Data:** In an experiment that tests the rate of photosynthesis in water plants, the water plants were placed at different distances from a light source. Bubbles of gas coming from the plants were counted to measure the rate. What can you say about how the rate is affected by the light?

Photosynthesis in Water Plants

Beaker number	Distance from light	Bubbles per minute
1	10 cm	45
2	30 cm	30
3	50 cm	19
4	70 cm	6
5	100 cm	1

22. **Making and Using Graphs:** Use the data from Question 21 to graph the relationship between photosynthesis rate and distance from light.
23. **Classifying:** Classify these common foods into a category of carbohydrates, lipids, or proteins: butter, bread, candy bar, cereals, cheese, cornstarch, fish, margarine, meats, pasta, peanut butter, shortening, sugar, vegetable oil.

24. **Concept Mapping:** Complete the events chain concept map to sequence the parts of matter from smallest to largest: element, atom, compound.

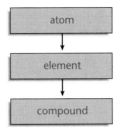

25. **Hypothesizing:** Make a hypothesis about what will happen to wilted celery when placed into a glass of plain water.

Performance Assessment

1. **Analyzing the Data:** In Activity 3-1 on page 77, you studied osmosis in an egg. Place 10 to 12 raisins in a glass of water and allow them to stand overnight. Explain your results.
2. **Making Observations and Inferences:** The MiniLAB on page 68 asked you to find out how enzymes work. Label two containers A and B. Make 1 cup of prepared instant pudding and divide it between the two containers. Add 1 teaspoon of fresh pineapple juice to container A and stir. Add 1 teaspoon of canned pineapple juice to container B and use a clean spoon to stir. What can you infer about differences between canned and fresh pineapple juice? What can you infer about your results in terms of enzyme activity?
3. **Designing an Experiment:** Design an experiment to show that respiration takes place in growing lima bean seeds.

Chapter 3 Review

Developing Skills

21. **Interpreting Data** The closer a plant is to the light, the faster its rate of photosynthesis.
22. **Making and Using Graphs** See page 88 for graph.
23. **Classifying** Carbohydrates: bread, candy bar, cereals, cornstarch, pasta, sugar

    Lipids: butter, margarine, peanut butter, shortening, vegetable oil

    Proteins: cheese, fish, meats
24. **Concept Mapping** See student page.
25. **Hypothesizing** Wilted celery will become crisp as water molecules move by osmosis into its cells to reach equilibrium.

Performance Assessment

1. Raisins are dehydrated, and water in the glass diffused into the raisins, causing them to become plump. Use the Performance Task Assessment List for Analyzing the Data in **PASC**, p. 27. P
2. An enzyme in the fresh pineapple will keep pudding from solidifying; the enzyme in the canned pineapple has been destroyed. Use the Performance Task Assessment List for Making Observations and Inferences in **PASC**, p. 17. P
3. The experiment should show that carbon dioxide is given off while the bean seeds are growing. Use the Performance Task Assessment List for Designing an Experiment in **PASC**, p. 23. P

Assessment Resources

 Reproducible Masters

Chapter Review, pp. 9, 10

Assessment, pp. 13-16

Performance Assessment, p. 17

Glencoe Technology

- **Chapter Review Software**
- **Computer Test Bank**
- **MindJogger Videoquiz**

Chapter Organizer

Section	Objectives/Standards	Activities/Features
Chapter Opener		Explore Activity: Observe seeds. p. 91
4-1 Cell Growth and Division (2 sessions, 1 block)*	1. **Describe** mitosis and explain its importance. 2. **Explain** differences between mitosis in plant and animal cells. 3. **Give** two **examples** of asexual reproduction. National Content Standards: (5-8) UCP1, UCP2, UCP3, A1, C2, C3, C5	Activity 4-1: Mitosis in Plant and Animal Cells, p. 96 MiniLAB: How does one cell become two? p. 97 Skill Builder: Sequencing, p. 98 Using Math, p. 98
4-2 Sexual Reproduction and Meiosis (3 sessions, 2 blocks)*	1. **Describe** the stages of meiosis and its end products. 2. **Name** the cells involved in fertilization and explain how fertilization occurs in sexual reproduction. National Content Standards: (5-8) UCP1, UCP5, A1, B1, C1, C2	Using Math, p. 100 Problem Solving: Chromosome Numbers and Meiosis, p. 101 Connect to Chemistry, p. 102 Skill Builder: Making and Using Tables, p. 103 Using Computers: Spreadsheet, p. 103
4-3 DNA (1½ sessions, 1 block)*	1. **Construct** and **identify** the parts of a model of a DNA molecule. 2. **Describe** how DNA copies itself. National Content Standards: (5-8) UCP1, UCP2, UCP5, A1, A2, B1, C1, C2, C3, E2, F5, G1, G2, G3	Using Technology: DNA Testing, p. 107 MiniLAB: Comparing RNA and DNA, p. 108 Skill Builder: Concept Mapping, p. 109 Using Math, p. 109 Activity 4-2: Making a Model of DNA, pp. 110-111
4-4 Science and Society Technology: Techniques in Cloning (½ session, ½ block)*	1. **Define** a clone in terms of inheritance of DNA. 2. **Identify** one benefit of cloning research. National Content Standards: (5-8) A1, A2, C2, E2, F4, F5, G1	Explore the Technology, p. 113 People and Science: Dr. Alfred L. Goldson, Radiation Oncologist, p. 114

* A complete Planning Guide that includes block scheduling is provided on pages 32T-35T.

Activity Materials

Explore	Activities	MiniLABs
page 91 pinto beans that have been soaked overnight, scalpel, paper towels, tap water, self-sealing plastic bag	page 96 prepared slide of an onion root tip, prepared slide of a whitefish embryo, microscope pages 110-111 six 8 1/2" × 11" sheets of construction paper—2 pieces of each color, scissors, 1 sheet of heavy paper, tape, safety goggles	page 97 colored tissue paper, construction paper, different colors of yarn, toothpicks, scissors, thread, glue, poster board page 108 construction paper, scissors

Need Materials? Call Science Kit (1-800-828-7777).

Teacher Classroom Resources

Reproducible Masters	Transparencies	Teaching Resources
Study Guide, p. 17 **Reinforcement,** p. 17 **Enrichment,** p. 17 **Activity Worksheets,** pp. 23-24, 27 **Lab Manual 8,** Mitosis **Concept Mapping,** p. 13	**Section Focus Transparency 13,** Higher and Higher **Teaching Transparency 7,** Mitosis	**Glencoe Life Science Interactive Videodisc** **Life Science CD-ROM** **Spanish Resources** **English/Spanish Audiocassettes** **Cooperative Learning Resource Guide** **Lab Partner** **Lab and Safety Skills** **Lesson Plans**
Study Guide, p. 18 **Reinforcement,** p. 18 **Enrichment,** p. 18 **Lab Manual 9,** Chromosomes	**Section Focus Transparency 14,** When Being Cross Is Good **Teaching Transparency 8,** Meiosis	
Study Guide, p. 19 **Reinforcement,** p. 19 **Enrichment,** p. 19 **Activity Worksheets,** pp. 25-26, 28 **Science Integration Activity 4,** The Effect of Radiation on Seeds **Critical Thinking/Problem Solving,** p. 8 **Cross-Curricular Integration,** p. 8 **Science and Society Integration,** p. 8 **Multicultural Connections,** pp. 11-12	**Section Focus Transparency 15,** New Cat **Science Integration Transparency 4,** DNA Up Close	**Assessment Resources** **Chapter Review,** pp. 11-12 **Assessment,** pp. 17-20 **Performance Assessment in the Science Classroom (PASC)** **MindJogger Videoquiz** **Alternate Assessment in the Science Classroom** **Performance Assessment** **Chapter Review Software** **Computer Test Bank**
Study Guide, p. 20 **Reinforcement,** p. 20 **Enrichment,** p. 20	**Section Focus Transparency 16,** Identical Twins	

Key to Teaching Strategies

The following designations will help you decide which activities are appropriate for your students.

L1 Level 1 activities should be within the ability range of all students, including those with learning difficulties.

L2 Level 2 activities should be within the ability range of the average to above-average student.

L3 Level 3 activities are designed for the ability range of above-average students.

LEP LEP activities should be within the ability range of Limited English Proficiency students.

LS These activities are designed to address different learning styles.

COOP LEARN Cooperative Learning activities are designed for small group work.

P These strategies represent student products that can be placed into a best-work portfolio.

GLENCOE TECHNOLOGY

The following multimedia resources are available from Glencoe.

The Secret of Life
What's in Stetter's Pond: The Basics of Life
In the Land of Milk and Money: Biotechnology
The Infinite Voyage Series
The Geometry of Life
National Geographic Society Series
STV: The Cell

Glencoe Life Science Interactive Videodisc
Mitosis and Meiosis: The Cycle of Life
Life Science CD-ROM

Teacher Classroom Resources

This is a representation of key blackline masters available in the Teacher Classroom Resources.

Teaching Aids

Section Focus Transparencies

Science Integration Transparencies

Teaching Transparencies

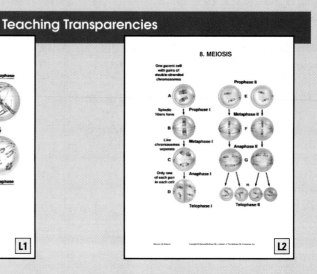

Meeting Different Ability Levels

Study Guide

Reinforcement

Enrichment Worksheets

Chapter 4 Cell Reproduction

Hands-On Activities

Science Integration Activity

Lab Manual

Assessment

Performance Assessment

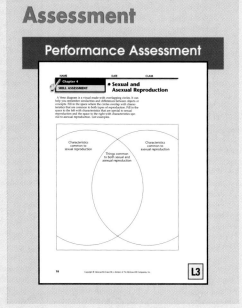

Enrichment and Application

Critical Thinking/Problem Solving

Cross-Curricular Integration

Science and Society Integration

Multicultural Connections

Concept Mapping

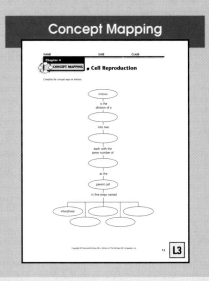

Cell Reproduction

CHAPTER OVERVIEW

Section 4-1 This section describes why cells divide, the cell cycle, and how body cells reproduce. Asexual reproduction is discussed.

Section 4-2 Sexual reproduction, the role of sex cells, and sex cell formation are presented in this section.

Section 4-3 This section discusses the chemical composition of DNA, how DNA replicates, and how DNA directs the production of proteins. The section ends with an explanation of mutations and why they occur.

Section 4-4 Science and Society The process of cloning with new techniques is presented.

Chapter Vocabulary

mitosis	meiosis
chromosome	fertilization
asexual	zygote
reproduction	DNA
sexual	gene
reproduction	RNA
sperm	mutation
egg	clone

Theme Connection

Stability and Change DNA controls all cell activities by directing the production of proteins in living organisms. Changes in the DNA result in evolutionary changes that are inherited.

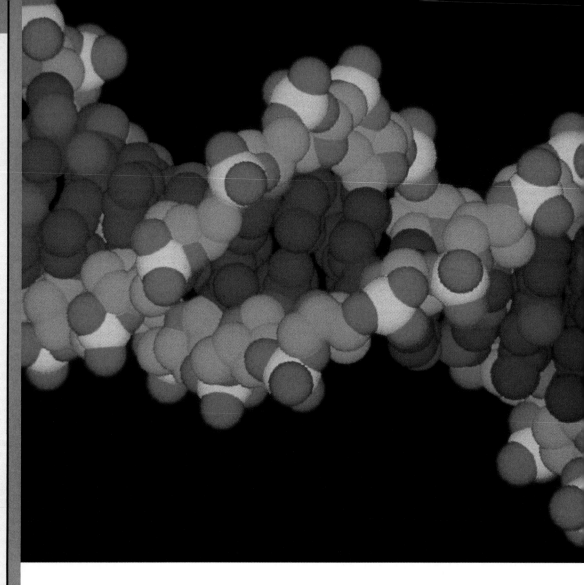

Previewing the Chapter

Section 4-1 Cell Growth and Division
- ▶ Why do cells divide?
- ▶ The Cell Cycle
- ▶ Steps of Animal Mitosis
- ▶ Asexual Reproduction

Section 4-2 Sexual Reproduction and Meiosis
- ▶ Sexual Reproduction
- ▶ The Importance of Sex Cells
- ▶ Meiosis

Section 4-3 DNA
- ▶ What is DNA?
- ▶ How DNA Copies Itself
- ▶ Genes
- ▶ Mutations

Section 4-4 Science and Society
  Technology: Techniques in Cloning

90

Learning Styles Look for the following logo for strategies that emphasize different learning modalities.

Kinesthetic	MiniLAB, pp. 97, 108; Making a Model, p. 102
Visual-Spatial	Explore, p. 91; Activity, p. 93; Visual Learning, pp. 93, 94, 100, 102; Demonstration, p. 95; Activity 4-1, p. 96; Across the Curriculum, p. 108
Interpersonal	Activity 4-2, pp. 110-111
Logical-Mathematical	Across the Curriculum, p. 95
Linguistic	Across the Curriculum, p. 94; Enrichment, pp. 95, 107; Using Science Words, p. 100; People and Science, p. 114

Chapter 4

Cell Reproduction

Do you know that your life hangs by a thread? Miles of threadlike DNA and protein condense to make up chromosomes in your cells, similar to the one to the left. When cells divide and organisms grow, chromosomes are the carriers of the information that shapes each new cell. In the following activity, find out what changes take place as a plant develops.

EXPLORE ACTIVITY

Observe seeds.

1. Carefully split open a pinto bean seed that has soaked in water overnight.
2. Observe both halves of the seed. Record your observations.
3. Place four other bean seeds in a moist paper towel in a self-sealing plastic bag.
4. Make observations for a few days.

Observe: Infer in your Science Journal what happened inside the seeds.

Previewing Science Skills

▶ In the **Skill Builders**, you will outline, make a table, and map concepts.

▶ In the **Activities**, you will observe, compare and contrast, infer, and make a model.

▶ In the **MiniLABs**, you will make a model and sequence events.

91

EXPLORE ACTIVITY

Purpose

IS **Visual-Spatial** Use the Explore Activity to introduce students to growth. As they read the chapter, they will understand that growth is the result of mitosis. **L1** **LEP**

Preparation

Soak the pinto bean seeds for 24 hours.

Materials

5 soaked pinto bean seeds, paper towels, plastic bag, and a hand lens for each group of 2 students

Teaching Strategies

The soaked seeds should split easily. If students have difficulty, forceps can be used to separate the seeds.

Observe

Students should observe the growth of a new plant from each seed.

✓Assessment

Performance Have students repeat the experiment using corn seeds and observe the new plant that grows from each seed. Use the Performance Task Assessment List for Carrying Out a Strategy and Collecting Data in **PASC,** p. 25. **P**

Prepare

Section Background

Every species has a characteristic number of chromosomes in each cell. The number varies with the species. A cat has 32 chromosomes, whereas a potato and a chimpanzee each have 48 chromosomes. In all sexually reproducing organisms, chromosomes occur in pairs. The two chromosomes of each pair are called homologous chromosomes.

Preplanning

- Obtain prepared onion root tip and whitefish embryo slides for Activity 4-1.
- To prepare for the Mini-LAB, obtain colored paper, poster paper, markers, toothpicks, yarn, thread, and glue.

1 Motivate

Bellringer

Before presenting the lesson, display **Section Focus Transparency 13** on the overhead projector. Assign the accompanying **Focus Activity** worksheet. [L1] [LEP]

Tying to Previous Knowledge

- Review the structure of the nucleus and the function of each of its parts from Chapter 2.
- Tell students that mitosis involves only the nucleus of the cell.

Science Words

mitosis
chromosome
asexual reproduction

Objectives

- Describe mitosis and explain its importance.
- Explain differences between mitosis in plant and animal cells.
- Give two examples of asexual reproduction.

Why do cells divide?

It's likely that each time you go to the doctor, a nurse measures your height and weight. Over the years, similar data collected from thousands of people have given the medical profession an idea of how people grow. Much of the growth happens because the number of cells in your body increases as you develop.

Constant Change

The fact is that you are constantly changing. You aren't the same now as you were a year ago or even a few hours ago. At this very moment, as you read this page, groups of cells throughout your body are growing, dividing, and dying. Worn-out cells on the palms of your hands are being replaced. Cuts and bruises are healing. Red blood cells are being produced in your bones at a rate of 2 to 3 billion per second to replace those that wear out. Muscles that you exercise are getting larger. Other organisms undergo similar processes. In **Figure 4-1**, a plant climbs a garden stake as the number of its cells increases. How do cells increase in number?

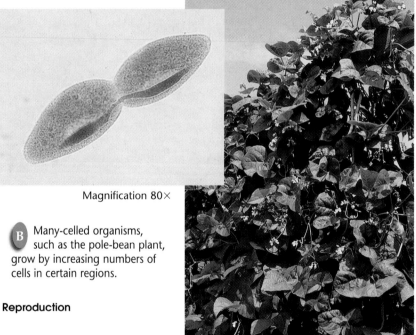

Figure 4-1

Cell division occurs in all organisms.

A A one-celled organism, such as a paramecium, reaches a certain size and then divides.

Magnification 80×

B Many-celled organisms, such as the pole-bean plant, grow by increasing numbers of cells in certain regions.

92 Chapter 4 Cell Reproduction

Program Resources

Reproducible Masters
Study Guide, p. 17 [L1]
Reinforcement, p. 17 [L1]
Enrichment, p. 17 [L3]
Activity Worksheets, pp. 23-24, 27 [L1]
Lab Manual 8
Concept Mapping, p. 13

Transparencies
Teaching Transparency 7
Section Focus Transparency 13 [L1]

The Cell Cycle

Organisms go through stages, or a life cycle, while they are alive, as illustrated in **Figure 4-2.** A simple life cycle is birth, growth and development, and death. Right now, you are in a stage in your life cycle called adolescence, a period of active growth and development. Cells also go through cycles. Most of the life of any cell is spent in a period of growth and development called *interphase.* During interphase, cells go through periods: rapid growth, growth and DNA synthesis, growth and preparation for division, and then cell division. In cell division, the nucleus divides and then the cytoplasm divides to form two new cells. Cell division is a continuous process. The cell cycle, as shown in **Figure 4-3,** is a series of events that takes place in a cell from one division to the next. Cells in your body that no longer divide, such as nerve and muscle cells, are always in interphase. Cells that actively divide, such as your skin cells, have a more complex cell cycle.

Mitosis

Cells divide in two steps. First, the nucleus of the cell divides, and then the cytoplasm divides. **Mitosis** is the process in which the nucleus divides to form two identical nuclei. Each of the two nuclei contains the same number and type of chromosomes as the original. The parent cell then undergoes division. Mitosis is described as a series of phases or steps. These steps have been named prophase, metaphase, anaphase, and telophase.

Figure 4-2

An organism undergoes many changes as it develops. These photographs show some of the stages of development of a red oak tree, starting as an acorn seed.

A Cell Cycle

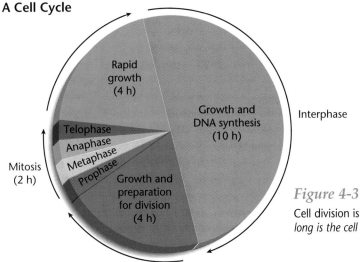

Figure 4-3

Cell division is a continuous process. *How long is the cell cycle represented in this graph?*

4-1 Cell Growth and Division **93**

2 Teach

Revealing Preconceptions

Many students may think that mitosis occurs in all the cells of an organism all the time. In some tissues, once the cells are formed, mitosis does not occur again. Red blood cells are formed, carry out their functions, and then are removed from the body. In plants, mitosis occurs in the growth regions and not throughout the whole plant.

Activity

Visual-Spatial Have students bring one of their baby pictures and a first- or second-grade picture to class. Identify the pictures by number and place them on the bulletin board. Then have the students identify the person in each picture. When all the pictures have been identified, discuss how the students have grown and changed from the time they were babies until the present. **L1** **LEP**

 Use Lab 8 in the **Lab Manual** as you teach this lesson.

Use **Teaching Transparency 7** as you teach this lesson.

Videodisc
STV: The Cell
How Cells Reproduce
Unit 3
Cell Reproduction

27458-29060

Visual Learning

Figure 4-3 How long is the cell cycle represented in this graph? *20 hours* **LEP** **IS**

Inclusion Strategies

Gifted Have students interview a physician, nurse, or head of a local special education program to find out about Down syndrome. They should find out what the latest research shows about detection, causes, and care. Students will want to ask about the characteristics and capabilities of a person with Down syndrome. **L3**

NATIONAL GEOGRAPHIC SOCIETY

Videodisc
STV: The Cell
How Cells Reproduce
Unit 3
Interphase

29163-29900
Mitosis

29909-33946

Figure 4-4

Each double-stranded chromosome is held together at a region called a centromere.

Centromere

Double-stranded chromosome

Figure 4-5

The steps of mitosis are shown in order on this page and the next.

A Interphase
During interphase, the nucleus can be clearly seen. The chromosomes cannot yet be seen, but are actively duplicating themselves.

Magnification 450×

B Prophase
During prophase, chromosomes become fully visible. The nucleolus and the nuclear membrane fade and disappear. Two small structures called centrioles move to opposite ends of the cell. Between the centrioles, threadlike spindle fibers begin to stretch across the cell.

94   Chapter 4 Cell Reproduction

Steps of Animal Mitosis

When a cell divides, the chromosomes in the nucleus play an important part. **Chromosomes** are structures in the nucleus that contain DNA. During *interphase*, you can't see chromosomes, but they are actively duplicating themselves. When the cell is ready to divide, the chromosomes condense and become visible. When the chromosomes appear, they are thick and doubled-stranded as in **Figure 4-4**. Each double-stranded chromosome is held together at a region called a centromere. Once chromosomes are double-stranded, the cell is ready to begin the process of division. Follow the steps of mitosis in the illustrations in **Figure 4-5** on these pages.

Magnification 450×

Magnification 450×

C Metaphase
In metaphase, the double-stranded chromosomes line up across the center of the cell. Each centromere becomes attached to a spindle fiber.

In most organisms, once the nucleus has divided, the cytoplasm also separates and two whole new cells are formed. In animal cells, the cytoplasm pinches in to form the new cells. The new cells then begin a period of growth, or interphase, again. They will take in water and other nutrients that they need to carry out cell processes.

Plant Cell Mitosis

Plant cells have rigid cell walls and do not pinch apart as animal cells do. Instead, a structure called a cell plate, shown in **Figure 4-6,** forms between the two new nuclei. New cell walls form along the cell plate. Plant cells do not have centrioles, but they do have spindle fibers during mitosis.

Magnification 450×

Magnification 450×

D **Anaphase**
As the process enters anaphase, each centromere divides. The two strands of each chromosome separate. Then, the separate strands begin to move away from each other toward opposite ends of the cell. *Why must the separated strands move to opposite ends of the cell?*

E **Telophase**
In the final step of mitosis, telophase, centrioles and spindle fibers start to disappear. The chromosomes stretch out and become harder to see. A nuclear membrane forms around each mass of chromosomes, and a new nucleolus appears in each new nucleus.

Figure 4-6

Cell plates form during plant cell mitosis.

Magnification 1000×

95

Across the Curriculum

Math It takes 15 minutes for certain embryo cells to divide. Have students calculate how many cells would be produced from one cell after 4 hours (65 536). **L1** **LS**

Activity 4-1

Mitosis in Plant and Animal Cells

Activity 4-1

Purpose

IS Visual-Spatial Students will observe the stages of mitosis in plant and animal cells and compare the processes and results. **L1**

Process Skills

observing, inferring, sequencing, comparing and contrasting, forming operational definitions

Time

40 minutes

Activity Worksheets, pp. 5, 23-24

Teaching Strategies

Alternate Materials If microscopes are not available for all students, a microprojector can be used for the class to work together.

Troubleshooting Students may have difficulty locating all phases in the onion root tip. You may want to place an onion root tip slide on the microprojector and point out the phases.

• Review the stages of mitosis before beginning this activity.

Answers to Questions

1. Many cells are smaller. Chromosomes can be seen in many of the cells.

2. Onion root tip cells are rectangular; whitefish embryo cells are almost round.

3. metaphase; anaphase

4. prophase

5. The whitefish embryo cells are much larger.

6. The whitefish chromosomes are smaller.

7. Cells in embryos and root tip undergo rapid and frequent cell division. As a result, the phases of mitosis can be seen easily.

Reproduction of body cells in plants and animals is accomplished by mitosis. In this activity, you will study prepared slides of onion root tip and the early whitefish embryo. These slides are used because they show rapidly dividing cells in the various stages of mitosis.

Whitefish cells
Magnification 320×

Onion root tip cells
Magnification 330×

Problem

How is mitosis in a plant cell different from mitosis in an animal cell?

Materials

• prepared slide of an onion root tip
• prepared slide of a whitefish embryo
• microscope

Procedure

1. Obtain prepared slides of onion root tip and whitefish embryo.
2. Set your microscope on low power and examine the onion root tip. Move the slide until you can see the area just behind the root tip. Turn the nosepiece to high power.
3. Use **Figure 4-5** on pages 94-95 to help you find a cell in interphase. Draw and label the parts of the cell you observe.
4. Repeat step 3 for prophase, metaphase, anaphase, and telophase.
5. Turn the microscope back to low power. Remove the onion root tip slide.
6. Place the whitefish embryo slide on the microscope stage under low power. Focus and find a region of dividing cells. Turn the microscope to high power.
7. Use **Figure 4-5** to help you find a cell in interphase. In a data table, draw and label the parts of the cell you observe.
8. Repeat step 7 for prophase, metaphase, anaphase, and telophase.

9. Return the nosepiece to low power. Remove the whitefish embryo slide from the microscope stage.

Analyze

1. How are the cells in the region behind the onion root tip different from those in the root tip?
2. What is the shape of the cells in the onion root tip and the whitefish embryo?
3. In which stage do the chromosomes move to the center of the cell and then move to opposite ends of the cell?
4. When do the spindle fibers appear in both the onion root tip and the whitefish embryo?

Conclude and Apply

5. **Compare** the sizes of the whitefish embryo cells and the onion root tip cells.
6. **Compare** the chromosomes of the whitefish embryo and the onion root tip.
7. Why are embryo cells and root tip cells used to study mitosis?

96 Chapter 4 Cell Reproduction

✔ Assessment

Performance To further assess students' understanding of the stages of mitosis, give each one a slip of paper listing a stage and have them describe what comes before and after that stage. Use the Performance Task Assessment List for Making Observations and Inferences in **PASC,** p. 17. **P**

Results of Mitosis

There are two important things to remember about mitosis. The first is that mitosis is the division of a nucleus. The second is that mitosis produces two new nuclei that have the same number of chromosomes as the original nucleus. Cells in your skin, like all cells in your body, each have 46 chromosomes. Each new skin cell produced by mitosis will also have 46 chromosomes. Cells in fruit flies have eight chromosomes. New fruit fly cells produced by mitosis will each have only eight chromosomes.

Asexual Reproduction

Your body forms two types of cells—body cells and sex cells. Skin, liver, bones, kidneys, lungs, and muscles are made up of different types of body cells. By far, your body has many more body cells than sex cells. The only sex cells that you have are the eggs or sperm in your reproductive organs.

Reproduction is the process by which an organism produces others of the same kind. Among living organisms, there are two types of reproduction—asexual reproduction and sexual reproduction. In **asexual reproduction,** new organisms are produced from one parent. You've just seen examples of this in the process of mitosis. Offspring produced by asexual reproduction have DNA that is identical to the DNA of the parent organism.

As demonstrated in **Figure 4-7,** several types of asexual reproduction are important in plants and animals. If you've ever grown a sweet potato in a jar of water, you've seen asexual reproduction take place. All the stems and leaves that grow out from the sweet potato have been produced by mitosis. New strawberry plants can be produced asexually from runners. Bacterial cells reproduce asexually by a process called fission. Fission is division of an organism into two equal parts.

Magnification 8×

Figure 4-7

Many organisms reproduce asexually. Hydra, shown above, and cattails on the left are examples.

4-1 Cell Growth and Division   97

For students who have mastered this section, use the **Reinforcement** and **Enrichment** masters.

4 Close

•MINI•QUIZ•

1. The process in which the nucleus of a cell divides to form two new nuclei is _____ . *mitosis*

2. New organisms are produced from one parent in _____ reproduction. *asexual*

Section Wrap-up

Review

1. process in which a cell nucleus divides into two nuclei, each of which has the same number of chromosomes; animal cells—have centrioles; plant cells—form a cell plate

2. budding and regeneration

3. **Think Critically** It must dissolve and the spindle fibers must form to prepare for mitosis.

USING MATH

Student answers will vary.

[(Present weight – Birth weight/Age) 5 years] + Present Weight

Skill Builder

Prophase
A. Chromosomes become visible.
B. Nucleolus and nuclear membrane fade, disappear.
C. Centrioles move to opposite ends of the cell.
D. Spindle fibers appear.

Metaphase
A. Double-stranded chromosomes line up.
B. Centromeres attach to spindle fibers.

Anaphase
A. Centromeres divide.
B. Two strands of chromosomes separate.
C. Separated strands move toward opposite ends of cell.

Telophase
A. Centrioles and spindle fibers disappear.
B. Chromosomes become harder to see.
C. Nuclear membrane forms in each cell.
D. Nucleolus appears in each new nucleus.

Budding and Regeneration

Budding is also a type of asexual reproduction in which a new organism grows from the body of the parent organism. Hydra can reproduce this way. When the bud on the adult becomes large enough, it breaks away to live on its own.

A few organisms can repair damaged or lost body parts by regeneration. During regeneration, a whole organism may develop from a piece of the organism, or two organisms may result when there is injury to one, as illustrated in **Figure 4-8**. Sponges, planaria, and sea stars can regenerate. If a sponge is cut into small pieces, a new sponge may develop from each piece. How do you think sponge farmers increase their "crop"?

Through cell division, organisms grow, replace worn-out or damaged cells, or produce whole new organisms. Fission and budding are types of asexual reproduction that result from mitosis. Replacement of lost body parts by regeneration is also a result of mitosis.

Magnification 10×

Figure 4-8

Some animals produce whole new body parts by regeneration. A planarian is a worm that can regenerate a head or a tail when cut in two.

Section Wrap-up

Review

1. What is mitosis and how does it differ in plants and animals?

2. Give two examples of asexual reproduction.

3. **Think Critically:** Why is it important for the nuclear membrane to dissolve in mitosis?

Skill Builder
Sequencing

Sequence the events in each stage of mitosis in animal cells. Begin with interphase. If you need help, refer to Sequencing in the **Skill Handbook.**

USING MATH

Find out how much you weighed at birth. Subtract that weight from what you weigh now. Divide that number by your age to find the average weight you have gained each year. Predict what you will weigh five years from now.

✓ Assessment

Oral How is mitosis different from cell division? *Mitosis is the division of the nucleus, and cell division is the division of the cytoplasm and organelles.* Use the Performance Task Assessment List for Making Observations and Inferences in **PASC,** p. 17. P

Sexual Reproduction 4•2 and Meiosis

Sexual Reproduction

In the last section, you learned that asexual reproduction occurs with only one parent. In **sexual reproduction,** a new organism is produced when sex cells from two parents combine. Each sex cell is produced by a different parent. The sex cell from the male parent is called the **sperm.** The sex cell from the female parent is called the **egg.** Sex cells like the ones in **Figure 4-9** are usually different in size from one another. Eggs are usually large and contain food material. Sperm are small with whiplike tails. The sperm head is almost all nucleus.

Production of Sex Cells

A human body cell has 23 pairs of chromosomes, but human sex cells have only 23 chromosomes. How does the chromosome number become reduced? How do sex cells form? The process of nuclear division that produces sex cells is called **meiosis.** Meiosis takes place in cells of reproductive organs in both plants and animals.

Science Words

sexual reproduction
sperm
egg
meiosis
fertilization
zygote

Objectives

- Describe the stages of meiosis and its end products.
- Name the cells involved in fertilization and explain how fertilization occurs in sexual reproduction.

Magnification 550×

Figure 4-9
The large cell is a human egg cell, or ovum. The smaller cells are human sperm.

Magnification 1100×

4-2 Sexual Reproduction and Meiosis **99**

Program Resources

 Reproducible Masters
Study Guide, p. 18 L1
Reinforcement, p. 18 L1
Enrichment, p. 18 L3
Lab Manual 9

 Transparencies
Teaching Transparency 8
Section Focus Transparency 14 L1

- Sexual reproduction is the production of offspring through meiosis and the fusion of sex cells or gametes. Such offspring are genetically different from both parents because in meiosis, genes are combined in new ways through genetic recombination. Genetic recombination produces variation between parents and offspring and gives some offspring a better chance of surviving in a changing environment.

- In animals, meiosis results in haploid egg and sperm cells. In plants, meiosis results in haploid spores that later lead to the production of egg and sperm cells.

- With the 23 pairs of chromosomes in the human body, there are more than 8 million different combinations possible for every human cell formed by meiosis.

1 Motivate

Bellringer

Before presenting the lesson, display **Section Focus Transparency 14** on the overhead projector. Assign the accompanying **Focus Activity** worksheet. L1 LEP

Tying to Previous Knowledge

Students will be familiar with cell division from having studied Section 4-1. Ask them to describe the stages of mitosis. Then tell the students that in meiosis the nucleus divides twice.

The zygote will have 188 chromosomes—94 from the egg and 94 from the sperm.

Using Science Words

 Linguistic Point out that *meiosis* comes from a Greek word for "diminution" or becoming less. Ask students what diminishes or becomes less in meiosis. *The chromosome number in each cell becomes less.*

Visual Learning

Figure 4-10 **Compare the number of chromosomes present in the different cells.** *Sex cells (eggs and sperm) have half the number of chromosomes present in the zygote and body cells.* **LEP**

GLENCOE TECHNOLOGY

 Videotape

The Secret of Life

In the Land of Milk and Money: Biotechnology

Find out about the applications of recombinant DNA technology.

Videodisc

The Infinite Voyage: The Geometry of Life

Chapter 3

Development of the Egg After Fertilization

How many chromosomes will be in a goldfish zygote if the egg of the goldfish contains 94 chromosomes?

The Importance of Sex Cells

In body cells, chromosomes are found in pairs. The 46 human chromosomes form 23 pairs of chromosomes. The pairs form because the chromosomes are alike. A cell that has two of every kind of chromosome is said to be *diploid*. Sex cells, on the other hand, contain only one chromosome from each matched pair. A sex cell with just one chromosome from each pair is *haploid*. *Haploid* means "single form." A human sex cell has 23 chromosomes, not 23 pairs of chromosomes. Therefore, it is haploid. For corn, the diploid number is 20, and the haploid number is 10. Usually, the haploid number of chromosomes is found only in sex cells of an organism.

Fertilization

What is so important about sex cells? Sexual reproduction starts with the formation of sex cells and ends when one sex

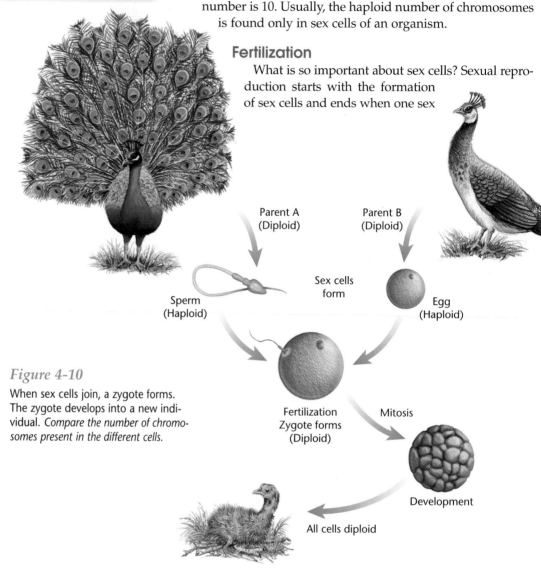

Figure 4-10
When sex cells join, a zygote forms. The zygote develops into a new individual. *Compare the number of chromosomes present in the different cells.*

Parent A (Diploid)

Parent B (Diploid)

Sperm (Haploid)

Sex cells form

Egg (Haploid)

Fertilization Zygote forms (Diploid)

Mitosis

Development

All cells diploid

100 Chapter 4 Cell Reproduction

Across the Curriculum

Language Arts Have students use library references to find the number of chromosomes in body cells of various plants and animals. Then have them find the number of chromosomes in the sex cells of each one. **L1**

cell joins with another and a new organism is begun. The joining of an egg and a sperm is called **fertilization.** The cell that forms in fertilization is called a **zygote.** If an egg with 23 chromosomes joins with a sperm that has 23 chromosomes, a zygote forms that has 46 chromosomes, or the diploid chromosome number for that organism. A zygote then begins to undergo mitosis and the organism develops. This process is illustrated in **Figure 4-10.**

Meiosis

Think what would happen if two cells formed through mitosis combined in sexual reproduction. The offspring would have twice as many chromosomes as its parent. In meiosis, there are two divisions of the nucleus, *meiosis I* and *meiosis II.* The different steps of each division have names like those in mitosis. In **Figure 4-11** on page 102, follow the steps of meiosis I.

Problem Solving

Chromosome Numbers and Meiosis

A horse and a donkey can mate to produce a mule. The mule is almost always sterile. Horses have a diploid chromosome number of 60. Donkeys have a diploid number of 66. Suggest an explanation for why the mule is sterile.

Solve the Problem:

1. **How many chromosomes would the mule receive from each parent?**
2. **What is the diploid number of the mule?**
3. **What would happen when meiosis occurs?**

Think Critically:

Why might a mule be sterile?

Horse

66
Donkey

60

Mule

101

Solve the Problem

1. 30 from the horse and 33 from the donkey
2. 63
3. Normal gametes would not form. There would not be pairs of chromosomes to pass on to the offspring.

Think Critically

Each sex cell could not have one-half the chromosomes of the original cell.

Using Science Words

In-vitro fertilization is a procedure that joins sperm and eggs outside the body. Fertilized eggs (actually two-day-old embryos) are implanted into the female.

NATIONAL GEOGRAPHIC SOCIETY

Videodisc

STV: The Cell

Meiosis

46998

Meiosis occurring in onion cells

46999

Genes on chromosomes

47008

Human chromosomes

47009

Cultural Diversity

Chromosome Number In 1887, Edouard-Joseph-Marie van Beneden, a Belgian scientist, discovered that each species has a fixed number of chromosomes. He also observed the formation of a haploid cell. In 1954, J. Hin Tijo and Albert Levan showed that each human body cell has 46 chromosomes.

Making a Model

Visual Learning

Use **Teaching Transparency 8** as you teach this lesson.

3 Assess

Check for Understanding

? FLEX Your Brain

Use the Flex Your Brain activity to have students explore MEIOSIS.

Activity Worksheets, p. 5

Reteach

Use an overhead transparency to show the stages of meiosis out of order. Have students identify each stage and place the stages in order.

Extension

For students who have mastered this section, use the **Reinforcement** and **Enrichment** masters.

Figure 4-11

In meiosis, there are two divisions of the nucleus—meiosis I and meiosis II. Meiosis begins with the pairing of like chromosomes. The steps of meiosis I are shown on this page.

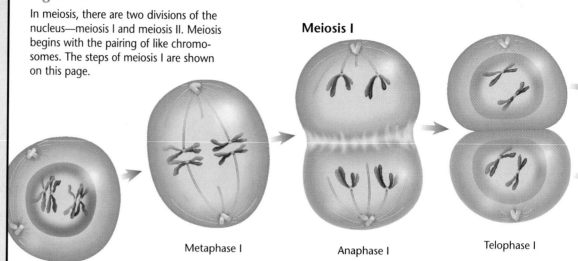

Meiosis I

Prophase I

Metaphase I

Anaphase I

Telophase I

Meiosis I

In *prophase I,* double-stranded chromosomes and spindle fibers appear. The nuclear membrane and the nucleolus disappear. Like chromosomes come together in matching pairs.

In *metaphase I,* the pairs of chromosomes line up in the center of the cell. Their centromeres become attached to the spindle fibers. In *anaphase I,* each double-stranded chromosome separates from its matching chromosome. Each one is pulled to opposite ends of the cell. Then in *telophase I,* the cytoplasm divides and two cells form. Each chromosome is still double-stranded.

Meiosis II

Meiosis II, as illustrated in **Figure 4-12,** then begins. In *prophase II,* the double-stranded chromosomes and spindle fibers reappear in each new cell. In *metaphase II,* the double-stranded chromosomes move to the center of the cell. There, the centromeres attach to spindle fibers. During *anaphase II,* the centromere divides, and the two strands of each chromosome separate and move to opposite ends of the cell. As *telophase II* begins, the spindle fibers disappear, and a nuclear membrane forms around the chromosomes at each end of the cell. Each nucleus contains only half the number of chromosomes that were in the original nucleus. A cell with 46 chromosomes at the beginning of meiosis I divides to produce cells that each have only 23 single-stranded chromosomes at the end of meiosis II.

102   Chapter 4 Cell Reproduction

When meiosis II is finished, the cytoplasm divides. Meiosis I forms two cells. In meiosis II, both of these cells divide into two cells. The two nuclear divisions result in four cells. Each of these four cells is a sex cell. Each sex cell has one-half the chromosomes of the original cell.

Figure 4-12

Figure 4-12
The steps of meiosis II are shown. *How many sex cells are finally formed?*

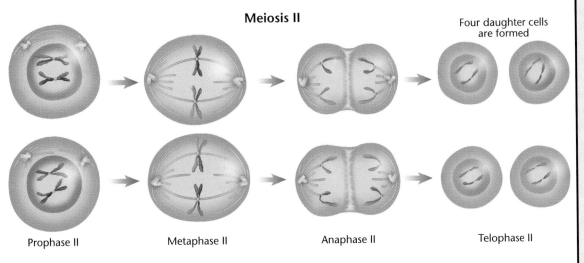

Meiosis II

Prophase II   Metaphase II   Anaphase II   Telophase II

Four daughter cells are formed

Section Wrap-up

Review

1. Distinguish between sexual and asexual reproduction.

2. What is a zygote, and how is it formed?

3. **Think Critically:** Plants grown from runners and leaf cuttings have exactly the same traits as the parent plant. Plants grown from seeds may vary from the parents in many ways. Suggest an explanation for this.

Skill Builder
Making and Using Tables
Make a table to compare mitosis and meiosis in humans. Vertical heads should include: *What type of cell* (Body or Sex), *Beginning cell* (Haploid or Diploid), *Number of cells produced, End-product cell* (Haploid or Diploid), and *Number of chromosomes in cells produced.* If you need help, refer to Making and Using Tables in the **Skill Handbook.**

Using Computers

Spreadsheet Make a data table that shows the steps of mitosis and the steps of meiosis. Use the data table to determine how the processes are similar and how they are different.

Skill Builder

Feature	Mitosis	Meiosis
What type of cell	Body cell	Sex cell
Beginning cell	Diploid	Diploid
Number of cells produced	Two	Four
End-product cell	Diploid	Haploid
Number of chromosomes produced	Same as parent–46	Half the parent cell number–23

Assessment

Performance Use this Skill Builder to assess students' abilities to compare and contrast meiosis and mitosis. Use the Performance Task Assessment List for Data Table in **PASC,** p. 37. **P**

---

4 Close

•MINI•QUIZ•

Use the Mini Quiz to check students' recall of chapter content.

1. In _____ reproduction, a new organism is produced when sex cells from two parents combine. *sexual*

2. The joining of an egg and a sperm is _____ . *fertilization*

3. The process of nuclear division that produces sex cells is _____ . *meiosis*

Section Wrap-up

Review

1. In sexual reproduction, a new organism is produced when cells from two parents combine. In asexual reproduction, a new organism is formed from one parent and is a duplicate of that parent genetically.

2. A zygote is a cell that forms in fertilization; it has a diploid number of chromosomes.

3. **Think Critically** Seeds are the result of sexual reproduction, which involves meiosis and fertilization. Plants grown from the same parent are the result of asexual reproduction.

Using Computers

The data sheet should reflect the information in the Skill Builder Table.

Prepare

Section Background

DNA molecules make up the chromosomes of cells and contain the instructions for making proteins. In cell reproduction, the DNA molecules in a cell's chromosomes are copied exactly. A DNA molecule can make an exact copy of itself because the double-stranded structure separates and each half is copied exactly. The process of mitosis begins after a cell makes exact copies of its DNA and chromosomes.

Preplanning

- Obtain the paper and tape for Activity 4-2.
- To prepare for the Mini-LAB, obtain colored construction paper and scissors.

1 Motivate

Bellringer

Before presenting the lesson, display **Section Focus Transparency 15** on the overhead projector. Assign the accompanying **Focus Activity** worksheet. L1 LEP

Tying to Previous Knowledge

Students are familiar with some template systems, such as keys and locks and peg-and-hole games. Ask for other examples. Use this knowledge to explain that DNA in the nucleus serves as a template for RNA.

Science Words

- DNA
- gene
- RNA
- mutation

Objectives

- Construct and identify the parts of a model of a DNA molecule.
- Describe how DNA copies itself.

4•3 DNA

What is DNA?

Have you ever sent a message to someone using a code? In order for him or her to read your message, that person had to understand the meaning of the symbols you used in your code. The chromosomes in the nucleus of a cell contain a code. This code is in the form of a chemical called deoxyribonucleic acid, or **DNA.** DNA is the master copy of an organism's information code. When mitosis takes place, the DNA code in the nucleus is copied and passed to the new cells. In this way, new cells receive the same coded information that was in the original cell. DNA controls the activities of cells with coded instructions. Every cell that has ever been formed in your body or in any other organism contains DNA.

History of DNA

Since the mid-1800s, it has been known that the nuclei of cells contain chemicals called nucleic acids. What does DNA look like? Scientist Rosalind Franklin discovered that the

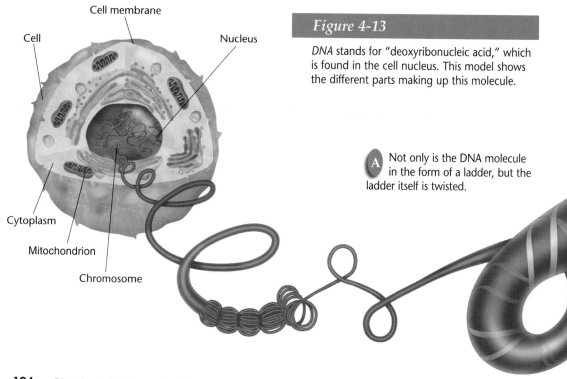

Cell membrane
Cell
Nucleus
Cytoplasm
Mitochondrion
Chromosome

Figure 4-13

DNA stands for "deoxyribonucleic acid," which is found in the cell nucleus. This model shows the different parts making up this molecule.

Ⓐ Not only is the DNA molecule in the form of a ladder, but the ladder itself is twisted.

Program Resources

 Reproducible Masters
Study Guide, p. 19 L1
Reinforcement, p. 19 L1
Enrichment, p. 19 L3
Cross-Curricular Integration, p. 8
Science and Society Integration, p. 8
Activity Worksheets, pp. 25-26, 28 L1
Science Integration Activities, pp. 7-8

Multicultural Connections, pp. 11-12
Critical Thinking/Problem Solving, p. 8

Transparencies
Section Focus Transparency 15 L1
Science Integration Transparency 4

DNA molecule was a strand of molecules in a spiral form. By using an X-ray technique, Dr. Franklin showed that the spiral was so large that it was probably made up of two spirals. As it turned out, the structure of DNA is similar to the handrails and steps of a spiral staircase. In 1953, using the work of Franklin and others, scientists James Watson and Francis Crick made a model of a DNA molecule.

A DNA Model

According to the Watson and Crick model, the sides (the "handrails") of the DNA molecule are made up of two twisted strands of sugar and phosphate molecules. The "stairs" that hold the two sugar-phosphate strands apart are made up of molecules called nitrogen bases. There are four kinds of nitrogen bases. These are adenine, guanine, cytosine, and thymine. In **Figure 4-13,** the bases are represented by the letters *A, G, C,* and *T.* The amount of cytosine in cells always equals the amount of guanine and the amount of adenine always equals the amount of thymine. This led to the hypothesis that these bases occur in pairs in the molecule. The Watson and Crick model shows that adenine always pairs with thymine, and guanine always pairs with cytosine. Like interlocking pieces of a puzzle, each base pairs up only with its correct partner.

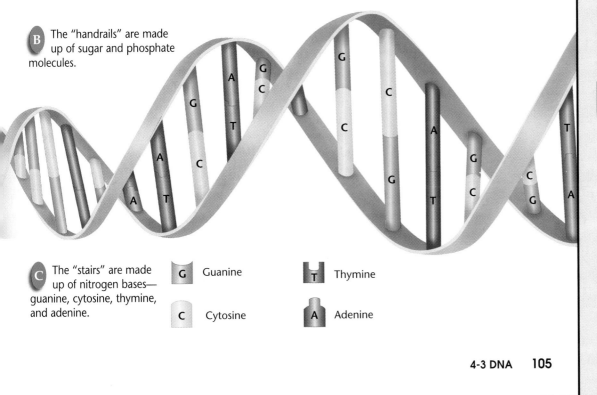

B The "handrails" are made up of sugar and phosphate molecules.

C The "stairs" are made up of nitrogen bases—guanine, cytosine, thymine, and adenine.

G Guanine **T** Thymine

C Cytosine **A** Adenine

4-3 DNA **105**

INTEGRATION
Chemistry

2 Teach

INTEGRATION
Chemistry

The DNA molecule is made up of five common elements: hydrogen, oxygen, carbon, nitrogen, and phosphorus. These elements are combined in different chemical groups and arranged in two chains.

Content Background

DNA, or deoxyribonucleic acid, is present in every cell of your body as well as the cells in all living organisms. It is so thin that it cannot be clearly visualized even through an electron microscope, some of which magnify more than 1 million times.

GLENCOE TECHNOLOGY

 Videodisc

The Infinite Voyage: The Geometry of Life
Chapter 2
DNA and the Double Helix

 Videotape

The Secret of Life
What's in Stetter's Pond: The Basics of Life
Discover how DNA may have evolved and the conditions under which primitive life-forms may have lived.

 Videodisc

STV: The Cell
DNA (computer graphic)

47000

Theme Connection

Scale and Structure In all of the cells of the human body, information is stored in the form of DNA, a complex biological molecule. DNA is responsible for directing all of the human body's functions.

105

Students are probably familiar with Morse code. Morse code uses only two symbols, the dot and the dash, in varying combinations to represent numbers and the letters of the alphabet. Anything that can be expressed in language can be expressed in Morse code. The DNA code has four symbols. The order of nitrogen bases, rather than the sequence of dots and dashes, expresses the information needed for life processes.

Concept Development

• Explain that knowing the base pairing rules allows you to predict the base sequence of a second strand of DNA if you know the sequence of the first strand.

• Make certain that students know that when they hear about the "code of life," they are hearing about the order of nitrogen bases in DNA.

Inquiry Question

Ask students to predict the base sequence of the second strand of DNA if the sequence of one strand is T G A A C T C G A T A C. *A C T T G A G C T A T G*

Student Text Question

What is the process by which DNA copies itself? *replication*

Figure 4-14
DNA uses itself as a pattern when it is copied.

Step 1 Enzyme separates DNA strand

Step 3 New strands of DNA produced

Enzyme

Original DNA molecule

Step 2 New bases in cytoplasm pair with original bases of DNA strands

How DNA Copies Itself

When chromosomes are doubled at the beginning of mitosis, the amount of DNA in the nucleus is also doubled. What is the process by which DNA copies itself? The Watson and Crick model can be used to show how this takes place. The two strands of DNA unwind and separate. Each strand then becomes a pattern on which a new strand is formed. **Figure 4-14** follows a strand of DNA as it produces two new DNA strands identical to the original DNA. The events that take place while DNA copies itself are given below.

Step 1. An enzyme breaks the bonds between the nitrogen bases. The two strands of DNA separate.

Step 2. The bases attached to each strand then pair up with new bases from a supply found in the cytoplasm. Adenine pairs with thymine. Cytosine pairs with guanine. The order of base pairs in each new strand of DNA will match the order of base pairs in the original DNA.

Step 3. Sugar and phosphate groups form the side of each new DNA strand. Each new DNA molecule now contains one strand of the original DNA and one new strand.

Integrating the Sciences

Chemistry The two strands of the DNA molecule are held together by chemical bonds between the bases. **Can any two molecules form a bond or are there restrictions on how chemicals can join together?** *There are restrictions.* **How might these affect the structure of DNA?** *Each base always joins with a specified base.*

Genes

Why is DNA important? All of the characteristics that you have are affected by the DNA that you have in your cells. It controls the color of your eyes, the color of your hair, and whether or not you can digest milk. These characteristics are called traits. How traits appear in you depends on the kinds of proteins your cells make. DNA stores the blueprints for making proteins.

Proteins are made of units called amino acids that are linked together in a certain order. A protein may be made of hundreds or thousands of amino acids. Changing the order of the amino acids changes the kind of protein made. The section of DNA on a chromosome that directs the making of a specific protein is called a **gene.** Genes control proteins that build cells and tissues and work as enzymes. Think about what might happen if an important protein couldn't be made in your cells.

DNA Fingerprinting

USING TECHNOLOGY

DNA Testing

In DNA testing, scientists isolate DNA from the blood or other biological material found at the scene of a crime. They also isolate DNA taken from the suspect involved. Both samples are treated with enzymes that cut the DNA into small fragments. The length of the fragments depends on the sequence of nitrogen bases in the DNA. The sequence of nitrogen bases differs from person to person.

A set of fingerprints?

The DNA fragments from both samples can be subjected to a technology that displays selected DNA fragments as bands. The markers on the bands from an individual can provide a DNA fingerprint. The probability that two people who are not identical twins would have the same set of marks is very small. The autoradiograph pictured below is the type of evidence that is presented to juries. Juries are so influenced by this kind of scientific data that legal experts insist that the evidence should be flawless before it is admitted.

Think Critically:

Is the probability of two persons having the same DNA small enough for a jury to convict a suspect? Explain your answer.

4-3 DNA   **107**

GLENCOE TECHNOLOGY

Videodisc
The Infinite Voyage: The Geometry of Life
Chapter 4
Master Genes and Their Influence

Chapter 5
Evolution, Molecular Genetics, and DNA Sequencing

Chapter 7
Selective Breeding

Community Connection

DNA Technology Invite a technician from a local law-enforcement laboratory to discuss DNA technology with the class. Have students prepare questions to ask the speaker.

Purpose

LS **Kinesthetic** Students will make a model of RNA to discover how it is different from DNA. **L1** **LEP**

Materials

colored construction paper, scissors

Teaching Strategies

A different color of construction paper will be needed for: deoxyribose, sugar, phosphate, adenine, quanine, cytosine, thymine, uracil, and ribose sugar.

Analysis

1. **DNA Base** **RNA Base**
   adenine   uracil
   guanine   cytosine
   thymine   adenine

2. **mRNA**   **tRNA**
   uracil   adenine
   guanine   cytosine

3. Uracil and ribose replace thymine and deoxyribose, respectively.

✓ Assessment

Oral Ask: **Why is DNA replication necessary for life?** *Without DNA replication, new cells would not receive the instructions for the proteins needed for life.* Use the Performance Task Assessment List for Making Observations and Inferences in **PASC,** p. 17. **P**

3 Assess

Check for Understanding

? FLEX Your Brain

Use the Flex Your Brain activity to have students explore DNA.

📁 **Activity Worksheets,** p. 5

MiniLAB

Comparing RNA and DNA

Procedure

1. Use construction paper and cut out shapes of sugars, phosphates, and bases to make a strand of DNA.
2. Separate the DNA strand. Use the left strand as the pattern for making a strand of RNA.
3. Find out what different parts you will need and cut those from the construction paper.
4. Make the RNA model.

Analysis

1. Look at the order of bases in the DNA strand. What is the order of bases in the RNA strand?
2. If the new strand of RNA that you have made were a piece of mRNA, what tRNA would link up with it in the cytoplasm?
3. Two new colors of construction paper will be needed to add RNA to the model. Why?

How does a cell know which proteins to make? The gene gives the directions for the order in which amino acids will be arranged. This order results in a particular protein.

RNA

In Chapter 2, you learned that proteins are made on ribosomes in cytoplasm. How does the code in the nucleus reach the ribosomes out in the cytoplasm? The codes for making proteins are carried from the nucleus to the ribosomes by a second type of nucleic acid, called ribonucleic acid or **RNA.** RNA is different from DNA in that it is made up of only one strand, and it contains a nitrogen base called uracil (U) in place of thymine and the sugar ribose. RNA is made in the nucleus on a DNA pattern.

Two different kinds of RNA are made from DNA in the nucleus—messenger RNA (mRNA) and transfer RNA (tRNA). Protein assembly begins as mRNA moves out of the nucleus and attaches to ribosomes in the cytoplasm. This process is illustrated in **Figure 4-15.** Pieces of tRNA pick up amino acids in the cytoplasm and bring them to the ribosomes. There, tRNA temporarily matches with mRNA and the amino acids become arranged according to the code carried by mRNA. The amino acids become bonded together, and a protein molecule begins to form.

Figure 4-15

RNA carries the code for a protein from the nucleus to the ribosome. There, its message is translated into a specific protein.

Across the Curriculum

History Have students study the history of DNA research and use poster board to draw and label a time line showing the events of DNA research since DNA was first removed from a cell nucleus in 1869. **P** **L1** **LS**

Mutations

Genes control the traits you inherit. Without correctly coded proteins, an organism can't grow, repair, or maintain itself. If a change occurs in a gene or chromosome, the traits of that organism are changed, as illustrated in **Figure 4-16.** Sometimes during replication, an error is made in copying a gene. Occasionally, a cell receives an entire extra chromosome. Outside factors such as X rays and chemicals have been known to change or break chromosomes. Any permanent change in a gene or chromosome of a cell is called a **mutation.** If the mutation occurs in a body cell, it may or may not be life threatening to the organism. If, however, a mutation occurs in a sex cell, then all the cells that are formed from that sex cell will have that mutation. Many mutations are harmful to organisms. Some mutations do not appear to have any effect on the organism. Many times, an organism with a mutation doesn't survive. But mutations also add variety to a species.

Figure 4-16

Albinism, resulting in the lack of pigments in skin, eyes, and hair, is an example of a mutation. *Is albinism generally a helpful or a harmful mutation? Explain.*

Section Wrap-up

Review

1. Which bases form pairs in a DNA molecule?

2. How does DNA make a copy of itself?

3. **Think Critically:** A single strand of DNA has the bases AGTAAC. Using letters, show what bases would match up to form a matching DNA strand from this pattern.

Skill Builder
Concept Mapping

Using a network tree concept map, show how DNA and RNA are alike and how they are different. If you need help, refer to Concept Mapping in the **Skill Handbook.**

USING MATH

There are four nitrogen bases that make up DNA. Groups of three bases code for certain amino acids. Each three-base combination is called a triplet. Determine the total number of possible triplet codes made from the four bases.

4-3 DNA   **109**

Skill Builder

Nucleic acids
- DNA
  - Double → contains → ACGT → found in → Nucleus
- RNA
  - number of strands
  - Single → contains → ACGU → found in → Cytoplasm and nucleus

✔ Assessment

Oral What are the two kinds of RNA, and how are they different? *messenger RNA and transfer RNA; mRNA copies a gene from DNA, and moves out of the nucleus to a ribosome, whereas tRNA picks up amino acids from the cytoplasm and brings them to the ribosomes so that proteins can be made.* Use the Performance Task Assessment List for Making Observations and Inferences in **PASC**, p. 17. **P**

Reteach

Have students make a drawing of DNA replication and protein synthesis.

Extension

📁 For students who have mastered this section, use the **Reinforcement** and **Enrichment** masters.

📁 Use these program resources as you teach this lesson.
Critical Thinking/Problem Solving, p. 8
Multicultural Connections, pp. 11-12

4 Close

•MINI•QUIZ•

Use the Mini Quiz to check students' recall of chapter content.

1. The chromosomes in the nucleus of a cell contain a chemical code called _____ . *DNA*

2. The part of a chromosome that directs the making of a specific protein is a(n) _____ . *gene*

3. Any permanent change in a gene or a chromosome is a(n) _____ . *mutation*

Section Wrap-up

Review

1. adenine and thymine, guanine and cytosine

2. The two strands of DNA separate and replicate identical strands. Each DNA molecule has one original strand and one new strand.

3. **Think Critically** T C A T T G

USING MATH

$4 \times 4 \times 4 = 64$

Purpose

LS **Interpersonal** Students will design and construct a model of DNA and demonstrate how DNA copies itself. **L1** **LEP** **COOP LEARN**

Process Skills

formulating a model, inferring, interpreting data, communicating, sequencing, forming a hypothesis

Time

one class period to brainstorm and make patterns, one class period to complete the activity

Materials

- Obtain construction paper in 6 different colors.
- Obtain a three-dimensional model of DNA from a biological supply company or use pictures of a DNA molecule such as the one on pages 104 and 105.
- Suggest that the groups use the same color for each of the molecules, for example: yellow for guanine, green for cytosine, purple for thymine, red for adenine, blue for sugar, and brown for phosphate.
- Students can model the sugars, phosphates, and bases from page 105 or pictures of other DNA molecules.
- Make sure that each group has 48 of each of the molecules that make up DNA. This will make a DNA molecule model that has 12 pairs that can be separated to copy itself.

Safety Precautions

Remind students to use care with scissors.

Activity 4-2

Design Your Own Experiment

Making a Model of DNA

How did James Watson and Francis Crick construct a three-dimensional model of DNA? Perhaps they brainstormed with pencils on paper napkins in restaurants. Or, maybe they tinkered in their laboratories with bits of wire, plastic, and string. In this lab, you will use colored construction paper to make a working model of DNA replication.

PREPARATION

Problem

What does a DNA molecule look like?

Thinking Critically

List the characteristics of the Watson and Crick model of the DNA molecule. How can these characteristics help you design and construct a DNA molecule model?

Objectives

- Design and construct a model of DNA.

- Demonstrate how the model copies itself.

Possible Materials

- 6 colors of construction paper (8½ × 11); 2 pieces of each color
- scissors
- heavy paper for patterns (1 sheet)
- tape

Safety Precautions

Use scissors carefully.

PLAN THE EXPERIMENT

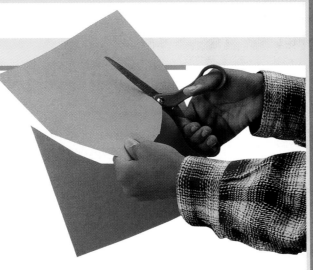

1. The six colors of construction paper correspond to the six different molecules that make up the DNA molecule.
2. As a group, decide what shapes you will use to represent each of the six molecules.
3. As a group, brainstorm a list of possible ways of representing the process of DNA replication.
4. Select the best idea. Then, as a group, list the steps you will follow. Be specific, describing exactly what you will do at each step.
5. Decide how you will identify the bonds that hold your molecule together and then demonstrate your model.

Check the Plan

1. Read over your entire experiment to make sure that all steps are in logical order.

2. Have you included all the parts of the DNA molecule in your model plan?
3. Does your model plan include the idea of new molecules being added to each half of the DNA molecule once it separates?
4. Will you have two molecules identical to the original when you have completed your modeling of DNA replication?
5. *Make sure your teacher approves your plan before you proceed.*

DO THE EXPERIMENT

1. Carry out the experiment as planned.
2. While the experiment is going on, write down any observations that you make in your Science Journal.

Analyze and Apply

1. **Compare** your models with those of other groups. Were

the molecules your group created the same as those of other groups?
2. Based on your observations of the DNA molecule model, **infer** why a DNA molecule seldom copies itself incorrectly.
3. **Explain** why models are useful to scientists.

Go Further

Use your DNA model to make a pattern for specific types of amino acids and proteins.

✔ Assessment

Performance Design an alphabet with four letters. How many words can you make using four letters? Use the Performance Task Assessment List for Model in **PASC**, p. 51. P

Possible Hypothesis

Most students will reflect that the DNA molecule has two sides connected with ladder-like pieces.

📁 **Activity Worksheets,** pp. 5, 25-26

PLAN THE EXPERIMENT

Possible Procedure

One possible procedure is to use the staircase model in the text.

Teaching Strategies

Tying to Previous Knowledge Most students know that the nucleus directs the activities of the cell, and the control center of the nucleus lies with the chromosomes. Chromosomes and their genes control the cells. Genes are actually under the control of the DNA code.

DO THE EXPERIMENT

Expected Outcome

Most models will be based on the Watson and Crick model in the text. P

Analyze and Apply

1. The model of each of the groups will probably be different from those of other groups.
2. Both A and T, and C and G pair, so DNA seldom copies itself incorrectly.
3. Models help to explain concepts visually.

Go Further

Have students find and use DNA code tables in a high school textbook to answer this problem.

Prepare

Section Background

- The original object of research on cloning was to find better ways to alter the genetic makeup of farm animals. The researchers hoped to discover ways to create herds of farm animals that produced more milk, more nutritious meat, or products such as prescription drugs or chemicals consumers might want.

- Dr. Wilmut was successful because he matched the phase of the cell cycle of the donor egg cell with the same phase in the mammary gland cell so that the cells were in synchrony. He accomplished this by starving the mammary gland cells so they would stop growing. This he felt was the key to making this experiment work.

- Dolly was the only lamb to survive from 277 eggs treated in the experiment. Other embryos grew after this treatment, but they did not survive the complete gestation period.

1 Motivate

Bellringer

Before presenting the lesson, display **Section Focus Transparency 16** on the overhead projector. Assign the accompanying **Focus Activity** worksheet. L1 LEP

2 Teach

Revealing Preconceptions

Students may think that clones are similar to identical

112

TECHNOLOGY:
4•4 Techniques in Cloning

Science Words

clone

Objectives

- Define a clone in terms of inheritance of DNA.
- Identify one benefit of cloning research.

How to Clone a Sheep

In February, 1997, Ian Wilmut, a Scottish scientist, stunned the world when he announced that his team of researchers had successfully cloned an adult Finn Dorset sheep. The researchers obtained an egg cell from one adult female sheep, took out its nucleus, and replaced that nucleus with the nucleus from a mammary gland cell of a different adult female sheep. The egg cell then divided and formed an embryo. Five months later, a newborn sheep named Dolly entered the world—the first clone of an existing adult female sheep.

What is a clone?

To scientists, a **clone** is an individual that is genetically identical to one of its parents. For many years, scientists have been able to produce genetically identical organisms in the laboratory by growing fertilized eggs to the 2-, 4-, or 8-cell embryo stage, then separating the cells. These cells then behave as if they were the original fertilized egg cell and

Figure 4-17

Cloning of an animal such as a sheep involves a series of steps.

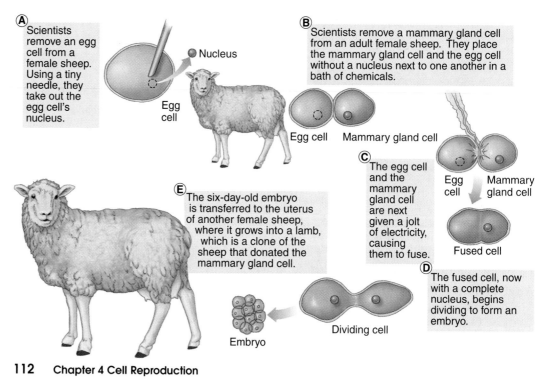

A Scientists remove an egg cell from a female sheep. Using a tiny needle, they take out the egg cell's nucleus.

B Scientists remove a mammary gland cell from an adult female sheep. They place the mammary gland cell and the egg cell without a nucleus next to one another in a bath of chemicals.

C The egg cell and the mammary gland cell are next given a jolt of electricity, causing them to fuse.

D The fused cell, now with a complete nucleus, begins dividing to form an embryo.

E The six-day-old embryo is transferred to the uterus of another female sheep, where it grows into a lamb, which is a clone of the sheep that donated the mammary gland cell.

Nucleus

Egg cell

Egg cell   Mammary gland cell

Egg cell   Mammary gland cell

Fused cell

Dividing cell

Embryo

112   Chapter 4 Cell Reproduction

---

twins. Although identical twins have the same DNA, they are not clones because a clone is defined as an individual who has the exact same DNA as its parent. Identical twins do not have the same DNA as either parent.

Content Background

Some researchers are exploring the potential of using the DNA molecule as the basis for future computers. Historians are exploring ways to use cloning to trace the migrations of humans from Africa to other parts of the world.

Program Resources

Reproducible Masters
Study Guide, p. 20 L1
Reinforcement, p. 20 L1
Enrichment, p. 20 L3

Transparencies
Section Focus Transparency 16 L1

continue growing into genetically identical siblings. However, these identical siblings are not clones. They grew from a fertilized egg cell that received half of its DNA from the mother and half from the father. A clone receives all of its DNA from just one parent. This process is illustrated in **Figure 4-17**.

Dr. Wilmut and Dolly

The Importance of Dolly

Why is the birth of Dolly important? Dolly amazed the scientific world because her growth was directed by the nucleus of a differentiated cell. A differentiated cell is a cell that is specialized to perform a specific function in the body. You know that you have skin cells, nerve cells, muscle cells, and so on. All of these differentiated cells have the exact same DNA as the undifferentiated fertilized egg that grew to become you. But at some point in your development, certain genes were turned on or off to instruct cells to become specialized skin, nerve, or muscle cells. How this occurs is not fully understood. Before Dolly, scientists generally accepted that once a cell became differentiated, it could not become any other kind of cell.

If scientists can figure out how a cell's DNA turns genes on or off to produce specific kinds of cells, they may eventually be able to use the nucleus of any body cell to grow new skin, nerve, or muscle cells—or any type of cell you need. Because these new cells will be genetically identical to your other cells, your body will not reject them.

Section Wrap-up

Review

1. Explain what a clone is.

2. How could information about how a clone is produced benefit people who have bone cancer?

 Visit the Chapter 4 Internet Connection at Glencoe Online Science, **www.glencoe.com/sec/science/life,** for a link to more information about cloning.

SCIENCE & SOCIETY

3 Assess

Check for Understanding

Ask students what they think the value of being able to clone an existing organism is. Ask them to identify a reason that cloning might be important in their own lives.

Reteach

Have students examine Figure 4-17 and explain in their own words the steps in the experiment that led to Dolly's birth.

Extension

For students who have mastered this section, use the **Reinforcement** and **Enrichment** masters.

4 Close

Ask students to summarize what they have learned about cloning. Have them brainstorm a list of the possible uses for cloning. Suggest that interested students visit a plant nursery or garden center and talk to a horticulturist about plant cloning.

Section Wrap-up

Review

1. A clone is an individual that is genetically identical to one of its parents.

2. Scientists could take healthy body cells from elsewhere on a patient with bone cancer, insert their nuclei into healthy egg cells, then grow new, healthy bone cells that can be used to replace the cancerous ones.

People and Science

Background

- As radiation passes through cells, it acts like a subatomic bullet that can knock electrons out of their proper orbits around atomic nuclei. When that happens, the chemical identity of the atoms, and that of the molecules those atoms make up, is changed dramatically. Cells that have been exposed to high-energy radiation cannot function or reproduce normally, and invariably die.

- In 1895, a German physicist, Wilhelm Conrad Röntgen (1845-1923), first observed a powerful and penetrating form of radiation, which he named "X rays." In 1995, in honor of the 100-year anniversary of that discovery, the American College of Radiology and the Radiologic Society of North America buried a time capsule in which they included materials and information about the most outstanding advances in the field of radiology during that period. The pioneering work of Dr. Goldson and his colleagues in developing intraoperative radiation therapy was included in the capsule.

Teaching Strategies

Linguistic Early development is a time of rapid cell division. Have students research articles about the effects of radiation on early development. Use the information acquired as the basis for a class discussion about the hazards of radiation exposure during pregnancy.

Career Connection

Career Path Radiologists must complete advanced medical training to become specialists.

People and Science

DR. ALFRED L. GOLDSON, *Radiation Oncologist*

On the Job

Q Dr. Goldson, tell us about the area of medicine in which you specialize.

A I'm a radiation oncologist, a cancer specialist. Radiation oncologists use radiation to treat cancer patients.

Q How does radiation kill cancer cells and destroy tumors?

A Radiation can damage any kind of cell. Cells are most vulnerable to damage from radiation when they are dividing—when the DNA is replicating and the cells are pulling apart into new daughter cells. The thing about cancer cells is that they usually divide faster and more often than normal, healthy cells—that's why you get a mass of cancer cells, a tumor—and so at any given time, more cancer cells are going through division than the normal cells around them are.

Q Describe your research on cancer and radiation.

A Right now, I'm involved in research on the early detection of breast and

Personal Insights

prostate cancer. One of the things I worked on in the past was to develop a technique called "intraoperative radiation therapy," in which a massive dose of radiation is administered directly to an inoperable cancer—one that can't be removed surgically—while the patient is right there on the operating table.

Q How do you feel about your accomplishments in cancer research?

A Here in the department of Radiation Oncology at Howard University, we've done a lot of things since the 1970s that have led to major technical advancements in treating cancer. We like to call ourselves "The Mouse that Roared"—a small department, but big on ideas!

Career Connection

While radiation oncologists use X rays to treat cancer, diagnostic radiologists are medical doctors who read X rays to diagnose injuries and disease. Contact the Department of Radiology at a local hospital and find out about:

- **X-ray technologists**
- **Radiation Therapists**

Make a brochure to interest your classmates in one of these health-care careers.

- High School: math and science
- College: (Pre-med) biology, chemistry, physics
- Four years of medical school, followed by a three-year residency in radiology.

Other health care professions in this area include X-ray technology (two-year certification program) and radiation therapy (three-year certification program).

For More Information

Students could contact the American Cancer Society for more information about cancer, its treatment, and careers in medicine and cancer research.

Summary

4-1: Cell Growth and Division

1. Cells divide to form new cells by mitosis and divisions of cytoplasm. There are four stages of mitosis.
2. Animal cells have centrioles. New animal cells form when the cytoplasm pinches in. Plant cells lack centrioles; their cytoplasm is separated by a cell plate following mitosis.
3. Asexual reproduction requires only one parent. Sexual reproduction requires two parents to produce new individuals.

4-2: Sexual Reproduction and Meiosis

1. Sexual reproduction involves the production of sex cells through the process of meiosis.
2. Sex cells from two parents fuse during fertilization to produce a zygote, a cell with the diploid chromosome number.
3. Meiosis is a division of nuclei in reproductive cells that results in sex cells, four cells with the haploid chromosome number.

4-3: DNA

1. DNA is made up of bases that appear in specific sequences, and it directs all activities of a cell. DNA can copy itself and is the framework on which RNA is made.
2. RNA assembles proteins in the cytoplasm based on instructions from genes in the nucleus.

3. Mutations are changes in an organism's DNA. Some are harmful, some have no effect, and some are beneficial.

4-4: Science and Society: Techniques in Cloning

1. Scientists used a differentiated cell from the mammary gland of an adult female sheep, fused with an egg cell without a nucleus, to create a clone of the sheep that contributed its mammary gland cell.
2. A clone is an individual that is genetically identical to one of its parents.

Key Science Words

a. asexual reproduction
b. chromosome
c. clone
d. DNA
e. egg
f. fertilization
g. gene
h. meiosis
i. mitosis
j. mutation
k. RNA
l. sexual reproduction
m. sperm
n. zygote

Reviewing Vocabulary

Match each phrase with the correct term from the list of Key Science Words.

1. nuclear structure containing DNA
2. new individual that is identical to a parent organism

Summary

Have students read the summary statements to review the major concepts of the chapter.

Reviewing Vocabulary

1. b	**6.** f
2. c	**7.** h
3. g	**8.** j
4. d	**9.** n
5. i	**10.** k

Assessment

Portfolio Encourage students to place in their portfolios one or two items of what they consider to be their best work. Examples include:

- MiniLAB mitosis models, p. 97
- Activity 4-2 model of DNA, p. 110
- Across the Curriculum, p. 108 P

Performance Additional performance assessments may be found in **Performance Assessment** and **Science Integration Activities.** Performance Task Assessment Lists and rubrics for evaluating these activities can be found in Glencoe's **Performance Assessment in the Science Classroom.**

GLENCOE TECHNOLOGY

Videodisc

Glencoe Life Science Interactive Videodisc
Side 1, Lesson 1
Mitosis and Meiosis: The Cycle of Life

11 to 14443

MindJogger Videoquiz

Chapter 4 Have students work in groups as they play the Videoquiz game to review key chapter concepts.

Chapter 4 Review

Checking Concepts

1. d	**6.** a
2. d	**7.** d
3. b	**8.** d
4. c	**9.** c
5. a	**10.** b

Understanding Concepts

11. A gene is a segment of DNA on a chromosome. Chromosomes are structures in the nucleus that contain DNA.

12. Offspring that are a result of asexual reproduction are identical genetically to the parent; offspring that are the result of sexual reproduction get half of their genetic material from each parent.

13. Spindle fibers pull chromosome strands to opposite poles.

14. Mitosis takes place in body cells; meiosis takes place in sex cells.

15. We notice the effects of harmful mutations more often, making us more aware of them.

Thinking Critically

16. A T C C G T C is matched by T A G G C A G.

17. U A G G C A G

18. No. In order for a mutation to be passed to offspring, the mutation must take place in a sex cell.

19. the copying of chromosomes in interphase; the separation of the copies at anaphase; the separation of two new cells at telophase

20. During meiosis, if one of the resulting eggs or sperm has both chromosome strands due to failure of the strands to separate, a zygote formed with that egg or sperm will have an extra chromosome.

116

---

3. segment of DNA that controls traits
4. deoxyribonucleic acid
5. formation of two new nuclei with identical chromosomes
6. fusion of an egg and a sperm
7. nuclear division that forms sex cells
8. a permanent change in a cell's DNA
9. cell that forms as a result of fertilization
10. nucleic acid that carries information for making proteins from the nucleus

Checking Concepts

Choose the word or phrase that completes the sentence.

1. _____ is a double spiral molecule with pairs of nitrogen bases.
a. RNA c. A protein
b. An amino acid d. DNA

2. RNA differs from DNA in that it contains _____.
a. thymine c. adenine
b. thyroid d. uracil

3. If a diploid tomato cell with 24 chromosomes undergoes meiosis, the sex cells produced will each have _____ chromosomes.
a. 6 b. 12 c. 24 d. 48

4. Chromosomes are doubled during _____.
a. anaphase c. interphase
b. metaphase d. telophase

5. During _____ in mitosis, double-stranded chromosomes separate.
a. anaphase c. metaphase
b. prophase d. telophase

6. The chromosome number in cells after mitosis is _____ the parent cell number.
a. the same as c. twice
b. half d. four times

116 Chapter 4 Cell Reproduction

---

7. Budding, fission, and regeneration are forms of _____.
a. mutations c. sexual reproduction
b. cell cycles d. asexual reproduction

8. _____ is any permanent change in a gene or a chromosome.
a. Fission c. Replication
b. Reproduction d. Mutation

9. Meiosis produces _____.
a. cells with the diploid chromosome number
b. cells with identical chromosomes
c. sex cells
d. a zygote

10. In the cell cycle, most of the life of any cell is spent in _____.
a. metaphase c. anaphase
b. interphase d. telophase

Understanding Concepts

Answer the following questions in your Science Journal using complete sentences.

11. What is the difference between genes and chromosomes?

12. What are the differences between offspring formed by sexual reproduction and asexual reproduction?

13. What do spindle fibers do during mitosis and meiosis?

14. In what cells do mitosis and meiosis take place?

15. Why is more known about harmful mutations than beneficial ones?

Developing Skills

21. Comparing and Contrasting

	Mitosis	Meiosis
Number of divisions	1	2
Numbers of cells produced	2	4
Numbers of chromosomes in parent cells	diploid	diploid
Numbers of chromosomes in resulting cells	diploid	haploid sex cells

22. Hypothesizing Incorrect division can result in an incorrect number of chromosomes.

Thinking Critically

16. If one strand of DNA had bases ordered ATCCGTC, what would be the bases of its other strand?
17. A strand of RNA matching the DNA strand ATCCGTC would have what base sequence?
18. A mutation takes place in a human skin cell. Will this mutation be passed on to the person's offspring? Explain your answer.
19. What processes in mitosis provide both new cells with identical DNA?
20. How could a zygote end up with an extra chromosome?

Developing Skills

If you need help, refer to the **Skill Handbook.**

21. **Comparing and Contrasting:** In a table, compare and contrast mitosis and meiosis as to: number of divisions, number of cells produced, number of chromosomes in parent cells and in sex cells.
22. **Hypothesizing:** Make a hypothesis about the effect of an incorrect mitotic division on the new cells produced.
23. **Making and Using Tables:** Compare and contrast DNA and RNA in a table.

24. **Concept Mapping:** Complete the events chain concept map of DNA synthesis.

DNA unwinds → New bases fill in → Two new strands of DNA form

25. **Sequencing:** Sequence the events that occur in sexual reproduction. Start with interphase in the parent cell and end with the formation of the zygote. Tell whether the number of chromosomes present at each stage is haploid or diploid.

Performance Assessment

1. **Events Chain:** Make a time line of the events involved in exploring DNA since it was first removed from a cell nucleus in 1869.
2. **Writing in Science:** Write a one-page paper about growth hormones called gibberellins and their use.
3. **Model:** Using the parts of the DNA model from Activity 4-2 and information in the MiniLAB on page 108 about RNA, make a model showing how a strand of RNA is made from DNA.

DNA and RNA

Nucleic acid	DNA	RNA
Number of strands	2	1
Type of sugar	Deoxyribose	Ribose
Letter names of bases	ACTG	ACGU
Where found?	Nucleus	Nucleus and cytoplasm

Chapter 4 Review

23. **Making and Using Tables**

Nucleic acid	DNA	RNA
Number of strands	2	1
Type of sugar	Deoxyribose	Ribose
Letter names of bases	ACTG	ACUG
Where found?	Nucleus	Nucleus and cytoplasm

24. **Concept Mapping** See student text.
25. **Sequencing** Order of events given for meiosis should reflect information on pages 102 and 103 and zygote information on page 100. Cell at the beginning of meiosis is diploid. Cells at end of meiosis are all haploid.

Performance Assessment

1. Students should research DNA and make a time line starting with its removal from a cell in 1869 and showing each event that has occurred since that time. Use the Performance Task Assessment List for Events Chain in **PASC,** p. 91. **P**
2. The report should include what happens to cells when growth hormones are used. Use the Performance Task Assessment List for Writing in Science in **PASC,** p. 87. **P**
3. Student models will be complex but should show mRNA and tRNA. Use the Performance Task Assessment List for Model in **PASC,** p. 51. **P**

Assessment Resources

 Reproducible Masters
Chapter Review, pp. 11-12
Assessment, pp. 17-20
Performance Assessment, p. 18

Glencoe Technology
- **Chapter Review Software**
- **Computer Test Bank**
- **MindJogger Videoquiz**

UNIT PROJECT 1

In Search of Cells

Even though cells make up ourselves and every living thing around us, we have not always known that cells are there. Some organisms, like protozoa, consist of only one cell, while other organisms, like people, consist of trillions of cells. While some cells, such as egg yolks, are large enough to be seen directly, most cells cannot be seen without using special equipment. Centuries ago, scientists speculated about the nature of life. They weren't sure what organisms were made up of. Scientists were like detectives, trying to reconstruct a crime based only on clues. As technologies such as the microscope developed, scientists finally were able to observe cells. Cell research has sometimes involved reconstructing the nature of cells by looking at the activities and materials that cells have produced, rather than looking at cells themselves. As technology has improved, researchers have been able to observe more and more details about cells. Technology has taken us from the limitations of the hand lens to being able to see atoms.

Cell Detectives

Many different scientists have contributed to our present knowledge of the cell. Explore these scientists' lives and research to see how they contributed to our current understanding of cells. A list of some cell researchers is given below. Working in groups, choose one individual or a pair of individuals to research.

Eduard Buchner

Jean-Baptiste Carnoy

Albert Claude

Jewel Plummer Cobb

Rosalind Franklin

Camillo Golgi

Ross Harrison and Alexis Carrel

Sir Hans Krebs

Rita Levi-Montalcini

Jacob Monod

Mikhail Sememovich Tswett

Rudolf Virchow

James Watson and Francis Crick

August Weismann

Objectives

LS **Linguistic** Students will research the cumulative nature of theory building and the feedback between knowledge and technology through a study of the development of the cell theory and the instruments that assisted in its evolution. **L1** **COOP LEARN**

Summary

In this project, students will research scientists who have contributed to our knowledge of the cell. Students will put together social, technological, and scientific factors that influence contributions to scientific knowledge. They will share this information in a creative manner and attempt to look at the influence that the changing state of technology has on the evolution or development of scientific knowledge about a particular subject.

Time Required

This project is designed to be worked on throughout Unit 1. One period is required to set up research groups and choose topics. At the end, one period is required to present scenarios. The rest of the work may be interspersed with other lessons over a period of two to three weeks.

Preparation

- Establish cooperative groups of 4 or 5 students.
- Be aware of sources on various scientists involved in cell research.
- Oversee materials for costumes and props.

Teaching Strategies

- Before beginning the cell research, show the class a folder. Tell them that you have the picture of an animal in the folder, and have them guess what animal it is. Use an unusual animal and give them some time to guess. They may not guess it at all. Now try the same thing, but give each group a different clue to the animal's identity. Have each group present its clue and then allow time to guess. If you have four or five groups, each with a different clue, students should be able to identify the animal this time. Discuss how particular clues influenced students' guesses. Compare this to the way scientific knowledge builds on previous ideas, modifying them as more information becomes available.
- Help students to recognize characteristics that are helpful in this approach to knowledge — open-mindedness, creativity, diverse thinking, methodical approach, and so on.

118

Procedure

For each scientist, research the following information:

1. What was already known about the cell by the time your researcher was alive?

2. What kind of equipment did he or she have to work with?

3. What were the limitations of the equipment available?

4. What was taking place socially or politically at the time your researcher was living? How did he or she dress? How did scientists communicate with each other at that time?

5. What contribution did your researcher make to the knowledge of cells?

6. How did this person arrive at his or her contribution?

Using Your Research

After you have completed your research, write a scenario using the people in your group. This scenario should answer the research questions listed above. Design costumes that reflect the lifestyle of the scientists researched. The period of time, social class, and geographic location should all influence the costumes you design. Props should include models of the equipment available at that particular time. Each group can present its scenario in chronological order. As a class, discuss how the scenarios show a history of cell research, and how they show a history of technological development as well.

Go Further

Knowledge of the cell is far from complete. More and more as technology improves, new questions are asked. Today, a great deal of cell research centers on looking for cures for various diseases that affect human life.

Questions, Questions

What makes cells become cancerous? How can HIV be stopped? Will gene therapy be effective?

Investigate the current research on muscular dystrophy, AIDS, cancer, or other diseases. How does this research make use of cell theory? What contributions might this research make to a better understanding of cells and how they work? Will you be the person in the next scenario?

Unit 1 Project **119**

- Help students determine which scientist(s) each group will select to investigate.

- Remind students that they will end their study of the cell with a chronological presentation, in period costume, with period instruments, and an explanation of how much they could know based on instruments of that period.

- Discuss what other instruments, besides microscopes, make modern knowledge of the cell possible (computers, cameras).

Go Further

Technology is now able to provide artificial tissues, organs, and limbs. Students may research these synthetic replacements to determine their limitations and uses. Students may wish to write a science fiction story about a time in the future when any human body part can be replaced with artificial parts. Would this be a good idea or not? What effect would this have on human lifestyle and life span? Would we have to rethink our definitions of robots and humans?

References

Brown, Walter V. and Eldridge M. Berthe. *Textbook of Cytology*. St. Louis: C.V. Mosby Co., 1974.

Curtis, Helena and N. Sue Barnes. *Biology*. New York: Worth Publishing, Inc., 1989.

de Duve, Christian. A *Guided Tour of the Living Cell*, Vols. 1 and 2. New York: Scientific American Library, W.H. Freeman and Co., 1984.

Watson, James, Bruce Alberts, Dennis Bray, Julian Lewis, Martin Ruff, and Ruth Roberts. *Molecular Biology of the Cell*, 2nd Ed. New York: Garland Publishing Co., 1989.

UNIT 2

Heredity and Evolution

In Unit 2, students will discover patterns of heredity and how inherited changes can influence evolutionary trends. Both simple Mendelian hereditary patterns and more complex patterns of multiple alleles and polygenic inheritance are explained. Students will investigate evidence for evolution, such as fossils, embryology, and DNA relationships. Darwin's theory of evolution by natural selection is presented as a primary mechanism of evolutionary change. The unit concludes with a discussion of how organisms are classified to reflect their evolutionary history.

CONTENTS

Chapter 5
Heredity   122

5-1   What is genetics?

5-2   Genetics Since Mendel

5-3   Human Genetics

5-4   Science and Society:
      The Human Genome
      Project

Chapter 6
Evolution   150

6-1   Mechanisms of
      Evolution

6-2   Evidence for Evolution

6-3   Primate Evolution

6-4   Science and Society:
      Extinction and
      Evolution

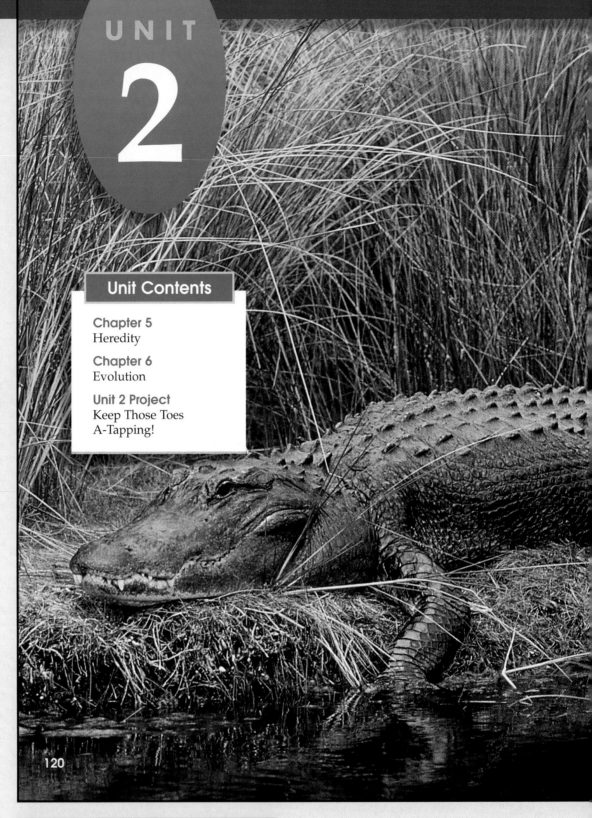

UNIT 2

Unit Contents

Chapter 5
Heredity

Chapter 6
Evolution

Unit 2 Project
Keep Those Toes
A-Tapping!

120

Science at Home

Patterns of Heredity Have each student select 100 pinto beans at random from a bag purchased at a grocery store. Count and keep a record of the number of dark marks making up the pattern on each bean. Have students determine the fewest number of dark marks, the largest number of dark marks, and the average number of dark marks per bean. **L1** **LEP**
COOP LEARN

Heredity and Evolution

What's Happening Here?

What do cockroaches, horse-shoe crabs, and crocodiles have in common? None of them has changed much since its first appearance on Earth. The American alligator sunning itself on the riverbank could be considered a living fossil because it still resembles crocodilian ancestors that lived 200 million years ago. Present-day crocodilians include alligators, caimans, and gavials. Are crocodilians so successful due to luck? Or is there some evolutionary principle at work? You'll learn about heredity and about the principles of evolution in this unit.

Science Journal

All organisms inherit certain characteristics from their parents. Baby crocodiles have sharp teeth and scaly skin, whereas rose plants have colorful flowers and thorny stems. In your Science Journal, describe characteristics of three organisms that you think were inherited from their parents.

121

INTRODUCING THE UNIT

What's Happening Here?

Ask students to make a hypothesis about how the organisms in the photo may be related. Tell them that alligators have remained almost unchanged for millions of years. Explain that in this unit, they will learn that similar characteristics among different organisms lead scientists to conclude that some fossil organisms are evolutionary ancestors of species alive today. Students will also learn how classification reflects these relationships.

Background

Alligator-like fossils 190 million years old indicate that the ancestors of today's alligators were nearly identical to the present-day species. Fossils almost identical to present-day species of cockroaches, horseshoe crabs, and coelacanths are also found in the fossil record. Although these organisms are similar to their ancestors, it does not mean that related species necessarily look alike. Many species look very different from their closest relatives. Remind students that there are reasons for these similarities and differences, and that they will explore these reasons in this unit.

Previewing the Chapters

Ask students to identify which photographs and illustrations in the unit show how similarities between organisms are transmitted from one generation to another and which show similarities between species over time.

Tying to Previous Knowledge

Stress the importance of the process of meiosis in determining genetic makeup and evolution.

Theme Connection

Stability and Change Display different kinds of fossils and encourage students to find similarities and differences between fossil organisms and living organisms. Have them explain how these similarities and differences illustrate relationships among organisms. *Students may hypothesize that similarities indicate that fossil species are ancestors of the living species.*

Cultural Diversity

Color Determination Some cultures have many terms for color. The English language has 11 basic color terms. Some cultures, for example New Guinea, have only two color terms—black and white. Cultures that recognize three color groups include black, white, and red in their classifications. Have students research how color is perceived by other cultures. **L2**

Chapter Organizer

Section	Objectives/Standards	Activities/Features
Chapter Opener		**Explore Activity:** Observe differences in human hairlines. p. 123
5-1 **What is genetics?** (1 session, ½ block)*	1. **Explain** how traits are inherited and explain Mendel's role in the history of genetics. 2. **Use** a Punnett square to predict the results of crosses. 3. **Explain** the difference between genotype and phenotype. **National Content Standards:** (5-8) UCP2, UCP3, A1, A2, C2, G1, G2, G3	**MiniLAB:** What are some common traits? p. 125 **Using Math,** p. 127 **Science Journal,** p. 128 **Skill Builder:** Observing and Inferring, p. 130 **Using Math,** p. 130 **Activity 5-1:** Expected and Observed Results, p. 131
5-2 **Genetics Since Mendel** (3 sessions, 2 blocks)*	1. **Explain** incomplete dominance. 2. **Compare** multiple alleles and polygenic inheritance, and give **examples** of both. **National Content Standards:** (5-8) UCP1, UCP2, A1, A2, C2, C4	**MiniLAB:** What are the distinguishing features of fingerprints? p. 133 **Activity 5-2:** Determining Polygenic Inheritance, pp. 134-135 **Skill Builder:** Comparing and Contrasting, p. 136 **Science Journal,** p. 136
5-3 **Human Genetics** (2 sessions, 1½ blocks)*	1. **Describe** two human genetic disorders. 2. **Explain** inheritance of sex-linked traits. 3. **Explain** the importance of genetic engineering. **National Content Standards:** (5-8) UCP1, UCP2, A1, C1, C2, C5, E2, F5	**Using Math,** p. 138 **Problem Solving:** Boy or Girl? p. 139 **Using Technology:** Karyotyping, p. 141 **Connect to Chemistry,** p. 142 **Skill Builder:** Concept Mapping, p. 143 **Science Journal,** p. 143
5-4 **Science and Society** **Technology: The** **Human Genome** **Project** (1 session, ½ block)*	1. **Describe** the goals of the Human Genome Project. 2. **Identify** some of the human diseases that are inherited. **National Content Standards:** (5-8) UCP1, E2, F5, G1	**Explore the Technology,** p. 145 **People and Science:** Sean B. Carroll, Geneticist, p. 146

* A complete Planning Guide that includes block scheduling is provided on pages 32T-35T.

Activity Materials

Explore	Activities	MiniLABs
page 123 No special materials are needed.	page 131 2 paper lunch bags, 100 red beans, 100 white beans pages 134-135 metric ruler, graph paper	page 133 washable ink pad, sheet of plain white paper, hand lens

Need Materials? Call Science Kit (1-800-828-7777).

Teacher Classroom Resources

Reproducible Masters	Transparencies	Teaching Resources
Study Guide, p. 21 **Reinforcement**, p. 21 **Enrichment**, p. 21 **Activity Worksheets**, pp. 29-30, 33 **Lab Manual 10**, Genetic Traits **Lab Manual 11**, 50:50 Chances **Concept Mapping**, p. 15	**Section Focus Transparency 17,** You Are What You Inherit **Teaching Transparency 9,** Genetics **Teaching Transparency 10,** Incomplete Dominance	**Glencoe Life Science Interactive Videodisc** **Life Science CD-ROM** **Spanish Resources** **English/Spanish Audiocassettes** **Cooperative Learning Resource Guide** **Lab Partner** **Lab and Safety Skills** **Lesson Plans**
Study Guide, p. 22 **Reinforcement**, p. 22 **Enrichment**, p. 22 **Activity Worksheets**, pp. 31-32, 34	**Section Focus Transparency 18,** Mixed-Up Genes	**Assessment Resources**
Study Guide, p. 23 **Reinforcement**, p. 23 **Enrichment**, p. 23 **Lab Manual 12**, Genes and Blood Clotting **Critical Thinking/Problem Solving**, p. 9 **Cross-Curricular Integration**, p. 9 **Science and Society Integration**, p. 9 **Multicultural Connections**, p. 13 **Technology Integration**, pp. 7-8 **Science Integration Activity 5**, The Law of Probability	**Section Focus Transparency 19,** Gene Game **Science Integration Transparency 5,** Testing Human Genes	**Chapter Review**, pp. 13-14 **Assessment**, pp. 32-35 **Performance Assessment in the Science Classroom (PASC)** **MindJogger Videoquiz** **Alternate Assessment in the Science Classroom** **Performance Assessment** **Chapter Review Software** **Computer Test Bank**
Study Guide, p. 24 **Reinforcement**, p. 24 **Enrichment**, p. 24	**Section Focus Transparency 20,** Mapping You	

Key to Teaching Strategies

The following designations will help you decide which activities are appropriate for your students.

L1 Level 1 activities should be within the ability range of all students, including those with learning difficulties.

L2 Level 2 activities should be within the ability range of the average to above-average student.

L3 Level 3 activities are designed for the ability range of above-average students.

LEP LEP activities should be within the ability range of Limited English Proficiency students.

LS These activities are designed to address different learning styles.

COOP LEARN Cooperative Learning activities are designed for small group work.

P These strategies represent student products that can be placed into a best-work portfolio.

GLENCOE TECHNOLOGY

The following multimedia resources are available from Glencoe.

The Secret of Life
Sex and the Single Gene: Cell Development
Tinkering with Our Genes: Genetic Medicine
Science and Technology Videodisc Series (STVS)
Plants & Simple Organisms
Microinjecting Polygenes

The Infinite Voyage Series
The Geometry of Life
National Geographic Society Series
STV: The Cell
STV: Human Body Vol. 3
Life Science CD-ROM

Teacher Classroom Resources

This is a representation of key blackline masters available in the Teacher Classroom Resources.

Teaching Aids

Section Focus Transparencies

Science Integration Transparencies

Teaching Transparencies

Meeting Different Ability Levels

Study Guide

Reinforcement

Enrichment Worksheets

Hands-On Activities

Science Integration Activity

The Law of Probability

L1

Lab Manual

Genetic Traits 10

L2

Assessment

Performance Assessment

The Case of the Wrinkled Peas

L3

Enrichment and Application

Critical Thinking/ Problem Solving

Heredity

Gene Therapy: Hope for Incurable Diseases

L2

Cross-Curricular Integration

Heredity

Eye Dominance and Heredity

L1

Science and Society Integration

Heredity

Science in the Designer Genes Business

L2

Technology Integration

Heredity and Evolution

Fixing Genetic Material

L1

Multicultural Connections

The Search for Super Rice

L2

Concept Mapping

Heredity

L3

Chapter 5

Heredity

CHAPTER OVERVIEW

Section 5-1 Genetics is defined within the context of the students' inheritance. Mendel's experiments and results are detailed.

Section 5-2 Additional genetic concepts developed since Mendel's original work are introduced. Incomplete dominance is described. Multiple alleles are contrasted with polygenic inheritance.

Section 5-3 Human genetics is explained in terms of Mendel's results. Some human genetic disorders and their causes are presented. Sex determination and sex-linked disorders are described.

Section 5-4 Science and Society The chapter ends with an overview of the goals of the human genome initiative.

Chapter Vocabulary

heredity	incomplete
allele	dominance
genetics	multiple alleles
dominant	polygenic
recessive	inheritance
Punnett	sex-linked
square	gene
genotype	pedigree
homozygous	genetic
heterozygous	engineering
phenotype	genome

Theme Connection

Stability and Change The theme of stability and change is further developed in this chapter. Homeostasis at the organism level is controlled by genes. This chapter introduces the basics of genetics and provides background for understanding evolution.

Previewing the Chapter

Section 5-1   What Is Genetics?
- ► What have you inherited?
- ► The Father of Genetics
- ► In Mendel's Garden
- ► Using a Punnett Square

Section 5-2   Genetics Since Mendel
- ► Incomplete Dominance
- ► Multiple Alleles
- ► Multiple Genes

Section 5-3   Human Genetics
- ► Genes and Health
- ► Sex Determination
- ► Sex-linked Disorders
- ► Why is genetics important?
- ► New Plants from Old

Section 5-4   Science and Society
Technology: The Human Genome Project

122

 Learning Styles

Look for the following logo for strategies that emphasize different learning modalities.

Visual-Spatial	Visual Learning, pp. 125-127, 129, 132, 133, 138-140, 144; People and Science, p. 146
Interpersonal	Explore, p. 123; Activity 5-2, pp. 134-135
Logical-Mathematical	MiniLAB, pp. 125, 133; Demonstration, p. 127; Enrichment, p. 128; Activity 5-1, p. 131; Using Technology, p. 141
Linguistic	Science Journal, p. 133

LS

Chapter

5

Heredity

Have you ever noticed how much organisms of the same species resemble each other? The zebras in the photo to the left have so many traits in common that they look almost identical. If you look closely, however, you will find small differences among individuals. Most traits found in organisms have at least two different forms, such as a straight hairline or a hairline with a widow's peak in humans. In the following activity, observe this genetic difference. Then in this chapter learn how you and your classmates received the genetic traits you have.

EXPLORE ACTIVITY

Observe differences in human hairlines.
1. Notice the two different kinds of hairlines in the small photo.
2. Record the name and kind of hairline of each of your classmates.

Observe: Do most of your classmates have the same kind of hairline as you? Record the percentage of each type of hairline in your Science Journal.

Previewing Science Skills

▶ In the Skill Builders, you will observe and infer, compare and contrast, and map concepts.

▶ In the Activities, you will predict, hypothesize, and collect and analyze data.

▶ In the MiniLABs, you will observe, and collect and analyze data.

123

Assessment Planner

Portfolio
Refer to page 147 for suggested items that students might select for their portfolios.

Performance Assessment
See page 147 for additional Performance Assessment options.
Skill Builders, pp. 130, 136, 143
MiniLABs, pp. 125, 133
Activities 5-1, p. 131; 5-2, pp. 134-135

Content Assessment
Section Wrap-ups, pp. 130, 136, 143, 145
Chapter Review, pp. 147-149
Mini Quizzes, pp. 136, 143

Group Assessment
Opportunities for group assessment occur with Cooperative Learning Strategies and Flex Your Brain Activities.

EXPLORE ACTIVITY

Purpose
LS **Interpersonal** Use the Explore Activity to introduce students to inheritance. Inform students that they will be learning more about inheritance and genetics as they read the chapter. **L1**

Preparation
Make a table on the chalkboard or on an overhead transparency to complete for your classes. Continue to add data to the table throughout the day.

Teaching Strategies
• Before the activity begins, you should clearly explain the difference between hairlines.
• Record data on the chalkboard for each class so that students can compare their results with the whole.

Observe
Answers will depend on the students' own hairlines. Most will have straight hairlines. Percentage of straight hairlines will likely approximate 60-90 percent.

✓ Assessment

Oral Have students locate and name a teacher in the building with a straight hairline and one with a widow's peak. Use the Performance Task Assessment List for Making Observations and Inferences in **PASC,** p. 17. **P**

Prepare

Section Background

- Mendel worked with mice before he worked with pea plants. The principles of genetics are the same for both plants and animals.
- Although the genotype may often be inferred from the phenotype, it is impossible to tell dominant homozygous and heterozygous genotypes apart by sight.

Preplanning

- To prepare for Activity 5-1, obtain 2 paper bags, 100 red beans, and 100 white beans for each pair of students.
- To prepare for the Mini-LAB, obtain paper and pencils.

1 Motivate

Bellringer

Before presenting the lesson, display **Section Focus Transparency 17** on the overhead projector. Assign the accompanying **Focus Activity** worksheet. L1 LEP

❔ FLEX Your Brain

Use the Flex Your Brain activity to have students explore INHERITANCE.

📁 **Activity Worksheets,** p. 5

5•1 What Is Genetics?

Science Words

- heredity
- allele
- genetics
- dominant
- recessive
- Punnett square
- genotype
- homozygous
- heterozygous
- phenotype

Objectives

- Explain how traits are inherited and explain Mendel's role in the history of genetics.
- Use a Punnett square to predict the results of crosses.
- Explain the difference between genotype and phenotype.

What have you inherited?

People have always been interested in why one generation looks like another. A new baby may look much like one of its parents. It may have eyes the same color as its father, or a nose like its mother. Eye color, nose shape, and many other physical features are types of traits that you may inherit from your parents. Every organism is a collection of traits, all inherited from its parents. **Heredity** (huh RED ut ee) is the passing of traits from parent to offspring. What controls these traits? As you will learn, traits are controlled by genes.

How Traits Are Inherited

Genes, as you learned in Chapter 4, are made up of DNA. They control all the traits that show up in an organism. Genes are found on chromosomes. When pairs of chromosomes separate into sex cells during meiosis, pairs of genes also separate from one another. As a result, each sex cell winds up with one form of a gene for each trait that an organism shows, as shown in **Figure 5-1.** If the trait is for hairlines, then the gene in one

Figure 5-1

An allele is one form of a gene. Alleles separate into sex cells during meiosis. In this example, the genes that control the trait for hairlines include *H*, the widow's peak hairline gene, and *h*, the straight hairline gene.

Ⓐ In a pair of like chromosomes, the genes that control a trait are located in the same position on each chromosome.

Ⓑ During meiosis, like chromosomes separate. Each chromosome now contains just one gene for the hairline trait, either *H* or *h*.

Ⓒ During fertilization, each parent donates one chromosome. This results in new pairs of chromosomes with two genes for the hairline trait.

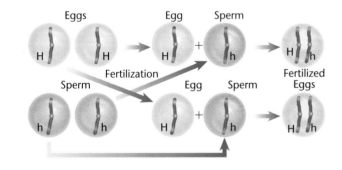

124 Chapter 5 Heredity

Program Resources

📁 **Reproducible Masters**
Study Guide, p. 21 L1
Reinforcement, p. 21 L1
Enrichment, p. 21 L3
Activity Worksheets, pp. 29-30, 33
Science Integration Activities, p. 9
Multicultural Connections, pp. 13-14
Concept Mapping, p. 15
Lab Manual 10, 11

📷 **Transparencies**
Section Focus Transparency 17 L1
Teaching Transparencies 9, 10

sex cell may control one form of the trait, such as a widow's peak. The gene for hairlines in the other sex cell may control the other form of the trait, such as a straight hairline. The different forms a gene may have for a trait are its **alleles** (uh LEELZ). The study of how traits are inherited through the actions of alleles is the science of **genetics.**

The Father of Genetics

The first recorded scientific study of how traits pass from one generation to the next was done by Gregor Mendel, a monk. Mendel was born in Austria in 1822. He learned how to grow plants as a child working in his family's orchard and farm. Mendel studied science and math and eventually became a priest and teacher. While teaching, Mendel took over a garden plot at his monastery. In 1856, he began experimenting with garden peas, like the plants shown in **Figure 5-2.** His observations of his father's orchard made him think that it was possible to predict the kinds of flowers and fruit a plant would produce. But something had to be known about the parents of the plant before such a prediction could be made. Mendel made extremely careful use of scientific methods in his research. After eight years of work on inheritance in pea plants, Mendel presented a paper detailing the results of his research to the Natural Science Society of Brünn, Austria. This paper was published by the Society in 1866 under the title, "Experiments with Plant Hybrids."

In 1900, three other scientists working in botany rediscovered Mendel's work. These other scientists had come to the same conclusions that Mendel had reached. Since that time he has been known as the Father of Genetics.

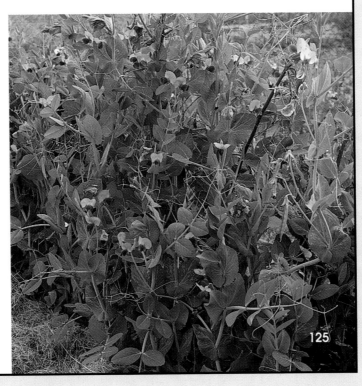

Figure 5-2

Mendel's work with garden peas such as these led to the science of genetics. *Why is Mendel called the Father of Genetics?*

125

125

Figure 5-3

Bees and other animals often visit only one species of flower at a time. *Why is this important to the plant species pollinated by animals?*

Figure 5-4

Mendel crossed tall and short pea plants in his garden.

A A cross between pure tall plants and pure short plants produced a generation of all tall plants.

In Mendel's Garden

Mendel chose ordinary garden peas, such as the ones you've eaten for dinner, for his experiments. Peas are easy to breed for pure traits. An organism that always produces the same traits in its offspring is called a *purebred*. Tall plants that always produce tall plants are purebred for the trait of tallness. Short plants that always have short offspring are purebred for the trait of shortness. In addition to height, Mendel studied six other traits of garden peas, shown in **Table 5-1**.

Dominant and Recessive Factors

In nature, insects such as the bee shown in **Figure 5-3** pollinate randomly as they go from flower to flower. In his experiments, Mendel became the pollinator himself. He crossed pea plants by taking pollen from the male reproductive structures of flowers of purebred tall plants and placing it on the female reproductive structures of flowers of pure short plants. This process is called cross-pollination. The results of his cross are shown in **Figure 5-4**. Notice that tall plants crossed with short plants produced all tall plants. It seemed as if whatever had caused the plants to be short had disappeared.

Mendel called the tall height form that appeared the **dominant** factor, because it dominated or covered up the short height form. He called the form that seemed to disappear the **recessive** factor. But what had happened to the recessive form?

Mendel allowed the new tall plants to self-pollinate. Then he collected the seeds from these tall plants and planted them. Both tall and short plants grew from these seeds. The recessive form had reappeared. Mendel saw that for every three tall plants there was one short plant, or a 3:1 ratio. He saw this 3:1 ratio often enough that he knew he could predict his results when he started a test. He knew that the *probability* was great that he would get that same outcome each time.

B When the first generation tall plants are crossed with each other, a second generation is produced with both tall plants and short plants.

Tall

Short

Tall

Tall

Parents

1st generation

Table 5-1

Traits Compared by Mendel							
Traits	**Shape of seeds**	**Color of seeds**	**Color of pods**	**Shape of pods**	**Plant height**	**Position of flowers**	**Flower color**
Dominant Trait	Smooth	Yellow	Green	Full	Tall	At leaf junctions	Purple
Recessive Trait	Wrinkled	Green	Yellow	Flat or constricted	Short	At tips of branches	White

Predictions Using Probability

Probability is a branch of mathematics that helps you predict the chance that something will happen. Suppose you and a friend want to go to different movies. You toss a coin to decide which movie you will see. Your friend chooses tails as the coin is in the air. What is the probability that the coin will land with the tail side of the coin facing up? You know there is one chance out of two possible outcomes that you will see your friend's choice of a movie. The probability of one side of a coin showing is one out of two, or 50 percent. You are dealing with probabilities.

Mendel also dealt with probabilities. One of the things that made his predictions accurate was that he worked with large numbers. He counted every plant and thousands of seeds. In fact, Mendel raised and studied almost 30 000 pea plants over a period of eight years. By doing so, Mendel increased his chances of seeing a repeatable pattern. Scientific research is based on repeatable results.

USING MATH

Suppose Mendel had 100 second generation pea plants. How many would he expect to be tall? How many would he expect to be short?

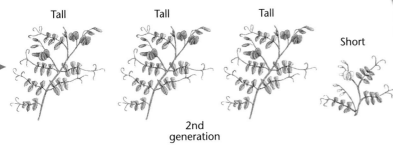

Tall   Tall   Tall   Short

2nd generation

C Mendel saw that in the second generation, there were three tall plants for every short plant, a ratio of 3:1.

Using a Punnett Square

Suppose you wanted to know what kinds of pea plants you would get if you pollinated a purple flower of a tall plant with pollen from a short plant that had white flowers. How could you predict what the offspring would look like without actually making the cross? A handy tool used to predict results in Mendelian genetics is the **Punnett square.** In a Punnett square, dominant and recessive alleles are represented by letters. A capital letter *(T)* stands for a dominant allele. A small letter *(t)* stands for a recessive allele. The letters are a form of shorthand. They show the genetic makeup, **genotype,** of an organism. Once you understand what the letters mean, you can tell a lot about the inheritance of a trait in an organism.

Alleles Determine Traits

Most cells in your body have two alleles for every trait. An organism with two alleles for a trait that are exactly the same is called **homozygous** (ho muh ZI gus). This would be written as *TT* or *tt.* An organism that has two different alleles for a trait is called **heterozygous** (het uh roh ZI gus). This condition would be written *Tt.* The purebred pea plants that Mendel used were homozygous for tall, *TT,* and homozygous for short, *tt.* The hybrid plants he produced were all heterozygous, *Tt.*

The physical trait that shows as a result of a particular genotype is its **phenotype** (FEE nuh tipe). Red is the phenotype for red flowering plants. Short is the phenotype for short plants. If you have brown hair, then the phenotype for your hair color is brown.

Figure 5-5

A Punnett square shows you all the ways in which alleles can combine. A Punnett square does not tell you how many offspring will be produced. By observing this Punnett square, you can see why all the plants in Mendel's first cross showed the dominant form of the trait in their phenotypes.

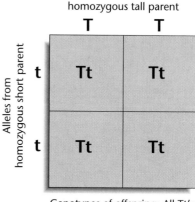

Alleles from homozygous tall parent

Genotypes of offspring: All Tt's
Phenotypes of offspring: All tall

Determining Genotypes and Phenotypes

In a Punnett square, the letters representing the two alleles from one parent are written along the top of the square. Those of the second parent are placed along the side of the square. Each section of the square is filled in like a multiplication problem with one allele donated by each parent. The letters that you use to fill in each of the squares represent the genotypes of offspring that the parents could produce.

The Punnett square in **Figure 5-5** represents the first type of cross-pollination experiment by Mendel. You can see that each homozygous parent plant has two alleles for height. One parent is homozygous for tall *(TT)*. The other parent is homozygous for short *(tt)*. The alleles inside the squares are the genotypes of the possible offspring. All of them have the genotype *Tt*. They all have tall as a phenotype. Notice that you can't always figure out a genotype just by looking at the phenotype. The combination of *TT* or *Tt* both produce tall plants when *T* is dominant to *t*. Study the sample problems below, then try Activity 5-1.

Figure 5-6

The color phenotype of this hibiscus flower is red. *Can you tell what the flower's color genotype is? Why or why not?*

Using Punnett Squares

Problem 1: Color in Peas

In peas, the color yellow is dominant to the color green. A homozygous yellow pea plant is crossed with a homozygous green pea plant. What will the genotypes and the phenotypes of all the possible offspring be?

Outcome:

Genotypes of all possible offspring: All **Yy**
Phenotypes of offspring: All yellow

Yellow parent (YY)

	Y	Y
y	Yy	Yy
y	Yy	Yy

Green parent (yy)

Problem 2: Wing Length in Fruit Flies

In fruit flies, long wings **(L)** are dominant to short wings **(l)**. Two heterozygous long-winged fruit flies (both **Ll**) are crossed. What are the possible genotypes of their offspring? What are the phenotypes?

Outcome:

Genotypes of all possible offspring: **LL**, **Ll**, and **ll**
Phenotypes of all possible offspring:
  LL and **Ll** = long wings
  ll = short wings

Long-winged mother (Ll)

	L	l
L	LL	Ll
l	Ll	ll

Long-winged father (Ll)

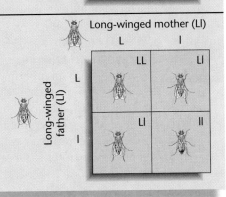

5-1 What Is Genetics?   **129**

---

4 Close

Demonstrate the alleles along a chromosome by showing students two long strings of the large plastic beads used as a toddler's toy. Explain that each different shape and color represents a different trait. When the two strands are brought together, you can demonstrate homozygous and heterozygous conditions.

Section Wrap-up

Review

1. Alleles are the forms of a gene that direct the formation of traits in an organism.

2. Genotype is the actual alleles an organism contains, and phenotype is the appearance of a trait in an organism. An example of a genotype is *Tt* for some plants. The phenotype is tall.

3. **Think Critically** weather forecasting, trends in populations, chances of winning the lottery

USING MATH

	B	b
b	Bb	bb
b	Bb	bb

50 percent are white, *bb*

50 percent are black, *Bb*

Table 5-2

Mendelian Inheritance

1. Traits are controlled by alleles on chromosomes.
2. An allele may be dominant or recessive in form.
3. When a pair of chromosomes separates during meiosis, the different alleles for a trait move into separate sex cells.

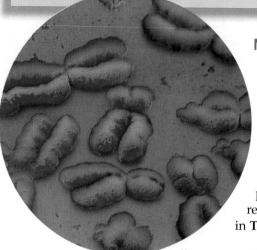

Mendel's Success

Gregor Mendel succeeded in describing how inherited traits are transmitted from parents to offspring. However, he didn't know anything about DNA, genes, or chromosomes. He did figure out that "factors" in the plants caused certain traits to appear. He also figured out that these factors separated when the plant reproduced. Mendel arrived at his conclusions by patient observation, careful analysis, and repeated experimentation. His work is summed up in **Table 5-2.**

Chromosome pairs
Magnification: 12 000×

Section Wrap-up

Review

1. How are alleles and traits related?

2. Differentiate between genotype and phenotype.

3. **Think Critically:** Give an example of how probability is used in everyday life.

Skill Builder
Observing and Inferring

Hairline shape is an inherited trait in humans. The widow's peak allele is dominant and the straight hairline allele is homozygous recessive. From your study of Mendel's experiments, infer how parents with widow's peaks could have a child without the trait. If you need help, refer to Observing and Inferring in the **Skill Handbook.**

USING MATH

Make a Punnett square showing a cross between two dogs. One dog is heterozygous for black coat, and the other dog is homozygous for white coat. Black is dominant to white. Use **B** for the dominant allele and **b** for the recessive allele. What percent of the offspring are expected to be white?

Skill Builder

Both parents would have to be heterozygous to produce children without the widow's peak trait.

✓ Assessment

Performance Use this Skill Builder to assess students' abilities to observe and infer how the hairline trait is inherited. Use the Performance Task Assessment List for Making Observations and Inferences in **PASC,** p. 17. **P**

Activity 5-1

Expected and Observed Results

Could you predict how many white flowers would result from crossing two heterozygous red flowers if you knew that white color was a recessive trait? Try this experiment to find out.

Problem

How does chance affect combinations of genes?

Materials

- 2 paper bags
- 100 red beans
- 100 white beans

Procedure

1. Place 50 red beans and 50 white beans into a paper bag. Place 50 red beans and 50 white beans into a second bag. Each bean represents an allele for flower color.
2. Label one of the bags "female" for the female parent. Label the other bag "male" for the male parent.
3. Without looking inside the bags, remove one bean from each bag. The two beans represent the alleles that combine when sperm and egg join.
4. Use a Punnett square to predict how many red/red, red/white, white/white combinations will be selected.

5. Make a data table and record the color combination of the beans each time you remove two beans. Then return them to their original bags.
6. Repeat step 5 ninety-nine times.
7. Count and record the total numbers of red/red, red/white, and white/white bean combinations in your data table.
8. Compile and record the class totals.

Analyze

1. Which combination occurred most often?
2. What was the ratio of red/red to red/white to white/white?
3. How did your predicted (expected) results compare with your observed (actual) results?

Conclude and Apply

4. Does chance affect allele combination? Explain.
5. What are the chances of selecting the same colors of alleles each time?
6. How do the results of a small sample **compare** with the results of a large sample?
7. **Hypothesize** how you could get predicted results to be closer to actual results.

Data and Observations   Sample Data

Beans	Red/Red	Red/White	White/White
Total	18	52	30
Class Total	376	751	373

5-1 What Is Genetics?   **131**

4. Each time an allele is chosen, there is a 50/50 chance it or its corresponding other allele will be chosen.
5. one chance in two
6. Small samples can sometimes give distorted results. The larger the sample, the more nearly accurate the results.
7. by increasing the size of the sample

Performance To further assess students' understanding of probability, have them repeat the activity using three different kinds of beans. Use the Performance Task Assessment List for Carrying Out a Strategy and Collecting Data in **PASC**, p. 25. **P**

Activity 5-1

Purpose

IS **Logical-Mathematical** Students will compare expected results with observed results using Punnett squares and red and white beans, and will investigate how the principles of heredity are related to chance. **L1**

Process Skills

predicting, observing, recording data, interpreting data, using numbers, making and using tables

Time

one class period

Teaching Strategies

- While one of each student pair places the beans in the bags, the other student should make the Punnett square to predict the color ratios.
- Emphasize the importance of completing all 100 trials. You might also challenge some students to do 1000 trials and have them compare the results of the two tests (100 vs. 1000).

Troubleshooting Explain to students that beans need to be returned to the bag rather than be kept out in order to keep the probabilities of choosing the color combinations the same throughout the activity.

📁 **Activity Worksheets,** pp. 5, 29-30

Answers to Questions

1. red/white
2. Results should show a generally close ratio of 1:2:1.
3. Answers will vary according to the individual numbers of the various color combinations chosen, but should reflect a generally close following of the expected results.

131

Prepare

Section Background

Many traits in humans are inherited by polygenic inheritance.

Preplanning

- For the MiniLAB, you will need tape, ballpoint pens, index cards, hand lenses, and ink pads with washable ink.
- For Activity 5-2, you will need metric rulers, graph paper, and pencils.

1 Motivate

Bellringer

Before presenting the lesson, display **Section Focus Transparency 18** on the overhead projector. Assign the accompanying **Focus Activity** worksheet. L1 LEP

Tying to Previous Knowledge

Explain that the concepts of incomplete dominance, multiple alleles, and polygenic inheritance do not displace Mendel's findings, but rather extend them.

2 Teach

Visual Learning

Figure 5-7 Ask students to explain how the inheritance pattern in four o'clock flowers differs from that of hibiscus. *Four o'clock flowers inherit color by incomplete dominance.* **Ask them if they can tell the genotype of a red four o'clock flower.** *Yes, it must be homozygous because if it were heterozygous, it would be pink.* LEP LS

132

5•2 Genetics Since Mendel

Science Words

incomplete dominance
multiple alleles
polygenic inheritance

Objectives

- Explain incomplete dominance.
- Compare multiple alleles and polygenic inheritance, and give examples of both.

Incomplete Dominance

When Mendel's work was rediscovered, scientists repeated his experiments. Mendel's results were true for peas and other plants again and again. However, when different plants were crossed, the results sometimes varied from Mendel's predictions. One scientist crossed pure red four o'clock plants with pure white four o'clocks. He expected to get all red flowers. But to his surprise, all the flowers were pink. Neither allele for flower color seemed dominant. Had the colors become blended like paint colors? He crossed the pink flowered plants with each other and red, pink, and white flowers were produced. These results are shown in **Figure 5-7**. The red and white alleles had not become "blended." Instead, the allele for white flowers and the allele for red flowers had resulted in an intermediate phenotype, a pink flower. **Incomplete dominance** is the production of a phenotype that is intermediate to those of the two homozygous parents. Flower color in four o'clock plants is inherited by incomplete dominance.

Figure 5-7

The diagrams show how color in four o'clock flowers is inherited by incomplete dominance.

Phenotypes: All pink
Genotypes: All RR′

Phenotypes: Red, pink, and white
Genotypes: RR, RR′, R′R′

Multiple Alleles

Mendel studied traits in peas that were controlled by just two alleles. However, many traits are controlled by more than two alleles. A trait that is controlled by more than two alleles is considered to be controlled by **multiple alleles.** One example of a trait controlled by multiple alleles is blood type in humans.

132   Chapter 5 Heredity

Program Resources

📁 **Reproducible Masters**
Study Guide, p. 22 L1
Reinforcement, p. 22 L1
Enrichment, p. 22 L3
Activity Worksheets, pp. 5, 31-32, 34

🖊 **Transparencies**
Section Focus Transparency 18 L1

In 1900, one scientist found three blood types in the human population. He called them A, B, and O. A and B alleles are both dominant alleles. When a person inherits one A and one B allele for blood type, both are expressed. The A and B alleles are considered to be co-dominant. Thus, a person with AB blood type shows both alleles in his or her phenotype. Both A and B alleles are dominant to the O allele, which is recessive.

A person with phenotype A blood inherited either the genotype AA or AO. A person with phenotype B blood inherited either genotype BB or BO. For a person to have type AB blood, an A allele must be inherited from one parent and a B allele must be inherited from the other parent. Finally, a person with phenotype O blood has inherited an O allele from each parent and has the genotype OO.

Multiple Genes

How many different shades of blue or brown eye color can you detect among your classmates? Eye color is an example of a single trait that is produced by a combination of many genes. **Polygenic** (pahl ih JEHN ihk) **inheritance** occurs when a group of gene pairs act together to produce a single trait. The effect of each allele may be small, but the combination produces a wide variety. You'll probably have trouble classifying all the different shades of blue or brown eyes in your class. **Figure 5-8** shows another trait that is inherited through a combination of gene pairs.

Figure 5-8

Fingerprints in your class will probably show varieties of the whorl (A), arch (B), and loop (C) patterns. There are many variations in these patterns. *Why are everyone's fingerprints unique?*

A Whorl B Arch C Loop

MiniLAB

What are the distinguishing features of fingerprints?

Every person has a unique set of fingerprints. Look at **Figure 5-8** to see some of the differences shown by fingerprints.

Procedure
1. Press the thumb and fingertips of one hand on a washable ink pad.
2. Separately place each fingertip on a sheet of blank paper and gently roll to produce a set of fingerprints.
3. Using a hand lens, observe your fingerprints to see if you can find a whorl, arch, or loop pattern.

Analysis
1. What patterns did you find?
2. Are fingerprints inherited as a simple Mendelian pattern, or as a more complex pattern?

MiniLAB

Purpose

LS **Logical-Mathematical** Students will interpret data to determine possible inheritance patterns of fingerprints. L1

Materials
tape, ballpoint pens, index cards, hand lenses, ink pads with washable ink

Teaching Strategies
Show students how to roll their fingertips from side to side on both the ink pad and the index card.

Analysis
1. Arches, loops, and whorls are common.
2. Fingerprints are more complex because each finger shows different patterns.

✓Assessment

Performance To further assess students' understanding of the inheritance of fingerprints, ask them to find out whether identical twins have identical fingerprints. *Fingerprints are partially controlled by embryonic environment, so even identical twins have different fingerprints.* Use the Performance Task Assessment List for Using Math in Science in **PASC**, p. 29. P

📁 **Activity Worksheets,** pp. 5, 34

Science Journal

Jumping Genes Have students write about Barbara McClintock and her work with genes that change position on chromosomes. Have them find out what organism she studied. L2 LS

Visual Learning

Figure 5-8 Why are everyone's fingerprints unique? *Fingerprints are different because they are controlled by multiple genes.* LEP LS

133

Activity 5-2

Purpose

 Interpersonal Design and carry out an experiment to determine polygenic inheri-tance. **L1** **COOP LEARN**

Process Skills

forming a hypothesis, mea-suring, recording data, calcu-lating, interpreting data, communicating, making and using tables, designing an experiment, separating and controlling variables

Time

one class period

Materials

metric rulers, graph paper, pencils

Possible Hypotheses

Student hypotheses will vary. Most students will hy-pothesize that the bar graph of class hand span data will look like a bell.

 Activity Worksheets, pp. 5, 31-32

Activity 5-2

Design Your Own Experiment
Determining Polygenic Inheritance

When several genes at different locations on chromosomes act together to produce a single trait, a wide range of phenotypes for the trait can result. By measuring the range of phenotypes and graphing them, you can determine if a trait is determined by polygenic inheritance. How would graphs differ if traits were inherited as simple dominant or recessive traits?

PREPARATION

Problem
How can the effect of polygenic inheritance be determined?

Form a Hypothesis
Based on the differences between the kinds of inheritance patterns you have studied, state a hypothesis about what a bar graph of the hand spans of stu-dents in your class would look like.

Objectives
• Design an experiment to determine what a bar graph of phenotypes for a polygenic trait will look like.

• Observe and measure the hand spans of students to the nearest centimeter.

Possible Materials
• metric ruler
• graph paper
• pencil

Safety Precautions
Always obtain the permission of any person included in an experiment on human traits.

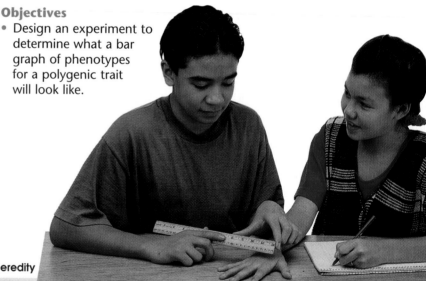

134   Chapter 5 Heredity

PLAN THE EXPERIMENT

1. As a group, agree upon and write out the hypothesis statement.
2. As a group, list the steps you need to take to test your hypothesis. Be specific; describe exactly what you will do at each step. List your materials.
3. Have another group read your plan to see if they could perform the experiment just as you have stated the steps.
4. If you need a data table, design one in your Science Journal so that it is ready to use as your group collects data.
5. Is there a control in your experiment?
6. How will you measure hand spans?

Check the Plan

1. Read over your entire experiment to make sure that all steps are in logical order.
2. Identify any constants, variables, and controls of the experiment.
3. *Make sure your teacher approves your plan before you proceed.*

DO THE EXPERIMENT

1. Carry out the experiment as planned.
2. While the experiment is going on, write down any observations that you make and complete the data table in your Science Journal.

3. **Graph** your results using a bar graph. Place the number of students on the *y*-axis and the measured intervals on the *x*-axis.
4. Using your graph, **estimate** how many genes might have been at work in determining the phenotypes you observed.

Analyze and Apply

1. **Compare** your results with other groups and determine if your results agree.
2. What was the range of measurements you made and what do you think accounted for the range?

Go Further

Find out if hair color is determined by polygenic inheritance. Collect data on hair color types and graph your results. Are the graphs similar?

PLAN THE EXPERIMENT

Possible Procedures

Students may choose too few of the many hand spans to measure. Most experimental designs will include measuring hand spans in one class. Others will choose to pool data from several classes.

Teaching Strategies

- Help students distinguish among range, median, mean, and mode.
- Results form a bell-shaped curve.

DO THE EXPERIMENT

Expected Outcome

Most students will collect enough data to graph a bell curve.

Analyze and Apply

1. Most groups will be in general agreement with a bell curve graph.
2. Answers will vary depending on observations.
3. Graphs will be bell shaped.
4. Students may estimate two to four genes.

Go Further

The graphs are similar because both are controlled by polygenic inheritance.

✔ Assessment

Performance Have students compare the inheritance of this pattern with that of the Explore Activity. Use the Performance Task Assessment List for Analyzing the Data in **PASC**, p. 27. **P**

Check for Understanding

Have a student write the heights of students on the chalkboard. Then ask students to graph the data (bell curve).

Reteach

Have students compare the three types of inheritance in this section by making a chart.

4 CLOSE

1. **Blood type in humans is controlled by _____ alleles.** *multiple*

2. **_____ occurs when a group of gene pairs act together to control a single trait.** *Polygenic inheritance*

3. **Pink four o'clock flowers occur as a result of _____ .** *incomplete dominance*

Section Wrap-up

Review

1. Multiple alleles—a single pair of genes that have more than two alleles; polygenic—multiple pairs of genes, each with two or more alleles.

2. A trait inherited by incomplete dominance appears when both alleles present are expressed.

3. **Think Critically** Cross pale gray offspring among themselves. If all three phenotypes are produced in the next generation, the trait is controlled by incomplete dominance.

Science Journal

Offspring are the result of combinations of different alleles from each parent.

136

Figure 5-9
Hair color, skin color, and eye color in humans are the results of the expression of more than one pair of genes.

Many human traits are controlled by polygenic inheritance, as shown in **Figure 5-9.** Height, weight, body build, and shape of eyes, lips, and ears are examples of some of these traits. It is estimated that skin color is controlled by three to six genes. Even more gene pairs may control the color of your hair and eyes.

Polygenic inheritance is, of course, not limited to human traits. Grain color in wheat, milk production in cows, and egg production in chickens are polygenic traits also.

The study of genes is no longer a simple look at a single trait controlled by one pair of alleles. Do you think Mendel would be amazed if he could see the amount of information that has come from his beginning work?

Section Wrap-up

Review

1. Compare multiple allele and polygenic inheritance.

2. Explain why a trait inherited by incomplete dominance is not a blend of two alleles.

3. **Think Critically:** A chicken that is purebred for black feathers is crossed with one that is purebred for white feathers. All the offspring produced have gray feathers. Is this a case of incomplete dominance? Explain.

Skill Builder
Comparing and Contrasting
Using the information on pages 132 and 133, compare and contrast multiple alleles and multiple genes to explain differences between the two concepts. Include examples of each. If you need help, refer to Comparing and Contrasting in the **Skill Handbook.**

Science Journal

In your Science Journal write a paragraph that explains why the offspring of two parents may not show much resemblance to either parent.

Skill Builder
Multiple alleles and multiple genes compare in that more than one gene is involved. They contrast in that in multiple alleles, more than two alleles exist that control a trait. In multiple genes, two or more gene pairs in an individual control a trait.

Assessment

Performance Have students prepare a table that lists types of inheritance. Use the Performance Task Assessment List for Data Table in **PASC,** p. 37. **P**

Human Genetics 5•3

Genes and Health

Sometimes a gene undergoes mutation that results in an unwanted trait. In Chapter 4, you learned that there are several types of mutations. Not all mutations are harmful, but some have resulted in genetic disorders among humans. Sickle-cell anemia and cystic fibrosis are two of the disorders that are the result of changes in DNA.

Recessive Genetic Disorders

Sickle-cell anemia is a homozygous recessive disorder in which red blood cells are sickle-shaped instead of disc-shaped, as shown in **Figure 5-10.** Sickle-shaped cells can't deliver enough oxygen to the cells in the body. In addition, the misshapen cells don't move through blood vessels easily. Body tissues may be damaged due to insufficient oxygen. Sickle-cell anemia is most commonly found in tropical areas and in a small percentage of African Americans. Many people with sickle-cell anemia die as children, but some patients live much longer. Sickle-cell anemia patients can be treated with drugs to increase the amount of oxygen carried by red blood cells, or by transfusions of blood containing normal cells.

Normal red blood cell

Sickled red blood cell

Magnification: 11 550×

Science Words

sex-linked gene
pedigree
genetic engineering

Objectives

- Describe two human genetic disorders.
- Explain inheritance of sex-linked traits.
- Explain the importance of genetic engineering.

Figure 5-10

Normal red blood cells maintain a disc shape. In sickle-cell anemia, red blood cells become misshapen. This occurs because the hemoglobin is not normal.

5-3 Human Genetics **137**

Section 5•3

Prepare

Section Background

- About 1000 simple recessive human disorders are presently known. Most human genetic disorders are recessive.
- Screening for sickle-cell anemia, phenylketonuria, and other serious genetic disorders is common at birth. Many states have laws that require doctors to screen for particular disorders.
- In 1990, the cystic fibrosis gene was "cured" in a test-tube culture. This advance gives an indication that genetic engineering may make cures of genetic disorders possible within students' lifetimes.

1 Motivate

Bellringer

Before presenting the lesson, display **Section Focus Transparency 19** on the overhead projector. Assign the accompanying **Focus Activity** worksheet. L1 LEP

Tying to Previous Knowledge

Most students will be aware of at least one person with a genetic disorder. This section bridges the mechanisms of inheritance in the previous two sections with practical applications in the students' own lives.

Program Resources

Reproducible Masters
Study Guide, p. 23 L1
Reinforcement, p. 23 L1
Enrichment, p. 23 L3
Cross-Curricular Integration, p. 9
Science and Society Integration, p. 9
Critical Thinking/Problem Solving, p. 9
Technology Integration, pp. 7-8
Lab Manual 12

Multicultural Connections, p. 13
Science Integration Activities, pp. 9-10

Transparencies
Section Focus Transparency 19 L1
Science Integration Transparency 5

You could use Punnett squares to demonstrate inheritance patterns discussed in this section.

USING MATH

about 1500 (30 000 ÷ 20 = 1500)

Student Text Questions

Is the organism that contributed these chromosomes male or female? *male, because there is a Y chromosome present* **What pair of sex chromosomes is in each of your cells?** *Students who are female should answer XX; males should answer XY.*

Visual Learning

Figure 5-11 Discuss with students the fact that most sex-linked genes are found on the X chromosome in humans. The SRC gene on the Y chromosome determines maleness. **LEP** **LS**

Teacher F.Y.I.

Point out the difference between sickle-cell trait and sickle-cell anemia. A person who has a single copy of the gene for this disorder has sickle-cell trait, and has both sickle-shaped and normal blood cells. These people are also carriers.

GLENCOE TECHNOLOGY

 Videotape

The Secret of Life

Use the videotape *Sex and the Single Gene: Cell Development* to explore the factors that determine an individual's sex.

USING MATH

If there are 30 000 white Americans in one town, about how many would carry the recessive allele for cystic fibrosis?

Figure 5-11

Sex in many organisms is determined by X (left) and Y (right) chromosomes. In the photograph, you can see how X and Y chromosomes differ from one another in shape and size.

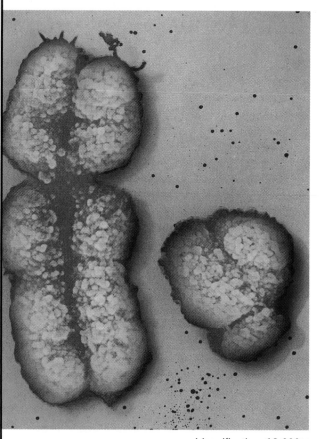

Magnification: 18 000×

138 Chapter 5 Heredity

Cystic fibrosis is another homozygous recessive disorder. In most people, a thin fluid is produced by the body to lubricate the lungs and intestinal tract. People with cystic fibrosis have thick mucus in these areas instead of a thin fluid. This mucus builds up in the lungs and digestive system of cystic fibrosis patients. Mucus in the lungs makes it hard to breathe, and eventually causes lung damage. Mucus in the digestive tract damages the pancreas. It also reduces or prevents the flow of digestive enzymes to the intestine. These enzymes are needed to break down food so that it can be absorbed by body cells.

Cystic fibrosis is the most commonly inherited genetic disorder among Caucasians. Nearly one white person in 20 carries a recessive allele for this disorder. In the United States, four white babies are born with this disorder every day. Patients with cystic fibrosis are often helped with antibiotics, special diets, and physical therapy to break up the thick mucus in their lungs.

Sex Determination

What determines the sex of an individual? Information on sex inheritance in many organisms, including humans, came from a study of fruit flies. Fruit flies have eight chromosomes in their cells. Among these chromosomes are either two X chromosomes, or an X chromosome and a Y chromosome. Several scientists concluded that the X and Y chromosomes contained genes that determined the sex of an individual. Females have two X chromosomes in their cells. Males have an X chromosome and a Y chromosome. You can see an X and a Y chromosome in **Figure 5-11.** Is the organism that contributed these chromosomes male or female?

Females produce eggs that have only an X chromosome. Males produce both X-containing sperm and Y-containing sperm. When an egg from a female is fertilized by an X sperm, the offspring is XX, a female. When an egg from a

Cultural Diversity

Malaria Protection People who have sickle-cell trait or sickle-cell anemia appear to have some protection against malaria. Use the information on the relationship between sickle-cell anemia and malaria to increase student awareness of health in different cultures. Point out that certain traits are found more often in (but are not limited to) certain racial or ethnic groups. As examples, Tay-Sachs occurs more frequently among Ashkenazi Jews (of Central European descent), sickle-cell anemia among African-Americans, and cystic fibrosis among caucasians in the United States.

female is fertilized by a Y sperm, the zygote produced is XY, and the offspring is a male. What pair of sex chromosomes is in each of your cells?

Sex-linked Disorders

Some inherited conditions are closely linked with the X and Y chromosomes that determine the sex of an individual. Scientists tell a story about a young boy who appeared to have normal intelligence, but couldn't be taught to pick ripe cherries on the family farm. His parents took him to a doctor. After observing him, the doctor concluded that the boy couldn't tell the difference between the colors red and green. Individuals who are red-green color-blind have inherited an allele on the X chromosome that prevents them from seeing these colors. Find out if you are color-blind by looking at **Figure 5-12.**

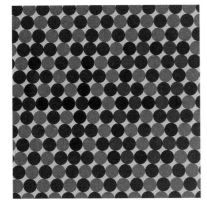

Figure 5-12
What number do you see in this pattern?

Problem Solving

Boy or Girl?

Probability refers to the chance that an event will occur. For example, if you flip a coin, it will land in only one of two ways—heads or tails. The probability of the coin landing heads or tails is one out of two, or one-half, or 50 percent. If you were to toss two coins at the same time, each coin still has only a one-half probability of landing with heads or tails up.

What is the probability that any single pregnancy will result in the birth of a girl?

Jennifer's mother is going to have a baby. Jennifer already has a brother, so she really hopes this baby is a girl. How would you determine the chance that this baby will be a girl?

Solve the Problem:
How can the outcome of each pregnancy be "predicted"?

Think Critically:
Does the fact that Jennifer's mother already has one boy and one girl affect the probability of the next child being a girl? Why or why not?

Solve the Problem
Each pregnancy has a 50/50 chance of being a girl regardless of the other children's sex.

Think Critically
No, each pregnancy is an independent event.

Visual Learning

Figure 5-12 What number do you see in this pattern? *People who have normal vision will see the number 48; people who are color blind will see a different number.* **LEP** **LS**

NATIONAL GEOGRAPHIC SOCIETY

Videodisc
STV: The Cell
Human chromosomes

47009
STV: The Human Body Vol. 3
Researcher studying genes

48819

139

140

Figure 5-13

Color blindness is a recessive sex-linked trait. Females who are heterozygous see colors normally, but their sons may be color-blind. When a woman with normal vision who carries one allele for color blindness (XX^C) marries a man with normal vision (XY), their children may have normal vision (XX, XY), be carriers (XX^C), or be color-blind (X^CY). *Why aren't the daughters of this marriage color-blind?*

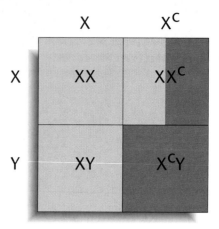

An allele inherited on a sex chromosome is a **sex-linked gene.** More males are color-blind than females. Can you figure out why? If a male inherits an X chromosome with the color-blind allele from his mother, he will be color-blind. Color-blind females are rare. **Figure 5-13** shows how males and females inherit the genes that result in color blindness.

Another sex-linked gene disease is hemophilia, a disorder in which blood does not clot properly. For people with this disease, a scrape can be life threatening. Like color blindness, males who inherit the X chromosome with the hemophilia allele will have the disorder. For a female to be a hemophiliac, she must inherit the defective allele on both X chromosomes. Hemophilia occurs in fewer than 1 in 7000 males. Females are usually carriers, having just one allele for hemophilia.

Pedigrees Trace Traits

How can you trace a trait through a family? A **pedigree** is a tool for tracing the occurrence of a trait in a family. Males are represented by squares and females by circles. A circle or

Figure 5-14

Pedigrees show the occurrence of a trait in a family. The icons in the key mean the same thing on all pedigree charts. Queen Victoria was a carrier of the recessive allele for hemophilia. *How many of her grandchildren on this chart were identified as hemophiliacs? How many were carriers?*

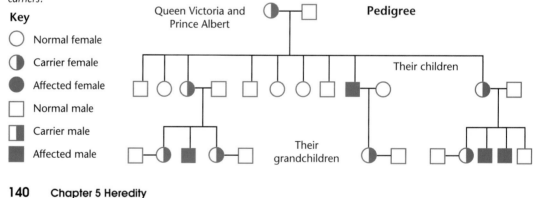

Key

○ Normal female

◐ Carrier female

● Affected female

□ Normal male

◧ Carrier male

■ Affected male

140 Chapter 5 Heredity

square that is completely colored or filled in shows that the person has the condition. Half colored circles or squares indicate carriers. A carrier has an allele for a trait, but does not show the trait. People represented by plain circles or squares are neither carriers nor affected by the trait.

You can see how the trait for hemophilia was inherited in Queen Victoria of England's family in the pedigree in **Figure 5-14.**

Why is genetics important?

If Mendel were to pick up a daily newspaper in any country today, he probably wouldn't believe his eyes. There is hardly a day that goes by anymore that there isn't a news article about the latest information on genetic research. The word *gene* has become a household word. In this chapter, you have learned

USING TECHNOLOGY

Karyotyping

Karyotyping is a process that allows scientists to study the chromosomes of an individual, sometimes even before birth.

Chromosomes are best seen during metaphase in mitosis when they are coiled. White blood cells divide frequently, so these cells are good to use for making a karyotype. The cells are fixed on microscope slides and dyed so that light and dark bands appear showing the position of DNA along the chromosomes. Then a photograph is taken of the chromosomes. The photograph is cut up into individual chromosomes. Scientists use the pattern of the stained bands, chromosome length, and the position of centromeres to identify pairs of matching chromosomes. Each pair is also numbered. The bands enable scientists to see if chromosomes have missing or duplicated parts.

Some genetic disorders can be identified just by looking at the karyotype of an individual. Sometimes this is done before birth when cells are taken from fluid around a fetus.

Think Critically:

Why would a missing part of a chromosome affect the traits an individual would have?

Karyotyping

USING TECHNOLOGY

Logical-Mathematical Karyotyping is a difficult concept for some students to understand. You may be able to find karyotypes from medical or college textbooks. You may also be able to acquire them from a local hospital or genetic counselor in your community. Use the karyotype to point out the sizes and shapes of some chromosomes, the location of the centromeres, and banding patterns, if they are obvious.

Think Critically

If part of a chromosome is missing, that means genes are missing. If the gene code is for important traits, the individual could have major difficulties.

GLENCOE TECHNOLOGY

 Videotape

The Secret of Life

Use the videotape *Tinkering with Our Genes: Genetic Medicine* to analyze both the benefits and ethical considerations of genetic research.

Videodisc

The Infinite Voyage: The Geometry of Life

Chapter 8

Hemophilia: The Effects of "Factor Eight"

Chapter 9

Hemophilia: Inserting a Gene That Can Cure

Chapter 10

Manipulating Genetic Engineering

141

CONNECT TO
CHEMISTRY

Answer PKU could be a recessive genetic disorder. Students could also relate the disorder to mistakes in coding DNA. If ignored, persons with PKU suffer retardation. If detected early in urine or blood samples, diet can control and prevent symptoms of PKU.

 INTEGRATION Chemistry

Understanding genes that control many traits of plants may help to produce more food from smaller land areas.

3 Assess

Check for Understanding

Ask students why genetics is important. They should respond with information about genetic disorders and sex-linked inheritance.

Reteach

Have students use a set of cutouts of X and Y chromosomes to model how a sex-linked disorder can be inherited.

Extension

For students who have mastered this section, use the **Reinforcement** and **Enrichment** masters.

142

CONNECT TO
CHEMISTRY

Information on genetic disorders has reached soft drink cans! People with the disease called phenylketonuria (PKU) have not inherited the ability to make an enzyme needed to break down phenylalanine. This chemical is present in certain artificially sweetened drinks. *Explain* how a person could be born with PKU.

 INTEGRATION Chemistry

that every trait that you have inherited is the result of genes expressing themselves. The same laws that govern inheritance of traits in humans govern the inheritance of traits in watermelons, wheat, and mice as well.

In this section on human genetics, you've seen that there are inherited disorders that affect the human population. You may even know someone with one of these disorders. Many genetic disorders are controlled by diet and preventive measures. Genetics is no longer something that you read about only in textbooks.

Knowing how genes are inherited causes some people to seek the advice of genetic counselors before having children. Couples who discover genetic traits that could lead to disorders in their children may elect to adopt healthy children instead.

Genetic Engineering

Through **genetic engineering,** scientists are experimenting with biological and chemical methods to change the DNA sequence that makes up a gene. Genetic engineering is already used to help produce large quantities of medicine, such as insulin, to meet the needs of people with a disease called diabetes. Genes can also be inserted into cells to change how those cells perform their normal functions, as shown in **Figure 5-15.** Genetic engineering research is also being used to find new ways to improve crop production.

Figure 5-15

Genetically engineered bacteria have been developed to clean up environmental problems such as oil spills. Such bacteria have alleles inserted into them that allow them to break down oil more efficiently than normal bacteria found in the environment.

New Plants From Old

One type of genetic engineering involves identifying genes that are responsible for desirable traits in one plant, then inserting these genes into other plants. In barley, for example, a single gene has been identified that controls several traits—including time of maturity, strength, height, and resistance to drought. Researchers plan to insert this gene into plants such as wheat, rice, and soybeans to create shorter, stronger, drought-resistant plants. Why are these traits of interest to scientists? Wheat, rice, and soybeans are subject to lodging, a condition in which plants fall over because of their own height and weight. If a new gene from the barley plant could result in shorter wheat or rice plants, then fewer plants would be lost, and crop yields would increase. This gene could even be used to develop lawn grass that stays a certain height and doesn't need to be watered as often! Research on genetic engineering in plants is done at centers such as the one shown in **Figure 5-16.**

Figure 5-16

Researchers at the International Rice Research Institute in the Philippines are using genetic engineering techniques to develop new strains of rice.

Section Wrap-up

Review

1. Describe two genetic disorders.

2. Explain why males are affected more often than females by sex-linked genetic disorders.

3. Describe the importance of genetic engineering.

4. **Think Critically:** Use a Punnett square to explain how a woman who is a carrier for color blindness can have a daughter who is color-blind.

Skill Builder
Concept Mapping

Use a network tree concept map to show how X and Y sex cells can combine to form fertilized eggs. Begin with female sex cells, each containing an X chromosome. Use two male sex cells. Indicate one with an X chromosome and one with a Y chromosome. If you need help, refer to Concept Mapping in the **Skill Handbook.**

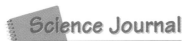

Science Journal

Hugo de Vries, Walter Sutton, Barbara McClintock, and Thomas Hunt Morgan each made discoveries in the field of genetics. Summarize their work in your Science Journal, and discuss the importance of their contributions.

SCIENCE & SOCIETY

Section 5•4

Prepare

Section Background

The Human Genome Project was headed by Nobel Laureate James Watson of Cold Spring Harbor Laboratory in New York from 1989 to 1992.

1 Motivate

Bellringer

Before presenting the lesson, display **Section Focus Transparency 20** on the overhead projector. Assign the accompanying **Focus Activity** worksheet. L1 LEP

Brainstorming

Have students imagine a book that has all of the information about humans written in code. Ask students: **If the code were a single word for each gene, how long would the book have to be?** *100 000 words*

2 Teach

Visual Learning

Figure 5-18 Why is it important to locate specific genes on such a map? *Just a few years ago, for instance, the gene for Huntington's disease, a deadly disorder resulting in deterioration of the nervous system, was unknown. Today, the exact location of the gene is known. This gives geneticists an attack point for searching for the exact cause and a possible cure for the disease.* LEP LS

Science Words

genome

Objectives

- Describe the goals of the Human Genome Project.
- Identify some of the human diseases that are inherited.

Figure 5-17

Computers are used to sequence the DNA of the human genome.

144 Chapter 5 Heredity

TECHNOLOGY:
5•4 The Human Genome Project

Tracking Human Genes

Think of the DNA that makes up the genes on your chromosomes. Strands of DNA are somewhat like a long railroad track. Just as railroad engineers use maps showing the various towns along rail routes, genetic engineers also are now able to locate particular genes along each chromosome. The resulting chromosome map representing all of the human genetic material, or **genome,** is a chart that shows the location of individual genes on a chromosome. A project called the *Human Genome Project* is aimed at sequencing all the DNA on all the human chromosomes.

There are more than 75 000 genes on the 46 chromosomes in each of your body cells, but only the first several thousand have been mapped so far. Much like a railroad engineer who stops at a specific station, scientists who study chromosomes are able to follow a piece of DNA and stop at the spot where a gene is located. Using these chromosome maps, they can identify genes responsible for specific traits.

DNA Sequencing

Sequencing DNA is a complicated matter. Tiny pieces of DNA are cut by enzymes. The pieces are inserted into bacteria, where they are reproduced as the bacteria reproduce. The DNA fragments are then placed on a gel in an electric field. After a while, the pieces of DNA line up like the ties on a railroad track, making a map, and the order of individual bases of DNA can be determined. This procedure is now automated.

Information From the Genome

Why would such gene maps be useful? More than 3000 human disorders are known to be inherited. Among these are cystic fibrosis, muscular dystrophy, and cancer. Already the Human Genome Project has identified genes that are responsible for Huntington's disease, and genes that identify people who may be at risk of developing breast cancer, colon cancer, high blood pressure, diabetes, and Alzheimer's disease. By examining a few drops of blood, scientists will soon

---

Program Resources

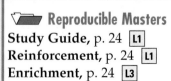

Reproducible Masters
Study Guide, p. 24 L1
Reinforcement, p. 24 L1
Enrichment, p. 24 L3

Transparencies
Section Focus Transparency 20 L1

---

be able to examine chromosomes and prepare a map of an individual's genes, shown in **Figure 5-18.** Developing a map for these genes is a first step toward diagnosing inherited disorders. Once a diagnosis is made, it may one day be possible to insert normal genes into human cells to correct genetic disorders. Trials are currently under way to do just that.

After examining a person's chromosome map, a physician may be able to tell if that person has inherited a gene that codes for a certain disorder. For example, a person at risk for developing heart disease might use this information to begin a program to reduce stress in his or her life, or to plan a diet that includes less fat. People with other disorders might decide not to have children as a result of this information. Both the methods and the data developed as part of the Human Genome Project are likely to benefit investigations of many other genomes, including a large number of commercially important plants and animals.

Emerging Technologies

Improving technology for DNA sequencing is one of the problems scientists are currently attacking. The Human Genome Project is also developing technology to find out if a particular gene is located on a chromosome and developing new mapping technologies and tags for specific genes. Every day new solutions are being worked out.

Figure 5-18

Studies of chromosomes can be used to produce maps such as the one shown above that identify where specific genes are located. *Why is it important to locate specific genes on such a map?*

Section Wrap-up

Review

1. What is the purpose of the Human Genome Project?

2. What would be the usefulness of having a complete map of the human genome?

 Visit the Chapter 5 Internet Connection at Glencoe Online Science, **www.glencoe.com/sec/science/life,** for a link to more information about the Human Genome Project.

SCIENCE & SOCIETY

Integrating the Sciences

Chemistry Review the chemical makeup of DNA with students. Point out that there are chemical techniques used to cut and sequence the DNA, and that machines are automated to read the DNA sequence based on chemical signals that are translated into electrical impulses recognizable by computers.

---

3 Assess

Check for Understanding

Ask questions 1 and 2 in the Section Review.

Reteach

Ask students how their lives might be affected by the human genome initiative.

Extension

For students who have mastered this section, use the **Reinforcement** and **Enrichment** masters.

4 Close

Discussion

Ask students how a greater knowledge of the human genome can improve lives of humans in the future.

Section Wrap-up

Review

1. The human genome initiative is aimed at mapping all the genes on all the human chromosomes.
2. Scientists will be able to examine all chromosomes and find genes that code for certain disorders.

Explore the Technology

Scientists in the Human Genome Project may be able to identify the gene that codes for producing insulin and eventually be able to create the gene and insert it into your pancreas cells, resulting in a cure for your disease.

People and Science

People and Science

Background

- Dr. Sean B. Carroll is a geneticist at the Howard Hughes Medical Institute at the University of Wisconsin–Madison. He and his colleagues discovered a set of genes in the buckeye butterfly that control wing shape and color patterns.

- The butterfly genes analyzed by Dr. Carroll's research team are similar to a set of well-studied genes in fruit flies that specify wing shape. In butterflies, however, these shape-determining genes are reactivated later in development to produce color patterns, as well. The distinctive color patterns of the world's 12 000 butterfly species can be viewed as slight variations on the same genetic theme.

Teaching Strategies

LS **Visual-Spatial** Obtain butterfly chrysalises from a biological supply company. Have students work in teams to care for the chrysalises until the butterflies emerge. When they do, have students record their observations of metamorphosis and make detailed drawings of the patterns on the butterflies' wings.

Career Connection

Career Path Research geneticists work in many different areas of biology, including botany, zoology, microbiology, and medicine.

- High School: math and science
- College: biology, chemistry, physics, and math
- Graduate School: Master's degree/Ph.D. in a specialized area of genetic research

SEAN B. CARROLL, *Geneticist*

On the Job

Q Dr. Carroll, how do genes direct the development of butterfly wings?

A The "blueprint" for the final shape and color pattern of a butterfly's wings is staked out by certain genes that are turned on and off in developing tissues while the butterfly is still a caterpillar. The same genes that are initially involved in constructing the wings are reactivated to make the wings' color patterns.

Q Is this discovery helpful in understanding how genes control development in other animals?

A Yes, we've learned that some basic genetic mechanisms for "making a body" are surprisingly similar in a great variety of animals, including humans.

Buckeye butterfly

Personal Insights

Q How does it feel to find out something new about how genes work?

A It's really like going through door after door after door. But until you actually get each door open and look inside, you have no idea what's in there. It's that journey that's really a blast!

Q Why do you think young people should consider becoming research scientists?

A You get to ask why—why and how. And that's fun! I think if you go into science for the same reason that people go into art or music or literature or sports—because they really enjoy it—that's a good reason, and that's enough.

Career Connection

Contact a natural history museum or the entomology department at your state university and ask about their insect collections. Be prepared to explain to your classmates how people in the following positions maintain and use these collections:

- **Museum curator**
- **Entomologist**

People trained in genetics also work for seed companies, nurseries, and agricultural research organizations, where they are involved in developing improved varieties of plants and breeds of farm animals. Some zoos employ geneticists to oversee captive-breeding programs for endangered species.

For More Information

For additional background information on Dr. Carroll's research, students can read "How to Make a Butterfly's Wing," *The New York Times*, July 5, 1994, p. C1 and "Gene Patterns Decorate Butterflies' Wings," *Science News*, July 9, 1994, p. 23.

Summary

5-1: What Is Genetics?

1. Traits are inherited through alleles passed from one generation to the next. Gregor Mendel determined the basic laws of genetics through controlled experiments.
2. The probability of types of offspring can be predicted in Punnett squares.
3. Genotype is the genetic makeup of an organism for a trait. Phenotype is the physical appearance that results from the genotype.

5-2: Genetics Since Mendel

1. In incomplete dominance, neither allele for a trait is completely dominant.
2. With multiple alleles, there are more than two alleles controlling a trait. In polygenic inheritance, more than one pair of alleles controls a trait.

5-3: Human Genetics

1. Some diseases are the result of genetic disorders.
2. Males express sex-linked traits more than females. Pedigrees may show patterns of inheritance in a family.
3. Genetic engineering may result in new solutions to many problems.

5-4: Science and Society: The Human Genome Project

1. The Human Genome Project seeks to locate and identify all the genes on human chromosomes.
2. Genome information may help pinpoint specific genetic disorders.

Key Science Words

a. allele
b. dominant
c. genetic engineering
d. genetics
e. genome
f. genotype
g. heredity
h. heterozygous
i. homozygous
j. incomplete dominance
k. multiple alleles
l. pedigree
m. phenotype
n. polygenic inheritance
o. Punnett square
p. recessive
q. sex-linked gene

Reviewing Vocabulary

Match each phrase with the correct term from the list of Key Science Words.

1. technology for changing a gene
2. gene located on the X chromosome
3. the allele that is hidden
4. different form of the same gene
5. a tool for predicting possible offspring
6. when a form intermediate between those of two parents appears
7. the study of heredity
8. more than one pair of alleles controls a trait
9. shows the pattern of gene inheritance in a family
10. all the genes in a species

Summary

Have students read the summary statements to review the major concepts of the chapter.

Reviewing Vocabulary

1. c	**6.** j
2. q	**7.** d
3. p	**8.** n
4. a	**9.** l
5. o	**10.** e

Assessment

Portfolio Encourage students to place in their portfolios one or two items of what they consider to be their best work. Examples include:

• MiniLAB, p. 125
• Enrichment, p. 128
• Inclusion Strategies, p. 141
P

Performance Additional performance assessments may be found in **Performance Assessment** and **Science Integration Activities**. Performance Task Assessment Lists and rubrics for evaluating these activities can be found in Glencoe's **Performance Assessment in the Science Classroom.**

GLENCOE TECHNOLOGY

MindJogger Videoquiz

Chapter 5 Have students work in groups as they play the Videoquiz game to review key chapter concepts.

Chapter 5 Review

1. a   **6.** b
2. c   **7.** d
3. a   **8.** a
4. c   **9.** a
5. b   **10.** b

Understanding Concepts

11. Multiple alleles means having more than two alleles, or types of genes, for a trait; the trait is controlled by one gene pair. Polygenic inheritance involves more than one gene pair controlling the trait. These gene pairs combine to produce the trait.

12. Mendel worked with a single trait, made controlled experiments, collected and analyzed data, reported on his experiments, and used large samples.

13. The blood types are A, B, AB, and O. A child with type O blood cannot have one or both parents with type AB blood; he or she can have parents with all other types as long as each has the O allele. A child with type AB blood cannot have one or both parents with type O; he or she cannot have both parents with type B or both parents with type A.

14. *Homozygous* means that both alleles for the trait are identical. *Heterozygous* means that the alleles for the trait are different.

15. *Dominant* means that the trait will show when there is a heterozygous genotype. It does not mean that it is the most common allele in a population. The most common may be a recessive allele.

Checking Concepts

Choose the word or phrase that completes the sentence.

1. _____ are located on chromosomes.
   a. Genes   c. Carbohydrates
   b. Pedigrees   d. Zygotes

2. Color blindness results from an allele that is _____.
   a. dominant
   b. on the Y chromosome
   c. on the X chromosome
   d. present only in females

3. _____ result(s) when two different alleles both appear in the phenotype.
   a. Incomplete dominance
   b. Polygenic inheritance
   c. Multiple alleles
   d. Sex-linked genes

4. During meiosis, _____ for a trait separate.
   a. proteins   c. alleles
   b. cells   d. pedigrees

5. The major job of genes is to control _____.
   a. chromosomes   c. cell membranes
   b. traits   d. mitosis

6. Blood type is inherited through _____.
   a. polygenic inheritance
   b. multiple alleles
   c. incomplete dominance
   d. recessive genes

7. Sickle-cell anemia is inherited through _____.
   a. polygenic inheritance
   b. multiple alleles
   c. incomplete dominance
   d. recessive genes

8. Eye color is a trait inherited through _____.
   a. polygenic inheritance
   b. multiple alleles

---

   c. incomplete dominance
   d. recessive genes

9. A normal female is produced if an egg unites with a sperm containing _____.
   a. an X chromosome
   b. XX chromosomes
   c. a Y chromosome
   d. XY chromosomes

10. Punnett squares are used to _____ the outcome of crosses of traits.
   a. dominate   c. assure
   b. predict   d. number

Understanding Concepts

Answer the following questions in your Science Journal using complete sentences.

11. Compare and contrast multiple allele inheritance and polygenic inheritance.

12. Explain the importance of Mendel's experimental methods.

13. Using an example, explain how blood typing could be used to detect the correct parent of a newborn if there was a mix-up in a nursery.

14. What is the difference between homozygous and heterozygous?

15. Explain why dominant traits might not necessarily be the traits that show up most frequently.

Thinking Critically

16. A pure black guinea pig is crossed with a pure white guinea pig. All the offspring are black. What are the probable genotypes of the parents and the offspring?

Thinking Critically

16. The parents are homozygous—one for black, one for white. All the offspring are heterozygous for black and white.

17. The dark gray feathers result from incomplete dominance of the black and white alleles, producing both black and white colors in feathers.

18. DNA is a chemical; a gene is a segment of DNA; an allele is a form of a gene specific for a trait; genes are located on chromosomes.

19. Finding where the genes are located may help predict the chances of having a child with the disease and eventually provide a cure.

20. The phenotype will show the dominant trait, whether the genotype is homozygous or heterozygous dominant, because the recessive gene does not show at all in the phenotype.

17. In breeding certain varieties of chickens, farmers found that chickens with black feathers always had chicks with black feathers, and chickens with white feathers always had chicks with white feathers. But mating a black chicken with a white chicken produced some chickens with dark gray feathers. How can this trait be explained?

18. Explain the relationship among DNA, genes, alleles, and chromosomes.

19. How will finding the location of disease-causing genes on chromosomes help to cure genetic diseases?

20. Explain how an organism that has a genotype *Gg* could have the same phenotype as an organism with the genotype *GG*.

Developing Skills

If you need help, refer to the **Skill Handbook.**

21. **Designing an Experiment:** Design an experiment to determine if a trait is transmitted by a dominant or recessive gene.

22. **Recognizing Cause and Effect:** A mutation in the gene for normal production of a plant's chlorophyll results in a plant with no chlorophyll. Explain the consequences.

23. **Classifying:** Classify the type of inheritance of each human trait that follows by making a chart: blood type, color blindness, eye color, height, hemophilia, sickle-cell anemia, weight.

24. **Interpreting Scientific Illustrations:** What genotypes and phenotypes would be produced if you crossed a red offspring with a pink offspring from **Figure 5-7B** on page 132?

25. **Concept Mapping:** On a separate sheet of paper, use the following terms to complete the network tree concept map below: phenotypes, recessive, genes, dominant.

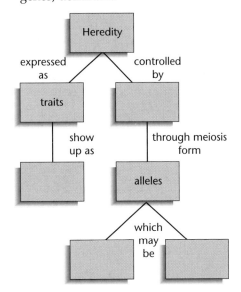

Performance Assessment

1. **Analyzing the Data:** Using **Table 5-1** and a Punnett square, show the expected color of peas when a plant that is heterozygous for green peas is crossed with one that is homozygous for yellow peas.

2. **Using Math in Science:** Make a Punnett square showing the possible offspring of a couple who have the blood phenotypes AB and O. What are the percentages and possible phenotypes of each of their children?

3. **Science Journal:** Conduct additional research on genetic screening. Use your information to write an essay in your Science Journal that describes why you think it would or would not be important to screen infants for additional genetic disorders.

Developing Skills

21. **Designing an Experiment** The student needs to describe an experiment in which crosses can be made between pure-bred organisms, similar to Mendel's experiments.

22. **Recognizing Cause and Effect** Because there is no chlorophyll, the plant cannot change sunlight into chemical energy, and therefore it will die. The mutation was harmful.

23. **Classifying** incomplete dominance: sickle-cell anemia; multiple alleles: blood type; polygenic inheritance: eye color, height, weight; sex-linked: hemophilia, color blindness

24. **Interpreting Scientific Illustrations** 50 percent *RR* and red; 50 percent *Rr* and pink

25. **Concept Mapping** See student page.

Performance Assessment

1. 50% are expected to be heterozygous for green peas, and 50% are expected to be homozygous for yellow peas. Use the Performance Task Assessment List for Analyzing Data in **PASC,** p. 27. **P**

2. The Punnett square will show half AO and half BO, so the offspring will have a 50% chance of being A and a 50% chance of being B. Use the Performance Task Assessment List for Using Math in Science in **PASC,** p. 29. **P**

3. The essay should say that genetic screening can be useful in providing early treatment for some genetic disorders. Use the Performance Task Assessment List for Science Journal in **PASC,** p. 103. **P**

Chapter Organizer

Section	Objectives/Standards	Activities/Features
Chapter Opener		**Explore Activity:** Model evolution by exploring camouflage coloring. p. 151
6-1 **Mechanisms of Evolution** (2 sessions, 2 blocks)*	1. **Compare** and **contrast** Lamarck's explanation of evolution with Darwin's theory of evolution. 2. **Explain** the importance of variations in organisms. 3. **Relate** how gradualism and punctuated equilibrium describe the rate of evolution. National Content Standards: (5-8) UCP2, UCP3, UCP4, A1, A2, B3, C1, C5, F2, F3, G1, G2, G3	**Problem Solving:** Why isn't Earth covered with pumpkins and pike? p. 155 **Using Math,** p. 156 **MiniLAB:** How does evolution occur? p. 157 **Activity 6-1:** Recognizing Variation in a Population, pp. 158-159 **Skill Builder:** Classifying, p. 161 **Science Journal,** p. 161
6-2 **Evidence for Evolution** (1 session, ½ block)*	1. **Describe** the importance of fossils as evidence of evolution. 2. **Explain** how relative and radioactive dating are used to date fossils. 3. **Give** examples of five types of evidence for evolution. National Content Standards: (5-8) UCP2, UCP4, A1, A2, B3, C2, C5, D1, D2, E2	**MiniLAB:** How are fossils made? p. 163 **Using Math,** p. 165 **Activity 6-2:** A Radioactive Dating Model, p. 167 **Connect to Chemistry,** p. 168 **Skill Builder:** Making and Using Tables, p. 170 **Using Computers:** Database, p. 170
6-3 **Primate Evolution** (1 session, ½ block)*	1. **Describe** the evidence that all primates evolved from a common ancestor. 2. **Describe** the ancestors of humans. 3. **Trace** the evolutionary history of humans. National Content Standards: (5-8) UCP4, B1, C1, E2	**Using Technology:** Bringing Bones to Life, p. 173 **Connect to Earth Science,** p. 174 **Skill Builder:** Concept Mapping, p. 175 **Science Journal,** p. 175
6-4 **Science and Society** **Issue: Extinction and Evolution** (1 session, ½ block)*	1. **Define** *extinction* and **identify** some causes of extinction. 2. **Define** *endangered species* and **identify** some causes of species endangerment. National Content Standards: (5-8) UCP4, C4, D2, F2, F3	**Connect to Earth Science,** p. 176 **Explore the Issue,** p. 177 **Science and Art:** Stone Age Art, p. 178 **Science Journal,** p. 178

* A complete Planning Guide that includes block scheduling is provided on pages 32T-35T.

Activity Materials

Explore	Activities	MiniLABs
page 151 sheet of newspaper classified ads; paper hole punch; 1 sheet each of white, unruled paper and black paper	pages 158-159 leaves, flowers, and seeds from one species of plant; metric ruler; magnifying glass; graph paper page 167 100 pennies, cardboard box with lid, graph paper	page 163 plaster of paris; small paper container; small object such as a seashell, key, or leaf; tap water

Need Materials? Call Science Kit (1-800-828-7777).

Teacher Classroom Resources

Reproducible Masters	Transparencies	Teaching Resources
Study Guide, p. 25 **Reinforcement**, p. 25 **Enrichment**, p. 25 **Activity Worksheets**, pp. 35-36, 39 **Lab Manual 13**, Overproduction	**Section Focus Transparency 21**, Change for Survival **Teaching Transparency 11**, Horse Evolution	**Glencoe Life Science Interactive Videodisc** **Life Science CD-ROM** **Spanish Resources** **English/Spanish Audiocassettes** **Cooperative Learning Resource Guide** **Lab Partner** **Lab and Safety Skills** **Lesson Plans**
Study Guide, p. 26 **Reinforcement**, p. 26 **Enrichment**, p. 26 **Activity Worksheets**, pp. 37-38, 40 **Science Integration Activity 6**, Geologic Time **Concept Mapping**, p. 17 **Cross-Curricular Integration**, p. 10 **Multicultural Connections**, pp. 15-16	**Section Focus Transparency 22**, Date with a Trilobite **Science Integration Transparency 6**, Fossil Signposts **Teaching Transparency 12**, Geologic Time Scale	**Assessment Resources** **Chapter Review**, pp. 15-16 **Assessment**, pp. 36-39 **Performance Assessment in the Science Classroom (PASC)** **MindJogger Videoquiz** **Alternate Assessment in the Science Classroom** **Performance Assessment** **Chapter Review Software** **Computer Test Bank**
Study Guide, p. 27 **Reinforcement**, p. 27 **Enrichment**, p. 27	**Section Focus Transparency 23**, A Brainy Idea	
Study Guide, p. 28 **Reinforcement**, p. 28 **Enrichment**, p. 28 **Lab Manual 14**, Variation **Critical Thinking/Problem Solving**, p. 10 **Science and Society Integration**, p. 10	**Section Focus Transparency 24**, Beware! Endangered Species	

Key to Teaching Strategies

The following designations will help you decide which activities are appropriate for your students.

L1 Level 1 activities should be within the ability range of all students, including those with learning difficulties.

L2 Level 2 activities should be within the ability range of the average to above-average student.

L3 Level 3 activities are designed for the ability range of above-average students.

LEP LEP activities should be within the ability range of Limited English Proficiency students.

LS These activities are designed to address different learning styles.

COOP LEARN Cooperative Learning activities are designed for small group work.

P These strategies represent student products that can be placed into a best-work portfolio.

GLENCOE TECHNOLOGY

The following multimedia resources are available from Glencoe.

The Secret of Life
It's in the Genes: Evolution
Gone Before You Know It: The Biodiversity Crisis
Science and Technology Videodisc Series (STVS)
Chemistry
Carbon-14 Dating
Earth & Space
Fossil CAT Scan

The Infinite Voyage Series
Life in the Balance
The Dawn of Humankind
The Keepers of Eden
The Great Dinosaur Hunt
Glencoe Life Science Interactive Videodisc
Evolution: Patterns of Change
Life Science CD-ROM

Teacher Classroom Resources

This is a representation of key blackline masters available in the Teacher Classroom Resources.

Teaching Aids

Section Focus Transparencies

Science Integration Transparencies

Teaching Transparencies

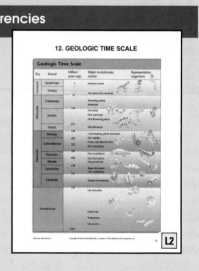

Meeting Different Ability Levels

Study Guide

Reinforcement

Enrichment Worksheets

Hands-On Activities

Science Integration Activity

Lab Manual

Assessment

Performance Assessment

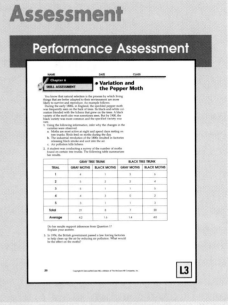

Enrichment and Application

Critical Thinking/ Problem Solving

Cross-Curricular Integration

Science and Society Integration

Multicultural Connections

Concept Mapping

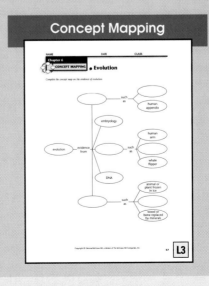

Chapter 6

Evolution

CHAPTER OVERVIEW

Section 6-1 Lamarck's explanation and Darwin's theory of evolution are compared and contrasted in relation to observed variation. Gradualism and punctuated equilibrium are discussed as modifications of Darwin's theory.

Section 6-2 The evidence for evolution by natural selection is presented. Relative and radioactive dating are compared.

Section 6-3 Evolution of primates from a common ancestor and the evolutionary history of humans are described.

Section 6-4 **Science and Society** Plant and animal extinctions are discussed in terms of evolution. Ways of saving endangered species are mentioned.

Chapter Vocabulary

species
evolution
natural
  selection
variation
gradualism
punctuated
  equilibrium
fossil
sedimentary
  rock
relative dating

radioactive
  element
homologous
vestigial
  structure
embryology
primate
hominid
Homo sapiens
extinction
endangered
  species

Theme Connection

Stability and Change Stability and change are discussed in this chapter. Theories of evolution and modifications are discussed along with indirect and direct evidence. The theme of systems and interactions is also developed in the fourth section. Plant and animal extinction rates are detailed.

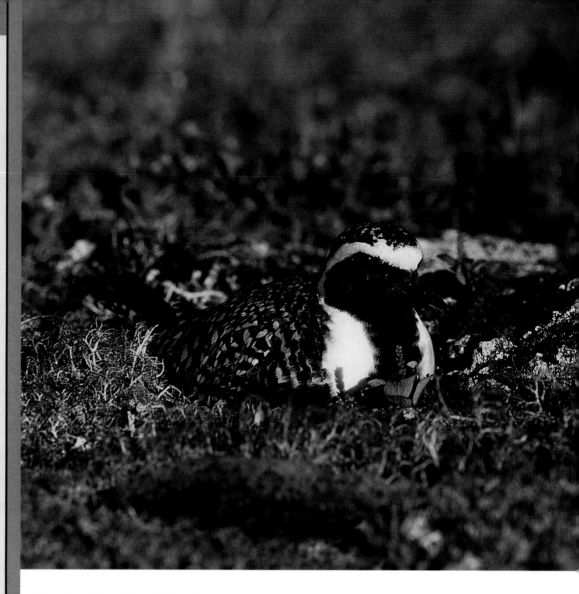

Previewing the Chapter

Section 6-1 Mechanisms of Evolution
- ► Early Thoughts About Evolution
- ► Evolution by Natural Selection
- ► Adaptation and Variation
- ► How fast does evolution occur?

Section 6-2 Evidence for Evolution
- ► Fossil Evidence
- ► The Fossil Record
- ► Other Evidence for Evolution

Section 6-3 Primate Evolution
- ► Primates
- ► Primate Classification
- ► Modern Humans

Section 6-4 Science and Society
Issue: Extinction and Evolution

150

Learning Styles Look for the following logo for strategies that emphasize different learning modalities. **LS**

Kinesthetic	Explore, p. 151; MiniLAB, pp. 157, 163; Activity, p. 171
Visual-Spatial	Visual Learning, pp. 153, 154, 156, 157, 160, 162, 164, 165, 166, 168, 169, 172, 174; Demonstration, pp. 155, 172; Across the Curriculum, p. 164; Activity, p. 168; Science and Art, p. 178
Interpersonal	Activity, p. 165; Science and Art, p. 178
Logical-Mathematical	Activities 6-1, pp. 158-159; 6-2, p. 167; Using Computers, p. 170; Science and Art, p. 178
Linguistic	Enrichment, p. 164; Science Journal, p. 173

LS

Chapter

6

Evolution

The golden plover is a shorebird that spends most of its time on the ground. It nests in areas of flat, open grasslands, with few trees. As you can see in the photograph, the golden plover has black-, gold-, and white-speckled feathers that enable it to blend in with its surroundings. Even golden plover eggs and chicks are camouflaged to disappear into the landscape.

Why is the golden plover speckled? Maybe these colors make it easier for the plover to survive in its particular environment. Like the plover, many animals blend in with their surroundings. You can use the classified ads from a newspaper to find out how camouflage works.

EXPLORE ACTIVITY

Model evolution by exploring camouflage coloring.

1. Spread a sheet of classified ads from a newspaper on the floor.
2. Using a paper hole punch, punch out 100 circles each from sheets of white paper, black paper, and classified ads.
3. Scatter all of the paper circles randomly on the spread-out sheets of classified ads. Pick up paper circles for 10 seconds.
4. Count each kind of paper circle; graph your data.
5. Repeat three times.

Observe: In your Science Journal, describe which paper circles were more difficult to find. What can you infer from this activity?

Previewing Science Skills

▶ In the **Skill Builders**, you will classify, make and use tables, and map concepts.

▶ In the **Activities**, you will collect and organize data, interpret data, formulate models, hypothesize, and experiment.

▶ In the **MiniLABs**, you will make and use models and infer.

151

EXPLORE ACTIVITY

Purpose

LS **Kinesthetic** Use the Explore Activity to show students that organisms must have adaptations, such as camouflage coloration, for survival. **L1** **LEP**

Preparation

For this activity, you will need lots of classified ads from the newspaper. Have the pieces of classified ads, black paper, and white paper cut before class.

Materials

newspaper classified ads, black and white construction paper, scissors, paper punch

Teaching Strategies

• In general, students should pick up far more of the white and black circles than the newsprint circles.

• If you find that students are picking up most of the circles within the time limit, add more circles or use less time.

Observe

Students should indicate that circles of classified ads were more difficult to find. You can infer that patterns can help conceal organisms from predators in their environment.

✔Assessment

Process Provide students with photographs of animals that exhibit concealing coloration or patterns and have them describe how each is beneficial to the animal. Use the Performance Task Assessment List for Making Observations and Inferences in **PASC**, p. 17. **P**

Assessment Planner

Portfolio

Refer to page 179 for suggested items that students might select for their portfolios.

Performance Assessment

See page 179 for additional Performance Assessment options.
Skill Builders, pp. 161, 170, 175
MiniLABs, pp. 157, 163
Activities 6-1, pp. 158-159; 6-2, p. 167

Content Assessment

Section Wrap-ups, pp. 161, 170, 175, 177
Chapter Review, pp. 179-181
Mini Quizzes, pp. 156, 170, 175

Group Assessment

Opportunities for group assessment occur with Cooperative Learning Strategies and Flex Your Brain Activities.

Prepare

Section Background

- Much attention has been given to Darwin's visit to the Galápagos, but he suspected the process of evolution before he ever began the tour on the *Beagle*.

- Gradualism and punctuated equilibrium are not so much competing hypotheses as they are attempts to explain the fossil evidence. Both have a place in the history of evolution.

Preplanning

- For Activity 6-1, obtain leaves, flowers, or seeds from one species of plant; metric ruler; magnifying glass; and graph paper for each student group.

1 Motivate

Bellringer

Before presenting the lesson, display **Section Focus Transparency 21** on the overhead projector. Assign the accompanying **Focus Activity** worksheet. L1 LEP

Tying to Previous Knowledge

Reinforce the differences between the terms *hypothesis* and *theory*. Explain that a theory is much more than a hypothesis and is an accepted explanation based on a large number of tests.

6•1 Mechanisms of Evolution

Science Words

species
evolution
natural selection
variation
gradualism
punctuated equilibrium

Objectives

- Compare and contrast Lamarck's explanation of evolution with Darwin's theory of evolution.
- Explain the importance of variations in organisms.
- Relate how gradualism and punctuated equilibrium describe the rate of evolution.

Early Thoughts About Evolution

On Earth today, there are millions of different types of organisms. Included among these are different species of plants, animals, bacteria, fungi, and protists. A **species** is a group of organisms whose members successfully reproduce among themselves. Have any of these species of organisms changed since they first appeared on Earth? Are they still changing today? Evidence from observation and experimentation shows that living things have changed through time and are still changing. Change in the hereditary features of a species over time is **evolution. Figure 6-1** shows how the horse has changed over time.

 A *Eohippus* was a small animal with four obvious toes.

Eohippus

Figure 6-1

The evolution of the horse can be traced back in time for at least 55 million years. Notice the change from several toes to a single hoof.

Mesohippus

B By 40 million years ago, *Mesohippus* had evolved, replacing *Eohippus.* Larger than *Eohippus, Mesohippus* also lost one toe during its evolution.

Merychippus

 C About 25 million years ago, *Mesohippus* was replaced by *Merychippus. Merychippus* was larger than *Mesohippus;* the two toes on the side were probably no longer functional.

152   Chapter 6 Evolution

Program Resources

Reproducible Masters

Study Guide, p. 25 L1
Reinforcement, p. 25 L1
Enrichment, p. 25 L3
Lab Manual 13
Activity Worksheets, pp. 5, 35-36, 39

Transparencies

Section Focus Transparency 21 L1
Teaching Transparency 11

In 1809, Jean Baptiste de Lamarck, a French scientist, proposed one of the first explanations as to how species evolve or change. Lamarck hypothesized that species evolve by keeping traits that their parents developed during their lives. Characteristics that were not used were lost from the species. According to Lamarck's explanation of evolution, if your Great Dane dog's ears were cropped when she was a puppy, her offspring would be born with cropped ears. Lamarck's explanation of evolution is often called the theory of acquired characteristics.

When you study genetics, you learn that genes on chromosomes control the inheritance of traits. The traits that are developed during an organism's life, such as cropped ears, are not inherited. After scientists collected large amounts of data on the inheritance of characteristics, Lamarck's explanation was rejected. The data showed that characteristics an organism develops during its lifetime aren't passed on to its offspring, as you can see from **Figure 6-2.**

Figure 6-2

Lamarck's explanation can be tested by experimentation. Female Great Danes with cropped ears don't give birth to puppies with cropped ears. Each puppy must have its ears cropped after birth to obtain the typical shape of a Great Dane's ears.

Evolution by Natural Selection

In the mid-1800s, Charles Darwin developed the theory of evolution still accepted today. At the age of 22, Darwin became the ship's naturalist aboard HMS *Beagle*. The *Beagle* was on a trip to survey the east and west coasts of South America. The ship sailed from England in December 1831. Darwin was responsible for recording information about all the plants and animals he observed during the journey.

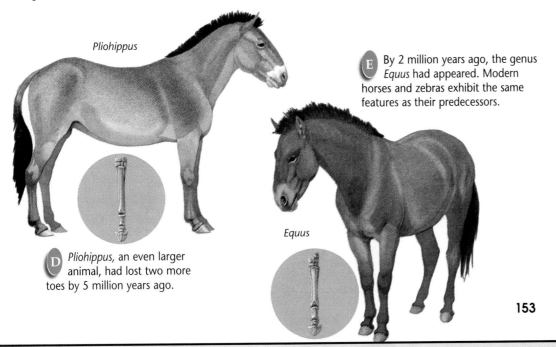

Pliohippus

E By 2 million years ago, the genus *Equus* had appeared. Modern horses and zebras exhibit the same features as their predecessors.

Equus

D *Pliohippus,* an even larger animal, had lost two more toes by 5 million years ago.

153

Figure 6-3

A map of Darwin's voyage is illustrated above. Some of the unique organisms he found in the Galápagos Islands included: a tree cactus, a marine iguana, a blue-footed booby, and a Hood Island saddleback tortoise.

Darwin's Observations

Darwin collected many plants, animals, and fossils from stops all along his route, as shown in **Figure 6-3.** He was amazed by the unique plants and animals he found in the Galápagos Islands. The Galápagos Islands are located in the Pacific Ocean, 600 miles west of the coast of Ecuador. The plants and animals Darwin saw in these islands must have come from Central and South America originally, yet in these 19 small islands, he observed many species that he had not seen before. He observed giant cactus trees, 14 species of finches, and huge land tortoises, but he saw few other reptiles, only nine mammal species, and no amphibians at all. Darwin was interested in the finches, however. He wondered how so many different, but closely related, species of finches could live on islands just a few miles apart.

For 20 years after the voyage, Darwin continued studying his collections. He thought about his observations and conducted further studies. He collected evidence of variations among species by breeding pigeons for racing. He also studied breeds of dogs and varieties of flowers. Darwin knew that people used artificial selection in breeding plants and animals. They chose parents that had traits they wanted in the offspring.

154   Chapter 6 Evolution

Principles of Natural Selection

Darwin's observations and experiments convinced him that individuals with traits most favorable for a specific environment survived and passed on these traits to their offspring. After extensive experiments all resulting in the same conclusions, this hypothesis eventually became the theory of evolution by **natural selection.** Natural selection means that organisms with traits best suited to their environments are more likely to survive. The factors Darwin identified that govern natural selection are:

1. Organisms produce more offspring than can survive.
2. Variations are found among individuals of a species.
3. Some variations enable members of a population to survive and reproduce better than others.
4. Over time, offspring of individuals with helpful variations make up more and more of a population.

Darwin wrote a book describing his theory of evolution by natural selection. His book, *On the Origin of Species by Means of Natural Selection*, was published in 1859. Although minor changes have been made to Darwin's theory as new information has been gathered, his theory is one of the most important concepts in the study of life science.

Problem Solving

Why isn't Earth covered with pumpkins and pike?

Assume that there are 70 seeds in one pumpkin and that this is the typical number for the species. The 70 seeds are planted and each seed grows into a plant that produces two pumpkins. All of their seeds are planted the following year. The number of seeds produced in each of three years can be calculated by multiplying the number of seeds planted times two pumpkins for each plant times 70 seeds in each pumpkin. The first year, 70 seeds are planted.

Year 1: $70 \times 2 \times 70 = 9800$
Year 2: $9800 \times 2 \times 70 = 1\,372\,000$
Year 3: $1\,372\,000 \times 2 \times 70 = 192\,080\,000$ seeds

The largest possible number of offspring produced by one individual is known as the biotic potential of a species.

Solve the Problem:

1. Assume that the ovaries of a pike contain 42 000 eggs, and that all the eggs are fertilized and hatch. Assume 50 percent are males and 50 percent are females, and all of the hatchlings survive and reproduce one year later. How many would there be a year later?
2. Should this process continue for just two years, there would be more than 880 million pike! How many would there be two years later?

Think Critically:

Why is the maximum rate of biotic potential never actually reached in nature?

155

Problem Solving

Solve the Problem

1. 21 000 females × 42 000 eggs = 882 000 000 offspring in the first generation
2. 441 000 000 females × 42 000 eggs = 18 522 000 000 000 offspring in the second generation

Think Critically

The maximum rate of biotic potential for the pike is never reached because not all of the eggs hatch, and many of the young are eaten by other organisms or die from environmental effects. Not all of the pumpkin seeds sprout. There is not enough food, water, space, and other necessities for all offspring to survive.

Discussion

Discuss the major distinctions between the methods used by Lamarck and Darwin to make their theories. Lamarck made a hypothesis, but did little testing. His theory was made with little research. Darwin's hypotheses were tested for years by experiments and observations before he ever reached his theory about evolution.

Demonstration

Visual-Spatial Show students photographs of different organisms. Let the students list the adaptations they think were selected for in each organism.

Use Lab 13 in the **Lab Manual** as you teach this lesson.

Figure 6-4

Variations may be beneficial, harmful, or neutral in a population.

B Variation causes some organisms to blend into their environments.

Adaptation and Variation

One of the points in Darwin's theory of evolution is that variations are found among individuals of a species. What are variations? A **variation** is the appearance of an inherited trait that makes an individual different from other members of the same species. Variations can be small, such as differences in the form of human hairlines, or large, such as an albino deer or a fruit without seeds. Variations are important in populations of organisms. A population is a group of organisms of one species that live in an area. If enough variations occur in a population as it produces new offspring, a new species may evolve from the existing species. It may take hundreds, thousands, or even millions of generations for a new species to evolve.

The Sources of Variations

Some variations are more helpful than others. An adaptation is any variation that makes an organism better suited to its environment. The variations that result in adaptation can be in an organism's color, shape, behavior, or chemical makeup. Camouflage is a protective adaptation that lets an organism blend into its environment, as shown in **Figure 6-4B.** An organism whose color or shape provides camouflage is more likely to survive and reproduce. These types of variations result from mutations, changes in an organism's DNA. Mutations are the source of variation among organisms.

A What would likely happen to an albino deer in its natural environment? Variations that result in a disadvantage, such as albinism, tend to decrease in a population over time.

What other factors bring about evolution? The movement of individuals into or out of a population brings in new genes and variations. Have you ever had an exchange student come to your school? The student probably brought new ideas, maybe a new style of dress, and even a new language. When new individuals come into an existing population, they can bring in new genes and variations in much the same way. The isolation of some individuals from others by geography and changes in climate can also result in evolutionary change, as you can see in **Figure 6-5.** Each of these factors affects how fast evolution occurs.

How fast does evolution occur?

How fast does evolution occur? Scientists are debating that question. Most scientists hypothesize that evolution occurs very slowly, perhaps taking tens or hundreds of millions of years. Other scientists hypothesize that evolution may occur very quickly, perhaps in a million years. As you study the evidence for evolution, you will see that evidence supports a combination of these two models.

Figure 6-5

The Tana River in Kenya separates two populations of giraffes. Over time, these two populations have become distinct. *Do these giraffes have different phenotypes? What does this tell you about their genotypes?*

 A The common giraffe is known as *Giraffa camelopardalis rothschildi.*

B This is the reticulated giraffe, *Giraffa camelopardalis reticulata.* The word *reticulated* means having lines or veins.

157

Mini**LAB**

How does evolution occur?

Assume that the diagrams below represent different species that evolved from the same ancestor.

Procedure
1. Determine which diagram might represent the simplest ancestral form.
2. Order the other organisms from simplest to most complex, using one structural change at a time.

Analysis
1. How are variations such as structural changes in a species similar to the model?
2. How does this model help illustrate how evolution occurs?

157

PREPARATION

Purpose

LS **Logical-Mathematical** Design and carry out an experiment to show the variation in a population of seeds. **L1**

COOP LEARN

Process Skills

forming a hypothesis, measuring in SI, using numbers, interpreting data, communicating, making and using tables, designing an experiment, separating and controlling variables

Time

one class period to plan the investigation and a second class period to complete the investigation

Materials

100 seeds per group, 1 metric ruler per group, graph paper for each student

Safety Precautions

Caution students not to put any seeds into their mouths.

Possible Hypotheses

Student hypotheses will vary. Possible hypotheses might be: "A sample of sunflower seeds will exhibit variations," or "A sample of sunflower seeds will exhibit variations in length."

📁 **Activity Worksheets,** pp. 5, 35-36

Activity 6-1

Design Your Own Experiment

Recognizing Variation in a Population

When you first see a group of plants or animals of one species, they may all look alike. However, when you look closer you will notice minor differences in each characteristic. Variations must exist in a population for evolution to occur. What kinds of variations have you noticed among species of plants or animals?

PREPARATION

Problem

How can you measure variation in a plant or animal population?

Form a Hypothesis

Make a hypothesis about the amount of variation there is in seeds, leaves, or flowers of one species of plant.

Objectives

- Observe, measure, and analyze variations in a population.
- Design an experiment that will allow you to collect data about variation in a population.

Possible Materials

- leaves, flowers, and seeds from one species of plant
- metric ruler

- magnifying glass
- graph paper

Safety Precautions

Do not put any seeds, flowers, or plant parts in your mouth.

PLAN THE EXPERIMENT

1. As a group, agree upon and write out the hypothesis statement.
2. As a group, list the steps you need to take to test your hypothesis. Be specific; describe exactly what you will do at each step. List your materials.
3. Decide what characteristic of seeds, leaves, or flowers you will study. For example, you could measure the length of seeds, the width of leaves, or the number of petals on the flowers of plants.
4. Design a data table in your Science Journal to collect data about one variation. Use the table to record the data your group collects as you complete the experiment.

Check the Plan
1. Read over your entire experiment to make sure that all steps are in logical order.

2. Identify any constants, variables, and controls of the experiment.
3. How many seeds, leaves, or flowers should you examine? Will your data be more accurate if you examine larger numbers?
4. Can you summarize the data in a graph or chart?
5. *Make sure your teacher approves your plan before you proceed.*

DO THE EXPERIMENT

1. Carry out the experiment as planned.
2. While the experiment is going on, write down any observations that you make and complete the data table in your Science Journal.

Analyze and Apply
1. **Compare** your results with those of other groups.

2. How did you determine the amount of variation present?
3. **Graph** your results placing the range of the variation on the *x*-axis and the numbers of each organism that had that measurement on the *y*-axis.
4. **Calculate** the mean and range of variation in your experiment.

Go Further

Design an experiment that would reveal variations in a collection of shells from a species of clam.

6-1 Mechanisms of Evolution   **159**

 Assessment

Performance Have students describe how they would measure variations in leaves or flowers of plants. Use the Performance Task Assessment List for Making Observations and Inferences in **PASC**, p. 17. **P**

3 Assess

Gradualism

Darwin hypothesized that the rate of evolution was steady, slow, and continuous. The model that describes evolution as a slow change of one species to another new species is known as **gradualism.** In this theory, there should be intermediate forms of all species. Look back at **Figure 6-1,** showing evolution of the horse. Fossil evidence shows intermediate forms between *Eohippus* and present-day horses. Horses appear to have evolved gradually over millions of years. There is fossil evidence like this to show the gradual evolution of many present-day species.

Figure 6-6

Evolution can occur slowly, as in gradualism, or rapidly, as in punctuated equilibrium.

Punctuated Equilibrium

But gradualism doesn't explain the evolution of some species, especially those in which few intermediate forms have been discovered. Another model, the **punctuated equilibrium** model, shows that rapid evolution of species can come about by the mutation of just a few genes. How fast is evolution by this model? New species could appear as quickly as every few million years, and sometimes even more rapidly than that. For example, bacteria that cause illness in humans can sometimes be killed by antibiotics such as penicillin. Penicillin has only been available for 50 years, yet some species of bacteria are now resistant to this drug. How did this happen so quickly? As in any population, some of the bacteria had mutations that enabled them to resist being killed by penicillin. When the drug was used to kill bacteria, the few individuals with this variation survived to reproduce. Over a short period of time, the entire population of bacteria became resistant to penicillin. The species had evolved quickly, an

160   Chapter 6 Evolution

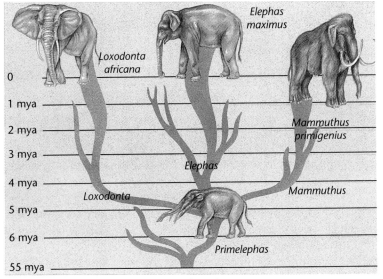

Figure 6-7
The evolution of the elephant illustrates how punctuated equilibrium usually occurs. Both present-day species of elephant evolved from a common ancestor that lived just 6 million years ago. Compare this rate of evolution to that of the horse. *Why does the evolution of the elephant fit the punctuated equilibrium model rather than the graduated model?*

4 Close

Students may know that antibiotics are not prescribed for illnesses caused by viruses because antibiotics are not effective against viruses. Ask students why it is best to prescribe antibiotics only when a specific bacterial disease has been diagnosed. They should know that physicians want to prevent the development of resistant strains of bacteria.

example of punctuated equilibrium. In this steplike pattern of evolution, mutations produce large changes in a short period of time. The fossil record also gives examples of this type of evolution, as you can see in **Figure 6-7.** Punctuated equilibrium and gradualism are compared in **Figure 6-6.**

Section Wrap-up

Review

1. Compare Lamarck's explanation of evolution with Darwin's theory.

2. How are variations important in a population?

3. Think Critically: Explain how the gradualism model of evolution differs from the punctuated equilibrium model.

Skill Builder
Classifying
Classify the following variations as being a part of an organism's shape, color, chemical makeup, or behavior: two species mate and don't produce offspring; two species of birds build nests in different places. If you need help, refer to Classifying in the **Skill Handbook.**

Science Journal

Go to the library and research the work of Alfred Wallace. In your Science Journal, write a paragraph that describes how the work of Alfred Wallace supported that of Charles Darwin.

Section Wrap-up

Review

1. Lamarck thought traits acquired by adults were passed to offspring. Darwin and Wallace thought that only certain traits that were inherited were passed to offspring.

2. Some variations are more advantageous for survival and reproduction in the environment than others. Those variations that confer an advantage to an organism are passed on to its offspring.

3. **Think Critically** Gradualism supports the hypothesis that an ancestral species slowly evolves to become another species. The punctuated equilibrium model explains that species tend to remain constant for millions of years and, suddenly, in the span of one or a few million years, become two or more species.

Skill Builder
chemical makeup; behavior

Assessment

Performance To assess students' abilities to classify organisms by their variations, ask them to classify two species of birds that eat different types of food. Use the Performance Task Assessment List for Making Observations and Inferences in **PASC,** p. 17. **P**

Science Journal
Alfred Wallace worked independently of Darwin, but came up with the same conclusions.

Prepare

Section Background

The fossil record is incomplete. Not all organisms become fossils.

Preplanning

- To prepare for the Mini-LAB, obtain plaster of paris, small paper cups, and small objects such as seashells.
- For Activity 6-2, obtain pennies, cardboard boxes, graph paper, and pencils.

1 Motivate

Bellringer

Before presenting the lesson, display **Section Focus Transparency 22** on the overhead projector. Assign the accompanying **Focus Activity** worksheet. L1 LEP

Tying to Previous Knowledge

Review the structure and function of DNA with students while they study the evidence for evolution.

2 Teach

Visual Learning

Figure 6-9 **Which of these would most likely be found in a layer of sedimentary rock?** *All of these fossils are found in sedimentary rock. The amber is sometimes found in stream beds.* LEP LS

Science Words

fossil
sedimentary rock
relative dating
radioactive element
homologous
vestigial structure
embryology

Objectives

- Describe the importance of fossils as evidence of evolution.
- Explain how relative and radioactive dating are used to date fossils.
- Give examples of five types of evidence for evolution.

Figure 6-8

Digging for fossils requires careful work. Paleontologists, scientists who study the past by examining fossils, sift tons of earth and stones to find tiny bones. They may use dental equipment such as dental picks and toothbrushes to remove dirt as they work to uncover larger bones.

162 Chapter 6 Evolution

6•2 Evidence for Evolution

Fossil Evidence

On a hot day in July 1975, in northern Texas, two people were strolling along the breezy shores of Lake Lavon. They came across some odd-looking rocks projecting from the muddy shore. They noticed the rocks seemed different from the surrounding limestone rocks. They took a few of the rocks to a local scientist specializing in reptiles and amphibians. The rocks were pieces of the skull of a fossil mosasaur, an extinct lizard that had lived in salt water.

A group of scientists returned to the site and carefully dug up the rest of the fossil mosasaur. This find provides evidence that about 120 million years ago, the northern Texas area—now more than 500 km from the Gulf of Mexico—was covered by a shallow sea. Fossils such as those found on the shores of Lake Lavon are examined by scientists called paleontologists, shown in **Figure 6-8**.

Program Resources

Reproducible Masters

Study Guide, p. 26 L1
Reinforcement, p. 26 L1
Enrichment, p. 26 L3
Cross-Curricular Integration, p. 10
Science Integration Activities, pp. 11-12
Concept Mapping, p. 17
Multicultural Connections, pp. 15-16
Activity Worksheets, pp. 5, 37-38, 40

Transparencies

Section Focus Transparency 22 L1
Teaching Transparency 12
Science Integration Transparency 6

Figure 6-9

Three examples of fossils are shown. *Which of these would most likely be found in a layer of sedimentary rock?*

 A This is an imprint fossil made by a leaf.

B This is the cast fossil of an ancient mollusk.

C An insect caught in resin that hardened over time is also a fossil.

Kinds of Fossils

The most abundant evidence for evolution comes from fossils like those found on the shore of Lake Lavon in Texas. **Fossils** are any remains of life from an earlier time, illustrated in **Figure 6-9.** Examples of fossils include:

1. the imprint of a leaf, feather, or organism in rock;
2. a cast made of minerals that filled in the hollows of an animal track, mollusk shell, or other parts of an organism;
3. a piece of wood or bone replaced by minerals;
4. an organism frozen in ice; and
5. an insect or other organism trapped in plant resin.

Sedimentary rock contains the most fossils. Sedimentary rock is a rock type formed by mud, sand, or other fine particles that settle out of a liquid. Limestone, sandstone, and shale are all examples of sedimentary rock. Fossils are found more often in limestone than in any other kind of sedimentary rock.

The Fossil Record

You learned that the mosasaur fossil found in Texas was 120 million years old. How was that date obtained? Scientists have divided Earth's history into eras and periods. These divisions make up the geologic time scale as shown in **Table 6-1.** Unique rock layers and fossils give information about the geology, weather, and life-forms of each time period. There are two basic methods for reading the record of past life. When these methods are used together, accurate estimates of the ages of certain rocks and fossils are made.

 INTEGRATION
Earth Science

MiniLAB

How are fossils made?

Find out how one kind of fossil is made.

Procedure

1. Pour a small amount of plaster of paris into a small paper cup or other paper container. Mix with a small amount of water.
2. Press a small object such as a seashell, key, or leaf into the mixture.
3. Carefully lift the object out of the container. Let the plaster dry for one to two days.
4. Tear the paper away from the container.

Analysis

1. What type of fossil have you made?
2. Infer whether a plant or an animal would be more likely to make this type of fossil. Explain your choice.

6-2 Evidence for Evolution   **163**

INTEGRATION
Earth Science

Fossils are usually found in sedimentary rock because it is the least disturbed kind of rock. Igneous rock rarely has fossils because the organisms were burned by the lava. Metamorphic rock has undergone changes due to pressure; this can also destroy fossils.

MiniLAB

Purpose

LS **Kinesthetic** Students will make a model that shows how fossil imprints are made. **L1** **LEP**

Materials

plaster of paris or clay, paper cup, seashell or other object

Teaching Strategies

Safety Precautions Have students wash their hands after handling plaster and organic materials.

Troubleshooting The plaster of paris will set more quickly if it is a thick paste and not watery.

Analysis

1. imprint
2. a plant; animals are more likely to make cast fossils

✓Assessment

Process Provide students with additional plaster or clay and allow them to make casts by putting the material into the imprints they made. Use the Performance Task Assessment List for Model in **PASC,** p. 51. **P**

📂 **Activity Worksheets,** pp. 5, 40

 Use **Science Integration Transparency 6** as you teach this lesson.

📂 Use **Concept Mapping,** p. 17, as you teach this lesson.

Integrating the Sciences

Earth Science Types of sedimentary rock include shale, limestone, sandstone, and conglomerate. Almost all fossils are found in sedimentary rock. Nearly 75 percent of Earth's surface is covered by sedimentary rock. In some places, the sedimentary layers are almost a mile thick. Many sedimentary rocks lack the necessary radioactive elements needed for radiometric dating.

Throughout the organism chapters of this book, geologic times are discussed in relation to the origin of organisms. Questions in the book do not require memorization of the eras. Some students may have difficulty comprehending the great spans of time. Use a time line clock to put the time in perspective. If a 12-hour period represents the total length of time since Earth was formed, humans appeared in the last fraction of a second. If a 4.5-m long time line is used, humans appeared in the last 0.1 cm. **LEP**

Visual Learning

Table 6-1 Allow students to discuss the evolutionary events shown in the table. Explain the vastness of geological time. A million years is only 0.03% of the total. **LEP** **LS**

Enrichment

LS **Linguistic** Have students use the library to find out how dinosaurs evolved and what might have caused their extinction. **L2**

Field Trip

Visit a natural history museum to observe the fossil collections and dioramas on display.

Inquiry Questions

Have students study the geologic time scale to answer the following questions: **During which period did simple invertebrates appear?** *the Cambrian period* **How many years ago did the first placental mammals appear?** *65 million years ago* **During which period did the first amphibians evolve?** *the Devonian period*

Table 6-1

Geologic Time Scale

Era	Period	Million years ago	Major evolutionary events	Representative organisms
Cenozoic	Quaternary	5	Humans evolve	
Cenozoic	Tertiary	65	First placental mammals	
Mesozoic	Cretaceous	144	Flowering plants dominant	
Mesozoic	Jurassic	213	First birds / First mammals / First flowering plants	
Mesozoic	Triassic	248	First dinosaurs	
Paleozoic	Permian	286	Cone-bearing plants dominant / First reptiles	
Paleozoic	Carboniferous	320	Great coal deposits form / First seed plants	
Paleozoic	Devonian	360	First amphibians	
Paleozoic	Silurian	408	First land plants / First jawed fish	
Paleozoic	Ordovician	438		
Paleozoic	Cambrian	505	Algae dominant / First vertebrates / Simple invertebrates	
	Precambrian	590 ... 3500	Life diversifies / Eukaryotes / Prokaryotes / Life evolves	

Across the Curriculum

Geography Have students use a geology text, geological map of the United States, or similar resource to describe the locations of different sedimentary rocks. **L1** **LS**

 Use **Teaching Transparency 12** as you teach this lesson.

Use **Science Integration Activities**, pp. 11-12, as you teach this lesson.

Figure 6-10

Fossils found in lower layers of sedimentary rock are usually older than fossils in upper layers.

A Paleontologists can date fossils by the age of the layer of rock where they occur.

B In the Grand Canyon, the Colorado River has cut through layers of sedimentary rock, exposing the layers.

Relative Dating

One method often used to determine the approximate age of a rock layer, or fossils within the layer, is to look at where the particular rock layer is. In undisturbed areas, older rock layers lie below successively younger rock layers, as shown in **Figure 6-10.** Fossils found in the lower layers of rock are older than those in upper layers. This method of dating fossils is known as **relative dating.** Relative dating can only estimate the age of a fossil.

Radioactive Dating

A method used to give a more accurate age to a rock layer or fossil is dating using radioactive elements. **Radioactive elements** give off radiation, a form of atomic energy. Uranium and a radioactive form of carbon are used in radioactive dating. Radioactive elements change to more stable products as they give off radiation. The radiation is given off at a constant rate, and the rate is different for each element. Scientists can measure how much of a radioactive element has changed. They can accurately determine the age of the rock by comparing the amount of stable product with the amount of radioactive element still present. For example, the radioactive element uranium changes to lead as it ages. Scientists can determine how old a fossil in a rock sample is by measuring the amounts of uranium and lead in the rock. The more lead there is, the older the rock, and by association, the older the fossil.

USING MATH

The Cenozoic era represents approximately 65 million years of Earth's 4.6-billion-year history. Approximately what percent of the total does this era represent?

Could relative dating be used to date fossils in rock layers that have been uplifted or disturbed in other ways? *No, radioactive dating would have to be used to accurately date the fossils.*

Visual Learning

Figure 6-11 **Why do we call these fossil fuels?** *These fuels are formed from the buried remains of ancient organisms; fossils are the evidence of ancient organisms.*

LEP

GLENCOE TECHNOLOGY

Videodisc

The Infinite Voyage: Life in the Balance

Introduction

Chapter 1
Understanding Mass Extinction

Chapter 2
Discovering Patterns of Extinction in Space

The Infinite Voyage: The Great Dinosaur Hunt

Chapter 2
The Evolution of Extinction

Chapter 11
Theories of Extinction

Fossils Show Evolution Occurred

Fossils are a record of organisms that lived in the past. But the fossil record is incomplete, much like a book with some pages missing. Because every living thing doesn't or can't become fossilized, the record will never be complete. By looking at fossils, scientists have determined that many simpler forms of life existed earlier in Earth's history, and more complex forms of life appeared later. The oldest fossil bacteria appeared 3.8 billion years ago. Simple invertebrates appeared in the Cambrian period, about 540 million years ago. The first land plants did not appear until the Silurian period, 438 million years ago. Dinosaurs ruled Earth during the Triassic and Jurassic periods, from 208 to 144 million years ago. The first mammals and birds did not appear until the Jurassic period, about 200 million years ago. **Figure 6-11** shows an artist's drawing of a typical scene of 300 million years ago. The fossil record gives scientists convincing evidence that living things evolved, but there are other types of evidence that support the theory of evolution.

Figure 6-11

In the Carboniferous period, 300 million years ago, amphibians and land plants ruled Earth. Most of the plants of this period eventually became peat, natural gas, petroleum, and coal. *Why do we call these fossil fuels?*

166

Use these program resources as you teach this lesson.
Cross-Curricular Integration, p. 10
Multicultural Connections, pp. 15-16

Activity 6-2

A Radioactive Dating Model

Determining the age of a rock using radioactive element measurement and detection requires expensive equipment. You can get a good idea, however, of how the process works by using a model that explains some of the mechanisms of the process.

Problem

How can a radioactive element be used to determine a fossil's age?

Materials

- 100 pennies
- cardboard box with lid
- graph paper
- pencil

Procedure

1. As a fossil ages, the amount of radioactive element decreases and the amount of stable element increases. Examine the graph below to see the decrease of a radioactive element over time.

2. Make a data table like the one shown.
3. Put 100 pennies face up in a cardboard box and replace the lid.
4. Shake the box for 10 seconds.
5. Take off the lid and take out all coins that are face down.
6. Record the number of coins that you take out.
7. Repeat steps 4 through 6 until all the coins have been removed.

Data and Observations   Sample Data

Trial #	Coins Left	# Removed
0	100	0
1	57	43
2	30	27
3	18	12
4	12	6
5	7	5
6	5	2
7	2	3
8	1	1

Analyze

1. What happens to the number of coins remaining after each trial?
2. Construct a graph of your results. Plot the number of coins remaining face up on the *y*-axis, and plot the trials on the *x*-axis. How does your graph compare with the graph to the left?
3. How does shaking the box represent the energy given off by radioactive elements when they become stable?

Conclude and Apply

4. **Describe** how this model is similar to the decay of a radioactive element.
5. How is this model unlike the decay of a radioactive element?
6. **Explain** why radioactive dating is considered more accurate than relative dating.

6-2 Evidence for Evolution   **167**

Assessment

Performance Have students design a model to mimic radiometric dating. Use the Performance Task Assessment List for Model in **PASC**, p. 51. **P**

Activity 6-2

Purpose

LS **Logical-Mathematical** Students will make and use a model to explain radioactive dating, and infer similarities between the model and radioactive dating. **L2**

COOP LEARN

Process Skills

measuring in SI, making models, communicating, sequencing, making and using graphs, forming operational definitions, interpreting data, observing and inferring

Time

45 minutes

Activity Worksheets, pp. 5, 37-38

Teaching Strategies

Troubleshooting Students may need to be reminded how to make a graph. Have them refer to the Skill Handbook.

Answers to Questions

1. The number decreases by about half.
2. This will be a graph similar to the one showing half-life in this activity.
3. Shaking the box requires energy and represents the energy given off when the elements change to become stable.
4. After a given period of time, about half the coins have changed because of the energy released.
5. The time for a given amount of a radioactive element to change to a stable element is exactly the same each time in the model.
6. Radioactive dating is more accurate because the rate of decay is exact. Fossil beds may be formed at different rates in different time periods.

167

CONNECT TO
CHEMISTRY

Scientists have produced charts that show the amino acid sequences of many proteins and enzymes in a wide variety of species. If two species of organisms have proteins or enzymes that have nearly all of the same amino acids in the same order, they are considered to be closely related. *Explain* what such a relationship indicates about the DNA of these organisms.

Other Evidence for Evolution

Besides fossils, what other evidence is there for evolution? Scientists have found more evidence by looking at similarities in chemical makeup, development, and embryological structure among organisms. You know that the functions of your arm, a dolphin's flipper, a bat's wing, and a bird's wing are all very different. Yet, as you can see in **Figure 6-12,** each of these structures is made up of the same kind of bones. Each has about the same number of muscles and blood vessels. Each of these limbs developed from similar tissues in the embryo. Body parts that are similar in origin and structure are called **homologous.** Homologous structures give evidence that two or more species share common ancestors.

Vestigial Structures

Vestigial structures also give evidence for evolution. A **vestigial structure** is a body part that is reduced in size and doesn't seem to have a function. Examples of vestigial organs in humans are the appendix and the muscles that move the ear. Manatees no longer have back legs, but they

Figure 6-12

A bird wing, bat wing, dolphin flipper, and human arm are homologous. Each has about the same number of bones, muscles, and blood vessels.

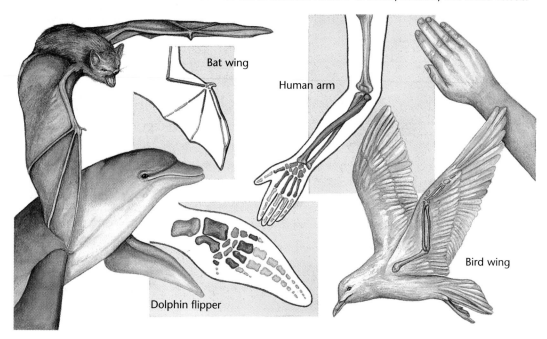

Bat wing

Human arm

Dolphin flipper

Bird wing

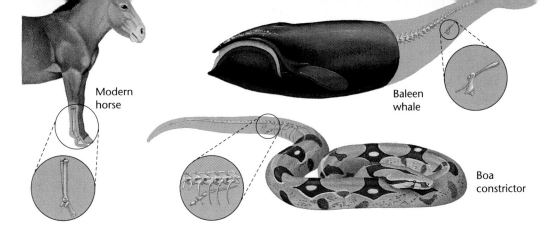

Modern horse

Baleen whale

Boa constrictor

retain the bones of the pelvis. Scientists think vestigial structures are parts that once functioned in an ancestor. **Figure 6-13** shows examples of other nonfunctioning structures.

Embryology

The study of the development of embryos is called **embryology.** An embryo is an organism in its earliest stages of development. Compare the embryos of the organisms in **Figure 6-14.** Do you see similarities? In the early stages of development, the embryos of fish, reptiles, birds, and mammals have a tail and gills or gill slits. Fish keep their gills, but the other organisms lose them as their development continues. In humans, the tail disappears, but fish, birds, and

Figure 6-13

Do whales or snakes have back legs? You can see that they don't, yet both animals have vestigial hip and leg bones where legs once existed. Horses once had four functional toes. Now they walk on just one toe; the others are no longer functional.

Fish   Lizard   Chicken   Rabbit

Figure 6-14

Similarities in the embryos of fish, lizards, chickens, and rabbits show evidence of evolution. Many of the same features are found in all of these organisms. *How can you tell which organism's embryo is which?*

6-2 Evidence for Evolution   **169**

4 Close

Figure 6-15

The DNA of humans and chimpanzees differs by less than one percent. *Why are chimpanzees often used as experimental animals in medical testing?*

lizards keep theirs. These similarities suggest an evolutionary relationship among all vertebrate species. This supports evidence from the fossil record that indicates aquatic, gill-breathing organisms evolved before air-breathing land vertebrates.

DNA

DNA is the molecule that controls heredity. Scientists can determine whether or not organisms are closely related by comparing their DNA. Organisms that are close relatives have similar DNA. By studying DNA, scientists have determined that dogs are the closest relatives of bears. You would probably not be surprised to learn that gorillas and chimpanzees, including the pygmy chimpanzee shown in **Figure 6-15,** also have DNA that is similar.

Genetic evidence also supports the view that primates all evolved from a common ancestor. Primates share many of the same proteins, including hemoglobin. Hemoglobin is a protein in red blood cells that carries oxygen. Many primates have hemoglobin that is nearly identical.

Section Wrap-up

Review

1. How is radioactive dating used to interpret the fossil record?

2. List five examples of evidence that support the theory of evolution.

3. **Think Critically:** Fossil leaves are found in the top layer, fossils of shells are found in the bottom layer, and fish bones are in the middle layer of three undisturbed beds of rock. List the fossil types from oldest to youngest.

Skill Builder
Making and Using Tables
Use **Table 6-1,** the Geologic Time Scale, to answer the following questions. During which periods did the first mammals and flowering plants appear? Which was the longest period of the Paleozoic era? If you need help, refer to Making and Using Tables in the **Skill Handbook.**

Skill Builder
Jurassic; Carboniferous

Primate Evolution 6•3

Primates

Monkeys, apes, and humans belong to the group of mammals called **primates.** The primates share several characteristics that led scientists to think all primates evolved from a common ancestor. All primates have opposable thumbs that allow them to reach out and bring food to the mouth, as shown in **Figure 6-16.** Having an opposable thumb allows you to cross your thumb across your palm and touch your fingers. Think of the problems you might have if you didn't have this type of thumb!

Primates also have binocular vision. Binocular vision permits a primate to judge depth or distance with its eyes. All primates have flexible shoulders and rotating forelimbs. These allow tree-dwelling primates to swing easily from branch to branch, and allow you to swing on a jungle gym. Each of these characteristics provides evidence that all primates share common ancestry.

Science Words

primate
hominid
Homo sapiens

Objectives

- Describe the evidence that all primates evolved from a common ancestor.
- Describe the ancestors of humans.
- Trace the evolutionary history of humans.

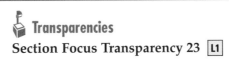

Figure 6-16

An opposable thumb allows tree-dwelling primates to hold onto branches. It also allows you to use your hand in many ways.

6-3 Primate Evolution   171

Prepare

Section Background

- Primates evolved during the Eocene. There are presently about 200 species of primates.
- Genetic studies indicate that the chimpanzee and gorilla are more closely related to humans than to any other primate.

Preplanning

Collect baseballs, keys, buttons, and pencils for the Motivate activity.

1 Motivate

Bellringer

Before presenting the lesson, display **Section Focus Transparency 23** on the overhead projector. Assign the accompanying **Focus Activity** worksheet. L1 LEP

Activity

LS **Kinesthetic** Have students look at **Figure 6-16** and handle various implements to develop an appreciation for the opposable thumb. Have some students tape their thumbs carefully to their hands and see what can be accomplished without the thumb. L1

Program Resources

📁 **Reproducible Masters**
Study Guide, p. 27 L1
Reinforcement, p. 27 L1
Enrichment, p. 27 L3

🗔 **Transparencies**
Section Focus Transparency 23 L1

Demonstration

 Visual-Spatial You can demonstrate binocular vision by having students cover first one eye and then the other. Show students that binocular vision works because a portion of each eye's visual field overlaps, enabling the brain to judge distance. **LEP**

Enrichment

Have students find out about the Leakey family and their work in East Africa. **L2**

GLENCOE TECHNOLOGY

Videodisc

The Infinite Voyage:
The Dawn of Humankind
Introduction

Chapter 2
Evolution of Early Man

Chapter 3
Evolution of Speech: Reasons and Physiological Variants

Chapter 4
Development of Modern Man

Chapter 6
Human Origins: Unexpected Finds in South Africa

Chapter 7
DNA Studies Create Controversy

Figure 6-17

Tarsiers and lemurs belong to a subgroup of primates called the prosimians, which means "before apes."

A Tarsier

B Lemurs

Primate Classification

Primates are divided into two major groups. The first group includes organisms such as lemurs and tarsiers, the prosimians, as shown in **Figure 6-17.** These animals are nocturnal, and have large eyes and excellent hearing. The second group, the higher primates, includes monkeys, apes, and humans.

Hominids

About 4-6 million years ago, our earliest ancestors branched off from the other higher primates. These ancestors, called **hominids,** were humanlike primates that ate both meat and vegetables and walked upright on two feet. Hominids shared some common characteristics with gorillas, orangutans, and chimpanzees, but a larger brain size separated them from these other great apes.

Out of Africa

In the early 1920s, Raymond Dart, a South African scientist, discovered a fossil skull in a quarry in South Africa. The skull had a small brain cavity but humanlike jaw and teeth. Dart named his discovery *Australopithecus. Australopithecus,* one of the earliest hominid groups discovered, means "southern ape." In 1974, an almost-complete skeleton of *Australopithecus* was discovered by an American scientist, Donald Johanson, and his colleagues. They named the fossil, estimated to be 2.9 to 3.4 million years old, Lucy, **Figure 6-18.** Lucy had a small brain but walked upright. Many scientists today think humans evolved in Africa from ancestors similar to Lucy.

Figure 6-18

The fossil remains of Lucy, a hominid, are estimated to be 2.9 to 3.4 million years old.

172   Chapter 6 Evolution

Cultural Diversity

Fossils in Africa Use the information about fossil sites in Africa to increase students' awareness of African geography and their appreciation of the rich heritage found there. Have students research sites such as Olduvai Gorge in Tanzania. **L2**

About 40 years after the discovery of *Australopithecus*, a discovery was made in East Africa by Louis, Mary, and Richard Leakey. The Leakeys discovered a fossil more similar to present-day humans than *Australopithecus*. They named this hominid *Homo habilis,* the "handy man," because they found simple stone tools near him. Scientists estimate *Homo habilis* to be 1.5 to 2 million years old.

Anthropologists suggest that *Homo habilis* gave rise to another species about 1.6 million years ago, *Homo erectus. Homo erectus* had a larger brain than *H. habilis*. This hominid migrated out of Africa about 1 million years ago. *Homo habilis* and *Homo erectus* both are thought to be direct ancestors of humans because they had larger brains and were more similar in form to humans than *Australopithecus*.

Modern Humans

Our species is named **Homo sapiens,** meaning "wise human." The fossil record shows that the human species, a very recent arrival on Earth, evolved about 300 000 years ago.

By about 125 000 years ago, two early groups of *Homo sapiens*, Neanderthal humans and Cro-Magnon humans, probably lived at the same time in parts of Africa and Europe.

Neanderthal humans had short, heavy bodies with thick massive bones, small chins, and heavy browridges, as you can see in **Figure 6-19.** The Neanderthal humans lived in family groups in caves and hunted mammoths, deer, and other large animals with well-made stone tools. For reasons that are not clear, Neanderthal humans disappeared from the fossil record about 35 000 years ago. Most scientists think Neanderthals were a side branch of human evolution, but not direct ancestors of modern humans.

Cro-Magnon fossils have been found in Europe, Asia, and Australia. These fossils are dated from 40 000 to about 10 000 years ago. The oldest recorded art dates from the caves of France where Cro-Magnon humans first painted bison, horses, and spear-carrying people. Cro-Magnon humans lived in caves, made stone carvings, cared for their elderly, and buried their dead, **Figure 6-20.** Standing about 1.6-1.7 m tall, the physical appearance of Cro-Magnon people was almost identical to that of modern humans. Cro-Magnon humans are thought to be direct ancestors of modern humans.

Figure 6-19

In this photograph, the skull of a Neanderthal, right, can be compared with the skull of a Cro-Magnon, left. *What differences can you see between these two skulls?*

174   Chapter 6 Evolution

Figure 6-20

This grave contained objects placed there by Cro-Magnon humans. Graves such as this, tools, and paintings on cave walls led scientists to hypothesize that Cro-Magnon humans had a well-developed culture.

Section Wrap-up

Review

1. Describe at least three kinds of evidence that suggests all primates shared a common ancestor.

2. What is the importance of *Australopithecus?*

3. Describe the differences among Neanderthal humans, Cro-Magnon humans, and modern humans.

4. **Think Critically:** Propose a hypothesis about why teeth represent the most abundant available fossils of hominids.

Science Journal

Write a story in your Science Journal about what a day would be like for you if you did not have thumbs.

Skill Builder

Concept Mapping

Using information in this section, make a concept map to show the sequence of human ancestors. Use the following terms: Neanderthal human, lemurs and tarsiers, *Homo habilis, Australopithecus,* modern *Homo sapiens,* Cro-Magnon human. If you need help, refer to Concept Mapping in the **Skill Handbook.**

Skill Builder

Concept maps should show an events chain in the following order: lemurs and tarsiers, *Australopithecus, Homo habilis,* Neanderthal human, Cro-Magnon human, and modern *Homo sapiens.*

✔ Assessment

Portfolio Use this Skill Builder to assess students' abilities to form a concept map of the sequence of human ancestors. Use the Performance Task Assessment List for Concept Map in **PASC,** p. 89. **P**

---

4 Close

•MINI•QUIZ•

Use the Mini Quiz to check students' recall of chapter content.

1. **Which order do monkeys, apes, and humans belong to?** *primates*
2. **About 4 to 6 million years ago, the _____ , our earliest ancestors, branched off from the other higher primates.** *hominids*
3. **Cro-Magnon people were almost identical to _____ .** *modern humans*

Section Wrap-up

Review

1. Homologous structures, similar DNA, and fossils suggest a common ancestry for primates.

2. *Australopithecus,* an early example of a hominid, had a small braincase, but humanlike jaws and teeth.

3. Neanderthal humans had short bodies with massive bones, and heavy browridges. Cro-Magnon humans were taller, invented art, made stone carvings, and buried their dead. Cro-Magnon people looked almost the same as modern humans.

4. **Think Critically** Teeth are the hardest parts of the organism and therefore the most likely to be fossilized.

Science Journal

The story should indicate activities that would be more difficult to perform.

Prepare

Section Background

The Endangered Species Handbook is published by the Animal Welfare Institute, P.O. Box 3650, Washington, D.C. 20007.

1 Motivate

Bellringer

Before presenting the lesson, display **Section Focus Transparency 24** on the overhead projector. Assign the accompanying **Focus Activity** worksheet. L1 LEP

Discussion

Ask students whether a minnow should prevent the building of a major dam that could supply water and electricity to thousands of people (snail darter story).

2 Teach

CONNECT TO

EARTH SCIENCE

Answer Most dinosaurs became extinct during the Cretaceous period. Hypotheses for extinction include meteor impact, climate change, and vegetational changes.

Content Background

Cheetahs have little genetic variability, and this is the apparent reason they are endangered.

ISSUE:

6•4 Extinction and Evolution

Science Words

extinction
endangered species

Objectives

- Define *extinction* and identify some causes of extinction.
- Define *endangered species* and identify some causes of species endangerment.

CONNECT TO

EARTH SCIENCE

Evidence that extinction is a natural event is found in the fossil record. Large-scale extinction events, such as extinction of the dinosaurs, have taken place in the past. Use **Table 6-1** to determine when these extinctions occurred. Use other references to *trace* what hypotheses scientists have made to explain why these extinctions happened.

The Nature of Extinction

By the time you finish reading this paragraph, one more plant or animal species may have joined the thousands that have become extinct. **Extinction,** the dying out of a species, is a natural event. Extinction occurs naturally because the environment is always changing. Through the slow process of natural selection, those species that are unable to change or adapt are naturally removed.

You know from the fossil record that many species such as the dinosaurs have become extinct throughout Earth's history. Evidence shows that there have been at least four mass extinctions during the 3.5 billion years that life has existed on Earth. Each of these mass extinctions, however, was followed by an explosion of new species that evolved rapidly to take advantage of the suddenly empty environment.

Endangered Species

Today, species are becoming extinct at an alarming rate. Human overpopulation and human activities such as illegal hunting, habitat destruction, and pollution are contributing to the extinction of many species worldwide. Two species in immediate danger of extinction are shown in **Figure 6-21.** There are fewer than 500 individuals of each of these species left in the wild. These are just two examples of **endangered species,** species with so few individuals left that they are in danger of becoming extinct. Since the 1500s, more than 500 species and subspecies have become extinct in the United States.

Efforts to halt the process of extinction included the passage of the Endangered Species Act of 1973. As a result of that law, the last remaining individuals of some species have been captured and contained in captive breeding programs. There were only 17 red wolves, *Canis rufus,* left in the world when the U.S. Fish and Wildlife Service began a captive breeding program. Now numbering only 60, the red wolf remains endangered. But captive breeding is extremely costly; nearly half of the money available for protecting all species is spent on just a dozen species.

Program Resources

 Reproducible Masters
Study Guide, p. 28 L1
Reinforcement, p. 28 L1
Enrichment, p. 28 L3
Science and Society Integration, p. 10
Critical Thinking/Problem Solving, p. 10
Lab Manual 14
Activity Worksheets, p. 5

Transparencies
Section Focus Transparency 24 L1

2 Points of View

Extinction Is Natural

Because extinction is a natural process, some people feel that we should not try to save endangered species. If species such as the California condor are dying out, it is because they have been unable to adapt to changing conditions; they are being selected against in their environment. Even species that are dying out due to human activity should be looked at as unable to adapt. Human beings are also part of nature; if elephants are killed off by people, their extinction is also due to natural selection.

Saving Endangered Species

Even though extinction is a natural process, most species endangered today are threatened by human actions rather than their changing environment. Water pollution in a stream, for example, can kill an entire population of pigtoe mussels before they have a chance to adapt. These freshwater mollusks are eaten by raccoons, muskrats, wading birds, and ducks. They are filter-feeders, removing pollutants and sediments from our drinking water. These are some of the reasons why scientists want to protect these endangered mussels. But other people think that we should protect all organisms, just because they exist.

A Siberian Tiger

Figure 6-21
Two of the most endangered species in the world are the Siberan tiger of Russia and the red wolf of the United States.

B Red Wolf

Section Wrap-up

Review

1. Explain why extinction could be viewed as a part of natural selection.

2. Mass extinctions have occurred several times in Earth's history. Why are scientists so concerned about the extinction of species today?

Explore the Issue

Should we just let endangered species die because they are unable to adapt? Or do human beings have an obligation to try to save as many of these species as we can, regardless of the cost?

SCIENCE & SOCIETY

---

3 Assess

Check for Understanding

? FLEX Your Brain

Use the Flex Your Brain activity to have students explore EXTINCTION.

🗂 **Activity Worksheets,** p. 5

Reteach
Have students list the effects of human activity in rain forests and the Arctic tundra.

Extension
🗂 For students who have mastered this section, use the **Reinforcement** and **Enrichment** masters.

4 Close

Ask students whether they think a species can be preserved when only 60 individuals are left. Remind them that species change through variations in individual organisms.

Section Wrap-up

Review

1. Extinction is a natural event that can open habitats to new species.

2. Today, most extinctions are the result of human interactions.

Explore the Issue

Some students may agree that humans should preserve species. Others may think preservation should be left up to nature. Still other students may think that extinction is not important.

177

Science & ART

Sources

- "Cave Art Find Is a Treasure Trove," *New York Times*, reported in *The Columbus Dispatch*. January 19, 1995, Columbus, Ohio, p. 8A.
- Hadingham, Evan. *Secrets of the Ice Age: The World of the Cave Artists*. New York: Walker and Co., 1979. Read about the myths and legends that surround cave art.
- Samachson, Dorothy and Joseph. *The First Artists*. New York: Doubleday, 1970. The authors describe major cave art collections in France and Spain, along with findings in Italy, Scandinavia, Russia, Africa, Australia, and the United States.

Background

Chauvet Cave also includes the skeletons of about 40 cave bears, extinct for the past 12 000 years. Some experts think the Stone Age artists chose this cave for their paintings because it had been used by the bears. A bear skull had been placed on a stone in the middle of the cave, suggesting an altar. Perhaps the artists believed their paintings captured the bears' power for themselves. The site also included flint knives, footprints, and traces of fireplaces. These clues will help archaeologists determine the purpose of cave art. The fact that the site was undisturbed for thousands of years excites researchers. So far, about 300 caves containing Stone Age (Paleolithic) paintings have been uncovered in Europe.

Teaching Strategies

- **IS Interpersonal** Discuss with students what the cave artists' work might tell us about their lives and their beliefs.

Stone Age Art

When did art begin? Radiocarbon dating has confirmed that artists began sketching on cave walls thousands of years ago. More than 300 cave paintings and engravings, some dating back 10 000 to 30 000 years, were found in December, 1994, in Chauvet Cave in south-central France. Many artists worked in the cave over thousands of years. These skilled artists used red, black, and yellow earth pigments and charcoal to draw the animals they saw daily. They included the woolly rhinoceros, mammoths, cave bears, and lions.

The murals also contain the first paintings found of a panther and owls, plus drawings or silhouettes of human hands.

Dropping Back in Time

Cave explorers, also called spelunkers, found Chauvet Cave after they noticed a draft coming from a pile of rubble. They cleared away the rubble, and lowered themselves into a cave that turned out to be the length of five football fields. Scientists believe that the original door to the cave had been blocked by geologic shifts over thousands of years.

First Findings

Cave paintings were first discovered in 1879 near Altamira, Spain. Another major collection of cave paintings was located near Lascaux, France, in 1940. That cave was opened to the public, but the change in the atmosphere faded the colorful paintings, which are 14 500 to 15 000 years old. In time, green fungus began to creep over them. To protect the paintings, the cave was closed to the public in 1963.

Painting Protection

Chauvet Cave is now open only to archaeologists, and an effort is being made to stabilize the climate inside the cave so the paintings will continue to be preserved. Many scientists believe that more cave art remains to be found. They hope that studying the paintings will help us learn how people first began to use symbols to express themselves.

Art in Chauvet Cave

*inter***NET** **CONNECTION**

Visit the Chapter 6 Internet Connection at Glencoe Online Science, **www.glencoe.com/sec/ science/life**, for a link to more information about cave art.

- **IS Visual-Spatial** Explore ways these paintings help describe the artists' position in human evolution. The paintings include human hands and show the existence of an opposable thumb.

- **IS Visual-Spatial** Discuss some pictures scientists might like to find as part of cave paintings. What pictures would best tell us about ancient life on Earth, including human life?

- **IS Logical-Mathematical** Ask students whether cave paintings would be considered fossils. Why or why not? Discuss the difference between fossils and artifacts.

- **IS Logical-Mathematical** Discuss how radioactive dating might be used to determine the age of cave paintings.

Chapter 6 Review

Summary

6-1: Mechanisms of Evolution

1. Hereditary features of a species of organism evolve or change over time.
2. Lamarck's explanation of acquired characteristics implied that characteristics parents developed during their lives were passed to their offspring. Darwin developed the natural selection theory of evolution, which states that organisms best adapted to their environments survive and reproduce.
3. Variations are differences in inherited traits among members of the same species.

6-2: Evidence for Evolution

1. Fossils are remains of life from the past; they are evidence of evolution.
2. The age of a fossil can be found by relative dating or radioactive dating.
3. Evidence for evolution is obtained by comparing embryology, homologous structures, chemical similarities, and vestigial structures.

6-3: Primate Evolution

1. Primates share several common characteristics including opposable thumbs, binocular vision, flexible shoulders, and rotating forearms.
2. Hominids are humanlike primates. The earliest known hominid is *Australopithecus*. *Homo habilis* is thought to be the direct ancestor of humans.

3. Modern humans, *Homo sapiens*, evolved 300 000 years ago.

6-4: Science and Society: Extinction and Evolution

1. Extinction is the dying out of a species due to changes in the environment.
2. Endangered species can be protected by captive breeding and by preserving their environments.

Key Science Words

a. embryology
b. endangered species
c. evolution
d. extinction
e. fossil
f. gradualism
g. hominid
h. homologous
i. *Homo sapiens*
j. natural selection
k. primate
l. punctuated equilibrium
m. radioactive element
n. relative dating
o. sedimentary rock
p. species
q. variation
r. vestigial structure

Reviewing Vocabulary

Match each phrase with the correct term from the list of Key Science Words.

1. remains of a once-living organism
2. structure with no obvious use
3. similar organisms that successfully reproduce
4. body structures that are similar in origin
5. change in hereditary feature over time

Chapter 6 Review   179

Chapter 6

Review

Summary

Have students read the summary statements to review the major concepts of the chapter.

Reviewing Vocabulary

1. e	**6.** q
2. r	**7.** b
3. p	**8.** f
4. i	**9.** a
5. c	**10.** k

 Assessment

Portfolio Encourage students to place in their portfolios one or two items of what they consider to be their best work. Examples include:

- Activity 6-1, pp. 158-159
- MiniLAB, p. 163
- Science Journal, p. 173 P

Performance Additional performance assessments may be found in **Performance Assessment** and **Science Integration Activities.** Performance Task Assessment Lists and rubrics for evaluating these activities can be found in Glencoe's **Performance Assessment in the Science Classroom.**

GLENCOE TECHNOLOGY

 Videodisc

Glencoe Life Science Interactive Videodisc
Side 1, Lesson 2
Evolution: Patterns of Change

14447 to 25820

MindJogger Videoquiz

Chapter 6 Have students work in groups as they play the Videoquiz game to review key chapter concepts.

Checking Concepts

1. b	**6.** c
2. b	**7.** d
3. d	**8.** a
4. c	**9.** b
5. a	**10.** b

Understanding Concepts

11. Giraffes with longer necks (variation) survived because they outlived shorter-necked individuals because they could eat leaves higher up on trees. This variation was passed on to offspring because it is a genetic trait.

12. Variations result from differences in genes. The resulting adaptations allow the best organisms to survive, passing along these genes. The poorer traits are eliminated because these genes are not passed on to offspring.

13. The organisms could not interact. If the environments were different, different variations would be naturally selected and different species could evolve.

14. Gradualism explains evolution as slow, steady, and continuous. Punctuated equilibrium shows rapid change in DNA and rapid evolution.

15. Relative dating compares the age of fossils found in different rock layers to each other. Radioactive dating is more accurate because it measures the actual amount of radioactive material present.

Thinking Critically

16. Lamarck: Ducks have webbed feet for swimming. The ducklings acquired the webbed feature. **Darwin:** A variation in a duck's feet was an advantage to survival in the environment. This

180

6. a difference in an inherited trait
7. organisms in danger of extinction
8. model of evolution showing slow change
9. the study of stages of an organism's early development
10. group containing monkeys, apes, and humans

Checking Concepts

Choose the word or phrase that completes the sentence.

1. _____ is the death of all members of a species.
   a. Evolution   c. Gradualism
   b. Extinction   d. Variation
2. The most accurate age of a fossil can be found using _____.
   a. natural selection
   b. radioactive elements
   c. relative dating
   d. camouflage
3. Homologous structures, vestigial structures, and fossils all provide evidence of _____.
   a. extinction
   b. species populations
   c. food choice
   d. evolution
4. A factor that controls natural selection is _____.
   a. inheritance of acquired traits
   b. unused traits become smaller
   c. organisms produce more offspring than can survive
   d. the size of an organism
5. A series of helpful variations in a species may result in _____.
   a. adaptation   c. extinction
   b. fossils   d. climate change

6. Organisms adapted to their environment are _____.
   a. extinct
   b. not reproducing
   c. surviving and reproducing
   d. forming fossils
7. The red wolf and the Siberian tiger are examples of _____.
   a. fossils   c. extinct species
   b. hominids   d. endangered species
8. Opposable thumbs and binocular vision are characteristics of _____.
   a. all primates   c. humans only
   b. hominids   d. monkeys and apes
9. The earliest known paintings were done by _____.
   a. *Australopithecus*
   b. Cro-Magnon humans
   c. *Homo sapiens*
   d. *Homo habilis*
10. A fossil has the same number of bones in its hand as a gorilla. This represents _____, one type of evidence for evolution.
    a. DNA
    b. homologous structures
    c. vestigial structures
    d. embryology

Understanding Concepts

Answer the following questions in your Science Journal using complete sentences.

11. Use Darwin's theory of natural selection to explain how the giraffe got its neck.

12. Describe what causes variations and how the resulting adaptations help explain evolution.

13. What could happen if members of a population became completely separated from each other?

was a feature that was passed on to the offspring.

17. Geographical isolation due to a volcanic eruption can separate a population. Two species may form.

18. Chameleons blend into their environment. This ability to change color is an adaptation for survival.

19. The layer of rock where it was found would be noted. Radioactive dating would be done. Comparisons to other fossils would be made.

20. An organism can become extinct because its habitat is destroyed, its food supply disappears or it is killed off. Tigers, carrier pigeons, and cheetahs are examples.

Developing Skills

21. Hypothesizing Ponds are not overpopulated by frogs because not all eggs survive. Some are eaten, the water dries up, and so on.

14. Compare gradualism and punctuated equilibrium.
15. Compare relative dating to radioactive dating of fossils. Which do you think is more accurate and why?

Thinking Critically

16. Explain how Lamarck and Darwin viewed a duck's webbed feet.
17. Using an example, explain how a new species of organism could evolve.
18. How is the coloration of chameleons an adaptation to their environment?
19. Describe the process a scientist would use to figure out the age of a fossil.
20. Explain how an organism could become extinct. Give an example.

Developing Skills

If you need help, refer to the **Skill Handbook.**

21. **Hypothesizing:** Frog eggs are common in ponds in spring. Make a hypothesis as to why ponds are not overpopulated by frogs in the summer. Use the ideas of natural selection to help you.
22. **Comparing and Contrasting:** Compare and contrast Cro-Magnon humans and Neanderthal humans.
23. **Observing and Inferring:** Observe the birds' beaks pictured below. Describe each. Infer the types of food each would eat and explain why.

24. **Concept Map:** Make an events chain map that lists the events that led Darwin to his theory of evolution. Include the following terms in your map: Darwin became naturalist on HMS *Beagle;* Darwin wrote *On the Origin of Species by Means of Natural Selection;* Darwin visited South America; Darwin bred pigeons; Darwin observed organisms in the Galápagos Islands; Darwin developed his theory of evolution by natural selection.
25. **Interpreting Data:** Interpret the following data. The chemicals present in certain bacteria were studied. Listed below are the chemicals found in each type. Each letter represents a different chemical. Use this information to determine which of the bacteria are closely related.

Chemicals Present	
Bacteria 1	A, G, T, C, L, E, S, H
Bacteria 2	A, G, T, C, L, D, H
Bacteria 3	A, G, T, C, L, D, P, U, S, R, I, V
Bacteria 4	A, G, T, C, L, D, H

Performance Assessment

1. **Display:** Make a collection of fossils found in your area. If you live in a city, visit a museum of natural history and include a report on the fossil history of the land the city is built upon.
2. **Poster:** Research the work of Alfred Wallace. On a world map, show the sites of his work and Darwin's. Compare and contrast the areas in which they worked.
3. **Model:** Using clay and a stone, shell, or other small object of your choice, make a model of a cast fossil. How does a cast fossil differ from an imprint fossil?

Chapter 6 Review

22. **Comparing and Contrasting** Neanderthals had short, thick bodies, heavy bones, small chins, and heavy browridges. Cro-Magnon humans looked similar to modern humans. They both lived in caves. Cro-Magnons made the first known cave paintings and buried their dead.
23. **Observing and Inferring** The hummingbird's beak is long and thin for sipping nectar. The bird of prey has a curved, sharp beak for eating prey. The duck has a wide, flat beak for dabbling in the mud. The pelican has a long beak with storage for scooping up food.
24. **Concept Map** Map should flow as follows: Darwin became naturalist on *HMS Beagle;* Darwin visited South America; Darwin observed organisms in the Galápagos Islands; Darwin bred pigeons; Darwin developed his theory of natural selection; Darwin wrote *On the Origins of Species by Means of Natural Selection.*
25. **Interpreting Data** Bacteria 2 and 4 have the same chemicals and are, therefore, most closely related.

Performance Assessment

1. Fossils will vary depending on locality. Use the Performance Task Assessment List for Display in **PASC,** p. 63. **P**
2. Wallace worked in Malaysia and South America; Darwin's voyage took him around the world. Use the Performance Task Assessment List for Poster in **PASC,** p. 73. **P**
3. A cast is made from material that fills an imprint. Use the Performance Task Assessment List for Model in **PASC,** p. 51. **P**

Assessment Resources

📁 Reproducible Masters
Chapter Review, pp. 15-16
Assessment, pp. 36-39
Performance Assessment, p. 20

Glencoe Technology
- 🔘 **Chapter Review Software**
- 🔘 **Computer Test Bank**
- 📼 **MindJogger Videoquiz**

UNIT PROJECT 2

Objectives

LS Logical-Mathematical Students will measure, record data, analyze statistics, and design a way to display data and conclusions about a human genetic characteristic. **L1**

COOP LEARN

Summary

In this project, students will collect data on foot length, and then find the mean, median, and mode for foot length in different age groupings for males and for females. The class will then display and discuss these statistics. Because this is a characteristic controlled by more than one gene, students should arrive at a bell-shaped curve on a graph. Through collection of large amounts of data, students will experience more accurate results than with a small sample.

Time Required

This project is designed to be worked on throughout Chapter 5. The students will need a period to plan and establish data charts. They will need time outside the class to collect their data. They will need another period to design and construct their statistical display. The transfer of data and statistical analysis may be done in an additional period, or may be interspersed with other lessons.

Preparation

- Establish cooperative groups of 4 or 5 students.
- Assemble materials: paper strips and pencils to make rulers, paper for charts, materials for display.

Teaching Strategies

- This project examines one trait governed by more than one gene. While the project can begin as you start Chapter 5, the principle of polygenic inheritance is covered in Section 5-2.

182

Keep Those Toes A-Tapping!

Genetics influences the way you look, the way you act, and your health. Some genetically influenced characteristics are obvious. Skin color, hair color, eye color, and height are frequently used to describe people. Other genetically influenced characteristics are not as noticeable. Do you have attached or free earlobes? Can you taste PTC? Do you have a "hitchhiker's thumb?" Because it is hard to study a number of genetic traits all at once, this project focuses on gathering data for only one characteristic—the length of feet.

Purpose

The length of the human foot is influenced by multiple genetic factors. To find out about this unique trait, you will first gather data, then analyze them, and finally design a method to display results.

Materials

- 10 people
- plastic metric ruler with raised markings
- calculator
- paper

Procedure

1. Work in a group. Each group will need to make a chart to record data. The chart will need to have horizontal heads that read *Age, Sex,* and *Foot length (left/right)* across the top. Down the side will be a number for each pair of feet you measure.

2. You can make a flexible ruler by placing a strip of paper over a plastic ruler with raised markings in centimeters. Rub over the ruler with your pencil to transfer the markings to your paper ruler.

3. Each group should measure the feet of *at least* ten people. These people should be a variety of ages. Include both males and females.

4. As a class, decide how you are going to divide the age groups. Are you going to make three-year groupings (ages 0-2, 3-5, etc.) or five-year groupings? Are you going to make smaller

groupings for young children, but increase the age increments to ten years for adults? Each group will now be assigned a particular age grouping and will make a chart for just that age group. Once these age-group charts are ready, each group will fill in the data they collected on the appropriate age charts.

5. Once everyone has recorded the data they measured, each group is ready to statistically analyze their particular age-group chart.

6. Calculate the mean, median, and mode of foot length for males and for females in a particular age group. The *mean* of male foot length is found by adding all the lengths of the male foot measurements in the age group and dividing by the number of male subjects in that age group. Repeat the process for female subjects.

7. The foot lengths for all males within the age group should be listed in order from shortest to longest. The *median* of male foot length can then be found by finding the length in the exact middle of the list. (If there is an even number, the median is calculated by averaging the two middle lengths.)

Go Further

Ask a local podiatrist for charts that list the average foot size for various age groups. See how close your results come to these published averages. Check with several local shoe stores to find out what sizes sell the most.

8. The *mode* is the foot length that appears most frequently in the data. The same process is applied to the female foot length data to find female median foot length and the mode for female foot length.

Using Your Research
As a class, decide how to display the data you have analyzed so that the differences between males and females and among different age groupings can be seen easily. Make construction paper feet to display your statistics. Try to come up with a creative way of showing how foot length changes with age. You might also want to show the variation in foot size within age groupings.

References
Burns, George W. *The Science of Genetics.* New York: Macmillan Co., 1969.

Gould, Stephen Jay. *Hen's Teeth and Horses' Toes.* New York: W.W. Norton and Co., 1983.

Leigh, Julia and David Savold. *The Day That Lightening Chased the Housewife and Other Mysteries of Science.* New York: Harper and Row—Perennial Library, 1989.

Sutton, Caroline. *How Do They Do That?* New York: Quill Press, 1982.

• Before dividing students into groups, show the class a picture of an individual from a magazine. Have the class describe this person using traits. Write the descriptive words on the board. Discuss how genetics affects some of the traits listed (pierced ears and salon-colored hair are not inherited). What other characteristics do genetics affect? A hitchhiker's thumb is one that bends outward; attached earlobes have no free-hanging lobe.

• Students should make their measurements from the top of the big toe to the heel.

• Discuss how large amounts of data make the outcome of an investigation reliable.

Go Further
What are some genetic traits that we can't observe just by looking at a person (e.g., blood type, ability to taste PTC, etc.)? The ability to taste phenylthiocarbamide (PTC) is determined by placing a test paper in the mouth. Assess students' abilities to analyze a genetic trait by having them use PTC test papers to see if they are "tasters" or "nontasters." Volunteers could also test their parents and other relatives and try to do pedigree analyses of the trait in their families. See page 140 for symbols used in drawing pedigrees. **CAUTION:** *When doing pedigrees, it is important to be sensitive to students who are adopted and may not know their genetic history.* Being a taster is determined by a dominant allele (T), so a taster may have the genotype TT or Tt. A nontaster can only have the genotype tt. By tracing how the alleles must be passed from generation to generation, students should be able to determine the genotypes for some of their relatives.

183

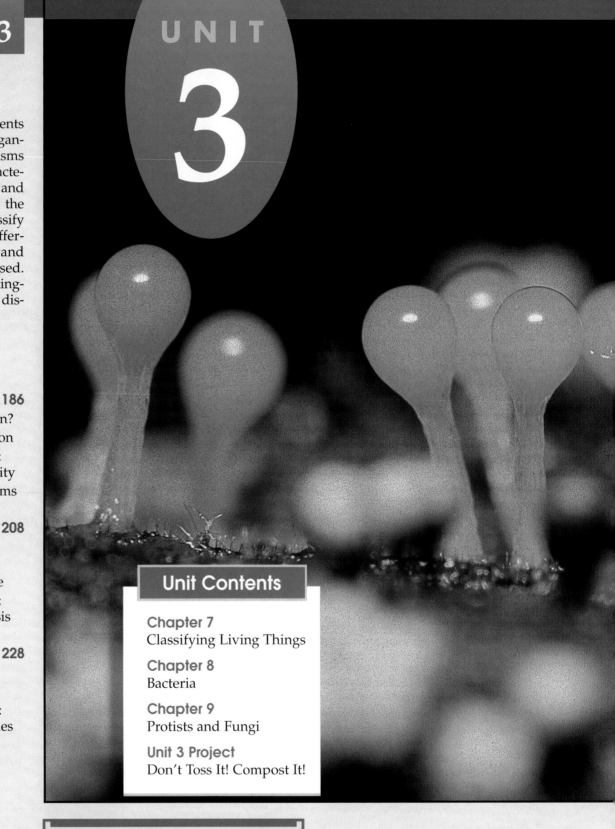

UNIT 3

Diversity of Life

Unit 3 introduces students to methods by which organisms are classified. Organisms in the kingdoms Archaebacteria, Eubacteria, Protista, and Fungi are compared, and the characteristics used to classify them are presented. Differences in cell structure and ecology are also discussed. The importance of each kingdom to human ecology is discussed.

CONTENTS

Chapter 7
Classifying Living Things   **186**

7-1   What is classification?

7-2   Modern Classification

7-3   Science and Society:
      Diminishing Diversity

7-4   Identifying Organisms

Chapter 8
Bacteria   **208**

8-1   Two Kingdoms of
      Bacteria

8-2   Bacteria in Your Life

8-3   Science and Society:
      Fighting Tuberculosis

Chapter 9
Protists and Fungi   **228**

9-1   Kingdom Protista

9-2   Kingdom Fungi

9-3   Science and Society:
      Monitoring Red Tides

U N I T

3

Unit Contents

Chapter 7
Classifying Living Things

Chapter 8
Bacteria

Chapter 9
Protists and Fungi

Unit 3 Project
Don't Toss It! Compost It!

Science at Home

Fun with Fungi Have students visit a supermarket and make lists of the kinds of mushrooms that are sold in the produce section, as well as the prices. Then have them research where each species of mushroom grows and whether it is collected in the wild or grown commercially. Ask students to compare the prices of the mushrooms with the ways they are collected. `L1` `LEP`

Diversity of Life

What's Happening Here?

Are these orange structures flowers? Actually, these structures are the fruiting bodies of an organism called a slime mold. Slime molds move amoebalike over decaying leaves on the forest floor. As the organism feeds, it grows, until suddenly drops appear on the upper surface. These drops develop into stalks with small, spore-containing heads. Elsewhere in the forest, a tree seems to sprout turkey tails, shown in the small photo. Slime molds and turkey tails are examples of protists and fungi, respectively, two of the kingdoms you will learn about in this unit.

Science Journal

Do you eat ice cream or take antibiotics when you are sick? If you do, then you have had contact with protists and fungi. Go to the library and look up *penicillin* and *carrageenan*. In your Science Journal, describe each item as a product of either a protist or a fungus, then explain whether you would find that product in ice cream or in antibiotics.

185

Theme Connection

Systems and Interactions Pose the following question. **Suppose a tree in a temperate forest is struck by lightning and killed. How would the nutrients and energy in the tree be released for use by other plants?** *Bacteria and fungi would attack the tree and begin its eventual decomposition. The products of decomposition are nutrients that can be used by other living plants.*

Cultural Diversity

Pathogens Have students make posters that show the geographical range for particular diseases. The maps may also indicate the kingdom to which each pathogen belongs and illustrations that show how the diseases spread. For example, people in tropical and subtropical regions are subject to malaria, which is caused by a protist. Mosquitoes are vectors for the disease. [L2]

INTRODUCING THE UNIT

What's Happening Here?

Have students compare the photograph of the slime mold fruiting bodies with that of the bracket fungus. Students may say that both are bright-colored, plantlike organisms. After students read the text, they should be able to indicate that both organisms reproduce by spores. Explain that the slime mold is a member of Kingdom Protista, and the fungus is a member of Kingdom Fungi.

Background

Introduce students to the idea that bacteria and fungi are useful organisms that decompose dead organic material and turn it back into organic compounds that can be recycled. Explain that without the function performed by bacteria and fungi, all life would eventually become extinct due to nutrients being locked away in other dead organisms.

Previewing the Chapters

Tell students that bacteria are sometimes used to clean up spilled oil. Ask them to explain how they think bacteria accomplish this task. Then encourage them to look through the chapters to become familiar with monerans, protists, and fungi.

Tying to Previous Knowledge

Some students will know that many diseases, such as pneumonia and strep throat, are caused by bacteria. They may even suggest that all bacteria cause disease. Use their preconception to explain that most of the monerans are useful in recycling materials. Explain that some members of Kingdom Protista and most of the members of Kingdom Fungi also perform this function.

Chapter Organizer

Section	Objectives/Standards	Activities/Features
Chapter Opener		Explore Activity: Use distinguishing features to classify organisms. p. 187
7-1 What is classification? (1 session)*	1. **Give examples** that show the need for classification systems. 2. **Describe** Aristotle's system of classification. 3. **Explain** Linnaeus's system of classification. National Content Standards: (5-8) UCP1, A1, A2, C1, C5, G1, G2, G3	Problem Solving: Classifying Unknown Organisms, p. 189 Activity 7-1: Classifying Seeds, pp. 190-191 Skill Builder: Observing and Inferring, p. 192 Using Math, p. 192
7-2 Modern Classification (2 sessions, 1½ blocks)*	1. **Name** the six kingdoms of living things. 2. **Identify** characteristics and members of each kingdom. 3. **List** the groups within each kingdom. National Content Standards: (5-8) UCP1, C1, E1, E2, C5	Using Technology: Beyond Appearances, p. 194 Science and History: Clarifying Classification, p. 19 Skill Builder: Concept Mapping, p. 196 Using Computers: Database, p. 196
7-3 Science and Society Issue: Diminishing Diversity (1 session, ½ block)*	1. **Identify** one negative effect of decreased species diversity. 2. **Explain** why the price of maintaining species diversity may be too high. National Content Standards: (5-8) UCP2, UCP3, UCP4, C4, F2	Explore the Issue, p. 199
7-4 Identifying Organisms (1 session, ½ block)*	1. **List** reasons scientific names are more useful to scientists than common names. 2. **Identify** the function of a dichotomous key. 3. **Demonstrate** how to use a dichotomous key. National Content Standards: (5-8) UCP1, A1, A2, C1, D1	MiniLAB: How are organisms named? p. 201 Skill Builder: Classifying, p. 203 Science Journal, p. 203 Activity 7-2: Using a Dichotomous Key, p. 204

* A complete Planning Guide that includes block scheduling is provided on pages 32T-35T.

Activity Materials

Explore	Activities	MiniLABs
page 187 insect collection, hand lens	pages 190-191 packet of 10 different kinds of seeds, hand lens, metric ruler, 2 sheets of paper page 204 No special materials are needed.	page 201 modeling clay

Need Materials? Call Science Kit (1-800-828-7777).

Teacher Classroom Resources

Reproducible Masters	Transparencies	Teaching Resources
Study Guide, p. 29 **Reinforcement**, p. 29 **Enrichment**, p. 29 **Activity Worksheets**, pp. 41-42 **Multicultural Connections**, pp. 17-18	**Section Focus Transparency 25,** Classified Information	**Glencoe Life Science Interactive Videodisc** **Life Science CD-ROM** **Spanish Resources** **English/Spanish Audiocassettes** **Cooperative Learning Resource Guide** **Lab Partner** **Lab and Safety Skills** **Lesson Plans**
Study Guide, p. 30 **Reinforcement**, p. 30 **Enrichment**, p. 30 **Lab Manual 15**, Classification **Concept Mapping**, p. 19	**Section Focus Transparency 26,** A Flower by Any Other Name **Teaching Transparency 13,** Life's Six Kingdoms **Teaching Transparency 14,** Classification Pyramid	**Assessment Resources**
Study Guide, p. 31 **Reinforcement**, p. 31 **Enrichment**, p. 31 **Science Integration Activity 7**, Human Impact on the Environment **Critical Thinking/Problem Solving**, p. 11 **Science and Society Integration**, p. 11	**Section Focus Transparency 27,** Lost Forever	**Chapter Review**, pp. 17-18 **Assessment**, pp. 47-50 **Performance Assessment in the Science Classroom (PASC)** **MindJogger Videoquiz** **Alternate Assessment in the Science Classroom** **Performance Assessment** **Chapter Review Software** **Computer Test Bank**
Study Guide, p. 32 **Reinforcement**, p. 32 **Enrichment**, p. 32 **Activity Worksheets**, pp. 43-44, 45 **Cross-Curricular Integration**, p. 11	**Section Focus Transparency 28,** All in the Name **Science Integration Transparency 7,** Living and Nonliving Classification	

Key to Teaching Strategies

The following designations will help you decide which activities are appropriate for your students.

L1 Level 1 activities should be within the ability range of all students, including those with learning difficulties.

L2 Level 2 activities should be within the ability range of the average to above-average student.

L3 Level 3 activities are designed for the ability range of above-average students.

LEP LEP activities should be within the ability range of Limited English Proficiency students.

LS These activities are designed to address different learning styles.

COOP LEARN Cooperative Learning activities are designed for small group work.

P These strategies represent student products that can be placed into a best-work portfolio.

GLENCOE TECHNOLOGY

The following multimedia resources are available from Glencoe.

The Secret of Life
What's in Stetter's Pond: The Basics of Life
Science and Technology Videodisc Series (STVS)
Plants & Simple Organisms
Simple Forms of Life in the Antarctic
Animals
Insect Museum

National Geographic Society Series
STV: Biodiversity
Glencoe Life Science Interactive Videodisc
Evolution: Patterns of Change
Life Science CD-ROM

Teacher Classroom Resources

This is a representation of key blackline masters available in the Teacher Classroom Resources.

Teaching Aids

Section Focus Transparencies

Science Integration Transparencies

Teaching Transparencies

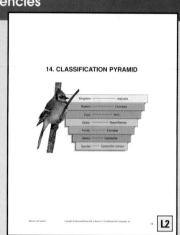

Meeting Different Ability Levels

Study Guide

Reinforcement

Enrichment Worksheets

Hands-On Activities

Science Integration Activity

Lab Manual

Assessment

Performance Assessment

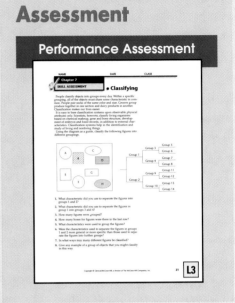

Enrichment and Application

Critical Thinking/Problem Solving

Cross-Curricular Integration

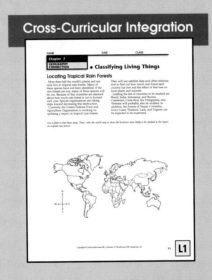

Science and Society Integration

Multicultural Connections

Concept Mapping

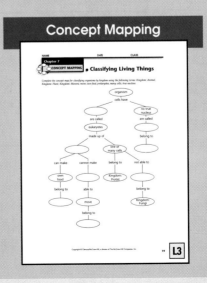

Classifying Living Things

CHAPTER OVERVIEW

Section 7-1 This section defines classification and explains its importance. The Linnaean system of binomial nomenclature is also explained.

Section 7-2 Classification of organisms into the six kingdoms is discussed in this section. Taxonomic units from phylum to species are also illustrated.

Section 7-3 *Science and Society* The importance of organism diversity is presented using the tropical rain forest as an example. Dangers due to rain forest destruction are discussed.

Section 7-4 The reasons scientific names are used are presented in this section. Students are given the opportunity to explore the use of a dichotomous key.

Chapter Vocabulary

classify	phylum
taxonomy	division
kingdom	class
binomial	order
nomenclature	family
genus	species
species	diversity
phylogeny	dichotomous key

Theme Connection

Stability and Change The theme of stability and change is used to explain how life scientists classify the millions of species of living things. The concept of evolutionary relatedness is emphasized. In addition, the theme of systems and interactions is further developed in the section that discusses the dangers of tropical rain forest and other habitat destruction.

Previewing the Chapter

Section 7-1   What is classification?
▶ Classifying
▶ Early History of Classification
▶ Scientific Naming

Section 7-2   Modern Classification
▶ Six-Kingdom System
▶ Prokaryotes and Eukaryotes
▶ Groups Within Kingdoms

Section 7-3   Science and Society
   Issue: Diminishing Diversity

Section 7-4   Identifying Organisms
▶ Common Names and Scientific Names
▶ Making and Using Dichotomous Keys

Learning Styles   Look for the following logo for strategies that emphasize different learning modalities. **LS**

Kinesthetic	Inclusion Strategies, pp. 194, 199; Using Computers, p. 196; MiniLAB, p. 201
Visual-Spatial	Activity 7-2, p. 204
Interpersonal	Activity 7-1, pp. 190-191
Logical-Mathematical	Explore, p. 187; Across the Curriculum, p. 189; Using Math, p. 192; Reteach, pp. 199, 202; Activity, p. 200
Linguistic	Across the Curriculum, p. 201; Science Journal, p. 203

Chapter 7

Classifying Living Things

How many plants and animals are discovered and described each year? You might think that the answer would be a very low number, or perhaps none at all! But you would be wrong. Life scientists discover, name, and describe hundreds of organisms every year. How do they decide if a certain plant belongs to the iris or orchid family of flowering plants? Or if an insect is more like a grasshopper or a beetle? Think about how scientists might make these distinctions.

EXPLORE ACTIVITY

Use distinguishing features to classify organisms.

1. Observe the organisms on the opposite page or in an insect collection in your class.
2. Notice the features that distinguish one group of organisms from the others.
3. Make a list of these distinguishing features.
4. Divide the organisms into two groups based on one feature.
5. Continue to divide the groups using different features until all the organisms have been separated.

Observe: In your Science Journal, describe how this activity models how biologists categorize living things.

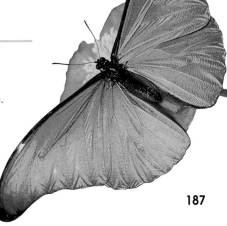

Previewing Science Skills

► In the **Skill Builders,** you will observe and infer, map concepts, and classify.

► In the **Activities,** you will model, classify, and observe.

► In the **MiniLAB,** you will classify and make models.

187

Prepare

Section Background

Linnaeus was able to classify only a few thousand organisms during his lifetime. Some scientists today conjecture that there may be 10 million different species. In this section, students will learn about classification from an historical perspective.

Preplanning

You will need a packet of ten different kinds of seeds, hand lens, metric ruler, and two sheets of paper for each student group for Activity 7-1.

1 Motivate

Bellringer

 Before presenting the lesson, display **Section Focus Transparency 25** on the overhead projector. Assign the accompanying **Focus Activity** worksheet. L1 LEP

? FLEX Your Brain

Use the Flex Your Brain activity to have students explore CLASSIFICATION.

📁 **Activity Worksheets,** p. 5

2 Teach

GLENCOE TECHNOLOGY

💿 **CD-ROM**

Life Science CD-ROM

Have students perform the interactive exploration for Chapter 7 to reinforce important chapter concepts and thinking processes.

Science Words

classify
taxonomy
kingdom
binomial nomenclature
genus
species

Objectives

- Give examples that show the need for classification systems.
- Describe Aristotle's system of classification.
- Explain Linnaeus's system of classification.

Classifying

When you go into a grocery store, do you usually go to one aisle to get milk, to another part of the store to get margarine, and to a third area to get yogurt? Most grocery stores group similar items, so the dairy products mentioned in the example above would be found together. When you place similar items together, you classify them. To **classify** means to group ideas, information, or objects based on similarities.

Classification is more a part of your life than you might think. Grocery stores, bookstores, and department stores group similar items together. In what other places is classification important?

Early History of Classification

More than 2000 years ago, Aristotle, a Greek philosopher, developed a system to classify living things. The science of classifying and naming organisms is called **taxonomy.** Aristotle began his system of taxonomy by dividing organisms into two large kingdoms, the plant and animal kingdoms. A **kingdom** is the largest of the taxonomic categories. Aristotle then divided the animal kingdom into smaller groups based on where animals live. Animals that live on land were in one group, animals that live in water were in another group, and animals that fly through the air were in a third group. The plant kingdom was also divided into three groups based on size and structure.

Figure 7-1

Aristotle's system of classification was unworkable for some kinds of organisms. Frogs live in water and on land. *Can you think of other organisms that don't fit into Aristotle's classification system? Why don't they fit?*

188 Chapter 7 Classifying Living Things

Program Resources

📁 **Reproducible Masters**
Study Guide, p. 29 L1
Reinforcement, p. 29 L1
Enrichment, p. 29 L3
Activity Worksheets, pp. 5, 41-42 L1
Multicultural Connections, pp. 17-18

 Transparencies
Section Focus Transparency 25 L1

Eventually, scientists began to criticize Aristotle's system because it had too many exceptions. Animals were classified according to where they lived, but what about frogs? Frogs, **Figure 7-1,** spend part of their lives in water and part on land. There was a problem with Aristotle's classification of plants, also. Trees aren't all related just because they're trees. A more logical classification system was needed.

Scientific Naming

Like Aristotle, Carolus Linnaeus, a Swedish physician and naturalist, created a system to classify organisms based on similarities in body structures and systems, size, shape, color, and methods of obtaining food.

Binomial Nomenclature

Linnaeus's system gives a two-word name to every organism. The two-word naming system is called **binomial nomenclature.** *Binomial* means "two names." The two-word species name is commonly called the organism's scientific name or Latin name. The first word of an organism's scientific name is the genus, and the second is the specific name. A **genus** is a group of different organisms that have similar characteristics. Together, the genus name and the specific name make up the scientific name of a particular species. A **species** is the smallest, most precise classification category. Organisms belonging to the same species can mate with one another to produce fertile offspring.

Problem Solving

Classifying Unknown Organisms

Laquitia and her parents were on a family vacation. They were driving through a national park in southern Arizona one evening just as the sun was setting. Suddenly, a tawny, heavily marked cat with a long tail ran across the road. The cat's spots included rings, speckles, slashes and bars. The area around the road was dense brush. Laquitia and her family were startled to see such a beautiful animal, and one they did not recognize.

Solve the Problem:

1. **What might be the important characteristics of the animal that would be needed to identify it?**
2. **Did Laquitia need other information to be able to determine the animal's species?**

Think Critically:

How would you begin to figure out what cat Laquitia saw?

189

Teacher F.Y.I.

The second word in a species name is officially referred to as the specific epithet.

Visual Learning

Figure 7-1 Can you think of other organisms that don't fit Aristotle's classification system? Why don't they fit? *Alligators, crocodiles, salamanders, penguins, and many other organisms do not fit Aristotle's system. Any organism that lives in more than one part of Earth, such as water and land, does not fit Aristotle's system.* **LEP**

Discussion

Ask students to list things that are classified. Ask what characteristics are used to group the items they list.

Problem Solving

Solve the Problem
1. length of tail, patterns of spots, and color
2. She needed to know that it was a cat, and the local habitat.

Think Critically
Students may suggest looking at a field guide to cats, mammals, or local fauna.

Across the Curriculum

Math Ask students how numbers are classified. Answers may include real and imaginary, whole numbers and fractions, decimals and percents, and rational and irrational. **LS**

3 Assess

Check for Understanding

Use question 23 in the Chapter Review and the key in Activity 7-2 to check for understanding of scientific naming and classification using characteristics.

Purpose

LS **Interpersonal** Students will design an experiment to test a hypothesis about seed classification. **L1** **COOP LEARN**

Process Skills

observing and inferring, classifying, forming operational definitions, communicating, making and using tables, comparing and contrasting, forming a hypothesis, designing an experiment, interpreting data, using numbers

Time

one class period

Materials

• Prepare packets of ten different kinds of seeds. Seeds that can easily be classified are lima beans, black-eyed peas, squash seeds, navy beans, peas, kidney beans, wrinkled corn seeds, popcorn, black beans, sunflower seeds, and pinto beans.

• Have all materials available as students make their plans.

Alternate Materials You may use objects other than seeds. Biological specimens should be used if at all possible.

Safety Precaution

Caution students not to put any seeds into their mouths. Use only edible seeds, not treated seeds.

Possible Hypotheses

Students may hypothesize that seeds may be classified by shape, size, or other characteristics. Any characteristic that is easily identifiable and consistent is appropriate.

📁 **Activity Worksheets,** pp. 5, 41-42

Activity 7-1

Design Your Own Experiment
Classifying Seeds

Scientists have developed classification systems to show how organisms are related. How do they determine what features they will use to classify organisms? Can you learn to use the same methods?

PREPARATION

Problem
How may seeds be classified?

Thinking Critically
What traits or physical features may be used to help classify various kinds of seeds?

Objectives
• Observe the seeds provided and notice their distinctive features.
• Classify seeds using your model.

Possible Materials
• packet of 10 different kinds of seeds
• hand lens
• metric ruler
• 2 sheets of paper

Safety Precautions 🧤☠️
Do not eat any seeds or put them in your mouth. Some kinds of seeds are poisonous.

PLAN THE EXPERIMENT

1. As a group, list the steps that you need to take to classify seeds. Be specific, and describe exactly what you will do at each step. List your materials.
2 Classify your seeds by making a model.
3. Make a key to identify your seeds.

Check the Plan

1. Read over your entire experiment to make sure that all steps are in logical order.
2. Identify the characteristics you will need

to divide the groups into ever smaller categories.
3. How will you check to see if your model works properly?
4. Have you checked to see that the characteristics you use are not repeated?
5. *Make sure your teacher approves your model before you proceed.*

DO THE EXPERIMENT

1. Carry out the experiment as planned.
2. While you are working, write down any observations that you make that would cause you to change your model.
3. Complete the key.

Analyze and Apply

1. **Compare** your key and model with those made by other groups.

2. Check your key by having another group use it.
3. In what ways can groups of different types of seeds be classified?
4. Why is it an advantage for scientists to use a standardized system to classify organisms? What observations did you make to support your answer?

Go Further

How would you classify a collection of wildflowers? How would this grouping be similar to the one for seeds?

✓ Assessment

Performance To further assess students' ability to classify, give them cutouts of ten different leaves and have them devise and describe a classification system for the leaves. Use the Performance Task Assessment List for Making and Using a Classification System in **PASC**, p. 49. P

Possible Procedures

Students may need help in understanding that some characteristics are inappropriate. Explain how a key works.

Teaching Strategies

• If students have difficulty dividing the seeds into groups, make sure they understand what *characteristics* are. Have them use a metric ruler and a hand lens as tools to assist them.

Troubleshooting Point out that highly variable characteristics such as color should be avoided.

DO THE EXPERIMENT

Expected Outcome

Students will have a useful dichotomous key if they are successful. P

Analyze and Apply

1. Comparisons may or may not result in agreement.
2. If the other group can key out seeds, the first group was successful.
3. size, color, shape, texture, how they are attached to the plant
4. Everyone would agree on the species name. Different classification systems could result in confusion and differing interpretations of scientific observations. Answers will vary but should be based on the observation that not every student classified the seeds in the same way.

Go Further

Classifying wildflowers would require knowing flower characteristics. The key should look similar.

Reteach

Have students describe an organism from its Latin name. Examples: *Musca domestica* (housefly), *Equus zebra* (zebra), and *Camelus dromedarius* (dromedary camel).

Extension

For students who have mastered this section, use the **Reinforcement** and **Enrichment** masters.

4 Close

•MINI•QUIZ•

1. The science of classifying and naming organisms is _____ . *taxonomy*

2. The two-word naming system created by Linnaeus is known as _____ . *binomial nomenclature*

3. A _____ is the largest taxonomic category. *kingdom*

Section Wrap-up

Review

1. The purpose is to arrange or group things according to similarities and differences.

2. Aristotle—two kingdoms: animals divided according to where they lived, plants divided according to size and structure; Linnaeus—groupings according to structures, systems, size, shape, color, food getting.

3. **Think Critically** Answers will vary. Stamps, clothing, food at a grocery, and organisms are all acceptable answers.

USING MATH

Logical-Mathematical The least number of characteristics needed is seven.

192

Figure 7-2

The photo on the left shows a dog, *Canis familiaris*. Other members of the genus *Canis* are the coyote, *Canis latrans* (middle), and the gray wolf, *Canis lupus* (right). Notice that they are all different species. *Why are they placed in the same genus?*

An example of a two-word, or species, name is *Canis familiaris*. This is a domesticated dog. Notice that the first word, the genus name, always begins with a capital letter. The second word, the specific name, begins with a lowercase letter. Both words in a scientific name are written in italics or underlined. Linnaeus chose the Latin language for his naming system because the meanings for Latin words are the same around the world. In Linnaeus's system, no two organisms have the same scientific name. Because of Linnaeus's system and the use of Latin, scientists around the world recognize the name *Canis familiaris* as a domesticated dog and not a gray wolf, *Canis lupus*. You can see the differences among a dog, a gray wolf, and a coyote in **Figure 7-2.**

Section Wrap-up

Review

1. What is the purpose of classification?

2. Compare and contrast Aristotle's and Linnaeus's classification systems.

3. **Think Critically:** List two examples of things that are classified based on their similarities.

Skill Builder

Observing and Inferring

What could you infer about the following species of oaks: *Quercus alba, Quercus rubra, Quercus suber, Quercus macrocarpa?* If you need help, refer to Observing and Inferring in the **Skill Handbook.**

USING MATH

What is the least number of characteristics needed to classify a group of eight different species to the level where there is only one species identified in each category?

192 Chapter 7 Classifying Living Things

Skill Builder

You could infer that they all have similar ancestors and are closely related because they have the same genus name.

✓ Assessment

Performance Assess students' abilities to observe and infer the differences and similarities among maples: *Acer rubrum, Acer saccharum,* and *Acer pennsylvanicum.* Use the Performance Task Assessment List for Making Observations and Inferences in **PASC,** p. 17. **P**

Modern Classification

Six-Kingdom System

How does the classification system used today by scientists differ from those of the past? Aristotle and Linnaeus developed their systems of classification using only characteristics of organisms that were easily observed. Scientists today also look at other types of traits to classify organisms. They look at the chemical makeup of the organism and its ancestors. Scientists find relationships among organisms by looking at similarities in genes and body structures. They also study fossils and the embryo of an organism as it develops. By studying all of these things, scientists can determine an organism's phylogeny. The **phylogeny** of an organism is its evolutionary history. Phylogeny tells scientists who the ancestors of an organism were and helps to classify it. Today, classification of many organisms is based on phylogeny.

The classification system most commonly used today separates organisms into six kingdoms. These kingdoms are Animal, Plant, Fungi, Protista, Eubacteria, and Archaebacteria. Organisms are placed into a kingdom based on several characteristics. These characteristics include: presence of a nucleus; single-celled or many-celled; ability to make food; and ability to move.

Prokaryotes and Eukaryotes

Kingdoms Archaebacteria and Eubacteria are made up of one-celled organisms that are simple in structure. Members of Kingdom Eubacteria include bacteria such as streptococcus,

Science Words

phylogeny
phylum
division
class
order
family

Objectives

- Name the six kingdoms of living things.
- Identify characteristics and members of each kingdom.
- List the groups within each kingdom.

Figure 7-3

Organisms are placed into kingdoms because they share many evolutionary characteristics.

 Amoebas are classified as protists.

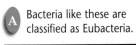 Bacteria like these are classified as Eubacteria.

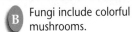 Fungi include colorful mushrooms.

7-2 Modern Classification   193

Section 7•2

Prepare

Section Background

- Taxonomy is the science of classifying. Today, taxonomy is usually based on inferred evolutionary relationships.
- Explain to students that the classification of an organism is always subject to change, based on new evidence.

1 Motivate

Bellringer

Before presenting the lesson, display **Section Focus Transparency 26** on the overhead projector. Assign the accompanying **Focus Activity** worksheet. [L1] [LEP]

Tying to Previous Knowledge

Use the information in the previous section to guide students in their understanding of the modern classification system.

2 Teach

Visual Learning

Figure 7-3 Ask students to name the two kingdoms not shown in the figure. *plants, animals* Show students photos of several different organisms and ask them to determine the kingdom of each. [LEP] [IS]

Program Resources

Reproducible Masters
Study Guide, p. 30 [L1]
Reinforcement, p. 30 [L1]
Enrichment, p. 30 [L3]
Concept Mapping, p. 19
Lab Manual 15

Transparencies
Teaching Transparencies 13 and **14**
Section Focus Transparency 26

194

USING TECHNOLOGY

Beyond Appearances ▼ ▲

Besides looking at an animal's physical traits, scientists now examine an animal's DNA to classify it. The more similar the DNA between two animals is, the more closely they are related. By using DNA information, scientists can tell which similar animals are actually separate species. They can also study how geographic separation and other factors affect populations. The information scientists gather helps zoos and wildlife programs conserve many species of animals.

Pandas Are Bears

Genetic studies are already changing the way some organisms are classified. By studying DNA, scientists have found that pandas really are bears, and not relatives of the raccoon as they once thought. The "Asiatic" lions in U.S. zoos are not truly from Asia. They have African genes as well. Some gray foxes living on islands off California's coast vary genetically from each other, but white rhinos in North and South Africa are genetically the same in spite of being separated by 2000 miles.

Think Critically:

How do scientists use DNA to classify organisms?

Giant panda

194 Chapter 7 Classifying Living Things

---

which causes strep throat, and cyanobacteria, one-celled plant-like organisms found in lakes and ponds. Archaebacteria and eubacteria don't have an organized nucleus surrounded by a membrane or membrane-bound organelles. These organisms are prokaryotes. *Prokaryote* means "before kernel," or before formation of a nucleus. All organisms other than eubacteria and archaebacteria are eukaryotes. Eukaryotes are organisms with a nucleus and organelles surrounded by membranes. *Eukaryote* means "true kernel," or true nucleus. **Figure 7-4** shows the differences between prokaryotes and eukaryotes. **Table 7-1** on page 195 gives examples and basic information about each of the six kingdoms. You will learn more about each kingdom in coming chapters.

Magnification: 24 000×

B A paramecium is an example of a eukaryotic cell.

Magnification: 110×

Groups Within Kingdoms

Suppose you go to a music store at the mall to buy a new CD. Will you look through all the CDs in the store until you find the one you're looking for? You don't have to search like that because the CDs are separated into categories of similar types of music such as rock, soul, classical, country, and jazz.

Figure 7-4

A prokaryotic cell has no nucleus. A nucleus is seen in the eukaryotic cell.

Table 7-1

Life's Six Kingdoms					
Archaebacteria	**Eubacteria**	**Protista**	**Fungi**	**Plant**	**Animal**
Prokaryotic	Prokaryotic	Eukaryotic	Eukaryotic	Eukaryotic	Eukaryotic
One-celled	One-celled	One- and many-celled	One- and many-celled	Many-celled	Many-celled
Some move	Some move	Some move	Don't move	Don't move	Move
Some members make their own food; others obtain it from other organisms.	Some members make their own food; others obtain it from other organisms.	Some members make their own food; others obtain it from other organisms.	All members obtain food from other organisms.	Members make their own food.	Members eat plants or other animals.

7-2 Modern Classification **195**

 Use these program resources as you teach this lesson.

Lab Manual 15

Concept Mapping, p. 19

Use **Teaching Transparencies 13** and **14** as you teach this lesson.

Theme Connection

Systems and Interactions The diversity of organisms seems immense to students. Explain that the systems and interactions among them create a unity that makes them understandable.

3 Assess

Check for Understanding

Ask students to identify the classification groups with the most members and with the fewest members. Ask them to explain.

Reteach

Play a game in which the object is to decide the lowest taxonomic category containing the organisms described. Example: **What level contains the bloodhound and poodle?** *species* **What level contains all the kinds of willow trees?** *genus* **What level contains all plants?** *kingdom*

Extension

For students who have mastered this section, use the **Reinforcement** and **Enrichment** masters.

4 Close

• MINI • QUIZ •

Use the Mini Quiz to check students' recall of chapter content.

1. The _____ of an organism is its evolutionary history. *phylogeny*
2. The classification system most commonly used today separates organisms into six major groups called _____ . *kingdoms*
3. Because bacteria don't have an organized nucleus or organelles, they are called _____ . *prokaryotes*
4. _____ are organisms with a nucleus and organelles surrounded by membranes. *Eukaryotes*

Section Wrap-up

Review

1. Archaebacteria: obtain energy through chemosynthesis; Eubacteria: one-celled prokaryotes—*streptococcus* bacteria; Protista: one- and many-celled eukaryotes—*Paramecium*; Fungi: one- and many-celled eukaryotes that can't move, obtain food from other organisms—mushroom; Plant: many-celled organisms that make their own food, can't move—oak tree; Animal: many-celled organisms that can move but can't make their own food—dog

2. Scientists classify organisms into smaller groups with more characteristics in common. A genus and species identify a specific organism.

3. **Think Critically** Protista

Using Computers

LS **Kinesthetic** The database will have categories of Kingdom, Phylum, Class, Order, Family, Genus, Species. Humans are Animalia, Chordata, Mammalia, Primates, Hominidae, Homo, *Homo sapiens*. Dogs are Animalia, Chordata, Mammalia, Carnivora, Canidae, Canis, *Canis familiaris*. Cats are Animalia, Chordata, Mammalia, Carnivora, Felidae, Felis, *Felis catus*.

✔ Assessment

Portfolio Use this Skill Builder to assess students' abilities to form a concept map of cell characteristics. Use the Performance Task Assessment List for Concept Map in **PASC,** p. 89. **P**

Figure 7-5

The classification of the blue jay shows its genus to be *Cyanocitta* and species to be *Cyanocitta cristata.* Each grouping shows organisms that have more in common. Each grouping also contains fewer organisms. When you reach the species level, a specific organism has been named—the blue jay.

Within each category, the CDs are divided by artists, and then into specific titles by the artist. Because of this classification system, you can easily find the CD you want.

Scientists classify organisms into groups in just the same way. Every organism can be placed into a kingdom. After being assigned to a kingdom, an organism can be classified into a **phylum,** the next smallest group. In the plant kingdom, the word **division** is used in place of phylum. Each phylum or division is separated into **classes.** Classes are separated into **orders,** and orders are separated into **families.** A genus is a group within a family. A genus can have one or more species.

Scientists go through this series of categories to name a specific organism the same way you went through a series of categories to find your compact disc. To understand how an organism is classified, look at the classification of the blue jay in **Figure 7-5.**

Section Wrap-up

Review

1. Name and describe a member of each kingdom.

2. Why are there smaller groups within each kingdom?

3. **Think Critically:** Classify as to kingdom a single-celled, eukaryotic organism that makes its own food.

Skill Builder
Concept Mapping
Use the following terms to make a network tree concept map. Provide linking words. *Cells, cell organelles, eukaryote, prokaryote, plants, animals, archaebacteria, eubacteria, no membrane-bound organelles, no nucleus, fungi, protists, organized nucleus.* If you need help, refer to Concept Mapping in the **Skill Handbook.**

Using Computers

Database Make a database that could be used to sort organisms based on any level of classification. Enter the classification information for humans, dogs, and cats.

Skill Builder

Clarifying Classification

Biologists estimate that there are between 5 and 30 million species of living things on Earth, of which only 1.5 million have been identified and classified. Aristotle was one of the first people to group living things into kingdoms. His two-kingdom system separated organisms into plants and animals.

Two Kingdoms to Three

Later classification systems further refined Aristotle's kingdoms by grouping organisms according to similarities in physical traits. When microscopes were invented, however, new organisms were discovered that did not really fit into either the plant or animal kingdoms. Scientists eventually constructed a third kingdom, Kingdom Protista, to hold all of the organisms that didn't fit elsewhere. As late as the 1960s, scientists disagreed about how to classify organisms such as bacteria and fungi. Botanists, scientists who study plants, placed fungi with the plants, whereas zoologists, who study animals, placed fungi with the protists.

Five Kingdoms to Six Kingdoms

Identifying organisms as either prokaryotes or eukaryotes led to a further change in the number of kingdoms. Organisms such as cyanobacteria, once called blue-green algae and classified as plants, then as protists, were reclassified as prokaryotes because they have no membrane surrounding their nuclear material. A five-kingdom system was proposed in 1969. In this system, cyanobacteria and the prokaryotic bacteria were reclassified into their own kingdom, Kingdom Monera. This system recognized the following five kingdoms: Monera, Protista, Fungi, Plant, and Animal. In the 1980s, scientists studying bacteria that lived in extreme environments, such as hot sulfur springs, discovered that these bacteria had different materials in their cell walls and cell membranes than other bacteria. These chemical differences led to the division of bacteria into two separate kingdoms, the Eubacteria and the Archaebacteria. Kingdoms Archaebacteria, Eubacteria, Protista, Fungi, Plant, and Animal are the six kingdoms recognized today. But taxonomy, like every science, is open to change. New research may result in additional kingdoms in the future.

Archaebacteria

Science Journal

Describe in your Science Journal what characteristics you would look for to classify an unknown cell into a particular kingdom. Explain how you would identify a cell as belonging to the eubacteria or to the protist kingdom.

domains suggested would include: Domain Bacteria (the eubacteria), Domain Archaea (the archaebacteria), and Domain Eukarya (all eukaryotes). The reason for this division is that most scientists think that the archaebacteria are more closely allied to eukaryotes than to eubacteria.

Teaching Strategies

Ask students the following questions:
- How is the six-kingdom system more accurate than the five-kingdom system?

- What does the existence of archaebacteria on Earth suggest about the possibility of life on other planets?
- Why are new species discovered today more likely to be described by their molecular structures rather than their physical traits?
- Which kingdom do you think is most diverse? Why?

Sources
- Campbell, Neil A. *Biology. Fourth Edition*. University of California, Riverside, The Benjamin/Cummings Publishing Company, Inc., 1996.
- Spangenburg, Ray, and Diane K. Moser. *The History of Science from the Ancient Greeks to the Scientific Revolution; The History of Science in the Eighteenth Century*. New York: Facts on File, 1993.

Background
- Aristotle was far ahead of the other scientists of his time. He used keen observations to classify more than 500 animals. He realized that dolphins, even though they lived in water, had placentas and were live-bearers, so he placed them with his land group. He also divided animals into those with red blood (the vertebrates) and those without (the invertebrates).
- By the end of the 17th century, scientists were using a genus name to describe a type of organism, such as *Rosa* to identify roses. A particular rose was then described using a long, descriptive phrase known as a *polynomial*.
- Linnaeus wrote a book, *Species Plantarum*, in 1753. He identified organisms by their genus name, then used the polynomial to identify each species. But in the margin of his book, he included a shorthand word for the species–which, used with the genus name, became the two-word description used today.
- Some scientists have suggested that instead of a six-kingdom system, organisms could be classified by a system including three *domains*. Domains would be higher taxonomic levels than kingdoms. The three

Prepare

Section Background

Tropical rain forests occur near the equator and may receive as much as 150-350 cm of rain per year.

1 Motivate

Bellringer

 Before presenting the lesson, display **Section Focus Transparency 27** on the overhead projector. Assign the accompanying **Focus Activity** worksheet. L1 LEP

Tying to Previous Knowledge

Make reference to the changes that have occurred in your particular biome due to human activity.

2 Teach

 NATIONAL GEOGRAPHIC SOCIETY

 Videodisc

STV: Biodiversity
Destroying Diversity
Unit 1
Habitat Destruction

00386-07361
Preserving Diversity
Unit 2

22598-44538

Science Words
species diversity

Objectives
- Identify one negative effect of decreased species diversity.
- Explain why the price of maintaining species diversity may be too high.

Figure 7-6

The northern spotted owl is a controversial bird because of its endangered status. This bird's habitat is old-growth forests, the area that loggers in the northwestern United States claim to be their most important area for harvesting timber.

198

Extinction Reduces Diversity

Are there any places on Earth that have escaped human intrusion? Places such as the old-growth forests of the northwestern United States and the rain forests of South America and Asia are large areas that are increasingly under pressure from humans. Individuals, communities, and international organizations are engaged in serious discussions about how best to protect and develop these areas. Why is there so much concern about these areas?

What is species diversity?

Rain forests, coral reefs, and similar environments are homes to hundreds of thousands of organisms. In a hectare of rain forest, for example, there may be 200 species of plants and more than 1000 species of animals. This great variety of plants, animals, and other organisms makes up **species diversity.** An ecosystem that has a high species diversity is more stable than one with fewer species.

How is species diversity changed?

Every minute, more than 20 hectares of rain forest are cleared by humans for farming or mining, or are cut for timber. Some areas of old-growth forests in the United States are protected from being cut because the northern spotted owl, an endangered species shown in **Figure 7-6,** lives in these areas. Protecting an entire forest because of one endangered species is one way to prevent a decrease in species diversity. But loggers and lumber companies still want to harvest the wood. Many organizations are at work to protect as many of the organisms as possible in areas of high species diversity.

However, human intrusion into undisturbed areas has in the past led to extinction of particular species, such as the Carolina parakeet and the passenger pigeon. Even though extinction is a natural process, today humans are causing extinction at a far greater rate than has ever occurred before.

Program Resources

📁 **Reproducible Masters**
Study Guide, p. 31 L1
Reinforcement, p. 31 L1
Enrichment, p. 31 L3
Critical Thinking/Problem Solving, p. 11
Science and Society Integration, p. 11
Science Integration Activities, pp. 13-14

📽 **Transparencies**
Section Focus Transparency 27 L1

2 Points of View

People have a right to use forest resources

Species diversity is important, but making a living and using the resources are more important. Science and technology can help to save many species. The livelihoods of people are more important than some decrease in diversity.

Science cannot replace extinct species

Species diversity outweighs human needs to use resources. A loss of species will result in ecosystems that are out of balance, and may even be in danger of collapsing. Losses of biological diversity are beyond the ability of science or technology to repair. Once a species becomes extinct, it cannot be brought back.

Section Wrap-up

Review

1. Why is the northern spotted owl an issue to people in North America?

2. How would you suggest that a consensus be reached between opposing parties in the species diversity debate?

interNET CONNECTION Visit the Chapter 7 Internet Connection at Glencoe Online Science, **www.glencoe.com/sec/science/life,** for a link to more information about biodiversity.

SCIENCE & SOCIETY

Figure 7-7

Rain forests are being cleared for farming and other uses at a rapid rate. *How can human needs for farming land be balanced with protection of other species?*

199

Prepare

Section Background

- Explain that a key usually contains the most general statements first, and progresses to more specific characteristics until a species is identified.

- Extinct organisms known only from fossils can also be classified using specialized dichotomous keys.

Preplanning

- For the MiniLAB, you will need to obtain clay and a Latin dictionary.

- For Activity 7-2, be sure to have paper and pencils available.

1 Motivate

Bellringer

Before presenting the lesson, display **Section Focus Transparency 28** on the overhead projector. Assign the accompanying **Focus Activity** worksheet. L1 LEP

Tying to Previous Knowledge

Make a comparison to the first two sections of the chapter where the classification system and groups within kingdoms are discussed.

2 Teach

Activity

LS **Logical-Mathematical** Pass out examples of taxonomic keys or field guides to the class. Ask students to describe the function of these books and when they might be used. Challenge them to key out a particular bird or other organism they know about. L1

7•4 Identifying Organisms

Science Words

dichotomous key

Objectives

- List reasons scientific names are more useful to scientists than common names.
- Identify the function of a dichotomous key.
- Demonstrate how to use a dichotomous key.

Common Names and Scientific Names

Have you heard anyone call the bird in **Figure 7-8A** a *Turdus migratorius*? In much of the United States, this bird is commonly called a robin, or a robin redbreast. However, people who live in England call the bird in **Figure 7-8B** a robin. In much of Europe, the same bird is called a redbreast. If you lived in Australia, you'd call the bird in **Figure 7-8C** a robin, or yellow robin. Are these the same species of bird? No, these birds are obviously very different from one another. Yet they all have the same or a similar common name. If you were talking informally to someone in your area who knew the same birds you did, you could use common names. But the common names for the different birds could be confusing if you were in another country or in another area of the same country. **Figure 7-9** gives examples of some common names that are confusing.

What would happen if life scientists used only common names when they tried to share information about the organisms they study? Many errors in understanding would probably result from the confusion. The system of binomial nomenclature developed by Linnaeus gives each bird a unique scientific name. The scientific name for the bird in **7-8A** is *Turdus migratorius*. The name for the bird in **7-8B** is *Erithacus rubecula*. The name for the bird in **7-8C** is *Eopsaltria australis.*

Figure 7-8

These three robins have the same common name, yet they are three different species.

200   Chapter 7 Classifying Living Things

Program Resources

Reproducible Masters
Study Guide, p. 32 L1
Reinforcement, p. 32 L1
Enrichment, p. 32 L3
Activity Worksheets, pp. 5, 43-44, 45 L1
Cross-Curricular Integration, p. 11

Transparencies
Section Focus Transparency 28

A

B

C

Functions of Scientific Names

Scientific names serve four basic functions. First, scientific names help scientists avoid errors in communication. A life scientist who studied yellow robins, *Eopsaltria australis,* would not be confused by information he or she read about *Turdus migratorius,* the American robin. Second, organisms with similar evolutionary histories are classified together. Because of this, you know that organisms that share the same genus name are more related than those that don't. Third, the scientific name gives descriptive information about the species. What can you tell from the species name *Turdus migratorius?* It tells you that this bird migrates from place to place. Fourth, the scientific names allow information about organisms to be organized and found easily and efficiently. Scientific names are used in guides to organisms called field guides. Scientists classify organisms by using more detailed lists of traits called **dichotomous** (di KAHT uh mus) **keys.**

Making and Using Dichotomous Keys

How would you go about finding out the name of an organism like the one pictured in **Table 7-2?** The easiest way would be to ask an expert on arthropods at a college in your area. However, no one knows, or is even expected to know, all of the arthropods or any other major taxonomic group. The first thing the expert might tell you is that the organism is some kind of mite or tick. But if you don't live near a college or natural history museum, what can you do? Should you give up? No, your next step should be to consult a field guide that contains descriptive words or photographs as an aid to identification. Field guides can be found in a library.

Figure 7-9

Common names can be misleading. Prairie dogs (A) are more closely related to squirrels than to dogs. Starfish (B) are not fish, yet seahorses (C) are fish.

MiniLAB

How are organisms named?

Give a scientific name to a new organism.

Procedure
1. Make a fictitious organism out of clay.
2. Give your organism a scientific name.
3. Make sure that your name is Latinized and infers information about the species.

Analysis
1. Present your organism to the class. Ask them to guess a name.
2. Why do scientists use Latin in giving organisms their names?

7-4 Identifying Organisms **201**

Use **Cross-Curricular Integration,** p. 11, as you teach this lesson.

Across the Curriculum

Languages Latin and Greek names usually describe the appearance or behavior of an organism. Have students look up scientific names and suggest how such names reflect the characteristics of the organism. **L1** **LS**

Revealing Preconceptions

Many students will try to skip steps in a key to identify an organism. Point out that skipping steps leads to misidentification.

Visual Learning

Figure 7-10 Use **Figure 7-10** to differentiate the features of mites and ticks used in their classification scheme. Students will be unable to successfully use **Table 7-2** if they are unable to identify the characteristics of the organisms. **LEP** **LS**

GLENCOE TECHNOLOGY

Videodisc

STVS: Animals
Disc 5, Side 1
Insect Museum (Ch. 22)

---

3 Assess

Check for Understanding

? FLEX Your Brain

Use the Flex Your Brain activity to have students explore CLASSIFICATION.

 Activity Worksheets, p. 5

Reteach

LS **Logical-Mathematical** Demonstrate how to use a key by making a transparency of a key. Show students photographs of an organism that can be identified using the key. Have them orally go through the steps of identification.

Extension

For students who have mastered this section, use the **Reinforcement** and **Enrichment** masters.

202

Figure 7-10

Notice the external features of mites and ticks used in identification. They all have eight legs and a body that is not divided into a cephalothorax and abdomen, as spiders are.

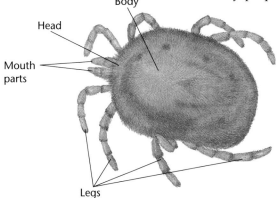

Body

Head

Mouth parts

Legs

Most field guides use descriptions, such as the external characteristics shown in **Figure 7-10**, illustrations of organisms, and information about the places they can be found to help in identification. Dichotomous keys used by scientists are arranged in steps with two descriptive statements at each step. Look at the dichotomous key for mites and ticks in **Table 7-2**. Notice that the descriptions are labeled 1a and 1b, 2a and 2b, and so on. To use the key, you must always begin with a choice from the first pair of descriptions. Notice that the end of each description is either the name of a species or directions to go on to another step. If you use the dichotomous key properly, you will eventually wind up with the correct name for your species.

Let's identify the soft tick in **Table 7-2**. Start at Step 1 of the key. Your tick is brown, so you go to Step 3. You measure your tick and find it is about 5 mm in length, so you go on to Step 4. Your tick is brown with an oval, flattened body, so you choose 4b. The dichotomus key tells you that your tick is an example of an *Ornithodoros* species, which are mammal soft ticks.

Table 7-2

Some Mites and Ticks of North America
1a If the animal is red, go to Step 2.
1b If the animal is brown or other than red, go to Step 3.
2a If the animal is smooth, globular, and somewhat elongated, it is a red freshwater mite, *Limnochares americana.*
2b If the animal is oval to rounded rectangle, and has dense velvety red hair, it is a velvet mite, *Trombidium* species.
3a If the animal is only 0.5 mm in length, it is a two-spotted spider mite, *Tetranychus uriticae.*
3b If the animal is larger than 0.5 mm in length, go to Step 4.
4a If the animal is reddish-brown, has a small shield of black-speckled gray near the head and has brown legs, it is an Eastern wood tick, *Dermacentor* species.
4b If the animal has a soft plate on the back, and is brown with an oval, flattened body, it is a mammal soft tick, *Ornithodoros* species.

Size: 5 mm

202   Chapter 7 Classifying Living Things

Integrating the Sciences

Earth Science Several minerals may share the same characteristic, but a mineral can be identified by a set of characteristics not shared by any other mineral. The Mohs scale of hardness is a reference scale that lists the hardness of ten minerals with 1 being the softest and 10 being the hardest.

A Size: 2 cm

B Size: 3 mm

C Size: 0.5 mm

Many kinds of field guides have been written. Field guides to plants, mushrooms, fish, butterflies, and every other kind of organism are available. You could find the names of species from all around the world by following the descriptions in the proper field guides, then checking your identifications by using a scientific dichotomous key.

Keys are useful in a variety of ways. It is important to know if a rock is igneous, metamorphic, or sedimentary when classifying fossils, for example. Minerals can be classified using a key that describes characteristics such as hardness, luster, color, streak, and cleavage. Why might you need to know several characteristics to classify a mineral, or a living thing?

Figure 7-11

Each of these animals is a different species of mite or tick.

INTEGRATION
Earth Science

Section Wrap-up

Review

1. List four reasons biologists use scientific names instead of common names in their communications.

2. Why can common names cause confusion?

3. What is the function of a dichotomous key?

4. **Think Critically:** Why would you infer that two species that look similar share a common evolutionary history?

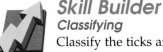

Skill Builder
Classifying
Classify the ticks and mites marked A, B, and C in **Figure 7-11,** using the dichotomous key in **Table 7-2.** If you need help, refer to Classifying in the **Skill Handbook.**

Select a field guide for grasses, trees, insects, or mammals. Write a book report describing the parts of the field guide. Select two organisms in the field guide that closely resemble each other. Compare them and explain how they differ, using labeled diagrams.

7-4 Identifying Organisms   203

Activity 7-2

Using a Dichotomous Key

Scientists who classify organisms have made many keys that allow you to identify an organism that may be unknown to you. Try this activity to see how it is done.

Purpose

Ⓛ Visual-Spatial Students will use a dichotomous key to interpret scientific illustrations.

L1

Process Skills

observing and inferring, comparing and contrasting, interpreting scientific illustrations, classifying

Time

30 minutes

📁 **Activity Worksheets,** pp. 5, 43-44

Teaching Strategies

Walk around the room as students are working. If they have the wrong answer, direct them to try again and to check their choices carefully.

Answers to Questions

1. seven different species

2. The title of the key specifies that it is a key to cats of North America.

3. Foxes are not cats.

4. The first pairs of descriptions are the most general. The following descriptions are more specific. The correct choice at the first pair of descriptions will lead to the correct identification.

Problem

How can a dichotomous key be used to identify native cats in the United States?

Materials

- paper and pencil

Procedure

1. Look at the cats pictured below.
2. Begin with Step 1 of the key to the right. Identify the cat labeled A.
3. On your paper, write the common and scientific names for the cat.
4. Use the same procedure to identify the species of the cat labeled B.

Analyze

1. According to the key, how many species of native cats reside in North America?

Conclude and Apply

2. How do you know that this key doesn't contain all the species of native cats in the world?
3. **Infer** why you couldn't identify a fox using this key.
4. **Explain** why it wouldn't be a good idea to begin in the middle of a key, instead of with the first step.

Key to Native Cats of North America

1a If the cat has a short tail, go to Step 2.

1b If the cat has a long tail, go to Step 3.

2a If the cat is distinctly mottled with long ear tufts tipped with black, and no cheek ruff, it is a lynx, *Felis lynx*.

2b If the cat has indistinct spots, short ear tufts, and a broad cheek ruff, it is a bobcat, *Felis rufus*.

3a If the cat has a plainly colored body, go to Step 4.

3b If the cat has a patterned body, go to Step 5.

4a If the cat is yellowish to tan above with white to buff below, it is a mountain lion, *Felis concolor*.

4b If the cat is either brown or black all over the body, it is a jaguarundi, *Felis yagouarundi*.

5a If the cat has a patterned body with tan and black, go to Step 6.

5b If the cat has black-bordered brown spots, tending to form lines on the body, it is an ocelot, *Felis pardalis*.

6a If the cat is large, spotted with black rosettes or rings in horizontal rows, it is a jaguar, *Felis onca*.

6b If the cat is small, with irregularly shaped spots and four dark-brownish stripes on the back and one on the neck, it is a margay, *Felis wiedii*.

A

B

204

✔ Assessment

Performance To further assess students' understanding of dichotomous keys, obtain a picture of another cat. Ask them to use the key to identify the cat. Use the Performance Task Assessment List for Making and Using a Classification System in **PASC**, p. 49. **P**

Chapter 7 | Review

Summary

7-1: What is classification?

1. To classify means to group ideas, information, or objects based on similarities. Taxonomy is the study of classifying and naming organisms.
2. Aristotle was the first to develop a system to classify organisms.
3. Linnaeus developed binomial nomenclature, a two-word naming system that gives every organism its own scientific name.

7-2: Modern Classification

1. The six kingdoms of living things are Archaebacteria, Eubacteria, Protista, Fungi, Plant, and Animal.
2. Organisms without a nucleus and membrane-bound cell organelles are prokaryotes. Eukaryotes have a nucleus and organelles.
3. Organisms are classified into these seven categories: kingdom, phylum, class, order, family, genus, and species. The genus and species names a specific organism.

7-3: Science and Society: Diminishing Diversity

1. Many species of plants and animals are lost each year due to human activities.
2. Some people think that progress may be more important than maintaining species diversity. Others argue that if diversity is not maintained, ecosystems will become unstable.

7-4: Identifying Organisms

1. Scientific names allow organisms with similar evolutionary histories to be classified together. Scientific names give descriptive information about the species.
2. Dichotomous keys are used to identify specific organisms.

Key Science Words

a. binomial nomenclature
b. class
c. classify
d. dichotomous key
e. division
f. family
g. genus
h. kingdom
i. order
j. phylogeny
k. phylum
l. species
m. species diversity
n. taxonomy

Reviewing Vocabulary

Match each phrase with the correct term from the list of Key Science Words.

1. to put similar organisms in groups
2. a group of similar species
3. the science of classification
4. system of naming organisms with two names
5. evolutionary history of organisms
6. tool for identifying organisms
7. subdivision of a class
8. largest group within the plant kingdom

Summary

Have students read the summary statements to review the major concepts of the chapter.

Reviewing Vocabulary

1. c	**6.** d
2. g	**7.** i
3. n	**8.** e
4. a	**9.** m
5. j	**10.** l

✓ Assessment

Portfolio Encourage students to place in their portfolios one or two items of what they consider to be their best work. Examples include:

- Skill Builder, p. 196
- Skill Builder, p. 203
- Dichotomous key for Activity 7-1, pp. 190-191 P

Performance Additional performance assessments may be found in **Performance Assessment** and **Science Integration Activities.** Performance Task Assessment Lists and rubrics for evaluating these activities can be found in Glencoe's **Performance Assessment in the Science Classroom.**

GLENCOE TECHNOLOGY

 Videodisc

Glencoe Life Science Interactive Videodisc
Side 1, Lesson 2
Evolution: Patterns of Change

14447 to 25820

MindJogger Videoquiz

Chapter 7 Have students work in groups as they play the Videoquiz game to review key chapter concepts.

Checking Concepts

1. b	**6.** c
2. d	**7.** c
3. a	**8.** d
4. d	**9.** b
5. d	**10.** d

Understanding Concepts

11. kingdom, phylum, class, order

12. Organisms that have similar or homologous structures and similar embryology are closely related and would be classified together.

13. Many species in the tropical rain forest have not yet been discovered and named. If we lose many more, the rain forest ecosystem may be upset, and we may be losing valuable sources of medicine and damaging our atmosphere.

14. The protist kingdom contains plantlike, animal-like and funguslike members. It seems as though organisms that did not quite fit into another kingdom were thrown into the protist kingdom.

15. The cell has a membrane-bound nucleus.

Thinking Critically

16. Binomial nomenclature is a two-word naming system for identifying specific organisms. The first word is the organism's genus; the second is its species. Written in Latin, it is important because scientists can identify a specific organism. In the case of disease-causing organisms, it is especially important to correctly identify the right organism.

17. Archaebacteria, Eubacteria, Protista, Fungi, Plantae, and Animalia are the kingdoms. Student exam-

---

9. variety of plants and animals

10. the most specific of an organism's scientific names

Checking Concepts

Choose the word or phrase that completes the sentence.

1. The group with the most members is a _____.
- a. family
- b. kingdom
- c. genus
- d. order

2. Organisms that are the most similar belong in a _____.
- a. family
- b. class
- c. genus
- d. species

3. The most complex organisms are _____.
- a. animals
- b. bacteria
- c. fungi
- d. protists

4. The closest relative of *Canis lupus* is _____.
- a. *Quercus alba*
- b. *Equus zebra*
- c. *Felis tigris*
- d. *Canis familiaris*

5. The scientific name is written as _____.
- a. *Genus Species*
- b. genus species
- c. genus Species
- d. *Genus species*

6. Mushrooms belong to the _____ kingdom.
- a. Animal
- b. Eubacteria
- c. Fungi
- d. Plant

7. The _____ of an organism is its evolutionary history.
- a. taxonomy
- b. biology
- c. phylogeny
- d. chemistry

8. The simplest eukaryotes are _____.
- a. animals
- b. fungi
- c. eubacteria
- d. protists

9. Trees and flowers are _____.
- a. animals
- b. plants
- c. fungi
- d. protists

---

10. Cells without an organized nucleus are _____.
- a. eukaryotes
- b. phylogeny
- c. species
- d. prokaryotes

Understanding Concepts

Answer the following questions in your Science Journal using complete sentences.

11. Two organisms of a family also belong to what other same classification groups?

12. How do embryology and similar structures help classify organisms?

13. Why is it important to maintain the species diversity found in tropical rain forests?

14. Members of Kingdom Protista are often called "misfits." Can you explain why?

15. You are examining a cell under the microscope. Explain how you know right away that this cell does not belong to Kingdom Archaebacteria.

Thinking Critically

16. Explain what binomial nomenclature is, and why it is important to scientists.

17. Name each of the six kingdoms, and identify a member of each kingdom.

18. Write a short dichotomous key to identify the members of your family. Use such things as sex, hair color, eye color, and weight in your key. Each individual's first and last name will make up the person's genus and species.

19. Discuss the relationship between tigers and lions, members of the genus *Felis*.

20. Scientific names often describe a characteristic of the organism. What does *Lathyrus odoratus* tell you about a sweet pea?

---

ples will vary.

18. Accept all logical keys. Students should write dichotomous keys.

19. As both tigers and lions belong in the same genus, they also have the same kingdom, phylum, class, order, and family. They differ in their species.

20. The name *odoratus* tells you that the sweet pea must have an odor.

Developing Skills

21. Comparing and Contrasting Kingdoms have a much larger number of organisms and therefore a greater variety of organisms than a genus.

22. Observing and Inferring Looking at its physical characteristics is the first step in identifying an organism. The appropriate identification keys will allow further classification.

Developing Skills

If you need help, refer to the **Skill Handbook.**

21. **Comparing and Contrasting:** Compare the number and variety of organisms in a kingdom and in a genus.
22. **Observing and Inferring:** You observe an organism you don't know. How would you go about determining its classification?
23. **Classifying:** Use the Key to Native Cats of North America on page 204 to identify the cats below.

24. **Making and Using Graphs:** Make a circle graph to show the number of species in each kingdom listed in the table below.

Number of Species of Organisms

Kingdom	Number of Species
Bacteria	10 000
Protista	51 000
Fungi	100 000
Plant	285 000
Animal	1 000 000

25. **Concept Mapping:** Starting with Aristotle, use information in Sections 7-1 and 7-2 to make an events chain concept map to show events leading to modern classification.

Performance Assessment

1. **Poster:** Make a poster of the six kingdoms. Draw organisms found in each kingdom, or make a collage using photographs.
2. **Making and Using a Classification System:** As a class, make a local field guide for the school grounds or a nearby park. Draw maps and give the location of plants. Identify the plants and animals by common name and as many scientific names as you can.
3. **Events Chain:** Make an events chain to explain how the cats in the activity on page 204 were identified. For additional help, turn to Concept Mapping in the Skill Handbook.

Assessment Resources

Reproducible Masters
Chapter Review, pp. 17-18
Assessment, pp. 47-50
Performance Assessment, p. 21

Glencoe Technology
Chapter Review Software
Computer Test Bank
MindJogger Videoquiz

23. **Classifying** Students should identify these cats: (A) bobcat; (B) jaguarundi; (C) jaguar.
24. **Making and Using Graphs** Students should find that 69.4% of the pie graph, or 250°, represents the animals; 19.8% of the graph, or 71°, represents the plants; 6.9% of the graph, or 25°, represents the fungi; 3.5% of the graph, or 13°, represents the protists; and 0.28% of the graph, or 1°, represents the monerans.
25. **Concept Mapping** Students' chain of events concept maps should begin with Aristotle. Linnaeus and his invention of binomial nomenclature should come next, followed by scientists studying phylogeny, cell type, similarities in genes and body structure, how embryos develop, and fossils. The concept map should end with the development of a five-kingdom system.

Performance Assessment

1. Posters will vary. Use the Performance Task Assessment List for Poster in **PASC**, p. 73.
2. Field guides will vary. Use the Performance Task Assessment List for Making and Using a Classification System in **PASC**, p. 49
3. Events should be listed as:
   • Read the first two statements.
   • Decide which statement is true.
   • Proceed according to directions at the end of the statement.
   • Continue choices until organism is identified. Use the Performance Task Assessment List for Events Chain in **PASC**, p. 91.

Chapter Organizer

Section	Objectives/Standards	Activities/Features
Chapter Opener		Explore Activity: Find out what bacteria look like. p. 209
8-1 Two Kingdoms of Bacteria (2 sessions, 1 block)*	1. **Describe** the characteristics of bacterial cells. 2. **Compare** aerobic and anaerobic organisms. **National Content Standards:** (5-8) UCP1, UCP5, A1, A2, C1, C2, C3, C4, C5, D1	MiniLAB: How quickly do bacteria grow? p. 212 Problem Solving: Controlling Bacterial Growth, p. 213 Skill Builder: Concept Mapping, p. 214 Using Math, p. 214 Activity 8-1: Observing Cyanobacteria, p. 215
8-2 Bacteria in Your Life (1 session, ½ block)*	1. **Identify** some ways bacteria are helpful. 2. **Explain** the importance of nitrogen-fixing bacteria. 3. **Explain** how some bacteria cause disease. **National Content Standards:** (5-8) UCP4, A1, A2, C1, C4, E2, F3, G1, G3	MiniLAB: How can you make yogurt? p. 217 Using Technology: Lyme Disease Vaccine, p. 218 Connect to Physics, p. 219 Skill Builder: Measuring in SI, p. 219 Using Computers: Spreadsheet, p. 219 Activity 8-2: Where are the most bacteria found? pp. 220-221
8-3 Science and Society Issue: Fighting Tuberculosis (1 session)*	1. **Describe** antibiotic resistance. 2. **Explain** the different opinions about how to deal with the problem of antibiotic resistance. **National Content Standards:** (5-8) A1, F1, F4, F5	Explore the Issue, p. 223 Using Math, p. 223 Science and Literature: Spoon River Anthology, p. 224 Science Journal, p. 224

* A complete Planning Guide that includes block scheduling is provided on pages 32T-35T.

Activity Materials

Explore	Activities	MiniLABs
page 209 plain yogurt, tap water, small dish, toothpicks, methylene blue, glass slide, coverslip, microscope	page 215 microscope, prepared slides of *Gloeocapsa* and *Anabaena*, living cultures of *Nostoc* and *Oscillatoria*, safety goggles, apron pages 220-221 sterile petri dishes with nutrient agar, incubator, cotton-tipped swabs, tape, tap water, safety goggles, apron	page 212 dried beans, distilled water, 10-mL graduated cylinder, 150-mL beaker, clock or watch with second hand, methylene blue, slide, coverslip, microscope page 217 liter of milk; large saucepan; stove; yogurt starter; tablespoon; spoon; clean, 1-liter thermos bottle with lid; refrigerator

Need Materials? Call Science Kit (1-800-828-7777).

Chapter 8 Bacteria

Teacher Classroom Resources

Reproducible Masters	Transparencies	Teaching Resources
Study Guide, p. 33 **Reinforcement**, p. 33 **Enrichment**, p. 33 **Activity Worksheets**, pp. 46-47, 50 **Lab Manual 16**, Bacteria **Lab Manual 17**, Shapes of Bacteria **Science and Society Integration**, p. 12 **Technology Integration**, pp. 9-10	**Section Focus Transparency 29**, Glued to the Tube **Science Integration Transparency 8**, Life Where You Least Expect It **Teaching Transparency 15**, Bacterial Shapes	**Glencoe Life Science Interactive Videodisc** **Life Science CD-ROM** **Spanish Resources** **English/Spanish Audiocassettes** **Cooperative Learning Resource Guide** **Lab Partner** **Lab and Safety Skills** **Lesson Plans**
Study Guide, p. 34 **Reinforcement**, p. 34 **Enrichment**, p. 34 **Activity Worksheets**, pp. 48-49, 51 **Lab Manual 18**, Bacterial Growth **Science Integration Activity 8**, Water Purification **Concept Mapping**, p. 21 **Critical Thinking/Problem Solving**, p. 12 **Cross-Curricular Integration**, p. 12 **Multicultural Connections**, pp. 19-20	**Section Focus Transparency 30**, It's a Small World **Teaching Transparency 16**, Plant Module Formation	**Assessment Resources** **Chapter Review**, pp. 19-20 **Assessment**, pp. 51-54 **Performance Assessment in the Science Classroom (PASC)** **MindJogger Videoquiz** **Alternate Assessment in the Science Classroom** **Performance Assessment** **Chapter Review Software** **Computer Test Bank**
Study Guide, p. 35 **Reinforcement**, p. 35 **Enrichment**, p. 35	**Section Focus Transparency 31**, Taking the Air	

Key to Teaching Strategies

The following designations will help you decide which activities are appropriate for your students.

L1 Level 1 activities should be within the ability range of all students, including those with learning difficulties.

L2 Level 2 activities should be within the ability range of the average to above-average student.

L3 Level 3 activities are designed for the ability range of above-average students.

LEP LEP activities should be within the ability range of Limited English Proficiency students.

LS These activities are designed to address different learning styles.

COOP LEARN Cooperative Learning activities are designed for small group work.

P These strategies represent student products that can be placed into a best-work portfolio.

GLENCOE TECHNOLOGY

The following multimedia resources are available from Glencoe.

Science and Technology Videodisc Series (STVS)
Plants & Simple Organisms
  Bacteriophage
  Bacteria in the Clouds
  Detecting Salmonella
  Using Bacteria to Detect Carcinogens
  Bacterial Digestion of Lignin

Bacterial Miners
Bacterial Waste Treatment
Animals
  Tick Research
Glencoe Life Science Interactive Videodisc
Decomposers: Earth's Natural Recyclers
Life Science CD-ROM

Teacher Classroom Resources

This is a representation of key blackline masters available in the Teacher Classroom Resources.

Teaching Aids

Section Focus Transparencies

GLUED TO THE TUBE

29 SECTION FOCUS TRANSPARENCY — L1

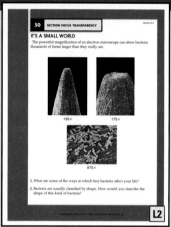

IT'S A SMALL WORLD

30 SECTION FOCUS TRANSPARENCY — L2

TAKING THE AIR

31 SECTION FOCUS TRANSPARENCY — L3

Science Integration Transparencies

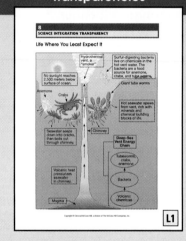

8 SCIENCE INTEGRATION TRANSPARENCY

Life Where You Least Expect It — L1

Teaching Transparencies

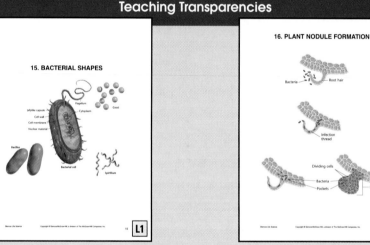

15. BACTERIAL SHAPES — L1

16. PLANT NODULE FORMATION — L2

Meeting Different Ability Levels

Study Guide

L1

Reinforcement

L2

Enrichment Worksheets

L3

208C

Hands-On Activities

Science Integration Activity

Lab Manual

Assessment

Performance Assessment

Enrichment and Application

Critical Thinking/ Problem Solving

Cross-Curricular Integration

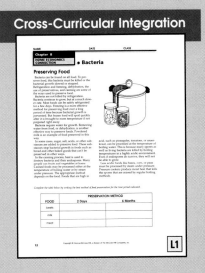

Science and Society Integration

Technology Integration

Multicultural Connections

Concept Mapping

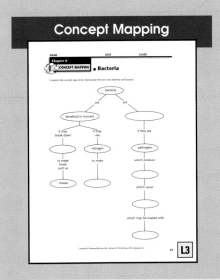

Bacteria

CHAPTER OVERVIEW

Section 8-1 Characteristics of bacteria are described in this section. Aerobic and anaerobic organisms are compared and contrasted.

Section 8-2 The ways that bacteria are important to humans and the environment are emphasized in this section. Examples of helpful and harmful bacteria are given.

Section 8-3 *Science and Society* Tuberculosis and its comeback are discussed as an example of a bacterial disease. The question of whether technology can solve the problem is asked.

Chapter Vocabulary

flagellum	pathogen
fission	antibiotic
aerobe	vaccine
anaerobe	toxin
saprophyte	endospore
nitrogen-fixing	antibiotic
bacteria	resistance

Theme Connection

Systems and Interactions The role of bacteria in maintaining a homeostatic balance is discussed. In addition to homeostasis, the ecological importance of bacteria is illustrated by using examples of niches filled by these organisms.

Previewing the Chapter

Section 8-1 Two Kingdoms of Bacteria
▶ What are bacteria?
▶ Eubacteria
▶ Archaebacteria

Section 8-2 Bacteria in Your Life
▶ Beneficial Bacteria
▶ Harmful Bacteria

Section 8-3 Science and Society
Issue: Fighting Tuberculosis

208

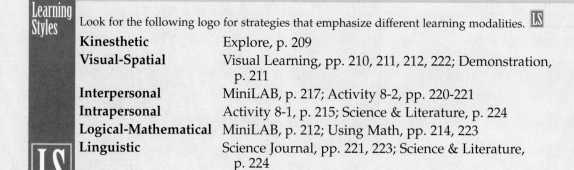

Learning Styles

Look for the following logo for strategies that emphasize different learning modalities. **LS**

Kinesthetic	Explore, p. 209
Visual-Spatial	Visual Learning, pp. 210, 211, 212, 222; Demonstration, p. 211
Interpersonal	MiniLAB, p. 217; Activity 8-2, pp. 220-221
Intrapersonal	Activity 8-1, p. 215; Science & Literature, p. 224
Logical-Mathematical	MiniLAB, p. 212; Using Math, pp. 214, 223
Linguistic	Science Journal, pp. 221, 223; Science & Literature, p. 224

Chapter 8

Bacteria

What would it be like to have millions of cells, like those shown on the previous page, living inside you? You really don't have to imagine such a situation because all of us have huge populations of them living in our small intestines all the time! These cells are bacteria called *Escherichia coli*. You can find out what some bacteria look like by doing the following activity. Then, in this chapter, learn more about these cells and why they are so important to the existence of life on Earth.

EXPLORE ACTIVITY

Find out what bacteria look like.

1. Mix a small drop of yogurt with ten drops of water in a small dish.
2. Add a drop of crystal violet. The crystal violet stains the bacteria so that you can see them.
3. After 30 seconds, dip a toothpick into the mixture and then touch the tip of the toothpick onto a glass slide. Let the slide dry.
4. Put a coverslip on top of the slide and observe the bacteria under low and then high magnifications using a microscope.

Observe: In your Science Journal, describe and draw what the bacteria look like.

Magnification 26 000×

Magnification 400×

Previewing Science Skills

▶ In the **Skill Builders**, you will **hypothesize**, **map concepts**, and **measure in SI**.

▶ In the **Activities**, you will **design an experiment**, **observe**, and **organize data**.

▶ In the **MiniLABs**, you will **research** and **infer**.

209

EXPLORE ACTIVITY

Purpose

LS **Kinesthetic** Use the Explore Activity to introduce monerans. Bacteria are living organisms. Explain that students will be learning about the characteristics of bacteria and cyanobacteria (the monerans). **L1**

Preparation

Be sure to purchase fresh, plain yogurt that is made with live cultures.

Materials

yogurt, methylene blue, toothpicks, slides, coverslips, microscopes, small dishes, water

Teaching Strategies

• You will probably want to check each microscope when students have the organism focused on high power to be certain that they are looking at the bacteria.

• Have the students make a drawing of what they see on the slide.

Observe

Yogurt bacteria are rod-shaped and are stained violet by methylene blue under the microscope.

✔ Assessment

Portfolio Have students design an experiment to determine whether yogurt bacteria are alive. They should place their experimental design in their portfolio. Use the Performance Task Assessment List for Designing an Experiment in **PASC**, p. 23. **P**

Assessment Planner

Portfolio
Refer to page 225 for suggested items that students might select for their portfolios.

Performance Assessment
See page 225 for additional Performance Assessment options.
Skill Builders, pp. 214, 219
MiniLABS, pp. 212, 217
Activities 8-1, p. 215; 8-2, pp. 220-221

Content Assessment
Section Wrap-ups, pp. 214, 219, 223
Chapter Review, pp. 225-227
Mini Quizzes, pp. 214, 219

Group Assessment
Opportunities for group assessment occur with Cooperative Learning Strategies and Flex Your Brain Activities.

Prepare

Section Background
- Leeuwenhöek was probably not the first person to observe bacteria, but he was the first to keep convincingly accurate records of his observations.
- Prokaryotes contain few cell organelles and none that are surrounded by membranes.

Preplanning
- Obtain dried beans for the MiniLAB.
- Obtain prepared slides of *Gloeocapsa* and *Anabaena* and living cultures of *Nostoc* and *Oscillatoria* for Activity 8-1.

1 Motivate

Bellringer
Before presenting the lesson, display **Section Focus Transparency 29** on the overhead projector. Assign the accompanying **Focus Activity** worksheet. L1 LEP

Tying to Previous Knowledge
This section should be bridged with Chapters 1 and 2. The discussion of the characteristics of living things and parts of a cell should be reviewed thoroughly.

Science Words
- **flagellum**
- **fission**
- **aerobe**
- **anaerobe**

Objectives
- Describe the characteristics of bacterial cells.
- Compare aerobic and anaerobic organisms.

What are bacteria?

In Chapter 2, you learned about two types of cells—prokaryotic cells and eukaryotic cells. Bacteria are classified as prokaryotes because they have no membrane-bound organelles, and their nuclear material consists of a single circular chromosome. Bacteria have cell walls and cell membranes and also contain ribosomes. Some bacteria contain chlorophyll, which enables them to use carbon dioxide, water, and sunlight to make their own food.

Types of Bacteria

Most bacteria don't make their own food. That means they have to rely on other organisms to provide them with food. These bacteria have to break down, or decompose, other living things to obtain energy. In nature, bacteria and fungi keep the world free of wastes by breaking them down. In doing so, they also release nutrients into the soil for use again. Bacteria are grouped into two kingdoms—Eubacteria and Archaebacteria, illustrated in **Figure 8-1**.

Eubacteria

When most people hear the word *bacteria,* they probably associate it with sore throats or other illnesses. However, very few bacteria cause illness. Most are important for other reasons. Bacteria are almost everywhere—in the air you breathe, the food you eat, and the water you drink. A shovelful of soil contains billions of them. Millions of bacteria live on and in your body. Most are beneficial to you.

Figure 8-1

Some bacteria are decomposers that break down living tissues. Others make their own food.

A Archaebacteria such as these *Halococcus* bacteria are found in salt lakes and on highly salted foods.

B Eubacteria called cyanobacteria make their own food.

Magnification 1000×

210   Chapter 8 Bacteria

Program Resources

Reproducible Masters
Study Guide, p. 33 L1
Reinforcement, p. 33 L1
Enrichment, p. 33 L3
Activity Worksheets, pp. 46-47, 50
Lab Manual 16, 17
Technology Integration, p. 9
Science and Society Integration, p. 12

Transparencies
Teaching Transparency 15
Section Focus Transparency 29 L1
Science Integration Transparency 8

Figure 8-2

Bacterial characteristics and shapes make bacteria different from other cells. Spirillum-, coccus-, and bacillus-shaped bacteria can be found in virtually any environment. *What common terms could be used to describe these cell shapes?*

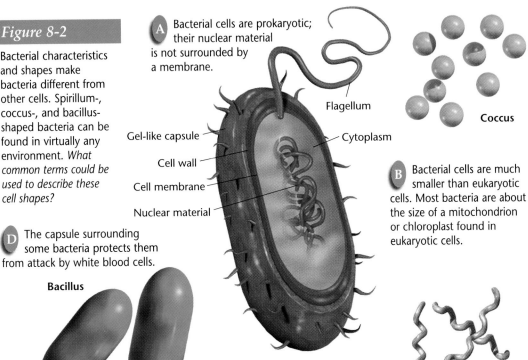

A Bacterial cells are prokaryotic; their nuclear material is not surrounded by a membrane.

Flagellum

Gel-like capsule

Cell wall

Cell membrane

Nuclear material

Cytoplasm

Coccus

B Bacterial cells are much smaller than eukaryotic cells. Most bacteria are about the size of a mitochondrion or chloroplast found in eukaryotic cells.

D The capsule surrounding some bacteria protects them from attack by white blood cells.

Bacillus

Bacterial cell

C The flagellum can help the bacterium to move in a liquid environment.

Spirillum

Bacterial Shapes

The bacteria that normally inhabit your home and body have three basic shapes—spheres, rods, and spirals. Sphere-shaped bacteria are called *cocci,* rod-shaped bacteria are called *bacilli,* and spiral-shaped bacteria are called *spirilla.* The general characteristics of bacteria can be seen in the bacillus shown in **Figure 8-2.** It contains cytoplasm, surrounded by a cell membrane and wall. The nuclear material is in the form of a single chromosome strand. Some bacteria have a thick, gel-like capsule around the cell wall. The capsule helps the bacterium stick to surfaces. How does a capsule help a bacterium to survive?

Many bacteria float freely in the environment on air and water currents, your hands, your shoes, and the family dog or cat. Many that live in moist conditions have a whiplike tail called a **flagellum** to help them move.

8-1 Two Kingdoms of Bacteria   **211**

Inclusion Strategies

Gifted Have students research botulism. They can collect articles from magazines, newspapers, etc. They should find out as much as they can about the moneran. As a group, they can problem solve and try to find a solution to help society cope socially with the disease. **L3**

Across the Curriculum

Math Ask students whether they'd rather have $10 every 20 minutes for 13 hours or one penny doubled every 20 minutes for 13 hours. Point out that the growth rate of bacteria is like that of the penny (*$400 by arithmetic growth or more than $5 trillion by exponential growth*). **L1**

MiniLAB

How quickly do bacteria grow?

Procedure

1. Obtain two or three dried beans.
2. Break them into halves and place the halves into 10 mL of distilled water in a glass beaker.
3. Observe how many days it takes for the water to become cloudy and develop an unpleasant odor.
4. Use methylene blue to dye a drop of water from the beaker and observe under the microscope.

Analysis

1. How long did it take for the water to become cloudy?
2. What did you observe on the slide that would make the water cloudy?
3. On what do you think the bacteria were feeding?

Cyanobacteria

A second type of eubacteria is known as cyanobacteria. Cyanobacteria are eubacteria that are producers. They make their own food using carbon dioxide, water, and energy from sunlight. Cyanobacteria contain chlorophyll and another pigment that is blue. This pigment combination gives cyanobacteria their common name, blue-green bacteria. But some cyanobacteria are yellow, black, or red. The Red Sea gets its name from red cyanobacteria.

All species of cyanobacteria are one-celled organisms. However, some of these organisms live together in long chains or filaments, as illustrated in **Figure 8-3.** Many are covered with a gel-like substance. This adaptation enables them to live in globular groups called colonies. Individual cells in colonies reproduce by fission. Cyanobacteria are important for food production in lakes and ponds. Because cyanobacteria make food from carbon dioxide, water, and the energy from sunlight, fish in a healthy pond can eat them and utilize the energy released from that food.

Have you ever seen a pond covered with smelly, green, bubbly slime? When large amounts of nutrients enter a pond, cyanobacteria increase in number and produce a mat-like growth called a bloom. The cyanobacteria die. Bacteria feed on them and use up all the oxygen in the water. As a result, fish and other organisms die.

Figure 8-3

Some species of cyanobacteria exist as filaments. The photograph shows *Oscillatoria,* which grows on damp soil and can make stones slick in damp areas.

Magnification 100×

212 Chapter 8 Bacteria

Magnification 9400×

Figure 8-4

In this color-enhanced electron micrograph, a bacterium is shown undergoing fission. Fission is the simplest form of asexual cell reproduction.

Reproduction

Bacteria reproduce by fission, as shown in **Figure 8-4. Fission** produces two cells with genetic material exactly like the parent cell's. Some bacteria also reproduce by a simple form of sexual reproduction. Two bacteria line up beside each other and exchange some DNA through a fine tube. This produces cells with different genetic material. As a result, the bacteria may have variations that give them an advantage for surviving in changing environments.

Most bacteria live in places where there is a supply of oxygen. Organisms that use oxygen for respiration are called **aerobes.** You are an aerobic organism. In contrast, some organisms, called **anaerobes,** can live without oxygen. For a few types of bacteria, oxygen is even deadly.

Problem Solving

Controlling Bacterial Growth

Many antibiotics are produced by molds and fungi. A scientist, while trying to grow bacteria in a petri dish, discovered a mold also growing on the plate. The area around this mold was free of bacteria. This event led to the discovery of penicillin, a common antibiotic. Streptomycin, another common antibiotic, was initially discovered in soil fungi. Soil containing this fungi was spread on agar in a petri dish that had bacteria previously growing on it. Again, this fungal growth prevented the bacteria from growing in the dish.

Advertisers proclaim that their mouthwash products can fight bad breath by killing the bacteria that cause it. How could you test a mouthwash sample for antibiotic action?

Solve the Problem:

1. Describe an experiment that you could do to determine if your mouthwash is effective at killing bacteria.
2. What controls would you use in your experiment?

Think Critically:

Read the ingredients label on a bottle of mouthwash. Which of the ingredients appear to be ones that are effective against bacteria? What other ingredients are present? What role do colored dyes play?

8-1 Two Kingdoms of Bacteria   **213**

INTEGRATION
Earth Science

Have students read about smokers and locate them on a globe or world map.

4 Close

Use the Mini Quiz to check students' recall of chapter content.

1. **The nuclear material of monerans consists of a single, circular _____ .** *chromosome*

2. **_____ produces two cells with genetic material exactly like the parent cell.** *Fission*

3. **Some organisms that die when they are exposed to oxygen are called _____ .** *anaerobes*

4. **All species of cyanobacteria are one-celled organisms, but some of these live together in long chains, or _____ .** *filaments*

Section Wrap-up

Review

1. small cells without membrane-bound organelles

2. Aerobic bacteria require oxygen to live; anaerobic bacteria can live without oxygen.

3. **Think Critically** Bacteria feeding on cyanobacteria probably removed too much of the oxygen and may have added poisonous materials to the water.

USING MATH

LM **Logical-Mathematical** 6 hours and 40 minutes (1 048 576 bacteria)

INTEGRATION
Earth Science

Figure 8-5

Bacteria that live in mineral hot springs are thought to be descendants of ancient bacterial forms because they are anaerobes. *What problems would bacteria have to overcome to live in such conditions as these?*

Section Wrap-up

Review

1. What are the characteristics of bacteria?

2. How do aerobic and anaerobic organisms differ?

3. **Think Critically:** A mat of cyanobacteria is found growing on a lake with dead fish floating along the edge. What has caused this to occur?

Skill Builder
Concept Mapping

Use an events chain concept map to show what happens in a pond when a bloom of cyanobacteria dies. If you need help, refer to Concept Mapping in the **Skill Handbook.**

214   Chapter 8 Bacteria

Skill Builder

The concept map should show the following events: the cyanobacteria bloom, then die; bacteria feed on them, using all the oxygen in the water; other organisms, such as fish, may die.

Archaebacteria

Certain kinds of anaerobic bacteria, Kingdom Archaebacteria, are thought to have existed for billions of years. They are found in extreme conditions, such as the hot springs shown in **Figure 8-5,** salty lakes, muddy swamps, the intestines of cattle, and near deep ocean vents where life exists without sunlight.

Archaebacteria are placed into groups, depending upon the way they obtain energy. The methanogens use carbon dioxide for energy and produce the methane gas that bubbles up out of swamps and marshes. The extreme halophiles live in salty environments; some of them require a habitat ten times saltier than seawater to grow. The last group of archaebacteria, **Figure 8-6,** is the extreme thermophiles that survive in hot areas.

Figure 8-6

Thermophiles derive energy by oxid ing sulfur compounds such as those escaping from this deep-sea vent.

USING MATH

Some bacteria may undergo fission every 20 minutes. Suppose that one bacterium is present at the beginning of a timed period. How long would it take for the number of bacteria to increase to more than 1 million?

✓ Assessment

Performance To further assess students' ability to observe cyanobacteria, set up a practical to have them identify organisms and/or answer questions about them. Use the Performance Task Assessment List for Making Observations and Inferences in **PASC,** p. 17.
P

Activity 8-1

Observing Cyanobacteria

You can obtain many species of cyanobacteria from ponds. When you look at these organisms under a microscope, you will find that they have many similarities, but also are different from each other in important ways. In this activity, you will compare and contrast species of cyanobacteria.

Problem
What do cyanobacteria look like?

Materials
- microscope
- prepared slides of *Gloeocapsa* and *Anabaena*
- living cultures of *Nostoc* and *Oscillatoria*

Safety Precautions

Wash your hands after handling cyanobacteria. Do not put cyanobacteria near your mouth. Some kinds produce toxins.

Procedure
1. Make a data table in your Science Journal. Indicate whether each cyanobacterium sample is in colony form or filament form. Write a *yes* or *no* for the presence or absence of each characteristic in each type of cyanobacterium.
2. Observe the prepared slides of *Gloeocapsa* and *Anabaena* under

the low and high power of the microscope. Notice the difference in the arrangement of the cells. Draw and label a few cells of each species of cyanobacterium.
3. Make a wet mount of each living culture. Observe under low and high power of the microscope. In your Science Journal, draw and label a few cells of each species of cyanobacterium.
4. Return all slides with living material to your teacher for disposal.
5. Wash your hands before continuing.

Analyze
1. How does the color of cyanobacteria compare with the color of leaves on trees? What can you infer from this?

Conclude and Apply
2. How can you tell by observing them through a microscope that cyanobacteria belong to Kingdom Eubacteria?
3. **Describe** the activities and general appearance of live cyanobacteria.

Data and Observations   Sample Data

Structure	Ana-baena	Gloe-ocapsa	Nostoc	Oscill-atoria
Filament or colony	filament	colony	filament	filament
Nucleus	no	no	no	no
Chloro-phyll	yes	yes	yes	yes
Gel-like layer	yes	yes	yes	yes

Oscillatoria

Spore

Heterocyst

Anabaena

Nostoc

Gloeocapsa

Purpose
[IS] Intrapersonal Students will observe and record the characteristics of cyanobacteria.
L1

Process Skills
observing, classifying, interpreting data, forming operational definitions

Time
50 minutes

 Activity Worksheets, pp. 5, 46-47

Teaching Strategies
Troubleshooting If students have trouble seeing the jelly-like layer, have them reduce the amount of light coming through the diaphragm of the microscope.

Answers to Questions
1. Cyanobacteria are bluer than leaves but contain chlorophyll. Therefore, they can undergo photosynthesis.
2. Cyanobacteria do not have visible nuclei and are small cells.
3. Cyanobacteria appear blue-green in color and may form slimy colonies in water.

▼ **Assessment**

Performance Have students draw, label, and describe different cyanobacteria in a logical manner in their Science Journals. Use the Performance Task Assessment List for Scientific Drawing in **PASC**, p. 55. **P**

Prepare

Section Background

Bacteria are necessary to many organisms because they break down food, and they are necessary to the environment because they break down dead material and release the nutrients for recycling.

Preplanning

- For the MiniLAB, obtain milk, yogurt starter, and a thermos bottle.
- Obtain petri dishes and nutrient agar for Activity 8-2.

1 Motivate

Bellringer

 Before presenting the lesson, display **Section Focus Transparency 30** on the overhead projector. Assign the accompanying **Focus Activity** worksheet. L1 LEP

Brainstorming

Have students brainstorm a list of words that come to mind when they think of the actions of bacteria.

GLENCOE TECHNOLOGY

 Videodisc

STVS: Plants & Simple Organisms
Disc 4, Side 1
Using Bacteria to Detect Carcinogens (Ch. 7)

Bacterial Waste Treatment (Ch. 10)

Science Words

- saprophyte
- nitrogen-fixing bacteria
- pathogen
- antibiotic
- vaccine
- toxin
- endospore

Objectives

- Identify some ways bacteria are helpful.
- Explain the importance of nitrogen-fixing bacteria.
- Explain how some bacteria cause disease.

Figure 8-7

Bacteria that break down proteins in milk are used in production of various kinds of cheese.

Beneficial Bacteria

Have you had any bacteria for lunch lately? Anytime you eat cheese or yogurt, you eat some bacteria. Bacteria break down substances in milk to make many everyday products. Cheese-making is illustrated in **Figure 8-7**. If you have eaten sauerkraut, you ate a product made with cabbage and a bacterial culture. Vinegar is also produced by a bacterium.

Uses of Bacteria

Bacteria called saprophytes (SAP ruh fitz) help maintain nature's balance. A **saprophyte** is any organism that uses dead material as a food and energy source. Saprophytes digest dead organisms and recycle nutrients so that they are available for use by other organisms. Without saprophytic bacteria, there would be layers of dead material deeper than you are tall spread over all of Earth.

The roots of plants such as peanuts and peas contain nitrogen-fixing bacteria in growths called nodules, illustrated in **Figure 8-8**. These **nitrogen-fixing bacteria** change nitrogen from the air into forms useful for plants and animals. Plants benefit by using the nitrogen, and animals benefit because they obtain nitrogen for making needed proteins. It is estimated that nitrogen-fixing bacteria save U.S. farmers several million dollars in fertilizer costs every year.

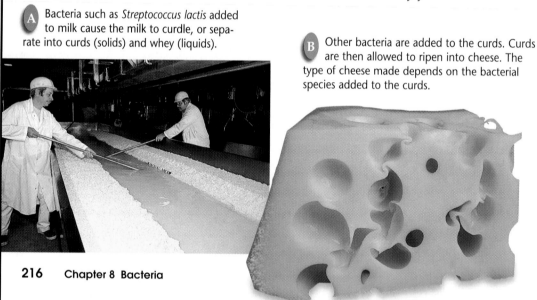

A Bacteria such as *Streptococcus lactis* added to milk cause the milk to curdle, or separate into curds (solids) and whey (liquids).

B Other bacteria are added to the curds. Curds are then allowed to ripen into cheese. The type of cheese made depends on the bacterial species added to the curds.

216   Chapter 8 Bacteria

Program Resources

 Reproducible Masters
Study Guide, p. 34 L1
Reinforcement, p. 34 L1
Enrichment, p. 34 L3
Cross-Curricular Integration, p. 12
Critical Thinking/Problem Solving, p. 12
Activity Worksheets, pp. 48-49, 51 L1
Lab Manual 18
Science Integration Activities, p. 15

Concept Mapping, p. 21
Multicultural Connections, pp. 19-20

Transparencies
Teaching Transparency 16
Section Focus Transparency 30

Figure 8-8

Root nodules, which form on the roots of peanuts, peas, and other legumes, contain nitrogen-fixing bacteria.

Bacteria — Root hair

A Root hairs curl in preparation for infection by the bacteria.

B Bacteria enter the roots through an "infection thread," which carries the bacteria into the root.

C Once inside "pockets" within the root, the bacteria produce a chemical that causes their host cells to grow and divide.

D These dividing cells form the nodule, which contains bacteria that change atmospheric nitrogen into more usable forms for plants and animals.

Infection thread

Dividing cells

Bacteria

Pockets

Nodule

Many industries rely on bacteria. Biotechnology is putting bacteria to use in making medicines, enzymes, cleansers, adhesives, and other products. Bacteria are extremely important in helping to clean up the extensive oil spills in Alaska, California, and Texas. The bacteria digest the oil.

Harmful Bacteria

Some bacteria are pathogens. A **pathogen** is any organism that produces disease. If you have ever had strep throat, you have had firsthand experience with a bacterial pathogen. Other pathogenic bacteria cause anthrax in cattle, and diphtheria, tetanus, and whooping cough in humans. Bacterial diseases in animals are usually treated effectively with antibiotics. An **antibiotic** is a substance produced by one organism that inhibits or kills another organism. Penicillin, a well-known antibiotic, works because it prevents bacteria from making cell walls. Without cell walls, bacteria cannot survive.

Some bacterial diseases can be prevented by vaccines. A **vaccine** is made from damaged particles from bacteria's cell walls, or from killed bacteria. Once injected, the white blood cells in the body learn to recognize the bacteria. If the particular bacteria then enter the body at a later time, the white blood cells immediately attack and overwhelm the bacteria. Vaccines have been produced that are effective against many bacterial diseases.

MiniLAB

How can you make yogurt?

Procedure

1. Bring a liter of milk almost to a boil in a saucepan. **CAUTION:** *Always be careful when using a stove or hot plate.*
2. Remove the pan from the burner and allow it to cool until it is lukewarm.
3. Add one or two heaping tablespoons of yogurt starter and stir.
4. Pour the mixture into a clean vacuum bottle and put on the lid.
5. Let stand for six hours and then refrigerate overnight.

Analysis

1. What do you think was in the yogurt starter?
2. Infer why you let the milk cool before adding the starter.

Discussion

Discuss bacteria as symbionts. (Bacteria in humans supply some vitamins and are useful in the digestion of foods high in fiber.)

MiniLAB

Purpose

IS **Interpersonal** Students will infer the presence of bacteria in yogurt starter.

L1 **COOP LEARN**

Materials

milk, saucepan, yogurt starter, thermos bottle

Analysis

1. bacteria
2. Yogurt culture is alive and excess heat can kill the bacteria. The bacteria culture grows well when lukewarm.

Assessment

Oral Ask students to formulate additional questions that they could test to extend the MiniLAB activity. Use the Performance Task Assessment List for Asking Questions in **PASC**, p. 19.

P

Activity Worksheets, pp. 5, 51

GLENCOE TECHNOLOGY

Videodisc

STVS: Plants & Simple Organisms

Disc 4, Side 1

Bacterial Digestion of Lignin (Ch. 8)

Bacterial Miners (Ch. 9)

USING TECHNOLOGY

Lyme Disease Vaccine

Lyme disease, a disease that causes swollen and painful joints, is caused by the bacterium *Borella burgdorferi*. The bacteria enter the body when an infected tick attaches itself to skin and feeds on the victim's blood. The bacteria reproduce in the victim's blood.

Help ahead?

Scientists are developing a vaccine to prevent this bacterial infection. The genes that control the production of the bacterium's surface protein are first duplicated. These genes are then inserted into different host bacteria, which produce quantities of the protein. The proteins are then inserted in an organism capable of developing Lyme disease. The organism's immune system reacts to the injected protein and destroys it. If the organism subsequently becomes infected with the disease-causing bacteria, the immune system "remembers" the surface proteins and destroys the bacteria. Such results encourage scientists working in the development of a vaccine to eliminate Lyme disease in humans.

Think Critically:

Why might a vaccine that contains surface proteins be useful in preventing a disease?

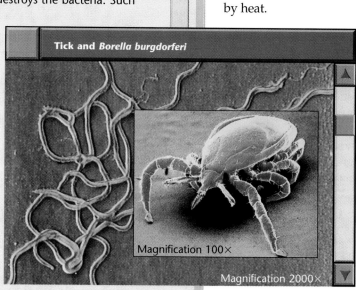

Tick and *Borella burgdorferi*

Magnification 100×

Magnification 2000×

218   Chapter 8 Bacteria

Some pathogens produce poisons. The poisons produced by bacterial pathogens are called **toxins.** Botulism, a type of food poisoning, is caused by a toxin that can cause paralysis and death. The bacterium that causes botulism is *Clostridium botulinum.* Many bacteria that produce toxins are able to produce thick walls around themselves and make themselves resistant to heat or drying. These thick-walled structures are called **endospores,** illustrated in **Figure 8-9.** Endospores can exist for more than 50 years and begin growing again. Botulism endospores require long exposure to heat to be destroyed. Once the endospores are in canned food, the bacteria can change back to regular cells and start producing toxins. Botulism bacteria are able to grow inside cans because they are anaerobes and do not need oxygen to live. Botulism toxins can be destroyed by heat.

Community Connection

Community Composting Many communities provide free or low-cost compost to citizens. Check to see whether your community maintains a compost pile.

Figure 8-9
Bacteria sometimes form endospores when conditions become unfavorable. Bacterial endospores are structures that can survive harsh winters, dry conditions, and heat. Endospores of botulism bacteria are sometimes found in canned foods where they can produce deadly toxins.

Magnification 15 000×

Pasteurization

Pasteurization, a process of heating food to a temperature that kills harmful bacteria, is used in the food industry. You are most familiar with it in pasteurized milk, but some fruit juices are also pasteurized. The process is named for Louis Pasteur, who first formulated the process for the wine industry in the 19th century in France.

Section Wrap-up

Review

1. Why are saprophytes helpful and necessary?

2. Why are nitrogen-fixing bacteria important?

3. Think Critically: Why is botulism associated with canned foods and not fresh foods?

Skill Builder
Measuring in SI
Air may have more than 3500 bacteria per cubic meter. How many bacteria might be in your classroom? If you need help, refer to Measuring in SI in the **Skill Handbook.**

CONNECT TO
PHYSICS

Many foods are "vacuum packed." What does this mean? *Research* how a vacuum helps prevent food spoilage.

Using Computers

Spreadsheet Create a spreadsheet that includes: Disease Name, Disease Organism, Transmission Route, and Symptoms. Enter information for the following diseases: whooping cough, tuberculosis, tetanus, diphtheria, and scarlet fever. Sort your data using Transmission Route.

8-2 Bacteria in Your Life   **219**

CONNECT TO
PHYSICS

Answer High heat and fast sealing are used to drive air from food in closed containers; cutting off oxygen makes most bacteria unable to reproduce.

4 Close

•MINI•QUIZ•

Use the Mini Quiz to check students' recall of chapter content.

1. **Bacteria called _____ help maintain nature's balance by decomposing other organisms.** *saprophytes*
2. **A(n) _____ is a drug used to kill bacteria.** *antibiotic*
3. **Some pathogens produce poisons called _____ .** *toxins*

Section Wrap-up

Review
1. They maintain homeostasis by recycling nutrients.
2. They change nitrogen from the atmosphere into a form that can be used by plants and animals.
3. **Think Critically** The bacteria that cause botulism are anaerobes.

Using Computers

Air—whooping cough, tuberculosis, diphtheria, scarlet fever

Objects—tetanus

Tetanus is transmitted by objects. Data will show that all listed diseases, except tetanus, are transmitted through the air. All listed diseases are caused by bacteria.

Skill Builder
Have students measure the classroom to find the length, width, and height in meters. Then have them multiply these measurements to obtain the volume in cubic meters. Multiply the volume by 3500 to obtain the number of bacteria per cubic meter.

 Assessment

Performance Assess students' abilities to measure in SI units by having them calculate the same situation for a room at home. Use the Performance Task Assessment List for Using Math in Science in **PASC,** p. 29. **P**

Interpersonal Students will design and carry out an experiment to determine where the highest populations of bacteria occur. **L1** **COOP LEARN**

Process Skills
communicating, observing and inferring, recognizing cause and effect, making and using tables, forming a hypothesis, designing an experiment, separating and controlling variables, interpreting data, using numbers

Time
2 class periods

Materials
In addition to materials listed on the student page, you may substitute boiled potatoes for nutrient agar.

Safety Precautions
Be sure that students observe the safety precautions listed on the student page.

Possible Hypotheses
Students may hypothesize that more bacteria may be found in air, on surfaces in the classroom, or on their own skin.

Activity Worksheets, pp. 5, 48-49

Activity 8-2

Design Your Own Experiment
Where are the most bacteria found?

Bacteria are so small that a microscope is needed to see them. On the other hand, bacterial colonies that contain millions of bacteria—all descendants from a single bacterium—can be seen without using a microscope. Each bacterium grows best at a particular temperature. Based on this information, design an experiment to determine where bacteria are found in the greatest concentrations.

PREPARATION

Problem
Where are the greatest concentrations of different kinds of bacteria found?

Form a Hypothesis
Based on the information you have learned about bacteria, state a hypothesis about where in your school you might be able to find the greatest concentrations of different bacteria.

Objectives
- Observe where different kinds of bacteria are found.
- Design an experiment that tests numbers of bacteria in particular locations.

Possible Materials
- sterile petri dishes with nutrient agar
- incubator
- cotton-tipped swabs
- tape
- water

Safety Precautions

The bacteria you collect have the potential for causing disease. *Be sure to close the petri dishes after putting a sample inside and tape them closed. Once the colonies begin to grow, do not open the petri dishes unless you have permission from your teacher. Wash your hands with soap and water when finished.* Your teacher will dispose of the petri dishes properly at the end of the experiment.

PLAN THE EXPERIMENT

1. As a group, decide upon and write out the hypothesis statement.
2. As a group, list the steps that you need to take to test your hypothesis. Be specific, describing exactly what you will do at each step and where you will sample for bacteria. Areas could include doorknobs, hands, etc. List your materials. *Have your teacher approve of your location and materials lists.*
3. Following your teacher's instructions, use a cotton-tipped swab to gently sweep the nutrient agar in petri dishes.
4. Prepare a data table in your Science Journal.

Check Your Plan

1. Read over your entire experiment to make sure that all steps are in logical order.
2. Identify any constants, variables, and controls of the experiment.
3. *Make sure your teacher approves your plan before you proceed.*

DO THE EXPERIMENT

1. Carry out the experiment as planned.
2. While the experiment is going on, write down any observations that you make and complete the data table in your Science Journal.

Analyze and Apply

1. **Compare** your results with those of other groups.

2. **Graph** your results using a bar graph, placing the number of kinds of bacteria on the y-axis and the locations where the samples were taken on the x-axis.
3. Where are the most kinds of bacteria found?
4. How would you find out what species of bacteria you sampled?

Go Further

Using the information you gained from this experiment, where would you search for additional types of bacteria?

Possible Procedures

Students may choose to swab various surfaces. They may also touch an object directly on the agar. Be sure students are able to summarize their data in a logical way. **P**

Teaching Strategies

Tying to Previous Knowledge Some students will infer that human skin is an excellent source of bacteria and swab different parts of their bodies.

Troubleshooting If fungi overrun the petri dishes, students may elect to estimate bacterial populations.

DO THE EXPERIMENT

Expected Outcome

Different surfaces will have different bacterial populations, as shown by numbers of individual colonies.

Analyze and Apply

1. Results may be different depending upon the objects sampled.
2. Graphs will vary depending on observations.
3. Most bacteria will be found on skin.
4. Students could use a dichotomous key.

Go Further

Student answers will depend on their results.

Science Journal

Linguistic Have students summarize their results in their Science Journal.

Assessment

Process Have students make pictures of the bacterial colonies on their petri dishes from different samples. Use the Performance Task Assessment List for Scientific Drawing in **PASC**, p. 55. **P**

Prepare

Section Background

M. tuberculosis is a typical moneran in many ways. The case study illustrates the interrelationship between parasites and their hosts.

1 Motivate

Bellringer

Before presenting the lesson, display **Section Focus Transparency 31** on the overhead projector. Assign the accompanying **Focus Activity** worksheet. [L1] [LEP]

Brainstorming

Brainstorm with students about microbes that they know can cause disease. Tell them which are viruses, bacteria, protists, etc. Lead them into a discussion of tuberculosis, a disease many will not know about.

2 Teach

Visual Learning

Figure 8-10 Use **Figure 8-10** to explain the effects of tuberculosis on the lungs. [LEP] [LS]

Science Words

antibiotic resistance

Objectives

- Describe antibiotic resistance.
- Explain the different opinions about how to deal with the problem of antibiotic resistance.

Figure 8-10

Mycobacterium tuberculosis attacks the lungs. This X ray shows blue areas on the lungs where the bacteria are growing.

ISSUE:
8•3 Fighting Tuberculosis

Past Wars with TB

Up until the 1950s, the United States waged a war against a tiny enemy called *Mycobacterium tuberculosis*. This bacterium produces a disease in humans called tuberculosis (TB). Tuberculosis is spread by close contact with other people who have advanced cases of the disease. **Figure 8-10** illustrates how TB attacks the lungs. If not treated properly, people can die of tuberculosis. By the 1980s, as a result of the development of effective antibiotics, death due to TB had become rare. The medical community in the United States thought the war had been won.

TB Makes a Comeback

Unfortunately, doctors and researchers fighting tuberculosis won the battle but not the war because tuberculosis has made a comeback. Between 1985 and 1991, the number of TB cases in the United States increased by 16 percent. In 1990 alone, more than 26 000 new cases were reported across the country. In areas where there are many homeless people and where public health programs have been reduced, tuberculosis infection is on the rise. Because the bacterium is easily passed among individuals, people living in places such as shelters have now become susceptible to TB.

Why has the number of TB cases increased, as illustrated in **Figure 8-11?** One critical factor is the AIDS epidemic. When the AIDS virus destroys a person's immune system, natural resistance to diseases such as TB is greatly impaired. It is estimated that more than 10 million Americans now harbor the tuberculosis bacterium and that half of current AIDS patients will develop TB.

Another troubling aspect of TB is that many strains are antibiotic resistant. **Antibiotic resistance** occurs when bacteria that are normally killed by a specific drug treatment mutate into those strains that are not affected by such treatments. How could this occur? After a few weeks of antibiotic treatments, some people feel better and therefore discontinue their prescriptions. If they stop taking antibiotics too soon, not all of the bacteria are killed. Sometimes antibiotic-resistant

---

Program Resources

 Reproducible Masters
Study Guide, p. 35 [L1]
Reinforcement, p. 35 [L1]
Enrichment, p. 35 [L3]

Transparencies
Section Focus Transparency 31 [L1]

bacteria survive and are then able to produce greater numbers of resistant populations of bacteria. Such actions can result in a strain of TB bacteria that no antibiotics can treat.

2 Points of View

Science Is the Answer

Some individuals believe that new technology is the answer to the current TB problem. Science and technology must produce new drugs to treat these new strains of bacteria that are causing tuberculosis once again. Discovering such drugs should be the major priority of the science community.

Society Is the Answer

An opposing view is one of social prevention. Instead of funding medical research for new treatments for tuberculosis, society should eliminate conditions enabling such mutant strains to develop. Poverty, lack of education, and poor public health programs must be addressed to prevent the crowding of homeless individuals into shelters. Public acceptance of the steps needed to prevent and control AIDS and the importance of proper treatment programs must also be increased. Some people support the position that preventive measures should be taken even without the permission of those who are most at risk.

Section Wrap-up

Review

1. Why does tuberculosis appear to be more of a problem in large cities and homeless shelters?

2. Infer how a bacterium becomes antibiotic resistant.

Explore the Issue

Should people be made to have treatment for such diseases as tuberculosis? Should society do more to help the health and environmental conditions of the poor in the United States? Give reasons for your answers.

Number of Tuberculosis Cases in the United States

Figure 8-11

The number of cases of tuberculosis in the United States is on the increase.

interNET CONNECTION

How much do you know about the history of tuberculosis in the United States? Visit the Chapter 8 Internet Connection at Glencoe Online Science, **www. glencoe.com/sec/science/ life**, for a link to more information about tuberculosis.

SCIENCE & SOCIETY

Across the Curriculum

Social Studies Have students analyze the effects tuberculosis has had on societies in the past, when it was known as "consumption." L2

Science Journal

Battling TB Have students write a one-page paper in their journal explaining why this battle with tuberculosis is difficult. LS

3 Assess

Check for Understanding

Have students answer questions 1 and 2 in the Section Wrap-up.

Reteach

Ask students to describe the actions of tuberculosis in the body.

Extension

For students who have mastered this section, use the **Reinforcement** and **Enrichment** masters.

4 Close

Have students look at prepared slides of tuberculosis bacteria in the microscope.

Section Wrap-up

Review

1. Tuberculosis is transmitted by close contact with infected people. Dense populations increase the risk.

2. When people stop taking prescribed medicine, antibiotic-resistant bacteria may cause a more dangerous reinfection.

Explore the Issue

Students will express a variety of answers, depending on their own viewpoints and values. Accept all answers that are accompanied by logical reasoning.

USING MATH

LS Logical-Mathematical

In 1984, there were 21 000 cases; in 1986, 22 000 cases; in 1985, approximately 21 500 cases.

Science & Literature

Sources

Masters, Edgar Lee. *Spoon River Anthology, An Annotated Edition.* Chicago: University of Illinois Press, 1992. Edited and annotated by John E. Hallwas.

Biography

- Edgar Lee Masters (1868-1950) was a poet, novelist, and lawyer who grew up in a small Illinois town. The towns he knew in his childhood became the basis for his most famous work, *Spoon River Anthology.*

- Too poor to attend college for more than one year, he studied law at his father's law firm. He eventually became a successful Chicago lawyer. He was a partner in renowned defense lawyer Clarence Darrow's firm from 1903 to 1911.

- While *Spoon River Anthology* is by far his best-known work, Masters also wrote plays, essays, novels, and biographies. His autobiography is entitled *Beyond Spoon River.*

Teaching Strategies

- Have students discuss any minor injuries they may have had that became infected. Ask whether any treatment was required, and if so, what kind.

- **Intrapersonal** Have students research tetanus to discover what toxin the bacterium produces and how an infection is treated.

- Ask students what type of vaccines they have had.

- Ask students what antibiotics they may have taken recently. Have them research how recent the development of that medication is. Possible approaches might be telephone calls to their doctors' offices or a class letter to a pharmaceutical company.

Spoon River Anthology
by Edgar Lee Masters

A hundred years ago, it was common for people to die from bacterial infections that can be little more than an inconvenience today.

One such case was immortalized in Edgar Lee Masters's anthology of poems, *Spoon River Anthology.* These poems are written in the voices of people who died in a small Illinois town in the late 1800s. While the town is fictitious, many of the people speaking in the poems were based on people Masters knew or knew about.

The poem "Charlie French" was inspired by a boy who died in 1881. A newspaper account of the tragedy read, in part:

> Died, on Friday evening, July 7, from the effects of a wound received from the accidental discharge of a toy pistol, Charlie Bell, aged 14 years. The incident occurred at Canton, July 4, producing lock-jaw a few days later…

The poem begins with the boy asking if anyone ever found out who caused the accident and describes all the fun of the Fourth of July celebration.

> *To have it all spoiled*
> *By a piece of cap shot under the skin of my hand.*
> *And all the boys crowding around me saying*
> *"You'll die of lock-jaw, Charlie, Sure."*
> *Oh Dear! Oh, dear!*
> *What chum of mine could have done it?*

Tetanus earned the name "lockjaw" because it results in muscle spasms that make the jaw clench. It's caused by toxins produced by the anaerobic bacterium *Clostridium tetani.*

Clostridium tetani is common in topsoil. It can enter the body through any wound but is most likely to infect through a puncture wound or a deep cut. Today, all children in the United States are routinely vaccinated against tetanus starting at two months of age.

Science Journal

In your Science Journal, hypothesize how a tetanus infection might have been caused by the sort of wound described in the poem and newspaper article quoted.

224 Chapter 8 Bacteria

- **Linguistic** Discuss or have students research how smallpox became the first viral disease apparently eliminated from humanity.

Science Journal

Tetanus bacteria is found in soil. The toy pistol and its ammunition may have had bacterial soil on it. Upon puncturing Charlie's skin, the bacteria released its toxins, causing paralysis.

Chapter 8 Review

Summary

8-1: Two Kingdoms of Bacteria

1. Bacteria are cells that contain DNA, ribosomes, and cytoplasm; they lack organelles. Most reproduce by fission.
2. Cyanobacteria are bacteria that make their own food. Most bacteria break down other living things to obtain energy.
3. The oldest types of bacteria are anaerobic, whereas more complex cells need oxygen and are called aerobes.

8-2: Bacteria in Your Life

1. Helpful bacteria aid in recycling nutrients, nitrogen fixation, and food production.
2. Some bacteria are harmful because they cause disease. Some bacterial diseases are anthrax, tetanus, and botulism.
3. A process like pasteurization keeps food free of harmful bacteria.

8-3: Science and Society: Fighting Tuberculosis

1. Antibiotic resistance is the evolutionary change of bacteria from strains that are mostly killed by a drug treatment to those strains that are not killed.
2. Some people believe that technology alone is enough to win the battle against tuberculosis, but others believe a social commitment is needed as well.

Key Science Words

a. aerobe
b. anaerobe
c. antibiotic
d. antibiotic resistance
e. endospore
f. fission
g. flagellum
h. nitrogen-fixing bacteria
i. pathogen
j. saprophyte
k. toxin
l. vaccine

Reviewing Vocabulary

Match each phrase with the correct term from the list of Key Science Words.

1. organism that decomposes dead organisms
2. structure by which some organisms move
3. heat-resistant structure in bacteria
4. substance that can prevent, not cure, a disease
5. any organism that produces disease
6. organism that must have oxygen to survive
7. organism that can survive without oxygen
8. harmful chemical produced by some pathogens
9. bacteria in roots that make nitrogen available to organisms
10. produces two cells with genetic material exactly like the parent's cells

Chapter 8 Review 225

Summary

Have students read the summary statements to review the major concepts of the chapter.

Reviewing Vocabulary

1. j	6. a
2. g	7. b
3. e	8. k
4. l	9. h
5. i	10. f

✓ Assessment

Portfolio Encourage students to place in their portfolios one or two items of what they consider to be their best work. Examples include:

- Flex Your Brain, p. 213
- Activity 8-2 experimental design and results, p. 221
- Chapter 8 Review, Performance Assessment Activities, p. 227 **P**

Performance Additional performance assessments may be found in **Performance Assessment** and **Science Integration Activities**. Performance Task Assessment Lists and rubrics for evaluating these activities can be found in Glencoe's **Performance Assessment in the Science Classroom.**

GLENCOE TECHNOLOGY

 Videodisc

Glencoe Life Science Interactive Videodisc
Side 1, Lesson 3
Decomposers: Earth's Natural Recyclers

25823 to 37733

MindJogger Videoquiz

Chapter 8 Have students work in groups as they play the Videoquiz game to review key chapter concepts.

Chapter 8 Review

Checking Concepts

Checking Concepts

1. a   **6.** a
2. b   **7.** b
3. a   **8.** c
4. d   **9.** d
5. a   **10.** a

Understanding Concepts

11. These cells are different from other kinds of bacteria.
12. It has been contaminated by exposure to the air.
13. Cyanobacteria make their own food by photosynthesis. They also have cell walls like plants.
14. Prokaryotes, of which bacteria are examples, have no nuclei or membrane-bound organelles.
15. Tuberculosis bacteria have developed antibiotic resistance.

Thinking Critically

16. Nitrogen would no longer be available in the form that plants could use; therefore, the plants would die unless fertilizer were added.
17. Bacteria can reproduce quickly, have means of moving, and can form endospores to survive extreme conditions.
18. Crops like beans, peas, and peanuts have bacteria that are nitrogen-fixing. These crops help to increase soil fertility.
19. spheres
20. Using fresh foods and keeping them refrigerated, washing hands and all surfaces and utensils, and properly cooking foods help to prevent food poisoning.

Checking Concepts

Choose the word or phrase that completes the sentence.

1. _____ is an example of bacteria that have developed antibiotic resistance.
a. Tuberculosis   c. Smallpox
b. Yellow fever   d. Cyanobacteria
2. Bacterial cells contain _____.
a. nuclei   c. mitochondria
b. DNA   d. no chromosomes
3. Bacteria that make their own food have _____.
a. chlorophyll   c. Golgi bodies
b. lysosomes   d. mitochondria
4. Most bacteria are _____.
a. anaerobic   c. many celled
b. pathogens   d. beneficial
5. Bacteria that are rod shaped are _____.
a. bacilli   c. spirilli
b. cocci   d. colonies
6. The structure that allows a bacterium to stick to surfaces is the _____.
a. capsule   c. chromosome
b. flagellum   d. cell wall
7. Blooms in ponds are caused by _____.
a. archaebacteria   c. cocci
b. cyanobacteria   d. viruses
8. Nutrients and carbon dioxide are returned to the environment by _____.
a. producers   c. saprophytes
b. flagella   d. pathogens
9. _____ is caused by a pathogenic bacterium.
a. Nitrogen fixation   c. Cheese
b. An antibiotic   d. Strep throat
10. Organisms that may find oxygen deadly are _____.
a. anaerobes   c. producers
b. aerobes   d. viruses

Understanding Concepts

Answer the following questions in your Science Journal using complete sentences.

11. Some scientists think that archaebacteria were among the first living things on Earth. What trait of archaebacteria seems to support this idea?
12. Milk produced by cows is free of bacteria, yet several hours later, it needs to be pasteurized. Why?
13. Why do you think the cyanobacteria used to be classified with the plants?
14. Explain why bacteria are prokaryotes.
15. Why are some antibiotic treatments used in the 1950s for tuberculosis no longer completely effective?

Thinking Critically

16. What would happen if nitrogen-fixing bacteria could no longer live on the roots of plants?
17. Why are bacteria capable of surviving in all environments of the world?
18. Farmers often rotate crops such as beans, peas, and peanuts with other crops such as corn, wheat, and cotton. Why might they make such changes?
19. The organism that causes bacterial pneumonia is called *Pneumococcus*. What is its shape?
20. What precautions can be taken to prevent food poisoning?

Developing Skills

21. Concept Mapping See reduced student page for answers.

22. Making and Using Graphs

Bacterial Reproduction Rates

Developing Skills

If you need help, refer to the **Skill Handbook.**

21. **Concept Mapping:** Use the following events chain to sequence the events that follow a pond bloom.

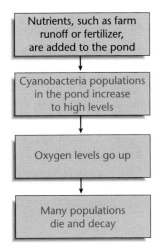

Nutrients, such as farm runoff or fertilizer, are added to the pond

↓

Cyanobacteria populations in the pond increase to high levels

↓

Oxygen levels go up

↓

Many populations die and decay

22. **Making and Using Graphs:** Graph the data of bacteria reproducing at different temperatures.

Bacterial Reproduction Rates

Temperature (°C)	Doubling Rate per Hour
20.5	2.0
30.5	3.0
36.0	2.5
39.2	1.2

23. **Interpreting Data:** What can you conclude about the effect of temperature on bacteria in question 22?
24. **Design an Experiment:** How could you decide whether a particular kind of bacteria can grow anaerobically?
25. **Comparing and Contrasting:** In a table, compare and contrast a virus, a bacterial cell, and a eukaryotic cell according to the structures they contain.

Performance Assessment

1. **Using Math in Science:** The Explore Activity on page 209 asked you to look at certain bacteria in yogurt. Assume that there are 50 000 bacteria in every drop of yogurt and there are 30 drops of yogurt in a milliliter. Calculate the number of bacteria in a 250-mL serving of yogurt.
2. **Designing an Experiment:** Look at the MiniLAB on page 217 again. Design an experiment that would test the effect of temperature on the final product—homemade yogurt. Carry out your experiment.
3. **Group Work:** Research the components of a successful compost heap. As a class project, set up a model composting project at your school, indicating the role of bacteria in the process.

Assessment Resources

Reproducible Masters
Chapter Review, pp. 19-20
Assessment, pp. 51-54
Performance Assessment, p. 22

Glencoe Technology
⦿ **Chapter Review Software**
⦿ **Computer Test Bank**
▭ **MindJogger Videoquiz**

Chapter 8 Review

23. **Interpreting Data** The most favorable temperature for growth is 30.5°C.
24. **Design an Experiment** Place the organism in an environment without oxygen and measure its growth rate.
25. **Comparing and Contrasting**

Virus or cell type	Structures	Nucleus
Virus	Nucleic acid and protein coat	No
Bacteria	Nucleic acid, cell wall, cytoplasm, ribosomes	No
Eukary-ote cell	cell membrane, cytoplasm, organelles, DNA in a nucleus	Yes

Performance Assessment

1. 50 000 bacteria/drop × 30 drops/mL = 5 × 250 mL = 375 000 000 bacteria = 375 million bacteria. Use the Performance Task Assessment List for Using Math in Science in **PASC,** p. 29. **P**
2. Experiments should hypothesize that high temperatures ruin the yogurt. Use the Performance Task Assessment List for Designing an Experiment in **PASC,** p. 23. **P**
3. Some bacteria act as saprophytes and help to decompose organic materials. Compost heaps will depend on local situations. Use the Performance Task Assessment List for Group Work in **PASC,** p. 97. **P**

Chapter Organizer

Section	Objectives/Standards	Activities/Features
Chapter Opener		Explore Activity: Dissect a mushroom to find out how it is put together. p. 229
9-1 **Kingdom Protista** (1½ sessions, 1 block)*	1. **Identify** the characteristics shared by all protists. 2. **Describe** the three groups of protists. 3. **Compare** and **contrast** the protist groups. National Content Standards: (5-8) UCP1, UCP4, A1, A2, C5	Using Math, p. 233 MiniLAB: What do slime molds look like? p. 237 Problem Solving: Puzzled About Slime, p. 238 Skill Builder: Making and Using Tables, p. 239 Using Computers: Spreadsheet, p. 239 Activity 9-1: Comparing Algae and Protozoa, p. 240
9-2 **Kingdom Fungi** (1½ sessions, 1 block)*	1. **Identify** the characteristics shared by all fungi. 2. **Classify** fungi into groups based on their methods of reproduction. 3. **Describe** the difference between the imperfect fungi and all other fungi. National Content Standards: (5-8) UCP1, UCP3, A1, A2, C2, C4, C5, D1, E2, F1, F3	Using Technology: Confocal Microscopy, p. 243 MiniLAB: How are spore prints made? p. 244 Skill Builder: Comparing and Contrasting, p. 245 Using Math, p. 245 Activity 9-2: Investigating Lichens, pp. 246-247
9-3 **Science and Society Technology: Monitoring Red Tides** (1 session, ½ block)*	1. **Identify** the causes and results of red tides. 2. **Describe** the technologies that are used to identify the locations of red tides and how such identification is useful to organisms. National Content Standards: (5-8) C3, C4, E2, F5, G1	Explore the Technology, p. 249 People and Science: Dr. Regina Benjamin, Family Practice Physician, p. 250

* A complete Planning Guide that includes block scheduling is provided on pages 32T-35T.

Activity Materials

Explore	Activities	MiniLABs
page 229 edible mushrooms	page 240 cultures of *Paramecium, Amoeba, Euglena, Spirogyra,* and *Micrasterias*; prepared slide of slime mold; 5 coverslips; microscope; dropper; 5 microscope slides pages 246-247 bits of live lichens of various forms, needles, tweezers, petri dishes with nutrient agar, sterile pond water, tape, microscope slide, coverslip, microscope, paper towels, apron, safety goggles	page 237 slime-mold growth medium page 244 edible mushrooms; sheet of white, unlined paper

Need Materials? Call Science Kit (1-800-828-7777).

Teacher Classroom Resources

Reproducible Masters	Transparencies	Teaching Resources
Study Guide, p. 36 **Reinforcement**, p. 36 **Enrichment**, p. 36 **Activity Worksheets**, pp. 52-53, 56 **Lab Manual 19**, Life in Pond Water **Lab Manual 20**, Molds **Lab Manual 21**, Yeasts **Concept Mapping**, p. 23 **Cross-Curricular Integration**, p. 13 **Multicultural Connections**, pp. 21-22	**Section Focus Transparency 32,** Stalking Wild Protists **Teaching Transparency 17,** Protists	**Glencoe Life Science Interactive** **Videodisc** **Life Science CD-ROM** **Spanish Resources** **English/Spanish Audiocassettes** **Cooperative Learning Resource** **Guide** **Lab Partner** **Lab and Safety Skills** **Lesson Plans**
Study Guide, p. 37 **Reinforcement**, p. 37 **Enrichment**, p. 37 **Activity Worksheets**, pp. 54-55, 57 **Lab Manual 22**, Lichens **Science Integration Activity 9,** Preparation of Carbon Dioxide **Critical Thinking/Problem Solving**, p. 13	**Section Focus Transparency 33,** Mushroom, Anyone? **Science Integration Transparency 9,** Whole or Sliced, Quartered or Diced **Teaching Transparency 18,** Zygote Fungus	**Assessment Resources** **Chapter Review**, pp. 21-22 **Assessment**, pp. 55-58 **Performance Assessment in the** **Science Classroom (PASC)** **MindJogger Videoquiz** **Alternate Assessment in the** **Science Classroom**
Study Guide, p. 38 **Reinforcement**, p. 38 **Enrichment**, p. 38 **Science and Society Integration**, p. 13	**Section Focus Transparency 34,** Stemming the Tide	**Performance Assessment** **Chapter Review Software** **Computer Test Bank**

Key to Teaching Strategies

The following designations will help you decide which activities are appropriate for your students.

L1 Level 1 activities should be within the ability range of all students, including those with learning difficulties.

L2 Level 2 activities should be within the ability range of the average to above-average student.

L3 Level 3 activities are designed for the ability range of above-average students.

LEP LEP activities should be within the ability range of Limited English Proficiency students.

LS These activities are designed to address different learning styles.

COOP LEARN Cooperative Learning activities are designed for small group work.

P These strategies represent student products that can be placed into a best-work portfolio.

GLENCOE TECHNOLOGY

The following multimedia resources are available from Glencoe.

Science and Technology Videodisc Series (STVS)
Plants & Simple Organisms
New Uses for Algae
Fungal Collection for Research
National Geographic Society Series
STV: The Cell

STV: Plants
Newton's Apple: Life Sciences
Glencoe Life Science Interactive Videodisc
Decomposers: Earth's Natural Recyclers
Life Science CD-ROM

Teacher Classroom Resources

This is a representation of key blackline masters available in the Teacher Classroom Resources.

Teaching Aids

Section Focus Transparencies

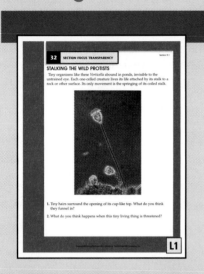

32 SECTION FOCUS TRANSPARENCY Section 9-1

STALKING THE WILD PROTISTS
Tiny organisms like these *Vorticella* abound in ponds, invisible to the untrained eye. Each one-celled creature lives its life anchored by its stalk to a rock or other surface. Its only movement is the springing of its coiled stalk.

1. Tiny hairs surround the opening of its cup-like top. What do you think they funnel in?

2. What do you think happens when this tiny living thing is threatened?

L1

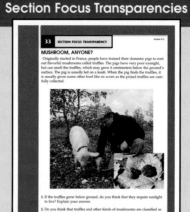

33 SECTION FOCUS TRANSPARENCY Section 9-2

MUSHROOM, ANYONE?
Originally started in France, people have trained their domestic pigs to root out flavorful mushrooms called truffles. The pigs have very poor eyesight, but can smell the truffles, which may grow 6 centimeters below the ground's surface. The pig is usually led on a leash. When the pig finds the truffles, it is usually given some other food like an acorn as the prized truffles are carefully collected.

1. If the truffles grow below ground, do you think that they require sunlight to live? Explain your answer.

2. Do you think that truffles and other kinds of mushrooms are classified as plants? Why or why not?

L2

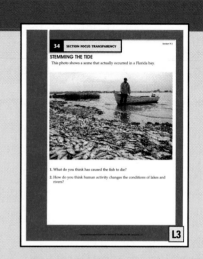

34 SECTION FOCUS TRANSPARENCY Section 9-3

STEMMING THE TIDE
This photo shows a scene that actually occurred in a Florida bay.

1. What do you think has caused the fish to die?

2. How do you think human activity changes the conditions of lakes and rivers?

L3

Science Integration Transparencies

9 SCIENCE INTEGRATION TRANSPARENCY

Whole or Sliced, Quartered or Diced

L1

Teaching Transparencies

17. PROTISTS

L1

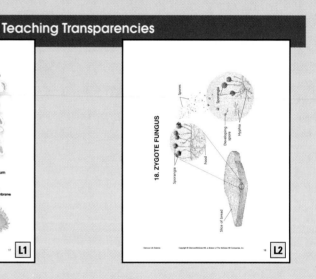

18. ZYGOTE FUNGUS

L2

Meeting Different Ability Levels

Study Guide

Chapter 9
STUDY GUIDE • **Kingdom Protista**

L1

Reinforcement

Chapter 9
REINFORCEMENT • **Kingdom Protista**

L2

Enrichment Worksheets

Chapter 9
ENRICHMENT • **Kingdom Protista**

Protists in a Jar

L3

Hands-On Activities

Science Integration Activity

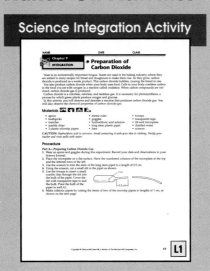

Preparation of Carbon Dioxide

L1

Lab Manual

Life in Pond Water 19

L2

Assessment

Performance Assessment

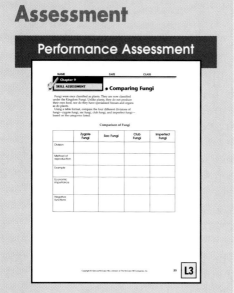

Comparing Fungi

L3

Enrichment and Application

Critical Thinking/ Problem Solving

Protists and Fungi

Fungi: A Source of Indoor Air Pollution

L2

Cross-Curricular Integration

Protists and Fungi

Dysentery and Malaria

L1

Science and Society Integration

Protists and Fungi

Choosing the Right Medicine

L2

Multicultural Connections

Awaking to Answers

L2

Concept Mapping

Protists and Fungi

L3

Chapter 9

Protists and Fungi

CHAPTER OVERVIEW

Section 9-1 Characteristics of all protists are first discussed in this section. Then the three groups of protists are described, compared, and contrasted with each other.

Section 9-2 In this section, fungi characteristics are identified. Fungi are then classified based on their methods of reproduction.

Section 9-3 Science and Society The effects of red tides and the technology used to track them are discussed.

Chapter Vocabulary

protist	sporangia
algae	ascus
protozoa	budding
pseudopod	basidium
cilia	lichen
hyphae	red tide
spore	

Theme Connection

Systems and Interactions/Stability and Change Evolutionary and ecological relationships of the protists and fungi are discussed.

Previewing the Chapter

Section 9-1   Kingdom Protista
► What is a protist?
► Plantlike Protists
► Animal-Like Protists
► Funguslike Protists

Section 9-2   Kingdom Fungi
► What are fungi?

Section 9-3   Science and Society
Technology: Monitoring Red Tides

228

Learning Styles

Look for the following logo for strategies that emphasize different learning modalities. **LS**

Kinesthetic	MiniLAB, p. 244
Visual-Spatial	Explore, p. 229; Visual Learning, pp. 231, 232, 234, 241, 248; Community Connection, p. 231; Demonstration, p. 234; MiniLAB, p. 237; Activity 9-1, p. 240
Interpersonal	Research Activity, p. 235; Community Connection, p. 243; Activity 9-2, pp. 246-247; People and Science, p. 250
Logical-Mathematical	Activity, p. 231; Across the Curriculum, pp. 237, 242; Using Math, p. 245
Linguistic	Across the Curriculum, p. 235

Chapter 9

Protists and Fungi

Do the mushrooms shown on the facing page look tasty? While many types of mushrooms cannot be eaten, mushrooms are sought after by gourmets for salads, casseroles, pizza, and other dishes. In the last chapter, you learned about members of the Kingdoms Archaebacteria and Eubacteria. Are members of the Kingdoms Protista and Fungi similar to the members of either kingdom of bacteria? In the following activity, find out something about mushrooms, which are fungi. Then in the chapter that follows, learn about the importance of protists and fungi.

EXPLORE ACTIVITY

Dissect a mushroom to find out how it is put together.

1. Examine an edible mushroom from the produce section of a grocery store using a hand lens.
2. Carefully pull the cap off and observe the gills on which will form hundreds of thousands of tiny reproductive structures called spores.
3. Use your fingers to pull the stalk apart lengthwise. Continue this process until the pieces are as small as you can get them.

Observe: How was the stalk of the mushroom connected to the rest of the organism? Describe the parts of the mushroom in your Science Journal.

Previewing Science Skills

▶ In the **Skill Builders,** you will compare and contrast and use variables, constants, and controls.

▶ In the **Activities,** you will observe, compare, and design an experiment.

▶ In the **MiniLABs,** you will make observations, predict, experiment, and estimate.

229

229

Prepare

Section Background

- Algae absorb much of the carbon dioxide produced by burning fossil fuels and release large amounts of oxygen to the atmosphere. Algae are the basis for most food chains in freshwater and marine habitats, and they are used to make industrial products.

- The population of diatoms may be so high in some ocean environments that a cubic meter of water may contain more than 35 000 000 of them.

Preplanning

- To prepare for the Mini-LAB, obtain slime mold cultures and culture dishes.

- Order cultures of protists needed for Activity 9-1.

1 Motivate

Bellringer

 Before presenting the lesson, display **Section Focus Transparency 32** on the overhead projector. Assign the accompanying **Focus Activity** worksheet. L1 LEP

Tying to Previous Knowledge

Bridge this section with what students learned in Chapter 7 about classification.

GLENCOE TECHNOLOGY

 Videodisc

STVS: Plants & Simple Organisms
Disc 4, Side 1
New Uses for Algae (Ch. 12)

Science Words

- protist
- algae
- protozoa
- pseudopod
- cilia

Objectives

- Identify the characteristics shared by all protists.
- Describe the three groups of protists.
- Compare and contrast the protist groups.

Radiolarian

Amoeba

Odonthalia

Euglena

Sporozoan

Laminaria

230 Chapter 9 Protists and Fungi

What is a protist?

Look at the organisms in **Figure 9-1.** Do you see any similarities among them? As different as they appear, all of these organisms belong to the protist kingdom. A **protist** is a single or many-celled organism that lives in moist or wet surroundings. All protists have a nucleus and are therefore eukaryotic. While there is a great variety among them, Kingdom Protista (pruh TIHS tuh) contains many different kinds of organisms that share more characteristics with each other than with members of any other kingdom. Some protists contain chlorophyll and make their own food, and others don't. Protists are plantlike, animal-like, and funguslike.

Evolution of Protists

Most scientists think that protists evolved from bacteria because protists, although simple, are more complex in structure than bacteria. Bacteria are thought to be the ancestors of animal-like and funguslike protists because these organisms can't make their own food. In contrast, plantlike protists probably evolved from cyanobacteria because both produce their own food.

Scientists hypothesize that protists, in turn, are the ancestors of the fungi, plant, and animal kingdoms. Plantlike protists, such as green algae, are the most probable ancestors of plants. Animal-like protists, the protozoa, are hypothesized to be the ancestors of animals because they have so many animal characteristics. Knowledge of the evolution of protists is incomplete because many lack hard parts and, as a result, there aren't many fossils of these organisms.

Figure 9-1

Kingdom Protista is made up of a diverse group of organisms. Some are one-celled and others are many-celled.

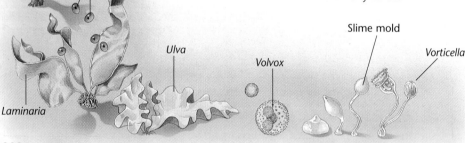

Ulva

Volvox

Slime mold

Vorticella

Program Resources

Reproducible Masters
Study Guide, p. 36 L1
Reinforcement, p. 36 L1
Enrichment, p. 36 L3
Cross-Curricular Integration, p. 13
Activity Worksheets, pp. 5, 52-53, 56 L1
Multicultural Connections, pp. 21-22
Lab Manuals 19, 20, 21
Concept Mapping, p. 23

Transparencies
Teaching Transparency 17
Section Focus Transparency 32 L1

Plantlike Protists

Plantlike protists are known as **algae.** Some species of algae are one-celled and others are many-celled. All algae can make their own food because they contain the pigment chlorophyll in their chloroplasts. Even though all algae have chlorophyll, not all of them look green. Many have other pigments that cover up their chlorophyll. Species of algae are grouped into six main phyla mainly according to their pigments and the form in which they store food.

Euglenas

Algae that belong to the phylum Euglenophyta have characteristics of both plants and animals. A typical euglena is shown in **Figure 9-2.** Like plants, these one-celled algae have chloroplasts and produce carbohydrate as food. When light is not present, euglenas require other food sources. Although euglenas have no cell walls, they do have a strong, flexible layer inside the cell membrane that helps them move and change shape. Many move by using flagella. Another animal-like characteristic of euglenas is that they have an adaptation called an eyespot that responds to light.

Diatoms

Diatoms, shown in **Figure 9-3,** belong to the phylum Chrysophyta (kruh SAHF uh tuh). *Chryso* means "golden brown." Diatoms are photosynthetic, one-celled algae that store food in the form of oil. They have a golden-brown pigment that masks the green chlorophyll.

Diatoms reproduce in extremely large numbers. When the organisms die, their small shells sink to the floor of the body of water and collect in deep layers. Ancient deposits made of diatoms are mined with power shovels and used in insulation, filters, and road paint. Diatom shells produce the sparkle that makes some road lines visible at night and the crunch you feel when you use toothpaste to brush your teeth.

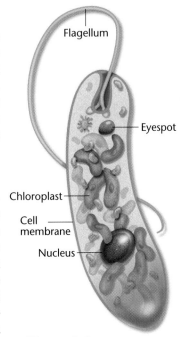

Figure 9-2

Euglenas are protists that have chloroplasts similar to plants. They can also move and sense light like animals.

Figure 9-3

The shells of diatoms contain silica, the main element in glass. The body of a diatom is like a small box with a lid. One half of the shell fits inside the other half. Diatom shells are covered with markings and pits that form patterns.

Magnification 130X

231

Community Connection

Pond Life Have students bring in pond or puddle water. Have them use a microscope and draw the protists observed. L1

LEP LS

Magnification 12 000×

Figure 9-4

Dinoflagellates are usually marine. Notice the groove that houses one of the two flagella that mark all members of this group.

Dinoflagellates

Phylum Pyrrophyta (puh RAHF uh tuh), the "fire" algae, contains species of one-celled algae called dinoflagellates that have red pigments. The name *dinoflagellate* means "spinning flagellates." One of the flagella moves the cell, and the other circles the cell, causing it to spin with a motion similar to a top. Dinoflagellates, shown in **Figure 9-4,** store food in the form of starch and oils.

Almost all dinoflagellates live in salt water. They are important food sources for many saltwater organisms.

Green Algae

Seven thousand species of green algae form the phylum Chlorophyta (kloh RAHF uh tuh), making it the most diverse group of protists. The presence of chlorophyll in green algae tells you that they undergo photosynthesis and produce food in the form of starch. **Figure 9-5** shows different forms of green algae.

Although most green algae live in water, others can live in many other environments, including trunks of trees and even on other organisms! Green algae can be one-celled or many-celled.

Figure 9-5

There are many different shapes among the species of green algae.

A *Chlamydomonas* is an example of a one-celled green alga.

B *Ulva,* also called sea lettuce, is a many-celled saltwater alga that grows in thin sheets.

Magnification 350×

Magnification 100×

C *Spirogyra* is a chainlike freshwater alga that has spiral-shaped chloroplasts.

D *Volvox* is a freshwater form that occurs in ball-shaped colonies. The colony rolls through the water using its flagella.

Magnification 125×

232 **Chapter 9 Protists and Fungi**

Red Algae

The red algae belong to the phylum Rhodophyta (roh DAHF uh tuh). *Rhodo* means red and describes the color of members of this phylum. If you've ever eaten pudding or used toothpaste, you have used something made with red algae. A carbohydrate called carrageenan found in red algae, such as Irish moss shown in **Figure 9-6,** is used to give toothpaste and pudding their smooth, creamy textures. Most red algae are many-celled. Some species of red algae can live up to 175 m deep in the ocean. Their red pigment allows them to absorb the limited amount of light that penetrates to those depths and enables them to produce a type of starch on which they live.

Figure 9-6

Carrageenan, from the red alga Irish moss, gives some puddings their smooth, creamy texture.

Brown Algae

Brown algae make up the phylum Phaeophyta (fee AHF uh tuh). Members of this phylum are many-celled and vary greatly in size. Kelp, shown in **Figure 9-7,** is an important food source for many fish and invertebrates. They form a dense mat of stalks and leaflike blades where small fish and other animals live.

People in many parts of the world eat brown algae. The thick texture of foods such as ice cream and marshmallows is produced by a carbohydrate called algin found in these algae. Brown algae are also used to make fertilizer. **Table 9-1** on page 234 summarizes the different phyla of plantlike protists.

USING MATH

Which is about 100 meters long: a baseball bat, the length of a football field, or the distance between two cities?

Figure 9-7

Giant kelp may be as much as 100 meters long and can form forests like this one located off the coast of California.

9-1 Kingdom Protista **233**

Table 9-1 Have students use **Table 9-1** to summarize the characteristics of plantlike protists. Ask questions based on the table and text.

Figure 9-8 Ask students to identify the two uses of an amoeba's pseudopods. *locomotion and food capture* [LEP] [LS]

Content Background

Of the 100 000 or so protozoans, only a few cause disease, but the illnesses they do cause can be severe and even fatal. Malaria, sleeping sickness, and dysentery are diseases produced by species of protozoans.

Demonstration

[LS] **Visual-Spatial** Use model plant and animal cells to illustrate the organelles in common between algae and plants and between protozoans and animals. [LEP]

NATIONAL GEOGRAPHIC SOCIETY

Videodisc

STV: The Cell

Amoeba digesting food

46973

Table 9-1

The Plantlike Protists			
Phylum	Example	Pigments	Other Characteristics
Euglenophyta Euglenas		Chlorophyll	One-celled alga that moves with a flagellum; has eyespot to detect light.
Bacillariophyta Diatoms		Golden Brown	One-celled alga with body made of two halves. Cell walls contain silica.
Dinoflagellata Dinoflagellates		Red	One-celled alga with two flagella. Flagella cause cell to spin. Some species cause red tide.
Chlorophyta Green Algae		Chlorophyll	One- and many-celled species. Most live in water; some live out of water, in or on other organisms.
Rhodophyta Red Algae		Red	Many-celled alga; carbohydrate in red algae is used to give some foods a creamy texture.
Phaeophyta Brown Algae		Brown	Many-celled alga; most live in salt water; important food source in aquatic environments.

Animal-Like Protists

One-celled, animal-like protists are known as **protozoa.** These complex organisms live in water, soil, and both living and dead organisms. Many types of protozoans are parasites. Remember that a parasite is an organism that lives in or on another organism. Protozoans contain special vacuoles for digesting food and getting rid of excess water. Four kinds of protozoan—sarcodines, flagellates, ciliates, and sporozoans—are recognized based on method of movement.

Figure 9-8

An amoeba constantly changes shape as it forms pseudopods to capture food and move from place to place. Many areas of the world have a species of amoeba in the water that causes dysentery. Dysentery causes a severe form of diarrhea.

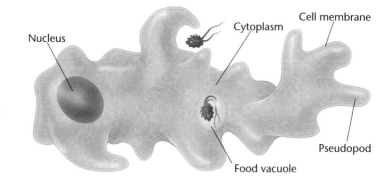

234 Chapter 9 Protists and Fungi

Integrating the Sciences

Earth Science The White Cliffs of Dover in England are made up primarily of the shells of sarcodines. Have students view foraminifera shells on a prepared slide. [L1]

Magnification 90×

Figure 9-9

Many of the saltwater rhizopods have skeletons made out of a chalk, calcium carbonate.

 The White Cliffs of Dover in England are made almost entirely of the shells of billions of rhizopods.

B Rhizopod shells come in a variety of shapes.

Rhizopoda

The first protozoans were probably similar to members of the phylum Rhizopoda. The amoeba shown in **Figure 9-8** is a typical species of this phylum. Rhizopods move about and feed using temporary extensions of their cytoplasm called **pseudopods.** The word *pseudopod* means "false foot." Pseudopods are footlike extensions of cytoplasm that the organism uses to move and to trap food. An amoeba extends the cytoplasm of a pseudopod on either side of a food particle such as a bacterium. Then the pseudopod closes and the particle is trapped. A vacuole forms around the food and it is digested. Members of the phylum Rhizopoda, as shown in **Figure 9-9,** are found in freshwater and saltwater environments, and certain types are found in animals as parasites.

Flagellates

Protozoans that move using flagella are called flagellates and belong to the phylum Zoomastigina (zoo mas tuh GEE nuh). All of the flagellates have one or many long flagella that whip through a watery environment, moving the organism along. Many species of flagellates live in fresh water, but some are parasites.

Trypanosoma, shown in **Figure 9-10,** is a flagellate that causes African sleeping sickness in humans and other animals. Another flagellate lives in the digestive system of termites. The flagellates are beneficial to the termites because they produce enzymes that digest the wood the termites eat. Without the flagellates, the termites would not be able to digest the wood.

Magnification 600×

Figure 9-10

Trypanosoma, responsible for African sleeping sickness, is spread by the tsetse fly in Africa. The disease causes fever, swollen glands, and extreme sleepiness.

9-1 Kingdom Protista   **235**

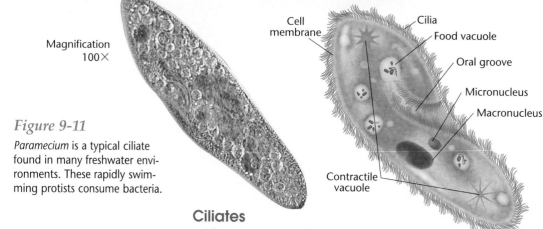

Figure 9-11

Paramecium is a typical ciliate found in many freshwater environments. These rapidly swimming protists consume bacteria.

Magnification 100×

Ciliates

The most complex protozoans belong to the phylum Ciliophora. Members of this phylum move by using cilia. **Cilia** are short, threadlike structures that extend from the cell membrane. Ciliates may be covered with cilia, or have cilia grouped in special areas of the cell. The beating of the cilia is organized so the organism can move swiftly in any direction.

A typical ciliate is *Paramecium*, shown in **Figure 9-11.** In the *Paramecium*, you can see another characteristic of the ciliates: each has two nuclei—a macronucleus and a micronucleus. The macronucleus controls the everyday functions of the cell. The micronucleus functions in reproduction. Paramecia usually feed on bacteria swept into the oral groove. Once the food is inside the cell, a food vacuole forms and the food is digested. Wastes are removed through the anal pore. As the name implies, a contractile vacuole contracts to remove excess water from the cell.

Sporozoans

The phylum Sporozoa contains only parasitic protozoans. Sporozoans have no way of moving on their own. All are parasites that live in and feed on the blood of humans and other animals, as shown in **Figure 9-12.**

Figure 9-12

Only female *Anopheles* mosquitoes spread the sporozoan that causes malaria. Malaria is spread when an infected mosquito bites a human. This disease causes many deaths each year.

236 Chapter 9 Protists and Fungi

Funguslike Protists

Funguslike protists include several small phyla of protists that have features of both protists and fungi. Slime molds and water molds are funguslike protists. Slime molds are much more attractive than their name sounds. Many are brightly colored. They form a delicate, weblike structure on the surface of their food supply. They obtain energy by decomposing organic materials.

Slime Molds

Slime molds have some protozoan characteristics. Many forms have beautiful colors. Examples of slime molds are illustrated in **Figure 9-13**. During part of their life cycle, the cells of slime molds move by means of pseudopods and behave like amoebas. Slime molds reproduce with spores the way fungi do. You will learn about reproduction in fungi in the next section.

Although most slime molds live on decaying logs or dead leaves in moist, cool, shady woods, one common slime mold is sometimes found crawling across lawns and mulch. It creeps along feeding on bacteria and decayed plants and animals. When conditions become less favorable, reproductive structures form on stalks, and spores are produced.

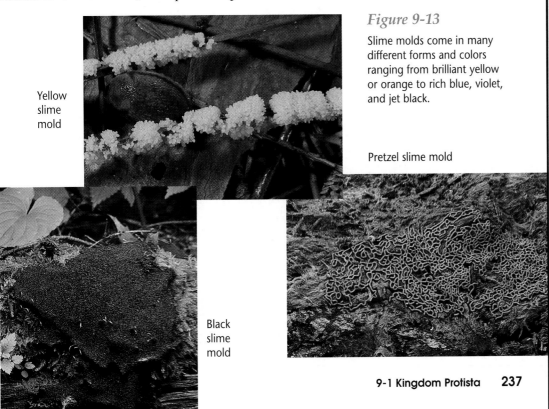

Yellow slime mold

Black slime mold

Pretzel slime mold

Figure 9-13

Slime molds come in many different forms and colors ranging from brilliant yellow or orange to rich blue, violet, and jet black.

9-1 Kingdom Protista   **237**

Problem Solving

Solve the Problem

1. Slime molds resemble amoebas in the way they move and feed. They appear to creep forward using streaming cytoplasm.

2. Slime molds resemble fungi in their methods of spore formation when conditions are unfavorable and in producing fruiting bodies that look like mushrooms.

Think Critically

Slime molds have characteristics of organisms in both protist and fungus kingdoms. They do not have cell walls of chitin. They move like amoebas and engulf large bacteria like the protists. Fungi, on the other hand, absorb nutrients rather than eat.

3 Assess

Check for Understanding

Use a culture of slime mold to illustrate the phases of its life cycle. It has a flagellate stage, a plasmodium stage, and a funguslike stage.

Reteach

Show students photographs or illustrations of various protists. Have students identify the life cycle phase of each.

Extension

For students who have mastered this section, use the **Reinforcement** and **Enrichment** masters.

Problem Solving

Puzzled About Slime

Slime molds, such as the scrambled egg slime mold, can be found covering moist wood as in the photograph shown below. They may be white or brightly colored red, yellow, or purple. If you looked at a piece of slime mold on a microscope slide, you would see that the cell nuclei move back and forth as the cytoplasm streams along. This helps the mold creep over the wood. At other times, when conditions are unfavorable, you can observe them to be dried structures that look like tiny mushrooms.

Solve the Problem:

1. How are slime molds like amoebas?
2. How are the reproductive structures of slime molds like fungi?

Think Critically:

Some books classify slime molds as fungi, but others classify them as protists. Why do most scientists classify slime molds as protists?

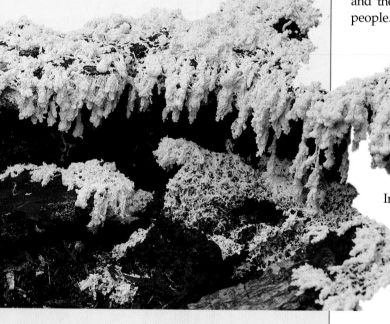

Water Molds and Downy Mildew

Water molds, downy mildews, and white rusts make up another phylum of funguslike protists. Most members of this large and diverse group live in water or moist places. Most water molds appear as fuzzy, white growths on decaying matter. They are called funguslike protists because they grow as a mass of threads over a plant or animal, digest it, and then absorb the organism's nutrients. Unlike fungi, however, they produce reproductive cells with flagella at some point in their reproductive cycles.

Some water molds are parasites of plants while others feed on dead organisms, such as fish. The most well-known member of this phylum is the water mold, pictured in **Figure 9-14A**, that caused the Irish potato famine in the 1840s. Potatoes were Ireland's main crop and the main food source for its people. When the potato crop became infected with the water mold, potatoes rotted in the fields, leaving many people with no food. Nearly one million people died in the resulting famine. Many others left Ireland and emigrated to the United States. This water mold continues to be a problem for potato growers, even in the United States.

Water molds have cell walls as do fungi, but their relatively simple cells are more like those of protozoans. **Figure 9-14B** shows a parasitic water mold that grows on decaying fish. If you have an aquarium, you may see water molds attack a fish and cause its death. Another important member of this phylum is a downy mildew that causes a serious disease in many plants.

Figure 9-14

The potato plant (A) and the fish (B) show the effects of the parasitic water molds.

A The Irish potato famine in the 1840s was the result of a water mold.

B A parasitic water mold growing on a fish will eventually kill it. Once the fish dies, the water mold will speed the decay of the fish.

Section Wrap-up

Review

1. What are the main characteristics of all protists?

2. Describe the characteristics of the three groups of protists.

3. Think Critically: Why aren't there many fossils of the different groups of protists?

Skill Builder
Making and Using Tables
Compare and contrast the protist groups. If you need help, refer to Making and Using Tables in the **Skill Handbook.**

Using Computers

Spreadsheet Use a spreadsheet to make a table that compares the characteristics of the four phyla of protozoans. Include phylum, example species, method of transportation, and other characteristics.

9-1 Kingdom Protista   **239**

Skill Builder
The table should include plantlike protists, animal-like protists, and funguslike protists. Suggest that students compare and contrast environment (habitat), reproductive cells, and whether free-living or parasitic.

 Assessment

Performance Assess students' abilities to make and use tables by having them write a statement in their journals comparing environments of slime molds and water molds. Use the Performance Task Assessment List for Analyzing the Data in **PASC**, p. 27. **P**

•MINI•QUIZ•

1. The most numerous of all the algae are the _____ . *diatoms*
2. Some _____ that cause red tides contain a nerve poison that can kill fish. *dinoflagellates*
3. One-celled animal-like protists are known as _____ . *protozoa*

Section Wrap-up

Review

1. Protists are simple organisms, mostly single-celled, that possess membrane-bound nuclei but don't have tissues or organs. Protists may be plantlike, animal-like, or funguslike.

2. Algae are plantlike protists with chloroplasts and cell walls. Protozoans are single-celled animal-like protists that have no cell walls and cannot make their own food. The funguslike protists are all consumers that produce structures similar to those of fungi.

3. Think Critically Protists are small and generally do not have any hard parts to be fossilized. The protists decay rapidly, leaving no trace of their existence.

Using Computers

Phylum—Sarcodina, Mastigophora, Ciliophora, Sporozoa; **Example Species**—amoeba, trypanosome, paramecium, plasmodium; **Method of Transportation**—pseudopod, flagella, cilia, none; **Other Characteristics**—some have hard shells, many are parasites, at least two nuclei, parasitic

Comparing Algae and Protozoans

Purpose

LS **Visual-Spatial** Observe the differences between algae and protozoa. L1

Process Skills

observing and inferring, defining operationally, classifying, making and using tables, comparing and contrasting

Time

50 minutes

📁 **Activity Worksheets,** pp. 5, 52-53

Teaching Strategies

Alternate Materials Other protozoans and algae may be used. It is best to use local materials from the same phyla, if available. Prepared slides may be used if living material is not available.

Troubleshooting Maintain separate cultures for each organism. Prevent students from using the same dropper in more than one culture.

Answers to Questions

1. *Paramecium*—cilia; *Amoeba*—pseudopoda; *Euglena*—flagella

2. *Spirogyra* and *Micrasterias* have no structures to enable them to move.

3. Only *Spirogyra* is many-celled; the rest of the protists are one-celled.

4. *Spirogyra, Micrasterias,* and *Euglena* make their own food. We know they make their own food because they contain chloroplasts and, therefore, undergo photosynthesis. Organisms that undergo photosynthesis are producers, capable of making their own food.

5. *Paramecium, Amoeba,* and *Euglena* are like animals because they obtain their food from other sources.

6. None had fungal characteristics.

240

Algae and protozoan cells have characteristics that are similar enough to place them within the same kingdom. However, the diversity of apparent forms within Kingdom Protista is great. In this activity, you can observe many of the differences that make organisms in Kingdom Protista so diverse.

Problem

What are the differences between algae and protozoans?

Materials

- cultures of *Paramecium, Amoeba, Euglena, Spirogyra,* and *Micrasterias*
- prepared slide of slime mold
- 5 coverslips
- microscope
- dropper
- 5 microscope slides

Procedure

1. Make a data table in your Science Journal for your drawings and observations.

2. Make a wet mount of the *Paramecium* culture.

3. Observe the wet mount first under low and then under high power. Draw and label the organism.

4. Repeat steps 2 and 3 with the other cultures. Return all preparations to your teacher and wash your hands.

5. Observe the slide of slime mold under low and high power. Record your observations.

Analyze

1. For each organism that could move, **label** the structure that enabled the movement.

2. Which protists could not move? Why?

3. **Describe** whether each protist had one cell or many cells.

Conclude and Apply

4. Which protists make their own food? **Explain** how you know that they make their own food.

5. **Identify** those protists with animal characteristics.

6. Which protists had fungal characteristics?

Magnification 22✕

Data and Observations   Sample Data

	Paramecium	Amoeba	Euglena	Spirogyra	Micrasterias
Drawing	Students' drawings should resemble *Paramecium* on page 236.	Students' drawings should resemble *Amoeba* on page 234.	Students' drawings should resemble *Euglena* on page 231.	Students' drawings should resemble *Spirogyra* on page 232.	Students' drawings should show a single-celled green alga with a flowerlike shape.

✓ **Assessment**

Performance To further assess students' abilities to identify algae and protozoans, have students make and view scrapings from fish tanks, making sure they wash their hands thoroughly afterward. Use the Performance Task Assessment List for Making Observations and Inferences in **PASC**, p. 17. P

Kingdom Fungi 9•2

What are fungi?

Do you think you can find members of Kingdom Fungi in a quick trip around your house or apartment? You can find fungi in your kitchen if you have mushroom soup or fresh mushrooms. Yeasts are a type of fungi used to make bread and cheese. You may also find mold, a type of fungus, growing on an old loaf of bread, or mildew, another fungus, growing on your shower curtain.

As important as fungi seem in the production of different foods, they are most important in their role as organisms that decompose or break down organic materials. Food scraps, clothing, dead plants, and animals are all made of organic material. Fungi work to decompose, or break down, all these materials and return them to the soil. The materials returned to the soil are then reused by plants. Fungi, along with bacteria, are nature's recyclers. They keep Earth from becoming buried under mountains of waste materials.

Characteristics of Fungi

Fungi were once classified as plants. But, unlike plants, fungi do not make their own food or have the specialized tissues and organs of plants, such as leaves and roots. Most species of fungi are many-celled. The body of a fungus is usually a mass of many-celled, threadlike tubes called **hyphae** (HI fee), as illustrated in **Figure 9-15.**

Fungi don't contain chlorophyll and therefore cannot make their own food. Most fungi feed on dead or decaying tissues. Organisms that obtain food in this way are called *saprophytes.* A fungus secretes enzymes to digest food outside of itself. Then the fungus cells absorb the digested food. Fungi that cause athlete's foot and ringworm are parasites. They obtain their food directly from living things.

Science Words

- hyphae
- spore
- sporangia
- ascus
- budding
- basidium
- lichen

Objectives

- Identify the characteristics shared by all fungi.
- Classify fungi into groups based on their methods of reproduction.
- Describe the difference between the imperfect fungi and all other fungi.

Figure 9-15

Hyphae usually grow underground. They may join to produce reproductive structures such as the scarlet cap mushrooms pictured.

Hyphae

 9-2 Kingdom Fungi **241**

Section 9•2

Prepare

Section Background

- Mycology is the study of fungi.
- The poison of the *Amanita* group of mushrooms acts like the venom of a rattlesnake.
- Many people are allergic to the spores produced by fungi.

Preplanning

- To prepare for the Mini-LAB, obtain mushrooms and paper towels.
- Obtain samples of lichens for Activity 9-2.

1 Motivate

Bellringer

Before presenting the lesson, display **Section Focus Transparency 33** on the overhead projector. Assign the accompanying **Focus Activity** worksheet. L1 LEP

Demonstration

Mix some yeast suspension with flour and water and put in a warm place. Ask students to predict what will happen over the course of the day. *The dough will rise as carbon dioxide is made by the yeast.*

Visual Learning

Figure 9-15 Invite students to explain why fungi were once classified as plants. Explain that unlike plants, fungi contain no chloroplasts or cellulose cell walls. LEP IS

Program Resources

Reproducible Masters
Study Guide, p. 37 L1
Reinforcement, p. 37 L1
Enrichment, p. 37 L3
Activity Worksheets, pp. 5, 54-55, 57 L1
Science Integration Activities, pp. 17-18
Critical Thinking/Problem Solving, p. 13
Lab Manual 22

Transparencies
Teaching Transparency 18
Section Focus Transparency 33 L1
Science Integration Transparency 9

2 Teach

Fungi grow best in warm, humid areas, such as tropical forests or the spaces between your toes.

Spores are reproductive cells that form new organisms without fertilization. The structures in which fungi produce spores are used to classify fungi into one of four phyla.

Zygote Fungi

The fuzzy black mold that you sometimes find growing on an old loaf of bread or perhaps a piece of fruit is a type of zygote fungus. Fungi that belong to this division, the phylum Zygomycota, produce spores in round spore cases called **sporangia** on the tips of upright hyphae, as illustrated in **Figure 9-16.** When each sporangium splits open, hundreds of spores are released into the air. Each spore will grow into more mold if it lands where there is enough moisture, a warm temperature, and a food supply.

Sac Fungi

Yeasts, molds, morels, and truffles are all examples of sac fungi. The spores of these fungi are produced in a little saclike structure called an **ascus.** The phylum Ascomycota is named for these sacs. The ascospores are released when the tip of an ascus breaks open.

Many sac fungi are well known by farmers because they destroy plant crops. Diseases caused by sac fungi are Dutch elm disease, apple scab, and ergot disease of rye.

Magnification 3×

Figure 9-16

In zygote fungi, spores are produced in round, black spore cases on the tips of upright hyphae, which turn black as they mature. The black fuzz you see on bread is actually a mass of mature spore cases. This photo shows the zygote fungus *Rhizopus* growing on a pear.

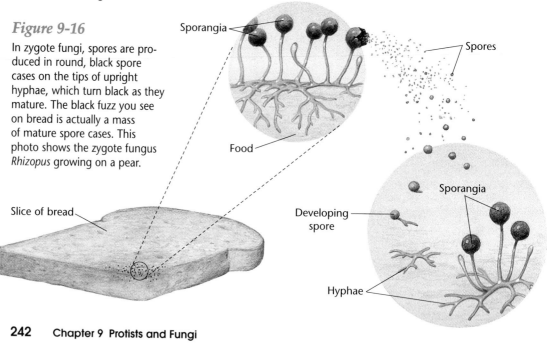

242   Chapter 9 Protists and Fungi

Across the Curriculum

Logical-Mathematical The fairy ring toadstool advances outward from the center at the rate of about 30 centimeters per year. One ring is estimated to be 150 years old. **How many meters in diameter is this ring?** *150 years × 30 cm/year × 2 radii = 900 cm = 9 meters* **P**

Yeast is an economically important sac fungus. Yeasts don't always reproduce by forming spores. They also reproduce asexually by budding, as illustrated in **Figure 9-17. Budding** is a form of asexual reproduction in which a new organism grows off the side of the parent. Yeasts are used in the baking industry. As yeasts grow, they use sugar for energy and produce alcohol and carbon dioxide as waste products. The carbon dioxide causes bread to rise.

Figure 9-17

Yeasts can reproduce by forming buds off the side of the parents. The bud pinches off and forms an identical cell.

Magnification 3400✕

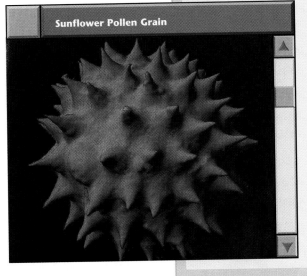

Sunflower Pollen Grain

USING TECHNOLOGY

Confocal Microscopy

Have you ever looked through a light microscope and been frustrated that you could only see two dimensions of the organism you were viewing? Do you think that you would be able to learn more about the organism if the image were sharpened and appeared in three dimensions?

The confocal microscope produces sharp images with light. Confocal microscopy works by focusing an intense beam of light on the specimen. When the light is reflected from the object, it passes through a mirror that splits the beam and allows the reflected light to travel through a tiny hole to a detector. The detector is usually a photographic plate or camera. To get the best photographs, a spinning disk with many small holes makes sure the light focuses on the entire object. The best confocal microscopes now also use lasers, electro-mechanical scanning, and computers to generate images that surpass anything you can see with the unaided eye or a light microscope.

*inter*NET CONNECTION

To find out more about confocal microscopes, visit the Chapter 9 Internet Connection at Glencoe Online Science, **www.glencoe.com/sec/science/life**.

9-2 Kingdom Fungi **243**

Community Connection

Fungal Medicine Have students work in fours to interview local physicians about the role of fungi in the health field. **L2**

COOP LEARN LS

USING TECHNOLOGY

Confocal microscopy has limited uses because the intense beam of light produces heat that can kill a specimen. Have students brainstorm possible uses of this technology.

Think Critically

The microscope could get a much better view of the inside of the fungus than a regular light microscope.

NATIONAL GEOGRAPHIC SOCIETY

Videodisc

Newton's Apple: Life Sciences
Mold
Chapter 3, Side A
Introduction

30609-31112
Can you eat mold?

33935-34165
Molds are fungi

31146-31738
Conditions for mold growth

32739-33307
Penicillin

35143

Figure 9-18

A mushroom is the spore-producing structure of a club fungus. Spores are contained in many club-shaped structures that line the gills of a mushroom cap. Gills are the thin sheets of tissue found under the mushroom cap.

MiniLAB

How are spore prints made?

Procedure

1. Obtain several mushrooms from the grocery store and let them age until the undersides look brown.
2. Remove the stems and arrange the mushroom caps with the gills down on a piece of unlined white paper.
3. Let the mushroom caps sit undisturbed overnight and remove them from the paper the next day.

Analysis

1. Draw and label a sketch of the results in your Science Journal.
2. Describe the marks on the page and what made them.
3. How could you estimate the number of new mushrooms that could be produced from one mushroom cap?

Club Fungi

The mushroom shown in **Figure 9-18** is a member of the phylum Basidiomycota. These fungi are commonly known as club fungi. The spores of these fungi are produced in a club-shaped structure called a **basidium.** The spores you observed on the gills of the mushroom in the Explore Activity at the beginning of this chapter were produced in the basidia.

Many of the club fungi are economically important. Rusts and smuts cause billions of dollars worth of damage to food crops each year. Cultivated mushrooms are an important food crop, but you should never eat a wild mushroom because many are poisonous.

Imperfect Fungi

The so-called imperfect fungi, phylum Deuteromycota, are species of fungi in which a sexual stage has never been observed. When a sexual stage of one of these fungi is observed, the species is immediately classified as one of the other three phyla. *Penicillium* is one example from this group. Penicillin, an antibiotic, is an important product of this fungus. Other examples of imperfect fungi are species that cause ringworm and athlete's foot.

Lichens

The colorful organisms in **Figure 9-19** are lichens. A **lichen** is an organism that is made of a fungus and either a green alga or a cyanobacterium. When two organisms live together, they often have a relationship in which both organisms benefit from living together. The cells of the alga live tangled up in the threadlike strands that make up the fungus. The alga gets a moist, protected place to live, and the fungus gets food from the alga. Lichens are an important food source for many animals.

Rocks tend to crumble as they erode. Lichens are important in the erosion process because they are able to grow on bare rock. When they grow on rock, lichens release acids as part of their metabolism. The acids break down the rock. As bits of rock accumulate and lichens die and decay, soil is formed. This soil supports the growth of other species. Organisms such as lichens that grow on bare rock are called pioneer species because they are the first organisms to arrive on the scene.

Earth scientists also use lichens to monitor pollution levels because many species of lichens quickly die when they are exposed to pollution. When the lichen species return to grow on tree trunks and buildings, it is an indication that the pollution has been cleaned up.

INTEGRATION
Earth Science

Figure 9-19

Lichens can grow upright, appear leafy, or as a crust on bare rock. All three forms may grow near each other. *Can you think of one way that lichens might be classified?*

British soldier lichen

Leafy lichen

Crusty lichen

Section Wrap-up

Review

1. How do fungi obtain food?

2. Why are Ascomycota called sac fungi?

3. **Think Critically:** If an imperfect fungus were found to produce basidia under some circumstances, how would the fungus be reclassified?

USING MATH

Of the 100 000 species of fungi, approximately 30 000 are sac fungi. From these data, estimate the percent of sac fungi as a part of the total fungi kingdom.

Skill Builder
Comparing and Contrasting
Organize information about fungi in a table. Use this information to compare and contrast the characteristics of the four divisions of fungi. If you need help, refer to Comparing and Contrasting in the **Skill Handbook.**

INTEGRATION
Earth Science

Wind, water, temperature changes, and living things are the main forces of erosion. Explain that all landforms—mountains, rivers, valleys, and plains, etc.—are the direct result of erosion forces.

4 Close

•MINI•QUIZ•

1. _____ are reproductive cells that form new organisms without fertilization. *Spores*

2. The spores of sac fungi are produced in a sac called a(n) _____ . *ascus*

3. The spores of Basidiomycota are produced in a club-shaped structure called a(n) _____ . *basidium*

4. The imperfect fungi are composed of fungi in which the _____ has never been observed. *sexual stage*

Section Wrap-up

Review

1. Some fungi get their food from decaying organisms, and others are parasites on living organisms.

2. Ascomycota produce spores in a saclike structure called an ascus.

3. **Think Critically** It would be reclassified as a club fungus belonging to division Basidiomycota.

USING MATH

LS Logical-Mathematical
30 000/100 000 = 30%

Skill Builder

Fungi Divisions	Example	Characteristics
Zygote fungi	Bread mold	Round spore cases on upright stalks
Sac fungi	Yeast, molds, morels, and truffles	Spores grow in a protective sac
Club fungi	Rusts and smuts	Spores grow in club-shaped structure
Imperfect fungi	Penicillin	No sexual stage has been observed
Lichens	British soldiers	Crusty, leafy, or upright; closely associated with algae

Assessment

Oral Why is it beneficial for fungi to produce both sexually and asexually? *Asexual reproduction allows for fungi to take full and rapid advantage of a food source. Sexual reproduction provides variety for surviving in different environmental conditions.*

Activity 9-2

Activity 9-2

PREPARATION

Purpose
IS **Interpersonal** Examine and classify lichen samples and determine how lichens are adapted to mutualism. **L1**
COOP LEARN

Process Skills
observing, classifying, inferring, interpreting data, communicating, forming operational definitions, forming a hypothesis, designing an experiment, making and using graphs, using numbers

Time
two or more class periods

Materials
- Lichens may be collected and occasionally misted with water or stored in a terrarium.
- Dry samples of reindeer moss (*Clasonia*) for several days prior to the activity.
- Place numbers on the lichen specimens to be used in the activity. The reindeer moss used for the microscope part of the activity should not be dried.

Safety Precautions
Be careful not to puncture yourself when using needles.

Possible Hypotheses
Students may suggest that the lichen cannot exist without both partners. They may also hypothesize that one partner can live without the other.

📁 **Activity Worksheets,** pp. 5, 54-55

PLAN THE EXPERIMENT

Possible Procedures
The petri dishes with nutrient agar can be used to grow the algae and fungi separately. If a very thin coating of sterile water is added to the agar's surface, the algae

246

Design Your Own Experiment
Investigating Lichens

Lichens grow on tree trunks, brick buildings, and bare rocks. They may be crusty growths, flattened leafy forms, or branching upright structures. Colors vary greatly. With all of these differences, all lichens include a cyanobacterium or alga in close association with a fungus. In this activity, you will discover whether the two organisms can grow separately.

PREPARATION

Problem
Can the cyanobacterium or alga and fungus of a lichen live separately?

Think Critically
Based on what you know about lichens, can the cyanobacterium or alga and fungus of a lichen live separately?

Objectives
- Examine and classify lichen samples.
- Determine whether the cyanobacterium or alga and fungus can live separately.

Possible Materials
- bits of live lichens of various forms
- needles
- tweezers
- petri dishes with nutrient agar
- sterile pond water
- tape
- microscope slide
- coverslip
- microscope
- paper towels

Safety Precautions

Be careful not to puncture yourself when using needles.

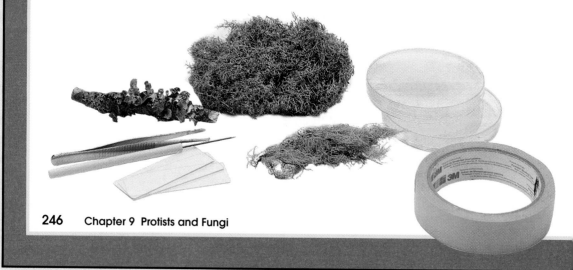

PLAN THE EXPERIMENT

1. As a group, agree upon and answer the Think Critically question.
2. As a group, list the steps you need to take. Be specific; describe exactly what you will do at each step. List your materials.
3. Your teacher will provide instructions on how to separate the fungus and alga or cyanobacterium correctly and safely and how to use a petri dish.
4. Design a data table in your Science Journal so that it is ready to use as your group collects data.

Check Your Plan

1. Read over your entire experiment to make sure that all steps are in logical order.

2. Identify the constants, variables, and controls of the experiment.
3. Do you have to run any tests more than one time?
4. Have you made provisions to summarize your data in a graph?
5. *Make sure your teacher approves your plan before you proceed.*

DO THE EXPERIMENT

1. Carry out the experiment as planned.
2. While the experiment is going on, write down any observations that you make and complete the data table in your Science Journal.

Analyze and Apply

1. **Compare** your results with those of other groups.
2. Did your lichen contain a cyanobacterium or an alga? How could you tell?

3. Did either or both of the organisms grow without the other?
4. **Graph** your results using a line on the same graph for each organism. Plot the number of cells or amount of growth on the *y*-axis and time on the *x*-axis.
5. Using your graph, **describe** the growth of both the cyanobacterium or alga and fungus.

Go Further

Based on the data of this experiment, propose another experiment that would expand your knowledge of organisms that live in permanent, close associations.

---

5. The algae are overgrown by the fungus and die, while the population of fungus cells increases rapidly.

Go Further

Student experiments should be redesigned to answer a different question about symbiosis in lichens.

Assessment

Performance To further assess students' understanding of lichens, provide other samples to classify. Use the Performance Task Assessment List for Making Observations and Inferences in **PASC**, p. 17. P

---

will grow on the medium. The fungi will grow on the medium with or without such a coating.

To remove the alga from the fungus, place a bit of lichen in a drop or two of water on the microscope slide. When teased with a needle, the lichen will release bits of the fungus and individual algal cells in the water. These can be taken up individually in a dropper and transferred to the agar surface.

Teaching Strategies

- Crustlike lichen samples may have different color and textures. Leaflike lichens will have a thin leathery feeling, and shrublike lichens will have a powdery texture.
- The reindeer moss that is soaked in water will be darker green and feel like a wet sponge.

DO THE EXPERIMENT

Expected Outcomes

Usually the fungi will overgrow the algae. If separated properly, some types of algae can also live without the fungi.

Analyze and Apply

1. Results will vary depending on experiment conducted and lichens used.
2. Answers will vary. Cyanobacteria have much smaller cells than algae.
3. Usually the fungus grows, but the algae die.
4.

Typical Graph

Prepare

Section Background

Review the concept of population explosions and what a satellite is before teaching this section.

1 Motivate

Bellringer

Before presenting the lesson, display **Section Focus Transparency 34** on the overhead projector. Assign the accompanying **Focus Activity** worksheet. L1 LEP

Tying to Previous Knowledge

Have students tell what they know about dinoflagellates from earlier in the chapter.

2 Teach

Demonstrate

Obtain a culture of dinoflagellates and set up a wet mount for students to view.

Discussion

Discuss with students the possible reasons for red tides. (Increased water temperatures, pollution, etc.)

Visual Learning

Figure 9-20 The dinoflagellates that cause red tides are a normal part of the marine phytoplankton. Ask students to list some of the damaging effects of such red tides. LEP IS

TECHNOLOGY:
9•3 Monitoring Red Tides

Science Words
red tide

Objectives
- Identify the causes and results of red tides.
- Describe the technologies that are used to identify the locations of red tides and how such identification is useful to organisms.

Figure 9-20
This California bay appears red due to an increase in the algal population.

What is a red tide?

Imagine a humpback whale dying and washing up on a beach. Then multiply this death by 14. Add to this grisly scene tons of dead fish littering beaches from Florida to Massachusetts. You don't have to imagine it because this event actually happened in 1987. The cause was a single species of the microscopic algae called dinoflagellates!

For reasons that are not completely understood, every once in a while the population of certain species of dinoflagellates increases rapidly. The population of algae may be so dense that red pigments in the algae cause the water to look red, as illustrated in **Figure 9-20**. This algal population explosion is known as **red tide.**

Each species of dinoflagellates releases a different kind of toxin. Fish and shellfish may eat the dinoflagellates that are a part of the plankton. The toxins are retained in the bodies of these organisms. People who then eat fish or shellfish that have become contaminated by the toxins can also become ill and sometimes die.

Underlying Causes

Red tides have several underlying causes. Red tides often occur where water becomes heated or where freshwater runoff creates a layer of warm surface water above a cold-water layer that is rich in nutrients. The upper warm layer appears unable to support such a large population of algae. However, dinoflagellates are able to swim into the lower layers and absorb nutrients at night and move to the upper layers to photosynthesize during the light hours. Sometimes the red tides spread for hundreds of

248 Chapter 9 Protists and Fungi

Program Resources

Reproducible Masters
Study Guide, p. 38 L1
Reinforcement, p. 38 L1
Enrichment, p. 38 L3
Science and Society Integration, p. 13

Transparencies
Section Focus Transparency 34

miles along a coastline. This occurs due to the dinoflagellates drifting with the current.

How are red tides monitored?

Scientists with the National Marine Fisheries Service laboratories sample marine waters and shellfish for the presence of dinoflagellates. When large numbers of the organisms are detected, they alert the public not to eat fish or shellfish from the affected areas. However, sampling methods are limited because they are able to observe only relatively small areas.

Within the last few years, satellites have been used to monitor red tides. Using infrared cameras, images such as the one in **Figure 9-21** can reveal the sea-surface temperatures. In one instance, satellites were able to track a warm-water mass harboring the dinoflagellates that cause red tide for more than three weeks. Such information could not be obtained in any other way. Some scientists have speculated that knowledge from such images may one day save the lives of hundreds of people and countless numbers of marine organisms.

Figure 9-21

In this satellite infrared image, warm water is being tracked from Maine into the Massachusetts Bay. Dinoflagellates, found in these warm waters, cause red tide to occur (shown red in the image).

Section Wrap-up

Review

1. What are the causes of red tides?

2. Why are satellites effective in monitoring the spread of red tides?

Explore the Technology

Suppose scientists find that over a long period of time, the number of red tides increases in an area. Hypothesize how the technology might be used to reduce the number of red tides.

SCIENCE & SOCIETY

9-3 Monitoring Red Tides   **249**

3 Assess

Check for Understanding

? FLEX Your Brain

Use the Flex Your Brain activity to have students explore EFFECTS OF RED TIDES.

🗁 **Activity Worksheets,** p. 5

Reteach

Show a film on the development of red tides.

Extension

🗁 For students who have mastered this section, use the **Reinforcement** and **Enrichment** masters.

4 Close

Have students make a model that shows how a satellite could take pictures of a large area.

Section Wrap-up

Review

1. Dinoflagellate populations increase dramatically when warm layers of water cover nutrient-rich layers of cold water.

2. Infrared images from satellites can detect temperature differences in water over very large areas.

Explore the Technology

Satellites might be used to search for pollution sources. Then polluters could be forced to clean up the pollution.

Background

- Dr. Benjamin grew up not far from Bayou La Batre. She signed on with the National Health Service Corps (now the National Health Service Loan Repayment Program), which paid her medical school tuition in return for a commitment that she spend several years practicing medicine in an "underserved" area that was in dire need of a doctor. The Corps sent her to Bayou La Batre. Rather than moving on after her years of service were completed, she decided to stay. In addition to her medical practice, Dr. Benjamin is currently serving a three-year term on the Board of Trustees of the American Medical Association.

- Fungi that would normally cause only minor infections and skin irritations in healthy people can cause very serious, even fatal, infections in people suffering from AIDS. One of the medical advances that has come out of AIDS research has been the discovery of a number of powerful and effective antifungal drugs. There are even antifungal drugs that can be administered in pill form.

Teaching Strategies

LS Interpersonal Divide the class into two teams. Have students in one team research the history of antibiotics and how their discovery and use changed the treatment of disease. Have the second team research the growing problem of antibiotic resistance. Challenge both teams to predict the fate of new antifungal drugs if they, too, are used indiscriminately in the years to come.

People and Science

DR. REGINA BENJAMIN, *Family Practice Physician*

On the Job

Q Dr. Benjamin, tell us about your practice in Bayou La Batre, Alabama.

A Bayou La Batre is a small, rural fishing village on Alabama's Gulf Coast, and I'm the only doctor in town. About 80 percent of my patients live below the poverty level. I see every kind of medical problem you can think of in my practice—children with colds and infections, work-related injuries like cuts and broken bones, diabetes, heart attacks, cancer … even shark bites.

Q What about fungal infections?

A Yes, I see a lot of skin infections caused by fungi—we get a lot of rain and humidity here. Most fungus infections on the skin appear as rashes. Fungal infections can be very difficult to treat, but because of AIDS research, we have a lot of new antifungal drugs that are proving to be very effective.

Personal Insights

Q Is antibiotic resistance a concern in your practice?

A Very much so. Like people in many places, people here were used to getting a shot of penicillin for everything. Sometimes it takes a lot of explaining to convince people to take all their medicine, or that every illness doesn't require an antibiotic.

Q What do you like best about being a doctor in a small town like Bayou La Batre?

A It's very rewarding. I feel like I'm making a difference here, that I am needed, that I'm contributing something.

Career Connection

Look in the State Government section of a phone book to locate the State Health Department. Contact them for information on careers in public health, including:

- **Public Health Nurse**
- **Epidemiologist**
- **Health Inspector**

Choose one of these careers and research the education and training required for it. Then report on your findings to your classmates.

250 Chapter 9 Protists and Fungi

Career Connection

Career Path Becoming a physician requires a bachelor's degree, four years of medical school, and usually three years of residency (supervised practice).

- High School: science and math
- College: biology and chemistry
- Medical school and residency

For More Information

Have students read the article about Dr. Regina Benjamin entitled "Poor Town Finds an Angel in a White Coat," from *The New York Times*, April 3, 1995. Information on a wide variety of health care careers can be found in *VGM's Handbook of Health Care Careers,* edited by Annete Selden, NTC Publishing Group, 1993.

Summary

9-1: Kingdom Protista

1. Protists are eukaryotic organisms. They can be one- or many-celled but do not have the complex organization found in plants and animals.
2. The protist kingdom is made up of a variety of organisms. Some are plantlike, others are animal-like, and others are funguslike.
3. Algae are photosynthetic and classified according to their pigments. Other protists, like flagellates and ciliates, are animal-like and have special means of locomotion. Funguslike protists include slime molds and water molds.

9-2: Kingdom Fungi

1. Fungi are saprophytes or parasites. They reproduce by spores. The structures in which a fungus produces spores are used to classify fungi into one of four phyla.
2. Zygote fungi produce spores in small round spore cases on the tips of upright hyphae. Bread mold is a zygote fungus. Sac fungi produce spores in small sacs called asci. Yeast is an example of a sac fungus. Mushrooms are club fungi. Club fungi produce spores in a club-shaped structure called a basidium. These structures line the gills of a mushroom cap.
3. Imperfect fungi are those for which no sexual stage has been observed. Penicillin is an important product of these fungi.

9-3: Science and Society: Monitoring Red Tides

1. Red tides are population explosions of dinoflagellates that can cause fish kills and illness or death in humans or other animals if affected animals are consumed.
2. Red tides are monitored by satellites. Satellite technology can capture images of high enough resolution to see the spread of red tides before they can be seen at the surface.

Key Science Words

a. algae
b. ascus
c. basidium
d. budding
e. cilia
f. hyphae
g. lichen
h. protist
i. protozoa
j. pseudopod
k. red tide
l. sporangia
m. spore

Reviewing Vocabulary

Match each phrase with the correct term from the list of Key Science Words.

1. footlike cytoplasmic extension
2. reproductive cell of a fungus
3. eukaryotic organism that is animal-like, plantlike, or funguslike
4. animal-like protist
5. plantlike protist
6. threadlike structures used for movement
7. population explosion of dinoflagellates

Chapter 9

Review

Summary

Have students read the summary statements to review the major concepts of the chapter.

Reviewing Vocabulary

1. j	**6.** e
2. m	**7.** k
3. h	**8.** f
4. i	**9.** l
5. a	**10.** g

✓ Assessment

Portfolio Encourage students to place in their portfolios one or two items of what they consider to be their best work. Examples include:

- Across the Curriculum, p. 235
- Across the Curriculum, p. 237
- MiniLAB spore prints, p. 244 **P**

Performance Additional performance assessments may be found in **Performance Assessment** and **Science Integration Activities.** Performance Task Assessment Lists and rubrics for evaluating these activities can be found in Glencoe's **Performance Assessment in the Science Classroom.**

GLENCOE TECHNOLOGY

 Videodisc

Glencoe Life Science Interactive Videodisc
Side 1, Lesson 3
Decomposers: Earth's Natural Recyclers

25823 to 37733

MindJogger Videoquiz

Chapter 9 Have students work in groups as they play the Videoquiz game to review key chapter concepts.

Checking Concepts

1. d	**6.** a
2. c	**7.** d
3. d	**8.** b
4. c	**9.** b
5. b	**10.** c

Understanding Concepts

11. Both have cell walls and chloroplasts.

12. Algae have chloroplasts and are able to make their own food.

13. Fungi help break down or decompose organic material and return nutrients to the soil.

14. Protozoans cannot make their own food and many have some method of movement.

15. Red tide is caused by a population explosion of dinoflagellates. The millions of these organisms with red pigments cause the water to look red. Some species of dinoflagellates have a nerve toxin that can kill fish and be absorbed by shellfish. Humans who eat fish or shellfish poisoned by this toxin can become ill.

Thinking Critically

16. no warmth, low humidity, and low moisture

17. named for their spiral-shaped chloroplasts.

18. one-celled—one-celled; colonial—many cells arranged in a group; chains—many cells arranged in a long filament; many-celled—organism has many different cells and is large

19. Fungi are made of soft parts that do not form fossils easily. They also live in warm, moist, humid areas that allow them to decompose rapidly. Because fungi have characteristics of plants but also their own unique characteris-

---

8. threadlike string of a fungus
9. contain spores in zygote fungi
10. organism made up of a fungus and an alga or a cyanobacterium

Checking Concepts

Choose the word or phrase that completes the sentence.

1. _____ (is) are examples of one-celled algae.
   a. Paramecia   c. Amoebas
   b. *Plasmodium*   d. Diatoms

2. Members of phylum Chrysophyta are _____ in color.
   a. green   c. golden-brown
   b. red   d. brown

3. Large numbers of _____ cause red tides.
   a. *Euglena*   c. *Ulva*
   b. diatoms   d. dinoflagellates

4. Brown algae belong to phylum _____.
   a. Rhodophyta   c. Phaeophyta
   b. Chrysophyta   d. Pyrrophyta

5. _____ moves by using cilia.
   a. *Amoeba*   c. *Plasmodium*
   b. *Paramecium*   d. *Euglena*

6. Funguslike protists can live well _____.
   a. on decaying logs
   b. in bright light
   c. on dry surfaces
   d. on metal surfaces

7. Decomposition is an important role of _____.
   a. protozoans   c. plants
   b. algae   d. fungi

8. Some _____ species are monitors of pollution levels.
   a. club fungus   c. slime mold
   b. lichen   d. imperfect fungus

---

9. _____ produce the spores in mushrooms.
   a. Sporangia   c. Asci
   b. Basidia   d. Hyphae

10. An example of an imperfect fungus is _____.
   a. mushroom   c. *Penicillium*
   b. yeast   d. lichen

Understanding Concepts

Answer the following questions in your Science Journal using complete sentences.

11. What characteristics do algae and plants have in common?

12. Explain why algae are plantlike.

13. How are fungi important to the environment?

14. How are protozoans like animals?

15. What are the causes and effects of a red tide like the one shown below?

Thinking Critically

16. What kind of environment is needed to prevent fungal growth?

17. Look at **Figure 9-5C** again. Why is *Spirogyra* a good name for this green alga?

---

tics, their origin is difficult to trace.

20. Fungi make enzymes that digest food outside of the organism. Then the nutrients are absorbed by the cells.

Developing Skills

21. Observing and Inferring *chloro*—green, *chryso*—gold, *phaeo*—brown, *pyrro*—fire, *rhodo*—red

22. Concept Mapping See student page.

23. Comparing and Contrasting Tables should show that Ascomycota reproduce by budding and making spores, example yeast; Zygomycota reproduce by producing spores in sporangia, example bread mold; Basidiomycota reproduce by making spores in basidia, example mushrooms.

24. Classifying Move by means of flagella—*Euglena*, dinoflagellates, *Volvox*, *Trypanosoma*; move by means of pseudopods—*Amoeba*, slime molds;

18. Compare and contrast one-celled, colonial, chain, and many-celled algae.
19. Discuss why scientists find it difficult to trace the origin of fungi. Why are fossils of fungi rare?
20. Explain the adaptations of fungi that enable them to get food.

Developing Skills

If you need help, refer to the **Skill Handbook.**

21. **Observing and Inferring:** Match the prefix of each alga, *Chloro-, Chryso-, Phaeo-, Pyrro-,* and *Rhodo-,* with the correct color: brown, green, fire, gold, and red.
22. **Concept Mapping:** Complete the following concept map on a separate sheet of paper.

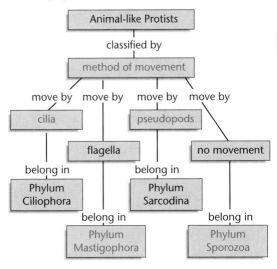

23. **Comparing and Contrasting:** Make a chart comparing and contrasting sac fungi, zygote fungi, and club fungi.

24. **Classifying:** Classify the following organisms based on their method of movement: *Euglena, Amoeba,* dinoflagellates, *Paramecium, Plasmodium,* slime molds, *Trypanosoma,* and *Volvox.*
25. **Observing and Inferring:** The figure below shows some of the structures and specialized adaptations of kelp, including the stipe, blade, holdfast, and gas floats. What purposes do these adaptations serve for kelp?

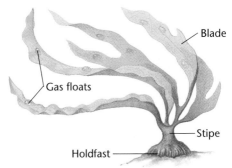

Blade

Gas floats

Stipe

Holdfast

Performance Assessment

1. **Designing an Experiment:** In your Science Journal, make a hypothesis and design an experiment to test what fruits allow *Penicillium* mold to grow. Conduct your experiment. Collect data and state a conclusion.
2. **Scientific Drawing:** In the MiniLAB on page 244, you made a spore print of the gills of a mushroom cap. Now carefully tease apart the top of the cap and make a wet-mount slide of an *extremely* small piece. Do the same with the stalk. In your Science Journal, draw and describe what you see, and compare the two.
3. **Making and Using a Classification System:** Develop a dichotomous key that you could use to identify the protozoan groups. If you need help, reread Section 7-4 of Chapter 7.

move by means of cilia—*Paramecium;* no movement—*Plasmodium*

25. **Observing and Inferring** Holdfast keeps the alga attached to one place; gas floats keep the alga upright and floating near the surface to receive plenty of light; stalk connects the blades

Performance Assessment

1. Experimental designs will vary. *Penicillium* will grow on most fruits. Use the Performance Task Assessment List for Designing an Experiment in **PASC,** p. 23. P
2. Drawings should show identifiable spores, basidia, and hyphae. Nuclei may or may not be present. Use the Performance Task Assessment List for Scientific Drawing in **PASC,** p. 55. P
3. Key to Identifying Protozoans
   1a If the organism moves, go to 2.
   1b If the organism drifts in the blood, it is a Sporozoan.
   2a If the organism has organelles that propel it through liquids, go to 3.
   2b If the organism's organelles for movement are pseudopods, it is a Sarcodine.
   3a If the organism has many small cilia, it is a Ciliate.
   3b If the organism has a single long flagellum, it is a Flagellate.

   Use the Performance Task Assessment List for Making and Using a Classification System in **PASC,** p. 49. P

Assessment Resources

 Reproducible Masters
Chapter Review, pp. 21–22
Assessment, pp. 55–58
Performance Assessment, p. 23

Glencoe Technology
- **Chapter Review Software**
- **Computer Test Bank**
- **MindJogger Videoquiz**

Objectives

LS **Kinesthetic** Students will observe a current concern, and apply and evaluate a technological solution to that concern. **L1** **COOP LEARN**

Summary

In this project, students will learn about composting as a means of organic waste disposal and will experiment with variables involved in the composting process to build an effective compost pile.

Time Required

A minimum of three weeks is required to produce compost. Two full periods are required to research and plan how and where to construct compost piles. One full period is needed at the end to evaluate the data collected from the compost piles. A few minutes will be needed daily to record observations, collect temperature data, and, when necessary, turn the compost piles.

Preparation

- Establish cooperative groups of 4 or 5 students.
- Obtain disposable plastic gloves.
- Help assemble materials that students decide they need, such as a glass jar, plastic aquarium, or window screens for compost piles; and soil, fertilizer, water, and organic wastes for composting.

Teaching Strategies

- Before students begin to research compost piles, ask them to think about the trash generated by the school. Discuss the kinds of trash, and the amount of trash. Take the class to the school dumpster. With students wearing disposable gloves, get a general idea of the different kinds of trash in the dumpster.

Don't Toss It, Compost It!

Have you ever thought about what happens to all the trash—paper cups, napkins, paper plates, and plastic spoons—after a football game or a dance? Of course, the first stop is a trash can or dumpster. This trash usually is picked up and trucked to a landfill. But our landfills are filling up, and we are running out of room for more.

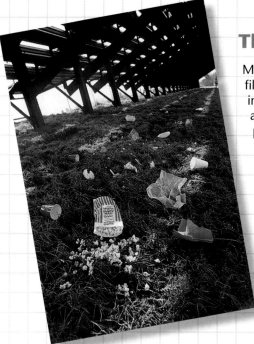

The Compost Solution

Many communities try to cut down on landfill problems through recycling and composting. Your community may already recycle aluminum cans, glass, plastic, and paper products. Your family can also compost kitchen scraps and garden wastes. A compost heap allows organic garbage to break down and decay—a natural process that occurs when living things die. As these organic materials are broken down by bacteria and fungi, nutrients are released into the environment and can be used by other living organisms. When you compost, you are taking part in a process that has been going on for millions of years.

How It Works

Bacteria and fungi start the composting process by breaking down large organic materials. This process releases heat as well as nutrients. The heat "cooks" and decomposes more material. A compost pile about 1 cubic meter in size works best to generate the heat required to make the process work. The pile needs to be turned or sifted regularly to keep material in contact with the warm center. The pile also needs to be kept moist and given plenty of air to keep the bacteria and fungi active.

Procedure

1. Work in groups to learn about composting. One group could set up a standard control compost pile. A small compost pile may be made in a large-mouthed glass jar or in a small, plastic aquarium. If outdoor composting is possible, you can make an enclosure using old window screens. Use four screens to make an open cube.

2. To build a control compost pile, alternate layers of soil (2 cm thick), a layer of kitchen or garden waste (4 cm thick), a sprinkle of fertilizer containing nitrogen, and a sprinkle of water. End with a top layer of soil. Do not add meat or fat—they can attract animals. Any type of grass clippings, leaves, or fruit or vegetable peels works well. Ash from a fireplace and sawdust are good additions also. Make sure that the soil you use is not sterilized potting soil purchased from a store. After the pile is completed, turn the compost once a week. This brings in oxygen and brings new material into the warm center. Your compost is ready when the materials have become uniformly brown and crumbly.

3. Each of the other student groups should design a compost pile that varies from the control pile in only one way. The amount of light, heat, or water the pile receives could be varied. The pile could be turned more or less frequently than the control pile.

4. Each group should hypothesize how one change will affect the compost. "If we wet the pile well, then compost will be made more quickly." "If we don't turn the pile each week, it will make compost more quickly."

Analyzing the Differences

Each group will record the changes in their compost pile before turning it by finding its temperature and observing its smell, texture, and appearance. At the end of the experiment, groups should decide which compost technique and recipe made good compost.

Go Further

Find out how your community's trash and garbage is processed. Does the community compost grass clippings, or are clippings bagged to be picked up with the trash? Almost one-quarter of the space in landfills in the United States is currently taken up by yard clippings! What advantages does composting have over landfills for organic materials?

- Broaden the discussion to include trash from homes. Is it different from school trash?

- Discuss places that have special trash problems—hospitals would be a good example of this, as would groceries and businesses using substances with radiation. Discuss ways that trash can be handled.

- Obtain permission to build one or more compost heaps on school grounds. Have students offer the school a proposal on what the compost can be used for—such as plantings on the school grounds.

Go Further

Some students may wish to do some research on trash solutions. Although landfills have improved and are constructed to prevent leakage of harmful chemicals into surrounding land and water, they are still anaerobic and slow. In some instances, they even stop the decay process. They must also be vented or methane gas may build up and cause problems. Other students may wish to try experiments growing plants using different mixtures of soil and compost to find how the soil-to-compost ratio affects plant growth.

References

Campbell, Stu. *Let It Rot! The Gardener's Guide to Composting.* Pownall, VT: Storey Communications, Inc., 1990.

"How to Build a Compost Heap." Rutgers Cooperative Extension of Morris County, CN 900, Morristown, NJ 07960.

Jones, Linda L. Cronin. "Strike It Rich with Classroom Compost." *The American Biology Teacher,* Vol. 54, No. 7, October, 1992, pp. 420-424.

Plants

In Unit 4, students are introduced to the plant kingdom. The unit is organized phylogenetically using a comparison of plant characteristics. After a thorough discussion of plant evolution, plant processes and relationships are discussed.

Chapter 10
Introduction to Plants   258

10-1 Characteristics of
Plants

10-2 Seedless Plants

10-3 Science and Society:
Cleaner Coal

Chapter 11
The Seed Plants   284

11-1 Seed Plants

11-2 Parts of Complex
Plants

11-3 Seed Plant
Reproduction

11-4 Science and Society:
Medical Treasures

Chapter 12
Plant Processes   312

12-1 Photosynthesis and
Respiration

12-2 Plant Responses

12-3 Science and Society:
Transgenic Crops

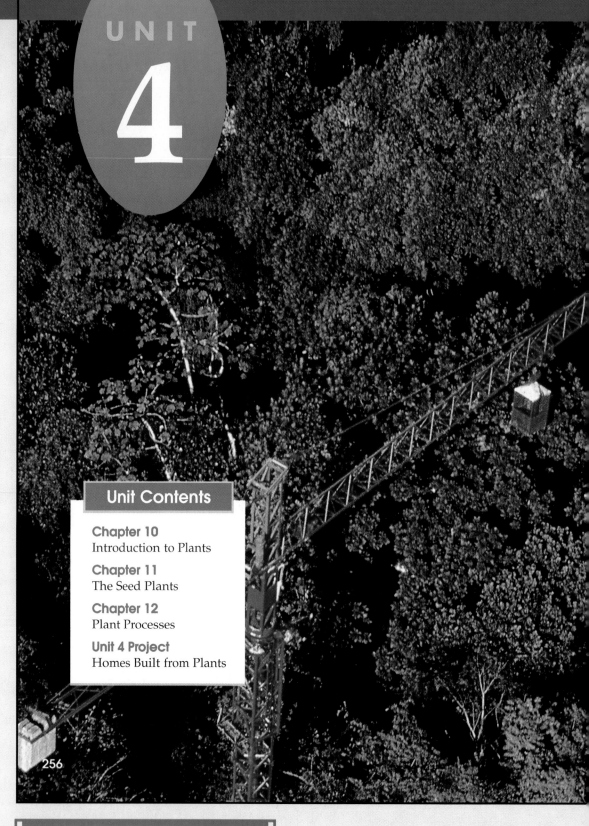

Unit Contents

Chapter 10
Introduction to Plants

Chapter 11
The Seed Plants

Chapter 12
Plant Processes

Unit 4 Project
Homes Built from Plants

256

Science at Home

Plant Growth Rates Have students grow plants in various places inside their homes—by a sunny window, in a dark closet, or by a window that receives indirect sunlight. Make sure each student grows the same type of plant in each location. Have students explain what happened to the plants. *Only plants that receive the appropriate amount of light thrive.*

L1

Plants

What's Happening Here?

The canopy of tropical rain forests contains many plants and animals found nowhere else on Earth. But how do you collect specimens 30 meters up in the air? Scientists of the Smithsonian Tropical Research Institute in Panama use a construction crane. The crane has a long girder with a gondola car at the tip. Researchers in the gondola car are lowered by cable to the exact spot in the canopy they wish to study. Some scientists are classifying new organisms, such as the flower shown in the smaller photograph, whereas others study how plants of the canopy respond to changes in carbon dioxide levels. You'll learn more about plant processes in this unit.

Science Journal

Plants take in carbon dioxide and release oxygen. Scientists tell us there are increasing levels of carbon dioxide in the air. In your Science Journal, describe how you think plants of the rain forest might respond to such increases in carbon dioxide levels.

257

Theme Connection

Scale and Structure Plant vascular tissues are important in determining the size to which plants may grow. Have students relate the scale of various plants to their phylogeny. *Mosses and liverworts grow only a few inches tall—they lack vascular tissue. Ferns grow taller but lack some cell types. Conifers and deciduous trees are some of the largest organisms on Earth, largely because of their vascular tissue.*

Cultural Diversity

Food for Thought People from different cultures rely on different plants for food. Most cultures rely on grains indigenous to their part of the world for their main source of carbohydrates. Have students prepare posters showing different plant parts that are used for food. *More familiar foods like peas, cinnamon, and tapioca may be presented as well as prickly pears and bamboo shoots.* L1 **COOP LEARN**

INTRODUCING THE UNIT

What's Happening Here?

Have students look at the photos and read the text. Ask them to explain what they observe. Point out to students that in this unit, they will study plant characteristics, plant processes, and the relationship of plants to other organisms.

Background

Most of the organisms in tropical rain forests live in the trees, especially in the top layer, or canopy. Working from the forest floor made it difficult for scientists to observe the organisms that never come down from the trees. Cranes that reach the canopy enable biologists to observe and collect organisms never seen before.

Previewing the Chapters

Encourage students to look through the unit to see which organisms are classified as plants. Ask them why they think most scientists don't classify algae as plants. Allow students to discuss which characteristics might be important in the classification of algae and plants. *Some algae are unicellular and none develop tissues. All plants are multicellular and develop tissues.*

Tying to Previous Knowledge

Lead students in a discussion that includes what they already know about plants. Many students will be able to identify common plants by leaf, seed, flower, or fruit. Then have students brainstorm a list of products they use that originate from plants. Be sure they identify nonfood items such as clothing or building materials in their lists.

Chapter Organizer

Section	Objectives/Standards	Activities/Features
Chapter Opener		**Explore Activity:** Determine what plant parts are edible. p. 259
10-1 **Characteristics of Plants** (2 sessions, 1 block)*	1. **List** the characteristics of plants. 2. **Describe** adaptations of plants that made it possible for them to survive on land. 3. **Compare** vascular and nonvascular plants. National Content Standards: (5-8) UCP1, UCP4, B1, C2, C5	**Connect to Chemistry**, p. 262 **Problem Solving:** What is in nature's medicine chest? p. 264 **Skill Builder:** Hypothesizing, p. 265 **Using Computers:** Spreadsheet, p. 265 **Science and Art:** The Art of Beatrix Potter, p. 266 **Science Journal**, p. 266
10-2 **Seedless Plants** (2 sessions, 2 blocks)*	1. **Describe** the life cycles of mosses and ferns. 2. **Compare** simple nonvascular plants with simple vascular plants. 3. **State** the importance of simple nonvascular and vascular plants. National Content Standards: (5-8) UCP1, UCP3, UCP4, A1, A2, C1, C2, C4, C5, D1, D2, E2, F5	**MiniLAB:** How much water can a moss hold? p. 268 **Connect to Earth Science**, p. 270 **MiniLAB:** How does water move in a plant? p. 271 **Activity 10-1:** The Effects of Acid Rain, pp. 272-273 **Using Technology:** The Ozone Connection, p. 276 **Skill Builder:** Concept Mapping, p. 277 **Using Math**, p. 277
10-3 **Science and Society** **Technology: Cleaner Coal** (1 session, ½ block)*	1. **Describe** the benefits and problems of burning coal. 2. **Explain** some of the methods for cleaning coal. National Content Standards: (5-8) A2, C2, E1, E2, F5	**Explore the Technology**, p. 279 **Activity 10-2:** Comparing Mosses, Liverworts, and Ferns, p. 280

* A complete Planning Guide that includes block scheduling is provided on pages 32T-35T.

Activity Materials

Explore	Activities	MiniLABs
page 259 No special materials are needed.	pages 272-273 moss plants; distilled water (pH 7); water with a pH of 5, 6, and 8 in different containers; metric ruler; hand lens; 10-mL graduated cylinder; apron; safety goggles; spray bottle page 280 live mosses, liverworts, and ferns with gametophytes and sporophytes; prepared slides of mosses, liverworts, and ferns; hand lens; forceps; dropper; microscope slide and coverslip; microscope; dissecting needle; pencil with eraser; safety goggles; apron	page 268 teaspoon, *Sphagnum* moss, 10 cm × 10 cm piece of cheesecloth, twist tie or string, balance, 100-mL graduated cylinder, tap water, 500-mL beaker, clock page 271 celery stalk with leaves, sharp knife, tall glass, tap water, red food coloring

Need Materials? Call Science Kit (1-800-828-7777).

Teacher Classroom Resources

Reproducible Masters	Transparencies	Teaching Resources
Study Guide, p. 39 **Reinforcement**, p. 39 **Enrichment**, p. 39 **Concept Mapping**, p. 25 **Critical Thinking/Problem Solving**, p. 14 **Cross-Curricular Integration**, p. 14 **Multicultural Connections**, pp. 23-24	**Section Focus Transparency 35**, Along a Rocky Coast	**Glencoe Life Science Interactive Videodisc** **Life Science CD-ROM** **Spanish Resources** **English/Spanish Audiocassettes** **Cooperative Learning Resource Guide** **Lab Partner** **Lab and Safety Skills** **Lesson Plans**
Study Guide, p. 40 **Reinforcement**, p. 40 **Enrichment**, p. 40 **Activity Worksheets**, pp. 58-59, 62, 63 **Lab Manual 23**, Vascular and Nonvascular Plants **Science Integration Activity 10**, Carbon Impressions	**Section Focus Transparency 36**, Getting Bogged Down **Science Integration Transparency 10**, Fossil Power Plants **Teaching Transparency 19**, Moss Life Cycle **Teaching Transparency 20**, Fern Life Cycle	**Assessment Resources** **Chapter Review**, pp. 23-24 **Assessment**, pp. 68-71 **Performance Assessment in the Science Classroom (PASC)** **MindJogger Videoquiz** **Alternate Assessment in the Science Classroom** **Performance Assessment** **Chapter Review Software** **Computer Test Bank**
Study Guide, p. 41 **Reinforcement**, p. 41 **Enrichment**, p. 41 **Activity Worksheets**, pp. 60-61 **Science and Society Integration**, p. 14	**Section Focus Transparency 37**, Ancient Forests, Today's Coal	

Key to Teaching Strategies

The following designations will help you decide which activities are appropriate for your students.

L1 Level 1 activities should be within the ability range of all students, including those with learning difficulties.

L2 Level 2 activities should be within the ability range of the average to above-average student.

L3 Level 3 activities are designed for the ability range of above-average students.

LEP LEP activities should be within the ability range of Limited English Proficiency students.

LS These activities are designed to address different learning styles.

COOP LEARN Cooperative Learning activities are designed for small group work.

P These strategies represent student products that can be placed into a best-work portfolio.

GLENCOE TECHNOLOGY

The following multimedia resources are available from Glencoe.

Science and Technology Videodisc Series (STVS)
Plants & Simple Organisms
City Trees
Super Trees
Sound of Thirsty Plants
Farming Indoors
Purifying with Plants
Ecology
Arctic Haze
Acid Rain and Plants

Hole in the Ozone
Greenhouse Effect
The Infinite Voyage Series
Secrets from a Frozen World
National Geographic Society Series
STV: Plants
STV: Rain Forest
GTV: Planetary Manager
Life Science CD-ROM

Teacher Classroom Resources

This is a representation of key blackline masters available in the Teacher Classroom Resources.

Teaching Aids

Section Focus Transparencies

Science Integration Transparencies

Teaching Transparencies

Meeting Different Ability Levels

Study Guide

Reinforcement

Enrichment Worksheets

Chapter 10 Introduction to Plants

Hands-On Activities

Science Integration Activity

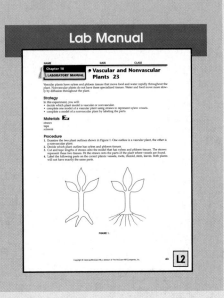

Lab Manual

Assessment

Performance Assessment

Enrichment and Application

Critical Thinking/ Problem Solving

Cross-Curricular Integration

Science and Society Integration

Multicultural Connections

Concept Mapping

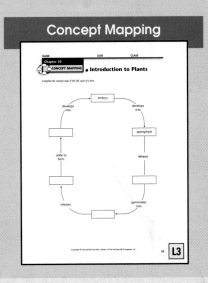

Chapter 10

Introduction to Plants

CHAPTER OVERVIEW

Section 10-1 Characteristics of plants are listed in this section. The adaptations of plants for survival on land are described. Differences between vascular and nonvascular plants are given.

Section 10-2 The life cycles of mosses and ferns are described. A comparison between nonvascular and vascular plants is made. Ecological importance of nonvascular and vascular plants is discussed.

Section 10-3 Science and Society Coal is the fossilized remains of plants such as mosses and ferns described in this chapter. This section details the technology used to remove sulfur dioxide and particulates.

Chapter Vocabulary

cellulose	alternation of
cuticle	generations
vascular	pioneer
plant	species
nonvascular	bog
plant	rhizome
rhizoid	frond
sporophyte	sorus
gametophyte	prothallus

Theme Connection

Stability and Change/Systems and Interactions The stability and change over time of plants from plantlike protists is detailed in this chapter. Emphasis is placed on the adaptations of plants for life on land. The systems and interactions of simple plants are also explained.

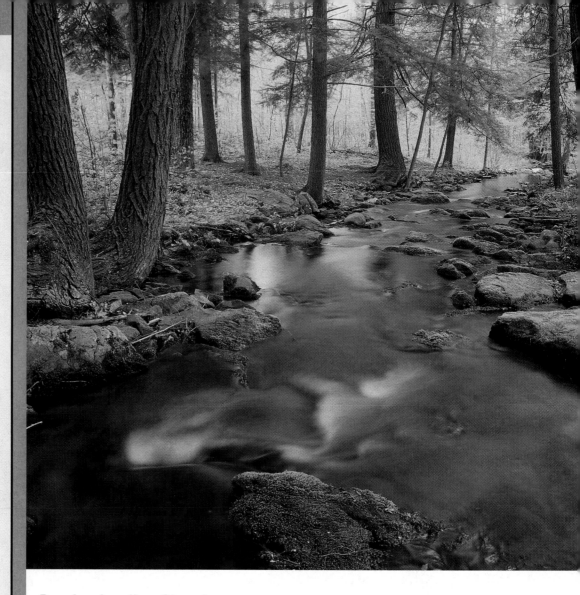

Previewing the Chapter

Section 10-1 Characteristics of Plants
▶ What is a plant?
▶ Origin and Evolution of Plants
▶ Adaptations to Land
▶ Classification of Plants

Section 10-2 Seedless Plants
▶ Seedless Nonvascular Plants
▶ Seedless Vascular Plants

Section 10-3 Science and Society
Technology: Cleaner Coal

258

Learning Styles

Look for the following logo for strategies that emphasize different learning modalities. **LS**

Kinesthetic	Explore, p. 259; Science & Art, p. 266
Visual-Spatial	Visual Learning, pp. 261, 264, 269, 270, 274, 275; Demonstration, pp. 261, 263, 274, 278; Science & Art, p. 266; MiniLAB, p. 271; Activity 10-2, p. 280
Interpersonal	Activity, p. 262; Activity 10-1, pp. 272-273
Logical-Mathematical	MiniLAB, p. 268; Activity, pp. 270, 278
Linguistic	Cultural Diversity, p. 264; Science & Art, p. 266; Science Journal, p. 276

LS

Chapter 10

Introduction to Plants

Plants are all around—in parks and gardens, by streams and on rocks, in houses, and even on dinner plates. Do you eat salads? Salads are made up of edible plants. What plants would you choose for a salad? Do you know what plant parts you would be eating? In the following activity, find out what plant parts are edible. Then in the chapter, learn about simple plant life.

EXPLORE ACTIVITY

Determine what plant parts are edible.

1. Make a list of five plants that you might eat during a typical day.
2. Decide which part of the plant you eat for each plant on your list.

Observe: In your Science Journal, list all the different parts of the plants you eat under two heads—Edible and Inedible. Explain your reasons for classifying each plant part as you did.

Previewing Science Skills

▶ In the Skill Builders, you will **hypothesize, map concepts,** and **measure in SI.**

▶ In the Activities, you will **design an experiment, observe,** and **organize data.**

▶ In the MiniLABs, you will **research** and **infer.**

259

EXPLORE ACTIVITY

Purpose

LS **Kinesthetic** Use the Explore Activity to introduce students to plant parts. Inform students that they will be learning more about simple plants and their parts in this chapter. L1 LEP

Materials

Ingredients for a salad might include lettuce, carrots, olives, tomatoes, and sunflower seeds.

Teaching Strategies

• Make a garden salad for the students; have them identify the plant parts as they eat the salad.

• You may wish to offer suggestions of additional vegetables if students do not produce a sufficiently diverse list.

Observe

Plant parts should include stems, roots, leaves, and flowers. Sample classifications will vary. For example, sweet potato *roots* are edible, whereas potato *stems* and leaves are generally considered to be inedible.

✓ **Assessment**

Portfolio Have students list additional plant parts that they eat in their portfolios. Use the Performance Task Assessment List for Science Portfolio in **PASC,** p. 105. P

Assessment Planner

Portfolio
Refer to page 281 for suggested items that students might select for their portfolios.

Performance Assessment
See page 281 for additional Performance Assessment options.
Skill Builders, pp. 265, 277
MiniLABs, pp. 268, 271
Activities 10-1, pp. 272-273; 10-2, p. 280

Content Assessment
Section Wrap-ups, pp. 265, 277, 279
Chapter Review, pp. 281-283
Mini Quizzes, pp. 265, 277

Group Assessment
Opportunities for group assessment occur with Cooperative Learning Strategies and Flex Your Brain Activities.

Prepare

Section Background

- The characteristic tissue formation of plants sets them apart from the plant-like protists.
- Fossils of tropical plants have been found in Antarctica, indicating that the continent was once located nearer the equator.
- Reproduction without the requirement of water was a major adaptation of land plants.
- Emphasize the fact that bryophytes do not contain roots, leaves, or stems because bryophytes have no vascular tissue.

1 Motivate

Bellringer

 Before presenting the lesson, display **Section Focus Transparency 35** on the overhead projector. Assign the accompanying **Focus Activity** worksheet. L1 LEP

Tying to Previous Knowledge

Discuss the basic relationship between plants and animals. Review the concepts of producer and consumer with students.

GLENCOE TECHNOLOGY

CD-ROM

Life Science CD-ROM

Have students perform the interactive exploration for Chapter 10 to reinforce important chapter concepts and thinking processes.

10•1 Characteristics of Plants

Science Words
cellulose
cuticle
vascular plant
nonvascular plant

Objectives
- List the characteristics of plants.
- Describe adaptations of plants that made it possible for them to survive on land.
- Compare vascular and nonvascular plants.

Figure 10-1

All plants are many celled and most contain chlorophyll. Grass, trees, and ferns all are members of Kingdom Plantae.

What is a plant?

Do you enjoy walking along nature trails in parks? Maybe you've taken off your shoes and wriggled your toes in the cool grass. Perhaps you climbed a tree and looked at the sky through green leaves. Or you may have walked down a nature trail to a low, damp place where ferns grow like the one shown in **Figure 10-1**. Grass, trees, and ferns all are members of the Plant Kingdom.

Now look at **Figure 10-2**. These organisms, mosses and liverworts, have characteristics that identify them as plants, too. What do they have in common with grass, trees, and ferns? What makes a plant a plant?

Characteristics of Plants

All plants are many celled and most contain the green pigment chlorophyll. Cell walls surround plant cells and provide structure. Most plants have roots or rootlike structures that hold them in the ground. As a result, most plants do not move around. Plants are made up of eukaryotic cells and have adapted successfully to life on land. In fact, they have adapted so well that they are found in nearly every environment on Earth. From frigid, ice-bound Antarctica to the hot, dry deserts of Africa, plants have evolved to survive even the most extreme conditions.

260

Program Resources

 Reproducible Masters
Study Guide, p. 39 L1
Reinforcement, p. 39 L1
Enrichment, p. 39 L3
Cross-Curricular Integration, p. 14
Critical Thinking/Problem Solving, p. 14
Multicultural Connections, p. 23
Concept Mapping, p. 25

 Transparencies
Section Focus Transparency 35 L1

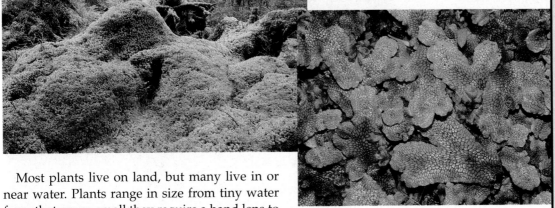

Most plants live on land, but many live in or near water. Plants range in size from tiny water ferns that are so small they require a hand lens to see, to the giant sequoia trees of the western United States, some of which are over one thousand years old and 100 m in height.

About 285 000 plant species have been identified, and scientists think there are many more still to be found, mainly in tropical rain forests. If you were asked to make a list of all the plants you could name, you probably would name vegetables, fruits, and field crops like wheat, rice, or corn. These plants are important food sources to humans and other consumers. Without green plants, life on Earth as we know it would not be possible.

Origin and Evolution of Plants

Where did the first plants come from? Like all life, early plants probably came from the sea, evolving from plantlike protists. What evidence is there that this is true? Both plants and green algae, a type of protist, have the same types of chlorophyll as well as carotenoids in their cells. Carotenoids are red, yellow, or orange pigments found in plastids and in all cyanobacteria.

Fossil Record

One way to understand the evolution of plants is to look at the fossil record. Unfortunately, plants usually decay before they form fossils. The oldest fossil plants are from the Silurian period and are about 420 million years old. Available fossils show that early plants were similar to the plantlike protists. Fossils of *Rhynia major* represent the earliest land plants known. This fossil plant is illustrated in **Figure 10-3**. These plants had no leaves, and their stems grew underground. Scientists hypothesize that these kinds of plants evolved into the simple plants of today.

Figure 10-2

Simple plants include mosses and liverworts like these.

Figure 10-3

Fossils of *Rhynia major* are representative of some of the earliest land plants known. These small plants had no true roots, stems, or leaves.

10-1 Characteristics of Plants **261**

Figure 10-4

Algae must have water to survive.

A Green algae produce their own food and move materials in and out through the cell membrane.

B If a pond completely dries up, the algae living in it will die.

Activity

LS Interpersonal Set up stations around the classroom that contain different plant parts. Have students write down the name of each plant part as they travel from station to station.

CONNECT TO
CHEMISTRY

Answer Cellulose is a tough material that enables plants to stand upright. By standing upright, plant leaves are able to catch more sunlight and make more food.

📂 Use **Critical Thinking/Probelm Solving**, p. 14, as you teach this lesson.

CONNECT TO
CHEMISTRY

Cellulose, found in plant cell walls, is an organic compound composed of many glucose units strung together. More than half the carbon in plants is in the form of cellulose. Raw cotton is over 90 percent cellulose. *Infer* which physical property you think makes cellulose ideal for helping plants survive on land.

Adaptations to Land

Imagine life for a one-celled green alga floating in a shallow pool. The water in the pool surrounds and supports the alga cell. It can make its own food through the process of photosynthesis, and materials move freely into and out of the cell membrane by osmosis, diffusion, and active transport. The alga has everything it needs to survive.

Now imagine a summer drought. The pool begins to dry up. Soon the alga lies on damp mud. What will happen to the alga? It can still make its own food, so it won't starve. As long as the ground stays damp, the alga can move materials in and out through the cell membrane. But it is no longer supported by the pool's water. What will happen to the alga if the land continues to dry up? The alga will lose water through the cell membrane by osmosis. Remember that in osmosis, water molecules diffuse through a membrane from an area of higher concentration to one of lower concentration. Water will diffuse out of the alga to the drier pond bottom. Without water, the alga will dry up and die, as illustrated in **Figure 10-4.**

Protection and Support

What adaptations would make it possible for plants to survive on land? Loss of water is a major problem plants face. What would help a plant conserve water? Plant cells have cell membranes as the protists do, but they also have rigid cell walls outside the membrane. Cell walls are made of

262 Chapter 10 Introduction to Plants

Community Connection

Attractive Plants Ask a florist or horticulturist to discuss the characteristics of plants that make them attractive and useful.

Across the Curriculum

Earth Science Ask students how the fossils of plants found in such places as Antarctica and high on mountains provide evidence of continental drift and mountain building.

Figure 10-5

Cell walls and waxy cuticles are adaptations that enable plants to survive on land.

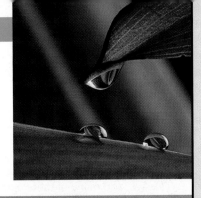

A Rain beads up on the leaves of some plants.

B But rain drips off the leaves of other plants.

Demonstration

LS **Visual-Spatial** Use a spray mister to mist water onto a plant leaf that has a thick cuticle. Allow students to observe how the water beads up and runs off. **LEP**

LS **Visual-Spatial** Place a stalk of celery in a glass of water to which some food coloring has been added. Once the colored liquid has reached the leaves, pull apart the stalk to show the parallel vascular tubes in the celery. **LEP**

cellulose, an organic compound made up of long chains of sugar molecules. Some woody plants are as much as 50 percent cellulose. Cell walls help prevent plant cells from drying out and provide structure and support. Most land plants also have a waxy, protective layer on stems and leaves called a **cuticle.** The next time it rains, go outside and see how raindrops bead up on the surface of most plant leaves or flower petals, as illustrated in **Figure 10-5.** The rain rolls right off because of the waxy cuticle. The cuticle is an adaptation that enabled plants to keep from drying out on land.

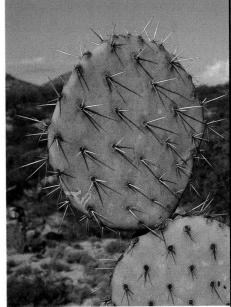

C Waxy cuticles prevent moisture loss from this prickly pear cactus.

Reproduction

Another problem plants faced in the move onto land was reproduction. Plants evolved from organisms that reproduced in water. As a matter of fact, these species were completely dependent on water for reproduction and survival. Simple plants still require water to reproduce. You will learn more about reproduction in simple plants in Section 10-2. More complex plants evolved ways to reproduce that do not require water. Reproduction in these plants is discussed in Chapter 11.

Life on Land

There are some advantages to life on land for plants. For one thing, there's more direct sunlight available for photosynthesis on land. Another advantage of life on land is the availability of carbon dioxide. There is much more carbon dioxide in the air than in the water. Carbon dioxide is a gas that plants use to produce food. All plants still need water to carry nutrients into and waste products out of their cells. They also need structures to support their weight on land. These supports eventually came in the form of stems and roots.

10-1 Characteristics of Plants **263**

Theme Connection

Stability and Change Point out the evolutionary changes necessary for plants, and subsequently animals, to invade the land.

Integrating the Sciences

Physics Have students conduct activities that explain how water pressure in plants provides some rigidity and support for plants. One such answer would be to run water through a hose and then pinch the walls to increase pressure. **L1**

Problem Solving

Think Critically

Some plants may become extinct before they can be studied. Political unrest can also slow the collection of plant materials for research.

---

3 Assess

264

Problem Solving

What is in nature's medicine chest?

People in all cultures have found ways to treat illnesses, and many of these cures involve plants. Various Native American cultures used willow bark to cure headaches. Heart problems were treated with foxglove in England, sea onions in Egypt, and seeds in Africa. In Peru, the bark of the cinchona tree was used to treat malaria. Scientists have found that many native cures are medically sound. Willow bark contains salicylates, the main ingredient in aspirin. Foxglove, shown below, and sea onions all contain digitalis or similar drugs. Cinchona bark contains quinine, an antimalarial drug.

Think Critically:

How might the destruction of the rain forests affect research for new drugs from plants?

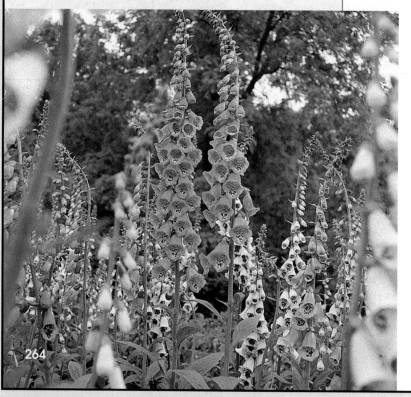

264

Classification of Plants

Today, plants usually are classified into major groups called divisions, as illustrated in **Figure 10-6.** A division is equivalent to the level phylum you've studied in the Protist Kingdom. The simplest plants—mosses and liverworts—are placed in the division Bryophyta (bri uh FITE uh).

Bryophytes

Bryophytes are small plants found in damp environments like the forest floor, the edges of ponds and streams, and near the ocean. Bryophytes are usually just a few cells thick, so they are able to absorb water directly through their cell walls. Plants in the more complex divisions of the Plant Kingdom have systems containing vascular tissue.

Vascular tissue is made up of long, tubelike cells in which water and nutrients are transported through the plant. Plants with vascular tissue are more than a few cells thick and are known as **vascular plants.** Most of the plants you know, such as trees, bushes, and flowers, are vascular plants. Plants without vascular tissue, the bryophytes, are **nonvascular plants.** Why is it an advantage for nonvascular plants to be located in damp areas?

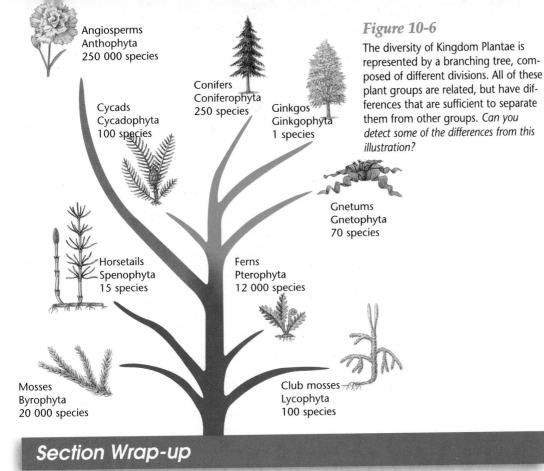

Angiosperms
Anthophyta
250 000 species

Conifers
Coniferophyta
250 species

Ginkgos
Ginkgophyta
1 species

Cycads
Cycadophyta
100 species

Gnetums
Gnetophyta
70 species

Horsetails
Spenophyta
15 species

Ferns
Pterophyta
12 000 species

Mosses
Byrophyta
20 000 species

Club mosses
Lycophyta
100 species

Figure 10-6

The diversity of Kingdom Plantae is represented by a branching tree, composed of different divisions. All of these plant groups are related, but have differences that are sufficient to separate them from other groups. *Can you detect some of the differences from this illustration?*

Section Wrap-up

Review

1. List the characteristics of plants.

2. Compare vascular and nonvascular plants.

3. **Think Critically:** If you left a board lying on the grass for a few days, what would happen to the grass underneath the board? Why?

Skill Builder
Hypothesizing

From what you have learned about adaptations necessary for life on land, make a hypothesis as to what types of adaptations land plants might have if they had to survive submerged in water. If you need help, refer to Hypothesizing in the **Skill Handbook.**

Using Computers

Spreadsheet Make a spreadsheet using the information in **Figure 10-6.** Title one column *Division* and another column *Number of Species.* Make a third column *Percent of Total* and complete the information for all three columns.

Using Computers

Division	Number of Species	Percent of Total
Anthophyta	250 000	88.1%
Coniferophyta	250	0.2%
Ginkgophyta	1	0.0%
Gnetophyta	70	0.0%
Cycadophyta	100	0.0%
Spenophyta	15	0.0%
Bryophyta	20 000	7.0%
Pterophyta	12 000	4.2%
Lycophyta	1 000	0.4%

Assessment

Performance Assess students' abilities to hypothesize about the adaptations necessary for land plants to live in water by having them devise a model for one of their suggested adaptations. Use the Performance Task Assessment List for Formulating a Hypothesis in **PASC,** p. 21. **P**

4 Close

•MINI•QUIZ•

Use the Mini Quiz to check students' recall of chapter content.

1. **All plants are many-celled and most contain the green pigment _____.** *chlorophyll*

2. **Plants range in size from tiny _____ to the giant _____.** *water ferns, sequoias*

3. **_____ are red, yellow, or orange pigments found in chloroplasts.** *Carotenoids*

4. **Plants with vessels that transport materials to other cells are called _____.** *vascular plants*

Section Wrap-up

Review

1. All plants are many-celled and most contain chlorophyll and carotenoids. Most plants have roots or rootlike fibers that hold them in the ground.

2. Vascular plants have tubes that carry water and food. Nonvascular plants get water and food by diffusion.

3. **Think Critically** The grass would turn yellow. Light is needed to keep up production of chlorophyll.

Skill Builder

Plants submerged in water would need more efficient chloroplasts because of less light. In addition, they would need ways to get additional carbon dioxide and ways to keep a balance between water inside cells and outside cells. **P**

265

Science & ART

Source

Beatrix Potter's Art. Paintings and drawings selected and introduced by Anne Stevenson Hobbs. New York: Frederick Warne/Penguin, 1989.

Biography

- Born in 1866, Beatrix Potter was raised in a prosperous Victorian English family. She was educated at home. This sheltered existence allowed her the time and isolation needed to become absorbed in her study of nature. Young Beatrix observed small pets and plants, read, drew, and painted almost obsessively.

- Potter's career as a story-teller and artist began when *The Tale of Peter Rabbit* was published in 1902. More than 20 books followed. With her income, Potter was able to buy a farm in England. At the age of 47, Beatrix married and made the farm her permanent home. She devoted the remainder of her life to farming, sheep-breeding, and preserving the countryside by buying tracts of land around her farm. Upon her death, Potter left over 4000 acres of land to her country.

Teaching Strategies

LS **Linguistic** Invite students to share the descriptive essays they wrote in their Science Journals.

LS **Visual-Spatial** Display a reproduction of one of Beatrix Potter's drawings of a plant, along with a real-life specimen of that plant. Have students compare the two and share their findings. **L1**

LS **Kinesthetic** Invite interested students to observe a plant in the same manner as Potter would and then draw a representational illustration.

The Art of Beatrix Potter

You may remember Beatrix Potter from books you read as a young child. She not only told stories that children and adults love, but she also illustrated them with wonderful pictures of animals and plants. In fact, Beatrix Potter is now almost as well known as an artist as she is a writer.

As a girl, Beatrix received training in drawing and painting. But she once said, "I don't want lessons. I want practice." And practice she did. In addition to an artist's feelings for life and living things, Beatrix had a scientist's sense of observation and detail. In fact, she had a scientific interest in all the natural world. Beatrix spent hours observing and studying animals and plants. She observed them in real life as well as in natural history books, collecting small pets such as mice and frogs, and studying mushrooms and lichens. Then she "copied" these natural specimens, as she called drawing something as realistically as possible.

When Beatrix grew older, she used this same technique to illustrate her children's books. But hundreds of pieces of her artwork—never before seen—were found after she died in 1943. Beatrix's interest in nature, including plants, became truly evident. Of all the illustrations discovered, more than 500 were of plants and fungi.

Flowering plants were often the subject of Beatrix's work. You can find roses, lilies, clematis, carnations, irises, foxgloves, and geraniums in the backgrounds of many of her book illustrations. Gardens—and the food plants that grow in them, such as onions—were also a favorite subject. So too were trees. According to Beatrix, many artists never really observed how a branch grows out from a tree trunk. "I did so many careful botanical studies in my youth," she wrote to a friend, "it became easy for me to draw twigs. And little details like that really add to the reality of a picture."

Science Journal

Choose a plant and observe it as closely and as carefully as possible. Then write a description of the plant in your Science Journal. If you like, draw a picture to accompany your descriptive essay.

LS **Visual-Spatial** Display a sample of Beatrix Potter's representational plant art with a sample from another artist (for example, Potter's *Waterlilies on Esthwaite Water*, 1906, with one of Claude Monet's waterlilies works). Have students compare and discuss the different approaches. **L1**

Other Works

The Complete Tales of Beatrix Potter. New York: Frederick Warne/Penguin, 1989.

Seedless Plants 10•2

Seedless Nonvascular Plants

If you were asked to name the parts of a plant, you probably would list roots, stems, leaves, and perhaps flowers. You may also know that many plants grow from seeds. But did you know that some simple plants have none of these parts? Nonvascular plants, the bryophytes, do not have roots, stems, or leaves. They do have rootlike fibers, stalks that look like stems, and leaflike green growths. Instead of growing from seeds, bryophytes grow from spores. The nonvascular plants include the mosses and liverworts. **Figure 10-7** shows some common types of nonvascular plants.

Mosses

A moss is a simple rootless plant with leaflike growths in a spiral around a stalk. Moss plants are held in place by rootlike filaments or threads made up of only a few long cells called **rhizoids.** One type of liverwort is a simple, rootless plant that has a flattened, leaflike body. Liverworts get their name because, to some people, they look like a liver. In the ninth century, liverworts were thought to be useful in treating diseases of the liver. The ending, *-wort*, means "herb," so the word *liverwort* means "herb for the liver." In liverworts, each rhizoid is made up of just one cell. Both mosses and liverworts grow in damp areas and range in size from 2 to 5 centimeters in height.

Of approximately 20 000 species of nonvascular plants, most are classified as mosses. Have you ever seen mosses growing on tree trunks, rocks, or the ground in damp or humid areas?

Science Words

rhizoid
sporophyte
gametophyte
alternation of
    generations
pioneer species
bog
rhizome
frond
sorus
prothallus

Objectives

- Describe the life cycles of mosses and ferns.
- Compare simple non-vascular plants with simple vascular plants.
- State the importance of simple nonvascular and vascular plants.

Figure 10-7

The seedless nonvascular plants include the mosses (left) and the liverworts (right). As a group, these plants are called bryophytes.

267

Section 10•2

Prepare

Section Background

- There are more species in the bryophyte division than in any other, except anthophytes, the flowering plants.
- Some mosses do have water-conducting cells, but they are not organized as vascular tissue.
- Rhizoids hold the plant to the ground, but do not conduct water to other cells in the plant as roots do.
- The ferns are the most diverse group of simple plants. They range in size from the tiny water fern to the tree fern, and in form from the maidenhair fern to the staghorn fern.

Preplanning

- To prepare for the MiniLAB on page 268, obtain *Sphagnum* moss and cheesecloth.
- The MiniLAB on page 271 requires food coloring and celery stalks.
- Moss plants and dilute hydrochloric or sulfuric acid are required for Activity 10-1.

1 Motivate

Bellringer

Before presenting the lesson, display **Section Focus Transparency 36** on the overhead projector. Assign the accompanying **Focus Activity** worksheet. L1 LEP

Tying to Previous Knowledge

Point out that students may know club mosses as ground pines from holiday decorations, and they know ferns from houseplants.

Program Resources

📁 **Reproducible Masters**
Study Guide, p. 40 L1
Reinforcement, p. 40 L1
Enrichment, p. 40 L3
Activity Worksheets, pp. 58-59, 62, 63 L1
Science Integration Activities, pp. 19-20
Lab Manual 23

📠 **Transparencies**
Teaching Transparencies 19, 20
Science Integration Transparency 10
Section Focus Transparency 36 L1

Figure 10-8
The leafy green gametophytes of moss support the stalks and capsules of the moss sporophytes. When the capsules mature and dry, they snap open and shoot spores into the air.

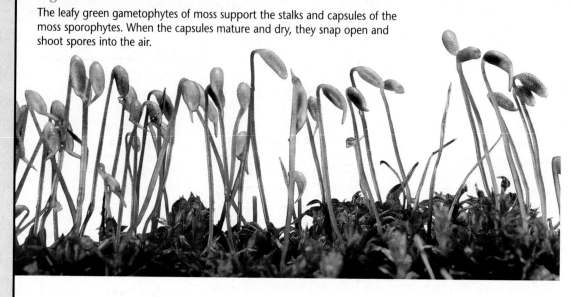

MiniLAB

How much water can a moss hold?

Procedure

1. Place a few teaspoons of *Sphagnum* moss on a piece of cheesecloth. Twist the cheesecloth to form a ball and tie it securely.
2. Weigh the ball.
3. Put 200 mL of water in a beaker and add the ball.
4. Predict how much water the ball will absorb.
5. Wait 15 minutes; remove the ball, draining off excess water.

Analysis

1. Weigh the ball and measure the amount of water left in the beaker.
2. In your Science Journal, calculate how much water the *Sphagnum* moss soaked up.

The Moss Life Cycle

A typical growth of moss plants looks like a soft, green carpet on the forest floor. Now look at the closeup of the moss plants in **Figure 10-8.** Can you see the structures sticking up out of the leaflike plants? These are the moss sporophytes. A **sporophyte** is the spore-producing diploid stage in the moss life cycle. It includes a stalk and a capsule in which numerous haploid spores are produced. Eventually, the capsule breaks open and the spores are scattered. When a haploid spore lands on wet soil or rocks, it germinates into a thin, threadlike green structure. Within a few days, small gametophyte moss plants begin to grow here and there along the thread. A **gametophyte** is the form of a moss plant that produces sex cells. Sometimes, a moss gametophyte produces only male or female sex cells, but often both types are produced. During a heavy dew or rain, the male sperm get splashed onto the female gametophyte. They swim to the female eggs. When the male and female sex cells unite, a diploid zygote forms. The zygote divides by mitosis to form an embryo, which in turn develops into a sporophyte, and the cycle begins again. In plants, this continual cycle, which alternates between the spore-producing phase and sex cell-producing phase, is called **alternation of generations.** In bryophytes, the most visible plant is the green, leaflike gametophyte. The life cycle of a moss, including the alternation of generations, is shown in **Figure 10-9.** Liverworts have a life cycle similar to that of the mosses.

268 Chapter 10 Introduction to Plants

Asexual Reproduction

In addition to the alternation of generations, both mosses and liverworts reproduce asexually. New moss plants can develop when a small piece of the parent plant breaks off. Liverworts develop small balls of cells within cuplike structures on the surface of the leaflike body. These balls of cells can be carried by water to new areas where they grow into new plants.

Alternate Generations

All plants have a life cycle in which the sporophyte and gametophyte alternate. In other words, all plants go through a diploid and a haploid stage. In nonvascular plants like mosses and liverworts, the sporophytes depend on the gametophytes for water and nutrients. In the more complex vascular plants, like tulips and oak trees, the sporophyte doesn't depend on the gametophyte for these things.

Figure 10-9

The life cycle of a moss alternates between gametophyte and sporophyte stages.

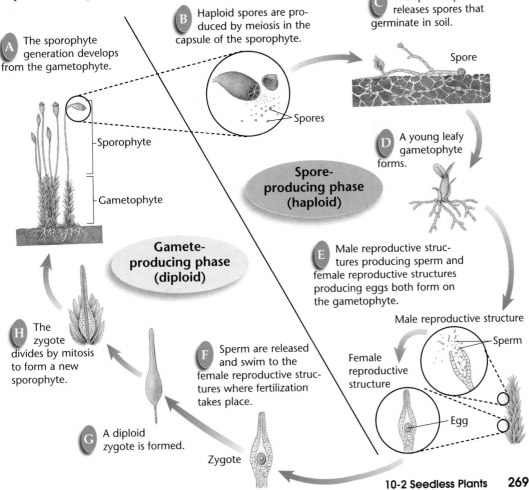

A The sporophyte generation develops from the gametophyte.

B Haploid spores are produced by meiosis in the capsule of the sporophyte.

Spores

C The spore capsule releases spores that germinate in soil.

Spore

D A young leafy gametophyte forms.

Spore-producing phase (haploid)

Gamete-producing phase (diploid)

Sporophyte

Gametophyte

E Male reproductive structures producing sperm and female reproductive structures producing eggs both form on the gametophyte.

Male reproductive structure

Sperm

Female reproductive structure

Egg

H The zygote divides by mitosis to form a new sporophyte.

F Sperm are released and swim to the female reproductive structures where fertilization takes place.

G A diploid zygote is formed.

Zygote

10-2 Seedless Plants 269

CONNECT TO
EARTH SCIENCE

*S*oil is a mixture of weathered rock and decaying organic matter (plant and animal). *Infer* what roles pioneer species such as lichens, mosses, and liverworts play in building soil.

Figure 10-10

Mosses are pioneer plants that often are among the first living things to inhabit a new environment like this lava field in Iceland. Over many years, the chemical actions of pioneer plants help to form soil.

Importance of Mosses and Liverworts

Mosses and liverworts are important in the ecology of many areas. Although mosses require moist conditions to grow and reproduce, many of them can withstand long periods of dryness. They are often among the first plants to grow in new environments, such as lava fields as shown in **Figure 10-10,** or disturbed environments, such as forests destroyed by fire.

When a volcano erupts, the lava covers the land and destroys the plants living there. After the lava cools, the spores of mosses and liverworts are carried to the new rocks by the wind and begin to grow wherever there is enough water. When they grow on rocks, their rhizoids can actually begin to penetrate tiny cracks in the rocks' surfaces. Weathering of rocks often begins when mosses release chemicals that begin to break down rocks. Organisms that are the first to grow in new or disturbed areas like these are called **pioneer species.** Pioneer plant species grow and die and begin to build up decaying plant material that provides nutrients for other less hardy plants. That is, these pioneer plants change the conditions in the environment so that other plants can grow there, too. Eventually, larger vascular plants can begin to grow, sending their roots down through the decaying plants and into the rocks underneath. As the roots grow, they crack the rocks apart even more, and the weathering process slowly turns the rocks into new soil.

Seedless Vascular Plants

A second and larger group of simple plants includes the seedless vascular plants, illustrated in **Figure 10-11.** These plants differ from mosses mainly because they have vascular tissue. They differ from other vascular plants because they produce spores rather than seeds. The vascular tissue in the seedless vascular plants is made up of long, tubelike cells. These cells carry water, minerals, and nutrients to all the cells throughout the plant. Why is having cells like these an advantage to a plant? Remember that bryophytes are only a few cells thick. Each cell absorbs water directly from its environment. As a result, these plants do not grow very big. Plants with vascular tissue, on the other hand, can grow bigger and thicker because each cell gets water and nutrients through the vascular tissue.

Types of Seedless Vascular Plants

Seedless vascular plants include the club mosses, spike mosses, horsetails, and ferns. Today, there are about 1000 species of club mosses, spike mosses, and horsetails. Ferns are more abundant, with at least 12 000 species known. During the Paleozoic era, there were many species that we now know of only from fossils. In the warm, moist Paleozoic era, some species of horsetails grew 15 meters tall. In contrast, simple vascular plant species today reach no more than 1 or 2 meters in height.

Figure 10-11

The seedless vascular plants include club mosses, spike mosses, horsetails, and ferns. All of these plants grow taller than mosses and liverworts, but none grow as tall as large trees.

Ferns

Club moss

Horsetails

Spike moss

10-2 Seedless Plants **271**

Purpose

IS **Interpersonal** In this activity, students will design an experiment to test the effects of acid rain on moss, conduct the experiment, and interpret data. **L1** **COOP LEARN**

Process Skills

designing an experiment, hypothesizing, observing and inferring, interpreting data, communicating, making and using tables, comparing and contrasting, recognizing cause and effect, separating and controlling variables, measuring in SI

Time

approximately 2-3 weeks from set-up to completion

Materials

Dilute hydrochloric acid or sulfuric acid can be used to lower pH. Test dilutions with pH paper to determine pH. Prepare enough solution to last for the entire time. You may use liverworts or other plants as a substitute if moss is unavailable.

Safety Precautions

Warn students to protect their eyes, skin, and clothes from acid solutions. If a spill on skin, eyes, or clothes occurs, flush the area for several minutes with water.

Possible Hypotheses

Students may suggest that a particular solution will kill or injure moss plants. Others may hypothesize that because moss are acid-loving, they will not be harmed. Similar hypotheses can be accepted.

📁 **Activity Worksheets,** pp. 5, 58-59

Activity 10-1

Design Your Own Experiment
The Effects of Acid Rain

Within the last fifty years, air pollution problems have grown in the United States. It has been found that air pollution from factories and automobile exhaust can result in air that contains high levels of sulfur dioxide and nitrogen oxides. These gases combine with water and eventually fall back to Earth as rain with a low or acidic pH. This is often called acid rain. What happens to plants growing in areas where air pollution is heavy?

PREPARATION

Problem

How does an acid solution affect plant tissues and plant growth?

Form a Hypothesis

Based on the information you have just learned, state a hypothesis about how the pH of water affects plant growth.

Objectives

- Observe the effects of different levels of pH on plants.
- Collect data that can be used to determine an answer to the problem.

Possible Materials

- moss plants
- distilled water, pH 7
- water in containers at pH 5, pH 6, pH 8
- metric ruler
- hand lens
- 10-mL graduated cylinder
- apron
- safety goggles
- spray bottle

Safety Precautions

Protect clothing and eyes from acid or base solutions.

PLAN THE EXPERIMENT

1. As a group, agree upon and write out the hypothesis statement. Which pH level is best for moss growth?
2. As a group, list the steps that you need to take to test your hypothesis. Be very specific, describing exactly what you will do at each step. List your materials.
3. Decide on the number of days you will collect data and how much water will be added to each plant.
4. Design a data table in your Science Journal, so that it is ready to use as your group collects data on the effects of the different pH treatments. Record your observations.

Check Your Plan

1. Read over your entire experiment to make sure that all steps are in logical order.
2. Identify any constants, variables, and controls of the experiment.
3. *Make sure your teacher approves your plan before you proceed.*

DO THE EXPERIMENT

1. Carry out the experiment as planned.
2. While the experiment is going on, write down any observations that you make and complete the data table in your Science Journal.

Analyze and Apply

1. **Compare** your results with those of other groups.
2. How was plant growth affected?

3. What evidence did you see of damaged tissue?
4. **Graph** your results using a line graph, placing amount of growth or tissue damage on the y-axis and time on the x-axis. You should have a different line on the graph for each pH solution tested.

Go Further

Compare the effects of pH on moss and a flowering plant.

Analyze and Apply

1. Depending on experimental procedures, groups may or may not report similar results.
2. Plant growth may be inhibited or discoloration may occur. The plants may die.
3. Discoloration is a result of damaged tissue.
4. Graphs will vary.

Go Further

Flowering plants will usually be affected more by low pH than moss.

 Assessment

Portfolio Have students take photographs or make large sketches of their experiment in progress and final phases. Encourage them to include these in their portfolio. Use the Performance Task Assessment List for Science Portfolio in **PASC**, p. 105. **P**

PLAN THE EXPERIMENT

Possible Procedures

Students will set up containers with moss plants. Each container will be sprayed with a different solution for the duration of the procedure. Spraying times will vary with experimental design and length of experiment. Once moss has been planted in separate containers, one may be used as the control and the other(s) as an experimental group(s). Students should be encouraged to water the control group with distilled water, pH 7. If there is only one experimental group, they may use water at pH 5, 6, or 8. If they have several experimental groups, they may be able to use water at all the pH levels available to determine the effects on the moss plants. Students will record both qualitative data (how the moss looks) and quantitative data (amount of growth, color change, or other measurable factors).

Teaching Strategies

Troubleshooting If students fail to obtain expected results, remind them that this often happens. Explain that they should try to explain why their results were unexpected and what may have caused the results.

DO THE EXPERIMENT

Expected Outcome

More moss may be damaged at high pH than low, since many species are adapted to acid conditions in bogs. Results may include discoloration of plants, inhibited growth of sporophytes, or even death.

Figure 10-12

Club mosses (A) and spike mosses (B) produce spores at the end of stems in structures similar to pinecones. Early photographers often used the dry, flammable spores of club moss as flash powder to provide light to take photographs.

Club Mosses and Spike Mosses

Look at the photographs of club mosses and spike mosses in **Figure 10-12.** Club mosses produce spores at the end of the stems in structures that look like tiny pinecones. The upright stems of club mosses have needlelike leaves. One kind of club moss, *Lycopodium,* is called the ground pine because it looks like a miniature pine tree. *Lycopodium* is found from arctic regions to the tropics, but never in large numbers. In some areas, *Lycopodium* is endangered because of overcollection for use in making holiday wreaths and other decorations.

Spike mosses look similar to club mosses. One species of spike moss, the resurrection plant, is adapted to desert conditions. When water isn't available, the plant curls up and looks dead. But when water falls in the desert, the resurrection plant unfurls its green leaves and begins making food again. The plant can go through this process many times if necessary.

Horsetails

Horsetails have a stem structure unique among the vascular plants. Their stems are jointed and have a hollow center surrounded by a ring of vascular tissue. Leaves grow in a spiral around each joint. In **Figure 10-13,** you can see these joints easily. If you pull on a horsetail stem, it will pop apart in sections. Spores grow in a structure at the tip of the stems, as in the club mosses. The stems of the horsetails contain silica, a gritty substance found in sand. In the days when pioneers traveled westward across the United States, horsetail stems were used to scour pots and pans. The common name of *Equisetum* today is still "scouring rush."

Figure 10-13

Horsetails contain silica and were used by pioneers in the United States to scrub pots and pans.

Ferns

The largest group of seedless vascular plants is the division Pterophyta (teh ruh FITE uh), which includes the ferns shown in **Figure 10-14.** There are about 12 000 living species of ferns, but like the other divisions of vascular plants, additional species are known only as fossils.

As Earth scientists have pieced together the geological history of Earth from clues left in rock layers, they have described a different climate during one part of the Paleozoic era. They have learned that during the Carboniferous period of the Paleozoic era, illustrated in **Figure 10-15,** the tallest plants were species of ferns that grew much taller than any fern species alive today. Species of tree ferns found today in tropical areas may reach 3 to 5 meters in height, but ancient tree ferns grew as high as 25 meters. The climate was tropical over much of Earth, with steamy swamps covering large areas. When the tree ferns and other plants died, many of them became submerged in water and mud before they could decompose.

Formation of Fuel

Over long periods of time, this plant material built up, became compacted and compressed, and eventually became coal. Today, a similar process is taking place in peat bogs. A **bog** is a poorly drained area with spongy, wet ground that is composed mainly of dead plants whose decay has been slowed by lack of oxygen. The principal plants in bogs are seedless plants such as bryophytes and ferns. Peat is one low-cost fuel that is being mined from many bogs. Scientists hypothesize that if the peat layers are buried by additional sediments, then compacted and compressed, the peat could eventually become coal, petroleum, or natural gas. This process takes millions of years.

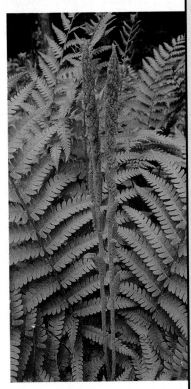

Figure 10-14

Most ferns produce spores in special structures on the leaves. The unusual cinnamon fern bears its spores on separate stalks.

Figure 10-15

The Carboniferous swamp forest contained many more species of club mosses, spike mosses, horsetails, and ferns than are alive today. These plants are believed to have formed the fuels now used.

10-2 Seedless Plants   **275**

Use **Teaching Transparency 20** as you teach this lesson.

3 Assess

Check for Understanding

Provide each student with a list of terms that can be used in an explanation of moss and fern life cycles. Have students write paragraphs that compare and contrast these two life cycles.

Reteach

Give pairs of students drawings of the stages of moss and fern life cycles on 3" × 5" notecards. Have students arrange the cards and label the drawings.

Extension

For students who have mastered this section, use the **Reinforcement** and **Enrichment** masters.

USING TECHNOLOGY

The Ozone Connection ▼ ▲

Ozone is a gas composed of three oxygen molecules linked together. You may have heard about a "hole in the ozone layer" and the damage it poses for plant and animal life, but how is the "hole" measured and what does ozone do? There is no hole, but the layer of ozone found above Earth's stratosphere is thinning. This ozone layer filters out much of the harmful ultraviolet radiation from the sun by absorption. Satellite imagery from the satellites placed into orbit by NASA measures the ozone layer. Can you see how people might think there was a hole in the ozone by looking at the image below taken by one of NASA's satellites? It has been found that another set of gases, called CFCs, is responsible.

If this harmful ultraviolet radiation were not absorbed by the ozone layer, catastrophe would result. It has been estimated that a one percent decrease in the thickness of the ozone layer would result in a tremendous decrease in plant growth over Earth, a further decrease in amphibian populations, an additional 50 000 cases of skin cancer, and 100 000 cases of cataracts in people.

Think Critically:

What might be the results of efforts to have an international ban on the production and use of CFCs by the end of the century?

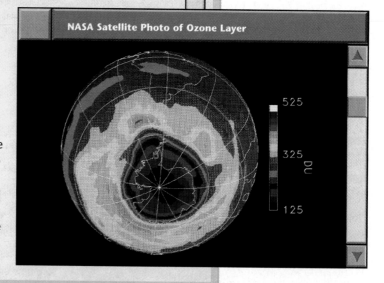

NASA Satellite Photo of Ozone Layer

525

325 DU

125

The Fern Life Cycle

In the life cycle of a fern, illustrated in **Figure 10-16,** the gametophyte and sporophyte do not depend on each other. Both forms of the fern plant can produce their own food. The fern plants that you may have seen growing in the woods are the sporophytes. A fern has leaves that grow above the ground from an underground stem called a **rhizome.** Roots grow from the rhizome to anchor the plant in the soil. The leaf of a fern is called a **frond.** Ferns produce spores in cases grouped together in structures called **sori** (*sing.,* sorus) on the lower sides of the mature fronds. The sori usually look like crusty rust, brown, or blackish-colored bumps. When a sorus opens, spores are released gradually as spore cases open. They land on damp soil or rocks and germinate into small, green, heart-shaped gametophyte plants. The gametophyte is called a

Science Journal

Life Cycles Have students describe, compare, and contrast the life cycles of mosses and ferns. **L1** **LS**

prothallus. The prothallus produces sex cells that unite to form the zygote. The zygote develops into the embryo, then sporophyte, as in the mosses. But in ferns, the gametophyte also produces new rhizomes, which can grow into the separate sporophyte fern plants you are familiar with.

Ferns have characteristics of both nonvascular and vascular plants. Like the bryophytes, ferns produce spores, yet they have vascular tissue like seed-bearing plants. Ferns have leaves, stems, and roots.

Figure 10-16

The fern's life cycle is similar in many ways to the life cycle of the moss. However, the sporophyte and gametophyte are both photosynthetic and capable of survival and growth without the other.

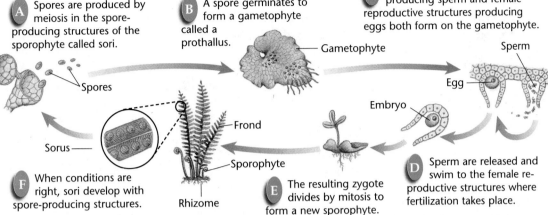

A Spores are produced by meiosis in the spore-producing structures of the sporophyte called sori.

B A spore germinates to form a gametophyte called a prothallus.

C Male reproductive structures producing sperm and female reproductive structures producing eggs both form on the gametophyte.

— Gametophyte

Sperm

Egg

Spores

Embryo

Sorus —

— Frond

— Sporophyte

Rhizome

D Sperm are released and swim to the female reproductive structures where fertilization takes place.

E The resulting zygote divides by mitosis to form a new sporophyte.

F When conditions are right, sori develop with spore-producing structures.

Section Wrap-up

Review

1. Describe the life cycle of mosses.

2. Explain the stages in the life cycle of a fern.

3. **Think Critically:** List ways simple plants affect your life each day. (HINT: Where do electricity and heat for homes come from?)

Skill Builder
Concept Mapping

Make a concept map showing how nonvascular and vascular plants are related. Include these terms in the concept map: *plant kingdom, bryophytes, liverworts, mosses, seedless plants, nonvascular plants, seedless vascular plants, club mosses, ferns, horsetails,* and *spike mosses*. If you need help, refer to Concept Mapping in the **Skill Handbook.**

USING MATH

There are approximately 8000 species of liverworts and 9000 species of mosses. Estimate what fraction of the bryophytes the mosses would be.

10-2 Seedless Plants **277**

Skill Builder

✔ Assessment

Portfolio Use this Skill Builder to assess students' abilities to form a concept map of the plant kingdom. Use the Performance Task Assessment List for Concept Map in **PASC,** p. 89. **P**

•MINI•QUIZ•

1. **The rootlike filaments that hold moss plants in place are called _____ .** *rhizoids*

2. **The continual cycle that alternates between the sporophyte and gametophyte of the moss plant is called _____ .** *alternation of generations*

3. **Club mosses, spike mosses, horsetails, and ferns are all included in the _____ vascular plants.** *seedless*

4. **The leaf of a fern is called a(n) _____ .** *frond*

Section Wrap-up

Review

1. Spores land on the ground and produce the upright, leafy gametophytes. Gametes are produced at the tops of the gametophytes. Sperm swim in water to reach the female gametes. When united, the gametes produce a zygote. The zygote grows into a spore-producing sporophyte that is dependent on the gametophyte.

2. A spore produces a heart-shaped gametophyte. The gametophyte produces gametes that unite to form a zygote. A zygote then produces an embryo that grows into a sporophyte.

3. **Think Critically** Simple plants such as mosses and ferns formed fossil fuels. These fuels are burned to produce electricity for homes, to power automobiles, and to provide heat.

USING MATH

5000 + 15 000 = 20 000 total bryophytes

15 000 ÷ 20 000 = 0.75 × 100 = 75%

TECHNOLOGY:
10•3 Cleaner Coal

Coal results from the fossilization of plant materials such as the mosses, liverworts, and ferns discussed in preceding sections of this chapter.

Preplanning

To prepare for Activity 10-2, obtain live mosses, liverworts, and ferns with gametophytes and sporophytes as well as prepared slides of these organisms.

1 Motivate

Bellringer

Before presenting the lesson, display **Section Focus Transparency 37** on the overhead projector. Assign the accompanying **Focus Activity** worksheet. L1 LEP

Tying to Previous Knowledge

Relate the burning of fossil fuels to the production of acid rain studied in Activity 10-1.

2 Teach

Demonstration

Visual-Spatial Use a map of the United States to illustrate the location of major coal deposits.

Activity

Logical-Mathematical Have students collect some rain or snow and determine its pH using pH paper.

Objectives

* Describe the benefits and problems of burning coal.
* Explain some of the methods for cleaning coal.

Figure 10-17

A cyclone collector uses gravity to reduce particle emissions.

Clean gas out

Gas and particles in

Collected dust out

Basic Cyclone Collector

Benefits and Problems of Burning Coal

In 1990, the United States obtained almost one-quarter of its energy from burning coal. Most coal is burned to produce electricity. Electrical utilities burn more coal than other fuels because coal is one-third to one-half as expensive as other fuels.

Coal is a dirty fuel because burning it releases many tons of particles and sulfur dioxide gas. In the 1970s, scientists found out for the first time that this sulfur dioxide gas was one cause of acid rain. With this knowledge, a focus was placed on learning to clean up the pollution created by burning fuel before more plants and fish died as a result.

Technologies for Cleaning Coal

Because the potential benefits of burning coal are so great, scientists have developed ways to reduce some of the problems. Fine particles and gases—two air pollutants—that are released by burning the coal are sources of air pollution requiring different solutions.

One way to reduce the particles released from burning coal is to allow them to be pulled into a collector by gravity. The device that accomplishes this is a cyclone collector, so named because it uses an air current to aid the settling of particles. Another way to reduce emissions is to introduce an electric field on a metal plate between the area where the coal is burned and the smokestack exit. The electrically charged particles of burned coal move toward the electric field. When they touch the metal, the particles lose their charge and fall into a collecting bin. The third way of removing particles is by filtering them into long bags that hang within the smokestack. By using these three methods, illustrated in **Figures 10-17, 10-18,** and **10-19,** some industries have lowered their emissions of particles by as much as 90 percent.

Gases are more difficult to clean than particles. The primary gas that scientists try to remove from coal smoke is sulfur dioxide. Within the past 15 years, considerable progress has

Program Resources

Reproducible Masters
Study Guide, p. 41 L1
Reinforcement, p. 41 L1
Enrichment, p. 41 L3
Activity Worksheets, pp. 60-61 L1
Science and Society Integration, p. 14

Transparencies
Section Focus Transparency 37 L1

Figure 10-18
Electrical fields are also used to collect dust particles.

Electrostatic Precipitator

Typical Bag Filter

Figure 10-19
Smokestacks contain bag filters for filtering out particles.

been made in achieving this goal. The main technology for cleaning sulfur dioxide is a scrubber. One of the most efficient methods is to channel the gas through water that contains limestone. In a chemical reaction, the limestone reacts with the sulfur dioxide, producing a "safer" compound, calcium sulfate. This chemical method for removing sulfur can be up to 90 percent efficient.

Section Wrap-up

Review

1. Why is coal used as the main energy source for generating electricity even though it is a dirty fuel?

2. What are the main problems with burning coal?

 Visit the Chapter 10 Internet Connection at Glencoe Online Science, **www.glencoe.com/sec/science/life,** for a link to more information about acid rain.

SCIENCE & SOCIETY

10-3 Cleaner Coal **279**

Activity 10-2

Comparing Mosses, Liverworts, and Ferns

Purpose

LS **Visual-Spatial** Students will observe the stages in the life cycles of mosses, liverworts, and ferns. **L1**

Process Skills
observing, classifying, comparing and contrasting

Time
50 minutes

Safety Precautions
Remind students not to eat any plant parts and be careful when using microscopes.

📁 **Activity Worksheets,** pp. 5, 60-61

Teaching Strategies
Alternate Materials Grow fern gametophytes by sprinkling the spores on a clay pot turned bottom up. First soak the pot overnight. Keep the pot in a plastic bag with water in the bottom to maintain a high humidity. Materials may be purchased from a biological supply house for little cost.

Troubleshooting Sometimes mold will overtake the fern gametophyte stages. Cultures should be discarded. Have additional specimens ready.

• If the moss spores are not liberated using water only, you may add an equal volume of glycerin to the water. This mixture will allow the spores to come out of the capsule more freely.

Answers to Questions
1. Rhizoids hold the plants to the ground.
2. Spores are found on the leaves of ferns, but in capsules on stalks of mosses and liverworts.
3. The gametophyte of the moss is a low-growing, leafy structure in a whorl around a stalk; that of the liverwort is a flat, leaflike

Mosses and liverworts make up the division of plants called bryophytes. Ferns make up the division pterophytes. Try this activity to observe the similarities and differences in these groups of plants.

Problem
How are mosses, liverworts, and ferns similar and different?

Materials

• live mosses, liverworts, and ferns with gametophytes and sporophytes
• prepared slides of mosses, liverworts, and ferns
• hand lens
• forceps
• dropper
• microscope slide and coverslip
• microscope
• dissecting needle
• pencil with eraser

Procedure
1. Examine prepared slides of sections of gametophytes and sporophytes of the three plants under the microscope.
2. Obtain a gametophyte of each plant. With a hand lens, observe the rhizoids, leafy parts, and stemlike parts, if any are present.
3. Obtain and use a hand lens to observe a sporophyte of each plant.
4. Locate the spore capsule on the moss plant. Remove the capsule and place it in a drop of water on the slide. Place a coverslip on the capsule. Use the eraser of a pencil to gently crush the capsule in order to release the spores. **CAUTION:** *Do not break the coverslip.* Observe the spores under low and high power.

5. Make labeled drawings of all observations and record your observations in a data table.

Data and Observations

Structure	Liverwort	Moss	Fern
Sporophyte			
Gametophyte			
Spores			

Analyze
1. What function do the rhizoids have?
2. Where are the spores formed on the moss? On a liverwort? On a fern?
3. Compare the gametophytes and sporophytes of each plant.

Conclude and Apply
4. What advantage is there to having spores shot out of capsules?
5. **Describe** the major difference between the life cycles of a moss and a fern.

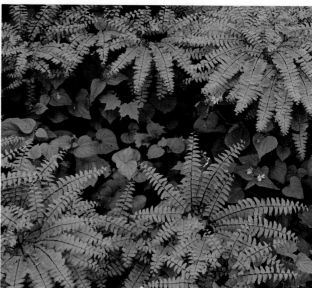

form. Sporophytes of mosses form a stalk with a capsule containing spores at the top. In liverworts, the sporophytes form umbrella-like structures with spores in cases underneath.
4. Shooting spores out of capsules helps to scatter them to other locations.
5. Gametophytes are the dominant stage in mosses and liverworts, but sporophytes are dominant in ferns.

✓ **Assessment**

Performance Assess students' understanding of the parts of mosses and liverworts by having them identify the parts of living specimens and describe a function for each. Use the Performance Task Assessment List for Making Observations and Inferences in **PASC**, p. 17. **P**

Chapter 10

Review

Summary

10-1: Characteristics of Plants

1. Plants are many-celled eukaryotes adapted to life on land.
2. Plants evolved from the plantlike protists.
3. Simple plants include nonvascular plants, the bryophytes, and vascular plants with vessels that conduct materials to plant cells.

10-2: Seedless Plants

1. Bryophytes—mosses and liverworts—reproduce through the process of alternation of generations, as do all plants.
2. Ferns, club mosses, and horsetails are seedless plants with vascular tissue.
3. Some nonvascular plants are pioneer species that help to break down rock into soil.

10-3: Science and Society: Cleaner Coal

1. Coal, a fuel composed of compressed plant material, can cause air pollution problems when burned.
2. Technologies to clean coal emissions include scrubbers and chemical methods to remove sulfur, one of the main components of air pollution produced by burning coal.

Key Science Words

a. alternation of generations
b. bog
c. cellulose
d. cuticle
e. frond
f. gametophyte
g. nonvascular plant
h. pioneer species
i. prothallus
j. rhizoid
k. rhizome
l. sorus
m. sporophyte
n. vascular plant

Reviewing Vocabulary

Match each phrase with the correct term from the list of Key Science Words.

1. plant that produces sex cells
2. wet ground made of dead, non-decaying plants
3. plant that produces spores
4. rootlike filament
5. underground fern stem
6. capsule made up of spore cases
7. plant containing vascular tissue
8. plant group including mosses and liverworts
9. the first plants to grow in new environments
10. waxy layer on stems and leaves

Chapter 10 Review   281

Summary

Have students read the summary statements to review the major concepts of the chapter.

Reviewing Vocabulary

1. f	**6.** l
2. b	**7.** n
3. m	**8.** g
4. j	**9.** h
5. k	**10.** d

Assessment

Portfolio Encourage students to place in their portfolios one or two items of what they consider to be their best work. Examples include:
- Skill Builder, p. 265
- Community Connection, p. 268
- Activity 10-2 drawings, p. 280 P

Performance Additional performance assessments may be found in **Performance Assessment** and **Science Integration Activities.** Performance Task Assessment Lists and rubrics for evaluating these activities can be found in Glencoe's **Performance Assessment in the Science Classroom.**

GLENCOE TECHNOLOGY

 MindJogger Videoquiz

Chapter 10 Have students work in groups as they play the Videoquiz game to review key chapter concepts.

Checking Concepts

1. d	**6.** a
2. b	**7.** a
3. a	**8.** c
4. d	**9.** b
5. d	**10.** d

Understanding Concepts

11. Burning coal produces pollutants that require scrubbers.
12. Alternation of generations appears to be an adaptation that assures that plants can survive periods of drought.
13. The sporophyte depends on the food produced by the gametophyte.
14. Land plant cells lose too much water and die.
15. Nonvascular plant cells absorb water directly from their surroundings. Vascular plant cells get water through vascular tissue.

Thinking Critically

16. Bogs are acidic; this kills the bacteria that normally cause decay.
17. Male gametes need water because they swim to female gametes during reproduction in simple plants.
18. *Sphagnum* moss retains water, thus allowing soil to hold water for plants growing in the garden soil.
19. Mosses absorb water directly into cells, so they must grow where it is wet.
20. Mosses begin to break down rocks, releasing the minerals that other plants need. These plants can then begin to grow on this new soil.

Developing Skills

21. **Concept Mapping** See student page.

282

Checking Concepts

Choose the word or phrase that completes the sentence.

1. Roots, cell walls of cellulose, and having many cells are examples of _____ characteristics.
   - a. moneran
   - b. protist
   - c. animal
   - d. plant
2. Plants probably evolved from _____.
   - a. animals
   - b. protists
   - c. fungi
   - d. monerans
3. _____ are plants only a few cells thick.
   - a. Bryophytes
   - b. Ferns
   - c. Horsetails
   - d. Vascular
4. _____ plants are plants with vessels.
   - a. Bryophyte
   - b. Moss
   - c. Nonvascular
   - d. Vascular
5. Nonvascular plants do not have _____.
   - a. zygotes
   - b. leaf-like structures
   - c. rhizoids
   - d. roots
6. Plants without vascular tissues are _____.
   - a. ferns
   - b. liverworts
   - c. moss
   - d. nonvascular
7. Ferns reproduce by forming _____.
   - a. spores
   - b. vascular tissue
   - c. seeds
   - d. flowers
8. _____ are seedless vascular plants.
   - a. Mosses
   - b. Liverworts
   - c. Horsetails
   - d. Oaks
9. Ferns do not have _____.
   - a. fronds
   - b. rhizoids
   - c. rhizomes
   - d. spores
10. All plants have a life cycle with _____.
    - a. seeds
    - b. fruits
    - c. flowers
    - d. alternation of generations

282 Chapter 10 Introduction to Plants

Understanding Concepts

Answer the following questions in your Science Journal using complete sentences.

11. Why are technologies needed to clean coal emissions?
12. Why is it an advantage for some plants to alternate generations?
13. Why are sporophytes dependent on the gametophytes in bryophytes?
14. What would happen if a land plant's waxy cuticle were destroyed?
15. Compare how the cells of nonvascular and vascular plants obtain water.

Thinking Critically

16. Human remains sometimes found in peat bogs are well preserved. Explain why this occurs.
17. Discuss the importance of water in reproduction of bryophytes and ferns.
18. Why is *Sphagnum* moss added to garden soil?
19. Explain why mosses are found on moist areas.
20. How do mosses change rocky areas so that other plants can take root?

22. **Classifying** Bryophytes: 3, 7, 9, 10; Seedless vascular: 1, 2, 4, 5, 6, 8
23. **Making and Using Graphs** See below.

Bryophytes 61%

Ferns 36%

Club mosses, spike mosses, and horsetails 3%

24. **Sequencing** Sporophyte produces haploid spores.
    Spores germinate into threadlike green structures.
    Gametophyte grows.
    Sex cells are produced.
    Sex cells unite.
    Zygote forms.
    Embryo develops.
    Embryo develops into sporophyte.

Developing Skills

If you need help, refer to the **Skill Handbook**.

21. **Concept Mapping:** Fill in the map showing the divisions of the Plant Kingdom.

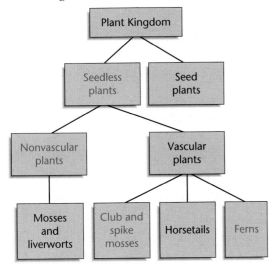

22. **Classifying:** Classify these plants in the correct group, bryophytes or seedless vascular: 1. club moss, 2. fern, 3. liverwort, 4. *Equisetum,* 5. horsetail, 6. *Lycopodium,* 7. moss, 8. spikemoss, 9. *Sphagnum,* 10. peat moss

23. **Making and Using Graphs:** Make a circle graph using the data below.

Simple Plant Species

Simple Plants	Number of Species
Bryophytes	20 000
Club mosses, spike mosses, and horsetails	1000
Ferns	12 000
Total	33 000

24. **Sequencing:** Put the steps of the moss life cycle in sequence starting with the sporophyte stage.
25. **Making and Using Tables:** Go for a walk in the woods and see how many different kinds of mosses there are. Make a table to record the locations and conditions under which each is growing.

Performance Assessment

1. **Model:** Develop a model of how coal is formed. Include a time line and a map of the parts of the U.S. where coal mines are located. Obtain samples of "soft" and "hard" coal and describe differences in their composition, origin, and formation.
2. **Design an Experiment:** Design an experiment using celery stalks with leaves, a sharp knife, water with food coloring, and a timer to determine the rate at which water moves up through vessels. Describe and record your experiment in your Science Journal.
3. **Graph from Data:** Graph the data from the MiniLAB on page 268 to compare the weight of the mass of *Sphagnum* before and after water is absorbed, or the amount of water in the beaker or jar before and after.

Chapter 10 Review

25. Answers will vary depending on location.

Performance Assessment

1. Student models will vary depending on materials used. Reference works will indicate where coal mines are located in the U.S. Use the Performance Task Assessment List for Model in **PASC**, p. 51. **P**
2. Student plans will vary considerably. Student records should show a plan for the experiment, materials, and how they would time the activity. Students will need to set time intervals as part of their experimental plan. One suggestion is that they put at least three celery stalks in colored water all at the same time. Then, at a given interval, one stalk can be removed and cut at a particular distance from the bottom of the stalk to see if the color has moved that far up the stalk. The second stalk can be removed and cut at a predetermined time interval. The third stalk should be left and the time checked when color first appears in the leaves. Use the Performance Task Assessment List for Designing an Experiment in **PASC**, p. 23. **P**
3. Data would be best illustrated with a bar graph. Use the Performance Task Assessment List for Graph from Data in **PASC**, p. 39. **P**

Assessment Resources

 Reproducible Masters
Chapter Review, pp. 23-24
Assessment, pp. 68-71
Performance Assessment, p. 24

Glencoe Technology
- **Chapter Review Software**
- **Computer Test Bank**
- **MindJogger Videoquiz**

Chapter Organizer

Section	Objectives/Standards	Activities/Features
Chapter Opener		Explore Activity: Dissect a seed to find out what is inside. p. 285
11-1 Seed Plants (2 sessions, 1 block)*	1. **List** the characteristics of seed plants. 2. **Describe** the main characteristics of gymnosperms and angiosperms and their importance. 3. **Compare** monocots and dicots. National Content Standards: (5-8) UCP1, UCP4, A1, B1, C1	Connect to Chemistry, p. 290 Skill Builder: Comparing and Contrasting, p. 291 Using Math, p. 291
11-2 Parts of Complex Plants (1½ sessions, 1 block)*	1. **Describe** the structures of roots, stems, and leaves. 2. **Describe** the functions of roots, stems, and leaves. 3. **Explain** some adaptations of roots, stems, and leaves to various environments. National Content Standards: (5-8) UCP1, UCP5, C1, C5, E2, F5	Using Technology: Skinning Your Fruit, p. 293 Skill Builder: Classifying, p. 295 Using Computers: Word Processing, p. 295
11-3 Seed Plant Reproduction (2½ sessions, 2 blocks)*	1. **Sequence** the stages in the life cycles of typical gymnosperms and angiosperms. 2. **Describe** the structure and function of the flower. 3. **Describe** methods of seed dispersal in seed plants. National Content Standards: (5-8) UCP1, UCP5, A1, A2, C1, C2, C5	MiniLAB: What's in a pinecone? p. 297 Connect to Physics, p. 297 Activity 11-1: Comparing Flowers and Seeds, p. 300 MiniLAB: How do plants disperse seeds? p. 301 Problem Solving: Are seed company claims true? p. 302 Skill Builder: Hypothesizing, p. 303 Science Journal, p. 303 Activity 11-2: Germination Rate of Seeds, pp. 304-305
11-4 Science and Society Issue: Medical Treasures (1 session, ½ block)*	1. **Describe** some plants that have value as medicines. 2. **Compare** and **contrast** views about preserving species that may or may not have value to humans. National Content Standards: (5-8) C5, F1, F4, G1	Explore the Issue, p. 307 People and Science: Mark Plotkin, Ethnobotanist, p. 308

* A complete Planning Guide that includes block scheduling is provided on pages 32T-35T.

Activity Materials

Explore	Activities	MiniLABs
page 285 lima bean or other large seed, small glass, tap water, scalpel, hand lens	page 300 apron, monocot and dicot flowers, monocot and dicot seeds, scalpel, forceps, iodine solution pages 304-305 seeds, distilled water, salt, potting soil, plant trays or plastic cups, seedling-warming cables, thermometer, 100-mL graduated cylinder, beakers, apron, safety goggles	page 297 hand lens; gymnosperm cone; large, white paper towel; drawing paper page 301 books on seed plants

Need Materials? Call Science Kit (1-800-828-7777).

Teacher Classroom Resources

Reproducible Masters	Transparencies	Teaching Resources
Study Guide, p. 42 **Reinforcement,** p. 42 **Enrichment,** p. 42 **Science and Society Integration,** p. 15	**Section Focus Transparency 38,** A Frozen Garden **Teaching Transparency 21,** Monocots and Dicots	**Glencoe Life Science Interactive Videodisc** **Life Science CD-ROM** **Spanish Resources** **English/Spanish Audiocassettes** **Cooperative Learning Resource Guide** **Lab Partner** **Lab and Safety Skills** **Lesson Plans**
Study Guide, p. 43 **Reinforcement,** p. 43 **Enrichment,** p. 43 **Lab Manual 24,** Root Structure and Function **Science Integration Activity 11,** Using Growth Rings as Indicators of Climate	**Section Focus Transparency 39,** Map "Root" Across the USA	
Study Guide, p. 44 **Reinforcement,** p. 44 **Enrichment,** p. 44 **Activity Worksheets,** pp. 64-65, 66-67, 68, 69 **Lab Manual 25,** Parts of a Fruit **Lab Manual 26,** Parts of a Seed **Lab Manual 27,** Plant Growth **Concept Mapping,** p. 27 **Critical Thinking/Problem Solving,** p. 15 **Multicultural Connections,** pp. 25-26	**Section Focus Transparency 40,** Bullseye! **Science Integration Transparency 11,** How Plants Hitch a Ride **Teaching Transparency 22,** Parts of a Flower	**Assessment Resources** **Chapter Review,** pp. 25-26 **Assessment,** pp. 72-75 **Performance Assessment in the Science Classroom (PASC)** **MindJogger Videoquiz** **Alternate Assessment in the Science Classroom** **Performance Assessment** **Chapter Review Software** **Computer Test Bank**
Study Guide, p. 45 **Reinforcement,** p. 45 **Enrichment,** p. 45 **Cross-Curricular Integration,** p. 15 **Technology Integration,** pp. 11-12	**Section Focus Transparency 41,** Life-Saving Plants	

Key to Teaching Strategies

The following designations will help you decide which activities are appropriate for your students.

L1 Level 1 activities should be within the ability range of all students, including those with learning difficulties.

L2 Level 2 activities should be within the ability range of the average to above-average student.

L3 Level 3 activities are designed for the ability range of above-average students.

LEP LEP activities should be within the ability range of Limited English Proficiency students.

LS These activities are designed to address different learning styles.

COOP LEARN Cooperative Learning activities are designed for small group work.

P These strategies represent student products that can be placed into a best-work portfolio.

GLENCOE TECHNOLOGY

The following multimedia resources are available from Glencoe.

Science and Technology Videodisc Series (STVS)
Chemistry
 Straw as Feed
 Oil from Wood
Plants & Simple Organisms
 Managing a Forest
 Bats as Pollinators
 Seed Banks
 New Grains
 Plant Clones
 Plant Chemical Repels Cockroaches

 Insecticide from Desert Plants
 Bitter Melon Cancer Treatment
 Rubber from Guayule
The Infinite Voyage Series
To the Edge of the Earth
National Geographic Society Series
STV: Plants
STV: Biodiversity
Life Science CD-ROM

Teacher Classroom Resources

This is a representation of key blackline masters available in the Teacher Classroom Resources.

Teaching Aids

Section Focus Transparencies

Science Integration Transparencies

Teaching Transparencies

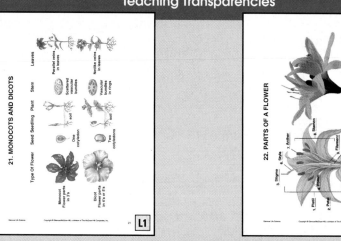

Meeting Different Ability Levels

Study Guide

Reinforcement

Enrichment Worksheets

Hands-On Activities

Science Integration Activity

Lab Manual

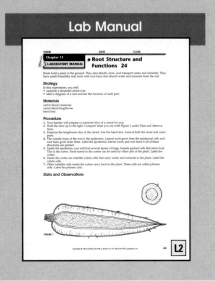

Assessment

Performance Assessment

Enrichment and Application

Critical Thinking/ Problem Solving

Cross-Curricular Integration

Science and Society Integration

Technology Integration

Multicultural Connections

Concept Mapping

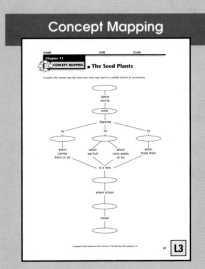

CHAPTER OVERVIEW

Section 11-1 The characteristics of seed plants, gymnosperms, and angiosperms are listed in this section. The importance of the gymnosperms and angiosperms is discussed. A comparison of evolution of gymnosperms and angiosperms is given.

Section 11-2 This section describes the structure and function of roots, stems, and leaves. Some of the unique adaptations of these plant parts are explained.

Section 11-3 Reproduction of gymnosperms and angiosperms is discussed in this section. The structure and function of the flower are described, and methods of seed dispersal are given.

Section 11-4 Science and Society The importance of plants for medicinal properties is discussed.

Chapter Vocabulary

gymnosperm	ovule
angiosperm	pollen grain
monocot	stamen
dicot	pistil
xylem	pistil
phloem	ovary
cambium	pollination
stomata	ethnobotanist
guard cell	

Theme Connection

Systems and Interactions/Stability and Change This chapter emphasizes the roles gymnosperms and angiosperms play in the environment. In addition, stability and change are used in a discussion of the phylogeny of these plants.

Previewing the Chapter

Section 11-1 Seed Plants
▶ What is a seed plant?
▶ Gymnosperms
▶ Angiosperms
▶ Origin and Evolution of Seed Plants
▶ Importance of Seed Plants

Section 11-2 Parts of Complex Plants
▶ Roots
▶ Stems
▶ Leaves

Section 11-3 Seed Plant Reproduction
▶ Gymnosperm Reproduction
▶ Angiosperm Reproduction

Section 11-4 Science and Society
  Issue: Medical Treasures

284

Learning Styles

Look for the following logo for strategies that emphasize different learning modalities. **LS**

Kinesthetic	Enrichment, p. 289; Demonstration, pp. 293, 306
Visual-Spatial	Explore, p. 285; Visual Learning, pp. 287, 288, 290, 292, 294, 296, 298, 299, 301; Demonstration, pp. 287, 288, 298; Activity, pp. 292, 293; Across the Curriculum, p. 289; MiniLAB, p. 297; Activity 11-1, p. 300
Interpersonal	People and Science, p. 308
Logical-Mathematical	Using Math, p. 291; Activity 11-2, pp. 304-305
Linguistic	Across the Curriculum, p. 298; MiniLAB, p. 301

Chapter 11

The Seed Plants

In Chapter 10, you learned about the different kinds of plants that do not produce seeds. However, most of the plants that you are familiar with do produce seeds. Look at the photo on the opposite page and notice the variety of seeds. Some seeds are found in cones, but most are produced by flowering plants. To find out more about seeds, try the activity below. Then in the chapter that follows, learn about the different parts of seed plants and how they reproduce.

EXPLORE ACTIVITY

Dissect a seed to find out what is inside.

1. Obtain a lima bean or other large seed.
2. If the seed is hard and dry, place it in water overnight.
3. Observe the seed and write your observations in your Science Journal.
4. Remove the paper-like covering from the seed and carefully pull apart the halves of the seed.
5. Look for the parts of the embryo plant on one side of the seed.

Observe: In your Science Journal, identify the different parts of the embryo plant.

Previewing Science Skills

▶ In the **Skill Builders**, you will compare and contrast, classify, and hypothesize.

▶ In the **Activities**, you will observe, analyze, collect and interpret data, and compare.

▶ In the **MiniLABs**, you will observe and identify.

285

EXPLORE ACTIVITY

Purpose

Visual-Spatial Use the Explore Activity to introduce students to the parts of seeds. Inform students that they will learn more about seeds and the plants that make seeds in this chapter. **L1** **LEP**

Preparation

Purchase lima beans from a grocery produce section.

Materials

Each student will need one lima bean seed.

Teaching Strategies

As you teach the parts of the seed, use the scientific names for the parts. The shell is a dried ovary; lima beans are seeds; seed leaves are cotyledons, etc.

Observe

Students should identify the embryo, its leaves, stem, and root. They should also indicate the presence of the seed covering and the stored food.

✓ Assessment

Performance To further assess students' understanding of seed parts, have them draw and label the embryo, stem, roots, and endosperm of the seed. Use the Performance Task Assessment List for Scientific Drawing in **PASC**, p. 55. **P**

Assessment Planner

Portfolio
Refer to page 309 for suggested items that students might select for their portfolios.

Performance Assessment
See page 309 for additional Performance Assessment options.
Skill Builders, pp. 291, 295, 303
MiniLABs, pp. 297, 301
Activities 11-1, p. 300; 11-2, pp. 304-305

Content Assessment
Section Wrap-ups, pp. 291, 295, 303, 307
Chapter Review, pp. 309-311
Mini Quizzes, pp. 291, 295, 303

Group Assessment
Opportunities for group assessment occur with Cooperative Learning Strategies and Flex Your Brain Activities.

Prepare

Section Background

- There are fewer than 1000 species of gymnosperms. Of these, about 550 are conifers, including pines, firs, junipers, and cedars.
- A single giant redwood may produce as much as 480 000 board feet of lumber.
- The economic value of seed plants to humans is practically incalculable. They are sources of food, shelter, and nourishment.

1 Motivate

Bellringer

 Before presenting the lesson, display **Section Focus Transparency 38** on the overhead projector. Assign the accompanying **Focus Activity** worksheet. L1 LEP

Demonstration

Line up several boxes of cereal. Have students identify the plants that produced the contents.

NATIONAL GEOGRAPHIC SOCIETY

 Videodisc

STV: Plants
What Is a Plant?
Unit 1, Side 1
Introduction

01061-0352

What is a seed plant?

Have you ever eaten vegetables like those shown in **Figure 11-1**? These vegetables include many varieties of potatoes, peppers, corn, and other foods. In the Philippines and Malaysia, coconut milk and banana flowers are used in many dishes. Corn and beans are staple foods in Mexico. All of these foods come from seed plants. What fruits and vegetables have you eaten today? If you had an apple, a peanut butter and jelly sandwich, and a glass of orange juice for lunch, you ate foods that came from seed plants.

Nearly all the plants you are familiar with are seed plants. Seed plants have roots, stems, leaves, and vascular tissue. There are more than 250 000 known species of seed plants in the world. What makes a seed plant different from simple plants is that it grows from a seed. A seed is the reproductive part of a plant that contains a plant embryo and stored food. Within a seed are all the parts needed to produce a new plant.

Most scientists classify seed plants into two major groups, the gymnosperms and angiosperms. Both of these groups produce seeds, but you will see that they have some important differences.

Gymnosperms

The oldest trees alive today are gymnosperms (JIHM nuh spurmz)—in fact, there's one bristlecone pine tree in the White Mountains of eastern California that is 4900 years old. **Gymnosperms** are vascular plants that produce seeds on the scales of female cones. The word *gymnosperm* comes from the Greek language and means "naked seed." Seeds of gymnosperms are not protected by a

Figure 11-1

Plants, like these being sold at a market in Ecuador, provide food for humans. *How are plants an important part of the world's food supply?*

Science Words

gymnosperm
angiosperm
monocot
dicot

Objectives

- List the characteristics of seed plants.
- Describe the main characteristics of gymnosperms and angiosperms and their importance.
- Compare monocots and dicots.

Program Resources

 Reproducible Masters
Study Guide, p. 42 L1
Reinforcement, p. 42 L1
Enrichment, p. 42 L3
Science and Society Integration, p. 15

 Transparencies
Section Focus Transparency 38 L1
Teaching Transparency 21

fruit. Gymnosperms do not produce flowers. Leaves of most gymnosperms are needlelike or scalelike. Most gymnosperms are evergreen plants that keep their leaves for several years.

Gymnosperms include four divisions of plants—the conifers, cycads, gingkoes, and gnetophytes. Of these, you are probably most familiar with the conifers—the pines, firs, spruces, cedars, and junipers, which make up the division Coniferophyta. This division contains the greatest number of species of gymnosperms. A few conifers produce male and female cones on separate trees, but most species produce both types of cones on the same tree. **Figure 11-2** shows examples of the four divisions of gymnosperms.

Figure 11-2

The gymnosperms include conifers, cycads, ginkgoes, and gnetophytes. Each of these plants is similar in that they generally produce naked seeds in cones.

 A Conifers are the largest, most diverse division of the gymnosperms. Most conifers are evergreen plants.

B Today, the ginkgoes are represented by only one living species.

C About 100 species of cycads exist today. The only present-day species that grows in the United States is found in Florida.

 D Most living gnetophytes can be found in the deserts or mountains of Asia, Africa, and Central or South America.

11-1 Seed Plants   287

2 Teach

Demonstration

LS **Visual-Spatial** Show students a pinecone, a rose, and a walnut. Ask students to explain what the plants that produce these structures all have in common. **LEP**

? FLEX Your Brain

Use the Flex Your Brain activity to have students explore SEED PLANTS.

📁 **Activity Worksheets,** p. 5

Visual Learning

Figure 11-1 How are plants an important part of the world's food supply? *Plants are a food source for humans as well as animals, which are in turn, consumed by humans.*

NATIONAL GEOGRAPHIC SOCIETY

💿 **Videodisc**
STV: Plants
What Is a Plant?
Unit 1, Side 1
Types of Plants

11765-18410

GLENCOE TECHNOLOGY

💿 **CD-ROM**

Life Science CD-ROM
Have students perform the interactive exploration for Chapter 11 to reinforce important chapter concepts and thinking processes.

Figure 11-3

Flowers of angiosperms come in an astonishing array of forms and colors. The orchid flowers shown here are only a few of the thousands of species of monocots. An even greater number of dicots exists.

Angiosperms

When you bite into a pear or an apple, you are eating part of an angiosperm (AN jee uh spurm). **Angiosperms** are vascular plants in which the seed is enclosed inside a fruit. A fruit is a ripened part of the plant where seeds are formed. If you eat a tomato, you'll find seeds inside. All angiosperms produce flowers, as shown in **Figure 11-3**. More than half of all known plant species are angiosperms.

Monocots and Dicots

There are two classes of angiosperms, or flowering plants: the monocots and the dicots. The terms *monocot* and *dicot* are shortened forms of the words *monocotyledon* and *dicotyledon*. The prefix *mono* means "one" and *di* means "two." A cotyledon is a seed leaf inside a seed. Therefore, **monocots** have one seed leaf inside their seeds and **dicots** have two. **Figure 11-4** compares the characteristics of monocots and dicots.

Monocots

Figure 11-4

The characteristics of monocots and dicots can be compared by observing differences in their plant parts.

A Monocots, such as these tiger lilies, are flowering plants with flower parts in threes. If you had cereal for breakfast this morning, you ate a bowlful of parts of monocot seeds. Cereal grains such as corn, rice, oats, and wheat are all monocots.

B In monocots, vascular bundles are scattered throughout the stems and show up as parallel veins in their narrow leaves.

 Seed

 Seedling

288 Chapter 11 The Seed Plants

Community Connection

Supermarket Survey Have students make a survey of the kinds of plants they can find at a supermarket. Discuss the list in class. Are more gymnosperms, monocots, or dicots on the list? **L1**

Origin and Evolution of Seed Plants

A seed is an adaptation that gives plants the ability to survive on land away from bodies of water. Gymnosperms and angiosperms today are found in tropical, temperate, and even extremely dry or cold environments.

Gymnosperms probably evolved from a group of plants that grew about 350 million years ago. Cycads and conifers have been found in the fossil record dated to the Paleozoic era, 200 million years ago. Flowering plants did not exist until the Cretaceous period, about 120 million years ago. Gaps in the fossil record prevent scientists from tracing the exact origins of flowering plants. After their appearance in the Cretaceous period, angiosperms evolved into many different forms. Today, they are the most common form of plant life on Earth.

Dicots

C Dicots, such as the *Lavatera* pictured, are flowering plants with flower parts in fours or fives. Examples of dicots include trees like oaks and maples, vegetables like lettuce and beans, fruits like watermelons and oranges, and many garden flowers.

D In dicots, vascular bundles occur in rings inside the stems. These bundles show up as branching, netlike veins in the broad dicot leaves.

Seed

Seedling

11-1 Seed Plants   **289**

Inquiry Question

Ask students to explain why scientists agree that seed plants evolved after mosses and ferns. *Seed plants are more complex. The fossil record contains fossils of mosses and ferns that are much older than any seed plant.*

Enrichment

�456 Kinesthetic Bring in several different kinds of seeds. Mix all the seeds together. Allow students to choose a seed and grow the plant. Have them identify the plant as a monocot, dicot, or gymnosperm as soon as the characteristics become visible. **L1 COOP LEARN**

Across the Curriculum

Art   Have students produce watercolor paintings or colored pencil drawings that are representative of particular plants. You may choose the style of the work or leave it up to the students. **L1 ⎎**

Inquiry Questions

- **Which groups of gymnosperms are not conifers?** *cycads, ginkgos, and gnetophytes*

- **Which group of seed plants is most fine furniture made from?** *Angiosperms (particularly hardwoods) because they have denser wood than gymnosperms.*

3 Assess

Check for Understanding

Ask students to compare the monocots and dicots according to their characteristics.

Reteach

Emphasize the important characteristics of the monocots and dicots again. Place several monocots and dicots around the room. Write a hint on an index card for each one. (Example: Notice the parallel veins in the leaves of this plant.) Have students identify them by using the hints.

Extension

📁 For students who have mastered this section, use the **Reinforcement** and **Enrichment** masters.

Throughout history, different cultures have depended on grains as the basis of their diets. *Identify* the complex carbohydrate that is common to many grains such as corn, beans, and rice.

Importance of Seed Plants

Imagine that your class is having a picnic in the park. You cover the wooden picnic table with a red-checked tablecloth and pass out paper cups and plates. You toast hot dogs and buns, eat potato chips, and drink apple cider. Perhaps you collect leaves or flowers for a science project. Later, you clean up and put leftovers in paper bags.

Now, let's imagine this scene if there were no seed plants on Earth. There would be no wooden picnic table and no pulp to make paper products such as cups, plates, and bags. Bread for buns, apples for cider, and potatoes for chips all come from plants. The tablecloth is made from cotton, a plant. Without seed plants, there would be no picnic!

Uses of Gymnosperms

On a global scale, conifers are the most economically important gymnosperms. As in **Figure 11-5,** most of the wood used for building construction and for paper production comes from conifers such as pines and spruces. Resin, a waxy substance secreted by conifers, is used to make chemicals found in soap, paint, varnish, and some medicines.

Figure 11-5

Gymnosperms such as pines (A) are commonly used in construction. (B) Resin, secreted by gymnosperms, is used to make household products such as paint and varnish.

290   Chapter 11 The Seed Plants

Cultural Diversity

Cultural Grains Use the Connect to Chemistry feature to broaden students' awareness of the fact that many cultures depend on a specific grain. Discuss some examples. **What grains does the United States depend on?** *corn and wheat* Have students find out about millet, sorghum, and amaranth.

Uses of Angiosperms

The most common plants on Earth are the angiosperms. They are important to all life because they form the basis for the diets of most animals. Grains such as barley and wheat and legumes such as peas and lentils were among the first plants ever grown by humans. Angiosperms also take in huge amounts of carbon dioxide for photosynthesis and release oxygen needed by other organisms. In addition to these important contributions, angiosperms are also the source of many of the fibers used in clothing such as flax and cotton fibers, shown in **Figure 11-6.** The production of medicines, rubber, oils, perfumes, pesticides, and some industrial chemicals also uses materials found in angiosperms.

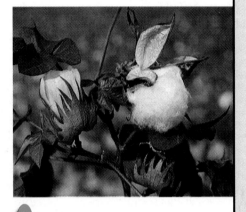

A

B

Figure 11-6

Cotton (A) is a flowering plant that yields long fibers that can be woven into cloth. A wide variety of cotton fabrics (B) can be produced. The differences are due to the tightness of the weave, the length of the fibers, and similar factors.

Section Wrap-up

Review

1. What are the characteristics of a seed plant?

2. Make a chart to compare the characteristics of gymnosperms and angiosperms.

3. **Think Critically:** The seed of the ginkgo has a fleshy covering that has an unpleasant smell, so only male trees are planted near homes. Why?

Skill Builder
Comparing and Contrasting

Monocots and dicots have some similarities and differences. Compare and contrast these two classes of plants. If you need help, refer to Comparing and Contrasting in the **Skill Handbook.**

USING MATH

How much time elapsed between the evolution of gymnosperms and the evolution of flowering plants?

11-1 Seed Plants   **291**

Skill Builder

Both monocots and dicots produce flowers. Monocots have one seed leaf inside their seeds, while dicots have two. Monocots have parallel veins in their leaves, and dicots have netlike veination. Monocots have flower parts in threes, and dicots have their flowering parts in fours or fives.

Assessment

Performance To assess students' abilities to compare and contrast monocots and dicots, have them give examples of monocots and dicots. Use the Performance Task Assessment List for Poster in **PASC**, p. 73. **P**

Prepare

Section Background

- The sweet potato is a fleshy root that contains stored food, but the Irish potato is a fleshy underground stem that contains stored food.
- Cactus spines are modified leaves that prevent water loss. Another modification of leaves is food storage, as seen in celery and rhubarb.

1 Motivate

Bellringer

Before presenting the lesson, display **Section Focus Transparency 39** on the overhead projector. Assign the accompanying **Focus Activity** worksheet. [L1] [LEP]

Activity

[LS] **Visual-Spatial** Give each student a piece of carrot, asparagus, and lettuce. Ask what part of a plant each comes from and what they have in common. Let the students eat their plant parts to see that each part stores food. [L1] [LEP]

Visual Learning

Figure 11-7 Use this illustration to explain to students that most plants have as much mass in the root system anchoring the plant as in the above-ground parts. [LEP] [LS]

Use Lab 24 in the **Lab Manual** as you teach this lesson.

11•2 Parts of Complex Plants

Science Words

xylem
phloem
cambium
stomata
guard cell

Objectives

- Describe the structures of roots, stems, and leaves.
- Describe the functions of roots, stems, and leaves.
- Explain some adaptations of roots, stems, and leaves to various environments.

Figure 11-7

The root system of a dandelion is longer than the plant is tall. When you pull up a dandelion, you often pull off the top portion of the plant. The root quickly produces new leaves and the dandelion problem remains.

Roots

Although there are differences between gymnosperms and angiosperms, they do share the same basic structures—roots, stems, and leaves. Imagine a large tree growing alone on top of a hill. What is the largest part? Chances are you named the trunk or the branches. But did you consider the roots? The root systems of most plants are as large or larger than the aboveground stems and leaves. Why are root systems, like the dandelion in **Figure 11-7,** so large?

Why are roots so important to plants? Water and minerals used by a plant enter by way of its roots. Roots have vascular tissue to move water and minerals from the ground up through the stems to the leaves. Roots also anchor plants in soil. If they didn't, plants could be blown away by wind or washed away by water. Each root system must support the plant parts that are above the ground—the stem, branches, and leaves of a tree, for example. In **Figure 11-7,** you can see the root system of a dandelion compared with the rest of the plant.

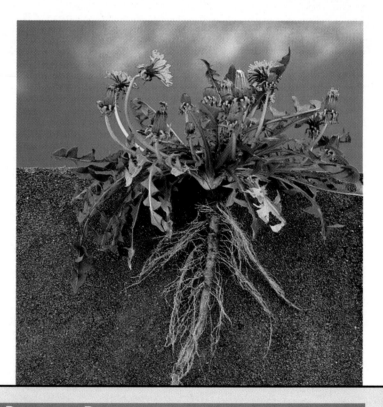

292

Program Resources

 Reproducible Masters
Study Guide, p. 43 [L1]
Reinforcement, p. 43 [L1]
Enrichment, p. 43 [L3]
Lab Manual 24
Science Integration Activities, pp. 21-22

Transparencies
Section Focus Transparency 39 [L1]

Roots also store food. When you eat carrots or beets, you eat roots made up of stored food. Root tissues may also perform special functions such as absorbing oxygen for the process of photosynthesis.

Stems

Did you know that the trunk of a tree is really its stem? Stems are the aboveground parts of plants that support leaves and flowers. The main functions of a stem are to support the plant and to allow movement of materials between leaves and roots. Some stems also have adaptations that allow them to store food. Potatoes and onions are underground stems with stored food. Sugarcane has an aboveground stem that stores large quantities of food. Stems of cacti are adapted to carry on photosynthesis and make food for the rest of the plant.

Peeling Citrus Fruit

11-2 Parts of Complex Plants   **293**

Use **Science Integration Activities,** pp. 21-22, as you teach this lesson.

NATIONAL GEOGRAPHIC SOCIETY

 Videodisc

STV: Plants
What Is a Leaf?
Unit 5, Side 3
Leaves—Vital to All

46016-52678

3 Assess

Check for Understanding

? FLEX Your Brain

Use the Flex Your Brain activity to have students explore PLANT PARTS.

 Activity Worksheets, p. 5

Reteach

Observe prepared slides of leaf cross sections under the microscope. Have students draw and label the parts. Ask students to explain the function of each leaf part.

Extension

For students who have mastered this section, use the **Reinforcement** and **Enrichment** masters.

294

Figure 11-8

The vascular tissue of seed plants includes xylem, phloem, and cambium. *Which of these tissues transports food throughout the plant?*

A Phloem transports sugar throughout the plant.

B Cambium produces additional xylem and phloem.

C Xylem transports minerals and water throughout the plant.

Wood

Bark

Vascular Tissue

Three main tissues make up the vascular system in the roots, stems, and leaves. **Xylem** (ZI lum) tissue is made up of tubular vessels that transport water and minerals up from the roots throughout the plant. **Phloem** (FLOH em) is a plant tissue made up of tubular cells. Phloem cells move food from leaves and stems, where it is made, to other parts of the plant for direct use or storage. Between xylem and phloem is the cambium. **Cambium** is a tissue that produces new xylem and phloem cells. All three tissues are illustrated in **Figure 11-8.**

Plant stems are either herbaceous or woody. Herbaceous stems are soft and green. Examples are the stems of peppers, corn, and tulips. Oak, birch, and other trees and shrubs have woody stems. Woody stems are hard and rigid.

Leaves

Have you ever rested in the shade of a tree's leaves on a hot, summer day? Leaves are the organs of the plant that usually trap light and make food through the process of photosynthesis.

Leaf Structure

Look at the structure of the leaf shown in **Figure 11-9.** The epidermis is a thin layer of cells that covers and protects both the upper and lower surfaces of a leaf. A waxy cuticle that protects the plant from wilting or drying out covers the epidermis of many leaves. Another adaptation of leaves is small pores in the leaf surfaces called **stomata.** Stomata allow carbon dioxide, water, and oxygen to enter and leave a leaf. The stomata are surrounded by **guard cells** that open and close the pores. The cuticle, stomata, and guard cells all are adaptations that enabled plants to survive on land by conserving water.

Leaf Cells

Inside a typical leaf are different layers of cells. One layer, the palisade layer, has rows of closely packed cells just below the upper epidermis. Palisade cells are packed with chloroplasts filled with chlorophyll. Most of the food made by leaves is made here. Between the palisade cells and the lower epidermis is a layer of loosely arranged cells, the spongy layer, and many air spaces. Xylem and phloem are located in this layer.

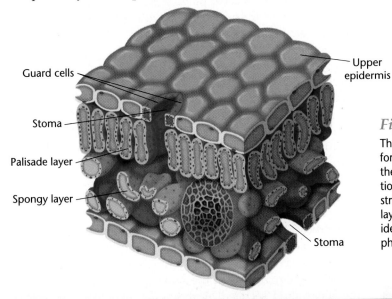

Guard cells

Stoma

Palisade layer

Spongy layer

Upper epidermis

Stoma

Figure 11-9

The layers of a leaf are adapted for photosynthesis. Notice that the palisade layer faces the direction from which the most light strikes the plant. The spongy layer provides the carbon dioxide needed by the other cells for photosynthesis.

Section Wrap-up

Review

1. What are the functions of roots, stems, and leaves?

2. What is the advantage of having most of the chloroplasts in cells toward the top of a leaf?

3. Think Critically: The cuticle and epidermis are transparent. How is this an advantage for a plant?

Skill Builder
Classifying
From what you have learned in this section, classify the following vegetables as roots, stems, or leaves: turnip, lettuce, asparagus, potato, and carrot. If you need help, refer to Classifying in the **Skill Handbook.**

Using Computers

Word Processing
Use a word processing program to make an outline of the structures and functions that are associated with roots, stems, and leaves.

Skill Builder
turnip—underground stem; lettuce—leaves; asparagus—stem; potato—underground stem; carrot—root

Assessment

Performance Assess students' abilities to classify vegetables by structure by giving them actual vegetables to classify. Use the Performance Task Assessment List for Making and Using a Classification System in **PASC,** p. 49. **P**

4 Close

•MINI•QUIZ•

Use the Mini Quiz to check students' recall of chapter content.

1. The _____ have vascular tissue to move water from the ground up to the leaves. *roots*

2. A _____ supports the above-ground parts of the plant. *stem*

3. Most of the food made by leaves is made in the _____ . *palisade layer*

Section Wrap-up

Review

1. Roots must carry water and minerals to other plant parts. Stems continue the transport of water and minerals to leaves. Leaves produce food from sunlight, water, and CO_2. Each part may be specialized for other functions, such as storing food or protecting the plant.

2. The top is where most sunlight hits, and therefore, where most photosynthesis takes place.

3. **Think Critically** Cuticles are transparent to allow light to reach the cells underneath for photosynthesis.

Using Computers

Answers will vary. For example, students may make a partial outline such as:

I. Roots
  A. Vascular tissue
    1. moves water
    2. moves minerals

Prepare

Section Background

Flowers that contain both stamens and pistils are perfect. Imperfect flowers may be either male (if they have only stamens) or female (if they have only pistils).

Preplanning

• To prepare for the MiniLAB on page 297, obtain cones with seeds.

• Activity 11-1 requires monocot and dicot flowers and seeds.

• To prepare for the MiniLAB on page 301, obtain ten different types of seeds.

• Activity 11-2 requires seeds, potting soil, plant trays, and seedling warming cables.

1 Motivate

Bellringer

Before presenting the lesson, display **Section Focus Transparency 40** on the overhead projector. Assign the accompanying **Focus Activity** worksheet. L1 LEP

Tying to Previous Knowledge

Anyone who has blown the fluff off a dandelion stem has acted as a seed dispersal mechanism. Ask students to recall any other times they have helped in the dispersal of seeds. One answer may be eating a watermelon.

Visual Learning

Figure 11-11 Aside from bats, what other animals might pollinate plants at night? *Hummingbird moths and some mammals.* LEP IS

11•3 Seed Plant Reproduction

Science Words

ovule
pollen grain
stamen
pistil
ovary
pollination

Objectives

• Sequence the stages in the life cycles of typical gymnosperms and angiosperms.

• Describe the structure and function of the flower.

• Describe methods of seed dispersal in seed plants.

Figure 11-10

Male cones of a Norway spruce tree release clouds of tiny pollen grains. These pollen grains unite with eggs to form zygotes.

296 Chapter 11 The Seed Plants

Gymnosperm Reproduction

Have you ever collected pinecones in the fall to use in making decorations? If you have, you've noticed that pinecones come in many different shapes and sizes. Many pine trees can be identified by looking at their cones.

Pine trees are typical gymnosperms. Each pine tree produces male cones and female cones on the sporophyte tree. Female cones consist of a spiral of woody scales on a short stem. Two **ovules,** the female reproductive part of the plant, are produced on top of each scale. As an ovule develops, it produces an egg cell, nutrient tissue, and a sticky fluid. **Pollen grains** containing sperm develop on the smaller male cone. Wind carries pollen grains to the female gymnosperm cones. Millions of pollen grains may be released in a great cloud from a single male cone. **Figure 11-10** illustrates the pollination process. Most of this pollen never reaches the female cones because the wind blows it onto other plants, water, and the ground. When a pollen grain does get blown between the scales of a female cone of the same species, it gets caught in the sticky fluid secreted by the ovule. A pollen tube grows from the pollen grain to the ovule. A sperm swims down the pollen tube and fertilizes the egg cell. As a result, a zygote forms that develops into an embryo.

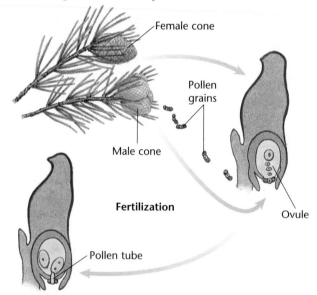

Female cone

Pollen grains

Male cone

Fertilization

Ovule

Pollen tube

Program Resources

 Reproducible Masters
Study Guide, p. 44 L1
Reinforcement, p. 44 L1
Enrichment, p. 44 L3
Activity Worksheets, pp. 5, 64-65, 66-67, 68, 69 L1
Critical Thinking/Problem Solving, p. 15
Multicultural Connections, pp. 25-26

Lab Manual 25, 26, 27
Concept Mapping, p. 27

Transparencies
Teaching Transparency 22
Science Integration Transparency 11
Section Focus Transparency 40 L1

Female cones of pine trees mature, open, and release their seeds during the fall or winter months. It may take a long time for seeds to be released from a pinecone. From the moment a pollen grain falls on the female cone until the time the seeds are released takes at least two years, depending on the species. Once seeds are released from a pinecone, they are carried away, eaten, or buried by animals. The buried seeds will germinate under favorable conditions and eventually grow into new pine trees.

Angiosperm Reproduction

Angiosperms all produce flowers. Flowers are important because they are the reproductive organs of angiosperms. When thinking about flowers, thoughts of bright colors and pleasing scents often come to mind. Although many such flowers do exist, other flowers are not brightly colored and do not smell at all. Have you ever looked at the flowers of wheat, rice, or grass? Why do you think there is such variety among flowers?

The Flower

The shape, size, and color of flowers tell you something about the life of the plant. Large flowers with bright-colored petals often attract insects and other animals, as shown in **Figure 11-11.** These insects pollinate the flowers while feeding on the flower's nectar and pollen. Flowers that aren't brightly

Figure 11-11

Some flowers bloom at night. Those that do are usually light colored or white, and they produce large amounts of scent molecules and nectar and pollen. *Aside from bats, what other animals might pollinate plants at night?*

11-3 Seed Plant Reproduction **297**

MiniLAB

What's in a pinecone?

1. Using a hand lens, look at the parts that make up a gymnosperm cone.
2. On a large paper towel, open the cone and note where the seeds are located.

Analysis

1. Make a drawing of the cone and seeds in your Science Journal.
2. Where are the seeds located?
3. Predict how this location is an advantage for the tree species.

CONNECT TO

PHYSICS

The scents produced by flowers are compounds that can evaporate quickly. Such compounds diffuse into the air as the kinetic energy of the air molecules strikes the scent molecules. *Infer* why you would be able to detect the odor of a group of flowers better on a windless day than a windy one.

Use **Critical Thinking/Problem Solving,** p. 15, as you teach this lesson.

---

2 Teach

Revealing Preconceptions

Emphasize that not all flowers are colorful. Many students will not recognize that trees such as oaks, pecans, and elms have flowers.

MiniLAB

Purpose

LS **Visual-Spatial** Students will observe pinecone structures and identify parts. **L1** **LEP**

Materials

cones with seeds, paper toweling

Analysis

1. Drawings will vary.
2. Seeds are found between woody scales.
3. Protection of the seeds ensures that more will survive.

✓ Assessment

Performance To further assess students' understanding of gymnosperm cones, have students place several cones in the refrigerator overnight, then examine them on removal and several hours later. Have students state how cone activity might be an advantage to gymnosperms. Use the Performance Task Assessment List for Making Observations and Inferences in **PASC,** p. 17. **P**

Activity Worksheets, pp. 5, 68

CONNECT TO

PHYSICS

Answer Scent molecules would remain closer together and closer to the source on a windless day.

Figure 11-12
The sepal and petals of a flower, such as this portulaca, may attract pollinators. *What are the male and female parts of the flower?*

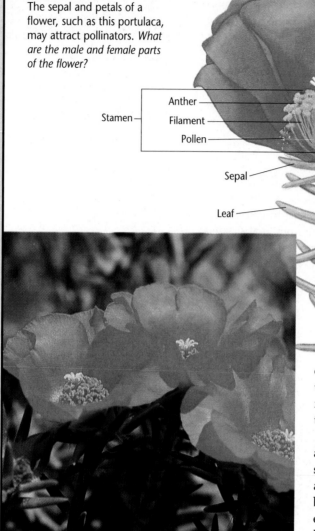

Petal
Stigma
Style
Pistil
Ovary
Ovules
Anther
Filament
Pollen
Stamen
Sepal
Leaf

colored often depend on wind for pollination. Their petals may be small, or the flowers may have no petals at all. Flowers that open at night often have strong scents.

Usually, the colored parts of a flower are the petals. Outside the petals are some small, leaflike parts called *sepals.* Sepals are easy to see when a flower is still a bud because they cover the bud. In some flowers, the sepals are as colorful as or even larger than petals.

Inside the flower are the reproductive organs of the plant. The **stamen** is the male reproductive organ of the flower. Stamens each consist of a filament and an anther where pollen grains form. The sperm develop in the pollen grain.

The **pistil** is the female reproductive organ of a flower. Pistils each consist of a sticky stigma where the pollen grains land, a long stalklike style, and an ovary. In angiosperms, the **ovary** is the swollen base of the pistil where ovules are formed. Eggs are produced as the ovule develops. You can see the parts of a typical flower in **Figure 11-12.**

298 Chapter 11 The Seed Plants

Across the Curriculum

Literature Have students work in teams to write a short story about a particular kind of flower and its pollination mechanism. The story should include a day in the life of a flower. **L1** **LS** **COOP LEARN** **P**

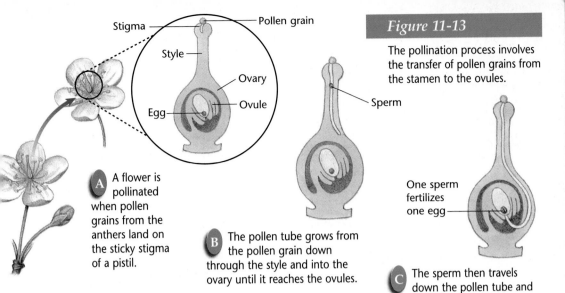

Stigma — Pollen grain

Style

Ovary

Egg — Ovule

(A) A flower is pollinated when pollen grains from the anthers land on the sticky stigma of a pistil.

(B) The pollen tube grows from the pollen grain down through the style and into the ovary until it reaches the ovules.

Figure 11-13

The pollination process involves the transfer of pollen grains from the stamen to the ovules.

Sperm

One sperm fertilizes one egg

(C) The sperm then travels down the pollen tube and fertilizes the egg. This zygote develops into the plant embryo part of the seed.

Development of a Seed

How does a seed develop? Follow **Figure 11-13** as this process is discussed. Pollen grains reach the ovule in a variety of ways. Pollen is carried by the wind or by animals such as insects, birds, and mammals. A flower is pollinated when pollen grains land on the sticky stigma. The pollen tube grows from the pollen grain down through the style and into the ovary until it reaches the ovules. The sperm then travels down the pollen tube and unites with, or fertilizes, the egg. This zygote develops into the plant embryo part of the seed. The transfer of pollen grains from the stamen to the stigmas is the process of **pollination.**

An embryo plant consists of cotyledons, stem, and root. Protected by a seed coat, the embryo also has a supply of stored food called the endosperm. After the seed germinates, the embryo will use the endosperm until it can begin to produce food on its own. You can see the parts of a seed in **Figure 11-14.**

Figure 11-14

The seed is a reproductive structure of a land plant that is capable of surviving unfavorable environmental conditions. The cotyledon stores food for the embryo plant, which will eventually germinate when conditions are favorable.

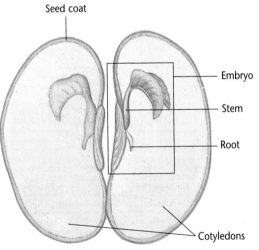

Seed coat

Embryo

Stem

Root

Cotyledons

11-3 Seed Plant Reproduction **299**

Enrichment

Visit an art museum to observe how many pieces of art contain flowers or other plant parts.

Visual Learning

Figure 11-13 Emphasize the process of pollination by having students trace the events using this illustration.
Figure 11-14 Review this illustration with students before they dissect seeds in Activity 11-1. **LEP** **LS**

📁 Use Labs 26 and 27 in the **Lab Manual** as you teach this lesson.

NATIONAL GEOGRAPHIC SOCIETY

💿 **Videodisc**
STV: Plants
What Is a Flower?
Unit 2, Side 1
Introduction

28889-32529
Reproduction

32529-39550
Pollination

39859-44901
From Flower to Seed

45047-48746

Activity 11-1

Activity 11-1

Comparing Flowers and Seeds

Purpose

LS Visual-Spatial Students will observe and analyze structures of seeds and flowers, and compare and contrast monocots with dicots. **L1**
LEP

Process Skills

observing and inferring, comparing and contrasting, classifying, forming operational definitions, measuring in SI

Time

50 minutes

Safety Precautions

Caution students to be careful when using sharp instruments such as a scalpel.

📁 **Activity Worksheets,** pp. 5, 64-65

Teaching Strategies

Troubleshooting Be sure the seeds have been soaked in water for at least 24 hours. Some students will find it difficult to orient the seeds properly for cutting. You should demonstrate where to make the cuts.

Answers to Questions

1. Monocots have flower parts in multiples of three, while dicots have them in multiples of four or five.
2. The anther has four pollen sacs; the stigma has three parts.
3. protects the embryo
4. cotyledons, stem, root, seed coat, and embryo
5. Monocots have flower parts in multiples of three. Dicot flower parts are in multiples of four or five.
6. The dicot has two cotyledons and the monocot has only one.
7. Answers will vary, but should include a comparison of structures, location, and numbers in seeds and flower.

300

You have read that monocots and dicots are similar because they are both groups of flowering plants. However, you have also learned that these two groups have major differences. Try this activity to compare and contrast monocot and dicot flowers and seeds.

Problem

How do monocot and dicot flowers and seeds compare?

Materials 🧤 🧹 ☠️

- monocot and dicot flowers
- monocot and dicot seeds
- scalpel
- forceps
- iodine solution

Procedure

1. Examine a monocot and dicot flower. Remove and count the outer row of sepals. Do the same with the petals.
2. Locate, remove, and observe the stamens and pistils of the flower.
3. Examine the two seeds.

Cut the seeds in half and observe the embryos.

4. Locate and observe the cotyledons in the seeds.
5. Test the seeds for starch by placing a drop of iodine on different parts of the seeds. A blue-black color indicates the presence of starch. **CAUTION:** *Iodine is poisonous and may stain or burn the skin.*

Analyze

1. **Compare** the numbers of stamens, petals, and sepals of monocot and dicot flowers.
2. **Describe** the anther and stigma of a flower.
3. What is the function of the outer covering and endosperm tissue of a seed?
4. What parts are the same in monocot and dicot flowers and seeds?

Conclude and Apply

5. **Distinguish** between a monocot and dicot flower.
6. How can you tell the difference between a monocot and a dicot by dissecting the seed?
7. **Compare** and **contrast** monocot and dicot flowers and seeds.

Data and Observations

Sample Data

	Number of Sepals	Number of Petals	Number of Cotyledons	Other Observations
Monocot	multiples of three	multiples of three	1	
Dicot	multiples of four or five	multiples of four or five	2	

300 Chapter 11 The Seed Plants

✓ **Assessment**

Performance Use germinated seeds to have students answer questions placed on index cards about plant parts. Use the Performance Task Assessment List for Making Observations and Inferences in **PASC,** p. 17.
P

Seed Dispersal and Germination

How do seeds get from the flower to the ground for germination? You know that seeds grow inside the ovary. A mature ovary is a fruit. Fruits may be either fleshy, like peaches or apples, or dry, like pecans and walnuts. Fleshy fruits are usually eaten by animals. After eating, these animals may then move to other locations where these seeds may germinate when released in the animals' feces. Other seeds are moved more directly by physical forces in the environment. Some seeds are able to float on top of water. These seeds, shown in **Figure 11-15**, are less dense than the water that they float on. Seeds of maple trees are dispersed by the wind. According to the law of physics, air moves around this type of seed in such a way that it remains in the air for a longer time than normal. In this way, the seeds use the motion of air currents to move them to new locations to germinate.

Figure 11-15

Seeds can be dispersed by various methods.

 A This coconut seed floats in the surf.

B Burdock seeds are dispersed by attaching to the fur of this cat. Seeds may also be dispersed while attached to feathers of birds and clothing.

C Other seeds, such as these milkweed seeds, are blown about by the wind. *Can you think of other ways seeds may be dispersed?*

11-3 Seed Plant Reproduction   **301**

 INTEGRATION
Physics

MiniLAB

How do plants disperse seeds?

Procedure
1. Make a list of ten different seeds, including those pictured in **Figure 11-15**.
2. Research each of the ten seeds to determine how they are dispersed.

Analysis
1. How are the seeds of each plant on your list dispersed—by wind, water, insects, birds, or mammals?
2. Identify features that tell you how each kind of seed is dispersed.

 Use **Science Integration Transparency 11** as you teach this lesson.

Visual Learning

Figure 11-15 Can you think of other ways that seeds may be dispersed? *Students may know that some seeds are transported through the digestive system of some animals, or they may suggest other ways.* **LEP** **LS**

301

Problem Solving

Are seed company claims true?

While purchasing seeds to plant in his vegetable garden, Juan noticed that each packet of seeds contained information about the number of seeds in the package, how deeply and how far apart they should be planted, and how quickly the seeds germinate. The packet of carrot seeds stated that there were approximately 200 seeds. They should be planted about two inches apart and a quarter of an inch deep. This particular packet of seeds said that the germination rate was 95 percent. Juan decided that he would test the claims of the seed company for the carrot seeds.

Think Critically:
What measurements should Juan make to determine whether the claims are true?

  Use Lab 25 in the **Lab Manual** as you teach this lesson.

When a seed reaches the soil, it may not germinate right away. Seeds often remain dormant until conditions are right for germination. Some seeds can remain dormant for hundreds of years, yet still be able to grow when environmental conditions become favorable. In 1982, seeds of the East Indian lotus germinated after 466 years!

Germination, as illustrated in **Figure 11-16,** is the development of a seed into a new plant. It begins when water is absorbed by seed tissues. Water signals the seed to begin growth. The food-storing endosperm provides the energy for the seeds of monocots to grow. Food is generally stored in the cotyledons of dicots. The energy supplies of many seeds can cause enough pressure to split rocks open when the roots begin to elongate and grow out of these seeds. Think about how much force must have been needed to split a rock. All of the energy came from stored energy in the seed, and later energy provided by photosynthesis. If the seed is a monocot, the seed containing the endosperm remains below the soil. In a dicot seed, such as a bean, the cotyledons are carried above ground by the elongating shoot. They also raise the seed above the surface. Eventually, as the food is used, the cotyledons shrivel and fall off of dicot plants. Then the plant, which now has leaves, takes over and makes all of its own food by photosynthesis.

Figure 11-16

In dicots, the seed and cotyledons may be raised above the soil. As the food is used up, the cotyledons shrivel and fall off.

Section Wrap-up

Review

1. Compare life cycles of angiosperms and gymnosperms.

2. Diagram the parts of a flower and label their functions.

3. List three methods of seed dispersal in plants.

4. **Think Critically:** Some conifers bear female cones on the top half of the tree and male cones on the bottom half. Why do you think this arrangement of male and female cones on trees is important?

Skill Builder
Hypothesizing
A corn plant produces thousands of pollen grains on top of the plant in flowers that have no odor or color. The pistils grow from the cob lower down on the plant. Explain how a corn plant is probably pollinated. If you need help, refer to Hypothesizing in the **Skill Handbook.**

Science Journal
Observe live specimens of several different types of flowers. Describe their structures. Include numbers of petals, sepals, and stamens, and describe the shape of the pistil in your Science Journal.

11-3 Seed Plant Reproduction   **303**

 Assessment

Performance Assess students' abilities to observe and infer how a flower is pollinated by giving each student a flower and having them identify how this particular specimen would be pollinated. Use the Performance Task Assessment List for Making Observations and Inferences in **PASC**, p. 17. **P**

Purpose

IS **Logical-Mathematical** Students will design and carry out an experiment to determine how germination rate varies with an environmental factor. **L1** **COOP LEARN**

Process Skills

designing an experiment, collecting and interpreting data, analyzing data, drawing conclusions, communicating, making and using tables, making and using graphs, comparing and contrasting, recognizing cause and effect, forming a hypothesis, separating and controlling variables, measuring in SI

Time

one full class period + 10 minutes for each of five additional days

Materials

- Students will plant seeds in shallow trays or plastic cups. They may use salt water solutions or warming cables in their setups, but not both.

Alternate Materials Use any fast germinating seeds. You may also substitute Epsom salt for table salt.

Safety Precautions

Have students review safety symbols and precautions on the student page.

Possible Hypotheses

Students may test germination rate with various solutions of salt, predicting that seeds will germinate at a higher rate with distilled water. They may predict that warmer soil will increase the rate of germination.

📁 **Activity Worksheets,** pp. 5, 66-67

Activity 11-2

Design Your Own Experiment
Germination Rate of Seeds

There are many environmental factors that affect the germination rate of seeds. Among these are soil temperature, air temperature, moisture content of soil, and salt content of soil. What happens to the germination rate when one of these variables is changed? Can you determine a way to predict the best conditions for seed germination?

Problem

How does seed germination vary with an environmental factor?

Form a Hypothesis

Based on your knowledge of seed germination, state a hypothesis about how germination rates vary with environmental factors.

Objectives

- Analyze the effect of an environmental factor on seed germination rate.
- Compare germination rates under different conditions.

Possible Materials

- seeds
- distilled water
- salt

- potting soil
- plant trays or plastic cups
- seedling warming cables
- thermometer
- graduated cylinder
- beakers

Safety Precautions

Be careful when using any electrical equipment to avoid shock hazards. Some kinds of seeds are poisonous. Do not place any seeds in your mouth.

304 Chapter 11 The Seed Plants

PLAN THE EXPERIMENT

1. As a group, agree upon and write out the hypothesis statement.
2. As a group, list the steps that you need to take to test your hypothesis. Be specific, describing exactly what you will do at each step. List your materials.
3. What variables will you be testing? What measurements will you take? What data will you collect? How often will you collect data?
4. If you need a data table, design one in your Science Journal so that it is ready to use as your group collects data.

Check Your Plan

1. Read over your entire experiment to make sure that all steps are in logical order.
2. Identify any constants, variables, and controls of the experiment.
3. Is it necessary to run any tests more than one time?
4. Will the data be summarized in a graph?
5. *Make sure your teacher approves your plan before you proceed.*

DO THE EXPERIMENT

1. Carry out the experiment as planned.
2. While the experiment is going on, write down any observations that you make and complete the data table in your Science Journal.

Analyze and Apply

1. **Compare** your results with those of other groups.
2. How did germination rates differ when the variable was altered?

3. **Graph** your results using a bar graph, placing germination rate on the *y*-axis and the environmental variables on the *x*-axis.
4. Using your graph, **estimate** the best range of one environmental condition to produce the best germination rate.

Go Further

What other experiments might be done to answer Analyze and Apply question 4 for other variables?

Go Further
Answers will vary. Other temperature ranges or chemical concentrations could be tested.

✓ Assessment
Oral Have students explain their results to the class. Use the Performance Task Assessment List for Oral Presentation in **PASC**, p. 71. **P**

Prepare

Section Background

Castor oil is a laxative made from the poisonous seeds of castor beans.

1 Motivate

Bellringer

Before presenting the lesson, display **Section Focus Transparency 41** on the overhead projector. Assign the accompanying **Focus Activity** worksheet. L1 LEP

Inquiry Question

Ask students what medicine they might take for a headache. When they suggest aspirin or an aspirin-substitute, point out that these medicines have similar compounds that were first obtained from plants.

2 Teach

Revealing Preconceptions

Most students will consider that medicines are synthetic chemicals. Point out that most have their origins in plants.

Demonstration

Kinesthetic Bring an aloe vera plant to class. If possible, obtain products that advertise aloe vera extract as an ingredient. Break off a leaf to show students the jelly-like sap of the plant. Let them feel the sap.

Science Words
ethnobotanist

Objectives
- Describe some plants that have value as medicines.
- Compare and contrast views about preserving species that may or may not have value to humans.

Figure 11-17
Ethnobotanists are attempting to understand the interactions of native peoples with plants. Such an understanding needs to be reached before the plants become extinct or the knowledge is lost.

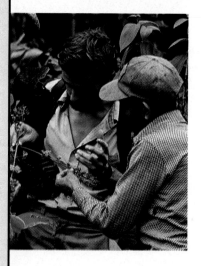

Treasures Found

When was the last time you had a headache? If you took an aspirin, you may have thought of that little tablet as a treasure when the pain subsided. Aspirin was originally developed from the bark of a willow tree. Now we have other medicines to get rid of that headache, but most of these other drugs were designed after the aspirin's molecular structure was determined. A few changes here and there, and new medicines are the result.

Other valuable medicines come from other plants. Many of the medicines for heart disease were developed from a plant called foxglove. Indian snakeroot yields a medicine that reduces high blood pressure. Even drugs for the treatment of cancer have been developed from the rosy periwinkle from Madagascar, Africa. As a matter of fact, most of the medicines used today by millions of people come directly from plants. Another large group of medicines is developed from by-products of compounds that were first derived from plants.

Hidden Treasures

A small but growing group of people are searching the less inhabited regions of the world, such as the rain forests of South America and Asia, for more of these hidden treasures. Pictured in **Figure 11-17,** an **ethnobotanist** is a person who studies the relationships between people of various cultures and the plants they use. Ethnobotanists have recently found that a high percentage of plants used by native peoples appear to be effective as medicines, such as the one shown in **Figure 11-18.** They are intrigued by the possibilities of drugs that may be found yet.

Perhaps there are plants that could yield cures for cancer, heart disease, AIDS, or any number of other diseases. However, time is running out for many species. Some scientists estimate that between one and 40 species are becoming extinct every year. Not only are plants becoming extinct, but cultures are as well. Many times, when native healers die, their knowledge dies with them.

306 Chapter 11 The Seed Plants

Program Resources

Reproducible Masters
Study Guide, p. 45 L1
Reinforcement, p. 45 L1
Enrichment, p. 45 L3
Cross-Curricular Integration, p. 15
Technology Integration, pp. 11-12

Transparencies
Section Focus Transparency 41 L1

2 Points of View

Preserving the Forests

Ethnobotanists are alarmed. They think destruction of the rain forest should be stopped and as many species as possible should be preserved. They also see a benefit in helping indigenous cultures to survive. Many ethnobotanists feel that the drug industry should be forced to pay the cultures involved for the research because it will eventually benefit the drug companies. Ethnobotanists see a value in preserving the possibility of new drugs and think that species should be protected because of their importance to the ecosystem.

Use of Forests for Survival

Drug companies see little reason to invest in studies by ethnobotanists. They feel that they can better use their profits by developing more synthetic drugs. One reason drug companies have reduced their search for medicinal plants is their belief that the expense in finding and developing these new medicines is greater than the potential profit.

Figure 11-18

A multitude of medicines have been developed from tropical plants, including a cancer-treatment drug made from this rosy periwinkle from Madagascar, an island off the coast of Africa.

Section Wrap-up

Review

1. What are some medicines that are derived from plants?

2. Why would an ethnobotanist travel to the rain forest?

Explore the Issue

Should drug companies be forced to spend some of their profits to fund the work of ethnobotanists? If yes, how would you address the concerns of drug companies? If no, how would you address the concerns of the ethnobotanists and environmentalists?

SCIENCE & SOCIETY

Section Wrap-up

Review

1. aspirin, cancer drugs, opiates, drugs to reduce high blood pressure

2. Ethnobotanists study the relationship between people and plants. Many native tribes are being absorbed into the mainstream of society, and their knowledge of plants is being lost. Ethnobotanists want to study these groups and their use of plants before it is too late.

Explore the Issue

Students' answers will vary. Answers that are supported by clear and logical thinking should be counted as correct.

307

Background

- It has been estimated that tropical rain forests are home to at least half of all the world's species of flowering plants. Nearly 50 major drugs have come from tropical forest plants, including quinine, curare, and codeine. Rain forest plants continue to be a rich source of many novel compounds that show promise as antiviral, antibacterial, and antifungal agents. The National Cancer Institute and many private pharmaceutical firms are expanding their efforts to find and test new plant compounds that might be effective against cancer and the AIDS virus.

- In addition to drugs, many food products come from rain forest plants. Chocolate is made from the seeds of the cacao tree. The seeds are packed into melon-shaped pods that grow directly out of the tree trunk. The Maya of Central America were the earliest cultivators of the cacao tree. They considered cacao seeds to be as valuable as gold and silver. The Aztecs called the thick, dark beverage they made from powdered cacao seeds "chocolatl," a word that combined the native word for water—"atle"—with a close approximation of the sound made when water and cocoa are whipped together vigorously.

Teaching Strategies

IS **Interpersonal** Have students organize a class field trip to a grocery store to search for foods and other products that come from rain forest plants. Students can create a classroom display of their findings. Challenge students to create recipes for a tropical feast that features foods from the forest. L1

308

People and Science

MARK PLOTKIN, *Ethnobotanist*

On the Job

Q **Dr. Plotkin, what is ethnobotany?**

A Ethnobotany is the study of how people use the plants in their environment. Most ethnobotanists work in the rain forest with native peoples to learn about the local plants that various tribes use for medicine, for food, and for building materials. My research has taken me to almost every Amazonian country, and to the rain forests of Central America, too.

Q **Could you describe your field work?**

A I often spend many weeks living in remote rain forest villages. I work closely with the village shaman, or medicine man. He takes me into the forest and I'll ask things like "What plant do you use to treat skin infections?" or "What do you use this plant for?" We collect plants and go back to the village where I put them into a plant press and dry them over a campfire. The dried plants are shipped back to this country where they are analyzed for compounds they contain that are effective in fighting disease.

Personal Insights

Q **How many of the world's plants have been studied for their possible medicinal value?**

A I'd say fewer than ten percent have been extensively studied. As rain forests and other natural areas are destroyed, we are losing our chance to learn the secrets of the plants that are disappearing. But there is still a lot out there that needs to be discovered, explored, and protected.

Career Connection

Contact an environmental organization such as the Nature Conservancy, the Sierra Club, or the World Wildlife Fund for information about preserving rain forests. Investigate how people involved in the following professions can help protect plant biodiversity, and report on your findings to your class:

- **Forester**
- **Plant Taxonomist**

308   Chapter 11 The Seed Plants

Career Connection

Career Path An ethnobotanist would typically have a master's or doctoral degree in plant science.

- High School: science and social studies
- College: plant biology, chemistry; anthropology
- Post-College: advanced degree in botany

For More Information

Vivid images of rain forest plants and human cultures are presented in Dr. Plotkin's book, *Tales of a Shaman's Apprentice*, Viking-Penguin, 1993. The National Wildlife Federation is a good source of reference materials and activity suggestions that focus on rain forests.

Summary

11-1: Seed Plants
1. Seed plants have stems, roots, and leaves and grow from seeds.
2. Gymnosperms produce seeds on cones; angiosperms produce seeds in flowers.
3. Seed plants are sources of oxygen, food, clothing, furniture, medicines, and other products.

11-2: Parts of Complex Plants
1. Roots absorb water and minerals, store food, and anchor the plant.
2. Stems transport food and water between roots and leaves and support the plant.
3. The palisade layer of leaves carries out photosynthesis. Xylem and phloem tissues are located in the spongy layer of the leaf.

11-3: Seed Plant Reproduction
1. Wind carries gymnosperm pollen to female cones where fertilization occurs.
2. In angiosperms, pollen reaches the stigma and travels to the ovary to fertilize the ovules.
3. Flowers contain male and female structures. The female pistils consist of the stigma, style, ovary, and ovules. The male stamens consist of the filament and anther, where pollen grains are formed.

11-4: Science and Society: Medical Treasures
1. Many plants such as foxglove and the Indian snakeroot have been used for a long time as medicines. Most medicines used today were originally found in plants. Ethnobotanists are carrying on a search for more medicines from plants.
2. The industry that makes medicines has reduced its search for drugs that come from plants because the expense in finding and developing them into new medicines is greater than the potential profit. Environmentalists state that species should be protected for their future potential as new drugs.

Key Science Words

a. angiosperm
b. cambium
c. dicot
d. ethnobotanist
e. guard cell
f. gymnosperm
g. monocot
h. ovary
i. ovule
j. phloem
k. pistil
l. pollen grain
m. pollination
n. stamen
o. stomata
p. xylem

Reviewing Vocabulary

Match each phrase with the correct term from the list of Key Science Words.

1. transfer of pollen grains to stigmas
2. cone-bearing plant
3. male part of a flower
4. small pore in leaf surface
5. person who studies human relationships with plants
6. flower-producing plant

Chapter 11 Review 309

Review

Summary

Have students read the summary statements to review the major concepts of the chapter.

Reviewing Vocabulary

1. m	6. a
2. f	7. k
3. n	8. p
4. o	9. g
5. d	10. j

✔ Assessment

Portfolio Encourage students to place in their portfolios one or two items of what they consider to be their best work. Examples include:
- MiniLAB, p. 297
- Across the Curriculum, p. 298
- Cultural Diversity, p. 307 P

Performance Additional performance assessments may be found in **Performance Assessment** and **Science Integration Activities**. Performance Task Assessment Lists and rubrics for evaluating these activities can be found in Glencoe's **Performance Assessment in the Science Classroom.**

GLENCOE TECHNOLOGY

▭ MindJogger Videoquiz

Chapter 11 Have students work in groups as they play the Videoquiz game to review key chapter concepts.

Checking Concepts

1. d	**6.** b
2. b	**7.** d
3. a	**8.** b
4. d	**9.** b
5. a	**10.** c

Understanding Concepts

11. Leaves are flattened and have a layer of cells containing many chloroplasts near the surface.

12. Gymnosperms are pollinated by the wind. To increase the chances of the wind blowing a pollen grain to a female cone, greater numbers must be produced.

13. Pollen sticks to the stigma on top of the style through which a tube forms, allowing the pollen to move down to the ovary.

14. One acceptable hypothesis is that natural selection favors colored petals since pollinators may be attracted by the colors. Other acceptable hypotheses may deal with chemical (scent) attractants.

15. A waxy cuticle protects the epidermis from water loss. Guard cells control the leaf's openings so water is not continually lost.

Thinking Critically

16. Resin prevents fungal and bacterial growth on the tree; insects cannot invade as easily.

17. Succulents can be found in areas that are hot or dry where plants must store water for long periods of time.

18. Water loss is prevented in conditions that are especially dry.

---

7. female part of a flower
8. vessel that transports water and minerals
9. plant with one embryo leaf
10. moves food from leaves and stems

Checking Concepts

Choose the word or phrase that completes the sentence.

1. Gymnosperms include _____.
  a. oaks   c. gladioli
  b. flowers   d. pines
2. Angiosperms produce _____.
  a. cones   c. needles
  b. flowers   d. woody scales
3. Colorful flowers are usually pollinated by_____.
  a. insects   c. clothing
  b. wind   d. birds
4. A(n) _____ is the sometimes sticky part of a flower.
  a. sepal   c. stamen
  b. ovary   d. stigma
5. The _____ contains food for the embryo.
  a. endosperm   c. stigma
  b. pollen grain d. root
6. The seed leaves of an embryo are _____.
  a. root hairs   c. guard cells
  b. cotyledons   d. stomata
7. _____ anchor plants in soil.
  a. Leaves   c. Stomata
  b. Stems   d. Roots
8. _____ produces new xylem and phloem.
  a. Epidermis   c. Pollen
  b. Cambium   d. An ovule
9. Most photosynthesis occurs in the _____.
  a. guard cells   c. epidermis
  b. palisade layer d. cuticle

19. Students can describe any brightly colored, large flower that they know: goldenrod, lilies, dandelions, and daisies are good examples.

20. Dry: acorn, almond, peanut; Fleshy: apple, cherry, orange

---

10. A(n) _____ provides waxy protection for the leaf.
  a. guard cell   c. cuticle
  b. epidermis   d. loose layer

Understanding Concepts

Answer the following questions in your Science Journal using complete sentences.

11. What adaptation in leaves increases photosynthesis?
12. Why do male pinecones produce so many pollen grains?
13. What adaptations of flowers help pollination?
14. Both petals and sepals of the gladiolus flower are brightly colored rather than just green. Make a hypothesis about why natural selection might have favored colored petals and sepals over plain green ones.
15. What prevents water loss in leaves?

Thinking Critically

16. What function does resin serve for the gymnosperm that secretes it?
17. Plants called succulents store large amounts of water in their leaves. In what kind of environment would you find succulents?
18. Why would cacti benefit from thick cuticles?
19. Use an example to describe a flower that is pollinated by insects.
20. Classify each fruit as dry or fleshy: acorn, almond, apple, cherry, orange, peanut.

Developing Skills

21. Comparing and Contrasting Roots—anchor in soil, absorb water and nutrients; stems—hold plant upright, transport water, food, and nutrients; leaves—make food for plant, transpire, move food to other plant parts; seeds—reproductive part of plant, produce new plant embryos

Developing Skills

If you need help, refer to the **Skill Handbook.**

21. **Comparing and Contrasting:** Compare the functions of roots, stems, leaves, and seeds in seed plants.

22. **Observing and Inferring:** Observe pictures or specimens of flowers and infer how each is pollinated. What features does each flower have to enable the pollination you suggest?

23. **Concept Mapping:** Fill in the following concept map to show the sequence of pollination and fertilization.

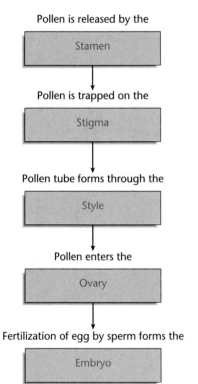

Pollen is released by the

Stamen

Pollen is trapped on the

Stigma

Pollen tube forms through the

Style

Pollen enters the

Ovary

Fertilization of egg by sperm forms the

Embryo

24. **Interpreting Data:** Interpret the data in the table below concerning stomata of these leaf types. What do these data tell you about where gas exchange occurs in each type?

Stomata (per mm²)

	Upper Surface	Lower Surface
Pine	50	71
Bean	40	281
Fir	0	228
Tomato	12	13

25. **Comparing and Contrasting:** Compare and contrast these structures of monocots and dicots: number of seed leaves, bundles arranged in stems, veins in leaves, flower parts.

Performance Assessment

1. **Scientific Drawing:** Use a key to wildflowers in your area. Identify at least ten kinds of wildflowers and make a sketch of each in your Science Journal. Describe the habitat for each flower and its blooming season.

2. **Design an Experiment:** Design an experiment to test the viability of seeds in the Problem Solving on page 302. Carry out your experiment. What percentage of seeds was viable?

3. **Scientific Drawing:** Collect three or four different kinds of flowers. Make wet mounts of the pollen from each flower, and observe under low and high power of the microscope. Describe the differences and make a labeled sketch of each type of pollen grain.

22. **Observing and Inferring** Pictures could include trumpet creeper flower pollinated by hummingbirds; goldenrod, clover, and dandelions pollinated by bees; ragweed, willows, corn, and grasses pollinated by wind.

23. **Concept Mapping** See student page.

24. **Interpreting Data** Pine and tomato have about equal numbers of stomata on both surfaces, indicating equal exchange on each leaf surface. Beans must use the lower stomata more; fir have no stomata on the upper surface, and therefore, all gas exchange occurs on the bottom.

25. **Comparing and Contrasting** Monocots have one seed leaf, scattered vascular bundles, parallel veins in leaves, and flower parts in threes. Dicots have two seed leaves, vascular bundles in rings, net veins in leaves, and flower parts in fours or fives.

Performance Assessment

1. Sketches and habitats will depend on student choices of wildflowers. Use the Performance Task Assessment List for Scientific Drawing in **PASC,** p. 55. P

2. Students should plant 100 seeds and derive the viability percentage. Accuracy would increase with a larger sample, e.g., 1000 seeds. Use the Performance Task Assessment List for Designing an Experiment in **PASC,** p. 23. P

3. Sketches will depend on flowers used. Use the Performance Task Assessment List for Scientific Drawing in **PASC,** p. 55. P

Assessment Resources

📁 **Reproducible Masters**

Chapter Review, pp. 25-26
Assessment, pp. 72-75
Performance Assessment, p. 25

Glencoe Technology

💿 **Chapter Review Software**
💿 **Computer Test Bank**
📼 **MindJogger Videoquiz**

311

Chapter Organizer

Section	Objectives/Standards	Activities/Features
Chapter Opener		**Explore Activity:** Observe the relationships between plants and water. p. 313
12-1 **Photosynthesis and Respiration** (3 sessions, 2 blocks)*	1. **Describe** the process of gas exchange in plants. 2. **Explain** the process and importance of photosynthesis. 3. **Describe** the process and importance of respiration and its relationship to photosynthesis. **National Content Standards:** (5-8) UCP1, UCP4, UCP5, A1, A2, B1, B3, C1, C3, E2	**Activity 12-1:** Stomata in Leaves, p. 316 **MiniLAB:** How can the use of carbon dioxide by plants be shown? p. 318 **Using Technology:** Pollution Cleanup, p. 319 **Connect to Physics,** p. 320 **MiniLAB:** How can respiration in yeast be shown? p. 321 **Using Math,** p. 323 **Skill Builder:** Comparing and Contrasting, p. 323
12-2 **Plant Responses** (1 session, 1 block)*	1. **Explain** the relationship between stimuli and tropisms in plants. 2. **Differentiate** between long-day and short-day plants. 3. **Explain** the relationship between plant hormones and responses. **National Content Standards:** (5-8) UCP1, A1, A2, C3	**Problem Solving:** The Forgotten Plant, p. 327 **Activity 12-2:** Plant Tropisms, pp. 328-329 **Skill Builder:** Comparing and Contrasting, p. 330 **Science Journal,** p. 330
12-3 **Science and Society Technology:** **Transgenic Crops** (1 session, ½ block)*	1. **Identify** two DNA technologies for improving crop production. 2. **Describe** the possible benefits of biotechnology. **National Content Standards:** (5-8) E1, E2, F5, G1	**Explore the Technology,** p. 333 **People and Science:** Flora Ninomiya, Horticulturist p. 334

* A complete Planning Guide that includes block scheduling is provided on pages 32T-35T.

Activity Materials

Explore	Activities	MiniLABs
page 313 1-gallon, resealable plastic bag; small potted plant; tap water	page 316 lettuce in dish of water, coverslip, microscope, microscope slide, salt solution, forceps pages 328-329 petri dish, tape, string, corn seeds, bean seeds, paper towels, tap water, ring stand, apron, safety goggles	page 318 safety goggles, 50 mL phenol red solution, 2 small beakers, plastic drinking straw, sprig of *Elodea*, bright light, stopwatch or clock with second hand page 321 safety goggles, 10 mL bromothymol blue, test tube, droppers, yeast suspension, sugar solution, stopwatch or clock with second hand

Need Materials? Call Science Kit (1-800-828-7777).

Teacher Classroom Resources

Reproducible Masters	Transparencies	Teaching Resources
Study Guide, p. 46 **Reinforcement**, p. 46 **Enrichment**, p. 46 **Activity Worksheets**, pp. 70-71, 74, 75 **Lab Manual 28**, Transpiration **Lab Manual 29**, Oxygen and Photosynthesis **Lab Manual 30**, Carbon Dioxide and Photosynthesis **Lab Manual 31**, Plant Respiration **Science Integration Activity 12**, Capillary Action **Science and Society Integration**, p. 16 **Multicultural Connections**, pp. 27-28	**Section Focus Transparency 42**, In the Conservatory **Teaching Transparency 23**, Gas Exchange in a Leaf	**Glencoe Life Science Interactive Videodisc** **Life Science CD-ROM** **Spanish Resources** **English/Spanish Audiocassettes** **Cooperative Learning Resource Guide** **Lab Partner** **Lab and Safety Skills** **Lesson Plans**
Study Guide, p. 47 **Reinforcement**, p. 47 **Enrichment**, p. 47 **Activity Worksheets**, pp. 72-73 **Lab Manual 32**, Tropisms **Concept Mapping**, p. 29 **Critical Thinking/Problem Solving**, p. 16 **Cross-Curricular Integration**, p. 16	**Section Focus Transparency 43**, Mum's the Word **Science Integration Transparency 12**, Seasons and the Seed Plants **Teaching Transparency 24**, Ethylene	**Assessment Resources** **Chapter Review**, pp. 27-28 **Assessment**, pp. 76-79 **Performance Assessment in the Science Classroom (PASC)** **MindJogger Videoquiz** **Alternate Assessment in the Science Classroom** **Performance Assessment** **Chapter Review Software** **Computer Test Bank**
Study Guide, p. 48 **Reinforcement**, p. 48 **Enrichment**, p. 48	**Section Focus Transparency 44**, Designer Plants	

Key to Teaching Strategies

The following designations will help you decide which activities are appropriate for your students.

L1 Level 1 activities should be within the ability range of all students, including those with learning difficulties.

L2 Level 2 activities should be within the ability range of the average to above-average student.

L3 Level 3 activities are designed for the ability range of above-average students.

LEP LEP activities should be within the ability range of Limited English Proficiency students.

LS These activities are designed to address different learning styles.

COOP LEARN Cooperative Learning activities are designed for small group work.

P These strategies represent student products that can be placed into a best-work portfolio.

GLENCOE TECHNOLOGY

The following multimedia resources are available from Glencoe.

Science and Technology Videodisc Series (STVS)
Plants & Simple Organisms
  Disease-Resistant Tomatoes
  Genetic Engineering in Barley
  New Grains
  Salt-Resistant Crops

National Geographic Society Series
Newton's Apple: Life Sciences
STV: Plants
Glencoe Life Science Interactive Videodisc
Photosynthesis and Cellular Respiration
Life Science CD-ROM

Teacher Classroom Resources

This is a representation of key blackline masters available in the Teacher Classroom Resources.

Teaching Aids

Section Focus Transparencies

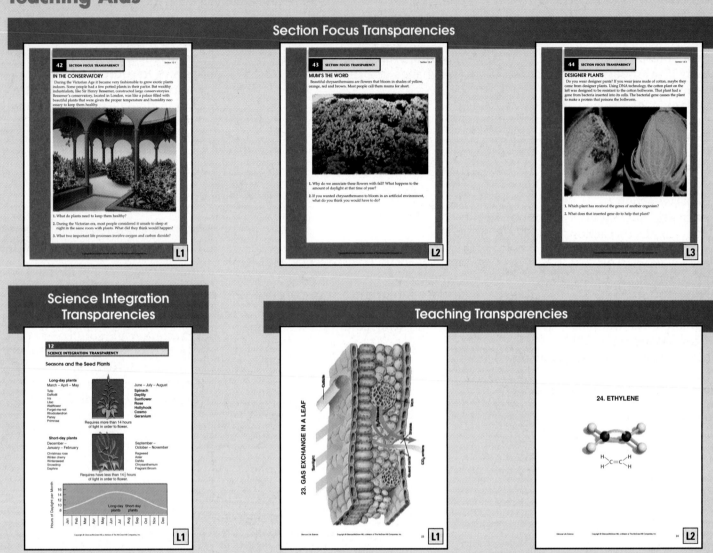

Meeting Different Ability Levels

Study Guide

Reinforcement

Enrichment Worksheets

Hands-On Activities

Science Integration Activity

Chapter 12
INTEGRATION • Capillary Action

L1

Lab Manual

Chapter 12
LABORATORY MANUAL • Transpiration 28

L2

Assessment

Performance Assessment

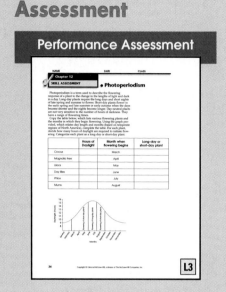

Chapter 12
SKILL ASSESSMENT • Photoperiodism

L3

Enrichment and Application

Critical Thinking/Problem Solving

Chapter 12
CRITICAL THINKING • Plant Processes

Night-Blooming Plants

L2

Cross-Curricular Integration

Chapter 12
MATH CONNECTION • Plant Processes

Symmetry

L1

Science and Society Integration

Chapter 12
SCIENCE & SOCIETY • Plant Processes

A Controversial Clearing

L2

Multicultural Connections

Chapter 12
MULTICULTURAL CONNECTIONS • Africa's Farmers Know Best

L2

Concept Mapping

Chapter 12
CONCEPT MAPPING • Plant Processes

L3

Chapter 12

Plant Processes

CHAPTER OVERVIEW

Section 12-1 The process and importance of photosynthesis and aerobic respiration are discussed in this section. Gas exchange is explained as a process that is needed in plants and animals.

Section 12-2 In this section, the relationship between stimuli and tropisms in plants is explained. The flowering response of plants to specific periods of darkness is explained and examples are given. Plant hormones are discussed in terms of the responses they produce in plants.

Section 12-3 Science and Society The usefulness of biotechnology to produce transgenic crops is discussed in this section.

Chapter Vocabulary

transpiration

photosynthesis

respiration

tropism

auxin

photoperiodism

long-day plant

short-day plant

day-neutral plant

biotechnology

transgenic crop

Theme Connection

Systems and Interactions/Energy Particular emphasis is placed on photosynthesis and respiration as homeostatic mechanisms and the role of energy in these processes.

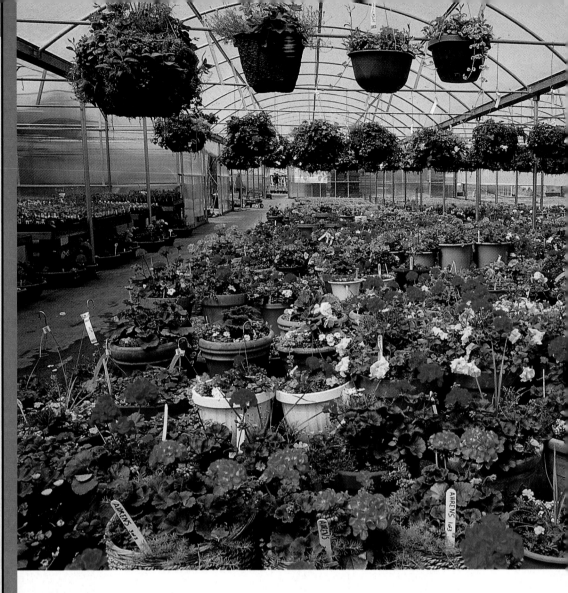

Previewing the Chapter

Section 12-1 Photosynthesis and Respiration
► Gas Exchange in Plants
► Photosynthesis
► Respiration

Section 12-2 Plant Responses
► What are plant responses?
► Plant Hormones
► Photoperiods

Section 12-3 Science and Society
Technology: Transgenic Crops

312

Learning Styles

Look for the following logo for strategies that emphasize different learning modalities.

Kinesthetic	People and Science, p. 334
Visual-Spatial	Visual Learning, pp. 315, 317, 319, 320, 321, 322, 324, 326, 331; Demonstration, pp. 315, 317; Activity 12-1, p. 316; MiniLAB, p. 318
Interpersonal	Activity 12-2, pp. 328-329
Logical-Mathematical	Explore, p. 313; MiniLAB, p. 321; Using Math, p. 323
Linguistic	Science Journal, p. 320

Plant Processes

Plants are similar in many ways to all other living things. They contain carbon, are made of cells, reproduce, and metabolize in ways similar to any other organism. What do you think would happen if the caretaker forgot to make water available to the plants on the opposite page? Of course, after a while they would begin to look unhealthy. Do the following activity to discover the relationship between plants and water. Find out how water enters and leaves a plant, and then you will be able to learn about plant processes in the chapter that follows.

EXPLORE ACTIVITY

Observe the relationships between plants and water.

1. Obtain a resealable plastic bag and a small potted plant from your teacher.
2. Water the plant; then place the plant and its pot in the plastic bag.
3. Seal the bag and place it in a sunny window.
4. Look at the plant at the same time every day for a few days.

Observe: In your Science Journal, describe what happens in the bag. If enough water is lost by a plant and not replaced, predict what will happen to the plant.

Previewing Science Skills

▶ In the **Skill Builders**, you will compare and contrast.

▶ In the **Activities**, you will observe, compare, interpret data, and make and use tables.

▶ In the **MiniLABs**, you will experiment and compare.

313

EXPLORE ACTIVITY

Purpose

LS **Logical-Mathematical** Use the Explore Activity to introduce students to plant processes. Inform students that they will be learning more about plant processes as they read the chapter. **L1**
COOP LEARN

Preparation

Obtain the materials needed at least two weeks in advance. The seedlings should be started in time for the plants to be large enough to show transpiration.

Materials

resealable bags and seedling plants in pots

Teaching Strategies

Have students record their hypotheses about what will happen when the plastic bags are placed over the seedlings.

Observe

Water was transported from the soil through the plant's stem and out the leaves. The plant will wilt or even die if too much water is lost.

✔ Assessment

Process Have students diagram the pathway of water using as much terminology as they know from previous chapters (i.e. roots, stems, vascular tissue, xylem, etc.). Use the Performance Task Assessment List for Scientific Drawing in **PASC**, p. 55. **P**

Assessment Planner

Portfolio
Refer to page 335 for suggested items that students might select for their portfolios.

Performance Assessment
See page 335 for additional Performance Assessment options.
Skill Builders, pp. 323, 330
MiniLABs, pp. 318, 321
Activities 12-1, p. 316; 12-2, pp. 328-329

Content Assessment
Section Wrap-ups, pp. 323, 330, 333
Chapter Review, pp. 335-337
Mini Quizzes, pp. 323, 330

Group Assessment
Opportunities for group assessment occur with Cooperative Learning Strategies and Flex Your Brain Activities.

Prepare

Section Background

- Photosynthesis accounts for more than 99 percent of the energy used by living organisms on Earth.
- Guard cells are usually the only cells in the leaf epidermis that contain chloroplasts.

Preplanning

- To prepare for the MiniLAB on page 318, obtain phenol red, straws, and *Elodea*.
- The MiniLAB on page 321 requires bromothymol blue, yeast solution, and sugar solution.
- Activity 12-1 involves lettuce, salt solutions, and microscopes.

1 Motivate

Bellringer

 Before presenting the lesson, display **Section Focus Transparency 42** on the overhead projector. Assign the accompanying **Focus Activity** worksheet. L1 LEP

Tying to Previous Knowledge

This section should be tied to Chapter 2, where the structure and function of chloroplasts and mitochondria are explained.

NATIONAL GEOGRAPHIC SOCIETY

 Videodisc
STV: Plants
Transpiration, artwork

53781

Science Words

transpiration
photosynthesis
respiration

Objectives

- Describe the process of gas exchange in plants.
- Explain the process and importance of photosynthesis.
- Describe the process and importance of respiration and its relationship to photosynthesis.

Figure 12-1

Stomata open when guard cells absorb water (left). They close when water is lost (right). *Would adding salt to the soil around a plant make the stomata open or close?*

Gas Exchange in Plants

When you breathe in, your lungs take in a mixture of gases called air. Air is made up of nitrogen, oxygen, and a small amount of carbon dioxide. On exhaling, you also release a mixture of gases, now with more carbon dioxide. Gas exchange is one of the ways living cells obtain raw materials and get rid of waste products.

Plants need water and carbon dioxide to survive. Plants absorb water and minerals through their roots. Water then travels up from the roots through the stem to the leaves. Water evaporates from leaf tissues and is released through stomata as water vapor. Carbon dioxide, a gas, gets into plants through stomata on leaf surfaces. Because leaves usually have more stomata (*sing.,* stoma) on the lower surface, more carbon dioxide reaches the spaces around the spongy layer, as shown in **Figure 12-2.** Water vapor is also found in the air spaces of the spongy layer.

Stomata

How does carbon dioxide enter a leaf? Each stoma is surrounded by two guard cells that control the size of the opening. Water moves into and out of guard cells by osmosis. When guard cells absorb water, they swell and the stoma opens, letting carbon dioxide in. Water vapor escapes during the process. When guard cells lose water, they relax and the stoma closes. **Figure 12-1** shows open and closed stomata.

Magnification 300×

Stoma
Guard cell

Magnification 300×

314   Chapter 12 Plant Processes

Program Resources

 Reproducible Masters
Study Guide, p. 46 L1
Reinforcement, p. 46 L1
Enrichment, p. 46 L3
Science and Society Integration, p. 16
Activity Worksheets, pp. 5, 70-71, 74, 75 L1
Science Integration Activities, pp. 23-24

Multicultural Connections, p. 27
Lab Manual 28, 29, 30, 31

Transparencies
Section Focus Transparency 42 L1
Teaching Transparency 23

Figure 12-2

Stomata are important to the gas-exchange process for photosynthesis to occur.

Sunlight

A A leaf's upper surface is covered by a cuticle.

Cuticle

Palisade layer

Guard cells

Stomata

Vein

CO₂ enters

H₂O leaves

B Carbon dioxide enters the leaf through the stomata. Oxygen also enters the stomata.

C The products of photosynthesis, water vapor and oxygen, exit through the stomata.

As shown above in **Figure 12-2,** light, water, and carbon dioxide all affect the opening and closing of stomata. Stomata usually are open during the day and closed at night. Less carbon dioxide enters and less water vapor escapes from the leaf when stomata are closed.

If you did the Explore Activity on page 313, you saw that water vapor condensed on the inside of the plastic bag. Loss of water vapor through stomata of a leaf is called **transpiration.** Far more water is lost through transpiration than is used during the food-making process of photosynthesis. During the day, stomata are open to allow necessary gases to be exchanged during photosynthesis.

Visual Learning

Figure 12-1 **Would adding salt to the soil around a plant make the stomata open or close?** *Salt would cause the water to leave the plant by osmosis. The decrease in water pressure would cause the stomata to close.* **LEP** **LS**

Figure 12-2 Make sure students understand that the function of the stomata is to allow the diffusion of carbon dioxide into the plant's photosynthesizing tissues. Water escapes from the plant when stomata are opened, but this is not their main function. **LEP** **LS**

2 Teach

Demonstration

LS **Visual-Spatial** You can model guard cells of a stoma by using the following demonstration. Blow up two balloons. Show students that when the balloons are blown up they remain straight. Now place cellophane tape along the side of two balloons as you blow them up. The balloons will bow outward. When they are put together, they will form an opening resembling a stoma. **LEP**

Use **Teaching Transparency 23** as you teach this lesson.

Use Lab 28 in the **Lab Manual** as you teach this lesson.

GLENCOE TECHNOLOGY

 CD-ROM

Life Science CD-ROM
Have students perform the interactive explorations for Chapter 12 to reinforce important chapter concepts and thinking processes.

NATIONAL GEOGRAPHIC SOCIETY

 Videodisc

Newton's Apple: Life Sciences
Plant Growth
Chapter 2, Side A
How do plants regulate water?

27281-28231

315

Activity 12-1 Stomata in Leaves

One of the interesting things about leaves is how stomata function to allow carbon dioxide gas into the leaf. Stomata are usually invisible without the use of a microscope. Try this activity to see some stomata for yourself.

Problem
Where are stomata in lettuce leaves?

Materials 🖐

• lettuce in dish of water
• coverslip
• microscope
• microscope slide
• salt solution
• forceps

Procedure

1. Remove a piece of an outer green leaf of the lettuce. Choose a leaf that is turgid (stiff) from absorbing water in the dish.
2. Bend the leaf back and use the forceps to strip off some of the transparent tissue covering the leaf. This is the epidermis. Prepare a wet mount of a small piece of this tissue.
3. Examine the preparation under low and then high power on the microscope. Draw and label the leaf section in your data table.
4. Note the location and spacing of the stomata. Count the number of stomata that are open.
5. Make a second wet mount of the lettuce leaf epidermis. This time, place a few drops of salt solution on the leaf. Predict what will happen to the stomata.
6. Examine the preparation under low and then high power on the microscope. Draw and label the leaf section in your data table.

7. Note the location and spacing of the stomata. Count the number of stomata that are open.

Data and Observations

Water Mount	Salt Solution Mount
Number of stomata open	Number of stomata open

Analyze

1. Describe the guard cells around a stoma.
2. How many stomata did you see in each leaf preparation?
3. How are guard cells different from the other cells of the leaf epidermis?
4. **Hypothesize** why a lettuce leaf becomes stiff in water.

Conclude and Apply

5. Were more stomata open or closed in the salt solution? Explain.
6. What can you **infer** about the function of stomata in a leaf?
7. How can you **conclude** that guard cells take part in photosynthesis?

316   Chapter 12 Plant Processes

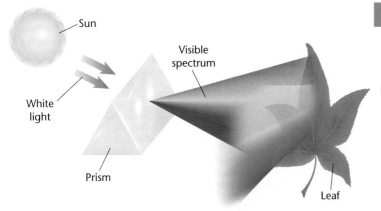

Sun

White light

Prism

Visible spectrum

Leaf

Figure 12-3

Light and chlorophyll are both essential parts of photosynthesis.

A As white light from the sun passes through a prism, it is separated into the colors of the visible spectrum. As white light strikes a green leaf, most of the colors, except green and yellow, are absorbed. Green and yellow are reflected by the leaf and are seen by the viewer.

Spring

Summer

Fall

B Deciduous trees, such as this sweet gum, change color in the autumn. Red pigments are usually masked by the green pigment chlorophyll during spring and summer. In the fall, the leaves lose their chlorophyll and the red pigments become visible.

Photosynthesis

Why aren't all the leaves of the trees in **Figure 12-3B** green? If you live in a place that has changing seasons, you know that the photograph on the far right was taken in the fall. Many trees and bushes change color as the days get shorter and the weather grows colder. Plants may change from green to red, brown, yellow, or orange. Some show all of these colors at the same time. These colors are the results of pigments such as the carotenoids that are present in plant leaves but are usually masked by the green color of chlorophyll. In the spring and summer, the green pigment chlorophyll is so abundant in the leaves that it covers up, or masks, all other pigments. In the fall, plant cells stop producing chlorophyll, and the other pigments become visible.

INTEGRATION Physics

INTEGRATION Physics

In physics, light is studied as energy in the form of packets and as waves. The packets of energy are called quanta. Shorter wavelengths of light provide greater amounts of energy.

12-1 Photosynthesis and Respiration **317**

Where does the energy for photosynthesis come from? *Light absorbed by the chlorophyll is the source of energy to drive the photosynthetic reaction.*

MiniLAB

Purpose

Visual-Spatial Students will relate the use of carbon dioxide by photosynthesis in plants by observing a chemical change. [L1]

COOP LEARN

Materials

25 mL of phenol red, two beakers, two straws, pieces of *Elodea*

Teaching Strategies

Test this before class to be sure that color changes are apparent.

Analysis

1. Carbon dioxide
2. Phenol red changes to orange when carbon dioxide is added.
3. When *Elodea* is added and light is provided, the plants begin photosynthesis and remove the added carbon dioxide, causing a change from orange to red in the beakers.

✓ Assessment

Performance To further assess students' understanding of laboratory methods used to detect carbon dioxide, see Performance Assessment, question 3, on p. 337. Use the Performance Task Assessment List for Designing an Experiment in **PASC**, p. 23. **P**

📁 **Activity Worksheets,** pp. 5, 74

MiniLAB

How can the use of carbon dioxide by plants be shown?

Procedure 🥽 🧪 🧤

1. Pour 25 mL of phenol red solution into a beaker, noting the original color.
2. Blow carefully through a straw. **CAUTION:** *Do not inhale.* Note the color.
3. Repeat this procedure in a second beaker to which you add a sprig of *Elodea*.
4. Place under a bright light for 15 minutes, observing every five minutes.

Analysis

1. What gas did you add to the solution?
2. Record in your Science Journal any observations you can make.
3. Relate the changes to photosynthesis.

As demonstrated in **Figure 12-3A** on page 317, white light from the sun passes through a prism. The prism then separates the white light into the colors of the visible spectrum. As white light strikes a green leaf, most of the wavelengths of light are absorbed except green and yellow. Green and yellow reflected back by the leaf are the colors that we see. Pigments within the plant, other than chlorophyll, trap energy from colors of light that chlorophyll does not absorb well. They give cells the colors red and yellow that you see in carrots, tomatoes, some fruits, and fall leaves.

The Food-Making Process

Chlorophyll is a pigment that traps light energy for plants to use in making food. **Photosynthesis,** illustrated in **Figure 12-4,** is the process in which plants use light energy to produce food.

What do plants need besides light to make food? During photosynthesis, a plant makes food from light energy, carbon dioxide, and water in a series of chemical reactions. Some of the light energy trapped in the chlorophyll is used to split water molecules into hydrogen and oxygen. Light energy is then used to join hydrogen and carbon dioxide together to form a new molecule of sugar. The sugar formed is glucose, the food a plant uses for maintenance and growth. The process is illustrated in the following equation:

$$6CO_2 + 6H_2O + \text{light energy} \longrightarrow C_6H_{12}O_6 + 6O_2$$

carbon dioxide water   chlorophyll sugar oxygen

Figure 12-4

During photosynthesis, carbon dioxide from the air, water in the soil, and energy from the sun react to form sugar and oxygen.

Energy from sunlight

Oxygen (O₂)

Carbon dioxide (CO₂)

Chlorophyll in leaves

Sugar (C₆H₁₂O₆)

Water (H₂O)

Theme Connection

Systems and Interactions This theme is reinforced by the discussion of photosynthesis and respiration. Ask students to explain what would happen to living things if either process were suddenly halted.

It takes six molecules of carbon dioxide (CO_2) and six molecules of water (H_2O) to form one molecule of sugar ($C_6H_{12}O_6$), plus six molecules of oxygen (O_2). That is, the reactants of photosynthesis are carbon dioxide and water, and the products of photosynthesis are sugar and oxygen.

What happens to these products in a plant? Plant cells make use of the products of photosynthesis in many ways. Some of the oxygen is used by the plant cells when breaking down food to release energy. The rest of the oxygen is released as a waste product and leaves the leaf through stomata. Sugar produced by plants is used for plant life processes such as growth, or it can be stored in the roots or stems of the plant as sugar, starch, or other products. When you eat beets, carrots, potatoes, or onions, you are eating stored food.

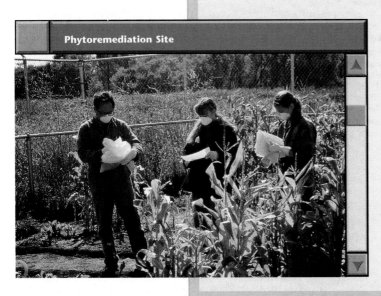

Phytoremediation Site

USING TECHNOLOGY

Pollution Cleanup

How are sites that have been polluted by industries that have gone out of business cleaned up? As pollution sites are reported, a list is made of those that are most polluted. Generally, the more polluted a site or the greater the danger that exists to the local population, the faster the site will be cleaned up. But the cost associated with such methods of cleaning up pollution sites is tremendous.

Studies are presently underway to test a cleanup method called *phytoremediation* in which plants do the cleaning. Different plants can live in different toxic areas. For instance, there are certain plants that can live in soil with high concentrations of arsenic, a toxic metal. These plants have been found to act as natural soil cleaners by absorbing some of the toxic materials in the soil. Then the plants can be harvested safely and treated as if they were ores. Just as metals are extracted from ores, the toxic metals can likewise be extracted from these plants. Interest in this method is increasing because of its advantages over other methods of cleanup.

Think Critically:

Why would it be advantageous to develop a method for harvesting plants as ore at polluted sites?

USING TECHNOLOGY

Have students use a dictionary to determine the meanings of *phyto-* and *remediation*. Then have them find other words that have these roots in them.

Think Critically
If plants can be found that accumulate toxic metals at a rate high enough that they can be harvested as ore, the method would be safer, cheaper, and more efficient than the cleanup methods presently used.

Discussion Question
Have students describe the chemical reactions that take place in photosynthesis and respiration. In photosynthesis, water splits into hydrogen and oxygen; oxygen escapes; and hydrogen is added to carbon dioxide, forming sugar. In respiration, oxygen is combined with sugar, and carbon dioxide, water vapor, and energy are released.

Visual Learning

Figure 12-4 Have students use the figure to trace the pathway of each of the reactants and products of photosynthesis in a plant. **LEP** **LS**

NATIONAL GEOGRAPHIC SOCIETY

 Videodisc

Newton's Apple: Life Sciences
Plant Growth
Chapter 2, Side A
Four ingredients of photosynthesis

20103-20376

Figure 12-5
In large cities with little green space, people will sometimes grow plants on top of buildings.

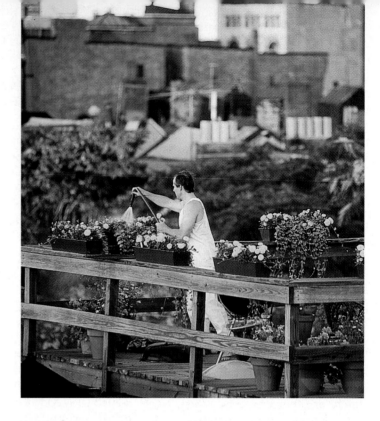

CONNECT TO
PHYSICS

Light energy can be described as waves or particles. Particles of light energy are called photons. One photon of light contains enough energy to excite a chlorophyll molecule. *Infer* what color of light is least used by a chlorophyll molecule.

Importance of Photosynthesis

Photosynthesis occurs only in producers—some bacteria, algae, and plants. In most algae and photosynthetic bacteria, photosynthesis occurs in all cells. However, in green plants, only certain cells are specialized for the purpose of photosynthesis.

Why is photosynthesis important to living things? An important role of photosynthesis is food production. Photosynthetic organisms are producers that provide food for nearly all the consumers on Earth. Second, through photosynthesis, which is occurring in the plants in **Figure 12-5**, plants remove carbon dioxide from the air and produce the oxygen that most organisms need to stay alive. As much as 90 percent of the oxygen entering our atmosphere today is a result of photosynthesis.

Respiration

Look at the photographs in **Figure 12-6.** Do these organisms have anything in common? Both of these organisms are similar in that they break down food to release energy.

320 Chapter 12 Plant Processes

Science Journal

Oxygen Supply In the temperate forests of North America, trees lose their leaves in the fall and winter. Do research to discover what keeps the world supplied with oxygen during this time. In your Science Journal, discuss what you learned about single-celled organisms that photosynthesize in marine biomes. **L2** **LS**

Respiration is the process by which organisms break down food to release energy. Respiration that uses oxygen to break down food chemically is called aerobic respiration. Cellular respiration occurs in the mitochondria of all eukaryotic cells. The overall chemical equation for aerobic respiration is:

$$C_6H_{12}O_6 + 6O_2 \longrightarrow 6CO_2 + 6H_2O + energy$$

| sugar | oxygen | carbon dioxide | water |

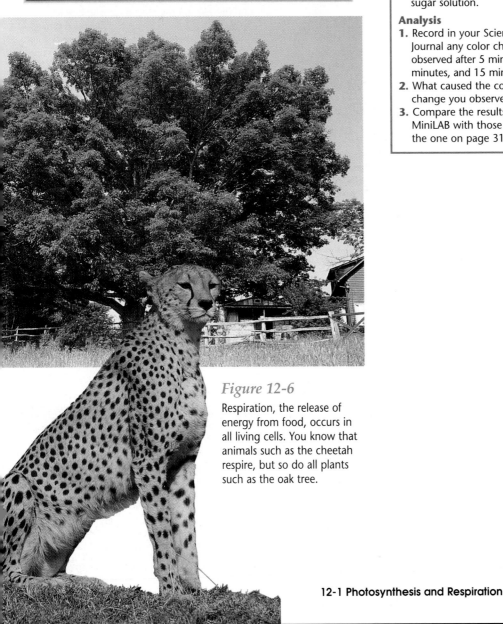

Figure 12-6

Respiration, the release of energy from food, occurs in all living cells. You know that animals such as the cheetah respire, but so do all plants such as the oak tree.

MiniLAB

How can respiration in yeast be shown?

Procedure

1. Pour 10 mL of bromothymol blue into a test tube.
2. Add 20 drops of yeast suspension and 10 drops of sugar solution.

Analysis

1. Record in your Science Journal any color change observed after 5 minutes, 10 minutes, and 15 minutes.
2. What caused the color change you observed?
3. Compare the results of this MiniLAB with those from the one on page 318.

MiniLAB

Purpose

Logical-Mathematical Students will record and analyze data that illustrate respiration in yeast. **L1**

COOP LEARN

Materials

10 mL bromothymol blue, 20 drops yeast suspension, 10 drops sugar solution, test tube

Teaching Strategies

Try the activity several times to gauge the amount of color change after 5, 10, and 15 minutes.

Analysis

1. The color gradually changes from blue to green over the 15-minute period.
2. Yeast carries on respiration, adding carbon dioxide to the solution.
3. Similar results should have occurred.

Assessment

Performance To further assess students' understanding of bromothymol blue as an indicator, design an experiment to test a hypothesis that states *Respiration takes place in plants only at night.* Use the Performance Task Assessment List for Designing an Experiment in **PASC**, p. 23. **P**

Activity Worksheets, pp. 5, 75

Science Journal

Forming an Operational Definition Have students write their own definitions of *photosynthesis* and *respiration*.

Visual Learning

Figure 12-6 Use this figure to emphasize that all organisms, including those that carry on photosynthesis, must respire. **LEP**

Is the equation for respiration familiar? How does it relate to the chemical reaction for photosynthesis? During photosynthesis, energy is stored in food. Photosynthesis occurs only in cells that contain chlorophyll, such as those in the leaves of plants. Respiration, as illustrated in **Figure 12-7**, occurs in all cells of all organisms. Respiration releases energy from food. **Table 12-1** compares the processes of photosynthesis and respiration.

Figure 12-7

Respiration takes place in the mitochondria in all cells.

B During these chemical reactions, energy is released. In this chemical reaction, oxygen and glucose are the reactants. *What are the products?*

Water

Energy

Carbon dioxide

Oxygen

A Glucose molecules are broken apart into simpler molecules. These then enter the mitochondrion, where they combine with oxygen molecules to form water and carbon dioxide.

Simpler substances

Glucose

Table 12-1

Comparing Photosynthesis and Respiration				
	Energy	**Raw Materials**	**End Products**	**In What Cells**
Photosynthesis	stored	water and carbon dioxide, plus energy	sugar, oxygen	cells with chlorophyll
Respiration	released	sugar, oxygen	water and carbon dioxide, plus energy	all cells

Importance of Respiration

Why do plant and animal cells need energy produced from the process of respiration? They use energy to build and repair their cells, tissues, and organs. Sugar produced during photosynthesis by plant cells is used to make cellulose to build cell walls. Plants also use energy from the food they make during photosynthesis for growth and reproduction, as illustrated in **Figure 12-8.**

Figure 12-8

Humans and other animals depend on the sugars produced by plants during photosynthesis. Animals use the sugars to produce energy through respiration.

A Some plants such as potatoes store extra food in special storage structures to be used later for other activities.

B Wheat is the most important source of food for a large portion of the world's population.

Section Wrap-up

Review

1. Explain how carbon dioxide and water vapor are exchanged in a leaf.

2. Why are photosynthesis and respiration important?

3. **Think Critically:** Humidity is caused by water vapor in the air. Why is humidity high in lush tropical rain forests?

Skill Builder
Comparing and Contrasting
Compare the raw materials and end products of photosynthesis and respiration and what happens to the energy involved in each process. If you need help, refer to Comparing and Contrasting in the **Skill Handbook.**

USING MATH

How many carbon dioxide molecules (CO_2) result from the respiration of six sugar molecules ($C_6H_{12}O_6$)?

Skill Builder
See **Table 12-1** at the bottom of student page 322.

Assessment

Portfolio Use this Skill Builder to assess students' abilities to compare and contrast photosynthesis and respiration. Use the Performance Task Assessment List for Analyzing the Data in **PASC,** p. 27. **P**

4 Close

•MINI•QUIZ•

Use the Mini Quiz to check students' recall of chapter content.

1. **Carbon dioxide and water vapor are gases that are exchanged by _____ through the stomata on leaf surfaces.** *diffusion*

2. **Loss of water vapor through stomata of a leaf is called _____ .** *transpiration*

3. **_____ is the process in which plants use energy from light to produce food.** *Photosynthesis*

4. **Aerobic respiration occurs in the _____ of cells.** *mitochondria*

Section Wrap-up

Review

1. Carbon dioxide and water vapor are exchanged by diffusion through the stomata on leaf surfaces.

2. Photosynthesis and respiration form a cycle of energy capture and use.

3. **Think Critically** Humidity is high in tropical rain forests because of the tremendous amount of water vapor transpired.

USING MATH

LS **Logical-Mathematical** For every sugar molecule respired there are six carbon dioxide molecules produced. So $6 \times 6 = 36$.

Prepare

Section Background

- Japanese scientists Yabuta and Sumiki discovered growth hormones called gibberellins during a study of rice plants that grew extremely tall because of "foolish rice disease," later found to be an overproduction of the hormone.
- Poinsettias can be made to flower by controlling the amount of darkness they receive several weeks prior to the time flowers are desired.

Preplanning

To prepare for Activity 12-2, obtain petri dishes, corn seeds, and bean seeds.

1 Motivate

Bellringer

Before presenting the lesson, display **Section Focus Transparency 43** on the overhead projector. Assign the accompanying **Focus Activity** worksheet. L1 LEP

Tying to Previous Knowledge

Review plant parts such as roots, stems, and leaves with students. Have them name the functions of each part.

Visual Learning

Figure 12-9 You may use **Figure 12-9** to introduce this section on plant responses. Point out that plants usually respond much more slowly than animals to stimuli, but thigmotropism can be a quick response. LEP IS

12•2 Plant Responses

Science Words

tropism
auxin
photoperiodism
long-day plant
short-day plant
day-neutral plant

Objectives

- Explain the relationship between stimuli and tropisms in plants.
- Differentiate between long-day and short-day plants.
- Explain the relationship between plant hormones and responses.

What are plant responses?

If you touch the leaflets of the *Mimosa pudica* plant in **Figure 12-9**, they will close up. Why does this happen? The leaflets fold up in response to the stimulus of being touched. A stimulus is anything in the environment that causes a change in the behavior of an organism. The response of a plant to a stimulus is a **tropism.** Most plants respond to stimuli by growing. Tropisms can be positive or negative. For example, plants might grow toward or away from a stimulus. Plants respond to stimuli by changing their behavior.

Tropisms

Touch is one stimulus that results in a change in a plant's behavior. The response to touch is *thigmotropism*, from the Greek word *thigma*, meaning "touch." Plants also respond to the stimuli of light, gravity, temperature, and amount of water. Some of the responses of plants are the result of plant hormones or chemical reactions in plant cells.

Did you ever see a plant that appeared to be leaning toward a window? Light is an important stimulus to plants. Plants respond to sunlight by enlarging or elongating the cells on the side of the plant opposite to the light source, causing the plant to bend in the direction of the light, or grow. The growth

Figure 12-9

The *Mimosa* plant's leaflets shown on the left respond to touch by folding up as demonstrated on the right. The response to touch is called *thigmotropism*.

324   Chapter 12 Plant Processes

Program Resources

Reproducible Masters

Study Guide, p. 47 L1
Reinforcement, p. 47 L1
Enrichment, p. 47 L3
Cross-Curricular Integration, p. 16
Activity Worksheets, pp. 5, 72-73 L1
Critical Thinking/Problem Solving, p. 16
Lab Manual 32
Concept Mapping, p. 29

Transparencies

Section Focus Transparency 43 L1
Teaching Transparency 24
Science Integration Transparency 12

response of a plant to light is called *phototropism*, illustrated in **Figure 12-10.** When the plant responds to light by growing toward it, the response is called a positive phototropism.

The growth response of an organism to gravity is called *gravitropism.* Plant roots tend to grow downward and the stems to grow upward. The downward growth of a plant's roots in response to the force of gravity is a positive gravitropism. Upward growth of the plant's stem away from the force of gravity is called negative gravitropism.

Plant Hormones

In garden centers and greenhouses, pots in which seedlings are grown are turned every few days. This is done because the seedlings lean toward the sunlight. As you have learned, this is an example of positive phototropism. But why do the seedlings respond to light in this way?

Auxin and Ethylene

Plant tropisms are controlled by plant hormones, which are chemical substances that affect growth. An **auxin** is a type of plant hormone. Auxins cause plant stems and leaves to exhibit positive phototropism. If light shines on a plant from

12-2 Plant Responses   **325**

2 Teach

Discussion

If possible, obtain a *Mimosa pudica* plant from a biological supply house. Touch one of the leaflets with a finger. Ask students to explain what they see.

Demonstration

Cut two stems of coleus. Apply a rooting hormone on one stem. Have students observe both stems for several days. Ask students to explain the results.

Use these program resources as you teach this lesson.
Concept Mapping, p. 29
Lab Manual 32

Integrating the Sciences

Physics Bring in a Venus's-flytrap plant. Touch the hairs to trigger the closing of the leaf. Ask students to relate the closing of the leaf to physics. *Pressure changes in the cells at the junction of the trap cause the leaf to close suddenly.*

Visual Learning

Figure 12-11 Use **Figure 12-11** to point out that hormones may be very simple molecules that cause a complex response in organisms. **LEP** **LS**

Use **Teaching Transparency 24** as you teach this lesson.

Figure 12-11

Ethylene, C_2H_2, is the plant hormone responsible for fruit ripening, such as these grapes found in a vineyard in Ontario, Canada.

one side, the auxin moves to the shaded side of the stem. The auxin causes cells on the shaded side of the stem to grow longer than cells on the lighted side. This causes the stem to curve toward the light.

Ethylene gas, a simple chemical compound of carbon and hydrogen, is another plant hormone. Many plants produce ethylene gas, illustrated in **Figure 12-11,** which causes fruits to ripen.

When fruits such as oranges, tomatoes, grapes, and bananas are picked before they have ripened, they stay green because their cells won't produce ethylene. Growers and shippers of fruit found out that it doesn't matter to the fruit where ethylene comes from for ripening. Growers now pick fruit before it is ripe. Shippers treat the picked fruit with ethylene gas during shipping. By the time fruit arrives at the store, it has ripened.

Figure 12-12

Long-day plants such as *Zinnia* and short-day plants such as primroses flower in response to specific periods of darkness.

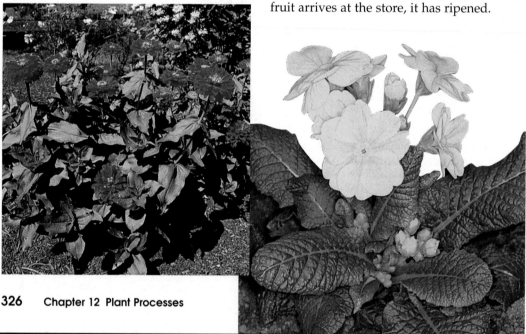

326 Chapter 12 Plant Processes

Community Connection

Forced Flowering Have students visit a commercial greenhouse to find out how plant growers encourage flowering of plants out of their normal season for flower production.

Cultural Diversity

Plant Foods Have students research the various plant foods eaten in different areas of the world. Ask them to relate the foods to the flowering patterns of the plants. **L2**

Photoperiods

Sunflowers bloom in the summer and cherry trees flower in the spring. Different plant species produce flowers at different times during the year. **Photoperiodism** is the flowering response of a plant to the change in the lengths of light and dark in a day.

The yearly motion of Earth around the sun is called a revolution. Earth is tilted during orbit and as a result, the seasons are produced. Photoperiods, the lengths of day and night, change with the changing seasons. For instance, have you noticed that during the winter, days are shorter than nights? Summers have the opposite photoperiod. Biologists who study photoperiodism in plants must be aware of these Earth science concepts.

Most plants require a specific period of darkness to flower. Plants that require short nights to flower are called **long-day plants.** You may be familiar with long-day plants such as spinach, lettuce, and potatoes. Similarly, plants that require long nights to flower are called **short-day plants.** Some short-day plants are poinsettias, strawberries, and ragweed. **Figure 12-12** shows both long-day plants and short-day plants.

Problem Solving

The Forgotten Plant

Jason's grandmother was preparing to visit her sick sister for a few weeks and needed Jason's help to get ready. Jason was so busy carrying out her luggage that on his way to the car, he knocked over a pot of geraniums that was growing on the patio. He knew that he should stop and pick up the pot, but he was in such a hurry that he kept going.

During the following two weeks, Jason completely forgot about the flowers. As he helped his grandmother unload the car upon her return, he noticed the pot on the way into the house. He picked up the pot, but to his surprise, the plant's stem had grown so that it was turning to the side. Jason admitted to his grandmother that he was responsible for what had happened to the plant. She told him not to worry about it because the plant would soon start to grow upright again.

Solve the Problem:

1. Explain why the plant grew as it did.
2. What hormone was likely responsible for the gravitropism?

Think Critically:

How could you prove that the plant growth was a normal response to a particular hormone?

Responses **327**

 Problem Solving

Solve the Problem

1. The plant was responding negatively to gravity and/or positively towards light.
2. auxin

Think Critically

You could perform an experiment by placing a higher concentration of plant hormone on one side of the plant than the other and observing the result.

📁 Use **Critical Thinking/ Problem Solving,** p. 16, as you teach this lesson.

3 Assess

Check for Understanding

❓ FLEX Your Brain

Use the Flex Your Brain activity to have students explore PHOTOPERIODISM.

📁 **Activity Worksheets,** p. 5

Reteach

Use different plants to demonstrate various tropisms. As an example, you might use a Venus's-flytrap to demonstrate thigmotropism or a plant that leans toward the light to demonstrate phototropism.

Extension

📁 For students who have mastered this section, use the **Reinforcement** and **Enrichment** masters.

Theme Connection

Systems and Interactions This theme is reinforced by the reaction of cells in plant organs to auxins and ethylene, two plant hormones. Have students look up the effects of other hormones such as gibberellins and cytokinins on plant tissues and organs. L1

Across the Curriculum

History Have students research the cause of "foolish rice disease" in Japanese rice plants. L2

Purpose
IS **Interpersonal** Students will design and carry out an experiment to observe plant tropisms in plant seedlings.
L1 **COOP LEARN**

Process Skills
observing and inferring, communicating, making and using tables, comparing and contrasting, recognizing cause and effect, forming operational definitions, forming a hypothesis, designing an experiment, separating and controlling variables, interpreting data, using numbers

Time
1 period to set up and then 10 minutes per day for up to two weeks

Materials
petri dish, tape, string, corn seeds, bean seeds, paper towel, water, ring stand

Safety Precautions
Some kinds of seeds are poisonous. Do not let any students put any seeds in their mouths.

Possible Hypotheses
Student hypotheses will vary. Possible hypotheses for this activity are "Corn seedling roots will show positive gravitropism;" "Corn seedling stems will show negative gravitropism;" and "Corn seedlings will not show any tropisms."

📁 **Activity Worksheets, pp. 5, 72-73**

Activity 12-2

Design Your Own Experiment
Plant Tropisms

A plant's response to a stimulus is called a tropism. Tropisms can be positive or negative. What kind of stimuli will plants respond to?

PREPARATION

Problem
How do plants respond to stimuli?

Form a Hypothesis
Based on your knowledge of tropisms, state a hypothesis about how the plant will respond to a stimulus.

Objectives
- Observe and analyze a plant response to a stimulus.
- Design an experiment that tests the effects of a variable, such as change in direction of light.

Possible Materials
- petri dish
- tape
- string
- corn seeds
- bean seeds
- paper towels
- water
- ring stand

Safety Precautions

Some kinds of seeds are poisonous. Do not put any seed in your mouth.

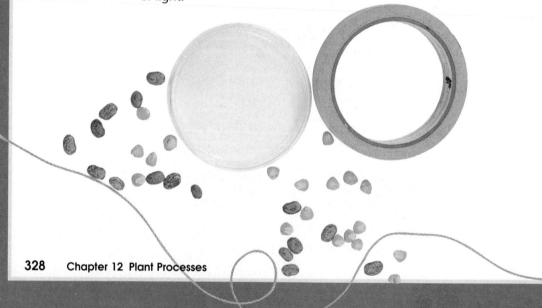

PLAN THE EXPERIMENT

Possible Procedures
Procedures will vary depending on the stimulus chosen by the student. In all cases, a control group in which the stimulus is not applied will be compared with an experimental group. Stimuli may be any of those discussed in the chapter or chosen by students.

PLAN THE EXPERIMENT

1. As a group, agree upon and write out the hypothesis statement.
2. As a group, list the steps that you need to take to test your hypothesis. Be specific, describing exactly what you will do at each step. List your materials.
3. Keep your seeds moist as you test for tropisms. Line the petri dish with paper towels before placing the seeds at equal distances around the outer edge of the petri dish. Tape the seeds to the paper towel and wet the towel.
4. Record the position of your seeds and devise a method to determine the effect of gravity on the seeds.
5. If you need a data table, design one in your Science Journal so that it is ready to use as your group collects data.

Check Your Plan

1. Read over your entire experiment to make sure that all steps are in logical order.
2. Identify any constants, variables, and controls of the experiment.
3. Is it necessary to run any tests more than one time?
4. Will the data be summarized in a graph? What kind of graph would be most useful?
5. *Make sure your teacher approves your plan before you proceed.*

DO THE EXPERIMENT

1. Carry out the experiment as planned.
2. While the experiment is going on, write down any observations that you make and complete the data table in your Science Journal.

Analyze and Apply

1. **Compare** your results with those of other groups.
2. **Identify** how the plants responded to the stimulus.
3. What name would you give to the movement you observed?
4. **Classify** the responses as positive or negative.

Go Further

Design an experiment to test the effect of another stimulus on a plant.

12-2 Plant Responses   **329**

Go Further

Students should produce an experimental design similar to the one for the original activity that will test a separate stimulus.

Assessment

Performance Students' performance of their experimental procedures should be the basis for assessment. Students' performance is more important than their results in this activity. Use the Performance Task Assessment List for Designing an Experiment in **PASC**, p. 23. **P**

Teaching Strategies
Tying to Previous Knowledge Review tropisms that are the result of stimuli from gravity, water, touch, etc. with students before they decide which stimulus they would like to test.

- **Troubleshooting** Be sure that students use a reasonable number of plants in each group.
- Be sure that students have a hypothesis that can be tested in the time available.
- If students decide to test a germinating seedling's response to gravity in a petri dish, be sure that the petri dish is kept vertical with seeds oriented in different directions.
- Make sure students understand that the same stimulus may elicit a positive tropism by one part of the plant while eliciting a negative tropism by another part.

DO THE EXPERIMENT

Expected Outcome
Outcomes will vary, depending on the stimulus chosen. For instance, plant stems and leaves react negatively to gravity but positively to light, while roots should have the opposite reactions.

Analyze and Apply
1. Groups with similar experimental designs should have similar results.
2. Depending on the stimulus, response may be positive or negative. Leaves and stems react positively to light, stems react negatively to gravity, etc.
3. The responses will be tropisms. The particular tropism depends on the experiment.
4. The tropism will be positive or negative, depending on the experiment.

329

4 Close

Section Wrap-up

Review

1. Phototropism: movement in response to light; gravitropism: response to gravity; thigmotropism: response to touch.
2. Long-day plants require short nights to produce flowers. Short-day plants require long nights to produce flowers. Day-neutral plants may produce flowers during a range of night lengths.
3. **Think Critically** Plant hormones are responsible for some tropisms.

Other plants, such as marigolds, corn, and other plants, shown in **Figure 12-13**, flower over a wide range of night lengths. Plants like these that aren't sensitive to the number of hours of darkness have a range of flowering times and are called **day-neutral plants.**

Fields of wildflowers bloom during late spring and summer, when bees and butterflies are most active and numerous. The photoperiodism of wildflowers may ensure that a plant produces its flowers at a time when there is an abundant population of pollinators. Photoperiodism also affects the distribution of flowering plants.

Figure 12-13

Examples of day-neutral plants are seen in this garden.

Section Wrap-up

Review

1. Describe the major forms of tropisms.
2. Compare three different photoperiod responses.
3. **Think Critically:** What is the relationship between plant hormones and tropisms?

Skill Builder
Comparing and Contrasting

Different plant parts exhibit positive and negative tropisms. Compare and contrast the responses of roots, stems, and leaves to light. If you need help, refer to Comparing and Contrasting in the **Skill Handbook.**

Skill Builder

Roots tend to grow away from light, while leaves and stems tend to grow toward light.

Science Journal

Imagine that you are a farmer in Costa Rica. For two years, you have tried unsuccessfully to grow strawberries. The berry plants grew very well both years, but they never produced any fruit. In your Science Journal, describe why you initially thought the berries should grow. Then explain what really happened.

Assessment

Performance Use this Skill Builder to assess students' abilities to compare and contrast the responses of roots, stems, and leaves to light by having them explain their results to you. Use the Performance Task Assessment List for Analyzing the Data in **PASC,** p. 27.

TECHNOLOGY:
Transgenic Crops 12•3

Using Biotechnology for Crop Modification

Within the last few years, a quick and precise method of improving crops has become available. This method, called **biotechnology,** is a process of moving DNA directly from one organism to another. Sometimes biotechnology uses DNA from organisms of the same species. At other times, DNA from one species is inserted into another species. A **transgenic crop** is one in which DNA from one kind of organism has been inserted into plants to improve the crop. The tomatoes in **Figure 12-14** are from transgenic plants. Improvements made include cotton plants that are no longer sensitive to herbicides, insect-resistant tobacco plants, and tomato plants whose fruit does not spoil quickly.

Science Words
biotechnology
transgenic crop

Objectives
• Identify two DNA technologies for improving crop production.
• Describe the possible benefits of biotechnology.

Figure 12-14

It is expected that in the future, more types of plants will be engineered to require less fertilizer, produce more fruit, or have the ability to grow in unfavorable conditions.

12-3 Transgenic Crops **331**

331

Figure 12-15

Two different technologies are used to produce transgenic crops.

A A particular bacterium, *Agrobacterium tumefaciens,* can cause plant diseases, but it can also be used to introduce DNA into a plant. The first thing that scientists do is remove the disease-causing genes from the bacterium. Then they insert known genes into the bacterium.

B After genes to be introduced have been decided on, scientists attach particular sequences of DNA to make the genes effective in only particular parts of a plant.

Transgenic Technologies

Transgenic crops are produced by two different technologies, as illustrated in **Figure 12-15.** The easiest method of producing transgenic crops is to insert DNA into plant cells using a virus or bacterium to transport the DNA. When the bacterium infects the plant, the DNA is transferred to the target plant cells, making them transgenic. Another method is to coat small metal balls with DNA and shoot them into the target cells of plants. This method was first used in 1987. The combination of these two methods gives scientists an unusual flexibility in modifying plants to do things that they wouldn't normally be able to do.

Benefits of Biotechnology

You may have noticed that many tomatoes in the grocery store are not red and ripe. This is because once tomatoes are ripened, they become mushy and spoil quickly. The first transgenic tomatoes that were approved for use had a gene inserted in them that slows the softening process. Therefore, the tomatoes could be left longer on the plant to become ripe and red, while maintaining their firmness for a longer period of time. There are many fruits such as peaches that could be altered in a similar way in the future.

Biotechnology has produced transgenic potatoes that produce their own insecticide. The insecticide is produced only in the leaves, which are attacked by the potato beetle. This modification to the plant enables the farmer to save considerable amounts of money by not applying insecticides to the plant, and it protects the environment. Also, there appear to be no insecticide residues on the potato when it reaches the market.

Scientists hope to produce transgenic crops in the future that can resist diseases. There appear to be some positive signs that this may happen soon. It is estimated that food production worldwide will have to triple in the next 40 years to meet the needs of the world's population. Biotechnology is one possible solution to this problem. If progress continues, only the imagination of trained scientists will limit the uses that can be made of the biotechnological approach to producing transgenic crops.

Figure 12-16
Potato beetle larvae.

Section Wrap-up

Review

1. What are two ways of producing transgenic crops?

2. What are the main benefits of the biotechnology used to produce transgenic crops?

Explore the Technology

How might biotechnology produce a transgenic crop that would improve your favorite food plant? In your Science Journal, describe the process that you would use to improve the plant if you were the scientist in charge.

*inter*NET
CONNECTION

Are foods from transgenic plants as safe as traditionally produced varieties? What might be advantages and disadvantages of using transgenic plants? Visit the Chapter 12 Internet Connection at Glencoe Online Science, **www.glencoe.com/ sec/science/life**, for a link to more information about transgenic plants.

SCIENCE & SOCIETY

Background

- According to Ms. Ninomiya, new varieties of roses for the cut-flower market are produced primarily by cross-pollination, whereby pollen from flowers on one plant is used to pollinate flowers on another. Test plants of new varieties are grown from the seeds that are produced.

- Sometimes mutations occur in plants that result in buds or other plant parts that develop differently from the norm. Most mutations have a negative outcome. But on rare occasions, a mutation results in improved flowers, fruit, or some other characteristic. Washington naval oranges are the result of a mutation that appeared on a sweet orange tree found growing in Brazil around 1820. The tree's owner noted the superior quality of the oranges growing on one branch, and multiplied the new variety, either by grafting or making cuttings from that branch. A half-century later, a missionary in Brazil sent a dozen naval orange trees to the U.S. commissioner of agriculture in Washington, D.C. Two of them were shipped to Riverside, California in 1873, and propagated vegetatively. The naval orange industry owes its beginnings to those two trees, one of which is still living.

People and Science

FLORA NINOMIYA, *Horticulturist*

On the Job

Q **Ms. Ninomiya, how did you get started in the nursery business, and what kind of nursery do you operate?**

A My grandfather established Ninomiya Nursery in 1913. It's a family business, and I'm the third generation of our family to be involved in growing flowers. We specialize in growing cut flowers, mostly roses. We have several hundred thousand rose plants growing in almost a million square meters of greenhouse space.

Q **Could you describe your duties in the nursery?**

A On a typical day, I would check to make sure that the watering of all the plants was going well, and that we didn't have any disease and insect problems. I would also check the flower-cutting process to monitor how much of each variety of flower we were cutting and make any necessary adjustments. Overall, it's my job to make sure that we produce good, salable flowers, and that we have enough crop for various holidays that come up. I also choose the new varieties of flowers that we are going

Personal Insights

to grow—there are new varieties coming along all the time.

Q **Are you using any new technologies in your flower-growing business?**

A Yes, we are just beginning to use the latest hydroponics technology in our greenhouses. With this new system, plants are grown in rocks, and there are computers that monitor the conditions and signal when the plants need water. Everything is watered automatically. Horticultural techniques are changing—I'm going to night school right now to learn more about computers!

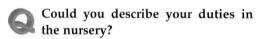

Career Connection

Choose one of these careers, and find out what educational course work and on-the-job training it requires. If possible, interview a person who works in the career you have chosen. Report the results of your investigation to your classmates.

Contact a vocational school and inquire about the following plant-related careers:

- **Turf Manager**
- **Arborist**
- **Plant Propagator**

334   Chapter 12 Plant Processes

Teaching Strategies

KS **Kinesthetic** After they research plant propagation techniques, have students make cuttings from a plant such as a Wandering Jew (*Zebrina pendula*), and root them in containers of water. Students can use the growing shoots in a classroom display about hydroponics. **L1**

Career Connection

Career Path A bachelor's degree in horticulture is usually required for a nursery management position like Flora Ninomiya's.

- High School: science and math
- College: plant science

Training for landscape designers, garden center managers, and similar plant-related careers can be obtained through two-year associate degree programs at vocational/technology schools.

For More Information

Your local County Extension Service can be a valuable source of information about plants, gardening, and plant-related issues.

Summary

12-1: Photosynthesis and Respiration

1. Carbon dioxide needed for photosynthesis enters leaves through stomata. Water vapor escapes during this process.
2. Chlorophyll in plants captures sunlight. Photosynthesis splits water molecules and makes glucose and oxygen, which are needed by all organisms.
3. Through respiration, energy stored in glucose becomes available for cell activities.

12-2: Plant Responses

1. Plants respond to stimuli by growing toward or away from light, gravity, or water.
2. Plants depend on a critical period of darkness in order to flower.
3. Many tropisms are controlled by hormones produced by plants.

12-3: Science and Society: Transgenic Crops

1. Transgenic crops are produced by two different biotechnologies. The easiest method is to insert DNA into the plant using a virus or bacterium. Another method is to coat small metal balls with DNA and shoot them into the cells.
2. The benefits of transgenic crops have been shown in such crops as tomatoes. Crops can be made that will produce fruit that will soften more slowly, resist disease, or produce their own insecticide.

Key Science Words

a. auxin
b. biotechnology
c. day-neutral plant
d. long-day plant
e. photoperiodism
f. photosynthesis
g. respiration
h. short-day plant
i. transgenic crop
j. transpiration
k. tropism

Reviewing Vocabulary

Match each phrase with the correct term from the list of Key Science Words.

1. a process of moving DNA directly from one organism to another
2. a plant hormone
3. flowering response of plant to light
4. using light to make glucose and oxygen
5. positive and negative plant responses
6. loss of water through stomata
7. plant that requires short nights to flower
8. plant that is not sensitive to hours of darkness
9. releasing energy from food
10. plant that has new DNA inserted into it

Chapter 12

Review

Summary

Have students read the summary statements to review the major concepts of the chapter.

Reviewing Vocabulary

1. b	**6.** j
2. a	**7.** d
3. e	**8.** c
4. f	**9.** g
5. k	**10.** i

✔ Assessment

Portfolio Encourage students to place in their portfolios one or two items of what they consider to be their best work. Examples include:

- MiniLAB assessment, p. 321
- Experiment design, pp. 328-329
- Science Journal, p. 330 P

Performance Additional performance assessments may be found in **Performance Assessment** and **Science Integration Activities.** Performance Task Assessment Lists and rubrics for evaluating these activities can be found in Glencoe's **Performance Assessment in the Science Classroom.**

GLENCOE TECHNOLOGY

 MindJogger Videoquiz

Chapter 12 Have students work in groups as they play the Videoquiz game to review key chapter concepts.

Chapter 12 Review

Checking Concepts

1. c
2. b
3. d
4. c
5. d
6. b
7. d
8. c
9. d
10. a

Understanding Concepts

11. Water is necessary to keep stomata open.

12. The raw materials of one process are the end products of the other.

13. Plants use energy from sunlight in photosynthesis.

14. Plants use energy to build and repair cells, tissues, and organs.

15. The grower knows that the fruit will be treated with ethylene gas during transit so that it will ripen by the time it arrives at the store.

Thinking Critically

16. If bananas were picked when they were ripe, they may be overripe by the time they arrive at a store.

17. (a)(b) negative gravitropism or positive phototropism (stem and tendrils up); (c) positive gravitropism (roots down); (d) positive phototropism (plant toward light)

18. Oxygen is a by-product of photosynthesis. Until organisms that used carbon dioxide for photosynthesis existed, little oxygen was available.

19. Crab-apple trees bloom in the spring, and sometimes in the fall, when the days are short and nights long. During these times, the increased dark periods may contribute to the trees' producing flowers.

20. Tropical countries have days and nights of about equal lengths for the entire year.

Checking Concepts

Choose the word or phrase that completes the sentence.

1. Stomata open to allow _____ the plant.
   a. sugar into
   b. sugar out of
   c. carbon dioxide into
   d. light into

2. A product of photosynthesis is _____.
   a. CO_2
   b. O_2
   c. H_2O
   d. H_2

3. Water, carbon dioxide, and energy are all products of _____.
   a. cell division
   b. photosynthesis
   c. growth
   d. respiration

4. A _____ plant needs short nights to flower.
   a. day-neutral
   b. short-day
   c. long-day
   d. nonvascular

5. Light, touch, and gravity are all examples of _____.
   a. tropisms
   b. growth behaviors
   c. responses
   d. stimuli

6. _____ is a plant's response to gravity.
   a. Phototropism
   b. Gravitropism
   c. Thigmotropism
   d. Hydrotropism

7. Stems, roots, and fruits are all plant parts that may respond to _____.
   a. tropisms
   b. sugar
   c. germination
   d. hormones

8. The primary reason leaves change colors is that the plant stops producing _____.
   a. hormones
   b. carotenoids
   c. chlorophyll
   d. pigments

9. The primary function of stomata is _____.
   a. photosynthesis
   b. to guard the interior cells
   c. to allow water to escape
   d. to permit carbon dioxide to enter the leaf

10. The products of photosynthesis are
   a. sugar and oxygen
   b. carbon dioxide and water
   c. chlorophyll and sugar
   d. carbon dioxide and oxygen

Understanding Concepts

Answer the following questions in your Science Journal using complete sentences.

11. What molecule is necessary to keep stomata open?

12. How do the raw materials and end products of photosynthesis and respiration compare?

13. Why is it important that plants grow toward the light?

14. Why do plants need energy?

15. Why would a grower intentionally pick unripened fruit?

Thinking Critically

16. Growers of bananas pick green bananas, then treat them with ethylene during shipping. Why?

17. Identify each response as positive or negative.
   a. stem grows up
   b. tendrils grow up
   c. roots grow down
   d. plant grows toward light

Developing Skills

21. **Hypothesizing** When water is limited and plants lose water through transpiration, guard cells close stomata to prevent further water loss.

22. **Observing and Inferring** Water plants that float on water have stomata on upper-leaf surfaces, whereas most land plants have more stomata on lower-leaf surfaces than on upper leaf surfaces.

18. Scientists who study sedimentary rocks and fossils suggest that oxygen did not occur on Earth until plantlike protists appeared. Why?

19. Explain why crab apple trees bloom in the spring, and sometimes in the fall, but not in the summer.

20. Why do day-neutral and long-day plants grow best in tropical countries?

Developing Skills

If you need help, refer to the **Skill Handbook.**

21. **Hypothesizing:** Make a hypothesis as to the action of leaf guard cells in desert plants.

22. **Observing and Inferring:** Based on your knowledge of plants, infer the amount and location of stomata in land and water plants.

23. **Classifying:** Classify each as short-day, long-day, or day-neutral plants by making a chart: corn, chrysanthemum, dandelions, dill, iris, poinsettia, red clover, spinach, sunflower, and tomato.

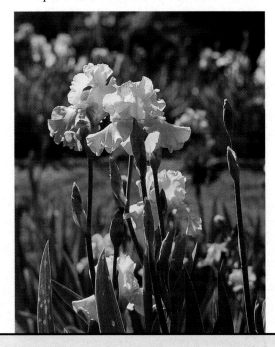

24. **Comparing and Contrasting:** Compare and contrast the action of each chemical on a plant: auxin, ethylene.

25. **Concept Mapping:** Complete the following concept map.

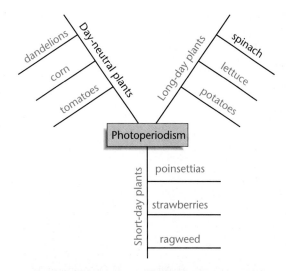

Performance Assessment

1. **Designing an Experiment:** Design a controlled experiment to test how well plant cuttings root with and without rooting hormone.

2. **Scientific Drawing:** Research the biotechnological methods of introducing DNA into plants. Write a paragraph in your Science Journal about how you could change a particular plant. Illustrate the differences as a *before drawing* and an *after drawing*.

3. **Designing an Experiment:** Design an experiment to show the rate of yeast respiration at three different temperatures.

Chapter 12 Review **337**

23. **Classifying** Short-day plants include iris, chrysanthemums, and poinsettias. Day-neutral plants include corn, tomatoes, and dandelions. Long-day plants include dill, red clover, spinach, and sunflowers.

24. **Comparing and Contrasting** auxin—stem grows toward light; ethylene—fruits ripen

25. **Concept Mapping** See student page.

Performance Assessment

1. Experimental designs should have a control group without hormone applications and one with hormones. Use the Performance Task Assessment List for Designing an Experiment in **PASC,** p. 23. P

2. The paragraph should describe the two methods listed in the Science and Society section on p. 332 and the drawing should reflect the descriptions. Use the Performance Task Assessment List for Scientific Drawing in **PASC,** p. 55. P

3. Experimental designs should have a control group at room temperature and a control to show respiration without yeast. Temperature variables should be within normal ranges for living organisms. Use the Performance Task Assessment List for Designing an Experiment in **PASC,** p. 23. P

Assessment Resources

Reproducible Masters
Chapter Review, pp. 27-28
Assessment, pp. 76-79
Performance Assessment, p. 26

Glencoe Technology
- **Chapter Review Software**
- **Computer Test Bank**
- **MindJogger Videoquiz**

UNIT
PROJECT
4

Homes Built from Plants

Objectives

IS **Kinesthetic** Students will make a model using information gained through observation and research. **COOP LEARN** **L1**

Summary

In this project, students will research various types of animal shelters and make a model of one. At least one group should be encouraged to build a life-size eagle's nest. They will also look at how the shelter reflects the needs of the animal and the materials available and how it is an adaptation to that environment for the animal involved. After building the model, students will have an opportunity to observe other animal home models.

Time Required

This project is designed to be worked on throughout Unit 4. One period is needed to allow groups to select an animal home to research. One period is needed for the model presentations. Actual building of models will have to be interspersed with lessons during the unit.

Preparation

- Establish cooperative groups of 4 or 5 students.
- Collect sources of information on animal homes.
- Plan on where students can collect materials for use in their models. Students will be using mostly plant materials.
- For larger models, such as a life-size eagle's nest, obtain permission from the school ar a local park where the nest can be constructed and be undisturbed.

ave you ever found an abandoned animal home? You may not have thought of it that way, but a bird's nest, a hornet's nest, an anthill, or a rabbit burrow is a carefully engineered home. Just like your home, these homes provide shelter and protection from the environment and predators. A nest is usually made up of twigs. Nest sizes differ. A hummingbird nest is the size of a quarter. Bald eagles' nests are large and may be 2 m in diameter. Eagles mate for life, and a nesting pair may return to the same nest for as long as 30 years. Each year, they add a little more to their nest, and it grows.

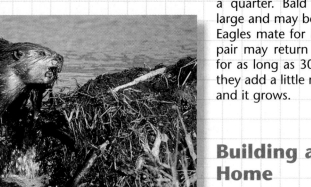

Building an Animal Home

Animals use all types of materials in their homes. Animals build with twigs, mud, grass, hair, fur, stone, wood, or anything else they can find. Ospreys, large birds that live near the water, may use discarded fishing line to build nests, and they frequently use seaweed. Some osprey nests have included bicycle tires and rubber gloves! Eagles, in contrast, are particular about what goes into their nests. Their nests are huge and made of sticks no less than one-half centimeter in diameter. Then they are lined with evergreens. These nests weigh hundreds of pounds. They are built to last many years. Built in trees, an eagle's nest is exposed to severe weather conditions.

Some animal homes are designed to blend into the environment. Prairie dog towns are basically underground. Other animals, like the magpie, a bird, like to collect shiny objects to "decorate" their homes.

Procedure

1. Work in groups to pick an animal home to research. What materials does the animal use? How big is the home? What kind of environment is involved? What is the actual building process the animal uses? How long is the home expected to last?

2. After researching a particular animal home, build a nest or burrow using the same materials and dimensions as the animal uses.

Using Your Research

Plan a display of your model homes that shows what is special about each animal home. Show how it fits into the particular environment in which it is found. How does it offer protection and shelter for the animal involved? Does the environment influence

the materials, and therefore the structure of the animal house? Think about the wind and rain that some structures have to withstand.

Go Further

Find out about animals that live in your area. What animals have you observed? If possible, go on a walk with a naturalist to learn more about the wildlife in your area. You may wish to focus particularly on bird nests. What types of materials do birds in your area use? Do they use twigs? Are the nests lined with feathers? Take pictures or make sketches of any nests you observe in your area. See whether you can decide what kind of bird is using the nest just by its size and makeup.

Unit 4 Project   **339**

Teaching Strategies

• Before having students choose animal homes to research, have them brainstorm some they are familiar with. List these on the chalkboard and have students analyze features common to all homes on the list. Help students to see that animal homes are affected by the materials available and the stresses of the environment. Draw from students the fact that plant materials and earth-related structures and materials (caves, soil burrows, and so on) are what animals draw on for building shelters. Discuss the needs that all living organisms have, bringing the focus onto the needs of shelter and protection, and how they help species survive. Discuss the effects of destruction of nesting sites on populations.

• Ask students about the specific shelter requirements of the animals they have named. For example, a reptile will seek out a shelter that helps it maintain a relatively stable body temperature. A desert dweller needs to have a shelter that provides a lower temperature and shade. An animal in a very cold climate might need a shelter that provides warmth.

Go Further

To assess students' understanding of the materials and engineering that go into animal-built shelters, have them construct nesting sites for various birds, such as a nest cone for mourning doves, or a bluebird box. Nesting materials can also be left out for the use of nearby birds. Hair from a hairbrush, string or yarn, and other supplies are useful to many birds. Interested students may also want to try constructing an animal burrow. A piece of clear plastic can be put on one side of the burrow model.

References

Boardman, Richard. *Animal Colonies: Development and Function Through Time.* New York: Van Nostrand Reinhold, 1973.

Cassidy, James, ed. *Book of North American Birds.* Pleasantville, NY: The Reader's Digest Association, Inc., 1990.

McDonnell, Janet. *Animal Builders.* Mankato, MN: Child's World, 1989.

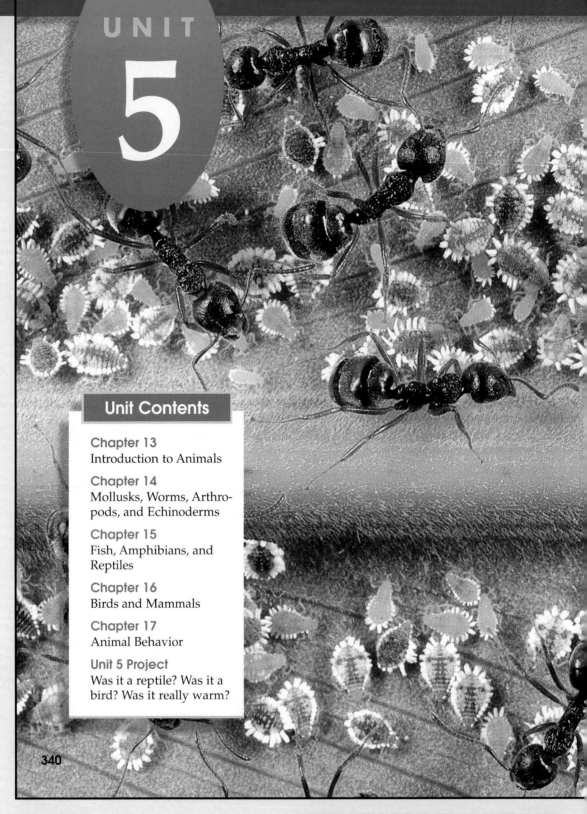

UNIT 5

Animals

Students examine animal characteristics and determine how they are used to classify animals into the different phyla. Characteristics of animals in the nine largest phyla are explored. At the conclusion of the unit, types of behavior and behavioral adaptations are presented.

CONTENTS

Chapter 13
Introduction to
Animals   342

13-1 What is an animal?

13-2 Science and Society:
      Animal
      Experimentation

13-3 Sponges and
      Cnidarians

13-4 Flatworms and
      Roundworms

Chapter 14
Mollusks, Worms,
Arthropods, and
Echinoderms   368

14-1 Mollusks

14-2 Segmented Worms

14-3 Arthropods

14-4 Science and Society:
      Pesticides and Insects

14-5 Echinoderms

Chapter 15
Fish, Amphibians,
and Reptiles   398

15-1 Fish

15-2 Amphibians

15-3 Science and Society:
      Amphibians and
      Ultraviolet Light

15-4 Reptiles

Chapter 16
Birds and Mammals   424

16-1 Birds

16-2 Mammals

16-3 Science and Society:
      California Sea Otters

Chapter 17
Animal Behavior   452

17-1 Types of Behavior

17-2 Behavioral Adaptations

17-3 Science and Society: Zoos and
      Captive Breeding

UNIT 5

Unit Contents

Chapter 13
Introduction to Animals

Chapter 14
Mollusks, Worms, Arthropods, and Echinoderms

Chapter 15
Fish, Amphibians, and Reptiles

Chapter 16
Birds and Mammals

Chapter 17
Animal Behavior

Unit 5 Project
Was it a reptile? Was it a bird? Was it really warm?

340

Science at Home

Pet Behavior Have students observe the behavior of a pet such as a dog, cat, or parakeet. Have them record how the animal reacts to people, food and other animals. Have students also observe a wild animal, such as a squirrel, in its environment. Have them compare the wild and domestic animals and share their observations with the class. **L1** **COOP LEARN**

Animals

What's Happening Here?

These ants aren't preparing to feast on the tiny aphids, but they do get food from them. Ants "farm" aphids for the sweet nectar they produce. The ants keep aphids in "pens" in underground burrows. In a similar relationship, a fish called the cleaner wrasse picks off and eats external parasites that annoy larger fish. The coral trout in this photograph pays visits to cleaner wrasse expressly to perform this service. Many relationships among animals benefit all the animals involved. In this unit, you will learn more about animals and their relationships.

Science Journal

African lions and spotted hyenas often fight over newly killed prey animals, such as gazelles and wildebeests. Researchers believed that lions killed such prey, but were robbed of their meal by hyenas. New evidence shows that it may be the lions that steal from hyenas. In your Science Journal, describe how you would determine which animal is the hunter.

341

Chapter Organizer

Section	Objectives/Standards	Activities/Features
Chapter Opener		Explore Activity: Identify which organisms are animals. p. 343
13-1 **What is an animal?** (½ session, ½ block)*	1. **Identify** the characteristics of animals. 2. **Determine** how the body plans of animals differ. 3. **Distingish** between invertebrates and vertebrates. National Content Standards: (5-8) UCP1, A1, C1, C5	Using Math, p. 345 Problem Solving: Identifying Animals, p. 346 Skill Builder: Concept Mapping, p. 347 Using Computers: Model, p. 347
13-2 **Science and Society Issue: Building Artificial Coral Reefs** (1 session, ½ block)*	1. **Discuss** the importance of coral reefs. 2. **Identify** some ways in which natural coral reefs are damaged. 3. **Describe** new technology for building artificial coral reefs. National Content Standards: (5-8) UCP1, UCP3, UCP4, UCP5, B2, C1, C4, C5, F2, F5	Internet Connection, p. 349
13-3 **Sponges and Cnidarians** (2 sessions, 1 1/2 blocks)*	1. **Identify** the structures that make up sponges and cnidarians. 2. **Describe** how sponges and cnidarians obtain food and how they reproduce. National Content Standards: (5-8) UCP1, UCP4, UCP5, A1, A2, B1, C1, C2, C3, D1	Connect to Chemistry, p. 351 MiniLAB: What are the adaptations of a sponge? p. 352 Activity 13-1: Observing a Cnidarian, p. 354 Skill Builder: Comparing and Contrasting, p. 356 Using Math, p. 356
13-4 **Flatworms and Roundworms** (2½ sessions, 2 blocks)*	1. **Compare** the body plans of flatworms and roundworms. 2. **Distinguish** between free-living and parasitic organisms. 3. **Identify** disease-causing flatworms and roundworms. National Content Standards: (5-8) UCP1, UCP5, A1, A2, C2, C3, E2, F5, G1	MiniLAB: How do planarians move? p. 358 Activity 13-2: Comparing Free-Living and Parasitic Worms, pp. 360-361 Using Technology: A Hookworm Vaccine, p. 362 Skill Builder: Sequencing, p. 363 Science Journal, p. 363 People and Science: Rebecca L. Johnson, Science Book Author, p. 364

* A complete Planning Guide that includes block scheduling is provided on pages 32T-35T.

Activity Materials

Explore	Activities	MiniLABs
page 343 No special materials are needed.	page 354 dropper, hydra culture, small dish, toothpick, *Daphnia* or brine shrimp, stereoscopic microscope pages 360-361 apron, culture dish with a planarian, prepared slide of a tapeworm, dropper, small piece of liver, small paintbrush, compound microscope, stereomicroscope, water, light source (small lamp)	page 352 *Grantia* sponge, small glass dish, hand lens, scissors, bleach, dropper, glass slide, coverslip, microscope page 358 dropper, planarians, watch glass, tap water, stereomicroscope

Need Materials? Call Science Kit (1-800-828-7777).

Teacher Classroom Resources

Reproducible Masters	Transparencies	Teaching Resources
Study Guide, p. 49 **Reinforcement**, p. 49 **Enrichment**, p. 49 **Cross-Curricular Integration**, p. 17	**Section Focus Transparency 45**, Into the Looking Glass **Teaching Transparency 25**, Animal Phylogeny	**Glencoe Life Science Interactive Videodisc** **Life Science CD-ROM** **Spanish Resources** **English/Spanish Audiocassettes** **Cooperative Learning Resource Guide** **Lab Partner** **Lab and Safety Skills** **Lesson Plans**
Study Guide, p. 50 **Reinforcement**, p. 50 **Enrichment**, p. 50 **Science and Society Integration**, p. 17	**Section Focus Transparency 46**, Coral Jewelry	
Study Guide, p. 51 **Reinforcement**, p. 51 **Enrichment**, p. 51 **Activity Worksheets**, pp. 76-77, 80 **Lab Manual 33**, Regeneration **Science Integration Activity 13**, Salt Concentration in Ocean Water **Concept Mapping**, p. 31 **Critical Thinking/Problem Solving**, p. 17	**Section Focus Transparency 47**, Harvesting Sponges **Teaching Transparency 26**, Cnidarians	**Assessment Resources** **Chapter Review**, pp. 29-30 **Assessment**, pp. 89-92 **Performance Assessment in the Science Classroom (PASC)** **MindJogger Videoquiz** **Alternate Assessment in the Science Classroom** **Performance Assessment** **Chapter Review Software** **Computer Test Bank**
Study Guide, p. 52 **Reinforcement**, p. 52 **Enrichment**, p. 52 **Activity Worksheets**, pp. 78-79, 81 **Concept Mapping**, p. 31 **Multicultural Connections**, pp. 29-30	**Section Focus Transparency 48**, Roundworms Underground **Science Integration Transparency 13**, A Look at Three Coral Reefs	

Key to Teaching Strategies

The following designations will help you decide which activities are appropriate for your students.

L1 Level 1 activities should be within the ability range of all students, including those with learning difficulties.

L2 Level 2 activities should be within the ability range of the average to above-average student.

L3 Level 3 activities are designed for the ability range of above-average students.

LEP LEP activities should be within the ability range of Limited English Proficiency students.

LS These activities are designed to address different learning styles.

COOP LEARN Cooperative Learning activities are designed for small group work.

P These strategies represent student products that can be placed into a best-work portfolio.

GLENCOE TECHNOLOGY

The following multimedia resources are available from Glencoe.

The Secret of Life
In the Land of Milk and Money: Biotechnology
Science and Technology Videodisc Series (STVS)
Animals
  Jellyfish

Blood Fluke Life Cycle
Nematodes
Ecology
  Evaluating Artificial Reefs
Life Science CD-ROM

Teacher Classroom Resources

This is a representation of key blackline masters available in the Teacher Classroom Resources.

Teaching Aids

Section Focus Transparencies

Science Integration Transparencies

Teaching Transparencies

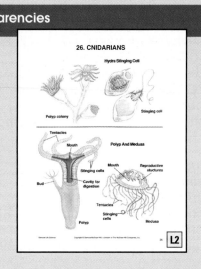

Meeting Different Ability Levels

Study Guide

Reinforcement

Enrichment Worksheets

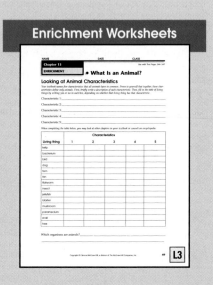

Chapter 13 Introduction to Animals

Hands-On Activities

Science Integration Activity

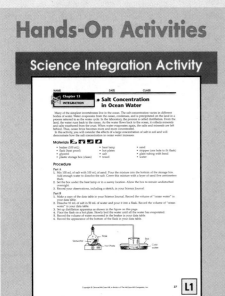

Salt Concentration in Ocean Water

L1

Lab Manual

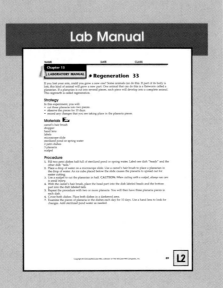

Regeneration 33

L2

Assessment

Performance Assessment

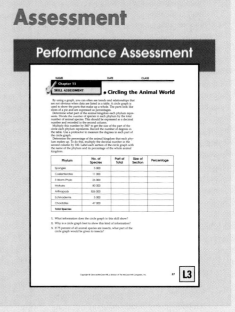

Circling the Animal World

L3

Enrichment and Application

Critical Thinking/ Problem Solving

Introduction to Animals

L2

Cross-Curricular Integration

Introduction to Animals

L1

Science and Society Integration

Introduction to Animals

L2

Multicultural Connections

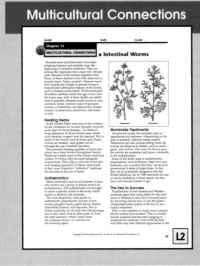

Intestinal Worms

L2

Concept Mapping

Introduction to Animals

L3

342D

Chapter 13

Introduction to Animals

CHAPTER OVERVIEW

Section 13-1 This section describes the characteristics that all animals have in common. The classification of animals as invertebrates or vertebrates and symmetry are discussed.

Section 13-2 Science and Society This section discusses the importance of coral reefs and identifies ways they are damaged. Technology for building artificial coral reefs using reef balls is described.

Section 13-3 This section introduces the animal phyla beginning with Porifera and Cnidaria. Their structure, methods of feeding, methods of reproduction, and origins are described.

Section 13-4 The flatworms and roundworms are described in detail. The body plan and way of life of each group is considered.

Chapter Vocabulary

vertebrate	hermaphro-
invertebrate	dite
radial	larva
symmetry	cnidarian
bilateral	tentacle
symmetry	polyp
sessile	medusa
filter feeder	free-living
collar cell	anus
regeneration	cyst

Theme Connection

Stability and Change Changing environments led to changes in complexity of animals inhabiting these environments. The groups of animals from invertebrates to vertebrates show a trend toward increasing complexity.

Previewing the Chapter

Section 13-1   What is an animal?
▶ Animal Characteristics
▶ Animal Classification

Section 13-2   Science and Society Technology: Building Artificial Coral Reefs

Section 13-3   Sponges and Cnidarians
▶ Sponges
▶ Cnidarians
▶ Origin of Cnidarians

Section 13-4   Flatworms and Roundworms
▶ Flatworms
▶ Roundworms

342

Learning Styles Look for the following logo for strategies that emphasize different learning modalities. LS

Kinesthetic	MiniLAB, p. 358
Visual-Spatial	Explore, p. 343; Demonstration, pp. 345, 351, 353; Reteach, p. 346; MiniLAB, p. 352; Activity, p. 359
Interpersonal	Activity, p. 345
Intrapersonal	Activity 13-1, p. 354
Logical-Mathematical	Across the Curriculum, pp. 345, 362; Using Math, p. 356; Activity 13-2, pp. 360-361
Linguistic	Across the Curriculum, p. 352; Using Science Words, p. 353; Reteach, p. 362; People and Science, p. 364

Introduction to Animals

Close your eyes and picture an animal. What comes to your mind? Did you imagine something with four legs and fur, or did you see an insect, a worm, or a person?

EXPLORE ACTIVITY

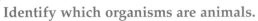

Identify which organisms are animals.

1. Look at the large photograph to the left.
2. On a sheet of paper, make a list of all the animals that you see in the photograph.
3. Identify the features that tell you these organisms are animals.
4. Now look at the smaller photograph below. Using your list of features of an animal, decide whether or not these organisms are animals. Explain your decision.

Observe: In your Science Journal, write a one-sentence definition of the word *animal*.

Previewing Science Skills

▶ In the **Skill Builders**, you will map concepts, sequence, and compare and contrast.

▶ In the **Activities**, you will observe, hypothesize, and collect data.

▶ In the **MiniLABs**, you will observe, describe, experiment, and predict.

343

EXPLORE ACTIVITY

Purpose

LS **Visual-Spatial** Use the Explore Activity to introduce the diversity of animals. Animal diversity can be seen in the chapter opening photograph. Explain to students that they will be learning more about the characteristics of animals. **L1** **LEP**

Teaching Strategies

• Ask students what animals are the most familiar to them. They probably will suggest dogs, cats, hamsters, fish, and other pets.

• Some students may not identify the organisms in the photograph, but they are fanworms. Fanworms are members of the annelid phylum, the same phylum to which earthworms belong.

Observe

Student answers will vary but should include that animals move around, consume their food, and reproduce.

✓ Assessment

Oral Ask students to apply their sentence description of an animal to a pet or zoo animal they are familiar with. If they cannot, suggest that they rethink their definition. Use the Performance Task Assessment List for Making Observations and Inferences in **PASC**, p. 17. **P**

Assessment Planner

Portfolio
Refer to page 365 for suggested items that students might select for their portfolios.

Performance Assessment
See page 365 for additional Performance Assessment options.
Skill Builders, pp. 347, 356, 363
MiniLABs, pp. 352, 358
Activities 13-1, p. 354; 13-2, pp. 360-361

Content Assessment
Section Wrap-ups, pp. 347, 349, 356, 363
Chapter Review, pp. 365-367
Mini Quizzes, pp. 347, 356, 363

Group Assessment
Opportunities for group assessment occur with Cooperative Learning Strategies and Flex Your Brain Activities.

Section 13•1

Prepare

Section Background

The branch of biology that deals with the study of animals is called zoology. Those scientists who study animals are zoologists. Zoologists group the animal kingdom into phyla. The nine largest phyla will be studied in this text. These phyla contain the majority of animal species.

Preplanning

Collect pictures of invertebrates and vertebrates for the bulletin board.

1 Motivate

Bellringer

Before presenting the lesson, display **Section Focus Transparency 45** on the overhead projector. Assign the accompanying **Focus Activity** worksheet. L1 LEP

Tying to Previous Knowledge

Review the characteristics of animals in Chapter 7. Make certain that students understand that animals are heterotrophic, have eukaryotic cells, and are many-celled.

Science Words

- vertebrate
- invertebrate
- radial symmetry
- bilateral symmetry

Objectives

- Identify the characteristics of animals.
- Determine how the body plans of animals differ.
- Distinguish between invertebrates and vertebrates.

Animal Characteristics

If someone asked you how the animal in **Figure 13-1** is different from a plant, what would you say? You might begin by saying that animals eat other living things, and they can move. Are these features enough to identify an organism as an animal? Let's look at all the characteristics that animals share.

1. Animals cannot make their own food. They depend on other living things in the environment for food. Some eat plants, some eat other animals, and some eat both plants and animals.
2. Animals digest their food. The proteins, fats, and carbohydrates in foods must be broken down into molecules small enough for their bodies to use.
3. Many animals move from place to place. Moving around lets them find food, mates, a better place to live, and escape from their enemies. Animals that move very slowly or not at all have adaptations that let them take care of these needs.
4. Animals have many cells. Different cells carry out different functions such as digesting food, getting rid of wastes, and reproducing.
5. Animal cells are eukaryotic. They each have a nucleus and organelles surrounded by membranes.

An animal is a many-celled eukaryotic organism that must find and digest its food. How does this description compare with the one you wrote for the Explore Activity on the previous page?

Figure 13-1

Animals depend on other living things for their food. *How do you get the nutrients you need to survive?*

344

Program Resources

 Reproducible Masters
Study Guide, p. 49 L1
Reinforcement, p. 49 L1
Enrichment, p. 49 L3
Cross-Curricular Integration, p. 17
Activity Worksheets, p. 5 L1
Multicultural Connections, pp. 29-30

 Transparencies
Section Focus Transparency 45 L1
Teaching Transparency 25

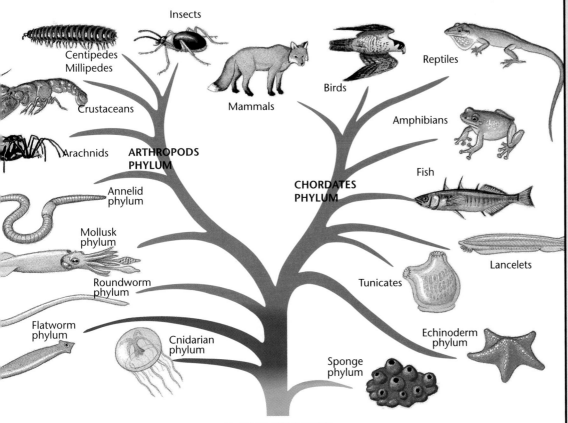

ARTHROPODS PHYLUM

Insects

Centipedes Millipedes

Crustaceans

Arachnids

Annelid phylum

Mollusk phylum

Roundworm phylum

Flatworm phylum

Cnidarian phylum

Mammals

Birds

Reptiles

Amphibians

CHORDATES PHYLUM

Fish

Lancelets

Tunicates

Echinoderm phylum

Sponge phylum

PROTIST ANCESTORS

Figure 13-2

This diagram shows the relationships among different groups in the animal kingdom.

Animal Classification

Scientists have identified and named more than 1 million species of animals. Some estimate that there are 3 to 30 million more to identify and name. Animals are classified into nine major phyla. The phyla are shown in **Figure 13-2.** All animals have the characteristics listed on page 344. When a scientist comes across a new animal, how does he or she begin to classify it?

Vertebrate or Invertebrate?

To classify an animal, first the scientist will look to see if the animal has a backbone. Animals with backbones are called **vertebrates.** Examples of vertebrate animals are fish, humans, birds, and snakes. About 97 percent of all animal species are invertebrates. **Invertebrates** are animals that don't have backbones. Sponges, jellyfish, worms, insects, and clams are all invertebrates.

USING MATH

In mathematics, a figure has *line symmetry* if it can be folded exactly in half. Draw a figure that has line symmetry and another figure that does not.

13-1 What is an animal?   **345**

Theme Connection

Scale and Structure Scientists classify animals on the basis of their body structure. They have either radial symmetry, bilateral symmetry, or no definite shape.

Across the Curriculum

Math Ask students to answer the following question: **A 1-cm³ sponge can filter 1 L of water in a 24-hour period. How many liters of water can a 2-m³ sponge filter in a 48-hour period?** *(2 m³ = 2 000 000; 2 000 000 L of water in 24 hours, 4 000 000 L of water in 48 hours)* **L1** **LS**

---

2 Teach

345

Where do you think the nerves and sense organs of animals with bilateral symmetry are located? *in the anterior end* Why? *It allows animals to sense the environment as they move forward.*

Problem Solving

Solve the Problem

2. No, because you would need to look at their cells to determine other characteristics.

Think Critically

To positively identify these organisms as animals, you would need to observe how they obtain food, whether they have many cells with different functions, and whether or not their cells are eukaryotic.

3 Assess

Check for Understanding

 FLEX Your Brain

Use the Flex Your Brain activity to have students explore CHARACTERISTICS OF ANIMALS.

📁 **Activity Worksheets,** p. 5

Reteach

🄛🄢 **Visual-Spatial** Have students make diagrams of a fish, a butterfly, and a dog and label the dorsal and ventral sides and the anterior and posterior ends.

Extension

📁 For students who have mastered this section, use the **Reinforcement** and **Enrichment** masters.

Problem Solving

Identifying Animals

It's summertime, and you and some friends are spending a few days at the beach. Your friends decide to do some fishing off a pier, but you don't enjoy fishing. You decide to go snorkeling instead. Putting on your flippers, face mask, and snorkel, you go exploring in the shallow water around the pier. You find lots of animals around the stones of the breakwater—clams, barnacles, a starfish, and lots of tiny fish. Attached to the stones, you also find several groups of small, brightly colored organisms that you have never seen before. You watch them for a while. They don't appear to move, except for waving gently in the water. Could these colorful organisms be animals?

Solve the Problem:

1. Compare your observations of the organisms below with the list of animal characteristics you have studied previously in this chapter.
2. Can you identify these organisms from your observations alone? Explain.

Think Critically:

How could you positively identify these organisms as animals?

346

Figure 13-3
Jellyfish have radial symmetry, butterflies have bilateral symmetry, and sponges are asymmetrical.

Radial symmetry

Bilateral symmetry

Asymmetry

Symmetry

After deciding whether or not a backbone is present, a scientist will look at the arrangement of the animal's body parts. This is called the animal's symmetry. Some animals have body parts arranged in a circle around a central point, the way spokes are arranged around the hub on a bicycle wheel. These animals have **radial symmetry.** Jellyfish have radial symmetry.

Bilateral Symmetry

Most animals have bilateral symmetry. Look in the mirror. Does your body look about the same on both sides? An animal with **bilateral symmetry** has its body parts arranged in the same way on both sides of its body. In Latin, the word *bilateral* means "two sides." Bilateral animals can be divided into right and left halves by drawing an imaginary line down the length of the body. Each half is a mirror image of the other half.

Some organisms have no definite shape and are called asymmetrical. There is no way their bodies can be divided into matching halves. Many sponges are asymmetrical. The three types of symmetry are shown in **Figure 13-3**. The parts of a bilaterally symmetrical animal are shown in **Figure 13-4**.

Once a scientist has determined whether an animal is an invertebrate or vertebrate and determined its symmetry, he or she will have to identify characteristics it has in common with other animal groups in order to classify it. You learned in Chapter 7 that animals in a group have similar characteristics because they descended from a common ancestor. The evolutionary history and relationships among animal phyla are represented by the diagram in **Figure 13-2**. It shows what are thought to be some of the major developments in animal evolution that led to the diversity we see today.

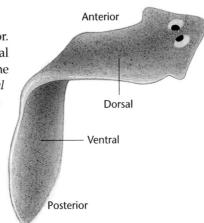

Anterior

Dorsal

Ventral

Posterior

Figure 13-4

Animals with bilateral symmetry have a definite front, or anterior, end. They also have a definite back, or posterior, end. The upper side of the animal is called the dorsal side, and the lower side is called the ventral side.

Section Wrap-up

Review

1. What are five characteristics of animals?

2. How are invertebrates different from vertebrates?

3. What are the types of symmetry? Give examples.

4. **Think Critically:** Radial symmetry is found among species that live in water. Why might radial symmetry be an adaptation uncommon among animals that live on land?

Skill Builder
Concept Mapping

Using the information on pages 344 to 347, make an events chain concept map showing the steps a scientist might use to classify a new animal. If you need help, refer to Concept Mapping in the **Skill Handbook.**

Using Computers

Model Use your computer to draw six fictitious animals. Determine the symmetry of each of the animals you have drawn.

Skill Builder

Students' events chain concept maps should begin with scientists determining whether the animal is a vertebrate or an invertebrate. The next step should be the scientist determining the animal's symmetry. The concept map should end with scientists comparing the organism's characteristics with those of other organisms. **P**

✓ Assessment

Oral Have students look at the animals you have posted on the bulletin board and choose one. Then ask them to use their concept maps to explain how they would classify their organisms. Use the Performance Task Assessment List for Concept Map in **PASC**, p. 89. **P**

---

4 Close

•MINI•QUIZ•

Use the Mini Quiz to check students' recall of chapter content.

1. **When a scientist finds a new animal, how does he or she begin to classify it?** *by looking to see whether it has a backbone*

2. **Are most animals invertebrates or vertebrates?** *invertebrates*

3. **What type of symmetry do most animals have?** *bilateral*

Section Wrap-up

Review

1. cannot make their own food, digest their food, most move, have many cells, have eukaryotic cells

2. Invertebrates do not have backbones; vertebrates have backbones.

3. radial symmetry—hydra; bilateral symmetry—dog, person; asymmetry—sponges

4. **Think Critically** An animal that lives in water may be sessile, move with the water, or just move up and down in the water column. Radial symmetry allows for all of these movements in water, but such an adaptation would be less useful on land.

Using Computers

Fictitious animals should show traits of animals. Students must be able to determine the symmetry of each one.

Prepare

Section Background

- Coral reefs date back to the Mesozoic era, 225 million years ago. Modern reefs may be as old as 2.5 million years.
- Coral reefs cover less than 0.2 percent of the ocean bottom, yet they capture about half of all the calcium that flows into the ocean every year. Coral reefs also release carbon dioxide into the atmosphere (about 100 million tons of carbon yearly).
- Coral reefs contain fewer species than tropical rain forests, but more phyla.
- About 10 percent of the fish caught worldwide for human consumption begin life near coral reefs.
- Reef balls range in size from 6 inches in diameter for use in home aquariums, up to 6 feet in diameter for use in the ocean.

1 Motivate

Bellringer

Before presenting this lesson, display **Section Focus Transparency 46** on the overhead projector. Assign the accompanying **Focus Activity** worksheet. L1 LEP

2 Teach

Brainstorming

Ask students to brainstorm a list of items that could be used as artificial reefs. Have them list the qualities that an item must have to be a good candidate for growing corals and other marine species.

Objectives

- Discuss the importance of coral reefs.
- Identify some ways in which natural coral reefs are damaged.
- Describe new technology for building artificial coral reefs.

Figure 13-5

Reef balls are built in molds. This newly-built reef ball will eventually be placed on the ocean bottom.

TECHNOLOGY: 13•2 Building Artificial Coral Reefs

Why are coral reefs important?

Coral reefs are the homes of beautiful tropical fish, sponges, and other colorful marine organisms. Coral reefs are built by small animals called corals. These animals give off chemicals that harden to form shells around their soft bodies. When thousands of these shells join, a coral reef forms. Offshore coral reefs provide homes for the young of commercially-important fish species. Scientists estimate that 35 000 to 60 000 different marine species can be found in and around healthy coral reefs. Coral reefs also help protect the shoreline from storms by breaking up heavy waves.

But coral reefs are easily damaged. Since the mid-1970s, coral reefs found offshore of 93 countries have been damaged or destroyed by water pollution from oil, gasoline, fertilizers, and soil. Careless boaters cause damage by dropping boat anchors on fragile reefs. Damaged or destroyed coral reefs do not protect the shore from storms, and without places to hide, small fish never have a chance to grow into larger fish.

What are artificial reefs?

Scuba divers have noticed that many more species of fish can be found around shipwrecks than above flat, sandy ocean bottoms. To improve fishing, some coastal communities have sunk ships, oil rigs, old tires, concrete pipes, and even steel drums offshore to create artificial reefs. But regular concrete dissolves in seawater, and metal rusts. Heavy storms sometimes move sunken ships around.

What are reef balls?

In 1993, a volunteer group in Georgia designed a new type of structure to repair or build new coral reefs. Called reef balls, these structures are made of concrete that is specially mixed so it does not dissolve in seawater. These structures have rough stony surfaces that provide many sites for

348 Chapter 13 Introduction to Animals

Program Resources

Reproducible Masters
Study Guide, p. 50 L1
Reinforcement, p. 50 L1
Enrichment, p. 50 L3
Science and Society Integration, p. 17

 Transparencies
Section Focus Transparency 46 L1

rough stony surfaces that provide many sites for tiny marine organisms to attach themselves. As you can see in **Figure 13-5,** reef balls have many holes in them. The holes are designed to allow water to flow through the ball, bringing nutrients to reef organisms. Reef balls are easy to place because they are molded around an inflated balloon. The balloon allows the concrete balls to float, so they can be towed behind a boat. At the reef site, the balloon can be deflated and removed. The balls sink slowly and can be placed exactly where they are needed—either to fill a hole in a living coral reef, or to create a new reef. Reef balls are heavier on the bottom; once in place they don't move, even during hurricanes.

Scientists are studying reef balls to see what organisms attach themselves to them, and how quickly organisms appear after the reef balls are put in place. Reef balls are now being used to repair damaged reefs and to build new coral reefs in Florida, Georgia, Texas, South Carolina, Virginia, Alabama, and Mississippi.

Figure 13-6
One month after being placed on the ocean bottom, these reef balls will soon host a variety of marine species.

Section Wrap-up

Review

1. Why are coral reefs important?

2. What kinds of human activities damage coral reefs?

3. What are some advantages of using reef balls instead of wrecked ships as artificial coral reefs?

Explore the Technology

Some coastal communities are using reef balls to grow young corals, which are then transplanted onto damaged areas on natural coral reefs. Smaller reef balls are being used in aquariums. Visit the Chapter 13 Internet Connection at Glencoe Online Science, **www.glencoe.com/sec/science/ life,** for a link to more information about coral reefs and reef balls.

SCIENCE & SOCIETY

13-2 Building Artificial Reefs   349

Section Wrap-up

Review

1. Coral reefs are important because they serve as nurseries for commercial fish species, break up storm waves, provide tourist destinations, and provide habitats for many different marine species.

2. Coral reefs are damaged by human activities such as water pollution from oil, gas, fertilizers, and soil erosion. They are also damaged by careless boaters who drop anchors on fragile coral reefs.

3. Reef balls do not move even during hurricanes; they permit the flow of nutrient-rich water through the structure; they can be placed precisely to repair damaged reef areas without harming the existing reef.

Concept Development

Invite an oceanographer or marine biologist to speak to the class about the importance of coral reefs and the activities that cause damage to them.

Activity

Have students contact petroleum companies in their area to find out how old oil rigs are being recycled as artificial coral reefs.

3 Assess

Check for Understanding

Ask questions 1–3 in the Section Review.

Reteach

Have students do some research to identify major coral reef systems of the world. Have them draw a map of the world and color in these coral reefs. From their maps, students should be able to list conditions that promote the growth of coral reefs, such as location in the tropics or location on the continental shelf.

Extension

For students who have mastered this section, use the **Reinforcement** and **Enrichment** masters.

4 Close

Divers and snorkelers visit coral reefs to look at the many kinds of organisms found there. Sometimes they collect pieces of coral, stand on coral reefs, or even drop boat anchors onto fragile reefs. All of these activities damage the reefs. Discuss with students whether coral reefs should be protected from any kind of human activity.

Prepare

Section Background

• Most sponges live in salt water. Fewer than 200 of all species live in fresh water. Nearly 80 percent of all life on Earth inhabits the oceans. Sea organisms make some of the most lethal poisons known.

• During periods of cold or dry weather, some freshwater sponges form specialized buds called gemmules. A gemmule is a cluster of sponge cells with a food supply surrounded by a protective covering. When conditions become favorable, the gemmule opens and the cells grow into a new sponge.

Preplanning

• Order *Grantia* sponges for the MiniLAB. Preserved *Grantia* sponges are available from biological supply companies.

• Order *Hydra* and *Daphnia* for Activity 13-1.

• Obtain modeling clay for the Section Wrap-up activity.

1 Motivate

Bellringer

Before presenting the lesson, display **Section Focus Transparency 47** on the overhead projector. Assign the accompanying **Focus Activity** worksheet. L1 LEP

Tying to Previous Knowledge

Review cell structure and function and tissue organization in Chapter 2 to help students understand differences in the body structures of sponges and cnidarians.

350

13•3 Sponges and Cnidarians

Science Words

- sessile
- filter feeder
- collar cell
- regeneration
- hermaphrodite
- larva
- cnidarian
- tentacle
- polyp
- medusa

Objectives

• Identify the structures that make up sponges and cnidarians.

• Describe how sponges and cnidarians obtain food and how they reproduce.

Sponges

Let's take a closer look at the different animal phyla. The least complex in body structure of the animal groups are sponges, cnidarians, flatworms, and roundworms.

Does the sponge in **Figure 13-7** look like an animal to you? Years ago, early scientists thought sponges were plants. However, sponges make up the group of animals with simple body plans and no body tissues, organs, or organ systems.

Characteristics of Sponges

All sponges live in water. Most are found in warm, shallow salt water near the coast, although some are found at ocean depths of 8500 m or more. A few species live in freshwater rivers, lakes, and streams. Sponges grow in many shapes, sizes, and colors. Some have radial symmetry, but most are asymmetrical. They may be smaller than a marble or larger than a compact car.

Adult sponges live attached to one place. They are often found with other sponges in colonies that never move, unless they are washed away by a strong wave. Organisms such as sponges that remain attached to one place during their lifetimes are **sessile.** Early scientists classified sponges as plants because they didn't move. As microscopes were improved, scientists observed that sponges couldn't make their own food, and so they were reclassified as animals.

350

Figure 13-7

Some species of sponges can grow to be larger than humans. *What characteristics make this sponge an animal?*

Program Resources

📁 **Reproducible Masters**
Study Guide, p. 51 L1
Reinforcement, p. 51 L1
Enrichment, p. 51 L3
Activity Worksheets, pp. 5, 76-77, 80 L1
Science Integration Activities, pp. 25-26
Concept Mapping, p. 31
Critical Thinking/Problem Solving, p. 17
Lab Manual 33

📦 **Transparencies**
Section Focus Transparency 47 L1
Teaching Transparency 26

Figure 13-8

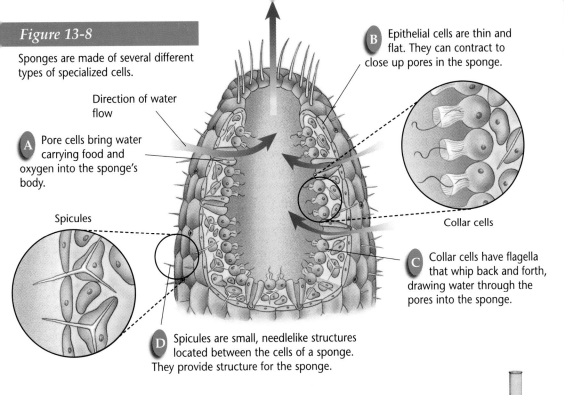

Sponges are made of several different types of specialized cells.

B Epithelial cells are thin and flat. They can contract to close up pores in the sponge.

Direction of water flow

A Pore cells bring water carrying food and oxygen into the sponge's body.

Spicules

Collar cells

C Collar cells have flagella that whip back and forth, drawing water through the pores into the sponge.

D Spicules are small, needlelike structures located between the cells of a sponge. They provide structure for the sponge.

The body of a sponge is covered with many small openings called pores. It is from the pores that sponges get their phylum name, Porifera. *Porifera* comes from a Latin word meaning "pore-bearing."

Figure 13-8 shows the body plan of a sponge. A sponge's body is a hollow tube closed at the bottom but with an opening in the top. The body wall has two cell layers made up of several different types of cells. There are cells that help a sponge get food, cells that digest food, cells that carry nutrients to all parts of the sponge, and cells that enable water to flow into the sponge.

Obtaining Food

Sponges filter food from water as it is pulled in through their pores. Bacteria, algae, protozoans, and other materials are filtered out of the water to be used as food. Organisms that obtain food this way are called **filter feeders.** Cells that line the inside of the sponge, called **collar cells,** help water move through the sponge. You can see in **Figure 13-8** that collar cells have flagella. The beating of the flagella in these cells moves water through the sponge, bringing oxygen to the cells and carrying away wastes.

CONNECT TO

CHEMISTRY

Spicules of "glass" sponges are composed of silica. Other sponges have spicules of calcium carbonate. *Relate* the composition of spicules to the composition of seawater.

13-3 Sponges and Cnidarians **351**

2 Teach

Visual Learning

Figure 13-7 **What characteristics make this sponge an animal?** *it has many cells, the cells are eukaryotic, cells have different functions, it must obtain food and digest it*

Figure 13-8 Ask students to study **Figure 13-8** to answer the following question: **What are the functions of the pore cells and the collar cells?** *Pore cells bring water containing food and oxygen into the sponge's body. Collar cells move the water through the sponge.* **LEP** **LS**

Demonstration

LS **Visual-Spatial** Mix soil and plant materials in a beaker of water. Pour the water mixture through a coffee filter into a container. Let students examine the contents of the filter and the container to see how the cells of a sponge might filter food particles from the water. **LEP**

CONNECT TO

CHEMISTRY

Answer Ocean water contains calcium and silica, which must be where sponges get their supplies of these elements for the formation of calcium carbonate and silica spicules.

Inclusion Strategies

Learning Disabled Have students select one type of animal described in this chapter. Have them design and construct a detailed, three-dimensional model of the animal and use the model to demonstrate the characteristics of animals.

Use these program resources as you teach this lesson.
Science Integration Activities, pp. 25-26
Critical Thinking/Problem Solving, p. 17

MiniLAB

What are the adaptations of a sponge?

Observe a *Grantia* sponge.

Procedure

1. Place the *Grantia* sponge in a small glass dish. Examine the sponge with a hand lens.
2. Make a drawing.
3. Insert the point of scissors into the opening of the sponge and cut lengthwise. Observe the inside of the sponge.
4. Add two drops of bleach to a microscope slide. **CAUTION:** *Be careful not to spill bleach on your hands or clothing or inhale.*
5. Put a small piece of sponge into the bleach. Add a coverslip. Observe the cells of the sponge.

Analysis

1. What traits of sponges did you observe?
2. How does water enter and leave the sponge?
3. Are spicules made of the same material as the rest of the sponge? Explain.
4. What is the function of the spicules?

The bodies of many sponges contain sharp, pointed structures called spicules. The soft-bodied sponges people use to take baths or to wash their cars have a skeleton of a fibrous material called spongin. Other sponges are supported by both spicules and spongin. Scientists classify sponges based on the kinds of materials that make up their skeletons. Two sponges are shown in **Figure 13-9.**

Reproduction

Sponges reproduce both asexually and sexually. They can reproduce asexually by forming buds. New sponges can also form from small pieces that break off the parent sponge. Sponge growers cut sponges into pieces, attach weights to them, and put them back into the ocean so the sponges can regenerate. **Regeneration** is the ability of an organism to replace body parts.

Sponges can reproduce sexually by egg and sperm. Some species of sponges have separate sexes, but most are hermaphrodites. A **hermaphrodite** (hur MAF ruh dite) is an animal that produces both sperm and eggs. Sperm are released in the water and carried by currents to other sponges where they fertilize the eggs. The fertilized egg develops into a young organism called a **larva** (*plural*, larvae). Larvae usually look very different from adults. Sponge larvae have cilia that allow them to swim about in the water. After a short time, the larvae settle down on objects where they will remain and grow into adult sponges.

Origin of Sponges

Sponges appeared on Earth about 600 million years ago in the Cambrian period. Their collar cells are similar to a type of colonial protozoan that is thought to be the ancestor of sponges. No other animal species is known to have evolved from sponges.

Figure 13-9

Sponges provide food for many fish, snails, and starfish. They are shelter for smaller organisms that live in their bodies. Researchers are collecting and testing sponges as possible sources of medicines.

352

Across the Curriculum

Language Arts Research and report on *Xestospongia muta,* a barrel sponge that grows like a giant vase off the Cayman Islands. Sponges of this type are being tested for potential medicines. L3 LS

Cnidarians

The colorful corals, flowerlike sea anemone, and iridescent Portuguese man-of-war you see in **Figure 13-10** belong to a group of animals called **cnidarians** (ni DAR ee uhnz). This name describes the stinging cells that all members of this phylum have. The word *cnidaria* is Latin for "stinging cells."

There are many different groups of cnidarians: the hydra group, the jellyfish, and the corals and sea anemones. Although many types of hydras live in fresh water, most cnidarians live in salt water. Most jellyfish live as individual organisms, but hydras and corals tend to form colonies. Corals, for example, live in colonies of polyps. Each polyp secretes a calcium carbonate shelter around its body. Over time, these calcium carbonate shelters join to those of neighboring corals to form a coral reef.

All cnidarians are radially symmetrical, with more complex bodies than those of sponges. They have two cell layers that are arranged into tissues and a digestive cavity where food is broken down. Most cnidarians have armlike structures called **tentacles** that surround the mouth. The tentacles are armed with the stinging cells that help the organism capture food, illustrated in **Figure 13-11.**

Figure 13-10

The corals, left, sea anemone, middle, and Portuguese man-of-war, above, all contain stinging cells characteristic of the phylum Cnidaria.

Figure 13-11

The stinging cells of cnidarians are located on their tentacles. *Why is this an appropriate location for these cells?*

A The stinging cell is a capsule that contains a coiled harpoon-like thread and poison. Each capsule has a hair that works like a trigger.

Sea anemone

Trigger

After firing

Nematocysts

Before firing

B When a small organism bumps into the trigger hair, the capsule explodes and shoots out its thread and poison. The poison paralyzes the organism until the tentacles wrap around it and pull it into the mouth.

13-3 Sponges and Cnidarians **353**

Community Connection

Field Trip Arrange a field trip to a freshwater pond to collect hydra. Invite a naturalist from the local park service to accompany the class to show students how to collect hydra.

Content Background

- Explain that budding, regeneration, and sexual reproduction may occur simultaneously in sponges.
- Explain that even though cnidarians have different body forms, they all have stinging cells, tentacles, two cell layers arranged into tissues, a digestive cavity, and radial symmetry.

Using Science Words

LS **Linguistic** *Polyp* is derived from the Greek word meaning "many feet." Medusa was the mythological Greek figure with writhing snakes instead of hair.

Demonstration

LS **Visual-Spatial** Dissect a preserved specimen of a sea anemone for students to see the tentacles, gastrovascular cavity, and basal disc.

Visual Learning

Figure 13-11 Why is this an appropriate location for these cells? *because the tentacles wave about and are the first body parts to touch food* **LEP** **LS**

GLENCOE TECHNOLOGY

 Videodisc
STVS: Animals
Disc 5, Side 1
Jellyfish (Ch. 2)

STVS: Ecology
Disc 6, Side 1
Evaluating Artificial Reefs (Ch. 11)

Activity 13-1

Activity 13-1

Observing a Cnidarian

The hydra has a body cavity that is a simple hollow sac. It is one of the few freshwater cnidarians. Try this activity to see how the tubelike body of the hydra reacts to food and other stimuli.

Purpose

IS **Intrapersonal** Students will observe the structure and method of feeding in a cnidarian. **L1** **LEP**

Process Skills

observing and inferring, identifying, using numbers, predicting, interpreting data, making and using tables

Time

40 minutes

Safety Precautions

Caution students to use extreme care when working with live animals.

Activity Worksheets, pp. 5, 76-77

Teaching Strategies

Troubleshooting Place several coverslips into the jar the day before the hydra are needed because they will probably cling to the coverslips. When you need hydra, remove the coverslips with forceps and place the coverslips directly into the dish that contains aquarium water.

• Assign one person to place the hydra in the dish under the stereoscopic microscope and the other one to record data.

Answers to Questions

1. The mouth is at the center of the tentacles.

2. Students who hypothesized that the hydra would react or contract should answer yes.

3. The tentacles stretched out to get the food.

4. When certain parts were touched, other parts may have responded by stretching or contracting.

5. The tentacles wrap around food and bring it into the mouth. Stinging cells in tentacles also stun prey.

Problem

How does a hydra react?

Materials

• dropper
• hydra culture
• small dish
• toothpick
• *Daphnia* or brine shrimp
• stereoscopic microscope

Procedure

1. Copy the data table and use it to record your observations.

2. Use a dropper to place a hydra into a dish along with some of the water in which it is living. **CAUTION:** *Be careful with living animals.*

3. Place the dish on the stage of a stereoscopic microscope. Bring the hydra into focus. Record the color of the hydra.

4. Identify and count the number of tentacles. Locate the mouth.

5. Observe the basal disk by which the hydra attaches itself to a surface.

6. Predict what will happen if the hydra is touched with a toothpick. Carefully touch the tentacles. Describe the reaction in the data table.

7. Drop a *Daphnia* or a small amount of brine shrimp into the dish. Observe how the hydra takes in food. Record your observations.

8. Return the hydra to the culture.

354 Chapter 13 Introduction to Animals

Data and Observations   Sample Data

Features	Observations
Color	green; varies with species
Number of tentacles	seven; varies with species
Reaction to touch	hydra should move away
Reaction to food	hydra should move toward food

Analyze

1. Where is the mouth of a hydra located in relation to the tentacles?

2. Did the hydra react to touch as you predicted?

3. How did the hydra react to food?

Conclude and Apply

4. When the hydra was touched, what happened to other areas of the animal?

5. **Describe** the advantages tentacles give hydra.

✓ Assessment

Performance To further assess students' abilities to observe the structure and method of feeding in a cnidarian, have them view a short video on cnidarians and describe the cnidarians' activities. Use the Performance Task Assessment List for Making Observations and Inferences in **PASC**, p. 17. **P**

Mouth   Tentacles   Mouth

Figure 13-12

The polyp form and medusa form are the two body plans of cnidarians.

Two Body Plans

Cnidarians have two different body plans. **Figure 13-12** shows these two plans. The **polyp** is shaped like a vase and is usually sessile. The **medusa** is bell-shaped and free-swimming. The hydra is an example of a polyp form, whereas the jellyfish is an example of a medusa form. Some cnidarians go through both polyp and medusa stages during their life cycles.

Medusa

Polyp

Cnidarians have a system of nerve cells called a nerve net. The nerve net carries impulses and connects all parts of the organism. This makes cnidarians capable of some simple responses and movements; for example, a hydra can somersault away from a threatening situation.

Reproduction

Cnidarians reproduce both asexually and sexually. Polyps reproduce asexually by producing buds that eventually fall off the parent and develop into new polyps, illustrated in **Figure 13-13.** Polyps can also reproduce sexually by producing either eggs or sperm. Sperm are released into the water and fertilize the eggs.

Medusa forms of cnidarians have both an asexual stage of reproduction and a sexual one. Free-swimming medusae produce eggs and sperm and release them into the water. The eggs are fertilized and develop into larvae. The larvae eventually settle down and grow into polyps. Young medusae bud off the polyp, and the cycle begins again.

Figure 13-13

Polyps like these hydra reproduce by budding.

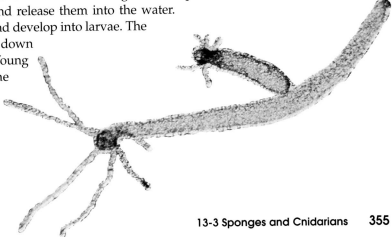

13-3 Sponges and Cnidarians   **355**

4 Close

Section Wrap-up

Review

1. Sponges have all the other characteristics of animals. They cannot make their own food, they have eukaryotic cells, they have many cells that have different functions, and they digest food.

2. Polyps reproduce asexually by budding, sexually by egg or sperm; medusas alternate between sexual and asexual reproduction.

3. **Think Critically** Corals secrete a hard skeleton around themselves which fossilizes easily. Fossil sponges might be rare except for fossilized spicules.

USING MATH

 Logical-Mathematical The sponge will move 15.625 mL of water in one minute.

Figure 13-14

The Great Barrier Reef consists of more than 2600 individual coral reefs. This reef complex stretches about 2000 km along the northeastern coast of Australia.

 INTEGRATION Earth Science

Origin of Cnidarians

Cnidarians were present on Earth during the Precambrian era more than 600 million years ago. Scientists think that the first form of cnidarian was the medusa. Polyps may have formed from larvae of medusae that became permanently attached to a surface. Most of the cnidarian fossils are fossils of corals.

Cnidarians and Coral Reefs

Many of the world's shallow seas are dominated by cnidarians. Coral reefs are found in warm waters of tropical seas between the latitudes of 30 degrees north and 30 degrees south. Besides corals, other cnidarians are found on these reefs. Sea anemones often attach themselves to dead coral, and many other organisms live on or near coral reefs. **Figure 13-14** illustrates the largest coral reef system in the world, the Great Barrier Reef of Australia.

Coral reefs protect beaches and shorelines from being washed away by ocean waves. The reefs are also areas of recreation for many people. The beautiful corals and large variety of sea life provide scuba divers and snorkelers with a wonderful view of life-forms in a reef.

Section Wrap-up

Review

1. Sponges are sessile organisms. Explain why a sponge is still considered to be an animal.

2. Describe the two body plans of cnidarians and their forms of reproduction.

3. **Think Critically:** Why are most fossils of cnidarians coral fossils? Would you expect to find a fossil sponge? Explain.

Skill Builder
Comparing and Contrasting
Compare and contrast the methods of obtaining food in sponges and cnidarians. If you need help, refer to Comparing and Contrasting in the **Skill Handbook.**

USING MATH

A sponge 1 cm in diameter and 10 cm tall can move 22.5 L of water through its body each day. Calculate the volume of water it will pump through its body in one minute.

Skill Builder
Sponges filter bacteria, algae, protozoans, and other materials out of the water. Collar cells help water move through the sponge. Cnidarians use the stinging cells on the tentacles to paralyze an organism, then the tentacles pull that organism into the mouth.

 Assessment

Performance Provide students with modeling clay. Have students make models of polyps and medusas. Ask them to demonstrate how these two body forms are similar. Use the Performance Task Assessment List for Model in **PASC**, p. 51. **P**

Flatworms and Roundworms

Flatworms

What kind of animal do you most likely think of when you hear the word *worm*? You probably think of an earthworm—the worm that crawls across pavement after a rain, or the worm used to bait a fishing hook. You probably wouldn't think immediately of tapeworms or any of the other many types of worms in the world. Just what is a worm? Worms are invertebrates with soft bodies and bilateral symmetry. They have three tissue layers organized into organs and organ systems. Worms live in many different environments. Some are very beautiful; others are not so attractive. There are flatworms, roundworms, and worms with segments. In this chapter, you will learn about flatworms and roundworms.

Types of Flatworms

As their name implies, flatworms have flattened bodies. They are members of the phylum Platyhelminthes (plat ih hel MIHN theez). Members of this phylum include planarians and tapeworms, as shown in **Figure 13-15.** Some flatworms are free-living but most are parasites. Remember, a parasite depends on another organism for food and a place to live. Unlike parasites, **free-living** organisms don't depend on one particular organism for food or a place to live. Most flatworms live in salt water, although there are a few species that live in fresh water.

Science Words

free-living
anus
cyst

Objectives

- Compare the body plans of flatworms and roundworms.
- Distinguish between free-living and parasitic organisms.
- Identify disease-causing flatworms and roundworms.

Figure 13-15

Planarians (A) are free-living flatworms. Tapeworms (B) are parasites.

13-4 Flatworms and Roundworms 357

Section 13•4

Prepare

Section Background

One of the world's serious health problems is caused by three species of blood fluke. Two to three hundred million people—mostly in Asia, Africa, and South America—are affected by these parasites.

Preplanning

- Order live planarians for the MiniLAB and Activity 13-2.
- Obtain prepared slides of tapeworms for Activity 13-2.

1 Motivate

Bellringer

Before presenting the lesson, display **Section Focus Transparency 48** on the overhead projector. Assign the accompanying **Focus Activity** worksheet. L1 LEP

Tying to Previous Knowledge

Review the definitions of *tissue, organs,* and *organ systems.* Review *bilateral symmetry* from Section 13-1. Discuss the presence or absence of tissues of organisms in this chapter—sponges, cnidarians, and now flatworms. Sponges do not have tissues, but cnidarians and flatworms do.

Program Resources

Reproducible Masters
Study Guide, p. 52 L1
Reinforcement, p. 52 L1
Enrichment, p. 52 L3
Activity Worksheets, pp. 5, 78-79, 81 L1
Concept Mapping, p. 31

Transparencies
Section Focus Transparency 48 L1

Purpose

IS Kinesthetic Students will observe planarians and determine how they move.
L1 LEP

Materials

culture of planarians, dropper, watch glass or clean jar lid, stereomicroscope

Teaching Strategies

• Planarians can be found in streams, rivers, and lakes and underneath rocks or leaves.

• Planarians may be placed in water from the classroom aquarium.

Troubleshooting If the planarians don't move, students may have to gently prod them with a small brush.

Analysis

1. Students should describe planarians as creeping or swimming.

2. Some students may infer that movement is by means of muscular undulations.

3. It does not depend on one particular organism for food or a place to live.

Assessment

Performance Have students feed the planarians pinhead-size pieces of liver and describe their behavior. Use the Performance Task Assessment List for Making Observations and Inferences in **PASC**, p. 17. **P**

Activity Worksheets, pp. 5, 81

Figure 13-16
Planarians have a simple body structure. *Why do you think these animals are called flatworms?*

Eyespot

Head

Cilia

Mouth

Pharynx

Digestive tract

Excretory system

Planarian

Mini LAB

How do planarians move?

Procedure

1. Use a dropper to transfer a planarian to a watch glass.
2. Add enough water so the planarian can move freely.
3. Place the glass under a stereomicroscope and observe.

Analysis

1. Describe how a planarian moves.
2. What body parts appear to be used in movement?
3. Explain why a planarian is described as a free-living flatworm.

Planarians

A planarian, like the one in **Figure 13-16,** is an example of a member of a free-living class of flatworms. It has a triangle-shaped head with two eyespots. There is one body opening, a mouth, on the ventral side of the body. A muscular tube called the pharynx connects the mouth and the digestive tract. A planarian feeds on small organisms and dead bodies of larger organisms. Most planarians live in fresh water, under rocks, or on plant material. Some live in moist places on land. Planarians vary in length from 3 mm up to 30 cm. Their bodies are covered with cilia. The cilia move the worm along in a slimy mucous track secreted by the underside of the planarian.

Planarians reproduce asexually simply by dividing in two. Planarians also have the ability to regenerate. A planarian can be cut in two, and each piece will grow into a new worm. Planarians reproduce sexually by producing eggs or sperm. Most are hermaphrodites, exchanging sperm with one another. They lay fertilized eggs that hatch in a few weeks.

Tapeworms

Tapeworms are parasitic members of the phylum Platyhelminthes. These worms use hooks and suckers to attach themselves to the intestine of a host organism, as you can see in **Figure 13-17.** Dogs, cats, humans, and other animals are hosts for tapeworms. A tapeworm doesn't have a mouth or a digestive system. Food that has already been digested by the host is absorbed from the intestine of the host by the worm. A tapeworm grows by producing new body segments immediately behind the head. Its ribbonlike body may grow to be 12 m long.

Each body segment of the tapeworm has both male and female reproductive organs. The segments produce eggs and sperm, and the eggs are fertilized in the segment. Once a segment is filled with fertilized eggs, it breaks off and passes out of the host's body. If a fertilized egg is eaten by another host, the egg hatches and develops into a new worm.

Tapeworm

Female reproductive organ

Hooks

Sucker

Segments

Male reproductive organ

Head

Figure 13-17

Tapeworms attach to an organism's intestine using hooks and suckers. Like most parasites, tapeworms produce huge numbers of eggs. *How do these eggs leave the host's body?*

Activity

Visual-Spatial Obtain specimens of tapeworms and *Ascaris* from a local veterinarian or a biological supply company. Have students observe and list the characteristics of each. [L1]

Enrichment

Have students use the library to research parasitic roundworms such as pinworms, whipworms, guinea worms, and filaria worms. [L2]

Visual Learning

Figure 13-16 Why do you think these animals are called flatworms? *because they are flattened in body shape*
Figure 13-17 How do these eggs leave the host's body? *The segment of the tapeworm's body that contains fertilized eggs breaks off and passes out in feces.* **LEP** LS

GLENCOE TECHNOLOGY

Videodisc
STVS: Animals
Disc 5, Side 1
Blood Fluke Life Cycle (Ch. 4)

Nematodes (Ch. 5)

Roundworms

If you own a dog, you've probably had to get medicine from your vet to protect your dog from heartworms. Heartworm disease in dogs is caused by roundworms. Roundworms make up the largest phylum of worms, the phylum Nematoda. It is estimated that there are more than a half million species of roundworms in the phylum Nematoda. Nematodes are found in soil, in animals, in plants, and in fresh water and salt water. Many are parasitic, but most are free-living.

Roundworms are slender and tapered at both ends. The body is a tube within a tube, with fluid in between. Unlike the organisms you have studied so far, nematodes have two body openings, a mouth and an anus. The **anus** is an opening at the end of the digestive tract through which wastes leave the body.

Figure 13-18

This free-living species of roundworm lives among cyanobacteria.

359

Cultural Diversity

Economics of Worms Have students research the economic effects of disease-causing roundworms in developing countries. [L2]

PREPARATION

Purpose

IS **Logical-Mathematical** Students will compare and contrast free-living and parasitic worms. **L1** **COOP LEARN**

Process Skills

comparing and contrasting, observing, recognizing cause and effect, communicating, classifying, making and using tables, forming operational definitions

Time

50 minutes

Materials

- Use prepared whole mount slides of tapeworms with heads and proglottids.
- Use a small brush to transfer the planaria from the culture to the culture dishes.
- Make sure the fed planaria are returned to a labeled dish so they will not be used again the same day.

Safety Precautions

Caution students to use extreme care when working with live animals.

📁 **Activity Worksheets,** pp. 5, 78-79

Activity 13-2

Design Your Own Experiment

Comparing Free-Living and Parasitic Worms

Observations of free-living and parasitic flatworms will help determine how each type of flatworm is adapted to its particular environment.

PREPARATION

Problem

How are the body parts of flatworms adapted to the environment in which they live?

Objectives

- Compare and contrast the body parts and functions of free-living and parasitic flatworms.
- Observe how these flatworms are adapted to their environments.

Possible Materials

- culture dish with a planarian
- prepared slide of a tapeworm
- dropper
- small piece of liver
- small paintbrush
- compound microscope
- stereoscopic microscope
- water
- light source, such as a lamp

Safety Precautions

Protect your clothing. Be careful when working with animals.

PLAN THE EXPERIMENT

1. As a group, make a list of possible ways you might design a procedure to compare and contrast types of flatworms. Your teacher will provide you with information on handling live flatworms.

2. In your group, choose one of the methods in your list from step 1. Now make a list of the steps you will need to take to follow the procedure. Be sure to describe exactly what you will do at each step. For example, if you want to compare the head of a tapeworm with the head of a planarian, state how you will do so.

3. Make a list of the materials that you will need to complete your experiment.

4. If you need a data table, design one in your Science

Journal so that it is ready to use as your group collects data.

Check the Plan

1. Read over your entire experiment to make sure that all steps are in logical order.

2. Will the data be summarized in a table?

3. *Make sure your teacher approves your plan before you proceed.*

DO THE EXPERIMENT

1. Carry out the experiment as planned.

2. While the experiment is going on, write down any observations that you make and complete the data table in your Science Journal.

Analyze and Apply

1. **Compare** and **contrast** parasitic and free-living worms.

2. Which body systems are better developed in free-living worms?

3. Which body system is more complex in parasitic worms?

Go Further

Suggest a reason for the lack of free movement in parasitic worms. What are two things that ensure success of parasitic worms?

13-4 Flatworms and Roundworms   361

Go Further

Parasitic worms do not have to move to find food once they have attached to a host. Two things that may ensure the success of parasitic worms are the ability to reproduce and to live off the host without killing the host.

Assessment

Performance To further assess students' understanding of the differences between free-living and parasitic worms, assign different cooperative groups and have them use their drawings and observations to explain the differences. Assess explanations for accuracy. Use the Performance Task Assessment List for Making Observations and Inferences in **PASC**, p. 17. **P**

PLAN THE EXPERIMENT

Possible Procedures
One possible procedure is to observe the body structures of both free-living and parasitic worms.

Teaching Strategies
Tying to Previous Knowledge Most students know that parasitic worms have body parts for attaching to a host and free-living worms have body parts for the environment in which they live.

DO THE EXPERIMENT

Expected Outcome
Student observations of the planaria may include those already listed: planaria move with a gliding motion; the most sensitive body parts are those near the head; the tubelike structure (pharynx) on the ventral side pulls food into the body. Planarians have flat bodies. The mouth is on the ventral side of the body. Student observations of the tapeworm will include body parts adapted for attaching to a host and a well-developed reproductive system.

Analyze and Apply

1. Parasitic worms have body parts adapted for attaching to a host. Free-living flatworms have no such parts.

2. Body systems that are better developed in free-living worms are the nervous system and digestive system.

3. In parasitic worms, the reproductive system is more complex.

361

3 Assess

Check for Understanding

? FLEX Your Brain

Use the Flex Your Brain activity to have students explore FLATWORMS AND ROUNDWORMS.

📁 **Activity Worksheets,** p. 5

Reteach

LS **Linguistic** Have each student write one short-answer question relating to flatworms and roundworms. Read the questions aloud. Ask students to respond as a group.

Extension

📁 For students who have mastered this section, use the **Reinforcement** and **Enrichment** masters.

USING TECHNOLOGY

A Hookworm Vaccine ▼ ▲

Scientists estimate that nearly 1 billion people on Earth are infected with hookworms. Hookworms use their sharp teeth to burrow into the wall of the small intestine. Blood delivers oxygen, water, iron, proteins, and many other substances to body cells. If the blood is being absorbed by hookworms rather than body cells, then the infected person may end up with iron deficiency, lethargy, weakness, protein malnutrition, and even mental retardation.

Preventing Hookworm Infection

Better sanitation and encouraging people to wear shoes are the traditional ways to prevent this parasite from spreading. But in many rural areas, hookworm is constantly reinfecting its victims. Recently, scientists have begun experimenting with vaccines to prevent the hookworm from burrowing into the skin and vaccines that prevent the hookworm from maturing once it is inside the host's body. Other scientists are working at the molecular level to find ways to interfere with the hookworm's life cycle.

inter**NET** CONNECTION

In what parts of the world are hookworm infections most prevalent? What difficulties arise in treating women of childbearing age? Visit the Chapter 13 Internet Connection at Glencoe Online Science, **www.glencoe.com/sec/ science/life,** for a link to more information about hookworms.

362 Chapter 13 Introduction to Animals

Types of Roundworms

Some roundworms are parasites of humans. The most common roundworm parasites of humans are *Ascaris,* hookworm, and *Trichinella.* Humans get hookworms by walking barefoot over dirt or through fields. Hookworm eggs hatch in warm, moist soil. Wearing shoes is the best protection against hookworms.

Ascaris is found in the intestines of pigs, horses, and humans. Eggs enter the host's body in contaminated food or water. They travel to the intestines where they mature and mate. Masses of worms can block the intestines and cause death if left untreated.

Trichinella worms, such as the one shown in **Figure 13-19,** cause the disease trichinosis. Humans can become infected when they eat undercooked pork that has *Trichinella* cysts. A roundworm **cyst** is a young worm with a protective covering. Trichinosis can be prevented by thoroughly cooking pork.

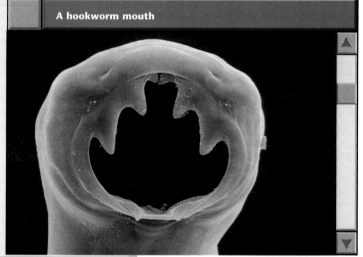

A hookworm mouth

Magnification: 377×

Across the Curriculum

Math Tapeworms as long as 9 m have been found in humans. Have students use a meterstick and measure 9 m to have an understanding of just how long tapeworms are. **L1** **LS**

Magnification: 75×

A **B**

Figure 13-19

Humans infected with *Trichinella* worms (A) are likely to suffer fever, vomiting, pain, and stiffness of skeletal muscles. *Ascaris* worms (B) can cause damage to the lungs, intestines, and even the brain.

The heartworms you treat your dog for enter the blood of a dog through a mosquito bite. They move to the heart, where they grow and reproduce. If the infection is left untreated, these roundworms can block the valves of the heart that lead to and from the lungs. Medicine can be given to dogs monthly to prevent heartworm disease.

Flatworms and roundworms are more complex than sponges and cnidarians. They have bilateral symmetry, three well-developed tissue layers, and organ systems. The roundworms have digestive systems with a mouth and an anus. Like sponges and cnidarians, these invertebrates probably evolved in the sea. They still live in moist environments.

Section Wrap-up

Review

1. Compare body plans of flatworms and roundworms.

2. Distinguish between a free-living flatworm and a parasitic flatworm.

3. What are three roundworms that cause diseases in humans? How can humans prevent infection?

4. **Think Critically:** Which organism is more complex, a cnidarian or a flatworm? Explain.

Skill Builder
Sequencing
Make an ordered list of the concepts found in Section 13-4. If you need help, refer to Sequencing in the **Skill Handbook.**

Science Journal

Write a public service announcement for your local television station informing the community about heartworm disease in dogs. Consult a veterinarian for information.

13-4 Flatworms and Roundworms **363**

363

Background

- *Science on the Ice: An Antarctic Journal* and *The Great Barrier Reef: A Living Laboratory* are two of Rebecca L. Johnson's books that present an in-depth look at how scientists study a variety of animals and other organisms in remote and challenging environments.

- Every year, the National Science Teachers Association (NSTA) and the Children's Book Council (CBC) put together an annotated bibliography of Outstanding Science Trade Books for Children. The books included in this list are evaluated by a review panel of educators and librarians appointed by the NSTA. As outstanding examples of good science writing, these books represent diverse subject areas and a variety of reading levels.

Teaching Strategies

LS Linguistic Challenge students to collect examples of science writing from as many different sources as possible, including newspapers, magazines, books, technical journals, and audio/visual materials. Create a classroom display of the students' collection and use it as the basis of a discussion on how science writers present information about science topics to different audiences.

People and Science

REBECCA L. JOHNSON, *Science Book Author*

On the Job

Q Ms. Johnson, what kind of books do you write?

A I write nonfiction science books for young adults. I especially like to write about scientists who do interesting research in unusual, often remote places. And since I prefer to write from firsthand experience, whenever possible I work directly with scientists in the field.

Q Where have you gone to do research for your books?

A All sorts of wonderful places. On Australia's Great Barrier Reef, I've worked with scientists who were investigating everything from genetics in giant clams to how corals respond to ultraviolet radiation. In Antarctica, I've had the opportunity to live in remote field camps with scientists who were studying penguins and seals. One of the most remarkable experiences I've had while gathering information for a book was exploring the ocean bottom at 700 meters down in a deep-diving research submersible.

Personal Insights

Q What's one of the most surprising things that you've learned about an animal in doing research for your books?

A The temperature of the seawater around Antarctica is below freezing, yet the fish that swim in those polar seas don't freeze because they have special proteins in their blood that act as a natural antifreeze.

Career Connection

Contact a local camera club or the art department of a vocational school or university to learn more about nature photography. Zoo exhibit designers learn about the natural environments of zoo animals so they can plan appropriate exhibits. Design a poster to interest your classmates in a career as a:

- **Nature Photographer**
- **Zoo Exhibit Designer**

364 Chapter 13 Introduction to Animals

Career Connection

Career Path Science writers can work for newspapers, television studios, magazines, textbook publishing companies, video production companies, health care organizations, advertising agencies, and in industry.

- High School: science, English
- College: science (all areas), journalism, creative writing, graphic design

A broad-based science background can be a distinct advantage in pursuing a career in science writing.

For More Information

In addition to being illustrated with photographs, science books and articles often contain considerable original artwork. Students may wish to contact the Guild of Natural Science Illustrators in Washington, DC, for more information about the field of natural science illustration and career opportunities as a scientific illustrator.

Summary

13-1: What is an animal?

1. Animals are many-celled eukaryotic organisms that must find and digest their food.
2. Invertebrates are animals that don't have backbones. Animals with backbones are vertebrates.
3. Animals that have body parts arranged the same way on both sides of the body have bilateral symmetry. Animals with body parts in a circle around a central point are radially symmetrical.

13-2: Science and Society: Building Artificial Coral Reefs

1. Built by animals called corals, coral reefs provide homes for marine organisms and protect the shoreline from storms.
2. Coral reefs are damaged by water pollution and other human activities.
3. New types of artificial coral reefs called reef balls can be built to repair damaged reefs or build new ones.

13-3: Sponges and Cnidarians

1. Sponges are considered the least complex animals in the animal kingdom. They are sessile and obtain food and oxygen by filtering water through pores.
2. Cnidarians are hollow-bodied animals with radial symmetry. Most have tentacles with stinging cells to obtain food. Jellyfish, hydras, and corals are cnidarians.

13-4: Flatworms and Roundworms

1. Flatworms belong to the phylum Platyhelminthes. They have bilateral symmetry. There are both free-living and parasite forms. Roundworms belong to the phylum Nematoda. They have a tube within a tube body plan and bilateral symmetry.
2. Free-living organisms don't depend on one particular organism for food or a place to live. Parasites depend on other organisms for food and a place to live.

Key Science Words

a. anus
b. bilateral symmetry
c. cnidarians
d. collar cell
e. cyst
f. filter feeder
g. free-living
h. hermaphrodite
i. invertebrate
j. larva
k. medusa
l. polyp
m. radial symmetry
n. regeneration
o. sessile
p. tentacle
q. vertebrate

Reviewing Vocabulary

Match each phrase with the correct term from the list of Key Science Words.

1. body parts the same on each side
2. a young worm enclosed in a covering
3. a young organism different from the adult
4. opening for digestive wastes

Summary

Have students read the summary statements to review the major concepts of the chapter.

Reviewing Vocabulary

1. b	6. i
2. e	7. q
3. j	8. l
4. a	9. h
5. o	10. p

✓ Assessment

Portfolio   Encourage students to place in their portfolios one or two items of what they consider to be their best work. Examples include:

- Skill Builder, p. 347
- Reteach, p. 355
- Performance Assessment, p. 356 **P**

Performance   Additional performance assessments may be found in **Performance Assessment** and **Science Integration Activities**. Performance Task Assessment Lists and rubrics for evaluating these activities can be found in Glencoe's **Performance Assessment in the Science Classroom.**

GLENCOE TECHNOLOGY

 MindJogger Videoquiz

Chapter 13   Have students work in groups as they play the Videoquiz game to review key chapter concepts.

Checking Concepts

1. d	**6.** c
2. a	**7.** b
3. b	**8.** c
4. a	**9.** c
5. c	**10.** a

Understanding Concepts

11. In bilateral symmetry, both halves of the organism are the same; flatworms are bilaterally symmetrical. Cnidarians have radial symmetry; all body parts are organized like the spokes of a wheel around a central point.

12. Budding involves cells or a new individual developing on and then breaking off the parent. Regeneration involves replacing lost parts through mitosis.

13. The tapeworm has hooks and suckers to hold onto the intestinal cells of its host; it has no digestive system because it gets already-digested food from its host.

14. dorsal—top side, ventral—bottom side, anterior—front end, posterior—back end

15. This cnidarian has caught its prey with the stinging cells in its tentacles. The tentacles are bringing the prey to its mouth.

Thinking Critically

16. Sponges have two cell layers; simple worms have three. Sponges have little cell organization; simple worms have tissues, organs, and organ systems. Sponges are asymmetrical; simple worms have bilateral symmetry.

17. Simple animals can reproduce in a variety of ways so they can respond to changing conditions. This allows them to survive in poor conditions.

---

5. attached, nonmoving animal
6. organism with no backbone
7. animal that has a backbone
8. cnidarian shaped like a tube or vase
9. an organism that makes both egg and sperm
10. used by cnidarians to capture food

Checking Concepts

Choose the word or phrase that completes the sentence.

1. Animal characteristics include all of the following except _____.
a. movement c. eukaryotic cells
b. digestion d. prokaryotic cells

2. All of these belong to the same group except _____.
a. fish c. jellyfish
b. hydras d. sea anemones

3. _____ is the opposite of dorsal.
a. Anterior c. Radial
b. Ventral d. Bilateral

4. Scientists classify sponges based on _____.
a. the material that makes up their skeletons
b. reproduction
c. their method of obtaining food
d. symmetry

5. Sponges are members of phylum _____.
a. Cnidaria c. Porifera
b. Nematoda d. Platyhelminthes

6. Sponges reproduce asexually by _____.
a. medusae c. budding
b. polyps d. eggs and sperm

7. The body plans of cnidarians are polyp and _____.
a. larva c. ventral
b. medusa d. bud

---

8. All are examples of cnidarians except _____.
a. corals c. planarians
b. hydras d. jellyfish

9. An example of a parasite is a _____.
a. sponge c. tapeworm
b. planarian d. jellyfish

10. Separate sexes are found in _____.
a. *Ascaris* c. sponges
b. planarians d. cnidarians

Understanding Concepts

Answer the following questions in your Science Journal using complete sentences.

11. Using examples, explain the difference between bilateral and radial symmetry.

12. Explain the difference between budding and regeneration.

13. Explain how the structure of a tapeworm and its lifestyle are related.

14. Identify the dorsal and ventral sides and the anterior and posterior ends of an animal.

15. Look at the photograph below. Explain what this photograph shows.

---

18. Sponges are filter-feeding omnivores that eat bacteria, algae, and protozoans. Hydra are carnivores that eat small animals they capture with their tentacles, such as *Daphnia*. Planarians are carnivores that eat smaller animals. Sponges absorb food, so they need pieces that are smaller than their cells. Hydra and planarians can digest larger particles.

Thinking Critically

16. Compare the body organization of a sponge to that of a simple worm.
17. What is the advantage to simple animals of having more than one means of reproduction?
18. List the types of food sponges, hydras, and planarians eat. Explain why the size of the food particles is different in each organism.
19. Compare and contrast the medusa and polyp body forms of cnidarians.
20. Why do scientists think the medusa stage was the first stage of the cnidarians?

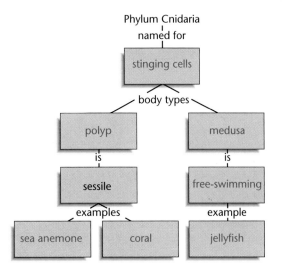

Developing Skills

If you need help, refer to the **Skill Handbook.**

21. **Sequencing:** Sequence the order of heartworm infection in a dog.
22. **Using Variables, Constants, and Controls:** Design an experiment to test the sense of touch in a planarian.
23. **Hypothesizing:** Make a hypothesis about cooking pork at high temperatures to prevent worms from developing, if present in the uncooked meat.
24. **Concept Mapping:** Complete the concept map of classification in the phylum Cnidaria.

Performance Assessment

1. **Lab Report:** Research tapeworms and other parasitic worms that live in humans. Find out how they are able to live in the intestines without being digested by the human host. Report your findings to the class.
2. **Invention:** Using the characteristics of an animal listed on page 344, invent a new animal. Choose an existing phylum and draw the animal, then write a description of how the animal moves, what it eats, and so on. Prepare a poster with your drawing and description and share it with the class.
3. **Slide Show:** Create a slide show of sponges and cnidarians found on a coral reef. You can make slides from pictures you have drawn or from borrowed photographs. Identify each organism and present the slide show to your class.

Chapter 13 Review 367

19. The medusa stage is bell-shaped and free-swimming. The polyp stage is tube-shaped and sessile.
20. The medusa is a moving form. The polyp stage develops from the medusa. More polyp fossils exist, possibly indicating later existence.

Developing Skills

21. **Sequencing** Infected mosquito bites dog—blood infected—to heart where the heartworms grow and reproduce—valves are blocked—death if untreated.
22. **Using Variables, Constants, and Controls** From their observations of planarians, students should be able to suggest an experiment that involves touching the planarian with natural objects and observing the movement.
23. **Hypothesizing** Cysts are destroyed by heat—the worm has no protection and is killed.
24. **Concept Mapping** See student page.

Performance Assessment

1. Some parasites are protected by a thick mucus. The hosts may not have the enzymes to break down the mucus. Use the Performance Task Assessment List for Lab Report in **PASC,** p. 47. P
2. Student descriptions should reflect information consistent with the characteristics of animals. Use the Performance Task Assessment List for Inventions in **PASC,** p. 45. P
3. Students should be able to identify each organism in their slide shows. Use the Performance Task Assessment List for Slide Show or Photo Essay in **PASC,** p. 77. P

Assessment Resources

Reproducible Masters
Chapter Review, pp. 29-30
Assessment, pp. 89-92
Performance Assessment, p. 27

Glencoe Technology
Chapter Review Software
Computer Test Bank
MindJogger Videoquiz

Chapter Organizer

Section	Objectives/Standards	Activities/Features
Chapter Opener		Explore Activity: Observe a scallop shell. p. 369
14-1 Mollusks (1½ sessions, 1 block)*	1. **Identify** the features of mollusks. 2. **Name** three classes of mollusks and identify a member of each. National Content Standards: (5-8) UCP1, UCP5, A1, B2, C1, C3, C5, F3	Using Math, p. 372 Skill Builder: Observing and Inferring, p. 373 Using Computers: Database, p. 373 Science and Art: Mollusks and the Art of the South Pacific Islanders, p. 374 Science Journal, p. 374
14-2 Segmented Worms (1 session, ½ block)*	1. **Describe** the traits of segmented worms. 2. **Describe** the structures and digestive process of an earthworm. 3. **Identify** the evolutionary relationship between segmented worms and mollusks. National Content Standards: (5-8) UCP1, UCP4, UCP5, A1, A2, C1, C2, C3, C5, F1	Problem Solving: How Do Tube Worms Eat? p. 377 Activity 14-1: Earthworms and Plant Growth, pp. 378-379 Skill Builder: Interpreting Scientific Illustrations, p. 381 Using Math, p. 381
14-3 Arthropods (2½ sessions, 2 blocks)*	1. **Identify** features used to classify arthropods. 2. **Relate** the structure of the exoskeleton to its function. 3. **Distinguish** between complete and incomplete metamorphosis. National Content Standards: (5-8) UCP1, UCP5, A1, A2, C1, C2, C3, C5, E2	Using Technology: A Fly in My Wound, p. 384 MiniLAB: What type of metamorphosis do fruit flies undergo? p. 385 Skill Builder: Sequencing, p. 388 Using Math, p. 388 Activity 14-2: Observing a Crayfish, p. 389
14-4 Science and Society Issue: Pesticides and Insects (1 session, ½ block)*	1. **Describe** the importance of pesticides in agriculture. 2. **Identify** the impact of pesticides on the environment. National Content Standards: (5-8) F4, F5, G1	Explore the Issue, p. 391
14-5 Echinoderms (1 session, ½ block)*	1. **Identify** the features of echinoderms. 2. **Describe** how sea stars get and digest food. National Content Standards: (5-8) UCP1, UCP5, A1, C1, C2, C3	MiniLAB: How do tube feet open clamshells? p. 393 Skill Builder: Observing and Inferring, p. 394 Science Journal, p. 394

* A complete Planning Guide that includes block scheduling is provided on pages 32T-35T.

Activity Materials

Explore	Activities	MiniLABs
page 369 scallop shells, hand lens	page 378-379 potting soil, sphagnum moss, 15-cm pots, small houseplants, earthworms, wire mesh, small pebbles, apron, safety goggles page 389 crayfish in small aquarium, stirrer, raw ground beef, antibacterial soap	page 385 banana, knife, glass jar, cheesecloth to cover mouth of jar, rubber band page 393 heavy books

Need Materials? Call Science Kit (1-800-828-7777).

Teacher Classroom Resources

Reproducible Masters	Transparencies	Teaching Resources
Study Guide, p. 53 **Reinforcement**, p. 53 **Enrichment**, p. 53 **Science Integration Activity 14,** Newton's Third Law of Motion **Concept Mapping**, p. 33	**Section Focus Transparency 49**, It's a Mollusk's Life **Science Integration Transparency 14,** Newton and the Cephalopods	**Glencoe Life Science Interactive Videodisc** **Life Science CD-ROM** **Spanish Resources** **English/Spanish Audiocassettes** **Cooperative Learning Resource Guide** **Lab Partner** **Lab and Safety Skills** **Lesson Plans**
Study Guide, p. 54 **Reinforcement**, p. 54 **Enrichment**, p. 54 **Activity Worksheets**, pp. 82-83 **Lab Manual 34**, Earthworm Anatomy **Critical Thinking/Problem Solving,** p. 18 **Cross-Curricular Integration**, p. 18 **Multicultural Connections**, pp. 31-32	**Section Focus Transparency 50,** Beyond the Fringe **Teaching Transparency 27**, Parts of an Earthworm	**Assessment Resources** **Chapter Review**, pp. 31-32 **Assessment**, pp. 93-96 **Performance Assessment in the Science Classroom (PASC)** **MindJogger Videoquiz** **Alternate Assessment in the Science Classroom** **Performance Assessment** **Chapter Review Software** **Computer Test Bank**
Study Guide, p. 55 **Reinforcement**, p. 55 **Enrichment**, p. 55 **Activity Worksheets**, pp. 84-85, 86 **Lab Manual 35**, Collect Insects **Lab Manual 36**, Grasshopper Anatomy **Lab Manual 37**, Metamorphosis **Technology Integration**, pp. 13-14	**Section Focus Transparency 51**, Legs, Legs, and More Legs **Teaching Transparency 28,** Metamorphosis	
Study Guide, p. 56 **Reinforcement**, p. 56 **Enrichment**, p. 56 **Science and Society Integration**, p. 18	**Section Focus Transparency 52**, Pests: Keep Off	
Study Guide, p. 57 **Reinforcement**, p. 57 **Enrichment**, p. 57 **Activity Worksheets**, p. 87	**Section Focus Transparency 53**, A Thorny Issue	

Key to Teaching Strategies

The following designations will help you decide which activities are appropriate for your students.

L1 Level 1 activities should be within the ability range of all students, including those with learning difficulties.

L2 Level 2 activities should be within the ability range of the average to above-average student.

L3 Level 3 activities are designed for the ability range of above-average students.

LEP LEP activities should be within the ability range of Limited English Proficiency students.

LS These activities are designed to address different learning styles.

COOP LEARN Cooperative Learning activities are designed for small group work.

P These strategies represent student products that can be placed into a best-work portfolio.

GLENCOE TECHNOLOGY

The following multimedia resources are available from Glencoe.

Science and Technology Videodisc Series (STVS)
Animals
Leeches
Conch Farming
Horseshoe Crab
March of the Spiny Lobster
Sounds of the Blue Crabs
Shrimp Farming
Tick Research

The Infinite Voyage Series
Insects: The Ruling Class
Glencoe Life Science Interactive Videodisc
Eyeing Evolution
Life Science CD-ROM

Teacher Classroom Resources

This is a representation of key blackline masters available in the Teacher Classroom Resources.

Teaching Aids

Section Focus Transparencies

Science Integration Transparencies

Teaching Transparencies

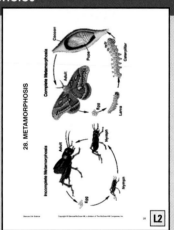

Meeting Different Ability Levels

Study Guide

Reinforcement

Enrichment Worksheets

Hands-On Activities

Science Integration Activity

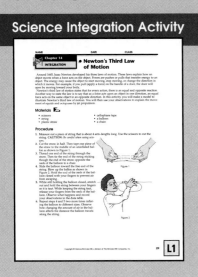

Chapter 14 INTEGRATION • Newton's Third Law of Motion

L1

Lab Manual

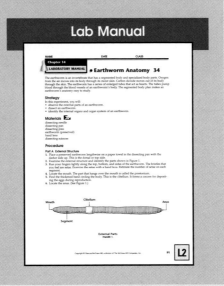

Chapter 14 LABORATORY MANUAL • Earthworm Anatomy 34

L2

Assessment

Performance Assessment

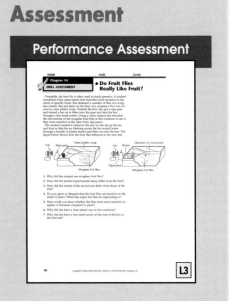

Chapter 14 SKILL ASSESSMENT • Do Fruit Flies Really Like Fruit?

L3

Enrichment and Application

Critical Thinking/ Problem Solving

Chapter 14 CRITICAL THINKING • Mollusks, Worms, Arthropods, and Echinoderms

Putting Worms to Work

L2

Cross-Curricular Integration

Chapter 14 HEALTH CONNECTION • Mollusks, Worms, Arthropods, and Echinoderms

Leeches and Medicine

L1

Science and Society Integration

Chapter 14 SCIENCE & SOCIETY • Mollusks, Worms, Arthropods, and Echinoderms

Better Ways to Battle Bugs

L2

Technology Integration

Unit 5 TECHNOLOGY • Animals

Building with Barnacles

L1

Multicultural Connections

Chapter 14 MULTICULTURAL CONNECTIONS • Attack of the Killer Caterpillars

L2

Concept Mapping

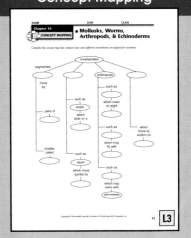

Chapter 14 CONCEPT MAPPING • Mollusks, Worms, Arthropods, & Echinoderms

L3

Mollusks, Worms, Arthropods, and Echinoderms

CHAPTER OVERVIEW

Section 14-1 This section describes the basic body plans of gastropods, bivalves, and cephalopods.

Section 14-2 A description of features of segmented worms is followed by a discussion of earthworms and leeches.

Section 14-3 The features used to classify arthropods are described. Examples of complete and incomplete metamorphosis are shown.

Section 14-4 Science and Society The importance of pesticides in agriculture, the impact of pesticides on the environment, and the possibility of farming without pesticides are explored.

Section 14-5 The features of echinoderms are presented in this section. How sea stars obtain and digest food is described.

Chapter Vocabulary

mollusk	Arthropoda
mantle	appendage
gill	exoskeleton
open	molting
circulatory	spiracle
system	metamor-
radula	phosis
closed	pesticide
circulatory	echinoderm
system	water-
setae	vascular
crop	system
gizzard	tube feet

Theme Connection

Stability and Change/Scale and Structure Comparisons of the adaptations that have evolved in the mollusks, segmented worms, and arthropods are shown.

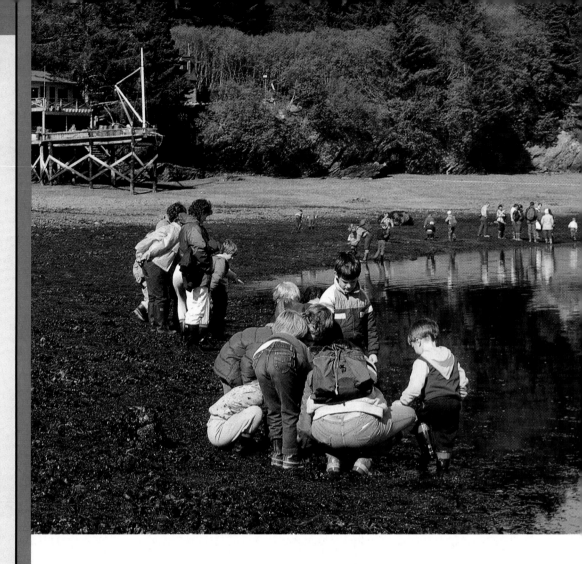

Previewing the Chapter

Section 14-1 Mollusks
► Features of Mollusks
► Classifying Mollusks
► Importance of Mollusks

Section 14-2 Segmented Worms
► Features of Segmented Worms
► Earthworms
► Leeches
► Marine Worms
► Evolution of Segmented Worms and Mollusks

Section 14-3 Arthropods
► Features of Arthropods
► Insects
► Metamorphosis
► Insect Adaptations
► Other Arthropods

Section 14-4 Science and Society
Issue: Pesticides and Insects

Section 14-5 Echinoderms
► Features of Echinoderms
► Classifying Echinoderms
► Importance of Echinoderms

368

Learning Styles Look for the following logo for strategies that emphasize different learning modalities. **LS**

Kinesthetic	Across the Curriculum, p. 383; Activity, p. 384; MiniLAB, p. 393
Visual-Spatial	Explore, p. 369; Visual Learning, pp. 371, 376, 380, 383, 393; Activity, p. 371; Science & Art, p. 374; Demonstration, pp. 376, 387; Reteach, p. 380; MiniLAB, p. 385; Activity 14-2, p. 389
Intrapersonal	Activity, p. 386
Logical-Mathematical	Activity 14-1, pp. 378-379
Linguistic	Across the Curriculum, pp. 372, 377; Science Journal, pp. 384, 394

Mollusks, Worms, Arthropods, and Echinoderms

If you've ever walked along a beach, you've probably seen lots of seashells. Seashells come in many different colors, shapes, and sizes. If you look closely, you will see that each shell has many rings or bands. In the following activity, find out what the bands tell you about the shell and the organism that made it. In the chapter that follows, you will learn about traits of other organisms with shells.

EXPLORE ACTIVITY

Observe a scallop shell.

1. Use a hand lens to observe a scallop shell.
2. Count the number of rings or bands on the shell. Count the large top point called the crown as one.
3. Compare the width of the bands. Do all of the scallops have the same number of bands?

Observe: What do you think the bands represent? Why do you think some are wider than others? Write down your observations in your Science Journal.

Previewing Science Skills

▶ In the **Skill Builders,** you will **observe and infer, interpret scientific illustrations,** and **sequence.**

▶ In the **Activities,** you will **observe, design an experiment,** and **collect** and **record data.**

▶ In the **MiniLABs,** you will **observe** and **record data.**

369

EXPLORE ACTIVITY

Purpose

LS **Visual-Spatial** Use the Explore Activity to introduce students to mollusks. Explain that they will be learning about mollusks and other complex invertebrates in the chapter. **L1** **LEP**

Preparation

Obtain one scallop shell, a hand lens, and a transparent metric ruler for each pair of students.

Teaching Strategies

Have each pair of students count the bands and answer the questions.

Observe

As a scallop grows, so does its shell. A new ring or band shows on the shell each year. The crown is one year's growth. Food supply, water temperature, oxygen content, pollutants, and the calcium carbonate in the water affect the width of the bands.

✔ Assessment

Oral Have students suggest ways in which the shell of a mollusk is an adaptation. *protection, conserves water, camouflage* Use the Performance Task Assessment List for Making Observations and Inferences in **PASC,** p. 17. **P**

Assessment Planner

Portfolio
Refer to page 395 for suggested items that students might select for their portfolios.

Performance Assessment
See page 395 for additional Performance Assessment options.
Skill Builders, pp. 373, 381, 388, 394
MiniLABs, pp. 385, 393
Activities 14-1, pp. 378-379; 14-2, p. 389

Content Assessment
Section Wrap-ups, pp. 373, 381, 388, 391, 394
Chapter Review, pp. 395-397
Mini Quizzes, pp. 381, 388

Group Assessment
Opportunities for group assessment occur with Cooperative Learning Strategies and Flex Your Brain Activities.

Prepare

Section Background

- The phylum Mollusca contains several classes. The three main classes are Gastropoda, the stomach-footed mollusks; Bivalvia, the two-shelled mollusks; and Cephalopoda, the head-footed mollusks. The bodies of all these mollusks are soft and unsegmented.

- Scientists have discovered mollusks living in hydrothermal vents in the ocean floor. They get their food from bacteria that live in their bodies. The bacteria use sulfide compounds to produce sugars that the mollusks use for food.

1 Motivate

Bellringer

Before presenting the lesson, display **Section Focus Transparency 49** on the overhead projector. Assign the accompanying **Focus Activity** worksheet. L1 LEP

Tying to Previous Knowledge

Most students will be familiar with mollusks such as oysters, clams, squid, garden snails, and octopuses.

2 Teach

370

Science Words

- mollusk
- mantle
- gill
- open circulatory system
- radula
- closed circulatory system

Objectives

- Identify the features of mollusks.
- Name three classes of mollusks and identify a member of each.

Features of Mollusks

The word *mollusk* comes from the Latin word meaning "soft," and describes the bodies of the organisms in the phylum Mollusca (mah LUS kah). **Mollusks** are soft-bodied invertebrates that usually have shells. They are found on land, in fresh water, and in salt water. Mollusks have bilateral symmetry and a fluid-filled body cavity that provides space for the body organs.

Mollusk Body Plan

All mollusks have a soft body usually covered by a hard shell. Covering the soft body is the mantle. The **mantle** is a thin layer of tissue that secretes the shell or protects the body if the mollusk does not have a shell. Between the soft body and the mantle is a space called the mantle cavity. In it are the **gills,** organs that exchange oxygen and carbon dioxide with the water. The body organs of mollusks are located together in an area called the visceral mass. The mantle covers the visceral mass. Finally, all mollusks have a muscular foot used for movement. These structures are illustrated in **Figure 14-1.**

The circulatory system of most mollusks is an **open circulatory system.** In this type of system, blood isn't always contained in vessels the way your blood is. Blood bathes a mollusk's organs directly in some areas of its body.

Classifying Mollusks

The classification of mollusks is based on whether or not the animal has a shell. Mollusks that have shells are classified by the kind of shell and by the kind of foot. In this section, you will learn about three classes of mollusks: gastropods, bivalves, and cephalopods.

Figure 14-1

All mollusks have the same basic body plan: a shell, foot, mantle, and visceral mass.

KEY

- Visceral mass
- Mantle
- Shell
- Foot

Snail

Clam

Squid

370 Chapter 14 Mollusks, Worms, Arthropods, and Echinoderms

Gastropods

The largest class of mollusks, the gastropods, includes snails, slugs, abalones, whelks, sea slugs, and conches. Look at the organism in **Figure 14-2A.** Can you see why these organisms are sometimes called univalves? Except for slugs, each member of the gastropods has a single shell. Many have a pair of tentacles with eyes at the tips. Gastropods obtain food by scraping and tearing it with the **radula,** a tongue-like organ with rows of teeth that works like a file. The radula is used for scraping algae and other food materials off of rocks.

Slugs and many snails are adapted to life on land. They move by muscular contractions of the foot. Glands in the foot secrete a layer of mucus for the foot to slide along in. Slugs live where it is moist. Slugs do not have shells, but they are protected by a layer of mucus.

Bivalves

Bivalves are mollusks that have a two-part shell joined by a hinge. Clams, oysters, and scallops are bivalve mollusks. These animals pull their shells closed with powerful muscles. To open their shells, they relax these muscles. The shell is made up of several layers made by the mantle. The smoothest layer is on the inside and protects the soft body.

Bivalves are well adapted to living in water. For protection, a clam burrows deep into the sand with its muscular foot. Mussels and oysters cement themselves to a solid surface or attach themselves with a strong thread. This keeps strong waves from washing them away. Scallops escape predators by rapidly opening and closing their shells. As the water is forced out, the organism moves rapidly in the opposite direction.

Figure 14-2

Although these animals look very different from one another, they are all mollusks.

A A snail is an example of a gastropod. It has one shell and moves along a mucus trail laid down on land.

C Cephalopods such as the octopus can swim or move by jet propulsion. *What does this organism have in common with gastropods and bivalves? How is it different?*

B A scallop is a bivalve. It moves by closing its two shells quickly.

371

USING MATH

Many snails move at a speed of about 8 cm per minute. If a typical snail does not stop to eat or rest, how far can it travel in an hour? *Haliotis* is a snail that travels at 48 m per hour. Which is faster: the speed of *Haliotis* or the speed of most other snail species? Which snail moves faster? How much faster?

INTEGRATION
Physical Science

Cephalopods

The class of mollusks with the most specialized and complex members are the cephalopods. The word *cephalopod* means "head-footed" and describes the body structure of these invertebrates. Squid, octopus, and the chambered nautilus belong to this class. Cephalopods have a large, well-developed head. Their "foot" is divided into many tentacles with strong suckers on each for capturing prey. Squid and octopuses have a well-developed nervous system and large eyes similar to human eyes. Unlike other mollusks, cephalopods have closed circulatory systems. In a **closed circulatory system,** blood containing food and oxygen is contained and transported in a series of vessels.

Adaptations of Cephalopods

All cephalopods live in oceans and have bodies adapted for swimming. They move quickly by jet propulsion, moving water in one direction while being propelled in another. A squid has a water-filled cavity between an outer muscular sheath and its internal organs. When the squid tightens its muscular sheath, water is forced out through an opening near the head. The jet of water sends it backward, and it can go elsewhere in a great hurry. A squid can swim at speeds of more than 60 m/s this way. It can briefly outdistance all but whales, dolphins, and the fastest fish. A squid can jump out of water and reach heights of almost 5 m above the ocean's surface. They can travel through the air as far as 15 m.

According to Newton's third law of motion, when one object exerts a force on a second object, the second one exerts a force on the first that is equal and opposite in direction. This

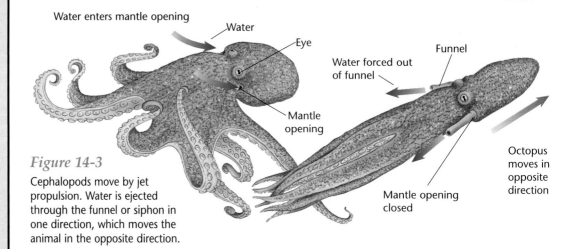

Figure 14-3
Cephalopods move by jet propulsion. Water is ejected through the funnel or siphon in one direction, which moves the animal in the opposite direction.

372   Chapter 14 Mollusks, Worms, Arthropods, and Echinoderms law is represented in the movement of a squid. Muscles exert force on water under the mantle. Water being forced out exerts force back, resulting in movement backward.

Squid cannot maintain their top speed for more than just a few pulses because a squid is inefficient, expending a lot of energy to move small amounts of fluid. As you can see in **Figure 14-3**, octopuses can also swim by jet propulsion, but they usually use their tentacles to creep more slowly over the ocean floor.

Importance of Mollusks

Mollusks provide food for fish, sea stars, and birds. Invertebrates, such as hermit crabs, use empty mollusk shells as shelter. Clams, oysters, snails, and scallops are used for food. Pearls are produced by many species of mollusks, but most are made by mollusks called pearl oysters, shown in **Figure 14-4**. Scientists are studying the nervous systems of the squid and octopus to understand how learning takes place and how memory works. Mollusks can also cause problems. For example, snails and slugs feed on plants and damage crops, and certain species of snails are hosts of parasites that infect humans.

Figure 14-4
Pearls are formed in many bivalve mollusks. Smooth mother of pearl is secreted by the mantle in layers around a grain of sand or other particle trapped between the mantle and the shell.

Section Wrap-up

Review

1. What features are used to classify mollusks?

2. Name the three classes of mollusks and identify a member from each class.

3. **Think Critically:** What adaptation makes a snail useful to have in an aquarium?

Skill Builder
Observing and Inferring
Observe the photographs of one-shelled mollusks and two-shelled mollusks in this section. Infer how bivalves are not adapted to life on land and gastropods are. If you need help, refer to Observing and Inferring in the **Skill Handbook.**

Using Computers

Database Use your computer to make a data table that compares and contrasts the features of one-shelled mollusks, two-shelled mollusks, and head-footed mollusks. Compare their methods of obtaining food, movement, circulation, and habitat. Use the data table to classify the mollusks shown in this section.

14-1 Mollusks **373**

Section Wrap-up

Review

1. kind of foot; presence or absence of shell; if present, kind of shell

2. gastropods—snails, slugs, abalones, whelks, sea slugs, and conches; bivalves—clams, oysters, and scallops; cephalopods—squid, octopuses, and chambered nautiluses

3. **Think Critically** The radula is the adaptation that a snail uses to scrape algae from the aquarium glass.

Skill Builder
Bivalves filter food from the water. They move slowly or remain attached to one place. Gastropods use their radulas and teeth to eat plants. Gastropods live on land and have a cavity in their bodies that serves as a lung.

Assessment

Process Ask students to compare in writing the muscular foot of bivalves and gastropods. *A gastropod's muscular foot spreads out under its body and a rippling motion moves it along. The foot of a bivalve is shaped and used differently. It is normally hidden inside the two shells. Some bivalves move by hooking the foot into the sand and pulling themselves along.* Use the Performance Task Assessment List for Writing in Science in **PASC,** p. 87. P

Using Computers				
Type of Mollusk	Nutrition	Movement	Circulation	Habitat
one-shelled mollusk	feed on plant material using radula	muscular contractions of foot, mucus layer	open circulatory system	land and water
two-shelled mollusk	use siphons to filter-feed	sessile or use foot to crawl, open and close shell	open circulatory system	water
head-footed mollusk	use tentacles with suckers to capture prey	jet propulsion or use tentacles	closed circulatory system	water P

Science & ART

Source

Price, Christine. *Made in the South Pacific*. E. P. Dutton: New York, 1979.

Background

Wood, stone, and metal are traditional materials used in art and jewelry making, as well as in the construction of homes, buildings, and a variety of other products. However, in places of the world where these materials may be scarce, bone and the shells of mollusks are often substituted. This can be clearly seen in many countries of the South Pacific, where mollusk shell is one of the most abundant resources. Since people began populating the islands of Polynesia, Micronesia, and Melanesia nearly 25 000 years ago, shell has been used to make tools, jewelry, canoes, and many other artifacts. Shell is an attractive medium for many reasons. It can be easily carved and fashioned into different shapes, and it comes in an array of colors. In the Solomon Islands off the eastern coast of New Guinea, for example, fishermen make bonito hooks in six different colors. The particular hook to use is chosen by observing the color of the smaller fish the bonito are feeding upon that day.

Teaching Strategies

Ⓛ Visual-Spatial Bring in a variety of seashells to display to students. Point out to students that mollusk shells come in many different sizes, shapes, and colors. Have students use seashell identification books to match the shells with particular mollusk species. L1

• Have students make observations on the characteristics of mollusk shells, such as texture, color, and thickness. Discuss how these characteristics are related to

Mollusks and the Art of the South Pacific Islanders

Mollusks have hard shells that are produced by chemical secretions from the animal's mantle and provide protection for its internal organs. Mollusks are important food sources in many parts of the world, particularly in places near the oceans. However, in countries that are bound by ocean waters, such as islands of the South Pacific, mollusks represent more than just food. In places like New Guinea, Tahiti, Fiji, and Samoa, mollusk shells are also important natural resources that are used in artwork, jewelry, tools, and other artifacts.

Shells as Tools

Because of their great abundance, mollusk shells have been useful to the seafaring peoples of the South Pacific Islands for thousands of years. For instance, in some parts of Micronesia, such as the Caroline and Gilbert Islands, mollusk shell is much more available than stone. The Gilbert Islanders, for example, once used the shells of giant clams to fashion the blades of their *adzes*, important tools in canoe building. Elsewhere, the blades are usually made from stone. The Gilbert Islanders also commonly used shells to decorate their wooden tackle boxes, necessities aboard any Micronesian canoe.

One of the more interesting uses of mollusk shell, still seen in some parts of the South Pacific today, was in the construction of fishhooks for catching the bonito, a large species of tropical fish. The making of the bonito hook, with the point and shank carved separately and lashed together with cord, was an art in itself. In fact, some bonito hooks were so beautiful, they were often passed down as family heirlooms. According to the lore of the fishermen, the shape and color of the bonito hook are important for attracting fish.

Map labels: Mariana Islands, Micronesia, Marshall Islands, Caroline Islands, Kiribati (Gilbert Islands), New Guinea, Melanesia, Solomon Islands, Polynesia, Samoa, Fiji, Tahiti

Science Journal

How are mollusk shells used today? What characteristics of mollusk shells make them useful? In your Science Journal, make a list of at least five common uses of mollusk shells.

374   Chapter 14 Mollusks, Worms, Arthropods, and Echinoderms

their usefulness in art, tool making, and jewelry making, and as building materials.

• You may want to present students with some information about the cultural habits and behaviors of South Pacific Islanders. However, how and when people arrived on the Pacific Islands is still a matter under debate. Discuss how the culture of the South Pacific Islanders is influenced by environment. Also point out culture has changed over time to what is observed today.

• You may wish to locate several books

about the artwork of South Pacific peoples to show students the extent of mollusk shell use.

Other Works

For more information about artwork in the South Pacific, try some of these other sources:

• Ambesi, Alberto Cesare. *Oceanic Art*. Hamlyn Publishing: London, 1970.
• Barrow, Thomas. *Art and Life in Polynesia*. Charles E. Tuttle: Rutland, VT, 1973.

Segmented Worms 14•2

Features of Segmented Worms

The worms you see crawling across sidewalks and driveways after a hard rain and the night crawlers you dig up for fishing belong to Phylum Annelida (uh NEL ud uh), the segmented worms. The tube-shaped bodies of all segmented worms are divided into many little sections. The word *annelid* means "little rings" and describes how the bodies of these worms look. Segmented worms are found in fresh water, salt water, and moist soil. Earthworms, leeches, and beautiful marine worms, shown in **Figure 14-5**, belong to this phylum.

Besides segments, members of this phylum have several other characteristics in common. Like mollusks, all segmented worms have a body cavity that holds their organs. On the outside of each segment are bristle-like structures called **setae** (*sing.*, seta) that help the worms move. Have you ever watched a bird try or tried yourself to pull an earthworm out of the ground? It's not that easy, is it? Segmented worms use their setae to hold on to the soil. Let's look more closely at the classes of segmented worms—earthworms, leeches, and marine worms.

Science Words
setae
crop
gizzard

Objectives
- Describe the features of segmented worms.
- Describe the structures and digestive process of an earthworm.
- Identify the evolutionary relationship between segmented worms and mollusks.

Figure 14-5

Segmented worms include three classes of organisms: earthworms, leeches, and a wide variety of marine worms. *What do these organisms have in common?*

A Common earthworms are usually 5 to 10 cm long. But this earthworm, the giant earthworm of Australia, is more than 3 m long.

B This painted leech from Malaysia waits on a leaf for a host to walk by.

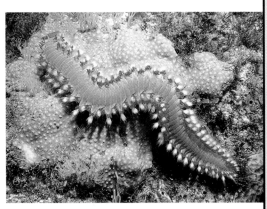

C Marine worms, class Polychaeta, have many setae on each segment. They include the free-moving bristleworms like this one, featherduster worms, and marine tube worms.

14-2 Segmented Worms **375**

Section 14•2

Prepare

Section Background
Earthworms are hermaphroditic. Fertilization occurs when two earthworms align their bodies next to each other and exchange sperm. The clitellum secretes a ring of mucus into which the egg and sperm are deposited as the clitellum moves forward. As this ring of mucus passes over the head of the worm, it closes into a cocoon from which the eggs later hatch into young.

Preplanning
Obtain live earthworms, potting soil, 15-cm pots, small house plants, sphagnum moss, wire mesh, and small pebbles for Activity 14-1.

1 Motivate

Bellringer
 Before presenting the lesson, display **Section Focus Transparency 50** on the overhead projector. Assign the accompanying **Focus Activity** worksheet. [L1] [LEP]

Tying to Previous Knowledge
Students have probably seen earthworms burrowing in the soil or crawling on sidewalks after a hard rain. Use their experiences with earthworms to discuss their observations about worms, where they live, and what they eat. Review flatworms and roundworms from Chapter 13.

Program Resources

Reproducible Masters
Study Guide, p. 54 [L1]
Reinforcement, p. 54 [L1]
Enrichment, p. 54 [L3]
Cross-Curricular Integration, p. 18
Activity Worksheets, pp. 5, 82-83 [L1]
Critical Thinking/Problem Solving, p. 18
Lab Manual 34
Multicultural Connections, pp. 31-32

Transparencies
Section Focus Transparency 50 [L1]
Teaching Transparency 27

375

2 Teach

Magnification: 5×

Figure 14-7

Segmented worms have circulatory, respiratory, excretory, digestive, muscular, and reproductive systems.

Brain — Hearts — Blood vessels — Crop — Gizzard — Mouth — Reproductive structures — Main nerve cord — Intestine — Waste removal tubes — Setae — Anus

Earthworms

The tubelike body of an earthworm has more than 100 segments. The segments aren't evident only on the outside; the body cavity is divided also. Each body segment, except for the anterior and posterior segments, has four pairs of setae. Earthworms move by using their setae, shown in **Figure 14-6,** and two sets of muscles in the body wall.

What Earthworms Eat

To obtain food, earthworms eat soil. When they burrow through soil, they are actually eating it. Earthworms get the energy they need to live from the bits of leaves and other organic matter found in the soil. The digestive system of an earthworm is made up of a crop, a gizzard, and an intestine. The soil eaten by an earthworm moves to the **crop,** which is a sac used for storage. Behind the crop is a muscular structure, the **gizzard,** which grinds the soil. Food then passes to the intestine, where it is broken down and absorbed by the blood. Undigested soil and waste materials leave the worm through the anus.

Earthworm Body Plan

Look at the body structure of an earthworm in **Figure 14-7.** Earthworms have two blood vessels along the dorsal and ventral sides of the body. The vessels meet in the anterior end where they are connected by five pairs of structures, called aortic arches, which pump blood through the body. Smaller vessels go into each body segment. An earthworm has a closed circulatory system; all the worm's blood is contained in blood vessels. Wastes are removed from earthworms by coiled tubes found in each segment. Earthworms have small, simple brains in the anterior segment. Nerves in each segment join to form a main nerve cord that connects to the brain. They respond to light, temperature, and moisture. Can you think of any times when they respond to moisture?

376 Chapter 14 Mollusks, Worms, Arthropods, and Echinoderms

Community Connection

Farming Earthworms Invite a farmer or someone from the agricultural extension service to discuss the importance of earthworms and how they are raised commercially.

An earthworm doesn't have gills or lungs. It lives in a thin film of water, where it exchanges oxygen and carbon dioxide by diffusion through its skin.

Like planarians, earthworms are hermaphrodites. Although each worm has both male and female reproductive structures, an individual worm can't fertilize its own eggs. It has to receive sperm from another earthworm in order to reproduce.

Leeches

Have you ever had to remove a leech from your body after swimming in a pond, lake, or river? Leeches belong to another class of segmented worms and are adapted to a lifestyle very different from earthworms. Their bodies are not as round or long as earthworms', and they don't have setae. Leeches feed on the blood of ducks, fish, and even humans. Two suckers, one at each end of the body, are used to attach to an animal. After the leech has attached itself, it cuts into the flesh and sucks out two to ten times its weight in blood. If you have ever had a leech attached to you, you probably didn't feel it. Leeches produce many chemicals, including an anesthetic. Why would producing an anesthetic be beneficial to a leech?

Problem Solving

How do tube worms eat?

Polychaetes are a class of segmented worms that live mainly in the ocean. They differ from earthworms in that they have tentacles, antennae, and specialized mouthparts. Each body segment has a pair of appendages called parapodia that assist in movement. Some polychaetes live in tubes made from sand or mud on the ocean floor.

Giant tube worms are related to polychaetes, but they have no eyes, mouths, or digestive systems. They were discovered in 1977. Some giant tube worms live in an area called the Galápagos Rift, deep beneath the ocean's surface near the Galápagos Islands. No sunlight reaches to these depths; thus, photosynthesis does not occur. However, the water is heated by volcanic activity along the rift. This volcanic activity also produces chemicals, such as sulfur compounds. These giant tube worms obtain energy and nutrients from bacteria that live inside their bodies. The bacteria release energy from the sulfur compounds that the tube worms absorb.

Solve the Problem:
1. **Infer why giant tube worms lack eyes.**

2. **Do you think giant tube worms have mouths? Why or why not?**
3. **Explain whether or not giant tube worms need digestive systems.**

Think Critically:
The segmented worms that live in the ocean are very different from earthworms. What adaptations do they have for living in the ocean?

377

Problem Solving

Solve the Problem
1. They live in darkness on the ocean floor.
2. No, because they do not eat.
3. They do not need digestive systems because they get energy and nutrients from bacteria inside their bodies.

Think Critically
Those that move have tentacles, antennae, and parapodia. They have specialized mouth parts for obtaining food. Some get their energy from bacteria that live inside their bodies.

Enrichment
Have students research and report on the medicinal leech *Hirudo medicinalis*. This leech produces the anticoagulant hirudin. L2

Inquiry Question
During dry periods, where are earthworms found in the soil? Explain. *Earthworms move deeper into soil where there is the necessary amount of moisture. They can't live in dry soil.*

Student Text Questions
- **Can you think of any times when they respond to moisture?** *Yes, when it rains, earthworms come out of the soil.*
- **Why would producing an anesthetic be beneficial to a leech?** *If a person cannot feel a bite, he or she would be less likely to pull the leech off.*

Across the Curriculum

History Have students research and report on the uses of leeches in medicine in the 1800s and through the 1900s. *They were used for bloodletting, applied to the temples for headaches, used to treat mental illness, gout, skin disease, and whooping cough.* L2

LS

GLENCOE TECHNOLOGY

 Videodisc
STVS: Animals
Disc 5, Side 1
Leeches (Ch. 7)

Purpose
LS **Logical-Mathematical** Students will determine whether the presence of earthworms in the soil affects the growth of plants. **L1** **COOP LEARN**

Process Skills
observing and inferring, communicating, recognizing cause and effect, forming a hypothesis, designing an experiment, separating and controlling variables, interpreting data

Time
two 45-minute periods and twice-weekly observations

Materials
- The number of variables in each group should be kept to a minimum. If this is not done, there is no way of knowing which variable caused the difference in results. The presence or absence of earthworms should be the only variable in each group.
- Earthworms may be obtained from a bait shop or from a biological supply company.
- A good soil mixture consists of two-thirds potting soil and one-third sphagnum moss or peat moss.

Safety Precautions
Caution students to keep their hands wet and to handle the earthworms gently.

Possible Hypotheses
Most student hypotheses will reflect increased plant growth and more healthy-looking plants in the soil that contains earthworms.

📁 **Activity Worksheets,** pp. 5, 82-83

Activity 14-1

Design Your Own Experiment
Earthworms and Plant Growth

Suppose you are digging up part of your school grounds to plant some flowers. The soil you are digging is heavy and thick and an orange-brown color. On the other side of the school, your classmates find soil that crumbles, is dark brown, and contains lots of earthworms. Where would you recommend planting the flowers? Does the presence of earthworms have any value to your garden?

PREPARATION

Problem
How does the growth of plants in an environment that contains earthworms compare with the growth of plants in one that doesn't?

Form a Hypothesis
Based on your reading and observations, state a hypothesis about the growth of plants in an environment with earthworms and one without.

Objectives
- Design an experiment that compares the growth of plants in two environments, one with earthworms and one without.
- Observe the growth of the plants for two weeks.
- Compare the results and explain any differences.

Possible Materials
- potting soil
- sphagnum moss
- 15-cm pots
- small houseplants
- earthworms
- wire mesh
- small pebbles

Safety Precautions
Protect clothing and eyes, and be careful when working with live animals. Alway keep your hands wet when handling earthworms. Dry hands cause mucus to be removed from earthworms.

PLAN THE EXPERIMENT

1. As a group, agree upon and write out the hypothesis statement.
2. As a group, identify what results would confirm or reject your hypothesis. Would you expect the plant without earthworms to die? Would you expect the plant with earthworms to grow taller than the plant without earthworms?
3. As a group, list the steps you will need to take to test your hypothesis. Be specific; describe exactly what you will do at each step. List your materials.
4. Prepare a data table in your Science Journal to record your observations.

Check the Plan

1. Read over your entire experiment to make sure that all steps are in logical order.
2. Identify all constants, variables, and controls

of the experiment.
3. Make sure you record your observations as the experiment proceeds. What data will you record?
4. Plan to examine plants to identify how earthworms have affected their growth. What measurements will you take?
5. *Make sure your teacher approves your plan before you proceed.*

DO THE EXPERIMENT

1. Carry out the experiment as planned.
2. While the experiment is going on, write down any observations that you make and complete the data table in your Science Journal.

Analyze and Apply

1. **Compare** your results with those of other groups.

2. **Compare** growth in the two sets of plants.
3. Describe what effect you think overwatering would have on the plants. How would it affect the earthworms?
4. Did your results support your hypothesis? Explain.

Go Further

Investigate the effect of soil pH on earthworms. Use potting soil, humus, and sphagnum moss in the pots with earthworms and plants.

14-2 Segmented Worms   **379**

Go Further

This will allow students the opportunity to examine the effect of a number of variables. The pH of each soil should be tested before placing the plants and the earthworms in the containers. A soil that is neither too acidic nor too basic should produce desirable results.

 Assessment

Oral Have students explain the relationship between earthworms and soil fertility. *Earthworms move organic matter throughout the soil, add nitrates to the soil, and dig tunnels that allow air and water to go deep into the soil.* Use the Performance Task Assessment List for Analyzing the Data in **PASC,** p. 27. **P**

PLAN THE EXPERIMENT

Possible Procedures
One possible procedure is to prepare two containers with a mixture of potting soil and sphagnum moss or peat moss. Plant two or three plants in each container. Add three to six medium-sized worms to one container. Both containers should receive adequate light and water. Students should observe the growth of each set of plants for three weeks.

Teaching Strategies
Tying to Previous Knowledge
Most students know that earthworms aerate the soil and the casts they produce fertilize the soil.

DO THE EXPERIMENT

Expected Outcome
Most results will show an increase in plant growth and healthier-looking plants in the container that has the earthworms.

Analyze and Apply
1. Student results should be similar to those of other groups.
2. Any improvements in plant growth should occur in the container with the earthworms. Earthworms aerate the soil, thus providing more oxygen for plant roots. They add nutrients to the soil through their waste products.
3. Overwatering could harm the plants by causing the growth of molds that would cause the plants to decay. It would eliminate oxygen in the soil so the earthworms and the plants would not have the oxygen they need.
4. Answers will be determined by student hypotheses.

379

3 Assess

Leeches and Medicine

Before the 1900s, physicians used leeches to drain blood from sick people. They thought that bleeding by leeches would drain away sickness and make the person well. The treatment usually caused patients to become weaker. Sometimes physicians used so many leeches on a patient that it even killed the person.

Today, leeches are used in microsurgery to keep blood flowing to reattached body parts. The tiny blood vessels in an ear, for example, can easily become blocked as blood begins to coagulate inside. Without a good flow of blood to the reattached body part, the tissue may die. To keep blood flowing, surgeons attach leeches to the microsurgery site, as you can see in **Figure 14-8.** Why? Because as the leeches feed on blood, chemicals in their saliva prevent the blood from coagulating. Leech saliva contains an anticlotting chemical that is being studied for use in patients with heart and circulatory diseases.

Figure 14-8

Using leeches to help blood flow in microsurgery patients sounds and looks odd, but leeches do not hurt the patient. *Why not?*

Marine Worms

There are more species of polychaetes, the marine worms, than of any other kind of annelid. More than 6000 known species of polychaetes include marine worms that float, burrow, build structures, or walk along the ocean bottom. Some polychaetes even produce their own light!

Ⓐ

Ⓑ

Figure 14-9

Polychaetes come in a wide variety of forms and colors. Bristleworms (A) move along the bottom of the sea. Fan worms (B) trap food in the mucus on their "fans." *How are these organisms similar to cnidarians and sponges?*

380 Chapter 14 Mollusks, Worms, Arthropods, and Echinoderms

Polychaetes, like other annelids, have segments with setae; the setae in these worms, however, occur in bundles.

Free-moving polychaetes, such as the bristleworm shown in **Figure 14-9A,** have a head with eyes, a tail, and paired fleshy outgrowths on their segments called parapodia, which help in feeding and locomotion. Sedentary polychaetes, such as the fan worm shown in **Figure 14-9B,** have specialized tentacles used for food gathering and respiration. Some polychaetes form tubes out of calcium carbonate, parchment, or mucus. Whenever these worms are startled, they retreat into their tubes.

Evolution of Segmented Worms and Mollusks

Scientists infer that mollusks and segmented worms share a common ancestor. They were the first of the animal groups to have bodies with space for the body organs. In both groups, an important stage of development is a structure that looks like a spinning top with cilia. This structure, a larva, is shown in **Figure 14-10.** This larva is the best evidence that mollusks and segmented worms share a common ancestor.

Figure 14-10
Mollusks and segmented worms have the same type of larva.

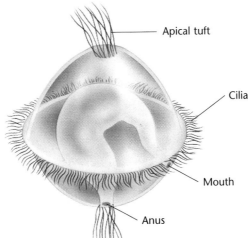

Apical tuft

Cilia

Mouth

Anus

Section Wrap-up

Review

1. What is the most noticeable feature of annelids?

2. Describe how an earthworm feeds and digests its food.

3. How are segmented worms related to mollusks?

4. **Think Critically:** Why do farmers promote the use of earthworms in the soil of their fields?

Skill Builder
Interpreting Scientific Illustrations
Use the diagram in **Figure 14-7** to list the body systems of an earthworm and organs in each system. If you need help, refer to Interpreting Scientific Illustrations in the **Skill Handbook.**

USING MATH

Suppose you dig up a small area of soil 10 cm wide, 10 cm deep, and 10 cm long and find six earthworms. Based on this sample, predict how many earthworms you would find in an area 10 m wide, 10 cm deep, and 10 m long.

14-2 Segmented Worms   **381**

Skill Builder
nervous system: brain, main nerve cord

reproductive system: reproductive structures, clitellum

circulatory system: aortic arches, blood vessels

digestive system: mouth, crop, gizzard, intestine, anus

excretory system: waste removal tubes

Assessment

Performance Use this Skill Builder to assess students' abilities to interpret scientific illustrations by having them point out external features in the photographs in Figure 14-7. Use the Performance Task Assessment List for Scientific Drawing in **PASC,** p. 55. **P**

Prepare

Section Background

- Animals with arthropod characteristics evolved more than 600 million years ago. In the course of evolution the bodies of arthropods came to have fewer segments. These segments have become fixed in number and more specialized in function.

- Millipedes and centipedes are often grouped as myriapods. *Myriapod* is Greek for "many feet."

- Centipedes can detach legs when attacked. The detached legs continue to wiggle while the centipede escapes.

Preplanning

- Obtain crayfish, several small aquaria, ground beef, and stirrers for Activity 14-2.

- For the MiniLAB, obtain small jars and ripe bananas.

1 Motivate

Bellringer

Before presenting the lesson, display **Section Focus Transparency 51** on the overhead projector. Assign the accompanying **Focus Activity** worksheet. [L1] [LEP]

Tying to Previous Knowledge

Students will be familiar with many species in the phylum Arthropoda. Many have eaten shrimp and lobsters. All will have some knowledge of insects and spiders. Use their knowledge of arthropods in identifying common characteristics.

Science Words

Arthropoda
appendage
exoskeleton
molting
spiracle
metamorphosis

Objectives

- Identify features used to classify arthropods.
- Relate the structure of the exoskeleton to its function.
- Distinguish between complete and incomplete metamorphosis.

Figure 14-11

The segments of some arthropods are fused together to form three regions: the head, the thorax, and the abdomen. *Can you see why these animals are named for jointed feet? Explain.*

Features of Arthropods

The arthropods make up the largest phylum of animals in the animal kingdom. They are adapted to almost every environment on Earth. Insects, shrimp, spiders, and centipedes are all members of the phylum **Arthropoda** (AR thruh pahd ah).

Arthropoda means "jointed foot" and describes the jointed appendages of arthropods. **Appendages** are structures that grow from the body. Your arms and legs are appendages. The jointed appendages of arthropods include legs, antennae, claws, and pinchers.

The bodies of arthropods are divided into segments like those of segmented worms. Because of this, scientists hypothesize that arthropods and segmented worms have a common ancestor. The bodies of some arthropods have many segments, whereas others have segments that are fused together to form body regions. **Figure 14-11** shows three regions formed by fused segments: the head, the thorax, and the abdomen. Arthropods have a body cavity and a digestive system with two openings, a mouth and an anus. They have a nervous system similar to that of annelids but with a larger brain.

Exoskeletons

All arthropods have an external covering called the **exoskeleton.** The exoskeleton covers, supports, and protects the body. The lightweight exoskeleton is made of protein and a carbohydrate called *chitin* (KITE un). This covering also keeps the animal's body from drying out.

An exoskeleton cannot grow as the animal grows. From time to time, the old exoskeleton is shed and replaced by a new one in a process called **molting.** The new exoskeleton is soft and takes a while to harden. During this time, the animal is not well protected from its predators.

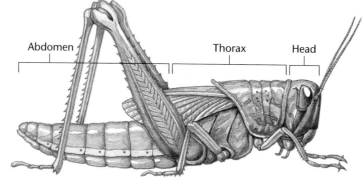

Abdomen   Thorax   Head

Program Resources

Reproducible Masters
Study Guide, p. 55 [L1]
Reinforcement, p. 55 [L1]
Enrichment, p. 55 [L3]
Activity Worksheets, pp. 5, 84-85, 86 [L1]
Technology Integration, pp. 13-14
Lab Manual 35, 36, 37

Transparencies
Section Focus Transparency 51 [L1]
Teaching Transparency 28

Insects

Insects make up the largest group of complex invertebrates. There are more than 700 000 classified species of insects, and scientists describe more each year. Why are insects so successful?

Insect Body Plan

Insects have three body regions: a head, a thorax, and an abdomen. With some insects, it is almost impossible to see where one region stops and the next one begins. The head has a pair of antennae, eyes, and a mouth. The thorax has three pairs of jointed legs and, in many species, one or two pairs of wings, as you can see in the insects shown in **Figure 14-12.** The abdomen is divided into 11 segments and has neither wings nor legs attached to it.

Insects are the only invertebrates that are able to fly. Most insects have either one or two pairs of wings. Flies have one pair. Some insects, such as silverfish and fleas, don't have wings. Flying enables insects to find new places to live, new food sources, and mates. It also helps them escape from their enemies.

The heads of most insects have simple and compound eyes and a pair of antennae. Simple eyes detect light and dark. Compound eyes contain many lenses and can detect colors and movement. The antennae are used for touch and smell.

Insects have open circulatory systems. The open circulatory system carries digested food to cells and removes wastes. Insect blood does not carry oxygen. Insects have openings called **spiracles** on the abdomen and thorax through which air enters and waste gases are expelled.

Figure 14-12

All insects have six legs, a head, a thorax, and an abdomen. They also have one or two pairs of wings, one pair of antennae, and a pair of compound eyes.

 A Stag beetle

C Dragonfly

B Leaf insect

383

383

USING TECHNOLOGY

A Fly in My Wound

For three hundred years, physicians treating wounds that would not heal noticed something odd. Patients who had wounds infested with fly larvae actually healed faster than those whose wounds were clean! Apparently, the larvae consumed the dead tissue and bacteria inside the wound, but left the healthy tissue alone.

Recently, many bacteria have evolved resistance to antibiotics. So physicians are looking for alternatives to antibiotics to treat patients. Enter the age of fly larvae therapy.

Maggot Therapy

Green bottle flies, *Lucilia sericata,* are grown to promote healing in persons with open sores. The larvae of these flies, called maggots, are placed in wounds that medicine has not been able to help. Early findings from a five-year study indicate that the maggots are able to help heal wounds that antibiotics cannot. Using fly larvae caused wounds to shrink about 20 to 25 percent a week. All the patients treated with maggot therapy healed completely within a month.

Think Critically:

What are the drawbacks to using maggot therapy in hospitals? Why would using maggots be considered a "back to nature" therapy?

Green Bottle Fly

Insects reproduce sexually. Females lay thousands of eggs, but only a fraction of the eggs develop into adults. Think about how overproduction ensures that each insect species will continue.

Metamorphosis

Many species of insects and other animals that hatch from fertilized eggs go through a series of changes in body form to become adults. This series of changes is called **metamorphosis.** Metamorphosis is controlled by chemicals secreted by the body of the animal. There are two kinds of metamorphosis, complete and incomplete, shown in **Figure 14-13.**

Complete Metamorphosis

Most insects, including butterflies, beetles, ants, bees, moths, and flies, develop through complete metamorphosis. The four stages of development are egg,

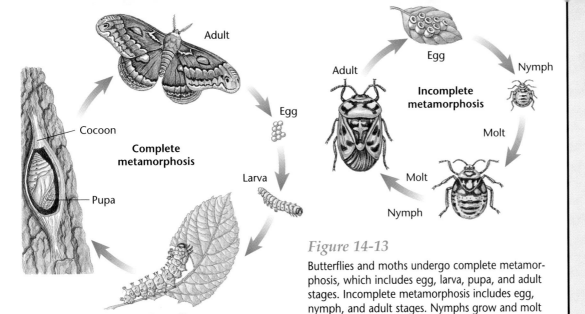

Figure 14-13

Butterflies and moths undergo complete metamorphosis, which includes egg, larva, pupa, and adult stages. Incomplete metamorphosis includes egg, nymph, and adult stages. Nymphs grow and molt several times before reaching the adult stage.

larva, pupa, and adult. A fertilized egg hatches into a worm-like larva stage. Maggots and caterpillars are larvae. The larva spends its time eating and growing before forming a shelter called a cocoon or chrysalis and going into a resting stage called the pupa. Inside, the larva changes and develops into an adult in a process that is little understood. The cocoon or chrysalis eventually opens and the adult comes out. The adults mate, the female lays eggs, and the cycle begins again.

Incomplete Metamorphosis

Grasshoppers, silverfish, lice, and crickets develop through incomplete metamorphosis. The three stages of development are egg, nymph, and adult. The fertilized egg hatches into a nymph that looks like a small adult, but without wings. A nymph molts several times before reaching the adult stage.

Insect Adaptations

Adaptations such as an exoskeleton, wings, and jointed appendages allow insects to live on land and to fly. Insects have short life spans; thus, natural selection takes place more quickly in insect populations than in organisms that take longer to reproduce. The small size of insects allows them to live in a wide range of environments and hide from their enemies. Many species of insects can live in the same area and not compete with one another for food.

14-3 Arthropods **385**

Cultural Diversity

Biological Controls Zhao Jinzao, a university vice-president in China, helped develop a technique for protecting crops from insects in the 1970s. He discovered that spiders could kill 80 to 90 percent of the harmful insects in cotton fields. Jinzao taught farmers to dig shallow pits every ten paces in their cotton fields and to throw straw into the pits. The straw protects the spiders from cold. Have students discuss the advantages of using spiders to control the insect population in crops.

Activity

[LS] Intrapersonal Have students look under rocks and logs for isopods. Isopods are land crustaceans. There are two easily distinguishable types. When disturbed, the pill bug curls into a ball. A sow bug does not curl into a ball. Isopods can be maintained in the classroom in a jar of moist soil with a slice of potato. [L1]

Other Arthropods

Although insects make up the largest group of arthropods worldwide, there are several other classes of arthropods: arachnids, centipedes, millipedes, and crustaceans. Examples of each class of arthropod are shown in **Figure 14-14** and **Figure 14-16** on page 388.

Arachnids

Spiders, scorpions, mites, and ticks are arachnids (uh RAK nudz). Arachnids are arthropods with two body regions, a head-chest region called the cephalothorax and an abdomen. They have four pairs of legs but no antennae. You can tell an arachnid from an insect because the arachnid has eight legs. Arachnids are adapted to kill prey with poison glands, stingers, or fangs. Appendages near the mouth are used to hold food.

A spider cannot chew its food. Using a pair of fangs, a spider injects poison into its prey; the poison paralyzes the prey. Then the spider releases enzymes that turn the prey into a liquid. The spider then sucks up the liquid. Oxygen and carbon dioxide are exchanged in structures called book lungs, so called because they look like the pages of a book. Spiracles on the abdomen allow oxygen and carbon dioxide to move into and out of the book lungs. You can see the book lungs of a spider in **Figure 14-15.**

Ticks and mites are arachnids that are parasites. Mites are tiny. House dust mites are found in the dust on floors and in bedding. There is even a type of mite that lives in the follicles of human eyelashes.

Figure 14-14

All arachnids have structures that are similar to those of insects.

A The spider is a common arachnid. Spiders have eight simple eyes used to sense light and darkness. This is a female wolf spider.

B A tick attaches to the skin of its host and feeds on its blood. This giant lion tick is found on lions and their prey.

386 Chapter 14 Mollusks, Worms, Arthropods, and Echinoderms

Across the Curriculum

Language Arts Have students prepare an illustrated report on the various hunting methods of spiders. Include the wolf spider and trap door spider. [L2]

Scorpions have a sharp stinger at the end of the abdomen that contains poison. Scorpions grab their prey with their pinchers and inject venom to paralyze them. The sting of a scorpion is painful to humans, but it usually is not fatal.

Centipedes and Millipedes

Even though centipedes and millipedes look like worms, you know they are not because worms don't have legs. Centipedes and millipedes make up two classes of arthropods. They have long bodies with many segments, exoskeletons, jointed legs, antennae, and simple eyes. They live on land in moist environments. Both reproduce sexually and make nests for their eggs. They stay with the eggs until they hatch.

Compare the centipede and millipede in **Figure 14-14C** and **D.** The centipede has one pair of jointed legs per segment, whereas the millipede has two pairs of legs per segment. Centipedes hunt for their food and have a pair of poison claws used to inject venom into their prey. Centipedes feed on snails, slugs, and worms. Their bites are painful to humans. Millipedes don't move as quickly as centipedes do. They feed on plants.

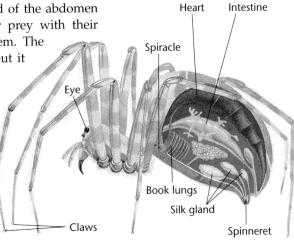

Figure 14-15

The external structures and book lungs of a spider are common to all arachnids.

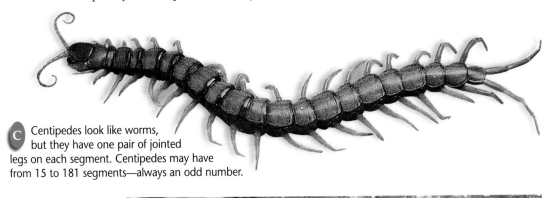

C Centipedes look like worms, but they have one pair of jointed legs on each segment. Centipedes may have from 15 to 181 segments—always an odd number.

D Millipedes are similar to centipedes, except they have two pairs of legs on each segment. A millipede may have more than 100 segments in its long abdomen.

387

Theme Connection

Stability and Change The body structures of centipedes and millipedes show the evolutionary link between arthropods and annelids. Annelids have nonspecialized segments with jointed appendages.

4 Close

Section Wrap-up

Review

1. jointed legs, body segments, exoskeleton

2. advantages: provides covering to prevent dehydration, supports body, protects body; disadvantages: body outgrows exoskeleton, animal vulnerable while new exoskeleton hardens

3. **Think Critically** Students may choose any insect to describe.

USING MATH

Answer about 8 times

Crustaceans

Crabs, crayfish, lobsters, shrimp, barnacles, pill bugs, and water fleas all belong to the class Crustacea. Crustaceans are arthropods that have one or two antennae and jaws called mandibles used for crushing food. Most crustaceans live in water, but some, like pill bugs, live on land in moist environments.

The blue crab in **Figure 14-16** is a typical crustacean. Although some crustaceans have three body regions, the blue crab has a head and thorax fused to form one region, the cephalothorax.

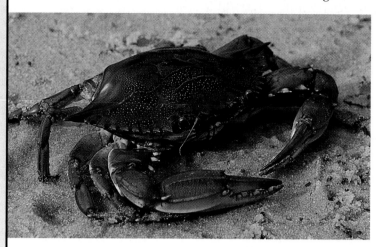

Figure 14-16

Blue crabs, crustaceans found along coastal waters of North America, are a source of food for humans and other animals.

Crustaceans have five pairs of legs. The first pair of legs are claws that catch and hold food. The other four pairs are walking legs. The five pairs of appendages on the abdomen are swimmerets. They help the crustacean move and are used in reproduction. The swimmerets also force water over the feathery gills that crustaceans use for breathing. If a crustacean happens to lose an appendage, it can regenerate the lost part.

Section Wrap-up

Review

1. What are three features of all arthropods?

2. What are the advantages and disadvantages of an exoskeleton?

3. **Think Critically:** Choose an insect you are familiar with and explain how it's adapted to its environment.

Skill Builder
Sequencing

Sequence the events in the complete metamorphosis of an insect and the events in incomplete metamorphosis. If you need help, refer to Sequencing in the **Skill Handbook**.

USING MATH

Of the major arthropod classes, 88% are insects, 7% are arachnids, 3% are crustaceans, 1% are centipedes and millipedes, and all others make up 1%. Show these data in a circle graph. In the number of species, insects outnumber all other arthropod species by how many times?

Skill Builder

Students should make a cycle concept map for each type of metamorphosis. Complete metamorphosis should show egg → larva → pupa → adult. The map of incomplete metamorphosis should show egg → nymph → adult. Each map should be made in a circular fashion to show the cycle continues.

Assessment

Content Ask students to write in their Science Journals how incomplete and complete metamorphosis are similar and how they are different. *Both are a series of changes. In complete metamorphosis, the young look very different from the adults.* Use the Performance Task Assessment List for Writing in Science in **PASC**, p. 87. P

Activity 14-2

Observing a Crayfish

A crayfish is a crustacean that has a fused head and thorax and a segmented body. It has a snout and eyes on movable eyestalks. Most crayfish have pincers. What do you think these pincers are for?

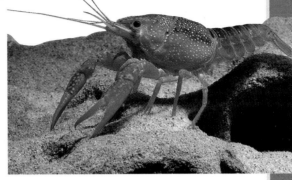

Problem
How does a crayfish use its appendages?

Materials
- crayfish in a small aquarium
- stirrer
- uncooked ground beef

Procedure
1. Copy the data table and use it to record your observations.
2. Your teacher will provide you with a crayfish in an aquarium. **CAUTION:** *Use care when working with live animals. Leave the crayfish in the aquarium while you do the activity.*
3. Touch the crayfish with the stirrer. How does the body feel?
4. Observe the crayfish moving in the water. How many body regions does it have? How many appendages does it have on each body region? What is the function of each appendage?

Record your answers in your data table.
5. Observe the compound eyes. On which body region are they located?
6. Drop a small piece of ground beef into the aquarium and observe how the crayfish eats.
7. Return the crayfish and aquarium to the place your teacher has assigned. Wash your hands.

Analyze
1. Describe the texture of the body of the crayfish.
2. How many body regions does it have?
3. How many appendages are located on each body region?
4. What structures does the crayfish use to get food?

Conclude and Apply
5. **Infer** how the location of the eyes is an advantage for the crayfish.
6. How does the structure of the claws aid in getting food?
7. What can you **infer** about the exoskeleton and protection?

Data and Observations   Sample Data

Body Region	Appendages	Function
Head	5 pairs including antennae and mouthparts	sense environment; hold and chew food
Thorax	5 pairs including claws and walking legs	capture food, defense, walking
Abdomen	5 pairs including swimmerets and tail	create water currents to help crayfish move; reproduction

14-3 Arthropods   **389**

Prepare

Section Background

The process in which levels of chemicals increase in animals higher on the food chain is called biological magnification.

1 Motivate

Bellringer

Before presenting the lesson, display **Section Focus Transparency 52** on the overhead projector. Assign the accompanying **Focus Activity** worksheet. `L1` `LEP`

2 Teach

Concept Development

Explain to students that many insects which damage crops are themselves preyed upon by other organisms. Many birds are insect eaters. When birds eat insects that have been sprayed, they ingest the pesticide, too. Describe how a pesticide can move through the food chain and accumulate in the top levels of organisms. Explain how DDT was linked to the reproductive problems of bald eagles, for example.

Use **Science and Society Integration,** p 18, as you teach this lesson.

3 Assess

Check for Understanding

Have students answer questions 1 and 2 in the Section Wrap-up.

Science Words

pesticide

Objectives

- Describe the importance of pesticides in agriculture.
- Identify the impact of pesticides on the environment.

Figure 14-17

Beneficial insects, such as this ladybird beetle, can reduce the need for pesticides. Ladybird beetles eat aphids, insects that harm crops.

Insect Pests

Most insects are harmless. In fact, many, such as the praying mantis, are beneficial. These insects eat harmful insects. Certain insects, such as bees, are necessary to help pollinate important farm crops, such as clover. Farmers and gardeners often find it necessary to destroy insects that damage crops and ornamental plants. Many farmers would not have successful crop yields without the use of pesticides. **Pesticides** are chemicals that kill undesirable plants and insects. It is estimated that for every $3 million spent on pesticides, about $12 million is returned in additional crops. In addition, the price of pesticides has increased much more slowly than the price of other farm expenses. This makes pesticides smart investments for farmers.

Environmental Damage

Although the use of pesticides is often beneficial to farmers, the environmental damage caused as a result of pesticide use is great. Studies show that most of the pesticides used in farm fields wind up in the soil, water, and air. Pesticides intended for general use kill beneficial insects along with the the harmful ones. Most of the nation's species of plants and animals are also affected. Birds and mammals either come in direct contact with the pesticide or feed on plants and animals that have been contaminated. It is estimated that there are about 45 000 human poisonings every year due to pesticides.

Our society has gained much from pesticides, but are we paying for them with our health? Can some alternative solution be reached? Use of chemicals that break down easily with the help of sunlight, water, and biological action may be an option. Use of natural predators, such as the ladybird beetle shown in **Figure 14-17,** and improved farming practices, such as the one shown in **Figure 14-18,** may also reduce the need for pesticides. Even if these solutions are able to cut the current problem in half, would we still have to continue to use pesticides? Can we live without them?

390 Chapter 14 Mollusks, Worms, Arthropods, and Echinoderms

Program Resources

Reproducible Masters
Study Guide, p. 56 `L1`
Reinforcement, p. 56 `L1`
Enrichment, p. 56 `L3`
Science and Society Integration, p. 18

Transparencies
Section Focus Transparency 52 `L1`

2 Points of View

Effects on Crops

The development and use of pesticides has enabled farmers worldwide to produce much more food per hectare than ever before. As the world's population increases, demands for quality food crops will increase as well. New ways of applying pesticides may reduce the amount needed and ensure that pesticides affect only the intended target—insect pests.

Effects on the Environment

Pesticides kill not only the intended pest, but other organisms as well. Even organisms that aren't killed may accumulate the pesticide in body tissues, leading to inability to reproduce or other harmful effects. Using natural checks and balances in farming will allow predators to keep insect pests under control.

Figure 14-18

Farmers can grow crops with little need for pesticides using the no-till method of farming. In no-till farming, the soil is not disturbed and natural predators of insect pests left in the soil help keep crop damage to a minimum.

Section Wrap-up

Review

1. Why are pesticides so important to our economy?

2. What are the current problems with pesticide use?

interNET CONNECTION Some insects are being used to control problem weeds, eliminating the need for chemical control. Visit the Chapter 14 Internet Connection at Glencoe Online Science, **www.glencoe.com/sec/science/life,** for a link to more information about biological control of insect pests.

SCIENCE & SOCIETY

14-4 Pesticides and Insects   391

391

Prepare

Section Background

- Phylum Echinodermata is made up of eleven classes, six of which are extinct and known only from fossils.

- The fossil record indicates that echinoderms date back to the Cambrian period over 500 million years ago. The fossil record also indicates that early adult forms were probably sessile and not free-living as are most modern forms.

- Echinoderms show an important evolutionary relationship to invertebrates in that they may share a common ancestor with the lower chordates.

Preplanning

For the MiniLAB, make sure students have their science books in class.

1 Motivate

Bellringer

Before presenting the lesson, display **Section Focus Transparency 53** on the overhead projector. Assign the accompanying **Focus Activity** worksheet. L1 LEP

Tying to Previous Knowledge

Even if students have not seen living echinoderms, they have probably seen dried sea stars and sand dollar skeletons. Use their knowledge of echinoderms to introduce this lesson.

Science Words

echinoderm
water-vascular system
tube feet

Objectives

- Identify the features of echinoderms.
- Describe how sea stars get and digest food.

14•5 Echinoderms

Features of Echinoderms

Unless you live near the ocean, you may not have seen an echinoderm, but most of you know what a sea star is. Sea stars, sea urchins, sand dollars, and sea cucumbers are all echinoderms (ih KI nuh durmz). **Echinoderms** are spiny-skinned invertebrates that live on the ocean bottom. The name *echinoderm* means "spiny-skin." The spiny part refers to the spines that cover the outside of these animals. Their bodies are supported and protected by an internal skeleton made of calcium carbonate plates. The plates are covered by the thin, spiny skin. Let's look more closely at one of the most well-known echinoderms, a sea star.

Water-Vascular System

Sea stars have a unique characteristic shared by all echinoderms, the **water-vascular system.** The water-vascular system is a network of water-filled canals. Thousands of tube feet are connected to this system. As water moves into and out of the water-vascular system, **tube feet** act like suction cups and help the sea star move and feed. Look at the tube feet shown in **Figure 14-19.** Another obvious trait of echinoderms is their radial symmetry. The bodies of sea stars, like many echinoderms, can be divided into five sections all located around a central point. **Figure 14-19** illustrates some of the other parts of a sea star.

Figure 14-19

Sea stars and all other echinoderms have a water-vascular system that allows them to move, eat, get oxygen, and get rid of waste.

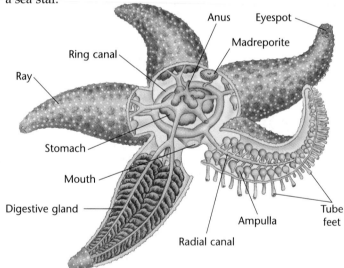

Anus · Eyespot · Madreporite · Ring canal · Ray · Stomach · Mouth · Digestive gland · Radial canal · Ampulla · Tube feet

Program Resources

 Reproducible Masters
Study Guide, p. 57 L1
Reinforcement, p. 57 L1
Enrichment, p. 57 L3
Activity Worksheets, pp. 5, 87 L1

Transparencies
Section Focus Transparency 53 L1

A Sea star

B Brittle star

C Sea urchin

Classifying Echinoderms

There are several classes of echinoderms: sea stars, brittle stars, sea urchins and sand dollars, and sea cucumbers. **Figures 14-20** and **14-21** show examples of each class.

Sea Stars

Sea stars have five or more arms arranged around a central point. These arms are lined with thousands of tube feet. A sea star uses its tube feet to open a clam. Once the shell is open slightly, the sea star turns its stomach inside out, pushes it through its mouth, and surrounds the soft body of the mollusk. The stomach secretes enzymes that digest the animal. Then when the meal is over, the sea star takes its stomach out of the clamshell and pulls it back inside its own body.

Sea stars reproduce sexually by releasing eggs and sperm into the water. Females can produce 200 000 eggs in one season. They can also repair themselves by regeneration. If a sea star loses an arm, it can grow a new one. If enough of the arm is lost, the arm itself can grow into a whole new sea star.

Brittle Stars

Brittle stars live hidden under rocks or litter on the ocean floor. They move much more quickly than sea stars and break off their arms as a defense when disturbed. Brittle stars can quickly regrow the lost parts.

Sea Urchins and Sand Dollars

Sea urchins and sand dollars have skeletons made of calcium carbonate plates, and both are covered with spines. Sand dollars are flat and covered with small, fine spines, whereas sea urchins are round and covered with longer spines.

Figure 14-20

Echinoderms include sea stars (A), brittle stars (B), and sea urchins (C). *What do these organisms have in common?*

Mini**LAB**

How do tube feet open clamshells?

Procedure
1. Hold your arm straight out, palm up.
2. Place a heavy book on your hand.
3. Have your partner time how long you can hold your arm up with the book on it.

Analysis
1. Describe how your arm feels after a few minutes.
2. If the book models the sea star and your arm models the clam, infer how a sea star successfully overcomes a clam to obtain food.

14-5 Echinoderms   **393**

3 Assess

Check for Understanding

Use the Flex Your Brain activity to have students explore ECHINODERMS.

Activity Worksheets, p. 5

Reteach

Have students explain why echinoderms have few predators.

Extension

For students who have mastered this section, use the **Reinforcement** and **Enrichment** masters.

4 Close

Ask students to explain how tube feet are used to open bivalves.

Section Wrap-up

Review

1. spiny skin, endoskeleton made of calcium plates, water-vascular system, tube feet, radial symmetry

2. Tube feet act like strong suction cups and help them move and get food.

3. **Think Critically** Sea stars graze on organisms that live among the coral.

Science Journal

Linguistic Journal entries should reflect the characteristics of the organism, its environment, way of life, and other information.

Sea Cucumbers

Sea cucumbers are soft-bodied echinoderms with a leathery covering. They have tentacles around the mouth and rows of tube feet on both the dorsal and ventral sides. When threatened, sea cucumbers may expel their internal organs. These organs regenerate in a few weeks.

Figure 14-21
A sea cucumber moves along the ocean bottom using tube feet.

Importance of Echinoderms

Echinoderms are providing scientists with information about how body parts regenerate. They are also important to the marine environment because they feed on dead organisms and help recycle materials. Many oyster and clam farmers don't find them as helpful. Sea stars feed on oysters and clams and destroy millions of dollars' worth of these organisms each year.

Scientists think that echinoderms are the invertebrate group that most closely resembles the chordates. Most have radial symmetry like some of the less complex invertebrates, but they also have complex body systems. Another reason scientists view them as advanced is the way an echinoderm embryo develops. It develops the same way the embryos of chordates do, and not like the embryos of less complex invertebrates. You will study chordates in the next chapter.

Section Wrap-up

Review

1. What features do all echinoderms have in common?

2. How do echinoderms move and get their food?

3. **Think Critically:** Why are sea stars common in coral reefs?

Skill Builder
Observing and Inferring
Observe the echinoderms pictured on pages 393 and 394, and infer why they are slow moving and live on the ocean floor. If you need help, refer to Observing and Inferring in the **Skill Handbook.**

Science Journal

Choose an echinoderm in this section and write about it in your Science Journal. Describe its appearance, how it gets food, where it lives, and other interesting facts.

Skill Builder

With the exception of the brittle star, the echinoderms pictured have no means of movement except for their tube feet. Therefore, they are not capable of rapid movement. Because they are not able to escape from predators, each has its own unique way of protecting itself. The sea star has few enemies; the sea urchin has long, pointed spines; the sea cucumber looks unappetizing as well as having other defenses.

The brittle star is capable of rapid movement, but spends most of its time hidden under debris.

 Assessment

Oral **Infer why echinoderms are bottom dwellers.** *Tube feet are on their ventral side.* Use the Performance Task Assessment List for Making Observations and Inferences in **PASC,** p. 17. **P**

Chapter 14 Review

Summary

14-1: Mollusks

1. Mollusks are soft-bodied invertebrates usually covered by a hard shell. Mollusks move using a muscular foot.
2. Mollusks with one shell are gastropods. Bivalves have two shells. Cephalopods have a foot divided into tentacles, and an internal shell.

14-2: Segmented Worms

1. Annelids, segmented worms, have a body cavity to hold internal organs. They also have setae to help them move.
2. Earthworms have a digestive system made of a mouth, crop, gizzard, intestine, and anus.
3. Mollusks and annelid worms develop from the same type of larva.

14-3: Arthropods

1. Arthropods are classified by number of body segments and appendages.
2. Arthropod exoskeletons cover, protect, and support the body.
3. Arthropods develop either by complete metamorphosis or incomplete metamorphosis.

14-4: Science and Society: Pesticides and Insects

1. Pesticides control populations of insects that damage food crops.
2. Pesticides can poison animals, including humans, that come in contact with them.

14-5: Echinoderms

1. Echinoderms are spiny-skinned invertebrates. They move by means of a water-vascular system.
2. Sea stars feed by using their tube feet to open the shells of mollusks.

Key Science Words

a. appendage
b. Arthropoda
c. closed circulatory system
d. crop
e. echinoderm
f. exoskeleton
g. gill
h. gizzard
i. mantle
j. metamorphosis
k. mollusk
l. molting
m. open circulatory system
n. pesticide
o. radula
p. setae
q. spiracle
r. tube feet
s. water-vascular system

Reviewing Vocabulary

Match each phrase with the correct term from the list of Key Science Words.

1. temporary food storage site
2. tongue-like organ in gastropods
3. bristles earthworms use to hold on to soil
4. structure in earthworms that grinds soil
5. arm, leg, or antenna
6. outer covering made of chitin
7. shedding of the exoskeleton
8. spiny-skinned invertebrate

Chapter 14 Review 395

Review

Summary

Have students read the summary statements to review the major concepts of the chapter.

Reviewing Vocabulary

1. d	6. f
2. o	7. l
3. p	8. e
4. h	9. q
5. a	10. i

✔ Assessment

Portfolio Encourage students to place in their portfolios one or two items of what they consider to be their best work. Examples include:

- Inclusion Strategies, p. 371
- Using Computers, p. 373
- Explore the Issue, p. 391 P

Performance Additional performance assessments may be found in **Performance Assessment** and **Science Integration Activities.** Performance Task Assessment Lists and rubrics for evaluating these activities can be found in Glencoe's **Performance Assessment in the Science Classroom.**

GLENCOE TECHNOLOGY

 Videodisc

Glencoe Life Science Interactive Videodisc
Side 2, Lesson 6
Eyeing Evolution

9310 to 18595

MindJogger Videoquiz

Chapter 14 Have students work in groups as they play the Videoquiz game to review key chapter concepts.

Checking Concepts

1. c	**6.** d
2. d	**7.** c
3. a	**8.** d
4. a	**9.** a
5. c	**10.** c

Understanding Concepts

11. Swimmerets are specialized appendages that help crustaceans move and are used in reproduction.

12. Sea stars regenerate. Cutting them apart causes the growth of more sea stars.

13. The exoskeleton is not fully hardened right after molting. Organisms usually hide at this time, until their exoskeleton is hard enough to protect them from predators.

14. An earthworm cannot take in oxygen from water-laden soil and must come to the surface for air.

15. Insects have wings, allowing them to go from place to place and to hover above a flower; hairs on their bodies pick up pollen. They also have simple and compound eyes that enable them to find flowers.

Thinking Critically

16. Gastropods use their radulas to tear plant materials and scrape algae; cephalopods use tentacles with strong suction cups to capture organisms and pry open shells; bivalves are filter feeders that get food by the action of sweeping cilia.

17. Clams burrow into the sand; oysters attach themselves by strong sticky threads to rocks; scallops eject water and move by rapidly opening and closing their shells; a

---

9. opening for air in some arthropods
10. structure that secretes the shell of mollusks

Checking Concepts

Choose the word or phrase that completes the sentence.

1. The _____ cover(s) the organs of mollusks.
   - a. gills
   - b. foot
   - c. mantle
   - d. visceral mass
2. An example of an annelid is a(n) _____.
   - a. snail
   - b. slug
   - c. octopus
   - d. earthworm
3. The organism with a closed circulatory system is a(n) _____.
   - a. earthworm
   - b. snail
   - c. oyster
   - d. slug
4. The largest phylum of animals is the _____.
   - a. arthropods
   - b. mollusks
   - c. annelids
   - d. echinoderms
5. Organisms with two body regions are _____.
   - a. insects
   - b. mollusks
   - c. arachnids
   - d. annelids
6. An example of an arthropod is a(n) _____.
   - a. snail
   - b. earthworm
   - c. slug
   - d. scorpion
7. Many _____ are organisms with radial symmetry.
   - a. annelids
   - b. mollusks
   - c. echinoderms
   - d. arthropods
8. The structures echinoderms use to move and open mollusk shells are _____.
   - a. calcium plates
   - b. arms
   - c. spines
   - d. tube feet

---

9. _____ body parts can be replaced by regeneration.
   - a. Sea star
   - b. Snail
   - c. Octopus
   - d. Slug
10. Of the following, the _____ is the only organism that doesn't sting or poison its food.
    - a. centipede
    - b. spider
    - c. millipede
    - d. scorpion

Understanding Concepts

Answer the following questions in your Science Journal using complete sentences.

11. Describe the adaptations of swimmerets.
12. Why is it unwise to cut apart sea stars and throw the parts back into a water area that is already overpopulated with them?
13. What problems do arthropods have immediately after molting?
14. Based on your observations of a segmented worm in Activity 14-1, explain why you see many earthworms on the surface of the soil after a heavy rain.
15. What structures do insects have that enable them to find food?

Thinking Critically

16. Describe how each class of mollusks obtains food.
17. Compare the ability of clams, oysters, scallops, and squid to protect themselves.
18. Compare an earthworm gizzard with teeth.
19. What evidence do scientists have that mollusks and annelids share a common ancestor?

---

squid ejects water and escapes by jet propulsion.

18. Grit in the earthworm gizzard grinds food much the way teeth break up and grind food in the mouth.

19. A space in the body to hold organs is present in both groups. The larva of both groups is similar.

20. It provides room in which to grow before the next molt.

20. After molting, but before the new exoskeleton hardens, an arthropod causes its body to swell by taking in extra water or air. How does this behavior help the arthropod?

Developing Skills

If you need help, refer to the **Skill Handbook.**

21. Classifying: Group the following animals into arthropod classes: spider, pill bug, crayfish, grasshopper, crab, silverfish, cricket, sow bug, tick, scorpion, shrimp, barnacle, butterfly.

22. Recognizing Cause and Effect: If all the earthworms were removed from a hectare of soil, what would happen to the soil? Why?

23. Observing and Inferring: Why are gastropods sometimes called univalves?

24. Classifying: The suffix *-ptera* means "wings." Find out the meaning of the prefix of each insect class and give an example of a member of each group.
DIPTERA   HOMOPTERA
ORTHOPTERA   HEMIPTERA
COLEOPTERA

25. Concept Mapping: Copy and complete the map below that describes the characteristics of arthropods.

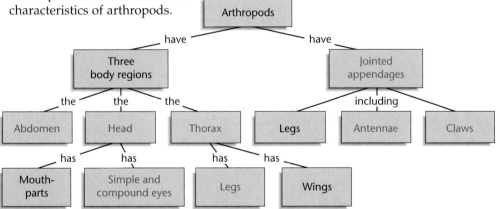

Performance Assessment

1. Scientific Drawing: Find out about the dance bees use to communicate with each other. Explain and diagram the dance for your class.

2. Display: Begin an insect collection of your own. There are many good books at your local library to guide you in how to collect and identify insects.

3. Slide Show/Photo Essay: Purchase some egg cases of a praying mantis or other insect from a biological supply house. Set up the eggs inside a lighted, ventilated screen box. Once the eggs hatch, take photographs or slides of all their activity. Make sure that you provide the proper food, water, and light for the arthropod you are photographing. When you have taken photographs of all stages of the arthropod's life cycle, prepare a slide show or photo essay of your results and present it to your class.

Chapter 14 Review   **397**

Assessment Resources

📂 Reproducible Masters
Chapter Review, pp. 31-32
Assessment, pp. 93-96
Performance Assessment, p. 28

Glencoe Technology
⊙ **Chapter Review Software**
⊙ **Computer Test Bank**
📼 **MindJogger Videoquiz**

Chapter 14 Review

Developing Skills

21. Classifying arachnids: spider, tick, scorpion; crustaceans: pill bug, crayfish, crab, sow bug, shrimp, barnacle; insects: grasshopper, silverfish, cricket, butterfly

22. Recognizing Cause and Effect The soil would become packed and hardened when the earthworms stopped aerating it. It would be less fertile because of lack of earthworm wastes.

23. Observing and Inferring Gastropods have only one shell, as compared to bivalves, which have two shells.

24. Classifying Diptera = two wings, flies; Orthoptera = straight wings, grasshoppers; Coleoptera = shielded wings, beetles; Homoptera = whole wings, leaf hoppers; Hemiptera = half wings, water bugs

25. Concept Mapping See student page.

Performance Assessment

1. Students may use library references to make drawings. Use the Performance Task Assessment List for Scientific Drawing in **PASC,** p. 55.

2. Students should obtain references on how to capture insects and prepare them for a collection. Use the Performance Task Assessment List for Display in **PASC,** p. 63.

3. Students may obtain help from a parent or other adult to take photographs. Use the Performance Task Assessment List for Slide Show or Photo Essay in **PASC,** p. 77.

Chapter Organizer

Section	Objectives/Standards	Activities/Features
Chapter Opener		Explore Activity: Model snake hearing. p. 399
15-1 Fish (1 session, 1 block)*	1. **Identify** the major characteristics of chordates. 2. **Explain** the differences between ectotherms and endotherms. 3. **Describe** the characteristics that identify the three classes of fish. National Content Standards: (5-8) UCP1, UCP4, UCP5, A1, A2, C1, C2, C3, C5, G1, G3	Using Math, p. 403 MiniLAB: How does a fish adjust to different depths? p. 404 Skill Builder: Sequencing, p. 405 Using Computers: Database, p. 405 Activity 15-1: Water Temperature and the Respiration Rate of Fish, pp. 406-407 People and Science: Nancy Aguilar, Ichthyologist, p. 408
15-2 Amphibians (1½ sessions, 1 block)*	1. **Describe** the adaptations that amphibians have for living in water and on land. 2. **Identify** the three kinds of amphibians and describe the characteristics of each. 3. **Describe** frog metamorphosis. National Content Standards: (5-8) UCP1, UCP4, A1, A2, C1, C2, C3, C5	MiniLAB: What are the adaptations of a frog? p. 410 Problem Solving: Red-Spotted Newts and Parasites, p. 411 Skill Builder: Sequencing, p. 412 Science Journal, p. 412 Activity 15-2: Metamorphosis in Frogs, p. 413
15-3 Science and Society Technology: Amphibians and Ultraviolet Light (1 session, ½ block)*	1. **Identify** the possible causes of the recent decline in amphibian populations. 2. **Explain** why amphibians are biological indicators. National Content Standards: (5-8) UCP2, C4	Explore the Technology, p. 415
15-4 Reptiles (1½ sessions, 1 block)*	1. **Identify** the adaptations that enable reptiles to live on land. 2. **Infer** why the early reptiles were so successful. 3. **Describe** the characteristics of the modern reptiles. National Content Standards: (5-8) UCP1, UCP4, UCP5, A1, C1, C3, C5, D2, E1, E2	Connect to Earth Science, p. 417 Using Technology: Treating Snakebites, p. 419 Skill Builder: Concept Mapping, p. 420 Using Math, p. 420

* A complete Planning Guide that includes block scheduling is provided on pages 32T-35T.

Activity Materials

Explore	Activities	MiniLABs
page 399 tuning fork, rubber-soled shoes	pages 406-407 goldfish, aquarium water, small fishnet, 600-mL beakers, ice water, stirring rod, thermometer, aquarium, apron	page 404 round balloon, bowl large enough to hold 2 such balloons, tap water
	page 413 4-L aquarium or jar, frog egg masses, lake or pond water, stereoscopic microscope, watch glass, small fishnet, aquatic plants, washed aquarium gravel, boiled lettuce	page 410 live frog, large jar to contain the frog, shoe box

Need Materials? Call Science Kit (1-800-828-7777).

Teacher Classroom Resources

Reproducible Masters	Transparencies	Teaching Resources
Study Guide, p. 58 **Reinforcement**, p. 58 **Enrichment**, p. 58 **Activity Worksheets**, pp. 88-89, 92 **Lab Manual 38**, Vertebrates **Concept Mapping**, p. 35 **Cross-Curricular Integration**, p. 19 **Science and Society Integration**, p. 19 **Science Integration Activity 15**, Sound Waves and Pitch	**Section Focus Transparency 54**, Going for a Ride **Science Integration Transparency 15**, Fish and Archimedes **Teaching Transparency 29**, Bony Fish	**Glencoe Life Science Interactive Videodisc** **Life Science CD-ROM** **Spanish Resources** **English/Spanish Audiocassettes** **Cooperative Learning Resource Guide** **Lab Partner** **Lab and Safety Skills** **Lesson Plans**
Study Guide, p. 59 **Reinforcement**, p. 59 **Enrichment**, p. 59 **Activity Worksheets**, pp. 90-91, 93 **Lab Manual 39**, Frog Anatomy **Concept Mapping**, p. 35	**Section Focus Transparency 55**, Spring Night on the Pond **Teaching Transparency 30**, Frog Metamorphosis	**Assessment Resources**
Study Guide, p. 60 **Reinforcement**, p. 60 **Enrichment**, p. 60	**Section Focus Transparency 56**, Now You See It	**Chapter Review**, pp. 33-34 **Assessment**, pp. 97-100 **Performance Assessment in the Science Classroom (PASC)** **MindJogger Videoquiz** **Alternate Assessment in the Science Classroom** **Performance Assessment** **Chapter Review Software** **Computer Test Bank**
Study Guide, p. 61 **Reinforcement**, p. 61 **Enrichment**, p. 61 **Concept Mapping**, p. 35 **Critical Thinking/Problem Solving**, p. 19 **Multicultural Connections**, pp. 33-34	**Section Focus Transparency 57**, What's the Difference?	

Key to Teaching Strategies

The following designations will help you decide which activities are appropriate for your students.

L1 Level 1 activities should be within the ability range of all students, including those with learning difficulties.

L2 Level 2 activities should be within the ability range of the average to above-average student.

L3 Level 3 activities are designed for the ability range of above-average students.

LEP LEP activities should be within the ability range of Limited English Proficiency students.

LS These activities are designed to address different learning styles.

COOP LEARN Cooperative Learning activities are designed for small group work.

P These strategies represent student products that can be placed into a best-work portfolio.

GLENCOE TECHNOLOGY

The following multimedia resources are available from Glencoe.

Science and Technology Videodisc Series (STVS)
Animals
Naming Fish
Studying Sharks
Nerve Regeneration in Garfish
Hump-Back Fish
Toadfish
Raising Super Fish
Sea Turtle Mystery
Alligator Courtship
Reptilian Sex Change
Rattlesnakes

Drugs from Snake Venom
National Geographic Society Series
STV: Biodiversity
STV: Rain Forest
Glencoe Life Science Interactive Videodisc
Structural Adaptations in Animals
Life Science CD-ROM

Teacher Classroom Resources

This is a representation of key blackline masters available in the Teacher Classroom Resources.

Teaching Aids

Section Focus Transparencies

Science Integration Transparencies

Teaching Transparencies

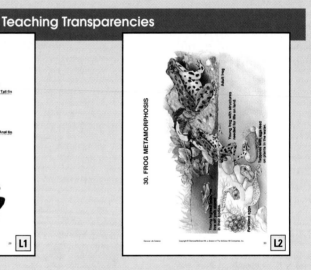

Meeting Different Ability Levels

Study Guide

Reinforcement

Enrichment Worksheets

Chapter 15 Fish, Amphibians, and Reptiles

Hands-On Activities

Science Integration Activity

Sound Waves and Pitch

Lab Manual

Vertebrates 38

Assessment

Performance Assessment

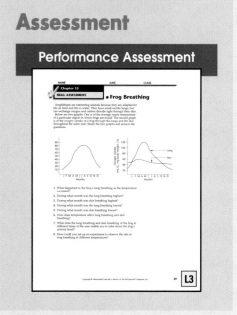

Frog Breathing

Enrichment and Application

Critical Thinking/Problem Solving

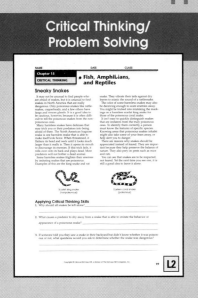

Sneaky Snakes

Cross-Curricular Integration

Fish, Amphibians, and Reptiles

Science and Society Integration

Something's Fishy

Multicultural Connections

Whose Beach Is It Anyway?

Concept Mapping

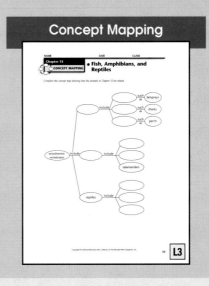

Fish, Amphibians, and Reptiles

Chapter 15

Fish, Amphibians, and Reptiles

CHAPTER OVERVIEW

Section 15-1 This section identifies the major characteristics of chordates. The structure of fish and three classes of fish are described. Examples of why fish are important are given.

Section 15-2 Amphibian adaptations for living in water and on land are examined. The characteristics of frogs, toads, and salamanders are described.

Section 15-3 *Science and Society* The reasons for the decline in amphibian populations are discussed.

Section 15-4 This section identifies the adaptations that enable reptiles to live on land and discusses why early reptiles were so successful. Modern reptiles are presented.

Chapter Vocabulary

chordate	scale
notochord	cartilage
dorsal hollow	amphibian
nerve cord	hibernation
gill slit	estivation
endoskeleton	biological
ectotherm	indicator
endotherm	reptile
fish	amniotic egg
fin	

Theme Connection

Stability and Change The theme of stability and change is introduced in a discussion of the chordates and vertebrates. The theme is developed through the discussion of adaptations that fish, amphibians, and reptiles have for living in their environments. The ecological importance of fish, amphibians, and reptiles is presented.

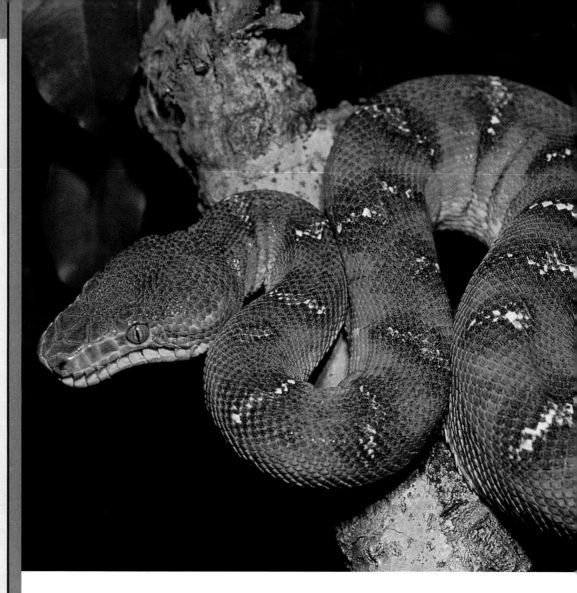

Previewing the Chapter

Section 15-1   Fish
- ▶ Vertebrate Characteristics
- ▶ What is a fish?
- ▶ Origins of Fish
- ▶ Jawless Fish
- ▶ Cartilaginous Fish
- ▶ Bony Fish

Section 15-2   Amphibians
- ▶ What is an amphibian?

- ▶ Origins of Amphibians
- ▶ Frogs, Toads, and Salamanders

Section 15-3   Science and Society
   Technology: Amphibians and Ultraviolet Light

Section 15-4   Reptiles
- ▶ What is a reptile?
- ▶ Origins of Reptiles
- ▶ Modern Reptiles

398

Learning Styles

Look for the following logo for strategies that emphasize different learning modalities. **LS**

Kinesthetic	MiniLAB, p. 404; People and Science, p. 408
Visual-Spatial	Demonstration, pp. 401, 402; Activity, p. 401; MiniLAB, p. 410; Activity 15-2, p. 413; Visual Learning, pp. 401, 403, 410, 414, 418
Interpersonal	Enrichment, p. 417
Logical-Mathematical	Activity 15-1, pp. 406-407
Linguistic	Science Journal, pp. 402, 412, 418
Auditory-Musical	Explore, p. 399

Fish, Amphibians, and Reptiles

How much do you know about reptiles? For example, do snakes have eyelids? Do they sting with their tongues? How can some snakes swallow animals that are larger than their own heads? If snakes have no ears, how do they hear? In the Explore Activity, find out how snakes hear. You'll learn more about snakes in the chapter that follows, and about other reptiles, fishes, and amphibians.

EXPLORE ACTIVITY

Model snake hearing.

1. Hold a tuning fork by the stem and tap it on a hard piece of rubber, such as the sole of a shoe.
2. Hold it next to your ear. What do you hear?
3. Now tap the tuning fork again. Press the base of the stem hard against your chin. What happens? Snakes receive similar vibrations from the environment.

Observe: Infer how a snake detects vibrations. In your Science Journal, predict how different animals use vibrations to hear.

Previewing Science Skills

▶ In the **Skill Builders**, you will **sequence** and **map concepts**.

▶ In the **Activities**, you will **hypothesize, observe, collect data, measure**, and **experiment**.

▶ In the **MiniLABs**, you will **compare** and **observe**.

399

Assessment Planner

Portfolio
Refer to page 421 for suggested items that students might select for their portfolios.

Performance Assessment
See page 421 for additional Performance Assessment options.
Skill Builders, pp. 405, 412, 420
MiniLABs, pp. 404, 410
Activities 15-1, pp. 406-407; 15-2, p. 413

Content Assessment
Section Wrap-ups, pp. 405, 412, 415, 420
Chapter Review, pp. 421-423
Mini Quizzes, pp. 412, 420

Group Assessment
Opportunities for group assessment occur with Cooperative Learning Strategies and Flex Your Brain Activities.

EXPLORE ACTIVITY

Purpose

LS **Auditory-Musical** Use the Explore Activity to introduce students to reptiles, one of the cold-blooded vertebrate classes they will study in this chapter. **L1**

Preparation

Before beginning this activity, obtain several tuning forks.

Materials

one tuning fork for every two students

Teaching Strategies

Advise students not to touch the prongs of the tuning fork after hitting it on the piece of rubber.

Observe

Sound is heard because vibrations traveled through the bones in the jaw and skull to the fluid in the inner ear. Here, the vibrations are translated into nerve impulses and interpreted as sound by the brain. In many snakes and salamanders, the jawbone is sensitive to ground vibrations.

✓Assessment

Portfolio Have students research and report on reptiles common to your area. Students should include habitat, type of food, and unusual habits of the reptiles in their report. Reports may be presented to the class. Use the Performance Task Assessment List for Oral Presentation in **PASC,** p. 71. **P**

399

Prepare

Section Background

Adaptations in fish are the result of millions of years of evolution by natural selection. The first known vertebrates were small, jawless, fishlike organisms called ostracoderms that appeared on Earth more than 500 million years ago. Scientists hypothesize that all other vertebrates can be traced back to the ostracoderms. The first modern fish appeared about 225 million years ago during the Triassic period.

Preplanning

• Obtain goldfish or guppies for Activity 15-1.
• Obtain balloons for the MiniLAB.

1 Motivate

Bellringer

Before presenting the lesson, display **Section Focus Transparency 54** on the overhead projector. Assign the accompanying **Focus Activity** worksheet. L1 LEP

Tying to Previous Knowledge

Vertebrates are animals that are familiar to students. Ask them to name as many kinds of vertebrates as possible. Then ask what characteristics these animals have in common.

Science Words

chordate
notochord
dorsal hollow nerve cord
gill slit
endoskeleton
ectotherm
endotherm
fish
fin
scale
cartilage

Objectives

• Identify the major characteristics of chordates.
• Explain the differences between ectotherms and endotherms.
• Describe the characteristics that identify the three classes of fish.

Figure 15-1

Tunicates and lancelets are both chordates.

A There are more than 2000 species of tunicates in the oceans. *How do you distinguish a tunicate from a sponge?*

400

Vertebrate Characteristics

Most of the groups of animals you've studied so far live in water. They are all invertebrates, and their bodies are adapted to aquatic life. Animals with backbones, the vertebrates, can be found in almost every environment—oceans, fresh water, land, and air. How are these organisms adapted to all these types of environments?

Traits of Chordates

Vertebrate animals belong to the phylum Chordata. The chordates are grouped into three subphyla, the largest being the vertebrates. The two small subphyla are made up of tunicates, also called sea squirts, and lancelets, pictured in **Figure 15-1**. Tunicates are sessile. They are filter feeders and live attached to objects in the sea. Lancelets also live in salt water. They can swim freely, but spend most of their time buried in the sand with only their heads sticking out. Lancelets are also filter feeders.

What do sea squirts and lancelets have in common with vertebrates such as whales, bears, fish, and humans? Actually, three things: at some time during their lives, all **chordates** have a notochord, a dorsal hollow nerve cord, and gill slits. The **notochord** is a flexible, rodlike structure along the dorsal side, or back, of the animal. In vertebrates, the notochord is eventually replaced by bones that make up a backbone in the adult. A **dorsal hollow nerve cord** is a tubular bundle of nerves that lies above the notochord. In most vertebrates, it develops into a spinal cord with a brain at the front end. **Gill slits** are paired openings located in the throat behind the mouth. Gill slits develop into gills in fish. Traces of gills can be seen in embryos of many vertebrates, including humans.

B The largest lancelets are only 8 cm in length. Lancelets are found in coarse gravel beds of crushed shells.

Program Resources

Reproducible Masters
Study Guide, p. 58 L1
Reinforcement, p. 58 L1
Enrichment, p. 58 L3
Cross-Curricular Integration, p. 19
Science and Society Integration, p. 19
Activity Worksheets, pp. 5, 88-89, 92 L1
Concept Mapping, p. 35
Lab Manual 38

Science Integration Activities, pp. 29-30

Transparencies
Section Focus Transparency 54 L1
Teaching Transparency 29
Science Integration Transparency 15

What is a vertebrate?

Vertebrates are named for the column of bones called vertebrae that encloses the dorsal nerve cord. The vertebrae, skull, and the rest of the internal skeleton is called the **endoskeleton.** *Endo-* means "within." An endoskeleton supports and protects internal organs and is the place where muscles are attached.

Most vertebrates are ectotherms. An **ectotherm** is an animal in which the internal body temperature changes with the temperature of its surroundings. Fish, amphibians, and reptiles are ectotherms. **Endotherms** are animals with constant internal body temperatures. Birds and mammals are endotherms. Do you have a constant internal body temperature? You are also an endotherm.

What is a fish?

Did you know that there are more differences among fish than among any other vertebrate class? That's because there are three separate classes of fish, but just one class each of amphibians, reptiles, birds, and mammals. Actually, there are more than 30 000 different species of fish—more than all other species of vertebrates. What is a fish, and how is it adapted to its environment?

Fish Adaptations

Fish are ectotherms that have many adaptations that allow them to live in water. Among these adaptations are gills, fins, and scales shown in **Figure 15-2.** All fish have gills for breathing. Gills consist of fleshy filaments that are filled with tiny blood vessels, so gills are bright red in color. To breathe, the fish takes in water through its mouth and passes the water over the gills. Oxygen in the water is taken up by blood vessels in the gill filaments. At the same time, carbon dioxide is released from the blood vessels and carried out of the fish's body. Most fish have **fins,** fanlike structures used for steering, balancing, and moving. The large tail fin moves back and forth to propel the fish through the water. Fish also have **scales,** hard, thin, overlapping plates that cover the skin and protect a fish's body. Most fish scales are made of bone. Fish are grouped into three classes: jawless fish, cartilaginous fish, and bony fish.

Figure 15-2

Fish have gills for taking in oxygen and scales for protection. Fins are used for steering, balancing, and moving.

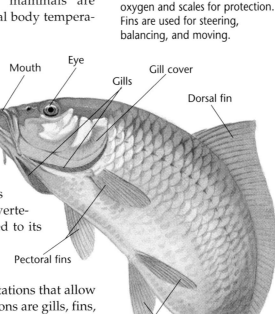

Mouth
Eye
Gills
Gill cover
Dorsal fin
Pectoral fins
Pelvic fins
Anal fin
Fin ray
Scales
Lateral line

2 Teach

Demonstration

Visual-Spatial Display preserved specimens of a tunicate, an amphioxus, and a frog so that students can observe characteristics of the three chordate subphyla.

Activity

Visual-Spatial Provide a skeleton of a fish (ectothermic vertebrate) and a cat (endothermic vertebrate). Have students observe the skeletons and determine how they are similar and how they are different. L1

Visual Learning

Figure 15-1 How do you distinguish a tunicate from a sponge? *A tunicate has a notochord, a dorsal hollow nerve cord, and gill slits at some time during its lifetime; a sponge does not.* LEP

GLENCOE TECHNOLOGY

 CD-ROM

Life Science CD-ROM
Have students perform the interactive exploration for Chapter 15 to reinforce important chapter concepts and thinking processes.

Use **Concept Mapping,** p. 35, as you teach this lesson.

Use **Science and Society Integration,** p. 19, as you teach this lesson.

Origins of Fish

Most scientists agree that fishes evolved from small, soft-bodied, filter-feeding organisms similar to present-day lancelets. The earliest fossils of fishes are those of jawless fish that lived 450 million years ago. Fossils of these early fish are usually found where ancient streams emptied into the sea. This makes it difficult to tell whether these fish ancestors evolved in fresh water or in salt water.

By 390 million years ago, a new group of fish, the placoderms, appeared. Placoderms flourished for 60 million years, then vanished from the fossil record. Early sharks evolved around the same time. They have continued to evolve into the shark species alive today. The first modern bony fishes appeared about 190 million years ago. Today, there are over 400 families of bony fish, the most successful class of fishes.

Jawless Fish

The lamprey in **Figure 15-3** belongs to the class of fish called Agnatha. This class includes the lampreys and hagfish. *Agnatha* is a Greek word that means "jawless." These jawless fish have round mouths and long, tubelike bodies covered with slimy skin with no scales. Fish in this class have very flexible bodies made of cartilage. **Cartilage** is a tough, flexible tissue that is not as hard as bone. Feel your ears and the tip of your nose. These are also made of cartilage.

Figure 15-3

There are three classes of fishes: Agnatha, the jawless fish; Chondrichthyes, the cartilaginous fish; and Osteichthyes, the bony fish.

A Lampreys are jawless fish that feed on the blood and body fluids of other fish.

B Skates are cartilaginous fish that feed on fish, mollusks, and crustaceans. The "wings" of skates and rays are modified pectoral fins.

C Rainbow trout are typical of the bony fishes. They have paired fins and a tail.

402 Chapter 15 Fish, Amphibians, and Reptiles

Hagfish live only in salt water, but many lampreys live in fresh water. Lampreys use their round mouths to attach to other fish by suction. They cut into the fish with toothlike structures and feed on its blood and body fluids.

Cartilaginous Fish

Sharks, skates, and rays are members of the class Chondrichthyes (kahn DRIHK thee eez). The word *chondros* is Greek for "cartilage" and *ichthyes* is Greek for "fish." Fish that belong to this class have skeletons made of cartilage like jawless fish. Unlike jawless fish, however, chondrichthyes have movable jaws and scales. The scales resemble vertebrate teeth and cause their skin to feel like fine sandpaper.

Like all animals, sharks and their relatives eat other organisms for food. Sharks are very efficient at finding and killing their food.

Bony Fish

Bony fish make up the class Osteichthyes (ahs tee IHK thee eez). The word *osteon* is Greek for "bone." Fish that belong to this class have skeletons made of bone. About 95 percent of all species of fish belong to this class.

The body structure of a typical bony fish, a tuna, is shown in **Figure 15-4.** The gills are covered and protected by a bony flap called the gill cover. Gill covers aid the fish in breathing. They open and close, moving water taken in through the mouth over the gills.

Most species of bony fish have separate sexes. To reproduce, the females release large numbers of eggs into the water. Males then swim over the eggs and release sperm. This behavior is called spawning. The joining of egg and sperm cells outside the female body is called external fertilization.

Figure 15-4

Although there are many different kinds of bony fish, they all share basic structures. *Why is the top fin called a dorsal fin?*

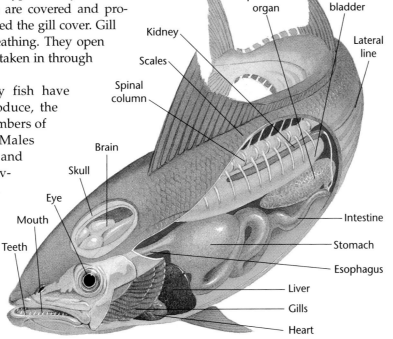

Reproductive organ
Swim bladder
Kidney
Lateral line
Scales
Spinal column
Brain
Skull
Eye
Mouth
Teeth
Intestine
Stomach
Esophagus
Liver
Gills
Heart

15-1 Fish **403**

INTEGRATION
Physics

(1) Submarines pump water into and out of special chambers to regulate the vertical force that causes the submarine to sink or rise. (2) Scuba divers use devices called buoyancy compensators to adjust to changes in water pressure.

Use **Science Integration Transparency 15** as you teach this lesson.

MiniLAB

Purpose
Kinesthetic Students will use a model to determine what happens when a fish fills or empties its swim bladder. [L1] [LEP] [COOP LEARN]

Materials
balloons, water, bowls

Teaching Strategies
The bowls need to be large enough to hold the balloons.

Analysis
1. The balloon filled with air floats. The balloon that has both air and water remains suspended in the water.
2. a swim bladder
3. A fish with a swim bladder filled with air would be near the surface; one with water and air, beneath the surface.

Assessment

Process Have students view goldfish in an aquarium and describe how they rise and fall in the water. Use the Performance Task Assessment List for Making Observations and Inferences in **PASC**, p. 17. [P]

Activity Worksheets, pp. 5, 92

INTEGRATION
Physics

MiniLAB

How does a fish adjust to different depths?
Find out what happens when a fish fills or empties its swim bladder.

Procedure
1. Fill a balloon with air.
2. Place it in a bowl of water.
3. Fill another balloon partially with water, then blow air into it until it is the same size as the balloon filled only with air.
4. Place the second balloon in the bowl of water.

Analysis
1. Compare the balloon filled with air to the balloon with air and water. Where is each balloon in the water?
2. What structure do these balloons model?
3. Compare where in the water (on the surface, below the surface) two fish would be if they had swim bladders similar to the two balloons.

Buoyancy

An important adaptation in most bony fish is the swim bladder, an air sac that helps control the buoyancy of the fish. Buoyancy is the ability of a liquid or a gas to exert an upward force on an object in it. Buoyancy was discovered by the Greek scientist Archimedes in 21 B.C. According to Archimedes' principle, the buoyant force on an object in a liquid is equal to the weight of the liquid displaced by the object. If the density of the object is greater than that of the liquid, the object will sink. If the density of the object is equal to the density of the liquid, the object will neither sink nor float. If the density of the object is less than the density of the liquid, the object will float.

The swim bladder in a fish allows it to adjust its density so it can rise or sink. Transfer of gases, mostly oxygen, between the swim bladder and the blood varies the inflation of the swim bladder and adjusts the density of the fish. As the swim bladder fills with gases, the fish becomes more buoyant and rises in the water. As the bladder deflates, the fish becomes less buoyant and sinks. Glands regulate the gas content in the swim bladder, enabling the fish to remain at a specific depth in the water with little effort. Deep-sea fishes often have oil in their swim bladders rather than gases. Some bottom-dwelling fish have no swim bladders.

Kinds of Bony Fish

Scientists who specialize in studying fish have organized bony fish into three groups: lobe-finned fish, lungfish, and ray-finned fish. Most fish are ray-finned fish. Examples of each group are shown in **Figure 15-5.**

Lobe-finned fish have fins that are lobelike and fleshy. It was thought that these organisms had been extinct for more than 70 million years. But in 1938, one individual of this group was caught in a net by some South African fishermen. Several living lobe-finned fish have been studied since. Lobe-finned fish are important because scientists think fish like these were the ancestors of the first land vertebrates.

Lungfish have both gills and lungs for breathing. This adaptation allows many to live in shallow waters that dry up in summer. When the water evaporates, lungfish burrow into the mud and cover themselves with mucus until water returns. Lungfish have been found along the coast of South America, Australia, and Africa.

Ray-finned fish have fins made of long, thin bones covered with skin. Yellow perch, tuna, salmon, swordfish, and eels are all ray-finned fish.

Figure 15-5

Bony fishes vary in appearance and behavior.

B The six species of lungfishes all grow to about 1.25 m. Lungfishes have long fins that are used to touch and sense their surroundings.

A The coelacanth is the only living representative of the lobe-finned fishes. Since its discovery in 1938, more than 60 of these fish have been caught near the Comoro Islands off the coast of Africa at depths of 200 to 300 m.

C This parrotfish, like other ray-finned fish, has fins that are supported by flexible, thin bones. Parrotfish have teeth that are fused into a parrotlike beak.

Section Wrap-up

Review

1. What are three characteristics of chordates?

2. Name the three classes of fish and give an example of each.

3. Explain the differences between ectotherms and endotherms.

4. **Think Critically:** Female fish lay thousands of eggs. Why aren't lakes and oceans overcrowded with fish?

Skill Builder
Sequencing
Suppose a fish is lying on the bottom of a lake. Sequence the events that must take place for the fish to rise to the surface of the lake. If you need help, refer to Sequencing in the **Skill Handbook.**

Using Computers

Database Use your computer to make a database of the characteristics of the three classes of fish. Use the database to identify fish.

15-1 Fish **405**

PREPARATION

Purpose
LS **Logical-Mathematical** Students will design and carry out an experiment to measure the effect of water temperature on fish. **L1** **COOP LEARN**

Time
one class period

Process Skills
observing, using numbers, interpreting data, communicating, comparing and contrasting, forming a hypothesis, recognizing cause and effect, designing an experiment, separating and controlling variables

Time
one class period

Materials
- Pet shops stock feeder fish that are inexpensive. If guppies are used, do not lower the temperature below 18°C.
- The starting temperature of the water is the temperature of the water in the aquarium or beaker.
- Students may notice that the number of times a fish's mouth opens is directly related to the openings of the gill coverings.

Safety Precautions
Students should handle fish with extreme care.

Possible Hypotheses
Most student hypotheses will reflect that the breathing rate of fish will increase as the temperature increases.

📁 **Activity Worksheets,** pp. 5, 88-89

Design Your Own Experiment
Water Temperature and the Respiration Rate of Fish

Last summer, it was very hot with few storms. One day after many sunny, windless days, you noticed that a lot of dead fish were floating on the surface of your neighbor's pond. What may have caused these fish to die?

PREPARATION

Problem
How does water temperature affect the respiration rate of fish?

Form a Hypothesis
Fish obtain oxygen from the water. State a hypothesis about how water temperature affects the respiration rate of fish.

Objectives
- Design and carry out an experiment to measure the effect of water temperature on the rate of respiration of fish.
- Observe the breathing rate of fish.

Possible Materials
- goldfish
- aquarium water
- 1 small fishnet
- 600-mL beakers
- container of ice water
- stirring rod
- thermometer
- aquarium

Safety Precautions

Protect your clothing. Use the fishnet to transfer fish into beakers.

PLAN THE EXPERIMENT

1. As a group, agree upon and write out the hypothesis statement. You might form a hypothesis that relates the amount of oxygen dissolved in water to water temperature, and how this would affect fish.
2. As a group, list the steps that you need to take to test your hypothesis. Be specific, and describe exactly what you will do at each step. List your materials.
3. How will you measure the breathing rate of fish? Will you count the number of times the fish opens its mouth or its gill covering?
4. Explain how you will change the water temperature in the beakers. Fish respond better to a gradual change in temperature than an abrupt change.
5. Record your data in a data table in your Science Journal.

Check the Plan
1. Read over your entire experiment to make sure that all steps are in logical order.
2. Identify any constants, variables, and controls of the experiment.
3. What data will your group collect? How many times will you run your experiment?
4. How will you measure the response of fish to changes in water temperature?
5. How will you relate the fish's behavior to the amount of oxygen in the water?
6. *Make sure your teacher approves your plan before you proceed.*

DO THE EXPERIMENT

1. Carry out the experiment as planned.
2. While the experiment is going on, write down any observations that you make and complete the data table in your Science Journal.

Analyze and Apply
1. **Compare** your results with those of other groups.
2. What were you measuring when you counted mouth or gill cover openings?
3. **Describe** how a decrease in water temperature affects respiration rate and behavior of the fish.
4. **Explain** how your results could be used to determine the kind of environment in which a fish can live.

Go Further

Fish may live in water that is totally covered by ice. How is this possible? What would happen to a fish if the water were to become very warm?

15-1 Fish   **407**

PLAN THE EXPERIMENT

Possible Procedures
One possible procedure is to place a fish in a beaker of aquarium water, lower the temperature with ice cubes, and count the gill cover openings as the temperature is lowered.

Teaching Strategies
Tying to Previous Knowledge Most students will have observed movement of fish in extremely cold weather and are aware that as the temperature decreases, movement decreases.

DO THE EXPERIMENT

Expected Outcome
Most students will observe a decrease in breathing rate with a decrease in temperature.

Analyze and Apply
1. Student results should be consistent with those of other groups.
2. The rate of breathing was measured by the number of mouth or gill cover openings.
3. As water temperature decreases, breathing rate decreases, and movements become slower.
4. Answers will vary but should include that fish need an environment in which water is not too warm nor too cold.

Go Further
Cold water contains much dissolved oxygen. Warm water has less oxygen, and the fish must breathe faster, thus using energy faster than it can take in oxygen. The fish could die.

Assessment

Performance Have students predict what will happen if they place a small drop of slightly diluted food coloring in front of the mouth of a goldfish or guppy. Use the Performance Task Assessment List for Formulating a Hypothesis in **PASC**, p. 21. **P**

407

Background

- Mudskippers are found in the tropics of the Indo-Pacific region, where they inhabit mangrove swamps and mudflats. As Ms. Aguilar stated, mudskippers spend considerable time out of the water. They carry around a limited supply of water trapped beneath the opercula covering their gill chambers, and use their stout fins to move about in search of food. Mudskippers are aggressive predators that prey on spiders, insects, isopods, crabs and other small crustaceans, and worms and mollusks that they actively dig out of the mud. When incoming tides flood the mudflats where mudskippers forage, they often climb up the spindly roots of mangrove trees.

- The region known as the Grand Banks off the coast of Newfoundland was once considered to be the world's richest cod fishing grounds. It's been said that 30 years ago, cod were so plentiful there that they could be caught simply by dipping a basket into the ocean. Today, the cod have all but disappeared due to overfishing and poor management of this rich natural resource. Dwindling fish populations are a growing concern worldwide.

Teaching Strategies

⬚ **Kinesthetic** Students can research the various species of fish that live in the lakes and rivers of their state. Then have them create life-sized models of each species and design a bulletin board that incorporates their models and graphically illustrates where each species is found, its particular habitat requirements, and other interesting data. ⬚

408

People and Science

NANCY AGUILAR, *Ichthyologist*

On the Job

Q Ms. Aguilar, tell us about your research on fish.

A I'm studying mudskippers, which inhabit mangrove swamps in the tropics. Mudskippers may spend up to 70 percent of their time *out of the water* foraging for food. Unlike other fish, they can tolerate being exposed to air for long periods of time. In my research, I'm trying to discover how mudskippers can survive so well out of the water.

Q What kind of experiments are you doing?

A When you take an "ordinary" fish out of water, its heart rate falls dramatically. But a mudskipper's heart rate stays pretty much the same when it's exposed to the air. The question is, why? What might that tell us about a mudskipper's metabolism and physiology? I've been studying mudskippers by putting them in small environmental chambers and implanting tiny wires just beneath their skin—a procedure that doesn't harm the fish, by the way. The wires are hooked up to a monitor that measures heart rate. I can then change the conditions inside the chamber—for example,

Personal Insights

increase the amount of oxygen or decrease the amount of water—and see how the fish respond.

Q Do you have any suggestions for young people interested in science as a career?

A In elementary and middle school, I didn't think classes like English or social studies were as important as science. But I've since discovered that that's not the case. Good writing skills, for example, are very important to me now in writing grant proposals in which I ask for research support. Subjects like English and math are important for anyone going into science.

Career Connection

Contact the nearest regional office of your state's Game, Fish and Parks Department, or a state-operated fish hatchery, and ask for information on the education and training required for the following careers that involve working with fish. Report on your findings to your classmates.

- **Hatchery Manager**
- **Stream Biologist**
- **Fisheries Manager**

Career Connection

Career Path There are many careers that involve working with and/or studying fish, in both marine and freshwater systems.

- High School: science and math
- College: biology; fishery science
- Post College: a master's degree or PhD is usually required for people interested in doing research on fish or aquatic systems, for fisheries managers, and for biologists who oversee the production of fish in hatcheries.

For More Information

Students may be interested in reading the article "The Cod That Disappeared," *New Scientist*, 16 September, 1995. State Game, Fish, and Parks Departments will often provide information, maps, and booklets about fish and other wildlife found in a particular region.

Amphibians 15•2

What is an amphibian?

The word *amphibian* comes from the Greek word *amphibios*, which means "double life." They are well named, for **amphibians** are ectothermic vertebrates that spend part of their lives in water and part on land. The frog, toad, and salamander shown in **Figure 15-6** are amphibians. What adaptations do these animals have that allow them to live both on land and in water? Amphibians have moist skin that is smooth, thin, and without scales. Oxygen and carbon dioxide are exchanged through the skin and the lining of the mouth. Amphibians also have small, simple, saclike lungs in the chest cavity to use for breathing.

Amphibian Adaptations

Because amphibians are ectotherms, their body temperatures change when the temperature of their surroundings changes. During the cold winter months, they become inactive. They bury themselves in mud or leaves until the temperature warms up. In the winter, the period of inactivity and lower metabolic needs is called **hibernation.** Amphibians that live in hot, dry environments have other adaptations. They become inactive and hide in the ground when temperatures become extreme. This kind of inactivity during the hot, dry summer months is called **estivation.**

A strong skeleton made of bone is another adaptation amphibians have for life on land. Organisms living in water are supported by the water. However, on land a stronger structure is needed to support the body.

Because their eggs lack shells and must be kept moist, most species of amphibians return to water to lay their eggs. The eggs hatch into larvae that live in water until they become adults and move to land.

Science Words

amphibian
hibernation
estivation

Objectives

* Describe the adaptations that amphibians have for living in water and on land.
* Identify the three kinds of amphibians and describe the characteristics of each.
* Describe frog metamorphosis.

Figure 15-6

Salamanders, toads, and frogs are types of amphibians. *Why must amphibians keep their skin moist?*

Marbled salamander

Blue poison-arrow frog

Fire-bellied toad

409

Program Resources

Reproducible Masters
Study Guide, p. 59 [L1]
Reinforcement, p. 59 [L1]
Enrichment, p. 59 [L3]
Activity Worksheets, pp. 5, 90-91, 93 [L1]
Concept Mapping, p. 35
Lab Manual 39

Transparencies
Section Focus Transparency 55 [L1]
Teaching Transparency 30

Section 15•2

Prepare

Section Background

* Biologists divide amphibians into three orders: Urodela, the salamanders; Anura, the frogs and toads; and Apoda, the wormlike caecilians. There are only about 250 species of salamanders in the world. Most are found in North America.
* Hell benders and mud puppies are two of the largest salamanders found in the United States. They reach lengths of 43 cm. Mud puppies have red external gills. Hell benders have skin that looks wrinkled and too large for their bodies.
* Caecilians are tropical amphibians that are not well known. The legless, wormlike creatures look, and to some extent, act like earthworms. They average 30 cm in length, have small eyes, and are often blind.

Preplanning

* Obtain live frogs for the MiniLAB.
* Obtain frog egg masses for Activity 15-2.

1 Motivate

Bellringer

Before presenting the lesson, display **Section Focus Transparency 55** on the overhead projector. Assign the accompanying **Focus Activity** worksheet. [L1] [LEP]

Tying to Previous Knowledge

Review the information in the previous section about lobe-finned fish. Ask students what adaptations would be necessary for a fish to survive on land.

2 Teach

410

MiniLAB

What are the adaptations of a frog?

Procedure

1. Obtain a live frog in a large jar.
2. Carefully observe the frog. Notice the position of its legs as it sits.
3. Observe the position of its mouth, eyes, nostrils, and ears.
4. Observe the color of its dorsal and ventral sides.
5. Take the frog out of the jar and put it in a box on the floor. Gently touch the back of the frog and observe how it jumps.
6. Return the frog to your teacher to place in an aquarium. Observe how the frog swims.

Analysis

1. Describe the adaptations the frog has for living in water.
2. What adaptations does it have for living on land?

Origins of Amphibians

The fossil record shows that fish were the first vertebrates on Earth. For about 150 million years, they were the only vertebrates. Then, as competition for food and space increased, some lobe-finned fish may have easily traveled across land searching for deeper water. The lobe-finned fish had lungs and bony fins that could have supported their weight on land. Amphibians are thought to have evolved from the lobe-finned fish about 350 million years ago.

Evolution favored the development of amphibians because there wasn't much competition on land. Almost free of predators, land provided a great supply of food in the form of insects, spiders, and other invertebrates. Amphibians were able to reproduce in large numbers, and many new species evolved. For 100 million years or more, amphibians were the dominant land animals.

Frogs, Toads, and Salamanders

Frogs and toads are amphibians that as adults have short, broad bodies with no neck, no tail, and four legs. The strong hind legs are longer than the front legs and are used for swimming and jumping. Bulging eyes and nostrils on top of the head let frogs and toads see and breathe while they are almost totally submerged in water. On spring nights, they make their presence known with loud, distinctive sounds.

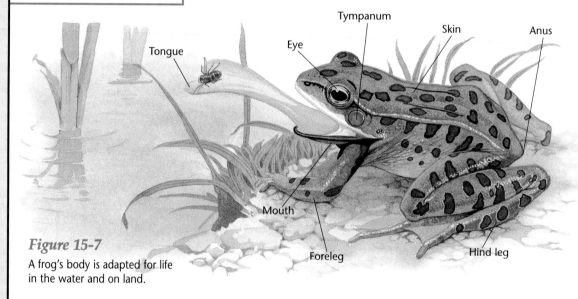

Figure 15-7
A frog's body is adapted for life in the water and on land.

410 Chapter 15 Fish, Amphibians, and Reptiles

You can see the external body structures of a frog in **Figure 15-7**. A frog's tongue is attached to the front of its mouth. When a frog sees an insect, it flips the loose end of its tongue out and the insect gets stuck in the sticky saliva on the tongue. Then the frog flips its catch back into its mouth. On each side of the head, just behind the eyes, frogs have round tympanic membranes. These membranes vibrate in response to sounds and are used by the frog to hear.

Toads, like frogs, breed in water, but they spend the rest of their time in the woods, fields, or in people's gardens. Toads have thick, warty skin that prevents them from drying out. Toads eat invertebrates such as spiders, earthworms, and caterpillars. Most toads have glands on their backs that secrete poison.

Salamanders and newts are amphibians often mistaken for lizards because of their long, slender bodies. The short legs of salamanders and newts appear to stick straight out from the sides of their bodies. Salamanders and newts can be found under dead leaves and rocks during the day, where they hide to avoid the drying heat of the sun. At night, they use their well-developed senses of smell and vision to feed on worms, crustaceans, and insects. Salamanders usually breed on land.

Problem Solving

Red-Spotted Newts and Parasites

Red-spotted newts, like other amphibians, lay their eggs in water. After completing metamorphosis, the young newts leave the pond and spend about six years on land, growing and maturing. Then they return to the pond to reproduce.

These newts often are infected with a parasite that is transmitted through the bite of a leech. Leeches that carry the parasites are found in ponds. When newts are analyzed for the presence of these parasites, scientists have found their blood may contain a million parasites per milliliter of blood—about the same as the number of red blood cells in the newt's blood! Yet the newts don't seem to suffer any illness as a result of this infection.

Solve the Problem:

1. **During what part of the red-spotted newt's life cycle do they become infected with the parasite?**
2. **What is the relationship between the parasites and leeches?**

Think Critically:

Newts spend about six years away from water. Why aren't newts killed by the parasites during this time?

411

Theme Connection

Stability and Change Amphibians have life cycles that include distinct changes in form. This pattern of change is called metamorphosis.

411

4 Close

Use the Mini Quiz to check students' recall of chapter content.

1. **How is hibernation different from estivation?** *Hibernation is inactivity in the winter. Estivation is inactivity during hot, dry, summer months.*

2. **What is a frog adaptation for catching insects?** *The tongue is attached to the front of the mouth.*

3. **Why are salamanders often mistaken for lizards?** *because of their long, slender bodies*

Section Wrap-up

Review

1. water—moist, smooth skin; bulging eyes and nostrils on top of the head; strong hind legs for swimming; land—lungs; strong skeleton; strong hind legs for jumping; moist, smooth skin

2. Frogs have short, broad bodies with no neck, no tail, and four legs; toads have thick, rough skin that keeps them from drying out; salamanders have smooth, moist skin and legs.

3. **Think Critically** They come out to hunt food when the air is moist.

Science Journal

 Linguistic Amphibians breathe through their skin, so they must keep their skin wet to survive.

Amphibian Metamorphosis

Recall that metamorphosis is a series of changes that a larva goes through to become an adult. Most frogs, toads, and salamanders have a two-stage life cycle: a larval stage that lives in water and an adult stage that lives on land. **Figure 15-8** shows the metamorphosis of a leopard frog.

The rate at which amphibians grow and the length of time they spend as larvae depend on the species, the water temperature, and available food. The less food and the cooler the temperature, the longer it takes for metamorphosis. Bullfrogs may not go through metamorphosis for a year or more after hatching, whereas spadefoot toads change into adults in about two weeks. Some amphibians, such as the mud puppy, never grow out of their larval stage. Mud puppies retain external gills even as adults.

Amphibians are unique animals in many ways. Their adaptations allow them to live both on land and in water.

Figure 15-8

Frogs undergo a series of changes from egg to adult.

Young legless tadpoles live off yolk stored in their bodies.

Fertilized eggs

Tadpoles with legs feed on plants in the water.

Young frogs with structures needed for life on land

Adult frog

Section Wrap-up

Review

1. List the adaptations of amphibians for living in water and their adaptations for living on land.

2. Describe the three kinds of amphibians.

3. **Think Critically:** Why do frogs and toads seem to suddenly appear after a rain?

Skill Builder
Sequencing
Sequence and describe each stage of frog metamorphosis. If you need help, refer to Sequencing in the **Skill Handbook.**

Science Journal

In your Science Journal, explain why it is important for amphibians to live in moist or wet environments.

Skill Builder
 Students should sequence the steps of frog development in this order: egg—legless tadpole—tadpole with legs—young frog with developing legs—adult frog on land. Descriptions should be similar to those in **Figure 15-8.** P

Assessment

Performance Assess students' understanding of metamorphosis by having them prepare a poster to compare frog metamorphosis with insect metamorphosis. Use the Performance Task Assessment List for Poster in **PASC,** p. 73. P

Metamorphosis in Frogs

When frogs reproduce, the female lays hundreds of jellylike eggs, and the male fertilizes them. Within a few weeks, these eggs hatch into tiny tadpoles that develop into frogs.

Problem
How long does it take for a tadpole to develop into a frog?

Materials
- 4-L aquarium or jar
- frog egg masses
- lake or pond water
- stereoscopic microscope
- watch glass
- small fishnet
- aquatic plants
- washed gravel
- boiled lettuce
- large rock

Procedure
1. Copy the data table in your Science Journal.
2. Obtain a mass of frog eggs and 4 L of lake or pond water from your teacher.
3. Prepare an aquarium with gravel, the rock, the pond water, and a few aquatic plants.
4. Place the egg mass in the aquarium. Use the fishnet to obtain a few eggs and place the eggs in a watch glass. The eggs should have the dark side up and the white yolk down. **CAUTION:** *Handle the eggs with care.*
5. Observe the eggs and record your observations in the data table.
6. Observe the eggs twice each week and record any changes that you observe.
7. When the tadpoles hatch, observe the stages of development twice a week. Identify the fin on the back, mouth, eyes, gill cover, gills, nostrils, and hind and front legs. Feed the tadpoles boiled lettuce. Observe how they feed.

Analyze
1. How long does it take for the eggs to hatch?
2. How long does it take for a tadpole to develop legs?
3. What organs do tadpoles have for breathing?
4. Which pair of legs appears first?

Conclude and Apply
5. What is the advantage of the jellylike coating around the eggs?
6. **Compare** the eyes of the young tadpoles with the eyes of the older ones.
7. **Explain** why plants are needed in the aquarium.
8. How long does it take for a tadpole to develop into a frog?

Data and Observations   Sample Data

Date	Observations
	Observations will vary according to the kind of frog eggs used. Students should note fishlike appearance of tadpoles, disappearance of the tail, and development of legs.

15-2 Amphibians   **413**

7. to add oxygen to the water and remove carbon dioxide
8. approximately two to four months, depending on the species

Assessment

Oral Have students describe the sequence of events in the development of frog eggs to adult frogs. Use the Performance Task Assessment List for Events Chain in **PASC**, p. 91. **P**

Purpose
IS **Visual-Spatial** Students will observe tadpole metamorphosis and use observations to study the life cycle of a frog. **L1**

Process Skills
observing and inferring, classifying, sequencing, forming operational definitions

Time
30 minutes on the first day; ten minutes per day of observation thereafter

Safety Precautions
Caution students to use extreme care when working with live animals.

Activity Worksheets, pp. 5, 90-91

Teaching Strategies
Alternate Materials Aged tap water can be used instead of pond water.

Troubleshooting The eggs must be handled carefully if they are to hatch. Make sure the temperature of the water remains about 20°C.

- Field-collected eggs can be shipped in January, February, and March. Peeper or toad eggs develop rapidly.
- Boil the lettuce leaves for ten minutes and then cool them.

Answers to Questions
1. answers will vary; 8 to 20 days
2. approximately four weeks, depending on the species
3. gills
4. the hind legs
5. It protects the eggs and keeps them from drying out.
6. The eyes of young are on each side of the head, as in fish. In older ones, they are nearer the top of the head.

Prepare

Section Background

Point out to students that clear cutting, acid rain, pesticides, introduction of amphibian-eating game fish to new locations, and ozone depletion are possible causes of the decline of amphibians. Because frogs, toads, and salamanders are biological indicators, it is important to know what is causing the decline.

1 Motivate

Bellringer

Before presenting the lesson, display **Section Focus Transparency 56** on the overhead projector. Assign the accompanying **Focus Activity** worksheet. L1 LEP

2 Teach

Visual Learning

Figure 15-9 Why are amphibians like these endangered? *due to habitat destruction, ozone depletion, acid rain, clear cutting, etc.* LEP IS

Brainstorming

Before beginning this lesson, have students brainstorm reasons for a worldwide decline in amphibian populations.

Concept Development

Students have already learned about the importance of species diversity in an ecosystem. Help them relate that concept to the use of key species as biological indicators in ecosystems.

414

Science Words

biological indicator

Objectives

• Identify the possible causes of the recent decline in amphibian populations.
• Explain why amphibians are biological indicators.

Figure 15-9

Three amphibians that are in danger of extinction are the Harlequin frog, the common toad, and the tiger salamander. *Why are amphibians like these endangered?*

A Harlequin frog

B Common toad

C Tiger salamander

414

TECHNOLOGY:
15•3 Amphibians and Ultraviolet Light

Where have all the frogs gone?

Where have all the amphibians gone? That's what scientists all over the world are wondering. There has been a decline in the populations of toads, frogs, newts, and salamanders in places as diverse as Costa Rica, Japan, and Australia. **Figure 15-9** illustrates the Harlequin frog, common toad, and tiger salamander, amphibians that are in danger of disappearing from Earth completely. Since 1990, scientists all over the world have been trying to find out why the populations of amphibians are declining so fast.

Amphibians as Biological Indicators

Amphibians are very sensitive to changes in their environment. Because they live on land and reproduce in water, they are directly affected by chemical changes in their environment, including soil erosion from land left bare after logging or water pollution from factories or power plants. Amphibians also absorb gases through their skin, making them susceptible to air pollutants as well. Because of their unique way of life, amphibians are considered to be **biological indicators**—species whose overall health reflects the health of their particular ecosystem as well. But air, soil, and water pollution aren't the only environmental problems affecting amphibians. Even amphibians living in remote, pollution-free environments are in decline. Some scientists now believe ozone depletion is affecting survival of amphibians worldwide.

Program Resources

Reproducible Masters
Study Guide, p. 60 L1
Reinforcement, p. 60 L1
Enrichment, p. 60 L3

Transparencies
Section Focus Transparency 56 L1

Cascade Frogs and Ultraviolet Light

The Cascade frog lives at high elevations (1220 m and above) in remote, undisturbed parts of the Cascade Mountains of California, Oregon, and Washington. There is no evidence of pollution in the area, yet eggs of this frog are dying. Researchers recently discovered why: eggs in the shallow, clear water of these mountains are subject to prolonged exposure to sunlight. As the ozone layer that shields Earth from the harmful rays of the sun becomes thinner, more ultraviolet rays are getting through. Apparently, the ultraviolet light damages the DNA of the eggs of the Cascade frogs. This damage is so severe that the frog eggs die. However, other amphibians exposed to this ultraviolet light are not affected. The Pacific tree frog, for example, is not suffering from a decline in population. Researchers have now discovered why.

When DNA is damaged by ultraviolet light, some organisms can repair the damage by producing an enzyme called photolyase. The eggs of the Pacific tree frog produce three times as much photolyase as the eggs of the Cascade frog, making them less likely to die as a result of excessive ultraviolet light. Both frogs are shown in **Figure 15-10.**

Figure 15-10

The Cascade frog (top) is declining in population whereas the Pacific tree frog (bottom) is not.

Section Wrap-up

Review

1. What are some of the proposed reasons for the decline of amphibians worldwide?

2. Why are amphibians considered to be biological indicators?

interNET **CONNECTION** Visit the Chapter 15 Internet Connection at Glencoe Online Science, **www. glencoe.com/sec/science/life,** for a link to more information about changes to amphibian populations.

SCIENCE & SOCIETY

---

3 Assess

Check for Understanding

Have students answer questions 1 and 2 in the Section Wrap-up.

Reteach

Have students recall from the previous section the characteristics of amphibians. Ask which of these can be affected by the problems discussed in this section. *All are affected.*

Extension

For students who have mastered this section, use the **Reinforcement** and **Enrichment** masters.

4 Close

Have students discuss why amphibians are more likely to be affected by acid rain and ozone depletion than reptiles, which also live in water. *In amphibians, respiration occurs through the skin, so they absorb poisonous gases and chemicals along with oxygen. They also lay their unshelled eggs in water.*

Section Wrap-up

Review

1. destruction of native habitats, pollution, acid rain, pesticides, ozone depletion
2. Their overall health reflects the health of their particular ecosystem.

Explore the Technology

Student answers will reflect that if ultraviolet light damages the DNA in frog eggs, it could damage the DNA in other organisms. Student answers concerning taking synthetic photolyase will vary. **P**

Prepare

Section Background

- Most lizards have four legs with five clawed toes on each foot, long tails, movable eyelids, and external ear openings. A few species, such as glass snakes and slow worms, have no legs. Lizards feed on anything they can catch and swallow.

- Snakes don't have legs, eyelids, outside ear openings, or bladders. They have a very flexible spine made up of 100 to 400 vertebrae, each attached to a pair of thin ribs. All snakes are meat-eaters.

- Turtles are the only reptiles with shells: the carapace, which covers the back; the plastron, which covers all or part of the underneath side; and the bridge, which connects the two. Turtles feed on insects, worms, fish, fruit, and other plant material.

1 Motivate

Bellringer

Before presenting the lesson, display **Section Focus Transparency 57** on the overhead projector. Assign the accompanying **Focus Activity** worksheet. **L1** **LEP**

Tying to Previous Knowledge

Most students are familiar with dinosaurs. Explain that dinosaurs were reptiles.

15•4 Reptiles

Science Words

reptile
amniotic egg

Objectives

- Identify the adaptations that enable reptiles to live on land.
- Infer why the early reptiles were so successful.
- Describe the characteristics of the modern reptiles.

Figure 15-11

Reptiles include lizards, snakes, turtles, crocodiles, and alligators.

A The common chameleon is a lizard that changes color quickly to match its surroundings.

B The coral snake is a poisonous snake found in open pine woods and near lakes.

C The painted turtle lives in slow-moving or still water.

What is a reptile?

How is a stegosaurus like a green iguana? Do you know that both of these animals are reptiles? A **reptile** is an ectothermic vertebrate with dry, scaly skin. The class Reptilia includes lizards, snakes, turtles, crocodiles, and alligators as well as extinct dinosaurs. Reptiles have many characteristics that are adaptations for life on land. A thick, dry, waterproof skin covered with scales prevents drying out and injury. With the exception of snakes, reptiles have four legs with claws that hold the body off the ground for moving quickly. The claws are used to dig, climb, and run.

Remember that even though amphibians live on land, they still need water for reproduction. Scientists hypothesize that, over time, some amphibians became less dependent on water and became the ancestors of reptiles.

Variety in Reptiles

Reptiles vary greatly in size, shape, and color. Giant pythons, 10 meters long, can swallow crocodiles whole. Some sea turtles have a mass of almost one metric ton and can swim faster than you can run. Three-horned lizards have movable eye sockets and tongues as long as their bodies. Reptiles live on every continent except Antarctica and in all the oceans except those in the polar regions. Some reptiles are shown in **Figure 15-11.**

416 Chapter 15 Fish, Amphibians, and Reptiles

Program Resources

 Reproducible Masters
Study Guide, p. 61 **L1**
Reinforcement, p. 61 **L1**
Enrichment, p. 61 **L3**
Multicultural Connections, pp. 33-34
Concept Mapping, p. 35
Critical Thinking/Problem Solving, p. 19

Transparencies
Section Focus Transparency 57 **L1**

Reptiles breathe with lungs. Even turtles that live in water must come to the surface to breathe air. A reptile's heart has three chambers. The lower chamber is partially divided. Oxygen-rich blood coming from the lungs is kept separated from blood that contains carbon dioxide returning from the body. This type of circulatory system provides a lot of oxygen to all parts of the body. Remember that oxygen is needed to release energy by cell respiration.

Development of the Amniotic Egg

One of the most important adaptations of reptiles for living on land is the way they reproduce. Unlike the eggs of most fish and amphibians, eggs of reptiles are fertilized internally, inside the body of the female. After fertilization, the female secretes a leathery shell around each egg, and then lays the eggs on land.

Figure 15-12 shows the structures in a reptile egg. The egg provides a complete environment for the embryo to develop in. This type of egg, called an **amniotic egg,** contains membranes to protect and cushion the embryo, and to help it get rid of wastes. It also contains a large food supply, the yolk, for the embryo to use as it develops. Small holes in the shell, called pores, allow oxygen and carbon dioxide to be exchanged. By the time it hatches, a young reptile looks like a small adult.

Origins of Reptiles

Reptiles first appeared in the fossil record nearly 280 million years ago. The earliest reptiles were called cotylosaurs. As these early reptiles no longer depended upon water for reproduction, they spread out across dry land. By the end of the Triassic period, reptiles began to dominate the land. Some reptiles even returned to the water to live, although they continued to lay their eggs on land. The Mesozoic era, which lasted from 225 million to 65 million years ago, is considered to be the Age of the Reptiles. Dinosaurs, descended from early reptiles, ruled Earth during this era, then died out.

Figure 15-12

The amniotic egg is one of the adaptations reptiles have for living on land. Young reptiles hatch from their eggs fully developed.

CONNECT TO

EARTH SCIENCE

Dinosaurs, reptiles that ruled Earth for 160 million years, died out by the end of the Cretaceous period. *Describe* what changes in the environment could have caused the extinction of the dinosaurs.

Community Connection

Live Snakes Invite a herpetologist to bring specimens of snakes to show the class. Have students handle snakes and then compare how snakes feel with how frogs and toads feel.

2 Teach

Content Background

The amniotic egg allows reptiles to live on land. It has four specialized membranes: the amnion, yolk sac, allantois, and chorion. It is named for the amnion, which surrounds the fluid in which the embryo floats. The yolk sac encloses the yolk, which is the food supply for the developing embryo. The allantois stores wastes produced by the embryo until the egg hatches. The chorion lines the outer shell enclosing the embryo and all other membranes.

Demonstration

Obtain a small turtle and place it in a terrarium for students to observe. Have the students use references to find out about care and feeding of turtles. *Be sure students wash their hands thoroughly after handling the turtle.*

Enrichment

LS **Interpersonal** Divide the class into cooperative groups and have them research the characteristics, habitats, and behaviors of reptiles that lived long ago, such as *Pteranodon, Elasmosaurus, Archelon, Tyrannosaurus,* and *Deinosuchus.*
COOP LEARN

CONNECT TO

EARTH SCIENCE

Answer Student answers should reflect the fact that dinosaurs were ectotherms.

Use **Multicultural Connections,** pp. 33-34, as you teach this lesson.

Modern Reptiles

Modern reptiles probably descended from cotylosaurs. Today there are three orders of reptiles: turtles; crocodiles and alligators; and lizards and snakes, shown in **Figure 15-13.**

Turtles

Turtles make up a successful order of animals. They can be found on almost every continent and in most of the world's oceans. The body of a turtle is covered by a hard shell on both top and bottom. Most turtles can withdraw into their shells for protection. Turtles have no teeth and use their beaks to feed on insects, worms, fish, and plants.

Crocodiles and Alligators

Crocodiles and alligators are among the world's largest living reptiles. They make up the order Crocodilia. Members of this order can be found in or near water in tropical climates.

Crocodiles have long, slender snouts and are aggressive. They can attack animals as large as cattle. Alligators are less aggressive. They have broad snouts and feed on fish, turtles, and waterbirds.

Figure 15-13

Reptiles include the turtles, crocodilians, and the lizards and snakes. *What traits do these animals have in common?*

A Turtles can be found nearly everywhere on Earth. Leatherback turtles roam the seas, feeding on jellyfish.

B Crocodilians include this gavial from India. Gavials are adapted to eat fish. They have more than 100 teeth in their slender jaws.

C This is an American five-lined skink, a type of lizard.

418

Lizards and Snakes

Lizards and snakes make up the largest group of reptiles. Lizards have movable eyelids, external ears, and legs with clawed toes on each foot. They feed on reptiles, insects, spiders, worms, and mammals. Snakes don't have legs, eyelids, or external ears.

Remember how you could feel the vibrations of the tuning fork in the Explore Activity on page 399? Snakes do not hear sound waves in the air. They "hear" vibrations in the ground. Sound waves in the ground are picked up by the lower jawbone and conducted to the bones of the snake's inner ear. From there, the vibrations are transferred to the snake's brain, where the sounds are interpreted.

Snakes are meat-eaters. Some snakes wrap around and constrict their prey. Others inject their prey with venom. Many snakes feed on rodents and help control their populations.

Australian Tiger Snake

USING TECHNOLOGY

Treating Snakebites

What would you do if you were bitten by an unknown snake? You'd probably want to know if the snake was poisonous or not, and if it was, which type of poisonous snake. Treatment of a snakebite usually involves injecting an antivenin, a chemical that neutralizes the snake's venom, or poison. But antivenin is only useful in treating the venom from a particular species of snake. If no one knows the species of the snake that bit you, your physician won't know which antivenin to use.

A Snake to the Rescue

Some snakes prey on other poisonous snakes, and yet they don't die from ingesting all those different types of venom. Researchers studying the Australian tiger snake have found out why. This snake produces an antivenin that neutralizes not only its own venom, but also the venom of a wide variety of snakes, including cobras, moccasins, rattlesnakes, and vipers. Scientists are now working out the molecular structure of this protein in hopes of learning how to duplicate its broad neutralizing power.

Think Critically:

Why would antivenin similar to that of the Australian tiger snake be extremely helpful in first-aid kits?

15-4 Reptiles   419

USING TECHNOLOGY

For more information on how and why antivenin is used, contact a local health facility.

Think Critically

Snakebite victims may not be able to identify their attacker, making it difficult to determine which antivenin to use. This antivenin would counteract the venom of a variety of snakes.

NATIONAL GEOGRAPHIC SOCIETY

 Videodisc
STV: Biodiversity
Alligator hatching, guarded by mother

46099

3 Assess

Check for Understanding

? FLEX Your Brain

Use the Flex Your Brain activity to have students explore REPTILES.

📁 **Activity Worksheets,** p. 5

Reteach

Prepare a quiz game with photographs or descriptions of lizards, turtles, alligators, crocodiles, snakes, frogs, and salamanders. Let students classify the animals as amphibians or reptiles and name a characteristic of each.

Extension

📁 For students who have mastered this section, use the **Reinforcement** and **Enrichment** masters.

Integrating the Sciences

Chemistry Snake venom contains poisonous proteins and toxins. Hemotoxins and neurotoxins attack red blood cells and the nervous system.

419

4 Close

•MINI•QUIZ•

Use the Mini Quiz to check students' recall of chapter content.

1. **Three characteristics of reptiles are** _____ . *ectothermic vertebrate; dry, scaly skin; lay eggs covered with a leathery shell*

2. **Reptiles breathe with** _____ . *lungs*

3. **Most reptile hearts have** _____ **chambers.** *three*

4. **The Age of Reptiles was the** _____ **era.** *Mesozoic*

Section Wrap-up

Review

1. a thick, dry, waterproof skin covered with scales; four legs with claws to hold the body off the ground (with the exception of snakes); three-chambered heart; lungs

2. Turtles are covered with a hard shell on both top and bottom. Most can withdraw into this shell for protection.

3. Early reptiles were able to live on land away from water; there was no competition from other large organisms; and they had many niches to fill.

4. **Think Critically** Predators think they are poisonous snakes and leave them alone.

Maternal Care in Reptiles

Most fish, amphibians, and reptiles show little or no care for their eggs once they have been laid. Turtles, for example, dig out a nest, deposit their eggs, cover the nest, and then leave. Turtles never see their own hatchlings and provide no care at all. The development of the amniotic egg gave reptiles the ability to emerge from the egg fully formed; thus, no maternal care is usually needed in reptiles. However, recently scientists have observed that crocodiles actually stay near their nests after laying their eggs. In fact, a few crocodile mothers have been photographed opening their nests in response to noises made by hatchlings. Then these crocodile mothers gather their babies up in their huge mouths and carry them to the safety of the river. The mother crocodile continues to keep watch over her hatchlings until they are big enough to protect themselves.

Figure 15-14
Unlike other reptiles, the Nile crocodile provides care for the hatchlings.

Section Wrap-up

Review

1. What adaptations do reptiles have for living on land?

2. How do turtles differ from other groups of reptiles?

3. Why were early reptiles so successful as a group?

4. **Think Critically:** Poisonous coral snakes and some harmless snakes have bright red, yellow, and black colors. How is this an advantage to nonpoisonous snakes?

Skill Builder
Concept Mapping

Make a concept map showing the major characteristics and orders of the reptile class. If you need help, refer to Concept Mapping in the **Skill Handbook.**

USING MATH

Many dinosaurs were very large. One large dinosaur, *Brachiosaurus*, was about 22 meters long and 12 meters high. The largest land mammal now living is the elephant. The average size of an elephant is 3 meters tall and 6 meters long. How does the elephant compare in height to the *Brachiosaurus*?

USING MATH

The elephant is one-fourth the height of *Brachiosaurus*.

Skill Builder

✓ Assessment

Performance Use this concept map to have students compare and contrast crocodiles and alligators. Have students identify these two groups from photographs. Use the Performance Task Assessment List for Concept Map in **PASC**, p. 89. **P**

Chapter 15 Review

Summary

15-1: Fish

1. The phylum Chordata includes lancelets, tunicates, and vertebrates; all have a notochord, dorsal hollow nerve cord, and gill slits.
2. Endothermic animals maintain a constant body temperature. The body temperature of an ectothermic animal changes with its environment.
3. Classes of fish include jawless fish, cartilaginous fish, and bony fish.

15-2: Amphibians

1. Amphibians are ectothermic vertebrates that live both in water and on land. The adaptations of amphibians include moist skin, mucous glands, and lungs.
2. Frogs live in wet areas and have jumping legs and sensory organs. Toads live in drier areas.
3. Frogs go through metamorphosis from egg, to tadpole, to adult. Legs develop, lungs replace gills, and the tail is lost.

15-3: Science and Society: Amphibians and Ultraviolet Light

1. Amphibian populations are declining due to habitat destruction, acid rain, pesticides, predation, and ozone depletion.
2. Amphibians are sensitive to pollution because they have no scales or shells for protection. They also absorb materials through their skin, including pollutants.

15-4: Reptiles

1. Reptiles are true land animals having dry, scaly skin. They lay amniotic eggs with leathery shells.
2. Early reptiles were successful because of their adaptations to living on land.
3. Reptile groups include turtles with tough shells, meat-eating crocodiles and alligators, and snakes and lizards.

Key Science Words

a. amniotic egg
b. amphibian
c. biological indicator
d. cartilage
e. chordate
f. dorsal hollow nerve cord
g. ectotherm
h. endoskeleton
i. endotherm
j. estivation
k. fin
l. fish
m. gill slit
n. hibernation
o. notochord
p. reptile
q. scale

Reviewing Vocabulary

Match each phrase with the correct term from the list of Key Science Words.

1. dorsal structure that becomes the backbone in vertebrates
2. ectothermic land and water animal
3. the internal skeleton
4. tough, flexible tissue
5. every vertebrate is also this
6. inactivity in winter

Chapter 15 Review 421

GLENCOE TECHNOLOGY

 Videodisc
Glencoe Life Science Interactive Videodisc
Side 2, Lesson 5
Structural Adaptations in Animals

11 to 9307

MindJogger Videoquiz
Chapter 15 Have students work in groups as they play the Videoquiz game to review key chapter concepts.

Chapter 15 Review

Checking Concepts

1. d	6. a
2. b	7. d
3. a	8. a
4. a	9. a
5. c	10. d

Understanding Concepts

11. Fish have gills, a streamlined body, and fins.
12. Lampreys and hagfish have no jaws and no scales.
13. Amphibians have moist skin with mucous glands; they respire through their skin and their lungs. Reptiles have dry, scaly skin to prevent water loss, and they lay amniotic eggs.
14. Frogs undergo metamorphosis. The eggs laid in water hatch into tadpoles. The tadpoles feed on plants while developing legs. Eventually metamorphosis is complete, and the adult moves onto land. The cycle then repeats itself.
15. They are adapted to living and feeding on the bottom and do not need to float to different depths.

Thinking Critically

16. Accept all logical answers with appropriate support.
17. The skin of amphibians absorbs oxygen, as well as any poisonous gases or chemicals in the area. Because of this, amphibians can be considered biological indicators—species that reflect the condition of the environment.
18. Tunicates and sponges are both sessile filter feeders.
19. All vertebrates have, at some time in their life, a notochord that becomes the vertebrae in vertebrates, a dorsal hollow nerve cord, and gill slits. They also all have an endoskeleton.

7. inactivity in summer
8. an ectotherm with dry skin
9. fanlike steering structure in fish
10. species that are sensitive to changes in the environment

Checking Concepts

Choose the word or phrase that completes the sentence.

1. Animals that have fins, scales, and gills are _____.
   a. amphibians   c. reptiles
   b. crocodiles   d. fish
2. Jawless fish do not have _____.
   a. cartilage   c. slimy skin
   b. scales   d. long tubelike bodies
3. An example of a cartilaginous fish is a
   _____.
   a. hagfish   c. perch
   b. tuna   d. goldfish
4. Most fish species belong to the _____ group.
   a. Osteichthyes   c. cartilaginous
   b. jawless   d. Chondrichthyes
5. A fish with both gills and lungs is a
   _____.
   a. shark   c. lungfish
   b. ray   d. perch
6. The group of vertebrates with lungs and moist skin is _____.
   a. amphibians   c. reptiles
   b. fish   d. lizards
7. A _____ is *not* a reptile.
   a. lizard   c. snake
   b. turtle   d. salamander
8. The group of reptiles that includes lizards also includes _____.
   a. snakes   c. crocodiles
   b. turtles   d. alligators

20. The amniotic egg provides the developing embryo with protection from drying out and from being eaten. An adequate supply of food is also provided. Reptile eggs can be laid anywhere on land, thus freeing reptiles from a reliance on water for reproduction.

9. Reptiles lay _____ eggs.
   a. amniotic   c. jelly-like
   b. brown   d. hard-shelled
10. At some time in their lives, all _____ have a notochord, dorsal hollow nerve cord, and gill slits.
    a. echinoderms   c. arthropods
    b. mollusks   d. chordates

Understanding Concepts

Answer the following questions in your Science Journal using complete sentences.

11. Describe the structural adaptations that allow fish to live in water.
12. Why are lampreys and hagfish not members of the class Chondrichthyes?
13. Compare the adaptations of amphibians and reptiles to life on land.
14. Describe the life cycle of a frog.
15. Suggest a reason why bottom-dwelling fish often do not have swim bladders.

Thinking Critically

16. Why do you think there are fewer species of amphibians on Earth than any other type of vertebrate?
17. Why are amphibians considered biological indicators?
18. In what ways are tunicates similar to sponges?
19. What features are common to all vertebrates?
20. Explain how the development of the amniotic egg led to the success of early reptiles.

Reptile eggs

Developing Skills

21. **Sequencing** Scales are present in fish, allowing them to move in water; skin with mucous glands is present in amphibians, which began to live on dry land; dry, scaly skin developed in reptiles that lived on land.

Developing Skills

If you need help, refer to the **Skill Handbook.**

21. **Sequencing:** Sequence the order in which these structures appeared in evolutionary history and explain what type of organism had this adaptation and why: skin has mucous glands; skin has scales; dry, scaly skin.

22. **Comparing and Contrasting:** Make a chart comparing the features of fish, amphibians, and reptiles.

23. **Designing an Experiment:** Design an experiment to find out the effect of water temperature on frog egg development.

24. **Making and Using Graphs:** Make a circle graph of the species of fish in the table below. What percent of the graph is accounted for by Osteichthyes?

Fish Species

Classes of Fish	# of Species
Jawless	45 species
Chondrichthyes	275 species
Osteichthyes	25 000 species

Amphibian eggs

25. **Concept Mapping:** Complete the concept map below describing the chordates.

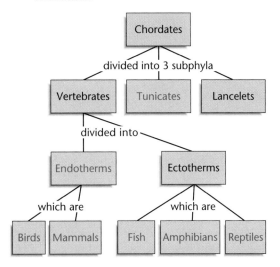

Performance Assessment

1. **Poster:** Obtain the eggs of an amphibian from a pond or a biological supply house. Keep the eggs in water of the correct temperature. Photograph the eggs each day until they hatch, then make a poster sharing the changes in the eggs daily.

2. **Conducting a Survey:** Many people are wary of reptiles. Write up questions about reptiles to find out how people feel about these animals. Give the survey to your classmates, then graph the results and share them with your class.

3. **Display:** Cut out pictures of fish from magazines and mount them on poster board. Then letter the names of each fish on 3" × 5" cards. Have your classmates try to match the names of the fish with their pictures. To make this activity more challenging, use only the scientific names of each fish on your cards.

22. **Comparing and Contrasting** Students' charts should list fish, amphibians, and reptiles down the left side of the chart. Across the top they should have the following categories: habitat, food, body features, and reproduction. Characteristics can be found in the text. Accept all logical charts.

23. **Designing an Experiment** Students should design experiments with a control—a set temperature—and groups with higher or lower temperatures—the variables.

24. **Making and Using Graphs** Students should find that 98.7% of the graph or 355° represents Osteichthyes, about 1.1% or 4° represents Chondrichthyes, and about 0.18% or 0.6° represents jawless fishes.

25. **Concept Mapping** See student page.

Performance Assessment

1. Posters should show egg, tadpole, and adult stages. Use the Performance Task Assessment List for Poster in **PASC,** p. 73. P

2. Student graphs will vary according to their survey audience. Use the Performance Task Assessment List for Conducting a Survey and Graphing the Results in **PASC,** p. 35. P

3. Students should find matching scientific names to fish relatively difficult. Suggest that students use simple names that describe the characteristics of each fish chosen. Use the Performance Task Assessment List for Display in **PASC,** p. 63. P

Assessment Resources

📁 **Reproducible Masters**

Chapter Review, pp. 33-34
Assessment, pp. 97-100
Performance Assessment, p. 29

Glencoe Technology

🔘 **Chapter Review Software**
🔘 **Computer Test Bank**
📼 **MindJogger Videoquiz**

Chapter Organizer

Section	Objectives/Standards	Activities/Features
Chapter Opener		**Explore Activity:** Model how a bird's gizzard works. p. 425
16-1 **Birds** (2 sessions, 1½ blocks)*	1. **Identify** the characteristics of birds. 2. **Identify** the adaptations birds have for flight. 3. **Explain** how birds reproduce and develop. **National Content Standards:** (5-8) UCP1, UCP4, UCP5, A1, A2, B2, C1, C3, C5, E2, F5	**MiniLAB:** How does oil affect the feathers of a bird? p. 428 **Activity 16-1:** Observing Contour and Down Feathers, p. 429 **Using Math,** p. 430 **Connect to Physics,** p. 431 **Using Technology:** Saving the Nene, p. 433 **Skill Builder:** Concept Mapping, p. 434 **Science Journal,** p. 434
16-2 **Mammals** (2 sessions, 1½ blocks)*	1. **Identify** the characteristics of mammals and explain how they enable mammals to adapt to different environments. 2. **Distinguish** among monotremes, marsupials, and placental mammals. 3. **Compare** reproduction and development in the three kinds of mammals. **National Content Standards:** (5-8) UCP1, UCP4, UCP5, A1, A2, C1, C2, C5	**MiniLAB:** What are the characteristics of hair? p. 436 **Activity 16-2:** Whale Insulation, pp. 438-439 **Problem Solving:** Bats and Sound Waves, p. 441 **Skill Builder:** Classifying, p. 445 **Using Math,** p. 445
16-3 **Science and Society** **Issue: California Sea** **Otters** (1 session, ½ block)*	1. **Explain** the economic importance of sea otters. 2. **Determine** whether the government should limit the range of California sea otters. **National Content Standards:** (5-8) C4, F2, G1	**Explore the Issue,** p. 447 **People and Science:** Tania Zenteno-Savin, Marine Biologist, p. 448

* A complete Planning Guide that includes block scheduling is provided on pages 32T-35T.

Activity Materials

Explore	Activities	MiniLABs
page 425 cracked corn, sunflower seeds, nuts, or other seeds; pea-sized gravel; mortar and pestle	page 429 bird's contour feather, bird's down feather, hand lens, scissors pages 438-439 test tubes, solid food shortening, plastic sandwich bags, plastic spoons, large beakers, ice water, heat source, thermometers, rubber bands, watch with a second hand	page 428 30 cm × 30 cm piece of lightweight cotton cloth, scissors, petroleum jelly or baby oil, two small bowls, tap water page 436 student's personal brush or comb, tap water, microscope slide and coverslip, microscope

Need Materials? Call Science Kit (1-800-828-7777).

Teacher Classroom Resources

Reproducible Masters	Transparencies	Teaching Resources
Study Guide, p. 62 Reinforcement, p. 62 Enrichment, p. 62 Activity Worksheets, pp. 94-95, 98 Lab Manual 40, Owl Pellets Science Integration Activity 16, Thermal Energy from Foods Concept Mapping, p. 37 Cross-Curricular Integration, p. 20 Science and Society Integration, p. 20 Multicultural Connections, pp. 35-36	Section Focus Transparency 58, Here, Hold This Teaching Transparency 31, Anatomy of a Bird	Glencoe Life Science Interactive Videodisc Life Science CD-ROM Spanish Resources English/Spanish Audiocassettes Cooperative Learning Resource Guide Lab Partner Lab and Safety Skills Lesson Plans
Study Guide, p. 63 Reinforcement, p. 63 Enrichment, p. 63 Activity Worksheets, pp. 96-97, 99 Critical Thinking/Problem Solving, p. 20	Section Focus Transparency 59, Dolphin Talk Science Integration Transparency 16, Mammal Sonar Teaching Transparency 32, Mammal Teeth	**Assessment Resources** Chapter Review, pp. 35-36 Assessment, pp. 101-104 Performance Assessment in the Science Classroom (PASC) MindJogger Videoquiz Alternate Assessment in the Science Classroom Performance Assessment Chapter Review Software Computer Test Bank
Study Guide, p. 64 Reinforcement, p. 64 Enrichment, p. 64	Section Focus Transparency 60, Otters Up Close and Personal	

Key to Teaching Strategies

The following designations will help you decide which activities are appropriate for your students.

L1 Level 1 activities should be within the ability range of all students, including those with learning difficulties.

L2 Level 2 activities should be within the ability range of the average to above-average student.

L3 Level 3 activities are designed for the ability range of above-average students.

LEP LEP activities should be within the ability range of Limited English Proficiency students.

LS These activities are designed to address different learning styles.

COOP LEARN Cooperative Learning activities are designed for small group work.

P These strategies represent student products that can be placed into a best-work portfolio.

GLENCOE TECHNOLOGY

The following multimedia resources are available from Glencoe.

The Secret of Life
Gone Before You Know It: The Biodiversity Crisis
Science and Technology Videodisc Series (STVS)
Animals
  Development of Walking in Chicks
  Songbird Study
  Sexing Birds
  Peregrine Falcon
  Black Bear Ecology
  How Bats Hear
  Temperature Regulation in Dogs

Kit Fox
Animal Models of Human Physiology
Ecology
  Bird Sanctuary
  Red Wolf
  Saving the Spotted Owl
National Geographic Society Series
STV: Biodiversity
Glencoe Life Science Interactive Videodisc
Structural Adaptations in Animals
Life Science CD-ROM

Teacher Classroom Resources

This is a representation of key blackline masters available in the Teacher Classroom Resources.

Teaching Aids

Section Focus Transparencies

58 SECTION FOCUS TRANSPARENCY Section 16-1

HERE, HOLD THIS

After a mother emperor penguin has laid an egg, she has quite an appetite. She's ready to go find fish. She leaves the egg on the capable feet of the father penguin. He will keep the egg warm in a fold of skin for up to two months until the baby penguin hatches. When she returns, both parents will take turns tending the new hatchling.

1. To what group of animals do penguins belong?
2. How do these animals care for their eggs compared with reptiles?
3. How does this egg compare to a reptile's egg?

L1

59 SECTION FOCUS TRANSPARENCY Section 16-2

DOLPHIN TALK

Did you know that some people have taught dolphins language? They show dolphins different colored flags inside a big tank. The dolphins respond to flag commands by swimming through hoops or throwing a ball. Other dolphins have been trained to get coins from salvaged shipwrecks underwater. But still other dolphins are swimming buddies to people who are recuperating from illnesses. These dolphins seem to have very good communication skills and show their human partners friendship and caring.

1. To which group of vertebrates do dolphins belong? How are they different from fish?
2. Dolphins are considered to be very intelligent. How do you think brain development in mammals compares to that of other vertebrates?

L2

60 SECTION FOCUS TRANSPARENCY Section 16-3

OTTERS UP CLOSE AND PERSONAL

An aquarium is the best place for people to watch playful California sea otters. The sea otter is an endangered species. All year long otters put on a show as they dive, roll, and eat their food in full view through the glass. As their natural habitat diminishes, the otters are sagest in Monterey Bay near the aquarium.

1. How is the sea otter adapted to its environment of the cool water of the Pacific?
2. Sea otters feast on abalone and sea urchin from the underwater kelp forest. People who fish for abalone and sea urchin for a living are in competition with the sea otter. Do you think the sea otter's habitat should be limited to help the people who fish for sea urchins and abalone? Explain your point of view.

L3

Science Integration Transparencies

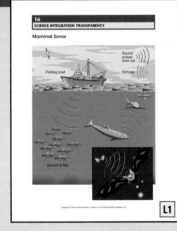

16 SCIENCE INTEGRATION TRANSPARENCY

Mammal Sonar

Sound pulses sent out
Echoes
Fishing boat
School of fish

L1

Teaching Transparencies

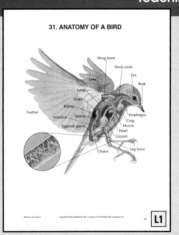

31. ANATOMY OF A BIRD

Wing bone, Vocal cords, Eye, Beak, Liver, Lung, Ovary, Kidney, Esophagus, Crop, Muscle, Heart, Gizzard, Feather, Intestine, Stomach, Eggshell gland, Cloaca, Leg bone

L1

32. MAMMAL TEETH

L2

Meeting Different Ability Levels

Study Guide

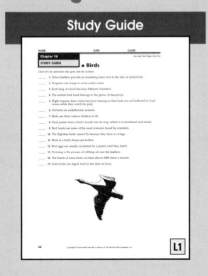

NAME DATE CLASS
Chapter 16 Use with Text Pages 424–434
STUDY GUIDE • **Birds**

Check (✓) the statements that agree with the textbook.

___ 1. Down feathers provide an insulating layer next to the skin of adult birds.
___ 2. Penguins use wings to swim under water.
___ 3. Each lung of a bird has four different chambers.
___ 4. The earliest bird fossil belongs to the genus *Archaeopteryx*.
___ 5. Flight requires keen vision but poor hearing so that birds are not bothered by loud noises while they watch for prey.
___ 6. All birds are endothermic animals.
___ 7. Birds use their contour feathers to fly.
___ 8. Food passes from a bird's mouth into its crop, where it is moistened and stored.
___ 9. Bird fossils are some of the most common found by scientists.
___ 10. The flightless birds cannot fly because they have no wings.
___ 11. Most of a bird's bones are hollow.
___ 12. Bird eggs are usually incubated by a parent until they hatch.
___ 13. Preening is the process of rubbing oil over the feathers.
___ 14. The hearts of some birds can beat almost 2000 times a minute.
___ 15. Some birds can digest food in less than an hour.

L1

Reinforcement

NAME DATE CLASS
Chapter 16 Use with Text Pages 424–434
REINFORCEMENT • **Birds**

Describe the four characteristics of birds.

1.
2.
3.
4.

Label the drawing of a contour feather with the names of its parts.

Write the name of the organ responsible for each step in the digestive process of a bird.

___ 9. where food is moistened and stored
___ 10. where food is partially digested
___ 11. where grain, seeds, and other foods are ground and crushed
___ 12. where the rest of digestion occurs

Answer the following questions.

13. How do the air sacs of birds help make the body lighter? _____
14. What is an amniotic egg? _____
15. What is the purpose of preening? _____
16. Compare and contrast contour feathers and down feathers. _____

L2

Enrichment Worksheets

NAME DATE CLASS
Chapter 16 Use with Text Pages 424–434
ENRICHMENT • **Birds**

Migration of Birds

Many animals, particularly birds, have a remarkable ability to migrate over thousands of miles of land and sea. Yet we know very little about how they do this. Clearly these animals must have special mechanisms of navigation and orientation.

The Atlantic golden plover flies over 8000 miles from the northeastern part of Canada, its arctic breeding ground, to an area of the southeastern part of South America. Perhaps it is even more extraordinary that the Pacific golden plover finds its way from the Western region of Alaska to the Hawaiian and Marquesan islands—relatively small pieces of land in the vast Pacific Ocean.

After much research, scientists have determined that these birds migrate by the sun and stars. Some species of birds cannot migrate successfully in cloudy weather. However, many other species continue to migrate under clouds and through fog. This posed another problem for scientists.

After many experiments, it was confirmed that by combining information from two sources, birds can orient themselves to a particular compass direction. These sources are the sun or constellation of stars, and their "internal clocks," which are related in some way to 24-hour periods. But even this does not explain birds' ability to migrate on a cloudy day or starless night.

There is much experimental evidence that birds can detect the earth's magnetic field and orient themselves with it. Very little is known about how birds detect magnetism. However, bits of magnetite, a magnetic iron ore, have been found in the head of pigeons. It is assumed at present that the magnetite is involved in some way in the pigeon's "magnetic on sense." Indeed, scientists have upset a pigeon's flight pattern by attaching a small magnet to the bird.

Much has been learned about navigation and orientation of birds and other animals. Yet scientists are far from understanding how birds can migrate with such accuracy for great distances year after year.

Question and Observations

1. Until lately recently how were birds thought to orient themselves on a long, migratory flight? _____
2. What recent discovery helps to explain how birds can fly in a particular compass direction? _____
3. How does a homing (messenger) pigeon find its way from the coast of France to its cage in the roof of a London apartment house? _____

L3

424C

Hands-On Activities

Science Integration Activity

Thermal Energy From Foods

Lab Manual

Owl Pellets 40

Assessment

Performance Assessment

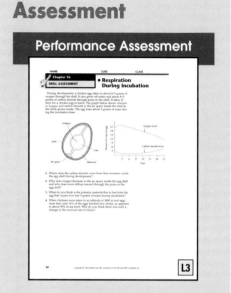

Respiration During Incubation

Enrichment and Application

Critical Thinking/ Problem Solving

Birds and Mammals

Cross-Curricular Integration

Birds and Mammals

Science and Society Integration

Birds and Mammals

Multicultural Connections

Saving the Puerto Rican Parrot

Concept Mapping

Birds and Mammals

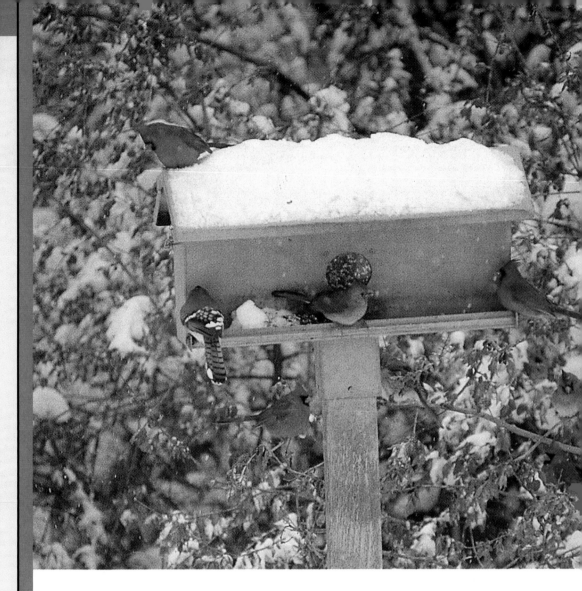

Chapter 16

Birds and Mammals

CHAPTER OVERVIEW

Section 16-1 This section describes the internal and external characteristics of birds and shows how these characteristics are adaptations for flight. Kinds of birds and the origin of birds are also discussed.

Section 16-2 This section builds on the concept of diversity through adaptation and classification. The orders of mammals are presented, and the section concludes with the origin of mammals.

Section 16-3 Science and Society The economic importance of the California sea otter is discussed. The Explore the Issue feature asks students whether the government should limit their range to the 402.5-km stretch where they presently live.

Chapter Vocabulary

incubate	monotreme
contour	marsupial
feather	placental
down feather	mammal
preening	gestation
mammal	period
mammary	placenta
gland	umbilical
herbivore	cord
carnivore	sea otter
omnivore	

Theme Connection

Stability and Change A discussion of adaptations of birds to life in the air and on land emphasizes the steps in the evolution of species that have moved from water to land environments. The theme of stability and change is continued in the discussion of mammal characteristics and origins. Both birds and mammals have developed homeostatic mechanisms for living on land.

424

Previewing the Chapter

Section 16-1 Birds
- ▶ Characteristics of Birds
- ▶ Adaptations for Flight
- ▶ Kinds of Birds
- ▶ Origin of Birds

Section 16-2 Mammals
- ▶ Characteristics of Mammals
- ▶ Classification of Mammals

- ▶ Diversity of Mammals
- ▶ Origin of Mammals

**Section 16-3 Science and Society
Issue: California Sea Otters**

424

424

Learning Styles Look for the following logo for strategies that emphasize different learning modalities.

Kinesthetic	Explore, p. 425; Activity, p. 427; MiniLAB, p. 428; Across the Curriculum, p. 441
Visual-Spatial	Demonstration, p. 427; Activity 16-1, p. 429; Activity, p. 431; MiniLAB, p. 436; Visual Learning, pp. 427, 428, 436, 440, 443, 447
Interpersonal	Science Journal, p. 428; People and Science, p. 448
Intrapersonal	Activity, p. 432
Logical-Mathematical	Activity 16-2, pp. 438-439; Enrichment, p. 441; Using Math, pp. 430, 445
Linguistic	Using Science Words, p. 428; Enrichment, p. 432; Science Journal, p. 434; Across the Curriculum, p. 437

Chapter
16

Birds and Mammals

Have you ever watched birds at a bird feeder? Birds eat one seed after another without stopping. They don't chew their food before swallowing because they don't have teeth. Many birds swallow small pebbles, stones, eggshells, and other hard materials, which are used to grind up seeds in a muscular sac called a gizzard. The activity below will show you how a gizzard works. In this chapter, you will learn about other characteristics of birds and about mammals.

EXPLORE ACTIVITY

Model how a bird's gizzard works.

1. Place some cracked corn, sunflower seeds, nuts, or other seeds and some gravel in a mortar.
2. Use a pestle and a churning motion to grind the seeds.
3. Describe the appearance of the seeds after grinding.

Observe: Describe in your Science Journal how a bird's gizzard helps it digest food .

Previewing Science Skills

▶ In the **Skill Builders**, you will **map concepts** and **classify**.

▶ In the **Activities**, you will **observe, infer, interpret data, hypothesize, measure,** and **compare.**

▶ In the **MiniLABs**, you will **observe, infer, compare,** and **describe.**

425

Assessment Planner

Portfolio
Refer to page 449 for suggested items that students might select for their portfolios.

Performance Assessment
See page 449 for additional Performance Assessment options.
Skill Builders, pp. 434, 445
MiniLABs, pp. 428, 436
Activities 16-1, p. 429; 16-2, pp. 438-439

Content Assessment
Section Wrap-ups, pp. 434, 445, 447
Chapter Review, pp. 449-451
Mini Quizzes, pp. 434, 445

Group Assessment
Opportunities for group assessment occur with Cooperative Learning Strategies and Flex Your Brain Activities.

EXPLORE ACTIVITY

Purpose
LS **Kinesthetic** Use the Explore Activity to introduce students to characteristics of birds. Explain that they will be learning more about the external and internal characteristics of both birds and mammals in this chapter. **L1**
COOP LEARN

Preparation
Purchase packaged bird seed and mix it with some corn and sunflower seeds. Collect gravel or pebbles.

Materials
one mortar and pestle, approximately one tablespoon of seeds, and one teaspoon of gravel for every three students

Teaching Strategies
• Assign one student to be the leader, one to get the materials and grind the seeds, and one to be the recorder.
• You may want to demonstrate how to use a mortar and pestle if students have not used them before.

Observe
The bird's gizzard uses hard materials such as small pebbles to grind the nuts and seeds the bird eats.

✔ Assessment

Performance Show students the internal structure of a bird. Ask which organ stores food (crop) and which organ grinds food (gizzard). **Ask students to infer how the two are different.** *The gizzard would have to be muscular to grind the food.* Use the Performance Task Assessment List for Making Observations and Inferences in **PASC,** p. 17. **P**

Prepare

Section Background

- Birds and mammals are warm-blooded, or endothermic, vertebrates. They produce heat to stay warm and lose heat to stay cool. Endothermic vertebrates have higher metabolic rates than do cold-blooded, or ectothermic, vertebrates. They have developed adaptations such as feathers, fur, body hair, and layers of fat for retaining heat. They sweat and pant to release body heat.

- Birds and mammals have complete four-chambered hearts. A four-chambered heart keeps oxygenated blood completely separate from deoxygenated blood.

Preplanning

- Obtain contour and down feathers for Activity 16-1.
- Obtain petroleum jelly, 15 × 15 cm pieces of cotton cloth, and bowls for the MiniLAB.

1 Motivate

Bellringer

Before presenting the lesson, display **Section Focus Transparency 58** on the overhead projector. Assign the accompanying **Focus Activity** worksheet. L1 LEP

Tying to Previous Knowledge

Compare reptile characteristics from Chapter 15 with the characteristics of birds. Point out the traits of reptiles that are modified in birds.

16•1 Birds

Science Words

incubate
contour feather
down feather
preening

Objectives

- Identify the characteristics of birds.
- Identify the adaptations birds have for flight.
- Explain how birds reproduce and develop.

Figure 16-1

Birds lay hard-shelled eggs in nests made out of many materials. *How are bird eggs similar to reptile eggs? How are they different?*

Characteristics of Birds

Did you see a bird today? Birds can be seen every day in backyards, flying above city streets, and in farmers' fields. What are the characteristics that make a bird a bird? Like reptiles, amphibians, and fish, birds are vertebrates. Unlike these animals, however, birds are endotherms, animals that maintain a constant internal body temperature. A bird's body temperature is about 40°C. You maintain a body temperature of 37°C.

Birds have feathers and scales. They are the only members of the animal kingdom that have feathers. All birds lay eggs enclosed in a hard shell. The parents keep the eggs warm, or **incubate** them, until they hatch. All birds have front legs that are modified into wings, whereas the two back legs support the body and usually have toes with claws. Birds have beaks instead of teeth.

Bird Eggs

If you've ever cooked scrambled eggs, you are familiar with a bird's egg. Like reptiles, all birds reproduce by laying an amniotic egg with a shell. Remember that an amniotic egg is a complete environment for a developing embryo. Fertilization is internal, and the shell is formed around the fertilized egg. Bird eggs have the same kind of membranes as reptile eggs to protect and nourish the embryo. Unlike a reptile's egg, a bird's egg has a hard shell made of calcium carbonate, the same chemical that makes up seashells, limestone, and marble. Birds also usually stay with their eggs to incubate them, and they care for their young. The length of time the eggs must be incubated varies with the species.

The female bird usually lays eggs in a nest, as you can see in **Figure 16-1**. Most birds lay from two to eight eggs at one time. All the eggs laid at one time are called a clutch.

Program Resources

 Reproducible Masters

Study Guide, p. 62 L1
Reinforcement, p. 62 L1
Enrichment, p. 62 L3
Cross-Curricular Integration, p. 20
Science and Society Integration, p. 20
Activity Worksheets, pp. 5, 94-95, 98 L1
Science Integration Activities, pp. 31-32
Concept Mapping, p. 37

Multicultural Connections, pp. 35-36
Lab Manual 40

Transparencies
Section Focus Transparency 58 L1
Teaching Transparency 31

Adaptations for Flight

People have always been fascinated by the ability of birds to fly. Flight requires a light, yet strong skeleton, wings, and feathers to help lift the bird off the ground. Keen senses, especially eyesight, and tremendous amounts of energy are also needed for flight.

Feathers

A bird's body is covered with two types of feathers, contour feathers and down feathers. Strong, lightweight **contour feathers** give birds their coloring and smooth, sleek shape. These are the feathers birds use to fly. The long contour feathers on the wings and on the tail help the bird steer and keep from tipping over. Have you ever watched ducks swimming in a pond on a freezing cold day, and wondered how they were able to keep warm? Soft, fluffy **down feathers** provide an insulating layer next to the skin of adult birds and cover the bodies of young birds. Feathers help birds maintain their constant body temperature. They grow in much the same way as your hair grows. The base of each feather shaft is rooted in a tiny follicle and receives nutrients from the bird's blood. When it is replaced, the feather either falls off or is pushed out by a new feather. As you can see in **Figure 16-2,** the shaft of the feather has many branches called barbs. Each barb has many cross branches called barbules that give the feather strength.

USING MATH

Count the number of different kinds of birds you observe outside at a certain time each day for a week. Graph your data.

Figure 16-2

Contour feathers are the feathers used for flight. Down feathers help insulate birds.

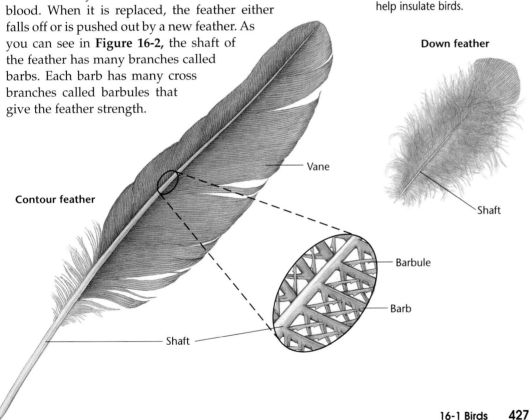

Down feather

Shaft

Contour feather

Vane

Barbule

Barb

Shaft

16-1 Birds **427**

Figure 16-3

Birds spend a lot of time preening. *Why is it important for this bird to preen its feathers?*

MiniLAB

How does oil affect the feathers of a bird?

The oil gland, or preen gland, secretes an oil that the bird squeezes out with its beak and spreads over its feathers and feet. Find out how the oil affects feathers.

Procedure

1. Cut two 15 × 15 cm pieces of cotton cloth.
2. Apply petroleum jelly or baby oil to one of the pieces.
3. Dip each piece in a bowl of water and pull it out.

Analysis

1. Compare what happens to the two pieces of cloth when placed in water.
2. What can you conclude about the importance of oil on bird feathers?

Figure 16-4

The hollow bones of birds are an adaptation for flight. *What advantage do the thin cross braces inside give bird bones?*

428   Chapter 16 Birds and Mammals

To care for its feathers, a bird has an oil gland located just above the base of its tail. Using its beak, the bird rubs oil from the gland over its feathers in a process called **preening.** The oil conditions the feathers and helps make them water repellent. The bird in **Figure 16-3** is preening.

Hollow Bones

Some of the adaptations birds have for flight can't be seen because they are internal. One of these is a bird's skeleton. Many of the bones in a bird's skeleton are fused for extra strength and more stability for flight. To make a bird lighter, most of its bones are hollow, with thin cross braces of bone inside for strength and support. The hollow spaces are filled with air. Why do you think it's important for a bird to have a light skeleton?

You can see the skeleton of a bird in **Figure 16-4.** A large sternum, or breastbone, supports the large chest muscles. The last bones of the spine support the tail feathers, which play an important part in steering and balancing during flight and landing.

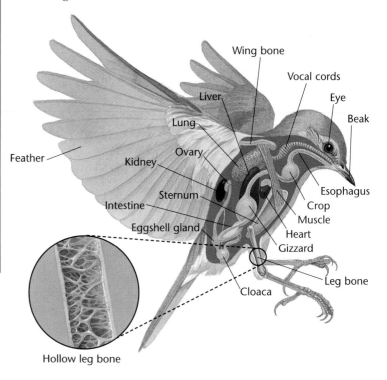

Wing bone
Vocal cords
Liver
Eye
Lung
Beak
Feather
Ovary
Kidney
Esophagus
Intestine
Sternum
Crop
Muscle
Eggshell gland
Heart
Gizzard
Cloaca
Leg bone
Hollow leg bone

Activity 16-1

Observing Contour and Down Feathers

Birds have two different kinds of feathers, including stiff contour feathers, which cover the body and wings, and fluffy down feathers that keep them warm. In this activity, you will observe the differences in contour and down feathers and how they are adapted for the jobs they do.

Problem
How do contour and down feathers differ?

Materials
- contour feather
- down feather
- hand lens
- scissors

Procedure
1. Use a hand lens to examine a contour feather. Find the shaft, vanes, and barbs. Use **Figure 16-2** on page 427 for help.
2. Hold the shaft by the thicker end and carefully bend the opposite end. Observe what happens when you release the bent end.
3. With the scissors, cut about 2 cm from the end of the shaft. Examine the cut end with the hand lens.
4. Separate the barbs near the center of the vanes. Use your fingers and gently rub the feather where you separated it. Observe what happens.

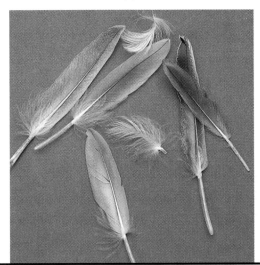

5. Draw the feather and label the shaft, vanes, and barbs.
6. Repeat step 1 with a down feather.
7. Draw and label the structure of a down feather.

Data and Observations   Sample Data

Contour Feather	Down Feather
Student drawings should resemble Figure 16-2	Student drawings should resemble Figure 16-2

Analyze
1. What happens when you release the bent end of the contour feather?
2. **Describe** how the cut end of the shaft appears.
3. What happens to the separated barbs when you rub them with your fingers?

Conclude and Apply
4. **Compare** the shape, filaments (barbs), and shaft of the down feather with those of the contour feather.
5. How is the structure of a contour feather adapted for flight?
6. How is the structure of a down feather adapted for insulation?

16-1 Birds   **429**

▸**Assessment**

Process Have students explain why down feathers of birds are used to make warm coats for humans. Ask them to examine some synthetic materials used in making warm coats and compare them with down feathers. Use the Performance Task Assessment List for Making Observations and Inferences in **PASC**, p. 17. **P**

Activity 16-1

Purpose
Ⓘ **Visual-Spatial** Students will observe the structure of a contour feather and a down feather. **L1**

Process Skills
measuring in SI, observing and inferring, interpreting data, comparing and contrasting, forming operational definitions

Time
50 minutes

📁 **Activity Worksheets,** pp. 5, 94-95

Teaching Strategies
Troubleshooting Some students may be allergic to feathers. Feathers can be kept in clear plastic bags and observed.
- Feathers may be obtained from a biological supply company or a craft store. Most feathers can be rinsed in warm water (no soap) and placed in a microwave oven for one to two minutes to dry.

Answers to Questions
1. It goes back to its previous position without breaking.
2. It is crisscrossed inside and has many air spaces.
3. They go back together.
4. Contour feathers are larger, with a more rigid shaft, and the barbs are hooked together. Down feathers are smaller, fluffier, and softer. The hairlike filaments are not hooked together, and the shaft is very flexible.
5. It is lightweight, flat, and spreads out to push the air. The locked barbs give it strength.
6. The fluffy feather traps air that is heated by the bird's body. It provides a layer of insulation.

429

INTEGRATION
Physics

USING MATH

Every 10 seconds, a crow beats its wings 20 times, a robin 23 times, a chickadee 270 times, and a hummingbird 700 times. Calculate how many times the wings of a crow beat in five minutes of flying time. The wings of a robin? A chickadee? A hummingbird?

Wings

A bird's most obvious adaptation for flight is its wings. Wings are attached to powerful chest muscles. By beating their wings, birds attain both *thrust* to go forward and *lift* to stay in the air. They beat their wings not just up and down but also back and forth.

If birds relied only on beating their wings for flight, they would tire quickly. Flying is made easier because wings provide lift without beating. In **Figure 16-5,** you can see the similarity in a bird's wing and an airplane wing. Both are curved on top and flat or slightly curved on the bottom. When a wing with this shape moves through the air, the air has a longer way to go around the curved upper surface than it does across the flatter bottom surface. The longer path taken by air moving over the upper surface reduces the air pressure there. This results in greater pressure on the lower surface of the wing. This difference in air pressure results in lift. A wing produces lift in proportion to its surface area. The larger the wing, the greater the lift. Birds with large wings can soar and glide for a long time. Once airborne, they can cover great distances without beating their wings.

Wings also serve important functions for birds that don't fly. Penguins use their wings to swim under water. While running or walking, ostriches and rheas balance with their wings.

Figure 16-5

Wings provide the upward force called lift for both birds and airplanes. Air moving over the curved upper surface of the wing reduces the air pressure there, resulting in an upward force. The amount of lift depends on wing area, the speed of air across the wing, and the shape and angle of the wing.

430 Chapter 16 Birds and Mammals

Getting Energy for Flight

Birds obtain the energy to fly from the food they eat. Because flying requires large amounts of energy, birds need to eat large amounts of food. An efficient digestive system breaks down the food quickly to supply birds with the energy they need. Some birds can digest food in less than an hour. It takes us more than a day to digest the food we eat!

Food first passes from the mouth into the crop, where it is moistened and stored. From there, it moves into the first part of the stomach, where it is partially digested. Then it moves into the gizzard. Small stones and grit in the gizzard grind and crush seeds and other foods. You saw how gravel helps break up seeds in the Explore Activity at the beginning of this chapter. Food then passes into the small intestine, where the rest of digestion occurs. Finally, nutrients are absorbed by the bloodstream.

Maintaining Body Temperature

Birds also need energy to maintain body temperature. Body heat is generated from the energy in food. Oxygen is needed to convert food to energy. To obtain oxygen, birds have efficient respiratory systems. As you can see in **Figure 16-6,** a bird has two lungs with balloonlike air sacs attached to each one. The air sacs are located in different parts of the body, including the hollow bones. These air sacs increase the amount of oxygen a bird can take in. Having these air sacs spread throughout the body also makes a bird lighter.

Birds have hearts with four chambers. Blood with oxygen is kept separate from blood without oxygen. Blood is circulated very quickly because birds have a rapid heartbeat rate. The hearts of some birds beat more than 1000 times in a minute. This rapid circulation supplies cells of the bird's body with the oxygen needed to release energy from food.

Figure 16-6

About 75 percent of the air inhaled by a bird passes into the air sacs. When a bird exhales, air with oxygen passes from the air sacs into the lungs. This means that a bird receives oxygenated air both when it inhales and when it exhales.

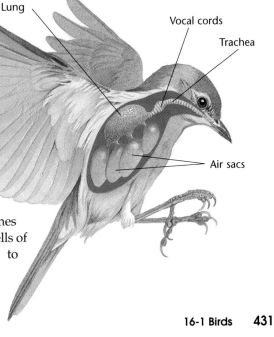

Lung

Vocal cords

Trachea

Air sacs

Theme Connection

Energy Because birds use large amounts of energy to fly, the crop and the gizzard help birds quickly obtain energy from food. The crop stores and releases food continuously. The gizzard grinds the food so that it is digested faster.

Activity

Visual-Spatial Use diagrams to show the internal and external structures of a bird. Point out the digestive tract, oil gland, and air sacs. Have students make a list of the bird's internal and external characteristics. **L1**

Inquiry Question

What is a disadvantage of being endothermic? *It takes more food to maintain body temperature.*

CONNECT TO

PHYSICS

Answer Ectothermic animals maintain body temperature from sunlight and warm objects in the environment. Endothermic animals generate body heat as a result of respiration. Students must also be aware that ectothermic animals do undergo respiration.

GLENCOE TECHNOLOGY

 Videodisc
STVS: Ecology
Disc 6, Side 1
Bird Sanctuary (Ch. 9)

Saving the Spotted Owl (Ch. 15)

Use **Science Integration Activities,** pp. 31-32, as you teach this lesson.

Kinds of Birds

The class Aves contains almost 9000 species of birds. Birds are classified into orders based on characteristic beaks, feet, feathers, and other physical features. Some common orders of birds are shown in **Figure 16-7.**

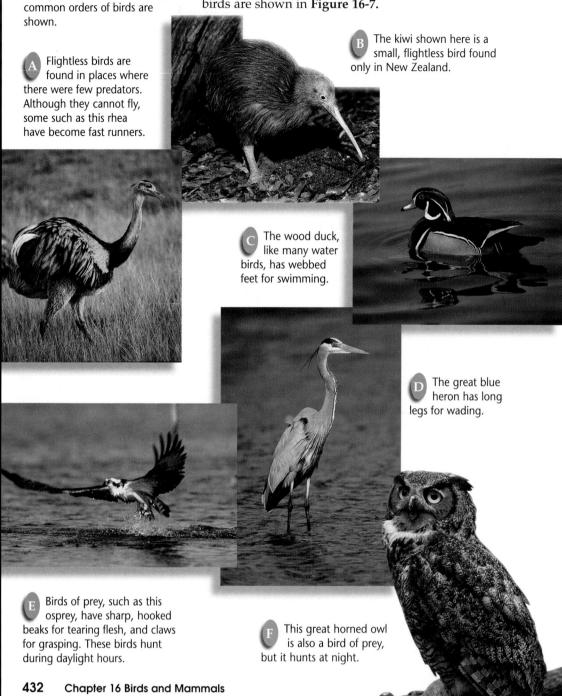

Figure 16-7

Examples of several of the most common orders of birds are shown.

A Flightless birds are found in places where there were few predators. Although they cannot fly, some such as this rhea have become fast runners.

B The kiwi shown here is a small, flightless bird found only in New Zealand.

C The wood duck, like many water birds, has webbed feet for swimming.

D The great blue heron has long legs for wading.

E Birds of prey, such as this osprey, have sharp, hooked beaks for tearing flesh, and claws for grasping. These birds hunt during daylight hours.

F This great horned owl is also a bird of prey, but it hunts at night.

432   Chapter 16 Birds and Mammals

Cultural Diversity

Earliest Bird Sankar Chatterjee is best known for his 1986 discovery of *Protoavis*, a 225 million-year-old fossil that may turn out to be the earliest known bird. Have students research the work of Chatterjee. **L2**

G There are two kinds of perching birds, insect eaters and seed eaters. Insect eaters, such as this nuthatch, have long pointed beaks for boring into wood to find insects.

H Seed-eating birds, such as this cardinal, have thick, strong beaks for cracking seeds.

Saving the Nene

The nene goose is an endangered bird found only on the volcanic islands of Hawaii. The nene is a small, land-dwelling goose. In the 1700s, there were approximately 25 000 of these geese in Hawaii. As settlers arrived, they brought with them cattle, sheep, pigs, goats, and rats that destroyed the nene's breeding grounds. By 1947, there were only 30 nene geese left in Hawaii.

Captive Breeding

In 1949, a captive-breeding program was begun in Hawaii to prevent extinction of the nene goose. How does captive breeding work? Scientists locate and capture the few remaining wild birds. These birds are housed in research centers, where they are fed, given medical attention, and protected from predators. The nene geese in the captive-breeding program mated and laid eggs, but only 24 goslings were raised. A captive-breeding program in England was more successful. Nearly 2000 captive-born nene geese have been released on the islands of Hawaii and Maui, although the nene goose is still an endangered species.

Think Critically:

Why is it important to reintroduce captive-born endangered animals to their native habitats?

Nene Geese

16-1 Birds   **433**

Across the Curriculum

History Pigeons have been associated with humans for many years. Over 4000 years ago in the Middle East, humans used pigeons as messengers or kept them for food. The first recorded use of pigeons as messengers was when Pharaoh Ramses III took the throne in 1200 B.C. Four birds were sent out in different directions to carry the message that a new Pharaoh had taken the throne.

NATIONAL GEOGRAPHIC SOCIETY

 Videodisc

STV: Biodiversity
Roseate spoonbill and egret: Florida

46084

3 Assess

Check for Understanding

? FLEX Your Brain

Use the Flex Your Brain activity to have students explore BIRDS.

📁 **Activity Worksheets,** p. 5

Reteach

Have students describe how the area near your school can be made into a more favorable habitat for birds.

Extension

📁 For students who have mastered this section, use the **Reinforcement** and **Enrichment** masters.

Use the Mini Quiz to check students' recall of chapter content.

1. **Birds keep their eggs warm, or _____ them.** *incubate*

2. **Feathers used for flight are _____ feathers.** *contour*

3. **The process of a bird rubbing oil over its feathers is _____ .** *preening*

4. **Soft, fluffy _____ feathers cover the bodies of young birds.** *down*

Section Wrap-up

Review

1. endotherms, feathers, lay eggs with a hard shell, incubate eggs

2. Contour feathers steer and keep the bird upright. The hollow bones in a bird's skeleton make the bird lighter, as do inflated air sacs.

3. **Think Critically** Reptiles are ectothermic and rely on warm weather for egg development. Birds are endothermic and incubate their eggs.

Science Journal

LS **Linguistic** Expressions such as "like water off a duck's back" are accurate, whereas "eats like a bird" is not. Other expressions are "eagle-eyes"—accurate; "wise as an owl"—not accurate; "happy as a lark"—not accurate; and "crazy as a loon"—not accurate. Students may write their own bird phrases and have classmates determine which are accurate and which are inaccurate.

Origin of Birds

Protoavis

Scientists have learned about the origins of most living things by studying their fossils. However, there are few fossils of birds to study. Those that do exist have served to convince scientists that birds developed from reptiles millions of years ago. Today, birds still have some characteristics of reptiles, including reptilelike scales on their feet and legs.

The oldest fossils thought to be birds are those of *Archaeopteryx* (ar kee AHP tuh rihks). Fossils of *Archaeopteryx* are about 150 million years old. Although scientists do not think *Archaeopteryx* was a direct ancestor of modern birds, it did have feathers and wings like today's birds. However, it also had solid bones, teeth, a long tail, and clawed front toes, just like a reptile.

Recently, scientists have discovered an older fossil that shows some characteristics of birds. *Protoavis* had two characteristics of birds: hollow bones and a well-developed breastbone with a keel. *Protoavis* lived about 225 million years ago, and is thought to be similar to birds. But as no fossil feathers have been found with *Protoavis*, scientists do not know if this animal was an ancestor of modern birds or a kind of ground-living dinosaur. **Figure 16-8** shows an artist's idea of what *Archaeopteryx* and *Protoavis* may have looked like.

Archaeopteryx

Figure 16-8

Archaeopteryx is considered a link between reptiles and birds. *Protoavis* may be an ancestor of birds.

Section Wrap-up

Review

1. List four characteristics shared by all birds.

2. How are a bird's feathers, air sacs, and skeleton adaptations for flight?

3. **Think Critically:** Explain why birds can reproduce in the Arctic but reptiles cannot.

Skill Builder
Concept Mapping

Make a network tree concept map that details the characteristics of birds. Use the following terms in your map: *birds, beaks, hollow bones, wings, eggs, adaptations for flight, feathers, air sacs.* If you need help, refer to Concept Mapping in the **Skill Handbook.**

Science Journal

There are many bird expressions in our language such as "like water off a duck's back" and "eats like a bird." In your Science Journal, write as many expressions as you can find and then determine which ones are accurate and which ones are not.

Skill Builder

Student concept maps should show a simple network tree beginning with birds, then adaptations for flight, which include hollow bones, feathers, and air sacs. Eggs, beaks, and wings should be on a level with adaptations for flight.

✓ Assessment

Performance Assess students' abilities to compare and contrast by having them examine beaks and feet of birds indigenous to your area. Have students make a model of a hypothetical bird's feet, and have them describe their bird's lifestyle. Use the Performance Task Assessment List for Model in **PASC,** p. 51. **P**

Mammals 16•2

Characteristics of Mammals

Cats, whales, moles, bats, horses, and people are all mammals. Mammals live almost everywhere, from tropical and arctic regions to deserts, woodlands, and oceans. What do you, a mole, and a whale have in common? Each species of mammal is adapted to its way of life, but all mammals share some characteristics. **Mammals** are endothermic vertebrates that have hair and produce milk to feed their young, as shown in **Figure 16-9.** Like birds, mammals provide care for their young.

Skin and Glands

Skin covers and protects the bodies of all mammals. Hair, horns, claws, nails, and hooves are produced by the skin. The skin contains many different kinds of glands. **Mammary glands,** which are characteristic of all mammals, produce the milk female mammals use to feed their young. Oil glands produce oil to lubricate and condition the hair and skin, and sweat glands help mammals stay cool. Many mammals have scent glands that are used for marking their territory, for attracting mates, and for defense.

Figure 16-9

Mammals, such as this harp seal, care for their young after they are born. *How do mammals feed their young?*

16-2 Mammals   435

Science Words

- mammal
- mammary gland
- herbivore
- carnivore
- omnivore
- monotreme
- marsupial
- placental mammal
- gestation period
- placenta
- umbilical cord

Objectives

- Identify the characteristics of mammals and explain how they enable mammals to adapt to different environments.
- Distinguish among monotremes, marsupials, and placental mammals.
- Compare reproduction and development in the three kinds of mammals.

Section 16•2

Prepare

Section Background

- The monotremes and marsupials make up only 5 percent of all mammals. Three species of monotremes—the duck-billed platypus and two species of echidnas—exist today. Monotremes live in Australia and on nearby islands in a wide range of habitats.

- The duck-billed platypus gets its name from its broad, flat, hairless snout, which resembles the bill of a duck.

- Most of the 250 species of marsupials live in Australia. Only one, the opossum, lives in North America.

Preplanning

- Obtain test tubes, shortening, plastic sandwich bags, plastic spoons, beakers, thermometers, rubber bands, a watch with a second hand, ice water, and a heating device for Activity 16-2.

- Set up microscopes and obtain microscope slides and cover slips for the Mini-LAB.

1 Motivate

Bellringer

🔲 Before presenting the lesson, display **Section Focus Transparency 59** on the overhead projector. Assign the accompanying **Focus Activity** worksheet. L1 LEP

Tying to Previous Knowledge

Have students compare the characteristics of birds and mammals. Both birds and mammals are endothermic, have four-chambered hearts, and four limbs for movement.

435

2 Teach

MiniLAB

What are the characteristics of hair?

Find out the characteristics of hair.

Procedure

1. Brush or comb your hair to remove a few loose hairs.
2. Take two hairs from your brush that look like they still have the root attached.
3. Make a wet mount slide of the two hairs, being sure to include the root.
4. Focus on the hairs with the low-power objective. Draw what you see.
5. Switch to the high-power objective and focus on the hairs. Draw what you see.

Analysis

1. Describe the characteristics of hair and root.
2. Infer how hair keeps an organism warm.

436

Hair

Nearly all mammals have hair on their bodies at some time during their lives. Many have thick fur that covers all or parts of their bodies; it traps air and helps to keep them warm. Others, such as whales and dolphins, have little hair. They rely on a thick layer of fat under their skin to keep them warm. Elephants and rhinoceroses have little hair. Porcupine quills and the spines of a hedgehog are modified hairs that protect them from their enemies. Sensory hairs, called whiskers, around the mouth of many mammals, such as cats, mice, and sea otters, help them keep in touch with their environment.

Teeth

Look at the mammals in **Figure 16-10.** Can you tell anything about what these animals eat by looking at their teeth? Almost all mammals have specialized teeth. Front teeth, called incisors, bite and cut. Next to the incisors, pointed canine teeth grip and tear. Premolars and molars shred, grind, and crush. Rabbits use their large incisors to cut off blades of grass. Animals like the rabbit that eat plants are called **herbivores.** Many mammals are herbivores. **Carnivores,** such as tigers, are flesh-eating animals. Some animals, like yourself, eat both plants and animals and are called **omnivores.** Scientists can tell what kind of food a mammal eats by examining its teeth.

Figure 16-10

Mammals have teeth specialized for the food they eat. *What do the teeth of the horse tell you about its chewing method?*

Body Systems

Mammals live active lives. Think of all the things you do each day! The body systems of mammals allow them to be active and to survive in many environments. All mammals have four-chambered hearts and networks of blood vessels. This type of heart allows blood coming from the lungs, full of oxygen, to be pumped directly to the body. Mammals have well-developed lungs made up of millions of microscopic sacs that increase the surface area for exchange of carbon dioxide and oxygen.

The nervous system is made up of a brain, spinal cord, and nerves. The brains in most mammals are larger than those of other animals the same size. Mammals are able to learn and remember more than other animals.

The digestive system of mammals varies according to the kind of food the animal eats. Carnivores have short digestive systems compared to those of herbivores. Meat is more easily digested than plant material. Herbivores need a long digestive system to help them break down the carbohydrate called cellulose found in plants.

Reproduction

All mammals reproduce sexually. Most mammals give birth to live young after a period of development inside an organ called the uterus. Mammal parents are protective. One or both parents care for the young after they are born. Many mammals are nearly helpless, and sometimes even blind, when they are born. They can't do much for themselves the first several days or even months. They just eat, sleep, and develop. The young of some mammals, such as antelope, deer, elephants, whales, and dolphins, must be well developed at birth to be able to move with their constantly moving parents. Those that live on land can usually stand by the time they are a few minutes old. Marine mammals can swim as soon as they are born.

During the nursing period, young mammals learn skills they need to survive. Among most kinds of mammals, the mother raises the young alone. In some species, males help with providing shelter, finding food, and protecting the young, as illustrated in **Figure 16-11**.

Figure 16-11

Both mother and father wolves care for their young after they are born. Mammals such as these Arctic wolves are taught how to hunt for their food by the parents.

16-2 Mammals **437**

Purpose

Logical-Mathematical Students will determine whether blubber (fat) is a good insulator. **L1** **COOP LEARN**

Process Skills

comparing and contrasting, observing and inferring, forming a hypothesis, interpreting data, separating and controlling variables, designing an experiment

Time

45 minutes

Materials

Alternate Materials If shortening is not available, students could use butter or margarine, or they could substitute pieces of Styrofoam or other insulating materials.

Safety Precautions

Make sure students use gloves if they are handling hot items.

Possible Hypotheses

Most student hypotheses will reflect that insulated test tubes will lose heat more slowly than uninsulated test tubes.

📁 **Activity Worksheets,** pp. 5, 96-97

Activity 16-2

Design Your Own Experiment
Whale Insulation

Mammals are endotherms that have adaptations, such as hair, which conserve and maintain body temperature. Whales are mammals that are nearly hairless, yet they live in cold seawater. How do whales maintain body temperature? Whales have a layer of fat called blubber beneath their skin. Blubber helps insulate the whale so it can survive in cold water.

PREPARATION

Problem

How effective is blubber in maintaining body temperature?

Form a Hypothesis

State a hypothesis about how blubber insulates a whale to help it maintain its body temperature.

Objectives

- Measure the rate of cooling of insulated and uninsulated test tubes.
- Compare the rates of cooling of insulated and uninsulated test tubes.
- Relate the results to the effectiveness of blubber as a means to maintain body temperature in whales.

Possible Materials

- test tubes
- solid food shortening
- plastic sandwich bags
- plastic spoons
- large beakers
- ice water
- heat source
- thermometers
- rubber bands
- watch with a second hand

Safety Precautions 🧤 🔥

Use gloves when working with ice or heated materials to protect your hands.

PLAN THE EXPERIMENT

1. As a group, agree upon and write out the hypothesis statement.
2. As a group, list the steps you need to take to test your hypothesis. Be specific; describe exactly what you will do at each step. List your materials.
3. How will you model the skin and blubber of a whale using test tubes, shortening, and plastic bags?
4. How will you determine the cooling rates of the insulated and uninsulated test tubes?
5. Prepare a data table in your Science Journal to record your data.

Check the Plan
1. Read over your entire experiment to make sure that all steps are in logical order.

2. Identify any constants, variables, and controls in the experiment.
3. How many test tubes will you need for the experiment?
4. What will you measure? How often will you take this measurement?
5. How will you show the relationship between insulation and maintenance of body heat?
6. *Make sure your teacher approves your plan before you proceed.*

DO THE EXPERIMENT

1. Carry out the experiment as planned.
2. While the experiment is going on, write down any observations that you make and complete the data table in your Science Journal.

Analyze and Apply
1. **Compare** your results with those of other groups.
2. How did the temperature change in each of the test tubes?

3. **Graph** your results using a line graph with the temperature on the *y*-axis and the time on the *x*-axis. Use a different color for each test tube.
4. How did the rate of cooling of the uninsulated test tube **compare** to the others? Explain.
5. **Infer** whether blubber is a good insulator for mammals that swim in cold seawater.

Go Further

Repeat this experiment using wool fabric or down feathers. Compare your results with the results of this experiment. Which material is a better insulator?

16-2 Mammals **439**

Go Further
Students may find other materials that also are good insulators.

Assessment
Performance Ask students to design and carry out an experiment that compares the insulating properties of polyester batting with that of Styrofoam peanuts. Use the Performance Task Assessment List for Designing an Experiment in **PASC**, p. 23. **P**

PLAN THE EXPERIMENT

Possible Procedures
Students may decide to fill three test tubes with hot water, covering one with shortening and a plastic bag, one with just a plastic bag, and one with a heavy layer of shortening and a plastic bag. After each bag is placed in a beaker full of ice water, they may measure temperatures in each tube every 5 minutes for 30 minutes.

Teaching Strategies
Demonstrate how to put several spoonfuls of shortening into a plastic bag and then place a test tube in it. Show students how to spread the shortening over the test tube by moving it with their fingers on the outside of the plastic bag.

DO THE EXPERIMENT

Expected Outcome
Students will observe that it takes about twice as long for the insulated test tube to lose heat and take on the temperature of the ice bath as it does for the uninsulated test tube.

Analyze and Apply
1. Student results should be similar.
2. Test tubes with insulation lost heat more slowly than those without insulation.
3. The graph will show a faster rate of cooling for the uninsulated test tube.
4. The rate of cooling of the uninsulated test tube was about twice that of the insulated tube. Insulation helps retain heat.
5. Blubber may be a good insulator because it is a fat, like shortening.

- Monotreme eggs are flexible and leathery like reptile eggs, not hard and smooth like bird eggs. But, unlike reptiles and like most birds, female monotremes incubate their eggs. Platypuses often curl up around their eggs. Echidnas carry their eggs around with them in a pouch that develops only during breeding season.

- A duck-billed platypus uses its bill to dig burrows in stream banks and to scoop up worms and other invertebrates from the bottoms of streams. The platypus reaches a length of about 46 cm and has thick brown fur, a broad flat tail, and webbed feet. The feet have claws, and the hind legs have sharp spines connected to poison glands used in defense.

Inquiry Question

What disadvantages do the developing young of monotremes have compared with the developing young of marsupials? *The developing young of marsupials are protected in their mother's pouch. The young of monotremes are left alone while the parents leave to find food, making the young more vulnerable to predators.*

GLENCOE TECHNOLOGY

Videodisc

STVS: Animals
Disc 5, Side 2
How Bats Hear (Ch. 18)

Figure 16-12

A duckbilled platypus is a mammal, yet it lays eggs. *Why is it classified as a mammal?*

Classification of Mammals

Mammals are classified into three groups based on how their young develop. The three groups are: monotremes, marsupials, and placental mammals.

Monotremes

Monotremes are mammals that lay eggs with tough, leathery shells. The female incubates the eggs. Mammary glands of the monotremes do not have nipples. When the young hatch, they nurse by licking milk from the skin and hair surrounding the female's mammary glands. The duckbilled platypus shown in **Figure 16-12** is a well-known monotreme.

Marsupials

The kangaroo in **Figure 16-13** is an example of a marsupial. **Marsupials** are pouched mammals that give birth to tiny, immature offspring. Kangaroos are born a few days after fertilization takes place and are about the size of a honeybee at birth. Immediately after birth, the young kangaroo crawls into the pouch on the female's abdomen and attaches to a nipple. It remains protected in the pouch while it feeds and develops more completely. After a few months, the young kangaroo, called a joey, is better developed and crawls out of the pouch.

The marsupials you are probably the most familiar with are kangaroos and opossums. Other marsupials are the koala, the Tasmanian devil, and the wallaby. Marsupials live primarily in Australia, Tasmania, and New Guinea.

Figure 16-13

Marsupials carry their developing young in a pouch on the outside of their bodies.

A A kangaroo and its joey keep watch for danger. The joey will return to its mother's pouch for protection.

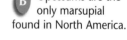

B Opossums are the only marsupial found in North America.

440 Chapter 16 Birds and Mammals

Integrating the Sciences

Earth Science Mammals diversified during the Cenozoic Era. Monotremes and marsupials developed in Australia and South America; placental mammals dominated North America. When North America and South America joined, placental mammals spread south. Australia remained geographically separate; it has the most diverse population of monotremes and marsupials in the world.

Visual Learning

Figure 16-12 Why is it classified as a mammal? *The duckbilled platypus has mammary glands.* **LEP** **LS**

Figure 16-14

A placental mammal's embryo develops inside a placenta in the uterus of a female.

Placental Mammals

In **placental mammals,** embryos develop inside the female in the uterus. The time during which the embryo develops in the uterus is called the **gestation period.** Gestation periods range from 16 days in hamsters to 650 days in elephants. Placental mammals are named for the **placenta,** a saclike organ developed by the growing embryo that attaches to the uterus, **Figure 16-14.** The placenta absorbs oxygen and food from the mother's blood. An **umbilical cord** attaches the embryo to the placenta. Several blood vessels make up the umbilical cord. The umbilical cord transports food and oxygen from the placenta to the embryo, and takes waste products away from the embryo. The mother's blood never mixes with the blood of the embryo.

Problem Solving

Bats and Sound Waves

Bats are acrobats of the night. They can fly around obstacles and can find insects to eat in complete darkness. Have you ever wondered how they do this? Some bats emit extremely high-pitched sounds through the mouth and nose when hunting for food. These sounds are usually too high pitched for humans to hear. Bats also make noises that people can hear, from whining sounds to loud twitters and squeaks. Bats can catch fast-flying insects or darting fish and at the same time avoid branches, wires, and other obstacles in a process called echolocation. The sound waves they send out travel in front of them, and this helps them locate objects.

Solve the Problem:

1. **What happens to a sound wave when it comes in contact with an object?**
2. **How does this enable bats to locate an object in the dark?**
3. **A bat has unusually large ears. How are the large ears an adaptation for hearing?**

Think Critically:

Explain what would happen to bats if they were allowed to search for prey in a soundproof room.

Across the Curriculum

Art Have students make a mobile that represents the orders of placental mammals listed on pages 442, 443, and 444. L1 LS

Problem Solving

Solve the Problem
1. Part of the wave is reflected back.
2. They locate the object from the reflected sound waves.
3. They catch the reflected sound waves.

Thinking Critically
They would have difficulty in a place that absorbs sound waves rather than reflects them.

Enrichment

LS **Logical-Mathematical** Have students research and report on gestation periods of mammals and the average number of offspring born to different species. Ask students to evaluate the data by comparing the size or activity of these organisms with the length of gestation. Some gestation periods are:

opossum	12 days
hamster	16 days
mouse	20 days
rabbit	31 days
dog	61 days
cat	63 days
guinea pig	68 days
cow	281 days
horse	336 days
camel	406 days
giraffe	442 days
whale	450 days
elephant	650 days	P

Inquiry Question

Do you think there is a connection between the size of mammals and their gestation periods? *Generally, the gestation periods of larger species are longer than those of smaller species.*

Use **Science Integration Transparency 16** as you teach this lesson.

Why are mammals important to people today? *We get meat, milk, hides, and fur from them. We make pets of mammals, use them for recreation, and use them in scientific research. They are a critical part of the food web.*

Diversity of Mammals

Monotremes make up one order of mammals. Scientists have divided marsupials into about seven orders. Placental mammals make up 18 other orders. Each order has adaptations that help identify it. Characteristics and examples of several of the most common orders of placental mammals are given in **Figure 16-15.** Additional examples of placental mammals are shown in **Figure 16-16** on page 444.

Figure 16-15

The major orders of placental mammals are shown here.

A Chiroptera
Bats, mammals that have front limbs adapted for flying, make up this order. Active at night, they feed on fruit, insects, or blood.

B Insectivora
Moles and shrews are insect-eating animals. Most have long skulls, narrow snouts, clawed feet, and are small in size.

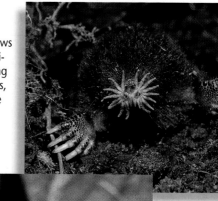

C Rodentia
Mice, rats, squirrels, beavers, porcupines, and gophers make up the largest order of mammals. They have chisel-like front teeth adapted for gnawing.

D Lagomorpha
Rabbits, hares, and pikas have long hind legs adapted for jumping and running. They have two pairs of upper incisors.

E Cetacea
Whales, dolphins, and porpoises have forelimbs that are modified into flippers for living in the ocean. They breathe through one or two blowholes on top of their heads. This order includes the largest mammals.

Cultural Diversity

Conservation To increase students' awareness of organisms indigenous to other countries, discuss protecting mammal wildlife. Mammals are perhaps the most exploited of all animals. African elephants, apes, pandas, manatees, dolphins, and rhinos are a few that are likely to become extinct within 50 years due to poaching and destruction of habitat. Have students bring articles from recent newspapers and science magazines about the exploitation of mammals and conservation efforts to save them. **L1**

F **Perissodactyla**
Horses, zebras, tapirs, and rhinoceroses have hooves with an odd number of toes. They are herbivores and have large, flat grinding molars. Their large skeletons are adapted for running.

G **Artiodactyla**
Deer, moose, camels, giraffes, pigs, and cows have hooves with an even number of toes. They are herbivores with large, flat molars and complex stomachs and intestines.

H **Carnivora**
Cats, dogs, bears, foxes, raccoons, otters, seals, sea lions, and walruses are carnivores with long, sharp canine teeth for tearing flesh. Most are carnivores, although a few may be omnivorous at times.

I **Proboscidea**
Elephants each have an elongated nose that forms a trunk and a pair of enlarged incisors that form tusks. Their skin is leathery.

J **Primates**
Humans, apes, and monkeys in this order have long arms with grasping hands and opposable thumbs. Their eyes face forward and they have large brains. Primates are omnivores.

443

Revealing Preconceptions

- Many people use the words *monkey* and *ape* interchangeably, but the words refer to different kinds of animals. The main differences between the two have to do with the structure of the skeleton, especially the skull, and the fact that most monkeys have tails, whereas apes do not.

- Camels do not store water in their humps. The humps contain fat, an energy-conserving adaptation that enables them to survive in the desert where food and water are scarce.

Inquiry Question

In what ways do retractable claws enable lions and other cats to survive?
They use their claws to climb trees and hold prey. They can be retracted, enabling the animals to walk and run faster.

Videodisc

STV: Biodiversity
Florida panther

46130
Humpback whale

46132
Manatees

46136
Mountain gorilla

46138

Visual Learning

Figure 16-15 See page 442. Look at the photographs and captions on this page and answer the questions. **Which order of mammals are plant eaters and have legs adapted for running and jumping?** *Lagomorpha* **Which order of mammals has the** **largest number of individuals?** *Rodentia* **Which order of mammals includes the largest mammals?** *Cetacea* **Which order of mammals have front limbs adapted for flying and are active at night?** *Chiroptera*

LEP **LS**

443

Content Background

- The Dermoptera, or gliding lemurs, have been placed in several different orders. Currently, scientists classify them in a separate order. More than half the gliding lemur's body is its tail. It looks like a furry kite as it spreads its limbs and leaps from tree to tree.

- Armadillos are the only members of the order Edentata that live in the United States. The armadillo's most distinguishing feature is its hardened skin, which is arranged in a pattern of either six or nine bands around the armadillo's body. The tough band acts like an armor to protect the animal from predators.

3 Assess

Check for Understanding

? FLEX Your Brain

Use the Flex Your Brain activity to have students explore MAMMALS.

📁 **Activity Worksheets,** p. 5

Reteach

Have students group the following characteristics under the headings Mammals, Birds, and Both Mammals and Birds: the most fully developed brains, endothermic, hollow bones, hair, feathers, shelled eggs, nurse their young, give birth to live young, extended parental care, outer ears, four-chambered heart, phylum Chordata.

Extension

📁 For students who have mastered this section, use the **Reinforcement** and **Enrichment** masters.

444

Figure 16-16

The orders of placental mammals also include these groups.

A **Sirenia**
Dugongs and manatees are strictly aquatic mammals. They have no back legs, and their front legs are modified into flippers. They graze on water plants in estuaries.

B **Edentata**
Giant anteaters, armadillos, and tree sloths such as this have no teeth and feed on insects.

C **Dermoptera**
Gliding lemurs have fangs and wide incisors.

D **Hyracoidea**
Hyraxes look like rabbits, but they have hooves and teeth like rhinoceroses. Hyrax species live on rocky slopes. They are herbivores.

E **Pholidota**
Pangolins are anteaters. They are covered with horny scales.

F **Tubulidentata**
Aardvarks dig up termites for food. They have only four or five teeth that are reduced to simple pegs without enamel.

444   Chapter 16 Birds and Mammals

Science Journal

Mammal Art Have students draw a mammal and label its parts, indicating the specialized adaptations that make it a mammal and help it survive. They should include information about how the mammal reproduces, what kind of skin and teeth it has, how it moves, how its lungs and heart function, etc. **L1**

Origin of Mammals

About 65 million years ago, dinosaurs became extinct. This opened up new habitats for mammals, and they began to branch out into many different species. Eventually, they became the most numerous animals on Earth. Some species died out, but others gave rise to modern mammals. Today, more than 4000 species of mammals have evolved from tiny, shrewlike creatures such as the one in **Figure 16-17** that lived 200 million years ago.

Mammals are important in maintaining a balance in the environment. Large carnivores such as lions help control populations of grazing animals. Bats help pollinate flowers, and some pick up plant seeds in their fur and distribute them. But mammals are in trouble today. As millions of acres of wildlife habitat are developed for shopping centers, recreational areas, and housing, many mammals are left without food, shelter, and space to survive.

Figure 16-17
Early insect-eating mammals may have looked like this.

Section Wrap-up

Review

1. Describe five characteristics of all mammals and explain how they allow mammals to survive in different environments.

2. Differentiate among placental mammals, monotremes, and marsupials.

3. Think Critically: Compare reproduction in placental mammals with that of monotremes and marsupials.

Skill Builder
Classifying

Classify the following animals into the three mammal groups: whales, echidnas, koalas, horses, elephants, opossums, kangaroos, rabbits, bats, bears, platypuses, and monkeys. Compare and contrast their characteristics. If you need help, refer to Classifying in the **Skill Handbook.**

USING MATH

The tallest land mammal is the giraffe at 5.6 m tall. Calculate your height in meters and determine how many of you it would take to be as tall as the giraffe.

16-2 Mammals   **445**

Skill Builder

Students should classify the organisms into three groups: placentals—horses, elephants, rabbits, bats, bears, monkeys; marsupials—koalas, opossums, kangaroos; monotremes—echidnas, platypuses. They should compare and contrast characteristics such as habitat, reproduction, nutrition, and locomotion. A chart or table would be the most useful method for students to use. **P**

 Assessment

Performance Assess students' abilities to classify another selection of mammals from Figures 16-15 and 16-16. Use the Performance Task Assessment List for Data Table in **PASC**, p. 37. **P**

Prepare

Section Background

Northern Pacific kelp forests are a habitat for all sorts of species. One is the sea urchin, which eats kelp. Sea otters hunt and eat sea urchins. Without otters, sea urchin populations can explode and devour so much kelp that they create "urchin barrens." Because kelp affects the waves and currents as well as providing habitat, otters are an important part of coastal ecosystems.

1 Motivate

Bellringer

Before presenting the lesson, display **Section Focus Transparency 60** on the overhead projector. Assign the accompanying **Focus Activity** worksheet. L1 LEP

2 Teach

Teacher F.Y.I.

Sea otters have been protected since the 1911 International Fur Seal Treaty. They also have been helped by the 1972 Marine Mammal Protection Act and the 1973 Endangered Species Act.

Content Background

California sea otters have only a 5 percent population growth rate, compared with a population growth rate of 20 percent a year for Alaskan sea otters.

SCIENCE & SOCIETY

ISSUE:
16•3 California Sea Otters

Science Words
sea otter

Objectives
- Explain the economic importance of sea otters.
- Determine whether the government should limit the range of California sea otters.

Figure 16-18

Sea otters choose rocks for smashing clams carefully and keep them for a while. Mother sea otters teach this behavior to their offspring.

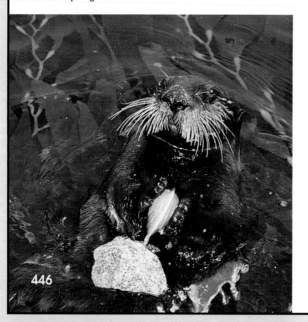

446

Sea Otters

Sea otters are placental mammals that belong to the order Carnivora. The size of basset hounds, they have a well-developed brain and good underwater vision, yet they rely mostly on touch to find their prey. Their whiskers are sensitive feelers and their front paws have pads. They live in kelp beds in the ocean along parts of the rocky coast of western North America and in a few coastal areas of Russia and Japan. Otters feed on animals that live in kelp beds, such as abalones, sea urchins, clams, and crabs. These food sources are also important cash crops for most area fishermen.

Sea otters are the only mammals, other than primates, known to use a tool for getting their food. **Figure 16-18** shows how a sea otter will carry a rock to the surface, lie on its back with the rock on its belly, and smash clams and other shellfish to get the meat inside. Every day, an otter must eat an amount of food equal to about 25 percent of its body weight. They eat so much because they are very lively and active. Sea otters clean and groom their fur with oils from skin glands. Air bubbles caught in their thick fur protect these skinny animals from the cold waters of the Pacific Ocean.

An Endangered Species

It is because of their fur that sea otters were hunted almost to extinction. By the 1900s, all that was left of the larger population of sea otters was a small band of survivors near Big Sur. Various laws have protected the otters, but today fewer than 2400 inhabit a 402.5-km stretch of shore south of San Francisco. California sea otters aren't increasing in numbers very quickly. Biologists are worried about the slow population growth. Pesticides and loss of habitat have been suggested as reasons for the slow growth.

Program Resources

2 Points of View

Limit the Range of Sea Otters

Abalones sell for high prices and sea urchins sell as delicacies in Japan. Sea urchins now surpass salmon as California's most valuable commercial fishery product. Commercial abalone and sea urchin fishermen believe the otters will destroy their industry. Some fishermen want to wipe out the otters completely. Most would like to limit otters to their current 402.5-km range, leaving the rest of the coast for fishing. They argue that the otters can be protected in this small range by controlling water pollution.

Allow Otters to Occupy More of Their Original Range

Biologists say complete recovery of the species depends on allowing the sea otters to reoccupy more of their original range. Northern kelp forests are a habitat for many species. The sea urchin is a big eater of kelp. Scientists have discovered that sea otters keep the urchins in check and help preserve the kelp forest habitat. The kelp-harvesting industry is a 50 million-dollar-a-year business in California. Without sea otters, the urchin populations explode and a great deal of kelp is destroyed. Kelp affects the waves and currents in the ocean, as well as providing habitats for many living things.

Figure 16-19

Kelp forests grow in the ocean. Sea urchins eat kelp, and the sea otters eat sea urchins. *How are each of these organisms classified?*

Section Wrap-up

Review

1. What are the characteristics of sea otters that make them economically important?

2. What is the greatest concern of biologists about the California sea otter population?

Explore the Issue

Should the government limit the range of the California sea otters to the 402.5-km stretch where they presently live? If yes, how can scientists increase their numbers in this area? If no, how will it affect commercial fishermen?

interNET
CONNECTION

Visit the Chapter 16 Internet Connection at Glencoe Online Science, **www.glencoe.com/sec/ science/life,** for a link to more information about sea otters and their habitat.

SCIENCE & SOCIETY

Explore the Issue

Yes answers should reflect that scientists can increase the sea otter numbers by forcing oil companies to move tanker lines much farther offshore, by stopping oil drilling in or adjacent to the otter range, and by controlling toxic runoff from agriculture. *No* answers should include that scientists believe that sea otters keep the urchins in check and help preserve the kelp forest upon which urchins depend for food. Commercial abalone and sea urchin fishermen believe otters will destroy their industry.

---

3 Assess

Check for Understanding

Have students answer questions 1 and 2 in the Section Wrap-up.

Reteach

Have students list the characteristics of sea otters. *Characteristics should include: small active mammals that consume large amounts of food (25 percent of their body weight each day), skinny, and thick fur.*

Extension

For students who have mastered this section, use the **Reinforcement** and **Enrichment** masters.

4 Close

Discuss the pros and cons of limiting the range of sea otters.

Section Wrap-up

Review

1. They use tools for smashing the sharp spines of sea urchins. By keeping sea urchins in check, they help preserve the kelp forest habitat upon which the urchins and other species depend.

2. Biologists are worried about the slow population growth due to pesticide use and loss of habitat.

Background

- Weddell seals spend about half their lives diving. The length of an average dive beneath the sea ice is about 20 minutes, but dives up to nearly an hour have been recorded. Weddells dive to 600m in search of the fish they eat. Northern elephant seals may be the deepest divers in the seal world. Scientists from the National Marine Mammal Laboratory in Seattle have attached microprocessor-controlled dive recorders to male elephant seals. Dives to more than 900 m were common, and one seal descended below 1500 m.

- As an adaptation for life in cold water, penguins and marine mammals have a thick layer of fat, called blubber, just under their skin. Blubber insulates these warm-blooded vertebrates and prevents heat loss to the environment. The layer of blubber on a Weddell seal's body is 10 to 15cm thick.

Teaching Strategies

IS **Interpersonal** Warm-blooded animals that live in cold climates exhibit characteristic adaptations to reduce heat loss to the environment. Have students work in pairs to design "hypothetical" polar mammals or birds that display physical and behavioral adaptations that would help conserve body heat. Have students present and describe their animals to the class. **L1**
COOP LEARN

People and Science

TANIA ZENTENO-SAVIN, *Marine Biologist*

On the Job

Q **Ms. Zenteno-Savin, tell us about your research on seals at the University of Alaska, Fairbanks.**

A I'm studying Weddell seals in Antarctica and northern elephant seals in California. Both these types of seals can dive to great depths and hold their breath for long periods of time. We know that when seals hold their breath, their hearts beat at a slower rate. I am studying several hormones in seals, trying to figure out if these hormones are involved in controlling heart rate in breath-holding seals, and if so, how they exert that control.

Q **What implications might your seal research have for people?**

A Physiologically, seals are not all that different from people. Gaining a better understanding of how hormones, breathing, and heart rate are related in seals might lead to a better understanding of the same thing in our bodies.

Q **What's it like to study seals in Antarctica?**

A For three months, I lived with several other researchers in a one-room hut

Personal Insights

on the sea ice, surrounded by colonies of Weddell seals. We often worked in subzero temperatures while we tagged, weighed, and measured seal pups, took blood samples for analysis, and studied young seals that were just beginning to dive.

Q **When did you first get interested in seals?**

A When I was in elementary school, I learned that seals are air-breathing mammals like us, but that they spend most of their lives in the water. The fact that seals can hold their breath for such a long time has always fascinated me.

Career Connection

Many large zoos and aquariums maintain seals, dolphins, and other marine mammals in captivity. Contact a zoo and ask for information on zoo-related careers working with marine mammals.

Create a poster to interest your classmates in one of the following careers:

- **Zookeeper**
- **Animal Trainer**
- **Zoo Educational Specialist**

Career Connection

Career Path Scientists who do research on marine mammals usually have a master's degree or Ph.D. in marine biology or a closely related field.

- High School: science and math
- College: biology and chemistry
- Post-College: Master's degree or Ph.D. in marine biology or ecology, wildlife biology, or animal physiology

For More Information

Students interested in careers working with animals at a zoo or aquarium could contact the American Zoo and Aquarium Association. To learn more about deep-diving animals, have students read "Going to Extremes," by Caroline Harding, *National Wildlife*, Aug./Sept. 1993, pp. 39-44.

Summary

16-1: Birds

1. Birds are endothermic animals that are covered with feathers and lay eggs. Their front legs are modified into wings.
2. The adaptations birds have for flight are: wings; feathers; a light, strong skeleton; and efficient body systems.
3. Birds lay eggs enclosed in hard shells. Most birds incubate their eggs until they hatch.

16-2: Mammals

1. Mammals are endothermic animals with hair. Female mammals have mammary glands that produce milk.
2. There are three groups of mammals. Mammals that lay eggs are monotremes. Mammals that have pouches for the development of their embryos are marsupials. Mammals whose offspring develop within a placenta inside a uterus are placental mammals.

16-3: Science and Society: California Sea Otters

1. Sea otters are carnivorous placental mammals that have well-developed brains and good underwater vision. They live in kelp beds and feed on abalones, sea urchins, clams, and crabs, destroying important commercial crops.
2. Sea otters keep sea urchins in check, helping to preserve the kelp forest habitat. Commercial abalone and sea urchin fishermen believe the otters will destroy their industry and want to limit the otters to their current 402.5-km range along the southern California coast.

Key Science Words

a. carnivore
b. contour feather
c. down feather
d. gestation period
e. herbivore
f. incubate
g. mammal
h. mammary gland
i. marsupial
j. monotreme
k. omnivore
l. placenta
m. placental mammal
n. preening
o. sea otter
p. umbilical cord

Reviewing Vocabulary

Match each phrase with the correct term from the list of Key Science Words.

1. plant eater
2. keep eggs warm
3. a mammal whose embryo develops inside a uterus
4. a feather birds use for flying
5. an animal that eats both plants and animals
6. a gland that produces milk
7. a feather that helps to insulate
8. activity of rubbing oil onto feathers
9. structure containing blood vessels that attaches the embryo to the placenta
10. pouched mammal

Chapter 16 Review 449

Summary

Have students read the summary statements to review the major concepts of the chapter.

Reviewing Vocabulary

1. e	**6.** h
2. f	**7.** c
3. m	**8.** n
4. b	**9.** p
5. k	**10.** i

 Assessment

Portfolio Encourage students to place in their portfolios one or two items of what they consider to be their best work. Examples include:
- Science Journal, p. 428
- Enrichment, p. 441
- Skill Builder, p. 445 P

Performance Additional performance assessments may be found in **Performance Assessment** and **Science Integration Activities.** Performance Task Assessment Lists and rubrics for evaluating these activities can be found in Glencoe's **Performance Assessment in the Science Classroom.**

GLENCOE TECHNOLOGY

Videodisc

Glencoe Life Science Interactive Videodisc
Side 2, Lesson 5
Structural Adaptations in Animals

11 to 9307

MindJogger Videoquiz

Chapter 16 Have students work in groups as they play the Videoquiz game to review key chapter concepts.

Checking Concepts

1. d	**6.** a
2. d	**7.** b
3. c	**8.** a
4. c	**9.** a
5. c	**10.** a

Understanding Concepts

11. To have enough energy to maintain body temperature and to fly, birds consume large quantities of food.

12. Placental mammals have a long period of development inside the mother. This provides food and protection for the embryo and a greater chance for survival.

13. Water birds have webbed feet for swimming. Shore birds have feet with long toes for wading through mud. Perching birds have claws for gripping onto branches. Birds of prey have clawed toes for catching prey.

14. Because the hollow bones and feathers of birds decompose easily, there are not many fossils to study.

15. The water helps support their weight.

Thinking Critically

16. Both birds and reptiles lay eggs. Reptiles lay large quantities of leathery-shelled eggs and do not incubate them. Birds lay fewer eggs at a time. The shells of bird eggs are hard, and the parents incubate them.

17. Hair traps air in land animals. The cetaceans probably have little hair because they live in water and are insulated with large amounts of fat.

18. Ducks have lighter wing bones than ostriches. Ostriches do not fly and

450

Checking Concepts

Choose the word or phrase that completes the sentence.

1. A(n) _____ is a flightless bird.
   a. duck   c. owl
   b. oriole   d. rhea
2. Wings of birds are not used for _____.
   a. flying   c. balancing
   b. swimming   d. eating
3. Birds crush and grind food in the
   _____.
   a. crop   c. gizzard
   b. stomach   d. small intestine
4. Omnivores are mammals that eat
   _____.
   a. plants
   b. animals
   c. plants and animals
   d. other animals
5. A _____ is an example of a marsupial.
   a. cat   c. kangaroo
   b. human   d. camel
6. Pouched mammals are called _____.
   a. marsupials   c. placental
   b. monotremes   d. chiropterans
7. _____ have mammary glands with no nipples.
   a. Marsupials   c. Placental mammals
   b. Monotremes   d. Lagomorphs
8. Teeth specialized for tearing are
   _____.
   a. canines   c. molars
   b. incisors   d. premolars
9. Bird eggs are different from reptile eggs in that they have _____.
   a. hard shells   c. albumen
   b. yolks   d. membranes
10. Humans and monkeys are _____.
    a. primates   c. edentates
    b. carnivores   d. cetaceans

Understanding Concepts

Answer the following questions in your Science Journal using complete sentences.

11. Why is it incorrect to tell someone who eats very little that he or she "eats like a bird"?

12. There are far more species of placental mammals than of marsupials or monotremes. Why do you think this is so?

13. Explain how the feet of birds are adapted to where they live.

14. Why is it difficult to study the origin of birds based on fossils?

15. Even dinosaurs were not as large as the biggest whales because their skeletons could not support that amount of weight. Infer how whales can grow to such a large size.

Thinking Critically

16. Discuss the differences in reproduction between birds and reptiles.

17. Give two reasons why whales have little hair.

18. Which type of bird would have lighter wing bones—ducks or ostriches? Explain.

19. What features of birds allow them to be fully adapted to life on land?

20. In what ways are birds and mammals similar?

Developing Skills

If you need help, refer to the **Skill Handbook.**

21. Classifying: You are a paleontologist studying fossil reptiles. One reptile appears to have hollow bones, a keeled

---

would be less likely to need this trait.

19. The amniotic egg, specialized beaks and feet, and wings allow land adaptation.

20. Both birds and mammals are endothermic and provide extended periods of care for their young. Both also can live in and adapt to nearly any environment found on Earth.

breastbone, and a short tail. How would you classify this animal?

22. **Observing and Inferring:** Look at the diagrams of animal tracks. Decide which track belongs to which type of mammal. Fill in the chart. Describe how each animal's foot is adapted to its environment.

Sample Data

Identifying Animal Tracks

Animal	Track	Adaptation
Bear	c	clawed toes
Beaver	a	webbed feet
Cheetah	g	clawed toes
Deer	b	feet for running, support
Horse	d	feet for running
Moose	e	feet for running, support
Raccoon	f	clawed toes

23. **Classifying:** You discover three new species of mammals. The traits of each species are as follows: Mammal 1—swims and eats plants; Mammal 2—flies and eats fruit; Mammal 3—runs and hunts. Classify each mammal into its correct group.

24. **Concept Mapping:** Complete the concept map describing the groups of mammals.

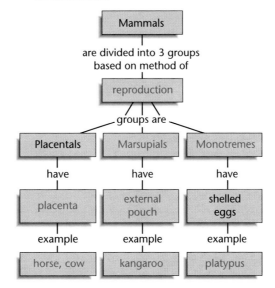

25. **Comparing and Contrasting:** Compare and contrast the teeth of herbivores, carnivores, and omnivores. How is each tooth type adapted to the animal's diet?

Performance Assessment

1. **Oral Presentation:** Research the birds in your area. Find out what they eat and the types of nests they build. Present your findings in a speech to your class.
2. **Letter:** Write to the National Wildlife Federation about endangered birds and mammals. Ask what is being done to save these animals.
3. **Song with Lyrics:** Write a song about the plight of California sea otters. Make sure you include two points of view in your song. Have a friend set your lyrics to music and perform it for the class.

Chapter 16 Review 451

Developing Skills

21. **Classifying** as a bird
22. **Observing and Inferring** a. beaver, b. deer, c. bear, d. horse, e. moose, f. raccoon, g. cheetah. Beavers' feet are adapted to living in water. The cheetah, raccoon, and bear have feet adapted for support and catching food. The deer and moose have feet adapted for support and running. The horse's hoof is adapted for running.
23. **Classifying** Mammal 1 belongs to the order Sirenia. Mammal 2 belongs to the order Chiroptera. Mammal 3 belongs to the order Carnivora.
24. **Concept Mapping** See student page.
25. **Comparing and Contrasting** Herbivores have large incisors for cutting off blades of grass. Carnivores have small incisors and large canine teeth to grip and tear food. Omnivores have canines and incisors to feed on both plants and animals.

Performance Assessment

1. Students should research common birds. Use the Performance Task Assessment List for Oral Presentation in **PASC**, p. 71. **P**
2. Students may decide to write about one particular animal. Use the Performance Task Assessment List for Letter in **PASC**, p. 67. **P**
3. Songs may be simple as long as the lyrics discuss the plight of California sea otters. Use the Performance Task Assessment List for Song with Lyrics in **PASC**, p. 79. **P**

Assessment Resources

 Reproducible Masters
Chapter Review, pp. 35-36
Assessment, pp. 101-104
Performance Assessment, p. 30

Glencoe Technology
- **Chapter Review Software**
- **Computer Test Bank**
- **MindJogger Videoquiz**

Chapter Organizer

Section	Objectives/Standards	Activities/Features
Chapter Opener		**Explore Activity:** Investigate a simple response. p. 453
17-1 Types of Behavior (1 session)*	1. **Distinguish** between innate and learned behavior. 2. **Recognize** reflex and instinctive actions and explain how they help organisms survive. 3. **Describe** and give examples of imprinting, trial and error, conditioning, and insight. National Content Standards: (5-8) UCP1, A1, A2, C2, C3, G1	**Using Math,** p. 455 **Connect to Chemistry,** p. 456 **MiniLAB:** How does insight help you solve problems? p. 458 **Skill Builder:** Observing and Inferring, p. 459 **Using Computers:** Spreadsheet, p. 459 **Activity 17-1:** Investigating Conditioning, pp. 460-461
17-2 Behavioral Adaptations (2 sessions)*	1. **Recognize** the importance of behavioral adaptations. 2. **Explain** how courtship behavior increases the chances of reproductive success. 3. **Evaluate** the importance of social behavior and cyclic behavior. National Content Standards: (5-8) UCP1, UCP3, UCP4, A1, B1, C2, C3, C5, E2, F5	**Using Math,** p. 464 **Science and Literature:** Honeybees, p. 465 **Science Journal,** p. 465 **MiniLAB:** How do earthworms respond to red light? p. 466 **Using Technology:** Artificial Insecticides, p. 467 **Problem Solving:** Abilities of Homing Pigeons, p. 468 **Skill Builder:** Using Variables, Constants, and Controls, p. 469 **Using Math,** p. 469
17-3 Science and Society Issue: Zoos and Captive Breeding (1 session)*	1. **Explain** the advantages of zoos. 2. **Determine** the advantages and disadvantages of captive breeding. National Content Standards: (5-8) A2, C4	**Explore the Issue,** p. 471 **Activity 17-2:** Observing Social Behavior in Ants, p. 472

* A complete Planning Guide that includes block scheduling is provided on pages 32T-35T.

Activity Materials

Explore	Activities	MiniLABs
page 453 clear piece of Plexiglas (~30 cm × 30 cm), wadded-up sheet of paper	pages 460-461 aquarium, glass cover for aquarium, thermometer, washed coarse gravel, tap water that has been aged 3 days, metric ruler, dish, food for guppies, fishnet, guppies, snails, freshwater plants, safety goggles, apron page 472 20 ants; large jar with wire, fine-mesh screen; soil; black paper; small, moist sponge; tap water; food (honey, sugar, bread crumbs) for ants; small can with lid; spoon or trowel; hand lens	page 458 chain made of 15 safety pins page 466 earthworm, moist paper towel, aluminum pie plate, flashlight, red cellophane, rubber band

Need Materials? Call Science Kit (1-800-828-7777).

Teacher Classroom Resources

Reproducible Masters	Transparencies	Teaching Resources
Study Guide, p. 65 Reinforcement, p. 65 Enrichment, p. 65 Activity Worksheets, pp. 100-101, 104 Science Integration Activity 17, Relationships Critical Thinking/Problem Solving, p. 21 Cross-Curricular Integration, p. 21	Section Focus Tansparency 61, Seeing Red? Teaching Transparency 33, Pavlov's Dogs	Glencoe Life Science Interactive Videodisc Life Science CD-ROM Spanish Resources English/Spanish Audiocassettes Cooperative Learning Resource Guide Lab Partner Lab and Safety Skills Lesson Plans
Study Guide, p. 66 Reinforcement, p. 66 Enrichment, p. 66 Activity Worksheets, p. 105 Lab Manual 41, Environment and Behavior Concept Mapping, p. 39 Science and Society Integration, p. 21 Multicultural Connections, pp. 37-38	Section Focus Transparency 62, Hide-'N-Seek Science Integration Transparency 17, Migration by Air, Land, and Sea Teaching Transparency 34, Learning, Chimp Style	**Assessment Resources** Chapter Review, pp. 37-38 Assessment, pp. 105-108 Performance Assessment in the Science Classroom (PASC) MindJogger Videoquiz Alternate Assessment in the Science Classroom
Study Guide, p. 67 Reinforcement, p. 67 Enrichment, p. 67 Activity Worksheets, pp. 102-103	Section Focus Transparency 63, Condor Comeback	Performance Assessment Chapter Review Software Computer Test Bank

Key to Teaching Strategies

The following designations will help you decide which activities are appropriate for your students.

L1 Level 1 activities should be within the ability range of all students, including those with learning difficulties.

L2 Level 2 activities should be within the ability range of the average to above-average student.

L3 Level 3 activities are designed for the ability range of above-average students.

LEP LEP activities should be within the ability range of Limited English Proficiency students.

LS These activities are designed to address different learning styles.

COOP LEARN Cooperative Learning activities are designed for small group work.

P These strategies represent student products that can be placed into a best-work portfolio.

GLENCOE TECHNOLOGY

The following multimedia resources are available from Glencoe.

Science and Technology Videodisc Series (STVS)
Animals
    March of the Spiny Lobster
    Migration of Killer Bees
Ecology
    Pheromone Rope
The Infinite Voyage Series
Insects: The Ruling Class
The Keepers of Eden

National Geographic Society Series
Newton's Apple: Life Sciences
STV: Rain Forest
Glencoe Life Science Interactive Videodisc
Structural Adaptations in Animals
Life Science CD-ROM

Teacher Classroom Resources

This is a representation of key blackline masters available in the Teacher Classroom Resources.

Teaching Aids

Section Focus Transparencies

61 SECTION FOCUS TRANSPARENCY   Section 17-1

SEEING RED?

Ever heard the phrase "seeing red"? It means getting so angry you're about to charge like a bull. But bulls don't really react one way or another to the color red. What do you think makes them charge?

1. If it isn't the color red, what do you think makes the bull charge?

2. Bulls don't have to be trained to charge at a cape. What kind of behavior is this? Reflex? Instinct? Conditioning? Explain your answer.

L1

62 SECTION FOCUS TRANSPARENCY   Section 17-2

HIDE-'N-SEEK

Can you find the chameleon? Sometimes it's not that easy. The chameleon can change its color to match the color of its surroundings.

1. How does changing color help the chameleon to avoid predators?

2. The chameleon survives by eating insects. How do you think its ability to change color might help it in this activity?

L2

63 SECTION FOCUS TRANSPARENCY   Section 17-3

CONDOR COMEBACK

Condors were disappearing from the face of the Earth, but then a few caring humans came to the rescue. By captive breeding, the number of condors has risen form 27 in 1987 to 68 in 1995.

1. What is captive breeding?

2. Why is it important to have enough males and females of a species for an experiment in captive breeding?

3. While the number of condors may have risen, some of their habitats have been destroyed. What effect could this have upon the condors?

L3

Science Integration Transparencies

17 SCIENCE INTEGRATION TRANSPARENCY

Migration by Air, Land, and Sea

L1

Teaching Transparencies

33. PAVLOV'S DOGS

L1

34. LEARNING, CHIMP STYLE

L2

Meeting Different Ability Levels

Study Guide

NAME   DATE   CLASS

Chapter 17
STUDY GUIDE   ● Types of Behavior

Complete the following crossword puzzle using words from the textbook.

Across

3. behavior that is modified by experience is called trial and _____

5. something inside an animal that causes it to act necessary for learning to occur

7. complex pattern of innate behavior, often having several parts

9. behavior that an organism is born with and which doesn't have to be learned

10. behavior that develops through experience

Down

1. a type of learning in which an animal forms a bond to another organism shortly after birth or hatching

2. the way an organism acts toward its environment

4. behavior that is modified when a response becomes associated with another stimulus

6. a form of reasoning that allows animals to use past experiences to solve new problems

8. an automatic response that does not involve the brain

Complete the following sentences using the terms below.

trial and error   conditioning   insight   reflex   learn
recognition   motivation   stimulus   instinct

1. Mating behavior of the male stickleback fish is an example of _____

2. Shivering is a _____ behavior.

3. Anything in the environment to which an organism responds is called a _____

4. Older quail will not crouch if a leaf falls because of their ability to _____

5. Imprinting is a learning pattern involving _____

6. Pavlov's dogs secreted saliva when a bell was rung even when no food was present because of _____

7. A hungry rat will be able to learn a maze if it receives food at the end. This is an example of _____

8. Riding a bike is learned by _____

9. Wolfgang Köhler's experiment showed that chimpanzees used _____ when they piled up boxes in order to reach bananas.

L1

Reinforcement

NAME   DATE   CLASS

Chapter 17
REINFORCEMENT   ● Types of Behavior

Tell which of the following examples are innate behavior and which are learned behavior. Write the word innate or learned in the space provided.

____ 1. Sowbugs move away from the light.

____ 2. A dog lies down when given a command to do so.

____ 3. A male stickleback fish protects its young.

____ 4. The pupil of a cat's eye increases in size in dim light.

____ 5. A seeing-eye dog leads a blind person.

____ 6. A dog barks twice for a bone.

____ 7. Spiders spin webs correctly on their first try.

____ 8. Birds fly south for the winter.

____ 9. A person rides a bicycle.

____ 10. A person shivers when cold.

Examples of types of behavior are given below. Identify each behavior and tell how it is helpful.

Behavior	Example	How it is helpful
	Fish will move to the place where they are fed when you tap on the tank.	
	A person taps the keys on a key ring until one works.	
	A duckling becomes socially attached to its mother.	
	A person sneezes.	

L2

Enrichment Worksheets

NAME   DATE   CLASS

Chapter 17
ENRICHMENT   ● Types of Behavior

Animal Behavior Research

The Austrian scientist Konrad Lorenz and the Dutch zoologist Niko Tinbergen were the founders of an approach to behavior called ethology. These two men shared the 1973 Nobel Prize in medicine and physiology with the German biologist Karl von Frisch. Ethology is the study of animal behavior in nature, and is concerned with innate behavior patterns. Instincts and reflexes are innate behaviors.

Learned behavior is acquired as the result of experience. There are several kinds of learned behavior. Habituation is a form of learning in which an animal learns not to respond to a stimulus. Animals may learn not to respond to stimuli that are repeated frequently and that are unimportant to them. Conditioning is another type of learned behavior. There are two kinds of conditioning. Ivan Pavlov, a Russian psychologist, knew that dogs respond to the smell of food by salivating. This

response is innate. Pavlov wanted to know if the salivation response could be caused by a less likely stimulus. Pavlov rang a bell at the same time a dog was given food. Eventually, the sound of the bell was enough to cause the dog to salivate. This type of conditioning is known as classical conditioning.

The American psychologist B.F. Skinner used pigeons to study another type of conditioning. He realized that if a certain behavior were followed by a reinforcement or reward, the behavior was more likely to be repeated in the future. To study this phenomenon, he invented the Skinner box. The box contained a typewriter-like key connected to a food supply. If the pigeon pecked at the key, it would be rewarded with food. This type of learning is called operant conditioning. If the response is reinforced with a reward, the behavior tends to be repeated.

Determine if the following behaviors are innate or learned. If the behavior is innate, determine if the behavior is an instinct or a reflex. If the behavior is learned, determine if the behavior is habituation, classical conditioning, or operant conditioning.

1. The first time a squirrel sees a nut, it tries to bury it.

2. Jane touches a hot curling iron and jerks her hand away.

3. John is startled by a loud noise outside his window. The same noise is repeated several times, but John no longer responds.

4. Cindy is teaching her dog tricks. When the dog rolls over at her command, the rewards it with praise.

5. A planaria moves away from bright light. It curls up when electrically shocked. When a bright light is closely followed by an electric shock, the planaria curls up. Eventually, the light alone causes the planaria to curl.

6. Kittens stalk imaginary prey.

L3

452C

Hands-On Activities

Science Integration Activity

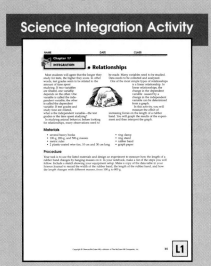

Chapter 17 · INTEGRATION — **Relationships**

L1

Lab Manual

Chapter 17 · LABORATORY MANUAL — **Environment and Behavior 41**

FIGURE 1. PILL BUGS (A) ROLL INTO A BALL (B) WHEN TOUCHED.   FIGURE 2.

L2

Assessment

Performance Assessment

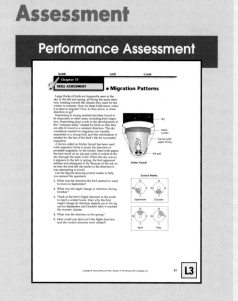

Chapter 17 · SKILL ASSESSMENT — **Migration Patterns**

L3

Enrichment and Application

Critical Thinking/ Problem Solving

Chapter 17 · CRITICAL THINKING — **Animal Behavior**

Sign Stimuli

L2

Cross-Curricular Integration

Chapter 17 · LIBRARY SCIENCE CONNECTION — **Animal Behavior**

Animals That Help People

L1

Science and Society Integration

Chapter 17 · SCIENCE & SOCIETY — **Animal Behavior**

Helping Sea Turtles

L2

Multicultural Connections

Chapter 17 · MULTICULTURAL CONNECTIONS — **Lemurs—Madagascar's Endangered Ghosts**

L2

Concept Mapping

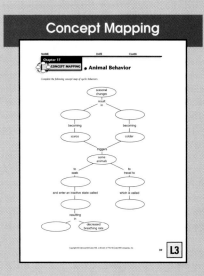

Chapter 17 · CONCEPT MAPPING — **Animal Behavior**

L3

452D

Animal Behavior

CHAPTER OVERVIEW

Section 17-1 This section discusses reflexes and instincts as innate behaviors. Kinds of learning, including imprinting, trial and error, conditioning, and insight, and examples of each are presented for students.

Section 17-2 This section describes territorial behavior, courtship behavior, social behavior, communication, and cyclic behavior.

Section 17-3 Science and Society The redesigning of zoos and captive breeding programs are discussed. The Explore the Issue feature asks whether captive breeding programs should be regulated.

Chapter Vocabulary

behavior	social
innate	behavior
behavior	society
reflex	communica-
instinct	tion
learning	pheromone
imprinting	cyclic
trial and error	behavior
motivation	circadian
conditioning	rhythm
insight	migration
territory	captive
aggression	breeding
courtship	
behavior	

Theme Connection

Stability and Change Complex patterns of behavior have evolved in animals to help them compete in a variety of environments.

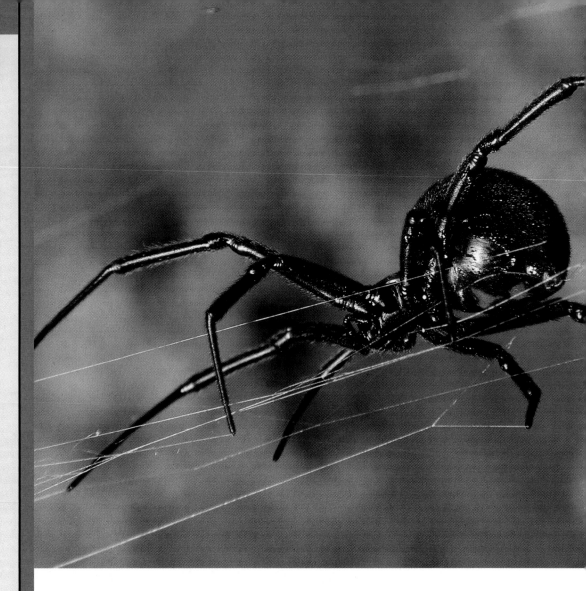

Previewing the Chapter

Section 17-1 Types of Behavior
▶ Innate Behavior
▶ Learned Behavior

▶ Social Behavior
▶ Communication
▶ Cyclic Behavior

Section 17-2 Behavioral Adaptations
▶ Instinctive Behavior Patterns
▶ Territorial Behavior
▶ Courtship Behavior

Section 17-3 Science and Society
Issue: Zoos and Captive Breeding

452

Chapter
17

Animal Behavior

A spider spins a web to catch its food. A bird makes a nest. Human infants cry when they are born and begin to focus on faces and smile when they are four weeks old. Do organisms learn these actions or do they occur automatically? In the activity, find out about things you do. Then in the chapter that follows, learn more about animal behavior.

EXPLORE ACTIVITY

Investigate a simple response.

1. Obtain a piece of clear Plexiglas and a wadded-up sheet of paper from your teacher.
2. Have a partner hold the Plexiglas a few centimeters from his or her face.
3. Stand about one meter away and gently toss the paper ball toward your partner's face.
4. Tell your partner not to blink and toss the paper ball again.

Observe: How does the tossed paper affect your partner's eyes? In your Science Journal, describe this behavior. Is this behavior learned or does it occur without learning?

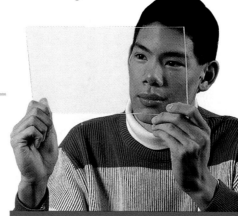

Previewing Science Skills

▶ In the **Skill Builders,** you will observe, infer, and use variables, constants, and controls.

▶ In the **Activities,** you will observe and collect and record data.

▶ In the **MiniLABs,** you will hypothesize, experiment, and record data.

453

Prepare

Section Background
- The scientific study of animal behavior as it occurs in the organism's natural environment is called *ethology*.
- Konrad Z. Lorenz, a founder of ethology, studied the behavior of birds and developed a theory of animal behavior that stressed its inherited aspects. Lorenz and two other ethologists, Karl von Frisch of Austria and Nikolaas Tinbergen of the Netherlands, received the 1973 Nobel Prize for their work.

Preplanning
- Obtain one 4-L aquarium, a thermometer, glass aquarium top, metric ruler, dish, food for fish, and a fish net for each group to conduct Activity 17-1. Purchase sand, water plants (*Elodea*), guppies, and snails for each group.
- Obtain safety pins for the MiniLAB.

1 Motivate

Bellringer
Before presenting the lesson, display **Section Focus Transparency 61** on the overhead projector. Assign the accompanying **Focus Activity** worksheet. L1 LEP

Tying to Previous Knowledge
Students are familiar with pet behaviors. Ask them to relate instances of feeding behavior and evidence of communication with other animals and with humans. Review the definitions of the terms *stimulus* and *response*.

17•1 Types of Behavior

Science Words
behavior
innate behavior
reflex
instinct
learning
imprinting
trial and error
motivation
conditioning
insight

Objectives
- Distinguish between innate and learned behavior.
- Recognize reflex and instinctive actions and explain how they help organisms survive.
- Describe and give examples of imprinting, trial and error, conditioning, and insight.

Innate Behavior

When you come home from school, does your dog run to meet you, barking and wagging its tail? After you play with him a while, does he sit at your feet and watch every move you make? Why does he do these things? Dogs are pack animals that generally follow a leader. They have been living with people for about 50 000 years. Dogs treat people as part of their own pack.

Behavior is the way an organism acts toward its environment. You may recall from Chapter 1 that anything in the environment to which an organism reacts is called a stimulus. You are the stimulus that causes your dog to bark and wag its tail. Your dog's reaction to you is a response. But how does your dog know when to wag and when to growl? Was this behavior learned or did your dog behave this way on his own?

Innate Behavior Is Inherited

Dogs and cats are quite different from one another in their behavior. They were born with certain behaviors, and they have learned others. A behavior that an organism is born with is an **innate behavior.** Such behaviors are inherited and they do not have to be learned.

Innate behavior patterns are usually correct the first time an animal responds to a stimulus. Kittiwakes are seabirds that nest on narrow ledges, as you can see in **Figure 17-1.** The chicks stand still as soon as they hatch. The chicks of a related bird, the herring gull, which nests on the ground, move around as soon as they can stand. Kittiwake chicks can't do this because one step could mean instant death. They hatch already knowing they must not move around.

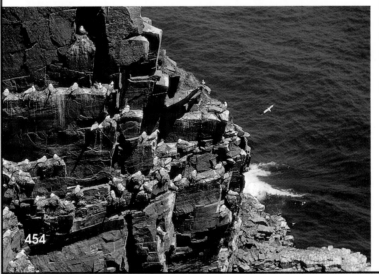

Figure 17-1

Some behaviors an animal is born with are necessary for survival. Kittiwakes, shown here, have behavioral adaptations for nesting on narrow ledges.

454

Program Resources

 Reproducible Masters
Study Guide, p. 65 L1
Reinforcement, p. 65 L1
Enrichment, p. 65 L3
Cross-Curricular Integration, p. 21
Activity Worksheets, pp. 5, 100-101, 104 L1
Science Integration Activities, pp. 33-34
Critical Thinking/Problem Solving, p. 21

Transparencies
Section Focus Transparency 61 L1
Teaching Transparency 33

Animals that have short life spans often have patterns of innate behavior. The lives of most insects, for example, are too short for the young to learn from the parents. In many cases, the parents have died by the time the young hatch. And yet, every insect reacts automatically to its environment. A moth will fly toward a light, and a cockroach will run away from it. They do not have to spend time learning what to do. Innate behavior allows animals to respond quickly to a stimulus in their environment without taking the time to choose a proper response.

Reflexes

The simplest innate behaviors are reflex actions. A **reflex** is an automatic response that does not involve the brain. Sneezing, shivering, yawning, jerking your hand away from a hot surface, and blinking your eyes when something is thrown toward you are all reflex actions. All animals have reflexes.

During a reflex, a message passes from a sense organ along the nerve to the spinal cord and back to the muscles. The message does not go to the brain. You are aware of the reaction only after it has happened. When you respond reflexively, you do not think about how you will respond. Your body reacts on its own, without your thought processes. A reflex is not the result of conscious thinking.

Instincts

An **instinct** is a complex pattern of innate behavior. Have you ever watched a spider spin a web? Spinning a web is complicated, and yet spiders spin webs correctly on the first try. **Figure 17-2** shows how sea turtles instinctively head for the sea as soon as they hatch. Unlike reflexes, instinctive behaviors may have several parts and take weeks to complete. Instinctive behavior begins when the animal recognizes a stimulus and continues until all parts of the behavior have been performed.

USING MATH

As many as two million spiders may live in one hectare of undisturbed meadow and catch as many as 500 insects per web per day. What is the total number of insects trapped in the webs of the spiders in one hectare per day?

Figure 17-2

Baby sea turtles move toward the ocean as soon as they hatch. *What stimulus do you think baby sea turtles are responding to when they do so?*

17-1 Types of Behavior **455**

2 Teach

USING MATH

one billion

Content Background

An example of instinctive behavior is found in the digger wasp. After the digger wasp comes from her underground pupa, she lives only a few weeks. In this time, she digs a hole for a nest, hunts prey to nourish her young, and lays her eggs. Her parents died several months before she emerged, so she was not taught these behaviors.

Activity

 Visual-Spatial Arrange pictures showing types of animal behavior on the bulletin board. Use pictures of behaviors associated with food-getting, protection, and reproduction. Have students list the kind of behavior—innate or learned—and the type—reflex, instinct, imprinting, trial and error, conditioning, or insight. L1

GLENCOE TECHNOLOGY

 Videodisc

STVS: Animals
Disc 5, Side 1
March of the Spiny Lobster (Ch. 11)

Use **Critical Thinking/ Problem Solving**, p. 21, as you teach this lesson.

Visual Learning

Figure 17-2 **What stimulus do you think baby sea turtles are responding to when they do so?** *Scientists disagree about this, but one explanation is that they head for the white edge of waves coming in to shore; turtles hatch at night and this may be the only light they see.*

LEP

Inclusion Strategies

Visually Impaired Allow students to collect crickets in an aquarium containing moist soil and lettuce or grass. Students can observe the crickets through a hand lens to note how their chirping sound is produced.

Figure 17-3
Newly hatched ruffed grouse chicks crouch down when something flies above their heads. Eventually they learn to crouch only when a predator flies overhead.

CONNECT TO
CHEMISTRY

When a salmon hatches, the chemical makeup of the water around it is imprinted in its memory. As an adult, the salmon follows streams with a chemical makeup that most closely corresponds to this memory. *Describe* why this imprinting is important for salmon survival.

Learned Behavior

All animals have both innate and learned behaviors. Learned behavior develops during an animal's lifetime. **Learning** is the result of experience or practice. You aren't born knowing how to play the piano; you must learn this behavior through practice. In animals, the more complex their brains, the more their behavior is the result of learning. Instinct almost completely determines the behavior of insects, spiders, and other arthropods. But fish, reptiles, amphibians, birds, and mammals all can learn.

Why is learning important for animals? Learning allows animals to respond to new situations. In changing environments, animals that have the ability to learn new behavior are more likely to survive. This is especially important in animals with long life spans because the longer an animal lives, the more likely it is that the environment in which it lives will change. Learning can modify instincts. For example, grouse and quail chicks leave their nest the day they hatch. They can run and find food, but they can't fly. When something moves above them, they crouch down and keep perfectly still until the danger is past, as you can see in **Figure 17-3.** They will crouch without moving even if the falling object is only a leaf. Older birds have learned that leaves will not harm them, but they, too, freeze when a hawk moves overhead.

Imprinting

Learned behavior includes imprinting, trial and error, conditioning, and insight. Have you ever seen young ducks following their mother? This is an important behavior because the adult bird has had more experience in finding food, escaping predators, and getting along in the world. **Imprinting** is a type of learning in which an animal forms a social attachment to another organism within a specific time period after birth or hatching.

Figure 17-4 shows Konrad Lorenz, an Austrian naturalist. Lorenz developed the concept of imprinting. Working with geese, he discovered that a gosling follows the first moving object it sees after hatching. It recognizes the moving object as its parent. It later recognizes similar objects as members of its own species. This behavior works well when the first moving object a young goose sees is an adult female goose. But goslings hatched in an incubator may see a human first and may imprint on him or her. Animals that become imprinted toward animals of another species never learn to recognize members of their own species.

Figure 17-4

Young goslings may imprint on any object or organism present during hatching. Konrad Lorenz is shown here with imprinted goslings.

Content Background

Explain that imprinting is an example of a behavior that is both innate and learned. Following and becoming attached to a moving object after birth or hatching is innate behavior. Recognizing that object is learned behavior.

Activity

LS **Kinesthetic** Assemble a large puzzle and then separate it into 12 or 14 sections (one section for every two students). Take the pieces of each section apart, place them in paper bags, and give one to each group. After each group has put its section of the puzzle together, have students fit their sections together. Discuss what type of behavior this is. *learned* **L1** **LEP**

COOP LEARN

Trial and Error

Can you remember when you learned to ride a bicycle? You probably fell many times before you learned to balance on the bicycle. But after a while you could ride without having to think about it. You have many skills that you have learned through trial and error. Skating and riding a bicycle are just a few.

Behavior that is modified by experience is called **trial and error** learning. Both invertebrates and vertebrates learn by trial and error. When baby chicks first learn to feed themselves, they peck at many spots before they get any food. As a result of trial and error, they soon learn to peck only at grain.

A particular stimulus may cause a different response in the same animal at different times. When your cat sees food, it will eat it if it's hungry, but if it has just eaten, it will ignore the food. For a hungry rat, shown in **Figure 17-5,** the motive to learn a maze may be the food at the end of the maze. **Motivation** is something inside an animal that causes the animal to act. It is necessary for learning to take place.

Figure 17-5

Perhaps you were motivated to learn to ride a bicycle because you wanted to ride with your friends. *Why does a rat learn how to find food inside a maze?*

17-1 Types of Behavior   **457**

GLENCOE TECHNOLOGY

💿 CD-ROM

Life Science CD-ROM

Have students perform the interactive exploration for Chapter 17 to reinforce important chapter concepts and thinking processes.

Inquiry Question

If a hungry rat eats a small amount of poisoned food that makes it ill, it will never eat poisoned food again. How can you explain this? *The rat's behavior changes because of trial and error learning.*

Visual Learning

Figure 17-5 **Why does a rat learn to find food inside a maze?** *motivation* **LEP** **LS**

Community Connection

Training Animals Have students interview a veterinarian or an animal trainer. Have them prepare a list of questions about what it is like to work in such a profession and what kind of schooling is required. **P**

457

MiniLAB

How does insight help you solve problems?

Procedure

1. Obtain 15 safety pins in a chain.
2. Make a hypothesis as to the fewest number of times you can open and close pins to break the chain into the following: 5 pins attached; 4 pins attached; 3 pins attached; 2 pins attached; and 1 alone. All pins must be closed at the end.
3. Test your hypothesis and compare it with those of your classmates.

Analysis
1. How was insight used in solving the problem?
2. What can you conclude about insight learning?

Conditioning

Animals often learn new behaviors by **conditioning.** In conditioning, behavior is modified so that a response previously associated with one stimulus becomes associated with another. Russian scientist Ivan P. Pavlov was the first person to study conditioning. He knew that the sight and smell of food made hungry dogs secrete saliva. Pavlov added another stimulus. He rang a bell when he gave the dogs food. The dogs began to connect the sound of the bell with food. Then Pavlov rang the bell without giving the dogs food. The dogs secreted saliva when the bell was rung even though he did not show them food. The dogs were conditioned to respond to the bell, shown in **Figure 17-6.**

American psychologist John B. Watson demonstrated that responses of humans can also be conditioned. In one experiment, he struck a metal object each time an infant touched a furry animal. The loud noise frightened the child. In time, the child became frightened by the furry animal when no sound was made.

Figure 17-6

Pavlov demonstrated conditioning in dogs in 1900.

B Ringing a bell each time he presented food to a dog established a connection between the bell and the food. The dog salivated when it smelled food and a bell was rung.

C Eventually the dog would salivate just at the sound of the bell. It had become conditioned to respond to the bell as if it were food.

A Pavlov noted that when dogs smell food, they salivate, a reflex.

458 Chapter 17 Animal Behavior

Insight

But how does behavior learned in the past help an animal when it is confronted by a new situation? Suppose you are given a new math problem to solve. Do you use what you have previously learned in math to solve the problem? If you have, then you have used a kind of learned behavior called insight. **Insight** is a form of reasoning that enables animals to use past experiences to solve new problems. In Wolfgang Kohler's experiments with chimpanzees, a bunch of bananas was placed too high for the chimpanzees to reach. Chimpanzees piled up boxes found in the room, climbed up on them, and reached the bananas, as you can see in **Figure 17-7**. Much of adult human learning also is based on insight.

Figure 17-7
Chimpanzees use insight to solve problems.

Section Wrap-up

Review

1. Compare innate behavior with learned behavior.

2. Compare a reflex with an instinct.

3. Describe the four types of learned behavior.

4. **Think Critically:** Use what you know about conditioning to explain how the term *mouthwatering food* came about.

Skill Builder
Observing and Inferring

Mammals have many ways of communicating. Observe a pet dog, cat, guinea pig, or hamster for a day. In your Science Journal, record the ways the mammal communicates its needs to humans or other mammals. Infer what each of the behaviors mean. If you need help, refer to Observing and Inferring in the **Skill Handbook.**

> **Using Computers**
>
> **Spreadsheet** Make a spreadsheet of the behaviors in this section. Sort the behaviors according to whether they are innate or learned behaviors. Then identify the type of innate or learned behavior.

Extension

📁 For students who have mastered this section, use the **Reinforcement** and **Enrichment** master.

4 Close

•MINI•QUIZ•

Use the Mini Quiz to check students' recall of chapter content.

1. **The way an organism acts toward its environment is _____ .** *behavior*
2. **_____ is something inside an animal that causes it to act.** *Motivation*
3. **Behavior modified by experience is _____ learning.** *trial and error*

Section Wrap-up

Review

1. Innate behavior is behavior an animal is born with. Learned behavior develops through experience.
2. A reflex is an automatic response that does not involve the brain. An instinct is a complex pattern of innate behavior.
3. Student responses should reflect information on pages 456-459.
4. **Think Critically** The response to the stimulus of food is production of saliva. Upon conditioning, the sight or smell of food can cause the production of saliva.

Using Computers			
Innate Behaviors		**Learned Behaviors**	
Type	**Example**	**Type**	**Example**
reflex	sneezing, blinking eyes	imprinting	ducklings following mother
instinct	spider spinning web, turtles running to sea	trial and error conditioning	riding a bike a specific response to a specific stimulus

Skill Builder

Student journal entries will be determined by the type of animal observed and the behavior observed. **P**

✓Assessment

Performance Have students write a description in their Science Journals of what kind of learning takes place when crows in a farmer's fields are no longer afraid of a scarecrow that has been in place for two weeks. Use the Performance Task Assessment List for Writing in Science in **PASC,** p. 87. **P**

Purpose
IS Interpersonal Students will demonstrate conditioning in fish. **L1** **COOP LEARN**

Process Skills
observing and inferring, experimenting, interpreting data, recognizing cause and effect, communicating, sequencing, making and using tables, forming a hypothesis, separating and controlling variables

Time
50 minutes; then 5 minutes each day for two to three weeks

Materials
• To age tap water, allow it to stand uncovered for three days.
• Possible stimuli for students to use are tapping on the tank, turning on a light, ringing a bell, or playing music.
• Caution students not to overfeed the fish.

Alternate Materials Small aquariums may be set up using 4-L glass jars or 3-L drink bottles.

Safety Precautions
Caution students to use extreme care when working with live animals.

Possible Hypotheses
Most student hypotheses will reflect that fish can be conditioned using a stimulus other than food.

📁 **Activity Worksheets,** pp. 5, 100-101

Activity 17-1

Design Your Own Experiment
Investigating Conditioning

How do fish locate and obtain their food? Can you predict how behaviors in fish can be modified?

PREPARATION

Problem
How can fish be conditioned to food?

Form a Hypothesis
Based on your observations, state a hypothesis about how fish can be conditioned to respond to a stimulus involving food.

Objectives
• Observe the behavior of fish in response to food.
• Design and carry out an experiment to demonstrate conditioning in fish.

Possible Materials
• aquarium
• glass cover for aquarium
• thermometer
• water heater for aquarium
• washed coarse sand
• tap water aged three days
• metric ruler
• dish
• food for guppies
• fish net
• guppies
• snails
• water plants

Safety Precautions 🥽 🐾
Protect clothing and eyes, and be careful when handling live animals.

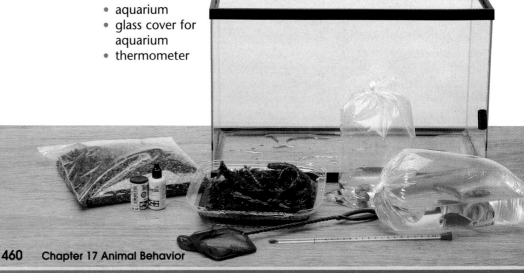

PLAN THE EXPERIMENT

1. As a group, agree upon and write out the hypothesis statement.
2. As a group, list the steps that you need to take to test your hypothesis. Be specific; describe exactly what you will do at each step. List your materials.
3. Identify how you will condition the fish. Will you feed them at the same time each day, or at the same place? What other stimulus will you condition the fish to respond to?
4. How will you determine that conditioning has taken place?
5. Prepare a data table in your Science Journal.

Check the Plan
1. Read over your entire experiment to make sure that all steps are in logical order.

2. Identify any constants, variables, and controls of the experiment.
3. How long will you conduct the conditioning experiment? How often will you feed the fish during that time?
4. Record the behavior of the fish in your data table.
5. *Make sure your teacher approves your plan before you proceed.*

DO THE EXPERIMENT

1. Carry out the experiment as planned.
2. While the experiment is going on, write down any observations that you make and complete the data table in your Science Journal.

Analyze and Apply
1. **Compare** your results with those of other groups.
2. **Describe** the stimulus you used to condition the fish.

3. **Explain** how long it took the fish to become conditioned. Was your hypothesis supported by your data?
4. Write a brief conclusion based on your data.

Go Further

Based on your data and observations, design an experiment using a stimulus to condition a family pet to sit up to ask for a treat.

17-1 Types of Behavior   461

Assessment

Performance Design an experiment to show how you could condition your dog to return a ball or a Frisbee to you. Carry out the experiment. *The dog could be rewarded with a treat for returning the ball or Frisbee.* Use the Performance Task Assessment List for Designing an Experiment in **PASC**, p. 23. **P**

PLAN THE EXPERIMENT

Possible Procedures
Fish may be trained to come to the top of the aquarium in response to whatever stimulus the group chooses to use instead of the presence of food.

Teaching Strategies
Tying to Previous Knowledge Most students have a knowledge of fish behavior and how to train pets. They may have trained a dog to sit up or to shake hands using food as the stimulus.

DO THE EXPERIMENT

Expected Outcome
Most results will reflect that the fish can be conditioned using a stimulus.

Analyze and Apply
1. Student results should be consistent with those of other groups.
2. Answers will vary.
3. The fish may be conditioned in two to three weeks. Answers will vary.
4. The fish were trained to come to the top of the aquarium in response to whatever stimulus the group used instead of the presence of food. Normally, the response would be produced by the presence of food.

Go Further

Students may design an experiment that would use a treat as a stimulus for sitting up.

Prepare

Section Background

- In territorial species, only individuals with territories mate. This helps ensure an adequate food supply for offspring. It also passes on the genes of the best-adapted individuals to the next generation.
- Animals use a variety of navigational devices to find their way. Honeybees use polarized light. Birds use the sun much as humans use a compass and map. Many birds migrate at night by observing the stars.
- One of the main characteristics of a society is the division of labor among its members.
- Some animal societies are organized into different levels of authority, such as a pecking order within a society of chickens.

Preplanning

Obtain earthworms, red cellophane, flashlights, and aluminum pans for the Mini-LAB.

1 Motivate

Bellringer

Before presenting the lesson, display **Section Focus Transparency 62** on the overhead projector. Assign the accompanying **Focus Activity** worksheet. L1 LEP

Tying to Previous Knowledge

Students will be familiar with aggressive and territorial behaviors of pets and other animals. Discuss familiar examples of territorial displays and threatening behaviors.

17•2 Behavioral Adaptations

Science Words

territory
aggression
courtship behavior
social behavior
society
communication
pheromone
cyclic behavior
circadian rhythm
migration

Objectives

- Recognize the importance of behavioral adaptations.
- Explain how courtship behavior increases the chances of reproductive success.
- Evaluate the importance of social behavior and cyclic behavior.

Instinctive Behavior Patterns

Many types of animal behavior are the result of complex patterns of innate behavior. Courtship and mating behavior within most animal groups is an example of an instinctive ritual that helps animals of one species recognize one another as possible mates. Animals also protect themselves and their sources of food by defending their territories. Instinctive behavior patterns such as these are the result of natural selection. Individual animals that did not show these behaviors died or failed to reproduce. Instinctive behavior, just like hair color, is inherited from an animal's parents.

Territorial Behavior

Many animals set up territories for feeding, mating, and raising young. A **territory** is an area that an animal defends from other members of the same species. Ownership of a territory is set up in different ways. Songbirds sing to set up territories. Sea lions bellow and squirrels chatter. Other animals leave scent marks. Some patrol the area to warn intruders. Why do animals defend their territories? Territories are areas that contain food, shelter, and potential mates. Animals need these things in order to survive. Defending territories is an instinctive behavior that improves the chances of survival for the offspring of an animal. **Figure 17-8** illustrates one way in which animals defend their territories.

Figure 17-8

Male sea lions patrol the area of the beach where their harem of females rests. Males ignore neighboring males with marked territories, but will attack and drive away any unattached male that tries to enter the territory.

462   Chapter 17 Animal Behavior

Program Resources

📁 **Reproducible Masters**
Study Guide, p. 66 L1
Reinforcement, p. 66 L1
Enrichment, p. 66 L3
Science and Society Integration, p. 21
Activity Worksheets, pp. 5, 105 L1
Concept Mapping, p. 39
Multicultural Connections, pp. 37-38
Lab Manual 41

🖊 **Transparencies**
Section Focus Transparency 62 L1
Teaching Transparency 34
Science Integration Transparency 17

Aggression

Have you ever watched a dog approach another dog eating a bone? What happens to the appearance of the dog with the bone? The hair stands up on its back, the lips curl, and the dog makes growling noises. This behavior is aggression. **Aggression** is a forceful act used to dominate or control another animal. Fighting and threatening are aggressive behaviors animals use to defend their territories, protect their young, or to get food.

Aggressive behaviors seen in birds include letting the wings droop below the tail feathers, taking another's perch, and thrusting the head forward in a pecking motion. These behaviors are intended to avoid physical contact. Fighting wastes valuable energy, and a missing feather or two can greatly reduce a bird's ability to fly.

Animals seldom fight to the death. They rarely use their teeth, beaks, claws, or horns to fight members of their own species. These structures are used for killing prey or for defense against members of another species. To avoid being killed, a defeated animal shows submission by crouching down or retreating, as shown in **Figure 17-9**.

Figure 17-9

Wolf puppies show their submission to adult males by making themselves as small as possible. *How does a pet dog respond when its owner reprimands it for poor behavior? Do you see any similarities in these two responses?*

Courtship Behavior

You have probably seen a male peacock spread the beautiful feathers on his lower back. The male frigate bird in **Figure 17-10** has a bright red pouch on his throat that takes about 25 minutes to blow up. A male sage grouse fans his tail, fluffs his feathers, and blows up his two air sacs. These are examples of a behavior that animals perform before mating. This type of behavior is called **courtship behavior.** Courtship behaviors allow male and female members of a species to recognize each other. They also allow males and females to be ready to mate at the same time. This helps ensure reproductive success.

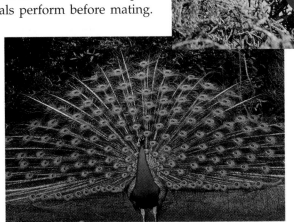

Figure 17-10

The male peacock (left) and the male frigate bird (above) are both trying to attract females with their displays.

17-2 Behavioral Adaptations   463

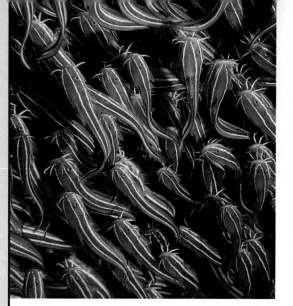

Figure 17-11
A predator of these fish has a hard time identifying just one fish to try to catch.

USING MATH

You can estimate the outside temperature by counting the number of times a cricket chirps. Use the equation

$$t = \frac{c}{4} + 37,$$

where *c* is the number of times a cricket chirps in one minute, and *t* is the temperature in degrees Fahrenheit. Find the temperature when a cricket chirps 80 times per minute.

In most species, the males perform courtship displays to attract a mate. Some courtship behaviors allow males and females to find each other across distances. Male fireflies produce different patterns in their flashes of light. A female of the same species recognizes the pattern and flashes back.

Social Behavior

Animals live together in groups for several reasons. One reason is that there is safety in large numbers, as you can see in **Figure 17-11.** A wolf pack is less likely to attack a herd of musk oxen than an individual musk ox. In some groups, large numbers of animals help keep each other warm. Penguins in Antarctica huddle together against the cold winds. Migrating animals in large groups are less likely to get lost than if they traveled alone.

Interactions among organisms of the same species are examples of **social behavior.** Social behaviors include courtship and mating, caring for the young, claiming territories, protecting each other, and getting food. These behaviors provide advantages for survival of the species.

Insects such as ants, bees, and termites live together in societies. A **society** is a group of animals of the same species living and working together in an organized way. Each member has a certain job. Usually, there is a female that lays eggs, a male that fertilizes the eggs, and workers that do all the other jobs in the society.

Some societies are organized by dominance. Wolves usually live together in packs. In a wolf pack, there is a dominant female. The top female controls the mating of the other females. If there is plenty of food, she mates and allows the others to do so. If food is scarce, she allows less mating, and usually she is the only one to mate.

Communication

In all social behavior, communication is important. **Communication** is an exchange of information. How do you communicate with the people around you? Animals in a group communicate with sounds and actions. Alarm calls, pheromones, speech, courtship behavior, and aggression are all forms of communication.

Honeybees
by Paul Fleischman

Poet Paul Fleischman wrote a book of poems in which he tried to re-create the "joyful noise" that insects make when communicating. Here is a selection from one of those poems, entitled "Honeybees." It is written for two people to read. One is the worker bee and reads the lines on the left. The other is the queen bee and reads the lines on the right.

Science Journal

Imagine you are a bee or some other animal with a structured social behavior. Write a poem or an essay from the animal's point of view.

Being a bee	*Being a bee*
	is a joy
is a pain	
	I'm a queen
I'm a worker	
I'll gladly explain.	*I'll gladly explain.*
	Upon rising, I'm fed
	by my royal attendants,
I'm up at dawn, guarding	
the hive's narrow entrance	
	I'm bathed
then I take out	
the hive's morning trash	
	then I'm groomed.
then I put in an hour	
making wax,	
without two minutes' time	
to sit and relax.	
	The rest of my day
	is quite simply set forth:
Then I might collect nectar	
from the field	
three miles north	
	I lay eggs,
or perhaps I'm on	
larva detail	
	by the hundred.

How did you react to that portion of the poem? Did the poet successfully describe the jobs workers and queens do? Try to read the poem out loud with a partner. Can you "hear" the buzz of the bees?

17-2 Behavioral Adaptations   465

Source

Fleischman, Paul. *Joyful Noise: Poems for Two Voices.* Harper and Row, 1988.

Biography

A West Coast writer, Paul Fleischman was born and raised in California and now resides in Pacific Grove. Fleischman is the author of more than 20 books for young readers, including a Newbery Honor Book, *Graven Images.*

Teaching Strategies

LS **Linguistic** Acquire the book *Joyful Noise* from your school or local library. Have students read the entire poem in pairs. Be sure they understand that lines of the poem directly across from each other should be read simultaneously; lines alone on one side of the poem should be read by that bee alone. Allow time for students to discuss their thoughts about the poem with their partners. [L2]

LS **Kinesthetic** Invite students to act out the bees' movements as they read the poem. [L1]

LS **Auditory-Musical** Allow students the opportunity to read the poem aloud to hear the sound of the lines. You may want to have one student read the queen bee's lines and the rest of the class read the worker bees' to more closely approximate the buzz of bees as well as the social structure of a hive. [L1]

LS **Linguistic** Have students discuss the difference in the way a poet such as Paul Fleischman writes about bees and the way a scientist would write about them. [L2]

LS **Linguistic** Invite students to share the poems or essays they wrote in their Science Journals.

Classics

• Teale, Edwin Way. *The Strange Lives of Familiar Insects.* Dodd, Mead, 1962. According to Paul Fleischman, this book inspired him to write *Joyful Noise.* For advanced readers.

• Mowat, Farley. *Never Cry Wolf.* Little, Brown, 1963. A tale of wolf—and human—behavior.

• Prince, J. H. *The Universal Urge.* Thomas Nelson, 1972. Describes courtship and mating among animals.

Other Works

Among Paul Fleischman's other works is *I Am Phoenix* (Harper, 1985), a collection of two-voice poems about birds.

INTEGRATION
Chemistry

Pheromones are chemical substances released by many kinds of animals to communicate with members of their species.

Visual Learning

Figure 17-12 If you gently brush a line of ants away from their trail, they go right back to it. Can you explain why? *They go right back to the trail because it contains pheromones laid down by other ants.* **LEP** **LS**

MiniLAB

Purpose
LS **Interpersonal** Students will observe the behavior of earthworms. **L1** **COOP LEARN**

Materials
earthworms, paper towels, pans, flashlights, red cellophane

Teaching Strategies
Make sure students' hands are wet when touching earthworms.

Analysis
1. The earthworm moves toward the red light because the red light is like night.
2. The response would be away from the bright light. Earthworms live in the soil where it is dark.

Assessment

Oral Have students predict the responses of a mole or shrew to this same experiment. Use the Performance Task Assessment List for Formulating a Hypothesis in **PASC**, p. 21. **P**

📁 **Activity Worksheets,** pp. 5, 105

466

INTEGRATION
Chemistry

Figure 17-12

Ants follow trails laid down by other ants to find and retrieve food. *Can you explain why?*

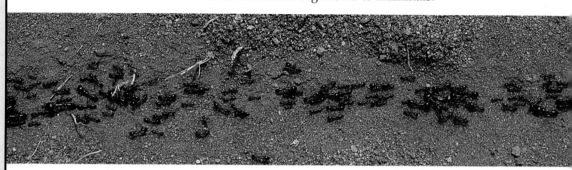

MiniLAB

How do earthworms respond to red light?

Procedure
1. Place an earthworm on a moist towel in a pan.
2. Cover a flashlight with red cellophane. Shine the red light on the anterior end of the earthworm. Record the response.

Analysis
1. Explain the response of the earthworm.
2. Predict the response of an earthworm to bright light. Explain your answer.

466 Chapter 17 Animal Behavior

Chemical Communication

Ants can sometimes be seen as in **Figure 17-12**, running along single file toward a piece of food. Male dogs stop frequently to urinate on bushes when you take them for walks. Both behaviors are based on chemical communication. The ants have laid down chemical trails that the others can follow, and the dog is letting other dogs know he has been there. In these behaviors, the animals are using pheromones to communicate. A **pheromone** is a chemical that is produced by one animal that influences the behavior of another animal of the same species. Animals that secrete pheromones range from one-celled organisms to mammals.

Both males and females use pheromones to establish territories, warn of danger, and attract mates. Certain ants, mice, and snails release alarm pheromones when injured or threatened. The odor warns other members of the species to leave the area.

Pheromones are powerful chemicals needed only in small amounts. They remain in the environment so the sender and the receiver do not have to be at a certain place in order to communicate. Because they linger in the environment, they may advertise the presence of an animal to predators as well as to the intended receiver of the message.

Sound Communication

Many insects communicate through sound. Male crickets rub a scraper on one forewing against a vein on the other forewing to produce chirping sounds. They use the sound to attract females and warn other males away from their territories. Each cricket species produces several calls that are different from those of other cricket species. The calls are often used to identify species. Male mosquitoes that are ready to mate use tiny hairs on their antennae to sense the buzzing sounds produced by females. The tiny hairs vibrate only to the frequency emitted by a female of the same species.

Inclusion Strategies

Learning Disabled Have students describe ways they have seen animals communicate.

Cultural Diversity

Hunting Behavior The lifestyle of hunting tribes revolved around the responses of the hunted animals. Have students select a Native American group and read about its relationship to the animals it depended on for food. **L2**

Vertebrates use a number of different forms of sound communication. Fish produce sounds by manipulating the air bladder. Rabbits thump the ground, gorillas pound their chests, and woodpeckers hammer hollow trees. Sound communication is useless in noisy environments. Seabirds that live on shorelines with pounding waves must rely on vision for communication.

Cyclic Behavior

What determines when an owl sleeps and when it wakes up? Animals show regularly repeated behaviors such as feeding in the day and sleeping at night or the opposite. Many reproduce every spring and migrate every spring and fall.

Cyclic behaviors are innate behaviors that occur in a repeating pattern. They are often repeated in

USING TECHNOLOGY

Artificial Insecticides

Japanese beetles and gypsy moths are two types of arthropod pests that attack plants. You may have seen people use traps like the one in the photograph to catch these pests. These traps are baited with artificial female pheromones. The pheromones lure the males of the species into the traps. Once inside, a sticky substance immobilizes the insects. Pheromone traps can be used to control at least 25 species of insects. Each insect species produces its own specific pheromone. Once the chemical structure of a particular species' pheromone is known, artificial pheromones can be produced and used to control individual insect pests. A Duke University organic chemist has developed methods for producing pheromones from a sugar called glucose. Artificial pheromones can also be released into the air at breeding time to confuse the insects so they cannot locate mates.

Think Critically:
What is a major advantage of using artificial pheromones as insecticides?

Japanese Beetle Trap

17-2 Behavioral Adaptations   467

Integrating the Sciences

Earth Science Some animals are very sensitive to changes in the environment that precede violent natural events, such as earthquakes and hurricanes. Discuss how an awareness of this behavior and the stimulus that causes it can be helpful to people.

Problem Solving

Abilities of Homing Pigeons

If a homing pigeon is placed into a dark box and transported to a strange, distant location, then released, there is a good chance it will soon show up in its home loft. Pigeons fitted with frosted lenses placed over their eyes to deprive them of visual landmarks will still return to the vicinity of their loft. Experiments have shown that pigeons can detect magnetic fields and sense changes in altitude as small as four millimeters. Pigeons also see ultraviolet light and hear extremely low-frequency sounds that come from wind blowing over ocean surf and mountain ranges thousands of kilometers away. What kind of behaviors do homing pigeons use to find their way back to their lofts?

Solve the Problem:

1. **Describe behaviors that may help homing pigeons find their way back to their lofts.**
2. **Are there learned behaviors that may help homing pigeons?**

Think Critically:

How would a pigeon that lost its vision find its way home?

468

response to changes in the environment. Behavior that is based on a 24-hour cycle is called a **circadian rhythm.** Animals that are active during the day are *diurnal.* Animals that are active at night are *nocturnal.*

Hibernation

Cyclic behaviors also occur over long periods of time. Hibernation is a cyclic response to cold temperatures and limited food supplies. During hibernation, an animal's body temperature drops to near that of its surroundings, and its breathing rate is greatly reduced. An animal in hibernation survives on stored body fat. The animal remains inactive until the weather becomes warm in the spring. Some mammals and many amphibians and reptiles hibernate.

Migration

Many birds and mammals move to new locations when the seasons change instead of hibernating. This instinctive seasonal movement of animals is called **migration.** Why do animals migrate? Most migrate in order to find food or reproduce in a better environment. Many species of birds fly hours or days without stopping. The blackpoll warbler flies nonstop a distance of more than 4000 km from North America to its winter home in South America. The trip takes nearly 90 hours. Arctic terns fly about 17 700 km from their breeding grounds in the Arctic to their winter home in the Antarctic. They return a few months later, traveling a distance of more than 35 000 km in less than a year. Gray whales swim from the cold Arctic waters to the waters off the coast of California. After the young are born, they make the return trip. **Figure 17-13** shows another animal that migrates.

Animals have many different behaviors. Some behaviors are innate and some are learned. Many are a combination of innate and learned behaviors. Appropriate behaviors help animals survive, reproduce, and maintain the species.

Figure 17-13

Wildebeests in Africa migrate in response to wet and dry seasons. These antelopes migrate more than 1600 kilometers following available water.

Section Wrap-up

Review

1. What are some examples of courtship behavior?

2. How are cyclic behaviors a response to stimuli in the environment?

3. Think Critically: How does migration help a species survive and reproduce?

Skill Builder

Using Variables, Constants, and Controls
Design an experiment to show that ants leave chemical trails to show other ants where food can be found. If you need help, refer to Using Variables, Constants, and Controls in the **Skill Handbook.**

USING MATH

Cicadas emerge from the ground every 17 years. The number of one type of caterpillar peaks every 5 years. If the peak cycles of the caterpillars and cicadas coincided in 1990, in what year will they coincide again?

17-2 Behavioral Adaptations **469**

Skill Builder

Accept all logical experiments. Students should identify a variable, constant, and control.

Assessment

Performance Assess students' abilities to use variables, constants, and controls by having them carry out their experiments. Use the Performance Task Assessment List for Designing an Experiment in **PASC,** p. 23. **P**

4 Close

•MINI•QUIZ•

Use the Mini Quiz to check students' recall of chapter content.

1. A forceful act used to dominate or control another animal is _____ . *aggression*

2. Courtship behaviors allow male and female members of a species to _____ . *recognize each other*

3. A group of animals of the same species living and working together in an organized way is a _____ . *society*

4. The exchange of information is _____ . *communication*

5. Animals that are active at night are said to be _____ . *nocturnal*

Section Wrap-up

Review

1. the bright red pouch on the throat of the male frigate bird; pheromones in gypsy moths

2. They are responses to changes in the environment, such as decrease in day length or temperature changes.

3. Think Critically Migration ensures that an animal lives and reproduces in a favorable environment so the likelihood of survival is increased.

USING MATH

The year these two cycles will again coincide is 2075.

470

SCIENCE & SOCIETY

ISSUE:
17•3 Zoos and Captive Breeding

Science Words

captive breeding

Objectives

- Explain the advantages of zoos.
- Determine the advantages and disadvantages of captive breeding.

Figure 17-14

The San Diego Wild Animal Park was one of the first zoos in America to place animals in outdoor exhibits built to resemble their natural habitats.

470

Redesigned Zoos

Twenty years ago, most zoos and aquariums kept their animals in small cages or behind glass windows, with little attempt to provide a natural habitat for the animals. But as more and more animals became endangered species in the wild, zoos changed from animal warehouses into places where captive animals live in open exhibits, eat balanced diets, and even reproduce and raise their young, as shown in **Figure 17-14.** Zoos now have a strong commitment to public education and can be powerful forces for wildlife conservation.

Captive Breeding

Captive breeding is the breeding of endangered species in captivity. Zoos use captive breeding to protect animals from extinction and to establish populations that can be released into suitable habitats. The last-known wild California condor was captured in 1987. Captive breeding at the San Diego Wild Animal Park has increased condor numbers from 21 to 68. Eight condors have even been released back into their previous habitat in California.

Zoo directors have learned two lessons from captive-breeding programs. One is to begin captive-breeding programs before wild populations are close to extinction. A second lesson zoo directors have learned is to start with enough male and female animals to prevent inbreeding. Inbreeding often leads to infertility, poor survival, and other genetic problems. To minimize these risks, the California condor program used elaborate, computer-based calculations to figure out which male should mate with each female.

Elephant populations in both Africa and Asia are declining due to habitat destruction, legalized shooting, and poaching. Modern elephant pens at zoos are expensive; few American zoos are willing to maintain bull elephants because they are dangerous to their keepers, to other elephants, and to themselves. There are only 17 sexually mature Asian bull elephants in accredited North American zoos. This leaves a narrow genetic base to guard against inbreeding.

---

2 Points of View

Disadvantages of Captive Breeding

Captive-breeding programs will not remain viable without breeding a surplus of animals. Some zoos sell surplus animals to dealers who sell them to other zoos, and sometimes to hunting camps where hunters pay to shoot exotic species. Captive-breeding programs won't solve the problem of habitat destruction. Some scientists feel that saving the last few members of a species is pointless if none of their habitat remains.

Advantages of Captive Breeding

Think of species that became extinct before captive-breeding programs could be established—the passenger pigeon, the Carolina parakeet, the Tasmanian wolf, and the dodo. Perhaps captive-breeding programs could have saved these animals from extinction.

Proponents of captive-breeding programs point out that there are already many species that survive only because self-sustaining captive populations were established before the animals became extinct in the wild. Examples are the California condor, the black-footed ferret, the Guam rail, and the Arabian oryx. Their survival in captivity has given conservation biologists time to try to find and protect habitats for the species to reestablish a wild population.

Figure 17-15

Elephants in zoos are hard to control, even with chained legs. *But how else can such species be saved from extinction?*

Visit the Chapter 17 Internet Connection at Glencoe Online Science, www.glencoe.com/sec/science/life, for a link to more information about survival plans for endangered species.

Section Wrap-up

Review

1. What evidence is there of increased public awareness of the plight of endangered species?

2. What two lessons have zoo directors learned from captive-breeding programs?

Explore the Issue

Should captive-breeding programs be regulated? If yes, how would the regulation be enforced? If no, then what can be done about the breeding of surplus animals?

SCIENCE & SOCIETY

Visual Learning

Figure 17-15 But how else can such species be saved from extinction? *by protecting and preserving natural habitat* **LEP** **LS**

animals to prevent inbreeding; and to start captive breeding before the wild population is in danger of extinction

Explore the Issue

Answers will vary but may include that rules could be established and people hired to see that they are enforced, and the surplus animals could be reestablished in the wild.

---

3 Assess

Check for Understanding

Have students answer questions 1 and 2 in the Section Wrap-up.

Reteach

Explain that in many parts of the world, prime animal habitat is being taken over for human settlement. Saving habitat is important for the survival of endangered animals. Captive breeding programs cannot prevent habitat destruction, but they can prevent many species from becoming extinct.

Extension

For students who have mastered this section, use the **Reinforcement** and **Enrichment** masters.

4 Close

Have students write a report stating their position for or against captive breeding programs. Suggest that they research the cost of maintaining zoos. Their reports should show that they have considered the conflict between preservation of animal populations and removing animals from their natural habitats.

Section Wrap-up

Review

1. the number of people who go to zoos, watch nature or wild animal programs on television, and attend exhibits or theme parks

2. to start with enough breeding male and female

Observing Social Behavior in Ants

Purpose

 Intrapersonal Students will observe the interaction of ants in a colony.

Process Skills

observing and inferring, classifying, forming operational definitions

Time

50 minutes and observation time each day

Safety Precautions

Caution students to handle ants with care. Some ants can bite and cause painful swelling. *Students who are allergic to other insect stings or bites should not handle ants.*

📁 **Activity Worksheets,** pp. 5, 102-103

Teaching Strategies

Alternate Materials Any glass container may be used.

Troubleshooting To prevent ants from crawling out of the jars, rub vegetable oil around the inside top of the jars.

- Collect the soil near the anthill. The ants can be collected at the same time. Try to find a queen for each colony.

- Have students discuss what they have observed about anthills.

Answers to Questions

1. Answers will vary. Usually, they take from one to five days.

2. Two or more ants may have been observed moving food or building a tunnel.

3. usually something sweet—answers will vary

4. The queen will look different from the workers.

5. a group of animals of the same species living and working together in an organized way

6. They live in colonies.

472

Ants live and work together in an organized way. Each member of an ant society has a certain job. In this activity you will observe ways in which ants are adapted to their environment and how they work together to survive.

Problem

How do ants live and work together to survive?

Materials

- 20 ants
- large jar with screen wire cover
- soil
- black paper
- small moist sponge
- water
- food—honey, sugar, bread crumbs
- small can with lid
- spoon or trowel
- hand lens

Procedure 🥽 🐾

1. Pour moistened, loose soil into a jar until it is ¾ full. Place the damp sponge on top of the soil.

2. Find an anthill. Use the spoon or trowel to place loose soil from the nest into the small can. Look for ants, small white eggs, cocoons, and larvae. Try to find a queen. The queen will be larger than the other ants. Cover. Place the ants in the refrigerator for a few minutes to slow their movements.

3. Gently place the ants in the jar. Add a small amount of each food in different areas. Place the screen wire cover on the jar.

4. Tape the black paper around the jar. Leave about 2 cm at the top uncovered.

472 Chapter 17 Animal Behavior

---

5. Take the paper off for a short period of time each day and observe the ants at work. Use a hand lens to observe individual ants.

6. Keep the sponge moistened and add food each day. Record your observations each day.

Data and Observations   Sample Data

Date	Number of ants visible	Other observations
		See teacher
		margin

Analyze

1. How long did it take for the ants to build tunnels after they were placed in the jar?

2. What did you observe that indicated the ants were working together?

3. What kinds of foods do ants prefer?

4. How do the ants in the jar differ?

Conclude and Apply

5. What is a society?

6. What evidence do you have that ants are social insects?

✓ **Assessment**

Performance Have students repeat the activity using a sugar substitute such as aspartame. Use the Performance Task Assessment List for Carrying out a Strategy and Collecting Data in **PASC**, p. 25. **P**

Summary

17-1: Types of Behavior

1. Behavior that an animal is born with is innate behavior. Other animal behaviors are learned through experience.
2. Reflexes are the simplest innate behaviors. An instinct is a complex pattern of innate behavior.
3. Learned behavior includes imprinting, in which an animal forms a social attachment soon after birth. Behavior modified by experience is learning by trial and error. Conditioning occurs when a response associated with one stimulus becomes associated with another. Insight uses past experiences to solve new problems.

17-2: Behavioral Adaptations

1. Behavioral adaptations such as defense of territory, courtship behavior, and social behavior help species of animals survive and reproduce.
2. Courtship behaviors allow males and females to recognize each other.
3. Interactions among members of the same species are social behaviors. Cyclic behaviors are behaviors that occur in repeating patterns.

17-3: Science and Society: Zoos and Captive Breeding

1. Zoos have a strong commitment to public education and wildlife conservation.
2. Captive-breeding programs should begin before the wild population is close to extinction.

Key Science Words

a. aggression
b. behavior
c. captive breeding
d. circadian rhythm
e. communication
f. conditioning
g. courtship behavior
h. cyclic behavior
i. imprinting
j. innate behavior
k. insight
l. instinct
m. learning
n. migration
o. motivation
p. pheromone
q. reflex
r. social behavior
s. society
t. territory
u. trial and error

Reviewing Vocabulary

Match each phrase with the correct term from the list of Key Science Words.

1. an inherited, not learned behavior
2. automatic response to a stimulus
3. forming a social bond to another organism after birth
4. the way an organism behaves toward its environment
5. change in behavioral response to a given stimulus
6. a defended area
7. using reasoning to solve a problem
8. an organized group of animals in a species
9. behaviors before mating
10. instinctive seasonal movement

Chapter 17 Review **473**

Summary

Have students read the summary statements to review the major concepts of the chapter.

Reviewing Vocabulary

1. j	**6.** t
2. q	**7.** k
3. i	**8.** s
4. b	**9.** g
5. f	**10.** n

✓ Assessment

Portfolio Encourage students to place in their portfolios one or two items of what they consider to be their best work. Examples include:
• Community Connection, p. 457
• Skill Builder, p. 459
• MiniLAB, p. 466 P

Performance Additional performance assessments may be found in **Performance Assessment** and **Science Integration Activities.** Performance Task Assessment Lists and rubrics for evaluating these activities can be found in Glencoe's **Performance Assessment in the Science Classroom.**

GLENCOE TECHNOLOGY

 Videodisc

Glencoe Life Science Interactive Videodisc
Side 2, Lesson 5
Structural Adaptations in Animals—Species Survival

6972 to 8746

MindJogger Videoquiz

Chapter 17 Have students work in groups as they play the Videoquiz game to review key chapter concepts.

Checking Concepts

1. c	**6.** d
2. a	**7.** a
3. d	**8.** b
4. c	**9.** c
5. a	**10.** a

Understanding Concepts

11. The behavior is imprinting. It is learned.

12. Reflex behaviors are automatic responses, so no learning, thinking, or insight is necessary.

13. Any example of trial and error is fine—tying shoes, writing, riding a bike, and so on.

14. Territories are expressed by marking with urine, chasing intruders away, making sounds (bird calls, howls, and so on), patrolling, or leaving scent markers.

15. conditioning: learned response associated with stimuli (Pavlov's dog bell with food); trial-and-error: learning by repeating (tying shoes); imprinting: attachment to another right after birth or hatching (goslings and mother goose)

Thinking Critically

16. Leaving the room when the bell rings is a learned conditioned response.

17. Migration allows organisms to survive changes in weather. The journey in many cases is long and stressful.

18. A habit is a learned behavior that has become automatic. A reflex is innate (not learned) behavior.

19. Behaviors that help get food, are protective, or are defensive help organisms to survive.

Checking Concepts

Choose the word or phrase that completes the sentence or answers the question.

1. An example of a reflex is _____.
a. writing c. sneezing
b. talking d. riding a bicycle

2. An instinct is an example of _____.
a. innate behavior
b. learned behavior
c. imprinting
d. conditioning

3. Nest building is an example of _____.
a. conditioning c. learned behavior
b. imprinting d. an instinct

4. The animals that depend least on instinct and most on learning are _____.
a. birds c. mammals
b. fish d. amphibians

5. _____ involves using reasoning to solve problems.
a. Insight c. Conditioning
b. Imprinting d. Trial and error

6. Something inside an animal that causes the animal to act is _____.
a. conditioning c. insight
b. imprinting d. motivation

7. Teeth, beaks, and claws are used for _____.
a. killing prey c. hurting other birds
b. fighting d. sending signals

8. All are examples of courtship behavior except _____.
a. fluffing feathers
b. taking over a perch
c. singing songs
d. releasing pheromones

9. An organized group of animals working specific jobs is a _____.
a. community c. society
b. territory d. circadian rhythm

10. The response of inactivity and slowed metabolism that occurs during cold conditions is _____.
a. hibernation c. migration
b. imprinting d. circadian rhythm

Understanding Concepts

Answer the following questions in your Science Journal using complete sentences.

11. What behavior does the photograph below show? Is this behavior innate or learned?

12. What are the advantages of reflex behaviors?

13. Give an example of something you learned by trial and error.

14. Describe the ways territoriality is expressed to other members of a species.

15. Compare conditioning, trial and error, and imprinting using examples of each.

Thinking Critically

16. Explain the type of behavior involved when the bell rings at the end of class.

17. Discuss the advantages and disadvantages of migration as a means of survival.

18. Explain how a habit, such as tying your shoes, is different from a reflex.
19. Use one example to explain how behavior increases an animal's chance for survival.
20. Hens lay more eggs in the spring when day length increases. How can farmers use this knowledge of behavior to their advantage?

Developing Skills

If you need help, refer to the **Skill Handbook.**

21. **Classifying:** Make a list of 25 things that you do regularly. Classify each as an innate or learned behavior. Which behaviors do you have more of?
22. **Observing and Inferring:** Make observations of a dog, cat, or bird for a week. Record what you see. How did the animal communicate with other animals and with you?
23. **Concept Mapping:** Complete the concept map below outlining the types of behavior.

24. **Using Variables, Constants, and Controls:** Design an experiment to get a specific response from an animal to a stimulus.
25. **Hypothesizing:** Make a hypothesis about how frogs communicate with each other. How could you test your hypothesis?

Performance Assessment

1. **Writing in Science:** View the film *Never Cry Wolf* (Disney). Write a summary of the behaviors of the wolves.
2. **Poster**: Draw a map showing the migration route of caribou, gray whales, or arctic wolves.
3. **Letter:** Write a letter to the local zoo or aquarium asking about their captive-breeding programs. Find out what species are successfully reproducing and what the zoo does with surplus animals. Report your findings to the class.

Assessment Resources

 Reproducible Masters
Chapter Review, pp. 37-38
Assessment, pp. 105-108
Performance Assessment, p. 31

Glencoe Technology
 MindJogger Videoquiz

---

20. A farmer can artificially lengthen the amount of 'daylight' hens are exposed to by using lights, thus stimulating hens to lay more eggs.

Developing Skills

21. **Classifying** Most behaviors will be learned, for example, brushing teeth or combing hair. Innate behaviors include sneezing, yawning, and other similar actions.
22. **Observing and Inferring** Observations will vary, but communication should be noted.
23. **Concept Mapping** See student page.
24. **Using Variables, Constants, and Controls** Any controlled experiment should be accepted. Be sure students identify the variable, constant, and control.
25. **Hypothesizing** Students will probably hypothesize that frogs communicate by using vocalizations. Accept all logical ways to test this hypothesis.

Performance Assessment

1. Students will point out both innate and learned behaviors. Use the Performance Task Assessment List for Writing in Science in **PASC,** p. 87. P
2. Student maps should indicate both spring and fall migration routes. Use the Performance Task Assessment List for Poster in **PASC,** p. 73. P
3. Student reports will vary depending on which zoos are contacted. Use the Performance Task Assessment List for Letter in **PASC,** p. 67. P

Was it a reptile? Was it a bird? Was it really warm?

Objectives

⟨LS⟩ Kinesthetic Students will apply a particular theory to research and construct a model. **L1** **COOP LEARN**

Summary

In this project, students will attempt to research a prehistoric life form based on a particular theoretical perspective. They will apply this research to build a model to represent a particular dinosaur. After doing this research and building the model, students will discuss the relationship between the apparent abilities of the animal and its body structure.

Time Required

This project is designed to be worked on throughout Unit 5. One to two periods are required to set up the project. Students will need time throughout the unit to work on research, construction, and presentation of dinosaur information.

Preparation

- Establish cooperative groups of 4 or 5 students.
- Make available valid information sources on dinosaurs and materials for dinosaur and diorama construction.

Dinosaurs are popular attractions in stories, as action figures, on TV shows, and in museum displays. We have lots of pieces of evidence from dinosaur times, but this evidence can be woven into a variety of theories about the nature of dinosaurs, dinosaur lifestyles, and the cause of dinosaurs' demise. Originally, most scientists accepted the idea that dinosaurs were cold-blooded reptiles, slow-moving, and not terribly bright. Today, however, new evidence is causing scientists to question some of these ideas. What are some of the current controversies?

Theories in Conflict

Scientists have many questions about dinosaurs.

- Were dinosaurs ectotherms or endotherms?
  - Were the dinosaurs more closely related to reptiles or to birds?
    - Were their bones solid like reptile bones or hollow inside like bird bones?
    - Were dinosaurs slow-moving, or could some of them move quickly?
  - Impressions of dinosaur skin show gland cells. Were these sweat glands or poison glands?
    - Did dinosaurs live in hot, humid areas or hot, dry areas?
- Are bones that are found together from one dinosaur or from many? How can scientists tell?
- Are the eggs that are found with dinosaur bones their dinner or their offspring?
- What caused dinosaurs to disappear? Was it geological processes such as volcanoes, or comets from space?

Using a Theory to Design a Dinosaur

Jack Horner from the Museum of the Rockies in Bozeman, Montana, thinks that dinosaurs were more closely related to birds than to reptiles. He has found that many dinosaur bones appear to be hollow. By using the CAT scan X ray, he has determined that the skull and other bones contain many holes for blood vessels similar to those found in warm-blooded animals like birds. These ideas are still controversial, but you can use Dr. Horner's theory to construct a dinosaur yourself.

Procedure

1. In a small group, look at books about dinosaurs and find one particular dinosaur that you can construct a model of. Make sure there is a diagram of the bone structure to give you guidance.

2. Decide what materials you will use. Straws are hollow and lightweight like bird bones. Use bendable straws. The straws can be joined with tape.

3. Study the environments in which your dinosaur seems to have lived. Build your dinosaur and a realistic diorama showing its environment and something about its lifestyle.

Using Your Research

Prepare a brief description to explain the model and diorama. Arrange a day for groups to share these projects with each other. Look for connections among the various forms researched. Can a food web be constructed to connect the various dinosaurs?

Go Further

Discuss whether or not you agree with Dr. Horner's theory. What differences in dinosaur lifestyle are involved in accepting or rejecting Dr. Horner's theory? Do you have any different theory about the nature of dinosaurs? What evidence currently supports this theory? What further evidence would be needed to support your theory?

Unit 5 Project   **477**

References

Herner, John R. and James Gorman. *Digging Dinosaurs.* New York: Workman Publishers, 1988.

Lessem, Don. "Skinning the Dinosaur." *Discover,* March 1989, pp. 38-44.

Robbins, Jim. "The Real Jurrasic Park." *Discover,* March 1991, pp. 52-59.

Trefil, James. "Whale Feet." *Discover,* May 1991, pp. 44-48.

Teaching Strategies

- Before students begin their research, ask them to talk about how dinosaurs are represented in stories, movies, TV shows, stuffed toys, etc. Discuss why people seem to have such an interest in dinosaurs, and how representations are not always scientifically based or accurate.

- Focus on bone structure and why animals stand on hind legs or need to be on all fours to support themselves.

- Discuss how bones are constructed to withstand stress. Consider getting input from a structural engineer as to how some structures are supported. Discuss how this relates to dinosaurs.

- Focus on differences between warm-blooded and cold-blooded animals and their needs. Discuss current research on why it is now thought that dinosaurs were warm-blooded instead of the long-held idea that they were more reptilian, or cold-blooded in nature.

Go Further

One major area of current research focuses on what caused the dinosaurs to become extinct. Students could look at the various theories suggested to explain the major extinctions of dinosaurs that appear in the fossil record. What kinds of evidence support each particular theory? What further evidence would be helpful in deciding among the various theories? Another area of controversy is the idea of remnant dinosaurs. The most famous example of this would be the Loch Ness monster.

477

UNIT 6

Ecology

In Unit 6, students are introduced to the fundamentals of ecology, including interaction of abiotic and biotic factors, characteristics of populations, the flow of energy, and the cycling of matter. Ecosystems and succession in ecosystems are explored. The unit concludes with a discussion of renewable and nonrenewable resources and how these resources are affected by human use.

CONTENTS

Chapter 18
**Life and
the Environment**   480

18-1 The Living
Environment and the
Nonliving Environment

18-2 Interactions Among
Living Organisms

18-3 Matter and Energy

18-4 Science and Society:
Bringing Back the
Wolves

Chapter 19
Ecosystems   508

19-1 How Ecosystems
Change

19-2 Land Environments

19-3 Science and Society:
Protecting Antarctica

19-4 Water Environments

Chapter 20
**Resources and
the Environment**   538

20-1 Natural Resources

20-2 Science and Society:
Recycling

20-3 Conservation and
Wildlife Protection

20-4 Maintaining a Healthy
Environment

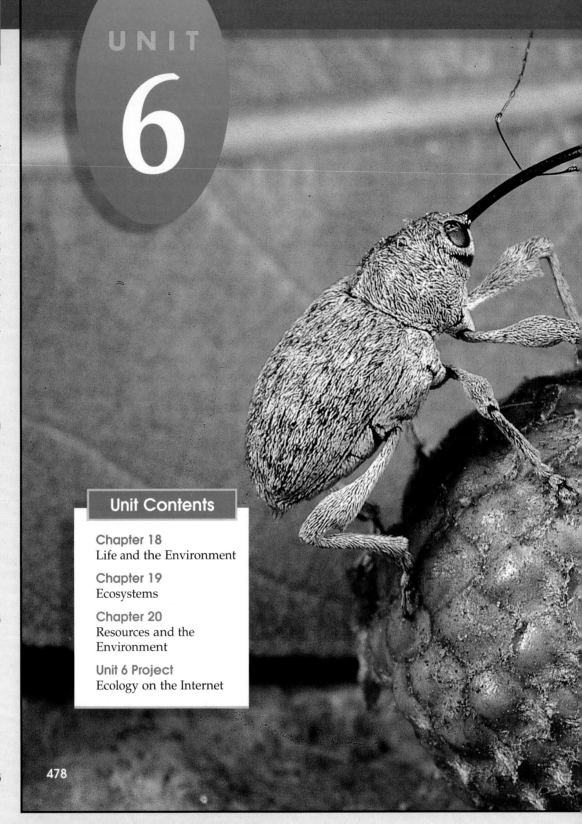

UNIT
6

Unit Contents

Chapter 18
Life and the Environment

Chapter 19
Ecosystems

Chapter 20
Resources and the
Environment

Unit 6 Project
Ecology on the Internet

478

Science at Home

Air Pollution Survey Have each student cover the top of a jar with plastic wrap and secure it with a rubber band. Then, use a tongue depressor to spread an even layer of petroleum jelly on the plastic wrap. Place the jars in different locations, and leave them undisturbed for at least 24 hours. Have students use hand lenses to observe the plastic wrap on each jar.

L1   **COOP LEARN**   **LEP**

Ecology

What's Happening Here?

Mighty oak trees grow from acorns, but not all acorns become trees. As acorns ripen and fall off the tree, many become food for bears, deer, and mice. Some acorns are attacked by acorn weevils such as the one in the large photograph. Weevils drill holes through the hulls to lay their eggs. Filbert worm moths use these holes to lay their own eggs. When their larvae hatch, they eat their way out. Green fungi invade. Maggots and springtails arrive to eat the fungi. What's in a nutshell? Life!

Science Journal

Many organisms have relationships that are not understood until it is too late. The dodo bird of the island of Mauritius was discovered in 1507, but by 1681, the last dodo had been killed. Soon people living in Mauritius noticed that the *Calvaria major* tree had become scarce. The dodo was known to eat the seeds of this tree. In your Science Journal, suggest a reason for the scarcity of this tree after the dodos died.

479

Theme Connection

Energy All living things require energy to live. Ask students to describe and give examples of how energy flows through an ecosystem. *Energy moves from one organism to another through a series of interactions called a food web.*

Cultural Diversity

Adapting to the Environment Some people living in desert areas wear loose-fitting, lightweight, light-colored garments that reflect sun and heat. They may live in tents that enable them to take advantage of the smallest breezes. A tundra, rain forest, or deciduous forest would call for different cultural adaptations. Have students construct a model to show one culture's food, clothing, and shelter.

Chapter Organizer

Section	Objectives/Standards	Activities/Features
Chapter Opener		Explore Activity: Determine the space available to an individual. p. 481
18-1 The Living Environment and the Nonliving Environment (1½ sessions, 1 block)*	1. **Identify** biotic and abiotic factors in an ecosystem. 2. **Describe** the characteristics of populations. 3. **Explain** the levels of biological organization. National Content Standards: (5-8) UCP1, A2, C4, D1	Connect to Earth Science, p. 484 Activity 18-1: Soil Composition, p. 485 Skill Builder: Observing and Inferring, p. 487 Science Journal, p. 487
18-2 Interactions Among Living Organisms (1½ sessions, 1 block)*	1. **Describe** the characteristics of populations. 2. **Identify** the types of relationships that occur among populations in a community. 3. **Compare** the habitat and niche of a species in a community. National Content Standards: (5-8) UCP1, A1, A2, C4, E2, F2, F5	Activity 18-2: Identifying a Limiting Factor, pp. 490-491 Using Technology: Monitoring Mayflies, p. 493 MiniLAB: How do nitrogen-fixing bacteria affect plants? p. 494 Skill Builder: Making and Using Tables, p. 495 Using Math, p. 495
18-3 Matter and Energy (2 sessions, 2 blocks)*	1. **Explain** how energy flows through ecosystems. 2. **Describe** the cycling of matter in the biosphere. National Content Standards: (5-8) UCP2, A1, B1, B3, C4	Problem Solving: Changes in Antarctic Food Webs, p. 499 MiniLAB: How does temperature affect the water cycle? p. 500 Skill Builder: Classifying, p. 501 Science Journal, p. 501
18-4 Science and Society Issue: Bringing Back the Wolves (1 session, ½ block)*	1. **Compare** and **contrast** the possible benefits and drawbacks of reintroducing wolves into Yellowstone National Park. National Content Standards: (5-8) C4, F2	Explore the Issue, p. 503 Science and Literature: Never Cry Wolf, p. 504 Science Journal, p. 504

* A complete Planning Guide that includes block scheduling is provided on pages 32T-35T.

Activity Materials

Explore	Activities	MiniLABs
page 481 meterstick	page 485 3 small paper cups containing freshly dug soil, newspaper, hand lens, metric scale, 250-mL beaker, tap water, jar with lid pages 490-491 bean seeds, small planting containers, soil, tap water, labels, trowel or spoon, aluminum foil, sunny window sill or other light source, refrigerator or oven	page 494 freshly picked legumes (alfalfa, beans, clover, vetch) and non-legumes, tap water, plastic tub, hand lens, prepared slide of nitrogen-fixing bacteria page 500 tap water, 250-mL beaker, heat source, watch glass to cover beaker, ice cubes

Need Materials? Call Science Kit (1-800-828-7777).

Teacher Classroom Resources

Reproducible Masters	Transparencies	Teaching Resources
Study Guide, p. 68 **Reinforcement**, p. 68 **Enrichment**, p. 68 **Activity Worksheets**, pp. 106-107 **Lab Manual 42**, Communities **Lab Manual 43**, Living Things in Soil **Multicultural Connections**, pp. 39-40	**Section Focus Transparency 64**, Natural Environments	**Glencoe Life Science Interactive Videodisc** **Life Science CD-ROM** **Spanish Resources** **English/Spanish Audiocassettes** **Cooperative Learning Resource Guide** **Lab Partner** **Lab and Safety Skills** **Lesson Plans**
Study Guide, p. 69 **Reinforcement**, p. 69 **Enrichment**, p. 69 **Activity Worksheets**, pp. 108-109, 110 **Lab Manual 44**, Limiting Factors **Lab Manual 45**, Living Space **Critical Thinking/Problem Solving**, p. 22 **Science and Society Integration**, p. 22	**Section Focus Transparency 65**, People, People Everywhere	*Assessment Resources* **Chapter Review**, pp. 39-40 **Assessment**, pp. 122-125 **Performance Assessment in the Science Classroom (PASC)** **MindJogger Videoquiz** **Alternate Assessment in the Science Classroom**
Study Guide, p. 70 **Reinforcement**, p. 70 **Enrichment**, p. 70 **Activity Worksheets**, p. 111 **Science Integration Activity 18**, Carbon Dioxide and Limestone **Concept Mapping**, p. 41 **Cross-Curricular Integration**, p. 22 **Technology Integration**, pp. 15-16	**Section Focus Transparency 66**, Food and Energy **Science Integration Transparency 18**, Phase Changes and the Water Cycle **Teaching Transparency 35**, Life and the Food Web/Energy Pyramid **Teaching Transparency 36**, Cycles	**Performance Assessment** **Chapter Review Software** **Computer Test Bank**
Study Guide, p. 71 **Reinforcement**, p. 71 **Enrichment**, p. 71	**Section Focus Transparency 67**, Endangered and Threatened Species	

Key to Teaching Strategies

The following designations will help you decide which activities are appropriate for your students.

L1 Level 1 activities should be within the ability range of all students, including those with learning difficulties.

L2 Level 2 activities should be within the ability range of the average to above-average student.

L3 Level 3 activities are designed for the ability range of above-average students.

LEP LEP activities should be within the ability range of Limited English Proficiency students.

LS These activities are designed to address different learning styles.

COOP LEARN Cooperative Learning activities are designed for small group work.

P These strategies represent student products that can be placed into a best-work portfolio.

GLENCOE TECHNOLOGY

The following multimedia resources are available from Glencoe.

Science and Technology Videodisc Series (STVS)
Animals
Temperature Regulation in Dogs
The Infinite Voyage Series
Secrets from a Frozen World
The Living Clock

National Geographic Society Series
STV: Rain Forest
GTV: Planetary Manager
Glencoe Life Science Interactive Videodisc
Ecosystems: Wetlands
Life Science CD-ROM

Teacher Classroom Resources

This is a representation of key blackline masters available in the Teacher Classroom Resources.

Teaching Aids

Section Focus Transparencies

Science Integration Transparencies

Teaching Transparencies

Meeting Different Ability Levels

Study Guide

Reinforcement

Enrichment Worksheets

Chapter 18 Life and the Environment

Hands-On Activities

Science Integration Activity

L1

Lab Manual

L2

Assessment

Performance Assessment

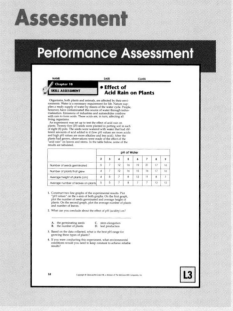

L3

Enrichment and Application

Critical Thinking/ Problem Solving

L2

Cross-Curricular Integration

L1

Science and Society Integration

L2

Technology Integration

L1

Multicultural Connections

L2

Concept Mapping

L3

Chapter 18

Life and the Environment

CHAPTER OVERVIEW

Section 18-1 The ecological units that make up the biosphere and the ways in which biotic and abiotic factors within the biosphere interact are discussed in this section.

Section 18-2 Characteristics of populations and how populations interact in communities are discussed.

Section 18-3 This section focuses on how energy flows through an ecosystem. All organisms need energy to carry on life processes. Also discussed is the cycling of materials—water, nitrogen, oxygen, and carbon—through the biosphere.

Section 18-4 Science and Society This section explores the possible benefits and drawbacks of reintroducing wolves into Yellowstone National Forest.

Chapter Vocabulary

biosphere	symbiosis
ecology	mutualism
abiotic factor	commensal-
biotic factor	ism
population	parasitism
community	habitat
ecosystem	niche
population	food chain
density	food web
limiting	ecological
factor	pyramid
carrying	water cycle
capacity	nitrogen
predation	cycle

Theme Connection

Systems and Interactions In learning about relationships among living and nonliving things in populations, communities, and ecosystems, students will discover ways in which all the living and nonliving things on Earth are interconnected.

480

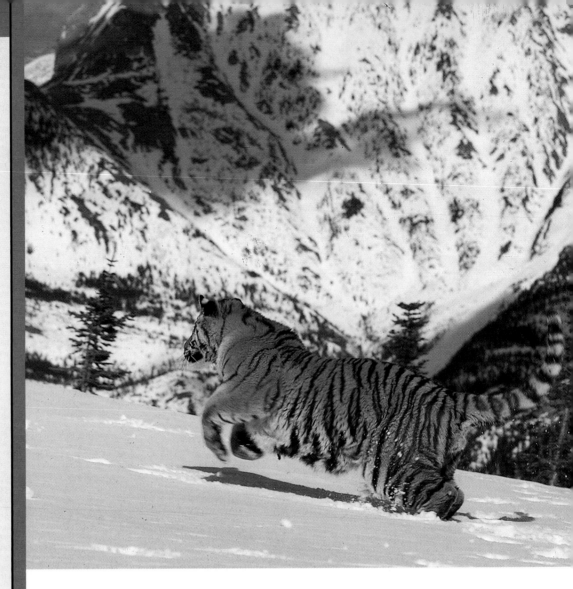

Previewing the Chapter

Section 18-1 The Living Environment and the Nonliving Environment
- ▶ The Biosphere
- ▶ Abiotic Factors
- ▶ Biotic Factors

Section 18-2 Interactions Among Living Organisms
- ▶ Characteristics of Populations
- ▶ Interactions in Communities

Section 18-3 Matter and Energy
- ▶ The Flow of Energy Through Ecosystems
- ▶ The Cycles of Matter

Section 18-4 Science and Society
Issue: Bringing Back the Wolves

480

Learning Styles Look for the following logo for strategies that emphasize different learning modalities. **LS**

Kinesthetic	Activity 18-1, p. 485; Reteach, p. 500
Visual-Spatial	Integrating the Sciences, p. 483; Activity, pp. 493, 497; Across the Curriculum, p. 493; MiniLAB, p. 494
Interpersonal	Activity 18-2, pp. 490-491; MiniLAB, p. 500
Intrapersonal	Activity, pp. 483, 492
Logical-Mathematical	Explore, p. 481; Science Journal, p. 489
Linguistic	Science Journal, p. 492; Cultural Diversity, p. 498; Science Journal, p. 501; Science & Literature, p. 504

Chapter 18

Life and the Environment

The Siberian tiger preys on wild pigs, deer, and other large mammals, and may travel for days in search of a meal or a mate. Tidepool anemones, shown below, feed on small fish and crabs, using their tentacles to capture prey that swims or floats by. A male tiger's home range covers an area of thousands of square kilometers. The anemone's living space is not much larger than its body. How does the number of organisms in an area affect each individual? You share your science classroom with other students. How much space is available to each student?

EXPLORE ACTIVITY

Determine the space available to an individual.

1. Use a meterstick to measure the length and width of the classroom.
2. Multiply the length times the width to find the area of the room in square meters.
3. Count the number of individuals in your class. Divide the number of square meters in the classroom by the number of individuals.

Observe: How much space does each person have? Determine the amount of space that each person would have if the number of individuals in your class doubled.

Previewing Science Skills

▶ In the **Skill Builders**, you will **observe and infer, make and use tables,** and **classify.**

▶ In the **Activities**, you will **observe, collect data, graph, hypothesize,** and **compare results.**

▶ In the **MiniLABs**, you will **observe** and **infer.**

481

Prepare

Section Background

- No living thing lives alone. Every living thing depends somehow upon certain other living and nonliving things.
- The word *ecology* comes from the Greek word *oikos,* meaning "house" or "place to live" and *-ology,* meaning "the study of."
- Ecologists learn about the biosphere by studying smaller and simpler ecological units within it.

Preplanning

Obtain soil for Activity 18-1.

1 Motivate

Bellringer

 Before presenting the lesson, display **Section Focus Transparency 64** on the overhead projector. Assign the accompanying **Focus Activity** worksheet. L1 LEP

Tying to Previous Knowledge

Review producers and consumers from Chapter 3 and classification and species from Chapter 7. Students will be familiar with organisms from their own environment. Ask how these organisms are interacting in their environment.

GLENCOE TECHNOLOGY

◉ CD-ROM

Life Science CD-ROM

Have students perform the interactive exploration for Chapter 18 to reinforce important chapter concepts and thinking processes.

18•1 The Living Environment and the Nonliving Environment

Science Words

- biosphere
- ecology
- abiotic factor
- biotic factor
- population
- community
- ecosystem

Objectives

- Identify biotic and abiotic factors in an ecosystem.
- Describe the characteristics of populations.
- Explain the levels of biological organization.

The Biosphere

Think of all the living organisms on Earth. There are millions of species of plants, animals, fungi, and bacteria. Where do all these organisms live? Living things can be found from 11 000 m below the surface of the ocean to the top of mountains 9000 m high. The part of Earth that supports living organisms is known as the **biosphere.** The biosphere, while it seems huge, is actually a small portion of Earth. It includes the topmost portion of Earth's crust, all the waters that cover Earth's surface, and the surrounding atmosphere. The thickness of the biosphere could be compared to the thickness of the skin of an apple.

Within the biosphere, there are many different environments. For example, red-tailed hawks are found in environments where tall trees border open grassland. The hawks nest high in the trees, and soar over the land in search of rodents and rabbits to eat. In environments with plenty of moisture, such as the banks of streams, willow trees provide food and shelter for birds, mammals, and insects. All organisms interact with the environment. The science of **ecology** is the study of interactions that take

Figure 18-1

The biosphere is the region of Earth that contains all living organisms. It extends from high in the atmosphere to the bottom of the deepest ocean. An ecologist is a scientist who studies relationships among organisms and between organisms and the physical features of the biosphere.

482

Program Resources

📁 **Reproducible Masters**
Study Guide, p. 68 L1
Reinforcement, p. 68 L1
Enrichment, p. 68 L3
Activity Worksheets, pp. 106-107 L1
Multicultural Connections, pp. 39-40
Lab Manual 42, 43

🔦 **Transparencies**
Section Focus Transparency 64 L1

place among organisms and between organisms and the physical features of the environment. Ecologists, such as the one in **Figure 18-1,** are the scientists who study the environment.

Abiotic Factors

A forest environment is made up of trees, birds, insects, and other living things that depend on one another for food and shelter. But these organisms also depend on factors that surround them such as water, soil, sunlight, temperature, and air. These factors, the nonliving, physical features of the environment, are called **abiotic factors.** Abiotic, *a* meaning "not" and *biotic* meaning "living," factors have effects on living things and often determine the organisms that are able to live in a certain environment. Some abiotic factors are shown in **Figure 18-2.**

Figure 18-2

Abiotic factors help determine which species can survive in an area.

A **Soil** Soil consists of minerals mixed with decaying bodies of dead organisms. It contains both living and nonliving components.

B **Light** Seasonal events, such as flowering in plants or migration of birds, are often triggered by a change in the number of hours of daylight.

C **Water** Many organisms live in water rather than air. Fish, clams, algae, and whales are examples of organisms that live in water.

D **Temperatures** Temperatures change with daily and seasonal cycles. Desert-dwelling tortoises are active only in the cool, early morning hours. During the hottest part of the day, they rest in the shade under rocks.

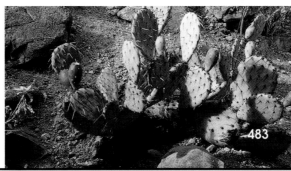

-483

Integrating the Sciences

Earth Science The biosphere is made up of water, land, and air. These three parts are called the hydrosphere, the lithosphere, and the atmosphere. Have students use Earth science reference books and draw how the three spheres intersect. L1 LS

Figure 18-3

Many wildflowers living on the forest floor flower and produce seeds early in the spring when they receive the maximum amount of sunlight. When the leaves are fully out on the trees, they receive little direct sun.

484

Water

Water is an important abiotic factor. The bodies of almost all organisms are 50 to 95 percent water. You learned in Chapter 3 that water is an important part of cytoplasm and of the fluid that surrounds cells. Respiration, photosynthesis, digestion, and other important life processes can take place only in the presence of water.

Soil

The type of soil in a particular location helps determine the type of plants and other organisms in that location. Soil type is determined by the relative amounts of sand, clay, and organic matter, or humus, the soil contains. Most soil is a combination of sand, clay, and humus. Humus is the decayed remains of dead organisms. The greater the humus content, the more fertile the soil.

Light and Temperature

The abiotic factors of light and temperature also impact the environment. As you learned in Chapter 12, through the process of photosynthesis, the radiant energy of sunlight is transformed into chemical energy that drives virtually all of life's processes. The availability of sunlight is a major factor in determining where green plants and photosynthetic protists live, as shown in **Figure 18-3**. For example, sunlight does not penetrate very far into deep water, so most algae are found living near the surface. Because little sunlight reaches the shady darkness of the forest floor, plant growth there is limited.

Activity 18-1

Soil Composition

Soil is more than minerals mixed with the decaying bodies of dead organisms. It contains other biotic and abiotic factors.

Problem
What are the components of soil?

Materials

- 3 small paper cups containing freshly dug soil
- newspaper
- hand lens
- scale
- beaker of water
- jar with lid

Procedure

1. Obtain 3 cups of soil sample from your teacher. Record the source of your sample in your Science Journal.
2. Pour one of your samples onto the newspaper. Sort through the objects in the soil and separate abiotic and biotic items. Use a hand lens to help identify the items. Describe your observations in your Science Journal.
3. Carefully place the second sample in the jar, disturbing it as little as possible. Quickly fill the jar with water and screw the lid on tightly. Without moving the jar, observe its contents for several minutes. Record your observations in your Science Journal.
4. Weigh the third sample. Record the weight in your Science Journal. Leave the sample undisturbed for several days, then weigh it again. Record the second weight in your Science Journal.

Analyze

1. What kinds of materials did you find in your soil sample?
2. What did you observe in step 3?
3. Did the soil weight change over time? If so, why?

Conclude and Apply

4. Did you **observe** organisms in your sample? What were they doing?
5. **Describe** the abiotic factors in your sample. What biotic factors did you **observe?**

18-1 The Living Environment and the Nonliving Environment **485**

Activity 18-1

Purpose

Kinesthetic Students will observe the components of soil. [L1] [LEP]

Process Skills
observing, measuring in SI, comparing and contrasting, forming operational definitions

Time
45 minutes

Safety Precautions
Have students wash their hands after handling the soil.

Activity Worksheets, pp. 5, 106-107

Teaching Strategies

- The soil should be from a single area, such as a garden, lawn, vacant field, or pond bank. Tell students the source of the soil.
- Samples should be freshly dug soil that has been disturbed as little as possible.
- The purpose of step 3 is to observe air bubbles rising from the soil.

Troubleshooting To avoid incorporating air when filling the jar with water, have students pour water into the jar from a beaker rather than the tap.

Answers to Questions

1. Student answers will vary.
2. bubbles of air rising from the soil sample
3. The soil weight changed because the water contained in the sample evaporated.
4. Students will observe worms, fungi, and other soil organisms.
5. Abiotic factors include air, water, and minerals. Biotic factors include decaying plants and animals, and soil organisms.

✔ Assessment

Performance Have students design an experiment that will allow them to compare the growth of organisms that live in different kinds of soil. They may sprinkle soil samples on moist bread or agar plates. Bacterial and mold colonies should be visible after a few days. Cover samples to avoid contamination. Use the Performance Task Assessment List for Designing an Experiment in **PASC,** p. 23. [P]

485

Biotic Factors

Mushrooms would not be able to grow without the decaying bodies of other organisms to feed on. Bees could not survive without pollen from flowers. Some species of owls and woodpeckers prefer to nest in the hollow trunks of dead trees. Abiotic factors do not provide everything an organism needs for survival. All organisms also depend on other organisms for food, shelter, protection, or reproduction. Living organisms in the environment are called **biotic factors.**

Levels of Biological Organization

You have already learned that the living world is highly organized. Atoms are arranged into molecules, which are in turn organized into cells. Cells form tissues, tissues form organs, and organs form body systems. Similarly, the biotic and abiotic factors studied by ecologists form layers of organization, as shown in **Figure 18-4.**

Figure 18-4

Organism—An organism is a single individual from a population.

Population—A population is all of the individuals of one species living in the same area at the same time. The individuals in a population are capable of breeding with one another.

Community—A community is made up of populations of different species that interact in some way.

Ecosystem—An ecosystem consists of communities and the abiotic factors that affect them.

Biosphere—The biosphere is the highest level of biological organization. It is made up of all the ecosystems on Earth.

486

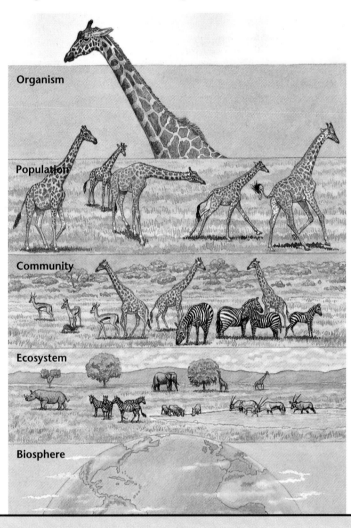

Theme Connection

Scale and Structure The biosphere contains many ecosystems that can be divided into smaller categories such as populations and communities.

Populations

Individual organisms of the same species that live in the same place and can produce young form a **population.** Members of several populations on a coral reef are seen in **Figure 18-5.** Members of populations of organisms compete with each other for food, water, mates, and space. The resources of the environment and how the organisms use these resources determine how large a population can be.

Communities

Most populations of organisms do not live alone; they live and interact with populations of other organisms. Groups of populations that interact with each other in a given area form a **community.** Populations of organisms in a community depend on each other for food, shelter, or for other needs.

Ecosystem

An **ecosystem** is made up of a biotic community and the abiotic factors that affect it. The rest of this chapter will discuss in more detail the kinds of interactions that take place between abiotic and biotic factors.

Figure 18-5

This coral reef is an example of an ecosystem. It contains hundreds of populations of organisms, as well as ocean water, sunlight, and other abiotic factors.

Section Wrap-up

Review

1. What is the difference between an abiotic factor and a biotic factor? Give at least five examples of each.

2. What is the difference between a population and a community? A community and an ecosystem?

3. **Think Critically:** Could oxygen in the atmosphere be considered an abiotic factor? Why or why not? What about carbon dioxide?

Skill Builder
Observing and Inferring
Each person lives in a population as part of a community. Describe your population and community. If you need help, refer to Observing and Inferring in the **Skill Handbook.**

 Science Journal

Write a paragraph in your Science Journal explaining how soil is influenced by biotic factors and by other abiotic factors.

Skill Builder
Students may describe themselves as humans, living in towns, inside houses, and as consumers, students, children, brothers, sisters, etc. **P**

Assessment

Portfolio Cut out pictures from magazines that can illustrate soil formation. Make a poster using the pictures. Be sure to include all stages. Use the Performance Task Assessment List for Poster in **PASC,** p. 73. **P**

4 Close

Section Wrap-up

Review

1. Biotic factors are living. Abiotic factors are nonliving. biotic—monerans, protists, fungi, plants, and animals; abiotic—soil, temperature, air, water, light, wind

2. An ecosystem is made up of a community and the abiotic factors that affect it. A population is all the organisms of one species living together in the same place at the same time. A community is all the populations that live in the same place at the same time.

3. **Think Critically** Living organisms interact with oxygen and carbon dioxide, which are considered abiotic factors.

Science Journal

Journal entries should reflect that decaying plants and animals return nutrients to the soil and that abiotic factors affect the soil.

Prepare

Section Background

• Animal population size can be influenced by the amount of food and space available. Population size is also controlled by environmental and biological factors. Natural populations cannot increase forever. An important factor influencing the size of a plant population is the amount of daily sunlight and the space available in which roots can grow and spread.

• *Symbiosis* comes from a Greek word meaning "living together."

Preplanning

• Obtain bean seeds for Activity 18-2.

• Obtain legume plants and nonlegume plants for the MiniLAB.

1 Motivate

Bellringer

Before presenting the lesson, display **Section Focus Transparency 65** on the overhead projector. Assign the accompanying **Focus Activity** worksheet. [L1] [LEP]

Tying to Previous Knowledge

Have students recall the needs of living things from Chapter 1. Knowing what living things need to grow and reproduce will be helpful in understanding limiting factors.

18•2 Interactions Among Living Organisms

Science Words

population density
limiting factor
carrying capacity
predation
symbiosis
mutualism
commensalism
parasitism
habitat
niche

Objectives

• Describe the characteristics of populations.
• Identify the types of relationships that occur among populations in a community.
• Compare the habitat and niche of a species in a community.

Characteristics of Populations

As shown in **Figure 18-6,** populations can be described according to certain characteristics. These include the size of the population, spacing (how the organisms are arranged in a given area), and density (how many individuals there are in a specific area). Suppose you spent several months observing a population of field mice living in a pasture. You would probably observe changes in the size of the population as older mice die, baby mice are born, and some mice wander away to make their homes elsewhere. The size of a population—the number of individual organisms it contains—is always changing, although some populations change more rapidly than others. The number of pine trees changes fairly slowly in a forest, but a forest fire, if left undisturbed, could quickly reduce the population of pine trees in the forest.

Figure 18-6

Populations have several characteristics that define them.

 Spacing A characteristic of populations is spacing. In some populations, such as the oak trees of an oak-hickory forest, individuals are spaced fairly evenly throughout the area.

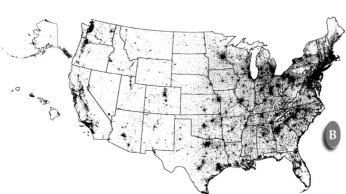

B **Density** Population density is higher in and around cities than in rural areas.

Program Resources

📁 **Reproducible Masters**
Study Guide, p. 69 [L1]
Reinforcement, p. 69 [L1]
Enrichment, p. 69 [L3]
Science and Society Integration, p. 22
Activity Worksheets, pp. 108-109, 110 [L1]
Critical Thinking/Problem Solving, p. 22
Lab Manual 44, 45

📠 **Transparencies**
Section Focus Transparency 65 [L1]

Population Density

At the beginning of this chapter, when you calculated how much space is available to each student in your classroom, you were measuring another population characteristic. The size of a population occupying an area of a specific size is called **population density.** The more individuals there are in a given amount of space, the more dense the population. For example, if there are 100 mice in an area of a square kilometer, the population density is 100 mice per km^2. The change in size of the entire human population over time is shown in **Figure 18-7.**

Figure 18-7

The size of the human population is increasing at a rate of about 1.6 percent per year. At the present time, it is about 5.6 billion. Scientists estimate that by the year 2000, Earth will contain 6 billion people.

Limiting Factors

Populations cannot continue growing larger and larger forever. In any ecosystem, there are limits to the amount of food, water, living space, mates, nesting sites, and other resources available. A **limiting factor** is any biotic or abiotic factor that restricts the number of individuals in a population. A limiting factor can also indirectly affect other populations in the community. For example, a drought might restrict the growth of seed-producing plants in a forest clearing. Fewer plants means that food may become a limiting factor for a mouse population that feeds on the seeds. Food may also become a limiting factor for hawks and owls that feed on the mice, as well as for the deer in **Figure 18-8.**

Competition is the contest among organisms to obtain the resources they need to survive and reproduce. As population density increases, so does competition among individuals.

Carrying Capacity

Suppose a population of robins continues to increase in size, year after year. At some point, food, nesting space, or other resources will become so scarce that some individuals may not

Figure 18-8

In many parts of the United States, deer populations have become large enough to exceed the environment's ability to produce adequate food. Individuals starve or, weakened from lack of food, fall victim to a variety of diseases.

489

Science Journal

Population Growth Use the *World Almanac* to find out what the world population size was for your state in 1950, 1960, 1970, 1980, and 1990. Plot the data on a line graph. Make a point showing what you think the population for 2000 will be. **L1** **LS**

Activity 18-2

Activity 18-2

PREPARATION

Purpose

IS **Interpersonal** Students will determine how abiotic factors affect the germination of seeds. **L1** **COOP LEARN**

Process Skills

observing and inferring, communicating, making and using tables, making and using graphs, comparing and contrasting, recognizing cause and effect, forming a hypothesis, designing an experiment, using numbers, separating and controlling variables, and interpreting data

Time

50 minutes for setup; 10 minutes every three to four days for data collection for a period of about two weeks

Materials

Square milk cartons may be used for planters.

Alternate Materials Other kinds of seeds such as pea seeds or flower seeds may be used.

Possible Hypotheses

Most student hypotheses will reflect that abiotic factors such as light, water, and temperature affect the germination of seeds.

📁 **Activity Worksheets,** pp. 5, 108-109

Activity 18-2

Design Your Own Experiment

Identifying a Limiting Factor

Organisms depend on many biotic and abiotic factors of their environment to survive. When these factors are limited or are not available to organisms, it can affect their survival. By experimenting with some of these limiting factors, you will see how organisms depend on all parts of their environment.

PREPARATION

Problem

How do abiotic factors such as light, water, and temperature affect the germination of seeds?

Form a Hypothesis

Based on what you have learned about limiting factors, make a hypothesis about how one specific abiotic factor may affect the germination of a bean seed. Be sure to consider factors that you can easily change.

Objectives

- Observe the effects of an abiotic factor on the germination and growth of bean seedlings.
- Design an experiment that demonstrates whether or not a specific abiotic factor limits the germination of bean seeds.

Possible Materials

- bean seeds
- small planting containers
- soil
- water
- labels
- trowel or spoon
- aluminum foil
- sunny window or other light source
- refrigerator or oven

Safety Precautions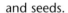

Wash hands after handling soil and seeds.

PLAN THE EXPERIMENT

1. As a group, agree upon and write out a hypothesis statement.
2. Decide on a way to test your group's hypothesis. Keep available materials in mind as you plan your procedure. List your materials.
3. Prepare a data table in your Science Journal.

Check the Plan

1. Read over your entire experiment to make sure that all steps are in logical order.
2. Identify any constants, variables, and controls in your experiment.
3. Be sure the factor you test is measurable. Record all data and observations in the data table.

4. Remember to test only one variable at a time and use suitable controls.
5. *Make sure your teacher has approved the plan before you proceed.*

DO THE EXPERIMENT

1. Carry out the experiment as planned.
2. While the experiment is going on, write down any observations that you make and complete the data table in your Science Journal.

3. **Graph** your results using a bar graph that compares the number of bean seeds that germinated in the experimental container with the number of seeds that germinated in the control container.

Analyze and Apply

1. **Compare** your results with those of other groups.
2. How did the abiotic factor you tested affect the germination of bean seeds?

Go Further

Design an experiment that tests the effects of several abiotic and biotic factors on bean seed germination.

 Assessment

Performance Have students design an experiment to test the effect of the same variable on another kind of seed. Use the Performance Task Assessment List for Designing an Experiment in **PASC**, p. 23. **P**

PLAN THE EXPERIMENT

Possible Procedures
Plant the same number of seeds in each of two planters and keep all conditions the same, except for the variable. Observe the results.

Teaching Strategies
Tying to Previous Knowledge Most students will have observed the germination of seeds and will be aware of the conditions that are necessary for germination. Ask students to identify them.

Troubleshooting Caution students not to plant the seeds too deep in the soil and not to overwater the planter.

DO THE EXPERIMENT

Expected Outcome
Most results will reflect that seeds need water and the proper temperature to germinate.

Analyze and Apply

1. Student results should be consistent with those of other groups using the same variable.
2. Student answers will vary based on the abiotic factor tested.
3. The graph will be determined by the number of seeds used in the experiment.

Go Further

The growth and development of bean seeds can be affected by many different factors. Students should choose one abiotic or biotic factor to test at a time. Although one factor is usually tested at a time, students may wish to choose an abiotic and a biotic factor to test together.

Figure 18-9

This graph shows how the size of a population increases until it reaches the carrying capacity of its environment. At first, growth is fairly slow, then speeds up as the number of adults capable of reproduction increases. Once the population reaches carrying capacity, its size remains fairly stable. *Why don't most populations achieve their biotic potential?*

Carrying capacity

Number of individual organisms

Time

Figure 18-10

When the moose population on an island in Lake Superior exceeded the island's carrying capacity, food became a limiting factor and many of the animals starved. Then wolves moved onto the island and began preying on the moose. The size of the prey population got smaller, finally leveling off at a number that was within the island's carrying capacity.

be able to survive or reproduce. When this happens, the environment has reached its carrying capacity. **Carrying capacity** is the largest number of individuals an environment can support and maintain for a long period of time. If a population begins to exceed the environment's carrying capacity, some individuals will be left without adequate resources and may die or be forced to move elsewhere.

Biotic Potential

What would happen if there were no limiting factors? A population living in an environment that supplies more than enough resources for survival will continue to grow, as more and more young are produced. A population's biotic potential is the number of individuals each female of a population can produce under the best possible conditions of plenty of food, water, ideal weather, and no disease or enemies. However, most populations never reach their biotic potential, or do so for only a short time. Eventually, the carrying capacity of the environment is reached and the population stops growing, as shown in **Figure 18-9.**

Interactions in Communities

Populations are regulated not only by the supply of food, water, and sunlight, but also by the actions of other populations.

The most obvious way one population can limit another is by **predation,** the feeding of one organism on another. Owls and hawks are predators that feed on mice, which are their prey. Predators are biotic factors that limit the size of prey populations. Because predators are more likely to capture old, ill, or extremely young prey, predation also helps maintain the health of a prey population by leaving only the strongest individuals to reproduce and pass their genes on to the next generation. **Figure 18-10** gives one example of the effects that predation can have on a population.

Symbiosis

There are many types of relationships between organisms in ecosystems. Many species of organisms in nature have close, complex relationships for survival. When two or more species live together, their relationship is

USING TECHNOLOGY

Monitoring Mayflies

When water pollutants such as raw sewage or chemicals enter a creek or a river, they are immediately diluted and begin to flow downstream. As a result, it is difficult to identify where the pollutants are coming from. Biologists can use population sizes of aquatic organisms to help them determine whether a stream is suffering ill effects from water pollution and help them identify the source of the pollutants.

Stream-dwelling invertebrates such as mayflies, caddis flies, stone flies, aquatic earthworms, and midges have differing tolerances to pollution levels. These organisms also tend to stay in one place. In a process called biomonitoring, biologists use comparisons of the population sizes of these species to evaluate the health of that part of the stream. Mayflies tend to be sensitive to pollutants. Large numbers of them indicate a healthy stream. Aquatic earthworms and midges can tolerate much higher pollutant levels. A stream with large populations of these organisms could be in trouble.

Think Critically:

The invertebrates used in these types of studies have short life spans, usually less than a year. Why is this trait useful in biomonitoring?

Mayfly Nymph

For more information on biomonitoring, contact the water division of the local environmental agency in your area.

Think Critically

A short life span means population size can change quickly. These populations quickly reflect changes in the health of the stream. If the organisms lived for many years, it might take longer to detect population changes.

Activity

Visual-Spatial Have students look at lichens with a hand lens. Then have them prepare a wet-mount slide with lichen pieces and observe the slide under the microscope. Ask: **How do the algae and fungus benefit each other?** *Algae make food for the fungus and the fungus provides a habitat and moisture for the algae.* L1

Use Lab 44 in the **Lab Manual** as you teach this lesson.

Across the Curriculum

Art Have students research symbiosis and make a drawing to illustrate a symbiotic relationship. L1 LS

Purpose

LS Visual-Spatial Students will compare the roots of a legume plant and a nonlegume plant. **L1**

Materials

legume plant (clover, alfalfa), nonlegume plant (geranium, coleus), prepared slide of nitrogen-fixing bacteria, beaker of water, hand lens

Teaching Strategies

A hand lens can be used to observe the nodules on the legume plant.

Analysis

1. The legume plant has nodules that contain nitrogen-fixing bacteria.
2. The bacteria make nitrogen available to the plant roots. The plant gives bacteria food.

Activity Worksheets, pp. 5, 110

Assessment

Performance Have students conduct research about symbiotic relationships. They should record the information they find in their Science Journals. Use the Performance Task Assessment List for Science Journal in **PASC**, p. 103. **P**

3 Assess

Check for Understanding

? FLEX Your Brain

Use the Flex Your Brain activity to have students explore SYMBIOSIS.

Activity Worksheets, p. 5

494

How do nitrogen-fixing bacteria affect plants?

Alfalfa, beans, and clover are able to thrive in nitrogen-poor soil because they have a symbiotic relationship with a certain kind of bacterium that lives in their roots and helps fix nitrogen.

Procedure

1. Closely wash then examine the roots of a legume plant and a nonlegume plant.
2. Examine a microscope slide of the nitrogen-fixing bacteria.

Analysis

1. What differences do you observe in the roots of the two plants?
2. How do the bacteria aid the plant? How does the plant aid the bacteria?

---

called a symbiotic relationship. **Symbiosis** is any close relationship between two or more different species. Not all relationships benefit one organism at the expense of another as in predation. Symbiotic relationships can be identified by the type of interaction between organisms, as shown in **Figure 18-11.** There are many types of symbiotic relationships between organisms. These are usually described by how each organism in the relationship is affected by the relationship.

A symbiotic relationship that benefits both species is called **mutualism.** An example of mutualism is the lichen. As you learned in Chapter 9, each lichen species is made up of a fungus and an alga or cyanobacterium. The fungus provides a protected living space for the alga, and the alga or bacterium provides the fungus with food.

In shallow tropical seas, brightly colored anemone fish find protection from predators by swimming among the stinging tentacles of sea anemones. The presence of the fish does not affect the anemone in any harmful or beneficial way. **Commensalism** is a symbiotic relationship that benefits one partner but does not harm or help the other.

Figure 18-11

There are many examples of symbiotic relationships in nature.

A The partnership between the desert yucca plant and the yucca moth is an example of mutualism. Both species benefit from the relationship. The yucca depends on the moth to pollinate its flowers. The moth depends on the yucca for a protected place to lay its eggs and a source of food for its larvae.

C Fleas and lice are parasites that feed on the blood of mammals and other vertebrates.

Magnification 54×

B Tropical orchids grow on the trunks of trees. The tree provides the orchid with a sunlit living space high in the forest canopy. This relationship is an example of commensalism because the orchid benefits from the relationship without harming or helping the tree.

494 Chapter 18 Life and the Environment

Cultural Diversity

Habitat Diversity Discuss and compare the habitats of different peoples on Earth. Show pictures of different cultures in distant parts of the world, such as Australia, Japan, Nigeria, the United Kingdom, India, Turkey, and others. Have students research the habitats of other peoples.

Parasitism is a symbiotic relationship that benefits the parasite and does definite harm to the parasite's partner, or host. Many parasites live on or in the body of the host, absorbing nutrients from the host's body fluids. Tapeworms live as parasites in the intestines of mammals. Mistletoe is a parasitic plant that penetrates tree branches with its roots.

Habitats and Niches

In a community, every species has a particular role. Each also has a particular place to live. The physical location where an organism lives is called its **habitat.** The habitat of an earthworm is soil. The role of an organism in the ecosystem is called its **niche.** The niche of an earthworm is shown in **Figure 18-12.** What a species eats, how it gets its food, and how it interacts with other organisms are all parts of its niche. As an earthworm moves through the soil, it takes the soil into its body to obtain nutrients. The soil that leaves the worm enriches the soil. The movement of the worm through soil also loosens it and aerates it, creating a better environment for plant growth.

Figure 18-12
Each organism in an ecosystem uses and affects its environment in a particular way.

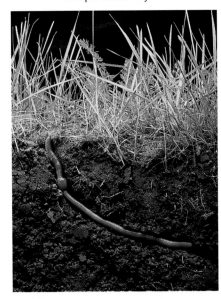

Section Wrap-up

Review

1. Describe how limiting factors can affect the organisms in a population.

2. Describe the difference between a habitat and a niche.

3. **Think Critically:** A parasite can obtain food only from its host. Most parasites slightly weaken but do not kill their hosts. Why?

Skill Builder
Making and Using Tables
Make a table that describes each of the following interactions among organisms: commensalism, mutualism, parasitism, and predation. Use + and − signs to indicate benefit and detriment, and 0 for no effect. Be sure to indicate this for each species in the relationship. If you need help, refer to Making and Using Tables in the **Skill Handbook.**

USING MATH

In a 12 m² area of weeds, there are 46 dandelion plants, 212 grass plants, and 14 bindweed plants. What is the population density per square meter of each species?

18-2 Interactions Among Living Organisms **495**

Skill Builder

INTERACTION	SPECIES	
	A	**B**
Commensalism	+	0
Mutualism	+	+
Parasitism	+	-
Predation	+	-

✔ Assessment

Performance Have students research data about population growth or decline of a threatened or endangered species. Have students put the data with number of individuals of the species on the vertical axis and year on the horizontal axis. Students should describe the growth curve. Use the Performance Task Assessment List for Graph from Data in **PASC**, p. 39. **P**

Reteach
Have students make a list of symbiotic relationships with which they are familiar and identify each.

Extension
For students who have mastered this section, use the **Reinforcement** and **Enrichment** masters.

4 Close

•MINI•QUIZ•

Use the Mini Quiz to check for understanding.
1. **The number of individuals per unit of living space is the _____ .** *population*
2. **What a species eats, how it gets its food, and how it interacts within a community is its _____ .** *niche*
3. **The place where an organism lives is its _____ .** *habitat*

Section Wrap-up

Review

1. A limiting factor such as food, living space, drought, and competition can cause a population to become smaller or limit the growth of a population.
2. The place where an organism lives is its habitat. The way an organism interacts with its surroundings is its niche.
3. **Think Critically** If a parasite kills its host, the parasite is likely to die because its food supply will disappear.

USING MATH

dandelion: 46/12 = 3.8; grass: 212/12 = 17.6; bindweed: 14/12 = 1.2

Prepare

Section Background

1 Motivate

Bellringer

Before presenting the lesson, display **Section Focus Transparency 66** on the overhead projector. Assign the accompanying **Focus Activity** worksheet. L1 LEP

Tying to Previous Knowledge

Review the definitions of *producers, consumers,* and *decomposers* from Chapter 3. Review the definition of energy as the ability to do work. Solar energy is transformed into chemical energy that is stored in food. Through chemical changes in cells, this energy is released for use by an organism. Review metabolism from Chapter 3.

496

18•3 Matter and Energy

Science Words

food chain
food web
ecological pyramid
water cycle
nitrogen cycle

Objectives

• Explain how energy flows through ecosystems.
• Describe the cycling of matter in the biosphere.

Figure 18-13

In any community, energy flows from producers to consumers. You will be able to follow several food chains in the pond ecosystem shown here.

The Flow of Energy Through Ecosystems

As you can see, life on Earth is not simply a collection of living organisms. Even organisms that appear to spend most of their time alone do interact with other members of their same species as well as with other organisms. Most of the interactions between members of different species are feeding relationships involving the transfer of energy from one organism to another. Energy moves through an ecosystem in the form of food. Recall from Chapter 3 that producers are organisms that capture energy from the sun and use it to fuel photosynthesis, thus producing chemical bonds in carbohydrates. Consumers are organisms that obtain energy when they feed on producers or other consumers. The transfer of energy does not end there. When organisms die, other organisms called decomposers obtain energy when they break down the bodies of the dead organisms. This transfer of energy through a community can be modeled by food chains, as seen in **Figure 18-13**, and food webs.

A The first link in any food chain is a producer. In this pond ecosystem, the producers are phytoplankton, algae, and a variety of plants, both aquatic and those on the shore.

B The second link of a food chain is usually an herbivore, an organism that feeds only on producers. Here, snails and small aquatic crustaceans are feeding on the algae and pond plants.

C The third link of a food chain is a carnivore, an animal that feeds on other animals. Some of the carnivores in this pond are bluegill, turtles, and frogs.

496

Program Resources

Reproducible Masters
Study Guide, p. 70 L1
Reinforcement, p. 70 L1
Enrichment, p. 70 L3
Cross-Curricular Integration, p. 22
Activity Worksheets, p. 111 L1
Science Integration Activities, pp. 35-36
Concept Mapping, p. 41
Technology Integration, pp. 15-16

Transparencies
Section Focus Transparency 66 L1
Teaching Transparencies 35, 36
Science Integration Transparency 18

Food Chains and Food Webs

A **food chain** is a simple way of showing how energy from food passes from one organism to another. The pond community pictured in **Figure 18-13** shows examples of many aquatic food chains. When drawing a food chain, arrows between organisms indicate the direction of energy transfer. An example of a pond food chain would be as follows.

phytoplankton → insects → bluegill → bass

Food chains usually have three or four links. Most have no more than five links. This is due to the decrease in energy available at each link. The amount of energy left by the fifth link is only a small portion of the total amount of energy available at the first link. This is because at each transfer of energy, a portion of the energy is lost as heat due to the activities of the organisms as they search for food and mates.

Single food chains are too simple to describe the many interactions among organisms in an ecosystem. There are actually many food chains in any ecosystem. A **food web** is a series of overlapping food chains, as seen in **Figure 18-14** on page 498. This concept provides a more complete model of the way energy moves through a community. Food webs are also more accurate models because they show the many organisms that feed on more than one level in an ecosystem.

D The fourth link of a food chain is a top carnivore, which feeds on other carnivores. Examples of these consumers in this pond are large fish such as crappies and bass.

E When organisms die in any ecosystem, bacteria and fungi, decomposers, feed on the dead organism, breaking down the remains of the organism.

497

2 Teach

Activity

LS **Visual-Spatial** Have pictures of plants, animals, and fungi from magazines mounted on construction paper. Give each group of four students seven to ten pictures. Have them develop a food chain using the pictures. Have them present their food chain to the class. Then have the teams use their food chains to make a food web. **L1** **COOP LEARN** **LEP**

GLENCOE TECHNOLOGY

Videodisc

GTV: Planetary Manager

Animal

47481

47482

47492

47493

47499

47508

47509

47510

Visual Learning

Figure 18-13 Encourage students to take time to study **Figure 18-13** closely. Carefully explain each feeding level in the illustration. You may wish to have students examine algae under the microscope as you discuss the beginning of the food web. **LEP** **LS**

Integrating the Sciences

Chemistry The processes of photosynthesis and chemosynthesis both produce food for organisms. In photosynthesis, the energy from sunlight is used by green plants to make glucose from carbon dioxide and water. In chemosynthesis, bacteria produce food and oxygen by using dissolved sulfur compounds that escape from lava.

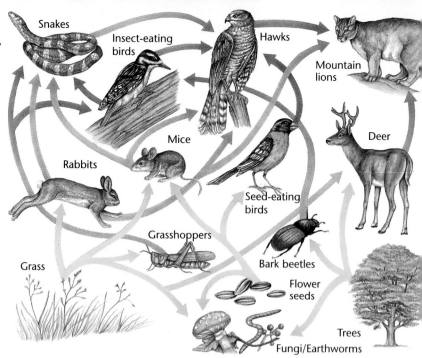

Figure 18-14

A food web includes many food chains. It provides a more complete and informative model of the complex feeding relationships in a community than a single food chain.

Ecological Pyramids

Virtually all the energy available to the biosphere comes from the sun, but producers capture and transform only a small fraction of the energy that reaches Earth's surface. When an herbivore eats a plant, some of the energy in the plant is transferred to the herbivore, but most of it is given off into the atmosphere as heat. The same thing happens when a carnivore consumes an herbivore. This transfer of energy can be modeled by an **ecological pyramid.** The base of an ecological pyramid represents the producers of an ecosystem. The next higher levels represent successive organisms in the food chain.

Figure 18-15

An energy pyramid illustrates that energy decreases at each successive feeding step. *Why aren't there more levels in an energy pyramid?*

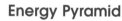

Energy Pyramid

The flow of energy from grass seeds to the hawk in **Figure 18-15** can be illustrated by an energy pyramid. An energy pyramid compares the energy available at each level of the food chain in an ecosystem. Just as most food chains have three or four links, a pyramid of energy usually has

498   Chapter 18 Life and the Environment

Integrating the Sciences

Chemistry Have students research biological magnification. Biological magnification is the process in which hazardous wastes become increasingly concentrated in successive levels of the food chain. It can be seen in environmental levels of lead. L1

Cultural Diversity

What do people eat? Have students research how people in different countries meet their food needs. Suggest that students choose a country and find information about the diets of the people living there. Have them consider whether the people are mainly first-, second-, or third-level consumers. L1 LS three or four levels. Only about ten percent of the energy available at each level of the pyramid is transferred to the next level. By the time the top level is reached, the amount of energy is greatly reduced.

The Cycles of Matter

The energy that is dispersed at each link in the food chain is constantly renewed by sunlight. But what about the physical matter that composes the bodies of living organisms? Matter on Earth is never lost or gained, but used over and over again. In other words, it is recycled. The carbon atoms present in your body right now have been on Earth since the planet was formed billions of years ago. They have been recycled untold billions of times. Many important materials that make up your body cycle through ecosystems. These materials are water, carbon, and nitrogen.

Problem Solving

Changes in Antarctic Food Webs

The food chain in the ice-cold Antarctic Ocean is based on phytoplankton—microscopic algae that float near the water's surface. The algae are eaten by tiny shrimplike krill, which are consumed by baleen whales, squid, and fish. The fish and squid are eaten by toothed whales, seals, and penguins. Humans have hunted baleen whales for the past 150 years, greatly reducing their numbers. Now that most countries have put a stop to whale hunting, there is hope that the population of baleen whales will increase. How will an increase in the whale population affect the food web?

Solve the Problem:
1. Draw a food web for the Antarctic.
2. Which organisms compete for the same source of food?

Think Critically:
1. Populations of seals, penguins, and krill-eating fish increased in size as populations of baleen whales declined. Why?
2. What might happen if the number of baleen whales increases, but the amount of krill does not?

18-3 Matter and Energy   **499**

Purpose

 Interpersonal Students will model what happens to water vapor as it rises in the atmosphere. [L1] **LEP**

COOP LEARN

Materials

Bunsen burner, ring stand with ring, wire gauze, beaker, watch glass that will cover the beaker opening, ice cube, water

Analysis

1. It evaporates.
2. It condenses back into a liquid.
3. Heat causes evaporation. Cold temperatures cause condensation.

 Assessment

Performance Have students diagram the water cycle in their Science Journals. Use the Performance Task Assessment List for Scientific Drawing in **PASC**, p. 55.
P

 Activity Worksheets, pp. 5, 111

INTEGRATION
Chemistry

Less than three percent of the water on Earth is fresh. Most of this water is frozen in the polar ice caps.

3 Assess

Check for Understanding

? FLEX Your Brain

Use the Flex Your Brain activity to have students explore CYCLING OF MATTER.

 Activity Worksheets, p. 5

500

INTEGRATION
Chemistry

MiniLAB

How does temperature affect the water cycle?

Find out what happens to water vapor as it rises in the atmosphere.

Procedure

1. Place a small amount of water in a beaker and heat it over a low flame.
2. Place a watch glass over the top of the beaker. Make sure it completely covers the beaker opening. Place an ice cube in the watch glass.

Analysis

1. What happens when the water at the bottom of the beaker is heated?
2. What happens at the top of the beaker, under the watch glass?
3. What does temperature have to do with these events?

500 Chapter 18 Life and the Environment

Figure 18-16

A water molecule that falls on the ground as rain can follow several paths through the water cycle. *Identify as many of these paths as you can in this diagram.*

The Water Cycle

Water molecules on Earth are on a constant journey, rising into the atmosphere, falling to land or the ocean as rain, and flowing into rivers and oceans. The **water cycle** involves the processes of evaporation, condensation, and precipitation.

When energy is added to a liquid, its molecules begin moving faster. The more energy the molecules absorb, the faster they move, until they are moving so fast they break free and rise into the atmosphere. The liquid evaporates, or changes from a liquid to a gas. The heat of the sun causes water on the surface of Earth to evaporate and rise into the atmosphere as gaseous water vapor.

As the water vapor rises, it encounters colder and colder air temperatures. As the molecules of water vapor become colder, they move more slowly. Eventually, the water vapor changes back into tiny droplets of water. It condenses, or changes from a gas to a liquid. These water droplets clump together to form clouds. When the droplets become large and heavy enough, they fall back to Earth as rain, or precipitation. This process is illustrated in **Figure 18-16.**

Reteach

 Kinesthetic Play the food web game. List producers, consumers, and decomposers in a food web on separate index cards. Give a card to each student. Give each producer a ball of colored yarn. Have producers roll the ball to their primary consumers, then each subsequent consumer until each chain is complete. This creates a food web.

Community Connection

Cleaning the Water Take students on a field trip to a local water treatment plant and have a plant engineer explain the water treatment process.

The Nitrogen Cycle

Nitrogen is an important element needed by organisms to make proteins. Even though nitrogen gas makes up 78 percent of the atmosphere, most living organisms cannot use nitrogen in this form. It has to be combined with other elements in a process called nitrogen fixation. You can see in **Figure 18-17** how nitrogen is converted into usable compounds by bacteria associated with plants. A small amount is changed into nitrogen compounds by lightning. The transfer of nitrogen from the atmosphere to plants, and back to the atmosphere or directly into plants again, is the **nitrogen cycle.**

Phosphorous, sulfur, and other elements needed by living organisms are also used and returned to the environment. Just as we save aluminum, glass, and paper products to be reused, the materials that organisms need to live are recycled continuously in the biosphere.

Figure 18-17

Nitrogen can be cycled from bacteria on plant roots to plants, then to animals, and directly back to plants again as a result of decomposition.

Atmospheric nitrogen converted by lightning

Animals eat plants

Animals and plants die and decompose

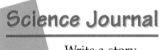

Bacteria on special plants fix nitrogen and change it to a usable form

Plants use nitrogen

Section Wrap-up

Review

1. How does the flow of energy through an ecosystem compare with the cycling of matter?

2. **Think Critically:** Using your knowledge of food chains and the energy pyramid, explain why there are fewer lions than gazelles on the African plains.

Skill Builder
Classifying

Look at the food web pictured in **Figure 18-14.** Classify each organism pictured as a producer, herbivore, omnivore, carnivore, or decomposer. If you need help, refer to Classifying in the **Skill Handbook.**

Science Journal

Write a story that describes the experiences of a single water molecule as it moves through the water cycle.

4 Close

•MINI•QUIZ•

Use the Mini Quiz to check students' recall of chapter content.

1. **A series of overlapping food chains is a(n)** _____ . *food web*

2. **The transfer of energy can be modeled by a(n)** _____ . *ecological pyramid*

3. **The** _____ **involves the processes of evaporation, condensation, and precipitation.** *water cycle*

Section Wrap-up

Review

1. Energy moves in one direction through the biosphere and is constantly renewed by the sun. Matter is recycled.

2. **Thinking Critically** There is not enough energy left at the top of an energy pyramid to support large numbers of top carnivores.

Skill Builder

snake—carnivore, insect-eating bird—carnivore, hawk—carnivore, mountain lion—carnivore, mouse—herbivore, rabbit—herbivore, seed-eating bird—herbivore, deer—herbivore, grasshopper—herbivore, bark beetle—herbivore, flower seed—producer, tree—producer, grass—producer, fungus—decomposer, earthworm—decomposer

✓ Assessment

Performance Put a picture of an underwater pond scene on the overhead projector. Have students identify the producers, consumers, and decomposers. Then have students diagram a possible food chain or web in their Science Journals. Use the Performance Task Assessment List for Making Observations and Inferences in **PASC**, p. 17. **P**

Science Journal

IS **Linguistic** Journal entries should describe a water molecule going through evaporation, condensation, and precipitation. **P**

Prepare

Section Background

- Wolves are expert hunters and prey mainly on large hoofed animals, such as caribou, deer, elk, and moose. Many people fear wolves. They believe that wolves attack human beings, and the animal's eerie howl frightens them. But wolves avoid people as much as possible.

- Wolves can destroy sheep, cows, and other livestock for food. Many hunters dislike wolves because they kill game animals. The hunters mistakenly think that wolves wipe out the game in certain areas.

1 Motivate

Bellringer

 Before presenting the lesson, display **Section Focus Transparency 67** on the overhead projector. Assign the accompanying **Focus Activity** worksheet. L1 LEP

2 Teach

GLENCOE TECHNOLOGY

 Videodisc

STVS: Animals
Disc 5, Side 2
Temperature Regulation in Dogs (Ch. 19)

ISSUE:
18•4 Bringing Back the Wolves

Objectives

- Compare and contrast the possible benefits and drawbacks of reintroducing wolves into Yellowstone National Park.

Figure 18-18

Wolves were viewed as dangerous, undesirable animals even before they began preying on livestock of ranchers.

The Gray Wolf in North America

The gray wolf is a large carnivore that preys on deer, elk, bison, and other mammals. The species was once common throughout North America, but the population began to decline back in the 1600s, when European settlers started moving into the wolves' habitat. As ranches and farms grew in size and number, the habitat of the wolves and their prey shrank in size. Prey populations were further reduced because the settlers killed large numbers of the animals the wolves depended upon for food. As their food supply dwindled, wolves began to prey on the easily captured cattle and sheep belonging to the ranchers.

By the early 1900s, wolves had almost disappeared from the lower 48 states of the United States. Wolves are still being killed, not only by hunters looking for fur pelts, but also by ranchers who believe wolves are preying on their livestock. Many states have designated the gray wolf an endangered species, which means that wolf hunting is now illegal in most of this country. Today, gray wolf populations are found primarily in Canada, Alaska, northern Michigan, Wisconsin, Montana, Idaho, and, for the first time in decades, in Wyoming's Yellowstone National Park.

Wolves Return to Yellowstone

Until 1995, Yellowstone had been without a wolf population for at least 70 years. Without these predators, prey populations of bison and elk reached record numbers. During hard winters, thousands of these animals starved. The presence of a predator would help keep elk, bison, and deer populations within the carrying capacity of park lands. So, in 1995, wildlife experts introduced into Yellowstone 14 wolves captured in Canada, with the hope that they will reproduce and repopulate the area. Wolf lovers and wildlife biologists are happy about restoring the park community to its original ecological balance. But neighboring ranchers in the region are worried about losing their livestock to wolves.

Program Resources

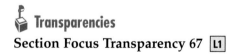 **Reproducible Masters**
Study Guide, p. 71 L1
Reinforcement, p. 71 L1
Enrichment, p. 71 L3

Transparencies
Section Focus Transparency 67 L1

2 Points of View

Wolves are an environmental and economic benefit.

Those in favor of bringing wolves back to Yellowstone estimate that the increase in visitors who want to see wolves in the wild will bring an extra $20 million a year into the area's economy. They also claim that there is so much prey available in the park that wolves will rarely venture out to neighboring ranches. Any rancher who loses livestock to a wolf will be paid for the loss, and ranchers will be permitted to shoot wolves caught in the act of killing livestock.

Wolves are an economic threat.

Ranchers point out that there are no fences around Yellowstone, and many of the newly introduced wolves have already been seen roaming outside park boundaries. The ranchers are worried about having to put time, energy, and money into guarding their herds against marauding wolves. Hunters wonder if wolves will so severely reduce populations of the elk and deer that recreational hunting of those animals in areas near the park will no longer be possible.

Figure 18-19

Another issue that can be raised is whether humans should allow wolves to be wolves simply for their own sake and without considering only their effects on the lives of humans.

Section Wrap-up

Review

1. Why was it difficult for the European settlers to accept the presence of the wolves?

2. What problems does the removal of a predator population cause?

 Visit the Chapter 18 Internet Connection at Glencoe Online Science, **www.glencoe.com/sec/science/life**, for a link to more information about wolves in Yellowstone National Park.

SCIENCE & SOCIETY

18-4 Bringing Back the Wolves   **503**

Explore the Issue

Answers will vary, but may include that there is so much prey available in the park that the wolves would rarely go to the ranches. Ranchers will be paid for their losses and allowed to shoot wolves caught in the act of killing livestock. The prey populations will die from starvation and disease. **P**

3 Assess

Check for Understanding

Have students answer questions 1 and 2 in the Section Wrap-up.

Reteach

Discuss the advantages of having the wolves: control of prey populations and the extra income for the area's economy. Have students determine whether the disadvantages to farmers and hunters are greater than the advantages.

Extension

For students who have mastered this section, use the **Reinforcement** and **Enrichment** masters.

4 Close

Red wolves (*Canis rufus*) have been reintroduced to parts of the southeastern U.S. Have students find out more about this species and the problems encountered with reintroduced populations. Some possible sources are *Newsweek*, "The Return of the Native," January 23, 1995, p. 53, and Dorothy Hinshaw Patent, *Gray Wolf, Red Wolf* (New York: Clarion, 1990).

Section Wrap-up

Review

1. They were competing with the wolves for both habitat and food.

2. Overpopulation of prey species, which can result in starvation. Prey can also damage other species in the community.

503

Science & Literature

Science & Literature

Source

Mowat, Farley. *Never Cry Wolf.* Bantam Books: New York, 1963.

Biography

Farley Mowat is recognized internationally as one of Canada's most brilliant authors. For nearly half a century, Mowat has written about the land, seas, and people of the Far North. Born in Belleville, Ontario, in 1921, Mowat always had a keen interest in nature, mainly due to the influence of his uncle, a professional ornithologist. Mowat has said of himself, "I am a Northern Man. I like to think I am a reincarnation of the Norse saga men and, like them, my chief concern is with the tales of men, and other animals, living under conditions of natural adversity."

Teaching Strategies

LS **Linguistic** Have students read the book *Never Cry Wolf* as an assignment for your unit on ecology or animals. Have students write short book reports after completing the book. You may wish to have students read the book on their own. Or you may want to read particular portions as a class to highlight certain issues and ideas.

• Show the movie version of *Never Cry Wolf* in class. You may wish to preview the movie and write a list of questions students can answer as they watch the film. Alternatively, students could write a list of 20-25 different facts they learn about wolves from watching the movie.

Science Journal

Obtain copies of the book and movie versions of *Never Cry Wolf* from your local library. Read the book first and then view the movie. In your Science Journal, compare the two versions. Why do you think books and movies like *Never Cry Wolf* are important for raising public awareness about environmental issues?

Never Cry Wolf
by Farley Mowat

When Canadian biologist Farley Mowat was left in the remote Canadian wilderness to study wolves, he didn't know what to expect of these mysterious and often misunderstood predators. Mowat was hired by the Canadian Wildlife Service to investigate and live among the wolves to help solve the country's growing "*Canis lupus* problem." It seemed that hunters were telling tales of packs of bloodthirsty wolves slaughtering caribou by the thousands and contributing to the extinction of this majestic herbivore. The government's answer was naturalist Mowat, who spent two summers and one winter studying Arctic wolves in their natural habitat.

In the book *Never Cry Wolf,* Farley Mowat details his year-long expedition learning about wolves and surviving on the frozen tundra of northern Canada. But this action-packed book is more than just a story of adventure and intrigue. It's also a story about a stunning scientific discovery. Following the publication of *Never Cry Wolf,* Mowat's conclusions about the habits and behaviors of wolves were criticized by the Canadian government and members of the scientific community. Instead of the fierce killers many people expected, Mowat discovered that wolves are gentle and skillful providers and devoted protectors of their young. Mowat also challenged the idea that wolves were causing the decline in the caribou population by showing that his wolf population fed almost exclusively on mice during the warmer summer months when mouse populations skyrocketed.

Filled with beautiful images of animals in their natural setting, *Never Cry Wolf* is also a book about one person's struggle to preserve a vanishing species. Mowat's heroic efforts to document never-before-seen behaviors in wild wolves has focused international attention on the many varieties of wolves that are rapidly going extinct in North America and elsewhere. In 1983, Mowat's epic book was made into an entertaining movie.

504 Chapter 18 Life and the Environment

• Discuss with students the biology, behavior, and distribution of wolves as they read the book or watch the film. Point out how the behavior and anatomy of wolves are closely tied with their environment. Also, discuss the current status of wolves as an endangered species.

Other Works

Farley Mowat has written several books that deal with animals in their natural environments. You may want to tell interested students about the following works: *People of the Deer, Owls in the Family, The Dog Who Wouldn't Be,* and *A Whale for the Killing.*

Summary

Summary

18-1: The Living Environment and the Nonliving Environment

1. Ecology is the study of interactions among organisms and between organisms and their physical environment.
2. The biosphere is the region of Earth where all life is found. It consists of biotic and abiotic factors.
3. Populations are made up of all the members of a species living in the same place at the same time. A community includes all the populations in an area.

18-2: Interactions Among Living Organisms

1. A population can be described by characteristics that include size, spacing, and density.
2. A limiting factor is a factor that restricts the number of individuals in a population. Carrying capacity is the number of individuals an environment can support.
3. Symbiosis is a close relationship between two or more species.
4. The physical location where an organism lives is its habitat. A niche is the role of an organism in an environment.

18-3: Matter and Energy

1. Food chains and food webs are models used to describe the feeding relationships in a community. An energy pyramid is used to describe the flow of energy through a community.
2. Energy is dispersed at each level of the food chain, but is replenished by the sun.

Matter is not replenished, but is recycled.

18-4: Science and Society: Bringing Back the Wolves

1. Gray wolves are predators that were once common throughout what is now the United States.
2. Efforts are being made to restore wolf populations to Yellowstone National Park. The issue is controversial, partly because ranchers worry about losing livestock to wolves.

Key Science Words

a. abiotic factor
b. biosphere
c. biotic factor
d. carrying capacity
e. commensalism
f. community
g. ecological pyramid
h. ecology
i. ecosystem
j. food chain
k. food web

l. habitat
m. limiting factor
n. mutualism
o. niche
p. nitrogen cycle
q. parasitism
r. population
s. population density
t. predation
u. symbiosis
v. water cycle

Reviewing Vocabulary

Match each phrase with the correct term from the list of Key Science Words.

1. any living thing in the environment
2. number of individuals of a species living in the same place at the same time

Chapter 18 Review   505

Summary

Have students read the summary statements to review the major concepts of the chapter.

Reviewing Vocabulary

1. c	6. f
2. r	7. i
3. b	8. k
4. o	9. l
5. v	10. g

✓ Assessment

Portfolio Encourage students to place in their portfolios one or two items of what they consider to be their best work. Examples include:
- Skill Builder, p. 487
- Science Journal, p. 501
- Explore the Issue, p. 503 [P]

Performance Additional performance assessments may be found in **Performance Assessment** and **Science Integration Activities**. Performance Task Assessment Lists and rubrics for evaluating these activities can be found in Glencoe's **Performance Assessment in the Science Classroom.**

GLENCOE TECHNOLOGY

 Videodisc

Glencoe Life Science Interactive Videodisc
Side 2, Lesson 9
Ecosystems: Wetlands—Exploring the Food Web

49728 to 50637

MindJogger Videoquiz

Chapter 18 Have students work in groups as they play the Videoquiz game to review key chapter concepts.

Checking Concepts

1. a
2. d
3. c
4. d
5. d
6. d
7. b
8. d
9. c
10. c

Understanding Concepts

11. Before wolves were reintroduced, one food chain might have grass or other producers as the first link, and elk as the second. After wolves are returned, the food chain would have a third level, the wolf.

12. A food chain includes only one organism at each level. A food web can have many different organisms at each level.

13. A parasite lives on or in the body of a living host, causing it harm. A predator kills its prey immediately.

14. A parasite benefits from the relationship, while the host is harmed. In a commensal relationship, one benefits and the other is neither helped nor harmed.

15. In a mutualistic relationship, both organisms benefit. Lichens, composed of algae and fungi, are an example of mutualism. Fungi provide a habitat for the algae, and the algae provide food for the fungi.

Thinking Critically

16. More energy is available from organisms lower on the food chain.

17. The virus benefits and the host is harmed.

18. The carrying capacity of the environment limits population size. Biotic potential is the largest size the population can reach under the best pos-

3. all parts of Earth inhabited by living organisms
4. job of a species in its environment
5. the cycle that includes evaporation, condensation, and precipitation
6. all the populations in an ecosystem
7. interaction of biotic and abiotic factors
8. series of overlapping food chains
9. where an organism lives in an ecosystem
10. model that compares amount of energy available at each level of a food chain

Checking Concepts

Choose the word or phrase that completes the sentence.

1. All are abiotic factors except _____.
   a. animals   c. sunlight
   b. air   d. soil
2. Coral reefs and oak-hickory forests are examples of _____.
   a. niches   c. populations
   b. habitats   d. ecosystems
3. A _____ is made up of all populations in an area.
   a. niche   c. community
   b. habitat   d. ecosystem
4. The number of individuals in a population occupying an area of a specific size describes the population's _____.
   a. clumping   c. spacing
   b. size   d. density
5. An example of a herbivore is a _____.
   a. wolf   c. tree
   b. moss   d. rabbit
6. The level of the food chain with the most energy contains _____.
   a. omnivores   c. decomposers
   b. herbivores   d. producers

7. _____ is a relationship in which one organism is helped and the other is harmed.
   a. Mutualism   c. Commensalism
   b. Parasitism   d. Symbiosis
8. Materials that are cycled in the biosphere include all of the following except _____.
   a. nitrogen   c. water
   b. sulfur   d. energy
9. A model that shows how energy is lost as it flows through an ecosystem is a
   _____.
   a. pyramid of biomass
   b. pyramid of numbers
   c. pyramid of energy
   d. niche
10. Returning wolves to Yellowstone adds a(n) _____ to the food web.
    a. producer   c. top carnivore
    b. herbivore   d. decomposer

Understanding Concepts

Answer the following questions in your Science Journal using complete sentences.

11. Describe two possible food chains that would apply to Yellowstone Park.
12. What is the difference between a food chain and a food web?
13. Compare parasitism and predation.
14. Compare parasitism and commensalism.
15. Using examples, explain some possible advantages of mutualism.

Thinking Critically

16. What would be the advantage to a human or other omnivore of eating a diet of organisms that are lower rather than higher on the food chain?

sible conditions. Carrying capacity usually prevents populations from reaching biotic potential.

19. They break down dead organisms and make the nutrients contained in their bodies available for living plants to use.

20. Answers will vary. Habitat is the student's home. Niche includes food, ways of getting food and water, relationships with other organisms, and daily activities.

Developing Skills

21. **Observing and Inferring** Water, light, and air are all abiotic factors; algae and snails are two different populations; goldfish, algae, and plants make a community.

17. Why are viruses considered parasites?
18. What does carrying capacity have to do with whether or not a population reaches its biotic potential?
19. Why are decomposers vital to the cycling of matter in an ecosystem?
20. Describe your own habitat and niche.

Developing Skills

If you need help, refer to the **Skill Handbook.**

21. **Observing and Inferring:** A home aquarium contains water, an air pump, a light, algae, a goldfish, and algae-eating snails. What are the abiotic factors in this environment? Which of these items would be considered a population? A community?
22. **Classifying:** Classify the following organisms as producer, herbivore, omnivore, or carnivore: cow, deer, grass, rabbit, human, wolf, green alga, and oak tree.
23. **Classifying:** Classify each event in the water cycle as the result of either evaporation or condensation.
    a. A puddle disappears after a rainstorm.
    b. Rain falls.
    c. A lake becomes shallower.
    d. Clouds form.
24. **Concept Mapping:** Use the following information to draw a food web of organisms living in a goldenrod field.
    Goldenrod sap is eaten by aphids. Goldenrod nectar is eaten by bees. Goldenrod pollen is eaten by beetles. Goldenrod leaves are eaten by beetles. Stinkbugs eat beetles. Spiders eat aphids. Assassin bugs eat bees.

25. **Making and Using Graphs:** Use the following data to graph the population density of a deer population over the years. Plot the number of deer on the y-axis and years on the x-axis. Propose a hypothesis to explain what might have happened to cause the changes in the size of the population.

Arizona Deer Population

Year	Deer per 400 hectares
1905	5.7
1915	35.7
1920	142.9
1925	85.7
1935	25.7

Performance Assessment

1. **Poster:** Use your own observations or the results of library research to develop a food web for a nearby park, pond, or other ecosystem. Make a poster display illustrating the food web.
2. **Oral Presentation:** Research the steps in the phosphorous cycle. Find out what role phosphorous plays in the growth of algae in ponds and lakes. Present your findings to the class.
3. **Writing in Science:** Use the results of library research to write a paper or present an oral report explaining how the lives of gray wolves in Canada or Alaska differ from the lives of red wolves living in the southeastern United States.

22. **Classifying** producer—grass, green alga, oak tree; herbivore—cow, deer, rabbit; carnivore—wolf; omnivore—human
23. **Classifying** (a) evaporation, (b) condensation, (c) evaporation, (d) condensation
24. **Concept Mapping**

25. **Making and Using Graphs** The deer population built up so high that the deer consumed all the available food. The next generations lost animals to starvation.

Performance Assessment

1. The poster should illustrate the concepts presented on pages 496 and 497. Use the Performance Task Assessment List for Poster in **PASC,** p. 73. [P]
2. Excess phosphorus causes an increase in the growth of algae. When the algae die, they remove much of the oxygen from the pond, affecting all other organisms in the pond. Use the Performance Task Assessment List for Oral Presentation in **PASC,** p. 71. [P]
3. The paper should include how the adaptations and food sources of gray wolves differ from those of red wolves. Use the Performance Task Assessment List for Writing in Science in **PASC,** p. 87.

Assessment Resources

Reproducible Masters
Chapter Review, pp. 39-40
Assessment, pp. 122-125
Performance Assessment, p. 32

Glencoe Technology
⊙ Chapter Review Software
⊙ Computer Test Bank
📼 MindJogger Videoquiz

507

Section	Objectives/Standards	Activities/Features
Chapter Opener		Explore Activity: Find out about some of the factors that define your part of the biosphere. p. 509
19-1 How Ecosystems Change (1 session, 1 block)*	1. **Describe** how ecosystems change over time. 2. **Explain** how new communities arise in areas that were bare of life. 3. **Compare** and **contrast** pioneer communities and climax communities. National Content Standards: (5-8) UCP3, UCP4, A1, A2, B1, C4, C5	Connect to Chemistry, p. 511 Skill Builder: Sequencing, p. 513 Science Journal, p. 513 Activity 19-1: Observing Succession, pp. 514-515
19-2 Land Environments (2 sessions, 1½ blocks)*	1. **Explain** how climate influences land environments. 2. **Describe** the six biomes that make up land environments on Earth. 3. **Compare** and **contrast** the adaptations of plants and animals found in each biome. National Content Standards: (5-8) UCP4, A1, A2, B3, C4, C5, D1, D3, E2, F5	Using Math, p. 518 Activity 19-2: Studying a Land Environment, p. 519 MiniLAB: Is there tundra or taiga in South America? p. 520 Using Technology: Ecosystems in Space? p. 523 Problem Solving: Saving the Rain Forests, p. 525 Skill Builder: Concept Mapping, p. 527 Using Computers: Database, p. 527
19-3 Science and Society Issue: Protecting Antarctica (1 session, ½ block)*	1. **Describe** the living and nonliving resources of the Antarctic. 2. **Explain** why the countries of the world are trying to agree about how to manage the scientific and mineral resources of Antarctica. National Content Standards: (5-8) D1, F2	Explore the Issue, p. 529
19-4 Water Environments (2 sessions, 1½ blocks)*	1. **Distinguish** between flowing freshwater and standing freshwater ecosystems. 2. **Describe** important seashore and deep ocean ecosystems. National Content Standards: (5-8) UCP1, A1, C4, D1, G1	MiniLAB: Freshwater Environments, p. 532 Skill Builder: Comparing and Contrasting, p. 533 Science Journal, p. 533 People and Science: Alfred Gonzalez, Salmon Habitat Researcher, p. 534

* A complete Planning Guide that includes block scheduling is provided on pages 32T-35T.

Activity Materials

Explore	Activities	MiniLABs
page 509 world map	pages 514-515 debris from a pond; boiled pond water; clean, 1-L jars with lids; microscope; dropper; microscope slides; coverslips; pond-life guide book page 519 graph paper, tape measure, notebook, thermometer, hand lens, field guides, binoculars, pencil	page 520 world map page 532 large, self-sealing freezer bag; dip net; pond; plastic cup; hand lens

Need Materials? Call Science Kit (1-800-828-7777).

Teacher Classroom Resources

Reproducible Masters	Transparencies	Teaching Resources
Study Guide, p. 72 **Reinforcement**, p. 72 **Enrichment**, p. 72 **Activity Worksheets**, pp. 112-113 **Concept Mapping**, p. 43 **Critical Thinking/Problem Solving**, p. 23	**Section Focus Transparency 68**, A Changing World	**Glencoe Life Science Interactive Videodisc** **Life Science CD-ROM** **Spanish Resources** **English/Spanish Audiocassettes** **Cooperative Learning Resource Guide** **Lab Partner** **Lab and Safety Skills** **Lesson Plans**
Study Guide, p. 73 **Reinforcement**, p. 73 **Enrichment**, p. 73 **Activity Worksheets**, pp. 114-115, 116 **Cross-Curricular Integration**, p. 23 **Science and Society Integration**, p. 23	**Section Focus Transparency 69**, Life in Different Climates **Science Integration Transparency 19**, Elevation and Climate **Teaching Transparency 37**, Seasons: Tilt Makes the Difference **Teaching Transparency 38**, Biome Map	**Assessment Resources** **Chapter Review**, pp. 41-42 **Assessment**, pp. 126-129 **Performance Assessment in the Science Classroom (PASC)** **MindJogger Videoquiz** **Alternate Assessment in the Science Classroom** **Performance Assessment** **Chapter Review Software** **Computer Test Bank**
Study Guide, p. 74 **Reinforcement**, p. 74 **Enrichment**, p. 74	**Section Focus Transparency 70**, National Parks	
Study Guide, p. 75 **Reinforcement**, p. 75 **Enrichment**, p. 75 **Activity Worksheets**, p. 117 **Science Integration Activity 19**, Waves, Currents, and Coastal Features **Multicultural Connections**, pp. 41-42	**Section Focus Transparency 71**, Water, Water Everywhere	

Key to Teaching Strategies

The following designations will help you decide which activities are appropriate for your students.

L1 Level 1 activities should be within the ability range of all students, including those with learning difficulties.

L2 Level 2 activities should be within the ability range of the average to above-average student.

L3 Level 3 activities are designed for the ability range of above-average students.

LEP LEP activities should be within the ability range of Limited English Proficiency students.

LS These activities are designed to address different learning styles.

COOP LEARN Cooperative Learning activities are designed for small group work.

P These strategies represent student products that can be placed into a best-work portfolio.

GLENCOE TECHNOLOGY

The following multimedia resources are available from Glencoe.

Science and Technology Videodisc Series (STVS)
Ecology
Preserving Duck Habitats
Acid Rain and Plants
Fish and Acid Rain
Purifying with Plants
The Infinite Voyage Series
Life in the Balance
To the Edge of the Earth
Secrets from a Frozen World

National Geographic Society Series
STV: Plants
STV: Biodiversity
STV: Rain Forest
Glencoe Life Science Interactive Videodisc
Ecosystems: Wetlands
Life Science CD-ROM

Teacher Classroom Resources

This is a representation of key blackline masters available in the Teacher Classroom Resources.

Teaching Aids

Section Focus Transparencies

Science Integration Transparencies

Teaching Transparencies

Meeting Different Ability Levels

Study Guide

Reinforcement

Enrichment Worksheets

Hands-On Activities

Science Integration Activity

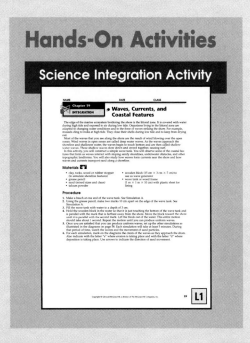

Assessment

Performance Assessment

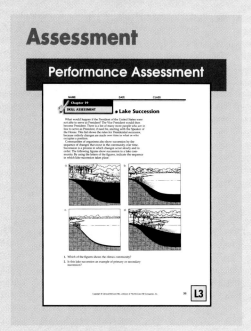

Enrichment and Application

Critical Thinking/Problem Solving

Cross-Curricular Integration

Science and Society Integration

Multicultural Connections

Concept Mapping

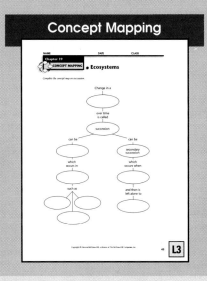

Chapter 19

Ecosystems

CHAPTER OVERVIEW

Section 19-1 Primary succession and secondary succession are differentiated in this section.

Section 19-2 This section discusses the factors that determine climate, identifies the location of six major land biomes, and describes the characteristics of each one.

Section 19-3 **Science and Society** The living and nonliving resources of Antarctica are presented. How the scientific and mineral resources of Antarctica may be protected is discussed.

Section 19-4 Characteristics of freshwater and saltwater environments are described.

Chapter Vocabulary

ecological
  succession
primary
  succession
pioneer
  community
secondary
  succession
climax
  community
precipitation
biome
tundra

permafrost
taiga
temperate
  deciduous
  forest
tropical rain
  forest
grassland
desert
plankton
estuary
intertidal
  zone

Theme Connection

Stability and Change Ecosystems change in ways that can be predicted. In the process of succession, each community affects the environment, often creating conditions that result in the community's replacement.

Systems and Interactions Organisms interact with their environment. Where an organism can live depends on the abiotic and biotic factors in the environment.

508

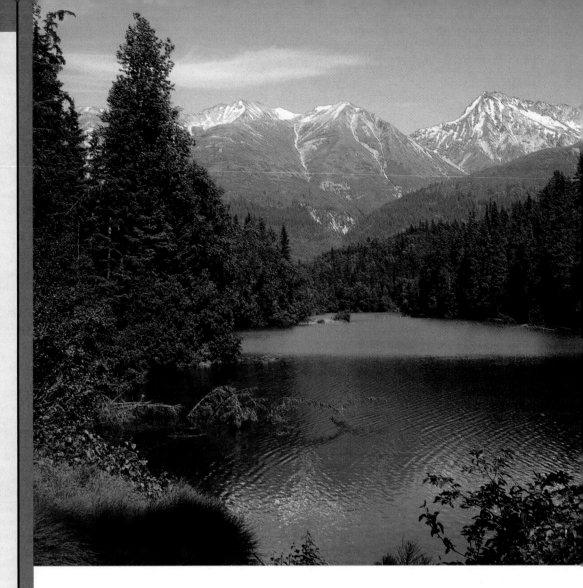

Previewing the Chapter

Section 19-1   How Ecosystems Change
▶ Ecological Succession

Section 19-2   Land Environments
▶ Factors that Determine Climate
▶ Land Biomes

Section 19-3   Science and Society Issue: Protecting Antarctica

Section 19-4   Water Environments
▶ Freshwater Biomes
▶ Saltwater Biomes

508

Learning Styles

Look for the following logo for strategies that emphasize different learning modalities. **LS**

Kinesthetic	MiniLAB, p. 532
Visual-Spatial	Explore, p. 509; Making a Model, p. 512; Activity, p. 517; Integrating the Sciences, p. 518
Interpersonal	Activity 19-1, pp. 514-515
Intrapersonal	Activity 19-2, p. 519; Activity, p. 525
Logical-Mathematical	Using Math, pp. 517, 518, 522; Integrating the Sciences, p. 517; MiniLAB, p. 520; Enrichment, p. 521
Linguistic	Enrichment, pp. 524, 526; Across the Curriculum, pp. 524, 531; Reteach, p. 526

508

Chapter 19

Ecosystems

Do you live in a part of the country with pine trees and mountain lakes? Or is your home a region of grassy prairie, sandy beach, or desert scrub? You know that all life is found in the biosphere. In this chapter, learn what kinds of ecosystems exist in the biosphere and find out some of the reasons why their biotic and abiotic factors differ.

EXPLORE ACTIVITY

Find out about some of the factors that define your part of the biosphere.

1. Locate your city or town on a globe or world map. Find your latitude.
2. Locate another city at the same latitude as yours, but on a different continent.
3. Locate a third city at a latitude very close to the equator.
4. Using references, compare average annual rainfall and average high and low temperatures for all three cities.

Observe: In your Science Journal, explain what effect latitude has on temperature and rainfall.

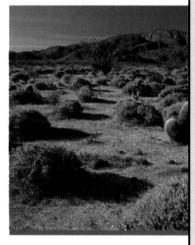

Previewing Science Skills

▶ In the **Skill Builders**, you will **sequence, map concepts,** and **compare and contrast.**

▶ In the **Activities**, you will **observe, collect data, predict,** and **compare results.**

▶ In the **MiniLABs**, you will **observe and infer.**

509

Assessment Planner

Portfolio
Refer to page 535 for suggested items that students might select for their portfolios.

Performance Assessment
See page 535 for additional Performance Assessment options.
Skill Builders, pp. 513, 527, 533
MiniLABs, pp. 520, 532
Activities 19-1, pp. 514-515; 19-2, p. 519

Content Assessment
Section Wrap-ups, pp. 513, 527, 529, 533
Chapter Review, pp. 535-537
Mini Quizzes, pp. 513, 527, 533

Group Assessment
Opportunities for group assessment occur with Cooperative Learning Strategies and Flex Your Brain Activities.

Prepare

Section Background

The most important cause of succession is the altering of the physical environment by the community itself. Communities tend to alter the area in which they live in such a way as to make it less favorable for themselves and more favorable for other communities.

Preplanning

Obtain pond water, debris from a pond, jars with lids, and pond life guidebooks for Activity 19-1.

1 Motivate

Bellringer

Before presenting the lesson, display **Section Focus Transparency 68** on the overhead projector. Assign the accompanying **Focus Activity** worksheet. `L1` `LEP`

Tying to Previous Knowledge

Find a decaying log with fungi, slime molds, lichens, mosses, ferns, and plant seedlings on it. Display the log and ask students how it illustrates succession.

19•1 How Ecosystems Change

Science Words

ecological succession
primary succession
pioneer community
secondary succession
climax community

Objectives

- Describe how ecosystems change over time.
- Explain how new communities arise in areas that were bare of life.
- Compare and contrast pioneer communities and climax communities.

Ecological Succession

Imagine hiking through a forest. Huge trees tower over the trail. You know it can take many years for trees to grow this large, so it's easy to think of the forest as something that has always been here. But this area has not always been covered with trees. Long ago it may have been a pond full of fish and frogs and surrounded by water-loving plants. As time passed, the decomposed bodies of plants and animals slowly filled in the pond, until it eventually became a lush green meadow full of grass and wildflowers. Gradually, over many more years, seeds blew in, trees began to grow, and a forest developed. The process of gradual change from one community of organisms to another is called **ecological succession.** The changes associated with succession usually take place in a fairly predictable order and involve animals, plants, and other organisms.

Figure 19-1

Stages in Primary Succession

 B Mosses and ferns gradually replace the lichens. These plants can grow even in extremely poor, thin soil. As they die, their decomposed bodies add humus to the soil. Insects and other small animals appear.

A Life on this bare rock begins with a pioneer community of lichens. These hardy organisms produce acids that help to break down the rock. The acids release chemicals and nutrients from the rock that can then be absorbed by the lichens. The decaying bodies of dead lichens contribute to soil formation.

510   Chapter 19 Ecosystems

Program Resources

Reproducible Masters

Study Guide, p. 72 `L1`
Reinforcement, p. 72 `L1`
Enrichment, p. 72 `L3`
Activity Worksheets, pp. 5, 112-113 `L1`
Critical Thinking/Problem Solving, p. 23
Concept Mapping, p. 43

 Transparencies

Section Focus Transparency 68 `L1`

Primary Succession

Think about conditions around an erupting volcano. Incredibly hot, molten lava flows along the ground, destroying everything in its path. As the lava cools, it forms new land. In Chapter 18, you learned how soil is formed from bare rock. Similar events happen to this newly formed land. Particles of dust and ash fall to the ground. The forces of weather and erosion break up the lava rock. A thin layer of soil begins to form. Birds, wind, and rain deposit more dust, along with bacteria, seeds, and fungal spores. Plants start to grow and decay. A living community has begun to develop.

Ecological succession that begins in a place that does not have soil is called **primary succession.** The first community of organisms to move into a new environment is called the **pioneer community,** as shown in **Figure 19-1.** Members of pioneer communities are usually hardy organisms that can survive drought, extreme heat and cold, and other harsh conditions. Pioneer communities change the conditions in their environments. These new conditions support the growth of other types of organisms that gradually take over.

CONNECT TO

CHEMISTRY

Freezing temperatures are harmful to living organisms because water is the main component of cells. As water freezes, it expands. *Explain* what happens to cells when ice crystals form inside them.

C As the soil layer thickens, its ability to absorb and hold water improves. Grasses, wildflowers, and other plants that require richer, more moist soil begin to take over. Butterflies, bees, and caterpillars come to feed on the leaves and flowers. When these plants die, they also enrich the soil, which will become home to earthworms and other large soil organisms.

D Thicker, richer soil supports the growth of shrubs and trees. More insects, birds, mammals, and reptiles move into the area. After hundreds or thousands of years of gradual change, what was once bare rock has become a forest.

19-1 How Ecosystems Change   511

2 Teach

CONNECT TO

CHEMISTRY

Answer The ice crystals take up extra room, piercing cell membranes and destroying the cells.

Revealing Preconceptions

Students may think that communities change because of time. Make sure they understand that communities change over time because organisms change the physical environment, making it suitable for other organisms.

Use **Concept Mapping,** p. 43, as you teach this lesson.

GLENCOE TECHNOLOGY

Videodisc

The Infinite Voyage: To the Edge of the Earth

Chapter 6
Exploring Life in the Canadian Arctic

Chapter 7
Eskimo Survival in the Arctic

Cultural Diversity

Lost Cities Abandoned cities and other cultural sites may become overgrown when the cultures that built them decline or are wiped out, or when the people leave the area. Then the natural processes of succession take over, and the cultural site may disappear in undergrowth. Have students research what happened in the Mayan and Incan sites in the rain forests of Central and South America. L2

Community Connection

Field Trip Take students on a field trip to see the stages of succession. Stages of succession can be seen in ponds, woods, road cuts, power-line clearings, and outcrops. Allow students to identify the different stages of succession.

Figure 19-2
The tangled growth of weeds and grasses in untended yards, on abandoned farms, and along country roadsides are the beginning stages of secondary succession.

Secondary Succession

What happens when a forest is destroyed by a fire or a city building is torn down? After a forest fire, nothing is left except dead trees and ash-covered soil. Once the rubble of a demolished building has been taken away, all that remains is bare soil. But these places do not remain lifeless for very long. The soil may already contain the seeds of weeds, grasses, and trees. More seeds are carried to the area by wind and birds. As the seeds germinate and plants begin to grow, insects, birds, and other wildlife move in. Ecological succession has begun again. Succession that begins in a place that already has soil and was once the home of living organisms is called **secondary succession,** shown in **Figure 19-2.**

Climax Communities

Succession involves changes in abiotic factors as well as biotic factors. You have already seen how lichens, mosses, and ferns change the environment by helping to form the rich, thick soil needed for the growth of shrubs and trees. Shrubs and trees also cause changes in abiotic factors. Their branches shade the ground beneath them, reducing the temperature. Shade also reduces the rate of evaporation, increasing the moisture of the soil. Sunlight, temperature, and moisture determine which species will grow in soil.

The redwood forest shown in **Figure 19-3** is an example of a community that has reached the end of succession. As long as the trees are not cut down or destroyed by fire or widespread disease, the species that make up the redwood community tend to remain the same. When a community has reached the final stage of ecological succession, it is called a **climax community.** Because primary succession begins in areas with no life at all, it can take hundreds or even thousands of years for a pioneer community to develop into a climax community. Secondary succession is a shorter process, but it still may take a century or more.

512   Chapter 19 Ecosystems

Comparing Pioneer and Climax Communities

As you have seen, pioneer communities are simple. They contain only a few species, and feeding relationships can usually be described with simple food chains. Climax communities are much more complex. They may contain hundreds of thousands of species, and feeding relationships usually involve complex food webs. Interactions among the many biotic and abiotic factors in a climax community create a more stable environment that does not change very much over time. Climax communities are the end product of ecological succession. A climax community that has been disturbed in some way will eventually return to the same type of community, as long as all other factors remain the same.

Figure 19-3

This forest of redwood trees is an example of a climax community. Redwoods live for hundreds of years. They create dense shade on the ground beneath them. Needles constantly fall from their branches. As the needles decompose, they form an acid soil that promotes the growth of young redwoods but prevents the growth of many other types of plants.

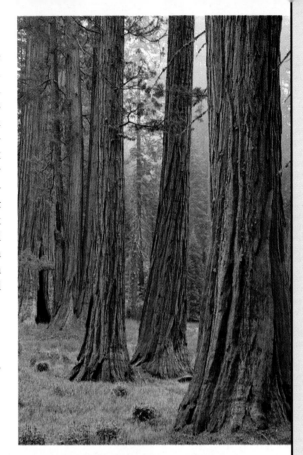

Section Wrap-up

Review

1. What is ecological succession?

2. What is the difference between primary and secondary succession?

3. **Think Critically:** What kind of succession will take place on an abandoned, unpaved country road? Why?

Skill Builder
Sequencing
Describe the sequence of events in primary succession. Include the term *climax community*. If you need help, refer to Sequencing in the **Skill Handbook.**

Science Journal

In your Science Journal, draw a food chain for a pioneer community of lichens and a food web for the climax community of an oak-maple forest. Write a short paragraph comparing the two communities.

PREPARATION

Purpose

IS **Interpersonal** Students will formulate a model of succession in a pond community.

L1 **COOP LEARN**

Process Skills

hypothesizing, formulating a model, inferring, observing, interpreting data, communicating, sequencing, making and using tables, forming operational definitions, forming a hypothesis, comparing and contrasting

Time

50 minutes, 10 minutes two times a week

Materials

- Find a pond that has large populations of protists. Collect pond water and plant materials with debris from the edges of the pond. Boil the pond water in a glass beaker for 5 to 10 minutes to sterilize it. Cover the pond water and allow it to cool. Spread the plant material on waxed paper and let it dry overnight.

Alternate Materials If pond water is not available, aquarium water and bottom debris from an aquarium may be used. Purchase protozoan cultures if necessary.

- Ordinary microscope slides and coverslips may crush some pond organisms, making their behavior difficult to observe. If so, depression slides can be used. Regular slides may also be used by raising the coverslip and using a toothpick to make a ring of petroleum jelly in which the drop of pond water is placed. If needed, methyl cellulose or strands of cotton can be used to slow the activity of microorganisms.

Activity 19-1

Design Your Own Experiment
Observing Succession

Succession in large ecosystems with long-lived organisms such as trees can take decades or centuries. Succession progresses much more rapidly in the smaller community of microscopic organisms that live in freshwater ponds. The short life cycles of these organisms make it possible to observe changes associated with succession in as little as two weeks.

PREPARATION

Problem

Do biotic and abiotic factors influence succession in a pond community?

Form a Hypothesis
Using what you know about primary and secondary succession, state a hypothesis about how a biotic or abiotic factor might influence succession in a pond community.

Objectives

- Observe changes in a pond community during succession.
- Design an experiment that demonstrates how a single biotic or abiotic factor influences succession in a pond community.

Possible Materials

- debris from a pond
- boiled pond water
- clean 1-L jars, with lids
- microscope
- dropper
- microscope slides
- coverslips
- pond life guide book

Safety Precautions

Wash hands after handling pond water.

Safety Precautions

Wash hands after handling pond water and pond debris.

Possible Hypotheses

Students may want to compare succession with and without light, at different temperatures, or with and without the presence of fish or aquatic plants.

 Activity Worksheets, pp. 5, 112-113

PLAN THE EXPERIMENT

1. As a group, agree upon and write out a hypothesis statement. Be sure your hypothesis tests only one variable. Some possibilities are comparing succession with and without light, at different temperatures, or with and without the presence of fish or aquatic plants.
2. As a group, list the steps you need to take to test your hypothesis. Be specific. Describe exactly what you will do at each step. List your materials.
3. Decide on how long you will observe your ecosystem. You will want to record the date each time you observe the organisms in your experiment. You will also want to record the number of slides you observed and the number of each type of organism. It is not necessary to know the names of each organism.

Clear drawings will identify them.
4. Design a data table in your Science Journal so it is ready to use as your group collects data.

Check the Plan
1. Read over your entire experiment to make sure that all steps are in logical order.
2. Identify any constants, variables, and controls in your experiment.
3. How much water will you put in each jar? How many drops will you observe?
4. *Make sure your teacher has approved the plan before you proceed.*

DO THE EXPERIMENT

1. Carry out the experiment as planned.
2. While the experiment is going on, write down any observations that you make and complete the data table in your Science Journal.

Analyze and Apply
1. **Compare** your results with those of other groups.
2. **Identify** how the biotic or abiotic factor you tested affects succession in the community.
3. How many different species did you observe?
4. **Compare** the species present. Were they similar? Different?

Go Further
Find out about the food and habitat preferences of one or more of the species you observed.

19-1 How Ecosystems Change   515

PLAN THE EXPERIMENT

Possible Procedures
Students must choose one variable to test and keep all others constant.

Teaching Strategies
Sampling techniques must be consistent. Samples should always be removed from the same part of the container (bottom, middle, or surface), and records should indicate from where the samples were taken.

DO THE EXPERIMENT

Expected Outcome
Results will vary with the type of experiment conducted. A normal progression of organisms is bacteria, paramecia, and rotifers. Paramecia may appear as soon as a week into the experiment.

Analyze and Apply
1. Students' results should be consistent with those of other groups. Variations in experimental design will vary students' results.
2. Answers will depend on the hypothesis and procedure.
3. Species types and numbers will vary, depending on the abiotic or biotic factor being tested.
4. Answers will vary.

Go Further

Have students use library references to identify the food and habitat preferences of the organisms.

Assessment

Performance Have students repeat the activity using different kinds of plants or plants grown from a more crowded pond. In such cases, the types and numbers of organisms they start with may be different from those they first observed in this activity. Succession will proceed at a different rate. Use the Performance Task Assessment List for Carrying Out a Strategy and Collecting Data in **PASC**, p. 25. [P]

515

Prepare

Section Background

Ecologists differ as to the number of biomes that should be recognized because biomes gradually merge into one another.

Preplanning

- Obtain thermometers, hand lenses, binoculars, and field guides for Activity 19-2.
- Obtain world maps for the MiniLAB on page 520.

1 Motivate

Bellringer

 Before presenting the lesson, display **Section Focus Transparency 69** on the overhead projector. Assign the accompanying **Focus Activity** worksheet. L1 LEP

Tying to Previous Knowledge

Review the concepts of adaptation and the conditions necessary for life from Chapter 1.

INTEGRATION
Earth Science

Climatic changes on Earth depend on how directly sunlight falls on the hemispheres as Earth revolves around the sun. The amount of sunlight falling on the hemispheres changes as Earth orbits the sun because of the tilt of Earth's axis (23.5°).

Science Words

- **precipitation**
- **biome**
- **tundra**
- **permafrost**
- **taiga**
- **temperate deciduous forest**
- **tropical rain forest**
- **grassland**
- **desert**

Objectives

- Explain how climate influences land environments.
- Describe the six biomes that make up land environments on Earth.
- Compare and contrast the adaptations of plants and animals found in each biome.

INTEGRATION
Earth Science

Figure 19-4

Because Earth is tilted on its axis, the angle of the sun's rays changes during the year. These changes create the seasons. When a region is tilted toward the sun, as the United States is during summer in the Northern Hemisphere, it is warmed to higher temperatures. During the winter, when the same area is tilted away from the sun, temperatures are colder. The tilt of Earth's axis does not have much of an effect on regions near the equator.

516 Chapter 19 Ecosystems

19•2 Land Environments

Factors that Determine Climate

What does a desert in Arizona have in common with a desert in Africa? They both have water-conserving plants with thorns, lizards, heat, little rain, and poor soil. How are the plains of the American west like the veldt of central Africa? Both regions have dry summers, wet winters, and huge expanses of grassland that support grazing animals such as elk and antelope. Many widely separated regions of the world have similar ecosystems. Why? Because they have similar climates. Climate is the general weather pattern in an area. The factors that determine a region's climate include temperature and precipitation.

Temperature

The sun supplies life on Earth not only with light energy for photosynthesis, but also heat energy for warmth. The temperature of a region is regulated primarily by the amount of sunlight that reaches it. In turn, the amount of sunlight is determined by an area's latitude and elevation.

Latitude

As **Figure 19-4** shows, not all parts of Earth receive the same amount of energy from the sun. When you conducted the Explore Activity at the beginning of this chapter, you probably concluded that temperature is affected by latitude. The nearer a region is to the north or south pole, the higher its latitude, the smaller the amount of energy it receives from the sun, and the colder its climate.

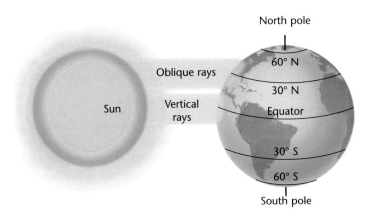

North pole

Oblique rays

Vertical rays

Sun

60° N
30° N
Equator
30° S
60° S

South pole

Figure 19-5

Because Earth is curved, rays of sunlight reaching higher latitudes near the poles are more spread out, and therefore weaker than the sunlight reaching lower latitudes near the equator. Climates near the equator are warm and those near the poles are cold.

Seasonal changes in sunlight also have an effect on the temperature of a climate. Because Earth is tilted on its axis, the angle of the sun's rays changes as Earth moves through its yearly orbit. During winter in the northern hemisphere, regions north of the equator are tilted away from the sun. Rays of sunlight are spread over a larger area, reducing their warming effect. As a result, winter temperatures are colder than summer temperatures.

Elevation

A region's elevation, or distance above sea level, also has an influence on temperature. Earth's atmosphere acts as insulation that traps some of the heat that reaches Earth's surface. At higher elevations, the atmosphere is thinner, so more heat escapes back into space. As a result, the higher the elevation, the colder the climate. The climate on a mountain will be cooler than the climate at sea level at the same latitude. Higher elevations affect plant growth as seen in **Figure 19-6.**

USING MATH

Earth is tilted at an angle of 23.5°. Without using a protractor, sketch an angle that measures about 23.5°. Then check your angle by measuring it with a protractor.

Figure 19-6

These Rocky Mountain bristlecone pines show the effects of higher elevations on plants. These trees are shaped by the wind and stunted by the cold, harsh conditions.

19-2 Land Environments   **517**

 Use **Science Integration Transparency 19** as you teach this lesson.

Integrating the Sciences

Earth Science Climate is the average temperature and precipitation over a period of time. Weather involves atmospheric changes on a daily basis. Have students record temperature and precipitation for a week and average the figures. Ask what kind of climate this represents. **L1** **LS**

Figure 19-7

Moist air moving into California from the Pacific Ocean is forced upward when it reaches the Sierra Nevada Mountains. As air rises, it loses its moisture in the form of rain or snow. By the time the air reaches Nevada and Utah, on the other side of the mountains, it is dry. This area is in the mountains' "rain shadow." It receives so little rain that it has become a desert.

518

Precipitation

Water is one of the most important factors affecting the climate of an area. **Precipitation** is the amount of water that condenses and falls in the form of rain, snow, sleet, hail, and fog. Differences in temperature have an important effect on patterns of precipitation.

Have you heard the expression, "Hot air rises"? Actually, hot air is pushed upward whenever cold air sinks. Cold air is more dense than hot air, so it tends to move toward the ground. This pushes warm air near Earth's surface upward. In warm tropical regions near the equator, the air, land, and oceans are constantly being heated by the direct rays of the sun. As the cooler air sinks, the warm air is pushed upward into the atmosphere. This warm air carries with it large amounts of water vapor from the oceans. When the air reaches a high enough altitude in the atmosphere, the water vapor it contains cools and condenses as rain. While the air rises, it also moves slowly toward either the north or south pole. The air loses virtually all of its moisture by the time it reaches a latitude of about 30°. Because of this pattern, deserts are common at latitudes near 30° in both the northern and southern hemispheres. Latitudes between 0° and 22° receive much larger amounts of rain.

The Rain Shadow Effect

The presence of mountain ranges also has an effect on rainfall patterns. As **Figure 19-7** shows, air that is moving toward a mountain range is forced upward by the shape of the land. As warm air is forced upward, it cools, condensing the water vapor it contains and creating rain or snow. By the time the air has passed over the mountains, it has lost its moisture. The region on the opposite side of the mountain range receives very little rain because it is in a "rain shadow" created by the mountains.

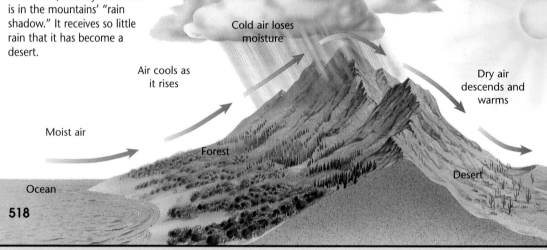

Cold air loses moisture

Dry air descends and warms

Air cools as it rises

Moist air

Forest

Desert

Ocean

Activity 19-2 · Studying a Land Environment

An ecological study includes observation and analysis of living organisms and the physical features of the environment.

Problem
How do you study an ecosystem?

Objectives
- Observe biotic and abiotic factors of an ecosystem.
- Analyze the relationships among organisms and their environment.

Materials
- graph paper
- tape measure
- notebook
- pencil
- thermometer
- hand lens
- binoculars
- field guides

Procedure
1. Choose a portion of an ecosystem near school or home as your area of study. You might choose to study a pond, a forest area in a park, a garden, or another area.
2. Decide the boundaries of your study area.
3. Using a tape measure and graph paper, make a map of your study area.
4. Using a thermometer, measure and record the air temperature in your study area.
5. Observe the organisms in your study area. Use field guides to identify them. Use a hand lens to study small organisms. Use binoculars to study animals you cannot get near. Also look for evidence, such as tracks or feathers, of organisms you do not see.
6. Record your observations in a table like the one shown. Make drawings to help you remember what you see.
7. Visit your study area as many times as you can and at different times of day for four weeks. At each visit, be sure to make the same measurements and record all observations. Note how biotic and abiotic factors interact.

Analyze
1. What type of ecosystem did you study?
2. Describe the abiotic factors of the ecosystem.
3. How many populations did you observe?

Conclude and Apply
4. **Identify** relationships among the organisms in your study area, such as predator-prey or symbiosis.
5. **Diagram** a food chain or food web for your ecosystem.
6. **Predict** what might happen if one or more abiotic factors were changed suddenly.
7. **Predict** what might happen if one or more populations were removed from the area.

Data and Observations

Sample Data

Date	Time of Day	Temperature	Organisms Observed	Observations and Comments
		Data will vary depending on ecosystem chosen.		

19-2 Land Environments   **519**

Assessment

Performance Have students alter the ecosystem in a way that is not harmful, such as placing shade over a part of the ecosystem. Have them observe how foreign materials affect the behavior of living things in the ecosystem. Use the Performance Task Assessment List for Carrying Out a Strategy and Collecting Data in **PASC**, p. 25. **P**

Purpose
IS **Intrapersonal** Students will conduct a study of an ecosystem, determine how the living and nonliving things interact, and determine the relationships that exist among the organisms that live there. **L1**

Process Skills
observing, classifying, communicating, defining operationally, recognizing cause and effect, predicting

Time
50 minutes for setup; four weeks to make observations

Activity Worksheets, pp. 5, 114-115

Teaching Strategies
- Arrange for students to check out binoculars, field guides, and other useful materials from the science classroom and library. Approve the ecosystems chosen by the students before they begin their observations.
- This may be the first independent study that students have undertaken at great depth and for such a dedicated length of time. Offer encouragement to students at regular intervals during this investigation.
- Offer students a set time each day to discuss any problems they may have concerning their method of study.

Answers to Questions
1-5. Answers will vary depending on the ecosystem chosen by each group.
6. Populations would be affected by changes in temperature, water or other abiotic factors. The entire ecosystem could be disrupted.
7. Other populations could be affected.

519

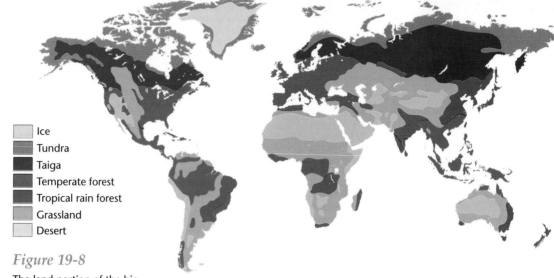

Ice
Tundra
Taiga
Temperate forest
Tropical rain forest
Grassland
Desert

Figure 19-8

The land portion of the biosphere can be divided into several biomes: tundra, taiga, temperate forest, tropical rain forest, grassland, and desert are the most commonly known. Biomes in different areas have the same kinds of abiotic factors. Different biotic factors fill the same niches in different biomes.

MiniLAB

Is there tundra or taiga in South America?

Procedure

1. Compare the latitudes where tundra is found in the northern hemisphere with the same latitudes in South America.
2. Compare the latitudes where taiga is found in the northern hemisphere with the same latitudes in South America.

Analysis

1. Are either of these biomes found in South America? Explain why or why not.

520   Chapter 19 Ecosystems

Land Biomes

Regions with similar climates tend to have ecosystems with climax communities of similar structure. Tropical rain forests are climax communities found near the equator, where temperatures are warm and rainfall is plentiful. Coniferous forests grow where winter temperatures are cold and there is moderate rainfall. Large geographic areas that have similar climates and ecosystems are called **biomes** (BI ohmz). The six most common biomes are mapped in **Figure 19-8.**

Tundra

At latitudes surrounding the north pole lies a biome that receives little precipitation, but is covered with ice most of the year. The **tundra** (TUN dra) is a cold, dry, treeless region, sometimes called a cold desert, where winters are six to nine months long. For some of those months, the land remains dark because the sun never rises above the horizon. For a few days during the short, cold summer, the sun never sets. Precipitation averages less than 25 cm per year, and winter temperatures drop to −40°C, so water in the tundra soil remains solidly frozen during the winter. During the summer, only the top few inches thaw. Below the thawed surface is a layer of permanently frozen soil called **permafrost.** The cold temperatures slow down the process of decomposition, so the soil is also poor in nutrients.

Tundra plants are resistant to drought and cold. They include species of lichens known as reindeer moss, true

 Use **Teaching Transparency 38** as you teach this lesson.

mosses, grasses, and small shrubs, as you can see in **Figure 19-9.** During the summer, mosquitoes, blackflies, and other biting insects are abundant. Many birds, including ducks, geese, various shorebirds, and songbirds, migrate to the tundra to nest during the summer. Hawks, snowy owls, mice, voles, lemmings, arctic hares, caribou, and musk oxen are also found there.

Taiga

Just below the tundra, at latitudes between about 50°N and 60°N, and stretching across Canada, northern Europe, and Asia, lies the world's largest biome. The **taiga** is a cold region of cone-bearing evergreen trees, including pines, firs, hemlocks, spruces, and cedars. This biome is also called the northern coniferous forest. Although the winter is long and cold, the taiga is warmer and wetter than the tundra. Precipitation is mostly snow and averages 35 to 40 cm each year. There is no permafrost. The ground thaws completely during the summer, making it possible for trees to grow. There are few shrubs and grasses, primarily because the forests of the taiga are so dense that little sunlight penetrates through the trees. Lichens and mosses grow on the forest floor.

Figure 19-9

Land is so flat in the tundra that water does not drain away. Because the permafrost also prevents water from soaking into the soil, part of the tundra becomes wet and marshy during the summer. Permafrost also prevents trees and other deep-rooted plants from growing in the tundra biome.

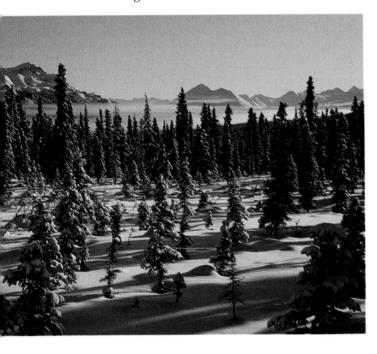

Figure 19-10

The climax community of the taiga is coniferous forest dominated by fir and spruce trees. Mammal populations include moose, black bears, lynx, and wolves.

19-2 Land Environments   521

Figure 19-11

The mild climate and rich soil of the temperate deciduous forest support a wide variety of organisms. Animal life includes deer, foxes, squirrels, mice, snakes, and a huge number of bird and insect species.

USING MATH

Make a graph that shows the average yearly precipitation in each of the land biomes.

Temperate Deciduous Forest

Temperate forests are found in both the northern and southern hemispheres, at latitudes below about 50°. Temperate regions usually have four distinct seasons each year. Precipitation ranges from about 75 to 150 cm and is distributed evenly throughout the year. Temperatures range from below freezing during the winter to 30°C or more during the warmest days of summer.

There are many coniferous forests in the temperate regions of the world, particularly in mountainous areas. However, most of the temperate forests in Europe and North America are dominated by climax communities of deciduous trees, which lose their leaves every autumn. These forests are called **temperate deciduous forests.** In the United States, they are found primarily east of the Mississippi River and include oak-maple and oak-hickory forests.

The loss of leaves in the fall signals a dramatic change in the life of the deciduous forest. Food becomes less abundant, and the leafless trees no longer provide adequate shelter for many of the organisms that live there during the warmer months. Some animals, particularly birds, migrate to warmer regions during the winter. Other organisms reduce their activities and their need for food by going into hibernation until spring. When spring does arrive, smaller plants benefit from the temporarily sunny forest floor and are able to grow before the trees leaf out again.

Layers of Vegetation

Forests form layers of vegetation, as illustrated in **Figure 19-12.** At the top of the forest is the canopy, which consists of the leafy branches of trees. The *canopy* shades the ground below and provides homes for birds, insects, mammals, and many other organisms.

Beneath the canopy and above the forest floor is the shrub layer or *understory*. The understory is made up of shorter plants that tolerate shade, along with organisms that depend on these plants for food and shelter.

The forest floor is dark and moist. It is home to many insects, worms, and fungi, as well as plants that can survive in dim light. Leaves, twigs, seeds, and the bodies of dead animals that fall to the forest floor either decompose or are eaten.

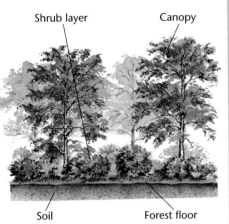

Figure 19-12

All forests are made up of layers with distinctly different biotic and abiotic factors.

Shrub layer
Canopy
Soil
Forest floor

USING TECHNOLOGY

Ecosystems in Space?

Humans have long been fascinated with the idea of setting up an ecosystem in which life continues indefinitely, even after the system is completely sealed off from the outside world. The possibilities of space flight have made this subject even more interesting. Humans traveling on extremely long journeys in outer space will not be able to bring with them all the food, water, and oxygen they need. It would be ideal if a spacecraft could be set up as a self-sustaining ecosystem in which all materials are recycled.

Keeping the Balance

The biggest problem in developing a self-sufficient ecosystem is determining exactly which species of plants, animals, protists, bacteria, and fungi should be included. Scientists have managed to develop stable, sealed ecosystems containing shrimp, algae, and dozens of species of bacteria and other microorganisms. These ecosystems are contained in glass spheres and have become known as microcosms, or "little worlds." Kept at room temperature and in dim, indirect sunlight, some of these microcosms are still healthy after eight years.

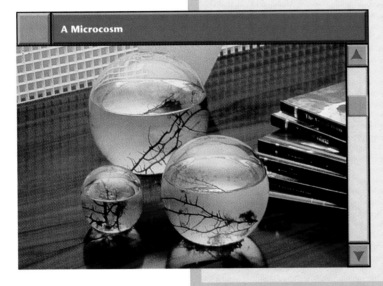

A Microcosm

Think Critically:

Why is it important to keep the sealed microcosms out of direct sunlight?

19-2 Land Environments   **523**

523

Tropical Rain Forest

The most important climax community in the equatorial regions of the world is the lush, green plant growth of the **tropical rain forest.** Rainfall averages 200 to 225 cm each year, and some areas receive as much as 400 cm of rain annually. Temperatures are warm and stable, never varying much above or below about 25°C. The abundant rainfall and high temperatures combine to create a hot, humid environment that can be compared to the atmosphere inside a greenhouse.

Plants

The highest part of the rain forest canopy is formed by the leaves and branches of trees that may reach 30 m to 40 m in height. A rain forest may contain more than 700 species of trees and over 1000 species of flowering plants. The canopy is so dense that it prevents much sunlight from filtering through to the regions below. Vines that are rooted in the soil grow up along tree trunks to reach the sun. Some types of plants such as bromeliads (bro ME lee adz) and orchids reach the light by anchoring themselves on tree trunks instead of in the soil. The understory is only dimly lit. Many of the plants growing

Figure 19-13

Tropical rain forests are lush, productive ecosystems containing more than half of all the species that exist on Earth. Parrots, monkeys, and an enormous number of insects live in the forest canopy. The peccary, a relative of the pig, roams the forest floor.

524

here have huge leaves that catch what little sunlight is available. The forest floor is almost completely dark, with few plants besides ferns and mosses. Many of the tallest trees have support roots that rise above the ground. Most plants have shallow roots that form a tangled mat at the soil surface.

Animals

The rain forest is home to a huge number of animals. It is estimated that 1000 hectares (about 2500 acres) of rain forest in South America contain thousands of insect species, including 150 different kinds of butterflies. The same patch of forest also contains dozens of species of snakes, lizards, frogs, and salamanders, and hundreds of varieties of brightly colored birds, including parrots, toucans, cockatoos, and hummingbirds. Tree-dwelling mammals include monkeys, sloths, and bats. Ocelots and jaguars are tropical cats that prowl the forest floor in search of small mammals such as pacas and agoutis, or piglike peccaries, shown in **Figure 19-13.**

Problem Solving

Saving the Rain Forests

Many of the world's rain forests are being destroyed for economic reasons. Logging and farming provide livelihoods for people living in these areas. When a section of rain forest is cleared, trees that can be used as lumber are removed and sold. The remaining plants are cut down and burned, the ash is used to fertilize the soil, and food crops are planted. After a couple of years, the soil becomes too poor to produce a harvest, so the land is abandoned and another patch of forest is cleared.

There are other ways in which people can make a living from the rain forest. Latex, a material used in surgical gloves, rubber bands, tires, and shoes, is the sap of rubber trees. Carefully tapping the trees provides a continual harvest without harming the forest. Many rain forest plants produce edible fruits, nuts, and oils that can be harvested year after year, without the need for clearing land. Harvesting these plants, rather than clearing land on which other crops can be grown for only a short time, could provide people with a sustainable income.

Solve the Problem:

Describe how competition between species is affecting the world's rain forests. What species are in competition? What are they competing for?

Think Critically:

Suppose a family could earn the same amount of money in two different ways. One is to clear several hectares of rain forest, sell the timber, and grow food crops for two years. The other is to harvest latex and edible fruits and nuts from a larger area of rain forest for four years. Which course of action would you recommend? Why? Give reasons why the family might choose the other method.

19-2 Land Environments **525**

Problem Solving

Solve the Problem

Humans are competing with rain forest species for living space and for resources.

Think Critically

Possible reasons for recommending the continuous harvest include these. At the end of four years, the source of income is still there; it has not been destroyed. The diversity of species in the rain forest has also been preserved. Possible reasons for choosing the one-time timber harvest might include that the annual income is lower for the continual-harvest option, and it might not provide the family with enough to live on.

Activity

IS **Intrapersonal** Have students prepare a chart of the major features of the three zones of the tropical rain forest: the canopy, understory, and forest floor. **L1**

Inquiry Questions

- **Infer why most rain forest animals live in trees.** *More food is available in the trees than on the ground.*

- **Why do few plants grow on the forest floor in the tropical rain forest?** *The broad leaves of the tall trees shade the forest floor, making it difficult for low-growing plants to receive enough sunlight to grow.*

Across the Curriculum

Geography Provide students with an outline map of North, South, and Central America. Have them color in regions that are tropical rain forests. Then have them devise a method for determining the approximate percentage of land covered by this biome in the western hemisphere.

Figure 19-14

Like many other grasslands, the prairies and plains of North America are hot and dry during the summer and cold and wet during the winter. They once supported huge herds of bison. Today, they are inhabited by pronghorn, gophers, ground squirrels, prairie chickens, and meadowlarks.

Grassland

Temperate and tropical regions that receive between 25 cm and 75 cm of precipitation each year and are dominated by climax communities of grasses are known as **grasslands,** illustrated in **Figure 19-14.** Most grasslands have a dry season, when little or no rain falls, which prevents the development of forests. There are grasslands on virtually every continent, and they are known by a variety of names. The prairie and plains of North America, the steppes of Asia, the veldts of Africa, and the pampas of South America are all grasslands.

Grass plants have extensive root systems, called sod, that absorb water when it rains and can withstand drought during long dry spells. The roots remain dormant during winter and sprout new stems and leaves when the weather warms in the spring. The soil is rich and fertile, and many grassland regions of the world are now important farming areas. Cereal grains such as wheat, rye, oats, barley, and corn, which serve as staple foods for humans, are types of grasses.

The most noticeable animals in grassland ecosystems are usually mammals that graze on the stems, leaves, and seeds of grass plants. Kangaroos graze in the grasslands of Australia. In Africa, common grassland inhabitants include wildebeests and zebras.

Desert

The **desert,** the driest biome on Earth, receives less than 25 cm of rain each year and supports little plant life. Some desert areas may receive no rain for years. When rain does come, it quickly drains away due to the sandy soil. Any water that remains on the ground evaporates rapidly, so the soil retains very little moisture.

Because of the lack of water, desert plants are spaced widely apart and much of the ground is bare. Some areas receive enough rainfall to support the growth of a few shrubs and small trees. Barren, windblown sand dunes are characteristic of the driest deserts, where rain rarely falls. Most deserts are covered with a thin, sandy or gravelly soil that contains little humus.

Adaptations of Desert Plants and Animals

Desert plants have developed a variety of adaptations for survival in the extreme dryness and hot and cold temperatures of this biome. Cactus plants, with their reduced, spiny leaves, are probably the most familiar desert plants of the western hemisphere. Certain species of euphorbias that look like cacti are found in African deserts. Both cacti and euphorbs have large, shallow roots that quickly absorb any water that becomes available.

Water conservation is important to all desert animals. Some, like the kangaroo rat, never need to drink water. They get all the moisture they need from the breakdown of food during digestion. Other adaptations involve behavior. Most animals are active only during the early morning or late afternoon, when temperatures are less extreme. Few large animals are found in the desert because there is not enough water or food to support them.

Figure 19-15

Desert organisms are adapted to hot, dry conditions.

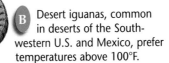

A The accordion-like sides of this barrel cactus can expand to store water following infrequent desert rains.

B Desert iguanas, common in deserts of the Southwestern U.S. and Mexico, prefer temperatures above 100°F.

Section Wrap-up

Review

1. Name two biomes that receive less than 25 cm of rain each year.

2. Compare the adaptations of tundra organisms to their environment with those of a desert organism.

3. Think Critically: Explain how the canopies of temperate deciduous forests and tropical rain forests are similar and different.

Skill Builder
Concept Mapping
Make a concept map of plant and animal species found in each land biome. If you need help, refer to Concept Mapping in the **Skill Handbook.**

Using Computers

Database Create a database of information on Earth's land biomes. Include data on temperature range, precipitation, limiting factors, and descriptions of climax communities.

19-2 Land Environments   **527**

Skill Builder
Students' concept maps should show each biome with extended branches that identify characteristics about that biome. Accept all logical concept maps.

Assessment

Portfolio Have students write a story with the setting being one of the biomes and the plot revolving around characteristics of that biome. Students can take turns reading stories to the class. **P**

---

4 Close

•MINI•QUIZ•

Use the Mini Quiz to check students' recall of chapter content.

1. _____ are large geographic areas with similar climates and climax communities. *Biomes*

2. **What are six major land biomes on Earth?** *tundra, taiga, temperate deciduous forest, tropical rain forest, grassland, desert*

3. **The amount of water that condenses and falls in the form of rain, sleet, hail, and fog is _____.** *precipitation*

4. **The permanently frozen soil of the tundra is _____.** *permafrost*

Section Wrap-up

Review

1. tundra and desert

2. Tundra plants are resistant to cold and drought. Desert plants are resistant to heat and drought. Many tundra organisms migrate during the coldest months. Desert animals have adaptations for water conservation.

3. **Think Critically** Tropical rain forest trees retain their leaves all year long, so the forest floor is always dark. Temperate deciduous trees lose their leaves during winter. Some plants have a chance to begin growing on the forest floor in early spring, before the trees grow new leaves. Both canopies limit growth on the forest floor.

Using Computers

Students can take data from the text. Each biome discussion includes data on temperature and precipitation ranges, climax communities, etc.

527

Prepare

Section Background

A number of topics that cannot be studied elsewhere on Earth can be studied in Antarctica. Antarctica is almost completely covered with snow and ice all year long. It has an effect on worldwide weather patterns.

1 Motivate

Bellringer

 Before presenting the lesson, display **Section Focus Transparency 70** on the overhead projector. Assign the accompanying **Focus Activity** worksheet. L1 LEP

2 Teach

GLENCOE TECHNOLOGY

 Videodisc

The Infinite Voyage: Secrets from a Frozen World
Introduction

Chapter 1
The Southern Ocean—A Rich Marine Ecosystem

Chapter 2
Krill: The Vital Link of the Food Chain

ISSUE:

19•3 Protecting Antarctica

Objectives

* Describe the living and nonliving resources of the Antarctic.
* Explain why the countries of the world are trying to agree about how to manage the scientific and mineral resources of Antarctica.

Figure 19-16

Antarctica provides researchers an area to investigate many topics that cannot be easily studied anywhere else on Earth. Some of these are the adaptations to the cold climate, and the movements of Earth's crust.

The Coldest Place on Earth

The continent of Antarctica is almost completely covered with snow and ice all year long. Winter is dark and long, with temperatures dipping to −90°C. The polar ice is 3 to 5 km deep, and is made up of about 70 percent of Earth's fresh water. During the winter, shelves of ice extend from the land out over the ocean, essentially doubling the size of the continent. The yearly freezing and thawing of this ice mass has important effects on worldwide weather patterns and is the force that drives ocean currents.

Life in the Antarctic

Only two percent of the Antarctic land mass is free of ice. Portions of this area receive no precipitation at all. Organisms found here include hundreds of species of lichens, along with mosses and two species of tiny flowering plants. Birds and seals use Antarctic shores as breeding grounds. Although the land is barren, the waters of the Antarctic Ocean teem with organisms, especially during spring and summer, when whales, fish, penguins, and other animals feed on masses of krill.

Antarctica's Mineral Riches

Antarctica is also a land of mineral riches. The largest coal field in the world may lie beneath its mountains. Gold and platinum have been found, and geologists believe the continent contains rich deposits of iron, lead, copper, and uranium. There are probably enormous quantities of oil beneath Antarctica's continental shelf. Many people are worried that exploiting these mineral reserves will have negative effects on the environment. In 1991, many nations of the world agreed to protect the Antarctic continent from development by declaring it a "reserve devoted to peace and science." The agreement bans all mining exploration for 50 years, limits visits by tourists, and sets up strict rules to protect the environment.

528 Chapter 19 Ecosystems

Program Resources

 Reproducible Masters
Study Guide, p. 74 L1
Reinforcement, p. 74 L1
Enrichment, p. 74 L3

Transparencies
Section Focus Transparency 70 L1

2 Points of View

Should we mine it?

Worldwide reserves of many valuable materials, especially oil, are declining. It can be argued that mining Antarctica's resources would supply humanity with needed raw materials and help improve the lives of people all over the world. At the same time, no one doubts that mining activities would have significant effects on Antarctica's fragile environment. Environmental and territorial issues would have to be resolved before mining could begin. Mining operations would bring more people into the area, as well as add to existing pollution problems that are already hampering scientific research.

Create an Antarctic Park

Many nations, particularly developing countries, have no claims to Antarctica and cannot afford to explore its mineral riches. Leaders in these and other nations would like to see Antarctica kept free of all claims and the region's mineral resources managed for the benefit of all humankind. Some groups would like to go even further by declaring the continent a park or nature preserve and banning all mining forever. Environmentalists point out that there is already competition between human scientists and other species of organisms for the little ice-free land that exists.

Section Wrap-up

Review

1. To which biome would you assign the ice-free portions of the Antarctic landmass? Why?

2. What are some of the limiting factors that restrict life in the Antarctic?

Explore the Issue

Should the nations of the world agree to keep Antarctica as an international park protected from mining operations and other types of development? If so, what kinds of limits should be placed on human activities in the area? If not, how much mineral exploration should be allowed, and how might conflicts over territory be resolved?

Figure 19-17

This map of Antarctica shows territorial claims and disputed areas.

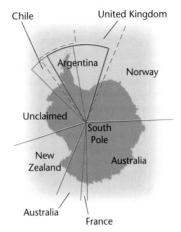

Antarctica

Chile — United Kingdom — Argentina — Norway — Unclaimed — South Pole — New Zealand — Australia — Australia — France

*inter*NET CONNECTION

Visit the Chapter 19 Internet Connection at Glencoe Online Science, **www.glencoe.com/sec/ science/life,** for a link to more information about Antarctica.

SCIENCE & SOCIETY

3 Assess

Check for Understanding

Have students answer questions 1 and 2 in the Section Wrap-up.

Reteach

Using a globe, review the location of Antarctica with students. Review with students the reasons scientists are interested in Antarctica.

Extension

For students who have mastered this section, use the **Reinforcement** and **Enrichment** masters.

4 Close

Have students prepare a list of ways that Antarctica is an asset to countries of the world.

Section Wrap-up

Review

1. tundra, because it receives little precipitation and has freezing temperatures all year long

2. cold temperatures, lack of soil, lack of water, covering of ice and snow

Explore the Issue

Student answers will vary. Some may feel that limits are necessary. Others may feel that people should be allowed to use whatever resources are available. Students may suggest various methods for determining ownership, such as lotteries, auctions, or other means. **P**

19•4 Water Environments

Prepare

Section Background

- More than 95% of Earth's water is found in oceans. The remaining 5% is fresh water.
- Salinity is the amount of dissolved salts present in water. As the surface of an ocean evaporates, the salinity level rises because there is more salt left in relation to the amount of remaining water. However, the salinity of ocean water remains stable because water with low salt levels is constantly flowing into the ocean water from rivers and in the form of rain.
- Limnology is the branch of biological science that has to do with lakes and ponds.

Preplanning

To prepare for the MiniLAB, obtain self-sealing plastic bags, pond muck, gravel, pond debris, plants, dip nets, and hand lenses.

1 Motivate

Bellringer

Before presenting the lesson, display **Section Focus Transparency 71** on the overhead projector. Assign the accompanying **Focus Activity** worksheet. L1 LEP

Tying to Previous Knowledge

From Chapters 8 and 9, review cyanobacteria, protists, and green algae and their importance in the food chains of marine and freshwater ecosystems.

Science Words

plankton
estuary
intertidal zone

Objectives

- Distinguish between flowing freshwater and standing freshwater ecosystems.
- Describe important seashore and deep-ocean ecosystems.

Figure 19-18

This swift-flowing mountain stream is fed by melting snow.

A The cold water, rapid current, and high oxygen content of this stream provide the kind of habitat required for fish such as fast-moving brook trout.

Freshwater Biomes

You've learned that temperature and precipitation are the most important factors determining which species can survive in a land environment. The limiting factors in water environments are the amount of salt in the water or salinity, dissolved oxygen, and sunlight. Fresh water contains little or no dissolved salts, and so has a low salinity. Earth's freshwater biomes include flowing water like these rivers and streams, as well as still or standing water, such as lakes and ponds.

Rivers and Streams

Flowing freshwater environments range from small, swiftly flowing streams to large, slow rivers. The faster a stream flows, the clearer its water tends to be, and the higher its oxygen content. Swift currents quickly wash loose particles downstream, leaving a rocky or gravelly bottom. The tumbling and splashing of swiftly flowing water mixes in air from the atmosphere, increasing the oxygen content of the water.

Most of the nutrients that support life in flowing-water ecosystems are washed into the water from land. In areas where the water movement slows down, such as wide pools in streams or large rivers, debris settles to the bottom. These environments tend to have higher nutrient levels and lower dissolved oxygen levels. They contain organisms such as freshwater mussels, minnows, and leeches that are not so well adapted for swiftly flowing water. They also tend to have more plant growth.

530

B With muscular, streamlined bodies, chinook salmon are well adapted to fast-moving turbulent waters.

Program Resources

📁 **Reproducible Masters**
Study Guide, p. 75 L1
Reinforcement, p. 75 L1
Enrichment, p. 75 L3
Activity Worksheets, pp. 5, 117 L1
Science Integration Activities, pp. 37-38
Multicultural Connections, pp. 41-42

📦 **Transparencies**
Section Focus Transparency 71 L1

Figure 19-19

Ponds and lakes differ in the types of communities inhabiting them.

A The warm, sunlit waters of this pond are home to a large variety of organisms. Plants and algae form the basis of a food web that includes snails, insects, frogs, snakes, turtles, and fish.

B The population density of the warm, shallow water of the lake shore is high. Population density is much lower in the colder, darker waters at the center of a deep lake.

Lakes and Ponds

A lake or pond forms when a low place in the land fills with rainwater, snowmelt, or water from a stream. The waters of lakes and ponds hardly move at all. They contain more plant growth than flowing-water environments contain.

Ponds are smaller shallow bodies of water. Because they are shallow, sunlight can usually penetrate all the way to the bottom, making the water warmer and promoting the growth of plants and algae. In fact, many ponds are almost completely filled with plant material, so that the only clear, open water is at the center. Because of the lush growth in pond environments, they tend to be high in nutrients.

Lakes are larger and deeper than ponds. They tend to have more open water because most plant growth is limited to shallow areas along the shoreline. In fact, organisms found in the warm, sunlit waters of the lakeshore are often similar to those found in ponds.

Floating in the warm, sunlit waters near the surface of freshwater lakes and ponds are microscopic algae, plants, and other organisms known as **plankton.** If you were to dive all the way to the bottom, you would discover there few, if any, plants or algae growing. Colder temperatures and lower light levels limit the types of organisms that can live in deep lake waters. Most lake organisms are found along the shoreline and in the warm water near the surface.

19-4 Water Environments   **531**

Figure 19-20
These Canada geese are swimming in an estuary of the Chesapeake Bay.

MiniLAB

Freshwater Environments

Procedure

1. Cover the bottom of a self-sealing freezer bag with about two centimeters of gravel, muck, and other debris from the bottom of a pond. If plants are present, add one or two to the bag. Use a dip net to capture small fish, insects, or tadpoles.
2. Carefully pour pond water into the bag until it is about two-thirds full. Seal the bag.
3. Keep the bag indoors at room temperature, out of direct sunlight.

Analysis

1. Using a hand lens, observe as many organisms as possible. Record your observations. After two or three days, return your sample to the original habitat.
2. Write a short paper describing the organisms in your sample ecosystem and explaining their interactions.

Saltwater Biomes

About 95 percent of the water on the surface on Earth contains high concentrations of salts. The saltwater biomes include the oceans, seas, and a few inland lakes, such as the Great Salt Lake in Utah.

Estuaries

Virtually every river on Earth eventually flows into the ocean. The area where a river meets the ocean contains a mixture of fresh water and salt water. These areas, called **estuaries,** are located near coastlines and border the land. Salinity changes with the amount of fresh water brought in by rivers and streams, and with the amount of salt water pushed inland by the tides.

Estuaries like the one in **Figure 19-20** are extremely fertile, productive environments because freshwater streams bring in tons of nutrients from inland soils. Nutrient levels in estuaries are higher than those in freshwater or other saltwater ecosystems. Estuarine organisms include many species of algae, a few salt-tolerant grasses, shrimp, crabs, clams, oysters, snails, worms, and fish. Estuaries serve as important nursery grounds for many species of ocean fish. They also provide much of the seafood consumed by humans.

Seashores

All of Earth's landmasses are bordered by ocean water. The fairly shallow waters along the world's coastlines contain a variety of saltwater ecosystems, all of which are influenced by the tides and by the action of waves. The gravitational pull of the sun and moon causes the tides to rise and fall twice each day. The height of the tides varies according to the phases of the moon, the season, and the slope of the shoreline. The **intertidal zone** is the portion of the shoreline that is covered with water at high tide and exposed to the air during low tide. Organisms living in the intertidal zone must not only be adapted to dramatic changes in temperature, moisture, and salinity, but also be able to withstand the force of wave action. Two kinds of intertidal zones are shown in **Figure 19-21.**

A **Sandy beaches**
Wave action keeps the sandy bottom in constant motion, and organisms that live on sandy shores, such as clams, crabs, and worms, burrow into the sand to avoid being washed away.

Open Ocean

Life abounds out in the open ocean, where there is no land. The ocean can be divided into life zones, based on the depth to which sunlight can penetrate into the water. The lighted zone of the ocean is the upper 200 m or so. It is the home of the plankton that make up the foundation of the food chain in the open ocean. Below about 200 m, where sunlight cannot reach, is the dark zone of the ocean. Animals living in this region either feed on material that floats down from the lighted zone, or they feed on each other.

Figure 19-21
Organisms living in intertidal zones have adaptations to survive in these changing environments.

B **Rocky shores**
Algae, mussels, barnacles, snails, and other organisms adapted for clinging to the rocks are typically found on rocky shores. These organisms must be able to tolerate the heavy force of breaking waves.

Section Wrap-up

Review

1. What are the similarities and differences of a lake and a stream?

2. What biotic or abiotic factor limits life on the floor of a tropical forest and the bottom of the deep ocean? Why?

3. **Think Critically:** Why do few plants grow in the waters of a swift-flowing mountain stream?

Skill Builder
Comparing and Contrasting
Compare and contrast the effects of (1) temperature in the tundra and desert and (2) sunlight in deep-lake and deep-ocean waters. If you need help, refer to Comparing and Contrasting in the **Skill Handbook.**

Science Journal

Write a paragraph in your Science Journal explaining how starting from the equator and moving toward the north pole is like climbing a mountain. Refer to abiotic factors in your explanation.

19-4 Water Environments **533**

533

People and Science

Background

- According to Mr. Gonzalez, populations of salmon throughout the United States are declining at an alarming rate. Many large rivers used by migrating salmon, such as the Columbia in Oregon and the Sacramento in California, are sites of hydroelectric dams that suck in large amounts of water to cool the turbines. In the process, huge numbers of young salmon are shredded and killed each season as they move downstream from the spawning grounds where they were hatched.

- The spawning migrations of salmon are one of the wonders of nature. Adult salmon spend their lives in the sea, but when ready to breed they return to the freshwater creeks and streams where they were hatched. The journey upstream for some salmon species may be more than 5000 kilometers. Once they reach their natal stream, female salmon spawn their eggs in shallow nests in the streambed, and males fertilize them. Salmon die after spawning.

Teaching Strategies

Have students create a bulletin board that illustrates the life cycle of a salmon. Students should be able to identify the fish's ecological requirements at each stage of its life cycle.

People and Science

Alfred Gonzalez, *Salmon Habitat Researcher*

On the Job

Q Mr. Gonzalez, tell us about your research on salmon habitat restoration.

A I am studying "stream enhancement techniques" that are aimed at improving the habitat for the coho salmon and other fish. My study site is located on a creek that was once prime salmon spawning habitat, but which was partially buried by a landslide in the 1980s. About ten years after the landslide, a variety of instream devices, such as logs for fish to hide under, were installed in the stream. Now I'm trying to determine if those devices are helping to restore the spawning areas that adult salmon use when they come upstream to breed.

Q How do you do that?

A Well, for example, I'm currently analyzing the effect of the instream devices on the size of the sand and gravel particles that settle out of the water onto the streambed. I collect samples from the streambed upstream and downstream from the devices, measure the size of the particles, and compare these data to those collected in previous years. If the devices work correctly,

Personal Insights

they will help increase the type of gravelly substrate that salmon prefer to use for their nests in which they lay their eggs.

Q What do you like best about the type of ecological research you do?

A I enjoy establishing my field research plan and collecting data out in the field. Being outdoors gives me a sense of freedom and adventure while pursuing science.

Career Connection

The Nature Conservancy is an organization that works to preserve the habitats of rare and endangered plants and animals, and has land and water conservation programs in all 50 states. Contact the Nature Conservancy office nearest you, and ask about job and volunteer opportunities.

- **Wildlife Refuge Manager**
- **Wetlands Ecologist**

Career Connection

Career Path High school: science and math

College: biology, chemistry, and environmental science

A graduate degree and/or special training is required for many environmentally and ecologically oriented jobs, which can range from waste management to soil conservation, and from parks and outdoor recreation to forestry.

For More Information

A good source of information for a wide variety of environmental careers is *The New Complete Guide to Environmental Careers* (1993), published by Island Press.

Summary

19-1: How Ecosystems Change
1. Ecological succession is the gradual change from one community to another.
2. Primary succession occurs in a place where there is no soil. It begins with organisms that make up a pioneer community.
3. Secondary succession begins in a place that has soil and was once the home of living organisms.
4. A community that has reached the final stage of ecological succession is a climax community.

19-2: Land Environments
1. Temperature and precipitation determine the climate of a region.
2. Large geographic areas with similar climate and climax communities are biomes.
3. Earth's land biomes include tundra, taiga, temperate forest, tropical rain forest, grassland, and desert.

19-3: Science and Society: Protecting Antarctica
1. Antarctica is the coldest, driest region on Earth.
2. Antarctica is an important site for scientific research and a land of rich mineral resources.
3. Most nations of the world are concerned about protecting Antarctica's environment. There is an active debate about how to manage the continent's resources.

19-4: Water Environments
1. Freshwater biomes include flowing water, such as rivers and streams, and standing water, such as lakes and ponds.
2. Saltwater biomes include estuaries, seashores, and the deep ocean.

Key Science Words

a. biome
b. climax community
c. desert
d. ecological succession
e. estuary
f. grassland
g. intertidal zone
h. permafrost
i. pioneer community
j. plankton
k. precipitation
l. primary succession
m. secondary succession
n. taiga
o. temperate deciduous forest
p. tropical rain forest
q. tundra

Reviewing Vocabulary

Match each phrase with the correct term from the list of Key Science Words.

1. one community of organisms replaces another
2. the first organisms to inhabit an area
3. region with similar climate and climax communities
4. permanently frozen soil
5. northern coniferous forest
6. equatorial region that receives large amounts of rainfall

Chapter 19 Review   535

Summary
Have students read the summary statements to review the major concepts of the chapter.

Reviewing Vocabulary

1. d	6. p
2. i	7. c
3. a	8. e
4. h	9. g
5. n	10. j

✔ Assessment

Portfolio Encourage students to place in their portfolios one or two items of what they consider to be their best work. Examples include:
- Skill Builder, p. 513
- Activity 19-1 results and answers, p. 515
- Explore the Issue, p. 529 P

Performance Additional performance assessments may be found in **Performance Assessment** and **Science Integration Activities**. Performance Task Assessment Lists and rubrics for evaluating these activities and other products generated throughout the chapter can be found in Glencoe's **Performance Assessment in the Science Classroom.**

GLENCOE TECHNOLOGY

Videodisc
Glencoe Life Science Interactive Videodisc
Side 2, Lesson 9
Ecosystems: Wetlands

38845 to 52792

MindJogger Videoquiz
Chapter 19 Have students work in groups as they play the Videoquiz game to review key chapter concepts.

Checking Concepts

1. d	**6.** c
2. b	**7.** c
3. b	**8.** a
4. d	**9.** d
5. b	**10.** a

Understanding Concepts

11. In both cases, the climate gets colder and vegetation changes.

12. The climax community of the taiga is made up of coniferous trees. The climax community of the temperate deciduous forest is deciduous trees.

13. A desert is a dry region with few plants. A grassland is a region dominated by grass plants; it receives more rainfall than a desert.

14. An estuary is a body of water made up of fresh water from a river or stream and salt water from the ocean. It is affected by tides but not waves. A seashore is on the coastline, is in contact only with salt water, and is affected by both tides and waves.

15. As air moves up the mountains, it cools, and the moisture in it condenses as rain or snow. When the air reaches the other side of the mountains, it no longer contains moisture, so the region on the other side of the mountain range gets much less precipitation.

Thinking Critically

16. The temperate forest soil contains more humus because the plants of the tropical forest absorb decayed material before it can become part of the soil.

7. dry region with poor soil
8. where fresh water mixes with salt water
9. between high and low tide lines
10. tiny organisms floating in the open ocean

Checking Concepts

Choose the word or phrase that best completes the sentence.

1. The climate of an area is determined by _____.
   a. plankton   c. limiting factors
   b. succession   d. abiotic factors
2. Tundra and desert are examples of _____.
   a. ecosystems   c. habitats
   b. biomes   d. communities
3. The _____ is a treeless, cold, and dry biome.
   a. taiga   c. desert
   b. tundra   d. grassland
4. All are examples of grasslands except _____.
   a. pampas   c. steppes
   b. veldts   d. estuaries
5. Mussels and barnacles have adapted to the wave action of the _____.
   a. sandy beach   c. open ocean
   b. rocky shore   d. estuary
6. The _____ contains the largest number of species of any biome.
   a. taiga
   b. temperate deciduous forest
   c. tropical rain forest
   d. grassland
7. A _____ is the end result of succession.
   a. pioneer community
   b. limiting factor
   c. climax community
   d. permafrost

8. Trees are part of the climax communities of each biome except _____.
   a. tundra   c. tropical rain forest
   b. taiga   d. grassland
9. Freshwater biomes include all except _____.
   a. lakes   c. rivers
   b. ponds   d. estuaries
10. All except _____ have flowing water.
    a. ponds   c. seashores
    b. rivers   d. streams

Understanding Concepts

Answer the following questions in your Science Journal using complete sentences.

11. How is traveling north from the equator similar to climbing a mountain?
12. What is the difference between taiga and temperate deciduous forest?
13. Compare desert and grassland.
14. What is the difference between an estuary and the seashore?
15. Explain how a mountain range affects precipitation in nearby regions.

Thinking Critically

16. Would a soil sample from a temperate deciduous forest contain more or less humus than soil from a tropical rain forest? Explain.
17. A grassy meadow borders an oak-maple forest. Is one of these ecosystems undergoing succession? Why?
18. Describe how ecological succession eventually results in the layers of vegetation found in forests.
19. In what ways would scientists and tourists visiting the ice-free areas of the Antarctic cause disturbances in the ecosystem?

17. The meadow is undergoing succession. Because a forest is already growing there, the climate must be appropriate for development of a forest.

18. Succession begins with small plants that grow low to the ground. As soil develops, it supports the growth of increasingly larger plants.

19. Accept reasonable answers. Some possibilities include the following. Human shelters might take up some of the space needed by birds, seals, or other animals for nesting sites. They might frighten the animals. Litter and waste left behind by humans might harm the animals.

20. The environment indoors is about the same temperature as the tropics.

20. Why do many tropical rain forest plants make good houseplants?

Developing Skills

If you need help, refer to the **Skill Handbook.**

21. **Interpreting Data:** Interpret the data given and decide which area of a stream has more species of living things and explain why. A riffle is a small wave. Find the total number of organisms in each area.

Stream Organisms

Organism	Number in Pools	Number in Riffles
Midgefly larva	0	1000
Caddisfly larva	2	70
Snail	2	12
Water penny	1	7
Mayfly larva	1	250
Stonefly larva	1	61
Horsefly larva	1	8
Water strider	8	1
Dragonfly larva	2	7
Water diving beetle	3	1
Whirligig beetle	5	2

22. **Concept Mapping:** Make a concept map for water environments. Include these terms: *saltwater ecosystems, freshwater ecosystems, intertidal zone, lighted zone, dark zone, lake, pond, river, stream, flowing water,* and *standing water.*

23. **Comparing and Contrasting:** Compare and contrast adaptations of organisms living in swiftly flowing streams and organisms living in the rocky intertidal zone.

24. **Making and Using Graphs:** Make a bar graph of the amount of rainfall per year in each biome.

Rainfall Amounts

Biome	Rainfall/Year
Deciduous forests	100 cm
Tropical rain forests	225 cm
Grasslands	50 cm
Deserts	20 cm

25. **Hypothesizing:** Make a hypothesis as to what would happen to succession in a pond if the pond owner removed all the cattails and reeds from around the pond edges every summer.

Performance Assessment

1. **Display:** Use your own observations or the results of library research to develop an exhibit that describes the important characteristics of one of Earth's land or water biomes. Include photographs or drawings of representative organisms and a possible food web.

2. **Poster:** Research the steps in succession that have taken place since the dramatic forest fires that burned much of Yellowstone National Park in 1988. Create a poster exhibit that describes those changes.

3. **Oral Presentation:** Use the results of library research to write a paper or present an oral report describing the experiences of people who have explored or conducted scientific research in the Antarctic.

2. Accept all reasonable responses. Use the Performance Task Assessment List for Poster in **PASC,** p. 73. [P]

3. Students' reports or presentations should be accompanied by a list of sources. Use the Performance Task Assessment List for Making Observations and Inferences in **PASC,** p. 17. [P]

Developing Skills

21. **Interpreting Data** Riffles have more organisms than pools. Oxygen is more available in places where water flows than in pools. Total number of organisms in pools: 26. Total number of organisms in riffles: 1419.

22. **Concept Mapping** Students should separate water ecosystems into marine and freshwater ecosystems. They should then give examples of each type.

23. **Comparing and Contrasting** Both types are adapted to avoid being washed away by the force of flowing water or waves. Their adaptations allow them to cling to hard surfaces or hide under rocks in crevices.

24. **Making and Using Graphs**

25. **Hypothesizing** Succession might slow down or stop altogether.

Performance Assessment

1. Students' projects should reflect information contained in the chapter. Make sure all references are included. Use the Performance Task Assessment List for Booklet or Pamphlet in **PASC,** p. 57. [P]

Chapter Organizer

Section	Objectives/Standards	Activities/Features
Chapter Opener		**Explore Activity:** Make a model to demonstrate how plant cover protects topsoil. p. 539
20-1 **Natural Resources** (2 sessions, 2 blocks)*	1. **Identify** natural resources and explain why they are important to living organisms. 2. **Distinguish** between renewable and nonrenewable resources. 3. **Describe** renewable and nonrenewable energy sources. **National Content Standards:** (5-8) UCP3, A1, B1, B3, D1, E2, F5	**MiniLAB:** How does mining for mineral deposits affect the environment? p. 541 **Using Technology:** Geothermal Power, p. 545 **Skill Builder:** Concept Mapping, p. 546 **Using Math,** p. 546 **Science and Art:** Trash Art, p. 547 **Science Journal,** p. 547
20-2 **Science and Society** **Technology: Recycling** (1 session, ½ block)*	1. **Identify** ways to recycle and reuse natural resources. **National Content Standards:** (5-8) C4, F5	**Explore the Technology,** p. 549
20-3 **Conservation and** **Wildlife Protection** (1½ sessions, 1 block)*	1. **Define** *conservation* and explain how it relates to natural resources. 2. **Describe** how renewable resources can be conserved and protected. **National Content Standards:** (5-8) A1, A2, C4, F2	**Activity 20-1:** Investigating Biodegradable Substances, pp. 552-553 **Using Math,** p. 555 **Skill Builder:** Comparing and Contrasting, p. 556 **Using Computers:** Database, p. 556
20-4 **Maintaining a Healthy** **Environment** (1½ sessions, 1 block)*	1. **Describe** the origins and effects of pollutants that contaminate air, land, and water. 2. **Explain** the causes of acid rain, ozone depletion, and the greenhouse effect. **National Content Standards:** (5-8) A1, A2, B1, D1, F2	**Problem Solving:** Indoor Air Pollution, p. 558 **MiniLAB:** Is acid rain falling on your town? p. 559 **Connect to Chemistry,** p. 560 **Activity 20-2:** Modeling the Greenhouse Effect, p. **Skill Builder:** Comparing and Contrasting, p. 564 **Using Math,** p. 564

* A complete Planning Guide that includes block scheduling is provided on pages 32T-35T.

Activity Materials

Explore	Activities	MiniLABs
page 539 moist sand, potting soil, small plastic basin or aluminum foil baking pan, clumps of moss, tap water, beaker, spray bottle	pages 552-553 clay pots, petri dishes, aluminum pie tins, potting soil, sand or gravel, labels, biodegradable and nonbiodegradable wastes, hand lens, microscope, apron page 561 2 clear plastic, 2-L soda bottles with the tops cut off and the labels removed; 2 thermometers; 4 cups of potting soil; masking tape; plastic wrap; rubber band; lamp with 100-watt lightbulb; chronometer or watch with second hand	page 541 chocolate-chip cookie or nut-filled brownie, paper plate, toothpick page 559 rain gauge, sample of rainwater, pH indicator paper, pH chart, tap water, distilled water, small beakers

Need Materials? Call Science Kit (1-800-828-7777).

Chapter 20 Resources and the Environment

Teacher Classroom Resources

Reproducible Masters	Transparencies	Teaching Resources
Study Guide, p. 76 **Reinforcement**, p. 76 **Enrichment**, p. 76 **Activity Worksheets**, p. 122 **Lab Manual** 46, Home Energy Needs **Cross-Curricular Integration**, p. 24	**Section Focus Transparency 72,** Products from Nature **Teaching Transparency 40,** pH Scale	**Glencoe Life Science Interactive Videodisc** **Life Science CD-ROM** **Spanish Resources** **English/Spanish Audiocassettes** **Cooperative Learning Resource Guide** **Lab Partner** **Lab and Safety Skills** **Lesson Plans**
Study Guide, p. 77 **Reinforcement**, p. 77 **Enrichment**, p. 77	**Section Focus Transparency 73,** Cutting Down on Trash	*(Assessment Resources)*
Study Guide, p. 78 **Reinforcement**, p. 78 **Enrichment**, p. 78 **Activity Worksheets**, pp. 118-119 **Critical Thinking/Problem Solving,** p. 24 **Science and Society Integration**, p. 24	**Section Focus Transparency 74,** In the Name of Progress	**Chapter Review**, pp. 43-44 **Assessment**, pp. 130-133 **Performance Assessment in the Science Classroom (PASC)** **MindJogger Videoquiz** **Alternate Assessment in the Science Classroom** **Performance Assessment** **Chapter Review Software** **Computer Test Bank**
Study Guide, p. 79 **Reinforcement**, p. 79 **Enrichment**, p. 79 **Activity Worksheets**, pp. 120-121, 123 **Lab Manual** 47, Acid Rain **Lab Manual** 48, Water Pollution **Science Integration Activity 20,** Venus—The Greenhouse Effect **Concept Mapping**, p. 45 **Multicultural Connections**, pp. 43-44	**Section Focus Transparency 75,** People and Nature **Science Integration Transparency 20,** Home Waste Water Hazards **Teaching Transparency 39,** Greenhouse Effect	

Assessment Resources

(see above column)

Key to Teaching Strategies

The following designations will help you decide which activities are appropriate for your students.

L1 Level 1 activities should be within the ability range of all students, including those with learning difficulties.

L2 Level 2 activities should be within the ability range of the average to above-average student.

L3 Level 3 activities are designed for the ability range of above-average students.

LEP LEP activities should be within the ability range of Limited English Proficiency students.

LS These activities are designed to address different learning styles.

COOP LEARN Cooperative Learning activities are designed for small group work.

P These strategies represent student products that can be placed into a best-work portfolio.

GLENCOE TECHNOLOGY

The following multimedia resources are available from Glencoe.

Science and Technology Videodisc Series (STVS)
Chemistry
  Solar House
  Solar Tower
  Wind Power
  Hydroelectric Power
  Geothermal Wells
  Oil from Wood
  Energy-Integrated Farm
The Infinite Voyage Series
Miracles by Design
Crisis in the Atmosphere

National Geographic Society Series
STV: Atmosphere
STV: Water
STV: Biodiversity
GTV: Planetary Manager
Glencoe Life Science Interactive Videodisc
Decomposers: Earth's Natural Recyclers
Life Science CD-ROM

Teacher Classroom Resources

This is a representation of key blackline masters available in the Teacher Classroom Resources.

Teaching Aids

Section Focus Transparencies

Science Integration Transparencies

Teaching Transparencies

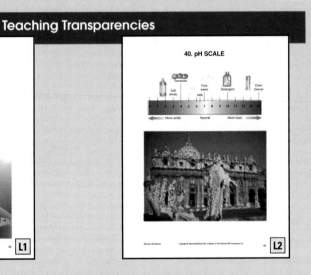

Meeting Different Ability Levels

Study Guide

Reinforcement

Enrichment Worksheets

Chapter 20 Resources and the Environment

Hands-On Activities

Science Integration Activity

Lab Manual

Assessment

Performance Assessment

Enrichment and Application

Critical Thinking/Problem Solving

Cross-Curricular Integration

Science and Society Integration

Multicultural Connections

Concept Mapping

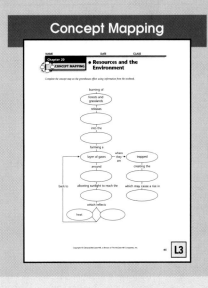

Resources and the Environment

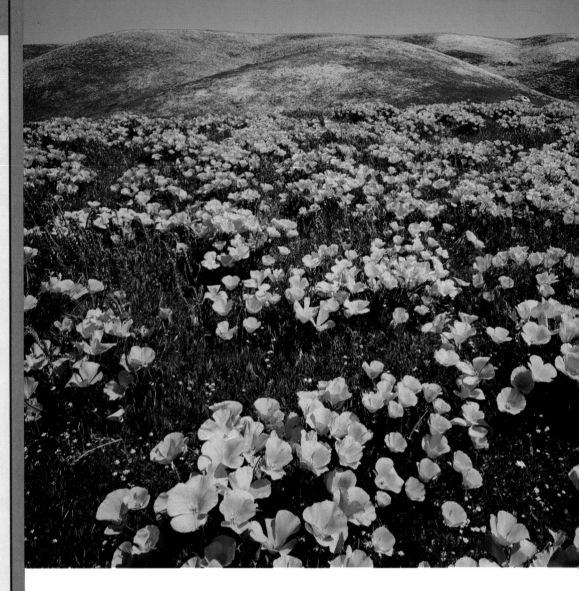

CHAPTER OVERVIEW

Section 20-1 This section defines natural resources and explains how renewable and nonrenewable resources are different. It then identifies energy sources for use now and in the future.

Section 20-2 Science and Society Solid waste is defined, and how it can be recycled and reused are discussed.

Section 20-3 This section describes conservation of energy, soil, wildlife, land, and minerals.

Section 20-4 Problems resulting from air, land, and water pollution are discussed.

Chapter Vocabulary

natural resources
renewable resource
nonrenewable resource
fossil fuel
photovoltaic (PV) cell
hydroelectric power
geothermal energy
nuclear energy
solid waste
recycling
soil depletion

erosion
soil management
extinction
endangered species
pollutant
smog
acid rain
ozone depletion
greenhouse effect
global warming
hazardous waste
groundwater

Theme Connection

Energy Even though the total amount of energy on Earth remains constant, the sources of usable energy are limited. Energy is often transformed from one form to another, such as fossil fuel energy into electricity and then into light energy.

Previewing the Chapter

Section 20-1 Natural Resources
▶ Earth's Resources
▶ Energy for Now and the Future

Section 20-2 Science and Society
Technology: Recycling

Section 20-3 Conservation and Wildlife Protection
▶ What We Can Do
▶ Energy Conservation

538

▶ Soil Conservation
▶ Wildlife Conservation
▶ Land Conservation
▶ Mineral Conservation

Section 20-4 Maintaining a Healthy Environment
▶ How are we affecting the environment?
▶ Air Pollution
▶ Land Pollution
▶ Water Pollution

Learning Styles

Look for the following logo for strategies that emphasize different learning modalities. **LS**

Kinesthetic	Explore, p. 539; MiniLAB, p. 541; Science & Art, p. 547; Inclusion Strategies, p. 551
Visual-Spatial	Visual Learning, pp. 541, 550
Interpersonal	Activity, pp. 542, 554; Activities 20-1, pp. 552-553; 20-2, p. 561; Science & Art, p. 547; MiniLAB, p. 559
Intrapersonal	Enrichment, pp. 543, 544; Science & Art, p. 547
Linguistic	Across the Curriculum, p. 551

LS

Chapter 20

Resources and the Environment

There's nothing quite so reliable as the ground beneath your feet. We walk on it, plant lawns and flowers on it, and pave it over when we want to drive cars on it. But topsoil is something we shouldn't take for granted. Topsoil is the loose, nutrient-rich, top layer of soil that is vital to the growth of everything from food crops to forests. Layers of vegetation growing on topsoil help protect it and keep it in place. What happens when topsoil is left unprotected?

EXPLORE ACTIVITY

Make a model to demonstrate how plant cover protects topsoil.

1. Use a mixture of moist sand and potting soil to create a miniature landscape in a plastic basin or aluminum foil baking pan. Include hills and valleys in the shape of your landscape.
2. Cover portions of your landscape with clumps of moss as a plant cover. Leave some sloping portions of your landscape without plant cover.
3. Simulate a rainstorm by pouring water from a beaker or spraying water from a spray bottle over your landscape.

Observe: What happens to the land that is not protected by plant cover?

Previewing Science Skills

▶ In the **Skill Builders,** you will compare and contrast and map concepts.

▶ In the **Activities,** you will measure, graph, model, calculate, and infer.

▶ In the **MiniLABs,** you will observe and infer.

539

Prepare

Section Background

- The two types of natural resources are nonrenewable and renewable. Nonrenewable resources, which include minerals (such as aluminum, tin, and copper) and fossil fuels (coal, oil, and natural gas), are present in limited supplies and are depleted by use. They are not replenished by natural processes within a reasonable period of time. Fossil fuels, for example, take millions of years to form.

- Renewable resources include trees in forests; fish in lakes, rivers, and the ocean; fertile agricultural soil; and fresh water in lakes and rivers. Nature replaces these resources, and they can be used forever as long as they are not overexploited in the short term.

Preplanning

Obtain chocolate chip cookies for the MiniLAB.

1 Motivate

Bellringer

Before presenting the lesson, display **Section Focus Transparency 72** on the overhead projector. Assign the accompanying **Focus Activity** worksheet. L1 LEP

Tying to Previous Knowledge

Review with the students the definitions of *habitat*, the *water cycle*, the *carbon cycle*, and the other cycles of matter from Chapter 18.

Science Words

natural resource
renewable resource
nonrenewable resource
fossil fuel
photovoltaic (PV) cell
hydroelectric power
geothermal energy
nuclear energy

Objectives

- Identify natural resources and explain why they are important to living organisms.
- Distinguish between renewable and nonrenewable resources.
- Describe renewable and nonrenewable energy sources.

Earth's Resources

A robin eats an earthworm and collects twigs for a nest. A fox burrows underground, preparing a den for pups that soon will be born. The leaves of a tree absorb carbon dioxide from the air. A sea otter stays afloat by wrapping itself in a strand of giant kelp while it uses a stone to hammer open the shell of a sea urchin. A group of young humans climbs aboard a bus for the ride to school.

What do these organisms have in common? They rely on Earth's natural resources for their survival. **Natural resources** are the parts of the environment used by living organisms for food, shelter, and all other needs. The natural resources people use include food, air, water, and the materials and energy needed to make everything from clothes and buildings to automobiles, books, computers, gasoline, and electricity.

Renewable Resources

Water moves from the surface of Earth to the atmosphere and back again via the water cycle, and oxygen is produced by plants during photosynthesis. Oxygen and water are examples of renewable resources. A **renewable resource** is a natural resource that is recycled or replaced by ongoing natural processes. The cotton used to make a pair of blue jeans is renewable because a new crop of cotton plants can be grown every year. Plants and animals such as those in **Figure 20-1**, water, and air are all renewable resources.

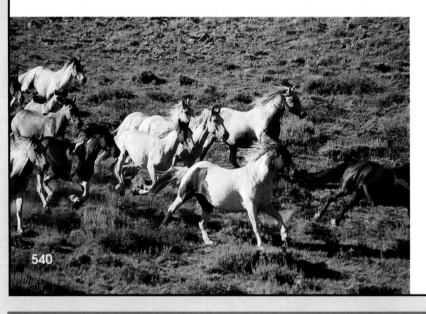

540

Figure 20-1

Renewable natural resources include food crops, forests, and wildlife such as this herd of wild mustangs.

Program Resources

Reproducible Masters
Study Guide, p. 76 L1
Reinforcement, p. 76 L1
Enrichment, p. 76 L3
Cross-Curricular Integration, p. 24
Activity Worksheets, pp. 5, 122 L1
Lab Manual 46

 Transparencies
Section Focus Transparency 72 L1
Teaching Transparency 40

A Aluminum is a metal, which is used to make beverage containers, window frames, and airplane parts.

Figure 20-2

Nonrenewable resources include many metals and minerals, as well as fossil fuels such as coal and oil.

B Lightbulbs contain a filament made of the metal tungsten, a nonrenewable substance.

Nonrenewable Resources

We do have to worry about running out of some natural resources. Whenever you take home a plastic grocery bag, drink from a plastic cup, paint a wall, or travel by car, bus, or airplane, you are using products made from a natural resource called petroleum. Petroleum, also known as crude oil, is a thick, greasy liquid formed from the bodies of organisms that died hundreds of millions of years ago. It is the raw material used to make fuel oil, gasoline, plastics, paints, chemicals, fertilizers, pesticides, and a huge number of other products that have become essential to modern life. Petroleum is an example of a nonrenewable resource. A **nonrenewable resource** is a natural resource that is available only in limited amounts and is not replaced by natural processes in a short period of time. Several products made from nonrenewable resources are shown in **Figure 20-2.**

Minerals are also nonrenewable. Minerals are nonliving solids formed in Earth's crust. They include diamonds; the graphite used in pencil leads; and the sand, gravel, and limestone used to make concrete. They also include metals, such as aluminum, copper, tin, iron, gold, silver, tungsten, and uranium. Phosphorus, an element important to the growth of plants and animals, is recycled so slowly in nature that it can be consumed before it is naturally replaced. As a result, it is considered nonrenewable. Topsoil is also considered nonrenewable because it takes hundreds of years to replace it.

MiniLAB

How does mining for mineral deposits affect the environment?

Procedure
1. Place a chocolate chip cookie or nut-filled brownie on a paper plate. Pretend the cake is Earth's crust and the nuts or chips are mineral deposits.
2. Use a toothpick to locate and dig up mineral deposits. Try to disturb the "land" as little as possible.
3. When mining is completed, do your best to restore the land to its original condition.

Analysis
1. Were you able to restore the land to its original condition? Describe the kinds of changes in an ecosystem that might result from a mining operation.
2. Compare the difficulty of mining deposits found close to the surface and those found deeper within Earth's crust.

Integrating the Sciences

Earth Science It will take millions of years for today's plants to become coal. There is a large supply of coal in the ground. Even though the United States burns more than 900 million metric tons of coal each year, coal supplies are so plentiful that there is enough to last hundreds of years.

Visual Learning

Figure 20-2 Ask students: **What is one way to preserve metal and mineral resources?** *recycling, restricting mining* LEP LS

---

2 Teach

MiniLAB

Purpose
LS **Kinesthetic** Students will "mine" a model mineral deposit and attempt to restore it to its original condition. L1 LEP

Materials
chocolate chip cookie or nut-filled brownie, paper plate, toothpick

Teaching Strategies
Oatmeal raisin cookies may also be used.

Analysis
1. Students will experience difficulty in the restoration process. This compares with how some forms of mining disrupt Earth's surface. An ecosystem could be permanently damaged from mining.
2. The ones closer to the surface are easier to remove than those found deeper within Earth's crust.

✓ Assessment

Oral Which do you think is more expensive: mining at the surface or deep within Earth's crust? Explain. *In general, surface mining is less expensive.* Use the Performance Task Assessment List for Making Observations and Inferences in **PASC**, p. 17. P

📁 **Activity Worksheet,** pp. 5, 122

🖊 Use **Teaching Transparency 40** as you teach this lesson.

541

Gas   Oil deposit   Water

Figure 20-3

Fossil fuels—coal, oil, and natural gas—are formed when organisms die and are buried by layers of rock and sediment.

Energy for Now and the Future

Imagine what life must have been like before the invention of the electric lightbulb or the internal combustion engine. You would probably have used candles for light at night, wood for cooking and heating, and horse-drawn wagons or walking for getting from one place to another. Today, we have electric lights, TV and radio, electric ranges, gas stoves, gasoline-powered automobiles, and jet-fueled airplanes. It takes energy to operate and manufacture all these modern conveniences. Where does all this energy come from?

Fossil Fuels

Right now, much of this energy comes from fossil fuels. **Fossil fuels** include coal, natural gas, and fuels made from oil, including gasoline, diesel fuel, jet fuel, heating oil, and kerosene. Fossil fuels are used to power everything from automobiles and lawn mowers to factories. They provide us with many of the conveniences we consider essential to modern life. However, the burning of fossil fuels is contaminating our air and water in ways that are harmful to the environment. There are other forms of energy that can provide some or all of the power we need without polluting the environment the way fossil fuels do. These alternatives include solar power and geothermal energy. Nuclear energy is another energy source but with pollution problems of its own.

Figure 20-4

There are a number of ways to capture solar energy and use it instead of fossil fuels to heat homes, heat water, generate electricity, and even cook food.

A This home has been designed to use passive solar heating. The floors and walls are made of heat-absorbing stone, tile, or brick. At night, the heat stored in the walls and floors keeps the building warm.

Solar Energy

You've probably heard many times that you should never leave a pet or a child in a parked car with the windows closed. Sunlight entering through the windows heats the interior of the car. This heat cannot escape through the closed windows, and the temperature inside can rise very rapidly to dangerous levels. Several ways solar energy is trapped and used are seen in **Figure 20-4.**

Solar Cells

Have you ever wondered how a solar-powered calculator works, or how the solar-powered satellites that orbit Earth generate their electricity? These devices use photovoltaic solar cells to turn sunlight directly into electric current. **Photovoltaic (PV) cells** are wafers made of the mineral silicon covered with thin layers of metals. When the sun's light hits these layers, a stream of electrons, or electricity, is created. PV cells are small, and they are easy to use because they don't have any moving parts. Presently, they are too expensive to use for generating large amounts of electricity, but improvements in the materials used to make them could bring prices down in the next decade or so. In the meantime, PV cells are being used to power a variety of commercial items, including flashlights, radios, watches, and toys.

B Active solar designs use a combination of solar collectors, pumps, and fans to heat indoor air and to heat water for household use.

Solar radiation — Solar collector
Water pipes
Thermostat
Heat
Hot water radiator
Pump Water storage tank

C A solar cooker is a foil-lined container topped with glass. If properly designed, it can reach temperatures of about 177°C and cook meals in much the same way as a Crock-Pot.

D In the Mojave Desert in southern California, an electric utility company is using a huge array of mirrors to focus sunlight on a water-filled tower. The steam produced by this system is used to generate electricity.

20-1 Natural Resources 543

543

Water and Wind Power

Water that is prevented from flowing downstream by a dam contains tremendous amounts of stored energy. When the water flows through openings in the dam, that stored energy is released. Hydroelectric power plants, located inside dams, use the energy of flowing water to turn gigantic turbines that generate enormous amounts of electricity. **Hydroelectric power,** or hydropower, is electricity produced by the energy of flowing water. Hydroelectric power plants like the one in **Figure 20-5A** are extremely efficient and they produce no pollution. They are also a renewable energy resource because the water in the reservoir behind the dam is continually replaced by Earth's water cycle.

The energy of wind can also be used to generate electricity. In regions with frequent winds, electricity is produced on "wind farms," which consist of large numbers of windmills as shown in **Figure 20-5B.** A windmill is a turbine that is rotated by the wind instead of by steam or water. Like hydropower, wind power is a nonpolluting, renewable energy source.

Figure 20-5

Water and wind power are actually indirect forms of solar power.

A When the water in this reservoir flows through narrow channels inside the dam, it turns the blades of turbines to generate electricity.

B Wind turning the turbine blades of the windmills on this "wind farm" also produces electricity.

Geothermal Energy

When a volcano erupts, lava and hot gases pour out of the ground. When a geyser erupts, steam and hot water spew out of the ground. Where does the energy for these eruptions come from? It comes from hot, molten rock that lies deep underground. Volcanoes erupt when molten rock is forced to the surface. Geysers are created when underground water that has been trapped in hot rock is heated to the boiling point and forced up through cracks and openings in the ground. The thermal energy contained in Earth's crust is called **geothermal energy.** In parts of the world where the crust is cracked or especially thin, this heat comes very close to the surface. In some

Integrating the Sciences

Physics The molecules that make up liquid water are closely packed, held together by weak chemical bonds. When water evaporates and turns into steam, the chemical bonds are broken, and the molecules spread farther apart. So, when liquid water turns to steam, it expands and its volume increases. If the expanding steam is forced through a narrow opening, its energy can be used to move physical objects, including the turbine blades of an electric generator. Ask students to use this information to describe how a whistling teakettle works.

Figure 20-6

Geothermal power plants use the thermal energy that lies deep in the crust of Earth to generate electricity.

of these areas, people have built geothermal power plants, which take advantage of Earth's heat to create steam that is used to generate electricity.

Geothermal Power

The Geysers, the largest geothermal power plant in the world, is located near Santa Rosa, California, in a canyon famous for its steamy geysers and pockets of boiling mud. The power plant began operation in 1960 with a single turbine, and now contains more than two dozen generating stations. It produces enough electricity for about two million people.

Wells drilled deep into the ground tap steam trapped in the hot rock. The steam is piped up to the surface, across the blades of giant turbines, then into huge cooling towers, where it condenses back into a liquid. The liquid water is then pumped underground to be reheated by geothermal energy.

Running out of Steam

During the 1980s, the flow of steam from the wells began to slow. There was not enough steam to operate all the turbines, so several were shut down. By the mid 1990s, it became necessary to pipe in water from a nearby lake and wastewater-treatment plant to add to the underground water supply. Apparently, the geothermal heat is a renewable resource, but the underground water is not.

Think Critically:

Why did the Geysers begin running out of steam? What happened to the underground water supply?

Geysers Power Plant

20-1 Natural Resources   **545**

USING TECHNOLOGY

For more information on geothermal power plants, contact a local gas or electric company.

Think Critically

The underground supply was being used up faster than it was being replaced. Some of the steam used to generate power escapes into the atmosphere. Not all of it can be recovered and returned underground.

3 Assess

Check for Understanding

? FLEX Your Brain

Use the Flex Your Brain activity to have students explore ENERGY SOURCES.

📁 **Activity Worksheets,** p. 5

Reteach

Have students list reasons renewable resources must be used wisely and nonrenewable resources must be conserved.

Extension

📁 For students who have mastered this section, use the **Reinforcement** and **Enrichment** masters.

4 Close

1. **Coal, natural gas, and oil are** _____ . *fossil fuels*

2. **The parts of the environment that supply living organisms with food, shelter, and other needs are** _____ . *natural resources*

3. **The thermal energy contained in Earth's crust is** _____ . *geothermal energy*

4. **A natural resource that is recycled or replaced by ongoing natural processes is** _____ . *renewable*

Section Wrap-up

Review

1. Natural resources are the raw materials that organisms use for survival.

2. Renewable resources can be replaced by nature. Examples are water, plants, animals, sunlight, and air. Nonrenewable resources cannot be replaced by nature. Examples are petroleum, diamonds, metals, phosphorus, and topsoil.

3. **Think Critically** Fossil fuels come from plants which fix energy from the sun.

USING MATH

(10 000 miles/30 mpg = 333.3 gallons)

(10 000 miles/25 mpg = 400.0 gallons – 333.3 = 66.7 gallons)

Figure 20-7

While nuclear energy is produced cleanly and cheaply, there are concerns about disposal of radioactive wastes and the safety of the reactors.

Nuclear Energy

Another form of energy that does not require burning fossil fuels takes advantage of the huge amounts of energy bound up in the nuclei of atoms. **Nuclear energy** is produced when billions of uranium nuclei are split apart in a nuclear fission reaction. This energy can be used to heat water to produce steam, which turns turbines that produce electricity. A primary advantage of nuclear power is that it does not contribute to air pollution. However, the mining of uranium does disrupt land ecosystems, and nuclear power plants produce radioactive waste materials. These materials, particularly spent fuel, remain radioactive for thousands of years. Safe disposal of nuclear wastes is a problem that has not yet been solved.

Section Wrap-up

Review

1. What are natural resources?

2. Compare and contrast renewable and nonrenewable resources, and give five examples of each.

3. **Think Critically:** Explain how fossil fuels could be considered a form of solar energy.

Skill Builder
Concept Mapping

Draw a concept map that shows the relationships among the following terms: *renewable resources, nonrenewable resources, fossil fuels, solar energy, nuclear energy, geothermal energy, wind power, hydroelectric power, passive solar, active solar, photovoltaic cells, gasoline, coal, oil,* and *natural gas.* If you need help, refer to Concept Mapping in the **Skill Handbook.**

USING MATH

Most cars in the United States are driven about 10 000 miles each year. If a car can travel 30 miles per gallon of gasoline, how many gallons will be used in a year? How many more gallons would be used by a car that gets 25 miles per gallon?

Skill Builder

Assessment

Oral Have students explain the concept map in the Skill Builder describing renewable and nonrenewable resources. Use the Performance Task Assessment List for Concept Map in **PASC,** p. 89. **P**

Trash Art

Does your art class have a major mural planned, but no paint and no money to buy paint? Has your theater group been rehearsing a new play that would be terrific, if only you had costumes and some scenery? Cans of leftover paint and discarded fabric and wood—just what you need—are probably buried in your local landfill, wasted and useless. But if your community has a recycling program for art supplies, these materials might be waiting in a warehouse for you to come and get them.

A Partnership of Environment and Art

One such group, Materials for the Arts, MFA, began in New York City in 1979, but the idea has since spread nationwide. People involved in a similar program in Los Angeles, for example, visit Hollywood studios, picking up old movie sets and other materials that used to end up in the landfills. Local theater groups are delighted to wander through MFA's warehouse of donated materials, selecting what they need for their own productions—at no charge.

They select from used computers and stereo equipment, poster board, choir robes, rubber chickens, or one-of-a-kind items such as a frozen-drink machine or a watch display case. These donated resources also help cultural groups furnish their offices and set up exhibitions.

In 1994-95, the New York City MFA diverted 428 tons of materials from the city's solid-waste stream. Donors receive tax credits for their castoffs.

Art from By-Products

Other organizations have also set up material exchanges and reuse programs. The Children's Museum of Boston and other museums, for example, raise money by selling bags of discarded industrial by-products that can be used for art projects. Young artists can fill their bags themselves with foam, wood, and plastic pieces of every size, color, and shape, plus paper in a rainbow of colors and textures and an intriguing collection of buttons, beads, and other objects.

So if you are looking for common or extraordinary art materials, contact your local solid waste-disposal agency. You might be able to rescue materials from the landfill as you express yourself!

Science Journal

In your Science Journal, list some benefits of recycling art materials—for the donating organization, for the receiving organization or individual, and for the community.

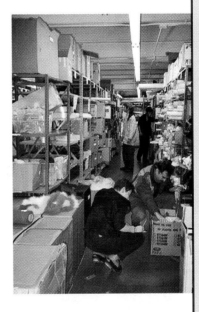

20-1 Natural Resources   547

Recycling Art Organizations

- Arts Clearinghouse, part of Materials for the Arts, Atlanta, 404/853-3261
- Children's Museum of Boston, Recycle Center, 617/426-8855
- Materials for the Arts, Los Angeles, 213/485-1097
- Materials for the Arts, New York City, 212/255-5924
- re:ART, Columbus, Ohio, 614/228-9400

Background

Many materials-exchange programs are city agencies, sponsored by the solid-waste authority, sanitation department, cultural affairs, conservation corps, and/or similar departments. Although several programs use the name Materials for the Arts, they are not connected. Programs offering similar services may have different titles. In 1994-95, the New York City MFA program listed 1085 active recipient arts organizations and 888 contributing businesses, companies, and individuals. This group offers a booklet titled "How to Start a Materials Donation Program." You can order a copy by calling 212/255-5924.

Teaching Strategies

Interpersonal Ask students to investigate whether your community collects recycled materials for the arts. Students might call a recycling hot line, art museums, arts groups, or relevant city agencies. Perhaps the local solid-waste department permits artists to rescue recycled materials at their own risk. L1

Kinesthetic Have students search the trash at school, at home, or at a business (with permission, and while wearing gloves and protective clothing) for clean, safely handled, nonfood items that might be used in art projects. Have them pool the materials they collect. Then ask them to select discarded materials and create works of art. Display their work in a school hallway to increase student awareness of reuse possibilities. Have the class post note cards, describing the sources of materials for their projects. If creating "trash art" is not feasible, ask students to suggest packaging materials that might lend themselves to artwork, such as foam "peanuts." L1

Intrapersonal Encourage students to identify and interview local artists who use recycled materials. Where do they find these materials? Why and how do they use them? L1

SCIENCE & SOCIETY
Section 20•2

Prepare

Section Background

- Recycling could stop the accumulation of glass, plastic, aluminum, and motor oil in the environment.
- Some plastics are melted down and spun into polyester fiber that is used to make sleeping bags, insulation, and fishing line. Plastics can be made into decay-resistant materials for picnic tables, waterfront decks, boat hulls, bathtubs, and automobile panels.

1 Motivate

Bellringer

 Before presenting the lesson, display **Section Focus Transparency 73** on the overhead projector. Assign the accompanying **Focus Activity** worksheet. L1 LEP

2 Teach

Brainstorming

Have students brainstorm a list of all the solid waste they as individuals add to landfills each day.

GLENCOE TECHNOLOGY

 Videodisc

The Infinite Voyage: Miracles by Design
Chapter 4
Biodegradable Plastic: The Miracle Material?

Science Words

solid waste
recycling

Objectives

- Identify ways to recycle and reuse natural resources.

Figure 20-8

Tires don't decay and are almost indestructible. Most tires discarded in the U.S. each year are dumped in landfills.

Beyond Garbage

How many garbage bins do you set beside the curb for pickup every week? Many communities ask people to separate newspapers, aluminum, glass, plastic, yard waste, and even magazines and office paper from the rest of their garbage. All of the material that's hauled away by the garbage truck is known as solid waste. **Solid waste** is unwanted, solid material that must be disposed of. The easiest way to get rid of it is to bury it. So, most of the solid waste we produce is dumped into landfills, which are really nothing more than enormous holes dug in the ground. As the hole is filled, the dumped waste is covered with soil. Out of sight, out of mind.

By now you've probably heard that we're running out of space to put landfills. Landfills disrupt the local ecosystem, they sometimes don't smell very good, and they often contain toxic materials that can pollute air and water. Not many people want a landfill for a neighbor. The obvious way to reduce our need for landfill space is to reduce the amount of waste we bury. We can do that by recycling, of course, and by reusing items we used to throw away. We can also stop using things we don't need.

Precycle and Reuse

Here's a riddle: What do you bring home from every shopping trip that you don't want, don't need, but have to pay for anyway? The answer: packaging. Most of the paper, plastic, and cardboard used to wrap new items for display on store shelves gets thrown in the trash as soon as the product is brought home. We pay to bring home garbage. You can avoid some of this senselessness by precycling, which means choosing products with less packaging, or with packaging made from recycled materials, whenever you have the choice. Precycling also means using recycled products, such as recycled paper, when they are available.

548 Chapter 20 Resources and the Environment

Program Resources

 Reproducible Masters
Study Guide, p. 77 L1
Reinforcement, p. 77 L1
Enrichment, p. 77 L3

 Transparencies
Section Focus Transparency 73 L1

Reuse means using the same item over and over again, without changing or reprocessing it. When you go on a picnic, you can take along reusable plastic plates instead of disposables. Instead of throwing away that empty margarine tub, you can wash it and use it to store leftovers, or your button collection, or those loose nails in the toolbox.

Recycle

If you can't reuse an item made from a natural resource, the next best thing is to recycle it. **Recycling** is reuse that requires changing or reprocessing of an item or resource. Used paper is turned into paper towels, bathroom tissue, newsprint, and stationery. Ranchers are using shredded paper instead of straw for bedding in the stalls of barns and stables.

Plastic soda bottles are melted down, spun into fibers, and woven into winter jackets and wall-to-wall carpeting. Grass clippings and leaves are turned into soil-enriching compost. Aluminum cans and glass bottles are melted and re-formed into more cans and bottles.

People in many communities have become so conscientious about recycling that recyclables are piling up, just waiting to be put to use. It would help if more people would purchase recycled products. We could also use more ideas about how to produce goods using recyclables. Can you think of some new possibilities?

Figure 20-9

Many products, including clothing, are made from recycled plastics. This jacket is actually made from recycled soda pop bottles.

Section Wrap-up

Review

1. Describe two advantages of reducing the amount of material we dump in landfills.

2. Compare and contrast precycling, recycling, and reusing.

Explore the Technology

Find out what kinds of recycled products are available in your community. Write an essay, create a poster, or design an advertising campaign that points out the advantages of these products and encourages people to use them.

*inter*NET
CONNECTION

Visit the Chapter 20 Internet Connection at Glencoe Online Science, **www.glencoe.com/sec/ science/life,** for a link to more information about Recycling.

SCIENCE & SOCIETY

Prepare

Section Background

Erosion is a natural process that has occurred since Earth was formed. Humans have little control over the erosion caused by glaciers, rivers, and oceans. A number of human activities, mainly farming and logging, have increased erosion. Scientists estimate that between 2 and 3 billion metric tons of soil are lost from farmlands in the United States each year.

Preplanning

Obtain petri dishes, petri dish lids, clay pots, potting soil, aluminum pie tins, scissors, metric rulers, and gravel or sand for Activity 20-1.

1 Motivate

Bellringer

Before presenting the lesson, display **Section Focus Transparency 74** on the overhead projector. Assign the accompanying **Focus Activity** worksheet. L1 LEP

Tying to Previous Knowledge

Ask students to list how they affect the environment each day. They should think of transportation, waste, and water.

Visual Learning

Figure 20-10 What are some other ways you can conserve energy? *Students may mention turning out lights and turning off televisions and radios when not being used. Lowering the thermostat in the winter and raising it in the summer are also examples.* LEP LS

20•3 Conservation and Wildlife Protection

Science Words

soil depletion
erosion
soil management
extinction
endangered species

Objectives

- Define *conservation* and explain how it relates to natural resources.
- Describe how renewable resources can be conserved and protected.

Figure 20-10

There are many actions individuals can take to save energy. Repairing a drafty window or door will reduce the amount of energy needed to heat your home. *What are some other ways you can conserve energy?*

What We Can Do

As the human population continues to grow, so will the demand for natural resources. Many Native American traditions teach people to think about how their actions will affect the next seven generations of their descendants. That means thinking of your children, grandchildren, and great-grandchildren. It means thinking about what Earth will be like 200 years from now. There are actions we can take to make sure we leave Earth's land, soil, minerals, wildlife, and other resources in good condition for the generations of humans who will live on this planet after us.

Energy Conservation

There are two ways to conserve nonrenewable resources: by using less or by using a renewable resource instead. You already know about some of the alternatives to nonrenewable fossil fuels. More and more home builders are designing houses that take advantage of sunlight, not only for warmth but also for lighting; well-placed windows and skylights can reduce the need to use electric lights during the day. The amount of electricity produced by solar, wind, and geothermal power plants will also increase. But the fact remains that most of the energy we use today is produced by fossil fuels.

In order for us to conserve those resources, we need to reduce our energy use. Recycling is one way to do that, as well as reducing energy use as shown in **Figure 20-10.** Making recycled aluminum cans uses only 10 percent of the energy required to make new cans. Recycling glass also saves energy.

Program Resources

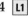 **Reproducible Masters**
Study Guide, p. 78 L1
Reinforcement, p. 78 L1
Enrichment, p. 78 L3
Science and Society Integration, p. 24
Activity Worksheets, pp. 5, 118-119 L1
Critical Thinking/Problem Solving, p. 24

Transparencies
Section Focus Transparency 74 L1

Figure 20-11

Soil management helps reduce the erosion of farmland.

A By plowing at right angles to the slope of the land, contour plowing reduces erosion by reducing the downhill flow of water.

B Strip-cropping involves alternating rows of plant crops, such as corn, with a cover crop such as wheat or hay.

C In areas with very steep hillsides, the downhill flow of water is reduced by creating flat areas, or terraces.

D No-till farming is a soil management method that involves planting crops in untilled, undisturbed soil.

Soil Conservation

Farmers take mature plants away from the land where they were grown. The crop is not left to decay and replenish the nutrients in the soil. This removal of soil nutrients is called **soil depletion.** Unless the nutrients are replaced, the soil will eventually become too poor for crop growth.

Farmers replenish the soil in several ways, including crop rotation and the addition of fertilizers. Crop rotation is the process of changing the crops that are grown in a field from one season to the next. Every few seasons, a nonfood crop is grown and tilled back into the soil to help replenish nutrients. Fertilizers include organic materials, such as compost, and inorganic chemicals made from petroleum. Other methods of soil conservation are shown in **Figure 20-11.**

Erosion

Another process that destroys soil fertility is erosion. **Erosion** is the wearing away of soil by wind and water. As you discovered when you conducted the Explore Activity at the beginning of the chapter, flowing water can pick up loose topsoil and wash it away. When a farmer plows a field and plants seeds, irrigation water or rainfall can carry the topsoil away. An important part of farming is soil management. As shown in **Figure 20-11, soil management** is the use of plowing methods that reduce soil erosion. These methods include contour plowing, strip-cropping, terracing, and no-till farming. Plowing, or tilling, loosens the soil, making it easier for seedlings to take root, but also making it easier for water and wind to carry the soil away.

20-3 Conservation and Wildlife Protection   551

2 Teach

Brainstorming

Ask students to brainstorm a list of things that are being done today to protect the environment. *They can name people or organizations that are involved in each example.*

Content Background

Explain that many people open a refrigerator and stand in front of the open door deciding what they want. This is an expensive habit. Money and energy can be saved by deciding what they want before opening the door and then closing it quickly.

NATIONAL GEOGRAPHIC SOCIETY

Videodisc

STV: Water
Water Quality
Unit 1, Side 1

00392-17833
Water Conservation
Unit 2, Side 1

17877-33972
Water Management
Unit 3, Side 1

34063-51042

Inclusion Strategies

Gifted Have students make a game that promotes wise use of resources or that raises other environmental issues. Game cards can give bonus moves or points for environmentally wise activities. L3

Across the Curriculum

Social Studies Drought and unsound agricultural practices contributed to the formation of the dust bowl in the United States. Have students use the library to learn about the dust bowl period in history and write a report on what the dust bowl was, where it occurred, its causes, and its effect on people. L1 LS P

Activity 20-1

PREPARATION

Purpose
LS **Interpersonal** Observe decomposition in nature using substances found in garbage and determine which materials are biodegradable and which are nonbiodegradable.
L1 **COOP LEARN**

Process Skills
observing and inferring, interpreting data, communicating, classifying, making and using tables, comparing and contrasting, forming operational definitions, forming a hypothesis

Time
50 minutes; 10 minutes two times a week for 4-6 weeks to collect data

Materials
- Divide the biodegradable and nonbiodegradable materials into approximately 2-cm sized squares for class use.
- Discuss differences between organic and inorganic materials and the role of bacteria and fungi in decomposition.
- Biodegradable substances may include fruits and vegetables, eggshells, bread, packaging materials, coffee grounds, leaves, nuts, water-soluble plastics, and newspapers. Nonbiodegradable substances may include glass, aluminum, tin, and plastic.
- Have students stir the soil to look at the items once every two or three days for four to six weeks.

Alternate Materials Milk cartons may be used instead of clay pots. Cartons or pots may be enclosed in a plastic bag and tied. Materials placed in soil must be well covered by the soil.

Activity 20-1

Design Your Own Experiment
Investigating Biodegradable Substances

People can take advantage of the recycling that goes on in nature to help reduce the amount of garbage that must be disposed of and to help replenish the soil that is used to grow food crops. A biodegradable substance is something that can be broken down by decomposer organisms found in nature. By encouraging the decomposition of organic, biodegradable materials found in our trash, we can produce compost. Compost is decaying organic matter, and it makes an excellent fertilizer.

PREPARATION

Problem
What kinds of materials are biodegradable?

Form a Hypothesis
Based on what you know about the process of decomposition in nature, state a hypothesis about what kinds of substances can be used to make compost. Be sure to think about conditions necessary for a material to decompose.

Objectives
- Distinguish between biodegradable and non-biodegradable substances.
- Observe the decomposition of biodegradable materials.

Possible Materials
- clay pots
- petri dishes
- aluminum pie tins
- potting soil
- sand or gravel
- labels
- biodegradable and non-biodegradable waste materials
- hand lens or microscope

Safety Precautions

Wash hands after handling soil or waste materials.

Safety Precautions
Caution students to wash their hands after handling soil and waste materials.

Possible Hypotheses
Most student hypotheses will reflect that organic materials are biodegradable and inorganic materials are not.

 Activity Worksheets, pp. 5, 118-119

PLAN THE EXPERIMENT

1. As a group, agree upon and write out a hypothesis statement.
2. As a group, list the steps you need to take to test your hypothesis. Be specific. Describe exactly what you will do at each step. List your materials.
3. Use the clay pots as the containers. The lids of the petri dishes will cover the clay pots. Use the pie pans underneath the pots. The pots can be placed in plastic bags if necessary.
4. When you plan your experiment be sure to identify and test only one variable at a time.
5. Prepare a data table in your Science Journal for all data and observations.

Check the Plan
1. Read over your entire experiment to make sure that all steps are in logical order.
2. Identify any constants, variables, and controls of the experiment.
3. *Make sure your teacher approves your plan before you proceed.*

DO THE EXPERIMENT

1. Carry out the experiment as planned.
2. While the experiment is going on, write down any observations that you make and complete the data table in your Science Journal.

Analyze and Apply
1. Which of the materials tested decomposed?
2. Which substances were biodegradable? Nonbiodegradable?
3. **Explain** how substances that are not biodegradable affect the environment.
4. **Describe** the kinds of organisms you observed.

Go Further

What kinds of ideas do your results give about what people can do to reduce the amount of garbage they produce? Make a plan for recycling the food wastes produced in the school cafeteria.

20-3 Conservation and Wildlife Protection   **553**

Assessment

Performance Design an experiment to compare different plastic bags and packaging materials that are supposed to be biodegradable. Some plastic bags break down when exposed to sunlight. Use the Performance Task Assessment List for Designing an Experiment in **PASC**, p. 23. P

PLAN THE EXPERIMENT

Possible Procedures
Moistened soil should be placed in the pots to within 1.5 cm from the top. The biodegradable materials should be placed in one container and the non-biodegradable materials should be placed in a second container. They should be covered with a thick layer of soil, and the petri dish lid should be placed tightly over the top and placed in an aluminum pie tin to which water is added.

Teaching Strategies
Tying to Previous Knowledge Most students have observed that molds and an unpleasant odor appear when foods decay. They are aware that bacteria and fungi are responsible for the decomposition.

DO THE EXPERIMENT

Expected Outcome
The organic materials will decompose and the inorganic materials will not.

Analyze and Apply
1. leaves, vegetables, fruit peelings, newspaper (organic materials)
2. the substances that decomposed; the substances that did not decompose
3. They pollute the environment.
4. fungi (molds) and the action of bacteria

Go Further

People can avoid using materials that are not recyclable and dispose of non-biodegradable materials in ways provided by the community. They can recycle materials and use biodegradable materials. Plans may include using the food wastes for feeding animals.

553

Figure 20-12

Species can become endangered because of pollution, habitat destruction, illegal hunting, or competition with introduced species.

A The destruction of habitat is the reason many species have become endangered, including the grizzly bear in the United States.

B When the pesticide DDT was sprayed on farmland, runoff from rainwater carried it into nearby streams, where it entered the food chain. DDT built up in the tissues of fish, which were eaten by ospreys. DDT buildup in the bodies of the ospreys caused them to produce eggs with shells so thin that the eggs broke open before the young could develop.

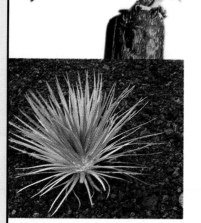

C Close to 200 plants that grow only in the Hawaiian Islands, including this silver sword plant, are endangered because plants introduced from other parts of the world compete with the natives for space.

Wildlife Conservation

A hundred years ago, flocks of up to a million passenger pigeons darkened the skies above North America. Today, there are none. The last member of the species died in a New York zoo in 1914. Human hunters killed them by the thousands until the species became extinct. **Extinction** is the dying out of an entire species. Extinction occurs naturally. During the course of evolution, many species that once lived on Earth have become extinct. But the large numbers of humans that now inhabit the planet have changed the environment in ways that have increased the rate of extinction.

An **endangered species** is a species that is in danger of becoming extinct unless action is taken to protect it. The bald eagle, the national bird of the United States, is an example of a species that was endangered for a time. In the early 1960s, there were only about 400 bald eagles left in the 48 lower states. Today, because of preservation efforts, the population has increased to approximately 5000.

Most species become endangered because of damage to their habitat by human activities, such as building roads, clearing forest for farms, or draining marshes to provide dry land for buildings. In other cases, pollution is to blame. The pesticide

DDT was used in the United States from the 1940s to the 1970s. This chemical provided farmers with relief from large populations of insect pests that were damaging crops. But DDT remains in the environment for many years after it is used, and is passed along the food chain from one organism to another. When it entered the bodies of large predatory birds, such as eagles, ospreys, falcons, and pelicans, the birds began to have difficulty reproducing. After DDT use was stopped, the bird populations began to recover. Bald eagles had recovered enough by 1995 to be removed from the list of endangered species. As **Figure 20-12** shows, a species may also become endangered because it is overhunted by predators or it is crowded out by a species that has been introduced from another part of the world.

Wildlife conservation involves determining what species are endangered and why, and devising methods of helping the species recover. Recovery may involve ridding the environment of a harmful pollutant, as was the case with DDT, preserving the habitat of a species, or even breeding a disappearing species in captivity and restoring it to the wild.

Land Conservation

When European settlers came to North America following Columbus's voyage, much of the continent was covered with forests. Settlers cut down many of these forests to clear land for farming, and to obtain fuel and lumber for building. About 100 years ago, people began to realize that if this kept up, we would eventually lose all of our forests. So the United States government began a conservation program that involved setting aside wilderness areas, including forests, that are protected from development.

USING MATH

Most eagles have a wingspan of about 180 centimeters; however, one of the largest, the harpy eagle, has a wingspan of 2.4 meters. How many times greater is the harpy eagle's wingspan than the average eagle's wingspan?

Figure 20-13

From the human perspective, wetlands may look like nothing more than a muddy waste. In the past, wetlands have been destroyed by filling them in with dirt and using the land for roads and buildings (left). Today, many remaining wetlands have been set aside as protected areas (right).

20-3 Conservation and Wildlife Protection **555**

Content Background

Deforestation is occurring worldwide. It is becoming serious in South America. The rain forests are home to millions of species of plants and animals. Many animals live in trees and do not move down to the ground. Discuss with students what happens to plants and animals as the forests are cleared for farms and ranches.

USING MATH

about 1.3 times

Discussion

Display a state map on the bulletin board and highlight state park and wildlife preserves. Ask how many students have visited these areas. Discuss with students how these areas are beneficial to humans.

3 Assess

Check for Understanding

? FLEX Your Brain

Use the Flex Your Brain activity to have students explore MINERAL RESOURCES. **P**

Activity Worksheets, p. 5

Reteach

Have students choose one of the following and write an essay based on what they have studied in this section: energy conservation, soil conservation, wildlife conservation, land conservation, and mineral conservation.

Extension

For students who have mastered this section, use the **Reinforcement** and **Enrichment** masters.

1. **A species that is in danger of becoming extinct unless action is taken to protect it is** _____ . *endangered*

2. **The wearing away of soil by wind and water is** _____ . *erosion*

3. _____ **is the use of plowing methods to reduce erosion.** *Soil management*

4. **The removal of soil nutrients is called** _____ . *soil depletion*

Section Wrap-up

Review

1. Soil depletion is the removal of soil nutrients. Crops are not left to decay and replenish the nutrients. Erosion is the wearing away of the soil by wind and water.

2. pollution, habitat destruction, overhunting, competition

3. **Think Critically** If there has not been enough time for new plant cover to grow, topsoil on slopes could be washed away.

Using Computers

Example entry: Organism: Manatee; Scientific name: *Trichechus mammatus;* Biome: tropical estuaries; Habitat: shallow, brackish water; Niche: first-order consumers of aquatic vegetation.

Many other countries also have made efforts to protect forests and wildlife habitat by developing national parks and protected areas. Worldwide, more than 8000 national parks and other protected areas, covering about 8.5 million square kilometers, an area slightly larger than the country of Brazil, have now been set aside. In many areas, including the United States, land that was once forested is being replanted. The replanting of a forest is called reforestation. Reforestation not only restores forest ecosystems, it also helps preserve topsoil because the roots of young trees help hold the soil in place.

Figure 20-14
This open-pit copper mine cuts into the Arizona landscape.

Mineral Conservation

Minerals are nonrenewable resources. Because we can't replace them, the way to conserve them is to use less and to recycle whenever we can. Another good reason for conserving natural resources, in addition to saving them for future generations of humans, is to avoid the need for mining. Most minerals—and fossil fuels—are obtained by digging mines into the earth. Mining disrupts ecosystems, destroys topsoil, and sometimes releases toxic substances into air and water as you can see in **Figure 20-14.** Recycling aluminum, steel, silver, and other minerals reduces the need for mining operations.

Section Wrap-up

Review

1. Compare and contrast soil depletion and erosion.

2. Describe four ways in which a species can become endangered.

3. **Think Critically:** What might happen to topsoil during the first rain after a forest fire? Why?

Skill Builder
Comparing and Contrasting
Compare and contrast biodegradable and nonbiodegradable materials, and give five examples of each. If you need help, refer to Comparing and Contrasting in the **Skill Handbook.**

Using Computers

Database Create an endangered species database. Include each organism's common and scientific names, biome in which it's found, habitat description, and reasons why it is threatened. Sort the database by biome to obtain a list of endangered species from each of the world's biomes.

Skill Builder
Biodegradable materials are broken down by natural processes; nonbiodegradable materials are not. Paper, wood, dead organisms, food scraps, and cotton cloth are biodegradable. Plastic bags, motor oil, aluminum cans, plastic containers, and plastic toys are not.

✓Assessment

Performance Have students make a concept map that compares biodegradable and nonbiodegradable materials. Use the Performance Task Assessment List for Concept Map in **PASC,** p. 89. **P**

Maintaining a Healthy Environment 20•4

How are we affecting the environment?

Safeguarding our air, land, and water requires more than conserving resources. It also means paying attention to how our use of those resources affects the health of the environment. Every action we take—from starting the car or revving up the lawn mower to painting a building, using bug spray, cutting down a tree, or turning on a light—has an effect on the environment.

Air Pollution

On almost any warm, sunny, windless day in any heavily populated area in the world, you can see a distinct layer of brown haze hanging in the air. The brown color comes from smoke, dust, and exhaust fumes that pollute the air. Wherever there are cars, trucks, airplanes, factories, furnaces, fireplaces, cookstoves, or power plants, there is air pollution. Although it also is caused by volcanic eruptions, forest fires, and the evaporation of chemicals such as paints and dry-cleaning fluid, most air pollution results from the burning of fossil fuels.

Smog

Burning wood or coal in a fireplace produces smoke and ash as well as heat. Burning gasoline not only provides the energy to run a car engine, but also creates a number of unwanted chemicals, including tiny particles of soot and a number of

Science Words

- pollutant
- smog
- acid rain
- ozone depletion
- greenhouse effect
- global warming
- hazardous waste
- groundwater

Objectives

- Describe the origins and effects of pollutants that contaminate air, land, and water.
- Explain the causes of acid rain, ozone depletion, and the greenhouse effect.

 INTEGRATION
Chemistry

Figure 20-15

Some important components of smog are nitrogen oxide, nitrogen dioxide, ozone, and soot particles. Tiny particles of soot suspended in the air obscure visibility and irritate the eyes and respiratory system.

557

Section 20•4

Prepare

Section Background

- The ozone layer is a layer of gases around Earth. Ozone, a molecule made of three oxygen atoms, traps excess ultraviolet radiation and prevents it from reaching Earth.

- More than 70% of Earth's surface is covered with water, but about 94% of it is in oceans and for the most part unavailable for human use. Only about 3% of water found on Earth is fresh water, and most is frozen in polar ice caps and glaciers.

Preplanning

- To prepare for the Mini-LAB, obtain rain samples, distilled water, and pH paper.
- For Activity 20-2, obtain plastic soda bottles, potting soil, plastic wrap, and masking tape.

1 Motivate

Bellringer

 Before presenting the lesson, display **Section Focus Transparency 75** on the overhead projector. Assign the accompanying **Focus Activity** worksheet. [L1] [LEP]

Tying to Previous Knowledge

Relate examples of humans' effects on the environment to each of the topics in this section. Humans have the ability to affect the entire biosphere in an adverse way. They also have the ability to solve problems and prevent problems from developing.

2 Teach

Problem Solving

Indoor Air Pollution

Pollution is often worse indoors than it is outdoors. For example, the air in a typical home or office contains not only pollutants carried in from outdoor air, but also pollutants that originate from items and activities inside the building. New paint and carpeting may give off fumes from chemicals used to make them. Pollutants are also contributed by woodstoves and gas ranges that are not vented to the outdoors. Carbon monoxide (CO), a gas made up of molecules with one carbon and one oxygen atom, is a by-product of the burning of wood, charcoal, and fossil fuels. When inhaled, CO replaces oxygen in the blood. CO poisoning causes nausea, fatigue, and even death if concentrations are high enough. Cigarette smoke, household cleaning products, and insect sprays also contribute to indoor air pollution.

Solve the Problem:

1. What gases might pollute the indoor air of a home with a woodstove that leaks smoke into the room?
2. Read the labels on containers of household bleach, bathroom cleanser, window cleaner, or other cleaning products. Why do some of them state: "Use only in a well-ventilated area" or "Avoid inhaling fumes"?

Think Critically:

1. Would it be a good idea to use a barbecue grill and charcoal briquettes to cook a meal indoors? Why or why not?
2. Would it be a good idea to use a kerosene stove to heat a cabin with all the doors and windows closed? Why or why not?

Across the Curriculum

Health Have students research sources of indoor air pollution. Have them study the effects of asbestos, formaldehyde, radon, cigarette smoke, and other substances found in houses. [L1]

gases, such as carbon monoxide, carbon dioxide, and oxides of nitrogen and sulfur. The unwanted by-products created when wood and fossil fuels are burned are examples of pollutants. Any substance that contaminates the environment and causes pollution is a **pollutant.** The brown haze shown in **Figure 20-15** on page 557 that hangs over cities and other heavily populated areas is **smog,** a form of air pollution that is created when sunlight reacts with pollutant chemicals produced by burning fuels.

Acid Rain

When you studied the water cycle, you learned that water evaporates from Earth's surface and rises into the air as water vapor. While in the atmosphere, it comes into contact with air pollutants. When it condenses and falls to the ground, it may bring pollutants with it.

Carbon dioxide gas mixes with water molecules during the formation of raindrops. This combination forms a weak acid solution. **Figure 20-16** shows common acid solutions. Acids are measured by a value called pH. Solutions with a pH below about 7.0 are acid. Because carbon dioxide and water mix to form a weak acid, most normal rainfall has a pH of 5.6.

Air pollutants can increase the acidity of rain. The sulfur oxides and nitrogen oxides released by the burning of fossil fuels react with water vapor in the atmosphere to form strong acids, including sulfuric acid and nitric

Figure 20-16

The pH scale shows whether a solution is acid or base. Acid rain can have damaging effects.

Tomatoes

Soft drinks

Pure water

Detergent

Oven cleaner

Milk

| 0 | 1 | 2 | 3 | 4 | 5 | 6 | 7 | 8 | 9 | 10 | 11 | 12 | 13 | 14 |

← More acidic   Neutral   More basic →

A Acid solutions have a pH below 7.0. Basic solutions have a pH greater than 7.0. Neutral solutions have a pH of 7.0.

B Exposure to acid rain has worn away many features of these statues. Many famous sculptures and buildings around the world are being damaged by acid rain.

acid. **Acid rain** is rain or snow with a pH below 5.6. It washes valuable nutrients from the soil, which can lead to the death of trees and other plants. When the runoff from acid rain flows into a lake or pond, it changes the pH of the water. Algae and microscopic plankton cannot survive, so fish and other organisms that depend on them for food also die off. A lake damaged by acid rain may look attractive because the water is very clear. But the water is clear because few organisms can survive in it.

Ozone Depletion

When ozone forms in the lower atmosphere, it is considered an air pollutant. But ozone also forms in the upper atmosphere. This upper ozone layer helps protect living organisms from exposure to harmful ultraviolet (UV) radiation from the sun. The ozone layer acts as a kind of sunscreen that protects living organisms from harmful ultraviolet (UV) radiation. When ultraviolet rays from the sun strike molecules of oxygen in the upper atmosphere, ozone is formed. Because the formation of ozone absorbs UV rays, they are prevented from reaching Earth's surface.

MiniLAB

Is acid rain falling on your town?

Use a rain gauge to collect samples of the rain that falls outside your home or school building.

Procedure
1. Dip a piece of pH indicator paper into the sample.
2. Compare the color of the paper with the pH chart provided. Record the pH of the rainwater.
3. Use separate pieces of pH paper to test the pH of tap water and distilled water. Record these values.

Analysis
1. Is the rainwater acid, basic, or neutral?
2. How does the pH of the rainwater compare with the pH of tap water? Distilled water?

20-4 Maintaining a Healthy Environment   **559**

MiniLAB

Purpose

IS **Interpersonal** Students will collect samples of rain and determine the pH of each sample. **L1** **COOP LEARN**

Materials
rain gauge, baby food jars, pH paper

Teaching Strategies
- Have students collect their samples from the same sources of precipitation.
- Have students use the same chart to determine the pH level of the samples.
- Have students test pH of rain on different days throughout the school year to detect variations.

Analysis
1. pH will be either acid or neutral. The pH of normal rain is 5.6. Student results may range from pH 4 to pH 7.
2. Tap water is about pH 6; distilled water is pH 7.

Assessment

Performance Set up several liquids of unknown pH. Have students test each with pH paper to determine whether they are acids or bases. Have them make a table that shows the pH of the unknown liquids and compares them with substances of known pHs. Use the Performance Task Assessment List for Analyzing the Data in **PASC**, p. 27. **P**

Activity Worksheets, pp. 5, 123

Integrating the Sciences

Earth Science When ultraviolet light from the sun strikes ozone molecules, they are split apart. The ozone quickly re-forms, releasing heat that escapes into space. The ozone layer protects living organisms from harmful radiation. Some substances, such as fluorocarbons, destroy the ozone layer.

Answer It became obvious to the world leaders that each CFC molecule released into the atmosphere could destroy many molecules of the protective ozone layer. If CFCs continued to be released, total destruction of protective ozone would be unavoidable.

Use **Multicultural Connections,** pp. 43-44, as you teach this lesson.

Content Background

An extreme example of the greenhouse effect at work has been observed on the planet Venus. Boiling-hot temperatures are common on the planet's surface both at night and during the day. This is not because the planet is so close to the sun, but because the atmosphere is dense with hydrocarbon vapors that trap the sun's reflected radiation.

Videodisc

The Infinite Voyage: Crisis in the Atmosphere

Introduction

Chapter 1
Historical Aspects of the Greenhouse Effect and Fossil Air

Chapter 3
Our Future Climate

Chapter 4
The Greenhouse Effect: Future and Past

560

Chlorofluorocarbon, or CFC, molecules contain atoms of chlorine, fluorine, and carbon. When they reach the ozone layer in the upper atmosphere, they are broken apart by the energy of ultraviolet light. The breakdown of CFCs releases chlorine atoms, which act as catalysts for a reaction that destroys ozone and produces ordinary oxygen. Because chlorine atoms are not changed during this reaction, each one continues to destroy ozone molecules. *Explain* how this information persuaded world leaders to agree that CFC use should be stopped.

Scientists have discovered that Earth's protective ozone layer is becoming thinner. The thinning of the ozone layer is called **ozone depletion.** It is being caused by air pollutants, but not pollutants that are created by the burning of fossil fuels. Chemicals known as chlorofluorocarbons (CFCs), used in the cooling systems of refrigerators, freezers, and air conditioners, are the primary cause of ozone depletion. These chemicals contain chlorine and fluorine molecules. When these molecules escape into the air, they rise in the atmosphere until they reach the level of the protective ozone layer. Once there, they react with ozone molecules, destroying them. The ozone layer is so important to the survival of life on Earth that world governments have agreed to stop producing CFCs by the year 2000.

Greenhouse Effect

Earth's atmosphere allows sunlight to penetrate to Earth's surface, but prevents much of the resulting heat from radiating back out into space. This heat-trapping feature of the atmosphere shown in **Figure 20-17** is known as the **greenhouse effect.** The greenhouse effect is essential to life on Earth. Without it, temperatures would be too cold to support life as we know it.

The heat-reflecting atmospheric gases responsible for the greenhouse effect are called greenhouse gases. The most important is carbon dioxide. Although carbon dioxide is a normal part of the atmosphere, it is also a by-product of the burning of fossil fuels. Over the past hundred years or so, humans have been burning larger and larger quantities of fossil fuels and releasing more and more carbon dioxide and other greenhouse gases into the atmosphere. These extra greenhouse gases could cause Earth to get warmer. The potential warming of Earth is called **global warming.**

Figure 20-17
The moment you step inside a greenhouse, you feel the results of the greenhouse effect. Heat trapped by the glass walls warms the air inside. Similarly, atmospheric greenhouse gases, particularly carbon dioxide, trap heat close to Earth's surface.

Carbon dioxide in atmosphere

Heat trapped near Earth's surface

Solar radiation

560 Chapter 20 Resources and the Environment

Integrating the Sciences

Physical Science About 45% of the sunlight reaching Earth is absorbed by Earth's surface. This energy rises from the surface as infrared radiation. Most infrared radiation cannot escape through the atmosphere. It is absorbed by gas molecules in the atmosphere such as carbon dioxide, water vapor, and ozone.

Activity 20-2

Modeling the Greenhouse Effect

You can create models of Earth with and without heat-reflecting greenhouse gases, then experiment with the models to observe the greenhouse effect.

Problem
How does the greenhouse effect influence temperatures on Earth?

Materials
- 2 clear, plastic, 2-liter soda bottles with tops cut off and labels removed
- 2 thermometers
- 4 cups of potting soil
- masking tape
- plastic wrap
- rubber band
- lamp with 100-watt lightbulb
- chronometer or watch with a second hand

Procedure
1. Copy the data table and use it to record your temperature measurements.
2. Put an equal volume of potting soil in the bottom of each container.
3. Use masking tape to affix a thermometer to the inside of each container. Place the thermometers at the same height relative to the soil. Shield each thermometer bulb by putting a double layer of masking tape over it.
4. Seal the top of one container with plastic wrap held in place with the rubber band.
5. Place the lamp with the exposed 100-watt bulb between the two containers and exactly 1 cm from each, as shown in the diagram. Do not turn on the light.
6. Let the apparatus sit undisturbed for five minutes, then record the temperature in each container.
7. Turn on the light. Record the temperature in each container every 2 minutes for the next 15 to 20 minutes.

8. Plot the results for each container on a graph.

Analyze
1. How did the temperature of each container change during the experiment?
2. What was the temperature difference between the two containers at the end of the experiment?

Conclude and Apply
3. What does the lightbulb represent in this experimental model? What does the plastic wrap represent?
4. **Describe** the ways in which this experimental setup is similar to and different from Earth and its atmosphere.

Data and Observations   Sample Data

Time	Open Container Temperature	Sealed Container Temperature
0 minutes	22°C	22°C
2 minutes	22°C	22°C
4 minutes	22°C	22.5°C
6 minutes	22°C	23°C

Tape
Thin cardboard
Soil
1 cm   1 cm

20-4 Maintaining a Healthy Environment   561

---

2. Answers will vary.
3. sunlight; greenhouse gases
4. Both sun and lightbulb represent a constant source of light and heat. The open container is unlike Earth because there is nothing to model heat-reflecting gases. The plastic wrap is similar to heat-reflecting greenhouse gases, but not quite the same because it is a solid, not a gas.

✓ Assessment

Oral Have students summarize the information in their graphs in writing. Make sure summaries are clear and concise. Use the Performance Task Assessment List for Written Summary of a Graph in **PASC**, p. 41.
P

Activity 20-2

Purpose
LS **Interpersonal** Students will make a model of the atmosphere and determine what happens when light shines on it. **L1** **COOP LEARN**

Process Skills
modeling, observing and inferring, interpreting data, communicating, comparing and contrasting, recognizing cause and effect, measuring in SI, making and using tables, making and using graphs

Time
50 minutes

Safety Precaution
Caution students to use extreme care in working with thermometers and 100-watt lightbulbs.

📁 **Activity Worksheets,** pp. 5, 120-121

Teaching Strategies
- Drafts in the room could affect experimental results. Close doors and windows if necessary.
- Have students work in teams of three to four students.
- Make sure the containers are set up so that the thermometers face outward, away from the light bulb. This will make it easier for students to read the temperatures.

Troubleshooting If the two thermometers did not read the same temperature when the experiment began, readings will need to be adjusted so both sets of data can be plotted on the same graph.

Answers to Questions
1. In most cases, students observe that the temperature increased, then leveled off; at the end of the experiment the open bottle was cooler than the closed bottle.

561

Demonstration

Bring a container of trash to class. Include a glass bottle, an aluminum can, a twist tie, plastic wrappers, paper, coffee filter with grounds, vegetable peelings, eggshells, and other household items. Wear plastic gloves and remove each piece from the container. Discuss which items are biodegradable and which items are not.

Content Background

Explain that not all hazardous wastes come from large industries. Many harmful chemicals, such as paint removers and furniture cleaners, come from households.

Activity

Many household cleaners contain toxic chemicals. Many new products are on the market that are just as effective but do not contain harmful chemicals. Have students investigate cleaning products in their homes to see what kinds of ingredients they contain. Be sure to warn students to take care when handling cleaning products or other chemicals. L1

Effects of Global Warming

Scientists disagree about whether or not the increase in atmospheric carbon dioxide and other greenhouse gases is actually warming up Earth. What might happen if the temperature of Earth did increase? Rainfall amounts could decrease, which could affect the production of food as well as affect the health of wildlife. A rise in temperature of just 5°C would begin melting the polar ice caps, which would raise sea levels and result in the flooding of coastal communities. There could also be an increase in the number and severity of storms and hurricanes. Many people feel that the possibility of global warming is a good reason to slow our consumption of fossil fuels.

Figure 20-18

Many communities hold special waste-disposal days when people can bring leftover paints, pesticides, cleaning solvents, and other hazardous materials to a special collection site. This helps keep hazardous materials out of the regular trash and landfills. Some communities collect and mix leftover paints to use in public beautification projects such as graffiti removal.

Land Pollution

As you learned earlier in this chapter, the human population is running out of land space for burying solid waste. For a long time, most people expected that biodegradable items disposed of in landfills would eventually decompose. But it turns out that they decompose very, very slowly. A newspaper thrown away in the 1950s was still readable when scientists dug it out of a landfill almost 40 years later. What about the nonbiodegradable substances we put in landfills? We can't expect them to break down at all. Any toxic materials that are disposed of in landfills can pollute the soil and surrounding waterways.

Waste materials that are harmful to human health or poisonous to living organisms in general are known as **hazardous wastes.** Examples include pesticides, medical wastes that may contain disease organisms, nuclear waste, chemicals used in industry, and even items you might have around the house, such as dead batteries, paints, and household cleaning products. If hazardous wastes are not properly contained, they may leak out of the landfill and into the surrounding soil and waterways. One way to prevent this is shown in **Figure 20-18.**

Integrating the Sciences

Earth Science The basis for recycling of materials for human use is that many resources are nonrenewable. While in many cases, recycling is not the cheapest alternative—a throwaway is often less expensive—in the long run, recycling is the wiser alternative.

Water Pollution

Many pollutants released into the air may pollute water as well. Acid rain starts as an air pollution problem but ends up affecting waterways because rain eventually flows into rivers and lakes. Water flowing over or through soil that contains hazardous wastes can pick up and carry some of those wastes along with it, as shown in **Figure 20-19.**

There are also more direct sources of water pollution. Wastewater from factories and sewage-treatment plants is released into streams, rivers, or sometimes directly into the ocean. In the United States and many other countries, this wastewater is treated to remove some or all pollutants before it is released into the environment. But in a large part of the world, wastewater treatment is not always possible.

Groundwater Pollution

Water pollution can also affect **groundwater,** which is water contained in the soil or trapped in underground pockets formed by nonporous rock. The water in these underground pools, or aquifers, comes from rainfall and runoff that soak down through the soil. If the water contacts waste materials as it makes its way underground, the aquifer may become contaminated. Groundwater contamination has already become a problem for rural residents and communities that draw their water supply from wells tapped into underground aquifers as shown in **Figure 20-20.**

Figure 20-19

When rain falls on roads and parking lots, it can wash oil and grease into nearby waterways. When rain falls on agricultural land that has been sprayed with pesticides, chemical residues may be washed into neighboring streams.

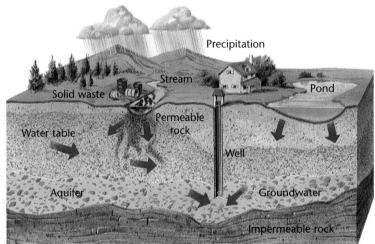

Figure 20-20

Pollutants leaking from storage containers can eventually reach and contaminate groundwater.

20-4 Maintaining a Healthy Environment **563**

Science Journal

Using Water Water usage is a growing concern in the world. By the year 2000, world demand for fresh water may be double what it was in the 1980s. Ask students to write their feelings about this and their suggestions for dealing with the problem in their journals. Ask them to use the library to investigate the topic. **L2**

4 Close

Section Wrap-up

Review

1. smog, acid rain, greenhouse effect, ozone depletion

2. the potential increase in the temperature of Earth's atmosphere due to the greenhouse effect

3. Think Critically Runoff from rainfall can carry these chemicals over the ground into streams or into the ground to underground aquifers.

USING MATH

A solution with a pH of 4 is 100 times more acidic than a solution with a pH of 6.

Skill Builder

Most air pollution is caused by the burning of fossil fuels. Water pollution can be caused by air pollution combined with acid rain. Water pollution is also caused by land pollution, when runoff carries pollutants from dumps and landfills into streams. Water

---

Figure 20-21

Oil spill cleanup involves using floating barriers, called booms, to contain as much oil as possible. Once contained, at least some of it can be skimmed off the water. Beach cleanup is just as messy. Oil-soaked sand has to be shoveled into bags by hand.

Ocean Pollution

Rivers eventually flow into the ocean, bringing their pollutants along. The areas of the ocean most seriously affected by pollution are those near the coastlines, where fresh water enters the sea or where factories, sewage-treatment plants, and shipping activities are concentrated. As is the case with freshwater pollution, the discharge of raw sewage into coastal waters can encourage the overgrowth of algae and plankton. Wastes produced by many of these species are poisonous to fish and wildlife.

Another well-known ocean pollution problem involves oil spills. About 4 million metric tons of oil are released into ocean waters every year. Some of that total comes from freshwater runoff and some comes from ships, including oil tankers that use ocean water to wash out empty tanks before refilling them. Some is also spilled by accident. One of the worst oil spills ever seen occurred during the Persian Gulf War, when a huge amount of crude oil was deliberately released into the water. Tens of thousands of seabirds died and salt marshes and other estuarine ecosystems were destroyed.

Section Wrap-up

Review

1. List four ways in which air pollution affects the atmosphere.

2. Describe global warming.

3. **Think Critically:** How can hazardous chemicals deposited in landfills affect groundwater? Surface streams?

Skill Builder
Comparing and Contrasting
Compare and contrast the causes and effects of air and water pollution. If you need help, refer to Comparing and Contrasting in the **Skill Handbook.**

USING MATH

The pH scale is based on the powers of 10. For example, a solution with a pH of 4 is ten times more acidic than a solution with a pH of 5. Similarly, a pH of 5 is ten times more acidic than a pH of 6. How many times more acidic is a solution with a pH of 4 than a solution with a pH of 6?

---

pollution also can be caused by direct release of pollutants into waterways. Air pollution causes smog, acid rain, the greenhouse effect, and ozone depletion. Water pollution causes overgrowth of algae and other products, which choke out the growth of other organisms. Oil spills damage ecosystems and destroy wildlife. Water pollution endangers people's drinking-water supplies.

Summary

20-1: Natural Resources

1. Natural resources are the parts of the environment that supply materials needed for the survival of living organisms.
2. Renewable resources are replaced by natural processes.
3. Nonrenewable resources cannot be replaced or are replaced slowly.
4. Energy sources include fossil fuels, solar energy, geothermal energy, and nuclear power.

20-2: Science and Society: Recycling

1. Solid waste is unwanted waste that is usually disposed of in landfills.
2. Solid waste can be reduced and resources can be conserved by precycling.
3. Resources can be reused, or recycled.

20-3: Conservation and Wildlife Protection

1. *Conservation* means "saving resources."
2. Soil management includes methods used to prevent soil depletion and erosion.
3. Extinction is the dying out of a species. An endangered species is in danger of extinction.
4. Land conservation includes setting aside protected lands for forests and wildlife.

20-4: Maintaining a Healthy Environment

1. Most air pollution is made up of waste products from the burning of fossil fuels.

2. The greenhouse effect is the warming of Earth by a blanket of heat-reflecting gases in the atmosphere.
3. Hazardous wastes buried in landfills can pollute the surrounding soil and water.
4. Wastes dumped on land or disposed of in landfills can pollute water. Acid rain can also pollute water.

Key Science Words

a. acid rain
b. endangered species
c. erosion
d. extinction
e. fossil fuel
f. geothermal energy
g. global warming
h. greenhouse effect
i. groundwater
j. hazardous waste
k. hydroelectric power
l. natural resource
m. nonrenewable resource
n. nuclear energy
o. ozone depletion
p. photovoltaic (PV) cell
q. pollutant
r. recycling
s. renewable resource
t. smog
u. soil depletion
v. soil management
w. solid waste

Reviewing Vocabulary

Match each phrase with the correct term from the list of Key Science Words.

1. natural resource that is not replaceable
2. coal, oil, and natural gas

Chapter 20 Review   565

Chapter 20

Review

Summary

Have students read the summary statements to review the major concepts of the chapter.

Reviewing Vocabulary

1. m	**6.** u
2. e	**7.** d
3. k	**8.** t
4. f	**9.** o
5. r	**10.** g

Assessment

Portfolio Encourage students to place in their portfolios one or two items of what they consider to be their best work. Examples include:
- Across the Curriculum, p. 551
- Flex Your Brain, p. 555
- Activity 20-2 results and answers, p. 561 **P**

Performance Additional performance assessments may be found in **Performance Assessment** and **Science Integration Activities.** Performance Task Assessment Lists and rubrics for evaluating these activities can be found in Glencoe's **Performance Assessment in the Science Classroom.**

GLENCOE TECHNOLOGY

 Videodisc

Glencoe Life Science Interactive Videodisc
Side 1, Lesson 3
Decomposers: Earth's Natural Recyclers

25823 to 37733

MindJogger Videoquiz

Chapter 20 Have students work in groups as they play the Videoquiz game to review key chapter concepts.

Checking Concepts

1. a   6. c
2. d   7. d
3. b   8. d
4. c   9. b
5. b   10. c

Understanding Concepts

11. Wastes produced by any combustion of fossil fuels, whether in cars or factories, include carbon monoxide, carbon dioxide, oxides of sulfur and nitrogen, and soot particles.
12. Sulfur and nitrogen oxides released during the combustion of fossil fuels react with atmospheric gases and sunlight to produce acids.
13. They contribute to global warming and destroy protective ozone.
14. It takes millions of years to produce what humans can consume in a century.
15. in areas of frequent high winds

Thinking Critically

16. Fossil fuels were formed from plants and animals. Wood is from a plant.
17. the desert region because sunshine is more constant there than at the poles
18. If on a slope, it might be eroded away because there are no longer any plant roots to secure it.
19. Yes. It is constantly replaced by natural processes and it contains natural resources.
20. Decomposers work to break down the waste materials into elements to be used again by living things.

3. energy source that takes advantage of the power of flowing water
4. heat energy from below the surface of Earth
5. reducing the use of natural resources by reusing an item after it has been reprocessed
6. removal of nutrients from soil caused by taking mature plants away from the area where they were grown
7. the dying out of a species
8. a combination of smoke, dust, and oxides of nitrogen and sulfur often characteristic of the air around large cities
9. a thinning of the layer of the atmosphere that absorbs ultraviolet radiation
10. an increase in the temperature of Earth due to an increase in the levels of carbon dioxide and certain other gases in the atmosphere

Checking Concepts

Choose the word or phrase that completes the sentence or answers the question.

1. Which of the following is not a fossil fuel?
   a. wood   c. natural gas
   b. oil   d. coal
2. Which of the following is not a renewable resource?
   a. wood   c. sunlight
   b. water   d. aluminum
3. Which of the following is not either directly or indirectly caused by the sun?
   a. hydroelectric power
   b. nuclear energy
   c. photosynthesis
   d. wind power

4. A renewable energy source found deep in the crust of Earth is _____.
   a. groundwater   c. geothermal energy
   b. solid waste   d. nuclear power
5. Methods used to prevent or reduce soil depletion and erosion are forms of _____.
   a. crop rotation   c. fertilizers
   b. soil management   d. terracing
6. Carbon dioxide is not a(n) _____.
   a. component of acid rain
   b. greenhouse gas
   c. ozone-depleting gas
   d. waste product from the burning of fossil fuels
7. Smog does not include _____.
   a. ozone   c. soot
   b. nitrogen oxides   d. acid rain
8. Ozone depletion is caused by _____.
   a. acid rain   c. carbon dioxide
   b. oxygen   d. chlorine molecules
9. Water pollution is not caused by _____.
   a. air pollution   c. solid-waste disposal
   b. ozone depletion   d. runoff from rainfall
10. _____ is/are not an example of a hazardous waste.
    a. Dead batteries   c. Soil
    b. Nuclear waste   d. Cleaning products

Understanding Concepts

Answer the following questions in your Science Journal using complete sentences.

11. What kinds of substances are found in automobile exhaust?
12. What causes acid rain?
13. How do CFCs affect the environment?
14. Why are fossil fuels considered non-renewable resources?
15. Where would wind power be a useful source of energy?

Developing Skills

21. **Interpreting Data** It is possible the wildflower is becoming endangered because of habitat destruction.

22. **Comparing and Contrasting** Strip-cropping helps reduce erosion caused by wind and water. Contour plowing reduces the downhill flow of water. Terracing reduces the downhill flow of water. No-till farming does not disturb the soil.

Thinking Critically

16. Why does the burning of wood produce similar pollutants to the burning of fossil fuels?
17. What would make a better location for a solar power plant—a polar region or a desert region? Why?
18. Why is it beneficial to grow another crop on soil after the major crop has been harvested?
19. Is garbage a renewable resource? Why or why not?
20. What is the importance of decomposers in a landfill?

Developing Skills

If you need help, refer to the **Skill Handbook.**

21. **Interpreting Data:** A certain species of wildflower grows only by the shores of a certain marsh in Minnesota. Recent observations indicate the population is about half the size it was last year. Portions of the marsh are being drained and filled with dirt to make room for a housing development. What could be happening to the wildflower?

22. **Comparing and Contrasting:** Compare and contrast the reasons for using crop rotation, strip-cropping, contour plowing, terracing, and no-till farming.

23. **Making and Using a Concept Map:** Make a concept map that shows the ways in which the air may become polluted. Use the following terms and events: burning of fossil fuels produces, nitrogen oxide, nitrogen dioxide, sulfur oxides, sunlight, ozone, carbon dioxide, carbon monoxide, escape of CFCs into atmosphere, smog, ozone depletion, and acid rain.
24. **Comparing and Contrasting:** Compare and contrast biodegradable and recyclable materials. Are all biodegradable substances recyclable? Are all recyclable materials biodegradable? Explain.
25. **Observing and Inferring:** Forests consume large amounts of atmospheric carbon dioxide. How might deforestation affect the greenhouse effect and global warming?

Performance Assessment

1. **Poster:** Use a poster to illustrate and describe three things students at your school can do to conserve natural resources.
2. **Conducting a Survey and Graphing the Results:** Develop a survey, in the form of a questionnaire, that asks families what kinds of items they recycle, how often, and where. Send the questionnaires home with students from your school. Set a deadline for their return. Organize the data and report your results to those who participated.
3. **Writing in Science:** Research the recovery of one or more species that were once endangered but are no longer considered at risk. Write a paper or give an oral report explaining what you learned.

Chapter 20 Review

23. **Making and Using a Concept Map** Concept maps should show the following logic: burning of fossil fuels produces nitrogen oxide, nitrogen dioxide, sulfur oxides, carbon dioxide, and carbon monoxide; oxides react with sunlight to produce smog and acid rain; escape of CFCs into the atmosphere causes ozone depletion.
24. **Comparing and Contrasting** A biodegradable material is one that can be broken down by natural decomposition processes. A recyclable material is one that can be reused if it is reprocessed in some way. A biodegradable substance is recyclable. A recyclable material is not necessarily biodegradable.
25. **Observing and Inferring** It might increase global warming because it increases the concentration of carbon dioxide in the atmosphere.

Performance Assessment

1. Accept all reasonable answers. Use the Performance Task Assessment List for Poster in **PASC**, p. 73. P
2. Student questionnaires should be well organized and ask questions based on information in the chapter. Data should be tallied in a logical manner. Use the Performance Task Assessment List for Oral Presentation in **PASC**, p. 71. P
3. The bald eagle, wild turkey, and bison are three examples. Students should include all sources. Use the Performance Task Assessment List for Writing in Science in **PASC**, p. 87. P

Assessment Resources

Reproducible Masters
Chapter Review, pp. 43-44
Assessment, pp. 130-133
Performance Assessment, p. 34

Glencoe Technology
- Chapter Review Software
- Computer Test Bank
- MindJogger Videoquiz

UNIT PROJECT 6

Objectives

IS Intrapersonal Students will search for, gather, and analyze information from the Internet about the ecological problems of their choosing. They will make presentations to the class using the information they obtained in their searches. **L1 COOP LEARN**

Summary

The Internet is a collection of data bases stored in hundreds of thousands of computers that are loosely linked together. Accessing the Internet requires a computer, modem, and a subscription with one of the commercial servers. Using the Internet, students can access scientists and universities, as well as private corporations throughout the world. After deciding on a topic about which they would like to learn more, students will collect data and prepare a report that contains text and graphics, and perhaps sound and photography.

Time Required

One-half class period will be needed to discuss the requirements of the project and to decide if students will work independently or with a partner. Each student will need approximately 3 hours on-line at the computer and an additional 3 to 5 hours to prepare a report. Allow an additional 2 to 3 days for student presentations.

Preparation

- If students will be using a computer lab or computers in the library, arrange access with the person in charge.
- Prepare a letter to parents telling them that students will be using Internet resources.
- Make a list of URLs (Internet addresses) that students can use as a starting point. Since URLs change often, it is especially important that you check the addresses before you provide them to students.
- If you are not familiar with the Internet, ask another faculty member for background help.

UNIT PROJECT 6

Ecology on the Internet

How do you use resources to find out more about a topic in which you are interested? In the past, you might have gone to the library and used a card catalog or periodical listing to find books, magazines, and newspaper articles that were related to your topic. Now, at most libraries, you can find these same book and article listings on a computer. Many libraries have put their holdings in one large database to make research less time consuming and more successful.

Within the last few years, a new and more powerful source for information retrieval has become widely available. This source of information is called the Internet. If you have a computer or have access to a school computer, you can do research on the Internet.

What is the Internet?

The Internet is made up of millions of computers all around the world that are linked to each other. The information available in these computers is varied and vast. You can find information on almost any imaginable topic. Using electronic mail, or E-mail, you can speak directly to an expert in a particular field or to a person in another country. Some computers store articles or entire books about particular subjects. Others contain thousands of photographs and illustrations.

Getting Started

To use the Internet, you must have a way to access it. Many schools have direct access to the Internet or have a connection through a local college or university. A phone line can also give you access by way of a commercial server. Each server differs slightly in the way it works. Time, practice, and perhaps some cooperative work with a few friends will help you locate the information you need. Reference books about using the Internet will be helpful, as will consulting with persons already proficient at using the Internet for research.

Beginning Your Search

For this project, you will choose an ecological problem that you wish to research and report about to your class. You may want to look again through Chapters 18-20 for some ideas. Some problems you might wish to consider are air quality, water quality, sound pollution, or landfills. Biodiversity, endangered or threatened species, and population growth are other possible topics. Pick a topic that isn't too broad or narrow.

Internet Yellow Pages

The Internet yellow pages contain the Internet addresses for locating information about any imaginable topic. However, because no one is in charge of the Internet, an Internet address may be available for a time, then the people managing it may decide to remove it.

Internet Search Tips

One important method for finding information is a search tool. A search tool allows you to type in a key word or words. The tool then searches thousands of sites on the Internet for those that contain your key words. After a few seconds, the search tool returns many addresses for other links related to your topic. The *gopher* is one type of search tool. Thousands of gopher servers are available on the Internet.

Using Your Research

After you've completed your search for information and gathered all of your data, you should be ready to prepare a presentation for your class. Your presentation should be put together using a multimedia computer program. Your presentation should consist of at least five different screens with text, graphics, and perhaps sound and photos. Design each screen so that it is appealing to the audience. Your report should include information from at least five sites on the Internet. Identify your sites and explain to the class how you got to them.

A word of caution. Not everything on the Internet is valid. Much of the information has not been edited, reviewed, or even verified as true. If you have questions about anything you find, ask a parent, teacher, or other adult to help you determine whether or not it is useful.

Teaching Strategies

- Remind students to take notes while they are online, just as they would for any other research project. URLs are important for navigating back to an area of interest.
- If students work in pairs, it is best to have a student who does not have Internet access at home work with one who does.
- Remind students that community libraries and university libraries provide Internet access. There may be a fee for logging onto the Internet.
- Explain that search engines may provide useful starting points. The Web Crawler is one example.
- Provide students with a rubric that explains what portion of their grade will be determined by each part of their Internet project.

Go Further

Once students have begun to use the Internet, they can use it as a vehicle to answer many different questions and complete other classroom exercises. You may wish to assign another problem that challenges them even further.

References

The following books—all 1995 copyrights—are available from Ventana Press, P.O. Box 13964, Research Triangle Park, North Carolina 27709-3964:

- *HTML Publishing on the Internet for Mac*
- *Internet Roadside Attractions*
- *Walking the World Wide Web*

In Unit 7, students are given an introduction to the systems of the human body and how they function. An understanding of how organs and systems interact is developed. Also included are discussions on human development from conception to birth. Interwoven throughout the unit are basic health concepts designed to give adolescents the knowledge needed to make informed health choices.

CONTENTS

Chapter 21
Bones, Muscles,
and Skin   572

21-1 The Skeletal System

21-2 Science and Society: Biomaterial—Speeding Bone Fracture Recovery

21-3 The Muscular System

21-4 Skin

Chapter 22
Nutrients and Digestion 598

22-1 Nutrition

22-2 Science and Society: Nutrients Combat Cancer

22-3 Your Digestive System

Chapter 23
The Circulatory System 624

23-1 Circulation

23-2 Science and Society: "Growing" Heart Valves

23-3 Blood

23-4 Your Lymphatic System

Chapter 24
Respiration and
Excretion   650

24-1 Your Respiratory System

24-2 Science and Society: Restricting Cigarette Advertising

24-3 Your Urinary System

Unit Contents

Chapter 21
Bones, Muscles, and Skin

Chapter 22
Nutrients and Digestion

Chapter 23
The Circulatory System

Chapter 24
Respiration and Excretion

Chapter 25
The Nervous and Endocrine Systems

Chapter 26
Reproduction and Growth

Chapter 27
Immunity

Unit 7 Project
On Your Mark!
Get Set! Go!

Chapter 25
The Nervous and
Endocrine Systems   676

25-1 Your Nervous System

25-2 Science and Society: Care and Treatment of Alzheimer's Patients

25-3 The Senses

25-4 Your Endocrine System

Chapter 26
Reproduction and Growth 702

26-1 Human Reproduction

26-2 Fertilization to Birth

26-3 Development After Birth

26-4 Science and Society: A New View of the Old

Chapter 27
Immunity   728

27-1 Disease

27-2 Your Immune System

27-3 Science and Society: Should a harmful virus be destroyed?

27-4 Noncommunicable Disease

The Human Body

What's Happening Here?

On a frigid winter morning, students blow "smoke" every time they exhale. Is this white vapor really smoke? When you inhale, the oxygen in the air diffuses into fingerlike sacs in your lungs, then into tiny blood vessels. The blood carries oxygen to all body cells, where it is exchanged for waste products such as water and carbon dioxide. In your lungs, blood cells release these products to the tiny sacs shown in the the large photograph. The next time you exhale, the water and carbon dioxide leave your body in a cloud of "smoke." You will learn more about your respiratory and circulatory systems in this unit.

Science Journal

As altitude increases, the gases in air spread out, so there is less oxygen for each breath a person takes. People who live high up in the mountains have adaptations to this thinner air. Name one such group, and in your Science Journal describe their adaptations to such thin air.

571

Chapter Organizer

Section	Objectives/Standards	Activities/Features
Chapter Opener		**Explore Activity:** Which muscles tire first? p. 573
21-1 **The Skeletal System** (2 sessions)*	1. **Identify** the five major functions of the skeletal system. 2. **Compare** and **contrast** movable and immovable joints. National Content Standards: (5-8) UCP1, A1, A2, C1	**Problem Solving:** Shape Influences Strength, p. 577 **Skill Builder:** Comparing and Contrasting, p. 579 **Using Math,** p. 579 **Activity 21-1:** Observing Bones, pp. 580-581
21-2 **Science and Society Technology: Biomaterial—Speeding Bone Fracture Recovery** (1 session)*	1. **Describe** the source of the hardness of bone. 2. **Compare** the standard method of bone fracture treatment with the biomaterial injection method. 3. **Discuss** some methods to prevent bone fractures. National Content Standards: (5-8) E2, F1, F5	**Explore the Technology,** p. 583
21-3 **The Muscular System** (2 sessions)*	1. **Describe** the major function of muscles. 2. **Compare** and **contrast** three types of muscles. 3. **Explain** how muscle action results in movement of body parts. National Content Standards: (5-8) UCP1, UCP3, A1, A2, B1, B2, C1	**Using Math,** p. 586 **MiniLAB:** How do muscle pairs work? p. 587 **Connect to Chemistry,** p. 588 **Skill Builder:** Sequencing, p. 588 **Science Journal,** p. 588 **Activity 21-2:** Observing Muscle, p. 589 **Science and Literature:** Frankenstein, p. 590 **Science Journal,** p. 590
21-4 **Skin** (1 session)*	1. **Compare** and **contrast** the epidermis and dermis of the skin. 2. **List** the functions of the skin. 3. **Discuss** how skin protects the body from disease and how it heals itself. National Content Standards: (5-8) UCP1, A1, C1, E2, F1, F5	**Using Technology:** Robot Skin, p. 592 **MiniLAB:** Do you always sweat? p. 593 **Connect to Earth Science,** p. 594 **Skill Builder:** Concept Mapping, p. 594 **Using Math,** p. 594

* A complete Planning Guide that includes block scheduling is provided on pages 32T-35T.

Activity Materials

Explore	Activities	MiniLABs
page 573 snap-type clothespin	pages 580-581 beef bone (cut in half and lengthwise), turkey bone (cut in half and lengthwise), hand lens, scalpel page 589 prepared slides of smooth, skeletal, and cardiac muscles; microscope; cooked turkey leg; dissecting pan or cutting board; 2 dissecting probes; hand lens	page 587 muscle diagram page 593 large, clear plastic food bag; masking tape; clock

Need Materials? Call Science Kit (1-800-828-7777).

Teacher Classroom Resources

Reproducible Masters	Transparencies	Teaching Resources
Study Guide, p. 80 **Reinforcement,** p. 80 **Enrichment,** p. 80 **Activity Worksheets,** pp. 124-125 **Lab Manual 49,** Analyzing Bones **Science Integration Activity 21,** Differences in a Species **Concept Mapping,** p. 47 **Cross-Curricular Integration,** p. 25 **Science and Society Integration,** p. 25	**Section Focus Transparency 76,** Putting It Together **Teaching Transparency 41,** Human Skeletal System	**Glencoe Life Science Interactive Videodisc** **Life Science CD-ROM** **Spanish Resources** **English/Spanish Audiocassettes** **Cooperative Learning Resource Guide** **Lab Partner** **Lab and Safety Skills** **Lesson Plans**
Study Guide, p. 81 **Reinforcement,** p. 81 **Enrichment,** p. 81 **Critical Thinking/Problem Solving,** p. 25	**Section Focus Transparency 77,** Finding Support	**Assessment Resources** **Chapter Review,** pp. 45-46 **Assessment,** pp. 143-146 **Performance Assessment in the Science Classroom (PASC)** **MindJogger Videoquiz** **Alternate Assessment in the Science Classroom** **Performance Assessment** **Chapter Review Software** **Computer Test Bank**
Study Guide, p. 82 **Reinforcement,** p. 82 **Enrichment,** p. 82 **Activity Worksheets,** pp. 126-127, 128 **Lab Manual 50,** Muscle Action **Multicultural Connections,** pp. 45-46	**Section Focus Transparency 78,** Robotics **Science Integration Transparency 21,** The Body's Lever Systems **Teaching Transparency 42,** Human Muscle System	
Study Guide, p. 83 **Reinforcement,** p. 83 **Enrichment,** p. 83 **Activity Worksheets,** p. 129	**Section Focus Transparency 79,** Peel It Off!	

Key to Teaching Strategies

The following designations will help you decide which activities are appropriate for your students.

L1 Level 1 activities should be within the ability range of all students, including those with learning difficulties.

L2 Level 2 activities should be within the ability range of the average to above-average student.

L3 Level 3 activities are designed for the ability range of above-average students.

LEP LEP activities should be within the ability range of Limited English Proficiency students.

LS These activities are designed to address different learning styles.

COOP LEARN Cooperative Learning activities are designed for small group work.

P These strategies represent student products that can be placed into a best-work portfolio.

GLENCOE TECHNOLOGY

The following multimedia resources are available from Glencoe.

Science and Technology Videodisc Series (STVS)
Human Biology
Arthritis Research
Tricycle for the Handicapped
Orthopedic Implants

National Geographic Society Series
STV: Human Body Vol. 2
STV: Human Body Vol. 3
STV: The Cell
Glencoe Life Science Interactive Videodisc
Exercise: Health Is Just a Heartbeat Away
Life Science CD-ROM

Teacher Classroom Resources

This is a representation of key blackline masters available in the Teacher Classroom Resources.

Teaching Aids

Section Focus Transparencies

Science Integration Transparencies

Teaching Transparencies

Meeting Different Ability Levels

Study Guide

Reinforcement

Enrichment Worksheets

Chapter 21 Bones, Muscles, and Skin

Hands-On Activities

Science Integration Activity

NAME   DATE   CLASS

Chapter 21
INTEGRATION   ● Differences in a Species

When you go to a museum and look at an exhibit of a vertebrate fossil, you are usually observing an animal's skeletal system. To use fossil dating efficiently, paleontologists first separate fossils into groups. The most useful group is called a species. A species is a group of organisms whose members can mate with one another to produce fertile offspring. Small differences in individuals may result in the development of a new species by a series of gradual changes. These changes can be traced from one geologic time division to another, if the fossil record is good.
In this activity, you will describe the variations present within a species. You will also describe a species in terms of one characteristic.

Materials
● graph paper   ● meter stick   ● pencils (colored)

Procedure
1. The species you will study is *Homo sapiens*, or your own species. You and your classmates are all members of this species.
2. Record all characteristics of the species you can. Record which of the characteristics you could measure and compare for all members of the species.
3. Make a copy of the data table in your Science Journal to record your height and the height of each of your classmates. Round off the height to the nearest decimeter (0.1 m). Measure and record the information in your data table. Find the average height.
4. Measure and record the heights of a class of younger students. Record the information in your data table. Find the average height.
5. Graph the Frequency (number of persons having the same height) on the vertical axis against the Height (m) on the horizontal axis. Use one color for your own class and a second color for the younger class. Remember to indicate what your colors represent.

Copyright © Glencoe/McGraw-Hill, a division of The McGraw-Hill Companies, Inc.

43   **L1**

Lab Manual

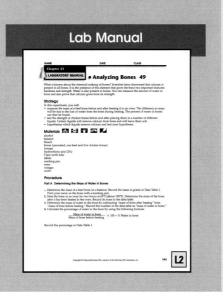

NAME   DATE   CLASS

Chapter 21
LABORATORY MANUAL   ● Analyzing Bones   49

What is known about the chemical makeup of bones? Scientists have discovered that calcium is present in all bones. It is the presence of this element that gives the bone two important features—hardness and strength. Water is also present in bones. You can measure the amount of water in bone and also prove that calcium gives bone its strength.

Strategy
In this experiment, you will:
● measure the mass of bone before and after heating it in an oven. The difference in mass will be due to the loss of water from the bone during heating. The percent of water in bones can then be found.
● test the strength of chicken bones before and after placing them in a number of different liquids. Certain liquids will remove calcium from bone and will leave them soft.
● hypothesize which liquids remove calcium and test your hypotheses.

Materials
alcohol
balance
bleach
bones (uncooked, one beef and five chicken bones)
forceps
hydrochloric acid (2%)
5 jars (with lids)
labels
marking pen
oven
vinegar
water

Procedure
Part A   Determining the Mass of Water in Bones
1. Determine the mass of a beef bone on a balance. Record the mass in grams in Data Table 1. Print your name on the bone with a marking pen.
2. Heat the bone in an oven for two hours at 65°C(about 150°F). Determine the mass of the bone after it has been heated in the oven. Record its mass in the data table.
3. Determine the mass of water in the bone by subtracting "mass of bone after heating" from "mass of bone before heating." Record this number in the data table as "mass of water in bone."
4. Calculate the percentage of water in the bone by using the following formula.

$$\frac{\text{Mass of water in bone}}{\text{Mass of bone before heating}} \times 100 = \% \text{ Water in bone}$$

Record this percentage in Data Table 1.

Copyright © Glencoe/McGraw-Hill, a division of The McGraw-Hill Companies, Inc.

141   **L2**

Assessment

Performance Assessment

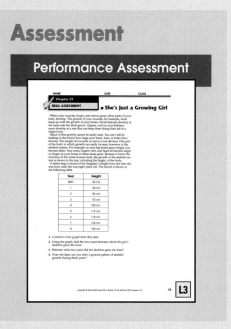

NAME   DATE   CLASS

Chapter 21
SKILL ASSESSMENT   ● She's Just a Growing Girl

When your muscles, bones, and nerves grow, other parts of your body develop. The growth of your muscles, for example, must keep up with the growth of your bones. Facial features develop at the same rate the skull grows. Organs, such as your kidneys, must develop at a rate that can keep them doing their job in a larger body.
Much of this growth cannot be easily seen. You can't tell by looking in the mirror how large your heart, liver, or brain have become. The length of a muscle or nerve is not obvious. One part of the body in which growth can easily be seen, however, is the skeletal system. For example, as your leg bones grow longer, you become taller. Your arms, fingers, feet, and head all become larger or longer as your bones in those areas grow. Because it forms the structure of the entire human body, the growth of the skeletal system is shown in the size, including the height, of the body.
A father kept a record of his daughter's height from the time she was born until she was eight years old. The record is shown in the following table.

Year	Height
Birth	42 cm
1	50 cm
2	86 cm
3	92 cm
4	104 cm
5	112 cm
6	118 cm
7	124 cm
8	128 cm

1. Construct a bar graph from this data.
2. Using the graph, find the two years between which the girl's skeleton grew the most.
3. Between what two years did her skeleton grow the least?
4. From the data, can you infer a general pattern of skeletal growth during these years?

Copyright © Glencoe/McGraw-Hill, a division of The McGraw-Hill Companies, Inc.

35   **L3**

Enrichment and Application

Critical Thinking/Problem Solving

NAME   DATE   CLASS

Chapter 21
CRITICAL THINKING   ● Bones, Muscles, and Skin

A Space Shuttle Ride for a Phantom Head

As you know, the skeleton protects the soft tissues in your body. But the skeleton may not be able to provide much protection from radiation. This is a concern for scientists at NASA who are currently studying the effects of radiation on astronauts.
NASA wants to study the effects of radiation on the human body because of the potential danger to space shuttle astronauts and future space-station workers. In outer space, there are cosmic rays that generally do not reach Earth's surface. Cosmic rays are radiation given off by the sun and other stars. Scientists know that radiation can cause serious biological damage. It disrupts the normal chemical processes of living cells, causing them to grow abnormally or die.
Recently NASA sent a human skull, which was donated to science, into space. This special skull has been aboard several shuttle flights. The skull is filled with plastic containing tiny meters and detectors that are able to measure different levels and kinds of radiation.
NASA wants to reproduce the effects of space travel on the human body as closely as possible. NASA's aim is to discover not only the amount of radiation that astronauts are exposed to while in space, but also the kinds of radiation. Radiation from particles, for example, can be much more damaging than radiation from other particles.

Through the use of this special skull, NASA has also been able to compare the levels and kinds of radiation that occur in different orbits. It is believed that the radiation in polar orbits has more energy than the radiation in orbits around the equator.
Although sending a human skull into space may seem eerie, astronauts may find comfort in knowing that it is accompanying them for their own good. NASA also plans to send a human torso into space. It will be used to test the effects of charged particles on the body's soft tissue.

Applying Critical Thinking Skills
1. Why would astronauts find comfort in having a human skull accompany them on their space shuttle flight? _____

2. How can scientists justify making a human skull part of their experiments? _____

3. What are some of the possible solutions that may protect astronauts from being exposed to dangerous levels of cosmic radiation? _____

Copyright © Glencoe/McGraw-Hill, a division of The McGraw-Hill Companies, Inc.

25   **L2**

Cross-Curricular Integration

NAME   DATE   CLASS

Chapter 21
FINE ARTS CONNECTION   ● Bones, Muscles, and Skin

Leonardo da Vinci

Leonardo da Vinci lived and painted during the Renaissance period in Italy. You may be familiar with his famous painting, the *Mona Lisa*. But da Vinci was not only a painter. He was an inventor, a scientist, and an architect, too. He recorded all of his observations, ideas, and sketches in a notebook. Da Vinci also studied the human body in detail and provided the first accurate pictures of human anatomy. By dissecting animal and human bodies, da Vinci learned how muscles, tendons, and bones work together to produce movement.

Find out more about Leonardo da Vinci's notebooks at the library. You may want to find the book The Notebooks of Leonardo da Vinci, *by Edward MacCurdy, Garden City Publishing Co., 1941–42. Look for da Vinci's drawings of human anatomy, especially bones and muscles. Choose one sketch, draw it to the space below, and label a few of the bones or muscles shown. Use an encyclopedia or anatomical reference to identify the bones and muscles you choose.*

Copyright © Glencoe/McGraw-Hill, a division of The McGraw-Hill Companies, Inc.

25   **L1**

Science and Society Integration

NAME   DATE   CLASS

Chapter 21
SCIENCE & SOCIETY   ● Bones, Muscles, and Skin

Tissue Replacement

For about 30 million Americans, movement means pain because the cartilage in their hips, knees, or other joints has worn away. With the cartilage missing, the bones that meet at their joints scrape against each other.
To reduce this pain, doctors might replace the cartilage with metal or plastic devices. Although about one million people a year undergo this procedure, the artificial cartilage often does not protect their joints and may have harmful side effects.
Alternatively, doctors might replace the entire joint. In 1992, more than 400 000 people received artificial hips, knees, or shoulder joints. These metal or plastic joints allow pain-free movement, but they last only 10 to 15 years and then must be replaced.
Now scientists are developing a method that will some day help people "regrow" their own cartilage. So far the process, called tissue engineering, has been successful in animals. Researchers take a small chunk of cartilage from the knee of a dog or rabbit and place it into a biodegradable framework that allows the cells to grow. Eventually, the piece is

replaced in the animal's knee. The framework dissolves, and the new cartilage begins to cushion the animal's movement.
Scientists have used this approach to grow new cartilage in cows, whose joints must support considerable weight. Some human trials are underway with more planned. "It's becoming more and more clear that materials like metal and plastic are not suitable replacements for the real thing," says Robert Langer, Germeshausen Professor at the Massachusetts Institute of Technology. "We are trying to understand how to take normal living cells and engineer them to regenerate new tissue. We're trying to get back to Mother Nature."
Researchers are also using tissue engineering to help burn patients grow their own new skin. In this way, doctors do not have to cause a wound by moving skin from one place on the patient's body to another. Tissue engineering shows great promise, but some basic challenges remain, such as how to encourage blood vessels and nerves to grow into new tissue and how to stop the tissue from growing too large.

You Decide
1. What other uses of tissue engineering do you predict for the future? What health problems might this process solve? _____

2. If you had arthritis or another joint problem, would you volunteer for tissue engineering to see if this process could help you grow new cartilage? Why or why not? _____

Copyright © Glencoe/McGraw-Hill, a division of The McGraw-Hill Companies, Inc.

25   **L2**

Multicultural Connections

NAME   DATE   CLASS

Chapter 21
MULTICULTURAL CONNECTIONS   ● Curing with Curare

In the humid air of the Amazon rain forest, a thousand shades of green form a lush canopy overhead. Vines coil up the trunks of giant trees as scarlet birds flit from branch to branch. A monkey howls and then is silent.
The ways of the forest hunter are successful and effective. Hunting game to eat and using plants for food, shelter, and medicine are all part of the forest technology that has been known to the indigenous (native) people of the South American rain forest for generations. Knowledge of what is *arihi a rima,*—"of the forest"—has kept people such as the Yanomamo of northern Brazil in harmony with the vast ecosystem in which they live.
But as progress and development invade the rain forest, indigenous people struggle to move forward without losing their cultural identity and health. Certain aspects of rain forest life have become known to people throughout the world. The treasure trove of medicinal plants once restricted to the rain forest can be used by everyone. One example is *curare*, the toxic substance used by the Yanomamo to make potent arrows for hunting. This powerful drug is used for medicinal purposes by physicians all over the world because of its effectiveness as a muscle relaxant during surgery.

Rain Forest Technology
The natural resource of the South American rain forest are extraordinary. Plants provide building materials for houses, abundant food, as well as a wealth of bark, seeds, leaves, and roots with a variety of properties. Many are used for healing. Some plants are used for their strong toxins. Two groups of trees and shrubs, the genus *Chondodendron* and the genus *Strychnos,* provide the compounds that yield the poison curare. The name curare comes from the word *wourari,* meaning poison.
Traditionally, the people in a Yanomamo village would be able to produce all their everyday items from the resources available in the rain forest. To go out hunting for game to eat meant making a bow and arrow from local materials. They make their bows from dense, hard palm wood. The bow is strung with long twisted cords of inner tree bark. Thin splinters of palm wood are used to make bow raised

points, the curare-poisoned tips of arrows. These arrows are about a foot long. They are weakened in several places by cutting them part of the way through. The Yanomamo usually make the arrow point in a bundle of 30 or 40 at once. They make the poison by first soaking the vine of the plant in hot water. Then other plant ingredients are added to make the mixture stick to the arrows. What results is a dark brown resin that has an aromatic odor. Traditionally, it is prepared to an earthenware pot or gourd. Then the dark mixture is painted onto the sharp arrow points. Longer lance-like points are used to kill larger game such as the tapir (a large hoofed animal), similar to the pig). Small bone barbs are used for hunting birds. Yanomamo boys become skilled hunters at about age ten. Being able to hunt is an important skill necessary for the well being of the community.

How Curare Works
If a hunter hits the target animal, such as a monkey, the arrow breaks along its weakened shaft. The monkey is unable to pull out the arrow point, and the curare enters its blood

Copyright © Glencoe/McGraw-Hill, a division of The McGraw-Hill Companies, Inc.

45   **L2**

Concept Mapping

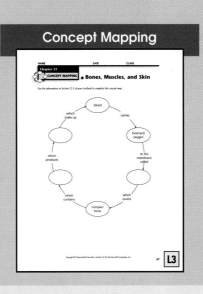

NAME   DATE   CLASS

Chapter 21
CONCEPT MAPPING   ● Bones, Muscles, and Skin

Use the information in Section 21-1 of your textbook to complete this concept map.

blood

which make up → carries

food and oxygen

which produces → to the membrane called

which contains → which covers

compact bone

Copyright © Glencoe/McGraw-Hill, a division of The McGraw-Hill Companies, Inc.

47   **L3**

572D

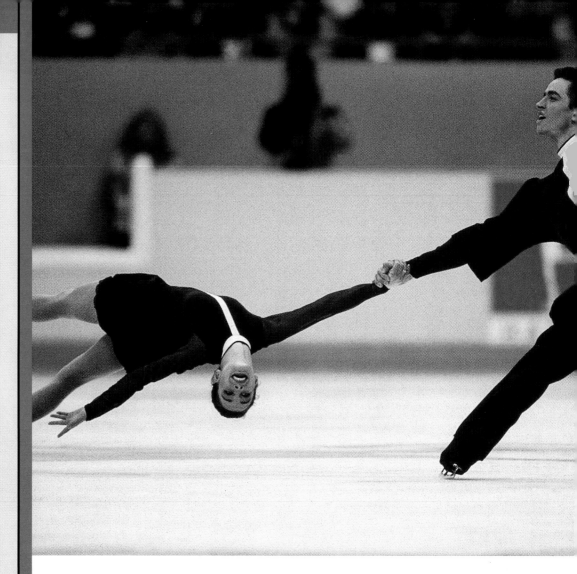

Chapter 21

Bones, Muscles, and Skin

CHAPTER OVERVIEW

Section 21-1 This section discusses the five major functions of the skeletal system. Details of the characteristics, composition, and development of bone and bone joints are also discussed.

Section 21-2 Science and Society This section discusses a new technology for the treatment of broken bones.

Section 21-3 The major functions and types of muscles are discussed. An explanation of how muscle action produces movement is given.

Section 21-4 This section discusses the tissue layers, structures, and five major functions of the skin.

Chapter Vocabulary

skeletal	voluntary
system	muscle
periosteum	involuntary
marrow	muscle
cartilage	skeletal
joint	muscle
ligament	tendon
immovable	smooth
joint	muscle
movable joint	cardiac
fracture	muscle
biomaterial	epidermis
osteoporosis	melanin
muscle	dermis

Theme Connection

Energy One of the themes in this text is energy. When the skeletal and muscular systems work to produce movement, energy is used, and work is done. Food is processed by the digestive system and converted into glucose. The chemical energy in glucose is used by muscles and transformed into mechanical energy.

Previewing the Chapter

Section 21-1 The Skeletal System
- ▶ A Living Framework
- ▶ Looking at Bone
- ▶ Bone Development
- ▶ Where Bones Meet

Section 21-2 Science and Society Technology: Biomaterial— Speeding Bone Fracture Recovery

Section 21-3 The Muscular System
- ▶ Moving the Human Body
- ▶ Levers—Your Body's Simple Machines
- ▶ Types of Muscle Tissue
- ▶ Muscles at Work

Section 21-4 Skin
- ▶ The Body's Largest Organ
- ▶ Skin at Work

572

Learning Styles

Look for the following logo for strategies that emphasize different learning modalities. **LS**

Kinesthetic	Explore, p. 573; Activity, p. 576; MiniLAB, p. 586; Science & Literature, p. 590; Demonstration, p. 592
Visual-Spatial	Demonstration, pp. 575, 578; Visual Learning, pp. 575, 576, 578, 583, 585, 593; Making a Model, p. 577; Activity 21-2, p. 589; MiniLAB, p. 593
Interpersonal	Activity 21-1, pp. 580-581
Logical-Mathematical	Discussion, p. 582
Linguistic	Across the Curriculum, p. 577; Science & Literature, p. 590

Bones, Muscles, and Skin

Like these figure skaters on the left, you might have experienced tired muscles after completing some strenuous physical activity. While you often feel your body's large muscles ache when they tire, your body also has many small muscles that can tire and ache after just a little bit of work.

EXPLORE ACTIVITY

Which muscles tire first?

1. Hold a snap-type clothespin between the thumb and forefinger of your writing hand.
2. Rapidly pinch the clothespin open and shut.
3. Count the number of times you can pinch the clothespin before your fingers become tired.
4. Record the number.
5. Repeat this activity with your other hand and compare the results.

Observe: Do small muscles that are exercised more (those in your writing hand) tire faster or slower than those in the hand that gets less exercise?

Previewing Science Skills

▶ In the Skill Builders, you will outline, sequence, and map concepts.

▶ In the Activities, you will observe and collect data.

▶ In the MiniLABs, you will observe and experiment.

573

EXPLORE ACTIVITY

Purpose

Ⓛ Kinesthetic Use the Explore Activity to introduce students to muscles. Inform students they will learn more about the muscular and skeletal systems as they read the chapter. L1 LEP

Preparation

Before you begin this chapter, have students bring in some wooden snap-type clothespins. Provide extras.

Materials

one snap-type clothespin

Teaching Strategies

Tell students to count only until the first sign of tiredness, not until exhaustion or cramping. Remind students that this is not a competition.

Observe

Students should observe that small muscles that are exercised tire more slowly than those in a hand that gets less exercise.

✓ **Assessment**

Performance Have students design another activity that tests the fatigue time of the right and left arm muscles. Use the Performance Task Assessment List for Designing an Experiment in **PASC**, p. 23 P

Assessment Planner

Portfolio
Refer to page 595 for suggested items that students might select for their portfolios.

Performance Assessment
See page 595 for additional Performance Assessment options.
Skill Builders, pp. 579, 588, 594
MiniLABs, pp. 586, 593
Activities 21-1, pp. 580-581; 21-2, p. 589

Content Assessment
Section Wrap-ups, pp. 579, 583, 588, 594
Chapter Review, pp. 595-597
Mini Quizzes, pp. 579, 588

Group Assessment
Opportunities for group assessment occur with Cooperative Learning Strategies and Flex Your Brain Activities.

Prepare

Section Background

- Bone tissue is composed chiefly of collagen and the mineral hydroxyapatite, $Ca_{10}(PO_4)_6(OH)_2$.
- In an embryo, the skeleton begins as a cartilaginous-membrane system. In the last months of the pregnancy, the cartilage and the membranes are replaced by bone.
- Bone tissue is constantly being reabsorbed and re-formed. Calcium and vitamin D are necessary for the deposition of new bone material.

Preplanning

To prepare for Activity 21-1, obtain a long beef bone and a long turkey bone. Boil the bones and have them cut in half along the length and width by a butcher.

1 Motivate

Bellringer

Before presenting the lesson, display **Section Focus Transparency 76** on the overhead projector. Assign the accompanying **Focus Activity** worksheet. L1 LEP

Tying to Previous Knowledge

Recall the role of the skeletal system in classifying vertebrate animals. Remind students of the function of the skeletal system in providing internal support and framework for the vertebrate body.

Science Words

- skeletal system
- periosteum
- marrow
- cartilage
- joint
- ligament
- immovable joint
- movable joint

Objectives

- Identify the five major functions of the skeletal system.
- Compare and contrast movable and immovable joints.

A Living Framework

The skull and crossbones flying from a pirate ship have long been a symbol of death. Thus, you might think that bones are dead structures made of rocklike material. It's true that a skeleton's bones are no longer living, but the bones in your body are very much alive. Each bone is a living organ made of several different tissues. Cells in these bones take in nutrients and expend energy. They have the same requirements as your other cells.

Major Functions of Your Skeletal System

All the bones together make up your **skeletal system,** which is the framework of your body. The human skeletal system has five major functions. First, it gives shape and support to your body, like the framework of the building, in **Figure 21-1.** Second, bones protect your internal organs: ribs surround the heart and lungs and a skull encloses the brain. Third, major muscles are attached to bone. Muscles move bones. Fourth, blood cells are formed in the red marrow of some bones. Bone marrow is a soft tissue in the center of many bones. Finally, the skeleton is where major quantities of calcium and phosphorus compounds are stored for later use. Calcium and phosphorus make bone hard.

Figure 21-1

The 206 bones of a mature adult support the body just as the steel girders support the sky-scraper.

Looking at Bone

As you study bones, you'll notice several characteristics. The differences in sizes and shapes are probably most obvious. Bones are frequently classified according to their shape. Bone shapes, shown in **Figure 21-2,** are genetically controlled and are also modified by the work of muscles that are attached to them.

Bone Structure

Upon close examination, you'll find that a bone isn't all smooth. There are bumps, edges, round ends, rough spots, and many pits and holes. Muscles and ligaments attach to some of the bumps and pits. Blood vessels and nerves enter and leave through the holes. Many internal and external characteristics of bone are seen in the picture of the humerus shown in **Figure 21-3.**

As you can see in **Figure 21-3,** the surface of the bone is covered with a tough, tight-fitting membrane called the **periosteum** (per ee AHS tee um). Small blood vessels in the periosteum carry nutrients into the bone. Cells involved in

Figure 21-2

The shape of a bone provides clues about what function it performs within the body. *Which of these bones supports the body?*

Femur — Long
Metacarpals — Short
Carpal — Irregular
Phalanges — Short
Scapula — Flat

Figure 21-3

Bone is made of layers of living tissue. Within the compact bone are circular structures called Haversian systems. When the bone is cut, you can see that these layers look like the growth rings of tree trunks.

Humerus
Cartilage
Periosteum
Marrow cavity
Spongy bone
Compact bone
Haversian system
Spongy bone
Blood vessels supplying living bone
Blood vessels and nerves in center
Compact bone
Periosteum

21-1 The Skeletal System   575

the growth and repair of bone are also found here. Under the periosteum is compact bone, a hard, strong layer of bone. Compact bone contains bone cells, blood vessels, and a protein base or scaffolding with deposits of calcium and phosphorus. The flexible protein base keeps bone from being too rigid and brittle or easily broken.

Spongy bone is found toward the ends of long bones like the humerus. Spongy bone is much less compact and has many small open spaces that make the bone lightweight. If all your bones were completely solid, you'd have a much greater mass. Long bones have large openings, or cavities. The cavities in the center of long bones and the spaces in spongy bone are filled with a fatty tissue called **marrow**. Marrow produces red blood cells at an incredible rate of more than 2 million cells per second. White blood cells are also produced in bone marrow, but in lesser amounts.

Cartilage

Notice in **Figure 21-3** that the ends of the bone are covered with a thick, smooth layer of tissue called **cartilage.** Cartilage does not contain blood vessels or minerals. It is flexible and is important at joints, where it absorbs shock and makes movement easier by reducing friction. In older people, cartilage sometimes wears away, resulting in a condition called arthritis. People with arthritis feel pain when they move.

Figure 21-4

As a person matures, the jawbone becomes larger. Osteoclasts break down bone on the inner part of the jaw. Osteoblasts add bone to the outer part. *Why is this important when your permanent teeth replace your baby teeth?*

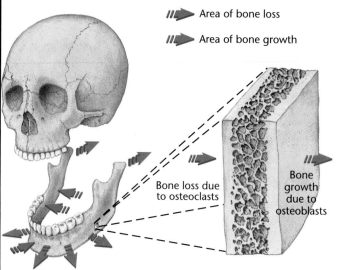

Area of bone loss

Area of bone growth

Bone loss due to osteoclasts

Bone growth due to osteoblasts

Bone Development

Before you were born, your skeleton was first made in the form of cartilage. Gradually, the cartilage was broken down and replaced by bone-forming cells called osteoblasts (AHS tee oh blastz). These cells deposit calcium and phosphorus that make bone tissue hard. At birth, your skeleton was made up of more than 300 bones. As you developed, some bones fused, or grew together.

Healthy bone tissue is dynamic. **Figure 21-4** shows how it is always being formed and re-formed. Osteoblasts are bone cells that build up bone. A second type of bone cell, called an osteoclast, breaks down bone tissue in other areas.

576 Chapter 21 Bones, Muscles, and Skin

This is a normal process in a healthy person. When osteoclasts break down bone, calcium and phosphorus are released. This process keeps the calcium and phosphorus in your bloodstream at about the same level.

Where Bones Meet

Think of the different actions you performed this morning. You opened your mouth to yawn, chewed your breakfast, reached for a toothbrush, and stretched out your arm to turn the doorknob as you walked out the door. All these motions were possible because your skeleton has joints. Any place where two or more bones meet is a **joint**. A **ligament** is a tough band of tissue that holds bones together at joints. Many joints, such as your knee, are held together by more than one ligament.

Problem Solving

Shape Influences Strength

When designing a building, architects know that certain shapes provide more support. Look closely at the shapes of the Haversian systems found in the cross section of a bone. Does the design of this system add to the overall strength of a bone? To find out, build two different support structures and see which is stronger.

Solve the Problem:

1. **Determine what geometric design is formed by Haversian systems.**
2. **Use two flat disks of modeling clay, plastic drinking straws, and two wooden boards to build the two different support structures. Insert the straws, in random order, in one clay disk. Lay a board on top. Put pressure on the board until the structure collapses.**
3. **Repeat the process, but this time arrange the straws in the same design as the Haversian systems.**

Think Critically:

Think about which structure withstood more pressure and how this relates to the strength of your bones.

1. What geometric shape provides strength for your bones?
2. Compact bone is made of a series of Haversian systems. What is the impact of several of these structures on the strength of the whole bone?
3. Suppose paper were wrapped around the straws in the second structure. What impact would this have on the strength of the structure? What part of an actual bone does the paper represent?

577

Types of Joints

Joints are classified as immovable or movable. Refer to **Figure 21-5** as you learn about different types of joints. An immovable joint allows little or no movement. The joints of the bones of your skull and pelvis are classified as immovable. A **movable joint** allows the body to make a wide range of movements. Gymnastics and working the controls of a video game require movable joints. There are several types of movable joints: pivot, ball-and-socket, hinge, and gliding joints. In a pivot joint, one bone rotates in a ring of another stationary bone. Turning your head is an example of a pivot movement. In a ball-and-socket joint, one bone has a rounded end that fits into a cuplike cavity on another bone. This provides a wider range of movement. Thus, your hips and shoulders can swing in almost any direction.

A third type of joint is a hinge joint. This joint has a back-and-forth movement like hinges on a door. Elbows, knees, and fingers have hinge joints. While hinge joints are less flexible than the ball-and-socket, they are more stable. They are not as easily dislocated, or put out of joint, as the shoulder. A fourth type of joint is a gliding joint, where one part of a bone

Figure 21-5
When a pitcher winds up, several types of joints are in action.

Shoulder

Ball-and-socket joint

Knee

Hinge joint

Skull

Arm

Vertebrae

Pivot joint

Gliding joint

578 Chapter 21 Bones, Muscles, and Skin

slides over another bone. Gliding joints move in a back-and-forth motion and are found in your wrists and ankles and between vertebrae. Gliding joints are the most frequently used joints in your body. You can't write a word, pick up a sock, or take a step without using a gliding joint.

Making a Smooth Move

Think about what happens when you rub two pieces of chalk together. Their surfaces begin to wear away. Without protection, your bones will also wear away at the joints. Recall that cartilage is found at the ends of the humerus. Cartilage helps make joint movements easier. It covers the ends of bones in movable joints. It reduces friction and allows the bones to slide over each other more easily. The joint is also lubricated by a fluid that comes from nearby capillaries. Pads of cartilage called disks are also found between the vertebrae. Here, cartilage acts as a cushion and prevents injury to your spinal cord. **Figure 21-6** shows how a damaged joint can be replaced.

Your skeleton is a living framework in the form of bones. Bones not only support the body but also supply it with minerals and blood cells. Joints are places between bones that enable the framework to be flexible and to be more than just a storehouse for minerals.

Figure 21-6

Some joints damaged by arthritis can be replaced with artificial joints. *How would this kind of surgery help a person stay active?*

Section Wrap-up

Review

1. What are the five major functions of a skeleton?

2. Name and give an example of a movable joint.

3. What are the functions of cartilage?

4. **Think Critically:** A thick band of bone forms around a healing broken bone. In time, the thickened band disappears. Explain how this extra amount of bone can disappear.

Skill Builder

Comparing and Contrasting

Compare and contrast the type and range of movement for a ball-and-socket joint and a gliding joint. If you need help, refer to Comparing and Contrasting in the **Skill Handbook.**

USING MATH

Calculate the percent of bones for each of these skeletal divisions. Remember the adult skeleton has 206 bones: skull—29 bones; vertebral column—26 bones; rib cage—25 bones; shoulder—4 bones; arms and hands—60 bones; hip—2 bones; legs and feet—60 bones.

21-1 The Skeletal System   **579**

Activity 21-1

Purpose

IS **Interpersonal** Design and carry out an experiment to identify and describe the parts of a beef and turkey bone. Identify the characteristics of each bone that make it adapted for movement on land or in the air. **L1**

COOP LEARN

Process Skills

observing, inferring, communicating, making and using tables, comparing and contrasting, forming a hypothesis, designing an experiment, separating and controlling variables, interpreting data

Time

45-50 minutes

Materials

Obtain the materials for each group. Have a butcher cut the bones in half lengthwise. If possible, provide some cross-sections of each bone type.

Safety Precautions

Caution students to use extreme care when using sharp instruments such as a scalpel. Care should be taken when handling the cut bones, since the edges may be sharp.

Possible Hypothesis

Students will most likely state that bones of flying animals will be lighter in mass than those of land animals.

📁 **Activity Worksheets,** pp. 5, 124-125

Activity 21-1

Design Your Own Experiment
Observing Bones

A skeleton aids movement. To move, animals must overcome the force of gravity. Land animals need skeletons that provide support against gravity. A flying animal needs a skeleton that provides support yet also allows it to overcome the pull of gravity and fly. Bones are adapted to the functions they perform. Find out if there is a difference between the bones of a land animal and those of a flying animal.

PREPARATION

Problem

Your task is to compare the bones of flying animals with those of animals that move on land. How does the type of movement affect the bone structures of these animals?

Form a Hypothesis

State a hypothesis that predicts how the bone structures of land-dwelling and flying animals are different. When making your hypothesis, consider the role gravity plays in hampering movement.

Objectives

- Design an experiment that reveals the difference between the bones of a land animal and those of a flying animal.
- Observe the differences in bone structure.

Possible Materials

- beef bone (cut in half and lengthwise)
- turkey leg bone (cut in half and lengthwise)
- hand lens
- scalpel

Safety Precautions

Use extreme care when using sharp instruments. Make sure to wash your hands with soap after completing this experiment.

580   Chapter 21 Bones, Muscles, and Skin

Inclusion Strategies

Gifted Domestic turkeys do not fly. Wild turkeys fly for short distances. Assign students the task of finding out why turkeys are such poor fliers. **L3**

PLAN THE EXPERIMENT

1. As a group, agree upon and write out the hypothesis statement.
2. As a group, list the steps that you need to take to test your hypothesis. Consider the following factors: How does mass affect the amount of force needed to move an object? What type of shape provides strength yet reduces mass?
3. List your materials.
4. Design a data table and record it in your Science Journal so that it is ready to use as your group collects data.

Check the Plan
1. Read over your entire experiment to make sure that all steps are in logical order.
2. Identify any constants, variables, and controls of the experiment.
3. *Make sure your teacher approves your plan before you proceed.*

DO THE EXPERIMENT

1. Carry out the experiment as planned.
2. While the experiment is going on, write down any observations that you make and complete the data table in your Science Journal.

Analyze and Apply
1. **Compare** your results with those of other groups.
2. In what ways were the different types of bone tissue similar?

3. **Infer** which type of bone would require more force to move. Explain why.
4. Does your data indicate any adaptations for flight in the bones? **Predict** the bone structure of birds known for their soaring and gliding flight.

Go Further

Find out whether bone structure or body shape is more important for movement of animals living in water.

Possible Procedures
One possible procedure is to compare the masses of the bones and also their internal structures.

Teaching Strategies
Tying to Previous Knowledge Students may be more familiar with chicken bones, but because they are less compact they do not reveal the internal structures as easily as beef or turkey bones.

DO THE EXPERIMENT

Expected Outcomes
Results will reveal that the beef bone is thicker, more compact, and has a marrow-filled center.

Analyze and Apply
1. Student results should be consistent with those of other groups.
2. Both bones have a periosteum, compact bone, and spongy bone.
3. The denser, more compact beef bone requires more force to move than the lighter-weight turkey bone.
4. The lighter weight of the turkey bones is an adaptation to flight. Birds known for soaring and gliding flight have very light bones.

Go Further

A sleek body shape provides a way for water animals to move quickly through the water. This may be more important than the bone structure because the water buoys up the animal's body.

Process Show students pictures of the skeletons of a penguin and a hawk. Have them identify which bird is probably a good flier. Use the Performance Task Assessment List for Scientific Drawing in **PASC**, p. 55. **P**

Prepare

Section Background

Discuss with the class that, although bones have strength and flexibility, they can fracture. The repair of a break can be accelerated using a newly developed biomaterial.

1 Motivate

Bellringer

Before presenting the lesson, display **Section Focus Transparency 77** on the overhead projector. Assign the accompanying **Focus Activity** worksheet. L1 LEP

Tying to Previous Knowledge

Have students recall the role of osteoclasts in bone development.

2 Teach

Discussion

LM **Logical-Mathematical** Compare the step-by-step process of the standard method for treating a bone fracture with the new method.

Science Words

- fracture
- biomaterial
- osteoporosis

Objectives

- Describe the source of the hardness of bone.
- Compare the standard method of bone fracture treatment with the biomaterial injection method.
- Discuss some methods to prevent bone fractures.

Bone Strength

Compact bone is as hard as some kinds of rocks. This hardness is due to minerals. The major mineral in bone is hydroxyapatite, which contains both calcium and phosphorus. The minerals calcium hydroxide and calcium carbonate are also found in bone. These minerals and protein fibers make bone both strong and flexible.

Broken Bones

Although bones are strong and flexible, they can break. A break in a bone is called a **fracture.** A simple fracture occurs when a bone has broken, but the broken ends do not cut through the skin. In a compound fracture, the broken ends of the bone stick out through the skin.

The standard treatment for bone fractures is to "set" the break. This means the ends of the bone are brought into contact with each other and aligned. A cast is put over the broken area to keep it from moving. During the healing process, shown in **Figure 21-7,** the periosteum starts to make new bone cells. A thick band of new bone forms around the break, acting like a built-in splint. Over time, the thickened band disappears as bone is reshaped with the help of osteoclasts. The whole healing process may take weeks or months.

Figure 21-7

A series of steps occurs as a bone heals.

A 1st few days
Spongy bone forms where the break occurs.

B 1-2 weeks
Blood vessels regrow between bone sections. Spongy bone fills in and hardens.

C 2-3 months
The bone is almost healed.

Program Resources

 Reproducible Masters
Study Guide, p. 81 L1
Reinforcement, p. 81 L1
Enrichment, p. 81 L3
Critical Thinking/Problem Solving, p. 25

 Transparencies
Section Focus Transparency 77 L1

Biomaterial Fracture Treatment

A new treatment may reduce the time in the hospital for hip, wrist, ankle, vertebrae and other fractures. This treatment involves the use of a newly developed biomaterial. **Biomaterials** are artificial materials that can be used to replace body parts. With the new treatment, the doctor first aligns the bones. Then the biomaterial is mixed into a soft paste and injected into the fractured bone area. **Figure 21-8** shows a fractured bone that could be treated with this technology.

In about 15 minutes, the paste hardens into a naturally occurring mineral of the human skeleton. After 12 hours, the biomaterial is as hard as the regular compact bone. Because the material is so bonelike, the normal repair process takes place more quickly. Bone cells, blood vessels, and bone tissue develop, and the healing is completed in as much as half the regular time. Trial testing of this new treatment has recently started in the United States.

Who is at risk?

Approximately 250 000 people over the age of 45 will break a hip this year. Seventy percent of all bone fractures in this age group are due to osteoporosis. **Osteoporosis** is a disease that breaks down bone, causing it to become brittle and weak. Some causes of this disease are a diet low in calcium, consuming too much caffeine, and getting too little exercise.

A simple fracture can be both a medical and financial problem. In addition to spending a long time recuperating, the patient may also spend much money for treatment. This new method of treatment may help ease both problems.

Figure 21-8

It normally takes several months for the bones to grow back together in a break such as this. The use of a biomaterial can speed this process. *Why is it important for it to form a material that is found naturally in the body?*

Section Wrap-up

Review

1. What makes bone hard?

2. What is the first step in both treatment methods for a fracture?

3. What can you do now as a teenager that will prevent bone fractures when you become a senior citizen?

Explore the Technology

If a person injures a hip, what would be the advantages of using the biomaterial treatment?

*inter*NET CONNECTION

Researchers have developed a new therapy for hard-to-heal bone fractures. Visit the Chapter 21 Internet Connection at Glencoe Online Science, **www. glencoe.com/ sec/science/life**, for a link to more information about gene therapy.

SCIENCE & SOCIETY

21-2 Biomaterial—Speeding Bone Fracture Recovery **583**

583

Prepare

Section Background

- All the movements of the body are the result of a unique specialization of muscle tissue known as contractility.
- A nerve impulse traveling to a muscle causes chemicals called neurotransmitters to be released at the junction of the nerve and muscle, causing the muscle to contract.
- Muscle cells respond to any stimulus by contracting to their maximum; there is no partial contraction. This phenomenon is known as the all-or-none principle.
- Muscle activity produces heat, which the circulatory system moves to other body tissues to help maintain the body temperature.

Preplanning

- To prepare for the Mini-LAB, obtain charts of muscle anatomy.
- For Activity 21-2, obtain the number of cooked turkey legs required and the slides of skeletal, smooth, and cardiac muscles.

1 Motivate

Bellringer

Before presenting the lesson, display **Section Focus Transparency 78** on the overhead projector. Assign the accompanying **Focus Activity** worksheet. L1 LEP

Tying to Previous Knowledge

Everyone has had sore, overworked muscles. Have students recall that the soreness often is in muscles that aren't conditioned.

584

21•3 The Muscular System

Science Words

muscle
voluntary muscle
involuntary muscle
skeletal muscle
tendon
smooth muscle
cardiac muscle

Objectives

- Describe the major function of muscles.
- Compare and contrast three types of muscles.
- Explain how muscle action results in movement of body parts.

Moving the Human Body

A driver uses a road map to find out what highway connects two cities within a state. You can use the picture in **Figure 21-9** to find out which muscles connect some of the bones in your body. A **muscle** is an organ that contracts and gets shorter. This contraction, or pull, provides the force to move your body parts. In the process, energy is used and work is done. Imagine how much energy is used by the more than 600 muscles in your body each day. No matter how still you might try to be, there is always movement taking place in your body.

Muscle Control

Muscles that you are able to control are called **voluntary muscles.** Your arm and leg muscles are voluntary. So are the muscles of your hands and face. You can choose to move them or not to move them. In contrast, **involuntary muscles** are muscles you can't consciously control. You don't have to decide to make these muscles work. They just go on working all day long, all your life. Blood gets pumped through blood vessels, and food is moved through your digestive system by the action of involuntary muscles. You can sleep at night without having to think about how to keep these muscles working.

Figure 21-9

Nearly 35-40 percent of your body's mass is muscle tissue. *Do you think most of your muscles are voluntary or involuntary?*

Program Resources

Reproducible Masters

Study Guide, p. 82 L1
Reinforcement, p. 82 L1
Enrichment, p. 82 L3
Activity Worksheets, pp. 5, 126-127, 128 L1
Multicultural Connections, pp. 45-46
Lab Manual 50

Transparencies

Section Focus Transparency 78 L1
Science Integration Transparency 21
Teaching Transparency 42

Levers—Your Body's Simple Machines

Your skeletal system and muscular system work together to move your body like a machine. A machine is a device that makes work easier. A simple machine does work with only one movement. The action of muscles on bones and joints often works like one type of simple machine called a lever. A lever is defined as a rigid bar that moves on a fixed point, known as a fulcrum. **Figure 21-10** shows a common lever being used to open a can of paint. In your body, bones are the bar, joints are the fulcrum, and muscles provide the effort force to move the body part.

Levers are classified into three types. Examples of the three types of levers are shown in **Table 21-1.**

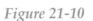

Figure 21-10

The screwdriver is used as a lever to open the can. You push down on the screwdriver and provide a force. The end of the screwdriver forces the lid up. *What is the fulcrum in this lever?*

Table 21-1

Types of Levers in the Human Body			
Lever Type	**First Class Lever** A first-class lever has the fulcrum between the load and the effort force.	**Second Class Lever** A second-class lever has the load between the fulcrum and the effort force.	**Third Class Lever** A third-class lever has the effort force between the fulcrum and the load.
Body Equivalent	In your body, the skull pivots on the top vertebra as neck muscles raise your head up and down.	Raising the body up on your toes occurs when the leg muscles pull on the leg and foot.	Your arm works like a lever when you pick up an object with your hand. The contracting biceps muscle is the force that moves the bones of your forearm at the elbow.

21-3 The Muscular System   **585**

2 Teach

Inquiry Question

Why aren't all muscles voluntary? *It would be impossible for a person to think about all the body functions that require muscle action—breathing, heart beating, stomach churning, etc.*

INTEGRATION
Physics

Levers are one kind of simple machine. Machines make work easier by decreasing the amount of force, transferring the amount of force, or changing the direction of force.

Use **Science Integration Transparency 21** as you teach this lesson.

NATIONAL GEOGRAPHIC SOCIETY

Videodisc
STV: Human Body Vol. 2
Muscular and Skeletal Systems
Unit 1, Side 1
Skeletal Muscles

21752-24658
How Muscles Work

24659-28863
Conclusion

28864-33550

Visual Learning

Figure 21-9 Do you think most of your muscles are voluntary or involuntary? *voluntary*

Figure 21-10 What is the fulcrum in this lever? *the rim of the can* **LEP** **LS**

Theme Connection

Energy Contracting muscles require the chemical energy from a simple sugar, glucose. The chemical energy is transformed into the mechanical energy of muscle activity. **What foods are readily converted to glucose within the body?** *Carbohydrates (sugars and starches) are oxidized into glucose for use in body activities.*

MiniLAB

Purpose

 Kinesthetic Students will observe the coordinated work of muscle pairs. **L1**
LEP

Materials
muscle anatomy chart

Teaching Strategies
Have students feel the muscles in their arms as they contract and relax.

Analysis
1. two—biceps and triceps
2. the biceps

Assessment

Process Have students use a chart of the muscular system and estimate the percent of body mass that is muscle tissue. Use the Performance Task Assessment List for Using Math in Science in **PASC**, p. 29. **P**

 Activity Worksheets, pp. 5, 128

Table 21-2

Types of Muscle		
Involuntary		**Voluntary**
Smooth muscle	**Cardiac muscle**	**Skeletal muscle**
Appearance: Smooth Location: Internal organs	Appearance: Striped Location: Only the heart	Appearance: Striped Location: Attached to bone
Small intestine	Heart	Biceps muscle

Types of Muscle Tissue

Three types of muscle tissue are found in your body: skeletal, smooth, and cardiac. **Skeletal muscles** are the muscles that move bones. They are attached to bones by thick bands of tissue called **tendons.** Skeletal muscles are the most numerous muscles in the body. When viewed under a microscope, skeletal muscle cells look striped, or striated. You can see the striations in **Table 21-2.** Skeletal muscles are voluntary muscles. You can control their use. You choose when to walk or not to walk. Skeletal muscles tend to contract quickly and tire easily.

The remaining two types of muscles, shown in **Table 21-2,** are involuntary. **Smooth muscles** are nonstriated, involuntary muscles that move many of your internal organs. Your intestines, bladder, and blood vessels are made of one or more layers of smooth muscles. These muscles contract and relax slowly.

USING MATH

Because of regular exercise during the track season, Hope was able to reduce her at-rest pulse rate from 76 beats per minute to 66. Estimate how many beats Hope is saving her heart muscle per day.

586 Chapter 21 Bones, Muscles, and Skin

 Use these program resources as you teach this lesson.
Multicultural Connections, pp. 45-46
Lab Manual 50

Use **Teaching Transparency 42** as you teach this lesson.

Cardiac muscle is found only in the heart. Like smooth muscle, cardiac muscle is also involuntary. As you can see from **Table 21-2**, cardiac muscle has striations like skeletal muscle. Cardiac muscle contracts about 70 times per minute every day of your life. You know each contraction as a heartbeat, but it isn't something you can control.

Muscles at Work

Skeletal muscle movements are the result of pairs of muscles working together. When one muscle of a pair contracts, the other muscle relaxes, or returns to its original length. For example, when the muscles on the back of your upper leg contract, they pull your lower leg back and up. Muscles always pull; they never push. When you straighten your leg, the back muscles relax and the muscles on the front of your upper leg contract. Compare how the muscles of your legs work with how the muscles of your arms work in **Figure 21-11**.

Muscle Action and Energy

When you straightened your leg, your muscles used energy. Muscles use chemical energy in the form of glucose. As the bonds in glucose break, chemical energy changes to mechanical energy and the muscle contracts. When the supply of glucose in a muscle is used up, the muscle becomes tired and needs to rest. During the resting period, the muscle is resupplied with glucose. Muscles also produce thermal energy when they contract. The heat produced by muscle contraction helps to keep your body temperature constant.

MiniLAB

How do muscle pairs work?

Find out which muscles are used to move your arm.

Procedure
1. Stretch your arm out straight. Bring your hand to your shoulder, then down again.
2. Using a muscle chart, determine which skeletal muscles in your upper arm enable you to perform this action.

Analysis
1. How many muscles were involved in this action?
2. Which muscle contracted to bring the forearm closer to the shoulder?

Figure 21-11

When the biceps of the upper arm contract, the lower arm moves upward. When the tricep muscles on the back of the upper arm contract the lower arm moves down.

Biceps contract

Triceps relax

Biceps relax

Triceps contract

21-3 The Muscular System   **587**

NATIONAL GEOGRAPHIC SOCIETY

Videodisc
STV: The Cell
Muscle-cell bands

46932
STV: Human Body Vol. 2
Smooth muscle

48671

3 Assess

Check for Understanding

? FLEX Your Brain

Use the Flex Your Brain activity to have students explore MUSCLE.

Activity Worksheets, p. 5

Reteach

Have students make flash cards with names of body organs that require muscle action. On the reverse side of each card, identify the type of muscle (skeletal, smooth, or cardiac) involved.

Extension

For students who have mastered this section, use the **Reinforcement** and **Enrichment** masters.

4 Close

CONNECT TO
CHEMISTRY

Muscle cells contain a protein called myoglobin that is similar to hemoglobin in red blood cells. Hemoglobin is a red pigment that carries oxygen. From this information, what can you *infer* about myoglobin?

Over a period of time, muscles can become larger or smaller, depending on whether or not they are used. You can see in **Figure 21-12** that skeletal muscles that do a lot of work, such as those in your writing hand or in the arms of a bricklayer, become large and strong. In contrast, if you just sit and watch TV all day, your muscles will become soft and flabby, and will lack strength. Muscles that aren't exercised become smaller in size.

Figure 21-12
The number of muscles in an adult does not increase with exercise and body building. The cells simply get larger.

Section Wrap-up

Review

1. What is the function of the muscular system?

2. Compare and contrast the three types of muscle.

3. Describe how a muscle attaches to a bone.

4. **Think Critically:** What happens to your upper arm muscles when you bend your arm at the elbow?

Skill Builder
Sequencing
Sequence the activities that take place when you bend your leg at the knee. If you need help, refer to Sequencing in the **Skill Handbook.**

Science Journal

Write a paragraph in your Science Journal identifying the three forms of energy involved in a muscle contraction. Describe what relationship this has to being a warm-blooded animal.

Observing Muscle

You learned that muscles can be identified by their appearance. In this activity, you will make observations to distinguish among the three types of muscle tissue.

Problem
What do different types of muscle look like?

Materials
- prepared slides of smooth, skeletal, and cardiac muscles
- microscope
- cooked turkey leg
- dissecting pan or cutting board
- dissecting probes (2)
- hand lens

Procedure
Part A
1. Copy the data table and use it to record your observations.
2. Using the microscope, first on low power and then on high power, observe prepared slides of three different types of muscle.
3. In the data table, draw each type of muscle that you observe.

Part B
4. Muscle tissue is made up of groups of cells held together in fibers, usually by a transparent covering called connective tissue. Obtain a piece of cooked turkey leg from your teacher. On the cutting board, use the two probes to tease the muscle fibers apart.
5. Use a hand lens to examine the muscle fibers and any connective tissue you see in the turkey leg.
6. Draw and measure five turkey leg fibers and describe the shape of these muscle fibers.

Data and Observations   Sample Data

Skeletal	Cardiac	Smooth
Student drawing should resemble photos on page 586.		
Length of fibers	Lengths will vary.	

Length of fibers: 1 cm | 2.5 cm | 1.5 cm | 0.75 cm | 1 cm

| Description of fibers | Fibers are long, tapered at ends. | |

Analyze
1. Which muscles on the slides have stripes, or striations?
2. What type of muscles do you expect to find in a turkey leg? Explain.
3. How are muscle fibers arranged in the prepared slides and in the turkey leg?

Conclude and Apply
4. **Predict** how the shape of a muscle fiber relates to its function.
5. Can you **conclude** that striations have anything to do with whether a muscle is voluntary or involuntary? Explain.

21-3 The Muscular System   589

Assessment

Performance To further assess students' understanding of muscle, have them examine a piece of turkey heart and describe the characteristics of the muscle. Use the Performance Task Assessment List for Making Observations and Inferences in **PASC**, p. 17. **P**

Purpose
IS **Visual-Spatial** Students will observe and identify parts of skeletal muscle and describe how muscle is attached to bone. **L1** **LEP**

Process Skills
observing, identifying, measuring, inferring

Time
45-50 minutes

Safety Precautions
Caution students to be careful handling the dissecting probe and to wash their hands when the activity is completed.

Activity Worksheets, pp. 5, 126-127

Teaching Strategies
Alternate Materials Cooked chicken legs can be used, but it may be more difficult to see certain parts.

Troubleshooting Caution students to use probes carefully when tearing the turkey tissue. They should spread out the tissue as finely as possible and notice the tapered ends of the muscle.

Answers to Questions
1. skeletal and cardiac muscles
2. Skeletal muscles; skeletal muscles are used to move the turkey's legs.
3. Accept all reasonable descriptions. Most will reply that the fibers lay side by side along the length of the bone.
4. The fiber is long, thicker in the center than it is at the ends. When a muscle contracts, it becomes even thicker in the center, but the ends stay about the same.
5. No. Both cardiac and skeletal are striated, but cardiac is involuntary and skeletal is voluntary.

Science & Literature

Science & Literature

Source

Mary Shelley, *Frankenstein, or, The Modern Prometheus,* 1818.

Biography

Mary Shelley (1797-1851) is best known for her classic book, *Frankenstein, or, The Modern Prometheus,* perhaps the most widely known horror novel. Mary Wollstonecraft Godwin Shelley, born in London in 1797, was the daughter of philosopher William Godwin and feminist writer Mary Wollstonecraft. With no formal training in writing, Mary began *Frankenstein* in 1816 at the age of nineteen. The classic tale of horror and romance has since been popularized by many motion pictures and television programs. Shelley wrote several other books during her lifetime, including her historical novel *Valperga: or, The Life and Adventures of Castruccio, Prince of Lucca* (1823), and *The Last Man* (1826), considered by many to be her best novel.

Teaching Strategies

LS **Linguistic** Have students discuss the analogy of the human body as a living machine.

LS **Kinesthetic** Have students construct simple marionette puppets of cardboard and yarn. Students will gain understanding of how the skeletal and muscular systems work by trying to simulate the major joints and muscle groups of the body.

LS **Linguistic** Initiate a discussion about famous athletes who can perform exciting and dazzling feats. Have students write a short essay about the athlete that most impresses them. Ask students to write about the reasons why they are impressed with the athlete's skills.

Frankenstein
by Mary Shelley

In this chapter, you've studied the structure of the bones and muscles of your own body. Many authors have written about the wondrous beauty and detailed complexity of the human form. In the classic book *Frankenstein,* author Mary Shelley pays tribute to the science of anatomy in the story of a scientist who creates a living being from lifeless body parts. In this excerpt from the book, Shelley details how Dr. Frankenstein slowly pieces together the Monster.

… As the minuteness of the parts formed a great hindrance to my speed, I resolved, contrary to my first intention, to make a being of a gigantic stature, that is to say, about eight feet in height, and proportionately large. After having formed this determination and having spent some months in successfully collecting and arranging my materials, I began.…

… His limbs were in proportion, and I had selected his features as beautiful. Beautiful!… His yellow skin scarcely covered the work of muscles and arteries beneath; his hair was of a lustrous black, and flowing; his teeth of a pearly whiteness; but these luxuriances only formed a more horrid contrast with his watery eyes, that seemed almost of the same colour as the dun white sockets in which they were set, his shriveled complexion and straight black lips.

Science Journal

Choose a living organism. Imagine you are a scientist trying to describe your organism to other scientists. Try to paint a vivid picture of the organism by describing its features in detail in your Science Journal.

Shelley paints a vivid picture of the Monster. In chilling detail, she describes its body form and facial features, which are both horrifying and yet beautiful because it is a working machine. Her description of the Monster's face creates the image of a technician piecing together a complex machine.

Many people agree that the human body is a sort of living machine. Mary Shelley's vivid descriptions have been compared to scientific writing. Which parts of this excerpt do you think would be similar to the writing of someone detailing the structure of the human body?

Classics

- Brooke, Peter, "Godlike Science/Unhallowed Arts: Language, Nature, and Monstrosity," in *The Endurance of Frankenstein,* George Levine and U.C. Knoepflmacher (eds.), University of California Press, 1979. This article examines Shelley's scientific-style writing.

- Gould, Steven Jay. *The Panda's Thumb.* New York: W.W. Norton, 1980. In this collection of essays, the author emphasizes the important evolutionary concept of adaptation.

Gould analyzes various controversies and questions in science and explains how all organisms, including humans, have evolved traits that enable them to survive and perform various behaviors in a given environment.

Other Works

Other works by Mary Shelley: "The Heir of Mondolfo" in *Seven Masterpieces of Gothic Horror,* edited by Robert and Donald Spector. New York: Bantam Books, 1963.

Skin

The Body's Largest Organ

Your skin is the largest organ of your body. Much of the information you receive about your environment comes through your skin. Skin is made up of two layers of tissue, the epidermis and the dermis. You can see in **Figure 21-13** that the **epidermis** is the surface layer of your skin. The cells on the top of the epidermis are dead. Thousands of these cells rub off every time you take a shower, shake hands, blow your nose, or scratch your elbow. New cells are constantly produced at the bottom of the epidermis that eventually replace the ones that are rubbed off. Cells in the epidermis produce the chemical melanin. **Melanin** (MEL uh nun) is a pigment that gives your skin color. The more melanin, the darker the color of the skin. Melanin increases when your skin is exposed to the ultraviolet rays of the sun. If someone has very few melanin-producing cells, very little color gets deposited. These people have less protection from the sun. They burn more easily and may develop skin cancer more easily.

The **dermis** is the layer of tissue under the epidermis. This layer is thicker than the epidermis and contains many blood vessels, nerves, and oil and sweat glands. Notice in **Figure 21-13** that fat cells are located under the dermis. This fatty tissue insulates the body. When a person gains too much weight, this is also where much of the extra fat is deposited.

Science Words
epidermis
melanin
dermis

Objectives
- Compare and contrast the epidermis and dermis of the skin.
- List the functions of the skin.
- Discuss how skin protects the body from disease and how it heals itself.

Figure 21-13

Hair, nails, and sweat and oil glands are all part of your body's largest organ.

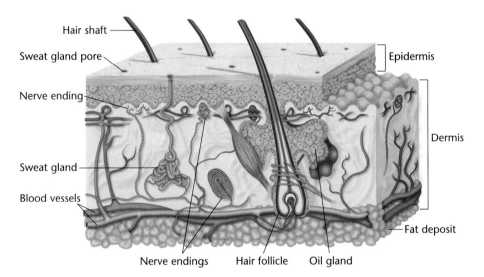

Hair shaft
Sweat gland pore
Epidermis
Nerve ending
Dermis
Sweat gland
Blood vessels
Fat deposit
Nerve endings   Hair follicle   Oil gland

21-4 Skin   591

Prepare

Section Background
- Fingerprints are ridges on the surface of the skin due to the ridges in the upper surface of the dermis.
- Hair develops from cells in the base of the hair follicle.

Preplanning
To prepare for the MiniLAB, obtain liter-sized, clear plastic bags and tape.

1 Motivate

Bellringer
 Before presenting the lesson, display **Section Focus Transparency 79** on the overhead projector. Assign the accompanying **Focus Activity** worksheet. [L1] [LEP]

Tying to Previous Knowledge
Recall the characteristics of mammals from Chapter 16. One characteristic is a hair-covered body. Humans lack this extensive hair covering.

NATIONAL GEOGRAPHIC SOCIETY

 Videodisc
STV: Human Body Vol. 3
Skin (close-up)

48771
Skin, magnified

48863

Program Resources

 Reproducible Masters
Study Guide, p. 83 [L1]
Reinforcement, p. 83 [L1]
Enrichment, p. 83 [L3]
Activity Worksheets, pp. 5, 129 [L1]

Transparencies
Section Focus Transparency 79 [L1]

Demonstration

 Kinesthetic Have students experience the cooling effect of evaporation of water from the skin. Each student should pat a wet cotton ball onto the wrist. Then have them gently blow on the wet surface. The evaporation removes body heat. Relate this experience to the evaporation of perspiration from the skin during hot weather. **LEP**

Videodisc
STV: Human Body Vol. 3
Immune System
Unit 1, Side 1
Introduction

00557-07510
First Defense—The Skin

07510-10160

USING TECHNOLOGY

Robot Skin

Think about all of the different objects you pick up in a day. Clues from the look and feel of the objects let you know if you need to use a gentle touch or a firm grip to pick up and hold them. Industrial robots are programmed to use a specific force to pick up an object. So long as they are built to handle a specific type of object, this works well. But, what if a robot had sense organs? Then it could adjust itself to doing different tasks.

Scientists are developing an artificial skin for robots that can sense the difference between a rock and a tomato and allow the robot to handle each correctly. The artificial skin is modeled after human skin. The outer layer is made of electrode sheets with gel between them. This layer reads the amount of pressure placed on an object. The inner layer consists of a layer of sensors between two rubber sheets. The sensors are made of a material that expands and contracts when placed in an alternating current. Thus, the sensors are sensitive to pressure and can detect very fine textures, friction, and surface irregularities.

Right now, scientists are looking for a way to make the artificial skin without making it too thick.

Think Critically:
Explain why artificial skin should not be too thick.

Robotic skin

Sensor layer · Inner layer · Outer layer · Rubber sheets · Electrode · Gel

---

Skin at Work

Your skin is not only the largest organ of your body, it also carries out several major functions. Among these functions are: protection, sensory response, formation of vitamin D, regulation of body temperature, and the excretion of wastes. Of these functions, the most important is that skin forms a protective covering over the body. As a covering, it prevents both physical and chemical injury as well as disease. Glands in the skin secrete fluids that damage or destroy some bacteria. Skin also prevents excess water loss from body tissues.

The skin also serves as a sensory organ. Specialized nerve cells in the skin detect and relay information about temperature, pressure, and pain. Some of the sensors are shown in **Figure 21-13.** Because of these sensors, you are

---

Community Connection

Industrial Robots Have students survey local manufacturing facilities to find out how many use industrial robots. Arrange for students to visit one facility to see how the robot is used. **L1**

Cultural Diversity

Tattoos A tattoo is a permanent mark or design put under the skin with indelible dyes. Have students research in which cultures tattoos are common and what they signify. **L2**

able to detect the softness of a cat, the sharp point of a pin, or the heat of a frying pan.

Vitamin D is essential for your good health. A third vital function of skin is the formation of vitamin D. Small amounts of this vitamin are produced in the epidermis in the presence of ultraviolet light from the sun. Vitamin D is needed for your body to absorb calcium from the food you eat.

Skin: Heat and Waste Exchange

Your skin plays an important role in helping to regulate your body temperature. Humans, unlike the fur-bearing animals in **Figure 21-14,** have very little hair to help them regulate body temperature. Hair is an adaptation that usually helps control body temperature. In humans, blood vessels in the skin can help release or hold heat. If the blood vessels expand or dilate, blood flow increases and heat is released. Less heat is released when the blood vessels constrict.

There are about 3 million sweat glands in the dermis. These glands also help regulate the body's temperature. When the blood vessels dilate, pores leading to the sweat glands in the skin open. Perspiration, or sweat, moves out onto the skin. Heat moves from inside the body to the sweat on the body's surface. The body then cools as the sweat evaporates. This system balances heat produced by muscle contractions.

The fifth function of the skin is to excrete wastes. Sweat glands release water, the salt sodium chloride, and a protein product called urea. If too much water and salt are released during periods of extreme heat or physical exertion, you might faint.

MiniLAB

Do you always sweat?
Find out if you sweat, even during periods of inactivity.

Procedure
1. Put your hand in a clear, plastic bag and tape it closed.
2. Sit quietly for 15 minutes. Observe what happens in the bag.

Analysis
1. Identify what formed inside the bag. Where did this substance come from?
2. Why is it necessary for this substance to form, even when you are inactive?

Figure 21-14

All animals must be able to control body temperature. Insulation of an animal's body occurs when the fur is raised. *What happens to fur when the animal's body temperature is high?*

593

593

4 Close

Have students infer how humans would receive information about environmental conditions if they did not have skin. Use this discussion to summarize the functions of the skin.

Section Wrap-up

Review

1. The epidermis is the thin outer layer of skin. It contains melanin. The dermis is under the epidermis. It is thicker and contains many blood vessels, nerves, and sweat glands.

2. protects, regulates body temperature, excretes waste, serves as sensory organ, forms vitamin D

3. As long as it is unbroken, skin prevents pathogens from entering the body. Skin prevents water loss.

4. **Think Critically** because they have lost the protection for body tissues and also from loss of body fluids

---

Figure 21-15
When intact, your skin protects you from bacteria and viruses that cause infection. Infections are a major problem if large areas of skin are injured.

CONNECT TO
EARTH SCIENCE

Find out about the effects of ultraviolet radiation on skin. *Infer* why mountain climbers risk becoming severely sunburned even in freezing temperatures.

Injury to the Skin

The epidermis may be very thin, but injury to large areas of the skin can cause death. If the epidermis is burned or rubbed away as in **Figure 21-15,** there are no longer any cells left that can divide to replace this lost layer. In severe cases of skin loss or damage, nerve endings and blood vessels in the dermis are exposed. Water is lost rapidly from the dermis and muscle tissues. Body tissues are exposed to bacteria and to potential infection, shock, and death.

Section Wrap-up

Review

1. Compare and contrast the epidermis and dermis.

2. List the five functions of skin.

3. How does skin help prevent disease in the body?

4. **Think Critically:** Why is a person who has been severely burned in danger of death from loss of water?

Skill Builder
Concept Mapping
Make an events chain concept map to show how skin helps keep body temperature constant. If you need help, refer to Concept Mapping in the **Skill Handbook.**

USING MATH

Find out if nails on your different fingers grow at different rates. Measure the length of each nail (base to tip). Measure the lengths again in a week. Calculate the amount of growth for each finger. Is there a difference?

Skill Builder
Students' sequences should approximate the following: body becomes hot or warm; blood vessels expand; sweat moves out onto skin; body cools as sweat evaporates.

Assessment

Process Have students construct models of skin with and without fur to determine the insulating qualities of fur. Use the Performance Task Assessment List for Model in **PASC,** p. 51. P

Review

Summary

21-1: The Skeletal System

1. Bones are living structures that protect, support, make blood cells, store minerals, and provide for muscle attachment.
2. Joints are classified as either movable or immovable.

21-2: Science and Society: Biomaterial —Speeding Bone Fracture Recovery

1. Broken bones heal when the broken ends contact each other and the periosteum forms a band of cells.
2. A new treatment using a biomaterial paste may speed the healing process.

21-3: The Muscular System

1. Muscle contracts to move bones and body parts.
2. Skeletal muscle is voluntary and moves bones. Smooth muscle is involuntary and controls movement in internal organs. Cardiac muscle is involuntary and causes contractions of the heart.
3. Skeletal muscles work in pairs. When one contracts, the other relaxes.

21-4: Skin

1. The epidermis is the outer layer of skin. The dermis is the inner layer and contains hair follicles, nails, nerves, sweat and oil glands, and blood vessels.
2. Protection, water retention, formation of vitamin D, and helping with body temperature are the skin's jobs.

Key Science Words

a. biomaterial
b. cardiac muscle
c. cartilage
d. dermis
e. epidermis
f. fracture
g. immovable joint
h. involuntary muscle
i. joint
j. ligament
k. marrow
l. melanin
m. movable joint
n. muscle
o. osteoporosis
p. periosteum
q. skeletal muscle
r. skeletal system
s. smooth muscle
t. tendon
u. voluntary muscle

Reviewing Vocabulary

Match each phrase with the correct term from the list of Key Science Words.

1. tough outer covering of bone
2. internal body framework
3. tissue or organ that contracts
4. a broken bone
5. voluntary muscle
6. involuntary heart muscle
7. outer layer of skin
8. skin pigment
9. place where bones meet
10. attaches muscle to bone

Chapter 21 Review **595**

Summary

Have students read the summary statements to review the major concepts of the chapter.

Reviewing Vocabulary

1. p	**6.** b
2. r	**7.** e
3. n	**8.** l
4. f	**9.** i
5. q	**10.** t

✔ Assessment

Portfolio Encourage students to place in their portfolios one or two items of what they consider to be their best work. Examples include:
- Skill Builder, p. 579
- Explore the Technology, p. 583
- Science Journal, p. 588 P

Performance Additional performance assessments may be found in **Performance Assessment** and **Science Integration Activities.** Performance Task Assessment Lists and rubrics for evaluating these activities can be found in Glencoe's **Performance Assessment in the Science Classroom.**

GLENCOE TECHNOLOGY

 Videodisc

Glencoe Life Science Interactive Videodisc
Side 2, Lesson 7
Health Is Just a Heartbeat Away

18598 to 27445

MindJogger Videoquiz

Chapter 21 Have students work in groups as they play the Videoquiz game to review key chapter concepts.

Chapter 21 Review

Checking Concepts

1. a	**6.** b
2. d	**7.** c
3. a	**8.** c
4. a	**9.** b
5. d	**10.** c

Understanding Concepts

11. Ligaments connect bones to bones; tendons connect muscles to bones.

12. Blood supply to bone might be cut off; broken bones would have a difficult time healing or would not heal at all.

13. Cartilage allows a joint to operate smoothly. When the cartilage is damaged or destroyed, the harder inner parts of the bones rub together, the movement is rough, and irritation of that joint results in pain and inflammation. Surgery would relieve the irritation and pain.

14. Bone repair is more natural and much quicker.

15. Failure to sweat when the body is overheated can result in heatstroke, which can lead to death.

Thinking Critically

16. If there is not enough exposure to sunlight, not enough vitamin D would be produced.

17. Sunblocks "mask" the skin cells and prevent full penetration of the sun's rays to the melanin of these cells.

18. Bones lacking calcium become too flexible and misshapen. They are also more easily broken.

19. Smooth muscle is made of individual, spindle-shaped cells in layers; skeletal muscle is made of striped fibers; cardiac muscle is a branching, fiber-like structure.

596

Checking Concepts

Choose the word or phrase that completes the sentence.

1. _____ bone is the most solid form of bone.
   a. Compact   c. Spongy
   b. Periosteum   d. Marrow
2. Blood cells are made in the _____.
   a. compact bone   c. cartilage
   b. periosteum   d. marrow
3. Minerals are stored in _____.
   a. bone   c. muscle
   b. skin   d. blood
4. _____ cover(s) the ends of bones.
   a. Cartilage   c. Ligaments
   b. Tendons   d. Muscle
5. Immovable joints are found _____.
   a. at the elbow   c. in the wrist
   b. at the neck   d. in the skull
6. The knees and fingers are examples of a _____ joint.
   a. pivot   c. gliding
   b. hinge   d. ball-and-socket
7. Vitamin _____ is made in the skin.
   a. A   c. D
   b. B   d. K
8. Dead cells are found on the _____.
   a. dermis   c. epidermis
   b. marrow   d. periosteum
9. A new biomaterial will be used to speed up _____ repair.
   a. blood   c. nerve
   b. bone   d. muscle
10. _____ helps retain fluids in the body.
    a. Bone   c. Skin
    b. Muscle   d. A joint

Understanding Concepts

Answer the following questions in your Science Journal using complete sentences.

11. Distinguish between the functions of ligaments and tendons.

12. What would result if a person had a disease of the periosteum?

13. In arthritis, cartilage is frequently damaged. Why might joint replacement surgery be helpful?

14. From a doctor's point of view, what are the advantages of using biomaterial in treating bone fractures?

15. Predict what would happen if a person's sweat glands didn't produce sweat.

Thinking Critically

16. When might skin not be able to produce enough vitamin D?

17. What effects do sunblocks have on melanin?

18. What would lack of calcium do to bones?

19. Using a microscope, how could you distinguish among the three muscle types?

20. What function of skin in your lower lip changes when a dentist gives you Novocaine for a filling in your bottom teeth? Why?

20. Novocaine numbs the nerves and your skin can no longer sense stimuli such as pressure or pain. Novocaine numbs nerves in the dermis.

Developing Skills

21. Observing and Inferring Accept all reasonable answers. Flexible joints are generally thought to be helpful during birth.

22. Designing an Experiment The experiment should involve comparing both resting and exercising heart rates.

23. Hypothesizing Sweat glands are not evenly distributed on the skin. This is known because, when exercising, sweat is found in certain areas (face, underarms) and not others.

Chapter 21 Review

Developing Skills

If you need help, refer to the **Skill Handbook.**

21. **Observing and Inferring:** The joints in the skull of a newborn baby are flexible, whereas those of a 17-year-old have grown together tightly. Infer why the infant's skull joints are flexible.

22. **Designing an Experiment:** Design an experiment to compare the heartbeat of athletes and nonathletes in your class.

23. **Hypothesizing:** Make a hypothesis about the distribution of sweat glands throughout the body. Are they evenly distributed?

24. **Concept Mapping:** Construct an events chain concept map to describe how a bone heals.

25. **Hypothesizing:** When exposed to direct sunlight, dark-colored hair absorbs more energy than light-colored hair. Given this information, make a hypothesis as to the most adaptive colors for desert animals.

CALVIN AND HOBBES

ALL THAT FUR MUST BE STRICTLY ORNAMENTAL.

Chapter 21 Review 597

Performance Assessment

1. **Display:** Find out the differences among first-, second-, and third-degree burns. A local hospital's burn unit or fire department are sources of information on burns. Display pictures of each type of burn and descriptions of treatments on a three-sided, free-standing poster board.

2. **Group Work:** How do tanning booths give you a suntan? As a class project, research the safety of this equipment. Present class results and assemble a safety pamphlet that reflects concepts from this chapter.

3. **Letter:** Write a letter to a local orthopedic surgeon asking about information on new treatments for knee injuries or back injuries.

---

24. Concept Mapping

Fracture occurs

↓

| Spongy bone forms at break |

1st few days

| Blood vessels regrow between sections, spongy bone fills in and hardens |

1-2 weeks

↓

| Thickened band of bone cells disappears; bone almost healed |

2-3 months

25. Hypothesizing Light colored hair will reflect more solar radiation than dark hair. Less heat will be absorbed.

Performance Assessment

1. First-degree burns involve the epidermis, second-degree burns reach the dermis, and third-degree burns extend through both the epidermis and dermis into the tissues below. Use the Performance Task Assessment List for Display in **PASC**, p. 63. **P**

2. Tanning booths use ultraviolet light to produce a tan over a series of visits. Failure to follow safety rules for exposure time to UV and faulty equipment can cause skin burns or disorders. Use the Performance Task Assessment List for Group Work in **PASC**, p. 97. **P**

3. Have students use the information to determine whether new treatment methods are as successful as older methods. Use the Performance Task Assessment List for Letter in **PASC**, p. 67. **P**

Assessment Resources

 Reproducible Masters

Chapter Review, pp. 45-46
Assessment, pp. 143-146
Performance Assessment, p. 35

Glencoe Technology

🔘 **Chapter Review Software**
🔘 **Computer Test Bank**
📼 **MindJogger Videoquiz**

Chapter Organizer

Section	Objectives/Standards	Activities/Features
Chapter Opener		**Explore Activity:** Make a model showing the length of your digestive tract. p. 599
22-1 **Nutrition** (2 sessions)*	**1. List** the six classes of nutrients. **2. Describe** the importance of each type of nutrient. **3. Explain** the relationship between diet and health. National Content Standards: (5-8) UCP1, UCP3, A1, A2, B1, B3, E2, F1, F5	**Connect to Chemistry,** p. 601 **Problem Solving:** A Menu for Endurance, p. 602 **Using Technology:** Bioengineered Foods, p. 606 **MiniLAB:** How much water? p. 607 **Skill Builder:** Making and Using Tables, p. 608 **Using Computers:** Spreadsheet, p. 608 **Activity 22-1:** Identifying Vitamin C Content, p. 609
22-2 **Science and Society** **Technology: Nutrients** **Combat Cancer** (1 session)*	**1. Describe** some added benefits of a diet rich in fruits and vegetables. **2. Explain** the advantages and disadvantages of antioxidants in the treatment of cancer. National Content Standards: (5-8) A1, E2, F1, F5	**Using Math,** p. 611 **Explore the Technology,** p. 611
22-3 **Your Digestive System** (2 sessions)*	**1. Distinguish** between mechanical and chemical digestion. **2. Name** the organs of the digestive system and describe what takes place in each. **3. Explain** how homeostasis is maintained in digestion. National Content Standards: (5-8) UCP1, A1, A2, B1, C1, C3, F1, G1	**MiniLAB:** How are fats emulsified? p. 615 **Skill Builder:** Observing and Inferring, p. 617 **Science Journal,** p. 617 **Activity 22-2:** Protein Digestion, pp. 618-619 **People and Science:** Georgia Larson, Clinical Dietician, p. 620

* A complete Planning Guide that includes block scheduling is provided on pages 32T-35T.

Activity Materials

Explore	Activities	MiniLABs
page 599 5 small index cards, 6.5 m of masking tape, meterstick	page 609 indophenol solution; graduated cylinder; glass marking pen; 10 identical test tubes; 10 dropping bottles each containing one of the following: water, orange juice, pineapple juice, apple juice, lemon juice, tomato juice, cranberry juice, carrot juice, lime juice, and mixed vegetable juice; safety goggles; apron pages 618-619 unflavored gelatin, dropper, 2 identical test tubes, test-tube rack, drinking glass, pepsin powder, cold tap water, graduated cylinder, glass marking pen, dilute hydrochloric acid, clock or stopwatch, goggles, apron	page 607 pan balance, 250-mL beaker, celery sticks, cookie sheet, conventional oven page 615 2 identical glasses, warm tap water, cooking oil, liquid dish soap, measuring spoons, stirrer

Need Materials? Call Science Kit (1-800-828-7777).

Teacher Classroom Resources

Reproducible Masters	Transparencies	Teaching Resources
Study Guide, p. 84 **Reinforcement,** p. 84 **Enrichment,** p. 84 **Activity Worksheets,** pp. 130-131, 134 **Lab Manual 51,** Testing for Carbohydrates **Lab Manual 52,** Testing for Proteins **Science and Society Integration,** p. 26 **Multicultural Connections,** pp. 47-48	**Section Focus Transparency 80,** Identifying the Nutrients in Foods	**Glencoe Life Science Interactive Videodisc** **Life Science CD-ROM** **Spanish Resources** **English/Spanish Audiocassettes** **Cooperative Learning Resource Guide** **Lab Partner** **Lab and Safety Skills** **Lesson Plans**
Study Guide, p. 85 **Reinforcement,** p. 85 **Enrichment,** p. 85	**Section Focus Transparency 81,** Nutrients: Natural and Supplements	**Assessment Resources**
Study Guide, p. 86 **Reinforcement,** p. 86 **Enrichment,** p. 86 **Activity Worksheets,** pp. 132-133, 135 **Lab Manual 53,** Digestion of Fats **Science Integration Activity 22,** The Breakdown of Starch **Concept Mapping,** p. 49 **Critical Thinking/Problem Solving,** p. 26 **Cross-Curricular Integration,** p. 26	**Section Focus Transparency 82,** Meeting Your Energy Needs **Science Integration Transparency 22,** Enzyme Catalysts at Work **Teaching Transparency 43,** Human Digestive System **Teaching Transparency 44,** Digestive Organs	**Chapter Review,** pp. 47-48 **Assessment,** pp. 147-150 **Performance Assessment in the Science Classroom (PASC)** **MindJogger Videoquiz** **Alternate Assessment in the Science Classroom** **Performance Assessment** **Chapter Review Software** **Computer Test Bank**

Key to Teaching Strategies

The following designations will help you decide which activities are appropriate for your students.

L1 Level 1 activities should be within the ability range of all students, including those with learning difficulties.

L2 Level 2 activities should be within the ability range of the average to above-average student.

L3 Level 3 activities are designed for the ability range of above-average students.

LEP LEP activities should be within the ability range of Limited English Proficiency students.

LS These activities are designed to address different learning styles.

COOP LEARN Cooperative Learning activities are designed for small group work.

P These strategies represent student products that can be placed into a best-work portfolio.

GLENCOE TECHNOLOGY

The following multimedia resources are available from Glencoe.

Science and Technology Videodisc Series (STVS)
Plants & Simple Organisms
  Disease-Resistant Tomatoes
  Genetic Engineering in Barley
Human Biology
  Insulin Pills
  Insulin Pump

The Infinite Voyage Series
A Taste of Health
National Geographic Society Series
STV: Human Body Vol. 1
Life Science CD-ROM

Teacher Classroom Resources

This is a representation of key blackline masters available in the Teacher Classroom Resources.

Teaching Aids

Section Focus Transparencies

80 SECTION FOCUS TRANSPARENCY — Section 22-1

IDENTIFYING THE NUTRIENTS IN FOODS

You body foods for two reasons. First, the energy you need to carry out your life functions and other activities is provided by the foods you eat. Second, the foods you eat supply your body with nutrients. Nutrients are substances in foods that provide your body with the materials it needs for growth, maintenance, and repair. Descriptions of some of the nutrients your body needs appear in the table shown.

Nutrient Type	Characteristics of Nutrient	Source (Plants or Animals)
Proteins	Used by the body for growth; are made of large molecules containing carbon, hydrogen, oxygen, and nitrogen	Animals and Plants
Sugars	Taste sweet; provide the body with energy	Plants
Starches	Often have a pasty or mushy texture; provide the body with energy	Plants
Cellulose	Cannot be digested by the body; have fibrous texture	Plants
Fats	Have a greasy or oily feel; observed in some foods after cooking as a layer of oil; provide the body with energy	Plants and Animals

1. What kinds of foods do you think are rich in sugars? In starches?
2. What kinds of foods do you think are rich in proteins? Give three examples.
3. Of the following, which foods do you think are rich in fats—butter, apples, beef, asparagus, lard? Explain.

L1

81 SECTION FOCUS TRANSPARENCY — Section 22-2

NUTRIENTS: NATURAL AND SUPPLEMENTS

Foods contain a variety of nutrients your body needs for development, growth, and repair. Eating a balanced diet should provide your body with all the nutrients it needs. However, many people choose to take vitamin or mineral supplements to make sure their body gets all the nutrients it needs.

1. What do the three drawings have in common?
2. Which is a better way to get nutrients—by eating a balanced diet or by taking nutrient supplements? Which way do you use to get your nutrients?
3. Why might a doctor advise some people to take a specific vitamin or mineral supplement in addition to eating a healthful diet? Give an example.

L2

82 SECTION FOCUS TRANSPARENCY — Section 22-3

MEETING YOUR ENERGY NEEDS

Think of all the ways you use energy each day. You use the energy provided by your body to think, walk, play sports, and even to sleep. You also use energy generated from Earth's energy resources when you ride a bus, play a radio, watch television, or cook your dinner.

1. What are some of the ways you use energy (electricity) in your everyday life? Where does this energy come from?
2. Why does your body need energy? Where does it get the energy it needs?
3. How is the way in which your body gets energy similar to the way in which energy is provided to a community by a power plant?

L3

Science Integration Transparencies

22 SCIENCE INTEGRATION TRANSPARENCY

Enzyme Catalysts at Work

L1

Teaching Transparencies

43. HUMAN DIGESTIVE SYSTEM

L1

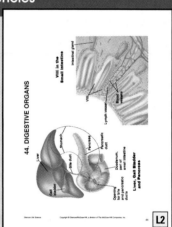

44. DIGESTIVE ORGANS

L2

Meeting Different Ability Levels

Study Guide

Chapter 22
STUDY GUIDE — Nutrition

L1

Reinforcement

Chapter 22
REINFORCEMENT — Nutrition

L2

Enrichment Worksheets

Chapter 22
ENRICHMENT — Nutrition

Do You Eat Too Much Fat?

L3

Hands-On Activities

Science Integration Activity

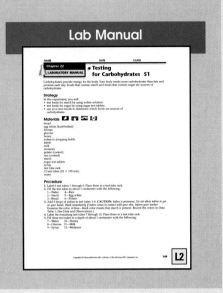

The Breakdown of Starch

Lab Manual

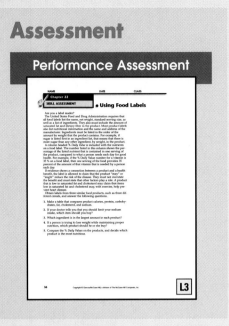

Testing for Carbohydrates 51

Assessment

Performance Assessment

Using Food Labels

Enrichment and Application

Critical Thinking/Problem Solving

Nutrients and Digestion

Sweeteners, Not Sweets, Spoil Your Appetite

Cross-Curricular Integration

Nutrients and Digestion

Should Cosmetic Use of Retin-A Be Allowed?

Science and Society Integration

Nutrients and Digestion

Advertising and Anorexia

Multicultural Connections

San Juan Agricultural Cooperative

Concept Mapping

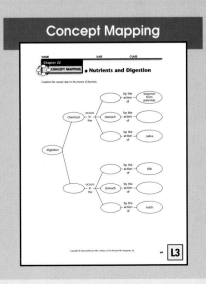

Nutrients and Digestion

Nutrients and Digestion

CHAPTER OVERVIEW

Section 22-1 This section focuses on the nutrients required by the body to carry on the activities of life. The six classes of nutrients and their functions within the body are also described.

Section 22-2 *Science and Society* Nutrients can have additional benefits beyond providing the substances needed for energy and growth. Cancer-combating chemicals, called antioxidants, found in foods are discussed.

Section 22-3 The process of digestion is introduced in this section. Each of the major organs of the digestive system is described and its specific function is detailed.

Chapter Vocabulary

nutrient	mechanical
carbohydrate	digestion
protein	chemical
amino acid	digestion
fat	saliva
vitamin	peristalsis
mineral	chyme
food group	villi
antioxidant	
digestion	

Theme Connection

Energy Energy is a major theme in this chapter. Nutrients contain chemical energy. Complex food compounds become simpler compounds that the body cells can use for metabolism. Chemical energy is transformed into heat and mechanical energy.

Previewing the Chapter

Section 22-1   Nutrition
▶ Why do you eat?
▶ Food Groups

Section 22-2   Science and Society
Technology: Nutrients Combat Cancer

Section 22-3   Your Digestive System
▶ Processing Food
▶ Enzymes—Nature's Chemists
▶ Where and How Digestion Occurs

598

Learning Styles

Look for the following logo for strategies that emphasize different learning modalities. **LS**

Kinesthetic	Enrichment, p. 603; Making a Model, p. 613
Visual-Spatial	Explore, p. 599; Demonstration, pp. 613, 614; Integrating the Sciences, p. 614; MiniLAB, p. 615
Interpersonal	Activity 22-1, p. 609
Logical-Mathematical	Across the Curriculum, pp. 603, 607; Enrichment, p. 605; MiniLAB, p. 606; Activity 22-2, pp. 618-619; People and Science, p. 620
Linguistic	Activity, p. 602; Science Journal, p. 604

Nutrients and Digestion

What does a paper fan have in common with your stomach? Both the fan and the walls of your stomach are made of folds. Because of the folds and its elastic walls, the stomach can expand to hold about 2 L of food or liquid. It is here that one part of the process that releases energy from food occurs. How long do you think it takes for the food to go through this process?

EXPLORE ACTIVITY

Make a model showing the length of your digestive tract.

1. Use index cards to make the following labels: Mouth (8 cm, 5-30 seconds), Pharynx and esophagus (26 cm, 10 seconds), Stomach (16 cm, 2-3 hours), Small intestine (4.75 m, 3 hours), and Large intestine (1.25 m, 2 days).
2. Place a piece of masking tape, 6.5 m long, on the classroom floor.
3. Beginning at one end of the tape, measure off and mark the lengths for each section of the digestive tract. Place each label next to its proper section.

Observe: What factors might alter the amount of time digestion takes?

Previewing Science Skills

▶ In the Skill Builders, you will compare and contrast, and observe and infer.

▶ In the Activities, you will predict, experiment, and collect data.

▶ In the MiniLABs, you will measure in SI and make models.

599

Assessment Planner

Portfolio
Refer to page 621 for suggested items that students might select for their portfolios.

Performance Assessment
See page 621 for additional Performance Assessment options.
Skill Builders, pp. 608, 617
MiniLABs, pp. 606, 615
Activities 22-1, p. 609; 22-2, pp. 618-619

Content Assessment
Section Wrap-ups, pp. 608, 611, 617
Chapter Review, pp. 621-623
Mini Quizzes, pp. 608, 617

Group Assessment
Opportunities for group assessment occur with Cooperative Learning Strategies and Flex Your Brain Activities.

EXPLORE ACTIVITY

Purpose
LS **Visual-Spatial** Use the Explore Activity to introduce students to the digestive system. Inform students that they will be learning about nutrients and the digestive system as they read the chapter. **L1** **LEP**

Materials
index cards, marking pens, masking tape, meterstick or tape

Teaching Strategies
• Discuss with students that the actual length and the amount of time food remains in the digestive tract varies among individuals and health conditions.
• Once the complete length of the digestive system has been marked off, encourage students to infer how something that long can fit inside their bodies.

Observe
Students should observe that the amount of time needed to digest food may depend on the type of food. Some foods go through several different chemical processes as they are digested or broken down into materials that are used by their bodies.

✔ Assessment

Process Have students infer what symptoms might occur in an infant if a birth defect causes constriction of a portion of the digestive tract. Use the Performance Task Assessment List for Making Observations and Inferences in **PASC,** p. 17. **P**

Prepare

Section Background

- Nutrients not immediately required for body activities are stored in cells in the form of glycogen or fat.
- The unique sequence of amino acids in a protein determines the shape of the molecule, and the shape of the protein molecule determines its function.
- Vitamins A, D, E, and K are fat soluble and are absorbed with the fats into cells.

Preplanning

- To prepare for the Mini-LAB, obtain fresh celery. Other fresh fruits or vegetables may be substituted for celery.
- To prepare for Activity 22-1, obtain the indophenol solution and the fruit and vegetable juices needed.

1 Motivate

Bellringer

Before presenting the lesson, display **Section Focus Transparency 80** on the overhead projector. Assign the accompanying **Focus Activity** worksheet. L1 LEP

Tying to Previous Knowledge

Have students discuss what is meant by a "balanced diet."

22•1 Nutrition

Science Words

- nutrient
- carbohydrate
- protein
- amino acid
- fat
- vitamin
- mineral
- food group

Objectives

- List the six classes of nutrients.
- Describe the importance of each type of nutrient.
- Explain the relationship between diet and health.

Why do you eat?

What to have for breakfast? You may base your decision on taste and amount of time available. A better factor might be the nutritional value of the food you choose. A chocolate-iced donut might be tasty and quick to eat but it provides few of the nutrients your body needs to carry out your morning activities. **Nutrients,** as shown in **Figure 22-1,** are substances in foods that provide energy and materials for cell development, growth, and repair.

Classes of Nutrients

There are six kinds of nutrients available in food: carbohydrates, proteins, fats, vitamins, minerals, and water. Carbohydrates, proteins, vitamins, and fats are all organic nutrients. In contrast, minerals and water are inorganic. They do not contain carbon. Foods containing carbohydrates, fats, and proteins are usually too complex to be absorbed right away by your body. These substances need to be broken down into simpler molecules before the body can make use of them. In contrast, minerals and water can be absorbed directly into your bloodstream. They don't require digestion or breakdown.

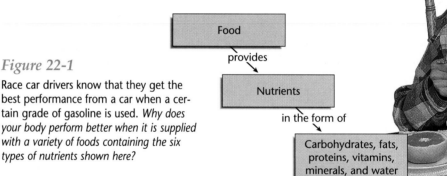

Figure 22-1

Race car drivers know that they get the best performance from a car when a certain grade of gasoline is used. *Why does your body perform better when it is supplied with a variety of foods containing the six types of nutrients shown here?*

Program Resources

 Reproducible Masters
Study Guide, p. 84 L1
Reinforcement, p. 84 L1
Enrichment, p. 84 L3
Science and Society Integration, p. 26
Activity Worksheets, pp. 130-131, 134 L1
Multicultural Connections, pp. 47-48
Lab Manual 51, 52

Transparencies
Section Focus Transparency 80 L1

Carbohydrates

Study the panels on several boxes of cereal. You'll notice that the number of grams of carbohydrates found in a typical serving is usually higher than the amounts of the other nutrients. That means that the major nutrient in the cereal is in the form of carbohydrates. **Carbohydrates** are the main sources of energy for your body. They contain carbon, hydrogen, and oxygen atoms. During cellular respiration, energy is released when molecules of carbohydrates break down in your cells.

Types of Carbohydrates

The three types of carbohydrates—sugar, starch, and cellulose—are shown in **Figure 22-2**. Sugars are simple carbohydrates. There are many types of sugars. You're probably most familiar with one called table sugar. Fruits, honey, and milk are sources of sugar. Your cells use sugar in the form of glucose. Starch and cellulose are complex carbohydrates. Starch is in foods such as potatoes and those made from grains such as pasta. Cellulose occurs in plant cell walls.

CONNECT TO
CHEMISTRY

There are many kinds of organic compounds, but only carbohydrates have a ratio of two hydrogen atoms to one oxygen atom. *Illustrate* the chemical formula for the most simple carbohydrate.

Figure 22-2

Carbohydrates are found in three forms: sugar, starch, and cellulose. While all three are made of carbon, hydrogen, and oxygen, their differences are due to chemical bonding and structure.

A The simple sugar, called glucose, is the basic fuel your body uses to carry out its life processes.

B Starch is stored in plant cells. During the digestive process, your body breaks down the complex starch molecules into the simple sugar glucose.

C Cellulose is a complex carbohydrate that is a major part of the strong cell walls in a plant. While your body cannot break down cellulose, its fiber is important in maintaining a smooth-running digestive system. Sources of cellulose include fresh fruits, vegetables, and grains.

22-1 Nutrition   **601**

2 Teach

Brainstorming

Are you what you eat? Have students consider this question before studying this section. Reconsider the question after completing this section.

Visual Learning

Figure 22-1 Why does your body perform better when it is supplied with a variety of foods containing the six types of nutrients shown here? *The variety provides all the nutrients your body needs.*

Figure 22-2 Samples of the three types of carbohydrates — simple sugars, starch, and cellulose—are given along with their basic molecular structures. Have students note the ratio of two hydrogen atoms and one oxygen atom for every carbon atom. **LEP** **LS**

CONNECT TO
CHEMISTRY

Answer The most basic carbohydrate would have the formula CH_2O (formaldehyde).

GLENCOE TECHNOLOGY

CD-ROM

Life Science CD-ROM
Have students perform the interactive exploration for Chapter 22 to reinforce important chapter concepts and thinking processes.

Problem Solving

A Menu for Endurance

Jackie is a long-distance runner. She plans to run in the Boston Marathon on Patriot's Day. This world famous race is more than 42 km. Like most runners, Jackie is conscious about eating foods that are filled with the right nutrients. The night before the marathon, Jackie must choose what to make for dinner. She wants a meal with a lot of complex carbohydrates. Her choices are:

(a) hamburger, french fries, chocolate cake, and a large glass of cola.

(b) spaghetti with tomato sauce, pasta and vegetable salad, whole-wheat roll, fresh fruit, and a large glass of water.

(c) fried chicken, mashed potatoes and gravy, broccoli with cheese sauce, roll and butter, and ice cream.

Solve the Problem:

1. Which menu should Jackie choose?
2. Name the carbohydrates found in this menu.

Think Critically:

1. Why does Jackie want to eat carbohydrates before the marathon?
2. Why won't a meal full of fats work as well?

602 Chapter 22 Nutrients and Digestion

Figure 22-3

Beans and fish are good sources of protein. The bean shown here is the winged bean, grown in the Philippines.

Proteins

Your body uses **proteins** for growth. As enzymes, they affect the rate of chemical reactions in your body. They are also involved in the replacement and repair of body cells. Proteins are large molecules that contain carbon, hydrogen, oxygen, and nitrogen. A molecule of protein is made up of a large number of subunits or building blocks called **amino acids**. Some sources of proteins are shown in **Figure 22-3.**

Essential Amino Acids

In Chapter 4, you learned that proteins are made according to directions supplied by genes that you inherit. Your body needs 20 different amino acids to be able to construct the proteins needed in your cells. Twelve of these amino acids can be made in your cells. The eight remaining amino acids are called essential amino acids. Your body doesn't have genetic instructions to construct them. Therefore, they have to be supplied through food you eat. Eggs, milk, and cheese contain all the *essential* amino acids. Beef, pork, fish, and chicken supply most of them. You might be surprised to know that whole grains such as wheat, rice, and soybeans supply many needed amino acids in addition to carbohydrates.

Fats

You may not think of fat as being a necessary part of your diet. In today's health-conscious society, the term *fat* has a negative meaning. However, **fats** are necessary because they provide energy and help your body absorb some vitamins. Fats are stored in your body in the form of fat tissue as shown in **Figure 22-4.** This tissue cushions your internal organs. You learned that carbohydrates are the main source of energy for your body. This is because people eat a large amount of carbohydrates. However, a gram of fat can release twice as much energy as a gram of carbohydrate. During this process, fat breaks down into smaller molecules called fatty acids and glycerol. Because fat is such a good storage unit for energy, any excess energy we eat is converted to fat until the body needs it.

Types of Fat

Fats are classified as unsaturated and saturated. Unsaturated fats, which come from plants, are usually liquid at room temperature. Corn, safflower, and soybean oils are all unsaturated fats. Some unsaturated fats are also found in poultry, fish, and nuts. Saturated fats are found in red meats and are usually solid at room temperature. Saturated fats have been associated with high levels of blood cholesterol. Cholesterol occurs normally in all your cell membranes. However, too much of it in your diet causes excess fat deposits to form on the inside walls of blood vessels. The deposits result in a cutoff of the blood supply to your organs and an increase in blood pressure. Heart disease and strokes may result from the cutoff of blood to the heart and brain.

Figure 22-4

Fat is stored in certain cells in your body.

A The cytoplasm and nucleus are pushed to the edge of the cell by the fat deposits.

B The first source of energy for your body comes from the sugars in carbohydrates. When the supply of sugars runs low, your body begins to use the reservoir of energy stored in body fat.

22-1 Nutrition **603**

Across the Curriculum

Mathematics Have students check the label of a food item for the grams of fat contained. Multiply this number by 9, the number of calories used when 1 g of fat is oxidized. Divide this number by the total number of calories in the food item. The calculation will give the percent of calories from fat. Recommendations are that no more than 30% of your calorie intake come from fat. **L1** **LS**

Enrichment

LS **Kinesthetic** Have students test for fats in foods by rubbing various foods on a piece of brown kraft paper (grocery bags). A translucent spot on the paper indicates the presence of a fat or oil. Moist foods produce only a wet spot that disappears after evaporation. **L1**

Discussion

Have students discuss whether milk is the perfect food. Milk does supply many of the nutrients, but it is not a perfect food and does lack certain essential vitamins and minerals.

GLENCOE TECHNOLOGY

Videodisc

The Infinite Voyage: A Taste of Health
Introduction

Chapter 1
Pima Indians: Old Traditions in Nutrition

Chapter 3
Developing Taste

Chaper 4
Consequences of a Fatty Diet

Chapter 5
Ancel Keys: Pioneer in Nutrition

Chapter 6
Genetic Links to Cholesterol

Students may have the mistaken notion that if vitamins are good for you, more is better. Some vitamins should not be taken in high doses. An excess of vitamin A can cause liver, skin, and muscle problems. Too much vitamin D can cause an imbalance of blood calcium, resulting in digestive and nervous system disorders.

Use these program resources as you teach this lesson.

Lab Manual 51, 52

Multicultural Connections, pp. 47-48

Science and Society Integration, p. 26

Figure 22-5
When taken in large doses, the excess amounts of water-soluble vitamins are excreted with urine. Excess amounts of fat-soluble vitamins accumulate in the body and may build up to harmful levels.

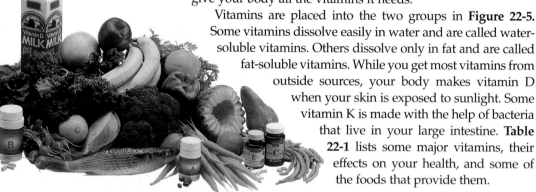

Vitamins

Those essential, organic nutrients needed in small quantities to help your body use other nutrients are called **vitamins.** For instance, vitamin D is needed for bone cells to use calcium. In general, vitamins promote growth and regulate body functions.

Most foods supply some vitamins, but no one food has them all. Although some people feel that taking extra vitamins is helpful, eating a well-balanced diet is usually good enough to give your body all the vitamins it needs.

Vitamins are placed into the two groups in **Figure 22-5.** Some vitamins dissolve easily in water and are called water-soluble vitamins. Others dissolve only in fat and are called fat-soluble vitamins. While you get most vitamins from outside sources, your body makes vitamin D when your skin is exposed to sunlight. Some vitamin K is made with the help of bacteria that live in your large intestine. **Table 22-1** lists some major vitamins, their effects on your health, and some of the foods that provide them.

Table 22-1

Vitamins		
Vitamin	**Body Function**	**Food Sources**
Water Soluble		
B (thiamine, riboflavin, niacin, B_6, B_{12})	growth, healthy nervous system, use of carbohydrates, red blood cell production	meat, eggs, milk, cereal grains, green vegetables
C	growth, healthy bones and teeth, wound healing	citrus fruits, tomatoes, green leafy vegetables
Fat Soluble		
A	growth, good eyesight, healthy skin	green/yellow vegetables, liver and fish liver oils, milk, yellow fruit
D	absorption of calcium and phosphorus by bones and teeth	milk, eggs, fish
E	formation of cell membranes	vegetable oils, eggs, grains
K	blood clotting, wound healing	green leafy vegetables, egg yolks, tomatoes

Science Journal

Vitamins and Health Following the discussion on vitamins, have students write a paragraph about why it might be a disadvantage to substitute vitamin pills for a proper diet. Their entries should indicate an understanding that other nutrients besides vitamins are needed for maintaining health. A well-balanced diet supplies the range of needed vitamins. L1 LS

Inclusion Strategies

Gifted Have students research a particular vitamin deficiency. They should include information about how the vitamin was discovered. L3 P

Table 22-2

Minerals		
Mineral	**Health Effect**	**Food Sources**
Calcium	strong bones and teeth, blood clotting, muscle and nerve activity	milk, eggs, green leafy vegetables
Phosphorus	strong bones and teeth, muscle contraction, stores energy	cheese, meat, cereal
Potassium	balance of water in cells, nerve impulse conduction	bananas, potatoes, nuts, meat
Sodium	fluid balance in tissues, nerve impulse conduction	meat, milk, cheese, salt, beets, carrots
Iron	carries oxygen in hemoglobin in red blood cells	raisins, beans, spinach, eggs
Iodine (trace)	thyroid activity, stimulates metabolism	seafood, iodized salt

Minerals

Inorganic nutrients that regulate many chemical reactions in your body are **minerals.** They are chemical elements such as phosphorus. About 14 minerals are used by your body for building cells, taking part in chemical reactions in cells, sending nerve impulses throughout your body, and carrying oxygen to body cells. Minerals used in the largest amounts in your body are presented in **Table 22-2.** Of the 14 minerals, calcium and phosphorus are used in the largest amounts for a variety of body functions. One of these functions is shown in **Figure 22-6.** Some minerals, called trace minerals, are required in only very small amounts. Copper and iodine are usually listed as trace minerals.

Figure 22-6

Calcium and phosphorus are used in the formation and maintenance of your bones. Look at the woman and the X ray of her spine. *What might happen if your diet is deficient in these minerals?*

22-1 Nutrition   605

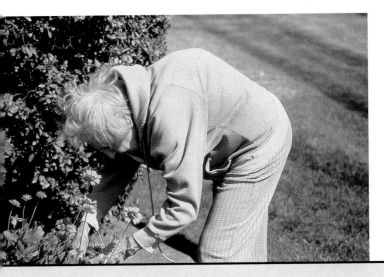

Visual Learning

Figure 22-6 What might happen if your diet is deficient in these minerals? *The bones would weaken and be susceptible to disease or fracture.* **LEP**
LS

Discussion

Explain why salt is iodized. Have students read a salt label to find out how the iodine is supplied.

Enrichment

LS **Logical-Mathematical** Have students keep track of all the foods and liquids they consume in one day. Estimate how much water was taken in and compare with the 2 L or so of water lost daily. L2

Content Background

• Discuss water in vegetables such as turnips and squashes. Many students have the impression that drinking water is the only source of water.

• Point out that water is necessary for chemical reactions in the body and that water is a by-product of the breakdown of all molecules in the process of respiration.

• Have students comment on the statement that where there is water, there is life.

Mini**LAB**

Purpose

LS Logical-Mathematical Students will calculate the water content of fresh celery. **L1 COOP LEARN**

Materials

pan balance, cut-up celery, 250-mL beaker, oven, oven mitt

Teaching Strategies

Safety Precautions Caution students to be careful when handling hot materials.

Troubleshooting Dry the fresh celery with a paper towel before determining mass.

Analysis

1. The water content will vary with the freshness of the celery used, but will be about 90% water.

2. Students may infer that the water content of fruits and vegetables is between 80 and 90 percent.

✔ Assessment

Performance Have students use their lab skills to determine the water content of fruits. Use the Performance Task Assessment List for Carrying Out a Strategy and Collecting Data in **PASC**, p. 25. **P**

USING TECHNOLOGY

Bioengineered Foods

The human population is growing. Some scientists warn that food production cannot keep up with this growth. A way to increase food supply is in the development of new bioengineered crops.

The Process

Traditionally, crossbreeding has been used to produce new varieties of plants. Now, through genetic engineering, new plants are developed in less time. In genetic engineering, a gene carrying the desired trait is transferred into a different species. For example, potatoes are susceptible to a type of virus. Scientists have transferred a gene that provides resistance to that virus into potatoes. Scientists are continuing this process on a variety of plants in hopes of improving their taste and nutritional value. New crops with higher levels of proteins and vitamins could improve the health of humans.

Some new crops are now available to consumers. The FDA has approved the sale of a new type of tomato. The new tomato can be picked when ripe and yet remains firm after harvest.

Think Critically:

Critics fear that new plants might harm the environment. How could this happen?

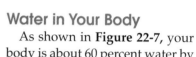
Inspecting crops for signs of the potato virus.

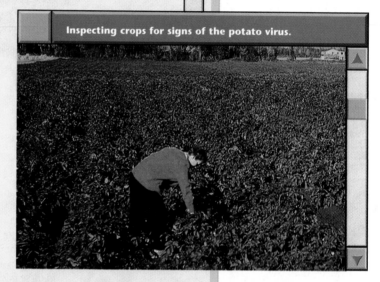

Water

You don't have to be lost in a desert to know how important water is for your body. Next to oxygen, water is the most vital factor for survival. You could live a few weeks without food, but only a few days without water. Most of the nutrients you have studied in this chapter can't be used by your body unless they are carried in a solution. This means that they have to be dissolved in water. Water enables chemical reactions to take place in cells.

Water in Your Body

As shown in **Figure 22-7**, your body is about 60 percent water by weight. This water is found in and around cells and in plasma and lymph. Water removes waste products from cells. Wastes dissolved in water leave your body as urine or perspiration. To balance water lost each day, you need to drink about 2 L of liquids. But don't think that you have to drink just water to keep your cells supplied. Most foods have more water in them than you realize. An apple is about 80 percent water, and many meats are as much as 90 percent water.

Water Loss

Your body also loses about 2 L of water every day through excretion, perspiration, and respiration. **Table 22-3** shows how water is lost from the body. The body is equipped to maintain its fluid content, however. When your body needs water, it sends messages to your brain. A feeling of thirst develops. Drinking a glass of water usually restores the body's homeostasis, and the signal to the brain stops.

Table 22-3

Water Loss	
Through	**Amount (mL/day)**
Exhaled air	350
Feces	150
Skin (mostly as sweat)	500
Urine	1800

Figure 22-7

About two-thirds of your body water is located within your body cells. Water helps maintain the cells' shapes and sizes. One-third of your body's water is outside of your body cells.

Water outside of body cells

Water inside of body cells

3 Assess

Check for Understanding

Have students review the vitamin table on page 604 to obtain a general concept of sources of each type of vitamin. Then have students infer the vitamins contained in some similar but unlisted fruits and vegetables.

Reteach

Have students make flash cards with names of fruits or vegetables. On the reverse side, identify the major vitamin content of each food. Students can test each other using the flash cards. **COOP LEARN**

Extension

For students who have mastered this section, use the **Reinforcement** and **Enrichment** masters.

Across the Curriculum

Math Students may use either a circle or bar graph to show the percent of water lost by various means, as shown in **Table 22-3**. L1 LS

4 Close

Figure 22-8

The pyramid shape reminds you that you should consume more servings from the bread and cereal group than from the meat and milk group. *Where should the least number of servings come from?*

Food Groups

Because no one food has every nutrient, you need to eat a variety of foods. Nutritionists have developed a simple system to help people plan meals that include all the nutrients required for good health.

Foods that contain the same nutrients belong to a **food group.** The food pyramid, in **Figure 22-8,** presents the basic food groups and serving suggestions. Eating a certain amount from each food group each day will supply your body with the nutrients it needs for energy and growth. Of course, most people eat foods in combined forms. Combinations of food contain ingredients from more than one food group and supply the same nutrients as the foods they contain. Examples of food group combinations include chili and macaroni and cheese.

Section Wrap-up

Review

1. List six classes of nutrients and give one example of a food source for each.

2. Describe a major function of each class of nutrient.

3. Discuss the relationship between your diet and your health.

4. Explain the importance of water in the body.

5. **Think Critically:** What foods from each food group would provide a balanced breakfast? Explain why.

Skill Builder
Making and Using Tables

Use the information in **Table 22-1** and **Table 22-2** to determine the vitamins and minerals needed for healthy bones. If you need help, refer to Making and Using Tables in the **Skill Handbook.**

Using Computers

Spreadsheet Use the format of **Table 22-2** and prepare a spreadsheet for a data table of the minerals. Use reference books to gather information about these following minerals and add them to the table: sulfur, magnesium, copper, manganese, cobalt, and zinc.

Skill Builder

vitamins C and D, calcium, and phosphorus

Assessment

Performance Assess students' abilities to make and use tables by having them use **Table 22-3** to find out how much water is lost through skin. Use the Performance Task Assessment List for Data Table in **PASC,** p. 37. P

Activity 22-1

Identifying Vitamin C Content

Vitamin C is found in a variety of fruits and vegetables. In some plants, the concentration is high; in others, it is low. Try this activity to test various juices and find out which contains the most vitamin C.

Problem
Which juices contain vitamin C?

Materials

- indophenol solution
- graduated cylinder
- glass marking pencil
- 10 test tubes
- test-tube rack
- 10 dropping bottles containing water, orange juice, pineapple juice, apple juice, lemon juice, tomato juice, cranberry juice, carrot juice, lime juice, mixed vegetable juice

Procedure

1. Make a data table like the one shown to record your observations.
2. Label the test tubes 1 through 10.
3. Predict which juices contain vitamin C. Record your predictions in your table.
4. Measure 5 mL of indophenol into each of the ten test tubes. **CAUTION:** *Wear your goggles and apron. Do not taste any of the juices.* Indophenol is a blue liquid that turns colorless when vitamin C is present. The more vitamin C in a juice, the less juice it takes to turn indophenol colorless.
5. Add 20 drops of water to test tube 1. Record your observations.
6. Begin adding orange juice, one drop at a time, to test tube 2.
7. Record the number of drops needed to turn indophenol colorless.
8. Repeat steps 6 and 7 to test the other juices.

Data Table and Observations

Test Tube	Juice	Prediction (yes or no)	Number of drops
1	water	Sample Data	
2	orange	yes	3
3	pineapple	no	4
4	apple	no	4
5	lemon	no	3
6	tomato	yes	5
7	cranberry	no	5
8	carrot	no	5
9	lime	yes	3
10	vegetable	yes	4

Analyze

1. What was the purpose of test tube 1?
2. Does the amount of vitamin C vary in fruit juices?
3. Which juice did not contain vitamin C?

Conclude and Apply

4. What is the best way to take in vitamins?
5. **Infer** why scurvy, a deficiency of vitamin C, is uncommon today.

22-1 Nutrition **609**

Activity 22-1

Purpose

IS **Interpersonal** Students will identify and compare juices that contain vitamin C. **L1**
COOP LEARN

Process Skills

observing, classifying, measuring, predicting, interpreting data, experimenting, communicating, comparing and contrasting

Time

50 minutes

Activity Worksheets, pp. 5, 130-131

Teaching Strategies

Troubleshooting Caution students not to mix up the droppers from the test-solution bottles.

Answers to Questions

1. control
2. Yes, the amount of vitamin C is influenced by whether vitamin C has been added; water has been added; or whether the juice is fresh, frozen, or canned.
3. Apple, cranberry, carrot, and vegetable; answers may vary if juices fortified with vitamins are used.
4. by eating a balanced diet
5. Fruits and vegetables containing vitamin C are included in diets. People consume a substantial variety of vitamin-fortified foods.

✓ Assessment

Performance To further assess students' abilities to identify vitamin C content, have them test for vitamin C in water that sat overnight with either shredded lettuce or shredded cabbage cores. Use the Performance Task Assessment List for Carrying Out a Strategy and Collecting Data in **PASC**, p. 25. **P**

609

Prepare

Section Background

Point out to students that a balanced diet having certain amounts from each food group provides the body with nutrients needed for energy and growth. An important additional benefit is that certain nutrients can reduce the risk of getting various diseases.

1 Motivate

Bellringer

Before presenting the lesson, display **Section Focus Transparency 81** on the overhead projector. Assign the accompanying **Focus Activity** worksheet. L1 LEP

Tying to Previous Knowledge

Recall the role of vitamins and minerals in nutrition and the effects on health if there is a deficiency.

2 Teach

Demonstration

Place an equal number of apple slices into each of two bowls. Rinse the apples in one bowl with lemon juice and the apples in the other bowl with water. After 30 minutes, note that the ascorbic acid in the lemon juice is an antioxidant. It prevents oxygen from reacting with the apple slices and keeps them from turning brown.

USING MATH

Percentages will depend on foods chosen.

22•2 Nutrients Combat Cancer

Science Words

antioxidant

Objectives

* Describe some added benefits of a diet rich in fruits and vegetables.
* Explain the advantages and disadvantages of antioxidants in the treatment of cancer.

A Good Habit to Have

How many times have you heard an adult say, "Eat your vegetables, they're good for you!" You know that vegetables and fruits contain vitamins and minerals necessary for energy and growth. In spite of this knowledge, you still might find them difficult to eat when you are looking at a plate containing the ones you like the least. Our eating habits are formed early in life. If the eating of fruits and vegetables was not encouraged when you were a child, it can be difficult for you to begin including these foods in your daily diet as an adult.

Recent Research

During the past several years, food scientists have learned that there are additional benefits to eating fruits and vegetables. Certain nutrients help lower the risk of getting various diseases. Recent animal studies have shown that some cancers may be prevented and possibly cured by the chemicals found in certain nutrients. Many of the cancer-combating chemicals are being tested to see how they aid cell defenses against the disease.

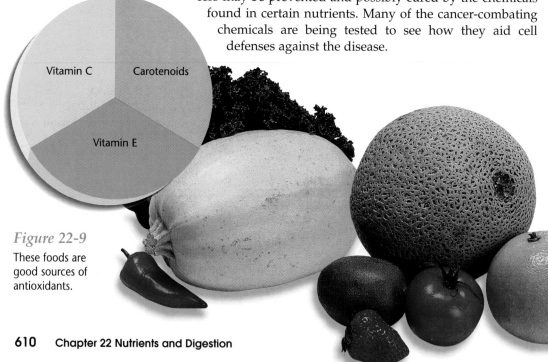

Figure 22-9
These foods are good sources of antioxidants.

Program Resources

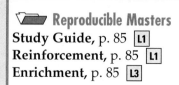 **Reproducible Masters**
Study Guide, p. 85 L1
Reinforcement, p. 85 L1
Enrichment, p. 85 L3

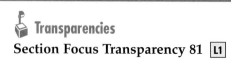 **Transparencies**
Section Focus Transparency 81 L1

Antioxidants

While scientists are looking at a variety of chemicals, one group being closely studied is known as antioxidants. **Antioxidants** are substances that prevent other chemicals from reacting with oxygen. Chemicals that enter the body from smoking or from pollutants in the air may cause cancer when they combine with oxygen. By preventing this reaction, the antioxidants can combat cancer. Antioxidants can also prevent cancer cells from repairing their damaged DNA. As shown in **Figure 22-9**, carotenoids, yellow-orange pigments found in carrots and squashes, are antioxidants; vitamins C and E are also antioxidants.

The Good and the Bad

Taking antioxidants with the drugs used to fight cancer seems to make these drugs more effective. Some doctors are prescribing antioxidants for their patients either to try to prevent cancer or to help fight an existing cancer condition. However, researchers warn that not all the cancer cells are always killed. Sometimes there are harmful side effects. For example, certain body functions may be disrupted. An additional problem is that antioxidants may also cause radiation treatments given to kill the cancer cells to be less effective. Tests are continuing to determine exactly what role antioxidants should play in cancer treatments. If laboratory tests show conclusively that antioxidants are successful disease fighters, you may hear, "Eat your antioxidants, they're good for you!"

Section Wrap-up

Review

1. What are some sources of antioxidants?

2. How do antioxidants react with cancer cells?

Explore the Technology

Many people think that if a small amount of a medicine is beneficial, then increasing the dosage will be even better. Why should a person not take large quantities of an antioxidant without the advice of a doctor? What dangers may result?

USING MATH

Food labels list nutritional information of packaged foods. Study the labels on five different kinds of canned or packaged foods. Make a bar graph and compare the percentage of vitamin C in each food.

*inter*NET CONNECTION

What information about antioxidants can be found on the nutrition information labels on packaged foods? Visit the Chapter 22 Internet Connection at Glencoe Online Science, **www.glencoe.com/sec/science/life**, for a link to more information about gene therapy.

SCIENCE & SOCIETY

Integrating the Sciences

Chemistry Oxidation is the reaction of a substance with oxygen. Sometimes this takes place rapidly, and heat and light are produced. An example is the burning of a fuel. In other cases, the process is slow and the heat is not noticeable. Examples are the rusting of iron and when food oxidizes as part of the metabolic processes. Have students research and list some oxidation reactions.

3 Assess

Check for Understanding

Have students answer questions 1 and 2 in the Section Wrap-up.

Reteach

Point out that painting iron prevents it from reacting with oxygen (rusting). This is analogous to antioxidants' preventing chemicals in the body from reacting with oxygen.

Extension

For students who have mastered this section, use the **Reinforcement** and **Enrichment** masters.

4 Close

Have students prepare a list of foods alleged to provide miraculous health benefits, but that are questionable. Example: Fish—a "brain food" that will make you smarter.

Section Wrap-up

Review

1. yellow-orange pigmented vegetables, foods with vitamins C and E
2. They prevent cancer cells from repairing damaged DNA, thereby hindering their growth and reproduction.

Explore the Technology

Answers will vary but may include the following. Large amounts of antioxidants may disrupt the normal functioning of body systems or interfere with other medicines taken. This could lead to a health disorder. **P**

Prepare

Section Background

Near the junction of the small and large intestines is a small, tube-like pouch called the appendix. It does not appear to have any function in the digestive process. The organ can become inflamed and require removal.

Preplanning

- To prepare for the Mini-LAB, obtain cooking oil and liquid dishwashing detergent. Optional materials include liquid hand soap and motor oil.
- To prepare for Activity 22-2, obtain the dilute hydrochloric acid (1%) solution and the pepsin powder.

1 Motivate

Bellringer

 Before presenting the lesson, display **Section Focus Transparency 82** on the overhead projector. Assign the accompanying **Focus Activity** worksheet. L1 LEP

Tying to Previous Knowledge

Recall from Chapter 1 that one feature of living things is that they use energy. Discuss with students that food is the source of energy for cell activities, but that it must be processed into usable molecules. This is the function of the digestive system.

22•3 Your Digestive System

Science Words

digestion
mechanical digestion
chemical digestion
saliva
peristalsis
chyme
villi

Objectives

- Distinguish between mechanical and chemical digestion.
- Name the organs of the digestive system and describe what takes place in each.
- Explain how homeostasis is maintained in digestion.

Processing Food

Like other animals, you are a consumer. The energy you need for life comes from food sources outside yourself. To keep the cells in your body alive, you take in food every day. Food is processed in your body in four phases: ingestion, digestion, absorption, and elimination. Whether it is a fast-food burger or a home-cooked meal, all the food you eat is treated to the same processes in your body. As soon as it enters your mouth, or is ingested, food begins to be broken down. **Digestion** is the process that breaks down food into small molecules so they can move into the blood. From the blood, food molecules are transported across the cell membrane to be used by the cell. Molecules that aren't absorbed are eliminated and pass out of your body as wastes.

The major organs of your digestive tract—mouth, esophagus (i SAH fuh guhs), stomach, small intestine, large intestine, rectum, and anus—are shown in **Figure 22-10.** Food passes *through* all of these organs. However, food *doesn't* pass through your liver, pancreas, or gallbladder. These three organs produce or store enzymes and chemicals that help break down food as it passes through the digestive tract.

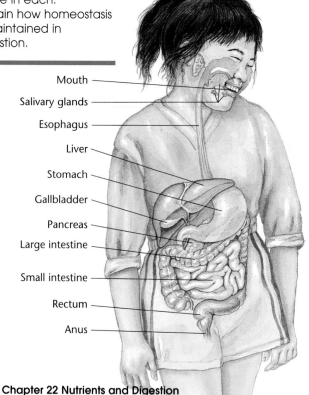

Mouth
Salivary glands
Esophagus
Liver
Stomach
Gallbladder
Pancreas
Large intestine
Small intestine
Rectum
Anus

Figure 22-10

The human digestive system can be described as a tube divided into several specialized sections. If stretched out, an adult's digestive system is between 6 to 9 m long.

Program Resources

 Reproducible Masters

Study Guide, p. 86 L1
Reinforcement, p. 86 L1
Enrichment, p. 86 L3
Cross-Curricular Integration, p. 26
Activity Worksheets, pp. 132-133, 135 L1
Science Integration Activities, pp. 43-44
Critical Thinking/Problem Solving, p. 26

Concept Mapping, p. 49
Lab Manual 53

Transparencies

Section Focus Transparency 82 L1
Science Integration Transparency 22
Teaching Transparency 43, 44

Enzymes—Nature's Chemists

Chemical digestion is only possible because of the actions of certain kinds of proteins. These proteins are called enzymes. Enzymes are molecules that speed up the rate of chemical reactions in your body. You can see in **Figure 22-11** that they speed up reactions without themselves being changed or used up. One way enzymes speed up reactions is by reducing the amount of energy necessary for a chemical reaction to begin.

Figure 22-11

A model of how enzymes work. *What happens to the enzyme after it is released?*

A The surface shape of an enzyme fits the shape of specific molecules called substrates.

B Enzyme and substrate join and the reaction occurs.

C Following the reaction, the enzyme and substrate separate. The enzyme is not changed by the reaction. The substrate is changed to a new chemical structure.

Enzymes in Digestion

A variety of enzymes are involved with the digestion of carbohydrates, proteins, and fats. Salivary amylase is a starch-degrading enzyme secreted in the mouth. This enzyme begins the process of breaking down the complex carbohydrate into simpler sugars. In the stomach, the enzyme pepsin causes complex proteins to break down into less complex proteins. In the small intestine, a number of other enzymes continue the process by breaking down the proteins into amino acids. The pancreas, an organ on the back side of the stomach, secretes several enzymes through a tube into the small intestine. Some continue the process of starch breakdown that started in the mouth. The resulting sugars are turned into glucose and used by the body's cells. Other enzymes are involved in the break-down of fats.

Other Enzyme Actions

Enzyme reactions are not only involved in the digestive process. Enzymes are also responsible for building your body. They are involved in the energy production activities of muscle and nerve cells. They are also involved in the blood-clotting process. Without enzymes, the chemical reactions of your body would not happen. You would not exist.

22-3 Your Digestive System   613

 Use these program resources as you teach this lesson.
Science Integration Activities, pp. 43-44
Cross-Curricular Integration, p. 26

 Use **Science Integration Transparency 22** as you teach this lesson.

2 Teach

INTEGRATION
Chemistry

A substance that changes the rate of a chemical reaction but is unchanged at the end of the reaction is called a catalyst. Enzymes are catalysts that function in living systems by speeding up reactions.

Demonstration

LS **Visual-Spatial** Substances called meat tenderizers contain enzymes. Often used is papain, an enzyme from papaya juice. Sprinkle some meat tenderizer on a small cube of beef. On another cube of beef, sprinkle some sugar. Heat both pieces in an oven at 250°F until brown. When cooled, examine both pieces and compare for tenderness. **LEP**

Making a Model

LS **Kinesthetic** Have students construct a three-dimensional model of an enzyme-molecule system. One such model may be similar to a lock-and-key system. **L1** **LEP**

Visual Learning

Figure 22-11 Use the model students make of how enzymes work (see above, Making a Model) to explain the three steps illustrated in the text. **What happens to the enzyme after it is released?** *It is unchanged and free to join with another molecule in another reaction.* **LEP** **LS**

613

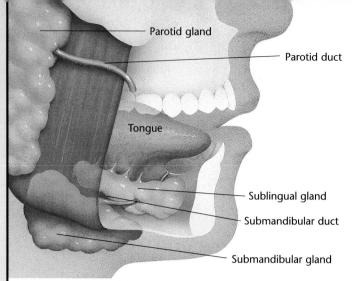

- Parotid gland
- Parotid duct
- Tongue
- Sublingual gland
- Submandibular duct
- Submandibular gland

Figure 22-12

About 1.5 L of saliva is produced by your body each day. *What happens in your mouth when you think about a food you like?*

- Muscles contract
- Food mass
- Muscles relax
- Food moves down

Figure 22-13

During peristalsis, muscles behind the food contract and push the food forward. Muscles in front of the food relax.

Where and How Digestion Occurs

Digestion is both mechanical and chemical. **Mechanical digestion** takes place when food is chewed and mixed in the mouth and churned in your stomach. **Chemical digestion** breaks down large molecules of food into different smaller molecules that can be absorbed by cells. Chemical digestion takes place in your mouth, stomach, and small intestine. Some digestive processes are both physical and chemical; for example, when bile acts on food. This process will be further discussed on the next page.

In Your Mouth

Mechanical digestion begins in your mouth. There, your tongue and teeth break food up into small pieces. Humans are adapted with several kinds of teeth for cutting, grinding, tearing, and crushing.

Some chemical digestion also starts in your mouth. As you chew, your tongue moves food around and mixes it with a watery substance called **saliva.** Saliva is produced by three sets of glands near your mouth that are shown in **Figure 22-12.** Saliva is made up mostly of water, but it also contains mucus and the enzyme salivary amylase. You learned that salivary amylase starts the breakdown of starch to sugar. Food that is mixed with saliva becomes a soft mass. The food mass is moved to the back of your tongue where it is swallowed and passes into your esophagus. Now the process of ingestion is complete, and the process of digestion has begun.

Passing Through the Esophagus

Your esophagus is a muscular tube about 25 cm long. Through it, food passes to your stomach in about four to ten seconds. No digestion takes place there. Smooth muscles in the walls of the esophagus move food downward by a squeezing action. These waves or contractions, called **peristalsis** (pe ruh STAHL sis), move food along throughout the digestive system. **Figure 22-13** shows how peristalsis works.

In Your Stomach

Your stomach is a muscular bag. When empty, it is some-what sausage-shaped with folds on the inside. As food enters from the esophagus, the stomach expands and the folds smooth out. Both mechanical and chemical digestion take place in the stomach. Mechanically, food is mixed by the muscular walls of the stomach and by peristalsis. Food is also mixed with strong digestive juices, which include hydrochloric acid and enzymes. The acid is made by cells in the walls of the stomach. The stomach also produces a mucus that lubricates the food, making it more slick. The mucus also protects the stomach from the strong digestive juices. Food moves through your stomach in about four hours. At the end of this time, the food has been changed to a thin, watery liquid called **chyme.** Little by little, chyme moves out of your stomach and into your small intestine.

In Your Small Intestine

Your small intestine may be small in diameter, but it is 4 m to 7 m in length. As chyme leaves your stomach, it enters the first part of your small intestine, called the duodenum. The major portion of all digestion takes place in your duodenum. A lot of different processes take place here at the same time. Digestive juices from the liver and pancreas are added to the mixture. Your liver, shown in **Figure 22-14,** produces a greenish fluid called bile. Bile is stored in a small sac called the gallbladder. The acid from the stomach makes large fat particles float to the top of the liquid. Bile physically breaks up these particles into smaller pieces, the way detergent acts on grease on dishes. This process is called emulsification. Although the bile physically breaks apart the fat into smaller droplets, the fat molecules are not changed chemically. Chemical digestion of carbohydrates, proteins, and fats occurs when the digestive juices from the pancreas are added.

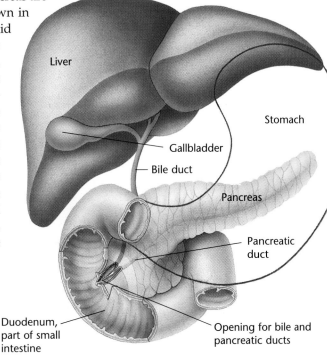

Figure 22-14

The liver, gallbladder, and pancreas are at the beginning of the small intestine.

Liver

Gallbladder

Bile duct

Stomach

Pancreas

Pancreatic duct

Duodenum, part of small intestine

Opening for bile and pancreatic ducts

 Use these program resources as you teach this lesson.
Concept Mapping, p. 49
Lab Manual 53
Critical Thinking/Problem Solving, p. 26

Use **Teaching Transparencies 43** and **44** as you teach this lesson.

616

You learned that the pancreas produces enzymes that help break down carbohydrates, fats, and proteins. Your pancreas also makes insulin. Insulin is a hormone that allows glucose to pass from the bloodstream into your body's cells. Without a supply of glucose, your body's cells have to use proteins and fats for energy.

In addition to insulin, the pancreas also produces another solution. This solution contains bicarbonate. Bicarbonate helps neutralize the stomach acid that is mixed with chyme.

Surface Area of the Small Intestine

Look at the wall of the small intestine in **Figure 22-15.** The walls of your small intestine are not smooth like the inside of a garden hose. Rather, they have many ridges and folds. These folds are covered with tiny, fingerlike projections called **villi.** Villi make the surface of the small intestine greater so that there are more places for food to be absorbed.

Food Absorption

By this time, chyme has become a soup of molecules that is ready to be absorbed through the cells on the surface of the villi. Peristalsis continues to move and mix the chyme. In addition, the villi themselves move and are bathed in the soupy liquid. Molecules of nutrients pass by diffusion, osmosis, or active transport into blood vessels in each villus. From there, blood transports the nutrients to all the cells of the body. Peristalsis continues to slowly force the remaining materials that have not been digested or absorbed into the large intestine.

Figure 22-15
Hundreds of thousands of densely packed villi give the impression of a velvet cloth surface. If the surface area of your villi could be stretched out, it would cover an area the size of a baseball diamond. *What would happen to a person's weight if the number of villi were drastically reduced? Why?*

616

In Your Large Intestine

As chyme enters the large intestine, it is still a thin, watery mixture. The main job of the large intestine, shown in **Figure 22-16**, is to absorb this water from the undigested mass. In doing so, large amounts of water are returned to the body and homeostasis is maintained. Peristalsis slows down somewhat in the large intestine. As a result, chyme may stay in the large intestine as much as three days. After the excess water is absorbed, the remaining undigested materials become more solid.

The Role of Bacteria

Bacteria that live in your large intestine feed on undigested materials like cellulose. This is a symbiotic relationship. The bacteria feed on cellulose, and in return they produce several vitamins that you need. Muscles in the rectum and anus control the release of solidified wastes from the body in the form of feces.

Food is processed in your digestive system for the purpose of supplying your body with raw materials for metabolism. These raw materials are in the form of nutrients. What is not digested is eliminated.

Figure 22-16

The large light colored area in this X ray is the large intestine.

Section Wrap-up

Review

1. Name in order the organs through which food passes as it moves through the digestive system.

2. Compare mechanical and chemical digestion.

3. How do activities in the large intestine help maintain homeostasis?

4. **Think Critically:** Crackers contain starch. Explain why a cracker held in your mouth for five minutes begins to taste sweet.

Skill Builder
Observing and Inferring
What would happen to food if the pancreas did not secrete its juices into the small intestine? If you need help, refer to Observing and Inferring in the **Skill Handbook.**

Science Journal

Write a paragraph in your Science Journal explaining what would happen to the digestion process if a person had a major portion of the stomach removed due to a disease.

•MINI•QUIZ•

Use the Mini Quiz to check students' recall of chapter content.

1. **Proteins that speed up digestive reactions are called _____ .** *enzymes*

2. **What causes the mechanical mixing of food in the stomach?** *peristalsis*

3. **What structures increase the surface area of the small intestine?** *villi*

Section Wrap-up

Review

1. mouth, esophagus, stomach, small intestine, large intestine, rectum, anus

2. Mechanical digestion involves cutting, grinding, tearing, and crushing by the teeth and tongue in the mouth and peristalsis by the walls of the stomach. Chemical digestion is the actual chemical breakdown of food into usable materials.

3. Water is removed and returned to the body.

4. **Think Critically** Complex carbohydrates in the cracker are broken down into simple sugars.

Science Journal

Journal entries should reflect that the digestive process is probably slowed somewhat because the stomach is not as large as it used to be. Also, patients are advised to eat small portions of food and to eat frequently. This puts less stress on the stomach's digestive activities.

Skill Builder
The food would be undigested and continue to be acidic from the stomach, possibly causing ulcers.

Assessment

Oral Assess students' abilities to infer by having them suggest what happens to digestion and nutrition in people who have lost their teeth. Use the Performance Task Assessment List for Making Observations and Inferences in **PASC,** p. 17. **P**

Activity 22-2

PREPARATION

Purpose

LS **Logical-Mathematical** Design and carry out an experiment to determine what happens to proteins in the digestive system. **L2** **COOP LEARN**

Process Skills

observing, measuring in SI, interpreting data, experimenting, inferring, communicating, recognizing cause and effect, forming a hypothesis, separating and controlling variables

Time

50-60 minutes

Materials

Obtain materials needed for each group. Have an alkaline solution, such as baking soda in water, available for use to test the inference in the Go Further activity.

Safety Precautions

Hydrochloric acid (HCl) is an irritant and corrosive. Vapors are harmful. Any spillage or contact with skin should be flushed with water. If a baking soda solution is used in the Go Further activity to neutralize the HCl, foaming will take place. Do not let the foam spill over the top of the test tube because this will obscure the results.

Possible Hypothesis

Most student hypotheses will reflect that an acidic environment is necessary for protein digestion by the enzyme pepsin.

📁 **Activity Worksheets,** pp. 5, 132-133

Activity 22-2

Design Your Own Experiment
Protein Digestion

You learned that proteins are large, complex, organic compounds necessary for living things to carry out their life processes. To be useful for cell functions, proteins must be broken down into their individual amino acids. The process of chemically breaking apart protein molecules involves several different factors, one of which is the presence of the enzyme pepsin in your stomach.

PREPARATION

Problem

Under what conditions will the enzyme pepsin begin the digestion of protein?

Form a Hypothesis

Formulate a hypothesis about what conditions are necessary for protein digestion to occur. When making your hypothesis, consider the various contents of the digestive juices that are found in your stomach.

Objectives

• Observe the effects of pepsin on gelatin.
• Design an experiment that tests the effect of a variable, such as the presence or absence of acid, on the activity of the enzyme, pepsin.

Possible Materials

• unflavored gelatin
• dropper
• 2 test tubes in a rack
• drinking glass

• pepsin powder
• cold water
• graduated cylinder
• marking pen
• dilute hydrochloric acid
• watch or clock

Safety Precautions

Always use care when working with acid. Wear goggles and an apron. Avoid contact with skin and eyes. Wash your hands thoroughly after pouring the acid.

Inclusion Strategies

Physically Challenged Place students who have difficulty manipulating glassware and acids in charge of formulating the hypothesis, observing, and recording the observations of the reaction.

PLAN THE EXPERIMENT

1. As a group, agree upon and write out the hypothesis statement.
2. As a group, list the steps you will need to take to test your hypothesis. Consider the following factors as you plan your experiment. Based on information provided by your teacher, how will you use the pepsin and the acid? How will the gelatin be prepared? How often will you make observations? Be specific, and describe exactly what you will do at each step.
3. List your materials.
4. Prepare a data table and record it in your Science Journal so that it is ready to use as your group collects data.

Check the Plan
1. Read over your entire experiment to make sure that all steps are in logical order.
2. Identify any constants, variables, and controls of the experiment.
3. Have you allowed for a control? How will the control be treated?
4. *Make sure your teacher approves your plan before you proceed.*

DO THE EXPERIMENT

1. Carry out the experiment as planned.
2. While the experiment is going on, write down any observations that you make and complete the data table in your Science Journal.

Analyze and Apply
1. **Compare** your results with those of other groups.
2. Did you **observe** a difference in the test tubes?
3. **Identify** the constants in this experiment.
4. Did the acid have any effect on the activity of the pepsin? How does this relate to the activity of this enzyme in the stomach?

Go Further

Infer why the activity of pepsin stops in the small intestine. Test your inference and see if the enzyme activity stops.

22-3 Your Digestive System **619**

✓ Assessment

Process Have students predict what would happen to the digestive process if a person took medicine that stopped the production of stomach acid because he or she had indigestion. Use the Performance Task Assessment List for Making Observations and Inferences in **PASC**, p. 17. **P**

PLAN THE EXPERIMENT

Possible Procedures
One possible procedure is to prepare the gelatin and pour an equal amount into each of two test tubes. When the gelatin sets, add 10 drops of acid to one of the test tubes. Sprinkle enough pepsin powder to cover the surface of the gelatin in both test tubes. Observe the gelatin every 10 minutes.

Teaching Strategies
Troubleshooting To prevent the gelatin from melting, place the test tubes in a glass of cold water. Students will need to note the amount of liquid after the acid is added in order to compare with the amount when the gelatin liquefies.

DO THE EXPERIMENT

Expected Outcome
Most results will reflect an increase in the amount of liquid in the acid test tube.

Analyze and Apply
1. Student results should be consistent with those of other groups.
2. Yes, the test tube with acid had more liquid than the one without acid.
3. gelatin and pepsin
4. Pepsin is active only in an acid medium. Acid is added to the food material from the walls of the stomach. This allows digestion to occur.

Go Further
The pepsin activity stops in the small intestine because something neutralizes the acid. The addition of an alkaline will stop the action of the acid.

Background

- A registered dietitian is one who has successfully passed the national Registration Examination for Dietitians. The exam is taken after a prospective dietitian has completed all of his or her educational and internship requirements. The letters *RD* behind a dietitian's name signify this professional qualification.

- In the United States, the profession of dietetics began with the founding of the American Dietetic Association in 1917 by Lenna Frances Copper. Historically, dietitians have played important roles during times of national crisis. For example, during the Great Depression of the 1930s, dietitians helped organize national food relief programs across the United States, and carried out public education programs on how to plan nutritionally adequate diets on limited budgets.

Teaching Strategies

LM **Logical-Mathematical** Using the food pyramid on page 608, have students design a balanced, varied, seven-day menu plan. Challenge students to come up with creative ways to include the minimum of five servings of fruits and vegetables each day. Display the menus in the classroom. **L1**

Career Connection

Career Path In addition to career opportunities in hospitals and other health care facilities, registered dietitians are also employed at fitness and wellness centers, as food service managers in schools and businesses, in industry as food researchers and marketing specialists, and as public health and nutrition educators.

GEORGIA LARSON, *Clinical Dietitian*

On the Job

Q Ms. Larson, as a registered dietitian who works in a hospital setting, what does your job entail?

A On a typical day, I spend most of my time performing nutritional assessments on newly admitted patients, and then designing diets for them that reflect their dietary, medical, even cultural needs.

Q How do you do a nutritional assessment?

A I start by asking questions about what and when a patient normally eats, recent changes in appetite or weight, illnesses, surgeries, and so on. I also ask specific questions about alcohol, sweets, and vitamin supplements. Then I put that information together with the patient's medical condition and design a diet that's right for him or her. Of course, some patients are too sick to answer questions. In those cases, I have to rely on medical histories and the results of lab tests on things like blood sugar, cholesterol, hemoglobin, kidney function, etc.

Personal Insights

Q Is there a common misconception many people have about nutrition?

A Yes, that there are "good" and "bad" foods. There aren't, really. Moderation is really the key to good nutrition. If you want a brownie, have one—just don't eat half the pan!

Q What do you enjoy most about your job?

A The variety. I interact with all kinds of people, of all ages and backgrounds. It's challenging to figure out and meet their needs.

Career Connection

Make a list of ten questions about what it's like to work as a registered dietician. Then contact a dietician at each of the following places and conduct an interview:

- **a wellness center**
- **a nursing home**
- **a weight-loss center**

Compile the results of your interviews and report on your findings to your class.

620 Chapter 22 Nutrients and Digestion

- High School: science, math, health, and home economics
- College: biology, chemistry, physiology, and nutrition science

Most colleges and universities require a nine-month or one-year postgraduate dietetic internship, although in some cases this requirement can be completed during the undergraduate course of study.

For More Information

Students can learn about educational requirements and careers in nutrition by contacting the American Dietetic Society.

Summary

22-1: Nutrition

1. Carbohydrates, fats, proteins, minerals, vitamins, and water are the six nutrients found in food.
2. Carbohydrates provide energy; proteins are needed for growth and repair; fats store energy and cushion organs. Vitamins and minerals regulate functions. Water makes up about 60 percent of body mass and is used for a variety of homeostatic functions.
3. Health is affected by the combination of foods that make up a diet.

22-2: Science and Society: Nutrients Combat Cancer

1. Certain nutrients, including those called antioxidants, may help prevent some types of cancers.
2. Antioxidants prevent chemicals that enter the body from reacting with oxygen. This may stop or prevent cancer from occurring.

22-3: Your Digestive System

1. The digestive system breaks down food mechanically by chewing and churning, and chemically, with the help of enzymes, into substances that cells can absorb and use.
2. Food passes through the mouth, esophagus, stomach, small intestine, large intestine, rectum, and out the anus.
3. Homeostasis is maintained by absorption of water in the large intestine.

Key Science Words

a. amino acid
b. antioxidant
c. carbohydrate
d. chemical digestion
e. chyme
f. digestion
g. fat
h. food group
i. mechanical digestion
j. mineral
k. nutrient
l. peristalsis
m. protein
n. saliva
o. villi
p. vitamin

Reviewing Vocabulary

Match each phrase with the correct term from the list of Key Science Words.

1. process of breaking down food
2. muscular contractions that move food
3. enzyme-containing fluid in the mouth
4. fingerlike projections in small intestine
5. starch and sugar
6. nutrient that forms a cushioning tissue
7. physical breakdown of food
8. inorganic nutrient
9. subunit of protein
10. liquid product that is the result of digestion

Have students read the summary statements to review the major concepts of the chapter.

Reviewing Vocabulary

1. f	**6.** g
2. l	**7.** i
3. n	**8.** j
4. o	**9.** a
5. c	**10.** e

✓ Assessment

Portfolio Encourage students to place in their portfolios one or two items of what they consider to be their best work. Examples include:

- Inclusion Strategies, p. 604
- Explore the Technology, p. 611
- Flex Your Brain, p. 616 P

Performance Additional performance assessments may be found in **Performance Assessment** and **Science Integration Activities.** Performance Task Assessment Lists and rubrics for evaluating these activities can be found in Glencoe's **Performance Assessment in the Science Classroom.**

GLENCOE TECHNOLOGY

 MindJogger Videoquiz

Chapter 22 Have students work in groups as they play the Videoquiz game to review key chapter concepts.

Chapter 22 Review

Checking Concepts

1. a   6. d
2. b   7. b
3. d   8. b
4. d   9. a
5. c   10. b

Understanding Concepts

11. Surface area is increased, making it easier for the enzymes to contact the substrates.
12. Ingestion is eating; digestion is breaking down of food; absorption is the taking in of nutrients by the blood and cells; elimination is ridding the body of wastes.
13. Cheese and butter are high in fat and can contribute to higher levels of blood cholesterol.
14. The pancreas and liver add enzymes to the small intestines.
15. A balanced diet gives the correct nutrients for all functions and reactions of the body.

Thinking Critically

16. Fat, because the gall bladder stores bile, which aids in the digestion of fats.
17. The stomach, because acids are produced there. Antacids are used to counteract an excess of acid.
18. Bile breaks fats down into smaller droplets, which have an increased surface area, so digestion occurs more quickly and easily. This is comparable to the action of soap on grease when washing dishes.
19. The body would retain those vitamins that are fat-soluble. They are stored in fat deposits. The water-soluble vitamins cannot be stored in fat deposits.

Checking Concepts

Choose the word or phrase that completes the sentence.

1. Most digestion occurs in the _____.
   a. duodenum   c. liver
   b. stomach   d. large intestine
2. The _____ makes bile.
   a. gallbladder   c. stomach
   b. liver   d. small intestine
3. Water is absorbed in the _____.
   a. liver   c. esophagus
   b. small intestine   d. large intestine
4. Food does not pass through the _____.
   a. mouth   c. small intestine
   b. stomach   d. liver
5. A substance that prevents chemicals from reacting with oxygen is_____.
   a. an enzyme   c. an antioxidant
   b. a fat   d. an antibiotic
6. The vitamin not used for growth is _____.
   a. A   b. B   c. C   d. K
7. In the _____, hydrochloric acid is added to the food mass.
   a. mouth   c. small intestine
   b. stomach   d. large intestine
8. The _____ produces enzymes that digest proteins, fats, and carbohydrates.
   a. mouth   c. large intestine
   b. pancreas   d. gallbladder
9. The food group containing yogurt and cheese is _____.
   a. dairy   c. meat
   b. grain   d. fruit
10. Carbohydrates are best obtained from _____.
    a. milk   c. meat
    b. grains   d. eggs

Understanding Concepts

Answer the following questions in your Science Journal using complete sentences.

11. What do chewing and the action of bile have in common in the digestive system?
12. Describe the differences among ingestion, digestion, absorption, and elimination.
13. Why should you limit the amount of cheese and butter in your diet?
14. List the structures that add enzymes to chyme in the small intestine.
15. Why is a balanced diet important to maintaining homeostasis?

Thinking Critically

16. What types of food would a person whose gallbladder has been removed need to limit in his or her diet? Explain your choice.
17. In what part of the digestive system do antacids work? Explain your choice.
18. Bile's action is similar to soap. Use this information to explain bile working on fats.
19. Vitamins are in two groups: water- or fat-soluble. Which of these might your body retain? Explain your answer.
20. Based on your knowledge of food groups and nutrients, discuss the meaning of the familiar statement: "You are what you eat."

20. If we eat nutritious food, the body will have good building blocks to use. If we eat a poor diet, the body will not have good building blocks to use.

Developing Skills

21. **Making and Using Graphs** Vitamin C and iron are given the greatest percent of RDA. A person on a fat-restricted diet could eat the product because the total fat is 5% of the RDA.

Developing Skills

If you need help, refer to the **Skill Handbook.**

21. **Making and Using Graphs:**
Recommended Dietary Allowances
(RDA) are made for the amounts of
nutrients people should take in to
maintain health. Prepare a bar graph
of the percent of RDA of each nutrient
from the product information listed
below.

Nutrient	Percent US RDA
Protein	2
Vitamin A	20
Vitamin C	25
Vitamin D	15
Calcium (Ca)	less than 2
Iron (Fe)	25
Zinc (Zn)	15
Total Fat 3.0 g	5
Saturated Fat 0.5 g	3
Cholesterol 0 mg	0
Sodium 60 mg	3

Which nutrients are given the greatest
percent of the Recommended Dietary
Allowance? Could a person on a fat-
restricted diet eat this product? Explain.

22. **Sequencing:** In a table, sequence the
order of organs through which food
passes in the digestive system. Indicate
whether ingestion, digestion, absorption, or elimination takes place in the
individual organs.

23. **Comparing and Contrasting:** Compare
and contrast the location, size, and
functions of the esophagus, stomach,
small intestine, and large intestine.

24. **Concept Mapping:** Make a concept
map showing the chain of events that
occur when fat is digested.

25. **Interpreting Data:** Using a label from a
food, list the nutrients and food groups
to which each belongs. How many
calories are provided by each type?

Performance Assessment

1. **Project:** Research the ingredients used
in diarrhea medications and in laxatives.
Make a chart showing where in the
digestive system these products act
and how they work.

2. **Consumer Decision—Making a Study:**
Collect labels from breakfast cereals.
Compare the amounts of vitamins and
minerals. Are the vitamins and minerals a
natural part of the ingredients? Were any
vitamins or minerals added to make the
cereal more healthful? Place all the cereals with added vitamins or minerals in
one group.

3. **Poster:** Use a paper plate and place mat.
Draw or cut out pictures of the foods you
would eat for one meal: breakfast, lunch,
or dinner. Label what vitamins and minerals are in each food.

2. Answers will vary depending on the
product chosen. Use the Performance
Task Assessment List for Consumer Decision-Making Study in **PASC,** p. 43. [P]

3. Answers will vary. Use the Performance
Task Assessment List for Poster in
PASC, p. 73. [P]

22. **Sequencing** Mouth: ingestion, mechanical and
chemical digestion; Esophagus: ingestion; Stomach:
mechanical and chemical
digestion; Small intestine:
chemical digestion and absorption; Large intestine:
absorption and elimina-
tion.

23. **Comparing and Con-
trasting** See question 22
for functions. Esophagus
is a tube between the
mouth and stomach,
26 cm. Stomach, 16 cm,
is a sac located between
the esophagus and small
intestine. The small intestine is located between
the stomach and large in-
testine, 4.75 m. The large
intestine, 1.25 m, is lo-
cated between the small
intestine and rectum.

24. **Concept Mapping**

Fatty food enters mouth - physically broken into smaller pieces

↓

Passes through esophagus

↓

Enters stomach and mixed with acidic digestive juices

↓

In duodenum, large fat particles float to top of chyme; bile breaks down these particles into smaller ones.

↓

Digestive juices from pancreas begin chemical digestion of fats. These are then used as nutrients by body cells.

25. **Interpreting Data** Answers will vary, depend-
ing on the label.

Performance Assessment

1. Products for diarrhea act
in the large intestine by
slowing down the con-
tractions. Laxatives also
act in the large intestine
by stimulating the intestinal wall, lubricating the
walls of the stool, or by
increasing the bulk of the
stool. Use the Perfor-
mance Task Assessment
List for Data Table in
PASC, p. 37. [P]

Chapter Organizer

Section	Objectives/Standards	Activities/Features
Chapter Opener		**Explore Activity:** Find out about the pumping activity of your heart. p. 625
23-1 Circulation (2 sessions)*	1. **Compare** arteries, veins, and capillaries. 2. **Trace** the pathway of blood through the chambers of the heart. 3. **Describe** pulmonary and systemic circulation. National Content Standards: (5-8) UCP1, A1, A2, B2, C1, C3, E2, F1, F5	MiniLAB: How hard does your heart work? p. 628 Using Technology: A New Heart Pump, p. 629 MiniLAB: What happens in a clogged artery? p. 630 Skill Builder: Concept Mapping, p. 632 Using Math, p. 632 Activity 23-1: Taking Blood Pressure, p. 633
23-2 Science and Society Technology: "Growing" Heart Valves (1 session)*	1. **Describe** some of the problems associated with animal-tissue or plastic replacement heart valves. 2. **Explain** the advantages of tissue-engineered replacement heart valves. National Content Standards: (5-8) UCP5, E2, F5	Explore the Technology, p. 635
23-3 Blood (2 sessions)*	1. **Describe** the characteristics and functions of the parts of blood. 2. **Explain** the importance of checking blood types before a transfusion is given. 3. **Describe** a disease and a disorder of blood. National Content Standards: (5-8) UCP1, A1, A2, B1, C1, F1, F4	Connect to Chemistry, p. 639 Problem Solving: The Baby Exchange, p. 640 Skill Builder: Making and Using Tables, p. 641 Using Math, p. 641 Activity 23-2: Comparing Blood Cells, pp. 642-643
23-4 Your Lymphatic System (2 sessions)*	1. **Describe** the functions of the lymphatic system. 2. **Explain** where lymph comes from. 3. **Explain** the role of lymph organs in fighting infections. National Content Standards: (5-8) UCP1, C1, F1	Skill Builder: Comparing and Contrasting, p. 645 Science Journal, p. 645 Science and Literature: Fantastic Voyage, p. 646 Science Journal, p. 646

* A complete Planning Guide that includes block scheduling is provided on pages 32T-35T.

Activity Materials

Explore	Activities	MiniLABs
page 625 stopwatch or clock with second hand	page 633 sphygmomanometer, stethoscope, table, 2 chairs, rubbing alcohol, cotton balls pages 642-643 prepared slides of human blood and blood of 2 other vertebrates (fish, amphibian, reptile, or bird), microscope, colored pencils	page 628 2 identical, plastic dishpans; water; 150-mL plastic cup; stopwatch or clock with second hand page 630 15-cm length of plastic tubing with a 1-cm diameter, medicine dropper, tap water, toothpick, solid shortening

Need Materials? Call Science Kit (1-800-828-7777).

Teacher Classroom Resources

Reproducible Masters	Transparencies	Teaching Resources
Study Guide, p. 87 **Reinforcement**, p. 87 **Enrichment**, p. 87 **Activity Worksheets**, pp. 136-137, 140, 141 **Lab Manual 54**, Heart Structure **Lab Manual 55**, Measuring Heartbeat **Lab Manual 56**, Blood Pressure **Science Integration Activity 23**, Preparation of Oxygen **Concept Mapping**, p. 51 **Multicultural Connections**, pp. 49-50	**Section Focus Transparency 83**, Clean Water In, Dirty Water Out **Science Integration Transparency 23**, Pipeline Pressure **Teaching Transparency 45**, Pulmonary Circulation **Teaching Transparency 46**, Systemic Circulation	**Glencoe Life Science Interactive Videodisc** **Life Science CD-ROM** **Spanish Resources** **English/Spanish Audiocassettes** **Cooperative Learning Resource Guide** **Lab Partner** **Lab and Safety Skills** **Lesson Plans**
Study Guide, p. 88 **Reinforcement**, p. 88 **Enrichment**, p. 88	**Section Focus Transparency 84**, Replacing Damaged Organs	**Assessment Resources** **Chapter Review**, pp. 49-50 **Assessment**, pp. 151-154 **Performance Assessment in the Science Classroom (PASC)** **MindJogger Videoquiz** **Alternate Assessment in the Science Classroom** **Performance Assessment** **Chapter Review Software** **Computer Test Bank**
Study Guide, p. 89 **Reinforcement**, p. 89 **Enrichment**, p. 89 **Activity Worksheets**, pp. 138-139 **Cross-Curricular Integration**, p. 27	**Section Focus Transparency 85**, Blood Banks	
Study Guide, p. 90 **Reinforcement**, p. 90 **Enrichment**, p. 90 **Critical Thinking/Problem Solving**, p. 27 **Science and Society Integration**, p. 27	**Section Focus Transparency 86**, What is a tonsillectomy?	

Key to Teaching Strategies

The following designations will help you decide which activities are appropriate for your students.

L1 Level 1 activities should be within the ability range of all students, including those with learning difficulties.

L2 Level 2 activities should be within the ability range of the average to above-average student.

L3 Level 3 activities are designed for the ability range of above-average students.

LEP LEP activities should be within the ability range of Limited English Proficiency students.

LS These activities are designed to address different learning styles.

COOP LEARN Cooperative Learning activities are designed for small group work.

P These strategies represent student products that can be placed into a best-work portfolio.

GLENCOE TECHNOLOGY

The following multimedia resources are available from Glencoe.

Science and Technology Videodisc Series (STVS)
Human Biology
  Modeling Blood Flow
  Measuring Blood Pressure
  Testing Heart Valves
  Heart Valve Balloon
The Infinite Voyage Series
The Champion Within
A Taste of Health

National Geographic Society Series
STV: Human Body Vol. 1
Glencoe Life Science Interactive Videodisc
Photosynthesis and Cellular Respiration
Life Science CD-ROM

Teacher Classroom Resources

This is a representation of key blackline masters available in the Teacher Classroom Resources.

Teaching Aids

Section Focus Transparencies

Science Integration Transparencies

Teaching Transparencies

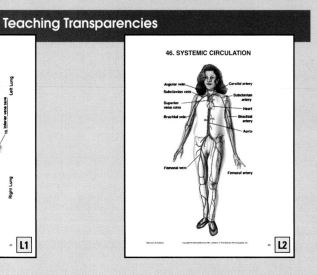

Meeting Different Ability Levels

Study Guide

Reinforcement

Enrichment Worksheets

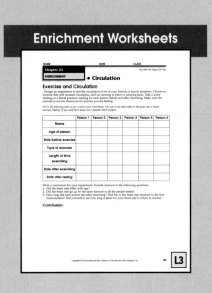

Hands-On Activities

Science Integration Activity

Preparation of Oxygen

L1

Lab Manual

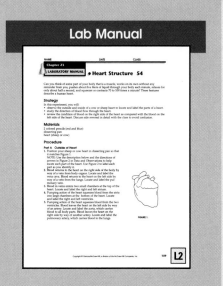

Heart Structure 54

L2

Assessment

Performance Assessment

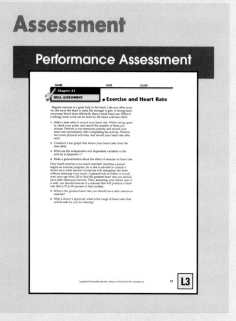

Exercise and Heart Rate

L3

Enrichment and Application

Critical Thinking/ Problem Solving

The Circulatory System

L2

Cross-Curricular Integration

Your Circulatory System

L1

Science and Society Integration

The Circulatory System

L2

Multicultural Connections

Looking into the Past Helps Modern Medicine

L2

Concept Mapping

The Circulatory System

L3

Chapter 23

The Circulatory System

CHAPTER OVERVIEW

Section 23-1 This section provides an overview of the major structures and functions of the circulatory system. The importance of the system is highlighted with a discussion of the diseases and disorders of the heart.

Section 23-2 Science and Society In this section new technology for replacing heart valves is discussed. Tissue-engineered valves are designed to reduce the problems associated with current replacement valves.

Section 23-3 This section focuses on blood, the flowing tissue of the circulatory system. The liquid and solid components of blood and their role in transporting materials are discussed.

Section 23-4 The lymphatic system is introduced, and its important role in fighting infections is detailed.

Chapter Vocabulary

atria
ventricle
pulmonary
  circulation
systemic
  circulation
coronary
  circulation
artery
vein
capillary
blood pressure

atherosclerosis
hypertension
heart murmur
plasma
hemoglobin
platelet
lymphatic
  system
lymph
lymphocyte
lymph node

Theme Connection

Stability and Change One of the themes in this text is stability and change. The circulatory system is a good example of a system that functions to maintain stability and respond to changes in the human life processes.

Previewing the Chapter

Section 23-1   Circulation
  ▶ Your Cardiovascular System
  ▶ Your Heart
  ▶ Blood Vessels
  ▶ Blood Pressure
  ▶ The Big Push
  ▶ Cardiovascular Disease

Section 23-2   Science and Society
Technology: "Growing" Heart Valves

Section 23-3   Blood
  ▶ Functions of Blood
  ▶ Parts of Blood
  ▶ Blood Clotting
  ▶ Blood Types
  ▶ Diseases and Disorders of Blood

Section 23-4   Your Lymphatic System
  ▶ Functions of Your Lymphatic System
  ▶ Lymphatic Organs
  ▶ A Disease of the Lymphatic System

Learning Styles   Look for the following logo for strategies that emphasize different learning modalities.

Kinesthetic	Activity, p. 627
Visual-Spatial	Visual Learning, pp. 627, 628, 632, 640, 644; Demonstration, pp. 627, 636, 638; MiniLAB, p. 630
Interpersonal	Explore, p. 625; Reteach, p. 640; Activity 23-2, pp. 642-643
Logical-Mathematical	MiniLAB, p. 628; Activity 23-1, p. 633

Chapter 23

The Circulatory System

You probably rarely think about how fast your heart beats. Unlike the girl below, usually you're unaware of its pumping action. Some situations change your awareness. You feel your heart pounding faster. You sense or "hear" the blood rushing through your body. Fear, stress, or anxiety causes your heart to beat faster. Strenuous activity also changes the rate of your heartbeat. In the following activity, find out how much exercise changes your heartbeat rate. Use your pulse rate, or number of heartbeats per minute, to track the change.

EXPLORE ACTIVITY

Find out about the pumping activity of your heart.

1. Find your pulse by placing your middle and index fingers over one of the carotid arteries in your neck. Don't press too hard.
2. Calculate your resting heartbeat rate. Count each beat of the carotid pulse rate for 15 seconds. Multiply the number of beats by four.
3. Jog in place for one minute. Take your pulse again.

Observe: How does your pulse rate change? What causes the change?

Previewing Science Skills

▶ In the **Skill Builders**, you will **map concepts, make and use tables**, and **compare** and **contrast**.

▶ In the **Activities**, you will **observe, experiment**, and **collect data**.

▶ In the **MiniLABs**, you will **compare, observe, infer,** and **interpret data**.

625

Assessment Planner

Portfolio
Refer to page 647 for suggested items that students might select for their portfolios.

Performance Assessment
See page 647 for additional Performance Assessment options.
Skill Builders, pp. 632, 641, 645
MiniLABs, pp. 628, 630
Activities 23-1, p. 633; 23-2, pp. 642-643

Content Assessment
Section Wrap-ups, pp. 632, 635, 641, 645
Chapter Review, pp. 647-649
Mini Quizzes, pp. 632, 641, 645

Group Assessment
Opportunities for group assessment occur with Cooperative Learning Strategies and Flex Your Brain Activities.

Prepare

Section Background

- The heart and the blood vessels make up the cardiovascular system. The action of the heart drives the blood through a closed system of tubes.
- Arteries have greater elasticity than veins in order to withstand the forceful movement of blood.
- Blood distributes materials for cell activities and functions in maintaining homeostasis within the cell.

Preplanning

- To prepare for Activity 23-1, obtain sphygmomanometers and stethoscopes.
- To prepare for the MiniLAB on page 628, obtain 2 large dishpans, a 150-mL plastic cup, a stopwatch or watch, water, and an apron for each laboratory team.
- To prepare for the MiniLAB on page 630, obtain 15 cm of clear plastic tubing (1 cm diameter), a medicine dropper, water, a toothpick, and solid shortening for each laboratory team.

1 Motivate

Bellringer

Before presenting the lesson, display **Section Focus Transparency 83** on the overhead projector. Assign the accompanying **Focus Activity** worksheet. L1 LEP

Tying to Previous Knowledge

Recall information in Unit 5 about ectotherms and endotherms. Relate the role of the circulatory system in maintaining a constant body temperature.

626

23•1 Circulation

Science Words

- atria
- ventricle
- pulmonary circulation
- systemic circulation
- coronary circulation
- artery
- vein
- capillary
- blood pressure
- atherosclerosis
- hypertension

Objectives

- Compare arteries, veins, and capillaries.
- Trace the pathway of blood through the chambers of the heart.
- Describe pulmonary and systemic circulation.

Your Cardiovascular System

With a body made up of trillions of cells, you seem quite different from a one-celled amoeba living in a puddle of water. But are you really that different? Even though your body is larger and made up of complex systems, the cells in your body have the same needs as the amoeba. Both of you need a continuous supply of oxygen and nutrients and a way to remove cell wastes.

An amoeba takes oxygen directly from its watery environment. Nutrients are distributed throughout its single cell by cytoplasmic streaming. In your body, a cardiovascular system distributes materials. *Cardio-* means "heart" and *vascular* means "vessel." Your cardiovascular system includes your heart, blood, and kilometers of vessels that carry blood to every part of your body as shown in **Figure 23-1.** It is a closed system because blood moves within vessels. The system moves oxygen and nutrients to cells and removes carbon dioxide and other wastes from the cells. Movement of materials into and out of your cells happens by diffusion. Recall from Chapter 3 that diffusion is when a material moves from a place of high concentration to an area of lower concentration.

Figure 23-1

The human circulatory system can be compared to a city's highway system. Goods are transported to individual homes or factories. Completed products and wastes are collected and removed. In a similar way, substances are transported throughout your body.

626   Chapter 23 The Circulatory System

Program Resources

 Reproducible Masters
Study Guide, p. 87 L1
Reinforcement, p. 87 L1
Enrichment, p. 87 L3
Activity Worksheets, pp. 5, 136-137, 140, 141 L1
Science Integration Activities, pp. 45-46
Lab Manual 54, 55, 56

Concept Mapping, p. 51
Multicultural Connections, pp. 49-50

Transparencies
Section Focus Transparency 83 L1
Teaching Transparencies 45, 46
Science Integration Transparency 23

Your Heart

Your heart is an organ made of cardiac muscle. It is located behind your sternum and between your lungs. Your heart has four cavities called chambers. The two upper chambers are the right and left **atria** (singular *atrium*). The two lower chambers are the right and left **ventricles.** During a single heartbeat, both atria contract at the same time. Then both ventricles contract at the same time. A valve separates each atrium from the ventricle below it so that blood flows only from an atrium to a ventricle. A wall prevents blood from flowing between the two atria or the two ventricles.

Pulmonary Circulation

Blood moves throughout your body in a continuous circulatory system. Scientists have divided the system into three sections. The beating of your heart controls blood flow through these sections. **Figure 23-2** shows blood movement through the section called pulmonary circulation. **Pulmonary circulation** is the flow of blood through the heart, to the lungs, and back to the heart. Use the illustrations in **Figure 23-3** to trace the path blood takes through this part of the circulatory system.

Figure 23-2

Pulmonary circulation moves blood between the heart and lungs.

C Oxygen-rich blood travels through the pulmonary vein and into the left atrium. The pulmonary veins are the only veins to carry oxygen-rich blood.

A Blood, high in carbon dioxide and low in oxygen, returns from the body to the heart. It enters the right atrium through the vena cava.

Superior vena cava

Aorta

Capillaries

Pulmonary vein

Right atrium

Pulmonary artery

Left atrium

Left ventricle

B The right atrium contracts, forcing the blood into the right ventricle. When the right ventricle contracts, the blood leaves the heart and goes through the pulmonary artery to the lungs. Here carbon dioxide diffuses out of the blood and oxygen diffuses into the blood.

Right lung

Inferior vena cava

Right ventricle

Left lung

D The left atrium contracts and forces the blood into the left ventricle. The left ventricle contracts, forcing the blood out of the heart and into the aorta.

23-1 Circulation   **627**

GLENCOE TECHNOLOGY

Videodisc

Glencoe Life Science Interactive Videodisc

Side 1, Lesson 4

Photosynthesis and Cellular Respiration

42444-43807

MiniLAB

Purpose

LS **Logical-Mathematical** The students will observe and record their rate of water bailing and compare it with the rate per minute of a pumping heart. **L1**
COOP LEARN

Materials

2 dishpans, water, 150-mL plastic cup, stopwatch

Teaching Strategies

To help students know whether they are moving too quickly or too slowly, have them first determine how many times they must bail water each second. Have them try to bail at a smooth, steady rate throughout the whole minute.

Analysis

1. Answers will vary depending on students' abilities, but most likely they will find it difficult to maintain the 70 cups per minute rate.

2. Students can stop or change arms. The heart must continue and not fatigue in order to maintain the body's activities.

✔ Assessment

Process Have students infer the tiring time for cardiac muscles and compare this with the tiring time for skeletal muscles. Use the Performance Task Assessment List for Making Observations and Inferences in **PASC**, p. 17. **P**

📁 **Activity Worksheets,** pp. 5, 140

Visual Learning

Figure 23-3 How does the rate change when a person stops exercising? *it gradually slows* **LEP** **LS**

Figure 23-3

The rate at which blood flows through the systemic system depends on how quickly the left ventricle contracts. *How does the rate change when a person stops exercising?*

Jugular vein
Subclavian vein
Superior vena cava
Brachial vein
Carotid artery
Subclavian artery
Heart
Aorta
Femoral artery
Femoral vein

MiniLAB

How hard does your heart work?

Procedure
1. Work with a partner. Fill a large dishpan with water. Put an empty dishpan next to it.
2. Use a 150-mL plastic cup. Fill it with water and pour it into the empty dishpan. Try to bail the water at a speed of 70 cups per minute. Have your partner count how many cups you bail per minute.

Analysis
1. Compare the number of cupsful you bail per minute to the 70 beats per minute of a resting heart.
2. What can you do when the muscles of your arm get tired? Explain why cardiac muscle in your heart cannot do the same.

Systemic Circulation

The final step of pulmonary circulation occurs when blood is forced from the left ventricle into the aorta. The aorta is the largest artery of your body and carries blood away from the heart. **Systemic circulation** moves oxygen-rich blood to all of your organs and body tissues except for the heart and lungs. It is the most extensive of the three sections of your circulatory system. **Figure 23-3** shows the major arteries and veins involved in systemic circulation. Once nutrients and oxygen are delivered by blood to your body cells and exchanged for carbon dioxide and wastes, the blood returns to the heart in veins. From the head and neck areas, blood returns through the superior vena cava. From your abdomen and the lower parts of your body, blood returns through the inferior vena cava. More information about arteries and veins will be presented later in this chapter.

Coronary Circulation

Your heart has its own blood vessels that supply it with nutrients and oxygen and remove wastes. These blood vessels are involved in coronary circulation. **Coronary circulation** is the flow of blood to the tissues of the heart. Whenever the coronary circulation is blocked, oxygen cannot reach the cells of the heart. The result is a heart attack.

📁 Use these program resources as you teach this lesson.
Science Integration Activities, pp. 45-46
Multicultural Connections, pp. 49-50
Lab Manual 55

🎮 Use **Teaching Transparencies 45** and **46** as you teach this lesson.

Blood Vessels

It wasn't until the middle 1600s that scientists confirmed that blood circulates only in one direction and that it is moved by the pumping action of the heart. They found that blood flows from arteries to veins. But they couldn't figure out how blood gets from arteries to veins. With the invention of the microscope, capillaries were seen.

Types of Blood Vessels

As blood moves out of the heart, it begins a journey through arteries, capillaries, and veins. **Arteries** are blood vessels that move blood away from the heart. Arteries, shown in **Figure 23-4** on the next page, have thick elastic walls made of smooth muscle. Each ventricle of the heart is connected to an artery, so with each contraction blood is moved from the heart into arteries.

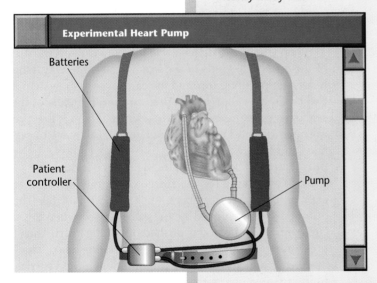

Experimental Heart Pump

Batteries

Patient controller

Pump

USING TECHNOLOGY

A New Heart Pump

Science has done much to increase the survival rate of heart transplant patients. But too often, the patient dies before a suitable donor heart is found. Biomedical engineers are working on several pumps that might help a patient's damaged heart continue to work until a donor heart is available.

A Pump for Future Patients

One potential pump might be made small enough to be implanted under the patient's ribs. This pump has a rotor that is suspended magnetically. A rotor is the rotating part of an electric motor. Most rotors are suspended by ball bearings or roller bearings. Both types of bearings need to be lubricated. Unfortunately, the lubricant might leak into the bloodstream and cause serious problems for the patient. A rotor that is magnetically suspended eliminates the need for lubricants. While this pump solves one problem, it creates another. Magnetically suspended rotors produce heat that might damage nearby body tissues.

Think Critically:
Pumps with bearings have more moving parts than the new pump. How does the number of moving parts affect the amount of wear and tear on the pump? How might this influence the life span of the pump?

 23-1 Circulation **629**

USING TECHNOLOGY

Artificial hearts have been developed as replacements for damaged hearts. Have students discuss why heart pumps might be more beneficial than artificial hearts.

Think Critically
Fewer moving parts mean less friction and less heat being generated. The pump should last longer.

GLENCOE TECHNOLOGY

 Videodisc

The Infinite Voyage: The Champion Within
Chapter 7
Combating Heart Attacks

The Infinite Voyage: A Taste of Health
Chapter 4
Consequences of a Fatty Diet

Chapter 6
Genetic Links to Cholesterol

Chapter 7
Arresting Heart Disease: A Case Study

NATIONAL GEOGRAPHIC SOCIETY

 Videodisc

STV: Human Body Vol. 1
Circulatory and Respiratory Systems
Unit 1, Side 1
The Heart's Central Role

4998-7737

Cultural Diversity

Diet Have students research the relationship of diets that are associated with certain cultures as well as being low in fats and the incidence of heart disease. L2

MiniLAB

Purpose
 Visual-Spatial The students will observe the restricted flow of water and make inferences about blood flow through blocked arteries. **L1** **LEP**
COOP LEARN

Materials
15-cm-length plastic tubing (1 cm diameter), medicine dropper, water, toothpick, solid shortening

Teaching Strategies
Have different teams insert different amounts of shortening to show varying degrees of blockage.

Analysis
1. The rate of flow of water through the clotted tube should be less.
2. An inference can be made that fatty deposits in an artery will reduce the blood flow.

✓ Assessment

Performance Have students design an instrument or a method that will remove the fatty deposit, but will not harm the blood vessel wall. Use the Performance Task Assessment List for Invention in **PASC**, p. 45. **P**

📁 **Activity Worksheets**, pp. 5, 141

Figure 23-4
Blood flows from arteries to capillaries to veins. Valves in veins help blood flow back toward the heart.

MiniLAB

What happens in a clogged artery?

Procedure
1. Use a 15-cm length of plastic tubing with a diameter of 1 cm.
2. With a medicine dropper, squirt water through the tube. Observe the flow of water through the tube.
3. Using a toothpick, insert a small glob of shortening into the tube.
4. Squirt water through the tube again.

Analysis
1. Compare the flow of water in each situation.
2. Infer what happens to the flow of blood in an artery that has a fatty deposit on its inside wall.

Veins are blood vessels that move blood to the heart. Veins have one-way valves to keep blood moving toward the heart. If there is a backward movement of the blood, the pressure of the blood closes the valves. Veins that are near skeletal muscles are squeezed when these muscles contract. This action helps blood move toward the heart. Blood in veins carries waste materials from cells and is therefore low in oxygen.

Capillaries are microscopic blood vessels that connect arteries and veins. The walls of capillaries are only one cell thick. Nutrients and oxygen diffuse to body cells through thin capillary walls. Waste materials and carbon dioxide move from body cells into the capillaries to be carried back to the heart.

Blood Pressure

When you pump up a bicycle tire, you can feel the pressure of the air on the walls of the tire. In the same way, when the heart pumps blood through the cardiovascular system, blood exerts a force called **blood pressure** on the walls of the vessels. This pressure is highest in arteries. There is less pressure in capillaries and even less in veins. As the wave of pressure rises and falls in your arteries, it is felt as your pulse. Normal pulse rates are between 65 and 80 beats per minute.

Blood pressure is measured in large arteries and is expressed by two numbers, such as 120 over 80. The first number is a measure of the pressure caused when the ventricles contract and blood is pushed out of the heart. Then blood pressure suddenly drops as the ventricles relax. The lower number is a measure of the pressure when the ventricles are filling up, just before they contract again.

630 Chapter 23 The Circulatory System

📁 Use these program resources as you teach this lesson.
Concept Mapping, p. 51
Lab Manual 56

The Big Push

You learned that a force is exerted on the inner walls of your blood vessels. The force is the result of blood being pumped through your body by the heart. Blood pressure is a measure of this force. To understand what a blood pressure reading means, you must first know more about pressure as explained in **Figure 23-5** and in the following text.

The total amount of force exerted by a fluid, such as blood or the gases in the atmosphere, depends on the area on which it acts. This is called pressure. In other words, pressure is the amount of force exerted per unit of area. This is written as:

$$P(\text{Pressure}) = \frac{F(\text{Force})}{A(\text{Area})}$$

Pressure Measurement

Scientists measure atmospheric pressure with a mercury barometer. At sea level, normal atmospheric pressure is 760 mm Hg. This means that the force of the atmosphere will raise a column of mercury (Hg) 760 millimeters in the barometer. Compare this to a normal blood pressure reading of 120 over 80 for a young adult. The first number is the systolic pressure (the pressure produced when the ventricles force blood from the heart). In this reading, the systolic pressure would raise a column of mercury 120 mm in a barometer. The second number is the diastolic pressure (the pressure at the end of the cardiac cycle). This pressure would raise a column of mercury 80 mm in a barometer.

Control of Blood Pressure

Special nerve cells in the walls of the aorta and the carotid arteries sense changes in your blood pressure. Messages are sent to the brain, and the amount of blood pumped by the heart is regulated. This provides for a regular, normal pressure within the arteries.

Figure 23-5

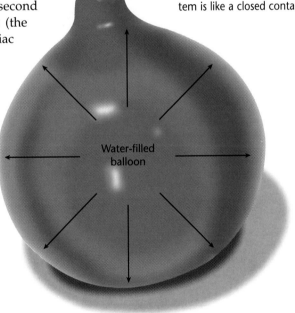

When pressure is exerted on a fluid in a closed container, the pressure is transmitted through the liquid in all directions. A balloon filled with water has the same amount of pressure pushing on all the inner surfaces of the balloon. Your circulatory system is like a closed container.

Water-filled balloon

23-1 Circulation **631**

 Use **Science Integration Tranparency 23** as you teach this lesson.

The term *pressure* is used in discussions of fluids, such as gases and liquids. Pressure is produced when a force is applied to a fluid. The greater the force, the greater the pressure on the fluid and the smaller the volume it takes up. Usually, the volume of a gas decreases by one-half when the pressure doubles.

Teacher F.Y.I.

The peak pressure of blood flow caused by the contraction of the ventricles is systolic pressure. The low pressure of blood flow produced when the heart relaxes is diastolic pressure.

3 Assess

Check for Understanding

Check students' understanding of the direction of blood flow in arteries (from the heart) and in veins (to the heart) and how the blood gets from the arteries to the veins (capillaries). Divide the class into groups of two students each. Hang a large paper on the wall for each pair. One student draws an outline of the body and the major vessels. The other labels the vessels.

Reteach

On a chart of the circulatory system, have students trace with their fingers the flow of blood from the heart to a major artery and back again.

Extension

For students who have mastered this section, use the **Reinforcement** and **Enrichment** masters.

4 Close

•MINI•QUIZ•

1. The lower chambers of the heart are the _____ . *ventricles*

2. What is the largest artery? *aorta*

3. What is coronary circulation? *flow of blood to tissues of the heart*

4. Another name for high blood pressure is _____ . *hypertension*

Section Wrap-up

Review

1. Students' answers should reflect information on pages 629 and 630.

2. Blood enters the right atrium and then moves into the right ventricle, which pumps blood to the lungs. From the lungs, blood enters the left atrium and then moves into the left ventricle, and out to the body through the aorta.

3. Students' answers should reflect information on pages 627 and 628.

4. Think Critically carbon dioxide

USING MATH

1 hour	=	4500
1 day	=	108 000
1 week	=	756 000
1 month (30 days)	=	3 240 000
1 year	=	38 880 000
80 years	=	3 110 400 000

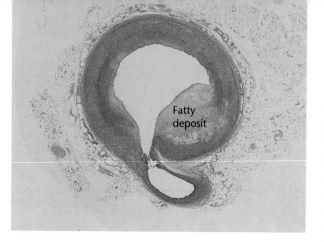

Fatty deposit

Figure 23-6

Atherosclerosis interferes with blood flow by blocking blood vessels with fatty substances. If an artery in the brain is clogged, a stroke occurs. *What happens if an artery in the heart is blocked?*

Cardiovascular Disease

Any disease or disorder that affects the cardiovascular system can seriously affect your health. Heart disease is the major cause of death in the United States. One leading cause of heart disease is **atherosclerosis,** a condition, shown in **Figure 23-6,** of fatty deposits on arterial walls. Eating foods high in cholesterol and saturated fats may cause these deposits to form. The fat builds up and forms a hard mass that clogs the inside of the vessel. As a result, less blood flows through the artery. If the artery is clogged completely, blood is not able to flow through.

Another disorder is high blood pressure, or **hypertension.** Atherosclerosis can cause hypertension. A clogged artery can cause the pressure within the vessel to increase. This causes the walls to lose their ability to contract and dilate. Extra strain is placed on the heart as it works harder to keep blood flowing. Being overweight as well as eating foods with too much salt and fat may contribute to hypertension. Smoking and stress also can increase blood pressure.

Section Wrap-up

Review

1. Compare and contrast the three types of blood vessels.

2. Explain the pathway of blood through the heart.

3. Contrast pulmonary and systemic circulations.

4. **Think Critically:** What respiration waste product builds up in blood and cells when the heart is unable to pump blood efficiently?

Skill Builder
Concept Mapping

Make an events chain concept map to show pulmonary circulation beginning at the right atrium and ending at the aorta. If you need help, refer to Concept Mapping in the **Skill Handbook.**

USING MATH

If a person's heart beats about 75 times per minute, calculate how many times it beats in one hour. Calculate the number of beats in one day, one week, one month, and one year. If the person lives to be 80 years old, how many times will the heart beat?

Skill Builder

Blood entering the right atrium is pumped into the right ventricle, which contracts and moves the blood to the lungs. From the lungs the blood enters the left atrium, which contracts and forces the blood into the left ventricle, which then contracts to move the blood to the body through the aorta.

✓Assessment

Performance Assess students' abilities to form a concept map of the flow of blood through the three types of vessels. Use the Performance Task Assessment List for Concept Map in **PASC,** p. 89. **P**

Taking Blood Pressure

Blood exerts a force on the walls of the blood vessels. This force is known as blood pressure. Blood pressure can be an indicator of a person's health. Try this activity to learn how to find a person's blood pressure.

Problem
How can you measure arterial pressure?

Materials
- sphygmomanometer
- stethoscope
- table
- chairs (2)
- alcohol
- cotton balls

Procedure
1. Make a table like the one shown. Use it to record your data.
2. Your partner should sit down on a chair with his or her arm on the table. One hand should face up.
3. Sit and face your partner. Place the blood pressure cuff on your partner's arm above the elbow. Align the arrows on the cuff. Close the cuff snugly.
4. Close the valve on the pump. Place the earpieces of the stethoscope in your ears. Place the round stethoscope head on your partner's brachial artery as shown in photograph. Squeeze the pump until the needle reaches 160 on the gauge. No blood will flow through this constricted area.

5. *Slowly* open the valve of the pump. **CAUTION:** *Do not open the valve quickly.* Observe the gauge. The needle will drop at a constant rate. Watch for the number on the gauge when you first hear a beat. This is the top, or systolic, number.
6. Keep listening for a heartbeat. Observe the needle on the gauge. Note the number on the gauge when you no longer hear a beat. This is the bottom, or diastolic, number.
7. Take the cuff off your partner's arm. Record your partner's blood pressure. Clean the earpieces of the stethoscope with alcohol and cotton balls. Trade places with your partner. **CAUTION:** *Do not do this activity on the same person more than one time per class period.*

Analyze
1. Describe what happened in your brachial artery during this activity.
2. As you listened to your partner's blood pressure, what did you notice about the needle on the gauge when you first heard the heartbeat rate and later when it stopped?

Conclude and Apply
3. How was your blood pressure different from your partner's?
4. What can you **infer** about a person who has low blood pressure?

Data and Observations

Sample Data

Blood Pressure	Example	Yours	Partner's
Systolic	120	110	120
Diastolic	80	70	70

23-1 Circulation **633**

Purpose
Logical-Mathematical Measure arterial pressure and interpret results of the data.
L1 COOP LEARN

Process Skills
interpreting data, communicating, making and using graphs, using numbers

Time
30-40 minutes

Activity Worksheets, pp. 5, 136-137

Teaching Strategies
Troubleshooting Remind students to close the valve on the pump. The pump should not be squeezed so the needle exceeds 160 on the gauge. The valve of the pump should be opened slowly.

Answers to Questions
1. The pressure in the brachial artery is temporarily closed off, then released.
2. The needle dropped on the gauge.
3. Pressures will probably be about the same unless other factors such as stress or health complications exist.
4. Accept all reasonable conclusions. Some will answer that this person will be tired, lacking in energy, and may appear pale.

Assessment

Performance To further assess students' understanding of blood pressure, have students graph the class data to find the average range of pressures. Use the Performance Task Assessment List for Graph from Data in **PASC**, p. 39. **P**

Prepare

Section Background

Discuss with the class that defective heart valves can cause various physical symptoms. Mechanical replacement valves may be rejected or cause infections and blood clots. Tissue-engineered valves may prevent these problems.

1 Motivate

Bellringer

 Before presenting the lesson, display **Section Focus Transparency 84** on the overhead projector. Assign the accompanying **Focus Activity** worksheet. L1 LEP

Tying to Previous Knowledge

Have students recall how blood pressure closes the one-way valves in veins.

2 Teach

Inquiry Question

How can a heart valve disorder cause dizziness after physical activity? *The brain is not receiving enough oxygen-enriched blood.*

GLENCOE TECHNOLOGY

 Videodisc
STVS: Human Biology
Disc 7, Side 1
Testing Heart Valves (Ch. 4)

Heart Valve Balloon (Ch. 6)

TECHNOLOGY:
23•2 "Growing" Heart Valves

Science Words
heart murmur

Objectives
- Describe some of the problems associated with animal-tissue or plastic replacement heart valves.
- Explain the advantages of tissue-engineered replacement heart valves.

Heart Valve Disease

As you can see in **Figure 23-7,** the heart has two sets of one-way valves that regulate the flow of blood within the heart's chambers. Blood passes through an atrioventricular or A-V valve as it moves between the atrium and ventricle in each half of the heart. Each half also has a semilunar valve. One is found where blood passes from the left ventricle into the aorta. The other is where blood moves from the right ventricle into the pulmonary artery. Sometimes a valve does not work properly and a stream or spurt of blood is able to move back through the valve. This defect is called a **heart murmur.** Many people are born with slight murmurs that do not affect them adversely. One cause for a heart murmur that does affect well-being may be an infection due to rheumatic fever. Symptoms of improperly functioning valves include shortness of breath, fatigue, and sometimes dizziness.

Pulmonary (semilunar) valve

Aortic (semilunar) valve

Mitral (A-V) valve

Tricuspid (A-V) valve

Figure 23-7

The characteristic "lub-dupp" sound of your heart is caused by the closing of the heart's valves. The "lub" is made as the ventricles contract and blood is forced back against the A-V valves. The "dupp" is made when blood is forced back against the semilunar valves.

634 Chapter 23 The Circulatory System

Program Resources

 Reproducible Masters
Study Guide, p. 88 L1
Reinforcement, p. 88 L1
Enrichment, p. 88 L3

Transparencies
Section Focus Transparency 84 L1

Standard Treatments

Severe heart murmurs may require surgery, like that performed in **Figure 23-8**, to replace damaged valves. In the United States alone, nearly 60 000 heart valves are replaced each year. Several methods are currently used to replace a valve. One technological device is a plastic ball-and-cage valve that allows blood to flow in only one direction. The pressure of blood forces the ball away from an opening. When the pressure is reduced, the ball falls back into the opening and seals it. Another method is to replace the valve with tissue from the blood vessels of a pig. The problem with these methods is that the body often rejects the materials. There are also risks of infections and the formation of blood clots. In addition, the artificial valve lasts only from ten to 20 years.

Growing a Valve

A promising new technique currently being tested is to "grow" a replacement valve within the heart. Heart muscle tissue is taken from the patient's heart and cultured. This means the heart cells are removed and grown in a laboratory. The cells that develop are placed into a material that serves as a framework for the growth of a living heart tissue valve. Then the damaged valve is surgically removed and the new cells in the framework material are placed in the valve area. In about six weeks, the framework material dissolves and the newly grown valve takes over the control of blood flow. Test results have so far been promising. But, more tests are necessary before the procedure will be used on humans.

Figure 23-8

Only a limited number of the 40 000 patients who require a heart transplant will be able to receive one. The new tissue-engineered valve could potentially help many of these patients.

Section Wrap-up

Review

1. Why will the new tissue-engineered valves be better than the plastic or pig tissue valves now being used?

2. Why will these new valves not be rejected by the body?

Explore the Technology

If this new technique is successful, what other body parts might be "grown" to replace damaged organs?

*inter*NET
CONNECTION

Visit the Chapter 23 Internet Connection at Glencoe Online Science, **www.glencoe.com/sec/ science/life,** for a link to more information about heart valves.

SCIENCE & SOCIETY

---

3 Assess

Check for Understanding

Have students answer questions 1 and 2 in the Section Wrap-up.

Reteach

Review with students the flow of blood through the heart and the function of the valves. The formation of tissue-engineered replacement valves is somewhat analogous to the process of bone development. Cartilage breaks down and is replaced by bone-forming cells.

Extension

For students who have mastered this section, use the **Reinforcement** and **Enrichment** masters.

4 Close

Have students prepare a list of ways that a patient would benefit from these new valves. Have them include health, economic, and social factors in the list.

Section Wrap-up

Review

1. The body will not reject the valves, infections and blood clots will be reduced, and the valves should not wear out.

2. The new valves will be made from the patient's own cells.

Explore the Technology

Answers will vary, but a list might include the sphincter valves in the digestive system or organs such as the nose or ear.

635

Prepare

Section Background

- Cells and cell fragments make up the formed elements in blood. The percent of these in total blood volume is about 46 percent.
- White blood cells can be classified into five groups—neutrophils, eosinophils, basophils, monocytes, and lymphocytes.
- Blood platelets are fragments of large cells called megakaryocytes that form in red bone marrow.
- A normal component of plasma is cholesterol, a lipid substance, necessary for cell activities. Excess cholesterol can cause fatty deposits in blood vessels.

Preplanning

To prepare for Activity 23-2, obtain the necessary microscope slides.

1 Motivate

Bellringer

Before presenting the lesson, display **Section Focus Transparency 85** on the overhead projector. Assign the accompanying **Focus Activity** worksheet. [L1] [LEP]

Demonstration

[LS] **Visual-Spatial** Fill two and one-half 2-L plastic soft drink bottles with red-colored water. The water represents the 5 L of blood in an adult.

23•3 Blood

Science Words

plasma
hemoglobin
platelet

Objectives

- Describe the characteristics and functions of the parts of blood.
- Explain the importance of checking blood types before a transfusion is given.
- Describe a disease and a disorder of blood.

Functions of Blood

Blood is a tissue consisting of cells, cell fragments, and liquid. Blood has many important functions. It plays a part in every major activity of your body. First, blood carries oxygen from your lungs to all body cells. It also removes carbon dioxide from your body cells and carries it to the lungs to be exhaled. Second, it carries waste products of cell activity to your kidneys to be removed. Third, blood transports nutrients from the digestive system to body cells. Fourth, materials in blood fight infections and help heal wounds. Anything that disrupts or changes any of these functions affects all the tissues of the body.

Blood makes up about eight percent of your body's total mass. If you weigh 45 kg, you have about 3.6 kg of blood moving through your body. The amount of blood in an adult would fill five one-liter bottles. If this volume falls, the body is sent into shock because blood pressure falls rapidly.

Parts of Blood

If you've ever taken a ride in a water slide at an amusement park, you have some idea of the twisting and turning travels of a blood cell inside a blood vessel. Surrounded by water, you travel rapidly through a narrow watery passageway, much like a red blood cell moves in the liquid part of blood.

Plasma

If you could examine blood closely, you would see that is it not just a red-colored liquid. Blood is a tissue made of red and white blood cells, platelets, and plasma. **Plasma** is the liquid part of blood and consists mostly of water. It makes up more than half the volume of blood. Nutrients, minerals, and oxygen are dissolved in plasma.

Figure 23-9

When blood is spun in a centrifuge, the cells separate out from the plasma. Each component plays a key role in body functions. The most dense part of blood settles to the bottom of a blood sample.

636   Chapter 23 The Circulatory System

Program Resources

 Reproducible Masters
Study Guide, p. 89 [L1]
Reinforcement, p. 89 [L1]
Enrichment, p. 89 [L3]
Cross-Curricular Integration, p. 27
Activity Worksheets, pp. 5, 138-139 [L1]

Transparencies
Section Focus Transparency 85 [L1]

Blood Cells

A cubic millimeter of blood has more than 5 million red blood cells. These disk-shaped blood cells contain **hemoglobin,** a chemical that can carry oxygen and carbon dioxide. Hemoglobin carries oxygen from your lungs to your body cells. Red blood cells also carry carbon dioxide from body cells to your lungs. Red blood cells have a life span of about 120 days. They are formed in the marrow of long bones at a rate of 2 to 3 million per second and contain no nuclei. About an equal number of old ones wear out and are destroyed in the same time period.

In contrast to red blood cells, there are only about 5000 to 10 000 white blood cells in a cubic millimeter of blood. White blood cells fight bacteria, viruses, and other foreign substances that constantly try to invade your body. Your body reacts to infection by increasing its number of white blood cells. White blood cells slip between the cells of capillary walls and out around the tissues that have been invaded. Here, they ingest foreign substances and dead cells. The life span of white blood cells varies from a few days to many months.

Circulating with the red and white blood cells are platelets. **Platelets** are irregularly shaped cell fragments that help clot blood. A cubic millimeter of blood may contain as many as 400 000 platelets. Platelets have a life span of five to nine days. **Figure 23-10** summarizes the solid parts of blood and their functions.

Figure 23-10

All blood cells and platelets originate in the red marrow of bones.

Magnification: 10 000×

A **White blood cells**
There are several types, sizes, and shapes of white blood cells. These cells destroy bacteria, viruses, and foreign substances.

B **Red blood cells**
These cells carry oxygen and carbon dioxide and are disk-shaped.

Magnification: 1970×

C **Platelets**
Platelets are small, irregularly shaped cell fragments that take part in the blood-clotting process.

637

Science Journal

CPR CPR is a life-saving technique. Have students write what the letters stand for and analyze the parts of the words to find out what they mean. Find out where they can learn CPR. *CPR stands for cardiopulmonary resuscitation. CPR is an emergency procedure that is used with mouth-to-mouth resuscitation when the heart has stopped beating.* Call the local American Red Cross for more information. **L1**

Visual-Spatial Put a thin layer of cotton fibers on the bottom of a wire sieve. Pour in some white glue. Have students notice how the glue is prevented from oozing through by the network of fibers. Relate this to a blood clot. The dried glue and fibers simulate a scab. **LEP**

Inquiry Question

Why would relatively minor cuts and bruises be hazardous to hemophiliacs? *Even minor cuts and bruises can cause major bleeding problems, which could result in death.*

Use **Cross-Curricular Integration,** p. 27, as you teach this lesson.

Blood Clotting

Everyone has had a cut, scrape, or other minor wound at some time. The initial bleeding is usually quickly stopped and the wounded area begins to heal. Most people will not bleed to death from a wound. The bleeding stops because platelets in your blood help prevent blood loss by making a blood clot. A blood clot is somewhat like a bandage. When you cut yourself, a series of chemical reactions causes threadlike fibers called fibrin to form a sticky net that traps escaping blood cells and plasma. This forms a clot and helps prevent further loss of blood. **Figure 23-11** shows the blood-clotting process that occurs after a cut. Some people have the genetic disease hemophilia. Their blood lacks one of the clotting factors that begins the clotting process.

Figure 23-11

A sticky blood clot seals the leaking blood vessel. Eventually, a scab forms, protecting the wound from further damage and allowing it to heal.

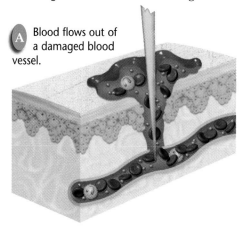

A Blood flows out of a damaged blood vessel.

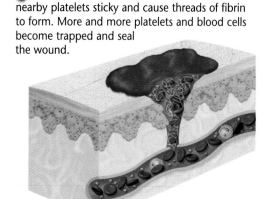

B Platelets stick to the area of the wound and release chemicals. The chemicals make other, nearby platelets sticky and cause threads of fibrin to form. More and more platelets and blood cells become trapped and seal the wound.

C The clot becomes harder. White blood cells destroy invading bacteria. Skin cells begin the repair process.

D The scab hardens and falls off after the wounded area has healed.

Integrating the Sciences

Chemistry When a person has a cut, substances released by broken platelets start chemical reactions that cause fibers to form. These fibers trap cells and a clot forms. Have students find out about the role of calcium and vitamin K in blood clotting. *Both calcium and vitamin K must be present in order for the chemical reactions in blood clotting to take place.*

Blood Types

Sometimes a person loses a lot of blood. This person may receive blood from another person through a blood transfusion. During a blood transfusion, a person receives blood or parts of blood. Doctors must be sure that the right type of blood is given. If the wrong type is given, the red blood cells of the person clump together. Clots form in the blood vessels and death results.

The ABO Identification System

Doctors know that there are four types of blood: A, B, AB, and O. Each type has a chemical identification tag called an antigen on its red blood cells. As shown in **Figure 23-12,** type A blood has A antigens. Type B has B antigens. Type AB has both A and B antigens on each blood cell. Type O has no A or B antigens.

Each type also has specific antibodies in the plasma of that blood. Antibodies are proteins that destroy or neutralize foreign substances, such as pathogens, in your body. The antibodies prevent certain blood types from mixing. Type A blood has antibodies against type B. If you mix type A blood with type B blood, type A red blood cells react to type B blood as if it were a foreign substance. The antibodies in type A respond by clumping the type B blood. Type B blood has antibodies against type A. Type AB has no antibodies, so it can receive blood from A, B, AB, and O. Type O has both A and B antibodies. **Table 23-1** lists the four blood types, what they can receive, and what types they can donate to.

Table 23-1

Blood Transfusion Possibilities		
Type	**Can receive**	**Can donate to**
A	O, A	A, AB
B	O, B	B, AB
AB	all	AB
O	O	all

Figure 23-12

The blood type you have depends on the presence or absence of certain antigens on your red blood cells and antibodies in your plasma. Your blood type is inherited and cannot be changed.

CONNECT TO

CHEMISTRY

Artificial blood substances have been developed to use in blood transfusions. They can carry oxygen and carbon dioxide. *Predict* what other properties they must have to be safe.

23-3 Blood **639**

Discussion

Survey the students to find out whether they know their own blood types. Discuss when it might be important for them to know their blood types.

CONNECT TO

CHEMISTRY

Answer Artificial blood must not cause clumping or be diseased.

Cultural Diversity

Blood as Food Some cultures consider blood to be a necessary part of their diet. Have students find out what necessary nutrients are supplied by blood (usually sodium and iron). [L1]

Across the Curriculum

History Have students research how blood transfusion experiments started in the early 1800s. [L2]

Teacher F.Y.I.

Treatment for the Rh problem, *erythroblastosis fetalis*, includes blood transfusion and exposure to fluorescent light.

Problem Solving

Solve the Problem
1. type O
2. yes

Think Critically
The babies could also have type A blood.

Visual Learning

Figure 23-13 How does this damage the affected tissues? *Built-up wastes are toxic to cells. Without oxygen, cell respiration cannot occur.* **LEP**

LS

---

3 Assess

Check for Understanding

? FLEX Your Brain

Use the Flex Your Brain activity to have students explore BLOOD.

📁 **Activity Worksheets,** p. 5

Reteach

LS **Interpersonal** Have each student, in a cooperative learning group, pick a different blood group. Let students decide who can give blood to whom or from whom they can receive blood. **COOP LEARN**

Extension

📁 For students who have mastered this section, use the **Reinforcement** and **Enrichment** masters.

Problem Solving

The Baby Exchange

Two mothers took home their new babies from the hospital on the same day. On the first day home when mother number one was removing the hospital name tag from her baby, she discovered the other mother's name was on the tag. The other mother was contacted but she was sure that she had the right baby. She did not want to give up the baby she had brought home from the hospital. Since the identity of the babies was disputed, the issue had to be decided in court. The blood types of all the parents and the two babies were studied. Mother number one and her husband both had blood type O. The baby they had at home had type B blood. Mother number two also had blood type O. The father had AB. Their baby was blood type O. Because blood types are inherited, what clues will help you solve this problem?

Solve the Problem:
1. **What is the only blood type the baby from family one could have?**
2. **Should the babies be exchanged?**

Think Critically:
Since A and B blood types are always dominant to blood type O, what other blood type could babies from family two have?

640

Community Connection

Blood Bank Visit a local blood bank and have students observe the various tests the collected blood undergoes before being sent to hospitals.

The Rh Factor

Just as antigens are one chemical identification tag for blood, the Rh marker is another. Rh blood type is also inherited. If the Rh marker is present, the person has Rh-positive (Rh+) blood. If it is not present, the person is said to be Rh-negative (Rh−). An Rh− person receiving blood from an Rh+ person will produce antibodies against the Rh+ factor.

A problem also occurs when an Rh− mother carries an Rh+ baby. Close to the time when the baby is about to be born, antibodies from the mother can cross the placenta and destroy the baby's red blood cells. If this happens, the baby can receive a blood transfusion before or right after birth. At 28 weeks and immediately after the birth, the mother can receive an injection that prevents the production of antibodies to the Rh factor. To prevent deadly consequences, blood groups and Rh factor are checked before transfusions and during pregnancies.

Diseases and Disorders of Blood

Blood, like other body tissues, is subject to disease. Because blood circulates to all parts of the body and performs so many vital functions, any disease of this tissue is cause for concern. Anemia is a disorder in which there are too few red blood cells, or too little hemoglobin in the red blood

---

cells. Because of this, body tissues can't get enough oxygen. They are unable to carry on their usual activities. Sometimes the loss of great amounts of blood or improper diet will cause anemia. Anemia can also result from disease or as a side effect of treatment for a disease. **Figure 23-13** illustrates another blood disease.

Leukemia is a disease in which one or more types of white blood cells are produced in increased numbers. However, these cells are immature and do not effectively fight infections. Blood transfusions and bone marrow transplants are used to treat this disease, but they are not always successful and death may occur.

Blood transports oxygen and nutrients to body cells and takes wastes from these cells to organs for removal. Cells in blood help fight infection and heal wounds. You can understand why blood is sometimes called the tissue of life.

Figure 23-13

Persons with sickle-cell anemia have deformed red blood cells. The sickle-shaped cells clog the capillaries of the person with this disease. Oxygen cannot reach tissues served by the capillaries and wastes cannot be removed. *How does this damage the affected tissues?*

Section Wrap-up

Review

1. What are the four functions of blood in the body?

2. Compare blood cells, plasma, and platelets.

3. Why is blood type checked before a transfusion?

4. Describe a disease and a disorder of blood.

5. **Think Critically:** Think about the main job of your red blood cells. If red blood cells couldn't pick up carbon dioxide and wastes from your cells, what would be the condition of your tissues?

Skill Builder

Making and Using Tables

Look at the data in **Table 23-1** on page 639 about blood group interactions. To which group(s) can type AB donate blood? If you need help, refer to Making and Using Tables in the **Skill Handbook.**

USING MATH

Calculate the ratio of the number of red blood cells to the number of white blood cells and to the number of platelets in a cubic millimeter of blood.

Skill Builder

A can receive from A and O.
B can receive from B and O.
AB can receive from A, B, AB, O.
O can receive from O.

Assessment

Performance Use this Skill Builder to assess students' abilities to use Table 23-1 to determine acceptable donors for someone with type AB blood. Use the Performance Task Assessment List for Analyzing the Data in **PASC**, p. 27. **P**

•MINI•QUIZ•

1. **What is blood?** *Blood is a tissue consisting of cells, cell fragments, and liquid.*

2. **Are there more red or white blood cells in the body?** *red blood cells*

3. **What type of blood group does a person have if the blood has no A or B antigens?** *O*

Section Wrap-up

Review

1. carries oxygen from the lungs to all body cells and removes carbon dioxide, carries wastes to kidneys, transports nutrients from the digestive system to body cells, has materials to fight infections and heal wounds

2. red cells—transport oxygen and carbon dioxide; white cells—fight infections; plasma—liquid part that carries dissolved nutrients, minerals, oxygen; platelets—help in blood clotting

3. to prevent blood cells from clumping

4. Leukemia is a disease in which large numbers of white blood cells are made; anemia is a disorder in which there is too little hemoglobin in the red blood cells.

5. **Think Critically** Respiration could not take place. Cells would have too much carbon dioxide and wastes and not enough oxygen needed to release energy from food. Wastes would become toxic and tissues would die.

USING MATH

500-1000 rbc for every wbc
12.5 rbc for every platelet

Activity 23-2

PREPARATION

Purpose

LS **Interpersonal** Design and carry out an experiment to observe and compare the characteristics of blood cells of selected vertebrates. **L1**

COOP LEARN

Process Skills

observing, classifying, communicating, inferring, comparing and contrasting, forming a hypothesis, designing an experiment, interpreting data, using numbers

Time

45-50 minutes

Materials

Obtain the necessary microscope slides for comparisons. Review procedures for working with prepared microscope slides to prevent damage to them.

Possible Hypotheses

Students will most likely state that the appearances will be similar and that red blood cells will be most numerous.

📁 **Activity Worksheets**, pp. 5, 138-139

Activity 23-2

Design Your Own Experiment
Comparing Blood Cells

Blood is an important tissue for all vertebrates. How do human blood cells compare with those of other vertebrates? Your task in this activity is to devise a method to compare and contrast blood cells of humans with those of two other vertebrates.

PREPARATION

Problem

What are some ways blood cells of vertebrates are alike or different?

Form a Hypothesis

Based on the information you have learned about human blood cells, state a hypothesis about the possible similarities and differences among vertebrate blood cells. Possible comparisons may include but not be limited to physical appearance and numbers present.

Objectives
- Observe the characteristics of red blood cells, white blood cells, and platelets.

- Design an experiment that compares human blood cells with those of other vertebrates.

Possible Materials
- prepared slides of human blood and blood of two other vertebrates (fish, amphibian, reptile, or bird)
- microscope
- colored pencils

Inclusion Strategies

Visually Impaired Place students who have difficulty seeing objects through a microscope in charge of forming the hypothesis and designing the experiment.

PLAN THE EXPERIMENT

1. As a group, agree upon and write out the hypothesis statement.
2. As a group, list the steps that you need to take to test your hypothesis. Consider the characteristics of each vertebrate that you compare. List your materials. Be sure that you can identify each kind of cell in the prepared slides.
3. Design a data table and record it in your Science Journal so that it is ready to use as your group collects data. Make sure that space is allowed for drawings of the cells that you observe.

Check the Plan
1. Read over your entire experiment to make sure that all steps are in logical order.

2. Identify any constants, variables, and controls of the experiment. Make sure you know how stains may have altered the appearance of the cells.
3. *Make sure your teacher approves your plan before you proceed.*

DO THE EXPERIMENT

1. Carry out the experiment as planned.
2. While the experiment is going on, write down any observations that you make and complete the data table.

Analyze and Apply
1. **Compare** your results with those of other groups.
2. Which type of blood cells are present in the greatest number?
3. How do red blood cells, white blood cells, and platelets **compare** among vertebrates?
4. Does each of the vertebrates studied have all three cell types?
5. Review the function of red blood cells. **Predict** what relationship you might find between the number of red blood cells and vertebrates with high rates of metabolism. Do your data support your answer?

Go Further

Describe how technology might design a new type of red blood cell to carry more oxygen.

23-3 Blood **643**

PLAN THE EXPERIMENT

Possible Procedures
One possible procedure is to observe the various types of blood cells and look for specific characteristics such as shape, number, and the presence or absence of a nucleus. A comparison data table should be prepared.

Teaching Strategies
• Remind students of the differences between red blood cells and white blood cells.
• Point out that white blood cells are colorless by nature, but are stained for slides to bring out details.

DO THE EXPERIMENT

Expected Outcome
Results will reveal that for the human, frog, and bird comparison, only the human red blood cells and platelets do not have a nucleus and that the white blood cells are similar in appearance.

Analyze and Apply
1. Student results should be consistent with those of other groups.
2. red blood cells
3. The red blood cells of the bird and frog are oval and have nuclei; human red blood cells are round without nuclei. All of the white blood cells are similar. The bird and frog platelets are alike in that they are spindle-shaped and have nuclei.
4. yes
5. Animals with high rates of metabolism should have more red blood cells.

Go Further
Answers will vary. A new red blood cell might have more hemoglobin in it or another substance that could carry more oxygen to cells.

✔ Assessment

Performance To further assess students' understanding of blood cells, obtain a slide of insect blood cells. Have students compare these blood cells with those of humans. Use the Performance Task Assessment List for Making Observations and Inferences in **PASC**, p. 17. **P**

1 Motivate

Bellringer

Before presenting the lesson, display **Section Focus Transparency 86** on the overhead projector. Assign the accompanying **Focus Activity** worksheet. L1 LEP

2 Teach

Activity

Collect magazine articles about research on HIV. Discuss how the lymphatic system is involved in HIV infection.

Visual Learning

Figure 23-15 How do muscles help move lymph? *As skeletal muscles contract they force lymph to move within the lymph vessels.* LEP LS

---

3 Assess

Check for Understanding

? FLEX Your Brain

Use the Flex Your Brain activity to have students explore LYMPH.

📁 **Activity Worksheets,** p. 5

Reteach

Have students construct a table to illustrate the functions of lymph nodes, the thymus, and the spleen.

Extension

📁 For students who have mastered this section, use the **Reinforcement** and **Enrichment** masters.

23•4 Your Lymphatic System

Science Words
lymphatic system
lymph
lymphocyte
lymph node

Objectives
* Describe the functions of the lymphatic system.
* Explain where lymph comes from.
* Explain the role of lymph organs in fighting infections.

Figure 23-14

Lymph is fluid that has moved from around cells into lymph vessels.

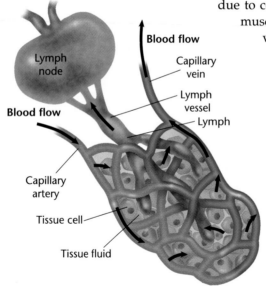

Blood flow
Lymph node
Blood flow
Capillary artery
Tissue cell
Tissue fluid
Capillary vein
Lymph vessel
Lymph

Functions of Your Lymphatic System

You have learned that blood carries nutrients and oxygen to cells. Molecules of these substances pass through capillary walls to be absorbed by nearby cells. Some of the water and dissolved substances that move out of your blood become part of a tissue fluid that is found between cells. Your **lymphatic system** collects this fluid from body tissue spaces and returns it to the blood through a system of lymph capillaries and larger lymph vessels. This system also contains cells that help your body defend itself against pathogens.

Lymphatic Organs

Once tissue fluid moves from around body tissues and into lymphatic capillaries, shown in **Figure 23-14,** it is known as **lymph.** Lymph enters the capillaries by absorption and diffusion. It consists mostly of water, dissolved substances, such as nutrients and small proteins, and **lymphocytes**, a type of white blood cell. The lymphatic capillaries join with larger vessels that eventually drain the lymph into large veins near the heart. There is no heartlike structure to pump the lymph through the lymphatic system. The movement of lymph is due to contraction of skeletal muscles and the smooth muscles in lymph vessels. Like veins, lymphatic vessels have valves that prevent the backward flow of lymph.

Lymph Nodes

Before lymph enters blood, it passes through bean-shaped structures throughout the body known as **lymph nodes.** Lymph nodes filter out microorganisms and foreign materials that have been engulfed by lymphocytes. When your body fights an infection, lymphocytes fill the nodes. They become inflamed and tender to the touch.

Tonsils are lymphatic organs in the back of the throat. They provide protection to the mouth and nose against pathogens. The thymus is a soft mass of tissue located behind

---

Program Resources

📁 **Reproducible Masters**
Study Guide, p. 90 L1
Reinforcement, p. 90 L1
Enrichment, p. 90 L3
Science and Society Integration, p. 27
Critical Thinking/Problem Solving, p. 27

 Transparencies
Section Focus Transparency 86 L1

the sternum; it produces lymphocytes that travel to other lymph organs. The spleen is the largest organ of the lymphatic system and is located behind the upper-left part of the stomach. Blood flowing through the spleen gets filtered. Here, worn out and damaged red blood cells are broken down. Specialized cells in the spleen engulf and destroy bacteria and other foreign substances.

A Disease of the Lymphatic System

As you read in Chapter 2, HIV is a deadly virus. When HIV enters a person's body, it attacks and destroys a particular kind of lymphocyte called helper T cells. Normally, helper T cells help produce antibodies to fight infections. With fewer helper T cells, a person infected with HIV is less able to fight pathogens. The infections become difficult to treat and often lead to death.

The lymphatic system, shown in **Figure 23-15,** collects extra fluid from body tissue spaces. It also produces lymphocytes that fight infections and foreign materials that enter your body. This system works to keep your body healthy.

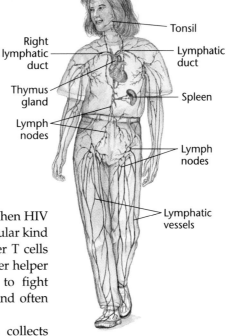

Figure 23-15

The Human Lymphatic System
How do muscles help move lymph?

Section Wrap-up

Review

1. Describe the role of your lymphatic system.

2. Where does lymph come from and how does it get into the lymphatic capillaries?

3. What happens when HIV enters the body?

4. **Think Critically:** When the amount of fluid in the spaces between cells increases, so does the pressure in these spaces. What do you infer will happen?

Skill Builder
Comparing and Contrasting
Compare and contrast your lymphatic system and your cardiovascular system. If you need help, refer to Comparing and Contrasting in the **Skill Handbook.**

Science Journal

An infectious microorganism gains entrance into your body. In your Science Journal, describe how the lymphatic system provides the body with protection against the microorganism.

Skill Builder
alike—transport fluids with cells, have vessels; some lymph vessels, like veins, have valves; unlike—type of fluid transported, different kinds of vessels, lymph fluid not pumped by a heart

Assessment

Portfolio Use this Skill Builder to assess students' abilities to compare and contrast the lymphatic and cardiovascular systems. Use the Performance Task Assessment List for Making Observations and Inferences in **PASC,** p. 17. P

•MINI•QUIZ•

1. **How does tissue fluid get into lymphatic capillaries?** *absorption and diffusion*

2. **_____ are structures in lymph vessels that prevent the backward flow of lymph.** *Valves*

3. **How does HIV cause the body to be less able to fight pathogens?** *It destroys helper T cells, which help produce antibodies to fight infections.*

Section Wrap-up

Review

1. Your lymphatic system collects fluid from body tissue spaces and returns it to the blood. This system also contains cells that help your body defend itself against pathogens.

2. Lymph comes from tissue fluids and gets into lymphatic capillaries by absorption and diffusion.

3. It attacks and destroys a particular kind of lymphocyte called helper T cells.

4. **Think Critically** The fluid will be forced out of the spaces and pass into lymph vessels.

Science Journal

Journal entries should reflect the roles of the lymph nodes, lymphocytes, and the spleen in providing protection against microorganisms. P

Science & Literature

Source

Asimov, Isaac. *Fantastic Voyage*, 1966.

Biography

Isaac Asimov (1920-1992) was born in Petrovichi, in the former Soviet Union. He came with his parents to the United States in 1923. After earning a Ph.D. in chemistry at Columbia University, he taught biochemistry and later became a professor at Boston University School of Medicine. Asimov wrote more than 450 books. He won many awards for his writing, including five Hugo Awards for science fiction.

Teaching Strategies

- Students may have expected the red blood cells to look red. Mention that Asimov explains elsewhere that red blood corpuscles look red when a large number are grouped together but are straw-colored individually.

- Students will recognize the milky, pulsating cell as a white blood cell, whose purpose is to ingest foreign substances. The cell moved by putting out extensions into which the mass of the cell flowed. Ask them why the white blood cells did not ingest *Proteus* with its crew as it would a bacterium of the same size. (*Asimov explained that although the vessel is small enough, its exterior is made of metal, and a white blood cell seeks only objects with a mucopolysaccharide cell wall.*) Discuss the other parts of the cell described by the author.

Fantastic Voyage
by Isaac Asimov

In this science fiction novel, a blood clot in the brain has caused a scientist defecting from his native country to lapse into a coma. Valuable information that governments are competing for is now beyond reach. He cannot be operated on from the outside, so a new method of miniaturization is used. It reduces in size by 1 million times a small submarine, called *Proteus*, and five passengers—one of whom is a brain surgeon. They are piloted through the scientist's circulatory system to the blood clot. The operation must be done quickly because the ship and passengers will return to normal size in 60 minutes. Read the author's description of their fantastic voyage in the bloodstream:

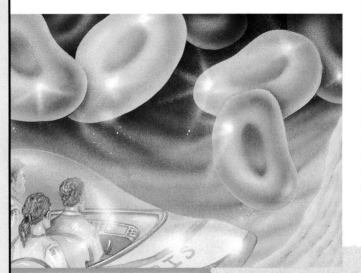

It was a vast, exotic aquarium they faced, one in which not fish but far stranger objects filled their vision. Large rubber tires, the centers depressed but not pierced through, were the most numerous objects. Each was twice the diameter of the ship, each an orange-straw color, each sparkling and blazing intermittently, as though faceted with slivers of diamonds.

Science Journal

What images does the description of traveling through the body in a miniature state bring to mind? In your Science Journal, describe what you see as you imagine moving through the body.

What were the objects described in the paragraph above? What color had you expected these objects to be?

646 Chapter 23 The Circulatory System

Classics

- Verne, Jules. *A Journey to the Center of the Earth.* Boston: H.L. Shepard, 1872. A young man tries to prove his courage by descending in a volcano to the center of the earth and backwards in time.

- Verne, Jules. *Twenty Thousand Leagues Under the Sea.* London: S. Low, Marsten, Searle and Rivington, 1885. This was another unusual voyage in which the author predicted many future inventions.

Other Works

Other works by Isaac Asimov: *The Bloodstream: River of Life* (nonfiction). New York: Collier, 1961. *Fantastic Voyage II: Destination Brain.* New York: Doubleday, 1987.

Summary

23-1: Circulation
1. Arteries carry blood from the heart; capillaries exchange food, oxygen, and wastes in cells; veins return blood to the heart.
2. Blood enters the right atrium, moves to the right ventricle, and then goes to the lungs by the pulmonary artery. Blood rich in oxygen returns to the left atrium, moves into the left ventricle, and then goes out through the aorta.
3. Pulmonary circulation is the path of blood to and from the heart and lungs. Circulation through the rest of the body is systemic circulation.

23-2: Science and Society: "Growing" Heart Valves
1. Tissue-engineered replacement valves may decrease the problems associated with present replacement valves.

23-3: Blood
1. Red blood cells carry oxygen; platelets form clots; white blood cells fight infection.
2. A, B, AB, and O blood types are determined by the presence or absence of antigens.
3. Anemia and leukemia are two diseases of the blood.

23-4: Your Lymphatic System
1. Lymphatic vessels return fluid to the circulatory system and fight disease.

2. Lymph structures filter blood, produce white blood cells that destroy bacteria, and destroy worn-out blood cells.

Key Science Words

a. artery
b. atherosclerosis
c. atria
d. blood pressure
e. capillary
f. coronary circulation
g. heart murmur
h. hemoglobin
i. hypertension
j. lymph
k. lymph node
l. lymphatic system
m. lymphocyte
n. plasma
o. platelet
p. pulmonary circulation
q. systemic circulation
r. vein
s. ventricle

Reviewing Vocabulary

Match each phrase with the correct term from the list of Key Science Words.

1. filters microorganisms
2. upper heart chambers
3. vessel connected to the heart ventricle
4. path of blood between heart and lungs
5. path of blood to the heart tissue
6. fatty deposit on artery walls
7. high blood pressure
8. liquid portion of the blood
9. blood vessel that connects arteries to veins
10. active in blood clot formation

Summary

Have students read the summary statements to review the major concepts of the chapter.

Reviewing Vocabulary

1. l	**6.** b
2. c	**7.** i
3. a	**8.** n
4. p	**9.** e
5. f	**10.** o

Assessment

Portfolio Encourage students to place in their portfolios one or two items of what they consider to be their best work. Examples include:
- MiniLAB Assessment, p. 628
- Activity 23-1 assessment graph, p. 633
- Science Journal, p. 645 [P]

Performance Additional performance assessments may be found in **Performance Assessment** and **Science Integration Activities.** Performance Task Assessment Lists and rubrics for evaluating these activities can be found in Glencoe's **Performance Assessment in the Science Classroom.**

GLENCOE TECHNOLOGY

 Videodisc

Glencoe Life Science Interactive Videodisc
Side 1, Lesson 4
Photosynthesis and Cellular Respiration

37736 to 49167

MindJogger Videoquiz

Chapter 23 Have students work in groups as they play the Videoquiz game to review key chapter concepts.

Checking Concepts

1. b   **6.** c
2. b   **7.** a
3. c   **8.** c
4. a   **9.** d
5. d   **10.** a

Understanding Concepts

11. if it had more hemoglobin
12. Valves help keep blood or lymph moving against gravity.
13. With anemia, a person's body tissues do not receive enough oxygen. This can result in a lack of energy.
14. A vein carries oxygen-rich blood from the lungs to the heart.
15. Coronary circulation is blood that circulates to heart tissue. Pulmonary circulation is bloodflow from the heart to the lungs and back to the heart.

Thinking Critically

16. oxygen-rich blood: aorta, coronary arteries, left atrium, left ventricle; carbon dioxide-full blood: coronary veins, inferior vena cava, right atrium, right ventricle, superior vena cava
17. A person might experience shortness of breath, fatigue, and sometimes dizziness.
18. The lymphatic system, like your veins, carries fluid away from body tissues and returns it to your circulatory system by way of your veins.
19. Blood and lymph move throughout the entire body.
20. These arteries are closest to the skin, where the pulse can be most easily felt.

Checking Concepts

Choose the word or phrase that completes the sentence.

1. Exchange of food, oxygen, and wastes occurs through the _____.
   a. arteries   c. veins
   b. capillaries   d. lymph vessels
2. Oxygen-rich blood first enters the _____.
   a. right atrium   c. left ventricle
   b. left atrium   d. right ventricle
3. Circulation to all body organs is _____.
   a. coronary   c. systemic
   b. pulmonary   d. organic
4. Blood is under great pressure in _____.
   a. arteries   c. veins
   b. capillaries   d. lymph vessels
5. Blood functions to _____.
   a. digest food   c. dissolve bone
   b. produce CO_2 d. carry oxygen
6. Infection is fought off by _____ cells.
   a. red blood   c. white blood
   b. bone   d. nerve
7. In blood, oxygen is carried by _____.
   a. red blood cells   c. white blood cells
   b. platelets   d. lymph
8. Clotting of blood requires _____.
   a. plasma   c. platelets
   b. oxygen   d. carbon dioxide
9. Type O blood has _____ antigen(s).
   a. A   c. A and B
   b. B   d. no
10. The largest filtering lymph organ is the _____.
    a. spleen   c. tonsil
    b. thymus   d. node

Understanding Concepts

Answer the following questions in your Science Journal using complete sentences.

11. What change in a human red blood cell would make it carry more oxygen?
12. What is the relationship between valves in veins and lymphatic capillaries and gravity?
13. Why would anemia cause fatigue?
14. Defend or refute this statement: Veins never carry oxygen-rich blood.
15. Distinguish between coronary and pulmonary circulation.

Thinking Critically

16. Identify the following as having oxygen-rich or carbon dioxide-full blood: aorta, coronary arteries, coronary veins, inferior vena cava, left atrium, left ventricle, right atrium, right ventricle, and superior vena cava.
17. What health problems are the result of improperly functioning heart valves?
18. Explain how the lymphatic system works with the cardiovascular system.
19. Why is cancer of the blood or lymph hard to control?
20. Pulse is usually taken at the neck or wrist. Why do you think this is so even though there are many arteries?

Developing Skills

If you need help, refer to the **Skill Handbook.**

21. **Concept Mapping:** Complete the events chain concept map showing how lymph moves in your body.

Tissue fluid around body cells moves into

↓

Lymphatic capillaries

↓

Lymphatic vessels

↓

Lymph nodes

↓

Blood vessels

↓

Blood moves through circulatory system and is filtered by the

↓

Spleen

22. **Comparing and Contrasting:** Compare the life span of the different types of blood cells.

23. **Interpreting Data:** Interpret the data listed below. Find the average heartbeat rate of four males and four females and compare the two averages.
Males: 72, 64, 65, 72
Females: 67, 84, 74, 67

24. **Designing an Experiment:** Design an experiment to compare the heartbeat rate at rest and after exercising.

25. **Hypothesizing:** Make a hypothesis to suggest the effects of smoking on heartbeat rate.

Performance Assessment

1. **Booklet:** With supervision, prepare a heart-healthy recipe. Check for recipes through the American Heart Association cookbooks. Bring a sample in to share with the class. Prepare copies of the recipe and explain why it is heart healthy. Put all the recipes together into a booklet.

2. **Pamphlet:** Prepare an informational pamphlet on heart transplants, their success, and what a heart-transplant patient has to do to remain healthy.

3. **Scientific Drawing:** Prepare a poster-sized drawing of the human heart and label its parts.

Developing Skills

21. **Concept Mapping** See student page.

22. **Comparing and Contrasting** red blood cells: 120 days; white blood cells: a few days to several months; platelets: 5 to 9 days

23. **Interpreting Data** Males = 68. Females = 73. In general, males' heartbeat rates are lower than females'.

24. **Designing an Experiment** An experiment should include trials for both resting and exercising heartbeat rates with different subjects.

25. **Hypothesizing** Cigarettes contain nicotine, which is a stimulant. Heartbeat rate will increase.

Performance Assessment

1. Students may want to distribute the booklet to other classes. Use the Performance Task Assessment List for Booklet or Pamphlet in **PASC,** p. 57. [P]

2. Pamphlet sections might include information on historical background, current status, role of drugs in preventing rejection, and expected life span after the operation. Use the Performance Task Assessment List for Booklet or Pamphlet in **PASC,** p. 57. [P]

3. Some students may want to use anatomy books for reference. Use the Performance Task Assessment List for Scientific Drawing in **PASC,** p. 55. [P]

Assessment Resources

 Reproducible Masters
Chapter Review, pp. 49-50
Assessment, pp. 151-154
Performance Assessment, p. 37

Glencoe Technology
🔘 **Chapter Review Software**
🔘 **Computer Test Bank**
📼 **MindJogger Videoquiz**

Chapter Organizer

Section	Objectives/Standards	Activities/Features
Chapter Opener		Explore Activity: Does physical activity change breathing rate? p. 651
24-1 **Your Respiratory System** (2½ sessions)*	1. **State** the functions of the respiratory system. 2. **Explain** how oxygen and carbon dioxide are exchanged in the lungs and in tissues. 3. **Trace** the pathway of air in and out of the lungs. 4. **Name** three effects of smoking on the respiratory system. National Content Standards: (5-8) UCP1, UCP3, UCP5, A1, A2, B1, B2, C1, C3, F1, F4	Science and Art: Breath Control in Singers, p. 655 Science Journal, p. 655 Connect to Chemistry, p. 656 MiniLAB: How do alveoli change surface area? p. 656 Using Math, p. 660 Skill Builder: Sequencing, p. 661 Science Journal, p. 661 Activity 24-1: The Effects of Exercise on Respiration, pp. 662-663
24-2 **Science and Society Issue: Restricting Cigarette Advertising** (1 session)*	1. **Describe** why tobacco use increased over the last century. 2. **Determine** what regulations are needed to control cigarette advertising. National Content Standards: (5-8) F1, F4, G1, G3	Explore the Issue, p. 665
24-3 **Your Urinary System** (2½ sessions)*	1. **Distinguish** between the excretory and urinary systems. 2. **Describe** how your kidneys work. 3. **Explain** what happens when urinary organs don't work. National Content Standards: (5-8) UCP1, UCP5, A1, C1, C3, E2, F1, F5	MiniLAB: What do kidneys look like? p. 667 Problem Solving: Frederick's Dilemma, p. 669 Using Technology: Kidney Stones, p. 670 Skill Builder: Concept Mapping, p. 671 Using Math, p. 671 Activity 24-2: Sweat Glands in the Skin, p. 672

* A complete Planning Guide that includes block scheduling is provided on pages 32T-35T.

Activity Materials

Explore	Activities	MiniLABs
page 651 stopwatch or clock with second hand	pages 662-663 clock or watch with second hand, drinking straws, 200 mL bromothymol blue solution, two 400-mL beakers, graduated cylinder, safety goggles, apron page 672 safety goggles, apron, 2% iodine solution, bond typing paper, hand lens	page 656 2 large sheets of paper, one of which is cut into 6 or more rectangles; tape; scissors page 667 kidney, scalpel, magnifying glass

Need Materials? Call Science Kit (1-800-828-7777).

Teacher Classroom Resources

Reproducible Masters	Transparencies	Teaching Resources
Study Guide, p. 91 **Reinforcement**, p. 91 **Enrichment**, p. 91 **Activity Worksheets**, pp. 142-143, 146 **Lab Manual 57**, Lung Capacity **Science Integration Activity 24**, Air **Critical Thinking/Problem Solving**, p. 28 **Cross-Curricular Integration**, p. 28 **Science and Society Integration**, p. 28	**Section Focus Transparency 87**, Water World **Science Integration Transparency 24**, Boyle's Law and Breathing **Teaching Transparency 47**, Respiratory System	**Glencoe Life Science Interactive Videodisc** **Life Science CD-ROM** **Spanish Resources** **English/Spanish Audiocassettes** **Cooperative Learning Resource Guide** **Lab Partner** **Lab and Safety Skills** **Lesson Plans**
Study Guide, p. 92 **Reinforcement**, p. 92 **Enrichment**, p. 92	**Section Focus Transparency 88**, Smoking and Health	**Assessment Resources** **Chapter Review**, pp. 51-52 **Assessment**, pp. 155-158 **Performance Assessment in the Science Classroom (PASC)** **MindJogger Videoquiz** **Alternate Assessment in the Science Classroom** **Performance Assessment** **Chapter Review Software** **Computer Test Bank**
Study Guide, p. 93 **Reinforcement**, p. 93 **Enrichment**, p. 93 **Activity Worksheets**, pp. 144-145, 147 **Concept Mapping**, p. 53 **Multicultural Connections**, pp. 51-52	**Section Focus Transparency 89**, Removing Liquid Wastes **Teaching Transparency 48**, Urinary System	

Key to Teaching Strategies

The following designations will help you decide which activities are appropriate for your students.

L1 Level 1 activities should be within the ability range of all students, including those with learning difficulties.

L2 Level 2 activities should be within the ability range of the average to above-average student.

L3 Level 3 activities are designed for the ability range of above-average students.

LEP LEP activities should be within the ability range of Limited English Proficiency students.

LS These activities are designed to address different learning styles.

COOP LEARN Cooperative Learning activities are designed for small group work.

P These strategies represent student products that can be placed into a best-work portfolio.

GLENCOE TECHNOLOGY

The following multimedia resources are available from Glencoe.

Science and Technology Videodisc Series (STVS)
Human Biology
  Nicotine and the Lungs
  Children and Smog

National Geographic Society Series
STV: The Cell
STV: Human Body Vol. 1
Glencoe Life Science Interactive Videodisc
Exercise: Health Is Just a Heartbeat Away
Life Science CD-ROM

Teacher Classroom Resources

This is a representation of key blackline masters available in the Teacher Classroom Resources.

Teaching Aids

Section Focus Transparencies

87 SECTION FOCUS TRANSPARENCY — Section 24-1
WATER WORLD
Scuba diving and snorkeling are activities that require special equipment that allow you to breathe while your body is in the water. This equipment is needed because unlike fish, which are adapted to breathe in oxygen dissolved in water, people breathe in oxygen from the air.

1. What special equipment is required for scuba diving? Why is this equipment important?
2. How do divers and fish differ in the way they get oxygen?

L1

88 SECTION FOCUS TRANSPARENCY — Section 24-2
SMOKING AND HEALTH
A burning cigarette gives off a variety of harmful chemicals in its smoke. These chemicals are air pollutants. The pollutants contained in cigarette smoke can be harmful to both smokers and nonsmokers. For this reason, many public places, such as restaurants and offices, have adopted a non-smoking policy to provide their patrons and employees with a smoke-free environment.

1. What are some of the health problems associated with cigarette smoking?
2. How do the health risks associated with cigarette smoking affect non-smokers?
3. How effective do you think the warning labels on cigarette packages are in encouraging people not to smoke?

L2

89 SECTION FOCUS TRANSPARENCY — Section 24-3
REMOVING LIQUID WASTES
Do you perspire heavily when you exercise? Perspiration is a liquid waste given off by the body. It contains mostly water and dissolved salts.

1. In addition to eliminating waste materials, in what other way is perspiring helpful to the body?
2. Why is it important to drink plenty of fluids when you are involved in activities that cause you to perspire heavily?
3. What other body system is involved in the removal of liquid wastes?

L3

Science Integration Transparencies

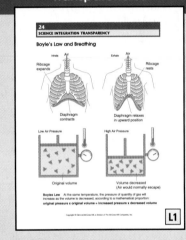

24 SCIENCE INTEGRATION TRANSPARENCY
Boyle's Law and Breathing

Boyle's Law At the same temperature, the pressure of gas will increase as the volume is decreased, according to a mathematical proportion:
original pressure × original volume = increased pressure × decreased volume

L1

Teaching Transparencies

47. RESPIRATORY SYSTEM

L1

48. URINARY SYSTEM

L2

Meeting Different Ability Levels

Study Guide

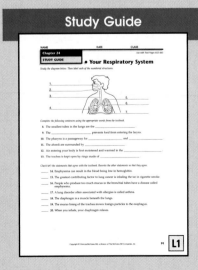

Chapter 24
STUDY GUIDE
Your Respiratory System

L1

Reinforcement

Chapter 24
REINFORCEMENT
Your Respiratory System

L2

Enrichment Worksheets

Chapter 24
ENRICHMENT
Your Respiratory System

L3

Hands-On Activities

Science Integration Activity

Chapter 24
INTEGRATION

● Air

L1

Lab Manual

Chapter 24
LABORATORY MANUAL

● Lung Capacity 57

L2

Assessment

Performance Assessment

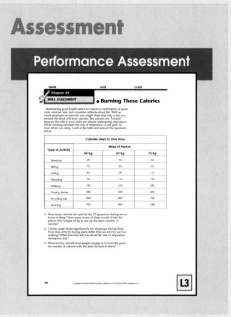

Chapter 24
SKILL ASSESSMENT

● Burning Those Calories

L3

Enrichment and Application

Critical Thinking/Problem Solving

Chapter 24
CRITICAL THINKING

● Respiration and Excretion

Additional Hazards of Cigarette Smoking

L2

Cross-Curricular Integration

Chapter 24
LANGUAGE ARTS CONNECTION

● Respiration and Excretion

Creative Writing

L1

Science and Society Integration

Chapter 24
SCIENCE & SOCIETY

● Respiration and Excretion

Should Asbestos Be Banned?

L2

Multicultural Connections

Chapter 24
MULTICULTURAL CONNECTIONS

● The Search for Compatible Kidneys

L2

Concept Mapping

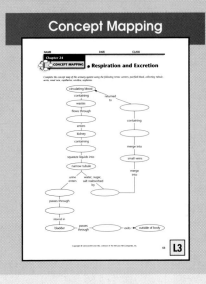

Chapter 24
CONCEPT MAPPING

● Respiration and Excretion

L3

Chapter 24

Respiration and Excretion

CHAPTER OVERVIEW

Section 24-1 This section details the basic anatomy and physiology of the respiratory system. The mechanics of breathing and a description of respiratory diseases and disorders are included.

Section 24-2 Science and Society The increased use of tobacco has brought about a rise in smoking-related health problems. The issue of regulating tobacco advertising is discussed.

Section 24-3 The organs of the urinary system and other excretory organs are introduced. Their role in maintaining homeostasis is explained.

Chapter Vocabulary

pharynx	passive
larynx	smoking
trachea	urinary
bronchi	system
alveoli	kidney
diaphragm	nephron
chronic	urine
bronchitis	ureter
emphysema	bladder
asthma	urethra

Theme Connection

Energy Energy transformation is a central theme in this text. Cells require oxygen in order to utilize nutrients. The cell respiration process provides the energy for all cellular activities. Chemical energy is transformed into heat and mechanical energy. Another theme developed in this chapter is homeostasis. The urinary system has an important function in maintaining the balance of liquids and various dissolved elements and compounds.

650

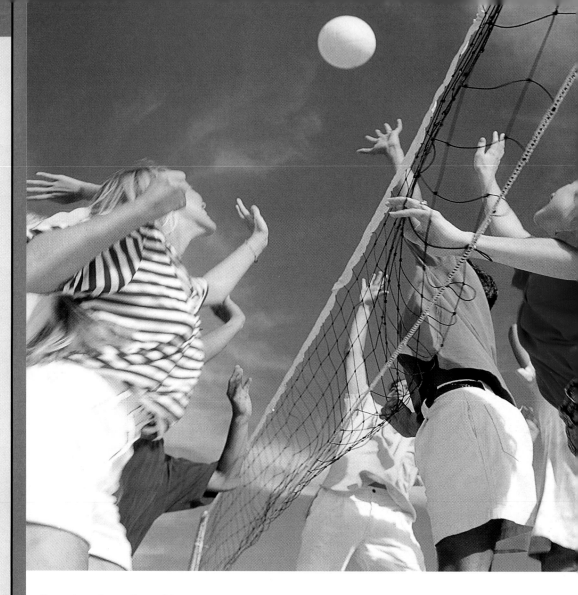

Previewing the Chapter

Section 24-1 Your Respiratory System
▶ Functions of Your Respiratory System
▶ Organs of Your Respiratory System
▶ How You Breathe
▶ A Life-Saving Maneuver
▶ Diseases and Disorders of the Respiratory System

Section 24-2 Science and Society
  Issue: Restricting Cigarette Advertising

Section 24-3 Your Urinary System
▶ Your Excretory System
▶ Organs of Your Urinary System
▶ Balancing Fluid Levels
▶ Diseases and Disorders of the Urinary System

650

Learning Styles

Look for the following logo for strategies that emphasize different learning modalities. **LS**

Kinesthetic	Activity, p. 654; Science & Art, p. 655
Visual-Spatial	Visual Learning, pp. 653, 657-660, 667; Demonstration, p. 659; MiniLAB, p. 667; Activity 24-2, p. 672
Interpersonal	Activity 24-1, pp. 662-663; Making a Model, p. 668
Logical-Mathematical	Explore, p. 651; MiniLAB, p. 656; Using Math, pp. 660, 671; Enrichment, p. 668; Across the Curriculum, p. 668
Linguistic	Science & Art, p. 655
Auditory-Musical	Science & Art, p. 655

Respiration and Excretion

You can live more than a week without food. You can live several days without water. But, you can live only several minutes without oxygen. Your body has the ability to store food and water. It cannot store much oxygen. It needs a continuous supply to keep your body cells functioning. Sometimes your body needs a lot of oxygen. Have you ever played volleyball so hard that it felt like your lungs would burst? How long did it take your breathing rate to return to normal? In the following activity, find out about one factor that can change your breathing rate.

EXPLORE ACTIVITY

Does physical activity change breathing rate?

1. Put your hand on your chest. Take a deep breath. Feel your chest move up and down slightly. Notice how your rib cage moves out and upward when you inhale.
2. Count your breathing rate for 15 seconds. Multiply this number by four to figure your breathing rate for one minute.
3. Jog in place for one minute and count your breathing rate again.

Observe: How did your breathing rate change? How long does it take for your breathing rate to return to normal? How is your breathing rate related to physical activity?

Previewing Science Skills

▶ In the **Skill Builders**, you will sequence and map concepts.

▶ In the **Activities**, you will design an experiment, observe, compare, and predict.

▶ In the **MiniLABs**, you will observe, compare, and describe.

EXPLORE ACTIVITY

Purpose

LS **Logical-Mathematical** Use the Explore Activity to discover how the respiratory system responds to physical activity. **L1**

Preparation

Have one stopwatch for every three students. Check with the school nurse to determine whether any student should not jog or engage in activities that stress the heart and respiratory system.

Materials

stopwatch, clock or watch with a second hand

Teaching Strategies

• Any student not able to participate in the physical part of this activity can be designated as a time-keeper or recorder.

• Review the reason for multiplying by four to obtain the pulse rate for one minute.

Observe

Students should observe that breathing rates increased after physical activity. Answers will vary on the length of time it takes for breathing rates to return to normal due to differences in physiology and fitness of students.

✓ Assessment

Performance Have students design another activity that will result in an increase in breathing rate. Do the activity and gather the data. Graph the results. Use the Performance Task Assessment List for Carrying Out a Strategy and Collecting Data in **PASC**, p. 25. **P**

Assessment Planner

Portfolio
Refer to page 673 for suggested items that students might select for their portfolios.

Performance Assessment
See page 673 for additional Performance Assessment options.
Skill Builders, pp. 661, 671
MiniLABs, pp. 656, 667
Activities 24-1, pp. 662-663; 24-2, p. 672

Content Assessment
Section Wrap-ups, pp. 661, 665, 671
Chapter Review, pp. 673-675
Mini Quizzes, pp. 661, 671

Group Assessment
Opportunities for group assessment occur with Cooperative Learning Strategies and Flex Your Brain Activities.

Prepare

Section Background

- The term *breathing* refers to the process of inhaling and exhaling air. *Respiration* refers to the chemical reaction that releases energy from glucose in body cells.
- The left lung is slightly smaller than the right lung to accommodate the heart.
- During one minute while the body is at rest, approximately 200 mL of oxygen is used by body cells; an equal amount of carbon dioxide is produced.

Preplanning

- To prepare for Activity 24-1, obtain the bromothymol blue solution.
- To prepare for the Mini-LAB, obtain one $8\frac{1}{2}" \times 14"$ sheet of paper and several $8\frac{1}{2}" \times 11"$ sheets of paper per laboratory team.

1 Motivate

Bellringer

Before presenting the lesson, display **Section Focus Transparency 87** on the overhead projector. Assign the accompanying **Focus Activity** worksheet. L1 LEP

Tying to Previous Knowledge

Recall from Chapter 23 the relationship of the circulatory system to the lungs. Remind students of the role of blood in providing oxygen to body cells.

Student Text Question

What do you think the plant produced when it was in the container? *The plant released oxygen.*

Science Words

- pharynx
- larynx
- trachea
- bronchi
- alveoli
- diaphragm
- chronic bronchitis
- emphysema
- asthma

Objectives

- State the functions of the respiratory system.
- Explain how oxygen and carbon dioxide are exchanged in the lungs and in tissues.
- Trace the pathway of air in and out of the lungs.
- Name three effects of smoking on the respiratory system.

Figure 24-1

Several processes are involved in how the body obtains, transports, and utilizes oxygen.

24•1 Your Respiratory System

Functions of Your Respiratory System

People have always known that air and food are needed for life. However, until about 225 years ago, no one knew why air was so important. At that time, a British chemist discovered that a mouse couldn't live in a container in which a candle had previously been burned. He reasoned that a gas in the air of the container had been destroyed when the candle burned. He also discovered that if he put a plant into the container, something necessary for life returned in eight or nine days. A mouse again could live in the container. Think about photosynthesis. What do you think the plant produced when it was in the container? It produced the gas for life that was later named oxygen.

Breathing and Respiration

People often get the terms *breathing* and *respiration* mixed up. Breathing is the process whereby fresh air moves into and stale air moves out of lungs. Fresh air contains oxygen, which passes from the lungs into your circulatory system. Blood then carries the oxygen to your individual cells. At the same time, your digestive system has prepared a supply of glucose in your cells from digested food. Now the oxygen plays a key role in the chemical reaction that releases energy from glucose. This chemical reaction, shown in the equation in **Figure 24-1**, is called respiration. Carbon dioxide is a waste product of respiration. At the end of this reaction, carbon dioxide wastes are carried back to your lungs in your blood. There, it is expelled

Breathing	Circulation	Respiration	Circulation	Breathing
Oxygen inhaled	Oxygen carried to body cells	Oxygen used in chemical reaction to release energy from glucose	Carbon dioxide removed from cells to lungs	Carbon dioxide exhaled

$$C_6H_{12}O_6 + 6O_2 \longrightarrow 6CO_2 + 6H_2O + Energy$$

(Glucose) + (Oxygen) ⟶ (Carbon dioxide) + Water + Energy

652   Chapter 24 Respiration and Excretion

Program Resources

📁 Reproducible Masters

Study Guide, p. 91 L1
Reinforcement, p. 91 L1
Enrichment, p. 91 L3
Cross-Curricular Integration, p. 28
Science and Society Integration, p. 28
Activity Worksheets, pp. 5, 142-143, 146 L1
Science Integration Activities, pp. 51-52

Lab Manual 57
Critical Thinking/Problem Solving, p. 28

🗄 Transparencies

Section Focus Transparency 87 L1
Teaching Transparency 47
Science Integration Transparency 24

Figure 24-2

Mount Everest is the highest place on Earth. Mountain climbers experience headaches, nausea, and shortness of breath while ascending to the top of mountains.

from your body in the stale air. Now think about why the first mouse was not able to stay alive in the container. **Figure 24-2** shows a place where the amount of oxygen in the atmosphere is reduced. How would this affect the ability of the human body to function?

Organs of Your Respiratory System

Your respiratory system is made up of body parts that help move oxygen into your body and carbon dioxide out of your body. The major structures and organs of your respiratory system are shown in **Figure 24-3.** These include your nasal cavity, pharynx (FER ingks), larynx, trachea, bronchi, bronchioles, and lungs. Air enters your body through two openings in your nose called nostrils or through your mouth. Once inside the nostrils, hair traps dust from the air. From your nostrils, air passes through your nasal cavity,

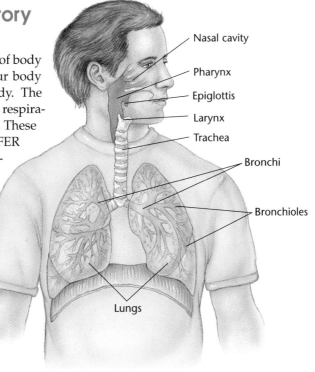

Nasal cavity
Pharynx
Epiglottis
Larynx
Trachea
Bronchi
Bronchioles
Lungs

Figure 24-3

Air can enter the body through both the nostrils and the mouth. *What is the advantage of having air enter through the nostrils?*

24-1 Your Respiratory System   **653**

- **What do you think could happen if you talk or laugh while eating?** *Food might enter the larynx and cause choking.*

- **Why is it necessary for the trachea to stay open all the time?** *If the trachea collapsed, it would be difficult or impossible to breathe.*

Revealing Preconceptions

When a person chokes on a piece of food, he or she often comments that "it went down the wrong tube." The pharynx serves as the passageway of both food into the esophagus and air into the trachea. The epiglottis is a small flap of tissue that normally closes over the tracheal opening when food is swallowed. Sometimes food or liquid gets past the epiglottis and goes into the trachea, triggering the choking reflex.

Activity

LS Kinesthetic Have students place their fingers on the front of their necks, tilt their heads backwards, and gently move their fingers up and down. Ask students whether they can feel the ridges that are the rings of cartilage around the trachea. **L1** **LEP**

Use these program resources as you teach this lesson.

Cross-Curricular Integration, p. 28

Science and Society Integration, p. 28

Science Integration Activities, pp. 47-48

Lab Manual 57

Critical Thinking/Problem Solving, p. 28

where it gets moistened and warmed. Glands that produce sticky mucus line the nasal cavity. The mucus traps dust, pollen, and other materials that were not trapped by the nasal hair. This helps filter and clean the air you breathe. Tiny hair-like structures, called cilia, move mucus and trapped material to the back of the throat where it can be swallowed.

Pharynx, Larynx, and Trachea

Warm, moist air now moves to the **pharynx,** a tubelike passageway for both food and air. At the lower end of the pharynx is a flap of tissue called the epiglottis. When you swallow, the epiglottis closes over your larynx. By doing this, food or liquid is prevented from entering your larynx. The food goes into your esophagus instead. What do you think could happen if you talk or laugh while eating?

The **larynx** is an airway to which your vocal cords are attached. Look at **Figure 24-4.** When you speak, muscles tighten or loosen your vocal cords. Sound is produced when air moves past, causing them to vibrate.

Below the larynx is the **trachea,** a tube about 12 cm in length. C-shaped rings of cartilage keep the trachea open and prevent it from collapsing. The trachea is lined with mucous membranes and cilia to trap dust, bacteria, and pollen. Why is it necessary for the trachea to stay open all the time?

Figure 24-4

Sound made by your vocal cords gets louder with increased air pressure. Pitch gets higher as muscular tension increases.

Epiglottis

Glottis closed

Vocal cords

Larynx

Trachea

Glottis open, showing inner trachea

Breath Control in Singers

How do you feel when you hear a great singer? Great singing is not just the result of talent. It also depends on training. Just as athletes train, serious singers spend hours learning and practicing breath control.

Recall that the vocal cords are attached to the larynx. When you sing, muscles tighten or loosen the cords. As you exhale, air moves up from the lungs, vibrating the vocal cords and resulting in sound. The more tightly the cords are stretched, the higher the pitch of the sound. The more relaxed the cords are, the lower the sound.

Vocal Training

Through special training, a singer learns how to expand the chest cavity so that the lungs fill with air from the bottom up, allowing the maximum amount of air in. The singer also learns to exhale air with controlled pressure. In this way, the singer can perform a long musical phrase in one breath.

Try it yourself. Inhale fairly quickly, as if in a brief musical pause, but inhale deeply, getting as much air into your lungs as possible. Then slowly exhale, controlling the release of air throughout. An opera singer, such as Kathleen Battle, shown here, must be able to control breathing in this way. Battle and other opera singers have great stamina. They need the stamina in order to sing for up to two hours and still be heard over a full-size orchestra by the hundreds of people in a huge auditorium.

To accomplish this control and stamina, singers exercise every day. They repeatedly loosen and tighten the diaphragm as well as muscles in the chest, throat, mouth, and jaw. Think of what you learned about the diaphragm and its role in breathing. Why would a strong diaphragm aid a singer's breath control?

The next time you listen to a singer perform beautifully, think about all of the physical training the singer has undergone as well. Then you can have a greater appreciation for both the physical ability and the artistic ability of the performer.

inter NET CONNECTION

Singers are not the only musicians who are trained in breath control. Visit the Chapter 24 Internet Connection at Glencoe Online Science, **www. glencoe.com/sec/science/ life**, for a link to more information about breath control.

24-1 Your Respiratory System 655

Background

An opera singer's ability to sing for an extended period of time louder than an orchestra in a large auditorium is called *technique*. Breath control is the mainstay of technique. It enables singers not only to sustain a long musical phrase but also to increase or decrease the volume for dramatic effect.

Teaching Strategies

LS Kinesthetic Have students place their hands on their ribs, close their mouths, and take a deep breath through their noses. Ask them to feel their lungs expanding with the air. Then have them say "aaaaaa" and let the air out slowly. Ask them to keep saying "aaaaaa" as long as they can. Students should feel their lungs deflate and note that they can say "aaaaaa" only as long as they still have air in their lungs.

LS Auditory-Musical Divide the class into small groups and distribute a balloon to each group. Ask students to visualize the balloon as a lung and its neck as a trachea. Have students blow up their balloons, then stretch the opening into a narrow slit. Students should note the sound. Then have them stretch the slit to make it longer and release a bit to make it shorter, noting the changes in pitch. Ask them to correlate the higher-pitched sounds and tauter stretch with the shorter vocal cords of females and the lower-pitched sounds and looser stretch with the longer vocal cords of males.

LS Auditory-Musical Play portions of an aria and other opera music for students to appreciate the voices and breath control of the singers.

LS Auditory-Musical Invite students who have studied voice or a wind instrument to perform for the class and demonstrate the breathing techniques they use.

LS Linguistic Invite students to share the essays they wrote in their Science Journals.

Classics

Suggested artists to play for students: Jessye Norman, Kathleen Battle, Kiri Te Kanawa; Placido Domingo, José Carreras, Luciano Pavarotti.

MiniLAB

Purpose

LS **Logical-Mathematical** Students observe and make inferences about the increased surface area provided by alveoli. **L1** **LEP**

Materials

one large piece of paper ($8\frac{1}{2}" \times 14"$), 3–4 smaller pieces of paper ($8\frac{1}{2}" \times 11"$, cut in half to measure $8\frac{1}{2}" \times 5\frac{1}{2}"$), scissors, tape

Teaching Strategies

You may want to use one color for the large sheet of paper and a different color for the smaller sheets.

Analysis

1. Answers will vary depending on student construction techniques, but will usually indicate at least a twofold increase in surface area.
2. The large piece of paper represents a lung; the smaller pieces represent alveoli sacs.
3. More gas can be exchanged because there is greater surface area within the same space.

✓ Assessment

Performance Have students calculate the surface area of a 10 × 20 cm cube and compare this with the surface area of one hundred 1 × 1 cm cubes that have the same volume as the larger cube. Use the Performance Task Assessment List for Using Math in Science in **PASC**, p. 29. **P**

📁 **Activity Worksheets,** pp. 5, 146

656

MiniLAB

How do alveoli change surface area?

Procedure

1. Make a cylinder out of a large sheet of paper. Tape it together.
2. Make cylinders out of small sheets of paper. Place them inside the large cylinder.
3. Unroll each cylinder. Place the small sheets next to each other in a rectangle. Lay the large sheet on top.

Analysis

1. Compare the surface area of the large sheet with all the small sheets put together.
2. What do the large sheet and small sheets represent?
3. How does this make gas exchange more efficient?

The Bronchi and the Lungs

At the lower end of the trachea are two short branches, called **bronchi,** that carry air into the lungs. Your lungs take up most of the space in your chest cavity. Within the lungs, the bronchi branch into smaller and smaller tubes. The smallest tubes are the bronchioles. At the end of each bronchiole are clusters of tiny, thin-walled sacs called **alveoli** (al VE uh li). As shown in **Figure 24-5,** lungs are actually masses of alveoli arranged in grape-like clusters. Capillaries surround the alveoli. The exchange of oxygen and carbon dioxide takes place between the alveoli and capillaries. This happens easily because the walls of the alveoli and the walls of the capillaries are only one cell thick. Oxygen diffuses through the walls of the alveoli and then through the walls of the capillaries into the blood. There the oxygen is picked up by hemoglobin in red blood cells and carried to all body cells. As this takes place, carbon dioxide is transported back from body cells in the blood. It diffuses through the walls of the capillaries and through the walls of the alveoli. Carbon dioxide leaves your body when you breathe out, or exhale.

Figure 24-5

There are about 300 million alveoli in each lung.

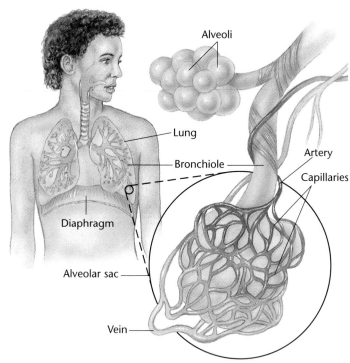

How You Breathe

Breathing is partly the result of changes in air pressure. Under normal conditions, a gas moves from an area of high pressure to an area of low pressure. **Figure 24-6** shows this. When you squeeze an empty plastic bottle, air rushes out. This happens because pressure outside the top of the bottle is less than inside the bottle. As you release your grip on the bottle, the pressure inside the bottle becomes less than outside the bottle. Air rushes back in.

Inhale and Exhale

Your lungs work in a similar way to the squeezed bottle. Your **diaphragm** is a muscle beneath your lungs that helps move air in and out of your body. It contracts and relaxes when you breathe. Like your hands on the plastic bottle, the diaphragm exerts pressure or relieves pressure on your lungs. Remember earlier when you felt your chest move up and down? When you inhale, your diaphragm, shown in **Figure 24-7,** contracts and moves down. The upward movement of your rib cage and the downward movement of your diaphragm cause the volume of your chest cavity to increase. Air pressure is reduced in your chest cavity. Air under pressure outside the body pushes into your air passageways and lungs. Your lungs expand as the air rushes into them.

When you exhale, your diaphragm relaxes and moves up to return to its dome shape. Your rib cage moves downward. These two actions reduce the size of your chest cavity. Your lungs also return to their original position. Pressure on your lungs is increased by these two actions. The gases inside your lungs are pushed out through the air passages.

Figure 24-6

If you squeeze a gas into a smaller space, you increase the pressure. If you provide a larger space, the pressure decreases.

Figure 24-7

Your lungs inhale and exhale about 500 mL of air with an average breath. This may increase to 2000 mL of air per breath when you do strenuous physical activity.

Inhale

Exhale

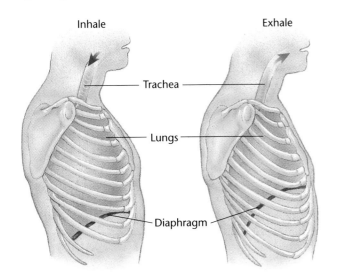

Inhale   Exhale

- Trachea
- Lungs
- Diaphragm

INTEGRATION
Physics

The total amount of force exerted by a gas depends on the size of its container. Pressure is the amount of force exerted per unit of area.

$$P = F/A$$

Boyle's law states that if you decrease the volume of a container of gas, the pressure of the gas will increase. If you increase the volume of the container, the pressure will decrease.

Enrichment

Have students research how dry air taken in through the nostrils has about 100 percent humidity by the time it reaches the alveoli. ▫ L2

Visual Learning

Figure 24-7 The position of the diaphragm during the process of inhaling and exhaling is shown. Compare this process with the squeezing of the plastic bottle in **Figure 24-6.** LEP IS

Discussion

Discuss with students ways singers practice their breathing in order to have greater breath control while singing. Refer to page 655.

Integrating the Sciences

Physical Science The rate at which oxygen or carbon dioxide will diffuse across a respiratory surface can be determined using Fick's law. Have students find out what Fick's law is. ▫ L2

657

Figure 24-8

The Heimlich maneuver is a technique used to save a person from choking.

A The rescuer stands behind the choking victim. She places a fist (thumb-side in) against the victim's stomach. The fist should be below the ribs and above the navel.

B With a sudden, sharp movement, the fist is thrust up and into the area below the ribs.

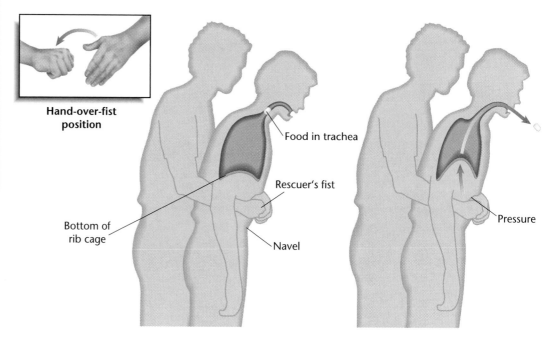

Hand-over-fist position

Food in trachea

Rescuer's fist

Bottom of rib cage

Navel

Pressure

A Life-Saving Maneuver

You learned that the epiglottis closes over your larynx to stop food from entering the trachea. Sometimes this process does not happen quickly enough. Each year, thousands of people die because food or other objects become lodged in the trachea. Air flow between the lungs and the mouth and nasal cavity is blocked. Death can occur in a matter of minutes.

Pressure Dislodges the Food

Rescuers use the Heimlich maneuver, **Figure 24-8,** to save the life of a choking victim. The theory behind this maneuver is to use pressure to force out the food. When the diaphragm is forced up, the volume of the chest cavity quickly decreases. Pressure is suddenly increased. Air is forced up in the trachea. There may be enough force to dislodge food or an object. The victim is able to breathe again.

Community Connection

CPR Arrange a visit to a Red Cross center or have a health specialist come to school and demonstrate both the Heimlich maneuver and the cardiopulmonary resuscitation (CPR) method to revive a person who is not breathing.

Diseases and Disorders of the Respiratory System

If you were asked to list some of the things that can harm your respiratory system, you would probably put smoking at the top. Many serious diseases are related to smoking. Being around others who smoke also can harm your respiratory system. Smoking, polluted air, and coal dust have been related to respiratory problems such as bronchitis, emphysema, cancer, and asthma.

Chronic Bronchitis

Bronchitis is a disease in which the bronchial tubes are irritated. This results in too much mucus being produced. Many cases of bronchitis clear up within a few weeks, but sometimes the disease will last for a long time. When bronchitis persists for a long time, it is called **chronic bronchitis.** Many cases of chronic bronchitis result from smoking. People who have chronic bronchitis cough often to try to clear the mucus from the airway, as **Figure 24-9** shows. However, the more a person coughs, the more the cilia and bronchial tubes can be harmed. When cilia are damaged, their ability to move mucus, bacteria, and dirt particles out of the lungs is impaired. If this happens, harmful substances, such as sticky tar from burning tobacco, build up in the airways. Sometimes scar tissue forms, impairing the functioning of the respiratory system.

Figure 24-9

Cilia help trap and move foreign matter.

A Coughing is a reflex that moves unwanted matter from respiratory passages.

B Hairlike cilia located around cancer cells in the respiratory passage are shown here. *How do damaged cilia weaken your body's defense against disease?*

LOAN OFFICER

659

3 Assess

Check for Understanding

Have students write a paragraph describing why air moves into the lungs when the ribs move upward and the diaphragm moves downward.

? FLEX Your Brain

Use the Flex Your Brain activity to have students explore RESPIRATION.

☞ **Activity Worksheets,** p. 5

Reteach

Use a lung demonstration apparatus to illustrate how the downward movement of the diaphragm causes reduced air pressure within the chest cavity.

Extension

☞ For students who have mastered this section, use the **Reinforcement** and **Enrichment** masters.

Emphysema

A disease in which the alveoli in the lungs lose their ability to expand and contract is called **emphysema.** Most cases of emphysema result from smoking. In fact, many smokers who are afflicted with chronic bronchitis eventually progress to emphysema. When a person has emphysema, cells in the bronchi become inflamed. An enzyme released by the cells causes the alveoli to stretch and lose their elasticity. As a result, alveoli can't push air out of the lungs. Less oxygen moves into the bloodstream from the alveoli. Blood becomes low in oxygen and high in carbon dioxide. This condition results in shortness of breath. As shown in **Figure 24-10,** some people with emphysema can't blow out a match or walk up a flight of stairs. Because the heart works harder to supply oxygen to body cells, people who have emphysema often develop heart problems as well.

Lung Cancer

Lung cancer is the leading cause of cancer deaths in men and women in the United States. Inhaling the tar in cigarette smoke is the greatest contributing factor to lung cancer. Once in the body, tar and other ingredients found in smoke are changed into carcinogens. These carcinogens trigger lung cancer by causing uncontrolled growth of cells in lung tissue. Smoking

Figure 24-10

Lung diseases can have major effects on breathing.

B A diseased lung cuts down on the amount of oxygen that can be delivered to body cells.

A Emphysema may take 20-30 years to develop. *Why does the long development time make it impossible to correct the damage?*

C A normal, healthy lung can exchange oxygen and carbon dioxide effectively.

also is believed to be a factor in the development of cancer of the mouth, esophagus, larynx, and pancreas. **Figure 24-11** shows one way people are reminded of the damage associated with smoking.

Asthma

Some lung disorders are common in nonsmokers. **Asthma** is a disorder of the lungs in which there may be shortness of breath, wheezing, or coughing. When a person has an asthma attack, the bronchial tubes contract quickly. Asthma is often an allergic reaction. An asthma attack can result from a reaction to breathing certain substances, such as cigarette smoke and plant pollen. Eating certain foods or stress also have been related to the onset of asthma attacks.

Except for certain bacteria, all living things would die without oxygen. Your respiratory system takes in oxygen and gets rid of carbon dioxide. This system also helps get rid of some pathogens. You can help keep your respiratory system healthy by avoiding smoking and breathing polluted air. Regular exercise helps increase your body's ability to use oxygen.

Figure 24-11

More than 80 percent of all lung cancer is related to smoking.

Section Wrap-up

Review

1. What is the main function of the respiratory system?

2. What happens when oxygen and carbon dioxide are exchanged in the lungs?

3. What causes air to move in and out of the lungs?

4. How does emphysema affect a person's alveoli?

5. **Think Critically:** How is the work of the digestive and circulatory systems related to the respiratory system?

Skill Builder
Sequencing
Sequence the pathway of air through the respiratory organs from the atmosphere to the blood and back to the atmosphere. If you need help, refer to Sequencing in the **Skill Handbook.**

Science Journal

Use library references to find out about a lung disease common among coal miners, stone cutters, and sand blasters. In your Science Journal, write a paragraph about the symptoms of this disease.

•MINI•QUIZ•

1. **Where is inhaled air moistened and warmed?** *nasal cavity*

2. **What keeps the trachea tube from collapsing?** *rings of cartilage*

3. **Masses of _____ are described as grapelike clusters.** *alveoli*

4. **When relaxed, the diaphragm has a _____ shape.** *dome*

Section Wrap-up

Review

1. supplies oxygen to the blood and removes carbon dioxide

2. Oxygen diffuses through the walls of the alveoli and then through the walls of the capillaries into the blood. Carbon dioxide diffuses in the opposite direction.

3. movement of the diaphragm and rib cage, and differences in pressure

4. Alveoli stretch and lose their elasticity.

5. **Think Critically** The digestive system provides food for respiration in cells. The circulatory system transports oxygen to break down food and carries respiration waste products to the lungs to be expelled.

Skill Builder
atmosphere—nasal cavity—pharynx—larynx—trachea—bronchi—bronchioles—alveoli—capillaries—blood—to all body cells—carbon dioxide comes back from body cells—capillaries—alveoli—and reverse of above P

✓ Assessment

Portfolio Use the Flex Your Brain activity to reinforce critical-thinking and problem-solving skills. In step 2, students might list how breathing feels, its function, what affects breathing rate, or things that interfere with breathing. Use the Performance Task Assessment List for Asking Questions in **PASC**, p. 19. P

Science Journal

Black lung is caused by inhalation of mineral dusts such as silica and coal dust. Fibrous tissue develops around the particles, replacing lung tissue. Less O_2-CO_2 exchange takes place.

PREPARATION

Purpose
LS Interpersonal Design and carry out an experiment to show how exercise affects the amount of carbon dioxide exhaled by the lungs. **L1**
COOP LEARN

Process Skills
communicating, hypothesizing, designing an experiment, observing and inferring, interpreting data, making and using tables, comparing and contrasting, recognizing cause and effect, and separating and controlling variables

Time
45-50 minutes

Materials
Obtain materials needed for cooperative groups.
Alternate Materials Limewater ($Ca(OH)_2$ saturated solution) may be used. CO_2 causes a cloudy white precipitate to form.

Safety Precautions
Caution students not to inhale through the straw. Do not let students share the same straw for this activity. Before the students exercise, make sure no student has a medical problem that the exercise would complicate.

Possible Hypotheses
Students will most likely state that it will take less time to cause a change in the bromothymol blue solution after exercising than before exercising.

📁 **Activity Worksheets,** pp. 5, 142-143

Activity 24-1

Design Your Own Experiment
The Effects of Exercise on Respiration

Breathing rate increases with an increase in physical activity. A bromothymol blue solution changes color when carbon dioxide is bubbled into it. Can you predict whether there will be a difference in the time it takes for the solution to change color before and after exercise?

PREPARATION

Problem
Your task in this activity is to compare the amount of carbon dioxide in exhaled air at rest and after activity. How will an increase in physical activity affect the amount of carbon dioxide exhaled?

Form a Hypothesis
State a hypothesis about how exercise will affect the amount of carbon dioxide exhaled by the lungs.

Objectives
- Observe the effects of the amount of carbon dioxide on the bromothymol blue solution.
- Design an experiment that tests the effects of a variable, such as the amount of carbon dioxide exhaled before and after exercise, on the rate at which the solution changes color.

Possible Materials
- clock or watch with second hand
- drinking straws
- bromothymol blue solution (200 mL)
- 400-mL beakers (2)
- graduated cylinder

Safety Precautions

Protect clothing from the solution. Wash hands after using the solution. **CAUTION:** *Do not inhale the solution through the straw.*

PLAN THE EXPERIMENT

Possible Procedures
- One possible procedure is to test for the amount of carbon dioxide exhaled from the lungs. Label the beakers A and B. Use beaker A for the resting test and beaker B for the activity test. Pour 100 mL bromothymol blue solution into each beaker.
- Test for exhaled carbon dioxide: Look at the clock and begin timing. Exhale through the straw into the bromothymol

---

Inclusion Strategies

Gifted Have students investigate why the bromothymol blue solution changes color. They can try other acid–base indicators such as litmus. **L3**

PLAN THE EXPERIMENT

1. As a group, agree upon and write out the hypothesis statement.
2. As a group, list the steps that you will need to take to test your hypothesis. Consider each of the following factors: How will you introduce the exhaled air into the bromothymol blue solution? How will you collect data on exhaled air before and after physical activity? What kind of activity is involved? How long will it go?
3. List your materials. Your teacher will provide instruction on safe procedures for using bromothymol blue.
4. Design a data table and record it in your Science Journal so that it is ready to use as your group collects data.

Check the Plan
1. Read over your entire experiment to make sure that all the steps are in logical order.
2. Identify any constants, variables, and controls of the experiment.
3. *Make sure your teacher approves your plan before you proceed.*

DO THE EXPERIMENT

1. Carry out the experiment as planned.
2. While the experiment is going on, write down any observations that you make and complete the data table in your Science Journal.

Analyze and Apply
1. What caused the bromothymol blue solution to change color? What color was it at the conclusion of each test?

2. **Compare** the time it took the bromothymol blue solution to change color before exercise and after exercise. Explain any difference.
3. What was the control?
4. Prepare a table of your data and **graph** the results.
5. Using your graph, **estimate** the time of color change if the time of your physical activity were twice as long.

Go Further

How could you determine the presence of any other gas exhaled from the lungs? Devise a test to see if water vapor is exhaled.

---

5. Estimates will vary, but should indicate that it would take less time for a color change if the physical activity were twice as long.

Go Further

Breathe on a cool, shiny metal or glass surface. Minute water droplets can be seen and felt. Paper treated with cobalt chloride will turn from blue to pink when it comes in contact with moisture.

Assessment

Performance To further assess students' abilities to analyze the effects of respiration, have them design an experiment to detect the water vapor exhaled in an hour/in a day. Use the Performance Task Assessment List for Designing an Experiment in **PASC**, p. 23. **P**

---

blue solution. **CAUTION:** *Do not inhale through the straw.* Continue exhaling for 15 seconds or until the bromothymol blue solution changes color. Note the color. Record the time it takes for the color change to occur. Carbon dioxide causes this change.
• Remember to exhale using the same force for both tests.

Teaching Strategies
• Experimental designs may vary. One possibility: Breathe into beaker A before exercising and then breathe into beaker B after three minutes of exercise.
• The carbon dioxide bubbled into the water forms carbonic acid, which causes the color change in the bromothymol blue solution.
• Remind students to blow quickly into the bromothymol blue solution after exercising.

DO THE EXPERIMENT

Expected Outcome
Results will reveal that more carbon dioxide is exhaled after exercising, and the bromothymol blue solution changes color more quickly.

Analyze and Apply
1. the presence of carbon dioxide; yellow
2. Bromothymol blue takes less time to change color after exercise. Exercise produces more carbon dioxide.
3. The control was breathing into beaker A, which contained the bromothymol blue solution, before exercising.
4. The table should contain all the data collected. The graph should accurately reflect the data.

Prepare

Section Background

The negative health aspects of tobacco use have been documented for many years. Tobacco products are required to have various health warnings on their packaging.

1 Motivate

Bellringer

Before presenting the lesson, display **Section Focus Transparency 88** on the overhead projector. Assign the accompanying **Focus Activity** worksheet. **L1** **LEP**

Tying to Previous Knowledge

Recall the effects of smoking on the circulatory and respiratory systems.

2 Teach

Activity

Have students look for tobacco product advertisements in magazines. Find out what techniques are used to promote the use of the product.

ISSUE:
24•2 Restricting Cigarette Advertising

Science Words

passive smoking

Objectives

- Describe why tobacco use increased over the last century.
- Determine what regulations are needed to control cigarette advertising.

Rise in Tobacco Use

Native Americans smoked tobacco in pipes long before Columbus sailed to the New World in 1492. Commercial production of tobacco began in North America in 1612. Tobacco was mostly smoked in pipes until the first cigarette-making machine was invented in the 1880s. This method of mass cigarette production caused the use of tobacco to increase greatly during the first half of the 20th century. By 1977, an average smoker in the United States consumed almost 13 000 cigarettes a year. With the rise in consumption came a rise in the number of people afflicted with smoking-related health problems. As early as the 1920s, doctors suspected a link between tobacco use and various types of cancer. Smoking was formally identified as a leading cause of cancers and other diseases in 1964 by the United States Surgeon General.

Smoking Restrictions

As evidence of the link between smoking and health problems mounted, many people wanted restrictions placed on cigarette use. They especially wanted use by teenagers regulated. As a result, it is illegal to sell tobacco products to individuals under the age of 18 in many states. Additional restrictions on all smokers include the banning of smoking on domestic air flights as shown in **Figure 24-12.** Smoking and nonsmoking areas are also found in many restaurants. Other public buildings prohibit smoking altogether. These efforts are to help prevent health problems in nonsmokers that are associated with passive smoking. Breathing in cigarette smoke-filled air is called **passive smoking.**

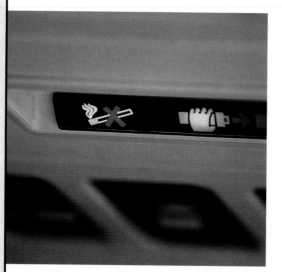

Figure 24-12

Smoking is not allowed on most airline flights because of the problems associated with passive smoking in a closely confined area.

664 Chapter 24 Respiration and Excretion

Program Resources

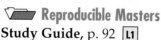 **Reproducible Masters**
Study Guide, p. 92 **L1**
Reinforcement, p. 92 **L1**
Enrichment, p. 92 **L3**

 Transparencies
Section Focus Transparency 88 **L1**

2 Points of View

▷ Limiting Advertising

Advertising is an effective way for a tobacco company to lure new smokers. In an effort to reduce the impact of advertising, laws have been passed restricting this kind of promotion. Since 1971, cigarette advertisements have been banned on radio and television. Recent public pressure has caused major professional sports stadiums to reposition the location of cigarette advertisements. Tobacco companies had placed large signs where they would appear during television broadcast coverage. Additional public pressure and threats of legal action have tried to influence the types of advertisements used by tobacco companies. Cartoon characters and social situations that were appealing to children and teenagers were targets for the protest, shown in **Figure 24-13**.

▷ Free Speech

Some people think that the protest groups have gone too far. They feel that the tobacco companies have the right to advertise when and where they want. Because the product is not illegal for adults to use, the rules against cigarette ads infringe on the companies' constitutional right of free speech. They also worry that if the government can regulate the cigarette advertisements, then it can impose more regulations on other companies.

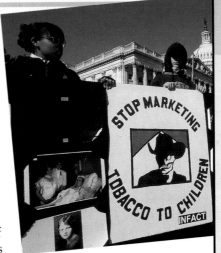

Figure 24-13

Many believe that tobacco companies should eliminate advertisements aimed at young people.

Section Wrap-up

Review

1. What is the major reason why cigarette use increased in the early 1900s?

2. What restrictions on smoking are presently enforced?

Explore the Issue

Should the tobacco companies be allowed to promote the use of their products through any kind of advertising they choose? If yes, how can the United States deal with the increased costs of health problems associated with tobacco use? If no, what actions should be taken to restrict cigarette advertising?

SCIENCE & SOCIETY

Across the Curriculum

Social Studies Have students trace the origin of tobacco ingestion to its use by Native Americans for special events. Christopher Columbus and Sir Walter Raleigh brought tobacco to Europe. [L2]
P

---

3 Assess

Check for Understanding

Have students answer questions 1 and 2 in the Section Wrap-up.

Reteach

Point out the success of restricting the advertisement of tobacco products in the United States. There has been an overall decrease in the number of smokers. However, the number of young people starting to smoke is rising.

Extension

📁 For students who have mastered this section, use the **Reinforcement** and **Enrichment** masters.

4 Close

Have students discuss the unglamorous aspects of smoking, such as bad breath. Ask students to discuss why a person might feel a need to start to smoke.

Section Wrap-up

Review

1. invention of cigarette rolling machines

2. cigarette purchases by persons under eighteen years of age, smoking in certain public buildings and on airline flights

Explore the Issue

Answers will vary, but may include the following. Yes: Eliminate subsidies to tobacco growers or have a special health tax on tobacco products. No: Ban all tobacco product advertising in publications usually read by young people **P**

665

Prepare

Section Background

Kidneys produce hormones that can bring about changes to increase blood pressure or increase formation of red blood cells.

Preplanning

- To prepare for the Mini-LAB, obtain animal kidneys from a butcher, scalpels, and magnifying glasses.
- To prepare for Activity 24-2, obtain the iodine solution. Test the bond paper to be used for reaction to the iodine. A small drop of iodine should cause a purple/black spot to appear.

1 Motivate

Bellringer

Before presenting the lesson, display **Section Focus Transparency 89** on the overhead projector. Assign the accompanying **Focus Activity** worksheet. L1 LEP

Tying to Previous Knowledge

Recall the function of a filter in a coffee-making machine. Only liquid is allowed to flow through. Have students compare this with the function of the urinary system.

24•3 Your Urinary System

Science Words

- **urinary system**
- **kidney**
- **nephron**
- **urine**
- **ureter**
- **bladder**
- **urethra**

Objectives

- Distinguish between the excretory and urinary systems.
- Describe how your kidneys work.
- Explain what happens when urinary organs don't work.

Figure 24-14

A flowchart shows how the urinary system is linked with the digestive, circulatory, respiratory systems, and skin to make up the excretory system.

Your Excretory System

Just as wastes, in the form of sewage or garbage, are removed from your home, your body eliminates wastes. You learned that undigested material is eliminated by your digestive system. The waste gas, carbon dioxide, is eliminated through the combined efforts of your circulatory and respiratory systems. Some salts are eliminated when you sweat. Together, these systems function as a part of your excretory system. If wastes aren't eliminated, you can become sick. Toxic substances build up and damage organs. If not corrected, serious illness or death occurs.

Another system is also involved in removing wastes. The organs of your urinary system are excretory organs. Your **urinary system** is made up of organs that rid your blood of wastes produced by the metabolism of nutrients and control blood volume by removing excess water produced by body cells. A specific amount of water in blood is important to maintain normal blood pressure, the movement of gases, and excretion of solid wastes. Your urinary system also balances certain salts and water that must be present in specific concentrations for cell activities to take place. **Figure 24-14** shows how the urinary system functions as a part of the excretory system.

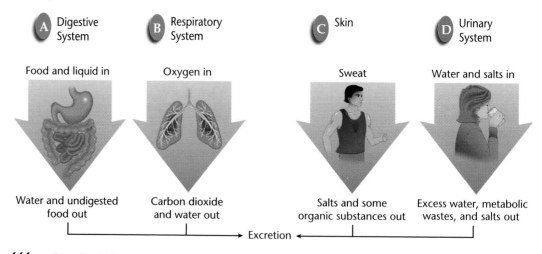

A Digestive System — Food and liquid in — Water and undigested food out

B Respiratory System — Oxygen in — Carbon dioxide and water out

C Skin — Sweat — Salts and some organic substances out

D Urinary System — Water and salts in — Excess water, metabolic wastes, and salts out

→ Excretion ←

Program Resources

 Reproducible Masters
Study Guide, p. 93 L1
Reinforcement, p. 93 L1
Enrichment, p. 93 L3
Activity Worksheets, pp. 5, 144-145, 147 L1
Concept Mapping, p. 53
Multicultural Connections, pp. 51-52

 Transparencies
Section Focus Transparency 89 L1
Teaching Transparency 48

Figure 24-15
The urinary system (A) is made of your kidneys, bladder, and the connecting tubes. The kidneys (B) are made up of many nephrons. A single nephron is shown in detail by the arrow.

Organs of Your Urinary System

The major organs of your urinary system, as shown in **Figure 24-15,** are two bean-shaped kidneys. Kidneys are located on the back wall of the abdomen at about waist level. The **kidneys** filter blood that has collected wastes from cells. All of your blood passes through your kidneys many times a day. In **Figure 24-15,** you can see that blood enters the kidneys through a large artery and leaves through a large vein.

The Filtering Unit

Each kidney is made up of about 1 million **nephrons,** the tiny filtering units of the kidney. Each nephron has a cuplike structure and a duct. Blood moves from the renal artery to capillaries in the cuplike structure. Water, sugar, salt, and wastes from your blood pass into the cuplike structure. From there, the liquid is squeezed into a narrow tubule. Capillaries that surround the tubule reabsorb most of the water, sugar, and salt and return it to the blood. These capillaries merge to form small veins. The small veins merge to form the renal veins, which return purified blood to your circulatory system. The liquid left behind flows into collecting tubules in each kidney. This waste liquid, or **urine,** contains excess water, salts, and other wastes not reabsorbed by the body. The average adult produces about 1 L of urine per day.

MiniLAB

What do kidneys look like?

Procedure
1. Use the kidney supplied by your teacher. Carefully cut the tissue lengthwise in half around the outline of the kidney.
2. Use a magnifying glass to observe the internal features of the kidney, or view the features in a model.
3. Compare the specimen or model with the kidney in **Figure 24-15.**

Analysis
1. What part makes up the major part of the kidney? Why is this part red in color?
2. How can the kidney be compared to a portable water-purifying system?

---

2 Teach

MiniLAB

Purpose
LS Visual-Spatial Students will observe the external and internal structures of a kidney. **L1** **LEP**

Materials
large animal kidneys, scalpels, magnifying glasses

Teaching Strategies
Safety Precautions Caution students about the use of sharp objects.

Troubleshooting Demonstrate the proper techniques for using the scalpels.

Analysis
1. the mass of nephron units; the color is due to the blood in the nephrons
2. Both serve as filtering mechanisms.

✓ Assessment

Oral Have students describe the evidence that two kidneys are more than enough to take care of the excretory functions. Use the Performance Task Assessment List for Making Observations and Inferences in **PASC,** p. 17. **P**

📁 **Activity Worksheets,** pp. 5, 147

Demonstration
Set up a diffusion experiment using a cellophane bag with a sugar solution placed in a container of water.

Visual Learning

Figure 24-15 Have students trace the pathway of the blood to, through, and from the kidney. **LEP** **LS**

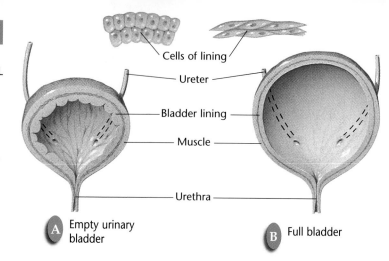

A Empty urinary bladder

B Full bladder

Urine Collection and Release

The urine in each collecting tubule drains into a funnel-shaped area of each kidney that leads to the ureters (YOOR ut urz). **Ureters** are tubes that lead from each kidney to the bladder. The **bladder** is an elastic muscular organ that holds urine until it leaves the body. **Figure 24-16** shows how the shape of the cells that make up the lining of the bladder change with the amount of urine stored in it. A tube called the **urethra** (yoo REE thruh) carries urine from the bladder to the outside of the body.

Water Loss Through Other Systems

Other parts of the excretory system also help your body maintain proper fluid levels. In addition to losing salt, an adult loses about 0.5 L of water each day through perspiration. When air is exhaled, you also lose water. You see this moisture when you see your breath on a cold day or breathe on a

Increased liquid intake

Hypothalamus

A Small amount of hormone produced

B Increased urine production

Figure 24-17

The amount of urine that you eliminate each day is determined by the level of a hormone. This hormone is produced by your hypothalamus.

668 Chapter 24 Respiration and Excretion

Across the Curriculum

Mathematics Calculate the ratio of body weight to amount of urine excreted. If an 11-kg child excretes 600 mL, what would you expect a 45-kg person to excrete? *2455 mL However, the amount is actually only around 1500 mL. The amount excreted by children is greater in proportion to their weight.* **L1** **IS**

cold window pane and notice a cloud or fog. Each day, about 350 mL of water is removed from your body through your respiratory system. A small amount of water is also expelled with the undigested material that passes out of your digestive system.

Balancing Fluid Levels

To maintain good health, the fluid levels within your body must be balanced and your blood pressure must be maintained. This happens because an area of your brain, the hypothalamus, is involved in controlling your body's homeostasis. It produces a hormone that regulates how much urine is produced. As **Figure 24-17** shows, if your brain detects too much water in your blood, it releases a lesser amount of the hormone. This signals the kidneys to return less water to the blood and increase the amount of urine excreted. If too little water is in the blood, more of the hormone is released and more water is returned to the blood. The amount of urine excreted decreases. At the same time, your brain sends a signal that causes you to feel thirsty. You drink liquids to quench this thirst.

Problem Solving

Frederick's Dilemma

Frederick was born with an unusual birth defect. He was born without sweat glands. Because of this, he finds it difficult to stay outdoors for any length of time, especially in the summer. Frederick is not usually able to enjoy a day at the beach or to participate in sports.

A local restaurant raised money to help Frederick. The money was used to develop a "cool suit." The suit resembles a space suit and fits under his clothing. Frederick uses a thermostat to control the pumping of a cool solution through the suit. The suit enables him to spend more time outside. He can even play sports such as baseball.

Solve the Problem:

1. **What function of the skin does the suit duplicate?**
2. **Describe how the suit could also be helpful during colder weather.**

Think Critically:

Explain how the circulating solution keeps Frederick cool.

669

Problem Solving

Solve the Problem

1. functions to regulate the body temperature
2. The suit can serve as an insulator against cold weather.

Think Critically

The circulating solution carries heat produced by Frederick away from his body, much as evaporation of sweat normally would.

Revealing Preconceptions

Students often do not recognize the need for drinking adequate amounts of water each day. Water is necessary for the proper functioning of the kidneys. Water leaves our bodies through perspiration from the skin, through water vapor in exhaled breath, and in urine.

USING TECHNOLOGY

Kidney Stones

Sometimes a small, solid particle may form in the area of the kidney where urine funnels into the ureter. Over time, more material such as calcium or uric acid may be added to the stone, causing it to increase in size. Small stones may remain in the kidney and not cause any problems. However, if a larger stone passes into a ureter, severe pain results. Traditionally, chemicals have been used to dissolve the stone or surgery was done to remove it.

A Nonsurgical Treatment

A new method of removal involves the use of a lithotripter. A lithotripter is a machine that produces shock waves. When these waves are focused on the kidney stone, it breaks into tiny pieces. These pieces are then carried out of the body with the urine. Earlier models of the lithotripter required the patient to be partially submerged in a tub of water. Newer machines use water-filled bags above and below the location of the kidneys as the patient lays flat on a table. Water in the bags transmits the shock waves to the kidney stone.

Think Critically:
What are some conditions necessary for the lithotripter to function successfully? What are some advantages of using the new lithotripter?

Shock Waves Breaking Up a Kidney Stone

Water-filled bags

Kidney stone

Water column

Shock wave generator

Diseases and Disorders of the Urinary System

What happens when someone's urinary organs don't work properly? Waste products that are not removed build up and act as poisons in body cells. Water that is normally removed from body tissues accumulates and causes swelling of the ankles and feet. Sometimes, fluids can also build up around the heart. The heart must work harder to move less blood to the lungs. Without excretion, there may also be an imbalance of salts. The body responds by trying to restore this balance. If the balance is not restored, the kidneys and other organs can be damaged.

Dialysis

Persons who have damaged kidneys may need to have their blood filtered by an artificial kidney machine in a process called dialysis, shown in **Figure 24-18.** During dialysis, blood from an artery is pumped through tubing that is bathed in a salt solution similar to blood plasma. Waste materials diffuse from the tube containing blood and are washed away by the salt solution. The cleaned blood is returned to a vein. A person with only one kidney can still function normally.

The urinary system is a purifying unit for the circulatory system. Wastes are filtered from blood as it passes through the kidneys. Some water, salts, and nutrients are reabsorbed to maintain homeostasis. Waste materials, dissolved in water, are eliminated from the body. This system helps to maintain the health of cells and, therefore, the entire body.

Figure 24-18
People who undergo dialysis must have the procedure done several days each week.

Section Wrap-up

Review

1. Describe the functions of the urinary system.

2. Explain how the kidneys remove wastes and keep fluids and salts in balance.

3. Compare the excretory and urinary systems.

4. What happens when urinary organs don't work?

5. Think Critically: Explain why reabsorption of certain materials in the kidneys is important.

Skill Builder
Concept Mapping
Using a network tree concept map, compare the excretory functions of the kidneys and the lungs. If you need help, refer to Concept Mapping in the **Skill Handbook.**

USING MATH

All 5 L of blood in the body pass through the kidneys in approximately five minutes. Calculate the average rate of flow through the kidneys in liters per minute.

24-3 Your Urinary System   **671**

Skill Builder

```
            Waste in
             blood
          /            \
    Lungs      are      Kidneys
      |      carried       |
    expel     to         expel
      |                    |
  Water and            Water
    CO₂                and salts
      |                    |
    from                  from
      |                    |
    Cells                Cells
```

Assessment

Portfolio Use this Skill Builder to assess students' abilities to form a concept map of the excretory functions of the lungs and kidneys. Use the Performance Task Assessment List for Concept Map in **PASC**, p. 89. **P**

4 Close

•MINI•QUIZ•

Use the Mini Quiz to check students' recall of chapter content.

1. **What are your main excretory organs?** *kidneys, lungs, skin*

2. **The filtering unit of the kidneys is the _____ .** *nephron*

3. **The _____ are the tubes that carry urine from the kidneys to the bladder.** *ureters*

4. **What is dialysis?** *a process by which blood from a person's artery is pumped through tubing that is bathed in a salt solution so that wastes are removed*

Section Wrap-up

Review

1. rids the body of wastes, controls blood volume by removing excess water, balances salts and water

2. Kidneys filter the blood to remove wastes, water, and salt. Necessary amounts of water, sugar, and salt are returned to the blood.

3. The excretory system includes all the organs that function to rid the body of a variety of wastes. The urinary system has organs that function to produce only one kind of waste, urine.

4. Waste products accumulate and act as poisons to the body.

5. Think Critically Many of the substances that are reabsorbed are needed by the body.

USING MATH

LM Logical-Mathematical 5 liters; 5 minutes = 1 liter per minute rate of flow

Sweat Glands in the Skin

Sweat glands are special structures in the skin that excrete waste. Try this activity to determine whether sweat glands are equally distributed on your body's surface.

Purpose

Ⓛ Visual-Spatial Design and carry out an experiment to determine the number and position of sweat glands. **L1**

COOP LEARN

Process Skills

experimenting, observing and inferring, analyzing, using numbers, comparing and contrasting, and predicting

Time

45-50 minutes

Safety Precautions

Students should pull up their sleeves during this experiment. Iodine is a poison and should not be put in the mouth or eyes.

📁 Activity Worksheets, pp. 5, 144-145

Teaching Strategies

Troubleshooting Obtain 2% tincture of iodine. Test the bond typing paper to ensure the starch/iodine reaction takes place.

Answers to Questions

1. Answers will vary due to individual differences. One test indicated 18 dots per centimeter.

2. Answers will vary due to individual differences. One test indicated two dots per centimeter.

3. Comparisons should show that there are fewer sweat glands on most other body parts.

4. The iodine reacted with the starch to form a purple/black compound wherever moisture from sweat glands touched the paper.

Problem

How are sweat glands distributed on the body?

Materials

- iodine solution
- cotton swabs
- bond typing paper
- hand lens

Procedure

1. Copy the data table and use it to record your observations.

2. Paint a 1 cm × 1 cm square on the skin of a palm of one hand with two percent tincture of iodine. **CAUTION:** *Iodine is a poison; do not get it in your mouth or eyes. Iodine can stain clothing.* Allow the iodine solution to dry on your skin for three to four minutes.

3. Using the thumb from the other hand, hold a 2 cm × 2 cm square piece of bond typing paper on the iodine spot for one minute.

4. Lift up the paper and examine it with a hand lens. You should see some purple/black dots. Each dot represents a sweat gland. Count the number of dots in 1 cm² of the paper. Record your data.

5. Select another spot, such as the back of the hand or forearm, to test for sweat gland distribution. Repeat steps 2 to 4.

Analyze

1. How many dots per centimeter did you observe on your palm?

2. How many dots per centimeter did you observe on the second spot tested?

3. Compare the results.

Conclude and Apply

4. Bond paper has starch in it. What caused the purple/black dots to appear on the paper?

5. From your observations, **compare** the distribution of sweat glands in different body areas. Why might they be more numerous on one part than another?

6. If this test were done on the palm of your hand after exercise, **predict** what you would expect to see.

Data and Observations

Sample Data

Body Location	Number of dots in 1 cm²
Palm of hand	18
Back of hand	2
Forearm	5
Other	results will vary

5. Sweat glands have greater distribution on the palm of the hand than on the back of the hand. Sweat glands may be more numerous on certain body parts in order to regulate body temperature or to form a protective film over the skin surface.

6. More dots would be seen because more sweat glands would be active and secreting moisture.

✓ Assessment

Performance To further assess students' abilities to use skills to locate the position of sweat glands, repeat the experiment to determine the distribution of sweat glands on the back of the neck. If there is a difference, infer an advantage to having more sweat glands on one surface than another. Use the Performance Task Assessment List for Carrying Out a Strategy and Collecting Data in **PASC**, p. 25. **P**

Summary

24-1: Your Respiratory System

1. Your respiratory system helps you take oxygen into your lungs and body cells and helps you remove carbon dioxide.
2. Inhaled air passes through the nasal cavity, pharynx, larynx, trachea, bronchi, bronchioles, and into the alveoli of the lungs.
3. Breathing results in part from the diaphragm's movement, which changes the pressure within the lungs.
4. Smoking causes many problems throughout the respiratory system.

24-2: Science and Society: Restricting Cigarette Advertising

1. Tobacco smoke has injurious effects on the body.
2. There is a controversy about how much control is needed over cigarette advertising.

24-3: Your Urinary System

1. Kidneys filter blood to remove wastes and keep sodium, water, and other chemicals in balance.
2. The kidneys are the major organs of the urinary system; they filter wastes from all of the blood in your body.
3. Parts of the digestive, circulatory, respiratory, and urinary systems work together as the excretory system.
4. When kidneys fail to work, dialysis may be used.

Key Science Words

a. alveoli
b. asthma
c. bladder
d. bronchi
e. chronic bronchitis
f. diaphragm
g. emphysema
h. kidney
i. larynx
j. nephron
k. passive smoking
l. pharynx
m. trachea
n. ureter
o. urethra
p. urinary system
q. urine

Reviewing Vocabulary

Match each phrase with the correct term from the list of Key Science Words.

1. nonsmoker breathing tobacco smoke
2. clusters of air sacs
3. where vocal cords are attached
4. branches of the trachea
5. muscle involved in breathing
6. disease of alveoli that have lost their elasticity
7. a cartilage-reinforced tube through which air moves to the bronchi
8. major urinary organ
9. tube from kidney to bladder
10. fluid waste

Chapter 24

Review

Summary

Have students read the summary statements to review the major concepts of the chapter.

Reviewing Vocabulary

1. k	**6.** g
2. a	**7.** m
3. i	**8.** h
4. d	**9.** n
5. f	**10.** q

 Assessment

Portfolio Encourage students to place in their portfolios one or two items of what they consider to be their best work. Examples include:

- Skill Builder, p. 661
- Explore the Issue, p. 665
- Across the Curriculum, p. 665 **P**

Performance Additional performance assessments may be found in **Performance Assessment** and **Science Integration Activities.** Performance Task Assessment Lists and rubrics for evaluating these activities can be found in Glencoe's **Performance Assessment in the Science Classroom.**

GLENCOE TECHNOLOGY

Videodisc

Glencoe Life Science Interactive Videodisc
Side 2, Lesson 7

Health Is Just a Heartbeat Away

18598 to 27445

MindJogger Videoquiz

Chapter 24 Have students work in groups as they play the Videoquiz game to review key chapter concepts.

Checking Concepts

1. b	**6.** d
2. c	**7.** a
3. a	**8.** b
4. b	**9.** c
5. a	**10.** c

Understanding Concepts

11. Air is warmed, filtered, and moistened in the nasal cavity.

12. The trachea must remain open for air to enter through it; the esophagus is soft and does not need to be open except when food is moving through.

13. The kidneys filter water, sugar, salt, and wastes from blood. Then most water, sugar, and salt is reabsorbed. Nitrogen waste is eliminated.

14. The respiratory system rids the body of carbon dioxide wastes and some water. Getting rid of wastes is the overall job of the excretory system.

15. Breathing involves the processes of inhaling and exhaling. Respiration is the process of releasing energy from foods.

Thinking Critically

16. When you inhale, the volume of your chest cavity increases. Air pressure is reduced in your chest cavity, and your lungs fill with air. When you exhale, the size of your chest cavity decreases. Pressure in your lungs increases and you breathe out.

17. Having many air sacs increases the surface area for gas exchange.

18. Cilia are destroyed and therefore cannot move particles out of the lungs; alveoli lose elasticity, decreasing total surface area of the lungs available for gas exchange.

674

Checking Concepts

Choose the word or phrase that completes the sentence.

1. When you inhale, your _____ contract(s) and move(s) down.
   a. bronchioles   c. nephrons
   b. diaphragm   d. kidneys

2. Air is moistened, filtered, and warmed in the _____.
   a. larynx   c. nasal cavity
   b. pharynx   d. trachea

3. Exchange of gases occurs between the _____ and capillaries.
   a. alveoli   c. bronchioles
   b. bronchi   d. trachea

4. The rib cage _____ when you exhale.
   a. moves up   c. moves out
   b. moves down   d. stays the same

5. _____ is a lung disorder that may occur as an allergic reaction.
   a. Asthma   c. Emphysema
   b. Chronic bronchitis   d. Cancer

6. A condition worsened by smoking is _____.
   a. arthritis   c. excretion
   b. respiration   d. emphysema

7. _____ are filtering units of the kidney.
   a. Nephrons   c. Neurons
   b. Ureters   d. Alveoli

8. Urine is temporarily held in the _____.
   a. kidneys   c. ureter
   b. bladder   d. urethra

9. About 1 L of water is lost per day through _____.
   a. sweat   c. urine
   b. lungs   d. none of these

10. All except _____ is(are) reabsorbed by blood after passing through the kidneys.
   a. salt   c. wastes
   b. sugar   d. water

19. If the kidneys stopped working, blood would accumulate the waste products of the body's organs. It would become poisonous, and the person would need immediate medical care.

20. Kidney stones in the ureter extend and block the tube and cause severe pain. Infections also may occur.

Understanding Concepts

Answer the following questions in your Science Journal using complete sentences.

11. Why is it better to breathe through your nose?

12. Why does the trachea have cartilage but the esophagus does not?

13. Explain how kidneys maintain homeostasis.

14. Why is the respiratory system considered part of the excretory system?

15. How are breathing and respiration different?

Thinking Critically

16. Compare air pressure in the lungs during inhalation and exhalation.

17. What is the advantage of the lungs having many air sacs instead of being just two large sacs, like balloons?

18. Explain the damage smoking does to cilia, alveoli, and lungs.

19. What would happen to the blood if the kidneys stopped working?

20. Explain why a kidney stone that becomes lodged in a ureter is painful.

Developing Skills

21. Making and Using Graphs

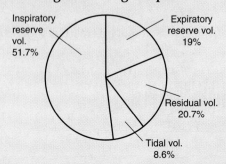

Inspiratory reserve vol. 51.7%

Expiratory reserve vol. 19%

Residual vol. 20.7%

Tidal vol. 8.6%

Developing Skills

If you need help, refer to the **Skill Handbook.**

21. **Making and Using Graphs:** Make a circle graph of total lung capacity.
    - Tidal volume (inhaled or exhaled during a normal breath) = 500 mL
    - Inspiratory reserve volume (air that can be forcefully inhaled after a normal inhalation) = 3000 mL
    - Expiratory reserve volume (air that can be forcefully exhaled after a normal expiration) = 1100 mL
    - Residual volume (air left in the lungs after forceful exhalation) = 1200 mL

22. **Interpreting Data:** Interpret the data below. How much of each substance is reabsorbed into the blood in the kidneys? What substance is totally excreted in the urine?

Materials Filtered by the Kidneys

Substance	Amount moving through kidney to be filtered	Amount excreted in urine
water	125 L	1 L
salt	350 g	10 g
urea	1 g	1 g
glucose	50 g	0 g

23. **Recognizing Cause and Effect:** Discuss how lack of oxygen is related to lack of energy.

24. **Hypothesizing:** Hypothesize the number of breaths you would expect a person would take per minute in each situation and give a reason for each hypothesis.
    - while sleeping
    - while exercising
    - while on top of Mount Everest

25. **Concept Mapping:** Make an events chain concept map showing what happens when urine forms in the kidneys. Begin with the phrase, *In the nephron....*

Performance Assessment

1. **Poster:** Make a poster showing the correct procedure for doing the Heimlich maneuver. Demonstrate the procedure for the class.

2. **Display:** Research the health problems associated with deteriorating asbestos found in the insulation used in older buildings. Prepare a report and display of your findings.

3. **Questionnaire/Interview:** Prepare a questionnaire that can be used to interview a doctor or nurse who works with lung cancer patients. Include questions on reasons for choosing the career, new methods of treatment, and the most encouraging or discouraging part of the job.

675

Chapter 24 Review

22. **Interpreting Data** water —124 liters; salt—340 grams; urea—all is excreted; glucose—50 grams

23. **Recognizing Cause and Effect** Insufficient oxygen results in a lack of energy because cells need oxygen to break down food to release energy. If oxygen isn't supplied, energy is not released.

24. **Hypothesizing** Number of breaths while sleeping is low because less oxygen is needed for rest. While exercising, breaths per minute increase to get oxygen to the muscles. On top of Mt. Everest there is less oxygen in the air; thus, more breaths per minute are taken.

25. **Concept Mapping** Events should be in this order: (1) In the nephron, water, sugar, salt, and wastes are filtered. (2) Capillaries reabsorb most of these substances and return them to the blood. (3) Waste-free blood is circulated throughout the body. (4) Excess waste is eliminated from the body.

Performance Assessment

1. Obtain literature from your local Red Cross chapter to ensure accuracy in information. Use the Performance Task Assessment List for Poster in **PASC,** p. 73. **P**

2. Invite a building contractor to discuss asbestos removal from older buildings. Use the Performance Task Assessment List for Writing in Science in **PASC,** p. 87. **P**

3. Students should prepare questionnaires and conduct interviews. Use the Performance Task Assessment List for Asking Questions in **PASC,** p. 19. **P**

Assessment Resources

 Reproducible Masters
Chapter Review, pp. 51-52
Assessment, pp. 155-158
Performance Assessment, p. 38

Glencoe Technology
- **Chapter Review Software**
- **Computer Test Bank**
- **MindJogger Videoquiz**

Chapter Organizer

Section	Objectives/Standards	Activities/Features
Chapter Opener		Explore Activity: Find a visual trick. p. 677
25-1 Your Nervous System (2 sessions)*	1. **Describe** the basic structure of a neuron and how an impulse moves. 2. **Compare** the central and peripheral nervous systems. 3. **Explain** the effects of drugs on the body. National Content Standards: (5-8) UCP1, UCP5, A1, A2, C1, C3, E2, F1, F5	Using Math, p. 682 Using Technology: Watching the Brain, p. 683 Skill Builder: Concept Mapping, p. 684 Using Computers: Flowchart, p. 684 Activity 25-1: Reaction Time, p. 685
25-2 Science and Society Issue: Care and Treatment of Alzheimer's Patients (1 session)*	1. **Describe** Alzheimer's disease and possible causes. 2. **Determine** the effects of the disease on patients and family. National Content Standards: (5-8) C3, F1	Explore the Issue, p. 687
25-3 The Senses (2 sessions)*	1. **List** the sensory receptors in each sense organ. 2. **Explain** what type of stimulus each sense organ responds to and how. 3. **Explain** the need for healthy senses. National Content Standards: (5-8) UCP1, A1, A2, B1, C1	MiniLAB: How is balance maintained? p. 690 MiniLAB: Who has a better sense of smell? p. 691 Activity 25-2: Investigating Skin Sensitivity, pp. 692-693 Skill Builder: Making and Using Tables, p. 694 Science Journal, p. 694
25-4 Your Endocrine System (2 sessions)*	1. **Explain** the function of hormones. 2. **Name** three endocrine glands and explain the effects of the hormones they produce. 3. **Explain** how a feedback system works. National Content Standards: (5-8) UCP1, A1, C1, C3, G1	Problem Solving: Why am I so tired? p. 696 Connect to Chemistry, p. 697 Skill Builder: Comparing and Contrasting, p. 697 Science Journal, p. 697 People and Science: Maureen Diggins, Researcher, p. 698

* A complete Planning Guide that includes block scheduling is provided on pages 32T-35T.

Activity Materials

Explore	Activities	MiniLABs
page 677 metric ruler	page 685 penny, meterstick pages 692-693 3" × 5" index cards, toothpicks, glue or tape, metric ruler	page 690 chalkboard, chalk, meterstick, stopwatch or clock with second hand page 691 aromatic foods, colognes, or household products

Need Materials? Call Science Kit (1-800-828-7777).

Teacher Classroom Resources

Reproducible Masters	Transparencies	Teaching Resources
Study Guide, p. 94 **Reinforcement,** p. 94 **Enrichment,** p. 94 **Activity Worksheets,** pp. 148-149 **Lab Manual 58,** Reaction Time **Multicultural Connections,** pp. 53-54	**Section Focus Transparency 90,** Message Center **Teaching Transparency 49,** The Reflex	**Glencoe Life Science Interactive Videodisc** **Life Science CD-ROM** **Spanish Resources** **English/Spanish Audiocassettes** **Cooperative Learning Resource Guide** **Lab Partner** **Lab and Safety Skills** **Lesson Plans**
Study Guide, p. 95 **Reinforcement,** p. 95 **Enrichment,** p. 95 **Science and Society Integration,** p. 29	**Section Focus Transparency 91,** Memory Loss and Forgetfulness	
		Assessment Resources
Study Guide, p. 96 **Reinforcement,** p. 96 **Enrichment,** p. 96 **Activity Worksheets,** pp. 150-151, 152, 153 **Lab Manual 59,** Parts of the Eye **Science Integration Activity 25,** Conduction of Heat **Concept Mapping,** p. 55 **Cross-Curricular Integration,** p. 29 **Technology Integration,** pp. 17-18	**Section Focus Transparency 92,** Extending the Senses **Science Integration Transparency 25,** The Eye's Built-In Optics	**Chapter Review,** pp. 53-54 **Assessment,** pp. 159-162 **Performance Assessment in the Science Classroom (PASC)** **MindJogger Videoquiz** **Alternate Assessment in the Science Classroom** **Performance Assessment** **Chapter Review Software** **Computer Test Bank**
Study Guide, p. 97 **Reinforcement,** p. 97 **Enrichment,** p. 97 **Critical Thinking/Problem Solving,** p. 29	**Section Focus Transparency 93,** Living with Diabetes **Teaching Transparency 50,** Endocrine Glands	

Key to Teaching Strategies

The following designations will help you decide which activities are appropriate for your students.

L1 Level 1 activities should be within the ability range of all students, including those with learning difficulties.

L2 Level 2 activities should be within the ability range of the average to above-average student.

L3 Level 3 activities are designed for the ability range of above-average students.

LEP LEP activities should be within the ability range of Limited English Proficiency students.

LS These activities are designed to address different learning styles.

COOP LEARN Cooperative Learning activities are designed for small group work.

P These strategies represent student products that can be placed into a best-work portfolio.

GLENCOE TECHNOLOGY

The following multimedia resources are available from Glencoe.

The Secret of Life
Tinkering with Our Genes: Genetic Medicine
Science and Technology Videodisc Series (STVS)
Human Biology
  Vision Diagnosis

National Geographic Society Series
STV: Human Body Vol. 2
Glencoe Life Science Interactive Videodisc
Medical Diagnostics: Tools of the Trade
Life Science CD-ROM

Teacher Classroom Resources

This is a representation of key blackline masters available in the Teacher Classroom Resources.

Teaching Aids

Section Focus Transparencies

90 SECTION FOCUS TRANSPARENCY   Section 25-1

MESSAGE CENTER

How often do you speak with your friends on the telephone? The telephone provides people with a way to communicate with others over both short and long distances. Your body also has a communication system. This system allows different parts of your body to interact with each other. It also allows you to respond to changes that take place outside your body.

1. What are some ways in which people communicate with each other?

2. Why is communication important?

3. Why is it important for different parts of your body to communicate with each other? What body system is responsible for carrying out this task?

L1

91 SECTION FOCUS TRANSPARENCY   Section 25-2

MEMORY LOSS AND FORGETFULNESS

Have you ever found yourself looking in strange places for an item you've misplaced, such as your keys, eyeglasses, or a book you need for school? Do you sometimes have difficulty remembering facts and dates you need to know for a test? Such problems are usually associated with forgetfulness. Forgetfulness is a type of short-term memory loss that is common for all people during some times of their lives.

1. What kinds of information do people forget?

2. When might forgetfulness become a problem?

3. How might you tell the difference between forgetfulness and a serious problem involving memory loss?

L2

92 SECTION FOCUS TRANSPARENCY   Section 25-3

EXTENDING THE SENSES

People use many tools and instruments that help to extend their senses. For example, some people with hearing impairments wear hearing aids that amplify sounds, or make them louder. Scientists also use many tools and instruments to help extend their senses. Some of these tools are shown below. Perhaps you've used some of these tools in your study of science or in other activities.

1. How many of the tools or instruments shown have you used? Which ones have you used?

2. What do the tools and instruments have in common?

3. How do the tools and instruments shown magnify objects? Why is this useful?

L3

Science Integration Transparencies

25 SCIENCE INTEGRATION TRANSPARENCY

The Eye's Built in Optics

L1

Teaching Transparencies

49. THE REFLEX

L1

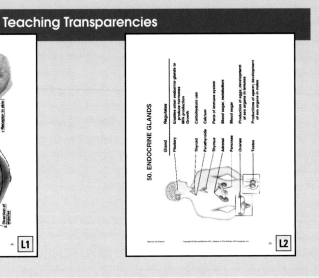

50. ENDOCRINE GLANDS

L2

Meeting Different Ability Levels

Study Guide

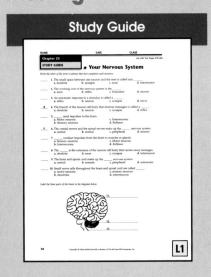

Chapter 25
STUDY GUIDE   • **Your Nervous System**

L1

Reinforcement

Chapter 25
REINFORCEMENT   • **Your Nervous System**

L2

Enrichment Worksheets

Chapter 25
ENRICHMENT   • **The Nervous System**

The Eyes Have It

L3

Chapter 25 The Nervous and Endocrine Systems

Hands-On Activities

Science Integration Activity

Conduction of Heat

Lab Manual

Reaction Time 58

Assessment

Performance Assessment

Optical Illusions

Enrichment and Application

Critical Thinking/Problem Solving

The Nervous and Endocrine Systems

Diseases of the Endocrine System

Cross-Curricular Integration

The Nervous and Endocrine Systems

Reading About Personal Experiences

Science and Society Integration

The Nervous and Endocrine Systems

Should Alzheimer's Patients Be Confined Against Their Will?

Technology Integration

The Human Body

Laser Eye Surgery

Multicultural Connections

Acupuncture: Doctors Get the Point

Concept Mapping

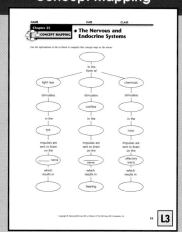

The Nervous and Endocrine Systems

676D

The Nervous and Endocrine Systems

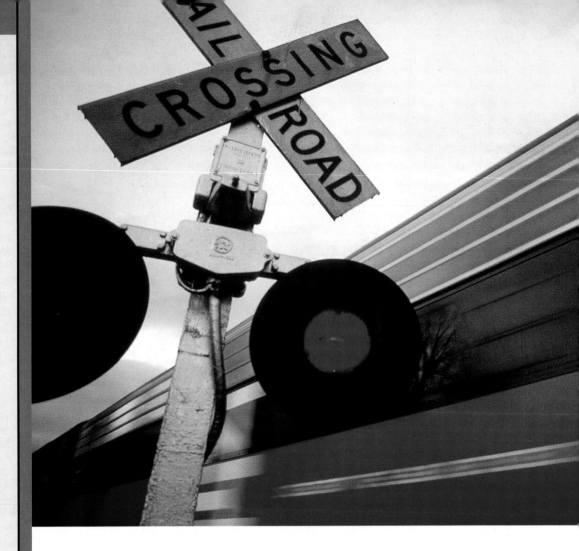

CHAPTER OVERVIEW

Section 25-1 This section centers on the functioning of the nervous system, the movement of impulses along neurons, and the two major divisions of the nervous system.

Section 25-2 Science and Society The symptoms and causes of Alzheimer's disease and the issues of care and treatment of patients are presented.

Section 25-3 This section describes the sense organs that enable the body to distinguish changes in its internal and external environments.

Section 25-4 The endocrine system is studied in this section. Its effect on tissues via its secretions, hormones, is described.

Chapter Vocabulary

neuron	cerebellum
dendrite	brain stem
axon	reflex
sensory neuron	Alzheimer's
interneuron	disease
motor neuron	retina
synapse	cochlea
central nervous	olfactory
system	cell
peripheral	taste bud
nervous	hormone
system	target
cerebrum	tissue

Theme Connection

Stability and Change A major concept of science is stability and change. This chapter on body regulation emphasizes the theme of homeostasis. The structures and functions of the nervous and endocrine systems are described. Their integrated efforts bring about the stable environment necessary for the healthy functioning of the body.

676

Previewing the Chapter

Section 25-1   Your Nervous System
  ▶ The Nervous System at Work
  ▶ Neurons
  ▶ The Central Nervous System
  ▶ The Peripheral Nervous System
  ▶ Drugs and the Nervous System

Section 25-2   Science and Society Issue: Care and Treatment of Alzheimer's Patients

Section 25-3   The Senses
  ▶ In Touch with Your Environment
  ▶ Vision
  ▶ Lenses: Refraction and Focus
  ▶ Hearing
  ▶ Smell
  ▶ Taste
  ▶ Touch, Pressure, Pain, and Temperature

Section 25-4   Your Endocrine System
  ▶ The Endocrine System
  ▶ A Negative-Feedback System

676

Learning Styles Look for the following logo for strategies that emphasize different learning modalities. ⬛

Kinesthetic	Demonstration, p. 679; Making a Model, p. 681; Reteach, p. 683; Activity 25-1, p. 685; Activity 25-2, pp. 692-693; MiniLAB, p. 690
Visual-Spatial	Explore, p. 677; Activity, p. 679; Visual Learning, pp. 679, 681, 682, 686, 689, 690; Demonstration, p. 689
Interpersonal	MiniLAB, p. 691; Reteach, p. 694; People and Science, p. 698
Linguistic	Across the Curriculum, pp. 680, 690

Chapter 25

The Nervous and Endocrine Systems

To survive, every organism must be able to detect what is happening around it. Sights or sounds may warn you of danger. Odors may help you find food. Sensations of hot and cold may protect you from fire or extreme temperatures. Your body interprets all of the sensations it receives to produce a picture of its surroundings. Sometimes the signals are interpreted differently from what they actually are. In the following activity, find out whether your eyes can trick you.

EXPLORE ACTIVITY

Find a visual trick.

1. Look at the figure at the bottom of the page.
2. Estimate the difference in height between pole A and pole C.
3. Use a metric ruler to measure the heights of poles A, B, and C.

Observe: What did you find out about their heights? How were you tricked by your eyes?

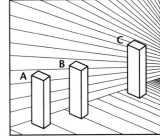

Previewing Science Skills

▶ In the **Skill Builders,** you will map concepts, make and use tables, and compare and contrast.

▶ In the **Activities,** you will experiment, predict, measure, and collect data.

▶ In the **MiniLABs,** you will experiment and collect and interpret data.

677

Prepare

Section Background

Inside a neuron at rest, the concentration of potassium ions is greater than that of sodium ions. A stimulus causes a brief reversal of the concentration of the ions. This produces a small electrical charge that travels from the excited area to the adjacent resting area. The entire action causes the impulse to travel along the nerve fiber.

Preplanning

To prepare for Activity 25-1, obtain pennies and a meterstick for each set of partners.

1 Motivate

Bellringer

 Before presenting the lesson, display **Section Focus Transparency 90** on the overhead projector. Assign the accompanying **Focus Activity** worksheet. L1 LEP

Tying to Previous Knowledge

Recall from Chapter 21 the movement of skeletal muscles. Indicate that coordinated body movements are accomplished by the nervous system.

GLENCOE TECHNOLOGY

CD-ROM

Life Science CD-ROM

Have students perform the interactive exploration for Chapter 25 to reinforce important chapter concepts and thinking processes.

Science Words

- neuron
- dendrite
- axon
- sensory neuron
- interneuron
- motor neuron
- synapse
- central nervous system
- peripheral nervous system
- cerebrum
- cerebellum
- brain stem
- reflex

Objectives

- Describe the basic structure of a neuron and how an impulse moves.
- Compare the central and peripheral nervous systems.
- Explain the effects of drugs on the body.

Figure 25-1

A neuron is made up of a cell body, dendrites, and an axon. *How does the branching of the dendrites allow for more impulses to be picked up by the neuron?*

25•1 Your Nervous System

The Nervous System at Work

It's your night to do the dishes. To make the job easier, you play your favorite CD. About halfway through, you turn around and come face to face with a hideous monster. You scream and drop a dish on the floor. Only then do you realize that the monster is your brother in a rubber mask. Suddenly, you're aware of being out of breath and your hands are shaking. Your knees feel wobbly, and your heart is racing. But then, after a few minutes, your breathing returns to normal and your heartbeat is back to its regular rate. What's going on?

Response to Stimuli

The scene shown on the next page is an example of how your body responds to changes in its environment and adjusts itself. Your body makes these adjustments with the help of your nervous system. Any change inside or outside your body that brings about a response is called a stimulus. Each day, you're bombarded by thousands of stimuli. Noise, light, the smell of food, and the temperature of the air are all stimuli from outside your body. A growling stomach is an example of an internal stimulus.

How can your body handle all these stimuli? Your body has internal control systems that maintain homeostasis. Breathing rate, heartbeat rate, and digestion are just a few of the activities that are constantly checked and regulated. Your nervous system and the endocrine system, a chemical control system described later in this chapter, are the main mechanisms by which homeostasis is maintained in your body.

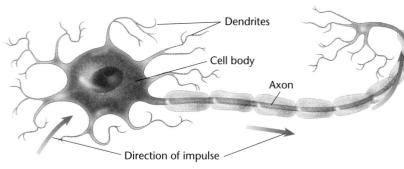

Dendrites

Cell body

Axon

Direction of impulse

Program Resources

 Reproducible Masters
Study Guide, p. 94 L1
Reinforcement, p. 94 L1
Enrichment, p. 94 L3
Activity Worksheets, pp. 148-149 L1
Lab Manual 58
Multicultural Connections, pp. 53-54

 Transparencies
Section Focus Transparency 90 L1
Teaching Transparency 49

Neurons

The working unit of the nervous system is the nerve cell, or **neuron**. The single neuron in **Figure 25-1** is made up of a cell body and branches called dendrites and axons. **Dendrites** receive messages and send them to the cell body. An **axon** carries messages away from the cell body. Any message carried by a neuron is called an impulse. Notice that the end of the axon branches. This allows the impulses to move to many other muscles, neurons, or glands.

Types of Neurons

Your skin and other sense organs are equipped with structures called receptors that respond to various stimuli. Three types of neurons—sensory neurons, motor neurons, and interneurons—then become involved with transporting impulses about the stimuli. As illustrated in **Figure 25-2, sensory neurons** receive information and send impulses to the brain or spinal cord. Once the impulses reach your brain or spinal cord, **interneurons** relay the impulses from the sensory to motor neurons. There are more interneurons in your body than either of the other two types of neurons. **Motor neurons** then conduct impulses from the brain or spinal cord to muscles or glands throughout your body.

Figure 25-2

In your nervous system, impulses travel a pathway. Together, the three types of neurons act like a relay team, moving impulses through your body from stimulus to response.

A When you saw the mask, sensory receptors in your eyes were stimulated.

B A message was sent to your brain by way of sensory neurons.

C Your brain sorted the information and determined a response.

E Your heart immediately started to pound and your breathing rate increased. You dropped the plate.

D The response was sent back along motor neurons to your muscles.

25-1 Your Nervous System   **679**

679

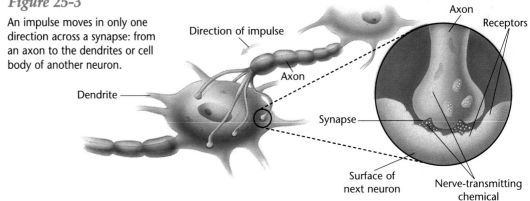

Figure 25-3

An impulse moves in only one direction across a synapse: from an axon to the dendrites or cell body of another neuron.

Direction of impulse
Dendrite
Axon
Axon
Receptors
Synapse
Surface of next neuron
Nerve-transmitting chemical

Synapse

Neurons don't touch each other. How does an impulse move from one neuron to another? To get from one neuron to the next, an impulse moves across a small space called a **synapse.** In **Figure 25-3,** you can see that when an impulse reaches the end of an axon, a nerve-transmitting chemical is released by the axon. This chemical diffuses across the synapse and starts an impulse in the dendrite or cell body of the next neuron. In this way, an impulse moves from one neuron to another.

The Central Nervous System

Figure 25-4

The brain and spinal cord (yellow) form the central nervous system, which sorts and interprets information from stimuli. All other nerves are part of the peripheral nervous system (green).

Spinal cord (CNS)
Brain
Brain
Cerebrum
Skull
Cerebellum
Spinal cord
Spinal cord
Spinal nerves (PNS)
Spinal nerve
Vertebra

Figure 25-4 shows how organs of the nervous system are grouped into two major divisions: the central nervous system (CNS) and the peripheral nervous system (PNS). The **central nervous system** is made up of the brain and spinal cord. The **peripheral nervous system** is made up of all the nerves outside the CNS, including cranial nerves and spinal nerves. These nerves connect the brain and spinal cord to other body parts.

Because you have a central nervous system with a brain, your body activities are coordinated. If someone pokes you in the ribs, your whole body reacts. Your neurons are adapted in such a way that impulses move in only one direction. Impulses in sensory neurons move from a receptor to the brain or spinal cord.

680 Chapter 25 The Nervous and Endocrine Systems

Your Brain

The brain is made up of approximately 100 billion neurons. You can see in **Figure 25-5** that the brain is divided into three major parts: cerebrum, cerebellum, and brain stem. The largest part of the brain, the **cerebrum,** is divided into two large sections called hemispheres. Here, impulses from the senses are interpreted, memory is stored, and the work of voluntary muscles is controlled. The outer layer of the cerebrum, the cortex, is marked by many ridges and grooves. The diagram also shows some of the tasks that sections of the cortex control.

A second part of the brain, the **cerebellum,** is behind and under the cerebrum. It coordinates voluntary muscle movements and maintains balance and muscle tone.

The **brain stem** extends from the cerebrum and connects the brain to the spinal cord. It is made up of the midbrain, the pons, and the medulla. The brain stem controls your heartbeat, breathing, and blood pressure by coordinating the involuntary muscle movements of these functions.

Your Spinal Cord

Your spinal cord is an extension of the brain stem. It is made up of bundles of neurons that carry impulses from all parts of the body to the brain and from the brain to all parts of your body. The spinal cord is about as big around as an adult thumb and 43 cm long. The CNS is protected by a bony cap called the skull, by vertebrae shown in **Figure 25-6,** and by three layers of membranes. Between some of these membranes is a fluid called cerebrospinal fluid. What purpose might this fluid serve?

Figure 25-5

Different areas of the brain control specific body activities.

Cerebrum · Tongue · Chewing · Motor · Sensory · Salivation · Swallowing · Smell · Pons · Medulla · Vision · Hearing · Cerebellum · Brain Stem

Spinal cord · Gray matter · White matter · Protective coverings · Spinal nerve · Spinal disk · Vertebra

Figure 25-6

Impulses travel to and from the brain by way of the spinal cord.

25-1 Your Nervous System 681

Figure 25-7

The divisions of the peripheral nervous system are shown. *What part of the PNS controls your breathing while you sleep?*

USING MATH

One of the longest spinal nerves extends from the spinal cord to muscles in the foot. Estimate the length of this nerve.

Figure 25-8

Your response in a reflex is controlled in your spinal cord, not in your brain.

The Peripheral Nervous System

Your brain and spinal cord are connected to the rest of your body by the peripheral nervous system. The PNS is made up of 12 pairs of cranial nerves from your brain and 31 pairs of spinal nerves from your spinal cord. These nerves link your central nervous system with all parts of your body. Spinal nerves are made up of bundles of sensory and motor neurons. For this reason, a single spinal nerve may have impulses going to and from the brain at the same time.

There are two divisions of the peripheral nervous system. The *somatic system* consists of the cranial and spinal nerves that go from the central nervous system to your skeletal muscles. The second division, the *autonomic system*, controls your heartbeat rate, breathing, digestion, and gland functions. When your salivary glands release saliva, your autonomic system is at work. Use **Figure 25-7** to help you remember these two divisions.

Reflexes

Have you ever jumped back from something hot or sharp? Then you've experienced a reflex. A **reflex** is an involuntary and automatic response to a stimulus. Usually, you can't control reflexes because they occur before you know what has happened. A reflex involves a simple nerve pathway called a reflex arc. **Figure 25-8** shows a reflex arc. As you reach for the pizza, some very hot cheese falls on your finger. Sensory receptors in your finger respond to the hot cheese, and an impulse is sent to the spinal cord. The impulse passes to an interneuron in the spinal cord that immediately relays the

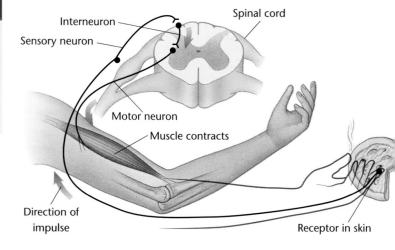

682 Chapter 25 The Nervous and Endocrine Systems

impulse to motor neurons. Motor neurons transmit the impulse to muscles in your arm. Instantly, without thinking, you pull your arm back in response to the burning food. This is a withdrawal reflex. A reflex allows the body to respond without having to think about what action to take. Reflex responses are controlled in your spinal cord, not in your brain. Your brain acts after the reflex to help you figure out what to do to make the pain stop.

Remember the rubber mask scare? What would have happened if your breathing and heartbeat rate didn't calm down within a few minutes? Your body system can't be kept in a state of continual excitement. The organs of your nervous system control and coordinate responses to maintain homeostasis within your body.

PET Image of a Human Brain

AWAKE

Watching the Brain

Scientists use positron emission tomography (PET) for research on brain functions and to diagnose brain disorders. A radioactive form of glucose is fed to or injected into the patient. The glucose is taken up by active cells in the brain, and positrons are given off. The positrons collide with electrons from the body and release gamma-ray energy. The path of the energy appears as an image on a color monitor. Scientists use the image to find out what areas of the brain are being stimulated.

Uses of PET

By comparing PET images formed when people perform different tasks, researchers have located the specific areas of the brain used for seeing, reading, hearing, speaking, and thinking. Other researchers are using the images to try to identify people who might be in the early stage of Alzheimer's disease. This disease is characterized by nerve cells that die or don't function properly.

Think Critically:

Epilepsy is a disorder of the nervous system that can result in convulsions. How can PET help scientists understand what happens when a person has an epileptic seizure?

Researchers are constantly looking for new noninvasive methods to study the brain. More precise knowledge can help scientists cure disorders throughout the body.

Think Critically
The images help scientists determine which areas of the brain are active during a seizure. This may lead to a better understanding of the cause and possibly a treatment or cure.

3 Assess

Check for Understanding

- Ask the question: **What might be the purpose of the extensive folding of the surface of the cerebrum?** *The folding provides a greater surface area to be contained within the enclosure of the skull.*

- Ask the question: **Why might a hard blow to the back of the head cause a vision disorder?** *The primary visual area of the brain is located at the lower rear side of the cerebrum.*

Reteach

LS **Kinesthetic** Have students make a papier-mâché or modeling clay model of the brain and its major parts. Use the model to review the parts and their functions.

Extension

For students who have mastered this section, use the **Reinforcement** and **Enrichment** masters.

4 Close

Section Wrap-up

Review

1. See Figure 25-1.

2. The central nervous system is made up of the brain and spinal cord. The peripheral nervous system is made up of cranial nerves and spinal nerves.

3. Sensory neurons receive information and send impulses to the spinal cord or brain. Motor neurons conduct impulses from the brain or spinal cord to muscles or glands throughout the body.

4. **Think Critically** Cocoa has caffeine, which is a stimulant that can cause sleeplessness.

Using Computers

Flowcharts should show an impulse moving from receptor to sensory neuron to interneuron in the spinal cord to motor neuron to muscle in the foot.

Figure 25-9

Caffeine, a substance found in cola, coffee, and other types of food and drink, can cause excitability and sleeplessness.

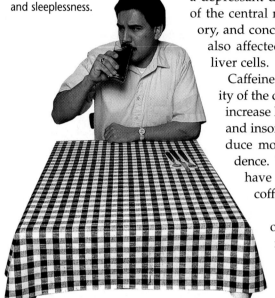

Drugs and the Nervous System

Many drugs, such as alcohol and caffeine, have a direct effect on your nervous system. When swallowed, alcohol passes directly through the walls of the stomach and small intestine into the circulatory system. Alcohol is classified as a depressant drug. A depressant slows down the activities of the central nervous system. Judgment, reasoning, memory, and concentration are impaired. Muscle functions are also affected. Heavy use of alcohol destroys brain and liver cells.

Caffeine is a stimulant, a drug that speeds up the activity of the central nervous system. Too much caffeine can increase heartbeat rate and cause restlessness, tremors, and insomnia. It can also stimulate the kidneys to produce more urine. Caffeine can cause physical dependence. When people stop taking caffeine, they can have headaches and nausea. Caffeine is found in coffee, tea, cocoa, and many soft drinks.

Think again about the mask scare. The organs of your nervous system control and coordinate responses to maintain homeostasis within your body. This task is more difficult when your body must cope with the effects of drugs.

Section Wrap-up

Review

1. Draw and label the parts of a neuron.

2. Compare the central and peripheral nervous systems.

3. Compare sensory and motor neurons.

4. **Think Critically:** During cold winter evenings, you often have several cups of hot cocoa. Explain why you have trouble falling asleep.

Skill Builder
Concept Mapping

Prepare an events chain concept map of the different kinds of neurons an impulse moves along from a stimulus to a response. If you need help, refer to Concept Mapping in the **Skill Handbook.**

Using Computers

Flowchart Create a graphic flowchart showing the reflex pathway of a nerve impulse when you step on a sharp object.

Skill Builder

The concept map should include the following: stimulus → sensory neurons → interneurons → motor neurons → response

 Assessment

Portfolio Use this Skill Builder to assess students' abilities to form a concept map of the chain of neurons involved between a stimulus and a response. Use the Performance Task Assessment List for Concept Map in **PASC,** p. 89. **P**

Activity 25-1

Reaction Time

Your body responds quickly to some kinds of stimuli. For example, reflexes allow you to react quickly without even thinking. Sometimes you can improve how quickly you react. Complete this activity to see if you can improve your reaction time.

Problem
How can reaction time be improved?

Materials
- penny
- meterstick

Procedure
Part A
1. Hold your right arm out with the palm down. Put a penny on the center of the back of your hand.
2. Tilt your hand so that the penny slides off. Try to catch the penny with your right hand.
3. Repeat steps 1 and 2 nine more times. In a table like the one shown, record how many times you catch the penny and how many times you drop the penny.
4. Repeat steps 1 through 3 with your left hand. Record your results.

Part B
1. Have a partner hold a penny about 0.5 m above the palm of your hand.
2. When your partner drops the penny, move your hand before the penny hits it.
3. Repeat steps 1 and 2 for different distances. Record your results in the table.
4. Repeat steps 1 through 3 with your other hand. Record your results.

Data and Observations   Sample Data

Part A				
	Right Hand		Left Hand	
Trial	Caught	Not caught	Caught	Not caught
1		✓		✓
2	✓			✓

Part B			
Trial	Distance above hand	Hit	Not hit
1	0.5 m	✓	
2	0.8 m		✓

Analyze
1. What was the stimulus in each activity?
2. What was the response in each activity?
3. What was the variable in each activity?

Conclude and Apply
4. **Compare** the responses of your writing hand and your other hand for both activities.
5. **Draw a conclusion** about how practice relates to stimulus-response time.

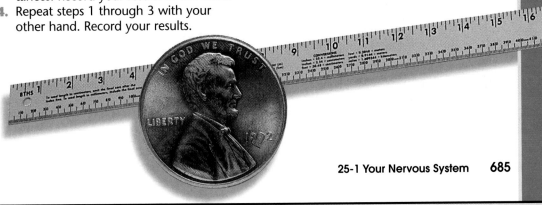

25-1 Your Nervous System   **685**

Activity 25-1

Purpose
KS **Kinesthetic** Observe reaction time using two different activities. **L1** **LEP** **COOP LEARN**

Process Skills
observing, communicating, using numbers, recognizing and using spatial relationships, measuring in SI, and interpreting data

Time
30 minutes

Materials
penny, meterstick

Activity Worksheets, pp. 5, 148-149

Teaching Strategies
In part B, students should keep their eyes on the penny as the partner releases it.

Answers to Questions
1. A—the penny sliding off the hand; B—dropping the penny
2. A—catching the penny; B—avoiding the penny
3. A—left hand if you are right-handed; B—distance
4. Answers will vary. The student will probably catch faster in Part A and avoid the penny more easily in Part B with the writing hand.
5. With practice, stimulus-response time will probably get better.

✔ Assessment

Performance To further assess students' abilities to measure and improve reaction time, have them formulate and test a hypothesis about how long it would take the non-writing hand to be trained to respond as the writing hand responds. Use the Performance Task Assessment List for Formulating a Hypothesis in **PASC**, p. 21. **P**

685

Prepare

Section Background

Not all cases of confusion and forgetfulness in older people are the result of Alzheimer's disease. An estimated 10 to 20 percent of older persons with these symptoms are treatable.

1 Motivate

Bellringer

Before presenting the lesson, display **Section Focus Transparency 91** on the overhead projector. Assign the accompanying **Focus Activity** worksheet. L1 LEP

Tying to Previous Knowledge

In Alzheimer's disease, the transmission of impulses from one neuron to the next is prevented.

2 Teach

Activity

Have students collect newspaper and magazine articles about new research dealing with the disease. L1

Visual Learning

Figure 25-11 How can support groups help ease the stress? *The groups may provide new suggestions on providing care, ways to relax, and a place to talk about the problems faced by caregivers.* LEP LS

SCIENCE & SOCIETY

ISSUE:

25•2 Care and Treatment of Alzheimer's Patients

Science Words

Alzheimer's disease

Objectives

• Describe Alzheimer's disease and possible causes.
• Determine the effects of the disease on patients and family.

What is Alzheimer's disease?

A definition of **Alzheimer's disease** is the failure of nerve cells in the brain to communicate. Researchers are attempting to determine what causes this communication failure. They know that Alzheimer's patients have low amounts of an enzyme that signals the production of a chemical called acetylcholine. Without this chemical, nerve impulses aren't carried from one neuron to the next. They have also found that the brains of Alzheimer's patients use far less oxygen than normal. Autopsies of Alzheimer's patients show major destruction of brain cells. This destruction results in such severe memory loss that, over a period of years, victims of this disease become very different people. In time, patients simply forget things. Items are misplaced. Later, they may not even recognize family members. Eventually, after losing both mental and physical abilities, they die. What causes this condition?

Figure 25-10

Alzheimer's disease is one of the leading causes of death among the elderly. Scientists are working on finding the cause and a cure.

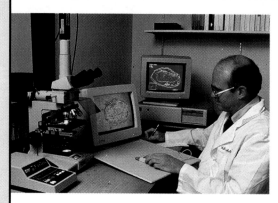

Cause and Diagnosis

For the scientist in **Figure 25-10,** finding the cause of Alzheimer's disease is difficult. Recently, there has been much research into both finding the cause of the disease and new ways to detect it. There is some evidence that the disease is inherited. Children of Alzheimer's patients show a greater chance of developing the disease than others. Some researchers believe that a defective chromosome may be the cause. Other evidence indicates that certain proteins in the cerebrospinal fluid may be the culprit. Diagnosing the disease has not been easy. There are several brain disorders that have similar symptoms. At present, the only way to positively identify whether a person has the disease is through an autopsy of the brain after death.

Program Resources

 Reproducible Masters
Study Guide, p. 95 L1
Reinforcement, p. 95 L1
Enrichment, p. 95 L3
Science and Society Integration, p. 29

Transparencies
Section Focus Transparency 91 L1

2 Points of View

Families Paying the Cost

It is estimated that there are about 4 million persons in the United States with Alzheimer's disease. The cost of this disease to families and society is high. During early stages of the disease, family members often care for the patient at home. Home care is much less expensive than the $18,000 to $30,000 per year charged by a nursing home. Most people's insurance will not pay for the long-term care that is needed. The federal programs of Medicare and Medicaid usually don't cover the costs incurred by the average patient. Medicaid won't pay until almost all of the patient's financial resources are gone. In addition to the financial burden, caregivers like those in **Figure 25-11** are under extreme emotional stress as they look after the patient 24 hours a day.

Finding Federal Funds

As the number of people living into their 70s, 80s, and 90s increases, the number afflicted with Alzheimer's also continues to rise. Predictions are that by the year 2050, there may be over 14 million victims. Some groups have suggested that federal funds should be made available to help pay for the care of these patients. They think that a patient or family should not have to become both emotionally and financially destitute before getting help.

Figure 25-11

Caregivers of Alzheimer's patients often suffer severe health problems because of the stress involved in taking care of the patient. *How can support groups help ease the stress?*

Section Wrap-up

Review

1. What are the possible causes of Alzheimer's disease?

2. Why is Alzheimer's disease hard to diagnose?

Explore the Issue

Should the federal government provide additional funds for Alzheimer's patients? If yes, how should the program be funded? Should taxes be raised or should funding for other programs be reduced? If no, then how can proper care be provided for these patients?

inter**NET** CONNECTION

What brain disorders have similar symptoms to those of Alzheimer's disease? How might physicians distinguish between these disorders? Visit the Chapter 25 Internet Connection at Glencoe Online Science, **www.glencoe.com/sec/science/life**, for a link to more information about the human brain.

SCIENCE & SOCIETY

25-2 Care and Treatment of Alzheimer's Patients　　**687**

Prepare

Section Background

- Sense organs receive stim-uli that are conveyed by nerve impulses to the brain, where they are interpreted.
- The senses of sight, hearing, balance, smell, and taste have highly specialized structures located in spe-cific areas of the body.
- Receptors of sensations of touch, pressure, pain, and temperature have less spe-cialized structures and are more generally located in the body.

Preplanning

- To prepare for Activity 25-2, obtain metric rulers, tooth-picks, glue, and index cards.
- To prepare for the MiniLAB on page 690, draw the two parallel lines on the chalk-board.
- To prepare for the second MiniLAB on page 691, ob-tain various liquid spices, flavor extracts, foods, and odorous substances. Also obtain a supply of 6-10 cot-ton balls per student.

1 Motivate

Bellringer

Before presenting the lesson, display **Section Focus Transparency 92** on the overhead projector. Assign the accompanying **Focus Activity** worksheet. L1 LEP

Tying to Previous Knowledge

Have students compare the sensory system of informa-tion processing with that of a telephone system. Have them brainstorm similarities and differences.

Science Words

- **retina**
- **cochlea**
- **olfactory cell**
- **taste bud**

Objectives

- List the sensory receptors in each sense organ.
- Explain what type of stimulus each sense organ responds to and how.
- Explain the need for healthy senses.

In Touch with Your Environment

Science fiction stories about space often describe energy force fields around spaceships. When some form of energy tries to enter the ship's force field, the ship is put on alert. Your body has an alerting system as well, in the form of sense organs. Your senses enable you to see, hear, smell, taste, touch, and feel what-ever comes into your personal territory. The energy that stimu-lates your sense organs may be in the form of light rays, heat, sound waves, chemicals, or pressure. Sense organs are adapted for capturing and transmitting these different forms of energy.

Vision

Think about the different kinds of objects you look at every day. It's amazing that, at one glance, you can see the words on this page, the color illustrations, and your classmate sitting next to you.

Light travels in a straight line unless something bends or refracts it. Your eyes are equipped with structures that bend light. As light enters the eye its waves are first bent by the cornea and then a lens. The lens directs the rays onto the retina (RET nuh). The **retina** is a tissue at the back of the eye that is sensitive to light energy. Two types of cells called rods and cones are found in the retina. Cones respond to bright light and color. Rods respond to dim light. They are used to help you detect shape and movement. Light energy stimulates impulses in these cells. The impulses pass to the optic nerve, which carries them to the brain. There the impulses are interpreted and you "see" what you are looking at.

Figure 25-12

Light moves through several transparent structures before striking the retina.

Choroid

Sclera

Retina

Optic nerve

Iris

Lens

Pupil

Virtreous humor

Ciliary muscles

Aqueous humor

Cornea

Conjunctiva

688 Chapter 25 The Nervous and Endocrine Systems

Program Resources

 Reproducible Masters

Study Guide, p. 96 L1
Reinforcement, p. 96 L1
Enrichment, p. 96 L3
Cross-Curricular Integration, p. 29
Activity Worksheets, pp. 150-151, 152, 153 L1
Science Integration Activities, pp. 49-50
Lab Manual 59

Concept Mapping, p. 55
Technology Integration, p. 17

Transparencies

Section Focus Transparency 92 L1
Science Integration Transparency 25

Lenses: Refraction and Focus

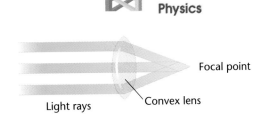

You know that light is refracted, or bent, when it passes through a lens. Just how it bends depends on the type of lens it passes through. A lens that is thicker in the middle and thinner on the edges is called a convex lens. The lens in your eye is convex. If you follow the light rays in **Figure 25-13**, you'll see that the lens causes parallel light rays to come together at a focus point. Convex lenses can be used to magnify objects. The light rays enter the eyes in such a way through a convex lens that the object appears enlarged. Magnifying lenses, hand lenses, microscopes, and telescopes all have convex lenses.

A lens that has thicker edges than the middle is called a concave lens. Follow the light rays in **Figure 25-13** as they pass through a concave lens. You'll see that this kind of lens causes the parallel light to spread out. A concave lens is used along with convex lenses in telescopes to allow distant objects to be seen clearly.

Correcting Vision

In an eye with normal vision, light rays are focused by the cornea and lens onto the retina. A sharp image is formed on the retina, and the brain interprets the signal as being clear. However, if the eyeball is too long from front to back, light from distant objects is focused in front of the retina. A blurred image is formed. This condition is called nearsightedness because near objects are seen clearly. To correct nearsightedness, eyeglasses with concave lenses are used. Concave lenses focus these images sharply on the retina. If the eyeball is too short from front to back, light from nearby objects is focused behind the retina. Again, the image appears blurred. Convex lenses correct this condition known as farsightedness. **Figure 25-14** shows how lenses are used to correct these vision problems.

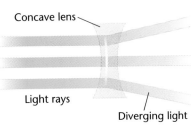

Figure 25-13

Light rays passing through a convex lens are bent toward the center and meet at a focal point. Light rays that pass through a concave lens are bent outward. They do not meet.

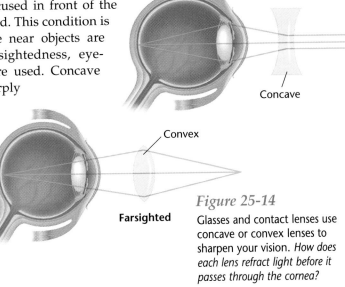

Figure 25-14

Glasses and contact lenses use concave or convex lenses to sharpen your vision. *How does each lens refract light before it passes through the cornea?*

---

2 Teach

INTEGRATION
Physics

A lens is made of transparent material that has at least one curved surface. Lenses bend light. A convex lens bends parallel rays of light so that they meet at one point called a focus. Convex lenses are converging lenses. A concave lens bends parallel rays of light so that they spread apart. Concave lenses are diverging lenses.

Demonstration

 Visual-Spatial Obtain a double convex lens to illustrate the refraction and focusing of light rays similar to that which occurs in the eye. **LEP**

Enrichment

Have students research the phenomenon of after-image, where an image is still perceived after its external cause is gone. **L2**

Visual Learning

Figure 25-14 How does each lens refract light before it passes through the cornea? *The concave lens spreads light apart first. The convex lens makes the light rays move closer together, in a parallel motion, before going through the lens of the eye.* **LEP**

GLENCOE TECHNOLOGY

 Videodisc
STVS: Human Biology
Disc 7, Side 1
Vision Diagnosis (Ch. 22)

Visual Learning

Figure 25-15 Why does spinning around make you dizzy? *Spinning makes the fluids in the inner ear send impulses to the brain that conflict with the actual position of the head. Dizziness results.* **LEP** **LS**

MiniLAB

Purpose

LS **Kinesthetic** The students will observe and make inferences concerning the factors that affect their abilities to maintain balance. **L1** **LEP**

Materials

chalkboard, chalk

Teaching Strategies

As one student stands between the lines, have a second student record how often the first student moves. Record whether the standing student tends to lean more to one side or the other.

Analysis

1. When the eyes are closed—when the eyes are open, a person can focus on a point to help the body remain still.
2. Answers will vary but could include such conditions as an inner ear infection or loud, continuous noise.

✔ Assessment

Performance To further assess students' understanding of balance, have them repeat the activity with feet apart or arms extended sideways. Use the Performance Task Assessment List for Carrying Out a Strategy and Collecting Data in **PASC**, p. 25. **P**

 Activity Worksheets, pp. 5, 152

690

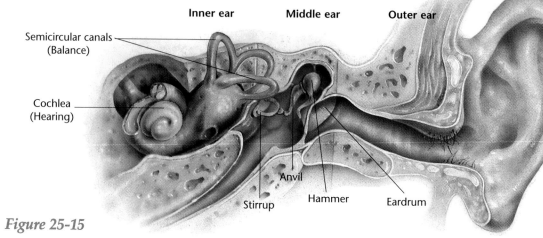

Inner ear   Middle ear   Outer ear

Semicircular canals (Balance)

Cochlea (Hearing)

Stirrup   Anvil   Hammer   Eardrum

Figure 25-15
Your ear responds to sound waves and to changes in the position of your head. *Why does spinning around make you dizzy?*

MiniLAB

How is balance maintained?

Procedure

1. Draw two parallel vertical lines on the chalkboard. Have a student stand between them, as still and straight as possible, for three minutes.
2. Observe how well balance is maintained.
3. Have the student close his or her eyes and repeat standing within the lines for three minutes.

Analysis

1. When was balance more difficult to maintain? Suggest an explanation.
2. What other factors might cause a person to lose the sense of balance?

Hearing

Sound energy is to hearing as light energy is to vision. When an object vibrates, it causes the air around it to vibrate, thus producing energy in the form of sound waves. When sound waves reach your ears, they stimulate nerve cells deep in your ear. Impulses are sent to the brain. The brain responds, and you hear a sound.

Figure 25-15 shows that your ear is divided into three sections: the outer, middle, and inner ear. Your outer ear traps sound waves and funnels them down the ear canal to the middle ear. The sound waves cause the eardrum to vibrate much like the membrane on a drum. These vibrations then move through three little bones called the hammer, anvil, and stirrup. The stirrup bone rests against a second membrane on an opening to the inner ear.

The Inner Ear

The **cochlea** (KOH klee uh) is a fluid-filled structure shaped like a snail's shell in the inner ear. When the stirrup vibrates, fluids in the cochlea also begin to vibrate. These vibrations stimulate nerve endings in the cochlea, and impulses are sent to the brain by the auditory nerve. Depending on how the nerve endings are stimulated, you hear a different type of sound. High-pitched sounds make the endings move differently from lower, deeper sounds.

Balance is also controlled in the inner ear. Special structures and fluids in the inner ear constantly adjust to the position of your head. This stimulates impulses to the brain, which interprets the impulses and helps you make the necessary adjustments to maintain your balance.

Across the Curriculum

Reading Find out how Helen Keller was able to learn to communicate even though she was deprived of sight and hearing at an early age. **L1** **LS**

 Use these program resources as you teach this lesson.

Cross-Curricular Integration, p. 29

Science Integration Activities, pp. 49-50

Lab Manual 59

Use Science Integration Transparency 25 as you teach this lesson.

Smell

A bloodhound is able to track a particular scent through fields and forest. Even though your ability to detect odors is not as sharp, your sense of smell is still important.

You can smell food because it gives off molecules into the air. Nasal passages contain sensitive nerve cells called **olfactory cells** that are stimulated by gas molecules. The cells are kept moist by mucous glands. When gas molecules in the air dissolve in this moisture, the cells become stimulated. If enough gas molecules are present, an impulse starts in these cells and travels to the brain. The brain interprets the stimulus. If it is recognized from previous experience, you can identify the odor. If you can't recognize a particular odor, it is remembered and can be identified the next time, especially if it's a bad one.

Taste

Have you ever tasted a new food with the tip of your tongue and found that it tasted sweet? Then when you swallowed it, you were surprised to find that it tasted bitter. **Taste buds** on your tongue are the major sensory receptors for taste. About 10 000 taste buds are found all over your tongue, enabling you to tell one taste from another.

Food: A Chemical Stimulus

Taste buds respond to chemical stimuli. When you think of food, your mouth begins to water with saliva. This adaptation is helpful because in order to taste something, it has to be dissolved in water. Saliva begins this process. The solution washes over the taste buds, and an impulse is sent to your brain. The brain interprets the impulse and you identify the taste.

Most taste buds respond to several taste sensations. However, there are areas of the tongue that seem more receptive to one taste than another. The four basic taste sensations are sweet, salty, sour, and bitter. **Figure 25-16** shows where these tastes are commonly stimulated on your tongue.

Smell and taste are related. When you have a head cold with a stuffy nose, food seems tasteless because it is blocked from contacting the moist membranes in your nasal passages.

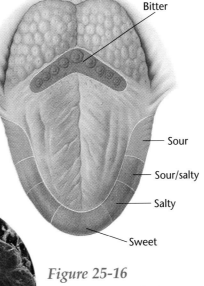

Bitter

Sour

Sour/salty

Salty

Sweet

Figure 25-16

Although your taste buds distinguish four separate taste sensations, scientists cannot determine differences among individual taste buds.

MiniLAB

Who has a better sense of smell?

Procedure
1. Design an experiment to test your classmates' abilities to recognize the odors of different foods, colognes, or household products.
2. Record their responses in a data table according to the sex of the individuals tested.

Analysis
1. Compare the numbers of correctly identified odors for both males and females.
2. What can you conclude about the differences between males and females in their ability to recognize odors?

MiniLAB

Purpose

IS Interpersonal Students will observe the difference in the sense of smell between males and females. L1 LEP
COOP LEARN

Materials
various spices, flavor extracts, and other odorous substances, cotton balls

Teaching Strategies
Safety Precautions Students should not use any substances with strong, noxious odors (e.g., ammonia).

Troubleshooting Place a small amount of the odorous substance on a cotton ball for each sniff test. Do not use the cotton ball for more than one person.

Analysis
1. Answers may vary, but more females should correctly identify specific odors.
2. Generally, females are more acutely aware of odors than are males.

Assessment

Performance Have students conduct a test of identifying odors while blindfolded. Use the Performance Task Assessment List for Carrying Out a Strategy and Collecting Data in PASC, p. 25. P

Activity Worksheets, pp. 5, 153

Use these program resources as you teach this lesson.
Concept Mapping, p. 55
Technology Integration, p. 17

Activity 25-2

PREPARATION

Purpose

LS **Kinesthetic** Design and carry out an experiment to determine the sensitivity of skin on various parts of the body by testing for the location of receptors in the skin. L1 LEP
COOP LEARN

Process Skills

observing and inferring, forming a hypothesis, designing an experiment, interpreting data, communicating, sequencing, making and using tables, comparing and contrasting, separating and controlling variables

Time

one class period to plan and hypothesize; one period to collect and analyze data

Materials

Obtain materials for cooperative groups. Prepare the test cards the day before the activity to allow glue to dry.

Safety Precautions

Caution students not to apply heavy pressure when using any contact points device.

Possible Hypotheses

Student hypotheses will vary. Possible hypotheses are: "The whole body is equally sensitive," "Nerve endings are various distances apart," and "The most sensitive parts of the body are the fingertips."

📁 **Activity Worksheets, pp. 5, 150-151**

Activity 25-2

Design Your Own Experiment
Investigating Skin Sensitivity

Your body responds to touch, pressure, and temperature. Not all parts of your body are equally sensitive to stimuli. Some areas are more sensitive than others. For example, your lips are sensitive to heat. This protects you from burning your mouth. Now think about touch. How sensitive is the skin on various parts of your body to touch? Which areas can distinguish the smallest amount of distance between stimuli?

PREPARATION

Problem

What areas of the body are most sensitive to touch?

Form a Hypothesis

Based on your experiences, state a hypothesis about which five areas of the body you believe to be more sensitive than others. Rank the areas from 5 (the most sensitive) to 1 (the least sensitive).

Objectives

- Observe the sensitivity to touch on various areas of the body.
- Design an experiment that tests the effects of a variable, such as the closeness of contact points, to determine which body areas can distinguish between the closest stimuli.

Possible Materials

- 3" × 5" index cards
- toothpicks
- glue or tape
- metric ruler

Safety Precautions

Do not apply heavy pressure when using any contact-points device.

692

Inclusion Strategies

Physically Challenged Any student who may not be able to manipulate the testing device may be in charge of formulating the hypothesis and the design of the experiment.

PLAN THE EXPERIMENT

1. As a group, agree upon and write out the hypothesis statement.
2. As a group, list the steps you need to take to test your hypothesis. Be specific in describing exactly what you will do at each step. Consider the following factors as you list the steps. How will you know that sight is not a factor? How will you use the card shown on the right to determine sensitivity to touch? How will you determine and record that one or both points of touch are felt? List your materials.
3. Design a data table in your Science Journal, so that it is ready to use as your group collects data.

Check the Plan

1. Read over your entire experiment to make sure that all steps are in order.
2. Identify any constants, variables, and controls of the experiment.
3. *Make sure your teacher approves the plan before you proceed.*

DO THE EXPERIMENT

1. Carry out the experiment as planned.
2. While the experiment is going on, write down any observations that you make and complete the data table in your Science Journal.

Analyze and Apply

1. **Compare** your results with those of other groups.
2. Which part of the body tested can distinguish between the closest stimuli?
3. Which part of the body is least sensitive?
4. Rank body parts tested from most to least sensitive. How did your test results **compare** with your hypothesis?
5. What do your answers to questions 2 and 3 indicate about the distribution of touch receptors in the skin?

Go Further

Plan an investigation to determine whether there are areas of the skin that are more sensitive to cold than others. Try it and see if your inference is correct.

5. Touch receptors are closer together in the skin of the fingertips and further apart on the back of the hand and forearm. Receptors in the palm and the back of the neck vary.

Go Further

Answers will vary, but may include inferences about the cheek, neck, and lips.

 Assessment

Performance To further assess students' understanding of skin sensitivity, repeat this activity on the lower part of the leg and on the foot. Use the Performance Task Assessment List for Designing an Experiment in **PASC**, p. 23. **P**

PLAN THE EXPERIMENT

Possible Procedures

One possible procedure is to predict what skin areas are more sensitive to touch. Rank the areas from most sensitive (5) to least sensitive (1) in a data table. Glue the toothpicks onto the card so that one side has a pair that are 1 mm apart; another side has a pair that are 3 mm apart. The other sides have pairs that are 5 mm and 10 mm apart. Label the card. With a partner's eyes closed, use the part of the card with toothpicks 1 mm apart and carefully touch the skin surface. Touch the fingertip, palm of hand, back of hand, forearm, and back of neck. Caution students not to apply heavy pressure. If the partner feels two points, record a plus (+) in the table. If the partner cannot feel both points, record a minus (-) in the table. Repeat using the 3-mm, 5-mm, and 10-mm sections of the card.

DO THE EXPERIMENT

Expected Outcome

Results will reveal that the more sensitive areas of the skin to touch are the fingertips, palms, and cheeks and the less sensitive areas are the back of the hand, forearm, and the back of the neck.

Analyze and Apply

1. Student results should be consistent with those of other groups.
2. fingertips and usually the palms
3. back of hand, forearm
4. Answers will vary. Students should answer honestly if predictions do not match outcomes.

? FLEX Your Brain

Use the Flex Your Brain activity to have students explore SENSES.

📁 **Activity Worksheets,** p. 5

Reteach

LS Interpersonal Have students make concept maps of the sensory organs.

Extension

📁 For students who have mastered this section, use the **Reinforcement** and **Enrichment** masters.

4 Close

•MINI•QUIZ•

1. **What is the function of the lens?** *to focus light rays onto the retina*

2. **What are the three small bones in the middle ear?** *hammer, anvil, stirrup*

Section Wrap-up

Review

1. eyes: rods and cones
   nose: olfactory cells

2. sound waves

3. The body responds to protect itself or maintain homeostasis.

4. Think Critically The brain could continue to function when injured, allowing you to carry out activities necessary for survival.

Science Journal

Entries should reflect knowledge of the type of skin receptors involved. **P**

Touch, Pressure, Pain, and Temperature

Figure 25-17
Many of the sensations picked up by receptors in the skin are stimulated by mechanical energy. Pressure, motion, and touch are examples.

How important is it to be able to feel pain inside your body? Several kinds of sensory receptors in your internal organs, as well as throughout your skin, respond to touch, pressure, pain, and temperature. These receptors pick up changes in touch, pressure, and temperature and transmit impulses to the brain or spinal cord. The body responds to protect itself or maintain homeostasis.

Your fingertips have many different types of receptors for touch. As a result, you can tell whether an object is rough or smooth, hot or cold, light or heavy. Your lips are sensitive to heat and prevent you from drinking something so hot that it would burn you. Pressure-sensitive cells in the dermis give warning of danger to a body part and enable you to move to avoid injury.

Your senses are adaptations that help you enjoy or avoid things around you. You constantly react to your environment because of information received by your senses.

Section Wrap-up

Review

1. What are the sensory receptors for the eyes and nose?

2. What type of stimulus do your ears respond to?

3. Why is it important to have receptors for pain and pressure in your internal organs?

4. **Think Critically:** The brain is insensitive to pain. What is the advantage of this?

Skill Builder
Making and Using Tables
Organize the information on senses in a table that names the sense organs and gives the type of energy to which they respond. If you need help, refer to Making and Using Tables in the **Skill Handbook.**

Science Journal

Write a paragraph in your Science Journal describing how you would feel as you held each of the following objects in your hands.

1. coarse sand from a beach
2. ice cube
3. silk blouse
4. snake

Skill Builder
Make sure the table lists sense organs for vision, hearing, smell, taste, touch, pressure, pain, and temperature. The energy listed should include light, heat, sound waves, chemicals, and pressure.

Assessment

Performance Assess students' abilities to make and use tables by having them write a statement comparing the different types of energy that stimulate the senses. Use the Performance Task Assessment List for Analyzing the Data in **PASC,** p. 27. **P**

Your Endocrine System

The Endocrine System

"The tallest man in the world!" and "the shortest woman in the land!" were common attractions in circuses of the past. These people became attractions because of their extraordinary height or lack of it. In most cases, their sizes were the result of a malfunction in their endocrine systems.

The endocrine system is the second control system of your body. Whereas impulses are control mechanisms of your nervous system, chemicals are the control mechanisms of your endocrine system. Endocrine chemicals called **hormones** are produced in several ductless glands throughout your body. As the hormones are produced, they move directly into your bloodstream. Hormones affect specific tissues called **target tissues.** Target tissues are frequently in another part of the body at a distance from the gland that affects them. Thus, the endocrine system doesn't react as quickly as the nervous system. **Table 25-1** shows the position of eight endocrine glands and what they regulate.

Table 25-1

Endocrine Glands

Gland	Regulates
Pituitary	Enables other endocrine glands to produce hormones Milk production Growth
Thyroid	Carbohydrate use
Parathyroids	Calcium
Thymus	Parts of immune system
Adrenal	Blood sugar; salt and water balance; metabolism
Pancreas	Blood sugar
Ovaries	Production of eggs; development of sex organs in females
Testes	Production of sperm; development of sex organs in males

Science Words

hormone
target tissue

Objectives

- Explain the function of hormones.
- Name three endocrine glands and explain the effects of the hormones they produce.
- Explain how a feedback system works.

Prepare

Section Background

- Glands that secrete substances such as oils, sweat, or mucus through ducts are exocrine glands.
- The hormones of the endocrine system regulate many of the body's metabolic functions, usually causing the rate of these functions to speed up or slow down.

1 Motivate

Bellringer

 Before presenting the lesson, display **Section Focus Transparency 93** on the overhead projector. Assign the accompanying **Focus Activity** worksheet. L1 LEP

Tying to Previous Knowledge

Recall from Section 25-1 how the nervous system responds to changes in environment and makes adjustments in the body.

Program Resources

 Reproducible Masters
Study Guide, p. 97 L1
Reinforcement, p. 97 L1
Enrichment, p. 97 L3
Critical Thinking/Problem Solving, p. 29

Transparencies
Section Focus Transparency 93 L1
Teaching Transparency 50

Problem Solving

Solve the Problem

1. pancreas
2. Glucose levels can be determined by testing the blood.
3. Those foods that are high in sugars.

Thinking Critically

Carrie's body cells are not able to use the glucose in her blood. Thus, her cells have no source of energy.

Enrichment

Have students research the function of the chemical messengers called prostaglandins. [L3]

3 Assess

Check for Understanding

?FLEX Your Brain

Use the Flex Your Brain activity to have students explore ENDOCRINE GLANDS.

📁 **Activity Worksheets,** p. 5

Reteach

Using a chart of the circulatory system, have students trace the pathway of a hormone to its target tissue.

Extension

📁 For students who have mastered this section, use the **Reinforcement** and **Enrichment** masters.

Problem Solving

Why am I so tired?

Carrie hasn't been feeling well lately. She complains about feeling tired all the time.

"She just doesn't have any get-up-and-go," her mother told the doctor. The doctor ordered some blood tests and questioned Carrie about her diet. After school one afternoon, Carrie went to the doctor's office to hear the results of the tests. The tests showed that her blood sugar was much too high. The doctor gave her strict instructions to follow detailing what she was allowed to eat every day. He told Carrie that she showed signs of diabetes.

Solve the Problem:

1. **What endocrine gland is involved in diabetes?**
2. **Why would the doctor use blood tests to diagnose diabetes?**
3. **What foods should Carrie avoid eating?**

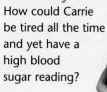

Think Critically:

How could Carrie be tired all the time and yet have a high blood sugar reading?

696 Chapter 25 The Nervous and Endocrine Systems

 Use Teaching Transparency 50 as you teach this lesson.

📁 Use **Critical Thinking/Problem Solving,** p. 29, as you teach this lesson.

The Pancreas— Playing Two Roles

You learned in Chapter 22 that the pancreas produces a digestive enzyme. This enzyme is released into the small intestine through ducts. The pancreas is also part of your endocrine system. This is because other groups of cells in the pancreas secrete hormones. One of these hormones, insulin, enables cells to take in glucose. Recall that glucose is the main source of energy for respiration in cells. Normally, insulin enables glucose to pass from the bloodstream through cell membranes. Persons who can't make insulin are diabetic. That means that insulin isn't there to enable glucose to get into cells.

A Negative-Feedback System

To control the amount of hormone an endocrine gland produces, the endocrine system sends chemical information back and forth to itself. This is a negative-feedback system, which is illustrated in **Figure 25-18.** It works much the way a thermostat works. When the temperature in a room drops below a certain level, a thermostat signals the furnace to turn on. Once the furnace has raised the temperature to the level set on the thermostat, the furnace shuts off. It will stay off until the thermostat signals again. In your body, once a target tissue responds to its hormone, the tissue sends a chemical signal back to the gland. This signal causes the gland to stop or slow down production of the hormone. When the level of the hormone in the bloodstream drops below a certain level, the endocrine gland is signaled to begin secreting the hormone again. In this way, the concentration of the hormone in the bloodstream is kept at the needed level.

Hormones secreted by endocrine glands go directly into the bloodstream and affect target tissues. The level of the hormone is controlled by a negative-feedback system. In this way, many chemicals in the blood and body functions are controlled.

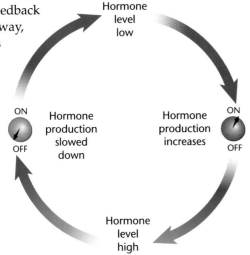

Figure 25-18
Many internal body conditions, such as hormone level and body temperature, are controlled by negative-feedback systems.

CONNECT TO
CHEMISTRY

The pituitary gland has been called the "master gland" because it controls so many other endocrine glands in the body. *Research* how the pituitary affects the kidneys and growth.

Section Wrap-up

Review

1. What is the function of hormones?

2. What is a negative-feedback system?

3. Choose one hormone and explain how it works.

4. **Think Critically:** Glucose passes from the bloodstream through cell membranes and into the cells. Glucose is required for respiration within cells. How would lack of insulin affect this process?

Skill Builder
Comparing and Contrasting
In what ways are the nervous system and endocrine system alike? If you need help, refer to Comparing and Contrasting in the **Skill Handbook.**

Science Journal

Imagine you are informed that you are diabetic and must give yourself daily insulin shots for the rest of your life. In your Science Journal, write about the different emotions you feel about this experience.

Skill Builder
Both systems send and receive messages. They also help to maintain homeostasis.

Assessment

Performance Write a paper explaining the similarities and differences of the nervous and endocrine systems. Use the Performance Task Assessment List for Writing in Science in **PASC,** p. 87. **P**

CONNECT TO
CHEMISTRY

Answer The pituitary gland controls when fluids are retained and released by the kidneys; it also controls growth in young animals and metabolism in adults.

4 Close

•MINI•QUIZ•

Use the Mini Quiz to check students' recall of chapter content.

1. **Endocrine glands do not have ____ .** *ducts*

2. **What is one of the hormones produced by the pancreas?** *insulin*

3. **What signals a gland to start secreting its hormone again?** *when the level of hormone in the bloodstream drops*

Section Wrap-up

Review

1. They are endocrine secretions that control activities in parts of the body not near to the gland.

2. A system that uses the level of hormone in the blood to signal when a gland should secrete the hormone again.

3. pancreas—secretes insulin; refer to Table 25-1 for other answers

4. **Think Critically** Insulin enables glucose to pass from the bloodstream into cells. Cells need glucose for respiration. Without insulin, cells can't respire.

Science Journal

Entries may reflect apprehension about having to take insulin or relief at knowing what was wrong.

Background

Both the U.S. Department of Agriculture and the U.S. Public Health Service monitor animal use in laboratories throughout the country. These agencies have strict regulations regarding the care and handling of laboratory animals. Inspectors check for cleanliness, and proper ventilation and temperature control in buildings where animals are housed. Animals must have access to food and fresh water at all times, and they cannot be unduly crowded. Laboratories in which live animals are maintained must also employ veterinarians to monitor the health of the animals and promptly treat their medical problems.

Teaching Strategies

IS Interpersonal Divide the class into groups, and assign an endocrine gland to each one. Challenge the groups to research their assigned gland, and make a list of the hormones it produces, the actions of those hormones, and their deficiency symptoms. Have each group report its findings to the class. **COOP LEARN**

Career Connection

Career Path Hormone researchers work in government and university laboratories, as well as in pharmaceutical companies that produce medicines and health care products for people and animals.

- High School: science and math
- College: biology, chemistry, physics, and math
- Graduate School: Master's degree/Ph.D. in endocrinology or physiology

People and Science

MAUREEN DIGGINS, *Researcher*

On the Job

Q Dr. Diggins, you teach full-time and also do research on hormones. What's the focus of your research?

A I'm interested in the relationship between body fat and hormones. I work with mice that have the "lethal yellow" mutation. Shortly after lethal yellow mice reach puberty, they get extremely fat, and when they do, they stop reproducing. Their excess fat apparently interferes with the hormones that control reproduction, and we're trying to figure out just how that happens.

Q How do you monitor what's happening to hormone levels in these fat mice?

A We perform what's called a radio-immunoassay on blood samples taken from the mice. This test "labels" hormones in the blood with a radioactive tracer, which makes it possible for us to measure, very accurately, how much of a hormone is present. Radioimmunoassays are used to measure hormone levels in people's blood in the same way.

Personal Insights

Q What's the greatest challenge in your job?

A Making time for teaching *and* research. I love teaching and it's a top priority. Compared to university professors, my research is fairly small-scale. But I do enough to keep active in my field, to experience the excitement of research, and to give my students a chance to experience it, too.

Career Connection

Trained laboratory technicians (or certified veterinary technicians) are usually responsible for the daily care and feeding of laboratory research animals. Contact each of the following educational facilities in your area for more information about these jobs.

- **Vocational/technical school**   • **Two-year college**

Write a one-page report describing the education and training required for such a position.

For More Information

Have students read "Melatonin," *Newsweek*, 7 August 1995, for more information on melatonin, a hormone with wide-ranging effects that has received considerable publicity.

Chapter 25 Review

Summary

25-1: Your Nervous System

1. A neuron is the basic unit of structure and function of the nervous system.
2. A stimulus is detected by sensory neurons. Impulses are carried to the interneurons and transmitted to the motor neurons. A response that is made automatically is a reflex.
3. The central nervous system contains the brain and spinal cord. The peripheral nervous system is composed of cranial and spinal nerves.

25-2: Science and Society: Care and Treatment of Alzheimer's Patients

1. A lack of an impulse-transmitting chemical in Alzheimer's patients causes memory loss.
2. Scientists think that heredity and a chemical imbalance are possible causes of Alzheimer's disease.

25-3: The Senses

1. The eyes respond to light energy, and the ears respond to sound waves.
2. Olfactory cells of the nose and taste buds of the tongue are stimulated by chemicals.

25-4: Your Endocrine System

1. Endocrine glands secrete hormones directly into the bloodstream.
2. Hormones affect specific tissues throughout the body.

Key Science Words

a. Alzheimer's disease
b. axon
c. brain stem
d. central nervous system
e. cerebellum
f. cerebrum
g. cochlea
h. dendrite
i. hormone
j. interneuron
k. motor neuron
l. neuron
m. olfactory cell
n. peripheral nervous system
o. reflex
p. retina
q. sensory neuron
r. synapse
s. target tissue
t. taste bud

Reviewing Vocabulary

Match each phrase with the correct term from the list of Key Science Words.

1. neuron carrying impulses from the brain
2. basic unit of nervous system
3. nerve cell between sensory and motor neurons
4. gap between neurons
5. division containing brain and spinal cord
6. an automatic response to stimuli
7. light-sensitive tissue of the eye
8. center for coordination of voluntary muscle action
9. inner-ear structure containing fluid
10. nerve cell that responds to gas molecules

Chapter 25

Review

Summary

Have students read the summary statements to review the major concepts of the chapter.

Reviewing Vocabulary

1. k   6. o
2. l   7. p
3. j   8. e
4. r   9. g
5. d   10. m

✓ Assessment

Portfolio Encourage students to place in their portfolios one or two items of what they consider to be their best work. Examples include:

- Brainstorming, p. 682
- Explore the Issue, p. 687
- Science Journal, p. 694 P

Performance Additional performance assessments may be found in **Performance Assessment** and **Science Integration Activities.** Performance Task Assessment Lists and rubrics for evaluating these activities can be found in Glencoe's **Performance Assessment in the Science Classroom.**

GLENCOE TECHNOLOGY

 Videodisc

Glencoe Life Science Interactive Videodisc
Side 2, Lesson 8
Medical Diagnostics: Tools of the Trade

27448 to 38842

MindJogger Videoquiz

Chapter 25 Have students work in groups as they play the Videoquiz game to review key chapter concepts.

Chapter 25 Review

Checking Concepts

1. d	**6.** c
2. b	**7.** a
3. c	**8.** c
4. c	**9.** c
5. a	**10.** d

Understanding Concepts

11. Taste is remembered. If something tastes bad and results in sickness, you remember it and do not taste the food again. Smell provides the same protection.

12. If a synapse were blocked, the chemicals that transmit nerve impulses would not be able to do this; therefore, these nerves could not be stimulated.

13. Once a certain level was reached, the gland would stop secreting its hormone until its level in the blood fell below a certain point.

14. The speed of the activities of the nervous system is more rapid than that of the endocrine system activities.

15. Stimulants speed up activities of the nervous system and can thereby cause excitability and sleeplessness.

Thinking Critically

16. Accept all reasonable responses; so messages do not get mixed; so that responses can be coordinated

17. Reflexes are automatic acts that occur without our thinking about them. Therefore, they happen very quickly and can shield our bodies from danger, such as sharp or hot objects.

Checking Concepts

Choose the word or phrase that completes the sentence.

1. Impulses cross synapses by way of _____.
  a. osmosis
  b. interneurons
  c. a cell body
  d. a chemical

2. The neuron structures that carry impulses to the cell body are _____.
  a. axons
  b. dendrites
  c. synapses
  d. nuclei

3. Neurons detecting stimuli in the skin and eyes are _____.
  a. interneurons
  b. motor neurons
  c. sensory neurons
  d. synapses

4. The _____ is the largest part of the brain.
  a. cerebellum
  b. brain stem
  c. cerebrum
  d. pons

5. The _____ controls voluntary muscle.
  a. cerebellum
  b. brain stem
  c. cerebrum
  d. pons

6. The _____ is the part of the brain that is divided into two hemispheres.
  a. pons
  b. brain stem
  c. cerebrum
  d. spinal cord

7. The _____ is (are) controlled by the somatic division of the PNS.
  a. skeletal muscles
  b. heart
  c. glands
  d. salivary glands

8. Ductless glands produce _____.
  a. enzymes
  b. target tissues
  c. hormones
  d. saliva

9. The _____ gland controls many other endocrine glands throughout the body.
  a. adrenal
  b. thyroid
  c. pituitary
  d. pancreas

---

10. The inner ear contains the _____.
  a. anvil
  b. hammer
  c. eardrum
  d. cochlea

Understanding Concepts

Answer the following questions in your Science Journal using complete sentences.

11. Discuss why taste and smell are protective.

12. What would happen if a chemical were blocking many synapses?

13. What would happen to an endocrine gland if the blood level of its hormone increased?

14. Contrast the speed of the activities of the nervous system with that of the endocrine system.

15. A drug label lists excitability and sleeplessness as possible side effects. Explain why the drug is a stimulant.

---

18. According to the level of hormone in the blood, target tissue sends a chemical signal back to the gland to stop or start hormone secretion.

19. Examples include giantism, dwarfism, or diabetes.

20. The skin on your face has more neurons or more closely arranged neurons than the skin on your back. The sense of touch is more pronounced on the face. You could check this out with a series of tests using something very lightweight, such as a feather or paint brush hair to test sensitivity on the face and neck.

Developing Skills

21. Classifying:
Sensory neurons—Location: sense organs and spinal cord; impulse direction: carry impulses to brain

Thinking Critically

16. Why is it helpful to have impulses move in only one direction in a neuron?
17. How are reflexes protective?
18. How is your endocrine system like the thermostat in a house?
19. Describe an example of a problem that results from improper gland functioning.
20. If a fly were to land on your face and another one on your back, which might you feel first? Explain how you would test your choice.

Developing Skills

If you need help, refer to the **Skill Handbook.**

21. **Classifying:** Classify the types of neurons as to their location and direction of impulse.
22. **Comparing and Contrasting:** Compare and contrast the structures and functions of the cerebrum, cerebellum, and brain stem. Include in your discussion the following functions: balance, involuntary muscle movements, muscle tone, memory, voluntary muscles, thinking, and senses.
23. **Concept Mapping:** Prepare a concept map showing the correct sequence of the structures through which light passes in the eye.
24. **Interpreting Scientific Illustrations:** Using the diagram of the synapse on page 680, explain how an impulse moves from one neuron to another.
25. **Observing and Inferring:** If an impulse traveled down one neuron, but failed to move on to the next neuron, what might you infer about the first neuron?

Performance Assessment

1. **Writing in Science:** Write a research paper on how hearing aids work. Explain why not all hearing-impaired persons can use hearing aids.
2. **Making Observations and Inferences:** On entering a hospital, the doctor notices that a patient's body movements are uncoordinated and he has difficulty maintaining his balance. What part of the brain might be injured?
3. **Photo Essay:** Choose one person to follow for a day. Take pictures of the person using all five senses. Make a photo essay of the pictures.

Interneurons — location: CNS; impulse direction: carry impulses from CNS to motor neurons
Motor neurons — location: muscles and glands; impulse direction: carry impulses from brain to muscles and glands

22. **Comparing and Contrasting** cerebrum — memory, senses, thinking; cerebellum—voluntary muscles, balance, muscle tone; brain stem—coordinating involuntary muscle movements, controls heartbeat, breathing, and blood pressure
23. **Concept Mapping** cornea → lens → retina → optic nerve → brain
24. **Interpreting Scientific Illustrations** A nerve-transmitting chemical is released from the axon of one neuron, diffuses across the synapse, and starts an impulse in the next neuron.
25. **Observing and Inferring** It may be lacking the nerve-transmitting chemical released by the axon.

Performance Assessment

1. A hearing aid has a microphone receiver that picks up sound which is amplified and transmitted to an earphone. If the nerve cells in the cochlea are damaged, the hearing aid will not be of use. Use the Performance Task Assessment List for Writing in Science in **PASC,** p. 87. P
2. The cerebellum. Use the Performance Task Assessment List for Making Observations and Inferences in **PASC,** p. 17. P
3. Display the photo essays on a class bulletin board. Use the Performance Task Assessment List for Slide Show or Photo Essay in **PASC,** p. 77. P

Assessment Resources

 Reproducible Masters
Chapter Review, pp. 53-54
Assessment, pp. 159-162
Performance Assessment, p. 39

Glencoe Technology
⊙ **Chapter Review Software**
⊙ **Computer Test Bank**
▭ **MindJogger Videoquiz**

Chapter Organizer

Section	Objectives/Standards	Activities/Features
Chapter Opener		**Explore Activity:** What is the proportion of boys to girls in the general population? p. 703
26-1 **Human Reproduction** (2 sessions, 1½ blocks)*	1. **Explain** the function of the reproductive system. 2. **Identify** the major structures of the male reproductive system. 3. **Describe** the functions of the major female reproductive organs. 4. **Explain** the stages of the menstrual cycle. National Content Standards: (5-8) UCP1, A1, A2, B1, C1, C2, C3	**Connect to Physics,** p. 705 **Skill Builder:** Sequencing, p. 708 **Using Math,** p. 708 **Activity 26-1:** Interpreting Diagrams, p. 709
26-2 **Fertilization to Birth** (2 sessions, 1½ blocks)*	1. **Describe** how an egg becomes fertilized. 2. **Identify** the major events in the stages of development of an embryo and fetus. 3. **Differentiate** between fraternal and identical twins. National Content Standards: (5-8) UCP1, A1, C1, C2, E2, F5	**Problem Solving:** When to Test, p. 712 **MiniLAB:** How does an embryo develop? p. 713 **Using Technology:** Prenatal Surgery, p. 714 **Skill Builder:** Making and Using Graphs, p. 715 **Science Journal,** p. 715
26-3 **Development After Birth** (2 sessions, 1 block)*	1. **State** the sequence of events of childbirth. 2. **Compare** the stages of infancy and childhood. 3. **Relate** adolescence to preparation for adulthood. National Content Standards: (5-8) UCP3, A1, A2, C3, F1	**Using Math,** p. 717 **MiniLAB:** What immunizations are given to babies and young children? p. 718 **Using Math,** p. 719 **Skill Builder:** Outlining, p. 720 **Science Journal,** p. 720 **Activity 26-2:** Average Growth Rate in Humans, p. 721
26-4 **Science and Society** **Technology: A New** **View of the Old** (1 session, 1/2 block)*	1. **Compare** the common view of aging with the new view of aging. 2. **Describe** factors that influence aging. National Content Standards: (5-8) F1, G1	**Explore the Technology,** p. 723 **People and Science:** Carol Ambrose, Nurse Midwife, p. 724

* A complete Planning Guide that includes block scheduling is provided on pages 32T-35T.

Activity Materials

Explore	Activities	MiniLABs
page 703 penny, pencil, paper	Page 709 No special materials are needed. page 721 graph paper, red and blue pencils	page 713 metric ruler, reference materials on human fetal development page 718 reference materials on childhood immunizations, student's personal immunization record

Need Materials? Call Science Kit (1-800-828-7777).

Teacher Classroom Resources

Reproducible Masters	Transparencies	Teaching Resources
Study Guide, p. 98 **Reinforcement**, p. 98 **Enrichment**, p. 98 **Activity Worksheets**, pp. 154-155 **Concept Mapping**, p. 57	**Section Focus Transparency 94**, The Reproductive Process **Teaching Transparency 51**, Reproductive Systems **Teaching Transparency 52**, The Menstrual Cycle	**Glencoe Life Science Interactive Videodisc** **Life Science CD-ROM** **Spanish Resources** **English/Spanish Audiocassettes** **Cooperative Learning Resource Guide** **Lab Partner** **Lab and Safety Skills** **Lesson Plans**
Study Guide, p. 99 **Reinforcement**, p. 99 **Enrichment**, p. 99 **Activity Worksheets**, p. 158 **Lab Manual 60**, Fetal Development **Science Integration Activity 26**, Chemical Reactions **Critical Thinking/Problem Solving**, p. 30	**Section Focus Transparency 95**, Gestation Periods	
		## Assessment Resources **Chapter Review**, pp. 55-56 **Assessment**, pp. 163-166 **Performance Assessment in the Science Classroom (PASC)** **MindJogger Videoquiz** **Alternate Assessment in the Science Classroom** **Performance Assessment** **Chapter Review Software** **Computer Test Bank**
Study Guide, p. 100 **Reinforcement**, p. 100 **Enrichment**, p. 100 **Activity Worksheets**, pp. 156-157, 159 **Cross-Curricular Integration**, p. 30 **Multicultural Connections**, pp. 55-56	**Section Focus Transparency 96**, Life Cycles **Science Integration Transparency 26**, The Human Yardstick	
Study Guide, p. 101 **Reinforcement**, p. 101 **Enrichment**, p. 101 **Science and Society Integration**, p. 30	**Section Focus Transparency 97**, Life Expectancy	

Key to Teaching Strategies

The following designations will help you decide which activities are appropriate for your students.

L1 Level 1 activities should be within the ability range of all students, including those with learning difficulties.

L2 Level 2 activities should be within the ability range of the average to above-average student.

L3 Level 3 activities are designed for the ability range of above-average students.

LEP LEP activities should be within the ability range of Limited English Proficiency students.

LS These activities are designed to address different learning styles.

COOP LEARN Cooperative Learning activities are designed for small group work.

P These strategies represent student products that can be placed into a best-work portfolio.

GLENCOE TECHNOLOGY

The following multimedia resources are available from Glencoe.

The Infinite Voyage Series
The Geometry of Life
The Champion Within

National Geographic Society Series
STV: Human Body Vol. 3
Life Science CD-ROM

Teacher Classroom Resources

This is a representation of key blackline masters available in the Teacher Classroom Resources.

Teaching Aids

Section Focus Transparencies

Science Integration Transparencies

Teaching Transparencies

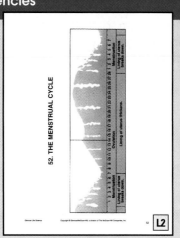

Meeting Different Ability Levels

Study Guide

Reinforcement

Enrichment Worksheets

Chapter 26 Reproduction and Growth

Hands-On Activities

Science Integration Activity

Chemical Reactions — L1

Lab Manual
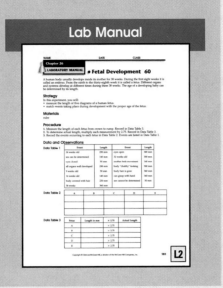

Fetal Development 60 — L2

Assessment

Performance Assessment

The Menstrual Cycle — L3

Enrichment and Application

Critical Thinking/Problem Solving

Reproduction and Growth — Identical Twins — L2

Cross-Curricular Integration

Reproduction and Growth — Babies—A Natural for Art — L1

Science and Society Integration

Reproduction and Growth — Living Longer — L2

Multicultural Connections

Growing Very Old, Very High Up — L2

Concept Mapping
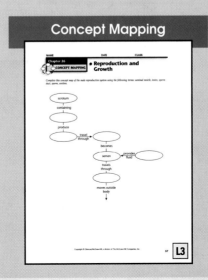

Reproduction and Growth — L3

702D

Reproduction and Growth

CHAPTER OVERVIEW

Section 26-1 In this section the structure and function of the major organs of the female and male reproductive systems are introduced. Special attention is given to the menstrual cycle.

Section 26-2 This section follows the development of a fertilized egg to the full-term fetus. The causes for multiple births are explored.

Section 26-3 This section opens with the details of childbirth. The development of a human is traced from infancy through adulthood. Characteristics of each stage are given.

Section 26-4 Science and Society This section deals with a new view of aging. Many persons over ninety are mentally alert and physically active.

Chapter Vocabulary

testis	pregnancy
sperm	embryo
semen	amniotic sac
ovary	fetus
ovulation	infancy
uterus	childhood
vagina	adolescence
menstrual cycle	adulthood
menstruation	centenarian
menopause	

Theme Connection

Energy A central theme in this text is that of energy. Energy drives many biological systems and is involved in the reproduction, growth, and development of organisms. The unique functioning of the human reproductive system is dependent on energy for cell division and the subsequent growth and differentiation of cells.

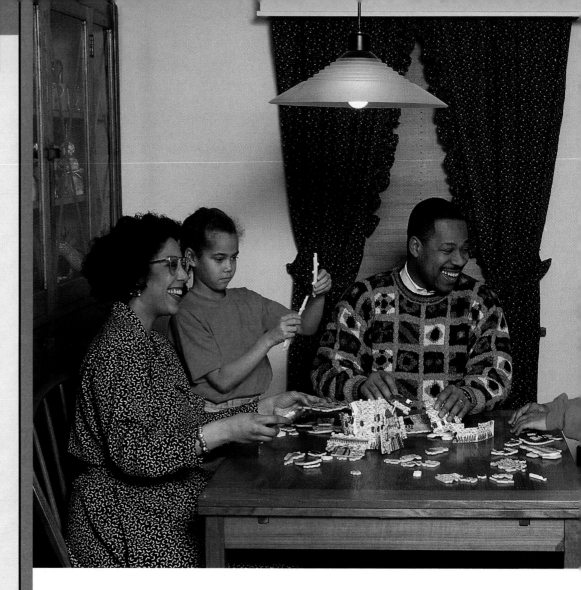

Previewing the Chapter

Section 26-1   Human Reproduction
▶ The Reproductive System
▶ The Male Reproductive System
▶ The Female Reproductive System
▶ The Menstrual Cycle

Section 26-2   Fertilization to Birth
▶ Fertilization
▶ A Biochemical Event

▶ Development Before Birth
▶ Multiple Births

Section 26-3   Development After Birth
▶ Childbirth
▶ Infancy and Childhood
▶ Adolescence
▶ Adulthood

Section 26-4   Science and Society
   Technology: A New View of the Old

702

Learning Styles

Look for the following logo for strategies that emphasize different learning modalities.

Visual-Spatial	Visual Learning, pp. 706, 707, 712, 714, 717, 718; Activity 26-1, p. 709; Demonstration, p. 712; Community Connection, p. 717
Intrapersonal	MiniLAB, p. 718
Logical-Mathematical	Explore, p. 703; MiniLAB, p. 713; Activity 26-2, p. 721
Linguistic	Across the Curriculum, p. 706

LS

Chapter 26

Reproduction and Growth

Some of your friends may come from families with both brothers and sisters. Some may come from families that have only boys or only girls. What do you think are the odds of a small family having all boys or all girls? Do the odds change as the family grows larger? What do you think is the proportion of boys to girls among children born in the general population?

EXPLORE ACTIVITY

What is the proportion of boys to girls in the general population?

1. Take a penny and toss it in the air. There is an equal chance that it will land heads or tails.
2. Toss the penny a hundred times.
3. Keep a record of the number of times it lands heads up and the number of times it lands tails up.
4. Keep a record of the order in which the penny lands.

Observe: Did your record reveal any patterns? Did you sometimes have five or more heads before a tails fell? How does this apply to families who have five girls or five boys? Infer the chances that a girl would be born if there were a sixth child.

Previewing Science Skills

▶ In the **Skill Builders**, you will **sequence, graph, and outline data**.

▶ In the **Activities**, you will **interpret and graph data**.

▶ In the **MiniLABs**, you will **interpret data** and **do research**.

703

Assessment Planner

Portfolio
Refer to page 725 for suggested items that students might select for their portfolios.

Performance Assessment
See page 725 for additional Performance Assessment options.
Skill Builders, pp. 708, 715, 720
MiniLABs, pp. 713, 718
Activities 26-1, p. 709; 26-2, p. 721

Content Assessment
Section Wrap-ups, pp. 708, 715, 720, 723
Chapter Review, pp. 725-727
Mini Quizzes, pp. 708, 715, 723

Group Assessment
Opportunities for group assessment occur with Cooperative Learning Strategies and Flex Your Brain Activities.

Prepare

Section Background

- As in all male mammals, the urethra carries both urine and semen outside the body. The release of urine through the urethra is controlled by sphincter muscles located above the opening of the vas deferens into the urethra. This system prevents urine and semen from mixing in the urethra.

- Occasionally, an egg from the ovary does not enter the opening of the oviduct but enters the abdominal cavity and becomes fertilized. An ectopic pregnancy occurs but does not continue because the tissues cannot sustain the embryo.

Preplanning

To prepare for Activity 26-1, obtain a pencil and sheet of paper for each group.

1 Motivate

Bellringer

Before presenting the lesson, display **Section Focus Transparency 94** on the overhead projector. Assign the accompanying **Focus Activity** worksheet. L1 LEP

Tying to Previous Knowledge

Recall from Chapter 4 cell division and the importance of meiosis to produce cells with 23 chromosomes in order to have a fertilized egg with the full complement of 46 chromosomes.

26•1 Human Reproduction

Science Words

- testis
- sperm
- semen
- ovary
- ovulation
- uterus
- vagina
- menstrual cycle
- menstruation
- menopause

Objectives

- Explain the function of the reproductive system.
- Identify the major structures of the male reproductive system.
- Describe the functions of the major female reproductive organs.
- Explain the stages of the menstrual cycle.

The Reproductive System

Reproduction is the process that continues life on Earth. Organisms that carry out sexual reproduction form eggs and sperm to transfer genetic information from one generation to the next. The number of offspring an organism has varies with the species. Humans, and other mammals, have fewer offspring than other animals. One determining factor in the number of offspring is the amount of care needed by the young animal. If you ever baby-sat, you know that babies and young children require almost constant attention. Unlike organisms such as spiders that can lay thousands of eggs resulting in thousands of offspring at one time as shown in **Figure 26-1,** most humans have only one or two babies at a time. This allows more individual attention for each offspring.

Figure 26-1

Animals that produce large numbers of eggs, such as spiders, right, have young that are usually able to take care of themselves from birth. Compare this to the ability of a human baby to care for itself. Imagine trying to care for these quintuplets!

Program Resources

📁 **Reproducible Masters**

Study Guide, p. 98 L1
Reinforcement, p. 98 L1
Enrichment, p. 98 L3
Activity Worksheets, pp. 5, 154-155 L1
Concept Mapping, p. 57

🔦 **Transparencies**

Section Focus Transparency 94 L1
Teaching Transparencies 51 and **52**

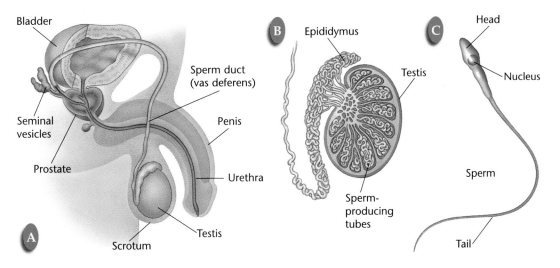

Figure 26-2

The structures of the male reproductive system are shown (A) with a close-up of a testis (B) and sperm (C). Once a male reaches puberty, the time when he sexually matures, he will produce about 300 million sperm per day.

The Male Reproductive System

You have read that most human body systems are alike in males and females. This is not the case for the reproductive system. Males and females each have structures specialized for their role in reproduction. **Figure 26-2** shows the structures of the male reproductive system. The external organs of the male reproductive system are the penis and scrotum. The penis is the male organ for reproduction and urination. The scrotum is a saclike pouch located behind the penis. Within the scrotum are the two testes. During puberty, the **testes** begin to produce the male reproductive cells, called **sperm,** and the male sex hormone, testosterone. Sperm are single cells with a head and tail. The tail moves the sperm, and the head contains genetic information. The external location of the scrotum helps regulate temperature for sperm production by holding the testes outside the male's body. As a result, the testes have a lower temperature, which is necessary for sperm production.

Sperm Movement

Many organs help in the production, transport, and storage of sperm inside the male body. After sperm are produced, they travel from the testes through tubes that circle the bladder. Behind the bladder, a gland called the seminal vesicle provides sperm with a fluid that gives them energy and helps them move. This mixture of sperm and fluid is called **semen.** Semen leaves the body through the urethra, the same tube that carries urine from the body. Semen and urine never mix. A muscle at the back of the bladder contracts to prevent urine from entering the urethra as sperm are ejected from the body.

CONNECT TO
PHYSICS

Shapes of objects are often related to their function. *Infer* why sperm have a streamlined head and an active tail.

Concept Development

Use a question box for questions that students may not want to ask in class. Use the questions to develop the content of lessons in the text.

Inquiry Question

Why is it not necessary for a human female to produce thousands of eggs at one time? *Eggs of humans are well protected and have a good chance of being fertilized.*

Teacher F.Y.I.

Sperm production is a continuous process from puberty throughout the life of a male. Oocyte production in females begins before birth. There is no production after birth. Each ovary may have nearly a million primary oocytes, cells from which eggs form.

CONNECT TO

PHYSICS

Answer The streamlined head of a sperm reduces friction as it moves through liquids. The active tail propels the sperm in a direction that can be against gravity.

NATIONAL GEOGRAPHIC SOCIETY

 Videodisc

STV: Human Body Vol. 3
Reproductive Systems
Unit 2, Side 2
The Female System

03087-07934
The Male System

07980-18202

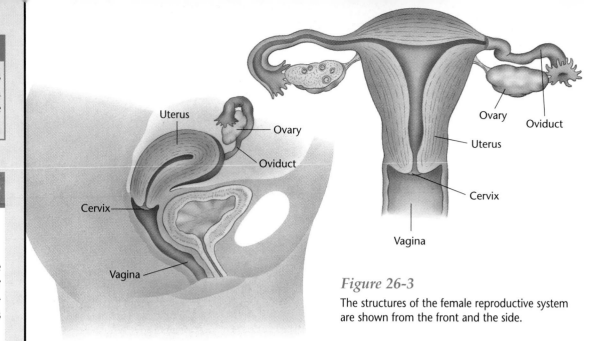

Figure 26-3

The structures of the female reproductive system are shown from the front and the side.

The Female Reproductive System

Eggs are the female reproductive cells. A female baby already has her total supply of eggs at birth. When the female reaches puberty, her eggs start to develop in **ovaries,** the female sex organs. Unlike the male's, most of the reproductive organs of the female are internal. The ovaries are located in the lower part of the body cavity. Each of the two ovaries is about the size and shape of an almond. **Figure 26-3** shows the structures of the female reproductive system.

Egg Movement

About once a month, an egg is released from an ovary. This process is called **ovulation.** The two ovaries take turns releasing the eggs. One month, the first ovary releases an egg; next month, the other ovary releases an egg and so on. When the egg is released, it enters the oviduct as shown in **Figure 26-4.** Sometimes, the egg is fertilized by a sperm. If fertilization occurs, it will happen in an oviduct. Cilia help sweep the egg through the oviduct to the uterus. The **uterus** is a hollow, pear-shaped, muscular organ with thick walls in which a fertilized egg develops. The lower end of the uterus is connected to the outside of the body by a muscular tube called the **vagina.** The vagina is also called the birth canal because a baby passes through this passageway when being born.

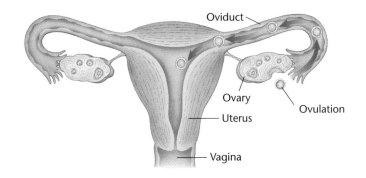
Oviduct

Ovary

Ovulation

Uterus

Vagina

The Menstrual Cycle

Before and after an egg is released from an ovary, as shown in **Figure 26-4,** the uterus undergoes certain changes during the menstrual cycle. The **menstrual cycle** is the monthly cycle of changes in the female reproductive system. This cycle of a human female averages 28 days. However, the cycle can vary in some individuals from 20 to 40 days. The changes include the maturing of an egg, the secretion of female sex hormones, and the preparation of the uterus to receive a fertilized egg, as shown in **Figure 26-5.** When an egg inside an ovary matures, the lining of the uterus thickens and prepares to receive a fertilized egg.

Three Phases

Generally, the 28-day cycle is divided into three phases. You can see in **Figure 26-5** that the first day of phase one starts when menstrual flow begins. Menstrual flow consists of blood and tissue cells from the thickened lining of the uterus. This monthly discharge is called **menstruation.** Menstruation usually lasts from four to six days and is a normal function of the female reproductive system.

The second phase begins after menstruation ends. Hormones cause the uterine lining to thicken again. One egg

Figure 26-4

The egg in the photograph has just been released from the ovary.

Magnification 1850×

Figure 26-5

The menstrual cycle begins on day one of menstruation. Ovulation occurs near day 14 of a regular 28-day cycle. Notice how the thickness of the uterine lining changes during the cycle.

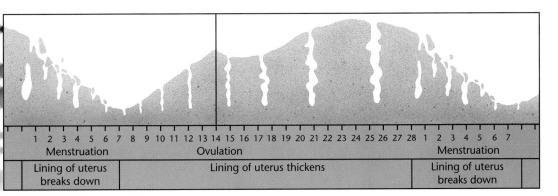

1 2 3 4 5 6 7 8 9 10 11 12 13	14 15 16 17 18 19 20 21 22 23 24 25 26 27 28	1 2 3 4 5 6 7
Menstruation	Ovulation	Menstruation
Lining of uterus breaks down	Lining of uterus thickens	Lining of uterus breaks down

NATIONAL GEOGRAPHIC SOCIETY

 Videodisc

STV: Human Body Vol. 3
Female reprod. organs 6: uterus (art)

48743

Female reprod. organs 7: vagina (art)

48744

3 Assess

Check for Understanding

? FLEX Your Brain

Use the Flex Your Brain activity to have students explore the REPRODUCTIVE SYSTEM.

 Activity Worksheets, p. 5

Reteach

Use diagrams to illustrate the pathway of the egg as it moves from the ovary to the uterus.

Extension

For students who have mastered this section, use the **Reinforcement** and **Enrichment** masters.

4 Close

•MINI•QUIZ•

1. Males produce _____ and females produce eggs. *sperm*

2. The passageway for both sperm and urine out of the male body is the _____ . *urethra*

3. Where is the egg usually fertilized? *in the upper third of the oviduct*

4. What is the normal length of the menstrual period? *4 to 6 days*

Section Wrap-up

Review

1. Its function is to reproduce the organism through eggs and sperm that carry genetic information from one generation to the next.

2. Sperm are produced in the testes, travel through tubes that encircle the bladder, are mixed with seminal fluid from the seminal vesicle, and leave the body via the urethra.

3. The ovaries produce the eggs; oviducts carry an egg to the uterus, where it will develop if fertilized; the vagina connects the uterus to the outside of the body and serves as the birth canal.

4. If the egg is not fertilized and does not implant in the uterus, the uterine lining is shed.

5. Think Critically If large amounts of blood are lost during the menstrual period, the body may have temporary iron deficiency because iron is a component of hemoglobin lost in menstrual blood.

USING MATH

50 years – 12 years = 38 years

38 years × 1 egg/month = 456 eggs

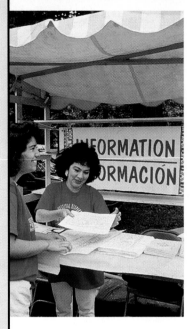

Figure 26-6

For most women menopause occurs sometime between the ages of 45 and 60.

---

in the ovary develops and matures until ovulation occurs. Ovulation occurs about day 14 of the cycle. Once the egg is released, it can be fertiziled for the next 24 to 48 hours as it moves through the oviduct. Pregnancy can occur if live sperm are in the oviduct 48 hours before or after ovulation.

During the third phase, hormones continue to increase the thickness of the uterine lining. In this way, the uterus is prepared for the arrival of a fertilized egg. If a fertilized egg does arrive, the uterus is ready to help support and nourish the developing embryo. If the egg is not fertilized, hormones decrease and the uterine lining deteriorates. Menstruation begins again. Another egg begins to mature and the cycle repeats itself.

Menopause

Menstruation begins when a girl reaches puberty and her reproductive organs have matured. For most females, the first menstrual period happens between ages eight and 13 and continues until age 45 to 55. Then, there is a gradual reduction of ovulation and menstruation. **Menopause** occurs when the menstrual cycle becomes irregular and eventually stops.

When the reproductive systems of males and females become mature, sperm and eggs are produced. The reproductive process allows for the species to continue.

Section Wrap-up

Review

1. What is the major function of a reproductive system?

2. Explain the movement of sperm through the male reproductive system.

3. List the organs of the female reproductive system and describe their functions.

4. Explain the cause of menstrual flow.

5. Think Critically: Adolescent females often require additional amounts of iron in their diet. Explain why.

Skill Builder

Sequencing

Sequence the movement of an egg through the female reproductive system. If you need help, refer to Sequencing in the **Skill Handbook.**

USING MATH

Usually, one egg is released each month during the reproductive years of a female. For a woman whose first menstruation starts at age 12 and ends at age 50, calculate the number of eggs released.

Skill Builder

ovary → oviduct → uterus → vagina

✓ Assessment

Performance Assess students' abilities to sequence by having them order the thicknesses of the uterus during a menstrual cycle using **Figure 26-5.** Use the Performance Task Assessment List for Events Chain in **PASC,** p. 91. P

Activity 26-1 · Interpreting Diagrams

Starting in adolescence, the hormone estrogen causes changes in the uterus. These changes prepare the uterus for the possibility of a fertilized egg embedding itself in the uterine wall.

Problem
What happens to the uterus during a female's monthly cycle?

Materials
- paper and pencil

Procedure
1. The diagrams below show what is explained in the previous section on the menstrual cycle.
2. Study the diagrams and their labels.
3. Use the information in Section 26-1 and the diagrams below to complete a table like the one shown.

Data and Observations
Sample Data

Days	Condition of uterus	What happens
1 - 6	breakdown of lining	menstruation
7 - 12	lining begins to thicken	egg matures in the ovary
13 - 14	lining is thicker	ovulation
15 - 28	lining thickens	egg moves to uterus

Analyze
1. What do the diagrams represent?
2. What do the pink- or red-colored parts of the diagrams represent?
3. How do these diagrams relate to **Figure 26-5**?
4. At what stage in the menstrual cycle is the uterine lining the most thickened?

Conclude and Apply
5. How long is the average menstrual cycle?
6. How many days does menstruation usually last?
7. How are the diagrams different?
8. On approximately what day in a 28-day cycle is the egg released from the ovary?
9. On what days does the lining of the uterus build up?
10. **Infer** why this process is called a cycle.
11. **Calculate** how many days before menstruation ovulation usually occurs.
12. **Interpret** the diagram to explain the menstrual cycle.

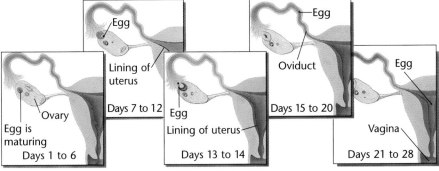

26-1 Human Reproduction 709

Purpose
LS **Visual-Spatial** Examine and interpret diagrams of the menstrual cycle. **L1**

Process Skills
observing, interpreting data

Time
40 minutes

📁 **Activity Worksheets,** pp. 5, 154-155

Teaching Strategies
Troubleshooting Stress the importance of examining labels and captions when interpreting diagrams.

Answers to Questions
1. the menstrual cycle in human females
2. the lining of the uterus wall and the time of menstruation
3. They explain what is shown in the diagram.
4. between days 21 and 28
5. 28 days
6. four to six days
7. Each shows the condition of the uterus at a different time during the menstrual cycle.
8. day 14
9. days 7 through 28
10. If fertilization does not occur, menstruation takes place, another egg is released, and the process is repeated.
11. about 14
12. You can see what is happening to the lining of the uterus each day of the cycle. The labels show the sequence of events taking place.

Assessment

Performance To further assess students' abilities to interpret diagrams, use this diagram to explain how fraternal twins form. Use the Performance Task Assessment List for Making Observations and Inferences in **PASC**, p. 17. **P**

Prepare

Section Background

- At the tip of a head of a sperm is a small sac filled with the enzyme hyaluronidase. This enzyme allows the sperm to penetrate the cell membrane of the egg.
- When fully developed, the placenta is about 2 cm in thickness and has a diameter of about 20 cm.
- The enlargement of the fetus and uterus eventually causes them to outgrow the space in the pelvis area. They then grow upward and displace such organs as the intestines and stomach.

Preplanning

To prepare for the MiniLAB, obtain a variety of reference materials from the library and one metric ruler per group.

1 Motivate

Bellringer

Before presenting the lesson, display **Section Focus Transparency 95** on the overhead projector. Assign the accompanying **Focus Activity** worksheet. [L1] [LEP]

Tying to Previous Knowledge

Have students recall from Chapter 22 the general need for good nutrition for the body and also relate this to the needs of the developing fetus. Draw attention to the government warnings on alcoholic beverages that indicate the dangers of alcohol consumption during pregnancy.

Science Words
pregnancy
embryo
amniotic sac
fetus

Objectives
- Describe how an egg becomes fertilized.
- Identify the major events in the stages of development of an embryo and fetus.
- Differentiate between fraternal and identical twins.

Fertilization

Before the invention of powerful microscopes, some people imagined a sperm to be a miniature person that grew in the uterus of a female. Others thought the egg contained a miniature individual that started to grow when stimulated by semen. In the latter part of the 1700s, experiments using amphibians showed that contact between an egg and sperm is necessary for life to begin development. With the formulation of the cell theory in 1839, scientists recognized that a human develops from a single egg that has been fertilized by a sperm. The uniting of a sperm with an egg is known as fertilization.

As you see in **Figure 26-7,** the process begins when sperm, deposited into the vagina, move through the uterus into the oviducts. Whereas only one egg is usually present, nearly 200 to 300 million sperm are deposited. Of that number, only one sperm will fertilize the egg. The nucleus of the sperm and the nucleus of the egg have 23 chromosomes each. When the egg and sperm unite, a zygote with 46 chromosomes is formed. Most fertilization occurs in an oviduct.

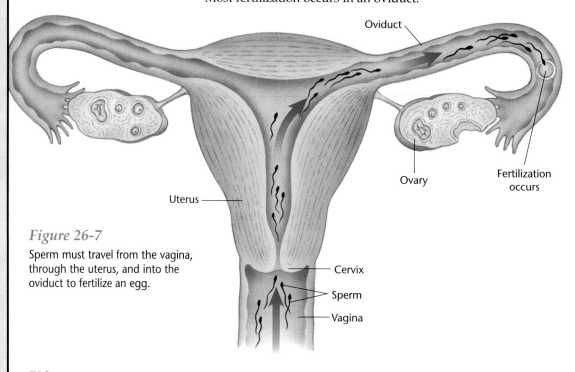

Figure 26-7
Sperm must travel from the vagina, through the uterus, and into the oviduct to fertilize an egg.

710 Chapter 26 Reproduction and Growth

Program Resources

 Reproducible Masters
Study Guide, p. 99 [L1]
Reinforcement, p. 99 [L1]
Enrichment, p. 99 [L3]
Activity Worksheets, p. 158 [L1]
Science Integration Activities, pp. 51-52
Lab Manual 60
Critical Thinking/Problem Solving, p. 30

Transparencies
Section Focus Transparency 95 [L1]

A Biochemical Event

Several hundred million sperm may be deposited in the vagina, but only several thousand actually reach the egg in the oviduct. During their travels, the sperm come in contact with chemical secretions in the vagina. It appears that this contact causes a change in the membrane of the sperm. The sperm then become capable of fertilizing eggs.

Although many sperm may actually reach the egg, only one will fertilize it. The one sperm that makes successful contact with the egg releases an enzyme from the saclike structure on its head. The enzyme has a direct effect on the protective egg-surface membranes. The structure of the membrane is disrupted, and the sperm head is now able to penetrate into the egg.

INTEGRATION
Chemistry

Figure 26-8
Only one sperm will fertilize an egg.

Magnification 1500×

Response of the Egg

Once a sperm has penetrated the egg, another series of chemical actions take place. These actions prevent further penetration by other sperm. Special chemical secretions are released onto the egg's surface, which prevent other sperm from binding. Other chemicals cause changes on the surface of the egg itself that prevent penetration. At this point, the nucleus of the successful sperm fuses with the nucleus of the egg. This fusion creates a new cell, called a zygote, that undergoes mitosis.

26-2 Fertilization to Birth **711**

711

Problem Solving

Solve the Problem

1. Amniocentesis, tissue sampling
2. when possible abnormalities are suspected or the mother has been injured

Think Critically

to determine the possibility of a sex-linked disorder such as hemophilia

Demonstration

Visual-Spatial Use Styrofoam balls to construct four- and eight-cell models to illustrate the earliest stages of growth of a zygote. Have students infer how these early stages develop into a hollow ball of cells. **LEP**

Inquiry Question

What is the difference between an embryo and a fetus? *During the first two months of pregnancy the unborn child is called an embryo; thereafter, it's called a fetus.*

Visual Learning

Figure 26-9 Have students find the umbilical cord and placenta. Explain that while the blood of the mother and embryo come close to each other, they never mix. Nutrients and wastes transfer between the blood vessels by diffusion. **LEP**

GLENCOE TECHNOLOGY

Videodisc

The Infinite Voyage: The Geometry of Life
Chapter 3
Development of the Egg After Fertilization

Problem Solving

When to Test

Expectant mothers want to know, "Will my baby be healthy?" Several tests can be done before a baby is born to test for birth defects. One test is known as amniocentesis. In this test, the doctor extracts a small amount of fluid from the amniotic sac. Examination of the fluid can reveal certain abnormalities. A similar test takes a tissue sample of the placenta. Analysis of the tissue shows evidence of certain diseases.

One of the most frequently used tests is ultrasound. A device generates sound waves that are transmitted through the mother's body. The sound waves produce photographic images of the fetus. Doctors observe the images to detect not only deformities, but also the size, sex, and rate of development of the fetus.

While each procedure has a benefit, each also has a risk. Some risks are to the mother, some are to the unborn fetus. Some doctors also think that too many tests may have an adverse effect on the development of the fetus.

Solve the Problem:

1. **Which procedures require entrance to the body to get materials for testing?**
2. **When do you think parents should request prenatal testing?**

Think Critically:

Under what conditions might it be important to know the sex of an unborn child?

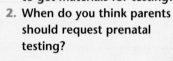

Ultrasound photo of a 23-week-old fetus

Integrating the Sciences

Physical Science Have students research the characteristics of ultrasonic waves. How are ultrasonic waves used in sonar? **L2**

Development Before Birth

The zygote moves along the oviduct to the uterus. During this time, the zygote is dividing and forming a ball of cells. After about seven days, the ball of cells is implanted in the wall of the uterus. The uterine wall has been thickening in preparation to receive a fertilized egg. Here the fertilized egg will develop for nine months until the birth of the baby. This period of time is known as **pregnancy.**

The Embryo

During the first two months of pregnancy, the unborn child is known as an **embryo.** In **Figure 26-9,** you can see how the embryo receives nutrients and removes wastes. Nutrients from the wall of the uterus are received by the embryo through villi. Blood vessels develop from the villi and form the placenta. The umbilical cord is attached to the embryo's navel and connects with the placenta. The umbilical cord transports nutrients and oxygen from the mother to the baby through a vein. Carbon dioxide and other wastes are carried through arteries in the umbilical cord back to the mother's blood. Other substances in the mother's blood can pass to the embryo. These include drugs, toxins, and disease organisms. Because these substances can harm the embryo, a mother should avoid harmful drugs, alcohol, and tobacco during pregnancy.

During the third week of pregnancy, a thin membrane begins to form around the embryo. This is called the amnion or the amniotic sac and is filled with a clear liquid called amniotic fluid. The amniotic fluid in the **amniotic sac** helps cushion the embryo against blows and can store nutrients and wastes. This sac attaches to the placenta.

During the first three months of development, all of an embryo's major organs form. A heart structure begins to beat and move blood through the embryo's blood vessels. At five weeks, the embryo is only as large as a single grain of rice, but there is a head with recognizable eye, nose, and mouth features. During the sixth and seventh weeks, tiny arms and legs develop fingers and toes.

Figure 26-9

By two months, the developing embryo is beginning to look human.

 A The protective membranes and fluid cushion the embryo from blows.

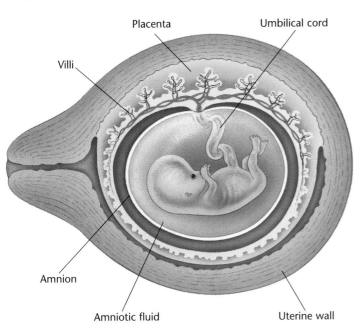

Placenta
Umbilical cord
Villi
Amnion
Amniotic fluid
Uterine wall

B The photograph shows an enlarged view of a two-month-old embryo. The actual size is approximately 2 cm.

MiniLAB

How does an embryo develop?

Procedure

1. Check the data in this table.

End of month	Length
3	8 cm
4	15 cm
5	25 cm
6	30 cm
7	35 cm
8	40 cm
9	51 cm

2. On a piece of paper, draw a line the length of the unborn baby at each date.
3. Using reference materials, find out what developmental events happen at each date.

Analysis

1. During which month does the greatest increase in length occur?
2. What size is the unborn baby when movement can be felt by the mother?

MiniLAB

Purpose

 Logical-Mathematical The students will observe data on the growth rate of a fetus and analyze changes in body proportions. L1

Materials

paper, pencil, metric ruler, reference materials

Teaching Strategies

Students can document other developmental occurrences at each month.

Analysis

1. from the end of the fourth month to the end of the fifth month
2. approximately 7-10 cm

Assessment

Performance To further assess students' understanding of fetus size, have them find and measure common household items that correspond to the fetus sizes. Set up a display of these items. Use the Performance Task Assessment List for Display in **PASC**, p. 63. P

📁 **Activity Worksheets,** pp. 5, 158

NATIONAL GEOGRAPHIC SOCIETY

 Videodisc

STV: Human Body Vol. 3

Embryo, 5 1/2 weeks (close-up)

48760

Fetus, 20 weeks

48769

USING TECHNOLOGY

Prenatal Surgery ▼ ▲

Until recently, babies born with certain deformities died soon after birth. One such deformity involves an opening in the diaphragm. The opening allows part of the intestine and other organs to protrude into the chest cavity. When this happens, the growth of the fetus's lungs is stunted, and the baby is unable to breathe when it is born. The usual procedure has been to operate on the baby immediately after birth. Unfortunately, this operation is successful only 20 percent of the time.

A New Procedure

Using new techniques, doctors are now able to operate on a fetus with this condition while it is still in the mother's uterus. A small incision is made in the mother's uterus. The fetus's chest is opened and the abdominal organs are returned to their proper place. Finally, a patch is placed over the hole in the diaphragm. The baby has a good chance of surviving after birth.

Prenatal surgery was unthinkable just a few years ago. Thanks to new methods, it is now performed to correct several abnormalities.

Think Critically:

Why are most prenatal surgeries done after at least the third month of development?

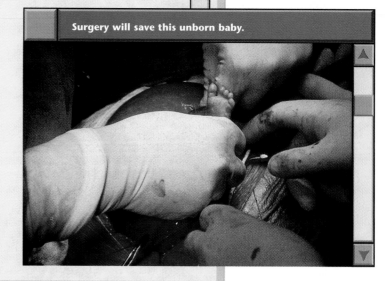
Surgery will save this unborn baby.

The Fetus

After the first two months of pregnancy, the developing baby is called a **fetus.** At this time, body organs are present. Around the third month, the fetus is 8-10 cm long, the heart can be heard beating, and the mother may feel the baby's movements within her uterus. The fetus sucks its thumb. By the fourth month, an ultrasound test can determine the sex of the fetus. By the end of the seventh month of pregnancy, the fetus is 30 to 38 cm in length. Fatty tissue builds up under the skin, and the fetus appears less wrinkled. By the ninth month, the fetus usually has shifted to a head-down position within the uterus. The head usually is in contact with the opening of the uterus to the vagina. The fetus is about 50 cm in length and weighs from 2.5 to 3.5 kg. The baby is ready for birth.

Multiple Births

Sometimes, two eggs leave the ovary at the same time. If both eggs are fertilized and both develop, twins are born. When two different eggs were fertilized by different sperm, the two babies are called fraternal twins. Fraternal twins may be two girls, two boys, or a boy and a girl. Fraternal twins do not always look alike. Not all twins are fraternal. Sometimes, a single egg splits apart shortly after it is fertilized. Each part of the egg then develops and forms an embryo. Because both children developed from the same egg and sperm, they are identical twins. Each has the same set of genes. Identical twins must be either two girls or two boys. You can see in **Figure 26-10** that one of these sets of twins looks exactly alike. Triplets and other multiple births may occur when either three or more eggs are produced at one time or an egg splits into three or more parts.

Figure 26-10

Pick out the fraternal twins and the identical twins.

Section Wrap-up

Review

1. What happens when an egg is fertilized?

2. What is one major event that occurs during the embryo and fetal stages?

3. Explain how twins are fraternal or identical.

4. **Think Critically:** If a mother is taking drugs, how can these harmful substances pass into the blood system of the embryo?

Skill Builder
Making and Using Graphs

Make a graph of the embryo and fetus sizes and the months of development using this data:
1 month = 0.5 cm, 2 months = 3 cm, 3 months = 8 cm, 4 months = 15 cm, 5 months = 25 cm, 6 months = 30 cm, 7 months = 35 cm, 8 months = 40 cm, 9 months = 50 cm. If you need help, refer to Making and Using Graphs in the **Skill Handbook.**

Science Journal

Write a poem in your Science Journal about how you would think, act, and feel if you were one half of a pair of twins.

Skill Builder

Students should make line or bar graphs. Time should be plotted on the *y*-axis and size on the *x*-axis.

Embryo and Fetus Sizes

Assessment

Performance Assess students' abilities to make and use graphs by having them make a statement about how the growth rate of a fetus compares with the growth rate of an embryo. Use the Performance Task Assessment List for Graph from Data in **PASC**, p. 39. **P**

•MINI•QUIZ•

1. **Of the millions of sperms deposited, how many will fertilize the egg?** *one*

2. **The _____ attaches the embryo to the placenta.** *umbilical cord*

3. **When does the embryo heart begin to beat?** *early in the first three months of development*

4. **The fetus usually has shifted to a head-down position by the _____ month.** *ninth*

Section Wrap-up

Review

1. Sperm and egg combine, producing a zygote with 46 chromosomes.

2. During the embryo stage, the amniotic sac and placenta develop and all the embryo's major organs form. During the fetal stage, the fetus continues to grow and develop.

3. Identical twins result when one fertilized egg splits and develops into two individuals. Fraternal twins are the result of two fertilized eggs.

4. **Think Critically** Since there is no direct connection between the fetus and the placenta of the mother, the exchange is due to diffusion between the capillaries of the umbilical cord and the placenta. **P**

Science Journal

Student answers will vary, but should reflect that not all thoughts, activities, and feelings would be the same as the other twin's. **P**

715

Prepare

Section Background

- The pressure and stretching of the uterine wall and an increase in progesterone causes contractions of the uterus.

- One of the most dramatic events for the newborn is the first breath. Air must be forced into the lungs because they are collapsed and the moist membranes are held together by surface tension.

- During the adolescent growth spurt, the hands and feet of teenagers may grow faster than the rest of the body. Eventually the arms, legs, and chest also grow, bringing the body into proportion.

Preplanning

- To prepare for Activity 26-2, obtain graph paper and red and blue pencils.

- To prepare for the Mini-LAB, obtain a list of the legally required immunizations for your state.

1 Motivate

Bellringer

Before presenting the lesson, display **Section Focus Transparency 96** on the overhead projector. Assign the accompanying **Focus Activity** worksheet. L1 LEP

Activity

Have students conduct a survey of parents with young children. Ask questions such as: What do you enjoy most about your children? How difficult is it to care for your children? L1

26•3 Development After Birth

Science Words

> infancy
> childhood
> adolescence
> adulthood

Objectives

- State the sequence of events of childbirth.
- Compare the stages of infancy and childhood.
- Relate adolescence to preparation for adulthood.

Childbirth

After nine months of developing within the mother, the baby is ready to be born. Like all newborn living things, the baby finds itself suddenly pushed out into the world. Within the uterus and amniotic sac, the baby was in a warm, watery, dark, and protected environment. In contrast, the new environment is cooler, drier, brighter, and not as protective.

Labor

The process of childbirth begins with labor, the muscular contractions of the uterus. In **Figure 26-11,** you can see one way some parents prepare for these labor contractions. As the contractions increase in strength and frequency, the amniotic sac usually breaks and releases its fluid. Over a period of hours, the contractions cause the opening of the uterus to widen to allow the baby to pass through. More powerful and frequent contractions push the baby out through the vagina into its new environment. Sometimes, the mother's pelvis is too small for the baby to fit through or the baby is in the wrong position for birth. In cases like this, the baby is delivered through an incision in the mother's uterus and abdomen. This surgery is called cesarean section.

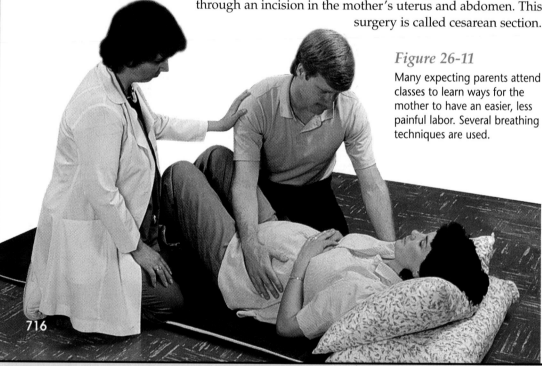

Figure 26-11

Many expecting parents attend classes to learn ways for the mother to have an easier, less painful labor. Several breathing techniques are used.

716

Program Resources

📁 **Reproducible Masters**

Study Guide, p. 100 L1
Reinforcement, p. 100 L1
Enrichment, p. 100 L3
Cross-Curricular Integration, p. 30
Multicultural Connections, pp. 55-56
Activity Worksheets, pp. 5, 156-157, 159 L1

 Transparencies

Section Focus Transparency 96 L1
Science Integration Transparency 26

Post Labor

At birth, the baby is still attached to the umbilical cord and placenta. The person assisting with the birth of the baby ties the cord and then cuts it. The baby takes its first breath of oxygen without the umbilical cord. It may also cry. Crying forces air into its lungs. The scar that later forms where the cord was attached is the navel. Soon after the baby's delivery, contractions expel the placenta from the mother's body.

Infancy and Childhood

The first four weeks after birth are known as the neonatal period. *Neonatal* means "newborn." In **Figure 26-12,** you can see that during this time, the baby adjusts to life outside of the uterus. Body functions such as respiration, digestion, and excretion are now performed by the baby rather than through the placenta. Unlike some other living things, the human baby must depend on others to survive. A newborn colt begins walking a few hours after its birth. The human baby is fed and has its diaper changed. It is not able to take care of itself.

Infancy

The next stage of development is **infancy,** the period from the neonatal stage to one year. It is a period of rapid growth and development for both mental and physical skills. An early skill is the ability to smile, which usually begins around six weeks of age. At four months, most babies can laugh, sit up when propped, and recognize their mother's faces. At eight months, the infant is usually able to say a few simple words, such as "mama" and "dada." One of the major events within the first year is the ability to stand unsupported for a few seconds.

USING MATH

Cotton diapers can be reused up to 100 times before having to be replaced. In your Science Journal, calculate how many diapers you would need for a year if an average of eight are used per day.

Figure 26-12

Development of the five senses occurs before birth. Newborns are attracted to bright colors and sweet tastes.

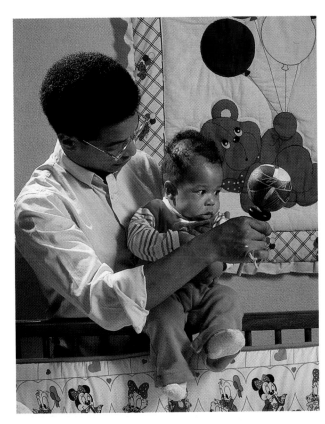

26-3 Development After Birth   **717**

2 Teach

USING MATH

Students should first determine how many diaper changes take place in a year. $365 \times 8 = 2920$. Then divide this total by 100 to find the number of diapers. (30)

Discussion

Discuss how the newborn infant must adjust to the new environment into which it has been born. Have students compare the conditions of the two environments.

Visual Learning

Figure 26-12 Based on the information presented in the caption for this photograph, have students develop a color scheme to use in decorating a newborn baby's room. Discuss which colors would be the most stimulating for the baby. **LEP** **LS**

Inclusion Strategies

Gifted Have students interview K-4 elementary school teachers to determine the physical and mental abilities of children at each level. Students can use a chart to present their findings on the developmental progression of individuals through childhood. Have them check their findings with references on childhood development. **L3**

Community Connection

Field Trip Arrange a visit to an X-ray lab to view X rays of bones of infants or young children and compare them with X rays of adults to show the difference in cartilage and bone ratios. **LS**

MiniLAB

What immunizations are given to babies and young children?

Find out whether you have all of your immunizations.

Procedure

1. Find out what immunizations are usually given to babies and young school children.
2. Compare these to your own immunization record.

Analysis

1. What boosters were you given?
2. Which vaccines are required for children to enter school?

Figure 26-13

Teens experience many changes in their physical and mental abilities. The changes sometimes result in a roller-coaster ride of emotions as they grow into young adults.

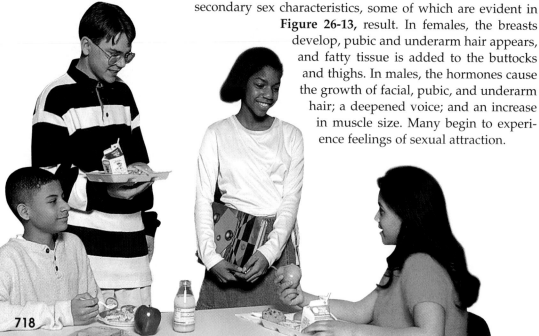

718

Childhood

After infancy is the **childhood** stage, which lasts until age 12. The physical growth rate for height and weight is not as rapid as in infancy. However, there is development of muscular coordination and mental abilities. By 18 months, the child is able to walk without help. Between two and three years, the child controls his or her bladder and bowel. At age three, the child can speak in simple sentences. By age five, many children can read a limited number of words. Throughout this stage, children develop their abilities to speak, read, write, and reason. At the same time, children also mature emotionally and learn how to get along with other people. Find out how old you were when you began to talk. What were your first words?

Adolescence

The next stage of development is adolescence; you are in this stage. **Adolescence** begins around ages 12 to 14. A part of adolescence is puberty. As you read earlier, puberty is the time of development when a person becomes physically able to reproduce. For girls, puberty occurs between ages 8 and 13. For boys, puberty occurs between ages 13 and 15.

During puberty, hormones that cause changes in the body are produced by the pituitary gland. The hormone FSH helps produce reproductive cells. LH helps with the production of sex hormones. As a result of the secretion of these hormones, secondary sex characteristics, some of which are evident in **Figure 26-13,** result. In females, the breasts develop, pubic and underarm hair appears, and fatty tissue is added to the buttocks and thighs. In males, the hormones cause the growth of facial, pubic, and underarm hair; a deepened voice; and an increase in muscle size. Many begin to experience feelings of sexual attraction.

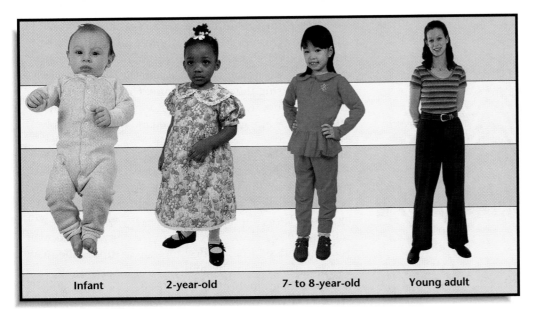

| Infant | 2-year-old | 7- to 8-year-old | Young adult |

The Growth Spurt

Adolescence is a time of your final growth spurt. Are you shorter or taller than your classmates? Because of differences in the time hormones begin functioning among individuals and between males and females, there are differences in boys' and girls' growth rates. Girls often begin their increase in growth between the ages of 11 and 13 and end at ages 15 to 16. Boys usually start their growth spurt at ages 13 to 15 and end at 17 to 18 years of age. Hormonal changes also cause underarm sweating and acne, requiring extra cleanliness and care. All of these physical changes can cause you to feel different or uncomfortable. This is normal. As you move through the period of adolescence, you will find that you will become more coordinated, be better able to handle problems, and gain improved reasoning abilities. You'll find that you enjoy spending more time with your friends.

Adulthood

The final stage of development is that of **adulthood.** It begins with the end of adolescence and extends to old age. Young adults are people in their 20s. Many young adults are completing an education, finding employment, and possibly marrying and beginning a family. This is when the growth of the muscular and skeletal system stops. You can see in **Figure 26-14** how body proportions change as you grow from infancy to adulthood.

Figure 26-14

You can compare body proportions of newborns, young children, teens, and adults with this chart. *How does the proportion of the head compared to the rest of the body change from baby to adult?*

USING MATH

At birth, the head of a newborn is about one-quarter of the total body length. Measure the head length and height of at least ten classmates. Calculate the ratio of the head length to body length for adolescents.

26-3 Development After Birth   **719**

USING MATH

Differential growth causes the head to grow less rapidly after birth, until the length of the head is approximately 1/8 of the body length.

GLENCOE TECHNOLOGY

 Videodisc

The Infinite Voyage Series: The Champion Within
Chapter 8
Physical Training and Senior Citizens

Use **Cross-Curricular Integration,** p. 30, as you teach this lesson.

3 Assess

Check for Understanding

? FLEX Your Brain

Use the Flex Your Brain activity to have students explore ADOLESCENCE.

Activity Worksheets, p. 5

Reteach

Have students find old photographs of children. Compare the physical development of adolescents 100 years ago with that of today's adolescents. Infer what factors might have caused puberty to begin later then than is common today.

Extension

For students who have mastered this section, use the **Reinforcement** and **Enrichment** masters.

Across the Curriculum

Health Have students find out about changes in the bones of older adults and the condition of kyphosis. **L2**

Cultural Diversity

Rites of Passage Have students find out about ceremonies in other societies, past or present, that mark the transition of young people to adults. **L2**

4 Close

Have students discuss the extension of the life span of humans due to improved health care.

Section Wrap-up

Review

1. Events include contractions, rupture of the amniotic sac, widening of the opening of the uterus to allow the baby to pass through that opening and the birth canal, and expulsion of the placenta.

2. Infancy is from age one month to one year and involves rapid growth and development of both mental and physical skills. Childhood is from age one year to age 12. There is development of muscular coordination and mental abilities.

3. Adolescence is a transition time. It is a time of physical growth to adult size and sexual maturity.

4. **Think Critically** Each person has his or her own rate and schedule of development.

Science Journal

Have students highlight the commonalities and differences between their lives and those of the people they interview.

Skill Builder

A. Neonatal
    1. first four weeks
  2. adjusts to life outside uterus
B. Infancy
  1. neonatal to one year
  2. rapid growth, development
C. Childhood
  1. ages 1 to 12
  2. development of muscular and mental abilities

Figure 26-15
Many senior citizens remain active both physically and intellectually.

People from age 30 to age 60 are in the stage of middle adulthood. During these years, physical strength begins to decline. Blood circulation and respiration become less efficient. Bones become more brittle, and the skin's elastic tissues are lost, causing the skin to become wrinkled. People in this group are busy with family and work commitments. They often care for aging parents as well as children.

Senior Citizens

Think about someone you know over age 65. How is that person like or different from you? Around this age, many people retire. In **Figure 26-15,** you see that they take up hobbies, travel, or volunteer at hospitals and community organizations. They are an active part of society. People over 75 years are placed in the oldest old-age group. While many in this group are mentally and physically active, others may need assistance in meeting their needs.

After birth, the stages of development of the human body begin with a baby making adjustments to a new environment. From infancy to adolescence, the body's systems mature, enabling the person to be physically and mentally ready for adulthood. During the next 50 years or longer, people live their lives making contributions to family, community, and society. Throughout the life cycle, people who care for their health enjoy a higher quality of life.

Section Wrap-up

Review

1. What are major events during childbirth?

2. Compare infancy with childhood.

3. How does the period of adolescence prepare you for the stage of adulthood?

4. **Think Critically:** Why is it hard to compare the growth and development of different adolescents?

Skill Builder
Outlining

Prepare an outline of the various life stages of human development from the neonatal period to adulthood. If you need help, refer to Outlining in the **Skill Handbook.**

Science Journal

Interview someone who's over 60 years of age. Find out what his or her life was like when he or she was your age. What does the person like to do now? Write a report in your Science Journal to share with the class.

D. Adolescence
  1. ages 12 to 18
  2. person becomes physically able to reproduce
E. Adulthood
  1. end of adolescence to old age
  2. several stages
    a. early adult—20s
    b. middle adult—30s to 50s
    c. young-old—65 to 74 years
    d. old-old—75 years +

✔ Assessment

Portfolio Use this Skill Builder to assess students' abilities to sequence the stages of human development. Use the Performance Task Assessment List for Events Chain in **PASC,** p. 91. **P**

Activity 26-2

Average Growth Rate in Humans

An individual's growth is dependent on both the effects of hormones and his or her genetic makeup.

Problem
Is average growth rate the same in males and females?

Materials
- graph paper
- red and blue pencils

Procedure
1. Construct a graph similar to graph A below. Plot mass on the vertical axis and age on the horizontal axis.
2. Plot the data given under Data and Observations for the average female growth in mass from ages 8 to 18. Connect the points with a red line.
3. On the same graph, plot the data for the average male growth in mass from ages 8 to 18. Connect the points with a blue line.
4. Construct a separate graph similar to graph B below. Plot height on the vertical axis and age on the horizontal axis.
5. Plot the data for the average female growth in height from ages 8 to 18. Connect the points with a red line. Plot the data for the average male growth in height from ages 8 to 18. Connect the points with a blue line.

Data and Observations

Age	Mass (kg)		Height (cm)	
	Female	Male	Female	Male
8	25	25	123	124
9	28	28	129	130
10	31	31	135	135
11	35	37	140	140
12	40	38	147	145
13	47	43	155	152
14	50	50	159	161
15	54	57	160	167
16	57	62	163	172
17	58	65	163	174
18	58	68	163	178

Analyze
1. Up to what age is average growth in mass similar in males and females?
2. Up to what age is average growth in height similar in males and females?

Conclude and Apply
3. When does the mass of females change most?
4. How can you explain the differences in growth between males and females?
5. **Interpret** the data to find whether the average growth rate is the same in males and females.

A   Average growth in mass

B   Average growth in height

26-3 Development After Birth   **721**

Activity 26-2

Purpose
Logical-Mathematical Graph the average growth rate in males and females, and determine whether there are differences in the average growth rates of males and females. **L1**

Process Skills
interpreting data, comparing and contrasting, using numbers, and making and using graphs

Time
30 minutes

Materials
graph paper, red and blue pencils

Activity Worksheets, pp. 5, 156-157

Teaching Strategies
Have students refer to Making and Using Graphs in the Skill Handbook for help in making a line graph.

Answers to Questions
1. up to about age 11
2. up to about age 12
3. Females increase the most in mass between the ages of 11 and 13.
4. Answers may vary slightly. Most students will say that girls enter adolescence at an earlier age than boys.
5. Average growth is the same until puberty. At puberty, girls have an earlier growth spurt than boys, but on the average boys grow taller than girls.

✔ Assessment

Performance To further assess students' understanding of growth, have them correlate the range of heights in their class to the statistical average. Use the Performance Task Assessment List for Using Math in Science in **PASC**, p. 29. **P**

721

Prepare

Section Background

Have students take a survey of parents, relatives, or friends to find out about persons over 80 years of age who are still active at work and in social activities. Compare numbers.

1 Motivate

Bellringer

Before presenting the lesson, display **Section Focus Transparency 97** on the overhead projector. Assign the accompanying **Focus Activity** worksheet. L1 LEP

Tying to Previous Knowledge

Osteoporosis is a disease characterized by the deterioration of bone. A diet lacking in calcium can cause osteoporosis because calcium is necessary for normal bone cell replacement.

2 Teach

Discussion

Students may not be familiar with the effects of diseases such as Alzheimer's disease, arthritis, osteoporosis, or diabetes. A brief discussion of each of these will help students.

SCIENCE & SOCIETY

TECHNOLOGY:
26•4 A New View of the Old

Science Words
centenarian

Objectives
- Compare the common view of aging with the new view of aging.
- Describe factors that influence aging.

inter NET CONNECTION

The American Association of Retired Persons is a non-profit organization dedicated to helping older Americans. Visit the Chapter 26 Internet Connection at Glencoe Online Science, **www.glencoe.com/sec/science/life**, for a link to more information about the AARP.

Figure 26-16

Predictions are that in about 40 years, the population of people 85 and over in the United States may reach 30 million.

722

Traditional Views

Imagine that you are over 90 years old. How would you describe yourself? When asked, some people describe the elderly as forgetful and physically impaired. They associate conditions such as arthritis, osteoporosis, diabetes, cancer, heart disease, and Alzheimer's disease with the aging process. They think that older people have less energy and mental ability than younger members of society. Limitations on hearing and sight also are considered common.

New Findings

The traditional view of the wellness of the elderly has changed somewhat with recent studies. When persons over 90 were studied, it was found that the majority of them were mentally alert and physically active. They suffered from fewer of the health problems that afflicted many people in their seventies and eighties. **Centenarians,** people a hundred years old and over, were found to be active in swimming, golf, tennis, bowling, and other physical activities. Some remained active in writing, art, reading, or even business operations.

Program Resources

Reproducible Masters
Study Guide, p. 101 L1
Reinforcement, p. 101 L1
Enrichment, p. 101 L3
Science and Society Integration, p. 30

Transparencies
Section Focus Transparency 97 L1

Longevity Genes

The study of the lives of people who are unusually long-lived has sought the causes for their good health. Researchers believe that certain genes are responsible for the aging process. However, genes are also responsible for a person's physical and mental abilities to deal with disease or injury. Scientists think that genes also have a role in how much functional reserve a person's organs have. Centenarians may have extensive areas of damaged brain cells, but they also have enough good cells in reserve to allow them to function rather normally. Obviously, good health habits, **Figure 26-17,** are also contributing factors to a good, long life.

A New View

The new findings of research on the physical and mental wellness of the very old has implications for health care. Traditionally, health care programs have assumed that more and more resources would be needed as a person grows older and older. However, it may be that because of the overall good health of people over 90, the health maintenance cost for a person in this age group may not be as staggering as was envisioned.

Figure 26-17

Researchers have found that an older person can do much to promote his or her health. This includes eating right, exercising regularly, monitoring stress, and not smoking.

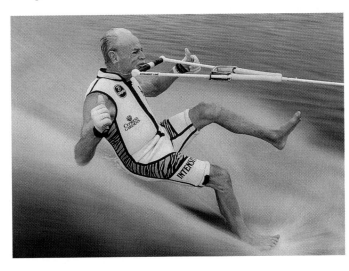

Section Wrap-up

Review

1. How do the old and new views of aging compare?

2. Describe how genes may affect the process of aging.

Explore the Technology

Health maintenance costs for the oldest of the old may be less than anticipated. What factors are responsible for this? What can you do now to help increase your years of good health?

SCIENCE & SOCIETY

Explore the Technology

The assumption that the cost of a person's medical care increases each year the person lives may not be true. The yearly medical bills of a 95-year-old person may not be higher than those of a 75- or 85-year-old person. In fact, the 95-year-old may be healthier because living that long could be due to following these guidelines: not smoking, eating nutritious foods, and staying mentally and physically active, as well as having the genes for long life. Students can eat nutritious food, get exercise, and avoid smoking.

Background

- The profession of midwifery (pronounced mid-WIF-ery) has existed for centuries. In the United States it received considerable public attention in the 1920s, when midwives were imported from England to work in rural Appalachia, where infant mortality and maternal death rates due to childbirth were especially high. Because these women often traveled by horseback into remote areas, they became known as "saddlebag midwives." Not only did they save the lives of countless mothers and babies, but they also provided much-needed public health and nutrition education.

- The are currently about 5000 certified nurse midwives practicing in the United States. The great majority are women, although the number of men entering the profession is on the increase. The professional organization to which nurse midwives belong is the American College of Nurse Midwives, which was founded in 1955. Ms. Ambrose is the current chairperson of Nurse Midwives of Color, an official committee of the ACNM that was formed to address the needs of midwives of color and to encourage sensitivity within the profession to ethnic and cultural issues.

Teaching Strategies

As a group, have students visit a local hospital to talk with a delivery room nurse or nurse midwife, and tour a delivery room. Afterwards, have them write an essay describing their experience.

People and Science

CAROL AMBROSE, *Nurse Midwife*

On the Job

Q Ms. Ambrose, what is a nurse midwife?

A Nurse midwifery is an advanced nursing specialty. A nurse midwife is a nurse with maternity nursing experience who has additional education and training in midwifery, that is, in the art of overseeing pregnancies and delivering babies. Nurse midwives work in a variety of settings, from hospitals to private practice.

Q As a nurse midwife who works in a private practice with a physician, could you describe a "typical" day?

A Well, a typical day for me would include regular office hours, during which I see pregnant clients and give prenatal care. Women also come for general reproductive health care, such as pap tests and breast exams. So there is a regular routine. But I'm also on call for patients who are close to delivering. When a woman does go into labor, I meet her at the hospital, take care of her during the labor process, and help her to deliver the baby. I then take care of both mother and baby afterward.

Personal Insights

Q What do you like best about your job?

A Developing close relationships with my patients, helping them through their entire pregnancy, and then being there to help deliver their babies. It's a real thrill to help someone you know well. And because there tends to be more family involvement with women who seek midwifery care, I not only know my patients, but I know their husbands and their other kids too!

Career Connection

In the business listings section of the phone book, look up the hospitals in your area. Choose one and contact its Education Department, Human Resources Department, or Career Information Office to learn more about the following health care careers, and ask about opportunities to "shadow" people in these positions for a day:

- **Delivery Room Nurse**
- **Phlebotomist**
- **Nurse Anesthetist**

724 Chapter 26 Reproduction and Growth

Career Connection

Career Path High School: science and math

College: Nursing degree

Graduate School: 1-2 year certification program in midwifery

For More Information

Students should contact the American College of Nurse Midwives in Washington, DC for more information about careers in nurse midwifery and schools that offer midwifery certification programs.

Summary

26-1: Human Reproduction

1. The reproductive system allows new organisms to be formed.
2. The testes produce sperm.
3. The female ovary produces an egg that can be fertilized.
4. When an egg is not fertilized, it disintegrates; the lining of the uterus is shed.

26-2: Fertilization to Birth

1. After fertilization, the zygote undergoes developmental changes to become an embryo and then a fetus.
2. The uterine muscles contract to push the baby out of the mother.
3. Twins occur when two eggs are fertilized or when a single egg splits after fertilization.

26-3: Development After Birth

1. Infancy is the stage of development from the neonatal period to one year. Childhood, which lasts to age 12, is marked by development of muscular coordination and mental abilities.
2. Adolescence is the stage of development when a person becomes physically able to reproduce. The final stage of development is adulthood.

26-4: Science and Society: A New View of the Old

1. Many conditions are associated with aging.
2. Certain genes are believed to be responsible for aging and for a person's adaptive ability and functional reserve.

Key Science Words

a. adolescence
b. adulthood
c. amniotic sac
d. centenarian
e. childhood
f. embryo
g. fetus
h. infancy
i. menopause
j. menstrual cycle
k. menstruation
l. ovary
m. ovulation
n. pregnancy
o. semen
p. sperm
q. testis
r. uterus
s. vagina

Understanding Vocabulary

Match each phrase with the correct term from the list of Key Science Words.

1. birth canal
2. male sex cells
3. egg-producing organ
4. release of egg from the ovary
5. nourishing fluid for sperm
6. ending of the menstrual cycle with age
7. sperm-producing organ
8. place where a fertilized egg develops into a baby
9. membrane that protects the unborn baby
10. name for the unborn child the first two months

Chapter 26 Review **725**

Chapter 26

Review

Summary

Have students read the summary statements to review the major concepts of the chapter.

Reviewing Vocabulary

1. s	6. i
2. p	7. q
3. l	8. r
4. m	9. c
5. o	10. f

✔ Assessment

Portfolio Encourage students to place in their portfolios one or two items of what they consider to be their best work. Examples include:
- Across the Curriculum, p. 711
- Think Critically in Section Wrap-up, p. 715
- Science Journal, p. 715 **P**

Performance Additional performance assessments may be found in **Performance Assessment** and **Science Integration Activities**. Performance Task Assessment Lists and rubrics for evaluating these activities can be found in Glencoe's **Performance Assessment in the Science Classroom**.

GLENCOE TECHNOLOGY

 MindJogger Videoquiz

Chapter 26 Have students work in groups as they play the Videoquiz game to review key chapter concepts.

Chapter 26 Review

Checking Concepts

1. c   **6.** c
2. b   **7.** d
3. a   **8.** c
4. a   **9.** b
5. b   **10.** a

Understanding Concepts

11. Sperm need to travel a great distance to fertilize the egg; therefore, many are needed so that one might make it to the egg.

12. Semen protects, nourishes, and helps the sperm move.

13. The lining provides a space for the zygote to attach.

14. Identical twins = one egg fertilized by one sperm: zygote splits after fertilization. Fraternal twins = two eggs fertilized by two different sperm.

15. The woman's body has a thickened uterine wall; the placenta develops to transport food and wastes through the umbilical cord; amniotic fluid cushions the embryo.

Thinking Critically

16. Both are paired organs and produce sex cells.

17. Meiosis occurs in the ovaries; ovulation occurs in the ovaries; fertilization occurs in the oviduct; implantation occurs in the uterus.

18. Embryo is the stage of development for the first two months; it is small and undergoes beginning development of all structures. The fetus is the developmental stage from two months until birth; it is more developed and its movement can be felt.

19. Mitosis continues to produce new cells.

Checking Concepts

Choose the word or phrase that completes the sentence.

1. The embryo develops in the _____.
a. oviduct   c. uterus
b. ovary   d. vagina

2. The monthly process of egg release is called _____.
a. fertilization   c. menstruation
b. ovulation   d. puberty

3. The union of an egg and sperm is _____.
a. fertilization   c. menstruation
b. ovulation   d. puberty

4. The egg is fertilized in the _____.
a. oviduct   c. vagina
b. uterus   d. ovary

5. Mental and physical skills rapidly develop during _____.
a. neonatal period   c. adulthood
b. infancy   d. adolescence

6. Puberty occurs during _____.
a. childhood   c. adolescence
b. adulthood   d. infancy

7. Sex characteristics common to males and females include _____.
a. breasts   c. increased fat
b. increased muscles   d. pubic hair

8. Growth stops during _____.
a. childhood   c. adulthood
b. adolescence   d. infancy

9. The period of development with three stages is _____.
a. infancy   c. adolescence
b. adulthood   d. childhood

10. The ability to reproduce begins at _____.
a. adolescence   c. childhood
b. adulthood   d. infancy

Understanding Concepts

Answer the following questions in your Science Journal using complete sentences.

11. Why are so many sperm released if only one is needed to fertilize an egg?

12. Why is semen necessary for sperm survival?

13. What is the purpose of the thickened uterine lining?

14. Explain the difference between identical and fraternal twins.

15. What features of a pregnant woman protect the developing child?

Thinking Critically

16. Explain the similar functions of the ovaries and testes.

17. Identify the structure in which each process occurs: meiosis, ovulation, fertilization, and implantation.

18. Describe the structural differences between embryo and fetus.

19. What kind of cell division occurs as the zygote develops?

20. Describe one major change in each stage of human development.

20. Neonatal: needs to be taken care of

Infancy: rapid growth and development of mental and physical skills: speak, stand

Childhood: development of mental and muscle coordination: walk, talk, read, reason

Adolescence: puberty—ability to reproduce; secondary sex characteristics and growth; increased reasoning

Adulthood: three stages; growth of skeleton and muscles stops

Developing Skills

If you need help, refer to the **Skill Handbook.**

21. **Classifying:** Classify each structure as female or male and internal or external: ovary, penis, scrotum, testes, uterus, and vagina.

22. **Concept Mapping:** Fill in the concept map of egg release.

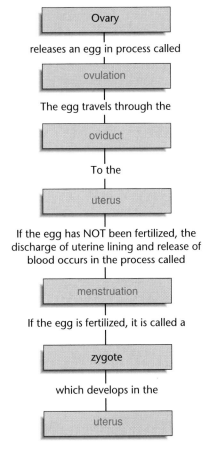

Ovary

releases an egg in process called

ovulation

The egg travels through the

oviduct

To the

uterus

If the egg has NOT been fertilized, the discharge of uterine lining and release of blood occurs in the process called

menstruation

If the egg is fertilized, it is called a

zygote

which develops in the

uterus

23. **Hypothesizing:** Make a hypothesis about the effects of raising identical twins apart from each other.

24. **Making and Using Graphs:** Use the data to make a graph showing the week of development versus size of the embryo. When is the fastest period of growth?

Embryo Size During Development

Week After Fertilization	Size
3	3 mm
4	6 mm
6	12 mm
7	2 cm
8	4 cm
9	5 cm

25. **Sequencing:** Sequence the steps involved in the birth process.

Performance Assessment

1. **Letter:** Find newspaper articles on the effects of smoking on the health of the developing embryo and newborn. Write a letter to the editor about why a mother's smoking is damaging to her unborn baby's health.

2. **Making a Model:** Put a soft fruit or vegetable (plum, peach, tomato) into a self-sealing plastic bag filled with water. Use this model to observe how the fetus is protected from shock by the amniotic fluid.

3. **Booklet or Pamphlet:** Make a booklet or pamphlet describing the special health care necessary for a pregnant woman.

Developing Skills

21. **Classifying** Female internal: ovary, uterus, vagina
   Male external: penis, scrotum, testes

22. **Concept Mapping** See student page.

23. **Hypothesizing** Students might hypothesize that they will be the same because of their genes or they might say the environment will have an effect, making them different from each other. Research now indicates that the genes rule.

24. **Making and Using Graphs** Students should make line or bar graphs with Week After Fertilization on the *x*-axis and Size on the *y*-axis. The most rapid period of growth occurs between weeks 7 and 8.

25. **Sequencing** Labor—amniotic sac breaks—baby moves out of uterus and vagina.

Performance Assessment

1. Responses will vary but should include information contained on page 712 of the student text. Use the Performance Task Assessment List for Letter in **PASC**, p. 67. P

2. Students should see that the fruit or vegetable is protected up to a point. A very hard blow damages it. Use the Performance Task Assessment List for Model in **PASC**, p. 51. P

3. Booklets can address diet, medical care, exercise, and other issues. Use the Performance Task Assessment List for Booklet or Pamphlet in **PASC**, p. 57. P

Assessment Resources

 Reproducible Masters
Chapter Review, pp. 55-56
Assessment, pp. 163-166
Performance Assessment, p. 40

Glencoe Technology
- **Chapter Review Software**
- **Computer Test Bank**
- **MindJogger Videoquiz**

Chapter Organizer

Section	Objectives/Standards	Activities/Features
Chapter Opener		**Explore Activity:** How are disease-causing organisms spread? p. 729
27-1 **Disease** (1 session, 1 block)*	1. **Describe** the work of Pasteur, Koch, and Lister in the discovery and prevention of disease. 2. **List** diseases caused by viruses and bacteria. 3. **Discuss** sexually transmitted diseases (STDs), their causes, and treatments. **National Content Standards:** (5-8) UCP1, A1, A2, B1, C3, F1, G1, G3	**Connect to Chemistry,** p. 732 **MiniLAB:** Are bacteria present in food? p. 733 **MiniLAB:** How fast do bacteria reproduce? p. 734 **Skill Builder:** Recognizing Cause and Effect, p. 735 **Science Journal,** p. 735 **Activity 27-1:** Microorganisms and Disease, p. 736
27-2 **Your Immune System** (2 sessions, 2 blocks)*	1. **Explain** the natural defenses your body has against disease. 2. **Describe** differences between active and passive immunity. 3. **Explain** how HIV affects the immune system. **National Content Standards:** (5-8) UCP1, A1, A2, C3, F1, G3	**Problem Solving:** Fighting TB, p. 739 **Skill Builder:** Comparing and Contrasting, p. 740 **Using Computers:** Flowchart, p. 740 **Science and History:** Ethics and the Hippocratic Oath, p. 741 **Science Journal,** p. 741 **Activity 27-2:** Stopping Growth of Microorganisms, pp. 742-743
27-3 **Science and Society Issue: Should a harmful virus be destroyed?** (1 session, ½ block)*	1. **Explain** the role of vaccines in preventing certain diseases. 2. **Discuss** whether a harmful virus should be destroyed. **National Content Standards:** (5-8) F4, G1	**Explore the Issue,** p. 745
27-4 **Noncommunicable Disease** (2 sessions, 1 block)*	1. **Identify** two noncommunicable diseases. 2. **Describe** the basic characteristics of cancer. 3. **Explain** what happens during an allergic reaction. **National Content Standards:** (5-8) A1, B1	**Using Math,** p. 748 **Using Technology:** Shots for Good Health, p. 749 **Skill Builder:** Making and Using Tables, p. 750 **Science Journal,** p. 750

* A complete Planning Guide that includes block scheduling is provided on pages 32T-35T.

Activity Materials

Explore	Activities	MiniLABs
page 729 dropper, peppermint food flavoring, cotton ball	page 736 6 fresh apples, rotting apple, alcohol, 6 self-sealing plastic bags, labels, marking pencil, paper towels, sandpaper, cotton ball, antibacterial soap, water pages 742-743 5 sterile petri dishes, 2-cm squares of filter paper, test chemicals (disinfectant, hydrogen peroxide, mouthwash, rubbing alcohol), transparent tape, labels, marking pencil, scissors, metric ruler, forceps, 4 small jars for the test chemicals, apron	page 733 four different food samples, 4 identical test tubes, tap water, droppers, methylene blue, mineral oil, warm water bath page 734 graph paper

Need Materials? Call Science Kit (1-800-828-7777).

Teacher Classroom Resources

Reproducible Masters	Transparencies	Teaching Resources
Study Guide, p. 102 Reinforcement, p. 102 Enrichment, p. 102 Activity Worksheets, pp. 160-161, 164, 165 Lab Manual 61, Communicable and Noncommunicable Diseases Lab Manual 62, Sexually Transmitted Diseases Concept Mapping, p. 59	Section Focus Transparency 98, Germs, Germs, Go Away Teaching Transparency 53, Koch's Postulates	Glencoe Life Science Interactive Videodisc Life Science CD-ROM Spanish Resources English/Spanish Audiocassettes Cooperative Learning Resource Guide Lab Partner Lab and Safety Skills Lesson Plans
Study Guide, p. 103 Reinforcement, p. 103 Enrichment, p. 103 Activity Worksheets, pp. 162-163 Critical Thinking/Problem Solving, p. 31 Cross-Curricular Integration, p. 31 Science and Society Integration, p. 31 Multicultural Connections, pp. 57-58 Science Integration Activity 27, Acids, Bases, and Indicators	Section Focus Transparency 99, Disease Statistics Teaching Transparency 54, Four Stages of Immunity Production	**Assessment Resources** Chapter Review, pp. 57-58 Assessment, pp. 167-170 Performance Assessment in the Science Classroom (PASC) MindJogger Videoquiz Alternate Assessment in the Science Classroom Performance Assessment Chapter Review Software Computer Test Bank
Study Guide, p. 104 Reinforcement, p. 104 Enrichment, p. 104	Section Focus Transparency 100, Diseases of the Past	
Study Guide, p. 105 Reinforcement, p. 105 Enrichment, p. 105 Lab Manual 63, Effects of Nicotine Lab Manual 64, Effects of Caffeine	Section Focus Transparency 101, Preventing Illness Through Safe Work Habits Science Integration Transparency 27, Poisons Around Us	

Key to Teaching Strategies

The following designations will help you decide which activities are appropriate for your students.

L1 Level 1 activities should be within the ability range of all students, including those with learning difficulties.

L2 Level 2 activities should be within the ability range of the average to above-average student.

L3 Level 3 activities are designed for the ability range of above-average students.

LEP LEP activities should be within the ability range of Limited English Proficiency students.

LS These activities are designed to address different learning styles.

COOP LEARN Cooperative Learning activities are designed for small group work.

P These strategies represent student products that can be placed into a best-work portfolio.

GLENCOE TECHNOLOGY

The following multimedia resources are available from Glencoe.

The Secret of Life
Nothing to Sneeze At: Viruses
The Infinite Voyage Series
The Living Clock

National Geographic Society Series
STV: Human Body Vol. 3
Life Science CD-ROM

Teacher Classroom Resources

This is a representation of key blackline masters available in the Teacher Classroom Resources.

Teaching Aids

Section Focus Transparencies

Science Integration Transparencies

Teaching Transparencies

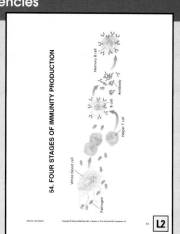

Meeting Different Ability Levels

Study Guide

Reinforcement

Enrichment Worksheets

Chapter 27 Immunity

Hands-On Activities

Science Integration Activity

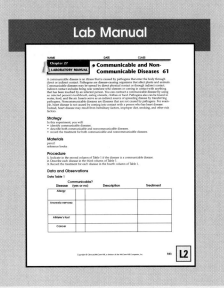

L1

Lab Manual

L2

Assessment

Performance Assessment

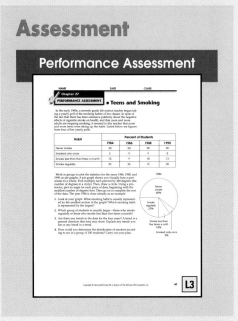

L3

Enrichment and Application

Critical Thinking/Problem Solving

L2

Cross-Curricular Integration

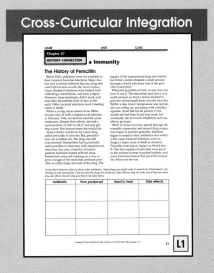

L1

Science and Society Integration

L2

Multicultural Connections

L2

Concept Mapping

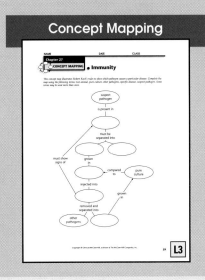

L3

Immunity

CHAPTER OVERVIEW

Section 27-1 This section describes the discovery of the causes of disease. There is a special focus on sexually transmitted diseases.

Section 27-2 The mechanisms of active and passive immunity are introduced. The action of the AIDS virus on the immune system is highlighted.

Section 27-3 Science and Society The role of vaccination in disease prevention is discussed. Students consider whether or not the last samples of disease-causing organisms should be destroyed.

Section 27-4 This section explores characteristics of noncommunicable diseases. There is a special emphasis on cancer.

Chapter Vocabulary

pasteurization
disinfectant
antiseptic
communicable
  disease
sexually
  transmitted
  disease
  (STD)
immune
  system
antigen
antibody
active
  immunity

passive
  immunity
lymphocyte
vaccination
noncommu-
  nicable
  disease
chronic
  disease
cancer
tumor
chemo-
  therapy
allergy
allergen

Theme Connection

Systems and Interactions One major theme of science is interaction among systems. This chapter focuses on interactions within the body. Natural defenses in the body aid in preventing disease. Vaccines also provide active immunity to the body systems.

Previewing the Chapter

Section 27-1 Disease
▶ Discovering Disease
▶ Keeping Clean
▶ Communicable Diseases
▶ Sexually Transmitted Diseases

Section 27-2 Your Immune System
▶ Natural Defenses
▶ Specific Defenses
▶ HIV and Your Immune System

Section 27-3 Science & Society
Issue: Should a harmful virus be destroyed?

Section 27-4 Noncommunicable Disease
▶ Chronic Disease
▶ Chemicals and Disease
▶ Cancer
▶ Allergies

Learning Styles

Look for the following logo for strategies that emphasize different learning modalities.

Kinesthetic	Explore, p. 729
Visual-Spatial	Demonstration, p. 731; Activity, p. 731; Activity 27-1, p. 736; Visual Learning, pp. 731, 732, 738, 739, 746
Interpersonal	MiniLAB, p. 733
Intrapersonal	Activity, pp. 732, 748
Logical-Mathematical	MiniLAB, p. 734; Activity 27-2, pp. 742-743
Linguistic	Across the Curriculum, p. 748

27

Immunity

Aaahh-choooo! What makes you sneeze? Some people have allergies. They sneeze when particles of dust or pollen get into their noses. Others sneeze when they have a cold. Have you noticed that when one person in your class gets a cold, others soon have it too? Allergies and colds are examples of your body's response to substances and organisms in the environment.

EXPLORE ACTIVITY

How are disease-causing organisms spread?

1. Work with a partner. Place a drop of peppermint food flavoring on a cotton ball. Pretend that the flavoring is a mass of cold viruses.
2. Use the cotton ball to rub an X over the palm of your right hand. Let it dry.
3. Shake hands with a classmate.
4. Have your classmate shake hands with another student.

Observe: How are some diseases spread? How many persons could be infected by your "virus?"

Previewing Science Skills

▶ In the Skill Builders, you will recognize cause and effect, compare and contrast, and make and use tables.

▶ In the Activities, you will experiment, hypothesize, and analyze data.

▶ In the MiniLABs, you will test and observe and calculate and graph results.

729

Assessment Planner

Portfolio
Refer to page 751 for suggested items that students might select for their portfolios.

Performance Assessment
See page 751 for additional Performance Assessment options.
Skill Builders, pp. 735, 740, 750
MiniLABs, pp. 733, 734
Activities 27-1, p. 736; 27-2, pp. 742-743

Content Assessment
Section Wrap-ups, pp. 735, 740, 745, 750
Chapter Review, pp. 751-753
Mini Quizzes, pp. 735, 740, 745, 750

Group Assessment
Opportunities for group assessment occur with Cooperative Learning Strategies and Flex Your Brain Activities.

EXPLORE ACTIVITY

Purpose
🅄 **Kinesthetic** Use the Explore Activity to introduce students to how disease can be spread. Inform students that they will be learning more about diseases and the immune system in this chapter. **L1** **COOP LEARN** **LEP**

Preparation
Before beginning this chapter, obtain the food flavoring and cotton balls. Orange, almond, or banana flavors can be used.

Materials
food flavorings, one cotton ball for each pair

Teaching Strategies
• Have students discuss how the mass of cold virus could have gotten on the hands in the first place.
• Have students infer whether the infection could be passed around on drinking glasses and handkerchiefs.
• **CAUTION:** *Following the activity, have students wash their hands thoroughly.*

Observe
Students should observe that the food flavoring can be spread to many persons. Anyone who can smell the odor of the flavoring on his or her hand is exposed.

✔Assessment

Performance Have students wash their hands with soap after all handshakes. Determine whether any flavoring remains on their hands. Discuss how washing affects the spread of disease. Use the Performance Task Assessment List for Making Observations and Inferences in **PASC**, p. 17. **P**

Prepare

Section Background

• Even after the work of Pasteur and Koch became known, the idea of diseases being caused by microorganisms was not accepted and was challenged many times.

• AZT and several other drugs are used alone or in combination to slow the multiplication of HIV.

Preplanning

• To prepare for the MiniLAB on page 733, obtain dairy products such as milk, sour cream, yogurt, or liquid from sour cream that is older than the sell-by date.

• To prepare for Activity 27-1, obtain a rotting apple or prepare one by scratching a fresh apple and rubbing dirt into the wounds. Allow this prepared apple to rot for several days before use.

1 Motivate

Bellringer

Before presenting the lesson, display **Section Focus Transparency 98** on the overhead projector. Assign the accompanying **Focus Activity** worksheet. L1 LEP

Discussion

Have students compare the population figures of their town with the 25 000 000 people who died of the plague. This is a graphic way of illustrating the effect disease has on history.

27•1 Disease

Science Words

pasteurization
disinfectant
antiseptic
communicable disease
sexually transmitted disease (STD)

Objectives

• Describe the work of Pasteur, Koch, and Lister in the discovery and prevention of disease.

• List diseases caused by viruses and bacteria.

• Discuss sexually transmitted diseases (STDs), their causes, and treatments.

Discovering Disease

"Ring around the rosie, A pocket full of posies, Ashes, Ashes, We all fall down."

You may not know that this rhyme is more than 600 years old. The rhyme is thought to be about a disease called the Black Death or the Plague. As shown in **Figure 27-1**, several times during the 14th century, the Plague spread through Europe. "Ring around the rosie" was a symptom of the disease—a ring around a red spot on the skin. People carried a pocket full of flower petals, "posies," and spices to keep away the stench of dead bodies. In spite of these efforts, 25 000 000 people "all fell down," and died from this terrible disease.

Causes of Disease

No one really knew what caused the Plague or how to stop it as it swept through Europe. It wasn't until about 150 years ago that scientists began to find out how some diseases are caused. Today, we know that diseases are caused by viruses, protozoans, and by harmful bacteria. These are known as pathogens.

The French chemist Louis Pasteur proved that harmful bacteria could cause disease. He developed a method of using heat to kill pathogens. **Pasteurization,** named for him, is the process of heating food to a temperature that kills most bacteria. Pasteur's work began the science of bacteriology.

Figure 27-1

Plague is caused by a bacterium that reproduces in fleas that live on rats. *What conditions in the towns of the 14th century encouraged the spread of this disease?*

730 Chapter 27 Immunity

Program Resources

 Reproducible Masters
Study Guide, p. 102 L1
Reinforcement, p. 102 L1
Enrichment, p. 102 L3
Activity Worksheets, pp. 5, 160-161, 164, 165 L1
Lab Manual 61, 62
Concept Mapping, p. 59

 Transparencies
Section Focus Transparency 98 L1
Teaching Transparency 53

Koch's Rules

Pasteur may have shown that bacteria caused disease, but there still was a problem. Being able to tell *which* specific organism caused a specific disease was the problem. It was a German doctor, Robert Koch, who first developed a way to isolate and grow, or culture, just one type of bacterium at a time. Koch developed a set of rules to be used for figuring out which organism caused a particular disease. These rules are illustrated in **Figure 27-2.**

Figure 27-2

The steps used to identify an organism as the cause of a specific disease are known as Koch's postulates.

1 In every case of a particular disease, the organism thought to cause the disease must be present.

2 The organism has to be separated from all other organisms and grown in a pure culture.

3 When the organism from the pure culture is injected into a test animal, it must cause the original disease.

4 Finally, when the suspect organism is removed from the test animal and cultured again, it must be compared with the original organism to see whether they are the same. Only when they match can you say that that organism is the pathogen that causes that disease.

27-1 Disease   **731**

2 Teach

Demonstration

LS **Visual-Spatial** Have students wash their hands and then lightly press their fingertips onto the surface of a sterile agar plate. Student petri dishes can be coded. Culture in a dark place, and have students observe the extent of the microorganism growth. **L1**

Activity

LS **Visual-Spatial** If possible, obtain microscope slides of stained bacteria to give students an idea of their size and number. **L1**

Visual Learning

Figure 27-1 **What conditions in the towns of the 14th century encouraged the spread of this disease?** *People threw both garbage and human wastes out into the streets. This was a breeding ground for both rats and bacteria.*

Figure 27-2 **Why is it not acceptable to use humans as the test animal in step 3?** *It is not ethically acceptable to knowingly give a disease to a human—even for medical or scientific testing.* **LEP** **LS**

GLENCOE TECHNOLOGY

CD-ROM

Life Science CD-ROM

Have students perform the interactive exploration for Chapter 27 to reinforce important chapter concepts and thinking processes.

Use **Concept Mapping**, p. 59, as you teach this lesson.

Use **Teaching Transparency 53** as you teach this lesson.

731

Figure 27-3

Antiseptics and strictly followed rules of cleanliness of modern operating rooms (right) have made surgical procedures safer than they once were (above).

732 Chapter 27 Immunity

Keeping Clean

Washing your hands before or after certain activities is an accepted part of your daily routine. This was not always true, even for doctors. Into the late 1800s, doctors, such as those in **Figure 27-3,** regularly operated in their street clothes and with bare hands. More patients died than survived as a result of surgery. Joseph Lister, an English surgeon, was horrified. He recognized the relationship between the infection rate and cleanliness in surgery. Lister dramatically reduced deaths among his surgical patients by washing their skin and his hands before surgery.

Disinfectants and Antiseptics

Lister used a disinfectant to reduce surgical infections. A **disinfectant** kills pathogens on objects such as instruments, floors, and bathtubs. Today, antiseptics are also used. An **antiseptic** kills pathogens on skin and prevents them from growing there for some time after. Doctors now wash their hands often with antiseptic soaps. They also wear sterile gloves to keep from spreading pathogens from person to person.

While the role of bacteria in causing disease was better understood and the use of disinfectants and antiseptics reduced the spread of disease, doctors did not understand why some diseases could not be controlled. Not until the late 1800s and early 1900s were viruses and their role in disease transmission discovered.

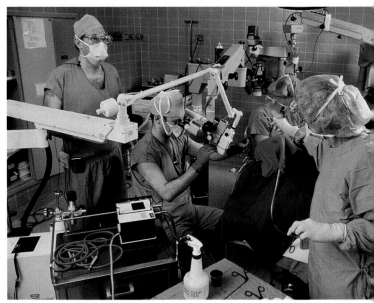

Communicable Diseases

A disease that is spread from one organism to another is a **communicable disease.** Communicable diseases are caused by agents such as viruses, pathogenic bacteria, protists, and fungi. Some diseases caused by these agents are listed in **Table 27-1.**

Table 27-1

Diseases and Their Agents			
Bacteria	**Protists**	**Fungi**	**Viruses**
Tetanus	Malaria	Athlete's foot	Colds
Tuberculosis	Sleeping	Ringworm	Influenza
Typhoid fever	sickness		AIDS
Strep throat			Measles
Bacterial	Mosquito	Virus	Mumps
pneumonia			Polio
Plague			Smallpox

Communicable diseases are spread through water and air, on food, by contact with contaminated objects, and by biological vectors. A biological vector is a carrier of a disease. Examples of vectors that spread disease are rats, flies, birds, cats, dogs, and the mosquito in **Figure 27-4.** To prevent the spread of disease, city water systems add chlorine to drinking water and swimming pools. Colds and many other diseases are spread through air. When you have a cold and sneeze, you hurl thousands of virus particles through the air. Each time you turn a doorknob, or press the button on a water fountain at school, your skin comes in contact with bacteria and viruses. Some dangerous communicable diseases are transmitted by sexual contact.

The spread of communicable diseases is closely watched by agencies in every country. In the United States, the Centers for Disease Control (CDC) in Atlanta, Georgia, monitors the spread of diseases throughout the country. The CDC also watches for diseases brought into the country.

Figure 27-4

How does the mosquito transport pathogens from infected people or objects to new hosts?

27-1 Disease 733

📁 Use Lab 61 in the **Lab Manual** as you teach this lesson.

MiniLAB

Are bacteria present in food? 🧤🥽

Methylene blue is used to detect bacteria. The faster the color fades, the more bacteria are present.

Procedure
1. Use the food samples provided by your teacher. Label four test tubes 1, 2, 3, and 4.
2. Fill three test tubes half full of the food samples.
3. Fill the fourth with water.
4. Add 20 drops of methylene blue and 2 drops of mineral oil to each tube.
5. Place the tubes into a warm-water bath.
6. Record the time and your observations.

Analysis
1. Compare how long it takes each tube to lose its color.
2. What was the purpose of tube 4?
3. Why is it important to eat and drink only the freshest food?

MiniLAB

Purpose

🔲 **Interpersonal** The students will observe the presence of some bacteria in all the samples. Spoiled foods will have more bacteria.

L1 COOP LEARN

Materials

methylene blue, test tubes, labels, samples of dairy products (milk, yogurt, cottage cheese, sour cream), pipet, mineral oil, warmwater bath, clock or watch

Teaching Strategies

To compare bacterial growth, you may have students use foods with and without added preservatives or raw and cooked hamburger meat.

Safety Precautions
- Methylene blue stains.
- Have students wash their hands with soap and water at the end of this activity.

Analysis
1. Length of time will vary with age of product used.
2. Tube 4 serves as a control.
3. By eating and drinking fresh foods, the contact with potentially harmful bacteria will be reduced.

✔ **Assessment**

Performance Have students repeat the activity using samples of the same foods, but with varying amounts of time past the expiration date. Use the data obtained to infer the varying levels of bacteria. Use the Performance Task Assessment List for Carrying Out a Strategy and Collecting Data in **PASC,** p. 25. **P**

📁 **Activity Worksheets,** pp. 5, 164

733

MiniLAB

Purpose
 Logical-Mathematical The students will observe the possible results if bacteria find favorable conditions and multiply rapidly. **L1**

Materials
paper, pencil, calculator

Analysis
1. At the five-hour point, there will be 32 768 bacteria. Students should label the vertical axis of the graph *thousands of bacteria* and the horizontal axis *time in hours.*
2. It is important to take all of any antibiotic that is prescribed. Failure to do so can leave a few bacteria that can begin dividing again and reinfect the body.

Assessment

Performance To further assess students' understanding of bacterial multiplication, have them multiply their totals by 10, by 100, and by 1000. Use the Performance Task Assessment List for Using Math in Science in **PASC**, p. 29. **P**

 Activity Worksheets, pp. 5, 165

3 Assess

Check for Understanding

? FLEX Your Brain

Use the Flex Your Brain activity to have students explore DISEASE.

 Activity Worksheets, p. 5

Figure 27-5
The bacterium that causes syphilis can be seen in this electron micrograph.

Magnification 19 000×

MiniLAB

How fast do bacteria reproduce?

Bacteria can divide every 20 minutes if conditions are favorable. Find out how many bacteria could be in your body after only a few hours.

Procedure
1. Make a chart like the one below.
2. Complete the chart up to the fifth hour.
3. Graph your data.

Analysis
1. How many bacteria are present after five hours?
2. Why is it important to take antibiotics promptly if you have an infection?

Time	Number of bacteria
0 hours 0 minutes	1
20 minutes	2
40 minutes	4
1 hour 0 minutes	8
20 minutes	16
40 minutes	32

Sexually Transmitted Diseases

Diseases transmitted from person to person during sexual contact are called **sexually transmitted diseases (STDs).** You can become infected with an STD by engaging in sexual activity with an infected person. STDs are caused by both viruses and bacteria. Some STDs are listed in **Table 27-2.**

Genital herpes causes painful blisters on the sex organs. Herpes can be transmitted during sexual contact or by an infected mother to her child during birth. The herpes virus, a latent virus, hides in the body for long periods and then reappears suddenly. There is no cure for herpes, and there is no vaccine to prevent it.

Gonorrhea and chlamydia are STDs caused by bacteria that may not produce any symptoms. When symptoms do appear, they may include painful urination, genital discharge, and sores. Penicillin and other antibiotics are used to treat these diseases. If left untreated, either disease can cause sterility, the inability to reproduce.

The strangely shaped bacterium that causes syphilis is seen in **Figure 27-5.** Syphilis has several stages. In stage 1, a sore appears on the mouth or genitals that lasts 10 to 14 days. Stage 2 may involve a rash, a fever, and swollen lymph glands. During stage 3, these symptoms may disappear. The victim often believes that the disease has gone away, but it hasn't. In Stage 4, syphilis may infect the cardiovascular and nervous systems. At this point, it is too late to treat syphilis. Nerve damage and death may result. Syphilis can be treated and cured with antibiotics only in the early stages.

Table 27-2

STDs	
Spread by virus	**Spread by bacteria**
Genital herpes	Gonorrhea
AIDS	Chlamydia
Genital warts	Syphilis

📁 Use Lab 62 in the **Lab Manual** as you teach this lesson.

HIV and AIDS

You learned that HIV is a latent virus that can exist in blood and body fluids. You can contract AIDS by having sex with an HIV-infected person, or by sharing an HIV-contaminated needle used to inject drugs. AIDS is also transmitted by contaminated blood transfusions. A pregnant female with AIDS may infect her child when the virus passes through the placenta or when nursing after birth. There is no vaccine to prevent AIDS and no medication to cure it. In the next section, more information will be given on how AIDS breaks down the immune system.

Sexually transmitted diseases are becoming more difficult to treat. Drugs that used to be effective in treating some of the diseases are now ineffective. For example, penicillin was used to treat syphilis. However, the organism that causes syphilis has become resistant to the antibiotic in some persons. New forms of the drugs must now be used, **Figure 27-6.**

Figure 27-6

New derivatives of penicillin and other classes of drugs are used to treat many diseases, including some STDs.

Section Wrap-up

Review

1. How did the discoveries of Pasteur, Koch, and Lister help in the battle against disease?

2. List a communicable disease caused by a virus, a bacterium, a protist, and a fungus.

3. What are STDs? How are they contracted and treated?

4. **Think Critically:** In what ways does Koch's procedure reflect the use of scientific methods?

Skill Builder
Recognizing Cause and Effect
How is not washing your hands related to the spread of disease? If you need help, refer to Recognizing Cause and Effect in the **Skill Handbook.**

Science Journal

Research the history of the building of the Panama Canal. Write a report in your Science Journal about the relationship between the disease, yellow fever, and the building of the canal.

Reteach
Review the concepts of microorganisms as causes of infections and the need to keep clean.

4 Close

•MINI•QUIZ•

1. **What kind of substances are used to kill microbes on people?** *antiseptics*

2. **STDs are caused by _____ .** *viruses, bacteria*

Section Wrap-up

Review

1. Pasteur—discovered bacteria could cause diseases; Koch—developed a way to isolate and grow one type of bacteria at a time; Lister—related infection to lack of cleanliness

2. Answers will vary. Possible answers are: virus—viral pneumonia; bacterium—strep throat; protist—sleeping sickness; fungus—athlete's foot.

3. They are sexually transmitted diseases. STDs are contracted by engaging in sexual activity with an infected person. Only those caused by bacteria can be treated with antibiotics.

4. **Think Critically** Answers will vary. Koch used strict isolation techniques and retested the suspected pathogen. He used controlled techniques.

Skill Builder
You can infer that not washing hands means that many viruses and bacteria accumulate on the skin and will be spread to other people and surfaces, increasing the chance of spreading disease.

✓ Assessment

Performance Assess students' abilities to recognize cause and effect by having them describe in writing the effect of sneezing on the spread of microorganisms. Use the Performance Task Assessment List for Writing in Science in **PASC**, p. 87. **P**

Science Journal

Student journal reports should include that building the canal was held up for years until mosquitoes causing malaria and yellow fever were brought under control. The canal was completed in 1914. **P**

Activity 27-1 Microorganisms and Disease

Purpose

IS **Visual-Spatial** Predict and explain how microbes spread. L1 **COOP LEARN**

Process Skills

observing and inferring, communicating, interpreting data, comparing and contrasting, recognizing cause and effect, separating and controlling variables, predicting

Time

30 minutes for initial setup; 15 minutes' observation time one week later

Safety Precautions

Have students wash their hands after handling microorganisms. Provide for safe disposal of apples.

📁 **Activity Worksheets,** pp. 5, 160-161

Teaching Strategies

Point out that brown areas found below the skin of apples (bruises) may be caused by bacteria.

Answers to Questions

1. Number 1 is a control and a standard for comparison.

2. It decayed.

3. The bruised and rubbed areas showed some decay.

4. Apple 5 decayed less than apples 2, 3, and 4.

5. Apple 6 did not decay. The skin was not broken, so pathogens could not enter the apple. Alcohol could have killed some.

6. Cleaning a wound removes pathogens, and infection is prevented.

7. Answers will vary with the hypotheses.

8. Pathogens can cause infection by entering damaged or cut surfaces of an organism.

Microorganisms are all around us. They are on the surfaces of everything we touch. Try this experiment to see how microorganisms are involved in spreading infections.

Problem

How do microorganisms cause infection?

Materials 🚫 🧤

- fresh apples (6)
- rotting apple
- alcohol
- self-sealing plastic bags (6)
- labels and pencil
- paper towels
- sandpaper
- cotton ball
- soap and water

Procedure

1. Label the plastic bags 1 through 6.
2. Put a fresh apple in bag 1 and seal it.
3. Rub the rotting apple over the entire surface of the remaining five apples. This is your source of microorganisms. **CAUTION:** *Always wash your hands after handling microorganisms.* Take one of the apples and put it in bag number 2.
4. Hold one apple 2 m above the floor and drop it. Put this apple into bag number 3.
5. Rub one apple with sandpaper. Place this apple in bag number 4.
6. Wash one apple with soap and water. Dry it well with a paper towel. Put this apple in bag number 5.
7. Use a cotton ball to spread alcohol over the last apple. Let it air dry and then place it in bag number 6.
8. Place all of the apples in a dark place for one week. Then wash your hands.
9. Write a hypothesis to explain what you think will happen to each apple.
10. At the end of the week, compare all the apples. Record your observations. **CAUTION:** *Give all apples to your teacher for proper disposal.*

736 Chapter 27 Immunity

Data and Observations   Sample Data

Apple	Observations
1	No change
2	Many brown spots, decay
3	Brown spots at breaks
4	Some brown, soft areas
5	Little or no decay
6	No change

Analyze

1. What was the purpose of apple number 1?
2. Explain any changes in apple number 2.
3. What happened on apples 3 and 4?
4. What effect did the soap and water have?

Conclude and Apply

5. Did you **observe** changes in apple number 6?
6. Why is it important to clean a wound?
7. Were your hypotheses supported?
8. **Relate** microorganisms and infection.

✓ Assessment

Performance To further assess students' understanding of microorganisms and disease, repeat the activity using hydrogen peroxide (3%) to clean an infected apple. Determine whether it stops bacterial growth. Use the Performance Task Assessment List for Carrying Out a Strategy and Collecting Data in **PASC**, p. 25. P

Your Immune System

Natural Defenses

In Section 27-1, harmful bacteria were discussed for their disease-causing properties. While most bacteria do not cause disease, some that do are able to enter your body through breaks in your skin. When this happens, your body mobilizes its defenses against the disease-causing intruders.

Your **immune system** is a complex group of defenses that your body has to fight disease. It is made up of cells, tissues, organs, and body systems that fight bacteria, viruses, harmful chemicals, and cancer cells.

Your Body's Lines of Defense

Several body systems are involved in maintaining your health. Your circulatory system contains white blood cells that engulf and digest foreign organisms and chemicals. These white blood cells constantly patrol the body, sweeping up and digesting bacteria that manage to get into the body. As you see in **Figure 27-7,** they slip between cells in the walls of capillaries to destroy bacteria. When the white blood cells cannot destroy the bacteria fast enough, a fever may develop. A fever helps fight pathogens by slowing their growth and speeding up your body's reactions.

Your respiratory system contains cilia and mucus that trap pathogens. When you cough, you expel trapped bacteria. In the digestive system, enzymes in the stomach, pancreas, and liver destroy pathogens. Hydrochloric acid in your stomach kills bacteria that enter your body on food.

All of these processes are general defenses that work to keep you disease-free. But if you do get sick, your body has another line of defense in the form of active and passive immunity.

Figure 27-7

White blood cells leave capillaries and engulf harmful bacteria in surrounding tissues.

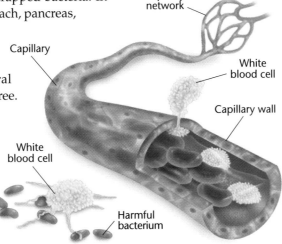

Capillary
network

Capillary

White blood cell

Capillary wall

White blood cell

Harmful bacterium

27-2 Your Immune System **737**

Science Words

- **immune system**
- **antigen**
- **antibody**
- **active immunity**
- **passive immunity**
- **lymphocyte**

Objectives

- Explain the natural defenses your body has against disease.
- Describe differences between active and passive immunity.
- Explain how HIV affects the immune system.

Specific Defenses

When your body fights disease, it is really battling proteins or chemicals that don't belong there. Proteins and chemicals that are foreign to your body are called **antigens.** Antigens are located on the surfaces of pathogens. When your immune system recognizes a foreign protein or chemical, as in **Figure 27-8,** it forms specific antibodies. An **antibody** is a protein made by an animal in response to a specific antigen. The antibody binds up the antigen, making it harmless. Antibodies help your body build defenses in two ways, actively and passively. **Active immunity** occurs when your body makes its own antibodies in response to an antigen. **Passive immunity** occurs when antibodies, which have been produced in another animal, are introduced into the body.

Figure 27-8

The response of your immune system to disease-causing organisms can be divided into four steps: recognition, mobilization, disposal, and immunity.

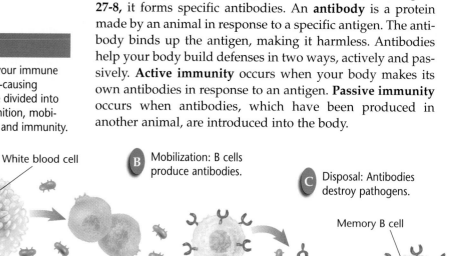

White blood cell

Pathogen

A Recognition: White blood cell surrounds pathogen and signals T cells. More T cells are produced. Helper T cells signal B cells.

B Mobilization: B cells produce antibodies.

Helper T cell

B cell

C Disposal: Antibodies destroy pathogens.

Memory B cell

Antibody

D Immunity: Some antibodies remain for future use.

Active Immunity

When your body is invaded by a pathogen, it immediately starts to make antibodies to inactivate the antigen. Once enough antibodies form, you get better. These antibodies stay on duty in your blood, becoming active when you again encounter the disease. Under these conditions, your body has active immunity to a particular disease.

Another way to develop active immunity to a particular disease is to be inoculated with a vaccine. A vaccine is a noninfectious form of the antigen and gives you active immunity against a disease without you actually getting the disease first. For example, suppose a vaccine for measles is injected into your body. Your body forms antibodies against the measles antigen. If you later encounter the virus, antibodies

📁 Use these program resources as you teach this lesson.
Cross-Curricular Integration, p. 31
Multicultural Connections, pp. 57-58
Science Integration Activities, pp. 53-54

necessary to fight that virus are already in your bloodstream ready to destroy the pathogen. Antibodies that immunize you against one virus may not guard against a different virus. As you grow older, you will be exposed to many more types of pathogens and will build a separate immunity to each one.

Passive Immunity

How is passive immunity different from active immunity? Recall that passive immunity is the transfer of antibodies into your body. For example, as a newborn, you were a bundle of passive immunity. You were born with all the antibodies that your mother had in her blood. However, these antibodies stayed with you only a few months. Passive immunity does not last as long as active immunity. Newborn babies lose their passive immunity in a few months. They need to be vaccinated to develop their own immunity.

You need to be vaccinated for protection from tetanus. Tetanus toxin is produced by a bacterium in soil. The toxin paralyzes muscles. Death can occur by suffocation when the muscles that control breathing become paralyzed. Tetanus antitoxin is produced in horses when they respond to injections of the tetanus antigen. Samples of antibodies made by horses are injected into humans. Because the antibodies provide limited immunity, booster shots are needed throughout life.

Problem Solving

Fighting TB

Annie Dodge Wauneka, the daughter of a Navajo chief, has spent much of her life fighting tuberculosis (TB). She was the first woman elected to the Tribal Council and was appointed chairperson of the health committee. TB killed many people on the reservation each year, so Wauneka decided to learn everything she could to fight it. She wrote a Navajo-English medical dictionary and got Navajo and non-Navajo doctors to work together on the disease. She also encouraged Navajo patients to trust non-Navajo doctors and to get early diagnoses. Her war against TB was so successful that she was awarded the Freedom Award, and the Navajo gave her the name "Warrior Who Scouts the Enemy."

Solve the Problem:

1. Why did Annie Dodge Wauneka choose to fight the spread of TB within her tribe?
2. Explain why the non-Navajo doctors were not successful in fighting TB on the reservation.

Think Critically:

Annie Dodge Wauneka was not a doctor. How was she able to help fight a disease without medical knowledge?

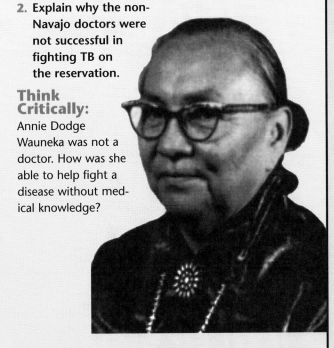

27-2 Your Immune System   739

Problem Solving

Solve the Problem

1. because the death rate from TB was high
2. They did not have the trust of the Navajos.

Think Critically

Wauneka helped with the social aspects of medical treatment. She brought together medical "experts" from both cultural traditions. She gave them ways to communicate with each other. She helped overcome the conflicts and distrust between native healers, patients, and non-Navajo doctors. **P**

Visual Learning

(See page 740.)
Figure 27-9 Why does this characteristic make the spread of AIDS more likely? *People don't know that they have the disease, so they can't begin medical treatment and may not take precautions that would avoid infecting other people.* **LEP** **IS**

3 Assess

Check for Understanding

Ask questions 1-3 and the Think Critically question in the Section Wrap-up.

Reteach

Use an anatomy chart or model to illustrate the various body systems involved in defense against microbes.

Extension

For students who have mastered this section, use the **Reinforcement** and **Enrichment** masters.

Use these program resources as you teach this lesson.
Science and Society Integration, p. 31
Critical Thinking/Problem Solving, p. 31

4 Close

Use the Mini Quiz to check students' recall of chapter content.

1. **How do disease-causing bacteria usually enter the body?** *through a break in the skin*

2. **What destroys pathogens in your digestive system?** *enzymes, also hydrochloric acid in the stomach*

3. **Substances made in response to antigens are called _____ .** *antibodies*

4. **What type of immunity does a newborn baby have?** *passive*

5. **HIV attacks cells called _____ .** *lymphocytes*

Section Wrap-up

Review

1. white blood cells, cilia, mucus, coughing, enzymes in the digestive system, hydrochloric acid, active immunity

2. Passive immunity does not last as long as active immunity.

3. HIV attacks lymphocytes, so your body cannot produce enough antibodies to attack the invading antigens.

4. **Think Critically** Antibodies formed after a vaccination to fight a specific type of antigen will protect only against that one antigen and not against any other antigens.

Using Computers

Use the textbook information to check the accuracy of the flowchart. See also the Skill Builder answer on this page.

Magnification 130 000×

Figure 27-9

A person can be infected with HIV (above) and yet not show any symptoms of the infection for several years. *Why does this characteristic make the spread of AIDS more likely?*

HIV and Your Immune System

HIV is different from other viruses. It attacks cells in the immune system called lymphocytes. **Lymphocytes** are white blood cells throughout the lymphatic system that chemically recognize antigens. They then produce antibodies and destroy invading antigens.

Because HIV destroys lymphocytes, the body is left with no way to fight invading antigens. The whole immune system breaks down. The body is unable to fight HIV, or any other pathogen. For this reason, people with AIDS die from other diseases such as pneumonia, cancer, or tuberculosis. The victim's body becomes defenseless.

When a microbe attacks your body, it must get past all of your natural defenses. If it gets past your skin or other defenses, it encounters your immune system—your last line of defense.

Section Wrap-up

Review

1. List natural defenses your body has against disease.

2. Why does passive immunity need to be renewed throughout life?

3. Explain how HIV attacks your immune system.

4. **Think Critically:** Several diseases have the same symptoms as measles. Why doesn't the measles vaccine protect you from all of these diseases?

Skill Builder
Comparing and Contrasting

Compare and contrast active and passive immunity. If you need help, refer to Comparing and Contrasting in the **Skill Handbook.**

Using Computers

Flowchart Use the information on pages 738 and 739 to create a flowchart that compares active and passive immunity.

Skill Builder

Compare: both confer immunity; contrast: in active immunity, the body makes its own antibodies, and in passive immunity, antibodies are introduced.

Assessment

Portfolio Use this Skill Builder to assess students' abilities to compare and contrast active and passive immunity by having them draw a concept map. Use the Performance Task Assessment List for Concept Map in **PASC,** p. 89. **P**

Ethics and the Hippocratic Oath

Doctors have always faced difficult decisions when treating seriously ill patients. But, the increasingly rapid pace of medical advances has made some decisions even more difficult for today's doctors. For example, is it ethical to use genetic engineering to give a developing fetus immunity to certain diseases? Should an elderly, very ill cancer patient receive chemotherapy? Where can doctors find ethical guidelines for their actions?

The First Oath

One of the earliest ethical guides for doctors was the Hippocratic Oath, written in 400 B.C. This oath has been revised several times, but one often-quoted version begins:

> *I swear by Apollo physician and Asclepius and Hygeia and Panaceia and by all the gods and goddesses. . . that I will fulfill. . . this oath.*

When the oath was written, the Greeks believed that the god Apollo drove the sun across the sky. They also believed that his son Asclepius had learned so much about healing from a centaur that he could bring the dead to life. The oath assumed these Greek gods were watching over human doctors.

In the centuries after the oath was written, most Christian, Jewish, and Islamic doctors focused on and agreed with that part of the oath that involved a doctor's duties to patients. Here the oath states that doctors are to use their skills "for the benefit of the sick" and "keep them from harm and injustice." Doctors are also to avoid all sexual contact with their patients and keep confidential all they know about their patients.

Current Practice

Over the years, the original oath has undergone many revisions. In fact, most of today's graduating medical students do not swear to uphold the Hippocratic Oath. A 1994 survey of six major medical schools found that only two used the oath—and in a modernized version. Other medical schools offered their students as many as four choices of oaths. At Harvard University, graduating medical students write their own oath every year.

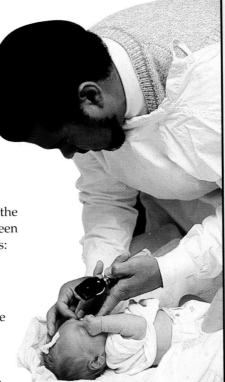

Science Journal

What should be included in an oath sworn by today's doctors? In your Science Journal, write four or five "rules" you think doctors should follow.

27-2 Your Immune System   **741**

Sources

- Dorman, John. "The Hippocratic Oath." *Journal of American College Health*, Vol. 44, 1 Sept. 1995.
- "Oath and Law of Hippocrates." *Harvard Classics*, Vol. 38, 1910.
- "Oath of Hippocrates." *Encyclopaedia Britannica*, Vol. 23, p. 889, 1985 ed.

Background

The Hippocratic Oath was never a law, only a code. It has two main sections. The first describes doctors' duties toward their medical school teachers. The second section covers duties to patients. Among other reflections of its time, it assumes only men will become physicians. It also warns physicians not to "use the knife." It's not clear whether that means they should leave surgery to those qualified to perform it or if surgery was considered unethical at that time. Hippocrates' famous quotation, "Above all, do no harm," was not part of his oath.

Teaching Strategies

Have students share the oaths they wrote and explain why they chose certain rules. Then discuss these ethical questions.

- Should people who abuse their health and who do not follow their doctors' orders still receive medical treatment for the problems that result?
- If doctors employed by companies discover health problems while examining employees, should they tell the companies? What if the health problem could affect others?
- If a patient will die soon, should the doctor violate confidentiality and tell the patient's family?

PREPARATION

Purpose

IS **Logical-Mathematical** Determine the chemicals that prevent growth of microorganisms. **L1** **COOP LEARN**

Process Skills

observing, hypothesizing, designing an experiment, predicting, interpreting data, communicating, recognizing cause and effect, comparing and contrasting, separating and controlling variables, and making models

Time

two class periods (one for setup, one for experiment)

Materials

Prepare nutrient agar plates, five for each group. Label four jars and four 2-cm squares of filter paper D, H, M, and A. Pour 50 mL of each of the following solutions into the jars: disinfectant, 3% hydrogen peroxide, mouthwash, and alcohol.

Alternate Materials Gelatin may be used instead of agar, but it may melt in warm temperatures. Bacterial growth is not as evident as on agar.

Safety Precautions

Caution students to handle forceps carefully. Do not allow students to open plates. Plates should be given to the teacher for proper disposal. Spray the cultures with a strong household disinfectant. Reseal dishes before disposal.

Possible Hypotheses

Student hypotheses will vary. Possible hypotheses are: "The chemicals will be equally effective in preventing microorganism growth" or "Hydrogen peroxide is most effective in preventing microorganism growth."

📁 **Activity Worksheets,** pp. 5, 162-163

Activity 27-2

Design Your Own Experiment
Stopping Growth of Microorganisms

Infections are caused by microorganisms. Without cleanliness, the risk of getting an infection from a wound is high. Earlier, you learned that disinfectants are chemicals that kill or remove disease organisms from objects. Antiseptics are chemicals that kill or prevent growth of disease organisms on living tissues. Your task is to devise a method to prevent the growth of microorganisms.

PREPARATION

Problem

What conditions do microorganisms need to grow? How can they be prevented from growing?

Form a Hypothesis

Based on your knowledge of disinfectants and antiseptics, state a hypothesis about methods that will prevent the growth of microorganisms.

Objectives

- Observe the effects of antiseptics and disinfectants on microorganism growth in petri dishes.
- Design an experiment that will test the effects of chemicals on microorganisms growing in contaminated petri dishes.

Possible Materials

- sterile petri dishes (5)
- filter paper (2-cm squares)

- test chemicals (disinfectant, hydrogen peroxide, mouthwash, alcohol)
- transparent tape
- pencil and labels
- scissors
- metric ruler
- forceps
- small jars for chemicals (4)

Safety Precautions

Handle the forceps carefully. On completion of the experiment, give your sealed petri dishes to your teacher for proper disposal.

Inclusion Strategies

Visually Impaired Students may require bright illumination and a magnifier to see the bacterial colonies. Do not remove the petri dish cover for better viewing.

PLAN THE EXPERIMENT

1. As a group, agree upon and write out a hypothesis statement.
2. As a group, list the steps that you will need to take to test your hypothesis. Consider what you learned about how infections are stopped. List your materials.
3. Design a data table and record it in your Science Journal so that it is ready to use as your group collects data.
4. Your teacher will give you instructions on procedures to use to conduct this experiment.

Check the Plan

1. Read over your entire experiment to make sure that all steps are in logical order.
2. Identify any constants, variables, and the control of the experiment.

3. *Make sure your teacher approves your plan before you proceed.*

DO THE EXPERIMENT

1. Carry out the experiment as planned.
2. While the experiment is going on, write down any observations that you make and complete the data table in your Science Journal.

Analyze and Apply

1. **Compare** your results with those of other groups.

2. How did you **compare** growth beneath and around each chemical-soaked square in the petri dishes?
3. **Interpret the data** to determine what substances appeared to be most effective in preventing microorganism growth and what substances appeared to be least effective.

Go Further

Design an experiment to determine whether sunlight has an effect on microorganism growth. Compare results with the results of the experiment with antiseptics and disinfectants.

27-2 Your Immune System **743**

Go Further

Sunlight does have a negative effect on many microorganisms. Results may vary depending on the amount of time the dish was exposed to sunlight.

 Assessment

Performance To further assess students' understanding of microorganism growth, have them repeat their experiment using different mouthwashes, antiseptics, and disinfectants. Use the Performance Task Assessment List for Carrying Out a Strategy and Collecting Data in **PASC**, p. 25. **P**

PLAN THE EXPERIMENT

Possible Procedures

Label the plates 1-5. Have students remove covers and rub a finger over the agar. Replace the cover and seal plate 5 with tape. Soak the 2-cm squares of filter paper in the test chemicals in the jars. Use forceps to remove the squares, and place one on the agar in each of the four remaining plates. Seal the dishes with tape and store them in a warm, dark place for 48 hours.

Teaching Strategies

Tying to Previous Knowledge Most students will have had experience with flavored gelatin and will recognize the semi-solid state of agar.

DO THE EXPERIMENT

Expected Outcome

Most experiments will show that the chemicals prevented microorganism growth. Disinfectants and hydrogen peroxide will have a more noticeable effect.

Analyze and Apply

1. Student results should be consistent with those of other groups.
2. Answers will vary, but students should be able to observe that there are fewer colonies beneath and around the chemical-soaked squares.
3. Answers will vary depending on the substances used. Household disinfectants and antiseptics (e.g., hydrogen peroxide) will have an effect. Some microorganisms are more susceptible to chemicals than others.

Prepare

Section Background

Some vaccination serums are made from toxins excreted as wastes by certain bacteria, such as those that produce diphtheria and tetanus. The toxins are known as exotoxins.

1 Motivate

Bellringer

Before presenting the lesson, display **Section Focus Transparency 100** on the overhead projector. Assign the accompanying **Focus Activity** worksheet. L1 LEP

Tying to Previous Knowledge

Review the functions of the blood. One of these is the transportation of materials in the blood to fight infections and heal wounds. Also review Jenner's work with the development of a smallpox vaccine as described on p. 36.

2 Teach

Enrichment

Research the nature of polio as a disease in the 1930s-1950s. Find out about the function of the iron lung. Also find out about the development of a polio vaccine by Jonas E. Salk and Albert B. Sabin. L2

Inquiry Question

What are toxins, and how are they used in vaccines? *See Section Background for answer.*

ISSUE:

27•3 Should a harmful virus be destroyed?

Science Words
vaccination

Objectives
- Explain the role of vaccines in preventing certain diseases.
- Discuss whether a harmful virus should be destroyed.

Protection from Disease

You learned that a vaccine can be given to a person to provide immunity from a disease. The process of giving a vaccine by injection or orally is called **vaccination.** Today, vaccines for several different diseases are available. Parents can now protect their children from polio, measles, mumps, diphtheria, whooping cough, chicken pox, and tetanus. Children can be protected, but only if they are vaccinated.

Diseases Are Returning

Thanks to intensive work by governments and the World Health Organization, smallpox, shown in **Figure 27-10,** has been completely eliminated from everywhere in the world. One estimate has polio becoming extinct by the year 2000. While this provides encouragement about the prospect of eradicating some diseases, the numbers of new cases of other diseases have increased. Dramatic rises in the number of new cases of measles, whooping cough, and tuberculosis have been reported in the United States. Several reasons for the return of these diseases can be cited. First, an increasing number of parents are neglecting to have their children vaccinated. Cost, lack of time, and a belief that the chance is very small that the child will actually get the disease all contribute to the apathy. Some parents fear that the vaccines will cause adverse reactions. Some don't realize that several diseases—measles, tetanus, and possibly chicken pox—require booster shots later in life.

With this in mind, a debate about what to do with samples of the disease-causing agents is going on. The debate is especially intense about samples of smallpox virus.

Figure 27-10

The last known case of smallpox was reported in Somalia in 1977.

Program Resources

 Reproducible Masters
Study Guide, p. 104 L1
Reinforcement, p. 104 L1
Enrichment, p. 104 L3

Transparencies
Section Focus Transparency 100 L1

2 Points of View

Keeping the Disease Agents

Although eliminated as a disease, live cultures of the smallpox virus are kept at the Centers for Disease Control, in Atlanta, Georgia. Additional cultures are kept at a similar laboratory in Russia. Scientists on one side of the argument want to keep the cultures. They reason that the live cultures might be useful in producing future vaccines or other medicines for diseases that are similar to smallpox. They also question the ethical right of humans to intentionally cause the extinction of a virus, even if it is harmful to people.

Destroying the Disease Agents

Scientists on the other side of the issue argue that all of the laboratory samples of the live smallpox virus should be destroyed. They fear that, even with all of the security precautions, such as those shown in **Figure 27-11**, taken to prevent the escape of the virus, an accident could happen. If the virus were to be released into an unprotected population, millions of people would die. The world's population is now considered unprotected because children are no longer vaccinated against smallpox. The extinction of a harmful virus is not as much of an issue as the public's safety.

Figure 27-11

Samples of many disease-causing organisms, such as the deadly Ebola virus, are kept in laboratories around the world. Special precautions are taken to protect the workers from becoming infected.

Section Wrap-up

Review

1. Why might parents think that their children would not get measles, even if they aren't vaccinated?

2. How does use of the polio vaccine prevent a polio epidemic from spreading across the United States?

Explore the Issue

Should samples of all disease agents be destroyed as soon as the disease is eliminated in the world? If yes, how would researchers be able to study the disease organism or find cures for similar diseases? If no, what measures can be taken to be sure laboratory workers do not become infected or terrorists do not steal the samples?

*inter*NET CONNECTION

The destruction and global eradication of all laboratory samples of the smallpox virus have been scheduled. Visit the Chapter 27 Internet Connection at Glencoe Online Science, **www.glencoe.com/sec/science/life**, for a link to more information about smallpox eradication.

SCIENCE & SOCIETY

27-3 Should a harmful virus be destroyed?   **745**

3 Assess

Check for Understanding

Ask questions 1 and 2 in the Section Wrap-up.

Reteach

Conduct a school survey to find how many students or members of their families have had measles, chicken pox, mumps, or whooping cough.

Extension

For students who have mastered this section, use the **Reinforcement** and **Enrichment** masters.

4 Close

•MINI•QUIZ•

Use the Mini Quiz to check recall of section content.

1. **The process of providing immunity by injecting a vaccine is called _____.** *vaccination*

2. **What disease has been successfully eliminated on our planet?** *smallpox*

Section Wrap-up

Review

1. the erroneous belief that the disease is no longer a threat to children

2. By using the vaccine, most people in the country are immune to the disease. So only the few who have not been vaccinated are at risk of contracting the disease.

Prepare

Section Background

- Noncommunicable diseases do not respond to antibodies and cause more deaths than communicable diseases.

- For some forms of cancer, less radical operations are being performed. Combined radiation and chemotherapy are used as postoperative treatment.

1 Motivate

Bellringer

Before presenting the lesson, display **Section Focus Transparency 101** on the overhead projector. Assign the accompanying **Focus Activity** worksheet. L1 LEP

Tying to Previous Knowledge

Recall the information in Chapter 23 concerning diseases and disorders of the heart. Review the role of diet and exercise in preventing heart disease.

2 Teach

Visual Learning

Figure 27-12 This figure shows an arthritic hand. Explain that this condition is often said to be brought on by an autoimmune process. Ask: **What does this mean?** *The immune system manufactures antibodies against its own body tissues, resulting in tissue damage.* LEP LS

27•4 Noncommunicable Disease

Science Words

- noncommunicable disease
- chronic disease
- cancer
- tumor
- chemotherapy
- allergy
- allergen

Objectives

- Identify two noncommunicable diseases.
- Describe the basic characteristics of cancer.
- Explain what happens during an allergic reaction.

Chronic Disease

Diseases and disorders such as diabetes, allergies, asthma, cancer, and heart disease are not caused by pathogens. These diseases are called **noncommunicable diseases** because they are not spread from one person to another. You can't "catch" them. Allergies, genetic disorders, lifestyle diseases, or chemical imbalances such as diabetes are not spread by sneezes or handshakes.

Some noncommunicable diseases are called **chronic diseases** because they last a long time. Some chronic diseases can be cured. Others cannot. Chronic diseases may result from improperly functioning organs, contact with harmful chemicals, or an unhealthy lifestyle. For example, your pancreas produces the hormone insulin. Diabetes is a chronic disease in which the pancreas cannot produce the amount of insulin the body needs.

Arthritis is a chronic disease resulting from a faulty immune system. The immune system begins to treat the body's own normal proteins as if they were antigens. The faulty immune system forms antibodies against the normal proteins in joints. You can see in **Figure 27-12** that the joints become distorted and movement becomes difficult and painful.

 Figure 27-12
Arthritis turns the body's immune system against itself. Joints may become severely deformed as these X rays show.

746   Chapter 27 Immunity

Program Resources

Reproducible Masters
Study Guide, p. 105 L1
Reinforcement, p. 105 L1
Enrichment, p. 105 L3
Lab Manual 63, 64

Transparencies
Section Focus Transparency 101 L1
Science Integration Transparency 27

Figure 27-13
In many farm communities, chemicals found in agricultural products, such as fertilizers and pesticides, seep into nearby lakes, rivers, and groundwater supplies. Eventually, the chemicals show up in the food supply of humans and other animals.

Chemicals and Disease

We are surrounded by chemicals. They are found in foods, cosmetics, cleaning products, pesticides, fertilizers, and building materials. Of the thousands of chemical substances used by consumers, less than two percent are harmful. Those chemicals that are harmful to living things are called toxins. Some chronic diseases are caused by toxins.

The Effects

The effects of a chemical toxin are determined by the amount that is taken into the body and how long the body is in contact with it. For example, a toxin taken into the body at low levels might cause cardiac or respiratory problems. Higher levels might even cause death. Some chemicals, such as asbestos, may be inhaled into the lungs over a long period of time. Eventually, the asbestos can cause chronic diseases of the lungs. Young children that eat, over a period of time, flakes of lead-based paints can suffer damage to their central nervous systems.

Manufacturing, mining, transportation, and farming produce waste products. These chemical substances may adversely affect the ability of the soil, water, and air to support life. Pollution, such as that caused by harmful chemicals, sometimes produces chronic diseases in humans. For example, the long-time exposure to carbon monoxide, sulfur oxides, and nitrogen oxides in the air may cause a number of diseases, including bronchitis, emphysema, and cancer of the lungs.

27-4 Noncommunicable Disease   747

**INTEGRATION
Chemistry**

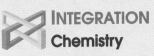

**INTEGRATION
Chemistry**

Many of the chemicals and other pollutants that people are exposed to adversely affect the health of both the world's environment and its living organisms. Chemicals released into the air damage the ozone layer and produce smog and acid precipitation. Agricultural chemicals build up in the water of lakes, rivers, and ponds. Scientists are working on understanding the cause-and-effect relationship between the pollutants and some chronic diseases.

Videodisc

STV: Human Body Vol. 3
Cancer cell surrounded by T cells (gray)

48820
Cancer cells growing in cilia

48827
Cells releasing granules—allergic reaction

48817
Patient being tested for allergies

48814
Woman receiving allergy shot

48818

Use Lab 63 and Lab 64 in the **Lab Manual** as you teach this lesson.

Use **Science Integration Transparency 27** as you teach this lesson.

Community Connection

EPA Have an employee of the Environmental Protection Agency explain how various pollutants are monitored in the local community.

USING MATH

Aside from the health reasons for not smoking, look at an economic one. In your Science Journal, calculate the amount of money a person will save in a week and in a year by not smoking. Assume that the person smokes a pack a day and the cost is $2.50 per pack.

Figure 27-14

Tobacco products have been directly linked to lung cancer.

Cancer

Cancer is a major chronic disease. **Cancer** results from uncontrolled cell growth. There are many different types of cancer, but most of them have these characteristics:

1. Uncontrolled cell growth results in large numbers of cells.
2. The large number of cells do not function as a part of the body.
3. The cells take up space and interfere with normal bodily functions.
4. The cells do not remain in one place, but travel throughout the body. In this way, cancer spreads and grows in many areas of the body.

A **tumor** is an abnormal growth anywhere in the body. If a tumor is near the surface of the body, you may feel it as a lump. But if it is deep inside the body, it may go undetected for years. There are two kinds of tumors, benign and malignant. Benign tumors are not cancerous. Malignant (muh LIHG nuhnt) tumors are cancerous and can spread.

Treatment

Treatment for cancer includes surgery to remove cancerous tissue, radiation with X rays to kill cancer cells, and chemotherapy. **Chemotherapy** is the use of chemicals to destroy cancer cells.

Cancers are complicated, and although **Table 27-3** gives some causes, no one fully understands how cancers form. Some scientists hypothesize that cancer cells form regularly in the body. However, the immune system destroys them. Only if the immune system fails or becomes overwhelmed does the cancer begin to expand. Still, warnings, such as that in **Figure 27-14,** emphasize that smoking, poor diet, and exposure to harmful chemicals stimulate some cancers to form.

Table 27-3

Causes of Cancer	
Carcinogens	**Oncogenes**
Substances that cause cancer: tars in tobacco smoke asbestos dust ultraviolet light radiation air pollution high-fat diet	Genes that cause a normal cell to become cancerous

Allergies

Have you ever broken out in an itchy rash after eating a favorite food? An **allergy** is an overly strong reaction of the immune system to a foreign substance. Many people have allergic reactions to cosmetics, shrimp, strawberries, and bee stings. Allergic reactions to some things such as antibiotics can even be fatal.

Allergens

Substances that cause the allergic response are called **allergens.** These are substances that the body would normally respond to as a mild antigen. Chemicals, dust, food, pollen, molds, and some antibiotics are allergens for some sensitive people. When you come in contact with an allergen, your immune system forms antibodies, and your body may react in many ways. When the body responds to an allergen, chemicals called histamines are released. Histamines

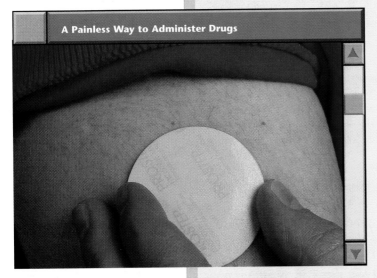

A Painless Way to Administer Drugs

27-4 Noncommunicable Disease   **749**

•MINI•QUIZ•

Use the Mini Quiz to check students' recall of chapter content.

1. **Can some chronic diseases be cured?** *yes*
2. **What are some chemicals that can cause chronic disease?** *household cleaners, car exhaust, cigarette smoke*
3. **The two kinds of tumors are _____ .** *benign, malignant*
4. **Substances that cause allergic responses are called _____ .** *allergens*

Section Wrap-up

Review

1. Answers may be: diabetes, allergies, asthma, cancer, heart disease.
2. uncontrolled cell growth, large number of cells that do not function as part of the body, cells interfere with normal bodily functions, cells travel throughout the body
3. **Think Critically** Joel's immune system seems to be reacting to the antibiotic. Joel should seek immediate medical care.

Science Journal

Journal entries will vary with each individual. The poems may reflect sadness, concern, and frustration. Students may also address the aspect of hope for recovery.

Figure 27-15
Some common substances such as cosmetics, foods, dust mites, and wool from sheep stimulate allergic responses in people. Hives are one kind of allergic reaction.

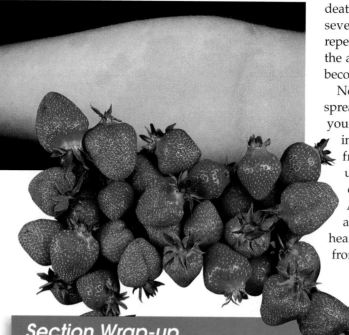

promote red, swollen tissues. Allergic reactions are sometimes treated with antihistamines. Pollen is an allergen that causes a stuffy nose, breathing difficulties, watery eyes, and a tired feeling in some people. Some foods cause blotchy rashes such as hives, shown in **Figure 27-15,** or stomach cramps and diarrhea. Most allergic reactions are minor. But severe allergic reactions can occur, causing shock and even death if not treated promptly. Some severe allergies are treated with repeated injections of small doses of the allergen, which allow the body to become less sensitive to it.

Noncommunicable diseases aren't spread from person to person. So, you don't have to worry about catching diabetes, cancer, or allergies from someone else. But, left untreated, certain noncommunicable diseases can be as deadly as AIDS or untreated syphilis. After all, most people die from chronic heart disease and cancer, and not from a virus or bacterium.

Section Wrap-up

Review

1. What are two noncommunicable diseases?
2. What are the basic characteristics of cancer?
3. **Think Critically:** Joel has an ear infection. The doctor prescribes an antibiotic. After taking the antibiotic, Joel breaks out in a rash and has difficulty breathing. What is happening to him? What should he do immediately?

Skill Builder
Making and Using Tables
Make a table that lists some chronic diseases and their treatments. Use the information in this section. If you need help, refer to Making and Using Tables in the **Skill Handbook.**

Science Journal

In your Science Journal, write a poem about how you would feel if you just learned that your best friend has a serious disease.

 Skill Builder

Chronic Disease	Treatment
Cancer	Radiation, chemotherapy
Allergies	Injections of allergens
Diabetes	Insulin

✓ Assessment

Performance Use this Skill Builder to assess students' abilities to make and use tables by having them classify other diseases and name the treatments given. Use the Performance Task Assessment List for Data Table in **PASC,** p. 37. **P**

Summary

27-1: Disease

1. Pasteur and Koch discovered that diseases are caused by microbes.
2. Communicable diseases caused by pathogenic bacteria, viruses, fungi, and protists can be passed from one person to another by air, water, food, and animal contact.
3. Sexually transmitted diseases (STDs) are passed during sexual contact.

27-2: Your Immune System

1. The purpose of the immune system is to fight disease.
2. Active immunity is long-lasting; passive immunity does not last.
3. AIDS is a communicable disease that damages the body's immune system so that it cannot fight any disease.

27-3: Science and Society: Should a harmful virus be destroyed?

1. Several diseases can be prevented if people are properly vaccinated.
2. Some scientists want to destroy the only remaining samples of the smallpox virus. Other scientists want to keep them.

27-4: Noncommunicable Disease

1. Causes of noncommunicable disease include genetics, chemicals, poor diet, and uncontrolled cell growth.
2. Chronic noncommunicable diseases include diabetes, cancer, arthritis, and allergies.

Key Science Words

a. active immunity
b. allergen
c. allergy
d. antibody
e. antigen
f. antiseptic
g. cancer
h. chemotherapy
i. chronic disease
j. communicable disease
k. disinfectant
l. immune system
m. lymphocyte
n. noncommunicable disease
o. passive immunity
p. pasteurization
q. sexually transmitted disease (STD)
r. tumor
s. vaccination

Reviewing Vocabulary

Match each phrase with the correct term from the list of Key Science Words.

1. causes an allergic reaction
2. disease spread through air, water, or contact
3. chemical that prevents pathogen growth on the skin
4. foreign protein attacked by the body
5. introduces antibodies for short-term immunity
6. white blood cell that fights pathogen
7. long-lasting, noncommunicable disease
8. uncontrolled cell division
9. use of chemicals to destroy cancer cells
10. introduces antigens for long-term immunity

Chapter 27 Review 751

Chapter 27

Review

Summary

Have students read the summary statements to review the major concepts of the chapter.

Reviewing Vocabulary

1. b	**6.** m
2. j	**7.** i
3. f	**8.** g
4. e	**9.** h
5. o	**10.** s

Assessment

Portfolio Encourage students to place in their portfolios one or two items of what they consider to be their best work. Examples include:

- Science Journal, p. 735
- Problem Solving, p. 739
- Explore the Issue, p. 745
- Using Technology, p. 749
  P

Performance Additional performance assessments may be found in **Performance Assessment** and **Science Integration Activities.** Performance Task Assessment Lists and rubrics for evaluating these activities can be found in Glencoe's **Performance Assessment in the Science Classroom.**

GLENCOE TECHNOLOGY

MindJogger Videoquiz

Chapter 27 Have students work in groups as they play the Videoquiz game to review key chapter concepts.

Chapter 27 Review

Checking Concepts

1. a   **6.** b
2. d   **7.** a
3. c   **8.** b
4. a   **9.** d
5. d   **10.** a

Understanding Concepts

11. His experiments helped make food safer to eat.

12. With arthritis, the immune system begins to treat the body's normal proteins as if they were antigens.

13. All of these activities put pathogens on your hands. If any of these pathogens get inside your body, you could get a disease.

14. Antibiotics kill bacterial STDs. Early treatment can prevent damage to body systems.

15. With active immunity, your body produces its own long-lasting antibodies when you get a disease or receive a vaccine. Passive immunity comes from several outside sources when the body does not produce antibodies to particular antigens.

Thinking Critically

16. In some cases, it is better to vaccinate. Some people may not produce antibodies to certain antigens and become ill.

17. Breast-fed babies receive immunity antibodies in their mothers' milk. These antibodies help fight diseases.

18. Your skin prevents pathogens from entering the body, your circulatory system has white blood cells that destroy the pathogens, the cilia and mucus of your respiratory system trap pathogens, and the digestive juices kill those pathogens that enter the stomach.

752

Checking Concepts

Choose the word or phrase that completes the sentence.

1. A pathogen causes a specific disease if it _____.
   a. is present in all cases of the disease
   b. does not infect other animals
   c. causes other diseases
   d. is treated with heat

2. Communicable diseases can be caused by _____.
   a. heredity   c. chemicals
   b. allergies   d. organisms

3. _____ is an example of an STD.
   a. Anthrax   c. AIDS
   b. Malaria   d. Pneumonia

4. _____ is caused by a virus.
   a. AIDS   c. Chlamydia
   b. Gonorrhea   d. Syphilis

5. Your body's defenses against pathogens include all of the following except _____.
   a. stomach enzymes
   b. skin
   c. white blood cells
   d. hormones

6. All of these are noncommunicable diseases except _____.
   a. allergies   c. asthma
   b. syphilis   d. diabetes

7. Lymphocytes are attacked by the virus that causes _____.
   a. AIDS   c. flu
   b. chlamydia   d. polio

8. _____ is a chronic joint disease.
   a. Asthma   c. Muscular dystrophy
   b. Arthritis   d. Diabetes

9. _____ are formed by the blood to fight invading antigens.
   a. Hormones   c. Pathogens
   b. Allergens   d. Antibodies

752   Chapter 27 Immunity

10. Cancer cells are destroyed by _____.
   a. chemotherapy
   b. antigens
   c. vaccines
   d. viruses

Understanding Concepts

Answer the following questions in your Science Journal using complete sentences.

11. How did Pasteur's experiments help society?

12. What makes arthritis different from other immune disorders?

13. Why is it important to wash your hands after using a rest room, petting a dog, handling a pet gerbil or parakeet, and before eating?

14. If a person gets a bacterial STD, why must it be treated promptly with antibiotics?

15. What's the difference between active and passive immunity?

19. Lymphocytes produce antibodies and destroy invading antigens.

20. Antigens are foreign proteins and chemicals that invade your body. Antibodies are formed by your body's immune system to destroy and remove the antigens. Antibiotics are drugs used to kill antigens when your own antibodies can't do the job.

Developing Skills

21. Making and Using Tables

Disease	Prevention
cancer	avoid smoking, alcohol; eat a balanced diet
diabetes	diet and/or insulin
tetanus	vaccination for tetanus; revaccinate if vaccination is over three years old
measles	prevent with vaccination

Thinking Critically

16. Which is better—to vaccinate people or to wait until they build their own immunity?
17. What advantage might a breast-fed baby have compared to a bottle-fed baby?
18. How does your body protect itself from antigens?
19. How do lymphocytes eliminate antigens?
20. Describe the differences among antibodies, antigens, and antibiotics.

Developing Skills

If you need help, refer to the **Skill Handbook.**

21. **Making and Using Tables:** Make a chart comparing the following diseases and their prevention: cancer, diabetes, tetanus, and measles.
22. **Concept Mapping:** Make a network tree concept map, comparing the defenses your body has against disease. Compare general defenses, active immunity, and passive immunity.
23. **Classifying:** Classify the following diseases as communicable or noncommunicable: diabetes, gonorrhea, herpes, strep throat, syphilis, cancer, and flu.
24. **Recognizing Cause and Effect:** Use a library reference to identify the cause of each disease as bacteria, virus, fungus, or protist: athlete's foot, AIDS, cold, dysentery, flu, pinkeye, pneumonia, strep throat, and ringworm.

25. **Making and Using Graphs:** Interpret the graph below showing the rate of polio cases. Explain the rate of cases between 1950 and 1965. What conclusions can you draw about the effectiveness of the polio vaccines?

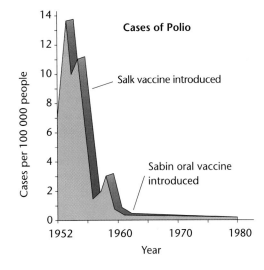

Performance Assessment

1. **Events Chain:** Design a flowchart (diagram) to show how a person with a cold could spread the disease to family members and to others.
2. **Research Project:** Write a report on Lyme disease, its causes, symptoms, and treatment.
3. **Letter:** Choose one side of the debate about whether or not the last cultures of the smallpox virus should be destroyed. Write a letter to the editor of a newspaper supporting the side you chose.

Chapter 27 Review   753

22. **Concept Mapping** See below.
23. **Classifying** communicable: gonorrhea, herpes, strep throat, syphilis, flu; noncommunicable: diabetes, cancer
24. **Recognizing Cause and Effect** *Students will need to consult references other than their texts.* bacteria: strep throat, pneumonia, pinkeye; virus: AIDS, cold, flu, pneumonia, pinkeye; fungus: ringworm, athlete's foot; protist: dysentery
25. **Making and Using Graphs** The rate of cases dropped right after the polio vaccine was introduced. The dramatic decline in polio cases was due to the newly discovered vaccines.

Performance Assessment

1. Have students compare their flowcharts to see how many people are infected by contact with one sick individual. Use the Performance Task Assessment List for Events Chain in **PASC**, p. 91. P
2. If possible, have students interview someone who has contracted the disease and include a transcript of the interview in the report. Use the Performance Task Assessment List for Writing in Science in **PASC**, p. 87. P
3. Responses will vary but should reflect information found in Section 27-3, pp. 744-745. Use the Performance Task Assessment List for Letter in **PASC**, p. 67. P

Concept Map of Immunity

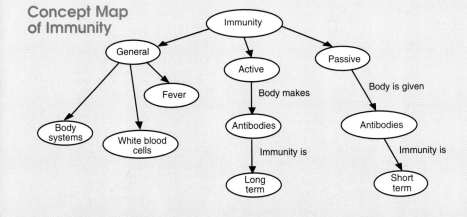

UNIT PROJECT 7

Objectives

Logical-Mathematical Students will practice recording observations, and analyzing the data produced through the use of graphs. **L1** **COOP LEARN**

Summary

In this activity, students will measure and record pulse and breathing rate in a resting state, a state of normal activity, and after exercise. They will also chart changes in performance produced by training for an athletic event. This information will then be analyzed and students will explain the results using their knowledge of various organ systems.

Time Required

This project is designed to be worked on throughout Unit 7. One period is required to set up the Olympic competition. One period a week is needed for training and for measuring performance. Students will need one period to analyze and explain their findings.

Preparation

- Establish cooperative groups of 4 or 5 students.
- Obtain ropes, balls, space for Olympic events, paper to record data, and graph paper.

On Your Mark! Get Set! Go!

The Olympics is exciting sports competition, showing the human body performing at its peak. Many organ systems must work together in an athletic performance. The brain receives input from the sensory organs and directs and coordinates the action. Nerves carry the brain's directions and stimulate the muscles to contract and to move the muscular and skeletal systems. In addition, nerves stimulate the circulatory and respiratory systems to speed the flow of blood with its increased supply of oxygen to the active muscles. The endocrine system also regulates performance through the hormones it releases. Adrenaline, for example, is released during times of stress and may give people greater strength, speed, and quickness than usual—certainly useful to athletes in a tough competition!

Classroom Olympics

How fast can you run the 100-m dash? Does training affect an athlete's ability to perform? For this project, you will plan, train, and participate in your own Olympic events. Working in groups, you will decide on physical activities that you would like to include in the classroom games. Each group is responsible for planning one event and demonstrating the event for the class. After the events are decided, each group will decide who will run the group's event and who will participate in the events planned by other groups.

Procedure for Training

1. Like Olympic athletes, the classroom athletes will need to train. Unlike Olympic athletes, you are competing against your own record. First, you want to look at your current performance level in your event. Have a partner record your level of achievement in the event so that you can compare your later performances to this baseline.

754

2. Now you are ready to train. Have your partner measure your body functions, such as pulse rate and breaths per minute, before you begin. Record these measurements—they will show the normal speed of your circulatory and respiratory systems. Sit quietly for 5 minutes and record the same measurements. These will show your resting rates. Now train in your event for 5 minutes and record these measurements again. Have your circulatory and respiratory systems slowed down or speeded up?

3. When you train, be consistent in the way you measure performance. For example, if you are doing a running event, time your runs on the same course each time. If you are jumping rope, you might measure how long you can jump rope without tripping up.

4. When your training time is finished, switch jobs and let your partner train while you record.

5. Although you will have some time to train in school, you may want to keep practicing at home—or at least do some conditioning. Athletes do conditioning exercises such as running to increase their strength and endurance in general, rather than practicing a particular skill.

Using Your Research
Within your group, analyze the data you have collected on

Go Further
You may wish to investigate the effect of learning on performance. Groups can design games or tasks that require learning a strategy and see how practice affects performance. One example of this kind of task would involve numbering small pieces of paper and having students pick them up in order.

performance. On the x-axis, record the number of each trial. On the y-axis, record your performance in seconds, minutes, meters, number of jumps, etc. Use another graph to chart the pattern of your pulse and breathing rates. Has your level of performance changed at all? Have your pulse and breathing rates changed at all as your training progressed? Think about the organ systems involved in your event and explain the changes or lack of change.

Teaching Strategies
Begin by asking students what sports events they like to play in or watch. Ask if anyone watches the Olympic competitions. What events do they particularly like? Are there any events that do not get much TV coverage on American stations? Discuss why some events are more heavily televised in the United States than others. Discuss what countries appear to have more success in certain events than other countries. For example, Russian gymnasts have been very successful, and winning Olympic cross-country skiers are frequently from just a few European countries. Why do athletes from these countries have such success? Factors can include things such as availability of snow for practice, cost of the sport, emphasis on training and support of athletes, and so on.

Go Further
Students can investigate Olympic records for various events. Why are records continually being broken? Are there any records that have not changed much over time? What factors influence how quickly an athletic record is likely to be broken?

Unit 7 Project 755

References
Mallon, Bill. *The Olympic Record Book.* Hamden, CT: Garland Pub., 1988.

Wallechinsky, David. *The Complete Book of the Olympics*, Rev. ed. New York: Penguin, 1988.

Care and Use of a Microscope

Arm
Supports the body

Low-power objective
Contains the lens with low-power magnification

Stage
Platform used to support the microscope slide

Stage clips
Hold the microscope slide in place

Coarse adjustment
Focuses the image under low power

Fine adjustment
Sharpens the image under high and low magnification

Eyepiece
Contains a magnifying lens you look through

Body
Connects the eyepiece to the revolving nosepiece

Revolving nosepiece
Holds and turns the objectives into viewing position

High-power objective
Contains the lens with the most magnification

Light source
Allows light to reflect upward through the specimen and the lenses

Base
Provides support for the microscope

Care of a Microscope

1. Always carry the microscope holding the arm with one hand and supporting the base with the other hand.
2. Don't touch the lenses with your finger.
3. Never lower the coarse adjustment knob when looking through the eyepiece lens.
4. Always focus first with the low-power objective.
5. Don't use the coarse adjustment knob when the high-power objective is in place.
6. Store the microscope covered.

Using a Microscope

1. Place the microscope on a flat surface that is clear of objects. The arm should be toward you.
2. Look through the eyepiece. Adjust the diaphragm so that light comes through the opening in the stage.
3. Place a slide on the stage so that the specimen is in the field of view. Hold it firmly in place by using the stage clips.

4. Always focus first with the coarse adjustment and the low-power objective lens. Once the object is in focus on low power, turn the nosepiece until the high-power objective is in place. Use ONLY the fine adjustment to focus with this lens.

Making a Wet-Mount Slide

1. Carefully place the item you want to look at in the center of a clean glass slide. Make sure the sample is thin enough for light to pass through.
2. Use a dropper to place one or two drops of water on the sample.
3. Hold a clean coverslip by the edges and place it at one edge of the drop of water. Slowly lower the coverslip onto the drop of water until it lies flat.
4. If you have too much water or a lot of air bubbles, touch the edge of a paper towel to the edge of the coverslip to draw off extra water and force air out.

Appendix B

SI/Metric to English Conversions

	When you want to convert:	To:	Multiply by:
Length	inches	centimeters	2.54
	centimeters	inches	0.39
	feet	meters	0.30
	meters	feet	3.28
	yards	meters	0.91
	meters	yards	1.09
	miles	kilometers	1.61
	kilometers	miles	0.62
Mass and Weight*	ounces	grams	28.35
	grams	ounces	0.04
	pounds	kilograms	0.45
	kilograms	pounds	2.2
	tons (short)	tonnes (metric tons)	0.91
	tonnes (metric tons)	tons (short)	1.10
	pounds	newtons	4.45
	newtons	pounds	0.23
Volume	cubic inches	cubic centimeters	16.39
	cubic centimeters	cubic inches	0.06
	cubic feet	cubic meters	0.03
	cubic meters	cubic feet	35.30
	liters	quarts	1.06
	liters	gallons	0.26
	gallons	liters	3.78
Area	square inches	square centimeters	6.45
	square centimeters	square inches	0.16
	square feet	square meters	0.09
	square meters	square feet	10.76
	square miles	square kilometers	2.59
	square kilometers	square miles	0.39
	hectares	acres	2.47
	acres	hectares	0.40
Temperature	Fahrenheit	5/9 (°F − 32)	Celsius
	Celsius	9/5 (°C + 32)	Fahrenheit

* Weight as measured in standard Earth gravity

Appendix C

Safety in the Classroom

1. Always obtain your teacher's permission to begin an investigation.
2. Study the procedure. If you have questions, ask your teacher. Be sure you understand any safety symbols shown on the page.
3. Use the safety equipment provided for you. Goggles and a safety apron should be worn when any investigation calls for using chemicals.
4. Always slant test tubes away from yourself and others when heating them.
5. Never eat or drink in the lab, and never use lab glassware as food or drink containers. Never inhale chemicals. Do not taste any substances or draw any material into a tube with your mouth.
6. If you spill any chemical, wash it off immediately with water. Report the spill immediately to your teacher.
7. Know the location and proper use of the fire extinguisher, safety shower, fire blanket, first aid kit, and fire alarm.
8. Keep all materials away from open flames. Tie back long hair and loose clothing.
9. If a fire should break out in the classroom, or if your clothing should catch fire, smother it with the fire blanket or a coat, or get under a safety shower. NEVER RUN.
10. Report any accident or injury, no matter how small, to your teacher.

Follow these procedures as you clean up your work area.

1. Turn off the water and gas. Disconnect electrical devices.
2. Return all materials to their proper places.
3. Dispose of chemicals and other materials as directed by your teacher. Place broken glass and solid substances in the proper containers. Never discard materials in the sink.
4. Clean your work area.
5. Wash your hands thoroughly after working in the laboratory.

Table C-1

First Aid	
Injury	**Safe Response**
Burns	Apply cold water. Call your teacher immediately.
Cuts and bruises	Stop any bleeding by applying direct pressure. Cover cuts with a clean dressing. Apply cold compresses to bruises. Call your teacher immediately.
Fainting	Leave the person lying down. Loosen any tight clothing and keep crowds away. Call your teacher immediately.
Foreign matter in eye	Flush with plenty of water. Use eyewash bottle or fountain.
Poisoning	Note the suspected poisoning agent and call your teacher immediately.
Any spills on skin	Flush with large amounts of water or use safety shower. Call your teacher immediately.

Safety Symbols

This textbook uses the safety symbols in **Table C-2** below to alert you to possible laboratory dangers.

Table C-2

Safety Symbols

DISPOSAL ALERT
This symbol appears when care must be taken to dispose of materials properly.

ANIMAL SAFETY
This symbol appears whenever live animals are studied and the safety of the animals and the students must be ensured.

BIOLOGICAL HAZARD
This symbol appears when there is danger involving bacteria, fungi, or protists.

RADIOACTIVE SAFETY
This symbol appears when radioactive materials are used.

OPEN FLAME ALERT
This symbol appears when use of an open flame could cause a fire or an explosion.

CLOTHING PROTECTION SAFETY
This symbol appears when substances used could stain or burn clothing.

THERMAL SAFETY
This symbol appears as a reminder to use caution when handling hot objects.

FIRE SAFETY
This symbol appears when care should be taken around open flames.

SHARP OBJECT SAFETY
This symbol appears when a danger of cuts or punctures caused by the use of sharp objects exists.

EXPLOSION SAFETY
This symbol appears when the misuse of chemicals could cause an explosion.

FUME SAFETY
This symbol appears when chemicals or chemical reactions could cause dangerous fumes.

EYE SAFETY
This symbol appears when a danger to the eyes exists. Safety goggles should be worn when this symbol appears.

ELECTRICAL SAFETY
This symbol appears when care should be taken when using electrical equipment.

POISON SAFETY
This symbol appears when poisonous substances are used.

SKIN PROTECTION SAFETY
This symbol appears when use of caustic chemicals might irritate the skin or when contact with microorganisms might transmit infection.

CHEMICAL SAFETY
This symbol appears when chemicals used can cause burns or are poisonous if absorbed through the skin.

Classification

Scientists use a six-kingdom system of classification of organisms. In this system, there are two kingdoms of organisms, Kingdoms Eubacteria and Archaebacteria, which are organisms that lack a true nucleus and membrane-bound organelles. The members of the other four kingdoms have cells each containing a nucleus and membrane-bound organelles. These kingdoms are Kingdom Protista, Kingdom Fungi, the Plant Kingdom, and the Animal Kingdom.

Kingdom Eubacteria

Cyanobacteria: one-celled prokaryotes; make their own food, contain chlorophyll;

Bacteria: one-celled prokaryotes; absorb food from their surroundings; are round, spiral, or rod shaped

Nostoc
450×

Kingdom Archaebacteria

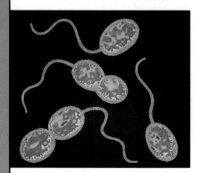

one-celled prokaryotes found in extreme environments; phyla found in salt ponds, hot sulfur springs, and deep ocean thermal vents

Thiocystis
6000×

Kingdom Protista

Phylum Euglenophyta: one-celled; can photosynthesize or take in food; most have one flagellum

Diatoms
63×

Phylum Bacillariophyta: most are one-celled; make their own food through photosynthesis; golden-brown pigments mask chlorophyll; diatoms

Phylum Dinoflagellata: one-celled; make their own food through photosynthesis; contain red pigments and have two flagella; dinoflagellates

Phylum Chlorophyta: one-celled, many-celled, or colonies; contain chlorophyll and make their own food; live on land, in fresh water, or in salt water; green algae

Volvox
100×

Phylum Rhodophyta: most are many-celled and photosynthetic; contain red pigments; most live in deep saltwater environments; red algae

Phylum Phaeophyta: most are many-celled and photosynthetic; contain brown pigments; most live in saltwater environments; brown algae

Amoeba
100×

Phylum Rhizopoda: one-celled; take in food; move by means of pseudopods; free-living or parasitic; sarcodines, amoebas

Phylum Zoomastigina: one-celled; take in food; have two or more flagella; free-living or parasitic; flagellates

Phylum Ciliophora: one-celled; take in food; have large numbers of cilia; ciliates

Phylum Sporozoa: one-celled; take in food; no means of movement; parasites in animals; sporozoans

Phylum Myxomycetes, Phylum Acrasiomycota: one- or many-celled; absorb food; change form during life cycle; cellular and plasmodial slime molds

Chocolate tube slime mold

Phylum Oomycota: live in water or on land; one- or many-celled parasites; absorb dead organic matter; cause diseases in plants and animals; water molds and mildews

Phylum Foraminifera: one-celled, mostly marine organisms that form shells using calcium carbonate; forams found as fossils in many sedimentary rocks

Kingdom Fungi

Division Zygomycota: many-celled; absorb food; spores are produced in sporangia; zygote fungi

Bread mold
5×

Division Ascomycota: one- and many-celled; absorb food; spores produced in asci; sac fungi

Division Basidiomycota: many-celled; absorb food; spores produced in basidia; club fungi

Scaly vase chanterelle

Division Deuteromycota: members with unknown reproductive structures; imperfect fungi

Lichens: organism formed by symbiotic relationship between an ascomycote or a basidiomycote and a green alga or a cyanobacterium; fungus provides protection and the alga or cyanobacterium provides food

Mountain wood fern

British soldier lichen

Plant Kingdom

Division Bryophyta: nonvascular plants that reproduce by spores produced in capsules; many-celled; green; grow in moist land environments; mosses and liverworts

Sphagnum moss

Spore Plants

Division Lycophyta: many-celled vascular plants; spores produced in cones; live on land; are photosynthetic; club mosses

Division Sphenophyta: vascular plants with ribbed and jointed stems; scalelike leaves; spores produced in cones; horsetails

Division Pterophyta: vascular plants with feathery leaves called fronds; spores produced in clusters of sporangia called sori; live on land or in water; ferns

Seed Plants

Division Ginkgophyta: deciduous gymnosperms; only one living species called the maidenhair tree; fan-shaped leaves with branching veins; reproduces with seeds; ginkgos

Ginkgo

Division Cycado-phyta: palmlike gymnosperms; large compound leaves; produce seeds in cones; cycads

Cycad

Division Coniferophyta: deciduous or evergreen gymnosperms; trees or shrubs; needle-like or scalelike leaves; seeds produced in cones; conifers

Pine forest, North Yukon, Canada

Pine flower

Division Gnetophyta: shrubs or woody vines; seeds produced in cones; division contains only three genera; gnetum

Welwitchia

Division Anthophyta: dominant group of plants; ovules protected at fertilization by an ovary; sperm carried to ovules by pollen tube; produce flowers and seeds in fruits; flowering plants

Apple blossoms

Apples

Phylum Porifera: aquatic organisms that lack true tissues and organs; they are asymmetrical and sessile; sponges

Azure sponge

Sea nettle jellyfish

Phylum Cnidaria: radially symmetrical organisms with a digestive cavity with one opening; most have tentacles armed with stinging cells; live in aquatic environments singly or in colonies; includes jellyfish, corals, hydra, and sea anemones

Phylum Platyhelminthes: bilaterally symmetrical worms with flattened bodies; digestive system has one opening; parasitic and free-living species; flatworms

Flatworm

Phylum Nematoda: round, bilaterally symmetrical body; digestive system with two openings; some free-living forms but mostly parasitic; roundworms

Scallop

Phylum Mollusca: soft-bodied animals, many with a hard shell; a mantle covers the soft body; aquatic and terrestrial species; includes clams, snails, squid, and octopuses

Phylum Annelida: bilaterally symmetrical worms with round, segmented bodies; terrestrial and aquatic species; well-developed body systems; includes earthworms, leeches, and marine polychaetes

Fanworm

Phylum Arthropoda: largest phylum of organisms that have segmented bodies with pairs of jointed appendages, and a hard exoskeleton; terrestrial and aquatic species; includes insects, crustaceans, spiders, and horseshoe crabs

Fiddler crab

Black and yellow argiope

Phylum Echinodermata: saltwater organisms with spiny or leathery skin; water-vascular system with tube feet; radial symmetry; includes starfish, sand dollars, and sea urchins

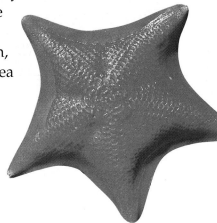

Bat star

Phylum Chordata: organisms with internal skeletons, specialized body systems, and paired appendages; all at some time have a notochord, dorsal nerve cord, gill slits, and a tail; include fish, amphibians, reptiles, birds, and mammals

Regal angelfish

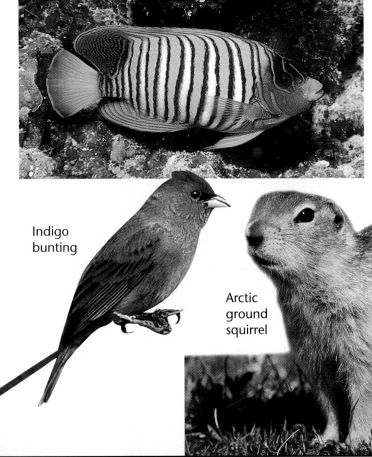

Indigo bunting

Arctic ground squirrel

Table of Contents

Organizing Information

Communicating 766
Classifying 766
Sequencing 767
Concept Mapping 767
Making and Using Tables 769
Making and Using Graphs 770

Thinking Critically

Observing and Inferring 773
Comparing and Contrasting 773
Recognizing Cause and Effect 774

Practicing Scientific Processes

Forming Operational Definitions 775
Forming a Hypothesis 775
Designing an Experiment to Test a Hypothesis 776
Separating and Controlling Variables 777
Interpreting Data 777

Representing and Applying Data

Interpreting Scientific Illustrations 779
Making Models 780
Measuring in SI 781
Predicting 783
Using Numbers 784

Organizing Information

Communicating

The communication of ideas is an important part of our everyday lives. Whether reading a book, writing a letter, or watching a television program, people everywhere are expressing opinions and sharing information with one another. Writing in your Science Journal allows you to express your opinions and demonstrate your knowledge of the information presented on a subject. When writing, keep in mind the purpose of the assignment and the audience with which you are communicating.

Examples Science Journal assignments vary greatly. They may ask you to take a viewpoint other than your own; perhaps you will be a scientist, a TV reporter, or a committee member of a local environmental group. Maybe you will be expressing your opinions to a member of Congress, a doctor, or to the editor of your local newspaper, as shown in **Figure 1.** Sometimes, Science Journal writing may allow you to summarize information in the form of an outline, a letter, or in a paragraph.

Figure 2

Classifying CDs.

Classifying

You may not realize it, but you make things orderly in the world around you. If you hang your shirts together in the closet or if your favorite CDs are stacked together, you have used the skill of classifying.

Classifying is the process of sorting objects or events into groups based on common features. When classifying, first observe the objects or events to be classified. Then, select one feature that is shared by some members in the group but not by all. Place those members that share that feature into a subgroup. You can classify members into smaller and smaller subgroups based on characteristics.

Remember, when you classify, you are grouping objects or events for a purpose. Keep your purpose in mind as you select the features to form groups and subgroups.

Example How would you classify a collection of CDs? As shown in **Figure 2,** you might classify those you like to dance to in one subgroup and CDs you like to

Figure 1

A Science Journal entry.

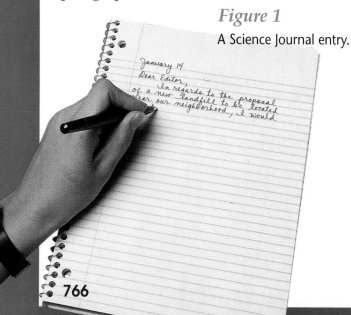

listen to in the next column. The CDs you like to dance to could be subdivided into a rap subgroup and a rock subgroup. Note that for each feature selected, each CD fits into only one subgroup. You would keep selecting features until all the CDs are classified. **Figure 2** shows one possible classification.

Figure 3

A recipe for bread contains sequenced instructions.

Sequencing

A sequence is an arrangement of things or events in a particular order. When you are asked to sequence objects or events within a group, figure out what comes first, then think about what should come second. Continue to choose objects or events until all of the objects you started out with are in order. Then, go back over the sequence to make sure each thing or event in your sequence logically leads to the next.

Example A sequence with which you are most familiar is the use of alphabetical order. Another example of sequence would be the steps in a recipe, as shown in **Figure 3.** Think about baking bread. Steps in the recipe have to be followed in order for the bread to turn out right.

Concept Mapping

If you were taking an automobile trip, you would probably take along a road map. The road map shows your location, your destination, and other places along the way. By looking at the map and finding where you are, you can begin to understand where you are in relation to other locations on the map.

A concept map is similar to a road map. But, a concept map shows relationships among ideas (or concepts) rather than places. A concept map is a diagram that visually shows how concepts are related. Because the concept map shows relationships among ideas, it can make the meanings of ideas and terms clear, and help you understand better what you are studying.

There is usually not one correct way to create a concept map. As you construct one type of map, you may discover other

Figure 4

Network tree describing U.S. currency.

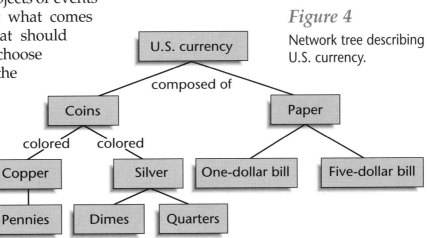

ways to construct the map that show the relationships between concepts in a better way. If you do discover what you think is a better way to create a concept map, go ahead and use the new one. Overall, concept maps are useful for breaking a big concept down into smaller parts, making learning easier.

Examples

Network Tree Look at the concept map about U.S. currency in **Figure 4.** This is called a network tree. Notice how some words are in rectangles while others are written across connecting lines. The words inside the rectangles are science concepts. The lines in the map show related concepts. The words written on the lines describe the relationships between concepts.

When you are asked to construct a network tree, write down the topic and list the major concepts related to that topic on a piece of paper. Then look at your list and begin to put them in order from general to specific. Branch the related concepts from the major concept and describe the relationships on the lines. Continue to write the more specific concepts. Write the relationships between the concepts on the lines until all concepts are mapped. Examine the concept map for relationships that cross branches, and add them to the concept map.

Events Chain An events chain is another type of concept map. An events chain map, such as the one describing a typical morning routine in **Figure 5,** is used to describe ideas in order. In science, an events chain can be used to describe a sequence of events, the steps in a procedure, or the stages of a process.

When making an events chain, first find the one event that starts the chain. This event is called the initiating event. Then,

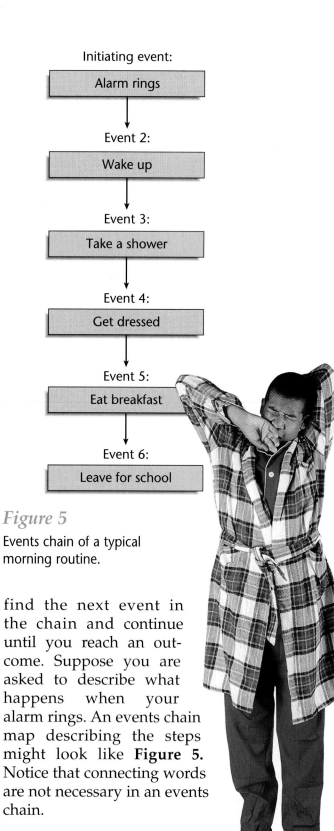

Initiating event:

Alarm rings

Event 2:

Wake up

Event 3:

Take a shower

Event 4:

Get dressed

Event 5:

Eat breakfast

Event 6:

Leave for school

Figure 5

Events chain of a typical morning routine.

find the next event in the chain and continue until you reach an outcome. Suppose you are asked to describe what happens when your alarm rings. An events chain map describing the steps might look like **Figure 5.** Notice that connecting words are not necessary in an events chain.

Cycle Map A cycle concept map is a special type of events chain map. In a cycle concept map, the series of events does not produce a final outcome. Instead, the last event in the chain relates back to the initiating event.

As in the events chain map, you first decide on an initiating event and then list each event in order. Because there is no outcome and the last event relates back to the initiating event, the cycle repeats itself. Look at the cycle map describing the relationship between day and night in **Figure 6.**

Figure 6

Cycle map of day and night.

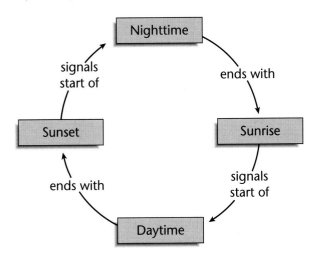

Spider Map A fourth type of concept map is the spider map. This is a map that you can use for brainstorming. Once you have a central idea, you may find you have a jumble of ideas that relate to it, but are not necessarily clearly related to each other. As illustrated by the homework spider map in **Figure 7,** by writing these ideas outside the main concept, you may begin to separate and group unrelated terms so that they become more useful.

Figure 7

Spider map about homework.

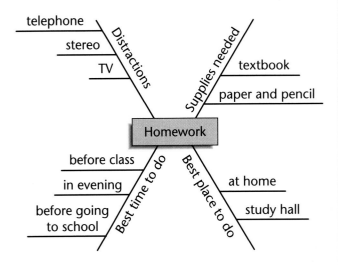

Making and Using Tables

Browse through your textbook and you will notice tables in the text and in the activities. In a table, data or information is arranged in a way that makes it easier for you to understand. Activity tables help organize the data you collect during an activity so that results can be interpreted more easily.

Examples Most tables have a title. At a glance, the title tells you what the table is about. A table is divided into columns and rows. The first column lists items to be compared. In **Figure 8,** the collection of recyclable materials is being compared in a table. The row across the top lists the specific characteristics being compared. Within the grid of the table, the collected data are recorded.

What is the title of the table in **Figure 8?** The title is "Recycled Materials." What is being compared? The different materials being recycled and on which days they are recycled.

Making Tables To make a table, list the items to be compared down in columns

Figure 8

Table of recycled materials.

Recycled Materials			
Day of Week	Paper (kg)	Aluminum (kg)	Plastic (kg)
Mon.	4.0	2.0	0.5
Wed.	3.5	1.5	0.5
Fri.	3.0	1.0	1.5

and the characteristics to be compared across in rows. The table in **Figure 8** compares the mass of recycled materials collected by a class. On Monday, students turned in 4.0 kg of paper, 2.0 kg of aluminum, and 0.5 kg of plastic. On Wednesday, they turned in 3.5 kg of paper, 1.5 kg of aluminum, and 0.5 kg of plastic. On Friday, the totals were 3.0 kg of paper, 1.0 kg of aluminum, and 1.5 kg of plastic.

Using Tables How much plastic, in kilograms, is being recycled on Wednesday? Locate the column labeled "Plastic (kg)" and the row "Wed." The data in the box where the column and row intersect are the answer. Did you answer "0.5"? How much aluminum, in kilograms, is being recycled on Friday? If you answered "1.0," you understand how to use the parts of the table.

Making and Using Graphs

After scientists organize data in tables, they may display the data in a graph. A graph is a diagram that shows the relationship of one variable to another. A graph makes interpretation and analysis of data easier. There are three basic types of graphs used in science—the line graph, the bar graph, and the circle graph.

Examples

Line Graphs A line graph is used to show the relationship between two variables. The variables being compared go on two axes of the graph. The independent variable always goes on the horizontal axis, called the x-axis. The dependent variable always goes on the vertical axis, called the y-axis.

Suppose your class started to record the amount of materials they collected in one week for their school to recycle. The collected information is shown in **Figure 9.**

You could make a graph of the materials collected over the three days of the school week. The three weekdays are the independent variables and are placed on the x-axis of your graph. The amount of materials collected is the dependent variable and would go on the y-axis.

After drawing your axes, label each with a scale. The x-axis lists the three weekdays. To make a scale of the amount of materials collected on the y-axis, look at the data values. Because the lowest amount collected was 1.0 and the highest

Figure 9

Amount of recyclable materials collected during one week.

Materials Collected During Week		
Day of Week	Paper (kg)	Aluminum (kg)
Mon.	5.0	4.0
Wed.	4.0	1.0
Fri.	2.5	2.0

Material Collected During Week

Figure 10

Graph outline for material collected during week.

Material Collected During Week

Figure 11

Line graph of materials collected during week.

was 5.0, you will have to start numbering at least at 1.0 and go through 5.0. You decide to start numbering at 0 and number by ones through 6.0, as shown in **Figure 10.**

Next, plot the data points for collected paper. The first pair of data you want to plot is Monday and 5.0 kg of paper. Locate "Monday" on the *x*-axis and locate "5.0" on the *y*-axis. Where an imaginary vertical line from the *x*-axis and an imaginary horizontal line from the *y*-axis would meet, place the first data point. Place the other data points the same way. After all the points are plotted, connect them with the best smooth curve. Repeat this procedure for the data points for aluminum. Use continuous and dashed lines to distinguish the two line graphs. The resulting graph should look like **Figure 11.**

Bar Graphs Bar graphs are similar to line graphs. They compare data that do not continuously change. In a bar graph, vertical bars show the relationships among data.

To make a bar graph, set up the *x*-axis and *y*-axis as you did for the line graph.

The data is plotted by drawing vertical bars from the *x*-axis up to a point where the *y*-axis would meet the bar if it were extended.

Look at the bar graph in **Figure 12** comparing the mass of aluminum collected over three weekdays. The *x*-axis is the days on which the aluminum was collected. The *y*-axis is the mass of aluminum collected, in kilograms.

Circle Graphs A circle graph uses a circle divided into sections to display data. Each section represents part of the whole. All the sections together equal 100 percent.

Suppose you wanted to make a circle graph to show the number of seeds that germinated in a package. You would count the total number of seeds. You find that there are 143 seeds in the package. This represents 100 percent, the whole circle.

You plant the seeds, and 129 seeds germinate. The seeds that germinated will make up one section of the circle graph, and the seeds that did not germinate will make up the remaining section.

Aluminum Collected During Week

Mass (Kg) — 4.0, 3.0, 2.0, 1.0

Monday   Wednesday   Friday

Day of Collection

Figure 12

Bar graph of aluminum collected during week.

To find out how much of the circle each section should take, divide the number of seeds in each section by the total number of seeds. Then multiply your answer by 360, the number of degrees in a circle, and round to the nearest whole number. The section of the circle graph in degrees that represents the seeds germinated is figured below.

$$\frac{129}{143} \times 360 = 324.75 \text{ or } 325 \text{ degrees (or } 325°)$$

Plot this group on the circle graph using a compass and a protractor. Use the compass to draw a circle. It will be easier to measure the part of the circle representing the non-germinating seeds, so subtract 325° from 360° to get 35°. Draw a straight line from the center to the edge of the circle. Place your protractor on this line and use it to mark a point at 35°. Use this point to draw a straight line from the center of the circle to the edge. This is the section for the group of seeds that did not germinate. The other section represents the group of 129 seeds that did germinate. Label the sections of your graph and title the graph as shown in **Figure 13.**

Figure 13

Circle graph of germinated seeds.

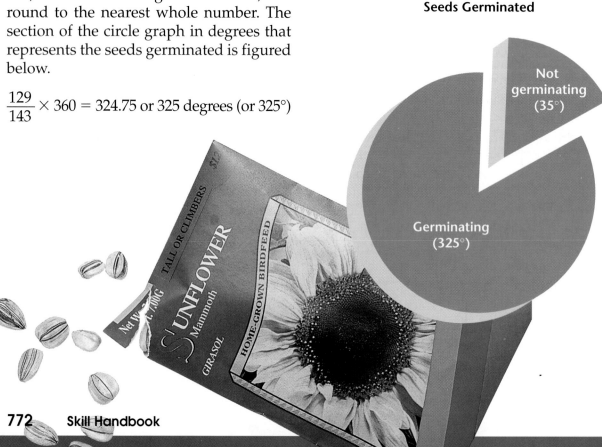

Seeds Germinated

Not germinating (35°)

Germinating (325°)

Thinking Critically

Observing and Inferring

Observing Scientists try to make careful and accurate observations. When possible, they use instruments such as microscopes, thermometers, and balances to make observations. Measurements with a balance or thermometer provide numerical data that can be checked and repeated.

When you make observations in science, you'll find it helpful to examine the entire object or situation first. Then, look carefully for details. Write down everything you observe.

Example Imagine that you have just finished a volleyball game. At home, you open the refrigerator and see a jug of orange juice on the back of the top shelf. The jug, shown in **Figure 14,** feels cold as you grasp it. Then you drink the juice, smell the oranges, and enjoy the tart taste in your mouth.

Figure 14

Why is this jug of orange juice cold?

As you imagined yourself in the story, you used your senses to make observations. You used your sense of sight to find the jug in the refrigerator, your sense of touch when you felt the coldness of the jug, your sense of hearing to listen as the liquid filled the glass, and your senses of smell and taste to enjoy the odor and tartness of the juice. The basis of all scientific investigation is observation.

Inferring Scientists often make inferences based on their observations. An inference is an attempt to explain or interpret observations or to say what caused what you observed.

When making an inference, be certain to use accurate data and observations. Analyze all of the data that you've collected. Then, based on everything you know, explain or interpret what you've observed.

Example When you drank a glass of orange juice after the volleyball game, you observed that the orange juice was cold as well as refreshing. You might infer that the juice was cold because it had been made much earlier in the day and had been kept in the refrigerator or you might infer that it had just been made, using both cold water and ice. The only way to be sure which inference is correct is to investigate further.

Comparing and Contrasting

Observations can be analyzed by noting the similarities and differences between two or more objects or events that you observe. When you look at objects or events to see how they are similar, you are comparing them. Contrasting is looking for differences in similar objects or events.

Figure 15

Table comparing the nutritional value of *Candy A* and *Candy B*.

Nutritional Value		
	Candy A	**Candy B**
Serving size	103 g	105 g
Calories	220	160
Total Fat	10 g	10 g
Protein	2.5 g	2.6 g
Total Carbohydrate	30 g	15 g

Example Suppose you were asked to compare and contrast the nutritional value of two candy bars, *Candy A* and *Candy B*. You would start by looking at what is known about these candy bars. Arrange this information in a table, like the one in **Figure 15.**

Similarities you might point out are that both candy bars have similar serving sizes, amounts of total fat, and protein. Differences include *Candy A* having a

higher calorie value and containing more total carbohydrates than *Candy B*.

Recognizing Cause and Effect

Have you ever watched something happen and then made suggestions about why it happened? If so, you have observed an effect and inferred a cause. The event is an effect, and the reason for the event is the cause.

Example Suppose that every time your teacher fed the fish in a classroom aquarium, she or he tapped the food container on the edge of the aquarium. Then, one day your teacher just happened to tap the edge of the aquarium with a pencil while making a point. You observed the fish swim to the surface of the aquarium to feed, as shown in **Figure 16.** What is the effect, and what would you infer to be the cause? The effect is the fish swimming to the surface of the aquarium. You might infer the cause to be the teacher tapping on the edge of the aquarium. In determining cause and effect, you have made a logical inference based on your observations.

Perhaps the fish swam to the surface because they reacted to the teacher's waving hand or for some other reason. When scientists are unsure of the cause of a certain event, they design controlled experiments to determine what causes the event. Although you have made a logical conclusion about the behavior of the fish, you would have to perform an experiment to be certain that it was the tapping that caused the effect you observed.

Figure 16

What cause-and-effect situations are occurring in this aquarium?

Practicing Scientific Processes

You might say that the work of a scientist is to solve problems. But when you decide how to dress on a particular day, you are doing problem solving, too. You may observe what the weather looks like through a window. You may go outside and see whether what you are wearing is warm or cool enough.

Scientists use an orderly approach to learn new information and to solve problems. The methods scientists may use include observing to form a hypothesis, designing an experiment to test a hypothesis, separating and controlling variables, and interpreting data.

Forming Operational Definitions

Operational definitions define an object by showing how it functions, works, or behaves. Such definitions are written in terms of how an object works or how it can be used; that is, what is its job or purpose?

Figure 17

What observations can be made about this dog?

Example Some operational definitions explain how an object can be used.
- A ruler is a tool that measures the size of an object.
- An automobile can move things from one place to another.

Or such a definition may explain how an object works.
- A ruler contains a series of marks that can be used as a standard when measuring.
- An automobile is a vehicle that can move from place to place.

Forming a Hypothesis

Observations You observe all the time. Scientists try to observe as much as possible about the things and events they study so they know that what they say about their observations is reliable.

Some observations describe something using only words. These observations are called qualitative observations. Other observations describe how much of something there is. These are quantitative observations and use numbers, as well as words, in the description. Tools or equipment are used to measure the characteristic being described.

Example If you were making qualitative observations of the dog in **Figure 17,** you might use words such as *furry, yellow,* and *short-haired.* Quantitative observations of this dog might include a mass of 14 kg, a height of 46 cm, ear length of 10 cm, and an age of 150 days.

Hypotheses Hypotheses are tested to help explain observations that have been made. They are often stated as *if* and *then* statements.

Examples Suppose you want to make a perfect score on a spelling test. Begin by thinking of several ways to accomplish this. Base these possibilities on past observations. If you put each of these possibilities into sentence form, using the words *if* and *then*, you can form a hypothesis. All of the following are hypotheses you might consider to explain how you could score 100 percent on your test:

> If the test is easy, then I will get a perfect score.

> If I am intelligent, then I will get a perfect score.

> If I study hard, then I will get a perfect score.

Perhaps a scientist has observed that plants that receive fertilizer grow taller than plants that do not. A scientist may form a hypothesis that says: If plants are fertilized, then their growth will increase.

Designing an Experiment to Test a Hypothesis

In order to test a hypothesis, it's best to write out a procedure. A procedure is the plan that you follow in your experiment. A procedure tells you what materials to use and how to use them. After following the procedure, data are generated. From this generated data, you can then draw a conclusion and make a statement about your results.

If the conclusion you draw from the data supports your hypothesis, then you can say that your hypothesis is reliable. *Reliable* means that you can trust your conclusion. If it did not support your hypothesis, then you would have to make new observations and state a new hypothesis—just make sure that it is one that you can test.

Example Super premium gasoline costs more than regular gasoline. Does super premium gasoline increase the efficiency or fuel mileage of your family car? Let's figure out how to conduct an experiment to test the hypothesis, "*if* premium gas is more efficient, *then* it should increase the fuel mileage of our family car." Then a procedure similar to **Figure 18** must be written to generate data presented in **Figure 19** on page 777.

These data show that premium gasoline is less efficient than regular gasoline. It took more gasoline to travel one mile (0.064) using premium gasoline than it does to travel one mile using regular gasoline (0.059). This conclusion does not support the original hypothesis made.

PROCEDURE

1. Use regular gasoline for two weeks.

2. Record the number of miles between fill-ups and the amount of gasoline used.

3. Switch to premium gasoline for two weeks.

4. Record the number of miles between fill-ups and the amount of gasoline used.

Figure 18

Possible procedural steps.

Figure 19

Data generated from procedure steps.

	Miles traveled	Gallons used	Gallons per mile
Regular gasoline	762	45.34	0.059
Premium gasoline	661	42.30	0.064

Separating and Controlling Variables

In any experiment, it is important to keep everything the same except for the item you are testing. The one factor that you change is called the *independent variable.* The factor that changes as a result of the independent variable is called the *dependent variable.* Always make sure that there is only one independent variable. If you allow more than one, you will not know what causes the changes you observe in the independent variable. Many experiments have *controls*—a treatment or an experiment that you can compare with the results of your test groups.

Example In the experiment with the gasoline, you made everything the same except the type of gasoline being used. The driver, the type of automobile, and the weather conditions should remain the same throughout. The gasoline should also be purchased from the same service station. By doing so, you made sure that at the end of the experiment, any differences were the result of the type of fuel being used—regular or premium. The type of gasoline was the *independent factor* and the gas mileage achieved was the *dependent factor.* The use of regular gasoline was the *control.*

Interpreting Data

The word *interpret* means "to explain the meaning of something." Look at the problem originally being explored in the gasoline experiment and find out what the data show. Identify the control group and the test group so you can see whether or not the variable has had an effect. Then you need to check differences between the control and test groups.

Figure 20

Which gasoline type is most efficient?

These differences may be qualitative or quantitative. A qualitative difference would be a difference that you could observe and describe, while a quantitative difference would be a difference you can measure using numbers. If there are differences, the variable being tested may have had an effect. If there is no difference between the control and the test groups, the variable being tested apparently has had no effect.

Example Perhaps you are looking at a table from an experiment designed to test the hypothesis: If premium gas is more efficient, then it should increase the fuel mileage of our family car. Look back at **Figure 19** showing the results of this experiment. In this example, the use of regular gasoline in the family car was the control, while the car being fueled by premium gasoline was the test group.

Data showed a quantitative difference in efficiency for gasoline consumption. It took 0.059 gallons of regular gasoline to travel one mile, while it took 0.064 gallons of the premium gasoline to travel the same distance. The regular gasoline was more efficient; it increased the fuel mileage of the family car.

What are data? In the experiment described on these pages, measurements were taken so that at the end of the experiment, you had something concrete to interpret. You had numbers to work with. Not every experiment that you do will give you data in the form of numbers. Sometimes, data will be in the form of a description. At the end of a chemistry experiment, you might have noted that one solution turned yellow when treated with a particular chemical, and another remained clear, like

Figure 21
Data.

water, when treated with the same chemical. Data, therefore, are stated in different forms for different types of scientific experiments.

Are all experiments alike? Keep in mind as you perform experiments in science that not every experiment makes use of all of the parts that have been described on these pages. For some, it may be difficult to design an experiment that will always have a control. Other experiments are complex enough that it may be hard to have only one dependent variable. Real scientists encounter many variations in the methods that they use when they perform experiments. The skills in this handbook are here for you to use and practice. In real situations, their uses will vary.

Representing and Applying Data

Interpreting Scientific Illustrations

As you read a science textbook, you will see many drawings, diagrams, and photographs. Illustrations help you to understand what you read. Some illustrations are included to help you understand an idea that you can't see easily by yourself. For instance, we can't see atoms, but we can look at a diagram of an atom and that helps us to understand some things about atoms. Seeing something often helps you remember more easily. Illustrations also provide examples that clarify difficult concepts or give additional information about the topic you are studying. Maps, for example, help you to locate places that may be described in the text.

Examples

Captions and Labels Most illustrations have captions. A caption is a comment that identifies or explains the illustration. Diagrams, such as **Figure 22,**

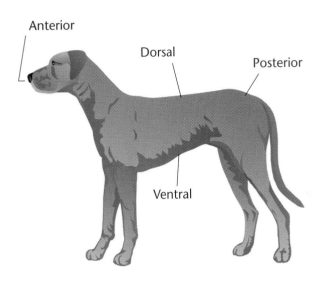

Figure 23

The orientation of a dog is shown here.

often have labels that identify parts of the organism or the order of steps in a process.

Learning with Illustrations An illustration of an organism shows that organism from a particular view or orientation. In order to understand the illustration, you may need to identify the front (anterior) end, tail (posterior) end, the underside (ventral), and the back (dorsal) side as shown in **Figure 23.**

You might also check for symmetry. A shark in **Figure 24** has bilateral symmetry. This means that drawing an imaginary line through the center of the animal from the anterior to posterior end forms two mirror images.

Radial symmetry is the arrangement of similar parts around a central point. An

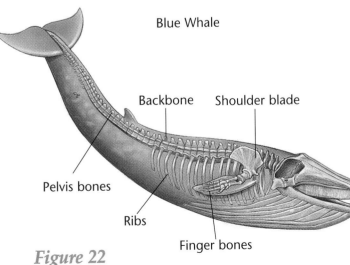

Figure 22

A labeled diagram of a blue whale.

Figure 24

A shark (A) illustrating bilateral symmetry and a pear (B) illustrating a longitudinal section and a cross section.

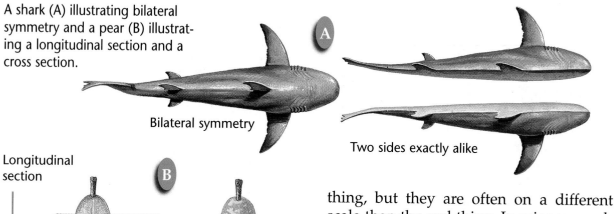

Bilateral symmetry

Two sides exactly alike

Longitudinal section

Cross section

object or organism such as a hydra can be divided anywhere through the center into similar parts.

Some organisms and objects cannot be divided into two similar parts. If an organism or object cannot be divided, it is asymmetrical. Regardless of how you try to divide a natural sponge, you cannot divide it into two parts that look alike.

Some illustrations enable you to see the inside of an organism or object. These illustrations are called sections. **Figure 24** also illustrates some common sections.

Look at all illustrations carefully. Read captions and labels so that you understand exactly what the illustration is showing you.

Making Models

Have you ever worked on a model car or plane or rocket? These models look, and sometimes work, much like the real thing, but they are often on a different scale than the real thing. In science, models are used to help simplify large or small processes or structures that otherwise would be difficult to see and understand. Your understanding of a structure or process is enhanced when you work with materials to make a model that shows the basic features of the structure or process.

Example In order to make a model, you first have to get a basic idea about the structure or process involved. You decide to make a model to show the differences in size of arteries, veins, and capillaries. First, read about these structures. All three are hollow tubes. Arteries are round and thick. Veins are flat and have thinner walls than arteries. Capillaries are small.

Now, decide what you can use for your model. Common materials are often best and cheapest to work with when making models. As illustrated in **Figure 25** on page 781, different kinds and sizes of pasta might work for these models. Different sizes of rubber tubing might do just as well. Cut and glue the different noodles or tubing onto thick paper so the openings can be seen. Then label each. Now you have a simple, easy-to-understand model showing the differences in size of arteries, veins, and capillaries.

What other scientific ideas might a model help you to understand? A model

Figure 25

Different types of pasta may be used to model blood vessels.

of a molecule can be made from gumdrops (using different colors for the different elements present) and toothpicks (to show different chemical bonds). A working model of a volcano can be made from clay, a small amount of baking soda, vinegar, and a bottle cap. Other models can be devised on a computer. Some models are mathematical and are represented by equations.

Measuring in SI

The metric system is a system of measurement developed by a group of scientists in 1795. It helps scientists avoid problems by providing standard measurements that all scientists around the world can understand. A modern form of the metric system, called the International System, or SI, was adopted for worldwide use in 1960.

The metric system is convenient because unit sizes vary by multiples of 10. When changing from smaller units to larger units, divide by 10. When changing from larger units to smaller, multiply by 10. For example, to convert millimeters to centimeters, divide the millimeters by 10. To convert 30 millimeters to centimeters, divide 30 by 10 (30 millimeters equal 3 centimeters).

Prefixes are used to name units. Look at **Figure 26** for some common metric prefixes and their meanings. Do you see how the prefix *kilo-* attached to the unit *gram* is *kilogram,* or 1000 grams? The prefix *deci-* attached to the unit *meter* is *decimeter,* or one-tenth (0.1) of a meter.

Examples

Length You have probably measured lengths or distances many times. The meter is the SI unit used to measure length. A baseball bat is about one meter long. When measuring smaller lengths, the meter is divided into smaller units called centimeters and millimeters. A centimeter is one-hundredth (0.01) of a meter, which is about the size of the width

Figure 26

Common metric prefixes.

Metric Prefixes			
Prefix	**Symbol**	**Meaning**	
kilo-	k	1000	thousand
hecto-	h	200	hundred
deka-	da	10	ten
deci-	d	0.1	tenth
centi-	c	0.01	hundredth
milli-	m	0.001	thousandth

Figure 27

Metric ruler showing centimeter and millimeter divisions.

of the fingernail on your ring finger. A millimeter is one-thousandth of a meter (0.001), about the thickness of a dime.

Most metric rulers have lines indicating centimeters and millimeters, as shown in **Figure 27.** The centimeter lines are the longer, numbered lines; the shorter lines are millimeter lines. When using a metric ruler, line up the 0-centimeter mark with the end of the object being measured, and read the number of the unit where the object ends, in this instance 4.5 cm.

Surface Area Units of length are also used to measure surface area. The standard unit of area is the square meter (m²). A square that's one meter long on each side has a surface area of one square meter. Similarly, a square centimeter, (cm²), shown in **Figure 28,** is one centimeter long on each side. The surface area of an object is determined by multiplying the length times the width.

Volume The volume of a rectangular solid is also calculated using units of length. The cubic meter (m³) is the standard SI unit of volume. A cubic meter is a cube one meter on each side. You can determine the volume of rectangular solids by multiplying length times width times height.

Liquid Volume During science activities, you will measure liquids using beakers and graduated cylinders marked in milliliters, as illustrated in **Figure 29.** A graduated cylinder is a cylindrical container marked with lines from bottom to top.

Liquid volume is measured using a unit called a liter. A liter has the volume of 1000 cubic centimeters. Because the prefix *milli-* means thousandth (0.001), a milliliter equals one cubic centimeter. One

Figure 29

A volume of 79 mL is measured by reading at the lowest point of the curve.

Figure 28

A square centimeter.

1 cm

1 cm

milliliter of liquid would completely fill a cube measuring one centimeter on each side.

Mass Scientists use balances to find the mass of objects in grams. You will use a beam balance similar to **Figure 30.** Notice that on one side of the balance is a pan and on the other side is a set of beams. Each beam has an object of a known mass called a *rider* that slides on the beam.

Before you find the mass of an object, set the balance to zero by sliding all the riders back to the zero point. Check the pointer on the right to make sure it swings an equal distance above and below the zero point on the scale. If the swing is unequal, find and turn the adjusting screw until you have an equal swing.

Place an object on the pan. Slide the rider with the largest mass along its beam until the pointer drops below zero. Then move it back one notch. Repeat the process on each beam until the pointer swings an equal distance above and below the zero point. Add the masses on each beam to find the mass of the object.

You should never place a hot object or pour chemicals directly onto the pan.

Instead, find the mass of a clean beaker or a glass jar. Place the dry or liquid chemicals in the container. Then find the combined mass of the container and the chemicals. Calculate the mass of the chemicals by subtracting the mass of the empty container from the combined mass.

Predicting

When you apply a hypothesis, or general explanation, to a specific situation, you predict something about that situation. First, you must identify which hypothesis fits the situation you are considering.

Examples People use prediction to make everyday decisions. Based on previous observations and experiences, you may form a hypothesis that if it is wintertime, then temperatures will be lower. From past experience in your area, temperatures are lowest in February. You may then use this hypothesis to predict specific temperatures and weather for the month of February in advance. Someone could use these predictions to plan to set aside more money for heating bills during that month.

Figure 30

A beam balance is used to measure mass.

Using Numbers

When working with large populations of organisms, scientists usually cannot observe or study every organism in the population. Instead, they use a sample or a portion of the population. To sample is to take a small representative portion of organisms of a population for research. By making careful observations or manipulating variables within a portion of a group, information is discovered and conclusions are drawn that might then be applied to the whole population.

Scientific work also involves estimating. To estimate is to make a judgment about the size of something or the number of something without actually measuring or counting every member of a population.

Examples Suppose you are trying to determine the effect of a specific nutrient on the growth of black-eyed Susans. It would be impossible to test the entire population of black-eyed Susans, so you would select part of the population for your experiment. Through careful experimentation and observation on a sample of the population, you could generalize the effect of the chemical on the entire population.

Here is a more familiar example. Have you ever tried to guess how many beans were in a sealed jar? If you did, you were estimating. What if you knew the jar of beans held one liter (1000 mL)? If you knew that 30 beans would fit in a 100-milliliter jar, how many beans would you estimate to be in the one-liter jar? If you said about 300 beans, your estimate would be close to the actual number of beans. Can you estimate how many jelly beans are on the cookie sheet in **Figure 31?**

Scientists use a similar process to estimate populations of organisms from bacteria to buffalo. Scientists count the actual number of organisms in a small sample and then estimate the number of organisms in a larger area. For example, if a scientist wanted to count the number of bacterial colonies in a petri dish, a microscope could be used to count the number of organisms in a one-square-centimeter sample. To determine the total population of the culture, the number of organisms in the square-centimeter sample is multiplied by the total number of square centimeters in the culture.

Figure 31

Sampling a group of jelly beans allows for an estimation of the total number of jelly beans in the group.

This glossary defines each key term that appears in **bold type** in the text. It also shows the page number where you can find the word used.

PRONUNCIATION KEY

a...b**a**ck (bak)	life (life)	u (+con)...flow**er** (flow ur)
ay...d**ay** (day)	oh...g**o** (goh)	sh...**sh**elf (shelf)
ah...f**a**ther (fahth ur)	aw...s**o**ft (sawft)	ch...na**t**ure (nay chur)
ow...fl**ow**er (flow ur)	or...**or**bit (or but)	g...**g**ift (gihft)
ar...**ca**r (car)	oy...c**oi**n (coyn)	j...**g**em (jem)
e...l**e**ss (les)	oo...f**oo**t (foot)	ing...s**ing** (sing)
ee...l**ea**f (leef)	ew...f**oo**d (fewd)	zh...vi**si**on (vihzh un)
ih...tr**i**p (trihp)	yoo...p**u**re (pyoor)	k...**c**ake (kayk)
i (i + con + e)...**i**dea	yew...f**ew** (fyew)	s...**s**eed, **c**ent (seed, sent)
(i dee uh)	uh...comm**a** (cahm uh)	z...**z**one, rai**s**e (zohn, rayz)

A

abiotic (ay bi AHT ihk) **factors:** nonliving physical features of the environment, including soil, water, temperature, air, light, wind, and minerals. (Chap. 18, p. 483)

acid rain: rain or snow with a pH below 5.6; formed when water vapor reacts with sulfur oxides and nitrogen oxides in the air to form strong acids. (Chap. 20, p. 559)

active immunity: long-lasting immunity caused by the body making its own antibodies in response to an antigen. (Chap. 27, p. 738)

active transport: movement of material through a cell membrane with the use of energy. (Chap. 3, p. 74)

adaptation: any characteristic of an organism that makes it better able to survive in its environment. (Chap. 1, p. 7)

adolescence: the stage of human development that begins around ages 12 to 14 and lasts until age 20. (Chap. 26, p. 718)

adulthood: the stage of human development that begins with the end of adolescence and extends into old age. (Chap. 26, p. 719)

aerobe: organism that must have oxygen to survive. (Chap. 8, p. 212)

aggression: a forceful act used to dominate or control another animal. (Chap. 17, p. 463)

AIDS: acquired immune deficiency syndrome, a fatal communicable disease caused by the HIV virus; destroys the body's ability to combat disease; spread by sexual contact, by use of contaminated needles, through the placenta, or by nursing from an AIDS-infected mother. (Chap. 2, p. 38)

algae: plantlike protists that contain chlorophyll and make their own food; they have no roots, stems, or leaves and live in or near water. (Chap. 9, p. 231)

allele: a different form a gene may have for a trait. (Chap. 5, p. 125)

allergen: a substance that causes an allergic response in the body. (Chap. 27, p. 749)

allergy: an overly strong reaction of the immune system to a foreign substance. (Chap. 27, p. 749)

alternation of generations: in some protists and in plants, a continuous cycle that alternates between sporophyte and gametophyte generations. (Chap. 10, p. 268)

alveoli (al VE uh li): in the lungs, clusters of tiny, thin-walled air sacs at the end of each bronchiole, where oxygen and carbon dioxide are exchanged. (Chap. 24, p. 656)

Alzheimer's disease: a disease of unknown cause in which nerve cells in the brain fail to communicate with each other, causing memory loss. (Chap. 25, p. 686)

amino (uh MEE noh) **acids:** the building blocks of proteins. (Chap. 22, p. 602)

amniotic egg: in reptiles, the leathery egg that provides a complete environment for the developing embryo. (Chap. 15, p. 417)

amniotic sac: a thin membrane that forms around an embryo, filled with amniotic fluid (a clear liquid that cushions the embryo). (Chap. 26, p. 713)

amphibian: ectothermic vertebrate that spends part of its life in water and part on land; examples are frogs, toads, and salamanders. (Chap. 15, p. 409)

anaerobe: organism that can survive without oxygen. (Chap. 8, p. 212)

angiosperm: a flower-producing vascular plant in which the seed is enclosed in a fruit, such as an apple. (Chap. 11, p. 288)

antibiotic: a substance produced by an organism that is used to kill or inhibit another organism. (Chap. 8, p. 217)

antibiotic resistance: in bacteria, the evolutionary change that occurs when bacterial strains that are mostly killed by a specific drug mutate into strains that are not affected by that drug. (Chap. 8, p. 222)

antibody: a protein made by the body in response to a specific antigen. (Chap. 27, p. 738)

antigens: proteins and chemicals that are foreign to the body. (Chap. 27, p. 738)

antioxidant: a substance that prevents chemicals that enter the body from reacting with oxygen; may stop or prevent cancer. (Chap. 22, p. 611)

antiseptic: a substance that kills pathogens on living tissue and helps prevent their regrowth. (Chap. 27, p. 732)

anus: an opening at the end of the digestive tract through which solid wastes leave the body. (Chap. 13, p. 359)

appendage: a structure growing from the body, such as an arm or leg. (Chap. 14, p. 382)

archaebacteria: prokaryotic, one-celled organisms that live as anaerobes in extreme environments, including salt ponds, hot sulfur springs, and deep ocean thermal vents. (Chap. 7, p. 193)

artery: a thick, elastic-walled vessel that moves blood away from the heart. (Chap. 23, p. 629)

Arthropoda (AR thruh pahd uh)**:** largest animal phylum, classified by the number of body segments and appendages; includes insects, shrimp, spiders, and centipedes. (Chap. 14, p. 382)

ascus: the small, saclike structure in which sac fungi produce spores. (Chap. 9, p. 242)

asexual reproduction: a type of reproduction in which a new organism is produced from one parent. (Chap. 4, p. 97)

asthma: a respiratory disorder in which the bronchial tubes contract rapidly, resulting in shortness of breath or coughing; caused by smoking, allergy, or stress. (Chap. 24, p. 661)

atherosclerosis (ath uh roh skluh ROH sus)**:** a circulatory system disorder in which fatty deposits form on artery walls and restrict blood flow. (Chap. 23, p. 632)

atria (AY tree uh)**:** the two upper chambers of the human heart. (Chap. 23, p. 627)

auxin: a plant hormone that causes plant stems and leaves to grow or bend toward light. (Chap. 12, p. 325)

axon: neuron structure that carries messages away from the cell body. (Chap. 25, p. 679)

B

basidium: the club-shaped, spore-producing structure of club fungi. (Chap. 9, p. 244)

behavior: the way an organism acts toward its environment. (Chap. 17, p. 454)

bilateral symmetry: describes animals with body parts arranged in the same way on both sides of their body. (Chap. 13, p. 347)

binomial nomenclature: the two-word naming system that provides every organism with its own scientific name. (Chap. 7, p. 189)

biogas: a mixture of gases that can be burned directly or converted to gas or liquid fuels. (Chap. 3, p. 84)

biogenesis: the theory that living things come only from other living things. (Chap. 1, p. 11)

biological indicator: species whose health reflects the health of its ecosystem. (Chap. 15, p. 414)

biomass: organic material from plants or animals that is used for energy. (Chap. 3, p. 84)

biomaterial: an artificial material that can be used to repair or replace body parts. (Chap. 21, p. 583)

biomes (BI ohmz)**:** large geographic areas that have similar climates and ecosystems. (Chap. 19, p. 520)

biosphere: all parts of Earth where life exists, including air, land, and water. (Chap. 18, p. 482)

biotechnology: a method of improving crops by moving DNA directly from one organism to another. (Chap. 12, p. 331)

biotic (bi AHT ihk) **factors:** living organisms in the environment, such as plants and animals. (Chap. 18, p. 486)

bladder: the bag-shaped, elastic, muscular organ that stores urine until it leaves the body; can hold up to 500 mL of urine. (Chap. 24, p. 668)

blood pressure: the force exerted by blood on the walls of the vessels; pressure is highest in the arteries and in young adults is normally 120 over 80. (Chap. 23, p. 630)

bog: low-lying, spongy, wet ground composed mainly of dead plants, whose decay has been slowed by lack of oxygen. (Chap. 10, p. 275)

brain stem: the part of the brain extending from the cerebrum to the spinal cord; controls and coordinates involuntary muscle movements such as heartbeats. (Chap. 25, p. 681)

bronchi: two short tubes that branch off the trachea and carry air into the lungs (*singular:* bronchus). (Chap. 24, p. 656)

budding: a type of asexual reproduction in which a new organism grows off the side of its parent. (Chap. 9, p. 243)

cambium (KAM bee um): vascular plant tissue that produces new xylem and phloem cells. (Chap. 11, p. 294)

cancer: a major chronic disease resulting from uncontrolled cell growth. (Chap. 27, p. 748)

capillary: a microscopic blood vessel that connects arteries and veins; nutrients, oxygen, and wastes are exchanged through their one-cell-thick walls. (Chap. 23, p. 630)

captive breeding: the breeding of an endangered species in captivity. (Chap. 17, p. 470)

carbohydrate: a nutrient that provides most of the body's energy; made up of carbon, hydrogen, and oxygen. (Chap. 22, p. 601)

cardiac muscle: involuntary muscle tissue found only in the heart. (Chap. 21, p. 587)

carnivore: an animal that eats only other animals. (Chap. 16, p. 436)

carrying capacity: the largest number of individuals that an environment can support and maintain over time. (Chap. 18, p. 492)

cartilage: a thick, tough, smooth flexible tissue that is harder than flesh but softer than bone (like your ears); it covers the ends of bones to allow movement and cushion shock. (Chap. 15, p. 402; Chap. 21, p. 576)

cell: the smallest unit of an organism that can perform life functions. (Chap. 1, p. 6)

cell membrane: the selectively permeable outer boundary of a cell that allows food and oxygen to move into the cell and wastes to leave it. (Chap. 2, p. 47)

cell theory: a major theory of life science: the cell is the basic unit of life; organisms are made up of one or more cells; and all cells come from other cells. (Chap. 2, p. 44)

cell wall: in plants, the rigid structure made of cellulose that surrounds a plant cell membrane and supports and protects it. (Chap. 2, p. 51)

cellulose: the organic compound that forms plant cell walls; made up of long chains of sugar molecules. (Chap. 10, p. 263)

centenarian: a person 100 years of age or older. (Chap. 26, p. 722)

central nervous system: part of the body's control system, including the brain and spinal cord; sorts and interprets information from stimuli. (Chap. 25, p. 680)

cerebellum: the part of the brain that coordinates voluntary muscle movements; maintains balance and muscle tone. (Chap. 25, p. 681)

cerebrum: the part of the brain that interprets impulses from the senses, stores memory, and controls the work of voluntary muscles. (Chap. 25, p. 681)

chemical digestion: chemical breakdown of food by fluids in the mouth, stomach, and small intestine; breaks down large molecules into smaller ones that can be absorbed by cells. (Chap. 22, p. 614)

chemotherapy: the use of chemicals to destroy cancer cells. (Chap. 27, p. 748)

childhood: the stage of human development from infancy to age 12 years. (Chap. 26, p. 718)

chloroplasts: plant cell organelles in which light energy is changed into chemical energy in the form of sugar during the process of photosynthesis. (Chap. 2, p. 51)

chordate: a member of the animal phylum Chordata that has a notochord, a dorsal hollow nerve cord, and gill slits at some time in its life cycle. (Chap. 15, p. 400)

chromatin: hereditary material in a cell's nucleus; chromatin coils into the form of chromosomes when a cell divides. (Chap. 2, p. 47)

chromosome: threadlike strands of DNA and protein in a cell nucleus that carry the code for the cell characteristics of an organism. (Chap. 4, p. 94)

chronic bronchitis: a respiratory disease in which too much mucus is produced in the bronchial tubes. (Chap. 24, p. 659)

chronic disease: a noncommunicable disease, such as diabetes or cancer, that lasts a long time. (Chap. 27, p. 746)

chyme (KIME): the thin, watery liquid in the digestive tract; partially digested food. (Chap. 22, p. 615)

cilia: short, hairlike structures that extend from the cell membrane and help tiny organisms move; found in respiratory passages. (Chap. 9, p. 236)

circadian rhythm: behavior on a 24-hour cycle. (Chap. 17, p. 468)

class: third-highest taxonomic category, below phylum (animals) or division (plants). (Chap. 7, p. 196)

classify: to group ideas, information, or objects, based on their similarities. (Chap. 7, p. 188)

climax community: the final stage of a community's ecological succession; it is stable, with a balance of abiotic and biotic factors. (Chap. 19, p. 512)

clone: an individual that is genetically identical to one of its parents (Chap. 4, p. 112)

closed circulatory system: a type of blood-circulation system in which vessels transport blood to the internal organs, as in humans. (Chap. 14, p. 372)

cnidarian (ni DAIR ee uhn)**:** a phylum of hollow-bodied animals with stinging cells. (Chap. 13, p. 353)

cochlea (KOH klee uh)**:** a sound-sensitive, fluid-filled structure in the inner ear; shaped like a snail's shell. (Chap. 25, p. 690)

collar cells: in sponges, the lining cells that help water move through the sponge by the beating of flagella. (Chap. 13, p. 351)

commensalism: a symbiotic relationship that benefits one partner but not the other. (Chap. 18, p. 494)

communicable disease: a disease that is transmitted from one organism to another by pathogenic bacteria, viruses, and fungi. (Chap. 27, p. 733)

communication: an exchange of information among animals by means of cries, movements, touch, speech, pheromones, and so on. (Chap. 17, p. 464)

community: all the populations of different species that live in the same place at the same time and interact with each other. (Chap. 18, p. 487)

compound light microscope: an instrument that uses light and convex lenses to magnify objects. (Chap. 2, p. 41)

conditioning: occurs when a response previously associated with one stimulus becomes associated with another stimulus. (Chap. 17, p. 458)

consumers: organisms that can't make their own food. (Chap. 3, p. 78)

contour feathers: on birds, the strong, lightweight feathers that are used for flight and that give birds their coloring and sleek shape. (Chap. 16, p. 427)

control: in an experiment, the standard for comparison. (Chap. 1, p. 15)

coronary circulation: the flow of blood to heart tissues. (Chap. 23, p. 628)

courtship behaviors: behaviors that help males and females of a species to recognize each other and prepare for mating. (Chap. 17, p. 463)

crop: in an earthworm, a sac in the digestive system that stores soil. (Chap. 14, p. 376)

cuticle: a waxy protective layer on the stems and leaves of land plants that helps the plant conserve water. (Chap. 10, p. 263)

cyclic behavior: innate behavior that occurs in a repeating pattern. (Chap. 17, p. 467)

cyst: a young, parasitic worm with a protective covering. (Chap. 13, p. 362)

cytoplasm: the gel-like substance inside the cell membrane that contains structures that carry out life processes. (Chap. 2, p. 48)

D

day-neutral plant: a plant that isn't very sensitive to the number of hours of darkness and thus can flower over a wide range of night lengths. (Chap. 12, p. 330)

dendrite: neuron structure that receives messages and sends them to the cell body. (Chap. 25, p. 679)

dermis: the inner layer of skin; it contains many blood vessels, nerves, and sweat and oil glands. (Chap. 21, p. 591)

desert: the driest biome on Earth; supports little plant life and receives less than 25 cm of rain yearly. (Chap. 19, p. 526)

development: all the changes undergone by living things as they grow. (Chap. 1, p. 8)

diaphragm (DI uh fram)**:** a muscle beneath the lungs that helps move air in and out of the body. (Chap. 24, p. 657)

dichotomous (DI kaht uh mus) **key:** a detailed list of traits used to identify a specific organism. (Chap. 7, p. 201)

dicot (DI kaht)**:** an angiosperm having two seed leaves inside its seed; for example, a maple tree. (Chap. 11, p. 288)

diffusion: the movement of molecules from areas of greater concentration to areas of lesser concentration. (Chap. 3, p. 72)

digestion: the process that breaks down food into small molecules that can be moved into the blood and used by the cell. (Chap. 22, p. 612)

disinfectant: a substance that kills pathogens on nonliving objects. (Chap. 27, p. 732)

division: second-highest taxonomic category in the

plant and fungi kingdoms (in the animal kingdom, *phylum* replaces division). (Chap. 7, p. 196)

DNA: *deoxyribonucleic acid*; a chemical in the nuclei of cells that codes and stores genetic information; consists of strands of molecules that control cell activities using coded instructions. (Chap. 4, p. 104)

dominant: the form of a trait that appears to dominate or mask another form of the same trait. (Chap. 5, p. 126)

dorsal hollow nerve cord: a tubular bundle of nerves that lies above the notochord in a chordate animal; the spinal cord in most vertebrates. (Chap. 15, p. 400)

down feathers: on birds, the soft, fluffy insulating feathers that cover their skin. (Chap. 16, p. 427)

echinoderm (ih KI nuh durm)**:** a spiny-skinned invertebrate that lives on the bottom of the ocean and that moves by means of a water-vascular system. (Chap. 14, p. 392)

ecological pyramid: a model used to describe the transfer of energy through a community. (Chap. 18, p. 498)

ecological succession: the process of gradual change from one community of organisms to another. (Chap. 19, p. 510)

ecology: the study of relationships between organisms and between organisms and their physical environments. (Chap. 18, p. 482)

ecosystem: a community interacting with the non-living (abiotic) parts of its environment. (Chap. 18, p. 487)

ectotherm: vertebrate animal whose internal body temperature changes with its environment. (Chap. 15, p. 401)

egg: in organisms that reproduce sexually, the sex cell (gamete) from the female parent. (Chap. 4, p. 99)

electron microscope: an instrument that bends beams of electrons in a magnetic field; magnifies objects too small to be seen with a light microscope. (Chap. 2, p. 42)

embryo: a fertilized egg during early growth; in humans, an unborn child during the first two months of pregnancy. (Chap. 26, p. 712)

embryology: the study of the development of embryos, which are the earliest stages of an organism's development. (Chap. 6, p. 169)

emphysema (em fuh SEE muh)**:** a respiratory disease in which the alveoli in the lungs lose their ability to expand and contract (elasticity), mostly caused by smoking. (Chap. 24, p. 660)

endangered species: a species with so few individuals left that it is in danger of extinction. (Chap. 6, p. 176; Chap. 20, p. 554)

endocytosis: a process by which cells transport a large body, such as a large protein molecule, through a cell membrane into the cytoplasm. (Chap. 3, p. 75)

endoplasmic reticulum (ER): a cell organelle consisting of folded membranes that move materials around within the cell. (Chap. 2, p. 48)

endoskeleton: the internal skeleton of an organism that supports and protects the internal organs and provides a frame for muscle. (Chap. 15, p. 401)

endospore (EN doh spor)**:** a thick-walled structure that some bacteria produce around themselves, especially for protection from heat and drought. (Chap. 8, p. 218)

endotherm: vertebrate animal that maintains a constant body temperature. (Chap. 15, p. 401)

enzyme (EN zim)**:** a protein that speeds chemical reactions in cells without being changed itself. (Chap. 3, p. 69)

epidermis: the surface or outer layer of your skin. (Chap. 21, p. 591)

equilibrium: condition in which molecules of a substance are spread evenly throughout a space. (Chap. 3, p. 72)

erosion: the wearing away of soil by wind, water, and ice. (Chap. 20, p. 551)

estivation: an adaptation for survival in hot, dry weather during which an animal becomes inactive and all body processes slow down. (Chap. 15, p. 409)

estuary: the areas where a freshwater river or stream meets the ocean; examples are salt marshes, deltas, and mud flats. (Chap. 19, p. 532)

ethnobotanist: a person who studies relationships between people from various cultures and their plants. (Chap. 11, p. 306)

eubacteria: prokaryotic, one-celled organisms found everywhere; reproduce by fission; include pathogenic bacteria and cyanobacteria. (Chap. 7, p. 193)

evolution: changes that occur over time in the hereditary features of a species of organisms. (Chap. 6, p. 152)

exocytosis: a process by which a cell moves large molecules out through the cell membrane. (Chap. 3, p. 76)

exoskeleton: on all arthropods, the hard, lightweight external covering that shields, supports, and protects the body. (Chap. 14, p. 382)

extinction: the dying out of an entire species. (Chap. 6, p. 176; Chap. 20, p. 554)

family: the fifth-highest taxonomic category, below an order. (Chap. 7, p. 196)

fat: a nutrient that provides energy; also cushions organs and helps the body absorb some vitamins. (Chap. 22, p. 603)

fermentation: a form of respiration that converts energy from glucose when the supply of oxygen is not sufficient; yeast and some bacteria use alcohol fermentation to release energy and CO_2. (Chap. 3, p. 81)

fertilization: in organisms that reproduce sexually, the fusion of an egg and a sperm (gametes). (Chap. 4, p. 101)

fetus (FEE tus): a developing embryo; in humans, the unborn baby after the first two months of pregnancy. (Chap. 26, p. 714)

filter feeder: an organism that obtains food and oxygen by filtering it from the water in which it lives. (Chap. 13, p. 351)

fin: fanlike structure on most fish; adapted for steering, balancing, and moving through the water. (Chap. 15, p. 401)

fish: an ectothermic vertebrate whose gills, fins, and scales adapt it to living in water. (Chap. 15, p. 401)

fission: a type of asexual reproduction used by bacteria in which one bacterium divides to form two cells with identical genetic material. (Chap. 8, p. 212)

flagellum (fluh JELL uhm): a whiplike tail on bacteria and some protists that helps them move through a moist environment. (Chap. 8, p. 211)

food chain: a model used to show how energy from food passes from one organism to another. (Chap. 18, p. 497)

food group: a group of foods containing the same nutrients; for example, the bread and cereal group. (Chap. 22, p. 608)

food web: a model used to describe a series of overlapping food chains. (Chap. 18, p. 497)

fossil fuel: a fuel created over time from the bodies of once-living organisms; coal, oil, and natural gas. (Chap. 20, p. 542)

fossils: remains of life from an earlier time. (Chap. 6, p. 163)

fracture: a break in a bone. (Chap. 21, p. 582)

free-living: describes an organism that finds its own food and place to live without depending on another organism; the opposite of a parasite. (Chap. 13, p. 357)

frond: the leaf of a fern. (Chap. 10, p. 276)

gametophyte: the sex cell-producing stage of a plant's life cycle. (Chap. 10, p. 268)

gasohol: a mixture of gasoline and ethanol. (Chap. 3, p. 84)

gene: the segment of DNA on a chromosome that directs the making of a specific protein, thus controlling traits that are passed to offspring. (Chap. 4, p. 107)

genetic engineering: biological and chemical methods to change a cell's DNA sequence to produce desirable traits or eliminate undesirable traits. (Chap. 5, p. 142)

genetics: the science of how traits are inherited through alleles passed from one generation to another. (Chap. 5, p. 125)

genome: a map of the location of individual genes on every chromosome of an individual. (Chap. 5, p. 144)

genotype: the genetic makeup of an organism for a trait. (Chap. 5, p. 128)

genus: a group of different organisms with similar characteristics; can have one or more species. (Chap. 7, p. 189)

geothermal energy: heat energy from below Earth's surface. (Chap. 20, p. 544)

gestation period: the time between fertilization and birth, during which the embryo develops in the uterus. (Chap. 16, p. 441)

gill: organ that exchanges carbon dioxide with water. (Chap. 14, p. 370)

gill slits: paired openings in the throat; in fish, they develop into gills for breathing underwater. (Chap. 15, p. 400)

gizzard: in an earthworm, a muscular digestive structure that grinds soil. (Chap. 14, p. 376)

global warming: the potential warming of Earth due to an increase in greenhouse gases in the atmosphere. (Chap. 20, p. 560)

Golgi (GAWL jee) **bodies:** cell organelles consisting of stacks of membrane-covered sacs that package and move proteins to the outside of the cell. (Chap. 2, p. 49)

gradualism: a model of evolution, showing a steady,

slow, and continuous change of one species into new species. (Chap. 6, p. 160)

grassland: a biome dominated by climax communities of grasses but that, due to a dry season, does not have forests. (Chap. 19, p. 526)

greenhouse effect: the process by which heat radiated from Earth's surface is trapped and reflected back to Earth by gases in the atmosphere, causing worldwide temperatures to increase. (Chap. 20, p. 560)

groundwater: water in the soil or trapped in underground rock pockets. (Chap. 20, p. 563)

guard cells: in a plant leaf, cells that surround the stomata to open and close them. (Chap. 11, p. 294)

gymnosperm: a vascular plant that produces seeds on the scales of female cones, such as pinecones. (Chap. 11, p. 286)

habitat: the physical location where an organism lives. (Chap. 18, p. 495)

hazardous waste: harmful or poisonous waste material. (Chap. 20, p. 562)

heart murmur: a defect in the heart caused by an improperly functioning valve. (Chap. 23, p. 634)

hemoglobin: a chemical in red blood cells that carries oxygen to tissues, carries carbon dioxide to lungs, and gives blood its red color. (Chap. 23, p. 637)

herbivore: an animal that eats only plants. (Chap. 16, p. 436)

heredity: passing of traits from parent to offspring. (Chap. 5, p. 124)

hermaphrodite: an animal that produces both sperm and eggs. (Chap. 13, p. 352)

heterozygous (het uh roh ZI gus)**:** an organism that has two different alleles for a trait. (Chap. 5, p. 128)

hibernation: an adaptation for winter survival during which an animal becomes inactive and all body processes slow down. (Chap. 15, p. 409)

homeostasis: the regulation of steady, life-maintaining conditions inside an organism or cell, despite changes in its environment. (Chap. 1, p. 7)

hominids: earliest humanlike primates; ate both meat and vegetables and walked upright on two feet. (Chap. 6, p. 172)

Homo sapiens: our own species, a hominid primate mammal; evolved 300 000 years ago. (Chap. 6, p. 173)

homologous: body parts in different species that are similar in origin and structure. (Chap. 6, p. 168)

homozygous (ho muh ZI gus)**:** an organism that has two identical alleles for a trait. (Chap. 5, p. 128)

hormones: chemicals secreted from endocrine glands that control specific body activities. (Chap. 25, p. 695)

host cell: a living cell in which a virus reproduces. (Chap. 2, p. 34)

hydroelectric power: energy source that uses the power of flowing water to produce electricity. (Chap. 20, p. 544)

hypertension (hi pur TEN chun)**:** a circulatory system disorder in which blood pressure is too high. (Chap. 23, p. 632)

hyphae (HI fee)**:** the mass of many-celled, threadlike tubes that form the body of a fungus. (Chap. 9, p. 241)

hypothesis: a prediction that can be tested. (Chap. 1, p. 15)

immovable joint: a juncture of two or more bones, made so that little or no movement occurs. (Chap. 21, p. 578)

immune system: a complex group of defenses that the body uses to fight disease. (Chap. 27, p. 737)

imprinting: a type of learned behavior in which an animal forms a social attachment to another organism soon after birth or hatching. (Chap. 17, p. 456)

incomplete dominance: the production of a phenotype in an offspring that is intermediate to the phenotypes of its two homozygous parents. (Chap. 5, p. 132)

incubate: to keep an egg or newborn animal warm; the heat helps eggs develop until they hatch and helps young to survive. (Chap. 16, p. 426)

infancy: the stage of human development from four weeks after birth to one year. (Chap. 26, p. 717)

innate behavior: behavior that an organism is born with and does not have to learn. (Chap. 17, p. 454)

inorganic compound: class of compounds made from elements other than carbon—for example, water. (Chap. 3, p. 70)

insight: a form of reasoning that enables animals to use past experiences to solve new problems. (Chap. 17, p. 459)

instinct: a complex pattern of innate behavior involving multiple actions. (Chap. 17, p. 455)

insulin: substance produced by the islets of Langerhans to help body cells use glucose in the blood for energy. (Chap. 4, p. 112)

interneurons: nerve cells throughout the brain and spinal cord that transmit impulses from sensory neurons to motor neurons. (Chap. 25, p. 679)

intertidal zone: the part of a shoreline that is submerged at high tide and exposed to air at low tide. (Chap. 19, p. 532)

invertebrate: animal without a backbone. (Chap. 13, p. 345)

involuntary muscle: a muscle that can't be consciously controlled, such as heart and digestive muscle. (Chap. 21, p. 584)

islets of Langerhans: cells in the pancreas that produce insulin. (Chap. 4, p. 112)

joint: any place where two or more bones meet; may be movable or immovable. (Chap. 21, p. 577)

kidney: the major organ of the urinary system; filters blood to produce the waste liquid called urine. (Chap. 24, p. 667)

kingdom: the highest and largest of the taxonomic categories; kingdoms include all other categories. (Chap. 7, p. 188)

larva: young organism that develops between egg and adult stages. (Chap. 13, p. 352)

larynx: airway to which the vocal cords are attached; the structure between the pharynx and trachea. (Chap. 24, p. 654)

law: a rule that describes a pattern in nature and what will happen under specific conditions. (Chap. 1, p. 17)

learning: the process of developing a behavior through experience or practice. (Chap. 17, p. 456)

lichen: an organism made up of a fungus and green algae or a cyanobacterium living in a mutualistic relationship. (Chap. 9, p. 244)

life span: the length of time an organism is expected to live. (Chap. 1, p. 8)

ligament: a tough band of tissue that holds bones together at joints. (Chap. 21, p. 577)

limiting factor: any biotic or abiotic factor that restricts the number of individuals in a population. (Chap. 18, p. 489)

long-day plant: a plant that requires short nights (long days) to flower. (Chap. 12, p. 327)

lymph (LIHMF): fluid in body tissues made up of water, dissolved substances, and lymphocytes. (Chap. 23, p. 644)

lymph nodes: bean-shaped structures throughout the body that filter microorganisms and foreign material from lymph before it returns to blood. (Chap. 23, p. 644)

lymphatic (lihm FAT ihk) **system:** collects fluid from body tissues and returns it to the blood through lymphatic capillaries and lymph vessels. (Chap. 23, p. 644)

lymphocyte (LIHM fuh site): a type of white blood cell that fights disease-causing antigens by engulfing and digesting them. (Chap. 23, p. 644; Chap. 27, p. 740)

lysosome (LI suh sohm): a cytoplasmic organelle that contains chemicals and digests wastes and worn-out cell parts. (Chap. 2, p. 50)

mammal: an endothermic vertebrate with insulating hair and mammary glands. (Chap. 16, p. 435)

mammary glands: glands in female mammals that produce milk for feeding young. (Chap. 16, p. 435)

mantle: in a mollusk, the thin layer of tissue that covers the soft body; it secretes chemicals that become a shell or protects the body if no shell exists. (Chap. 14, p. 370)

marrow: a red or yellow fatty tissue in certain bones that produces red and white blood cells. (Chap. 21, p. 576)

marsupial: a mammal with a pouch on its abdomen for carrying and nursing its young. (Chap. 16, p. 440)

mechanical digestion: physical breakdown of food into smaller particles by chewing (in the mouth) and by churning (in the stomach). (Chap. 22, p. 614)

medusa: a bell-shaped cnidarian that is free swimming. (Chap. 13, p. 355)

meiosis: the division of the cell nucleus to produce sex cells (gametes). (Chap. 4, p. 99)

melanin (MEL uh nun): a pigment produced by the epidermis that gives skin its color. (Chap. 21, p. 591)

menopause: in older females, the period of years when the menstrual cycle becomes irregular and eventually stops. (Chap. 26, p. 708)

menstrual (MEN strul) **cycle:** in females, the hormonally regulated cycle of egg production and menstruation. (Chap. 26, p. 707)

menstruation (men STRAY shun): in females, the monthly shedding of the lining of the uterus that takes place if fertilization of an egg has not occurred. (Chap. 26, p. 707)

metabolism: all of the chemical activities of an organism that enable it to live, grow, and reproduce. (Chap. 3, p. 78)

metamorphosis: the changes in body form during the life cycle—for example: egg, larva, pupa, adult. (Chap. 14, p. 384)

migration: the instinctive seasonal movement of animals, such as birds flying south for the winter. (Chap. 17, p. 468)

minerals: inorganic nutrients that regulate many chemical reactions in the body, such as building cells and sending nerve impulses. (Chap. 22, p. 605)

mitochondria: cell organelles that break down food molecules and release energy. (Chap. 2, p. 49)

mitosis: the process by which a nucleus divides into two identical cells, each containing the same number and type of chromosomes as the parent cell. (Chap. 4, p. 93)

mollusk: a soft-bodied invertebrate that usually has a hard shell, such as a snail. (Chap. 14, p. 370)

molting: the periodic shedding and replacing of the old, outgrown body covering, such as skin or an exoskeleton. (Chap. 14, p. 382)

monocot (MAHN uh kaht): an angiosperm having a single seed leaf inside its seed; for example, an orchid. (Chap. 11, p. 288)

monotreme: a mammal that lays eggs having a tough, leathery shell; for example, the duckbilled platypus. (Chap. 16, p. 440)

motivation: some stimulus within an animal that causes it to act. (Chap. 17, p. 457)

motor neuron: nerve cell that conducts impulses from the brain or spinal cord to muscles or glands throughout the body. (Chap. 25, p. 679)

movable joint: a place where two or more bones meet or are attached and allow a wide range of movement. (Chap. 21, p. 578)

multiple alleles: having more than two alleles that control a trait. (Chap. 5, p. 132)

muscle: an organ that can relax and contract to allow movement. (Chap. 21, p. 584)

mutation: any permanent change in an organism's genetic material (DNA). (Chap. 4, p. 109)

mutualism: a symbiotic relationship that benefits both partners. (Chap. 18, p. 494)

natural resources: parts of the environment that organisms need for survival, such as air, water, soil, and food. (Chap. 20, p. 540)

natural selection: Darwin's theory of evolution, which says that organisms best adapted to their environment are more likely to survive and pass their traits to their offspring. (Chap. 6, p. 155)

nephron (NEF rahn): the tiny filtering unit of the kidney. (Chap. 24, p. 667)

neuron: nerve cell that carries impulses throughout the body; the basic unit of the nervous system. (Chap. 25, p. 679)

niche: the role of an organism in the ecosystem. (Chap. 18, p. 495)

nitrogen cycle: the continuous movement of nitrogen from the atmosphere, to plants, and back to the atmosphere (or directly into plants) again. (Chap. 18, p. 501)

nitrogen-fixing bacteria: bacteria that change nitrogen from the air into nitrogen compounds that are useful to plants and animals. (Chap. 8, p. 216)

noncommunicable disease: a disease that cannot be spread from one organism to another; may be caused by genetics, chemicals, poor diet, or uncontrolled cell division. (Chap. 27, p. 746)

nonrenewable resource: a natural resource that is available only in limited amounts and that cannot be replaced by nature in a short period of time. (Chap. 20, p. 541)

nonvascular plant: a plant lacking vascular tissue; it absorbs water directly through its cell membranes. (Chap. 10, p. 264)

notochord: a flexible, rodlike structure along the dorsal side (back) of a chordate animal; the backbone in vertebrates. (Chap. 15, p. 400)

nuclear energy: energy produced from the splitting apart of uranium nuclei by a nuclear fission reaction. (Chap. 20, p. 546)

nucleus: the structure inside a cell that directs the cell's activities; contains chromatin; chromosomes in a dividing cell. (Chap. 2, p. 47)

nutrient: a substance in food that produces energy and materials for life activities. (Chap. 22, p. 600)

olfactory cell: nerve cells in the nasal passages that respond to chemical stimuli in air. (Chap. 25, p. 691)

omnivore: an animal that eats both plants and animals. (Chap. 16, p. 436)

open circulatory system: a type of blood-circulation system that lacks vessels but instead bathes internal organs in blood, such as in mollusks. (Chap. 14, p. 370)

order: the fourth-highest taxonomic category, below a class. (Chap. 7, p. 196)

organ: a structure made up of different types of tissues that work together to do a specific job; for example, the heart. (Chap. 2, p. 54)

organelles: in eukaryotic cells, the structures within the cytoplasm that break down food, move wastes, and store materials. (Chap. 2, p. 48)

organic compound: class of compounds in living organisms: carbohydrates, lipids, proteins, and nucleic acids. (Chap. 3, p. 68)

organism: a living thing that is made of one or more cells, uses energy, moves, responds to its environment, adjusts, reproduces, adapts, and has a life span. (Chap. 1, p. 6)

osmosis: the diffusion of water through a cell membrane. (Chap. 3, p. 73)

osteoporosis: a disease that breaks down bone, making it brittle and weak. (Chap. 21, p. 583)

ovary: in angiosperms, the swollen base of the pistil, where ovules form; in female animals, the organ that produces ova, or egg cells. (Chap. 11, p. 298; Chap. 26, p. 706)

ovulation (ahv yuh LAY shun): in human females, the monthly process of egg release from an ovary. (Chap. 26, p. 706)

ovule (OHV yewl): the female reproductive part of a plant that produces the eggs. (Chap. 11, p. 296)

ozone depletion: the thinning of Earth's ozone layer. (Chap. 20, p. 560)

parasitism: a symbiotic relationship that benefits the parasite but harms the parasite's partner. (Chap. 18, p. 495)

passive immunity: short-term immunity caused by antibodies introduced into the body from an outside source. (Chap. 27, p. 738)

passive smoking: the breathing in of cigarette smoke-filled air by nonsmokers. (Chap. 24, p. 664)

passive transport: movement of material across a cell membrane without the use of energy. (Chap. 3, p. 74)

pasteurization: heating food long enough to kill most bacteria. (Chap. 27, p. 730)

pathogen: any organism that produces disease. (Chap. 8, p. 217)

pedigree: a diagram that shows the occurrence of a trait in a family. (Chap. 5, p. 140)

periosteum (per ee AHS tee um): a tough, tight-fitting membrane that covers the surface of bones. (Chap. 21, p. 575)

peripheral nervous system: part of the control system, including cranial nerves and spinal nerves that connect the brain and spinal cord to other body parts. (Chap. 25, p. 680)

peristalsis (pe ruh STAHL sis): muscular contractions that move food through the digestive system. (Chap. 22, p. 614)

permafrost: in tundra areas of Earth, the nutrient-poor soil that remains permanently frozen underneath the top few inches of soil. (Chap. 19, p. 520)

pesticide: a chemical that kills undesirable plants or animal pests. (Chap. 14, p. 390)

pharynx (FER ingks): a tubelike passageway for both food and air, located between the nasal cavity and the esophagus. (Chap. 24, p. 654)

phenotype (FEE nuh tipe): a physical trait that shows as a result of an organism's particular genotype. (Chap. 5, p. 128)

pheromone: a powerful chemical produced by an animal to influence the behavior of another animal of the same species. (Chap. 17, p. 466)

phloem (FLOH em): vascular plant tissue made up of tubular cells that move food from leaves and stems to other parts of the plant for use or storage. (Chap. 11, p. 294)

photoperiodism: the flowering response of a plant to changes in the length of day and night. (Chap. 12, p. 327)

photosynthesis (foht oh SIHN thuh sus): chemical reaction used by producers, such as green plants, to produce food; light energy is used to produce chemical energy, converting carbon dioxide and water into sugar and oxygen. (Chap. 12, p. 318)

photovoltaic (PV) cell: a solar cell that turns sunlight directly into electric current. (Chap. 20, p. 543)

phylogeny: the evolutionary history of an organism. (Chap. 7, p. 193)

phylum: the second-highest taxonomic category in the animal kingdom. (Chap. 7, p. 196)

pioneer community: the first community to inhabit a new environment. (Chap. 19, p. 511)

pioneer species: the first plants to grow in a new or disturbed area; their decay creates material in which other plant species grow. (Chap. 10, p. 270)

pistil (PIHS tul): the female reproductive organ of a flower. (Chap. 11, p. 298)

placenta: in placental mammals, the saclike organ developed by a growing embryo; attaches to the uterus; absorbs food and oxygen from the mother's blood. (Chap. 16, p. 441)

placental mammal: a mammal whose young develop inside the female in the uterus. (Chap. 16, p. 441)

plankton: microscopic algae, plants, and other organisms that float in lakes, ponds, oceans, and other bodies of water. (Chap. 19, p. 531)

plasma: the liquid part of blood, made mostly of water but also containing dissolved nutrients, minerals, and oxygen. (Chap. 23, p. 636)

platelet: an irregularly shaped cell fragment that helps blood clot. (Chap. 23, p. 637)

pollen grain (PAHL un GRAYN): the male reproductive part of a plant that contains the sperm. (Chap. 11, p. 296)

pollination: the process that transfers pollen grains from the stamen to the stigmas. (Chap. 11, p. 299)

pollutant: any substance that contaminates the environment; mostly made up of waste products from burning fossil fuels. (Chap. 20, p. 558)

polygenic (pahl ih JENH ink) **inheritance:** occurs when groups of gene pairs act together to produce a specific trait. (Chap. 5, p. 133)

polyp: a vase-shaped cnidarian that usually is sessile. (Chap. 13, p. 355)

population: organisms of one species that live in the same place at the same time and that can produce offspring. (Chap. 18, p. 487)

population density: the number of individuals per unit of living space. (Chap. 18, p. 489)

precipitation: the amount of moisture that condenses and falls as rain, sleet, hail, or snow, or that forms as fog. (Chap. 19, p. 518)

predation: the feeding of one organism on another organism. (Chap. 18, p. 493)

preening: a behavior of birds in which a bird uses its beak to rub oil over its feathers to condition them and help make them water repellent. (Chap. 16, p. 428)

pregnancy: the period of time between fertilization and birth (in humans, nine months). (Chap. 26, p. 712)

primary succession: the development of new communities in areas that do not have any soil, such as new volcanic islands. (Chap. 19, p. 511)

primates: the group of mammals to which monkeys, apes, and humans belong and which share several characteristics, such as opposable thumbs. (Chap. 6, p. 171)

producers: green plants that make their own food by photosynthesis. (Chap. 3, p. 78)

protein: a nutrient made up of amino acids; used throughout the body for growth and to replace and repair cells. (Chap. 22, p. 602)

prothallus: the structure in ferns that produces gametes. (Chap. 10, p. 276)

protists: members of the Kingdom Protista; some are plantlike, others are animal-like, and others fungus-like; simple one-celled or many-celled organisms. (Chap. 9, p. 230)

protozoa (proht uh ZOH uh): one-celled animal-like protists; many are parasites. (Chap. 9, p. 234)

pseudopod (SEWD uh pahd): a footlike cytoplasmic extension used by some organisms to move and to trap food. (Chap. 9, p. 235)

pulmonary (PUL mo ner e) **circulation:** the path of blood from the heart, to the lungs, and back to the heart. (Chap. 23, p. 627)

punctuated equilibrium: a model of evolution, showing the rapid change of a species caused by the mutation of just a few genes. (Chap. 6, p. 160)

Punnett square: a tool that shows how genes can combine; used to predict the probability of types of offspring. (Chap. 5, p. 128)

radial symmetry: describes animals with body parts arranged in a circle around a central point, similar to a bicycle wheel. (Chap. 13, p. 346)

radioactive element: an element that gives off radiation, a form of atomic energy. (Chap. 6, p. 165)

radula: in gastropods, a tongue-like organ with rows of teeth that scrape and tear food. (Chap. 14, p. 371)

recessive: the form of a trait that seems to disappear in a population but can reappear depending on the way the alleles combine. (Chap. 5, p. 126)

recycling: the reusing of an item or resource after it has been changed or reprocessed. (Chap. 20, p. 549)

red tide: an algal population explosion that causes the water to look red and can cause fish kills and illness. (Chap. 9, p. 248)

reflex: the simplest type of innate behavior; an automatic, involuntary response to a stimulus. (Chap. 17, p. 455; Chap. 25, p. 682)

regeneration: the ability of an organism to replace body parts; a type of asexual reproduction in which a whole new organism grows from just a part of the parent organism. (Chap. 13, p. 352)

relative dating: estimating the age of a fossil by comparing it to younger fossils in the rock layers above and to older fossils in the rock layers below. (Chap. 6, p. 165)

renewable resource: a natural resource that is constantly being recycled or replaced by nature. (Chap. 20, p. 540)

reptile: an ectothermic vertebrate that has dry, scaly skin and that lays eggs covered with a leathery shell; examples are lizards, snakes, and turtles. (Chap. 15, p. 416)

respiration: the process by which organisms break down food to release energy. (Chap. 12, p. 321)

response: the reaction of an organism to a stimulus. (Chap. 1, p. 7)

retina: light-sensitive tissue at the back of the eye. (Chap. 25, p. 688)

rhizoid: rootlike filament containing only a few long cells; it holds moss plants in place. (Chap. 10, p. 267)

rhizome: the underground stem of a fern. (Chap. 10, p. 276)

ribosome: a cell organelle on which protein is made. (Chap. 2, p. 48)

RNA: *ribo*nucleic *a*cid; carries codes for making proteins. (Chap. 4, p. 108)

saliva: the watery substance produced in the mouth that begins the chemical digestion of food. (Chap. 22, p. 614)

saprophyte (SA proh fite): any organism that uses dead material as a food and energy source. (Chap. 8, p. 216)

scales: hard, thin, overlapping plates that cover and protect a fish's body. (Chap. 15, p. 401)

scientific methods: problem-solving procedures used by scientists; define the problem, make a hypothesis, test the hypothesis, analyze the results, and draw conclusions. (Chap. 1, p. 14)

sea otter: a carnivorous placental mammal with a well-developed brain and good underwater vision. (Chap. 16, p. 446)

secondary succession: a type of ecological succession that occurs in a place that has soil and was once the home of living organisms. (Chap. 19, p. 512)

sedimentary rock: a type of rock formed when fine particles like mud and sand settle out of water and become cemented together. (Chap. 6, p. 163)

semen: the mixture of sperm and a fluid that nourishes the sperm and helps them move. (Chap. 26, p. 705)

sensory neuron: nerve cell that transmits stimuli from receptors to the brain or spinal cord. (Chap. 25, p. 679)

sessile: describes organisms, such as trees, that remain attached to one place during their lifetime. (Chap. 13, p. 350)

setae (SEE tee): in a segmented worm, bristle-like structures on the outside of the body that help it grip soil and move. (Chap. 14, p. 375)

sex-linked gene: an allele inherited on a sex chromosome. (Chap. 5, p. 140)

sexual reproduction: a type of reproduction in which a new organism is produced by combining sex cells from two parents. (Chap. 4, p. 99)

sexually transmitted disease (STD): a disease transmitted from person to person during sexual contact. (Chap. 27, p. 734)

short-day plant: a plant that requires long nights (short days) to flower. (Chap. 12, p. 327)

skeletal muscles: voluntary muscles that work in pairs and move bones. (Chap. 21, p. 586)

skeletal system: the body's network of bones, which form a rigid frame to support the body, protect internal organs, generate red blood cells, and store calcium and phosphorus. (Chap. 21, p. 574)

smog: a brown-colored air pollution that occurs when sunlight reacts with waste products released by the burning of fuels. (Chap. 20, p. 558)

smooth muscles: involuntary muscles that move many internal organs, such as the stomach, intestines, and blood vessels. (Chap. 21, p. 586)

social behavior: interactions between organisms of the same species, including mating, caring for young, getting food, and claiming territory. (Chap. 17, p. 464)

society: a group of animals of the same species living and working together in an organized way. (Chap. 17, p. 464)

soil depletion: the removal of soil nutrients due to the taking away of mature plants from where they were grown. (Chap. 20, p. 551)

soil management: the use of plowing methods to prevent soil depletion and erosion. (Chap. 20, p. 551)

solid wastes: unwanted, solid materials that can be recycled, burned, buried, or dumped. (Chap. 20, p. 548)

sorus: structure that produces spores on the underside of a fern frond. (Chap. 10, p. 276)

species: a group of organisms whose members successfully reproduce among themselves. (Chap. 6, p. 152; Chap. 7, p. 189)

species diversity: the great variety of plants, animals, and other organisms on Earth. (Chap. 7, p. 198)

sperm: the gamete or reproductive cell from the male parent produced in the testes. (Chap. 4, p. 99; Chap. 26, p. 705)

spiracles: in an arthropod, openings in the abdomen and thorax through which air enters and waste gases leave. (Chap. 14, p. 383)

spontaneous generation: the theory that nonliving things produce living things; for example, that frogs arise spontaneously from mud. (Chap. 1, p. 10)

sporangia: the round spore cases of zygote fungi. (Chap. 9, p. 242)

spore: a reproductive cell that forms new organisms without fertilization; in fungi, ferns, and some protists. (Chap. 9, p. 242)

sporophyte: the spore-forming stage of a plant's life cycle. (Chap. 10, p. 268)

stamen (STAY mun): the male reproductive organ of a flower. (Chap. 11, p. 298)

stimulus: anything an organism responds to, such as sound, light, heat, vibration, odor, movement, hunger, thirst, and so on. (Chap. 1, p. 7)

stomata: small pores in the surface of a plant leaf that allow carbon dioxide, water, and oxygen to enter and leave. (Chap. 11, p. 294)

symbiosis: a close relationship between two organisms that live together. (Chap. 18, p. 494)

synapse: the small gap between two neurons across which impulses can travel. (Chap. 25, p. 680)

systemic (sihs TEM ihk) **circulation:** the flow of blood from the heart to all the body tissues (except lungs and heart) and back to the heart. (Chap. 23, p. 628)

taiga: a cold biome below the tundra; it has coniferous forests and is warmer and wetter than the tundra. (Chap. 19, p. 521)

target tissue: specific tissue affected by hormones. (Chap. 25, p. 695)

taste buds: the major taste receptors; tissue located on the tongue that responds to chemical stimuli. (Chap. 25, p. 691)

taxonomy (tak SAHN uh mee): the science of classifying and naming organisms. (Chap. 7, p. 188)

technology: the use of scientific knowledge to improve the quality of human life or to solve problems. (Chap. 1, p. 24)

temperate deciduous forest: climax communities of deciduous trees, which lose their leaves in the fall. (Chap. 19, p. 522)

tendon: a thick band of tissue that attaches a muscle to a bone. (Chap. 21, p. 586)

tentacles: the armlike structures that surround the mouths of some organisms and help them to capture food. (Chap. 13, p. 353)

territory: an area that an animal defends from other members of the same species. (Chap. 17, p. 462)

testis: organ in males that produces sperm and testosterone. (Chap. 26, p. 705)

theory: a description of nature, based on many observations; subject to change when evidence changes. (Chap. 1, p. 17)

tissues: groups of similar cells that do the same sort of work; for example, all muscle tissue contracts. (Chap. 2, p. 54)

toxin: a poison produced by disease-causing organisms (pathogens). (Chap. 8, p. 218)

trachea: a cartilage-reinforced tube that carries air to the bronchi. (Chap. 24, p. 654)

transgenic crop: a crop of plants that has new DNA inserted into it to improve yield, disease resistance, and so on. (Chap. 12, p. 331)

transpiration: in plants, loss of water vapor through the stomata of a leaf. (Chap. 12, p. 315)

trial and error: behavior that is modified by experience. (Chap. 17, p. 457)

tropical rain forest: hot, humid, equatorial biome; it contains the largest number of species. (Chap. 19, p. 524)

tropism: the response of a plant to a stimulus. (Chap. 12, p. 324)

tube feet: in echinoderms, the structures connected to the water-vascular system that help them move and feed. (Chap. 14, p. 392)

tumor: an abnormal growth of tissue. (Chap. 27, p. 748)

tundra (TUN dra): a very cold, dry, treeless biome where the sun is barely visible during the six to nine months of winter. (Chap. 19, p. 520)

U

umbilical cord: in placental mammals, a bundle of blood vessels connecting the placenta to the embryo; transports nutrients and oxygen from the placenta to the embryo and carries away wastes. (Chap. 16, p. 441)

ureters (YOOR ut urz): tubes that lead from each kidney to the bladder. (Chap. 24, p. 668)

urethra (yoo REE thruh): a tube that carries urine from the bladder to the outside of the body. (Chap. 24, p. 668)

urinary system: a system of excretory organs that rids blood of wastes, excess water, and excess salts. (Chap. 24, p. 666)

urine: waste liquid collected by the kidneys; contains water, salts, and other wastes. (Chap. 24, p. 667)

uterus: in females, the pear-shaped, hollow, muscular organ in which a fertilized egg develops into a baby; also called the womb. (Chap. 26, p. 706)

V

vaccination: administration of a weakened virus to develop immunity against a disease. (Chap. 27, p. 744)

vaccine: a solution made from damaged virus or bacteria particles or from killed or weakened viruses or bacteria; can prevent, but not cure, many viral and bacterial diseases. (Chap. 2, p. 36; Chap. 8, p. 217)

vagina: in females, the passageway that leads from the uterus to outside of the female's body; also called the birth canal because offspring pass through it when being born. (Chap. 26, p. 706)

variable: in an experiment, the factor tested. (Chap. 1, p. 15)

variation: the occurrence of an inherited trait that makes an individual different from other members of the same species. (Chap. 6, p. 156)

vascular plant: a plant containing vascular tissue made up of tubelike cells that transport food and water through the plant. (Chap. 10, p. 264)

vein: a vessel with one-way valves that moves blood toward the heart, carrying wastes. (Chap. 23, p. 630)

ventricles (VEN trih kulz): the two lower chambers of the human heart. (Chap. 23, p. 627)

vertebrate: animal with a backbone. (Chap. 13, p. 345)

vestigial structure: a body part that is reduced in size and that has no obvious use. (Chap. 6, p. 168)

villi (VIHL i): tiny, fingerlike projections on the inner surface of the small intestine. (Chap. 22, p. 616)

virus: a microscopic particle made of either a DNA or an RNA core and covered with a protein coat; it infects host cells in order to reproduce. (Chap. 2, p. 32)

vitamins: organic nutrients that promote growth, regulate body functions, and help the body use other nutrients. (Chap. 22, p. 604)

voluntary muscle: a muscle you can control, such as arm and leg muscles. (Chap. 21, p. 584)

W

water cycle: the continuous movement of water in the biosphere through evaporation, condensation, and precipitation. (Chap. 18, p. 500)

water-vascular system: in echinoderms, the network of water-filled canals to which thousands of tube feet are connected. (Chap. 14, p. 392)

X

xylem (ZI lum): vascular plant tissue made up of tubular vessels that transport water and minerals from the roots up through the plant. (Chap. 11, p. 294)

Z

zygote: in organisms that reproduce sexually, the cell that forms in fertilization. (Chap. 4, p. 101)

This glossary defines each key term that appears in **bold type** in the text. It also shows the page number where you can find the word used.

abiotic factor/factor abiótico: rasgos físicos inanimados que a menudo determinan los organismos que pueden sobrevivir en cierto ambiente. (Cap. 18, pág. 483)

acid rain/lluvia ácida: lluvia o nieve con un pH menor de 5.6, que lava nutrientes valiosos del suelo lo que puede conducir a la muerte de árboles y otras plantas. (Cap. 20, pág. 559)

active immunity/inmunidad activa: tipo de inmunidad que ocurre cuando el cuerpo fabrica sus propios anticuerpos en respuesta a un antígeno. (Cap. 27, pág. 738)

active transport/transporte activo: tipo de transporte celular que requiere energía para mover materiales a través de una membrana celular. (Cap. 3, pág. 74)

adaptation/adaptación: cualquier característica heredada que posee un organismo y que lo hace más apto para sobrevivir en sus alrededores. (Cap. 1, pág. 7)

adolescence/adolescencia: etapa de desarrollo que comienza alrededor de los 12 a 14 años, cuando una persona es capaz de producir progenie. (Cap. 26, pág. 718)

adulthood/edad adulta: etapa final de desarrollo que comienza al terminar la adolescencia y se extiende hasta la vejez. (Cap. 26, pág. 719)

aerobe/aerobio: organismo que usa oxígeno para llevar a cabo la respiración. (Cap. 8, pág. 212)

aggression/agresión: acto forzado que se usa con el propósito de dominar o controlar a otro animal. (Cap. 17, pág. 463)

AIDS/SIDA: Síndrome de Inmunodeficiencia Adquirida. Enfermedad causada por el VIH (Virus de Inmunodeficiencia Humana). El SIDA permite que muchos organismos causen enferme-dades en el cuerpo, las cuales conducen, tarde o temprano, a la muerte. (Cap. 2, pág. 38)

algae/alga: protista que parece una planta, contiene clorofila en sus cloroplastos y fabrica su propio alimento. (Cap. 9, pág. 231)

allele/alelo: formas diferentes que puede tener un gene para un rasgo. (Cap. 5, pág. 125)

allergen/alérgeno: sustancia que causa una reacción alérgica en el cuerpo. (Cap. 27, pág. 749)

allergy/alergia: potente reacción del sistema inmunológico a una sustancia extraña. (Cap. 27, pág. 749)

alternation of generations/alternación de generaciones: ciclo continuo que alterna entre la fase productora de esporas y la fase productora de células sexuales en los musgos. (Cap. 10, pág. 268)

alveoli/alvéolos: manojos de pequeños sacos de paredes delgadas ubicados en el extremo de cada bronquiolo. (Cap. 24, pág. 656)

Alzheimer's disease/enfermedad de Alzheimer: fracaso de las células nerviosas en el encéfalo para comunicarse. Ocasiona tal destrucción de las células nerviosas, que resulta en la pérdida paulatina de la memoria. (Cap. 25, pág. 686)

amino acid/aminoácido: unidad básica de la cual están compuestas las moléculas proteicas. (Cap. 22, pág. 602)

amniotic egg/huevo amniótico: tipo de huevo que contiene membranas que protegen y amortiguan el embrión, además de ayudarlo a deshacerse de los desperdicios. También contiene un abastecimiento grande de alimento, para el embrión. (Cap. 15, pág. 417)

amniotic sac/bolsa amniótica: bolsa pegada a la placenta que contiene el fluido amniótico que ayuda a proteger al embrión contra golpes y que puede almacenar nutrientes y desperdicios. (Cap. 26, pág. 713)

amphibian/anfibio: vertebrado de sangre fría que

pasa parte de su vida en agua y parte sobre tierra. (Cap. 15, pág. 409)

anaerobe/anaerobio: organismo que no necesita oxígeno, el cual incluso puede ser mortal para algunos de ellos. (Cap. 8, pág. 212)

angiosperm/angiosperma: planta vascular en la cual la semilla está rodeada dentro del fruto. (Cap. 11, pág. 288)

antibiotic/antibiótico: sustancia producida por un organismo que inhibe o mata otro organismo. (Cap. 8, pág. 217)

antibiotic resistance/resistencia a los antibióticos: ocurre cuando las bacterias que, cierto tratamiento de drogas normalmente mata, desarrollan mutaciones o cepas a las cuales no afecta el tratamiento. (Cap. 8, pág. 222)

antibody/anticuerpo: proteína que fabrica un animal en respuesta a un antígeno específico. (Cap. 27, pág. 738)

antigen/antígeno: proteína y sustancia química extraña para el cuerpo ubicada sobre las superficies de los patógenos. (Cap. 27, pág. 738)

antioxidant/antioxidante: sustancia que evita que otras sustancias químicas reaccionen con el oxígeno. (Cap. 22, pág. 611)

antiseptic/antiséptico: sustancia química que mata los patógenos sobre la piel. (Cap. 27, pág. 732)

anus/ano: abertura en el extremo del canal digestivo a través de la cual salen del cuerpo los desperdicios del animal. (Cap. 13, pág. 359)

appendage/apéndice: estructura que crece del cuerpo de un animal. (Cap. 14, pág. 382)

archaebacteria/arquebacterias: organismos unicelulares procarióticos que viven como anaeroblos en condiciones ambientales extremas, incluídas las lagunas salobres, los manantiales de aguas termales sulfurosas y las fisuras termales oceánicas profundas. (Cap. 7, pág. 193)

artery/arteria: vaso sanguíneo de paredes gruesas y elásticas que transporta sangre fuera del corazón. (Cap. 23, pág. 629)

Arthropoda/Arthropoda: el filo más grande de animales en el reino animal e incluye los insectos, los camarones, las arañas y los ciempiés. (Cap. 14, pág. 382)

ascus/ascus: estructuras que parecen sacos en donde producen esporas los hongos como las levaduras, los mohos, las morelas y las trufas. (Cap. 9, pág. 242)

asexual reproduction/reproducción asexual: tipo de reproducción en la cual un solo progenitor produce un nuevo organismo, como por ejemplo, en la mitosis. La progenie producida posee DNA idéntico al del organismo progenitor. (Cap. 4, pág. 97)

asthma/asma: trastorno de los pulmones en que la persona puede sentirse corta de aliento, resollos asmáticos o tos. A menudo, el asma es una reacción alérgica. (Cap. 24, pág. 661)

atherosclerosis/aterosclerosis: acumulación de depósitos grasos en las paredes arteriales. Es una de las causas principales de las enfermedades cardíacas. (Cap. 23, pág. 632)

atria/aurículas: las dos cavidades superiores del corazón. (Cap. 23, pág. 627)

auxin/auxina: tipo de hormona vegetal que causa el fototropismo positivo de las hojas y de los tallos de una planta. (Cap. 12, pág. 325)

axon/axón: parte de la neurona que transmite mensajes desde el cuerpo celular. (Cap. 25, pág. 679)

basidium/basidio: estructura en forma de basto en la cual producen esporas los hongos basidiomicetos. (Cap. 9, pág. 244)

behavior/comportamiento: manera en que actúa un organismo hacia su ambiente. (Cap. 17, pág. 454)

bilateral symmetry/simetría bilateral: animal cuyas partes corporales están distribuidas de la misma forma a ambos lados de su cuerpo. (Cap. 13, pág. 347)

binomial nomenclature/nomenclatura binaria: sistema de clasificación de Linneo que usa dos términos, o nombre científico, para nombrar cada organismo. (Cap. 7, pág. 189)

biogas/biogas: mezcla de gases que se pueden almacenar y transportar como el gas natural. (Cap. 3, pág. 84)

biogenesis/biogénesis: teoría que establece que los seres vivos provienen de otros seres vivos y no de seres inanimados. (Cap. 1, pág. 11)

biological indicator/índice biológico: especie de animales cuya salud general refleja la salud de sus ecosistemas particulares. (Cap. 15, pág. 414)

biomass/biomasa: material orgánico proveniente de plantas o animales que se usa como fuente energética. (Cap. 3, pág. 84)

biomaterial/biomaterial: material artificial que puede usarse para reemplazar partes del cuerpo. (Cap. 21, pág. 583)

biome/bioma: extensa área geográfica que tiene climas y ecosistemas similares. (Cap. 19, pág. 520)

biosphere/biosfera: la parte de la Tierra que sostiene organismos vivos. (Cap. 18, pág. 482)

biotechnology/biotecnología: método rápido y preciso de mejorar las cosechas en el cual el DNA se mueve directamente desde un organismo hasta otro. (Cap. 12, pág. 331)

biotic factor/factor biótico: cualquier organismo vivo en un ambiente. (Cap. 18, pág. 486)

bladder/vejiga: órgano muscular elástico que almacena la orina hasta que sale del cuerpo. (Cap. 24, pág. 668)

blood pressure/tensión arterial: fuerza ejercida sobre las paredes de los vasos sanguíneos a medida que la sangre es bombeada a través del sistema cardiovascular. (Cap. 23, pág. 630)

bog/pantano: área de drenaje malo con tierra fangosa y esponjosa compuesta principalmente de plantas muertas cuya descomposición ha sido retardada debido a la falta de oxígeno. (Cap. 10, pág. 275)

brain stem/bulbo raquídeo: parte del encéfalo que se extiende desde el cerebro y conecta el encéfalo con la médula espinal. Esta compuesto del encéfalo medio, el puente de Varolio y la médula. (Cap. 25, pág. 681)

bronchi/bronquios: dos ramificaciones cortas en el extremo bajo de la tráquea que llevan aire a los pulmones. (Cap. 24, pág. 656)

budding/gemación: una forma de reproducción asexual en la cual un nuevo organismo crece de un lado del progenitor. (Cap. 9, pág. 243)

cambium/cambium: tejido que produce nuevas células de xilema y de floema. (Cap. 11, pág. 294)

cancer/cáncer: enfermedad crónica que resulta del crecimiento descontrolado de las células. (Cap. 27, pág. 748)

capillary/capilar: vaso sanguíneo microscópico que conecta las arterias y las venas. (Cap. 23, pág. 630)

captive breeding/reproducción en cautiverio: apareamiento y reproducción en cautiverio de especies en peligro de extinción para proteger a los animales de la extinción. (Cap. 17, pág. 470)

carbohydrate/carbohidrato: la fuente principal de energía del cuerpo que contiene átomos de carbono, hidrógeno y oxígeno. (Cap. 22, pág. 601)

cardiac muscle/músculo cardíaco: tipo de músculo involuntario que solo se encuentra en el corazón. (Cap. 21, pág. 587)

carnivore/carnívoro: animal que come la carne de otros animales, como por ejemplo, el tigre. (Cap. 16, pág. 436)

carrying capacity/capacidad de carga: el mayor número de individuos que un ambiente puede soportar y mantener durante un largo período de tiempo. (Cap. 18, pág. 492)

cartilage/cartílago: capa lisa y gruesa de tejido que cubre los extremos de los huesos, en donde absorbe choques y facilita el movimiento, reduciendo la fricción. (Cap. 21, pág. 576)

cartilage/cartílago: tejido flexible fuerte que no es tan duro como el hueso. (Cap. 15, pág. 402)

cell/célula: son las unidades más pequeñas de los organismos y que llevan a cabo las funciones vitales. (Cap. 1, pág. 6)

cell membrane/membrana celular: estructura flexible que forma el límite externo de la célula y que permite que solo ciertos materiales entren y salgan de la célula. (Cap. 2, pág. 47)

cell theory/teoría celular: establece que todos los organismos están hechos de una o más células; que las células son las unidades estructurales y funcionales básicas de todos los organismos y que todas las células provienen de otras células existentes. (Cap. 2, pág. 44)

cell wall/pared celular: estructura rígida fuera de la membrana celular que apoya y protege las células vegetales. (Cap. 2, pág. 51)

cellulose/celulosa: compuesto orgánico hecho de cadenas largas de moléculas de azúcar, del cual están formadas las paredes celulares de las plantas. (Cap. 10, pág. 263)

centenarian/centenario(a): persona que pasa de los 100 años de edad. (Cap. 26, pág. 722)

central nervous system/sistema nervioso central: uno de los dos sistemas principales en que se divide el sistema nervioso. Está compuesto por el encéfalo y la médula espinal. (Cap. 25, pág. 680)

cerebellum/cerebelo: la segunda parte del encéfalo, ubicado detrás y debajo del cerebro, coordina los movimientos de los músculos voluntarios, mantiene el balance y tono muscular. (Cap. 25, pág. 681)

cerebrum/cerebro: la parte más grande del encéfalo, la cual está dividida en dos secciones grandes llamadas hemisferios. (Cap. 25, pág. 681)

chemical digestion/digestión química: digestión que rompe las moléculas grandes de alimento en moléculas más chicas que pueden ser absorbidas por las células. (Cap. 22, pág. 614)

chemotherapy/quimioterapia: uso de sustancias químicas para eliminar células cancerosas. (Cap. 27, pág. 748)

childhood/niñez: etapa que le sigue a la lactancia y que dura hasta los 12 años. (Cap. 26, pág. 718)

chloroplast/cloroplasto: organelos en las células vegetales que convierten la energía luminosa en energía química en una forma de azúcar. (Cap. 2, pág. 51)

chordate/cordado: cualquier miembro del filo animal Chordata con notocordio, cordón nervioso dorsal hueco y hendiduras branquiales en algún momento durante su vida. (Cap. 15, pág. 400)

chromatin/cromatina: heliografía genética en forma de filamentos largos para las operaciones de la célula, la cual es una forma de material hereditario. (Cap. 2, pág. 47)

chromosome/cromosoma: estructuras en el núcleo que contienen DNA. (Cap. 4, pág. 94)

chronic bronchitis/bronquitis crónica: tipo de bronquitis que persiste por largo período de tiempo. (Cap. 24, pág. 659)

chronic disease/enfermedad crónica: tipo de enfermedad no infecciosa que dura mucho tiempo. (Cap. 27, pág. 746)

chyme/quimo: líquido delgado y acuoso en el cual se convierten los alimentos después de unas cuatro horas de haber sido consumidos. (Cap. 22, pág. 615)

cilia/cilios: estructuras cortas que parecen hilos y se extienden de la membrana celular. (Cap. 9, pág. 236)

circadian rhythm/ritmo circadiano: comportamiento basado en un ciclo de 24 horas. (Cap. 17, pág. 468)

class/clase: grupo en el cual se divide el filo o división. (Cap. 7, pág. 196)

classify/clasificar: significa agrupar ideas, información u objetos basándose en sus semejanzas. (Cap. 7, pág. 188)

climax community/comunidad clímax: comunidad que ha alcanzado la etapa final de sucesión ecológica. (Cap. 19, pág. 512)

clone/clone: célula u organismo que exhiben un DNA genético idéntico (Cap. 4, pág. 112)

closed circulatory system/sistema circulatorio cerrado: tipo de sistema en el cual la sangre que contiene alimento y oxígeno se encuentra dentro de vasos sanguíneos. (Cap. 14, pág. 372)

cnidarian/celentéreo: describe las células urticantes que poseen todos los miembros del filo. (Cap. 13, pág. 353)

cochlea/cóclea: estructura en forma de caracol llena de líquido dentro del oído interno. (Cap. 25, pág. 690)

collar cell/célula del cuello: célula que forra el interior de las esponjas y la cual ayuda a mover el agua a través del animal. (Cap. 13, pág. 351)

commensalism/comensalismo: relación simbiótica en la cual solo uno de los organismos se beneficia y el otro ni se beneficia ni se perjudica. (Cap. 18, pág. 494)

communicable disease/enfermedad contagiosa: enfermedad que se transmite de un organismo a otro. (Cap. 27, pág. 733)

communication/comunicación: intercambio de información. (Cap. 17, pág. 464)

community/comunidad: grupo de poblaciones que interactúan entre sí en un área determinada. (Cap. 18, pág. 487)

compound light microscope/microscopio de luz compuesto: microscopio que permite que la luz pase a través de un objeto y luego a través de una o más lentes. (Cap. 2, pág. 41)

conditioning/acondicionamiento: modificación del comportamiento de modo que una respuesta antes asociada con un estímulo, se pueda asociar con otro estímulo. (Cap. 17, pág. 458)

consumer/consumidor: organismos que no pueden fabricar su propio alimento. (Cap. 3, pág. 78)

contour feather/pluma de contorno: tipo de plumas fuertes y livianas que les dan a las aves sus bellos coloridos y sus formas lisas y suaves. (Cap. 16, pág. 427)

control/control: estándar contra el cual se comparan los resultados de una prueba. (Cap. 1, pág. 15)

coronary circulation/circulación coronaria: flujo de sangre hacia los tejidos del corazón. (Cap. 23, pág. 628)

courtship behavior/comportamiento de cortejo: comportamiento de mutuo reconocimiento entre las hembras y los machos. (Cap. 17, pág. 463)

crop/buche: saco que usan los gusanos para almacenar la tierra que se comen. (Cap. 14, pág. 376)

cuticle/cutícula: capa cerosa protectora en los tallos y hojas de las plantas la cual les ayuda a conservar la humedad. (Cap. 10, pág. 263)

cyclic behavior/comportamiento cíclico: comportamiento innato que ocurre en un patrón repetitivo. (Cap. 17, pág. 467)

cyst/quiste: gusano parasítico joven con una cubierta protectora. (Cap. 13, pág. 362)

cytoplasm/citoplasma: material gelatinoso dentro de la membrana celular y afuera del núcleo que contiene una gran cantidad de agua y muchas sustancias químicas y estructuras que llevan a cabo los procesos vitales de la célula. (Cap. 2, pág. 48)

D

day-neutral plant/planta de día neutro: planta que no es sensible al número de horas de oscuridad para florecer. (Cap. 12, pág. 330)

dendrite/dendrita: parte de la neurona que recibe mensajes y los transmite al cuerpo celular. (Cap. 25, pág. 679)

dermis/dermis: capa de tejido debajo de la epidermis. (Cap. 21, pág. 591)

desert/desierto: el bioma más seco de la Tierra. Recibe menos de 25 cm de lluvia al año y sostiene poca vegetación y fauna. (Cap. 19, pág. 526)

development/desarrollo: todos los cambios por los que pasa un organismo a medida que crece. (Cap. 1, pág. 8)

diaphragm/diafragma: músculo ubicado debajo de los pulmones, el cual se contrae y relaja y, además, ayuda a mover el aire dentro y fuera del cuerpo. (Cap. 24, pág. 657)

dichotomous key/clave dicotómica: sistema que usan los científicos para clasificar organismos mediante el uso de una lista de características más detalladas. (Cap. 7, pág. 201)

dicot/dicotiledónea: tipo de angiosperma que contiene dos cotiledones dentro de sus semillas. (Cap. 11, pág. 288)

diffusion/difusión: movimiento de moléculas desde un área de mucha concentración a otra de menor concentración. (Cap. 3, pág. 72)

digestion/digestión: proceso que descompone los alimentos en moléculas más pequeñas que pueden entrar en la sangre, en donde son transportadas a través de la membrana celular para uso de las células. (Cap. 22, pág. 612)

disinfectant/desinfectante: sustancia química que mata los patógenos. (Cap. 27, pág. 732)

division/división: reemplaza el filo en los reinos de las plantas y de los hongos. (Cap. 7, pág. 196)

DNA/DNA: ácido desoxirribonucleico; código químico que contienen los cromosomas en el núcleo de una célula. (Cap. 4, pág. 104)

dominant/dominante: factor que domina u oculta a otro factor. (Cap. 5, pág. 126)

dorsal hollow nerve cord/cordón nervioso dorsal hueco: manojo de nervios que yace sobre el notocordio. (Cap. 15, pág. 400)

down feather/plumón: tipo de pluma suave y esponjosa que provee a las aves adultas una capa de aislamiento cerca de la piel y que cubre el cuerpo de las aves jóvenes. (Cap. 16, pág. 427)

E

echinoderm/equinodermo: invertebrado con piel espinosa que vive en el fondo del océano. (Cap. 14, pág. 392)

ecological pyramid/pirámide ecológica: modelo que representa la transferencia de energía accesible en la biosfera. (Cap. 18, pág. 492)

ecological succession/sucesión ecológica: proceso de cambio gradual de una comunidad de organismos a otra. (Cap. 19, pág. 510)

ecology/ecología: el estudio de las interacciones entre organismos y entre los organismos y los rasgos físicos de su ambiente. (Cap. 18, pág. 482)

ecosystem/ecosistema: está compuesto de una comunidad biótica y de los factores abióticos que la afectan. (Cap. 18, pág. 487)

ectotherm/ectotermo: animal cuya temperatura corporal interna cambia con la del ambiente. (Cap. 15, pág. 401)

egg/huevo: célula sexual de la progenitora o madre. (Cap. 4, pág. 99)

electron microscope/microscopio electrónico: microscopio que usa un campo magnético para doblar los rayos de luz. Puede ampliar las imágenes hasta 1 000 000 de veces. (Cap. 2, pág. 42)

embryology/embriología: el estudio del desarrollo de los embriones. (Cap. 6, pág. 169)

embryo/embrión: nombre que se le da al bebé nonato durante los dos primeros meses de desarrollo. (Cap. 26, pág. 712)

emphysema/enfisema: enfermedad en la cual los alvéolos pulmonares pierden su capacidad de expandirse y contraerse, haciendo que la persona se sienta corta de aliento. (Cap. 24, pág. 660)

endangered species/especie en peligro de extinción: especies con tan pocos miembros vivos que se encuentran a punto de desaparecer. (Cap. 6, pág. 176; Cap. 20, pág. 554)

endocytosis/endocitosis: proceso mediante el cual las células transportan dentro del citoplasma un cuerpo grande. (Cap. 3, pág. 75)

endoplasmic reticulum/retículo endoplasmático: membrana con pliegues que mueve materiales a través de la célula y que se extiende desde el núcleo hasta la membrana celular y ocupa mucho espacio en algunas células (Cap. 2, pág. 48)

endoskeleton/endoesqueleto: esqueleto interno de un animal vertebrado que apoya y protege los órganos internos y al cual se adhieren los músculos. (Cap. 15, pág. 401)

endospore/endoespora: pared gruesa que rodea a muchas bacterias que producen toxinas. (Cap. 8, pág. 218)

endotherm/de sangre caliente: animal cuya temperatura corporal interna se mantiene constante. (Cap. 15, pág. 401)

enzymes/enzimas: tipo de proteína que acelera las reacciones químicas en las células sin ser alteradas ellas mismas. (Cap. 3, pág. 69)

epidermis/epidermis: capa superficial de la piel. (Cap. 21, pág. 591)

equilibrium/equilibrio: estado en el cual las moléculas de una sustancia se encuentran esparcidas uniformemente a través de un espacio. (Cap. 3, pág. 72)

erosion/erosión: agotamiento del suelo por acción del viento y del agua. (Cap. 20, pág. 551)

estivation/estivación: período de inactividad durante tiempo extremadamente caluroso y seco en que los anfibios se vuelven inactivos y se esconden en la tierra. (Cap. 15, pág. 409)

estuary/estuario: ambiente muy fértil y productivo en donde un río desemboca en el océano que sirve de vivero para muchas especies de peces oceánicos. (Cap. 19, pág. 532)

ethnobotanist/etnobotánico(a): persona que estudia las relaciones entre la gente de varias culturas y las plantas que estas usan. (Cap. 11, pág. 306)

eubacteria/eubacterias: organismos unicelulares procarióticos que se encuentran en muchos lugares y que se reproducen mediante la fisión; incluyen las bacterias patogénicas y las cianobacterias. (Cap. 7, pág. 193)

evolution/evolución: cambio en los rasgos hereditarios de una especie a lo largo del tiempo. (Cap. 6, pág. 152)

exocytosis/exocitosis: proceso en que una célula mueve los desperdicios hacia la membrana celular, en donde se juntan con ella y el contenido es expulsado fuera de la célula. (Cap. 3, pág. 76)

exoskeleton/exoesqueleto: cubierta corporal externa que apoya y protege el cuerpo del animal. (Cap. 14, pág. 382)

extinction/extinción: cuando mueren todos los miembros de una especie. Ocurre naturalmente debido a que el ambiente cambia y los organismos incapaces de cambiar o adaptarse, perecen. (Cap. 6, pág. 176) (Cap. 20, pág. 554)

family/familia: grupo en el cual se separa el orden. (Cap. 7, pág. 196)

fat/grasa: sustancia que provee energía al cuerpo y le ayuda a absorber algunas vitaminas. (Cap. 22, pág. 603)

fermentation/fermentación: una forma de respiración que convierte energía de la glucosa cuando hay una cantidad insuficiente de oxígeno. (Cap. 3, pág. 81)

fertilization/fecundación: la unión de un huevo y un espermatozoide. (Cap. 4, pág. 101)

fetus/feto: nombre con que se conoce al bebé nonato después de los dos primeros meses de embarazo. (Cap. 26, pág. 714)

filter feeder/alimentador por filtración: organismos que obtienen su alimento filtrándolo del agua a través de sus poros. (Cap. 13, pág. 351)

fin/aleta: estructura en forma de abanico que usan los peces para cambiar de dirección, equilibrarse y moverse. (Cap. 15, pág. 401)

fish/pez: animal de sangre fría que posee muchas adaptaciones que le permiten vivir en el agua. (Cap. 15, pág. 401)

fission/fisión: proceso de reproducción que produce dos células con material genético idéntico al de la célula progenitora. (Cap. 8, pág. 212)

flagellum/flagelo: cola de las bacterias que parece un látigo y les ayuda en la locomoción. (Cap. 8, pág. 211)

food chain/cadena alimenticia: manera simple de mostrar cómo la energía de los alimentos pasa de un organismo a otro. (Cap. 18, pág. 497)

food group/grupo de alimentos: alimentos que contienen los mismos nutrientes. (Cap. 22, pág. 608)

food web/red alimenticia: serie de cadenas alimenticias sobrepuestas. (Cap. 18, pág. 497)

fossil/fósil: cualquier resto de vida de una época pasada. (Cap. 6, pág. 163)

fossil fuel/combustible fósil: tipo de combustible que incluye el carbón, el gas natural y los combustibles que se hacen del petróleo. (Cap. 20, pág. 542)

fracture/fractura: cualquier rotura en un hueso. Puede ser simple o compuesta. (Cap. 21, pág. 582)

free-living/de vida libre: organismo que no depende de otro para su alimentación o morada. (Cap. 13, pág. 357)

frond/fronda: hoja de un helecho. (Cap. 10, pág. 276)

gametophyte/gametofito: forma de una planta de musgo que produce células sexuales. (Cap. 10, pág. 268)

gasohol/gasohol: mezcla de gasolina y etanol. (Cap. 3, pág. 84)

gene/gene: sección de DNA en el cromosoma que dirige la fabricación de proteínas específicas. (Cap. 4, pág. 107)

genetic engineering/ingeniería genética: ciencia que experimenta con métodos biológicos y químicos para cambiar la secuencia del DNA que compone un gene. (Cap. 5, pág. 142)

genetics/genética: ciencia que estudia cómo se heredan los rasgos a través de los alelos. (Cap. 5, pág. 125)

genome/genoma: representa todo el material genético humano y el cual muestra la ubicación de los genes individuales en un cromosoma. (Cap. 5, pág. 144)

genotype/genotipo: muestra la composición genética de un organismo. (Cap. 5, pág. 128)

genus/género: un grupo de diferentes organismos que poseen características parecidas. (Cap. 7, pág. 189)

geothermal energy/energía geotérmica: energía térmica de la corteza que se puede aprovechar como fuente energética. (Cap. 20, pág. 544)

gestation period/período de gestación: tiempo durante el cual el embrión se desarrolla en el útero materno. (Cap. 16, pág. 441)

gill/branquia: órgano que intercambia oxígeno y dióxido de carbono con el agua en los moluscos. (Cap. 14, pág. 370)

gill slits/hendiduras branquiales: aberturas ubicadas en la garganta detrás de la boca de los cordados. (Cap. 15, pág. 400)

gizzard/molleja: estructura muscular de los gusanos ubicada detrás del buche que muele la tierra que estos comen. (Cap. 14, pág. 376)

global warming/calentamiento global: calentamiento potencial de la Tierra causado por la quema de combustibles fósiles y la liberación de dióxido de carbono y otros gases. (Cap. 20, pág. 560)

Golgi body/cuerpo Golgi: pilas de sacos cubiertos de membranas que empacan y mueven proteínas afuera de las células. (Cap. 2, pág. 49)

gradualism/gradualismo: modelo que describe la evolución como un cambio lento de una especie en otra especie nueva. (Cap. 6, pág. 160)

grassland/pradera: región tropical y templada en la cual domina la comunidad clímax de hierbas. (Cap. 19, pág. 526)

greenhouse effect/efecto de invernadero: retención en la atmósfera terrestre del calor proveniente del sol. Hace que aumenten las temperaturas terrestres. (Cap. 20, pág. 560)

groundwater/agua subterránea: agua atrapada en depósitos subterráneos que se formaron de roca no porosa. (Cap. 20, pág. 563)

guard cell/célula guardiana: célula alrededor del estoma que abre y cierra los poros de la hoja para ayudarla a conservar agua. (Cap. 11, pág. 294)

gymnosperm/gimnosperma: planta vascular que produce semillas en las escamas de los conos femeninos. (Cap. 11, pág. 286)

habitat/hábitat: lugar en donde vive un organismo. (Cap. 18, pág. 495)

hazardous waste/desperdicio peligroso: material de desecho dañino para la salud de los seres humanos o venenoso para los organismos vivos. (Cap. 20, pág. 562)

heart murmur/soplo al corazón: defecto congénito del corazón. (Cap. 23, pág. 634)

hemoglobin/hemoglobina: sustancia química que contienen los glóbulos rojos, la cual puede transportar oxígeno y dióxido de carbono. (Cap. 23, pág. 637)

herbivore/herbívoro: animal que come plantas. (Cap. 16, pág. 436)

heredity/herencia: transmisión de características de los progenitores a la progenie. (Cap. 5, pág. 124)

hermaphrodite/hermafrodita: animal que produce tanto espermatozoides como huevos. (Cap. 13, pág. 352)

heterozygous/heterocigoto: organismo que tiene dos alelos distintos para una característica. (Cap. 5, pág. 128)

hibernation/hibernación: período de inactividad durante el invierno en que algunos animales se entierran para protegerse del frío. (Cap. 15, pág. 409)

homeostasis/homeostasis: regulación del ambiente interno de un organismo para mantener las condiciones que le permiten vivir. (Cap. 1, pág. 7)

hominid/homínido: primates que parecían humanos y que comían tanto plantas como animales y caminaban erguidos. (Cap. 6, pág. 172)

Homo sapiens/Homo sapiens: el nombre de nuestra especie. (Cap. 6, pág. 173)

homologous/homólogo: estructuras parecidas en organismos de diferentes especies. (Cap. 6, pág. 168)

homozygous/homocigoto: organismo que tiene dos alelos idénticos para una característica. (Cap. 5, pág. 128)

hormone/hormona: sustancia química endocrina producida en varias glándulas sin conductos a través de todo el cuerpo. (Cap. 25, pág. 695)

host cell/célula huésped: célula en la cual se reproduce un virus. (Cap. 2, pág. 34)

hydroelectric power/potencia hidroélectrica: tipo de potencia eléctrica que se produce aprovechando la energía del aqua que fluye. Las plantas de energía hidroeléctrica son extremadamente eficientes y no producen contaminación. (Cap. 20, pág. 544)

hypertension/tensión alta: trastorno cardiovascular que puede ser causado por la aterosclerosis. (Cap. 23, pág. 632)

hyphae/hifa: masa filamentosa multicelular que compone el cuerpo de un hongo. (Cap. 9, pág. 241)

hypothesis/hipótesis: predicción que se puede probar. (Cap. 1, pág. 15)

immovable joint/articulación fija: tipo de articulación que permite poco o nada de movimiento. (Cap. 21, pág. 578)

immune system/sistema inmunológico: grupo complejo de defensas que posee el cuerpo para combatir enfermedades. (Cap. 27, pág. 737)

imprinting/impronta: tipo de aprendizaje en el cual un animal forma un lazo social con otro organismo dentro un período específico de tiempo después de nacido el animal. (Cap. 17, pág. 456)

incomplete dominance/dominancia incompleta: producción de un fenotipo intermedio a los de dos padres homocigotos. (Cap. 5, pág. 132)

incubate/incubar: acción de mantener abrigado un huevo hasta que salga la cría del cascarón. (Cap. 16, pág. 426)

infancy/lactancia: período de rápido crecimiento y desarrollo mental y físico del bebé desde el nacimiento hasta que cumple un año. (Cap. 26, pág. 717)

innate behavior/comportamiento innato: comportamiento heredado con el cual nace un animal. (Cap. 17, pág. 454)

inorganic compound/compuesto inorgánico: compuesto que no contiene carbono. (Cap. 3, pág. 70)

insight/discernimiento: forma de razonamiento que permite a los animales hacer uso de experiencias pasadas para resolver nuevos problemas. (Cap. 17, pág. 459)

instinct/instinto: patrón complejo de comportamiento innato. (Cap. 17, pág. 455)

insulin/insulina: sustancia química que produce un cuerpo sano y la cual usan las células para convertir la glucosa sanguínea en energía. (Cap. 4, pág. 112)

interneuron/interneurona: neurona que retransmite los impulsos desde las neuronas sensoriales hasta las neuronas motoras. (Cap. 25, pág. 679)

intertidal zone/zona entre la marea baja y la alta: porción de la costa cubierta de agua durante la marea alta y expuesta al aire durante la marea baja. (Cap. 19, pág. 532)

invertebrate/invertebrado: animal sin columna vertebral. (Cap. 13, pág. 345)

involuntary muscle/músculo involuntario: músculo que no puedes controlar conscientemente. (Cap. 21, pág. 584)

islets of Langerhans/islas de Langerhans: masas celulares en el páncreas que secretan la insulina. (Cap. 4, pág. 112)

J

joint/articulación: cualquier lugar en donde se unen dos o más huesos. (Cap. 21, pág. 577)

K

kidney/riñón: órgano que filtra la sangre y la limpia de los desperdicios que esta ha recogido de las células. (Cap. 24, pág. 667)

kingdom/reino: la categoría taxonómica más grande. (Cap. 7, pág. 188)

L

larva/larva: huevo fecundado que se desarrolla en un organismo joven. (Cap. 13, pág. 352)

larynx/laringe: vía respiratoria a la cual están adheridas las cuerdas vocales. (Cap. 24, pág. 654)

law/ley: descripción confiable de algo en la naturaleza, la cual está basada en muchas observaciones. (Cap. 1, pág. 17)

learning/aprendizaje: comportamiento que resulta de la experiencia o de la práctica. (Cap. 17, pág. 456)

lichen/liquen: organismo compuesto de un hongo y un alga verde o una cianobacteria. (Cap. 9, pág. 244)

life span/duración de vida: lapso de tiempo que se espera que viva un organismo. (Cap. 1, pág. 8)

ligament/ligamento: banda dura de tejido que mantiene unidos los huesos a las articulaciones. (Cap. 21, pág. 577)

limiting factor/factor limitativo: cualquier factor biótico o abiótico que limita el número de individuos en una población. (Cap. 18, pág. 489)

long-day plant/planta de día largo: planta que requiere noches cortas para florecer. (Cap. 12, pág. 327)

lymph/linfa: fluido corporal que entra a los capilares linfáticos después de moverse alrededor de los tejidos corporales. (Cap. 23, pág. 644)

lymph node/ganglio linfático: estructura que filtra los microorganismos y los materiales extraños de la linfa antes de que esta entre a la sangre. (Cap. 23, pág. 644)

lymphatic system/sistema linfático: sistema que recoge el fluido entre los tejidos corporales y lo devuelve a la sangre a través de un sistema de capilares y vasos linfáticos más grandes. (Cap. 23, pág. 644)

lymphocyte/linfocito: glóbulo blanco del sistema linfático que reconoce químicamente los antígenos. Puede producir anticuerpos y destruir los antígenos invasores. (Cap. 27, pág. 740)

lymphocyte/linfocito: un tipo de glóbulo blanco. (Cap. 23, pág. 644; Cap. 27, pág. 740)

lysosome/lisosoma: organelos en el citoplasma que contienen sustancias químicas que digieren desperdicios y partes celulares desgastadas, además de descomponer los alimentos. (Cap. 2, pág. 50)

M

mammal/mamífero: vertebrado de sangre caliente que tiene pelo y cuya hembra produce leche para amamantar a las crías. (Cap. 16, pág. 435)

mammary gland/glándula mamaria: tipo de glándula característica de todos los mamíferos, produce la leche que usan las madres de los mamíferos para amamantar a las crías. (Cap. 16, pág. 435)

mantle/manto: capa fina de tejido que secreta la concha y protege el cuerpo de los moluscos. (Cap. 14, pág. 370)

marrow/médula: tejido grasoso que rellena las cavidades centrales de los huesos largos y los espacios en el tejido óseo esponjoso y que produce glóbulos rojos. (Cap. 21, pág. 576)

marsupial/marsupio: animal con una bolsa abdominal en donde protege y amamanta a sus crías inmaduras y pequeñas. (Cap. 16, pág. 440)

mechanical digestion/digestión mecánica: digestión que tiene lugar cuando se mastican y se mezclan los alimentos en la boca y son revueltos en el estómago. (Cap. 22, pág. 614)

medusa/medusa: animal de vida libre de forma acampanada. (Cap. 13, pág. 355)

meiosis/meiosis: proceso de división del núcleo que produce células sexuales. (Cap. 4, pág. 99)

melanin/melanina: pigmento que le da color a la piel. (Cap. 21, pág. 591)

menopause/menopausia: ocurre cuando el ciclo menstrual se vuelve irregular y, a la postre, cesa. (Cap. 26, pág. 708)

menstrual cycle/ciclo menstrual: ciclo mensual de cambios en el sistema reproductor femenino. (Cap. 26, pág. 707)

menstruation/menstruación: descarga mensual que consiste en células sanguíneas y tejidos de la cubierta interior gruesa del útero. (Cap. 26, pág. 707)

metabolism/metabolismo: conjunto total de todas las actividades químicas de un organismo, que le permiten mantenerse vivo, crecer y producir progenie. (Cap. 3, pág. 78)

metamorphosis/metamorfosis: cambios por los que pasan muchos insectos y otros animales desde el huevo fecundado hasta ser adultos. (Cap. 14, pág. 384)

migration/migración: movimiento instintivo de ciertos animales de mudarse a lugares nuevos cuando cambian las estaciones, en lugar de entrar en estado de hibernación. (Cap. 17, pág. 468)

mineral/mineral: nutriente inorgánico que regula muchas reacciones químicas en el cuerpo. (Cap. 22, pág. 605)

mitochondria/mitocondria: organelos en donde se descomponen las moléculas de alimentos y se libera energía. (Cap. 2, pág. 49)

mitosis/mitosis: proceso mediante el cual una célula se divide formando dos células idénticas. (Cap. 4, pág. 93)

mixture/mezcla: una combinación de dos o más sustancias en la cual las sustancias individuales retienen sus propiedades. (Cap. 3, pág. 67)

mollusk/molusco: invertebrado de cuerpo blando, generalmente, con concha, simetría bilateral y una cavidad llena de fluido. (Cap. 14, pág. 370)

molting/muda: proceso mediante el cual un animal reemplaza su exoesqueleto. (Cap. 14, pág. 382)

monocot/monocotiledónea: tipo de angiosperma que contiene un cotiledón dentro de sus semillas. (Cap. 11, pág. 288)

monotreme/monotrema: mamífero que pone huevos con cáscara fuerte y correosa. (Cap. 16, pág. 440)

motivation/motivación: algo dentro de un animal que hace que este actúe. (Cap. 17, pág. 457)

motor neuron/neurona motora: neurona que conduce impulsos desde el encéfalo o la médula espinal hasta los músculos o las glándulas a través de todo el cuerpo. (Cap. 25, pág. 679)

movable joint/articulación móvil: articulación que permite que cuerpo realice una amplia gama de movimientos. (Cap. 21, pág. 578)

multiple alleles/alelos múltiples: cuando una característica es controlada por más de dos alelos. (Cap. 5, pág. 132)

muscle/músculo: órgano que se contrae y se encoge. La contracción provee la fuerza para mover las partes del cuerpo. (Cap. 21, pág. 584)

mutation/mutación: cualquier cambio permanente en un gene o cromosoma celular. (Cap. 4, pág. 109)

mutualism/mutualismo: relación simbiótica en la cual ambas especies se benefician. (Cap. 18, pág. 494)

natural resource/recurso natural: partes del ambiente que usan los organismos vivos. (Cap. 20, pág. 540)

natural selection/selección natural: establece que los organismos cuyos rasgos los hacen más aptos para sus ambientes sobreviven para pasar esas características a su progenie. (Cap. 6, pág. 155)

nephron/nefrón: unidad diminuta de filtración de los riñones. (Cap. 24, pág. 667)

neuron/neurona: elemento básico estructural del sistema nervioso. (Cap. 25, pág. 679)

niche/nicho: papel de un organismo en su ecosistema. (Cap. 18, pág. 495)

nitrogen cycle/ciclo del nitrógeno: transferencia de nitrógeno de la atmósfera a las plantas y de regreso a la atmósfera o directamente a las plantas nuevamente. (Cap. 18, pág. 501)

nitrogen-fixing bacteria/bacteria nitrificante: bacterias que convierte el nitrógeno del aire en una forma útil para las plantas y los animales. (Cap. 8, pág. 216)

noncommunicable disease/enfermedad no infecciosa: enfermedad que no se transmite de una persona a otra. (Cap. 27, pág. 746)

nonrenewable resource/recurso no renovable: recurso natural accesible solo en cantidades limitadas y que no se puede reemplazar mediante procesos naturales en un corto tiempo. (Cap. 20, pág. 541)

nonvascular plant/planta no vascular: planta que carece de tejido vascular. (Cap. 10, pág. 264)

notochord/notocordio: estructura flexible en forma

de bastón a lo largo del lado dorsal o espalda, de los animales cordados. (Cap. 15, pág. 400)

nuclear energy/energía nuclear: energía que se produce del rompimiento de billones de núcleos de uranio en una reacción de fisión nuclear. (Cap. 20, pág. 546)

nucleus/núcleo: el organelo más grande en el citoplasma de una célula eucariota, el cual dirige todas las actividades de la célula. (Cap. 2, pág. 47)

nutrient/nutriente: sustancia en los alimentos que provee energía y materiales para el desarrollo, crecimiento y reparación de las células. (Cap. 22, pág. 600)

olfatory cell/célula olfativa: célula nerviosa nasal sensible a las moléculas de gas. (Cap. 25, pág. 691)

omnivore/omnívoro: animal que come tanto plantas como animales. (Cap. 16, pág. 436)

open circulatory system/sistema circulatorio abierto: sistema circulatorio que no posee vasos sanguíneos. (Cap. 14, pág. 370)

order/orden: grupo en el cual se separa la clase. (Cap. 7, pág. 196)

organ/órgano: conjunto de tejidos que funcionan juntos para realizar una función particular. (Cap. 2, pág. 54)

organelle/organelo: estructura dentro del citoplasma de células eucariotas, que realiza una o más labores específicas. (Cap. 2, pág. 48)

organic compound/compuesto orgánico: son los compuestos que contienen carbono. (Cap. 3, pág. 68)

organism/organismo: seres vivos que poseen ciertas características que no poseen los seres inanimados. (Cap. 1, pág. 6)

osmosis/ósmosis: difusión del agua a través de una membrana celular. (Cap. 3, pág. 73)

osteoporosis/osteoporosis: enfermedad en que los huesos se desintegran y se vuelven quebradizos. (Cap. 21, pág. 583)

ovary/ovario: base hinchada del pistilo en donde se forman los óvulos en las angiospermas. (Cap. 11, pág. 298)

ovary/ovario: órgano sexual femenino en donde comienzan a madurar los óvulos cuando la mujer llega a la pubertad. (Cap. 26, pág. 706)

ovulation/ovulación: proceso en que se libera un óvulo cada mes de uno de los ovarios. (Cap. 26, pág. 706)

ovule/óvulo: parte reproductora femenina de la planta, la cual se produce en la parte superior de cada escama. (Cap. 11, pág. 296)

ozone depletion/agotamiento de la capa de ozono: adelgazamiento de la capa de ozono que protege la Tierra de los rayos ultravioletas dañinos del sol. Es causado por el uso de clorofluorocarbonos. (Cap. 20, pág. 560)

parasitism/parasitismo: relación simbiótica que beneficia al parásito y perjudica al huésped. (Cap. 18, pág. 495)

passive immunity/inmunidad pasiva: la que ocurre cuando se introducen en el cuerpo los anticuerpos producidos en otros animales. (Cap. 27, pág. 738)

passive smoking/fumar indirectamente: el respirar aire lleno del humo de cigarrillo. (Cap. 24, pág. 664)

passive transport/transporte pasivo: movimiento de partículas a través de la membrana celular mediante la difusión, en el cual la célula no utiliza energía para mover los materiales. (Cap. 3, pág. 74)

pasteurization/pasteurización: proceso de calentar los alimentos hasta una temperatura que mata la mayoría de las bacterias. (Cap. 27, pág. 730)

pathogen/patógeno: cualquier organismo causante de enfermedades. (Cap. 8, pág. 217)

pedigree/árbol genealógico: una herramienta que se usa para seguir la ocurrencia de una característica en una familia. (Cap. 5, pág. 140)

periosteum/periósteo: membrana fuerte y apretada que cubre la superficie de los huesos. (Cap. 21, pág. 575)

peripheral nervous system/sistema nervioso periférico: uno de los dos sistemas principales en que se divide el sistema nervioso. Está compuesto de todos los nervios por fuera del sistema nervioso central. (Cap. 25, pág. 680)

peristalsis/peristalsis: ondas o contracciones que bajan el alimento por el esófago y a través del sistema digestivo. (Cap. 22, pág. 614)

permafrost/permagel: capa de suelo permanentemente congelada debajo de la capa superficial del suelo de la tundra. (Cap. 19, pág. 520)

pesticide/pesticida: sustancia química que mata insectos y plantas indeseadas. (Cap. 14, pág. 390)

pharynx/faringe: pasaje en forma de tubo por donde pasa el aire y los alimentos. (Cap. 24, pág. 654)

phenotype/fenotipo: característica física observable como resultado de un genotipo particular. (Cap. 5, pág. 128)

pheromone/feromona: sustancia química producida por un animal, la cual influye en el comportamiento de otro animal de la misma especie. (Cap. 17, pág. 466)

phloem/floema: tejido vegetal compuesto de células tubulares. Transporta alimento desde las hojas y los tallos hasta otras partes de la planta, en donde es usado o almacenado. (Cap. 11, pág. 294)

photoperiodism/fotoperiodismo: respuesta de floración de una planta estimulada por el cambio en la longitud de horas de luz y de oscuridad. (Cap. 12, pág. 327)

photosynthesis/fotosíntesis: proceso en que las plantas usan energía luminosa para producir alimentos. (Cap. 12, pág. 318)

photovoltaic (PV) cell/pila fotovoltaica: disco hecho del mineral silicio cubierto con capas finas de metales. Se usa para generar electricidad utilizando la energía solar. (Cap. 20, pág. 543)

phylogeny/filogenia: historia de la evolución de un organismo. (Cap. 7, pág. 193)

phylum/filo: el grupo más pequeño después del reino. (Cap. 7, pág. 196)

pioneer community/comunidad pionera: primera comunidad de organismos que se mudan a un nuevo ambiente. (Cap. 19, pág. 511)

pioneer species/especies pioneras: organismos que son los primeros en crecer en áreas nuevas o que han sido alteradas. (Cap. 10, pág. 270)

pistil/pistilo: órgano reproductor femenino de la flor. (Cap. 11, pág. 298)

placenta/placenta: órgano en forma de saco que desarrolla el embrión en crecimiento y el cual se adhiere al útero. (Cap. 16, pág. 441)

placental mammal/mamífero placentario: animal cuyas crías se desarrollan dentro del útero de la madre. (Cap. 16, pág. 441)

plankton/plancton: plantas y algas microscópicas y otros organismos que flotan cerca de la superficie en aguas cálidas y soleadas. (Cap. 19, pág. 531)

plasma/plasma: parte líquida de la sangre que consiste principalmente en agua. Contiene nutrientes, minerales y oxígeno disueltos. (Cap. 23, pág. 636)

platelet/plaqueta: fragmentos celulares de formas irregulares que ayudan en la coagulación de la sangre. (Cap. 23, pág. 637)

pollen grain/grano de polen: parte reproductora masculina que contiene los espermatozoides. (Cap. 11, pág. 296)

pollination/polinización: transferencia de los granos de polen desde el estambre hasta los estigmas. (Cap. 11, pág. 299)

pollutant/contaminante: cualquier sustancia que contamina el ambiente y contribuye a la contaminación. (Cap. 20, pág. 558)

polygenic inheritance/herencia poligénica: ocurre cuando un grupo o un par de genes actúan juntos para producir una sola característica. (Cap. 5, pág. 133)

polyp/pólipo: animal que tiene forma de jarrón y que generalmente es sésil. (Cap. 13, pág. 355)

population/población: grupo de organismos individuales de la misma especie que viven en el mismo lugar y que pueden producir crías. (Cap. 18, pág. 487)

population density/densidad demográfica: el tamaño de una población que ocupa un área de tamaño específico. (Cap. 18, pág. 489)

precipitation/precipitación: agua que se condensa y cae en forma de lluvia, nieve, cellisca, granizo y neblina. (Cap. 19, pág. 518)

predation/predación: cuando un organismo se alimenta de otro. (Cap. 18, pág. 493)

preening/arreglarse las plumas con el pico: proceso mediante el cual las aves se frotan el aceite de sus glándulas sobre las plumas para hacerlas resistentes al agua. (Cap. 16, pág. 428)

pregnancy/embarazo: período de nueve meses de desarrollo del huevo fecundado hasta el nacimiento del bebé. (Cap. 26, pág. 712)

primary succession/sucesión primaria: sucesión ecológica que comienza en un lugar que no tiene suelo. (Cap. 19, pág. 511)

primate/primate: grupo de mamíferos que tienen pulgares oponibles, visión binocular y hombros flexibles. (Cap. 6, pág. 171)

producer/productor: organismo que fabrica su propio alimento. (Cap. 3, pág. 78)

protein/proteína: molécula grande que contiene carbono, hidrógeno, oxígeno y nitrógeno y la cual usa el cuerpo para el crecimiento. (Cap. 22, pág. 602)

prothallus/prótalo: gametofito que produce células sexuales que se juntan formando el cigoto. (Cap. 10, pág. 277)

protist/protista: organismo unicelular o multicelular que vive en lugares húmedos o mojados. (Cap. 9, pág. 230)

protozoa/protozoario: protista unicelular y complejo parecido a una planta que vive en agua, tierra y en organismos vivos y muertos. (Cap. 9, pág. 234)

pseudopod/seudópodo: extensión temporal del citoplasma, o pata falsa, de los Sarcodinos, las cuales usan para moverse y alimentarse. (Cap. 9, pág. 235)

pulmonary circulation/circulación pulmonar: flujo de sangre a través del corazón, hasta los pulmones y de regreso al corazón. (Cap. 23, pág. 627)

punctuated equilibrium/equilibrio puntuado: modelo que muestra que la evolución rápida de una especie puede resultar debido a la mutación de unos cuantos genes. (Cap. 6, pág. 160)

Punnett square/cuadrado de Punnett: herramienta útil para predecir los resultados en la genética de Mendel. (Cap. 5, pág. 128)

radial symmetry/simetría radiada: animal cuyas partes corporales están distribuidas en círculo alrededor de un punto central. (Cap. 13, pág. 346)

radioactive element/elemento radiactivo: aquél que despiden radiación, una forma de energía atómica. (Cap. 6, pág. 165)

radula/rádula: órgano que usan los gastrópodos para obtener alimento. (Cap. 14, pág. 371)

recessive/recesivo: factor que parece desaparecer o que es ocultado por uno dominante. (Cap. 5, pág. 126)

recycling/reciclar: proceso de usar nuevamente algún artículo que requiere un cambio o nuevo procesamiento del recurso. (Cap. 20, pág. 549)

red tide/marea roja: coloración rojiza de las aguas superficiales del océano causada por una explosión en la población de algas. (Cap. 9, pág. 248)

reflex/reflejo: respuesta automática que tienen todos los animales y que no involucra el cerebro. (Cap. 17, pág. 455)

reflex/reflejo: respuesta involuntaria y automática a un estímulo. (Cap. 25, pág. 682)

regeneration/regeneración: capacidad de un orga-

nismo para reemplazar partes corporales. (Cap. 13, pág. 352)

relative dating/datación relativa: método de datación de fósiles en el cual se estima la edad de un fósil mediante la comparación de las capas de rocas. (Cap. 6, pág. 165)

renewable resource/recurso renovable: : recurso natural reciclado o reemplazado mediante procesos naturales continuos. (Cap. 20, pág. 540)

reptile/reptil: vertebrado de sangre fría con piel seca y escamosa. (Cap. 15, pág. 416)

respiration/respiración: proceso de descomposición de los alimentos para liberar energía. (Cap. 12, pág. 321)

response/respuesta: reacción de un organismo a un estímulo. (Cap. 1, pág. 7)

retina/retina: tejido sensible a la energía luminosa en la parte posterior del ojo. (Cap. 25, pág. 688)

rhizoid/rizoide: filamento parecido a raíces o hilo compuesto de unas cuantas células largas, que anclan los musgos al suelo. (Cap. 10, pág. 267)

rhizome/rizoma: tallo subterráneo de donde crecen las hojas de los helechos hacia la superficie y las raíces en sentido opuesto para anclar la planta al suelo. (Cap. 10, pág. 276)

ribosome/ribosoma: estructura en el citoplasma en donde las células fabrican sus propias proteínas. (Cap. 2, pág. 48)

RNA /RNA: ácido ribonucleico; ácido nucleico que transporta los códigos para fabricar proteínas desde el núcleo hasta los ribosomas. (Cap. 4, pág. 108)

saliva/saliva: sustancia acuosa producida por tres conjuntos de glándulas cerca de la boca y la cual se mezcla con los alimentos. (Cap. 22, pág. 614)

saprophyte/saprofito: cualquier organismo que usa materiales muertos como fuentes alimenticias y energéticas. (Cap. 8, pág. 216)

scale/escama: placa dura, delgada y sobrepuesta que cubre la piel de los peces y los protege. (Cap. 15, pág. 401)

scientific methods/métodos científicos: procedimientos organizados que se usan para resolver problemas. (Cap. 1, pág. 14)

sea otter/nutria de mar: mamífero placentario del orden Carnivora. (Cap. 16, pág. 446)

secondary succession/sucesión secundaria: sucesión que comienza en un lugar que ya tiene suelo y el cual fue la morada de organismos vivos. (Cap. 19, pág. 512)

sedimentary rock/roca sedimentaria: tipo de roca formada de barro, arena y otras partículas finas que quedan al secárseles el agua y la cual contiene la mayor parte de los fósiles. (Cap. 6, pág. 163)

semen/semen: mezcla de espermatozoides y fluido producida en la vesícula seminal. (Cap. 26, pág. 705)

sensory neuron/neurona sensorial: recibe información y envía impulsos al encéfalo o a la médula espinal. (Cap. 25, pág. 679)

sessile/sésil: organismos como las esponjas, que permanecen anclados a un lugar durante toda su vida. (Cap. 13, pág. 350)

setae/seta: estructura parecida a una cerda que facilita la locomoción de un gusano. (Cap. 14, pág. 375)

sex-linked gene/gene ligado al sexo: un alelo heredado en un cromosoma del sexo. (Cap. 5, pág. 140)

sexual reproduction/reproducción sexual: tipo de reproducción en que las células sexuales de dos progenitores se combinan para producir un nuevo organismo. (Cap. 4, pág. 99)

sexually transmitted disease (STD)/enfermedad transmitida sexualmente: enfermedad transmitida de una persona a otra mediante el contacto sexual. Causada por virus o bacterias. (Cap. 27, pág. 734)

short-day plant/planta de día corto: planta que requiere noches largas para florecer. (Cap. 12, pág. 327)

skeletal muscle/músculo del esqueleto: tipo de músculo que mueve los huesos. (Cap. 21, pág. 586)

skeletal system/sistema óseo: armazón del cuerpo compuesta por todos los huesos del esqueleto. (Cap. 21, pág. 574)

smog/smog: forma de contaminación del aire que cuelga sobre las ciudades y otras áreas densamente pobladas. (Cap. 20, pág. 558)

smooth muscle/músculo liso: músculo involuntario no estriado que mueve muchos de los órganos internos. (Cap. 21, pág. 586)

social behavior/comportamiento social: interacciones entre organismos de la misma especie. (Cap. 17, pág. 464)

society/sociedad: grupo de animales de la misma especie que viven y trabajan juntos en una forma organizada. (Cap. 17, pág. 464)

soil depletion/desgaste del suelo: eliminación de nutrientes del suelo. (Cap. 20, pág. 551)

soil management/manejo del suelo: uso de métodos de arado del terreno para disminuir su erosión. (Cap. 20, pág. 551)

solid waste/desperdicio sólido: material sólido indeseado que se debe desechar. (Cap. 20, pág. 548)

sorus/soro: conjunto de estructuras en grupos en el cual producen esporas los helechos. (Cap. 10, pág. 276)

species/especie: grupo de organismos cuyos miembros se reproducen exitosamente entre sí. (Cap. 6, pág. 152)

species/especie: la categoría de clasificación más pequeña. (Cap. 7, pág. 189)

species diversity/diversidad de especies: la gran diversidad de plantas, animales y otros organismos que viven en un ambiente específico. (Cap. 7, pág. 198)

sperm/espermatozoide: célula reproductora masculina producida en los testículos. (Cap. 26, pág. 705)

sperm/espermatozoides: célula sexual del progenitor o padre. (Cap. 4, pág. 99)

spiracle/espiráculo: abertura en el abdomen y el tórax de los insectos, a través de la cual entra el aire y salen los gases de desecho. (Cap. 14, pág. 383)

sponteneous generation/generación espontánea: idea errónea de que los seres vivos provenían de materia inanimada. (Cap. 1, pág. 10)

sporangia/esporangios: esporas producidas dentro de cubiertas redondas en las puntas de las hifas. (Cap. 9, pág. 242)

spore/espora: célula reproductora que forman nuevos organismos sin ayuda de la fecundación. (Cap. 9, pág. 242)

sporophyte/esporofito: etapa diploide de producción de esporas en el ciclo de vida de los musgos. (Cap. 10, pág. 268)

stamen/estambre: órgano reproductor masculino de la flor. (Cap. 11, pág. 298)

stimulus/estímulo: cualquier cosa a la cual responde un organismo, como el sonido. (Cap. 1, pág. 7)

stomata/estoma: pequeños poros en la superficie de las hojas de las plantas, que permiten que el dióxi-

systemic circulation/circulación sistémica: transporta sangre rica en oxígeno a todos los órganos y tejidos corporales, excepto a los pulmones y al corazón. (Cap. 23, pág. 628)

taiga/taiga: región fría de árboles coníferos siempre verdes. (Cap. 19, pág. 521)

target tissue/tejido asignado: tipo de tejido que se ve afectado por las hormonas. (Cap. 25, pág. 695)

taste bud/papila gustativa: receptor principal para el sentido del gusto. (Cap. 25, pág. 691)

taxonomy/taxonomía: la ciencia de clasificar y nombrar organismos. (Cap. 7, pág. 188)

technology/tecnología: el uso del conocimiento científico para resolver problemas cotidianos o mejorar la calidad de la vida. (Cap. 1, pág. 24)

temperate deciduous forest/bosque deciduo de zonas templadas: comunidad clímax de árboles deciduos, los cuales pierden sus hojas en el otoño. (Cap. 19, pág. 522)

tendons/tendones: bandas de tejido grueso que unen los huesos a los músculos del esqueleto. (Cap. 21, pág. 586)

tentacle/tentáculo: estructura que parece un brazo. Rodea la boca en la mayoría de los celentéreos. (Cap. 13, pág. 353)

territory/territorio: área que un animal defiende de otros miembros de la misma especie. (Cap. 17, pág. 462)

testes/testículos: órganos ubicados dentro del escroto. Producen espermatozoides y testosterona. (Cap. 26, pág. 705)

theory/teoría: explicación de cosas o eventos que se basa en muchas observaciones. (Cap. 1, pág. 17)

tissue/tejido: grupo de células similares que realiza el mismo tipo de función en los organismos multicelulares. (Cap. 2, pág. 54)

toxin/toxina: veneno que producen los patógenos bacteriales. (Cap. 8, pág. 218)

trachea/tráquea: un conducto de unos 12 cm de largo ubicado debajo de la laringe y forrado con membranas mucosas y cilios para atrapar el polvo, las bacterias y el polen. (Cap. 24, pág. 654)

transgenic crop/cosecha transgénica: tipo de cosecha en que el DNA de una clase de organismo se injerta en plantas para mejorar la cosecha. (Cap. 12, pág. 331)

transpiration/transpiración: pérdida del agua a través del estoma de una hoja. (Cap. 12, pág. 315)

trial and error/comportamiento de tanteos: comportamiento que se puede modificar por la experiencia. (Cap. 17, pág. 457)

tropical rain forest/bosque pluvial tropical: la comunidad clímax más importante en las regiones ecuatoriales del mundo. (Cap. 19, pág. 524)

tropism/tropismo: respuesta positiva o negativa de una planta a un estímulo. (Cap. 12, pág. 324)

tube feet/patas tubulares: estructuras en los equinodermos que actúan como ventosas y los ayudan a alimentarse y a moverse. (Cap. 14, pág. 392)

tumor/tumor: crecimiento anormal en cualquier parte del cuerpo. (Cap. 27, pág. 748)

tundra/tundra: región extremadamente fría, sin árboles y seca. (Cap. 19, pág. 520)

umbilical cord/cordón umbilical: conecta el embrión a la placenta y transporta oxígeno y alimentos desde la placenta al embrión y saca los desechos que produce el embrión. (Cap. 16, pág. 441)

ureter/uréter: conductos que unen cada riñón con la vejiga. (Cap. 24, pág. 668)

urethra/uretra: conducto por donde sale la orina de la vejiga. (Cap. 24, pág. 668)

urinary system/sistema urinario: compuesto por los órganos que limpian la sangre y que controlan el volumen sanguíneo eliminando el exceso de agua que producen las células. (Cap. 24, pág. 666)

urine/orina: desecho líquido que contiene un exceso de agua, sales y otros desperdicios que no fueron absorbidos por el cuerpo. (Cap. 24, pág. 667)

uterus/útero: órgano muscular hueco con forma de pera y paredes gruesas en donde se desarrolla el óvulo fecundado. (Cap. 26, pág. 706)

vaccination/vacunación: proceso de administrar una vacuna. (Cap. 27, pág. 744)

vaccine/vacuna: se hace de partículas de virus dañados los cuales ya no pueden causar enfermedad. (Cap. 2, pág. 36)

V

vaccination/vacunación: proceso de administrar una vacuna. (Cap. 27, pág. 744)

vaccine/vacuna: se hace de partículas de virus dañados los cuales ya no pueden causar enfermedad. (Cap. 2, pág. 36)

vaccine/vacuna: sustancia hecha de partículas dañadas de las paredes celulares de bacterias o de bacterias muertas. (Cap. 8, pág. 217)

vagina/vagina: tubo muscular en el extremo inferior del útero que lo conecta al exterior del cuerpo. (Cap. 26, pág. 706)

variable/variable: factor que se prueba en un experimento. (Cap. 1, pág. 15)

variation/variación: aparición de un rasgo heredado que diferencia a un individuo de otros miembros de la misma especie. (Cap. 6, pág. 156)

vascular plant/planta vascular: planta con tejido vascular. (Cap. 10, pág. 264)

vein/vena: vaso sanguíneo con válvulas de una sola vía que transporta sangre hacia el corazón. (Cap. 23, pág. 630)

ventricles/ventrículos: las dos cavidades inferiores del corazón. (Cap. 23, pág. 627)

vertebrate/vertebrado: animal con columna vertebral. (Cap. 13, pág. 345)

vestigial structure/estructura vestigial: parte corporal de tamaño reducido que parece no tener ninguna función. (Cap. 6, pág. 168)

villi/microvellosidad: pequeñitas proyecciones parecidas a dedos que cubren los numerosos pliegues y arrugas del intestino delgado. (Cap. 22, pág. 616)

virus/virus: trozo de material hereditario cubierto de proteína que infecta y se reproduce solo dentro de una célula viva. (Cap. 2, pág. 32)

vitamin/vitamina: nutriente orgánico esencial necesario en pequeñas cantidades para ayudar al cuerpo a usar otros nutrientes. (Cap. 22, pág. 604)

voluntary muscle/músculo voluntario: músculo que puedes controlar conscientemente. (Cap. 21, pág. 584)

W

water cycle/ciclo del agua: viaje continuo del agua en la Tierra. (Cap. 18, pág. 500)

water-vascular system/sistema vascular acuífero: red de canales llenos de agua que es una característica distintiva de los equinodermos. (Cap. 14, pág. 392)

X

xylem/xilema: tejido compuesto de vasos tubulares que transportan agua y minerales desde las raíces a través de toda la planta. (Cap. 11, pág. 294)

Z

zygote/cigoto: célula que resulta de la fecundación. (Cap. 4, pág. 101)

Index

The Index for *Glencoe Life Science* will help you locate major topics in the book quickly and easily. Each entry in the Index is followed by the numbers of the pages on which the entry is discussed. A page number given in **boldface type** indicates the page on which that entry is defined. A page number given in *italic type* indicates a page on which the entry is used in an illustration or photograph. The abbreviation *act.* indicates a page on which the entry is used in an activity.

Aardvark, *444*
Abiotic factors, **483**-485, *483, 484, act.* 485
Acetylcholine, 686
Acid rain, *act.* 272-273, 558-**559,** *559, act.* 559
Acquired characteristics, Lamarck's theory of, 153, *153*
Active immunity, **738**-739
Active solar design, *543*
Active transport, **74,** *74*
Active viruses, 34, *34*
Adaptations, **7,** 156-157
  of amphibians, 409, *410, act.* 410
  of birds, 427-431, *427, 428, act.* 429, *430*
  of fish, 401, *401*
  of plants, 262-263, *262, 263*
  of sponges, *act.* 352
Adenine, 105, *106*
Adenovirus, *33*
Adolescence, **718**-719, *718*
Adrenal gland, 695
Adulthood, **719**-720, *720, 722*-723, *722, 723*
Advertising, of cigarettes, 665, *665*
Aerobes, **212**
Aerobic respiration, 321
African sleeping sickness, 235, *235*
Aged, 720, *720, 722*-723, *722, 723*
Aggression, **463,** *463*

Agnatha, 402-403, *402*
Agriculture, soil management in, 551, *551*
Aguilar, Nancy, 408, *408*
AIDS, **38**-39, 58, 222, 735, 740
AIDS virus (HIV), *32, 34, 36, 38, act.* 38-39, 58, 222, 645, 735, 740, *740*
Air pollution, 557-562
  acid rain, 558-559, *559, act.* 559
  greenhouse effect and, 560-562, *560, act.* 561
  indoor, 558
  ozone depletion, 276, *276,* 414-415, 559-560
  smog, 557-558, *557*
Albinism, *111*
Alcohol, 684
  in fermentation, *81,* 243
  as fuel, 84
Algae, **231**-234, *231, 232, 233, 234*
  brown, 233, *233, 234, 234*
  green, 232, *232, 234, 234, 262, 262*
  lichens and, 244
  protozoa vs., *act.* 240
  red, 233, *233, 234, 234,* 248
Alleles, **125,** *124,* 128, *128,* 129
  multiple, **132**-133
Allergens, **749**-750, *750*
Allergies, **749**-750, *750*
Alligators, 418-419
Altamira cave paintings (Spain), 178

Alternation of generations, **268,** 269
Alveoli, **656,** *656, act.* 656
Alzheimer's disease, 683, *683,* **686**-687, *686, 687*
Ambrose, Carol, 724, *724*
Amino acids, 107, 110, **602**-603
Amniotic egg, **417,** *417, 426, 426*
Amniotic sac, **713,** *713*
Amoebas, *193, 234,* 626
Amphibians, **409**-415, *409*
  adaptations of, 409, *410, act.* 410
  frogs. *See* Frogs
  metamorphosis of, 412, *412, act.* 413
  reproduction of, 409
  salamanders, 409, *409,* 411, 412, *414*
  toads, 409, *409,* 410, 411, 412, *414*
  ultraviolet light and, 414-415, *414, 415*
Amylase, salivary, 613
Anaerobes, **212,** 218
Anaphase, *95,* 102, *102, 103*
Anemia, 641
Angiosperms, **288**
  dicots, 288, *289, 303*
  monocots, 288, *288*
  origin and evolution of, 289
  parts of, 292-295, *292, 294, 295*
  reproduction of, 297-298, *297, 298*
  uses of, 291, *291*
Animal(s), *342, act.* 343, *344*

amphibians, 409-415, *409, 410, act.* 410, *411, 412, act.* 413, *414, 415*

arthropods, 382-391, *382, 383, 384, 385, act.* 385, *386, 387, 388, act.* 389, *390, 452*

behavior of. *See* Behavior

captive breeding of, *431, 433,* 470-471, *470, 471*

characteristics of, 344

classification of, 195, 344-347, *345, 346, 347*

cnidarians, 353-356, *353, act.* 354, *355, 356*

in desert, 527, *527*

echinoderms, 392-394, *392, 393, act.* 393, *394*

experimentation on, 348-349

fish, 400-408, *400, 401, 402, 403, act.* 404, *405, act.* 406-407, *456, 464*

flatworms, *347,* 357-363, *357, 358, act.* 358, *359, 360, act.* 360-361, *361*

habitats of, 495, 534, *554, 555, 556*

identifying, *act.* 187, 189, *189,* 200-204, *200, 201, 202, 203, act.* 204, *346*

invertebrates, 345, 346

mammals. *See* Mammals

mitosis in, 94-95, *94-95, act.* 96

mollusks, *368, act.* 369, 370-374, *370, 371, 372, 373, 381*

reptiles, *398, act.* 399, 416-420, *416, 417, 418, 419, 420*

roundworms, 359, *359,* 362-363, *362, 363*

segmented worms, 375-381, *375, 376, act.* 378-379, *380, 381, act.* 466, 495, *495*

sponges, 350-352, *350, 351, 352, act.* 352

symmetry of, 345-347, *346, 347*

in tropical rain forests, *524, 525*

vertebrates, 345, 400-401

wildlife conservation, 554-555, *554*

See also individual animals

Animal cells, 46-50

cell membrane of, *46, 47, 71,* 102. *See also* Cell transport

cytoplasm in, 48-50, *act.* 48

endoplasmic reticulum in, 48-49, *48*

Golgi bodies in, 49, *49, 75*

lysosomes in, 50

mitochondria in, 49-50, *49*

nucleus of, 47, *47*

plant cells vs., *act.* 53

Animal-like protists, 230, 234-236, *234, 235, 236*

Annelids. *See* Segmented worms

Ant(s), 466, *466, act.* 472

Antarctica, 364, 499, *499,* 528-529, *528*

Anteaters, *444*

Anthrax, 217

Antibiotic(s), 75, 160-161, 213, **217,** 244, 734, 735, *735*

Antibiotic resistance, **222-**223, 250

Antibodies, 639, **738-**739

Antigens, **738**

Antihistamines, 750

Antioxidants, *610,* **611**

Antiseptics, **732,** *732, act.* 742-743

Anus, **359**

Anvil, 690, *691*

Aorta, 628

Appendages, **382**

Aquifers, 563

Arachnids, 386-387, *386, 387, 452*

Archaebacteria, 193-194, 210, *214*

Archaeopteryx, 434, 434

Archimedes, 404

Aristotle's system of classification, 188-189, *188,* 196

Art

of Beatrix Potter, 266

mollusks in, 374

painting from observation, 26, *26*

singing, 655, *655*

Stone Age, 178, *178*

trash, 547, *547*

Arteries, **629,** *630, act.* 630

Arthritis, 746, *746*

Arthropoda (arthropods), **382-**391

arachnids, 386-387, *386, 387, 452*

centipedes, 387, *387*

crustaceans, 388, *388, act.* 389

features of, 382, *382*

insects, 383-385, *383, 384, 385,* 390-391, *390*

millipedes, 387, *387*

Artificial blood, 639

Artificial insecticides, 467, *467*

Artificial joints, *579*

Artificial skin, 592, *592*

Ascaris, 362, 363

Ascus, **242**

Asexual reproduction, **97-**98, *97, 98,* 269, 355, *355,* 358

Asimov, Isaac, 646

Aspirin, 306

Asthma, **661**

Atherosclerosis, **632,** *632*

Atoms

combining, 65-66, *66*

structure of, 64, *64*

Atria, **627,** *627*

Australian tiger snake, 419, *419*

Australopithecus, 172, 172

Autonomic system, 682, *682*

Auxin, **325**

Axon, **679,** 680, *680*

Babies, 717, *717*

Bacilli, 211, 211

Bacteria, *208,* 210-213, *act.* 215

antibiotics and, 75, 160-161, 213, 217, 222-223, 244, 250

appearance of, *act.* 209

beneficial, 216-217, *216, 217, act.* 217

in biotechnology, 332, *332*

characteristics of, *211*

classification of, 193-194, *193,* 196, *196*

controlling growth of, 213, *213, act.* 742-743

cyanobacteria, 212, *212 act.* 215

as decomposers, 210, *210, 497*

in deep-sea vents, 214, *214*

digestion and, 617

diseases caused by, 217-218, *218,* 222-223, *222, 223, 224,* 730-731, *730, 731, 733, 734, 734*

eubacteria, 193-194, 210-213

evolution of, 160-161

in food, *act.* 733

genetically engineered, *142*
harmful, 217-219, *219*, 730-731, *730, 731*, 733, 734, *734*, 737, *737*
identifying, 731, *731*
location of, *act.* 220-221
nitrogen-fixing, **216,** *act.* 495
reproduction of, 97, 212-213, *212, act.* 212, *act.* 734
shapes of, 211, *211*
structure of, 210
types of, 210, *210*
white blood cells and, 737, *737*
Bacterial cells, 52, *52*
Bacteriologist, 14
Bacteriophage, *33, 34*
Balance, 690, *690, act.* 690
Ball-and-socket joint, 578, *578*
Barb, 427
Basidium, **244**
Bats, 441, *441, 442,* 445
Beagle (research ship), 153-154, *154*
Beaver, *436*
Bees, 465, *465*
Beetle, *383, 390,* 467, *467*
Behavior, *452, act.* 453, **454-**473
    aggressive, 463, *463*
    communication, 464-467, *465, 466, act.* 466, *467*
    courtship, **463-**464, *463*
    cyclic, **467-**469, *469*
    innate, **454-**455, *454, 455*
    instinctive, 455, *455,* 462
    learned, 456-461, *456, 457, 458, act.* 458, *act.* 460-461
    social, **464,** *464, act.* 472
    territorial, 462-463, *462, 463*
Benjamin, Regina, 250, *250*
Bicarbonate, 82, 616
Biceps, *587*
Bilateral symmetry, *346,* **347,** *347*
Bile, 615
Binocular vision, 171
Binomial nomenclature, **189,** 192, *192,* 200-201
Biodegradable substances, *act.* 552-553, *552*
Biogas, **84-**85
Biogenesis, **11,** *11, act.* 11
Biological indicators, **414**
Biomass, **84-**85, *84, 85*
Biomaterial, **583,** *583*

Biomes, **520-**527, *520*
    deserts, *520,* 526-527, *527*
    freshwater, 530-531, *530, 531, act.* 532
    grasslands, *520,* 526, *526*
    saltwater, 532-533, *532, 533*
    taiga, *520, act.* 520, 521, *521*
    temperate deciduous forests, *520,* 522-523, *522, 523*
    tropical rain forests, 198, *199,* 261, 308, *342, 520,* 524-525, *524*
    tundra, 520-521, *520, act.* 520, *521*
Biosphere, **482-**483, *482, 486, act.* 509
Biotechnology, **331-**333, *331, 332*
Biotic factors, **486-**487, *486, 487*
Biotic potential, **492,** *492*
Birds, 426-434
    adaptations for flight, 427-431, *427, 428, act.* 429, *430*
    body temperature of, 426, 431, *431*
    captive breeding of, 433, *433*
    characteristics of, 426, *426*
    classification of, *197,* 200, *200*
    courtship behavior of, 463, *463*
    eggs of, 426, *426*
    feathers of, 426, 427-428, *427, 428, act.* 428, *act.* 429
    flightless, 430, *432*
    gizzard of, *act.* 425, 431
    hollow bones of, 428, *428*
    homing pigeons, 468, *468*
    insect eating, *433*
    kinds of, *424, 432, 432-433*
    migration of, 468-469
    origin of, 434, *434*
    pesticides and, 555
    preening of, 428, *428, act.* 428
    of prey, *432*
    reproduction of, 426, *426,* 433, *433*
    respiration of, 431, *431*
    seed eating, *433*
    wings of, 430, *430*
Birth, 714, 716-717, *716*
Birth canal (vagina), 706, *706, 710,* 711
Birth defects, 712, *712,* 714, *714*
Bivalves, *act.* 369, 371, *371, 373*
Bladder, *667,* **668,** *668,* 705
Blood, 68, *68,* 636-643
    artificial, 639

clotting of, 638, *638*
    diseases and disorders of, 640-641, *641*
    parts of, 636-637, *636, 637*
    Rh factor in, 640
Blood cells, 574, 637, 641
    comparing, *act.* 642-643
    red, 637, *637,* 641, *641,* 645
    white, 637, *637,* 641, 737, *737*
Blood pressure, **630-**631, *631, act.* 633
Blood types, 133, 639-640, *639*
Blood vessels, 629-630, *630, act.* 630
    arteries, 629, *630, act.* 630
    capillaries, 629, 630, *630,* 644, 656, *656,* 667, 737, *737*
    disorders of, 632, *632*
    veins, 629, 630, *630*
Blubber, *act.* 438-439
Bog, **275**
Bone(s), 575-583, *575, act.* 580-581
    of birds, 428, *428*
    compact, 576, 582
    development of, 576-577, *576*
    fractures in, 582-583, *582, 583*
    joints and, 577, 578-579, *578*
    shape of, 575, *575,* 577
    spongy, 576, 582
    structure of, 575-576, *575*
Bone marrow, 574, **576,** 637
Bony fish, 403-405, *403, 405*
Botanist, 14
Botulism, 218, *219*
Brain, 681, *681,* 683, *683*
Brain stem, **681,** *681*
Breathing
    control of, 655
    physical activity and, *act.* 651
    process of, 657, *657*
    respiration vs., 652-653
    See also Respiration
Breeding, captive, 433, *433,* **470-**471, *470, 471*
Brine shrimp, *act.* 18-19
Bristleworms, *380,* 381
Brittle stars, 393, *393*
Bronchi, **656,** 659
Bronchitis, **659,** *659*
Brown algae, 233, *233, 234, 234*
Bryophytes, 264, 267-270, *267, 268,*

act. 268, *269, 270*
BST (bovine somatotrophin), 24-25
Budding, 98, **243,** *243, 355*
Buoyancy, 404, *act.* 404
Butterflies, *385*

C

Cactus, *527*
Caffeine, 684, *684*
Calcium, 574, 576, 577, 605, *605*
Cambium, **294,** *294*
Camel, *443*
Cancer, **748**
    causes of, 748
    lung, 660-661, *661,* 748
    medicines for, 306, *307*
    nutrients and, 610-611, *610*
    skin, 591
    smoking and, 660-661, *661,*
      664-665, *748*
Capillaries, 629, **630,** *630,* 644, 656,
    656, 667, 737, *737*
Captive breeding, 433, *433,* **470-**
    471, *470, 471*
Carbohydrates, 69, **601**
    cellulose, 262-263, 601, *601,* 617
    glucose, 66, 80, *81,* 112-113, *322,*
      601, *601,* 696
    sugar, 318-319, *318,* 321, 322,
      323, 601, *601*
Carbon dioxide, 263
    exchange in plants, 314-315,
      315, act. 316
    in fermentation, *81,* 243
    greenhouse effect and, 560, *560*
    in photosynthesis, *act.* 82-83,
      291, 318, *318, act.* 318
    in respiration, 80, *act.* 82-83, 314,
      656
Carbonic acid, 82
Carboniferous period, *166,* 275,
    275
Carcinogens, 660, 748, *748*
Cardiac muscle, *586,* **587,** *act.* 589
Cardiovascular disease, 632, *632,*
    634-635, *634, 635*
Cardiovascular system, 626-635.
    See also Circulatory system;
    Heart
Careers
    bacteriologist, 14

botanist, 14
clinical dietitian, 620, *620*
ecologist, 14, *482,* 483
ethnobotanist, 306-307, 308, *308*
family practice physician, 250,
    250
geneticist, 14, 146, *146*
hormone researcher, 698, *698*
horticulturist, 334, *334*
ichthyologist, 408, *408*
life scientist, 14, 16
marine biologist, 448, *448*
nurse midwife, 724, *724*
radiation oncologist, 114, *114*
salmon habitat researcher, 534,
    534
science book author, 364, *364*
zoologist, 14
Carnivore, **436,** *443,* 496-497,
    502-503, *502, 503*
Carotenoids, 261
Carrageenan, 233, *233*
Carroll, Sean B., 146, *146*
Carrying capacity, 489, **492,** *492*
Cartilage, **402,** *575,* **576,** 579
Cartilaginous fish, 403
Cave paintings, 178, *178*
Cell(s), **6,** *act.* 31, 43-57, *act.* 63
    animal, 46-50, *46, 47, 48, act.* 48,
      49, act. 53, *71,* 75, 102
    bacterial, 52, *52*
    collar, **351,** *351*
    cycle of, 93, *93*
    differences among, 52, *52*
    division of, *43, act.* 91, 92-98, *92,*
      93, 94-95, act. 96, 97, *act.* 97
    energy in, 78-83, *78, 79, act.*
      82-83
    eukaryotic, 45, *46,* 194, 195, *195,*
      260
    host, **34**
    nucleus of, 45, 47, *47*
    organization of, 54-57, *55, 56,*
      act. 56-57
    plant, 51, *51, act.* 53
    prokaryotic, 45, *45,* 52, *52,* 194,
      195, *195*
    sex, 99-101, *99, 100*
    surface area and volume of, 50,
      50
    transplanted, 112-113, *113*
Cell membrane, *46,* **47,** *71,* 102.

    See also Cell transport
Cell plate, 95, *95*
Cell theory, 43-**44,** *44*
Cell transport, 71-77
    active, **74,** *74*
    diffusion, 72, *72, act.* 72, 73, *73*
    osmosis, 73, *73, act.* 77, 262
    passive, **74**
Cell wall, **51,** *51*
Cellular respiration, 79-80, *act.*
    82-83, 321, 322, *322*
Cellulose, 262-**263,** 601, *601,* 617
Celsius scale, 23, *23*
Centenarian, **722-**723
Centipedes, 387, *387*
Central nervous system (CNS),
    680-681, *680, 681*
Centrioles, *94,* 95
Centromere, 94, *94,* 102
Cephalopods, *371,* 372-373, *372*
Cerebellum, **681,** *681*
Cerebrospinal fluid, 681
Cerebrum, **681,** *681*
Cesarean section, 716
Cetaceans, *442*
CFCs (chlorofluorocarbons), 560
Chambered nautilus, 372
Chameleon, *416*
Chauvet Cave (France), 178, *178*
Cheese-making, 216, *216*
Chemical(s), 747, *747*
Chemical communication, 466,
    466, 467, 467
Chemical digestion, **614**
Chemotherapy, **748**
Chicken pox, 36
Childhood, **718,** *act.* 718
Chimpanzees, 170, *170, 443,* 459,
    459
Chiroptera, *442*
Chitin, 382
Chlamydia, 734
Chlamydomonas, 232
Chlorofluorocarbons (CFCs), 560
Chlorophyll, 51, 78, 231, 260, *260,*
    261, 295, 317, *317,* 318, *318*
Chloroplasts, **51,** *51,* 78, *79,* 231
Choking, 658, *658*
Cholesterol, 603, 632
Chondrichthyes, 403
Chordates (vertebrates), 345,
    400-401

Chromatin, **47,** *47*
Chromosomes, *90,* **94,** *94-95, 97*
  genetics and, 124, *124-125*
  in meiosis, 102, *102, 103*
  number of, 100, 101
  pairing of, 100, *102*
  X and Y, 138-140, *138, 140*
Chronic bronchitis, **659,** *659*
Chronic diseases, **746,** *746*
Chrysalis, 385, *385*
Chyme, **615,** 616, 617
Cicadas, 469
Cigarette advertising, 664-665, *664, 665*
Cilia, **236,** *236,* 654, 659, *659*
Ciliates, 236, *236*
Cinchona tree, 264, *264*
Circadian rhythm, **468**
Circulation
  coronary, **628**
  pulmonary, **627,** *628*
  systemic, **628,** *628, act.* 628, 629, *629*
Circulatory system, *act.* 625, 626-647, *626*
  blood, 68, *68,* 636-643, *636, 637, 638, 639, 641, act.* 642-643
  blood pressure and, 630-631, *631, act.* 633
  blood vessels, 629-630, *630, act.* 630, 632
  cardiovascular disease and, 632, *632,* 634-635, *634, 635*
  closed, **372**
  heart, *act.* 625, 627-628, *627, 628, act.* 628, 629, *629*
  lymphatic system, 644-645, *644, 645*
  open, **370**
Citrus fruits. *See* Fruit
Clam, 371
Class, **196**
Classification, *act.* 187, **188-205**
  of amoebas, *193*
  of animals, 195, 344-347, *345, 346, 347*
  Aristotle's system of, 188-189, *188,* 196
  of bacteria, 193-194, *193,* 196, *196*
  of birds, *197,* 200, *200*
  dichotomous keys in, 201-204,

act. 204
  DNA and, 194
  of dogs, 192, *192*
  five-kingdom system of, 193-197, *193, 194, 195, 197*
  of frogs, *188,* 189
  of mammals, 440-441, *440, 441*
  of mollusks, 370-374, *370, 371, 372, 373*
  naming organisms, 189, 192, *192,* 200-204, *200, 201, act.* 201, *202, 203, act.* 204
  of pandas, 194, *194*
  of plants, *act.* 190-191, 195, 264, *265*
  of primates, 172-173, *172*
  of seeds, *act.* 190-191
  taxonomy and, 188-192, *188, act.* 190-191
  of unknown organisms, 189, *189*
  of viruses, 33, *33*
  of wolves, 192, *192*
Cleanliness, 732, *732*
Climate, 516-518
  Earth's tilt on axis and, *516,* 517
  elevation and, 517, *517*
  latitude and, *516,* 517
  precipitation, 518, *518*
  rain shadow effect and, 518, *518*
  temperature, 517
Climax community, **512-**513, *513*
Clone, **112**
Closed circulatory system, **372**
Clotting, 638, *638*
Club fungi, 244, *244*
Club moss, 271, *271,* 274, *274*
Cnidarians, **353-**356, *353, act.* 354, *355, 356*
Coal
  burning of, 278-279, *278, 279*
  as fossil fuel, 542
Cocci, 211, *211*
Cochlea, **690,** *690*
Cocoon, 385, *385*
Coelacanth, *405*
Colds, 36
Collar cells, **351,** *351*
Color blindness, 139-140, *140*
Commensalism, **494,** *494*
Communicable diseases, **733,** *733*
Communication, **464-**467, *465, 466, act.* 466, *467*

Community, *486,* **487**
  climax, **512-**513, *513*
  energy flow through, 496-501, *496-497*
  interactions in, 492-495, *492, 494, 495, act.* 495
  pioneer, *510,* **511,** 513
Compact bone, 576, 582
Compost, 552
Compound(s), 66-67, *66*
  inorganic, **70,** *70*
  organic, **68-69**
Compound light microscope, **41-**42, *41*
Concave lens, 689, *689*
Conditioning, **458,** *458, act.* 460-461
Cones
  of eye, 688, *688*
  in plant reproduction, 287, 296-297, *296, act.* 297
Confocal microscopy, 243, *243*
Conifer, 287, *287,* 289. *See also* Gymnosperms
Coniferous forests, 521, *521*
Conservation, 550-556
  of energy, 550, *550*
  of land, 555-556, *555*
  of minerals, 556, *556*
  of soil, 551, *551*
  of wildlife, 554-555, *554*
Consumers, **78,** 496-497, *496-497*
Contour feathers, **427,** *427, act.* 429
Contour plowing, 551, *551*
Control, **15**
Controlled experiment, 10, *10*
Convex lens, 689, *689*
Coral, 353, *353, 356*
Coral reef, 353, 356, *356,* 487, *487*
Coronary circulation, **628**
Cotton, 291, *291*
Cotyledon, 288, 299, *299,* 303, *303*
Cotylosaurs, 417
Coughing, 659, *659*
Courtship behavior, **463-**464, *463*
Crab, 388, *388*
Crayfish, 388, *act.* 389
Crick, Francis, 104, 105, 108
Critical thinking, 20, *21*
Crocodiles, 418-419, *418,* 420, *420*
Cro-Magnon humans, 173, 174, *174, 175*
Crop, **376**

transgenic, **331**-333, *331, 332*
Crustaceans, 388, *388, act.* 389
Cuticle, **263,** *263,* 294
Cyanobacteria, 193, 213-215, *213, act.* 215
Cycads, **287,** *287,* 289
Cycles
    of cells, 93, *93*
    of matter, 499-501, *500, 501*
    menstrual, **707,** *707, 708*
    nitrogen, **501,** *501*
    water, **500,** *500, act.* 500
Cyclic behavior, **467**-469, *469*
Cyclone collector, 278, *278*
Cyst, **362**
Cystic fibrosis, 75, 138
Cytoplasm, **48**-50, *act.* 48
Cytosine, 105, 106

Dairy industry, 24-25, *24, 25*
Dandelion, *292*
Dart, Raymond, 172
Darwin, Charles, 153-155, *154*
Data analysis, 16
Dating
    fission track, 12, *12*
    radioactive, 12, *12,* 165, *act.* 167
    relative, **165,** *165*
Day-neutral plants, **330,** *330*
DDT, 555
Deciduous forests, *520,* **522**-523, *522, 523*
Decomposers, 210, *210,* 496, *497*
Dendrite, **679,** 680, *680*
Deoxyribonucleic acid. *See* DNA
Depressant, 684
Dermis, **591,** *591*
Desert, *520,* **526**-527, *527*
Desertification, 484
Development, **8,** *8*
    of bones, 576-577, *576*
    embryonic, 712-713, *713, act.* 713
    human, 712-723, *713, act.* 713, *716, 717, 718, act.* 718, *719, act.* 721, *722, 723*
    of seeds, 299-303, *303.*
    See also Growth
Diabetes, 112-113, *112,* 696, 746
Diagrams, interpreting, *act.* 709

Dialysis, 670-671, *671*
Diaphragm, 654, **657**
Diatoms, 231, *231,* 234, *234*
Dichotomous keys, **201**-204, *act.* 204
Dicots, **288,** *289, 303*
Diet. *See* Nutrients
Diffusion, **72,** *72, act.* 72, 73, *73,* 262
Digestion, **612**-619
    bacteria and, 617
    chemical, **614**
    enzymes in, 613, *613*
    mechanical, **614**
    of proteins, *act.* 618-619
Digestive system, 612, *612*
Digestive tract, *598*
    esophagus, 614, *614*
    large intestine, 617, *617*
    model of, *act.* 599
    mouth, 614, *614*
    small intestine, 615-616, *615*
    stomach, 615
Diggins, Maureen, 698, *698*
Dinoflagellates, 232, *232,* 234, *234,* 248, *249*
Dinosaurs, 176, 416, 417
Diphtheria, 217
Diploid cell, 100
Diseases, 730-736
    animal experimentation and, 348-349
    bacterial, 217-218, *218,* 222-223, *222, 223,* 224, 730-731, *730, 731,* 733, 734, *734*
    of blood, 640-641, *641*
    cardiovascular, 632, *632,* 634-635, *634, 635*
    chemicals and, 747, *747*
    chronic, **746,** *746*
    communicable, **733,** *733*
    fungi as cause of, 242, 250, 733
    Koch's rules for identifying, 731, *731*
    of nervous system, 683, *683,* 686-687, *686, 687*
    noncommunicable, **746**-750, *746, 747, 748, 749, 750*
    prevention of, 732, *732, act.* 742-743. *See also* Immunity
    protozoa as cause of, *234,* 235, *235,* 236, *236*
    of respiratory system, 659-661,

659, 660, 661
    roundworms as cause of, 362-363, *363*
    sexually transmitted (STDs), **734**-735, *734*
    of urinary system, 670-671, *670, 671*
    viral, 36, 37, *act.* 38-39, 58, *58,* 733, 744-745, *744*
    See also individual diseases
Disinfectants, **732,** *act.* 742-743
Diversity
    extinction and, 198-199, *198, 199*
    of mammals, 442, *442-444*
    species, **198**-199, *198, 199*
Division, **196**
DNA (deoxyribonucleic acid), 69, *90,* **104**-111, *104*
    in asexual reproduction, 97
    classification and, 194
    as evidence for evolution, 170, *170*
    models of, *104,* 105, *105, act.* 108-109
    replication of, 106, *106*
    RNA vs., *act.* 110
    sequencing of, 144, 145
    testing of, 106, *106*
    of transgenic crops, 332
Dogs, classification of, 192, *192*
Dominance, incomplete, **132,** *132*
Dominant traits, **126,** 127
Doolittle, Bev, 26
Dorsal hollow nerve cord, **400**
Dover (England), White Cliffs of, *235*
Down feathers, **427,** *427, act.* 429
Downy mildew, 238, 239
Dragonfly, *383*
Drugs
    nervous system and, 684, *684*
    from plants, 264, 306-307, *306, 307,* 308
Duck, *432*
Duck-billed platypus, 440, *440*
Duodenum, 615
Dysentery, *234*

E. coli, 208-209
Ear, 690, *690*

Eardrum, 690, *691*

Earth
motions of, 327
tilt on axis, *516*, 517

Earthworms, 375, 376-377, *376, act. 378-379, act. 466, 495, 495, 511*

Ebola virus, 58, *58*

Echinoderms, **392**-394, *392, 393, act. 393, 394*

Ecological pyramid, **498**

Ecological succession, **510**-512
primary, *510-511,* **511**
secondary, **512,** *512*

Ecologist, 14, *482,* 483

Ecology, **482**

Ecosystems, *486,* **487,** 488-495, 508-535, *act.* 509
changes in, 510-515, *510, 511, 512, 513, act.* 514-515
energy flow through, 496-501, *496-497, 498, 499, 500, 501*
land, 516-527, *517, 518, act.* 519, *520, act.* 520, *521, 522, 524, 525, 526, 527*
in space, 523, *523*
water, 530-533, *530, 531, 532, act.* 532, *533*

Ectotherms, **401**

Egg, **99,** *99*
amniotic, **417,** *417, 426, 426*
of bird, 426, *426*
in human reproduction, 706, 707, *707,* 710, 711, *711*

Elderly, 720, *720,* 722-723, *722, 723*

Electricity, energy sources for, 542-546, *542, 543, 544, 545, 546*

Electron, 64, *64*

Electron microscopes, **42**-43, *42*

Elements, 65, *65*
radioactive, **165**

Elephants, *161, 344, 436, 443,* 470, *470*

Elevation, 517, *517*

Embryo
human, **712**-713, *713, act.* 713
plant, 299, *299*

Embryology, **169,** *169*

Emphysema, **660,** *660*

Emulsification, 615, *act.* 615

Endangered species, **176, 554**
as biological indicators, 414-415, *414, 415*
captive breeding of, 433, *433,* 470-471
diminishing diversity and, 198-199, *198, 199*
extinction and, 176-177, *177*
ozone depletion and, 414-415, *414, 415*
protection of, 446-447, *446, 447,* 502-503, *503,* 554-555, *554*
reintroduction of, 502-503, *502, 503*

Endangered Species Act of 1973, 176

Endocrine glands, 695

Endocrine system, 695-697, *697*

Endocytosis, **75,** *76*

Endoplasmic reticulum (ER), **48**-49, *48*

Endoskeleton, **401**

Endosperm, 299

Endospores, **218,** *219*

Endotherms, **401,** 438

Energy
as basic need, 8, *8*
from biomass, 84-85, *84, 85*
in cells, 78-83, *78, 79, act.* 82-83
conservation of, 550, *550*
flow through ecosystems, 496-501, *496-497, 498, 499, 500, 501*
geothermal, **544**-545, *545*
nuclear, **546,** *546*
releasing, 79-81, *81*
solar, 542, 543, *543*
sources of, 542-546, *542, 543, 544, 545, 546*
water, 544, *544*
wind, 544, *544*

Energy pyramid, **498**-499, *498*

Environment, 482-505, 557-564
abiotic factors in, 483-485, *483, 484, act.* 485
biosphere, 482-483, *482, 486, act.* 509
biotic factors in, 486-487, *486, 487*
carrying capacity of, 489, 492, *492*
land, 516-527, *517, 518, act.* 519, *520, act.* 520, *521, 522, 524, 525, 526, 527*
limiting factors in, 489, *act.* 490-491
living space in, *act.* 481
pesticides and, 391
pollution of. *See* Pollution
population density in, 489, *489*
senses and, 688
water, 530-533, *530, 531, 532, act.* 532, *533*

Enzymes, **69,** *act.* 68, 613, *613*

Epidermis, 294, **591,** *591*

Epiglottis, 654

Equilibrium, **72**
punctuated, **160**-161, *160, 161*

Equisetum, 274

Erosion, **551,** *551*

Esophagus, 614, *614*

Estivation, **409**

Estuaries, **532,** *532*

Ethanol, 84, *84*

Ethics, 741

Ethnobotanist, **306**-307, 308, *308*

Ethylene, 326, *326*

Eubacteria, 196

Euglena, 231, *231,* 234, *234*

Eukaryotic cells, 45, *46,* 194, 195, *195,* 260

Evolution, **152**-179, *act.* 157
adaptation and variation and, 156-157, *156, 157, act.* 158-159
of bacteria, 160-161
of chimpanzees, 170, *170*
Darwin's theory of, 153-155
DNA and, 170
of elephants, *161*
embryology and, 169, *169*
evidence for, 162-170, *162, 163, 165, act.* 167, *168, 169, 170*
extinction and, 176-177, *177*
of giraffes, *157*
gradualism, 160, *160*
homologous structures and, 168, *168*
of horses, *152-153,* 160
of humans, 170-175, *170, 171, 172, 173, 174, 175*
model of, *act.* 151
of mollusks, 381, *381*
by natural selection, 153-155
of plants, 261, *261,* 289
of protists, 230

punctuated equilibrium, 160-161, *160, 161*
  of segmented worms, 381, *381*
  speed of, 157-161, *160, 161*
  theory of acquired characteristics, 153, *153*
  vestigial structures and, 168, *169*
Excretory system, 666. *See also* Urinary system
Exhalation, 657, *657*
Exocytosis, 75-**76,** *76*
Exoskeleton, **382**
Experimentation, 15-16, *15*
  on animals, 348-349
  biogenesis, 11, *11,* act. 11
  controlled, 10, *10*
  origin of life, 12, *12*
  spontaneous generation, 10, *10*
Extinction, **176-**177, *177,* **554**
  diminishing diversity and, 198-199, *198, 199*
  wildlife conservation and, 554-555, *554*
  See also Endangered species
Eye, 136, *136,* 688-689, *688, 689*

Fahrenheit scale, 23
Family, **196**
Fan worms, *380,* 381
Farming, soil management in, 551, *551*
Farsightedness, 689
Fats, **603,** *603,* 615, act. 615, 632
Fat-soluble vitamins, 604
Fatty acids, 603
Feathers, 426, 427-428, *427, 428,* act. 428, act. 429
Feedback systems, negative, 696-697, *697*
Female reproductive system, 706-708, *706, 707, 708*
Fermentation, 80-**81,** *81,* 243, act. 321
Ferns, 271, *271,* 275-277, *275, 277,* act. 280, 510
Fertilization, 100-**101,** *100,* 710-711, *710, 711*
Fertilizers, 233, 552, 747, *747*
Fetus, 712, *712,* **714,** *714*

Fibroblasts, 54
Filter feeders, **351,** *351*
Fin, **401,** *401*
Fingerprints, *133,* act. 133
Fireflies, 464
Fireworms, *375*
Fish, 400-408, *400,* **401**
  adaptations of, 401, *401*
  behavior of, 456
  bony, 403-405, *403, 405*
  buoyancy of, 404, act. 404
  cartilaginous, 403
  jawless, 402-403, *402*
  origins of, 402
  respiration of, 400, 401, *401,* 403, act. 406-407
  social behavior of, *464*
  vertebrate characteristics of, 400-401, *400*
Fission, reproduction by, 97, **212,** *212*
Fission track dating, 12, *12*
Five-kingdom system of classification, 193-197, *193, 194, 195, 197*
Flagellates, 235, *235*
Flagellum, **211,** *211,* 231, 232, *232, 351*
Flatworms, *347,* 357-363, *357, 358,* act. 358, *359, 360,* act. 360-361, *361*
Fleischman, Paul, 465
Flies, 384, *384,* 385
Flight, 430-431, *430*
Flightless birds, 430, *432*
Flower(s), *288, 289,* 297-298, *297, 298,* act. 300, 312
Flowering plants. *See* Angiosperms
Flu (influenza), 36
Fluid levels, in body, 668, *668, 669,* 670
Food
  absorption of, 616
  bacteria in, act. 733
  bioengineered, 606, *606*
  choking on, 658, *658*
  digestion of, 612-619
  from photosynthesis, 318-319, *318,* 323, *323*
  from seed plants, 286, *286*
  taste and smell of, 691, *691*
Food chain, 496-497, **497**

Food groups, **608,** *608*
Food poisoning, 218, *219*
Food web, **497,** *498, 499, 499*
Forests, 198, *198, 199,* 261, 275, 307, *342*
  coniferous, 521, *521*
  conservation of, 555-556
  temperate deciduous, *520,* **522-**523, *522, 523*
  tropical rain, 198, *199,* 261, 308, *342, 520,* **524-**525, *524*
Fossil(s), **163-**170
  dating, 165, *165,* act. 167
  digging for, 162, *162*
  kinds of, 163, *163*
  making, act. 163
  plants, 261, *261*
  reconstructing organisms from, 173, *173*
Fossil fuels, **542,** *542,* 550, 558, 560
Fossil record, 163-166, *165, 166,* 261
Fox, *443*
Foxglove, 306
Fractures, bone, **582-**583, *582, 583*
Franklin, Rosalind, 104
Fraternal twins, 715, *715*
Free-living organisms, **357,** *357, 359*
Freshwater biomes, 530-531, *530, 531,* act. 532
Frigate bird, 463, *463*
Frogs, 409, *409,* 410-411, *410,* act. 410
  classification of, *188,* 189
  metamorphosis of, 412, *412,* act. 413
  ultraviolet light and, *414,* 415, *415*
Frond, **276**
Fruit, 288, *289*
  plant hormones and, 326, *326*
  seed dispersal and, 301
  skinning, 293, *293*
Fruit flies, act. 385
Fuel
  alcohol as, 84
  fossil, **542,** *542,* 550, 558, 560
  from plants, 84-85, *84, 85,* 275, *275*
Fungi, 8, *193,* 195, 228, act. 229, 241-247

characteristics of, 241-242, *241*
club, 244, *244*
diseases caused by, 242, 250, 733
imperfect, 244
lichens, 244-245, *245,*
  act. 246-247, *510*
sac, 242-243, *243*
spores, 242, *242, act.* 244
zygote, 242, *242*
Funguslike protists, 237-239, *237,*
  act. 237, *238, 239*

Galápagos Islands, 154, *154,* 377
Gallbladder, 615, *615*
Gametophyte, **268,** *268,* 269, *269,*
  277, *277*
Garbage. *See* Waste
Gas, natural, 542
Gas exchange, in plants, 314-316,
  314, 315, act. 316
Gasohol, **84,** *84*
Gastropods, 371, *371*
Gavial, *418*
Gene(s), **107,** 110, 111
  health and, 137-138, *137*
  multiple, 133-136, *act.* 134-135
  sex-linked, 139-**140,** *140*
Gene therapy, 37
Generations, alternation of, **268,**
  269
Genetic disorders, 137-138, *137,*
  139-140, *140*
Genetic engineering, **142-**143, *142,*
  143
Geneticist, 14, 146, *146*
Genetics, 124, **125-**147
  dominant traits in, 126, 127
  human. *See* Human genetics
  importance of, 142
  incomplete dominance in,
    132, *132*
  Mendel's studies of, 125-127,
    125, 126-127, 130
  multiple alleles in, 132-133
  plant, 125-131, *125, 126-127, 129,*
    act. 131
  polygenic inheritance and,
    133-136, *act.* 134-135
  probability in, 127

Punnett square in, 128-129, *128,*
  129
recessive traits in, 126, 127, *act.*
  131, 137-138, *137*
theory of acquired characteris-
  tics in, 153, *153*
Genome, **144-**145, *145*
Genotype, **128,** 129, *129*
Genus, **189,** 197
Geologic time scale, *164*
Geothermal energy, **544-**545, *545*
Germination, 302-303, *303, act.*
  304-305
Gestation period, **441**
Geysers, 544, 545, *545*
Gill(s), **370,** 400, 401, *401,* 403
Gill slits, **400**
Ginkgoes, 287, *287*
Giraffe, evolution of, *157*
Gizzard, **376,** *act.* 425, 431
Glands
  adrenal, 695
  endocrine, 695
  mammary, **435**
  parathyroid, 695
  pituitary, 695, 697
  sweat, 669, *act.* 672
  thyroid, 695
Gliding joint, 578-579, *578*
Global warming, **560,** *560,* 562
Glucose, 66, 80, *81,* 112-113, *322,*
  601, *601,* 696. *See also*
  Carbohydrates
Glycerol, 603
Gnetophytes, 287, *287*
Goldson, Alfred L., 114, *114*
Golgi bodies, **49,** *49,* 75
Gonorrhea, 734
Gonzalez, Alfred, 534, *534*
Goose, 433, *433,* 456, *457*
Gradualism, **160,** *160*
Grafting, of skin, 54, *54*
Grasslands, *520,* **526,** *526*
Gravitropism, 325
Gray wolf, 502-503, *502, 503*
Great Barrier Reef (Australia),
  356, *356*
Green algae, 232, *232,* 234, *234,*
  262, *262*
Greenhouse effect, **560-**562, *560,*
  act. 561

Groundwater, **563,** *563*
Growth, 8, *8*
  of bacteria, 213, *213, act.* 742-743
  cellular, *43, act.* 91, 92-98, *92, 93,*
    94-95, act. 96, *97*
  rate of, in humans, *act.* 721
  stopping, *act.* 742-743.
  See also Development
Growth spurt, 719, *719*
Guanine, 105, 106
Guard cells, **294**
Gymnosperms, **287,** *287*
  origin and evolution of, 289
  parts of, 292-295, *292, 294, 295*
  reproduction of, 296-297, *296,*
    act. 297
  seeds of, 287
  uses of, 290, *290*

Habitat, **495,** 534, *554,* 555, 556
Hagfish, 403
Hair, 136, *136,* 436, *act.* 436, 593
Hairlines, *act.* 123, *124-125*
Hammer, 690, *691*
Haploid cell, 100
Haversian systems, *575,* 577
Hazardous wastes, **562,** *562*
Health, and genes, 137-138, *137.*
  See also Diseases; Nutrients
Health insurance, 687
Hearing, *act.* 399, 690, *690*
Heart, *act.* 625, 627-628, *627, 628,*
  act. 628, 629, *629*
Heart disease, 632. *See also*
  Cardiovascular disease
Heart murmur, **634-**635
Heart pump, 629, *629*
Heart transplant, *635*
Heart valve disease, 634-635, *634,*
  635
Heartworms, 359, 363
Hedgehog, 436
Heimlich maneuver, 658, *658*
Helmont, Jan Baptist van, 10
Hemoglobin, **637**
Hemophilia, *140,* 141, 638
Herbaceous stems, 294
Herbivore, **436,** *496*
Heredity, *act.* 123, **124,** *124-125.*

See also Genetics

Hermaphrodite, **352,** 358

Heron, *432*

Herpes, 734

Heterozygous organism, **128,** *act.* 131

Hibernation, **409,** 468, 522

Hinge joint, 578, *578*

Hippocratic Oath, 741

Histamines, 750

History
of classification systems, 188-189, *188,* 196-197
ethics and Hippocratic Oath in, 741
viruses through, 58, *58*

HIV (AIDS virus), *32, 34, 36, 38, act.* 38-39, *58,* 222, 645, 735, 740, *740*

Hives, 750, *750*

Homeostasis, **7,** 669, 671, 678, 684, 694

Hominids, **172,** *173*

Homo habilis, 173

Homo sapiens, **173**-175, *174, 175. See also* Human development

Homologous structures, **168,** *168*

Homozygous organism, **128**

Honeybees, 465, *465*

Hooke, Robert, 43

Hookworms, 362, *362*

Hormones, 24-25, **695,** 696-697, 698, 718
plant, 325-326, *326*

Horses
evolution of, *152-153,* 160
teeth of, *436*

Horsetails, 271, *271,* 274, *274*

Host cell, **34**

Human development, 716-723, *719*
adolescence, 718-719, *718*
adulthood, 719-720, *720,* 722-723, *722, 723*
childbirth, 714, 716-717, *716*
childhood, 718, *act.* 718
embryonic, 712-713, *713, act.* 713
growth rate of, *act.* 721
growth spurt in, 719, *719*
infancy, 717, *717*
senior citizens, 720, *720,* 722-723, *722, 723*

Human evolution, 170-175, *170,*

171, 172, 173, 174, 175

Human genetics, 133, *133, act.* 133, 137-141
eye color and, 136, *136*
hair color and, 136, *136*
hairlines and, *act.* 123, *124-125*
health and, 137-138, *137*
Human Genome Project and, 144-145, *145*
sex determination and, 138, *138,* 139, *702, act.* 703
sex-linked disorders and, 139-140, *140*
skin color and, 136, *136*

Human Genome Project, 144-145, *145*

Human reproduction, 704-715, *704*
birth in, 714, 716-717, *716*
embryonic development in, 712-713, *713, act.* 713
fertilization in, 710-711, *710, 711*
sex determination in, 138, *138,* 139, *702, act.* 703

Human reproductive system, 704-708
female, 706-708, *706, 707, 708*
male, 705, *705*

Human respiration, *act.* 651, 652-665, *652, act.* 662-663

Humerus, 575, *575,* 576, 579

Humus, *510*

Hydra, *97, 353, act.* 354, 355, *355*

Hydroelectric power, **544,** *544*

Hydroxyapatite, 582

Hypertension, **632**

Hyphae, **241,** *241*

Hypothalamus, *668*

Hypothesis, **15**

Hyrax, *444*

Ibuprofen, 75, *75*

Identical twins, 715, *715*

Iguana, *527*

Immovable joint, **578,** *578*

Immune system, **737**-740, *737, 738*

Immunity, 729, 737-751
active, **738**-739
passive, **738,** 739. *See also* Vaccine

Immunization, *act.* 718

Imperfect fungi, 244

Imprinting, **456,** *457*

Incisor, 436

Incomplete dominance, **132,** *132*

Incubation, **426**

Infancy, **717,** *717*

Infection, 594, *594*
causes of, *act.* 729, 730-731, *730, 731, 732, 732, act.* 736, *act.* 742-743, 744-745, *744*
prevention of, 732, *732, act.* 742-743

Influenza, 36

Information gathering, 15

Ingestion, 614

Inhalation, 657, *657*

Inheritance, polygenic, **133**-136, *act.* 134-135

Inherited traits. *See* Genetics

Innate behavior, **454**-455, *454, 455*

Inorganic compounds, **70,** *70*

Insect(s), 383-385, *383, 384, 385,* 390-391, *390*
adaptations of, 385
body plan of, 383, *383*
communication among, 465, *465,* 466, *466*
as disease carriers, 733, *733*
metamorphosis of, 384-385, *385, act.* 385
pesticides and, 390-391, *391,* 467, *467,* 747, *747*
social behavior of, 464, *act.* 472. *See also individual insects*

Insecticides, 747, *747*
artificial, 467, *467*

Insight, *act.* 458, **459,** *459*

Instinct, **455,** *455,* 462

Insulation
in humans, 591, 593, *593*
in whales, *act.* 438-439

Insulin, **112,** 616, 696

Insurance, 687

Interactions, in communities, 492-495, *492, 494, 495, act.* 495

International System of Units (SI), 22-23, *act.* 22, 23

Interneurons, **679**

Interphase, 93, 94, *94*

Intertidal zone, **532**

Intestines

large, 617, *617*
small, 615-616, *615*
Invertebrates, **345**
Involuntary muscles, **584**
Iodine, 605
Irish potato famine, 238, *239*
Iron, 605
Islets of Langerhans, **112**-113

Janssen, Zacharias, 41
Jawless fish, 402-403, *402*
Jellyfish, *346*, 353, *353*, 355
Jenner, Edward, 36
Jet propulsion, *371*, 372-373, *372*
Joey, 440, *440*
Johanson, Donald, 172
Johnson, Rebecca L., 364, *364*
Joint, **577**, 578-579, *578*
artificial, *579*

Kangaroos, 440, *440*
Karyotyping, 141, *141*
Kelp, *233*, 447, *447*
Kidney, **667**-668, *667*, *act.* 667, 670-671, *671*
Kidney stones, 670, *670*
Kingdom, **188**, 193-197, *193*, *194*, *195*, *197*
Kiwi, *432*
Koch, Robert, 731
Koch's rules for identifying diseases, 731, *731*
Kohler, Wolfgang, 459

Labor, 716, *716*
Lactic acid, 81
Ladybird beetle, *390*
Lakes, 531, *531*
Lamarck, Jean Baptiste de, 153
Lamprey, 402, *402*, 403
Lancelets, 400, *400*
Land environments, 516-527
Antarctica, 364, 499, *499*, 528-529, *528*

biomes, 520-527, *520*, *act.* 520, *521*, *523*, *524*, *525*, *526*, *527*
climate and, 516-518, *516*, *517*, *518*
conservation of, 555-556, *555*
studying, *act.* 519
Land pollution, 319, *319*, 562, *562*
Large intestine, 617, *617*
Larson, Georgia, 620, *620*
Larva, **352**, *381*, 384, *384*, 385, *385*
Larynx, **654**
Latent viruses, 35, *35*
Latitude, *516*, 517
Law, **17**
Leaf insect, *383*
Leakey, Louis, 172-173
Leakey, Mary, 172-173
Leakey, Richard, 172-173
Learning, **456**-461, *456*, *457*, *458*, *act.* 458, *act.* 460-461
Leaves, 294-295, *295*, 314-315, *314*, *315*, *act.* 316
Leeches, *375*, 377, 380, *380*
Leeuwenhoek, Anton Van, 41
Lemurs, 172, *172*, 444
Lenses, 689, *689*
Lescaux cave paintings (France), 178
Leukemia, 641
Levers, 585, *585*
Lichens, **244**-245, *245*, *act.* 246-247, *510*
Life
biogenesis and, 11, *11*, *act.* 11
features of, 6-8
organization of, 6
origins of, 11-13, *13*
spontaneous generation of, 10, *10*
Life functions, 6, *6*
Life science, 14, *14*
Life span, **8**
Ligament, **577**
Light
as environmental factor, *483*, 484, *484*
lenses and, 689, *689*
photons of, 320
plant response to, 324-325, *325*
sunlight, 8, *8*, *483*, 484, *484*, *516*, 517

ultraviolet, 414-415, 559
vision and, 688, *688*
white, *317*, 318
Lightning, 501
Limiting factor, **489**, *act.* 490-491
Linnaeus, Carolus, 189
Linnaeus's system of nomenclature, **189**, 192, *192*, 200-201
Lipids, 68
Lister, Joseph, 732
Literature
Fantastic Voyage, 646
Frankenstein, 590
"Honeybees," 465
The Lives of a Cell, 86
Never Cry Wolf, 504
Spoon River Anthology, 224
Lithotripter, 670, *670*
Liver, 615, *615*
Liverworts, *261*, 267, *267*, 269, 270, *270*, *act.* 280
Living space, *act.* 481
Living things, basic needs of, 8-9, *8*, *9*
Lizards, 416, *416*, *418*, 419
Lobe-finned fish, 404, *405*, 410
Lobster, 388
Long-day plants, *326*, **327**
Lorenz, Konrad, 456
"Lucy," 172, *172*
Lung(s), 656, *656*, 657, *657*
Lung cancer, 660-661, *661*, *748*
Lungfish, 404, *405*
Lycopodium, 274
Lyme disease, 218, *218*
Lymph, **644**, *644*
Lymph nodes, **644**-645
Lymphatic organs, 644, *644*
Lymphatic system, **644**-645, *644*, *645*
Lymphocytes, **644**, **740**
Lysosomes, **50**

Maggots, 384, *384*, 385
Malaria, *236*
Male reproductive system, 705, *705*
Mammals, **435**-447
body systems of, 437

characteristics of, 435-437, *435, 436, 437*

classification of, 440-441, *440, 441*

diversity of, 442, *442-444*

hair of, *436, act. 436*

marine, 437, *438, act. 438-439, 448*

marsupials, 440, *440*

maternal care in, 435, *435, 437, 437, 440, 440*

monotremes, 440, *440*

origin of, 445, *445*

placental, **441,** *441, 442-444*

reproduction of, 437, 441, *441*

respiration of, 437

teeth of, 436, *436, act.* 436.
See also individual mammals

Mammary glands, **435**

Manatee, *444*

Mantle, **370,** *370*

Marine mammals, 437, *438, act. 438-439,* 448

Marine worms, *375, 377, 380-381, 380*

Marrow, 574, **576,** 637

Marsupials, **440,** *440*

Masters, Edgar Lee, 224

Maternal care
in mammals, 435, *435, 437, 437, 440, 440*
in reptiles, 420, *420*

Matter
cycles of, 499-501, *500, 501*
nature of, 64-70

Mayflies, 493, *493*

Measles, 36

Measurement, 22-23, *act.* 22, *23*

Mechanical digestion, **614**

Medical insurance, 687

Medicine
ethics in, 741
painless administration of shots in, 749, *749.*
See also Drugs

Medusa, **355,** *355*

Meiosis, **99,** 101-103, *102, 103*

Melanin, **591**

Mendel, Gregor, 125-127, *125, 126-127,* 130

Menopause, **708**

Menstrual cycle, **707,** *707, 708*

Menstruation, **707**

Mesozoic era, 417

Messenger RNA (mRNA), 110

Metabolism, **78**

Metamorphosis, **384**
of amphibians, 412, *412, act.* 413
of insects, 384-385, *385, act.* 385

Metaphase, *94,* 102, *102, 103*

Methanol, 84

Microorganisms
destruction of, 744-745, *744*
disease and, *act.* 729, 730-731, *730, 731, 732, 732, act.* 736, *act.* 742-743, 744-745, *744*
stopping growth of, *act.* 742-743.
See also individual organisms

Microscopes, 40-43, *40, act.* 42
compound light, **41**-42, *41*
confocal, 243, *243*
early, 40-41
electron, **42**-43, *42*
stereoscopic light, 42

Migration, **468**-469, *469*

Mildew, downy, 238, 239

Miller, Stanley L., 12

Millipedes, 387, *387*

Mimosa pudica plant, 324, *324*

Minerals, **605**
in Antarctica, 528-529
conservation of, 556, *556*
as nonrenewable resources, 541, *541*
as nutrients, 9, 574, 576, 577, 605, *605*

Mining
diseases caused by, 747
environment and, *act.* 541
mineral conservation and, 556, *556*

Mistletoe, 495

Mites, 202, *202, 203,* 386

Mitochondria, **49**-50, *49, 322*

Mitosis, **93**-97, *94-95, act.* 96, 711

Mixture, **67,** *67*

Models

of digestive tract, *act.* 599

of DNA, *104,* 105, *105, act.* 108-109

of evolution, *act.* 151

of greenhouse effect, *act.* 561

Molar, 436

Mole, *442*

Mollusks, *368, act.* 369, **370**-374, *381*
bivalves, *act.* 369, 371, *371, 373*
body plan of, 370, *370*
cephalopods, *371,* 372-373, *372*
evolution of, 381, *381*
gastropods, 371, *371*
importance of, 373

Molting, **382**

Monocots, **288,** *288*

Monotremes, **440,** *440*

Mosquito, 733, *733*

Mosses, *261,* 267-270, *267, 268, act.* 268, *269, 270, 271, 271, 274, 274, act.* 280, *510*

Moths, *385,* 467

Motivation, **457**

Motor neuron, **679,** *679,* 683

Mouth
breathing through, *653*
digestion in, 614, *614*

Movable joint, **578,** *578*

Mowat, Farley, 504

Mucus, 75, 654

Mudskippers, 408

Multiple alleles, **132**-133

Multiple genes, 133-136, *act.* 134-135

Mumps, 36

Muscles, *act.* 573, **584**-589, *584*
bones and, 574
cardiac, 586, **587,** *act.* 589
involuntary, **584**
skeletal, *act.* 573, 574, **586,** *586, act.* 589
smooth, **586,** *586, act.* 589
voluntary, **584**
at work, 587-588, *587, act.* 587, *588*

Mushrooms, *193,* 195, *228, act.* 229, 244, *244*

Mutation, 38, **111,** *111,* 160-161, 698
Mutualism, **494,** *494*
Myoglobin, 588

Names, of organisms. *See* Nomen-clature
Nasal cavity, 653-654
Natural gas, 542
Natural resources, 538-556, **540**
    nonrenewable, **541,** *541, act.* 541
    protection of, *538, act.* 539
    renewable, **540,** *540*
Natural selection, 153-**155**
Neanderthal humans, 173, 174, *174*
Nearsightedness, 689
Needs, basic, 8-9, *8, 9*
Negative feedback systems, 696-697, *697*
Nematodes, 359, *359,* 362-363, *362, 363*
Nephron, **667,** *667*
Nerve net, 355
Nervous system, 678-687
    central, 680-681, *680, 681*
    diseases of, 683, *683,* 686-687, *686, 687*
    drugs and, 684, *684*
    neurons in, *678, 679, 679,* 683
    peripheral, 680, 682-683, *682*
    reaction time of, *act.* 685
    synapses in, 680, *680*
Neurons, *678,* **679,** *679,* 683
Neutrons, 64, *64*
Newts, 411, *411*
Niche, **495,** *495*
Ninomiya, Flora, 334, *334*
Nitrogen bases, 105, 106
Nitrogen cycle, **501,** *501*
Nitrogen-fixing bacteria, **216,** *act.* 495
No-till farming, 551, *551*
Nomenclature
    binomial, **189,** 192, *192,* 200-201
    common names, 200-201, *200, 201*
Noncommunicable diseases, **746-**750, *746, 747, 748, 749, 750*
Nonrenewable resources, **541,** *541,*

act. 541
Nonvascular plants, **264,** 267-270, *267, 268, act.* 268, *269, 270*
Nostrils, 653, *653*
Notochord, **400**
Nuclear energy, **546,** *546*
Nucleic acids, 69
Nucleolus, 47, *47*
Nucleus, **47**
    of atom, 64, *64*
    of cell, 45, 47, *47*
Nutrients, **600-**611, *600*
    in bioengineered foods, 606
    cancer and, 610-611, *610*
    carbohydrates, 69, 601, *601*
    classes of, 600
    fats, 603, *603,* 615, *act.* 615, 632
    in food groups, 608, *608*
    minerals, 9, 574, 576, 577, 605, *605*
    proteins, 602-603, *602, act.* 618-619
    vitamins, 593, 604, *604, act.* 609
    water, 9, *9,* 606-607, *607, act.* 607

Observation, 26, *26*
Ocean
    bacteria in, *4,* 214, *214*
    coral reefs in, 353, 356, *356,* 487, *487*
    life in, 533
    pollution of, 564, *564*
    red tide in, 248-249, *248, 249*
    shorelines of, 532, *533*
Octopus, *371,* 372, *372*
Oil, 542, *542*
Oil spills, 564, *564*
Olfactory cells, **691**
Omnivore, **436**
On the Origin of Species by Means of Natural Selection (Darwin), 155
Oparin, Alexander I., 11
Open circulatory system, **370**
Opossum, *440*
Opposable thumb, 171, *171*
Optical illusion, *act.* 677
Order, **196**
Organ(s), **54-**55, *55*
Organ system, 55

Organelles, **48-**50, *48, 49*
Organic compounds, 68-69
Organisms, **6,** *486*
    community of, *486,* 487
    free-living, **357,** *357, 359*
    heterozygous, **128,** *act.* 131
    homozygous, **128**
    identifying, *14, act.* 187, 189, *189,* 200-204, *200, 201, 202, 203, act.* 204, 346
    naming, 189, 192, *192,* 200-204, *200, 201, act.* 201, *202, 203, act.* 204
    population of, *486,* 487
    sensitive to pollution, 493, *493*
    sessile, **350**
    unknown, 189, *189*
Osmosis, **73,** *73, act.* 77, 262
Osprey, *432*
Osteichthyes (bony fish), 403-405, *403, 405*
Osteoblasts, 576, *576*
Osteoclasts, *576,* 577, 582
Osteoporosis, **583**
Ovaries, **298,** *298,* 695, **706,** *706*
Oviduct, 706, 710, *710*
Ovulation, **706,** 707, *707, 708*
Ovule, **296,** 299
Owl, *432*
Oxygen, 9, 656
    in photosynthesis, 318-319, *318,* 320, *320*
Oyster, 371
Ozone, 276
Ozone depletion, 276, *276,* 414-415, 559-**560**
Ozone layer, 276, *276,* 414-415, 559-560

Pain, 694, *694*
Painless shots, 749, *749*
Paleozoic era, 271, 275
Palisade layer, 295, *295*
Pancreas, *615,* 616, 695, 696
Pandas, classification of, 194, *194*
Pangolin, *444*
Paramecium, *43,* 236, *236*
Parapodia, 377, 381
Parasites, 411, *494,* 495

Parasitism, *494,* **495**
Parathyroid glands, 695
Parrotfish, *405*
Passive immunity, **738,** 739
Passive smoking, **664,** *664*
Passive solar heating, *542*
Passive transport, **74**
Pasteur, Louis, 11, 219, 730
Pasteurization, 219, **730**
Pathogens, **217**-218. *See also individual pathogens*
Pavlov, Ivan P., 458, *458*
Peacocks, 463, *463*
Pearl, *373*
Peat bog, 275
Pedigree, **140,** *140*
Penguins, 430
Penicillin, 160-161, 213, 217, 244, 734, 735, *735*
Penis, 705
Periosteum, **575**-576, 582
Peripheral nervous system (PNS), **680,** 682-683, *682*
Peristalsis, **614,** *614,* 615, 616, 617
Permafrost, **520,** *521*
Permeability, 80
    selective, 71, *71*
Perspiration, 593, *act.* 593, 606-607, 669, *act.* 672
Pesticides, **390**-391, *391,* 467, *467,* 555, 747, *747*
PET (positron emission tomography), 683, *683*
Petals, 297-298, *298*
Pharynx, **654**
Phenotype, **128,** 129, *129, act.* 134-135
Pheromone, **466,** 467
Phloem, **294,** *294,* 295
Phosphorus, 574, 576, 577, 605, *605*
Photon, 320
Photoperiodism, *326,* **327,** 330, *331*
Photosynthesis, 78, *78, 79, act.* 82-83, 291, 317-320, *317,* **318,** *318, act.* 318, 320, 322, 323, *323,* 496
Phototropism, 324-325, *325*
Photovoltaic (PV) cells, **543**
Phylogeny, **193**
Phylum, **196**
Phytoremediation, 319, *319*
Pigeons, 468, *468*

Pika, *442*
Pinecones, 296-297, *296, act.* 297. *See also* Gymnosperms
Pioneer community, *510,* **511,** 513
Pioneer species, **270,** *270*
Pistil, **298,** *298*
Pituitary gland, 695, 697
Pivot joint, 578, *578*
Placenta, **441,** 717
Placental mammals, **441,** *441, 442*-444
Placoderms, 402
Plague, 730, *730*
Planarians, *98,* 357, *357,* 358, *358, act.* 358, 360, *act.* 360-361
Plankton, **531**
Plant(s), *258,* 260-335
    active transport in, 74, *74*
    adaptations of, 262-263, *262, 263*
    cell mitosis in, 95, *95, act.* 96
    characteristics of, 260-261, *260, 261*
    classification of, *act.* 190-191, 195, 264, *265*
    conservation of, 554-555, *554*
    day-neutral, **330,** *330*
    in desert, 526-527, *527*
    earthworms and, *act.* 378-379, *511*
    edible parts of, *act.* 259
    flowers of, *288, 289,* 297-298, *297, 298, act.* 300, *312*
    fossils of, 261, *261*
    fuel from, 84-85, *84, 85,* 275, *275*
    gas exchange in, 314-316, *314, 315, act.* 316
    genetically engineered, 143, *143*
    genetics of, 125-131, *125, 126-127, 129, act.* 131
    hormones in, 325-326, *326*
    illustrating, 266
    living in water, 261
    long-day, *326,* **327**
    medicines from, 264, 306-307, *306, 307,* 308
    movement of water in, *act.* 271
    nitrogen cycle and, 501, *501*
    nitrogen-fixing bacteria and, 216, *act.* 495
    nonvascular, **264,** 267-270, *267, 268, act.* 268, 269, 270
    origin and evolution of, 261,

    261, 289
    photoperiodism in, *326, 327, 330, 331*
    photosynthesis in, 78, *78, 79, act.* 82-83, 291, 317-320, *317, 318, act.* 318, 320, 322, 323, *323,* 496
    pioneer, **270,** *270*
    pollution and, 319, *319,* 320, *320*
    reproduction of, *act.* 91, 263, 268, 269, 296-305, *296, 297, act.* 297, *298, 299, 303, act.* 304-305
    responses of, 324-330, *324, 325, act.* 328-329
    roots of, 292-293, *292,* 526
    seed. *See* Seed plants
    seedless, 267-277, *267, 268, act.* 268, 269, 270, 271, 274, 275, 277
    short-day, *326,* **327**
    stems of, 293-294, *294*
    in temperate deciduous forests, 522-523, *523*
    transgenic crops, 331-333, *331, 332*
    in tropical rain forests, 524-525, *524, 525*
    vascular, **264**
    water and, *act.* 271, *act.* 313, 314-315, *314, 315*
Plant cells
    animal cells vs., *act.* 53
    parts of, 51, *51*
Plantlike protists, 230, 231-234, *231, 232, 233, 234*
Plasma, **636,** *636,* 639
Platelets, **637,** *637,* 638, *638*
Platypus, 440, *440*
Plotkin, Mark, 308, *308*
Plowing, 551, *551*
Polio, 36
Pollen grains, **296,** *296,* 299
Pollination, 297-**299,** *297, 299*
Pollutant, **558**
Pollution
    acid rain, *act.* 272-273
    of air, 557-562, *557, 559, act.* 559, *560, act.* 561
    from burning coal, 278-279, *279*
    damage to ozone layer and, 276, *276,* 414-415, 559-560

as danger to wildlife, 554-555, *554*

of land, 319, *319*, 562, *562*

monitoring organisms sensitive to, 493, *493*

of ocean, 564, *564*

plants and, 319, *319*, 320, *320*

of water, 563-564, *563*, *564*

Polychaetes, *375*, 377, 380-381, *380*

Polygenic inheritance, **133**-136, *act.* 134-135

Polyps, 353, **355,** *355*

Ponds, 531, *531*

Population, *486*, **487**

carrying capacity and, 489, 492, *492*

characteristics of, 488-492, *488*

limiting factors of, 489, *act.* 490-491

Population density, **489,** *489*

Porcupine, 436

Pores, 351, *351*

Porifera, 351

Positron emission tomography (PET), 683, *683*

Potassium, 605

Potato famine, 238, *239*

Potter, Beatrix, 266

Precipitation, **518,** *518*

Precycling, 548

Predation, *492*, **493**

Preening, **428,** *428*, *act.* 428

Pregnancy, **712**-714, *713*

Premolar, 436

Prenatal surgery, 714, *714*

Pressure, 631, *631*, 694, *694*

Primary succession, *510-511*, **511**

Primates, **171**-175, *171*, *172*, *173*, *174*, *175*, *443*

Probability, in genetics, 127

Problem solving, **14**-16, *15*

Producers, **78,** 320, 496, *496*

Prokaryotic cells, 45, *45*, 52, *52*, 194, 195, *195*

Prophase, *94*, 102, *102*, *103*

Proteins, 69, **602**

amino acids, 107, 110, 602-603

digestion of, *act.* 618-619

as nutrients, 602-603, *602*, *act.* 618-619

Prothallus, **276**-277

Protists, *193*, 195, 196, **230**-240, *230*

as agents of disease, 733

animal-like, 230, 234-236, *234*, *235*, *236*

evolution of, 230

funguslike, 237-239, *237*, *act.* 237, *238*, *239*

plantlike, 230, 231-234, *231*, *232*, *233*, *234*

Protoavis, 434, *434*

Protons, 64, *64*

Protozoa, **234**-236, *234*, *235*, *236*, *act.* 240

Pseudopod, **235**

Puberty, 718, *718*

Pulmonary circulation, **627,** *628*

Punctuated equilibrium, **160**-161, *160*, *161*

Punnett square, **128**-129, *128*, *129*

Pupa, 385, *385*

Quinine, 264

Rabbits, 436, *442*

Rabies virus, *33*, 36

Radial symmetry, **346,** *346*

Radioactive dating, 12, *12*, 165, *act.* 167

Radioactive elements, **165**

Radioactive waste, 546

Radula, **371**

Rain, 518

acid, *act.* 272-273, 558-**559,** *559*, *act.* 559

Rain forests, 198, *199*, 261, 308, *342*, *520*, **524**-525, *524*

Rain shadow effect, 518, *518*

Rainbow trout, 402

Ray-finned fish, 404, *405*

Reaction time, *act.* 685

Recessive traits, **126,** 127, *act.* 131, 137-138, *137*

Recycling, 547, *547*, 548-**549,** *548*, *549*, 550, *act.* 552-553, 556

Red algae, 233, *233*, 234, *234*, 248

Red blood cells, 637, *637*, 641, *641*, 645

Red tide, **248**-249, *248*, *249*

Redi, Francesco, 10

Redwood trees, *513*

Reflex, **455, 682**-683, *682*, *act.* 685

Reflex arc, 682

Reforestation, 556

Regeneration, 98, *98*, **352**

Relative dating, **165,** *165*

Renewable resources, **540,** *540*

Reporting, 16

Reproduction, 7, *7*

of amphibians, 409

asexual, **97**-98, *97*, *98*, 269, 355, *355*, 358

of bacteria, 97, 212-213, *212*, *act.* 212, *act.* 734

of birds, 426, *426*, 433, *433*

captive breeding, 433, *433*, 470-471, *470*, *471*

cellular, *43*, *act.* 91, 92-98, *92*, *93*, *94-95*, *act.* 96, *97*

of cnidarians, 355, *355*

of fish, 403

by fission, 97, 212, *212*

of flatworms, 358, 359

human. *See* Human reproduction

of mammals, 437, 441, *441*

meiosis, 99, 101-103, *102*, *103*

of plants, *act.* 91, 263, 268, 269, 296-305, *296*, *297*, *act.* 297, *298*, *299*, 303, *act.* 304-305

by regeneration, 98, *98*, 352

of reptiles, 417, *417*

sexual, *98*, **99**-103, *100*, *102*, *103*

of sponges, 352

of viruses, 34-35, *34*, *35*

Reptiles, *398*, *act.* 399, **416**-420

alligators, 418-419

crocodiles, 418-419, *418*, 420, *420*

hearing of, *act.* 399

lizards, 416, *416*, *418*, 419

maternal care in, 420, *420*

origins of, 417

reproduction of, 417, *417*

snakes, *398*, *act.* 399, 416, *416*, *417*, 419, *419*

turtles, 416, *416*, 417, 418, *418*, 420, 455, *455*

variety in, 416-417, *416*

Research, *act.* 5, *17*, 534, 698

Resources. *See* Natural resources
Respiration, **321**-323, *321*, *act.* 321
   aerobic, 321
   of amphibians, 409
   of birds, 431, *431*
   breathing vs., 652-653
   cellular, 79-80, *act.* 82-83, 321,
     322, *322*
   effects of, *act.* 662-663
   of fish, 400, 401, *401*, 403,
     act. 406-407
   of humans, *act.* 651, 652-665,
     652, *act.* 662-663
   importance of, 323, *323*
   of mammals, 437
   photosynthesis vs., 322
   physical activity and, *act.* 651
   of reptiles, 417
Respiratory system
   diseases and disorders of,
     659-661, *659*, *660*, *661*
   functions of, 652-653, *652*
   organs of, 653-656, *653*, *654*, *656*
Response, **7**
   of plants, 324-330, *324*, *325*,
     act. 328-329
   simple, *act.* 453
   stimulus and, 7, *7*, 324, 454, 678,
     679
Retina, **688,** *688*
Reusing, 548
Revolution, 327
Rh factor, 640
Rhea, *432*
Rhinoceros, 436, *443*
Rhizoid, **267**
Rhizome, **276**
Rhynia major, 261, *261*
Ribonucleic acid (RNA), 69, **110,**
   110, *act.* 110
Ribosomes, **48**-49, *48,* 110
Rivers, 530
RNA (ribonucleic acid), 69, **108,**
   108, *act.* 110
Robot skin, 592, *592*
Rocks
   dating, 12, *12*
   sedimentary, **163,** *165*
Rod(s), of eye, 688, *688*
Rodents, *442*
Roots, 292-293, *292*, 526
Roundworms, 359, *359*, 362-363,

362, 363
Rusts, 238, 244

S

Sac fungi, 242-243, *243*
Safety, 22-23
Salamanders, 409, *409*, 411, 412,
   414
Saliva, **614,** *614*
Salivary amylase, 613
Salmon, 456
Salt, *act.* 18-19, *66*, *67*, 670, 671
Saltwater, *act.* 63, 68, *68*
Saltwater biomes, 532-533, *532*, *533*
San Diego Wild Animal Park,
   470, *470*
Sand dollars, 393
Saprophyte, **216,** 241
Sarcodines, 235, *235*
Saturated fats, 603, 632
Scales, **401,** *401*
Scallops, *act.* 369, 371, *371*
Scanning electron microscope
   (SEM), *42*, 43
Schleiden, Matthias, 43
Schwann, Theodor, 43-44
Science, 14
Scientific methods, **14**-16, *15*,
   act. 18-19
Scientific units, 22-23, *act.* 22, *23*
Scorpion, 386, 387
Scrotum, 705
Scrubber, 279
Sea anemones, 353, *353*
Sea cucumbers, 394, *394*
Sea lettuce, *232*
Sea lions, *462*
Sea otters, **446**-447, *446*
Sea stars, *392*, 393, *393*
Sea urchins, 393, *393*
Seashells, *368*, *act.* 369, 370,
   370, 374
Seashores, 532, *533*
Seasons, *516*, 517
Secondary succession, **512,** *512*
Sedimentary rock, **163,** *165*
Seed(s), *284*, 286
   classifying, *act.* 190-191
   development of, 299-303, *303*
   dispersal of, 301, *301*, *act.* 301

   dissecting, *act.* 285
   flowers vs., *act.* 300
   germination of, 302-303, *303,*
     act. 304-305
   parts of, 299, *299*
   planting, 302
Seed plants, 284-305, *284*
   food from, 286, *286*
   importance of, 290-291, *290*, *291*
   origin and evolution of, 289
   parts of, 292-295, *292*, *294*, *295*
   reproduction of, *act.* 91, 296-305,
     296, *297*, *act.* 297, *298*, *299*,
     303, *act.* 304-305
   See also Angiosperms;
     Gymnosperms
Seedless plants
   nonvascular, 267-270, *267*, *268*,
     act. 268, *269*, *270*
   vascular, 271-277, *271*, *274*, *275*,
     277
Segmented worms, 375-381
   earthworms, 375, 376-377, *376*,
     act. 378-379, *act.* 466, 495,
     495, *511*
   evolution of, 381, *381*
   features of, 375, *375*
   leeches, 375, 377, 380, *380*
   marine worms, 375, 377, 380-381,
     380
Selective permeability, 71, *71*
Semen, **705**
Seminal vesicle, 705
Senior citizens, 720, *720*, 722-723,
   722, *723*
Senses, 688-694
   environment and, 688
   hearing, *act.* 399, 690, *690*
   smell, 691, *act.* 691
   taste, 691, *691*
   touch, *act.* 692-693, 694, *694*
   vision, 171, *act.* 677, 688-689,
     688, *689*
Sensory neuron, **679,** *679*
Sepals, 298, *298*
Sessile organisms, **350**
Setae, **375,** *375*, *376*
Sex cells, 99-101, *99*, *100*
Sex determination, 138, *138*, 139,
   702, *act.* 703
Sex-linked genes, 139-**140,** *140*
Sexual reproduction, *98*, **99**-103,

100, 102, 103

Sexually transmitted diseases (STDs), **734**-735, *734*

Sharks, 402, 403

Shelley, Mary, 590

Short-day plants, *326*, **327**

Shrimp, *act.* 18-19, 388

SI (International System of Units), 22-23, *act.* 22, *23*

Sickle-cell anemia, 37, 137, *137, 641*

Skates, *402, 403*

Skeletal muscle, *act.* 573, 574, **586,** *586, act.* 589

Skeletal system, **574**-583, *574, 580*
 bones in, 575-583, *575, 576, 578, act.* 580-581, *582, 583*
 exoskeleton, 382
 functions of, 574
 healing fractures in, 582-583, *582, 583*
 joints in, 577, 578-579, *578*

Skin, 591-594, *591*
 artificial, 592, *592*
 color of, 136, *136*
 grafting of, 54, *54*
 injury to, 594, *594*
 as insulation, 591, 593, *593*
 sensitivity of, *act.* 692-693, 694, *694*

Skin cancer, 591

Skin patch, 749, *749*

Skull, 681

Slime molds, 237, *237, act.* 237, 238, *238*

Sloth, *444*

Slug, 371

Small intestine, 615-616, *615*

Smallpox, 36, 58

Smell, 691, *act.* 691

Smog, 557-**558,** *557*

Smoking, 659, 660-661, *661,* 664-665, *664, 748*

Smooth muscle, **586,** *586, act.* 589

Snail, 371, *371, 372*

Snake(s), *398, act.* 399, 416, *416, 417, 419, 419*

Snakebites, 419

Snow, 518

Social behavior, **464,** *464, act.* 472

Society, **464**

Sod, 526

Sodium, 605

Sodium bicarbonate, 82

Sodium chloride, *66, 67, 67*

Soil
 conservation of, 551, *551*
 earthworms and, *act.* 378-379, *511*
 as environmental factor, *483, 484, act.* 485
 pollution in, 319, *319*
 protection of, *538,* 539, *act.* 539

Soil depletion, **551**

Soil management, **551,** *551*

Solar collector, *543*

Solar cooker, *543*

Solar energy, *542,* 543, *543*

Solid waste, 547, *547,* **548**-549, *548, 549*

Solutions, 68

Somatic system, 682, *682*

Sori, **276**

Sound communication, 466-467

Sound waves, 441, 690, *690*

South Pacific Islanders, 374

Space, ecosystems in, 523, *523*

Spallanzani, Lazzaro, 10

Spawning, 403

Species, **152, 189**
 endangered. *See* Endangered species
 pioneer, **270,** *270*

Species diversity, **198**-199, *198, 199*

Sperm, **99,** *99,* 102, **705,** *705,* 710-711, *710,* 711, *711*

Spicules, *351,* 352

Spiders, 386-387, *386, 387, 452*

Spike moss, 271, *271, 274, 274*

Spinal cord, 681, *681, 682*

Spindle fibers, 95, *95,* 102

Spiracles, **383**

Spirilla, 211, *211*

Spirogyra, 232

Spleen, 645

Sponges, 350-352, *350, 351, 352, act.* 352

Spongy bone, 576, 582

Spongy layer, 295, *295*

Spontaneous generation, **10,** *10*

Sporangia, **242,** *242*

Spore, **242,** *242, act.* 244

Sporophyte, **268,** 269, *269,* 276,

277, 277

Sporozoans, 236, *236*

Squid, 372, 373

Squirrels, *442*

Stamen, **298**

Starch, 601, *601*

STDs (sexually transmitted diseases), **734**-735, *734*

Stereoscopic light microscope, 42

Stimulant, 684

Stimulus, **7,** *7,* 324, 454, 678, *679*

Stirrup bone, 690, *691*

Stomach, 615

Stomata, **294,** 314-315, *314, 315, act.* 316

Streams, 530, *530*

Streptococcus, 193, *216*

Streptomycin, 213

Strip-cropping, 551, *551*

Succession, **510**-512
 primary, *510-511,* **511**
 secondary, **512,** *512*

Sugar, 318-319, *318,* 321, *322, 323,* 601, *601. See also* Carbohydrates

Sulfur dioxide, 278, 279

Sunlight, 8, *8, 483,* 484, *484, 516,* 517

Surface area, of cells, 50, *50*

Suspensions, 68, *68*

Sweat, 593, *act.* 593, 606-607, 669, *act.* 672

Sweat glands, 669, *act.* 672

Swim bladder, 404, *act.* 404

Symbiosis, 493-**494,** *494*

Symmetry, 345-347, *346*
 bilateral, **346**-347, *347*
 radial, **345,** *346*

Synapse, **680,** *680*

Syphilis, 734, *734,* 735

Systemic circulation, **628,** *628, act.* 628, 629, *629*

Tadpoles, *act.* 413

Taiga, *520, act.* 520, **521,** *521*

Tapeworms, 357, *357,* 358-359, *act.* 360-361, *361,* 495

Target tissues, **695**

Tarsiers, 172, *172*

Taste, 691, *691*

Taste buds, **691,** *691*
Taxonomy, **188**-192, *188, act.* 190-191
Technology, **24**
  AIDS vaccine, 38-39
  centenarian studies, 722-723
  classification using DNA, 194
  for cleaner burning of coal, 278-279, *278, 279*
  confocal microscopy, 243, *243*
  in dairy industry, 24-25, *25*
  DNA testing, 107, *107*
  ecosystems in space, 523, *523*
  energy from biomass, 84-85, *84, 85*
  fission track dating, 12, *12*
  fly larvae therapy, 384, *384*
  geothermal power, 545, *545*
  "growing" heart valves, 634-635
  hookworm vaccine, 362
  Human Genome Project, 144-145, *145*
  ibuprofen, 75, *75*
  karyotyping, 141, *141*
  lithotripter for kidney stones, 670, *670*
  Lyme disease vaccine, 218
  monitoring organisms sensitive to pollution, 493, *493*
  ozone layer, 276, *276,* 414-415
  phytoremediation, 319, *319*
  positron emission tomography (PET), 683, *683*
  reconstruction from fossils, 173, *173*
  recycling, 548-549, *548, 549*
  saving endangered birds, 433, *433*
  skin grafts, 54, *54*
  skinning fruit, 293, *293*
  snakebite treatment, 419
  transgenic crops, 331-333, *331, 332*
  transplanted cells to treat diabetes, 112-113, *113*
Teeth, 436, *436, act.* 436
Telophase, *95,* 102, *102, 103*
Temperate deciduous forests, *520,* **522**-523, *522, 523*
Temperature
  of birds, 426, 431, *431*
  determinants of, 517
  diffusion and, *act.* 72
  as environmental factor, *483,* 484
  freezing point, 511
  respiration rate of fish and, *act.* 406-407
  sensing, 694
  water cycle and, 500, *act.* 500
  of whales, *act.* 438-439
Temperature scales, 23, *23*
Tendons, **586**
Tentacles, **353,** *353*
Termites, 235
Terracing, 551, *551*
Territorial behavior, 462-463, *462, 463*
Territory, **462**
Testes, 695, **705,** *705*
Tetanus, 217, 224, 739
Theory, **17**
  of acquired characteristics, 153, *153*
  cell, 43-**44,** *44*
  of evolution, 153-155
Thigmotropism, 324, *324*
Thinking, critical, 20, *21*
Thomas, Lewis, 86
Thumb, opposable, 171, *171*
Thymine, 105, 106
Thymus, 695
Thyroid gland, 695
Ticks, 202, *202, 203,* 386, *386,* 387
Time
  geologic, *164*
  reaction, *act.* 685
Tissues, **54,** *55*
Toads, 409, *409,* 410, 411, 412, *414*
Tobacco mosaic virus, *33*
Tonsils, 645
Topsoil. *See* Soil
Touch, *act.* 692-693, 694, *694*
Toxin, **218,** 739, 747, *747*
Trace minerals, 605
Trachea, **654**
Traits, *act.* 125
  alleles and, 128, *128*
  dominant, **126,** 127
  recessive, **126,** 127, *act.* 131, 137-138, *137*
  tracing, *140,* 141.
  See also Genetics

Transfer RNA (tRNA), 110
Transgenic crops, **331**-333, *331, 332*
Transmission electron microscope (TEM), 42-43, *42*
Transpiration, **315,** *315*
Transplantations, 112-113, *113,* 635
Transport. *See* Cell transport
Trash. *See* Waste
Trash art, 547, *547*
Trees, in climax communities, 512, *513. See also* Forests
Trial and error, **457,** *457*
Triassic period, 417
Triceps, *587*
Trichinella worms, 362, *363*
Trichinosis, 362
Tropical rain forests, 198, *199,* 261, 308, 342, 520, **524**-525, *524*
Tropism, **324**-325, *324, 325, act.* 328-329
Trout, 402
Trypanosoma, 235, *235*
Tube feet, **392,** *392, act.* 393
Tube worms, 377
Tuberculosis, 222-223, *222, 223,* 739
Tumor, **748**
Tuna, 403, *403*
Tundra, **520**-521, *520, act.* 520, *521*
Tunicates, 400, *400*
Turtles, 416, *416,* 417, 418, *418,* 420, 455, *455*
Twins, 714-715, *715*

Ultrasound, 712, *712,* 749
Ultraviolet light, 414-415, 559
Ulva, 232
Umbilical cord, **441,** 712, 717
Units of measure, 22-23, *act.* 22, *23*
Unsaturated fats, 603
Uranium, 546
Urea, 593
Ureters, **668**
Urethra, **668,** *668,* 705
Urinary system, **666**-671, *666*
  diseases and disorders of, 670-671, *670, 671*
  fluid levels in, 668, *668,* 669, 670
  organs of, 667-668, *667, act.* 667, *668*

Urine, **667,** 668, *668*
Uterus, **706,** *706,* 707, *707,* 710, *710*

Vaccination, **744**
Vaccine, **36,** *36,* act. 38-39, **217,** 218, 362, 738
Vacuoles, **50**
Vagina, **706,** *706,* 710, *710,* 711
Variable, **15**
Variation, **156**-157, *156, 157, act.* 158-159
Vascular plants, **264,** 271-277, *271, 274, 275, 277*
Vascular tissue, 294, *294*
Vegetation. *See* Plant(s)
Veins, **629,** 630, *630*
Ventricles, **627,** *627,* 628
Vertebrates, **345,** 400-401
Vestigial structures, **168,** *169*
Villi, **616,** *616*
Viral diseases, 36, 37, *act.* 38-39, 58, *58,* 733, 744-745, *744*
Virchow, Rudolph, 44
Viruses, **32**-39
    active, 34, *34*
    characteristics of, 32-33, *32*
    classification of, 33, *33*
    destruction of, 744-745, *744*
    helpful, 37
    in history, 58, *58*
    latent, 35, *35*
    reproduction of, 34-35, *34, 35*
    size of, 43.
    See also individual viruses
Vision, 171, *act.* 677, 688-689, *688, 689*
Vitamins, 593, **604,** *604, act.* 609
Vocal cords, 654, *654,* 655
Vocal training, 655
Volcanoes, 544
Volume, of cells, 50, *50*
Voluntary muscles, **584**
Volvox, 232

Waste
    in cells, 46, 47, 49, 50, 80
    elimination of, 593, 606, 617, 666-672
    hazardous, **562,** *562*
    radioactive, 546
    recycling, 547, *547,* 548-549, *548, 549*
    solid, 547, *547,* **548**-549, *548, 549*
Water, 70, *70*
    chemical formula for, 66, *66*
    diffusion of (osmosis), 73, *73,* 262
    as environmental factor, *483,* 484
    groundwater, **563,** *563*
    levels in body, 668, *668,* 669, 670
    as nutrient, 9, *9,* 606-607, *607, act.* 607
    plants and, *act.* 271, *act.* 313, 314-315, *314, 315, act.* 316
    plants living in, 261
Water cycle, **500,** *500, act.* 500
Water environments, 530-533
    freshwater, 530-531, *530, 531, act.* 532
    saltwater, 532-533, *532, 533*
Water molds, 238-239, *239*
Water pollution, 563-564, *563, 564*
Water power, 544, *544*
Water-soluble vitamins, 604
Water-vascular system, **392,** *392*
Watson, James, 104, 105, 108
Watson, John B., 458
Waxy cuticle, 294
Wetlands, *555*
Whales, *act.* 438-439, *442,* 469
Whiskers, 436
White blood cells, 637, *637,* 641, 737, *737*
White Cliffs of Dover, *235*
White light, *317,* 318
White rusts, 238
Whooping cough, 217
Wildebeest, *469*
Wildlife conservation, 554-555, *554*
Wind power, 544, *544*
Wings, 430, *430*

Withdrawal reflex, 683
Wolves
    aggressive behaviors of, 463, *463*
    care of young by, *437*
    classification of, 192, *192*
    extinction of, 176, *177*
    re-introduction of, 502-503, *502, 503*
    social behavior of, 464
    teeth of, *436*
Wood alcohol, 84
Wood duck, *432*
Woody stems, 294
Worms
    flatworms, *347,* 357-363, *357, 358, act.* 358, *359, 360, act.* 360-361, *361*
    roundworms, 359, *359,* 362-363, *362, 363*
    segmented, 375-381, *375, 376, act.* 378-379, *380, 381, act.* 466, 495, *495*

X chromosome, 138-140, *138, 140*
Xylem, **294,** *294,* 295

Y chromosome, 138-140, *138, 140*
Yeast, 81, *81,* 243, *243, act.* 321
Yellowstone National Park, 502-503, *503*
Yogurt, *act.* 217
Yucca plant, *494*

Zenteno-Savin, Tania, 448, *448*
Zoo(s), 470-471, *470, 471*
Zoologist, 14
Zygote, *100,* **101,** 710, 711, 712
Zygote fungi, 242, *242*

Art Credits

vi, Morgan-Cain and Associates; xii, Pond & Giles; xiii, Chris Forsey; xv, (t) Jim Jobst, (b) Morgan-Cain and Associates; xx, Michael Woods; xxi, Barbara Hoopes-Amber; xxv, Morgan-Cain and Associates; 10, Pond & Giles; 11, Chris Forsey; 13, 15, 21, 29, 32, 33, 34, 35, Morgan-Cain and Associates; 41, David Fischer; 44, Morgan-Cain and Associates; 46, 47, 48, 49, Felipe Passalacqua; 50, Morgan-Cain and Associates; 51, Felipe Passalacqua; 52, Morgan-Cain and Associates; 53, David Fischer; 61, 64, 65, 66, 67, 71, 73, 74, 76, 79, 80, 89, 93, 94, 95, 96, Morgan-Cain and Associates; 100, (l, r, b) Michael Maydak, (c) Morgan-Cain and Associates; 102, 103, 104-105, 106, 108, 112, 117, 124, Morgan-Cain and Associates; 126, 127, Pond & Giles; 128, Morgan-Cain and Associates; 129, 132, Pond & Giles, 140, 149, Morgan-Cain and Associates; 152-153, Michael Woods; 154, David Reed; 160, Morgan-Cain and Associates; 161, Laurie O'Keefe; 164, Morgan-Cain and Associates, (inset) Dave Reed; 165, David Fischer; 166, Barbara Hoopes-Amber; 167, Morgan-Cain and Associates; 168, Laurie O'Keefe; 169, (t) Michael Woods, (b) Pond & Giles; 171, Bob Giuliani; 181, Laurie O'Keefe; 188, Cende Hill; 195, Biruta Hansen, Cende Hill; 196, Morgan-Cain and Associates; 202, 203, Pond & Giles; 211, Morgan-Cain and Associates; 215, Felipe Passalacqua; 217, 223, 224, 227, Morgan-Cain and Associates; 230, Felipe Passalacqua; 231, Morgan-Cain and Associates; 234, (t) Felipe Passalacqua, (b) Morgan-Cain and Associates; 236, Morgan-Cain and Associates; 241, 242, David Fischer; 253, (l) Morgan-Cain and Associates, (r) David Fischer; 261, Jean Helmer; 265, Pond & Giles, Morgan-Cain and Associates; 269, 277, Pond & Giles; 278, 279, 283, Morgan-Cain and Associates; 288, 289, 294, Pond & Giles; 295, Morgan-Cain and Associates; 296, 298, 299, 303, Pond & Giles; 311, 315, 317, Morgan-Cain and Associates; 318, David Ashby; 321, 322, 326, Morgan-Cain and Associates; 332, David Ashby, Morgan-Cain and Associates; 337, Morgan-Cain and Associates; 345, Dave Genris; 346, Cende Hill; 347, Felipe Passalacqua; 351, Sarah Woodward; 353, 355, Cende Hill; 358, 359, Sarah Woodward; 367, Morgan-Cain and Associates; 370, Laurie O'Keefe; 372, Sarah Woodward; 374, Morgan-Cain and Associates; 376, 381, Sarah Woodward; 382, 385, David Ashby; 387, Michael Woods; 392, Sarah Woodward; 397, Morgan-Cain and Associates; 401, 403, 410, Michael Woods; 412, Felipe Passalacqua; 417, David Fischer; 423, Morgan-Cain and Associates; 427, Pond & Giles; 428, Michael Woods; 430, David Ashby; 431, Michael Woods; 434, Chris Forsey; 436, Laurie O'Keefe; 445, Rolin Graphics; 451 (r) Morgan-Cain and Associates, (l) Laurie O'Keefe; 458, Laurie O'Keefe; 459, David Ashby; 472, Barbara Hoopes-Amber; 475, Morgan-Cain and Associates; 483, Pond & Giles; 486, David Ashby; 488, 489, 492, Morgan-Cain and Associates; 496-497, Barbara Hoopes-Amber; 498, Laurie O'Keefe; 499, David Ashby; 500, 501, Felipe Passalacqua; 507, Cende Hill; 510-511, Chris Forsey; 516-517, Morgan-Cain and Associates; 518, Chris Forsey; 520, Morgan-Cain and Associates; 523, Chris Forsey; 524, Barbara Hoopes-Amber; 529, 542, 543, 559, 560, 561, Morgan-Cain and Associates; 563, Chris Forsey; 574, David Fischer; 575, (t, m) Felipe Passalacqua, (b) Chris Forsey; 576, Pond & Giles; 578, David Fischer, Morgan-Cain and Associates; 582, Morgan-Cain and Associates; 584, David Fischer; 585, Morgan-Cain and Associates; 586, Chris Forsey; 587, David Fischer; 591, 592, 600, 601, Morgan-Cain and Associates; 604, Pond & Giles; 607, Morgan-Cain and Associates; 608, Pond & Giles; 610, Morgan-Cain and Associates; 612, Laurie O'Keefe; 613, Morgan-Cain and Associates; 614, (t) David Fischer, (b) Tonya Hines; 615, 616, Tonya Hines; 626, 627, 628, Laurie O'Keefe; 629, 630, 631, Morgan-Cain and Associates; 634, Laurie O'Keefe; 638, 639, 644, Morgan-Cain and Associates; 645, 646, Laurie O'Keefe; 649, 652, Morgan-Cain and Associates; 653, Laurie O'Keefe; 654, David Fischer; 656, Laurie O'Keefe; 657, (t) Felipe Passalacqua, (b) Laurie O'Keefe; 658, 666, Morgan-Cain and Associates; 667, Laurie O'Keefe; 668, (t) David Fischer, (b) Laurie O'Keefe; 670, 677, 678, Morgan-Cain and Associates; 679, (l) Morgan-Cain and Associates, (r) David Ashby; 680, (t) Gina Lapurga, (b) Michael Courtney; 681, Michael Courtney; 682, (t) Morgan-Cain and Associates, (b) Michael Courtney; 688, Michael Courtney; 689, Morgan-Cain and Associates; 690, Jim Jobst; 691, Michael Courtney; 695, 697, Morgan-Cain and Associates; 700, Michael Courtney; 705, 706, Sarah Woodward; 707, (t) Sarah Woodward, (b) Tonya Hines; 709, Tonya Hines; 710, 713, Sarah Woodward; 721, 727, Morgan-Cain and Associates; 730, Chris Forsey; 731, Cende Hill; 732, Chris Forsey; 733, Felipe Passalacqua; 737, 738, Morgan-Cain and Associates; 756, David Fischer; 757, 767, 768, 769, 771, 772, Morgan-Cain and Associates; 779, Rolin Graphics; 780, (t) Barbara Hoopes-Amber, (b) David Ashby; 782, Morgan-Cain and Associates.

Photo Credits

Cover NHPA; **vi** Biophoto Associates/Photo Researchers, Inc.; **vii** (t) Science Photo Library/Photo Researchers, Inc., (bl) Visuals Unlimited/ Barbara Gerlach, (br) Matt Meadows; **viii** (tl) John Cancalosi/Tom Stack & Associates, (tr) John Reader/Science Photo Library/Photo Researchers, Inc., (b) Stephen J. Krasemann/DRK Photo; **ix** Elaine Shay; **x** (t) Visuals Unlimited/David Newman, (b) Robert Holland/DRK Photo; **xi** (t) Runk/Schoenberger from Grant Heilman, (c) Animals Animals/Zig Leszczynski, (b) Animals Animals/Prestige Pictures/ Oxford Scientific Films; **xii** Visuals Unlimited/Don W. Fawcett; **xiii** Larry Ulrich/DRK Photo; **xiv** (t) Duomo/William R. Sallaz, (b) Dr. Tony Brain/Science Photo Library/Photo Researchers, Inc.; **xv** CNRI/Science Photo Library/Photo Researchers, Inc.; **xvi** (tl) Ron Chapple/FPG International, (tr) D.W. Fawcett/Photo Researchers, Inc., (b) Doug Martin; **xvii** Jeff Smith/Fotosmith; **xviii xix** Matt Meadows; **xx** (l) John Cancalosi/DRK Photo, (r) Matt Meadows; **xxi** Mark Burnett; **xxii** (t) Mark Burnett, (c) StudiOhio, (b) Animals Animals/Ken Cole; **xxiii** (t) Mark Burnett, (b) Aaron Haupt; **xxiv** (t) John Gurche, (b) NASA/Photo Researchers, Inc.; **xxvi** (t) Rebecca L. Johnson, (b) Alfred Gonzales; **xxvii** (l) Greenwich Workshop, (r) SSEC/University of Wisconsin; **2-3** ZEFA-Bach/The Stock Market; **3** Earth Scenes/Zig Leszczynski; **4-5** Tom Vezo; **5** Matt Meadows; **6** Norbert Wu/The Stock Market; **7** (t) Frank Rossotto/The Stock Market, (b) Jane Burton/Bruce Coleman, Inc.; **8** (t) Animals Animals/Alfred B. Thomas, (b) Kenneth Murray/Photo Researchers, Inc.; **9** Clem Haagner/ABPL/Photo Researchers, Inc.; **12** Visuals Unlimited/Fermilab; **13** Earth Scenes/E.R. Degginger; **14** (tl) Visuals Unlimited/Kjell B. Sandved, (bl) Animals Animals/Herb Segars, (r) Randy Morse/Tom Stack & Associates; **15** (l) Visuals Unlimited/David M. Phillips, (r) Doug Martin; **16** Mark Burnett; **17 18 19** Matt Meadows; **20** S. Nielsen/DRK Photo; **23** (t) Doug Martin, (l) Glencoe photo, (c) Rick Weber, (r) Matt Meadows; **24** Lynn M. Stone/ DRK Photo; **25** Jane Latta/Photo Researchers, Inc.; **26** Greenwich Workshop; **30-31** Garden Matters; **31** Mark Burnett; **32** (l) Omikron/ Photo Researchers, Inc., (r) Institut Pasteur/CNRI/Phototake, NYC; **33** (tl) Tektoff-RM/CNRI/Science Photo Library/Photo Researchers, Inc., (tr) M. Wurtz/Biozentrum, University of Basel/Science Photo Library/ Custom Medical Stock Photo, (bl) Photo Researchers, Inc., (br) Lee D. Simon/Photo Researchers, Inc.; **36** (l) J.S. Reid/Custom Medical Stock Photo, (r) Glencoe photo; **38** Lennart Nilsson/Boehringer Ingelheim International GmbH; **40** (l) Visuals Unlimited/Mike Eichelberger, (r) George E. Jones/Photo Researchers, Inc.; **42** (l) Alfred Pasieka/Science Photo Library/Photo Researchers, Inc., (r) Dr. Tony Brian/Science Photo Library/Photo Researchers, Inc.; **43** (t) Jonathan Eisenback/Phototake, NYC, (b) M. Abbey/Photo Researchers, Inc. **45** (t) Roland Birke/ OKAPIA/Photo Researchers, Inc., (b) Visuals Unlimited/W. Banaszewski; **46** Biophoto Associates/Photo Researchers, Inc.; **47** Dr. Dennis Kunkel/Phototake, NYC; **48** Visuals Unlimited/Fred E. Hossler; **49** (t) Visuals Unlimited/D.W. Fawcett, (b) Visuals Unlimited/K.R. Porter, D. Fawcett; **51** Biophoto Associates/Photo Researchers, Inc.; **52** CNRI/Science Photo Library/Photo Researchers, Inc.; **54** Hans Halberstadt/Photo Researchers, Inc.; **56** Aaron Haupt; **57** Matt Meadows; **58** CDC/Phototake, NYC; **62-63** Aaron Haupt; **63** Mark Burnett; **64** (l) E.R. Degginger/Photo Researchers, Inc., (r) Keith Kent/Science Photo Library/Photo Researchers, Inc.; **65** Mark Burnett; **66** Matt Meadows; **67** Mark Burnett; **68** (l) Steve Lissau, (r) Matt Meadows; **70** Ted Clutter/Photo Researchers, Inc.; **71 72** Matt Meadows; **73** Mark Burnett; **75** StudiOhio; **77** Matt Meadows; **78** Gay Bungarner/Photo/NATS; **80** Aaron Haupt; **81** Doug Martin; **82 83** Matt Meadows; **84** Charles Fitch/FPG International; **85** (t) Georg Gerster/Photo Researchers, Inc., (b) Earth Scenes/John Eastcott & Yva Momatiuk; **86** (l) Science Photo Library/Photo Researchers, Inc., (r) SSEC/University of Wisconsin; **90-91** Scott Camazine/Photo Researchers, Inc.; **91** StudiOhio; **92** (l) Ken Wagner/Phototake, NYC, (r) Ann Reilly/Photo/NATS; **93** (t) Visuals Unlimited/Karlene V. Schwartz, (c) Michael P. Gadomski/Photo Researchers, Inc., (b) Hermann Eisenbeiss/Photo Researchers, Inc.; **94** (t) Biophoto Associates/Photo Researchers, Inc., (c r b) M. Abbey/Photo Researchers, Inc.; **95** (l c) M. Abbey/Photo Researchers, Inc., (r) Visuals Unlimited/R. Calentine; **96** (l) J.L. Carson/Custom Medical Stock Photo, (r) Animals Animals/E.R. Degginger; **97** (l) Visuals Unlimited/Walt Anderson, (r) Biophoto Associates/Photo Researchers, Inc.; **98** Visuals Unlimited/T.E. Adams; **99** (l) David M. Phillips/Photo Researchers, Inc., (r) G. Schatten/Photo

Researchers, Inc.; **101** (l, b) Grant Heilman, (r) Mark E. Gibson/The Stock Market; **107** Peter Menzel; **109** Alvin E. Staffan/Photo Researchers, Inc.; **110 111** StudiOhio; **113** Glencoe photo; **114** Jeff Fearing; **118** (t) Visuals Unlimited/Kevin Collins, (b) Gary Retherford/Photo Researchers, Inc.; **119** Visuals Unlimited/SIU; **120-121** Stephen J. Krasemann/DRK Photo; **121** David M. Dennis/Tom Stack & Associates; **122-123** M.P. Kahl/DRK Photo; **123** Matt Meadows; **125** Jane Grushow from Grant Heilman; **126** Runk/Schoenberger from Grant Heilman; **129** George H. Harrison from Grant Heilman; **130** Matt Meadows; **131** Biophoto Associates/Photo Researchers, Inc.; **133** Robert Wallis/SABA Press Photos; **137** Jackie Lewin, Royal Free Hospital/Science Photo Library/Photo Researchers, Inc.; **138** Biophoto Associates/Photo Researchers, Inc.; **139** (l) David M. Dennis/Tom Stack & Associates, (r) Suzanne Szasz/Photo Researchers, Inc.; **141** Biophoto Associates/Photo Researchers, Inc.; **142** Larry LeFever from Grant Heilman; **143** Nigel Cattlin/Holt Studio International/Photo Researchers, Inc.; **144** Peter Menzel; **145** Philippe Plailly/Science Photo Library/Photo Researchers, Inc.; **146** (t) University of Wisconsin-Madison News Service, photo by Jeff Miller, (b) Visuals Unlimited/Barbara Gerlach; **148** MAK-One; **150-151** Rod Planck/Photo Researchers, Inc.; **151** Aaron Haupt; **153** Doug Allemand/Jacana/ Photo Researchers, Inc.; **154** (tl) Christian Grzimek/Okapia/Photo Researchers, Inc., (tr c) Wolfgang Kaehler, (b) Kenneth W. Fink/Photo Researchers, Inc.; **155** Aaron Haupt; **156** (l) Richard P. Smith/Tom Stack & Associates, (r) Visuals Unlimited/John Gerlach; **157** (l) Joe McDonald/Tom Stack & Associates, (r) Mark Boulton/Photo Researchers, Inc.; **158** Jeremy Woodhouse/DRK Photo; **159** Matt Meadows; **162** Richard Nowitz/Phototake, NYC; **163** (l) John Cancalosi/Tom Stack & Associates, (c) David M. Dennis/Tom Stack & Associates, (r) John Cancalosi/DRK Photo; **165** Visuals Unlimited/D. Long; **170** Tim Davis/Photo Researchers, Inc.; **171** M. Loup/ Jacana/Photo Researchers, Inc.; **172** (tl) Steve Kaufman/DRK Photo, (tr, b) Tom McHugh/Photo Researchers, Inc.; **173** John Gurche; **174** John Reader/Science Photo Library/Photo Researchers, Inc.; **175** RIA-Novosti/Sovfoto; **177** (t) Barbara von Hoffman/Tom Stack & Associates, (b) Mark Newman/Tom Stack & Associates; **178** Jean-Marie Chauvet/ Sygma; **182** (l) Mark Burnett, (r) Hickson & Associates; **183** (t) Ken Frick, (b) Mark Burnett; **184 185** David M. Dennis/Tom Stack & Associates; **186-187** Matt Meadows; **187** Visuals Unlimited/Alex Kerstitch; **189** Animals Animals/Ken Cole; **190 191** StudiOhio; **192** (l) Renee Lynn/Photo Researchers, Inc., (c) Stephen J. Krasemann/DRK Photo, (r) Robert Winslow/Tom Stack & Associates; **193** (l) Institut Pasteur/CNRI/ Phototake, NYC, (c) Rod Planck/Tom Stack & Associates, (r) M.I. Walker/Photo Researchers, Inc.; **194** Tom & Pat Leeson/DRK Photo; **195** (l) Science Photo Library/Custom Medical Stock Photo, (r) David M. Phillips/Photo Researchers, Inc.; **196** S. Nielsen/DRK Photo; **197** Larry Ulrich/DRK Photo; **198** G.C. Kelly/Photo Researchers, Inc.; **199** Dr. Morley Read/Science Photo Library/Photo Researchers, Inc.; **200** (l) Stephen J. Krasemann/DRK Photo, (c) Animals Animals/Robert Maier, (r) John Cancalosi/DRK Photo; **201** (l) Animals Animals/Don Enger, (c) Index Stock/Phototake, NYC, (r) Dr. Paul A. Zahl/Photo Researchers, Inc.; **204** (l) Stephen J. Krasemann/DRK Photo, (r) Jeff Lepore/Photo Researchers, Inc.; **207** (t) Victoria Hurst/Tom Stack & Associates, (c) Animals Animals/Ken Cole, (b) Animals Animals/Peter Weimann; **208-209** David M. Phillips/Photo Researchers, Inc.; **209** Michael Abbey/Photo Researchers, Inc.; **210** (l) Alfred Pasieka/Science Photo Library/Photo Researchers, Inc., (r) Visuals Unlimited/R. Calentine; **212** (t) CNRI/Science Photo Library/Photo Researchers, Inc., (b) Visuals Unlimited/Hal Beral; **213** (t) Tom E. Adams/Peter Arnold, Inc., (b) Matt Meadows; **214** Visuals Unlimited/WHOI, D. Foster; **216** (l) David R. Frazier/Photo Researchers, Inc., (r) John Colwell from Grant Heilman; **218** (l) Microworks/CNRI/Phototake, NYC, (r) Spielman/CNRI/ Phototake, NYC; **219** Alfred Pasieka/Science Photo Library/Photo Researchers, Inc.; **220 221** Matt Meadows; **222** CNRI/Science Photo Library/Photo Researchers, Inc.; **224** Bill Binzen/The Stock Market; **228-229** Matt Meadows; **229** Mark Burnett; **231** Jan Hinsch/Science Photo Library/Photo Researchers, Inc.; **232** (t) Visuals Unlimited/David Phillips, (l, r) Runk/Schoenberger from Grant Heilman, (ct) Biophoto Associates/Photo Researchers, Inc., (cb) Carolina Biological Supply/ Phototake, NYC; **233** (t) Andrew J. Martinez/Photo Researchers, Inc., (b) Gregory Ochocki/Photo Researchers, Inc.; **235** (t) Kevin Schafer/Tom Stack & Associates, (c) Alfred Pasieka/Science Photo Library/Photo

PERIODIC TABLE OF THE ELEMENTS

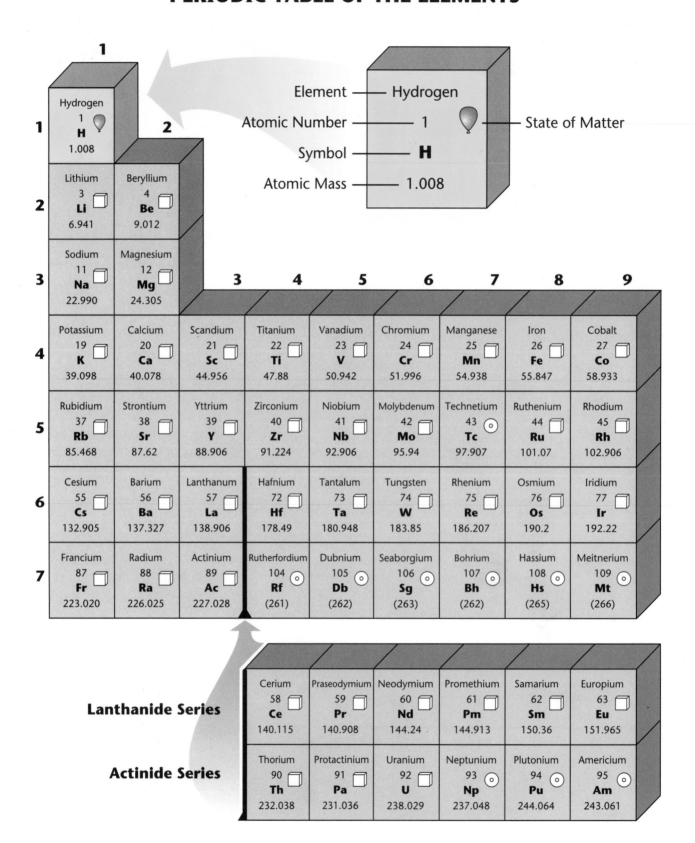